Flora of North America

Flora of North America

North of Mexico

Edited by FLORA OF NORTH AMERICA EDITORIAL COMMITTEE

VOLUME 3

Magnoliophyta: Magnoliidae and Hamamelidae

MAGNOLIALES (Magnolia order)

LAURALES (Laurel order)

PIPERALES (Pepper order)

ARISTOLOCHIALES (Birthwort order)

ILLICIALES (Star-anise order)

NYMPHAEALES (Water-lily order)

RANUNCULALES (Buttercup order)

PAPAVERALES (Poppy order)

HAMAMELIDALES (Witch-hazel order)

URTICALES (Nettle order)

LEITNERIALES (Corkwood order)

JUGLANDALES (Walnut order)

MYRICALES (Bayberry order)

FAGALES (Beech order)

CASUARINALES (She-oak order)

NEW YORK OXFORD · OXFORD UNIVERSITY PRESS · 1997

Oxford University Press

Oxford New York
Athens Auckland Bangkok Bogotá
Bombay Buenos Aires Calcutta Cape Town
Dar es Salaam Delhi Florence Hong Kong Istanbul
Karachi Kuala Lumpur Madras Madrid Melbourne
Mexico City Nairobi Paris Singapore
Taipei Tokyo Toronto Warsaw

and associated companies in

Berlin Ibadan

Published by Oxford University Press, Inc.,
198 Madison Avenue, New York, New York 10016

Library of Congress Cataloging-in-Publication Data
(Revised for volume 3)
Flora of North America north of Mexico
edited by Flora of North America Editorial Committee.
Includes bibliographical references and indexes.
Contents: v. 1. Introduction—v. 2. Pteridophytes and gymnosperms—
v. 3. Magnoliophyta: Magnoliidae and Hamamelidae
ISBN 0-19-511246-6 (v. 3)
1. Botany—North America.
2. Botany—United States.
3. Botany—Canada.
1. Flora of North America Editorial Committee.
QK110.F55 1997 581.97 92-30459

Printing (last digit): 9 8 7 6 5 4 3 2 1
Printed in the United States of America
on acid free paper

Contents

FOUNDING MEMBER INSTITUTIONS

Flora of North America Association

Arnold Arboretum
Jamaica Plain, Massachusetts

Biosystematics Research Institute
Ottawa, Canada

Canadian Museum of Nature
Ottawa, Ontario

Carnegie Museum of Natural
History
Pittsburgh, Pennsylvania

Field Museum of Natural History
Chicago, Illinois

Fish and Wildlife Service
United States Department of the
Interior
Washington, D.C.

Harvard University Herbaria
Cambridge, Massachusetts

Hunt Institute for Botanical
Documentation
Carnegie Mellon University
Pittsburgh, Pennsylvania

Jacksonville State University
Jacksonville, Alabama

Jardin Botanique de Montréal
Montréal, Quebec

Kansas State University
Manhattan, Kansas

Missouri Botanical Garden
St. Louis, Missouri

New Mexico State University
Las Cruces, New Mexico

New York State Museum
Albany, New York

Northern Kentucky University
Highland Heights, Kentucky

The New York Botanical Garden
Bronx, New York

The University of British Columbia
Vancouver, British Columbia

The University of Texas
Austin, Texas

Université de Montréal
Montréal, Quebec

University of Alaska
Fairbanks, Alaska

University of Alberta
Edmonton, Alberta

University of California
Berkeley, California

University of California
Davis, California

University of Idaho
Moscow, Idaho

University of Illinois
Urbana-Champaign, Illinois

University of Iowa
Iowa City, Iowa

University of Kansas
Lawrence, Kansas

University of Michigan
Ann Arbor, Michigan

University of Oklahoma
Norman, Oklahoma

University of Ottawa
Ottawa, Ontario

University of Southwestern
Louisiana
Lafayette, Louisiana

University of Western Ontario
London, Ontario

University of Wyoming
Laramie, Wyoming

Utah State University
Logan, Utah

For their support of the Flora of North America Project,
we gratefully acknowledge and thank:

National Science Foundation
The Pew Charitable Trusts
The Caleb C. and Julia W. Dula Foundation
The Surdna Foundation
The David and Lucile Packard Foundation
National Fish and Wildlife Foundation
ARCO Foundation
The William and Flora Hewlett Foundation
Edward Chase Garvey Memorial Foundation
Waste Management, Inc./Environmental Affairs Department
The Mellon Foundation
The Geraldine R. Dodge Foundation
Chevron
Union Pacific Railroad Company

Project Staff

Helen K. Jeude
Technical Editor

Deborah L. Kama
Database Manager (–1996)

John Myers
Illustrator

Gina Otterson
Departmental Secretary

Bruce D. Parfitt
Managing Editor (–1995)

Anne Keats Smith
Map Editor

Judith Unger
Project Coordinator

Alan T. Whittemore
Bryophyte and Vascular Plant Specialist

Yevonn Wilson-Ramsey
Illustration Editor and Illustrator

James L. Zarucchi
Managing Editor (1996–)

Eleanor Zeller
Clerk

Acknowledgments

Flora of North America is fortunate to have a tremendously skilled and dedicated staff. Although every person in the Organizational Center has contributed in many ways to the completion of this volume, Helen Jeude (technical editor), Anne Keats Smith (map editor), and Yevonn Wilson-Ramsey (illustration editor and illustrator) deserve special recognition. They have lived and breathed Magnoliidae and Hamamelidae for the past three years, and it is their careful work, attention to detail, and commitment to presenting the work of authors, reviewers, and editors in the best possible way that accounts for the quality that has been achieved here. Alan Whittemore helped significantly with the preparation of several major treatments, in addition to those he authored.

J. B. Phipps edited Magnoliales, Hamamelidales, Fagales, and Casuarinales; Grady Webster edited Laurales; John Thieret edited Piperales, Aristolochiales, Illiciales, and Nymphaeales; Bruce Parfitt edited Ranunculaceae (I edited the other families in Ranunculales); Robert Kiger edited Papaverales; Leila Shultz edited Urticales; and David Boufford edited Leitneriales, Juglandales, and Myricales. The editorial process involves a tremendous number of steps and includes intensive review at each step (e.g., sometimes by as many as fifty-five regional reviewers). We are all grateful to the authors, reviewers, and editors for the excellent work that they have done and for their patience with the process. Theodore Barkley, John Packer, John Strother, and John Thieret were both "early" and "final" reviewers for every manuscript. John Strother worked through many manuscripts more than twice.

Flora of North America owes its start to those who attended an organizing meeting on 30 April and 1 May 1982 at the Missouri Botanical Garden: G. Argus, D. Bates, W. Burger, S. Hatch, N. Holmgren, T. Jacobsen, M. Johnston, R. Kiger, J. Massey, J. McNeill, R. Mecklenburg, N. Morin, J. Phipps, D. Porter, J. Reveal, S. Shetler, F. Utech, and G. Webster.

The project has been designed, in large part, by the Editorial Committee. This committee has changed its makeup slightly over the years, but overall it has been a remarkably stable group. By 1986 the Editorial Committee consisted of G. Argus, T. Barkley, D. Boufford, L. Brouillet, D. Henderson, R. Kiger, B. MacBryde, J. McNeill, N. Morin, J. Packer, J. Phipps, L. Shultz, R. Spellenberg, J. Strother, J. Thieret, G. Webster, and D. Whetstone. Since then, Henderson, MacBryde, and Whetstone have resigned from the committee and have been replaced by R. Hartman, J. Fay, and M. Moore. James Estes, M. Johnston, D. Murray, A. Smith, G. Straley, and R. Thompson were added to the committee as permanent members. In 1991, bryophytes were added to the project, and the following bryologists were added to the Editorial Committee: W. Buck, M. Crosby, J. Engel, M. Hicks, D. Horton, N. Miller, B. Murray, W. Reese, R. Stotler, B. Thiers, and D. Vitt.

In 1994 an Informatics Committee was established, its members including T. Barkley, D. Boufford, L. Brouillet, R. Hartman, N. Morin, B. Murray, L. Shultz, and A. Smith, of the Editorial Committee, and new members F. Bisby, J. Schnase, and J. Wiersema. Taken together, the Editorial Committee and the Informatics Committee form the Flora of North America Organization, on behalf of which a Management Committee facilitates policy decisions.

The *Flora of North America* project is coordinated from the Organizational Center, located at the Missouri Botanical Garden. Support by the Missouri Botanical Garden for this Center has been instrumental in the success of the project, and we are grateful to Peter H. Raven for committing institutional resources and for contributing his time and energy to *Flora of North America*. The institutions at which editors work provide more than half of the overall cost of the project through in-kind support. We are grateful to all of the foundations and individuals who have financially supported *Flora of North America*. In addition to the staff (listed on p. viii), undergraduate interns, some of whom were funded by the Research Experiences for Undergraduates program of the National Science Foundation, have contributed to the project. From 1992 through 1995 they were Lena C. Hileman, Nicholas Lewin, Amy Prosser, and Georgina Witt.

Members of the Flora of North America Project Advisory Panel were: F. A. Almeda Jr., California Academy of Sciences; D. M. Bates, Liberty Hyde Bailey Hortorium; T. P. Bennett, Academy of Natural Sciences; A. Bouchard, Jardin Botanique de Montréal; W. C. Burger, Field Museum of Natural History; T. Duncan, University of California; C. G. Gruchy, National Museum of Natural Sciences, Canada; V. L. Harms, University of Saskatchewan; J. Hickman, University of California; A. G. Jones, University of Illinois; R. W. Kiger, Hunt Institute for Botanical Documentation; M. M. Littler, National Museum of Natural History; J. Massey, University of North Carolina, Chapel Hill; G. A. Mulligan, Vascular Plant Herbarium, Ottawa; G. B. Ownbey, University of Minnesota; J. G. Packer, University of Alberta; D. H. Pfister, Harvard University Herbaria; G. T. Prance, The New York Botanical Garden; R. F. Scagel, University of British Columbia; R. L. Shaffer, University of Michigan; R. F. Thorne, Rancho Santa Ana Botanical Garden; and B. L. Turner, University of Texas.

Members of the Flora of North America Project Database Consulting Group were: Guy Baillargeon, Biosystematics Research Centre, Ottawa; Chris Beecher, NAPRALERT, Chicago; Warren U. Brigham, Illinois Natural History Survey; Christian Burks, Genbank, Los Alamos; Theodore J. Crovello, California State University; Thomas Duncan, University of California; Janet Gomon, National Museum of Natural History, Washington D.C.; Ronald L. Hartman, Rocky Mountain Herbarium; Maureen Kelley, BIOSIS, Philadelphia; Robert W. Kiger, Hunt Institute for Botanical Documentation; Kenneth M. King, EDUCOM, Washington, D.C.; Robert Magill, Missouri Botanical Garden; Jim Ostell, National Library of Medicine, Bethesda; David Raber, Case Ware Inc., Costa Mesa; Beryl Simpson, University of Texas; Frederick Springsteel, University of Missouri; Kerry S. Walter, Center for Plant Conservation, Jamaica Plain.

In 1994, a new Advisory Group was established. Members are: John M. Briggs, Kansas State University; Warren U. Brigham, Illinois Natural History Survey; Jack Dangermond, Environmental Systems Research Institute, Inc.; Janet Gomon, National Museum of Natural History; Abraham Silberschatz, University of Texas; Steven Young, Environmental Protection Agency.

We are deeply grateful to these people and institutions for their hard work and continuing support on behalf of *Flora of North America*.

                                                                                    N. R. M.

Taxonomic Reviewers

Loran C. Anderson
Florida State University
Tallahassee, Florida

Peter W. Ball
University of Toronto
Mississauga, Ontario

William T. Barker
North Dakota State
University
Fargo, North Dakota

Rupert C. Barneby
The New York Botanical
Garden
Bronx, New York

A. Linn Bogle
University of New
Hampshire
Durham, New Hampshire

David E. Boufford
Harvard University Herbaria
Cambridge, Massachusetts

David M. Brandenburg
Neward, Ohio

Christopher Brayshaw
Royal British Columbia
Museum
Victoria, British Columbia

Nancy C. Coile
Florida Division of Plant
Industry
Gainesville, Florida

Raymond Cranfill
University of California
Berkeley, California

Thomas S. Elias
United States National
Arboretum
Washington, D.C.

L. L. Gaddy
Walhalla, South Carolina

James W. Hardin
North Carolina State
University
Raleigh, North Carolina

Robert R. Haynes
University of Alabama
Tuscaloosa, Alabama

C. Barre Hellquist
North Adams State College
North Adams,
Massachusetts

James Henrickson
California State University
Los Angeles, California

Jeri W. Higginbotham
Jacksonville State University
Jacksonville, Alabama

Carl S. Keener
Pennsylvania State
University
University Park,
Pennsylvania

Elbert L. Little
Arlington, Virginia

Karen L. Lu
Humboldt State University
Arcata, California

Paul J. M. Maas
University of Utrecht
Utrecht, Netherlands

Wayne E. Manning
Lewisburg, Pennsylvania

Jim R. Massey
University of North
Carolina
Chapel Hill, North Carolina

Christopher A. Meacham
University of California
Berkeley, California

Michael R. Mesler
Humboldt State University
Arcata, California

David C. Michener
Matthaei Botanical Gardens
Ann Arbor, Michigan

Cornelius H. Muller
University of California
Santa Barbara, California

Eliane M. Norman
Stetson University
DeLand, Florida

Robert Ornduff
University of California
Berkeley, California

Gerald B. Ownbey
University of Minnesota
St. Paul, Minnesota

George K. Rogers
University of West Indies
Bridgetown, Barbados

Leila M. Shultz
Harvard University Herbaria
Cambridge, Massachusetts

Richard W. Spellenberg
New Mexico State
University
Las Cruces, New Mexico

Stephen A. Spongberg
Arnold Arboretum
Jamaica Plain,
Massachusetts

Guy Sternberg
Starhill Forest
Petersburg, Illinois

Tod F. Stuessy
Natural History Museum
Los Angeles, California

Robert L. Wilbur
Duke University
Durham, North Carolina

Dennis W. Woodland
Andrews University
Berrien Springs, Michigan

Regional Reviewers

ALASKA

Robert Lipkin
Alaska Natural Heritage Program
University of Alaska
Anchorage, Alaska

Mary Stensvold
U.S.D.A. Forest Service, Alaska Region
Sitka, Alaska

NORTHWESTERN UNITED STATES

Kathleen E. Ahlenslager
Colville National Forest
Colville, Washington

Edward R. Alverson
The Nature Conservancy
Eugene, Oregon

Kenton L. Chambers
Oregon State University
Corvallis, Oregon

SOUTHWESTERN UNITED STATES

H. David Hammond
Flagstaff, Arizona

Charles T. Mason Jr.
University of Arizon
Tucson, Arizona

Donald J. Pinkava
Arizona State University
Tempe, Arizona

Teresa Prendusi
U.S.D.A. Forest Service, Southwestern
  Region
Albuquerque, New Mexico

CALIFORNIA

Roxanne L. Bittman
California Department of Fish &
  Game
Sacramento, California

Barbara Ertter
University of California
Berkeley, California

James R. Shevock
U.S.D.A. Forest Service
San Francisco, California

WESTERN CANADA

William T. Cody
Centre for Land & Biological
  Resources Research
Ottawa, Ontario

Vernon L. Harms
University of Saskatoon
Saskatoon, Saskatchewan

ROCKY MOUNTAINS

Bonnie Heidel
Montana Natural Heritage Program
Helena, Montana

Douglass M. Henderson (deceased)
University of Idaho
Moscow, Idaho

Tim Hogan
University of Colorado Museum
Boulder, Colorado

J. Stephen Shelly
U.S.D.A. Forest Service, Northern
  Region
Missoula, Montana

Stanley L. Welsh
Brigham Young University
Provo, Utah

NORTH CENTRAL UNITED STATES

William T. Barker
North Dakota State University
Fargo, North Dakota

Ralph E. Brooks
Black & Veatch
Overland, Kansas

Anita F. Cholewa
University of Minnesota
St. Paul, Minnesota

Theodore S. Cochrane
University of Wisconsin
Madison, Wisconsin

Neil A. Harriman
University of Wisconsin
Oshkosh, Wisconsin

Robert B. Kaul
University of Nebraska
Lincoln, Nebraska

Gary E. Larson
South Dakota State University
Brookings, South Dakota

Deborah Q. Lewis
Iowa State University
Ames, Iowa

Ronald L. McGregor
University of Kansas
Lawrence, Kansas

Lawrence R. Stritch
Midewin National Tallgrass Prairie
Wilmington, Illinois

Connie Taylor
Southeastern Oklahoma State
  University
Durant, Oklahoma

George A. Yatskievych
Missouri Botanical Garden
St. Louis, Missouri

SOUTH CENTRAL UNITED STATES

David E. Lemke
Southwest Texas State University
San Marcos, Texas

Jackie M. Poole
Texas Parks & Wildlife Department
Austin, Texas

Robert C. Sivinski
New Mexico Energy, Mineral, and
*   Natural Resources Department*
Santa Fe, New Mexico

EASTERN CANADA

Jacques Cayouette
Centre for Land & Biological
*   Resources Research*
Ottawa, Ontario

William J. Crins
Ontario Ministry of Natural Resources
Bracebridge, Ontario

Harold Hinds
University of New Brunswick
Fredrickton, New Brunswick

John K. Morton
University of Waterloo
Waterloo, Ontario

Anton A. Reznicek
University of Michigan
Ann Arbor, Michigan

GREENLAND

Bent Fredskild
University of Copenhagen
Copenhagen, Denmark

Geoffrey Halliday
University of Lancaster
Lancaster, England

NORTHEASTERN UNITED STATES

Ray Angelo
New England Botanical Club
Cambridge, Massachusetts

Steven E. Clemants
Brooklyn Botanic Garden
Brooklyn, New York

Tom S. Cooperrider
Kent State University
Kent, Ohio

Les Mehrhoff
University of Connecticut
Storrs, Connecticut

Edward G. Voss
University of Michigan
Ann Arbor, Michigan

SOUTHEASTERN UNITED STATES

Sidney McDaniel
Institute for Botanical Exploration
Mississippi State, Mississippi

R. Dale Thomas
Northeast Louisiana University
Monroe, Louisiana

Lowell E. Urbatsch
Louisiana State University
Baton Rouge, Louisiana

Alan S. Weakley
The Nature Conservancy
Raleigh, North Carolina

Thomas F. Wieboldt
Virginia State University
Blacksburg, Virginia

B. Eugene Wofford
University of Tennessee
Knoxville, Tennessee

FLORIDA

Walter S. Judd
University of Florida
Gainesville, Florida

Richard P. Wunderlin
University of South Florida
Tampa, Florida

Contributors

Tim A. Atkinson
Burlington, North Carolina

William T. Barker
*Department of Animal & Range
  Sciences
North Dakota State University
Fargo, North Dakota*

Kerry A. Barringer
*Brooklyn Botanic Garden
Brooklyn, New York*

A. Linn Bogle
*Plant Biology Department
University of New Hampshire
Durham, New Hampshire*

Allan J. Bornstein
*Department of Biology
Southeast Missouri State University
Cape Girardeau, Missouri*

David E. Boufford
*Harvard University Herbaria
Cambridge, Massachusetts*

D. E. Brink
Chugiak, Alaska

George F. Buddell II
*Herbarium
Department of Biological Science
Northern Kentucky University
Highland Heights, Kentucky*

Curtis Clark
*Biological Sciences Department
California State Polytechnic University
Pomona, California*

Bryan E. Dutton
*Brooklyn Botanic Garden
Brooklyn, New York*

Frederick B. Essig
*Department of Biology
University of South Florida
Tampa, Florida*

Dennis Festerling Jr.
Grand Rapids, Michigan

Bruce A. Ford
*Herbarium
University of Manitoba
Winnipeg, Manitoba*

John J. Furlow
*Herbarium
Ohio State University
Columbus, Ohio*

Lisa O'Rourke George
Castle Rock, Colorado

Gary L. Hannan
*Department of Biology
Eastern Michigan University
Ypsilanti, Michigan*

C. Barre Hellquist
*Department of Biology
North Adams State College
North Adams, Massachusetts*

Richard J. Jensen
*Department of Biology
Saint Mary's College
Notre Dame, Indiana*

George P. Johnson
*Department of Biological Sciences
Arkansas Tech University
Russellville, Arkansas*

John T. Kartesz
*Department of Biology
University of North Carolina
Chapel Hill, North Carolina*

Robert B. Kaul
*School of Biological Sciences
University of Nebraska
Lincoln, Nebraska*

Carl S. Keener
*Department of Biology
Pennsylvania State University
University Park, Pennsylvania*

Robert W. Kiger
*Hunt Institute for Botanical
  Documentation
Carnegie Mellon University
Pittsburgh, Pennsylvania*

Robert Kral
*Herbarium, General Biology
  Department
Vanderbilt University
Nashville, Tennessee*

Donald H. Les
*Department of Ecology &
  Evolutionary Biology
University of Connecticut
Storrs, Connecticut*

Henry Loconte
*Wavering Place Gardens & Nursery
Eastover, South Carolina*

Paul S. Manos
*Department of Botany
Duke University
Durham, North Carolina*

Frederick G. Meyer
Takoma Park, Maryland

Susan Meyer
*Shrub Sciences Laboratory
U.S.D.A. Forest Service
Provo, Utah*

Nancy P. Moreno
Department of Biology
Wiess School of Natural Sciences
Houston, Texas

C. H. Muller
Department of Biological Sciences
University of California
Santa Barbara, California

David F. Murray
University of Alaska Museum
Fairbanks, Alaska

Kevin C. Nixon
L. H. Bailey Hortorium
Cornell University
Ithaca, New York

Gerald B. Ownbey
Herbarium
University of Minnesota
St. Paul, Minnesota

Bruce D. Parfitt
Department of Biology
University of Michigan
Flint, Michigan

Marilyn M. Park
Wyoming, Michigan

James S. Pringle
Royal Botanical Gardens
Hamilton, Ontario

Gwynn W. Ramsey
Herbarium
Biology Department
Lynchburg College
Lynchburg, Virginia

Donald G. Rhodes
Herbarium
Department of Biological Sciences

Louisiana Tech University
Ruston, Louisiana

Susan Sherman-Broyles
State University of New York City
Cortland, New York

Leila M. Shultz
Harvard University Herbaria
Cambridge, Massachusetts

Ernest Small
Biosystematics Research Institute
Central Experimental Farm
Ottawa, Ontario

Daniel D. Spaulding
Wellington, Alabama

Kingsley R. Stern
Chico Herbarium
Department of Biological Sciences
California State University
Chico, California

Donald E. Stone
Organization for Tropical Studies
North American Office
Durham, North Carolina

John W. Thieret
Northern Kentucky University
Department of Biological Science
Highland Heights, Kentucky

Henk van der Werff
Herbarium
Missouri Botanical Garden
St. Louis, Missouri

Michael A. Vincent
Department of Botany
Miami University
Oxford, Ohio

Michael J. Warnock
Department of Life Sciences
Sam Houston State University
Huntsville, Texas

R. David Whetstone
Department of Biology
Jacksonville State University
Jacksonville, Alabama

Alan T. Whittemore
Missouri Botanical Garden
Flora of North America Project
St. Louis, Missouri

John H. Wiersema
U.S.D.A./Systematic Botany &
Mycology
BARC-West
Beltsville, Maryland

Karen L. Wilson
Royal Botanic Gardens Sydney
Sydney, Australia

Thomas K. Wilson
Department of Botany
Miami University
Oxford, Ohio

B. Eugene Wofford
Herbarium, Department of Botany
University of Tennessee
Knoxville, Tennessee

J. A. Woods
Chugiak, Alaska

Richard P. Wunderlin
Herbarium
Biology Department
University of South Florida
Tampa, Florida

Introduction

Nancy R. Morin, *Convening Editor*

Scope of the Work

Flora of North America North of Mexico is a synoptic floristic account of the plants of North America north of Mexico: the continental United States of America (including the Florida Keys and Aleutian Islands), Canada, Greenland (Kalâtdlit-Nunât), and St. Pierre and Miquelon. The flora is intended to serve both as a means of identifying plants within the region and as a systematic conspectus of the North American flora. Taxa and geographical areas in need of further study also are identified in the flora.

Flora of North America North of Mexico will be published in thirty (30) volumes. Volume 1 contains background information that is useful for understanding patterns in the flora. Volume 2 contains treatments of ferns and gymnosperms. Families in volumes 3–26, the angiosperms, are arranged according to the classification system of A. Cronquist (1981). Bryophytes will be covered in volumes 27–29. Volume 30 will contain the cumulative bibliography and index.

The first two volumes were published in 1993; second impressions of both volumes, including minor emendations, were released shortly thereafter. The bibliographic citation for the flora is: Flora of North America Editorial Committee, eds. 1993 +. *Flora of North America North of Mexico*. 3 + vols. New York and Oxford.

Volume 3 contains treatments of 32 families, 128 genera, and 741 species. For additional statistics, please refer to Table 1.

Contents · General

The published flora includes accepted names, literature citations, selected synonyms, identification keys, summaries of habitats and geographic ranges, descriptions, chromosome numbers, phenological information, and other biological observations. Economic uses, weed status, and conservation status are provided from specified sources. Each volume contains a bibliography and an index to the taxa included in the volume. A comprehensive, consolidated bibliography and comprehensive index will be published in the last volume. The treatments, written and reviewed by experts from throughout the systematic botanical community, are based on original observations of herbarium specimens and, whenever possible, on living plants. These observations are supplemented by critical reviews of the literature.

Table 1. *Statistics for Volume 3 of Flora of North America.*

Family	Total Genera	Total Species	Endemic Genera	Endemic Species	Introduced Genera	Introduced Species	Species and Infraspecies of Conservation Concern
Magnoliaceae	2	9	0	9	0	0	1
Annonaceae	3	12	2	10	0	1	3
Canellaceae	1	1	0	0	0	0	0
Calycanthaceae	1	2	0	2	0	0	0
Lauraceae	9	13	0	7	1	1	3
Saururaceae	2	2	0	1	0	0	0
Piperaceae	2	9	0	0	1	4	0
Aristolochiaceae	3	28	1	22	0	2	4
Illiciaceae	1	2	0	1	0	0	1
Schisandraceae	1	1	0	1	0	0	0
Nelumbonaceae	1	2	0	0	0	1	0
Nymphaeaceae	2	17	0	8	0	2	2
Cabombaceae	2	2	0	0	0	0	0
Ceratophyllaceae	1	3	0	1	0	0	0
Ranunculaceae	22	285	1	184	5	30	50
Berberidaceae	8	33	1	23	1	5	5
Lardizabalaceae	1	1	0	0	1	1	0
Menispermaceae	4	5	1	2	0	0	0
Papaveraceae	17	63	5	28	4	12	9
Fumariaceae	4	23	0	17	1	3	3
Platanaceae	1	3	0	0	0	0	0
Hamamelidaceae	3	5	1	3	0	0	0
Ulmaceae	4	19	1	7	0	3	1
Cannabaceae	2	3	0	0	0	2	0
Moraceae	7	18	1	2	4	13	0
Urticaceae	8	21	0	2	2	7	0
Leitneriaceae	1	1	1	1	0	0	1
Juglandaceae	2	17	0	12	0	0	2
Myricaceae	2	8	1	6	0	0	0
Fagaceae	5	97	1	62	0	1	11
Betulaceae	5	33	0	25	0	1	3
Casuarinaceae	1	3	0	0	1	3	0
Totals	128	741	17	436	21	92	99

Basic Concepts

Our goal has been and continues to be to make the flora as clear, concise, and informative as practicable so that it can be an important resource for both botanists and nonbotanists. To this end, we are attempting to be consistent in style and content from the first volume to the last. Readers may assume that a term has the same meaning each time it appears and that, within groups, descriptions may be compared directly with one another. Any departures from consistent usage will be explicitly noted in the treatments (see also "References" below).

Treatments are intended to reflect current knowledge of taxa throughout their ranges worldwide, and classifications are therefore based on all available evidence. Where notable differences of opinion about the classification of a group occur, appropriate references are mentioned in the discussion of the group.

Documentation and arguments supporting significantly revised classifications are published separately in botanical journals before publication of the pertinent volume of the flora. Similarly, all new names, names for new taxa, and new combinations are published prior to their use in the flora. No nomenclatural innovations will be published intentionally in the flora. Journals and series in which papers relevant to *Flora of North America North of Mexico* have been or may be published include *Brittonia, Canadian Journal of Botany, North American Flora, Novon, Systematic Botany, Systematic Botany Monographs,* and *Taxon,* among others.

Contents of Treatments

Taxa treated in full include native species, native species thought to be recently extinct, hybrids that are well established (or frequent), and waifs or cultivated plants that are found frequently outside cultivation and give the appearance of being naturalized. Taxa mentioned only in discussions include waifs or naturalized plants now known only from isolated old records and some nonnative, economically important, or extensively cultivated plants, particularly when they are relatives of native species. Excluded names and taxa are listed at the ends of appropriate sections, e.g., species at the end of genus and genera at the end of family.

Treatments are intended to be succinct and diagnostic but adequately descriptive. Characters and character states used in the keys are repeated in the descriptions. Descriptions of related taxa at the same rank are directly comparable.

With few exceptions, taxa are presented in taxonomic sequence. If an author is unable to produce a classification, the taxa are arranged alphabetically, and the reasons are given in the discussion.

Treatments of hybrids follow that of one of the putative parents. Hybrid complexes are treated at the ends of their genera, after the descriptions of species.

We have attempted to keep terminology as simple as accuracy permits. Common English equivalents have been used in place of Latin or Latinized terms or other specialized terminology whenever the correct meaning could be conveyed in approximately the same space, e.g., "pitted" rather than "foveolate," but "striate" rather than "with fine longitudinal lines." Specialized terms that are used are defined in the generic or family descriptions and, in some cases, are illustrated.

References

Authoritative general reference works used for style are *The Chicago Manual of Style, for Authors, Editors, and Copywriters,* ed. 13 (University of Chicago Press 1982); *Webster's New Geographical Dictionary* (Merriam-Webster 1988); and *The Random House Dictionary of the English Language,* ed. 2, unabridged (S. B. Flexner and L. C. Hauck 1987). *B-P-H/S. Botanico-Periodicum-Huntianum/Supplementum* (G. D. R. Bridson and E. R. Smith 1991) has been used for abbreviations of titles of serials; and *Taxonomic Literature,* ed. 2 (F. A. Stafleu and R. S. Cowan 1976–1988) and its supplements by F. A. Stafleu and E. A. Mennega (1992 +) have been used for abbreviations of titles of books.

Graphic Elements

All genera, and approximately one out of three species, are illustrated. Illustrated taxa are marked with an "F" (for figure) following the accepted name statement. The illustrations may be of typical or of unusual species, or they may show diagnostic traits or complex structures. Most illustrations have been drawn from herbarium specimens selected by the authors. In some cases living material or photographs have been used. Data on specimens that were used and parts that were illustrated have been recorded. This information, together with the archivally preserved original drawings, is deposited in the Missouri Botanical Garden Library and is available for scholarly study.

Specific Information in Treatments

Keys

A key to families of Magnoliophyta will be published separately. Keys are included in each volume for all ranks below families if two or more taxa are treated. For dioecious species, keys are designed for use with either staminate or pistillate plants. Keys are also designed to facilitate identification of taxa that flower before leaves appear. More than one key may be given, and for some groups tabular comparisons may be presented in addition to keys.

Nomenclatural Information

Basionyms, with author and literature citation, are given for accepted names. Synonyms in common use are listed in alphabetical order, without literature citations.

Common names in vernacular use are given in the appropriate language. In general, such names have not been created for use in the flora. Those preferred by governmental or conservation agencies are listed if known.

The last names of authors of taxonomic names have been spelled out. The conventions of *Authors of Plant Names* (R. K. Brummitt and C. E. Powell 1992) have been used as a guide for including first initials to discriminate individuals who share surnames.

If only one infraspecific taxon within a species occurs in the flora area, nomenclatural information (literature citation, basionym with literature citation, relevant synonyms) is given for the species, as is information on the number of infraspecific taxa in the species and their distribution worldwide, if known. A description and detailed distributional information are given only for the infraspecific taxon.

Descriptions

Character states common to all taxa are treated in the description of the taxon at the next higher rank. For example, if corolla color is yellow for all species treated within a genus, that character state is given in the generic description. Characters used in keys are repeated in the descriptions. Characteristics are given as they occur in plants from the flora area. Notable characteristics that occur in plants from outside the flora area are given in square brackets or are included in a brief discussion at the end of the description. In families with one genus and one or more species, the family description is given as usual, the genus description is condensed, and the species description is as usual.

In reading descriptions of vascular plants, the reader may assume, unless otherwise noted, that: the plant is green, photosynthetic, and reproductively mature; a woody plant is perennial; stems are erect; roots are fibrous; leaves are simple and petiolate. Because measurements and elevations are almost always approximate, modifiers such as "about," "circa," or " ± " are usually omitted.

Arrangements of elements within descriptions of taxa are from base to apex, proximal to distal, abaxial to adaxial. General features such as growth form, persistence, habit, and nutrition are given first. For a particular structure or organ system, description of parts follows the order: presence, number, position/insertion, arrangement, orientation, connation, adnation, coherence, adherence. Features of a whole organ follow the order: color, odor, symmetry, architecture, shape, dimensions (length, width, thickness, mass), texture, base, margin, peripheral region or sides, central area, apex, surface, vestiture, internal parts, exudates. Unless otherwise noted, dimensions are length × width. If only one dimension is given, it is length or height. All measurements are given in metric units. Measurements usually are based on dried specimens but these should not differ significantly from the measurements actually found in fresh or living material.

Chromosome numbers generally are given only if published, documented counts are available from North American material or from an adjacent region. No new counts are published intentionally in the flora. Chromosome counts from nonsporophyte tissue have been converted to the $2n$ form. A literature reference for each reported chromosome number is available in the Flora of North America database (see below). The base number ($x =$) is given for each genus. This represents the lowest known haploid count for the genus unless evidence is available that the base number differs.

Flowering time and often fruiting time are given by season, sometimes qualified by early, mid, or late or by months. Elevation generally is rounded to the nearest 100 m; elevations between 0 and 100 m are rounded to the nearest 10 m. Mean sea level is shown as 0 m, with the understanding that this is approximate. Elevation often is omitted from herbarium specimen labels, particularly for collections made where the topography is not remarkable, and therefore elevation is sometimes not known for a given taxon.

The term "introduced" is defined broadly to refer to plants that were released deliberately or accidentally into the flora and that now exist as wild plants in areas in which they were not recorded as native in the past. The distribution of nonnative plants is often poorly documented and may be ephemeral. The nature of introduced populations is discussed as far as understood.

If a taxon is globally rare or if its continued existence is threatened in some way, the words "of conservation concern" appear before the statements of elevation and geographic range, and a "C" is shown after the accepted name statement. Criteria for taxa of conservation concern are based on The Nature Conservancy's designations of global rank (G-rank), G1 and G2:

G1 Critically imperiled globally because of extreme rarity (5 or fewer occurrences or fewer than 1000 individuals or acres) or because of some factor(s) making it especially vulnerable to extinction.

G2 Imperiled globally because of rarity (5–20 occurrences or fewer than 3,000 individuals or acres) or because of some factor(s) making it very vulnerable to extinction throughout its range.

Taxa thought to have become extinct during the period of permanent European settlement, i.e., the past 500 years, are included in the flora. Treatments of such taxa have been reviewed by Larry Morse of The Nature Conservancy. Additional comments were provided by Ken Berg, Bureau of Land Management; Peggy Olwell, U.S. National Park Service; and Chris Topik, U.S.D.A. Forest Service.

Range maps are given for each species or infraspecific taxon. The maps are generalized and, in order to represent the probable range of a taxon, parts of states or povinces may be shaded even though documentation of occurrence there may be lacking. Occurrences in states or provinces listed in distribution statements are documented by specimens. We have assumed that details such as "northeastern Florida" are apparent on the map; consequently, directional qualifiers are not given in the list of territories, provinces, and states. Taxa that occur only in the flora area are indicated as endemic by an "E" after the accepted name statement. Authors are expected to have seen at least one specimen documenting each state record and have been urged to examine as many specimens as possible from throughout the range of each taxon. Additional information about distribution may be given in the discussion.

Distributions are stated in the following order: Greenland; St. Pierre and Miquelon; Canada (provinces and territories in alphabetic order); United States (states in alphabetic order); Mexico (11 northern states may be listed specifically, in alphabetic order); West Indies; Bermuda; Central America (Guatemala, Belize, Honduras, El Salvador, Nicaragua, Costa Rica, Panama); South America; Europe, or Eurasia; Asia (including Indonesia); Africa; Pacific Islands; Australia; Antarctic.

Discussion

The discussion section includes information on taxonomic problems and interesting biological phenomena. Statements of economic uses supplied and documented by the author(s), Native American medicinal plants based on D. E. Moerman (1986), and weed status determined in consultation with weed specialist Robert H. Callihan are given in order to make this information more easily available to users of the *Flora of North America North of Mexico*. Weediness is indicated by codes W1 and W2 at the end of the accepted name statement. (The codes are references to publications that provide basic authority for the designation of weediness: D. T. Patterson et al. 1989 [W1] and R. H. Callihan et al. 1995 [W2].) Authors may provide additional discussions on the various aspects. Toxicity, if known, also is mentioned in the discussion, and pertinent literature is cited. (Please see "Caution" below.)

Selected References

Major references used in preparation of a treatment or containing critical information about a taxon are cited after the discussion. These, and other works that are referred to briefly in the discussion or elsewhere, are included in the bibliography at the end of the volume and in the consolidated bibliography in the last volume of the flora.

The Database

Data contained in printed volumes of the *Flora of North America North of Mexico*, additional supporting data, authority files, more precise maps, and other useful information are available on CD-ROM and online at http://www.fna.org.

CAUTION

The Flora of North America Editorial Committee does not encourage, recommend, promote, or endorse any of the folk remedies, culinary practices, or various utilizations of any plant described within these volumes. Information about medicinal practices and/or ingestion of plants, or of any part or preparation thereof, has been included only for historical background and as a matter of interest. Under no circumstances should the information contained in these volumes be used in connection with medical treatment. Readers are strongly cautioned to remember that many plants in the flora are toxic or can cause unpleasant or adverse reactions if used or encountered carelessly.

Flora of North America

1. MAGNOLIACEAE Jussieu

· Magnolia Family

Frederick G. Meyer

Trees or shrubs, deciduous or evergreen, aromatic. **Pith** homogeneous or diaphragmed. **Leaves** alternate, simple, petiolate; stipules early or tardily deciduous, at first surrounding stem, adnate on adaxial side of petiole (free in *Magnolia grandiflora*), often ochreate, leaving persistent annular scar around node. **Leaf blade** pinnately veined, unlobed (or evenly 2–10-lobed in *Liriodendron*), margins entire. **Inflorescences** terminal, solitary flowers (often paired in *Magnolia ashei*), pedunculate; spathaceous bracts 2 (*Magnolia*) or 1 (*Liriodendron*). **Flowers:** perianth hypogynous, segments imbricate; tepals deciduous, 6–18, in 3 or more whorls of 3, ± similar or outer tepals sepaloid, inner tepals petaloid; stamens numerous, hypogynous, free, spirally arranged; filaments very short to 1/2 length of anthers; anthers introrse, latrorse, or extrorse, longitudinally dehiscent; connective with distal appendage; pistils numerous, superior, spirally arranged on elongate receptacle (torus), stalked or sessile, free or ± concrescent, 1-locular; placentation marginal, placenta 1; ovules 1–2; style 1, short and recurved (*Magnolia*) or large and winglike (*Liriodendron*); stigma 1, terminal or terminal decurrent (*Magnolia*) or recurved (*Liriodendron*). **Fruits** conelike syncarps consisting of aggregates of coalescent, woody follicles (follicetums, as in *Magnolia*) or apocarps consisting of aggregates of indehiscent samaras (samaracetums, as in *Liriodendron*). **Seeds** 1–2 per pistil, arillate, endosperm oily (*Magnolia*), or without aril, adherent to dry endocarp (*Liriodendron*).

Genera ca. 6(–12), species ca. 220 (2 genera, 9 species in the flora): mostly in Asia, the Pacific Islands, and the Western Hemisphere.

Magnoliaceae are pollinated by beetles.

Herbarium material of *Magnolia* is usually incomplete and inadequate for critical study. Collections should include material of the stipules, spathaceous bracts, a full complement of stamens, and all tepals to facilitate identification of *Magnolia* species.

SELECTED REFERENCES   Canright, J. E. 1960. The comparative morphology and relationships of the Magnoliaceae. III. Carpels. Amer. J. Bot. 47: 145–155.   Demuth, P. and F. S. Santamour Jr. 1978. Carotenoid flower pigments in *Liriodendron* and *Magnolia*. Bull. Torrey Bot. Club 105: 65–66.   Hardin, J. W. and K. A. Jones. 1989. Atlas of foliar surface features in woody plants, X. Magnoliaceae of the United States. Bull. Torrey Bot. Club 116: 164–173.   Nooteboom, J. P. 1985. Notes on Magnoliaceae. Blumea 31: 65–121.   Praglowski, J. 1974. Magnoliaceae Juss. Taxonomy by J. E. Dandy. World Pollen Spore Fl. 3: 1–48.   Sargent, C. S. 1890–1902. The Silva of North America. . . . 14 vols. Boston and New York. Vol. 1, pp. 1–20.   Spongberg,

S. A. 1976. Magnoliaceae hardy in temperate North America. J. Arnold Arbor. 57: 250–312.   Wood, C. E. Jr. 1958. The genera of the woody Ranales in the southeastern United States. J. Arnold Arbor. 39: 296–346.

1. Leaf blade entire, base deeply cordate or auriculate, or cuneate to abruptly narrowed or rounded, apex obtuse or acute to acuminate; stipules adnate on petiole or rarely free, early deciduous; tepals petaloid, usually spreading, creamy white, rarely greenish or yellow to orange-yellow, outermost tepals sepaloid, sometimes reflexed, greenish; anthers introrse or latrorse; follicles persistent, coalescent; seeds with brightly colored aril, extruded from follicles and suspended by funiculi . . . . .1. *Magnolia*, p. 4
1. Leaf blade evenly 2–10-lobed, base rounded to shallowly cordate or truncate, apex broadly truncate or notched; stipules free, erect, leafy, tardily deciduous; tepals petaloid, tip recurved, greenish yellow with feathered orange band near base, outermost tepals sepaloid, reflexed, green; anthers extrorse; samaras caducous, forming elongate spindle-shaped dry cone, indehiscent; seeds without aril, adherent to dry endocarp . . . . .2. *Liriodendron*, p. 10

1. MAGNOLIA Linnaeus, Sp. Pl. 1: 535. 1753; Gen. Pl. ed. 5, 240. 1754 · [For Pierre Magnol (1638–1715), professor and director of the botanical garden at Montpellier, France]

Kobus Nieuwland; *Tulipastrum* Spach

Trees or shrubs, deciduous or evergreen. **Pith** homogeneous or diaphragmed. **Leaves** distinctly alternate or sometimes crowded in terminal whorl-like clusters; stipules early deciduous, free or adnate to and proximal on petiole. **Leaf blade:** base deeply cordate or auriculate or cuneate to abruptly narrowed or rounded, margins entire, apex obtuse or acute to acuminate; surfaces abaxially chalky white or green to glaucous, pubescent or glabrous. **Flowers** protogynous, appearing with or before leaves; tepals 9–15, petaloid, usually spreading, creamy white, rarely greenish, yellow, or orange-yellow, outermost tepals sepaloid, sometimes strongly reflexed, greenish; stamens on elongate torus, early deciduous; filaments white or purple, very short; anthers introrse or latrorse. **Follicles** persistent, coalescent, forming conelike aggregate, abaxially dehiscent. **Seeds** with red, pink, or orange oily aril, extruded from follicles and suspended by funiculi. $x = 19$.

Species ca. 120 (8 in the flora): temperate and tropical regions, Western Hemisphere, Asia (Himalayas, China, Japan, Taiwan, Malaysia, and Indonesia).

All species of *Magnolia* in the flora are cultivated. Many of the Asiatic taxa are also grown in the flora area. Numerous hybrids and cultivars from those taxa have been introduced to horticulture.

SELECTED REFERENCES Callaway, D. J. 1994. The World of Magnolias. Portland.   Coker, W. C. 1943. *Magnolia cordata* Michaux. J. Elisha Mitchell Sci. Soc. 59: 81–88.   Heiser, C. B. 1962. Some observations on pollination and compatibility in *Magnolia*. Proc. Indiana Acad. Sci. 72: 259–266.   Johnson, D. L. 1989. Species and Cultivars of the Genus *Magnolia* (Magnoliaceae) Cultivated in the United States. M.S. thesis. Cornell University.   Johnson, D. L. 1989b. Nomenclatural changes in *Magnolia*. Baileya 23: 55–56.   McDaniel, J. C. 1966. Variations in the sweet bay magnolias. Morris Arbor. Bull. 17: 7–12.   Millais, J. G. 1927. Magnolias. London.   Rockwell, H. C. 1966. The Genus *Magnolia* in the United States. M.S. thesis. West Virginia University.   Santamour, F. S. Jr. 1969b. Cytology of *Magnolia* hybrids. I. Morris Arbor. Bull. 20: 63–65.   Thien, L. B. 1974. Floral biology of *Magnolia*. Amer. J. Bot. 61: 1037–1045.   Thien, L. B., W. H. Heimermann, and R. T. Holman. 1975. Floral odors and quantitative taxonomy of *Magnolia* and *Liriodendron*. Taxon 24: 557–568.   Tobe, J. D. 1993. A Molecular Systematic Study of Eastern North American Species of *Magnolia* L. Ph.D. thesis. Clemson University.   Treseder, N. G. 1978. Magnolias. Boston.   Vázquez-G., J. A. 1990. Taxonomy of the Genus *Magnolia* in Mexico and Central America. M.S. thesis. University of Wisconsin.   Vázquez-G., J. A. 1994. *Magnolia* (Magnoliaceae) in Mexico and Central America: A synopsis. Brittonia 46: 1–23.

1. Leaf blade deeply cordate to auriculate at base.
  2. Leaf blade abaxially chalky white, sometimes pale green to glaucous, pubescent; foliar buds, twigs, and follicles pubescent.
    3. Trees to 32 m; flowers solitary; follicetums globose-ovoid, 5–8 × 5–7 cm; leaf blade 50–110 cm; stamens (300–)350–580; pistils 50–80. . . . . 1. *Magnolia macrophylla*
    3. Trees to 12 m; flowers solitary or often paired; follicetums cylindric to nearly ovoid, 2.5–6.5 × 1.5–4 cm; leaf blade 17–56 cm; stamens 170–350; pistils 20–50 . . . . 2. *Magnolia ashei*
  2. Leaf blade abaxially green or glaucous, glabrous; foliar buds, twigs, and follicles glabrous.
    4. Leaf blade rhombic-obovate to obovate-spatulate or oblanceolate, usually more than 25 cm, gradually tapered from broadest part to base; stamens 100–200, 8–14 mm; follicetums 5.5–10 × 2.5–5 cm . . . . 3. *Magnolia fraseri*
    4. Leaf blade predominantly pandurate to broadly rhombic-spatulate, usually less than 25 cm, abruptly tapered from broadest part to base; stamens 83–137(–150), 4.5–8(–10.5) mm; follicetums 4–6 × 2.5–3.5 cm . . . . 4. *Magnolia pyramidata*
1. Leaf blade cuneate to abruptly narrowed or rounded at base.
  5. Leaf blade abaxially chalky white to glaucous, glabrous or silky-pubescent, somewhat leathery. . 6. *Magnolia virginiana*
  5. Leaf blade abaxially green to red-brown, glabrous, felted, or pilose, thin- to thick-leathery or not leathery.
    6. Trees evergreen; leaf blade thick-leathery, abaxially glabrous or red-brown felted, adaxially lustrous green; flowers 15–30(–45) cm across, strongly lemony fragrant. . . . . 7. *Magnolia grandiflora*
    6. Trees deciduous; leaf blade thin, not leathery, abaxially pale green to whitish, glabrous or pilose, adaxially green; flowers 5.5–11 cm across, malodorous or aromatic.
      7. Leaves usually in terminal whorl-like clusters; leaf blade cuneate to long-tapered at base; flowers malodorous, 5.5–11 cm across; tepals creamy white, 8–12 cm, spreading, outermost tepals reflexed, greenish; follicetums 6–10 × 2–3.5 cm; follicles long-beaked. . . . . 5. *Magnolia tripetala*
      7. Leaves not in whorl-like clusters; leaf blade cuneate to truncate or rounded, slightly oblique at base; flowers aromatic, 6–9 cm across; tepals greenish, glaucous, occasionally yellow to orange-yellow, usually less than 8 cm, erect, outermost tepals reflexed, greenish; follicetums 2–7 × 0.8–2.7 cm; follicles short-beaked . . . . 8. *Magnolia acuminata*

1. **Magnolia macrophylla** Michaux, Fl. Bor.-Amer. 1: 327. 1803 · Bigleaf magnolia [E] [F]

Magnolia michauxiana de Candolle

Trees, deciduous, single-trunked, to 15(–32) m. **Bark** yellowish to gray, smooth. **Pith** homogeneous. **Twigs and foliar buds** silky-pubescent. **Leaves** crowded in terminal whorl-like clusters; stipules (9–)11–17 × 3–6.5 cm, abaxially pilose, glandular. **Leaf blade** broadly elliptic to obovate-oblong, 50–110 × 15–30 cm, base truncate to deeply cordate or auriculate, apex acute to short-acuminate or obtuse; surfaces abaxially chalky white, sometimes pale green to glaucous, pilose, adaxially deep green, glabrous. **Flowers** solitary, fragrant, 35–40(–50) cm across; spathaceous bracts 2, outer bract abaxially rusty gray, inner bract thinner, glabrous; tepals creamy white, glandular, innermost whorl purple-blotched at base, outermost segments strongly reflexed, greenish; stamens (300–)350–580, 12.5–24.5 mm; filaments white; pistils 50–80. **Follicetums** globose-ovoid, 5–8 × 5–7 cm; follicles short-beaked, distally appressed silky-pubescent. **Seeds** ± ovoid, 10–12 mm, pointed, aril orange-red. $2n = 38$.

Flowering spring. Alluvial woods and sheltered valleys, piedmont; 150–300 m; Ala., Ark., Ga., Ky., La., Miss., N.C., Ohio, Tenn., Va.

The disposition of *Magnolia macrophylla* and its close relative *M. ashei* has been perplexing since *M. ashei* was described. Some investigators have treated them as geographic varieties or subspecies, and this has some questionable merit. In the foliar state *M. macrophylla* is hardly, if at all, distinguishable from *M. ashei,* but in other morphologic details of flower and fruit, they are readily distinguished. They also differ in the floral odors, which are distinct and chemically different (L. B. Thien et al. 1975). *Magnolia macrophylla* and *M. ashei* are allopatric. *Magnolia macrophylla* is a much larger, usually single-trunked tree of the piedmont with a wider distribution, larger leaves, more stamens, larger stipules, and both filiform and flagelliform trichomes on the leaves. The follicetum is nearly globose-ovoid, with more pistils and larger seeds. *Magnolia macrophylla* produces the largest leaves and flowers of any species of the genus.

In Arkansas *Magnolia macrophylla* was known from a single disjunct locality in Clay County, where only two trees were recorded in 1981 (R. B. Figlar 1981). A survey in 1995 failed to locate the species in the same site.

This handsome tree is occasionally cultivated. A close relative, *M. dealbata* Zuccarini, occurs in Mexico.

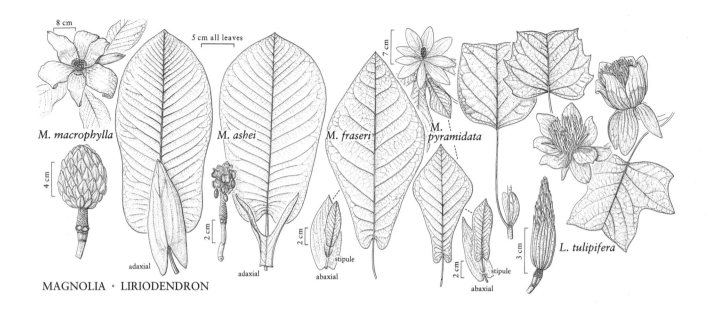

MAGNOLIA · LIRIODENDRON

The largest known tree of *Magnolia macrophylla*, 32 m in height with a trunk diameter of 53 cm, is recorded from Daniel Boone National Forest, Tight Hollow, Kentucky (American Forestry Association 1994).

The Cherokee tribe used *Magnolia macrophylla*, mainly the bark, as an analgesic, antidiarrheal, gastrointestinal aid, respiratory aid, and toothache remedy (D. E. Moerman 1986).

2. Magnolia ashei Weatherby, Rhodora 28: 35. 1926   · Ashe's magnolia   C E F

Magnolia macrophylla Michaux subsp. *ashei* (Weatherby) Spongberg; *M. macrophylla* var. *ashei* (Weatherby) D. L. Johnson

Trees, deciduous, often multitrunked, to 10(–12) m. **Bark** dark gray, smooth. **Pith** homogeneous. **Twigs and foliar buds** silky-pubescent. **Leaves** crowded in terminal whorl-like clusters; stipules 6–8 × 3–4 cm, abaxially pilose, glandular. **Leaf blade** broadly elliptic to obovate-oblong, (17–)25–46(–56) × (10–)15–30(–40) cm, base truncate to deeply cordate or auriculate, apex acute to short-acuminate; surfaces abaxially chalky white or pale green to glaucous, pilose, adaxially deep green, glabrous. **Flowers** solitary or often in pairs on adjacent twigs, fragrant, 15–38(–50) cm across; spathaceous bracts 2, outer bract abaxially rusty gray, inner bract glabrous; tepals creamy white, glandular, innermost whorl purple-blotched at base, outermost segments strongly reflexed, greenish; stamens 170–350, 13–20 mm; filaments white; pistils 20–50. **Follicetums** cylindric to nearly ovoid, 2.5–6.5 ×

1.5–4 cm; follicles short-beaked, distally appressed silky-pubescent. **Seeds** lenticular to somewhat globose, 8–10 mm, aril orange-red. $2n = 38$.

Flowering spring. Woodlands, ravines, and bluffs; coastal plain; of conservation concern; 0–50 m; Fla.

Magnolia ashei, the rarest species of *Magnolia* in the flora, is limited to six counties in the Florida panhandle; it is in danger of extirpation because of habitat disturbance. *Magnolia ashei* differs from *M. macrophylla* in being a smaller, often multitrunked tree with smaller leaves, fewer stamens and pistils, smaller seeds, smaller stipules, filiform trichomes, and smaller, nearly glabrous, cylindric follicetums. The flowers are often borne in pairs. *Magnolia ashei* flowers at an early age (three to four years from seed); it is a desirable small tree in cultivation.

3. Magnolia fraseri Walter, Fl. Carol., 159. 1788   · Mountain magnolia   E F

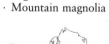

Magnolia auricularis Salisbury; *M. auriculata* W. Bartram

Trees, deciduous, single-trunked, to 25(–32.6) m. **Bark** gray to brownish, smooth. **Pith** homogeneous. **Twigs and foliar buds** glabrous. **Leaves** crowded in terminal whorl-like clusters; stipules (5.4–)7.6–9.5(–10) × 2–6 cm, abaxially glandular. **Leaf blade** rhombic-obovate to obovate-spatulate or oblanceolate, broadest near middle, gradually tapering to base, 20–30(–60) × 8–16(–27) cm, base deeply cordate or auriculate to somewhat truncate, apex obtuse to acute or somewhat acuminate; surfaces abaxially strongly glaucous, glabrous, adaxially deep green, glabrous. **Flowers** fragrant, 16–22 cm across; spathaceous bracts 2, abaxially glandular; tepals creamy white, the outermost greenish; stamens 100–200, 8–14 mm; filaments white; pistils 50–90. **Follicetums** ellipsoid, 5.5–10 × 2.5–5 cm, glabrous; follicles recurved, long-beaked, glabrous. **Seeds** lenticular, 7–10 mm, aril red. 2*n* = 38.

Flowering spring. Rich woods and coves; 300–1520 m; Ga., Ky., N.C., S.C., Tenn., Va., W.Va.

Magnolia fraseri, a tree of the Appalachian Mountains and upper piedmont, and its close congener *M. pyramidata*, of the lower piedmont and coastal plain, are sometimes confused taxonomically, but they differ in a series of good characters. Some investigators have treated them as geographic varieties or subspecies of *M. fraseri*. *Magnolia fraseri* differs from *M. pyramidata* in being a much larger tree with wide-speading branches, different leaf shape, larger flowers, more numerous stamens, and larger follicetums. In their present geographic distribution, these taxa are allopatric. In some localities *M. fraseri* may be abundant, as in the area near Sugar Grove, Smythe County, Virginia, where hundreds of specimens occur. It is a desirable flowering tree, occasionally cultivated.

The largest known tree of *Magnolia fraseri*, 32.6 m in height with a trunk diameter of 94 cm, is recorded from Great Smoky Mountains National Park, Tennessee (American Forestry Association 1994).

4. **Magnolia pyramidata** W. Bartram, Travels Carolina, 408. 1791 · Pyramid magnolia [E] [F]

Magnolia auriculata W. Bartram var. *pyramidata* (W. Bartram) Nuttall; *M. fraseri* Walter subsp. *pyramidata* (W. Bartram) E. Murray; *M. fraseri* var. *pyramidata* (W. Bartram) Pampanini; *M. macrophylla* Michaux var. *pyramidata* (W. Bartram) Nuttall

Trees, deciduous, single-trunked, to 11.9 m. **Bark** gray, smooth. **Pith** homogeneous. **Twigs and foliar buds** glabrous. **Leaves** crowded in terminal whorl-like clusters; stipules 5.4–7 × 2.5–3 cm, abaxially glandular. **Leaf blade** predominantly pandurate to broadly rhombic-spatulate, broadest above middle, abruptly tapering to base, 18–25(–30) × 7.8–14 cm, base deeply cordate to auriculate, or somewhat truncate, apex acute to short-acuminate; surfaces abaxially glaucous, glabrous, adaxially dull deep green. **Flowers** fragrant, 12–18 cm across; spathaceous bracts 2, abaxially glandular; tepals creamy white; stamens 83–137 (–150), 4.5–8(–10.5) mm; filaments white; pistils 36–60. **Follicetums** ellipsoid, 4–6 × 2.5–3.5 cm, glabrous; follicles recurved, short-beaked, glabrous. **Seeds** lenticular, 7–8 mm, aril red. 2*n* = 38.

Flowering spring. Rich woods and river bluffs, mostly coastal plain, sometimes lower piedmont; 0–120 m; Ala., Fla., Ga., La., Miss., S.C., Tex.

Confined largely to the coastal plain, *Magnolia pyramidata* differs from the allopatric *M. fraseri* in being a smaller tree with a narrower, pyramidal habit; *M. pyramidata* is very local and nowhere abundant. Morphologically, *M. pyramidata* differs from *M. fraseri* in the pandurate leaf blades, smaller flowers and stipules, fewer stamens and pistils, and smaller follicetums. *Magnolia pyramidata* is occasionally cultivated, but it is less hardy than *M. fraseri*.

The largest known tree of *Magnolia pyramidata*, 11.9 m in height with a trunk diameter of 69 cm, is recorded from Newton County, Texas (American Forestry Association 1994).

5. **Magnolia tripetala** (Linnaeus) Linnaeus, Syst. Nat. ed. 10, 2: 1082. 1759 · Umbrella-tree [E]

Magnolia virginiana Linnaeus var. (δ) *tripetala* Linnaeus, Sp. Pl. 1: 536. 1753; *Kobus tripetala* (Linnaeus) P. Parmentier; *Magnolia frondosa* Salisbury; *M. umbrella* Desrousseaux; *M. umbrella* var. *tripetala* (Linnaeus) P. Parmentier

Trees, deciduous, often multi-trunked, to 15 m. **Bark** gray, smooth. **Pith** homogeneous. **Twigs and foliar buds** glabrous. **Leaves** crowded in terminal whorl-like clusters; stipules (4–)6.6–9.4(–10) × 2.6–3.7 cm, abaxially red glandular, sparsely pilose. **Leaf blade** elliptic-oblong to narrowly obovate, or oblanceolate, (10–)26–57(–70) × (7.2–)10–30 cm, thin, broadest near middle, base cuneate to long-tapered, apex very short to long-acuminate or abruptly acute, rarely apiculate; surfaces abaxially densely pilose, especially on midvein, adaxially green, glabrous. **Flowers** malodorous, 5.5–11 cm across; spathaceous bracts 2, abaxially glandular; tepals spreading, creamy white, outermost whorl sepaloid, reflexed, greenish; stamens 81–103(–115), 8–17 mm; filaments purple; pistils (45–)53–66(–73). **Follicetums** cylindric to ovoid-cylindric, 6–10 × 2–3.5 cm; follicles long-beaked, glabrous. **Seeds** lenticular to nearly ovoid, 9–12 mm, aril deep pink to red. 2*n* = 38.

Flowering spring. Rich woods and ravines, mainly in uplands, rarely coastal plain; 0–1065 m; Ala., Ark., Fla., Ga., Ind., Ky., Md., Miss., N.C., Ohio, Okla., Pa., S.C., Tenn., Va., W.Va.

In Virginia *Magnolia tripetala* is a disjunct in the coastal plain.

The malodorous flowers of *Magnolia tripetala* are uniquely associated with this species. The tree is occasionally cultivated. Both filiform and flagelliform trichomes occur on the leaves. Sometimes cylindric trichomes also occur.

The largest known tree of *Magnolia tripetala*, 15.2 m in height with a trunk diameter of 87 cm, is recorded from Bucks County, Pennsylvania (American Forestry Association 1994).

6. **Magnolia virginiana** Linnaeus, Sp. Pl. 1: 535. 1753 · Sweet-bay, swamp-bay, laurier doux [E]

Magnolia australis Ashe; *M. australis* var. *parva* (Ashe) Ashe; *M. fragrans* Rafinesque 1817, not Salisbury 1796; *M. glauca* (Linnaeus) Linnaeus; *M. glauca* var. *pumila* Nuttall; *M. virginiana* subsp. *australis* (Sargent) E. Murray; *M. virginiana* var. *australis* Sargent; *M. virginiana* var. *glauca* Linnaeus; *M. virginiana* var. *grisea* Linnaeus; *M. virginiana* var. *parva* Ashe

Large shrubs or trees, evergreen to partly evergreen or deciduous, multitrunked to 10 m, or single-trunked to 28 m. **Bark** dark gray, smooth. **Pith** diaphragmed. **Twigs and foliar buds** silky-pubescent, sometimes glabrous. **Leaves** distinctly alternate, not in terminal whorl-like clusters; stipules 3–5.5(–6) × 0.3–0.5 cm, abaxially brownish puberulent, red-glandular. **Leaf blade** oblong to elliptic, ovate to obovate, 6–22 × 2.6–7 cm, somewhat leathery, base cuneate, apex obtuse to acute or rounded to somewhat acuminate; surfaces abaxially chalky white to glaucous, glabrous or densely silky-pubescent, adaxially dull green to lustrous. **Flowers** fragrant, 5–8 cm across; spathaceous bracts 2, outer bract abaxially silky-pubescent, inner bract nearly glabrous, red-glandular; tepals creamy white, red-glandular, outermost segments reflexed, greenish; stamens (32–)63–90(–102), 5.5–11 mm; filaments white; pistils (9–)19–33(–50). **Follicetums** ellipsoid to nearly globose, 2–5.5 × 1.5–3 cm; follicles short-beaked, glabrous. **Seeds** somewhat globose to lenticular, 5 mm, aril red. $2n = 38$.

Flowering spring. Swamps, bays, low wet woods, savannas; chiefly in coastal plain and lower piedmont; 0–540 m; Ala., Ark., Del., D.C., Fla., Ga., La., Md., Mass., Miss., N.J., N.Y., N.C., Pa., S.C., Tenn., Tex., Va.

The most widely distributed species of *Magnolia* in the flora, *M. virginiana* occurs in two growth forms: deciduous and multitrunked northward, and typically single-trunked and evergreen in the southern range. Where these forms overlap geographically in North Carolina and adjacent areas, intergradation occurs, and the identification of these intergrades is difficult, if not impossible. It has been impossible to pinpoint their occurrence in the zone of overlap. Herbarium specimens cannot be trusted to resolve this dilemma. Both filiform and flagelliform trichomes occur on the leaves, but these are without taxonomic significance. Some in-vestigators have treated these habital variants as geographic varieties or as subspecies, but infraspecific taxa are not recognized here. Without other defining characteristics and no clear geographic correlation, infraspecific taxa have little significance or taxonomic value in *M. virginiana*. A thorough field study is needed to clarify the taxonomy of this otherwise well-known plant.

Magnolia virginiana is widely cultivated. It was the first magnolia known in Europe, dating from 1688 in England. A few cultivars of both the deciduous and evergreen forms are now grown in cultivation. *Magnolia virginiana* is a parent of several hybrids, including the first known magnolia hybrid, *M.* ×*thompsoniana* (Loudon) C. de Vos (= *M. virginiana* × *M. tripetala*), dating to 1808. Other hybrids include the so-called Freeman hybrids of *M. grandiflora* × *M. virginiana* and *M. virginiana* × *M. hypoleuca* with its cultivar 'Nimbus'.

The largest known tree of *Magnolia virginiana* (the evergreen southern form), 28 m in height with a trunk diameter of 1.4 m, is recorded from Union County, Arkansas (American Forestry Association 1994).

The Houma and Rappahannock tribes used decoctions of leaves, twigs, and bark of *Magnolia virginiana* to treat colds and chills, to warm the blood, and as a hallucinogen (D. E. Moerman 1986).

7. **Magnolia grandiflora** Linnaeus, Syst. Nat. ed. 10, 2: 1082. 1759 · Southern magnolia, bull-bay, laurier tulipier [E]

Magnolia ferruginea Z. Collins ex Rafinesque; *M. foetida* (Linnaeus) Sargent; *M. lacunosa* Rafinesque; *M. virginiana* Linnaeus var. *foetida* Linnaeus

Trees, evergreen, single-trunked, to 37 m. **Bark** gray, rough, thick, furrowed in thick plates. **Pith** diaphragmed. **Twigs and foliar buds** densely red- or white-hairy. **Leaves** distinctly alternate, not in terminal whorl-like clusters; stipules 2, free, 4.5–13 × 1.5–3.5 cm, abaxially densely brown-silky, sometimes deeply notched. **Leaf blade** narrowly to broadly elliptic or oblanceolate, (7.5–)13–20(–26) × (4.5–)6–10(–12.5) cm, thick-leathery, base narrowly cuneate, apex abruptly tapered and acute to short-acuminate, rarely obtuse; surfaces abaxially glabrous to densely red-brown felted, adaxially bright green, lustrous, glabrous. **Flowers** strongly lemony fragrant, 15–30(–45) cm across; spathaceous bracts 2, leathery, outer bract abaxially brown to grayish pilose, deeply notched, smaller, inner bract abaxially appressed brown to grayish pilose, shallowly notched, larger; tepals creamy white; stamens (179–)213–383(–405), 16–29 mm; filaments purple; pistils (45–)55–81(–89). **Follicetums** cylindric to somewhat obovoid, 7–10 × 3.5–5 cm; follicles beaked, sparsely to densely silky-villous. **Seeds** lenticular to narrowly ellipsoid, (9–)12–14 mm, adaxially slightly grooved, aril red. $2n = 114$.

Flowering spring. Wooded dunes, hammocks, river bottoms, mesic woods, and ravine slopes; coastal plain; 0–120 m; Ala., Ark., Fla., Ga., La., Miss., N.C., S.C., Tex.

Magnolia grandiflora (a hexaploid) is highly variable, especially the leaves, which range from glabrous to densely red-brown felted on the abaxial surface. It is the only magnolia species in the flora with free stipules, and the inner spathaceous bract is unique among *Magnolia* taxa in the flora. Curled filiform trichomes occur on the abaxial leaf surface. In the wild, hybrids with *M. virginiana* (a diploid) have been reported but not confirmed by the present author. The compatibility of these taxa is well known from the Freeman hybrid, a highly sterile tetraploid growing at the U.S. National Arboretum in Washington, D.C. In crosses using the hexaploid *M. grandiflora,* this parent is dominant and nearly masks the other parent.

Magnolia grandiflora is an escape, and it naturalizes in the tidewater area of Virginia and locally elsewhere beyond its natural range in the southeastern United States. It ranks among the noblest of North American broadleaved trees and is cultivated widely in the United States and in many other countries. A large number of cultivars have been introduced to horticulture.

Southern magnolia (*Magnolia grandiflora*) is the state tree of both Louisiana and Mississippi.

The largest known tree of *Magnolia grandiflora,* 37.2 m in height with a trunk diameter of 1.97 m, is recorded from Smith County, Mississippi (American Forestry Association 1994).

The Choctaw and Koasati tribes used the bark of *Magnolia grandiflora* as dermatological and kidney aids (D. E. Moerman 1986).

8. **Magnolia acuminata** (Linnaeus) Linnaeus, Syst. Nat. ed. 10, 2: 1082. 1759 · Cucumber-tree [E]

Magnolia virginiana Linnaeus var. (ε) *acuminata* Linnaeus, Sp. Pl. 1: 536. 1753; *Kobus acuminata* (Linnaeus) Nieuwland; *Magnolia acuminata* var. *alabamensis* Ashe; *M. acuminata* var. *aurea* (Ashe) Ashe; *M. acuminata* subsp. *cordata* (Michaux) E. Murray; *M. acuminata* var. *cordata* (Michaux) Seringe; *M. acuminata* var. *ludoviciana* Sargent; *M. acuminata* var. *ozarkensis* Ashe; *M. acuminata* var. *subcordata* (Spach) Dandy; *M. cordata* Michaux; *Tulipastrum acuminatum* (Linnaeus) Small; *T. acuminatum* var. *aureum* Ashe; *T. acuminatum* var. *flavum* Small; *T. acuminatum* var. *ludovicianum* (Sargent) Ashe; *T. acuminatum* var. *ozarkense* (Ashe) Ashe; *T. americanum* Spach; *T. americanum* var. *subcordatum* Spach; *T. cordatum* (Michaux) Small

Trees, deciduous, single-trunked, to 30 m. **Bark** dark gray, furrowed. **Pith** homogeneous. **Twigs and foliar buds** silvery-pubescent. **Leaves** distinctly alternate, not in terminal whorl-like clusters; stipules 3.2–4.3 × 1.4–1.6 cm, abaxially pilose. **Leaf blade** broadly ovate-elliptic, oblong to oblong-obovate, rarely somewhat rotund, (5–)10–25(–40) × 4–15(–26) cm, base cuneate to truncate or broadly rounded, often somewhat oblique, apex acuminate; surfaces abaxially pale green to whitish, pilose to nearly glabrous, adaxially green, glabrous or rarely scattered pilose. **Flowers** slightly aromatic, 6–9 cm across; spathaceous bracts 2, abaxially silky-pubescent; tepals erect, strongly glaucous to greenish or sometimes yellow to orange-yellow, outermost tepals reflexed, much shorter, green; stamens (50–)60–122(–139), 5–13 mm; filaments white; pistils (35–)40–45(–60). **Follicetums** oblong-cylindric, often asymmetric, 2–7 × 0.8–2.7 cm; follicles short-beaked, glabrous. **Seeds** heart-shaped, somewhat flattened to somewhat globose, 9–10 mm, smooth, aril reddish orange. $2n = 76$.

Flowering spring. Rich woods, slopes, and ravines, often along streams; 0–1400 m; Ont.; Ala., Ark., Fla., Ga., Ill., Ind., Ky., La., Md., Miss., Mo., N.Y., N.C., Ohio, Okla., Pa., S.C., Tenn., Va., W.Va.

The vernacular name, cucumber-tree, alludes to the resemblance of the follicetum to the young fruit of cucumber. It is the only magnolia species in the flora that occurs naturally in Canada.

Studies of *Magnolia acuminata* have failed to reconcile the nature of variation in this widespread species. In an attempt to settle differences in variation patterns, J. W. Hardin (1954) recognized four infraspecific taxa in *M. acuminata*. Later (1972, 1989) Hardin abandoned his earlier views for a more conservative stance, stating that variation in *M. acuminata* lacked any consistent pattern or geographic correlation. This is the view taken here—no infraspecific taxa are accepted for *M. acuminata* at this time. Its flowers are normally greenish and glaucous or sometimes yellow to orange-yellow, less showy than those of other magnolias in the flora. In southern areas, trees with yellow to orange-yellow flowers (originally described by Michaux as *M. cordata*) occur in North Carolina, Georgia, Alabama, and perhaps elsewhere, together with trees that bear normal greenish flowers. Both filiform and flagelliform trichomes occur on the leaves; cylindric trichomes also occur.

Magnolia acuminata is of value to horticulturists because no other species of the genus has yellow tepals. *Magnolia acuminata* contains major quantities of xanthophyll lutein-5,6-epoxide and, in smaller amounts, α carotene-5,6-epoxide. Although this carotenoid occurs randomly throughout populations of *M. acuminata,* often it is masked by chlorophyll and not visibly expressed. Sometimes the carotenoid pigment shows through, as in the hybrid *M. acuminata* × *M. denudata* 'Elizabeth'. In that cross the *M. acuminata* parent tree was a nondescript plant with greenish flowers; yet out of this hybrid came 'Elizabeth', a stunning plant with light canary yellow flowers, a result completely unexpected. A thorough field study of *M. acuminata* is clearly warranted, and further investigation of the carotenoid flower pigments is needed to clarify the taxonomy of this widespread tree.

The largest known tree of *Magnolia acuminata*, 29.6 m in height with a trunk diameter of 1.26 m, is recorded from a specimen cultivated in Waukon, Iowa (American Forestry Association 1994).

The Cherokee and Iroquois tribes used *Magnolia acuminata*, largely the bark, as an analgesic, antidiarrheal, gastrointestinal aid, anthelmintic, toothache remedy, and for various other uses (D. E. Moerman 1986).

2. LIRIODENDRON Linnaeus, Sp. Pl. 1: 535. 1753; Gen. Pl. ed. 5, 239. 1754 · [Greek *lirion*, lily, and *dendron*, tree]

Tulipifera Miller

Trees, deciduous. **Pith** diaphragmed. **Leaves** distinctly alternate, not in false terminal whorl; stipules tardily deciduous on summer shoots, free. **Leaf blade** evenly (2–)4–6(–10)-lobed, base rounded to shallowly cordate or truncate, apex broadly truncate or notched; surfaces abaxially glaucous, adaxially lustrous, smooth. **Flowers** protogynous, appearing with the leaves; tepals (7–)9, petaloid, tip recurved, greenish yellow with feathered orange band or blotch near base, outermost tepals sepaloid, reflexed, green; stamens on short torus, tardily deciduous, whorled; filaments 1/3–1/2 length of extrorse anthers. **Samaras** deciduous, indehiscent, in elongate, spindle-shaped, dry cone. **Seeds** adherent to dry endocarp. *x* = 19.

Species 2 (1 in the flora): North America, e Asia.

SELECTED REFERENCES   Berry, E. W. 1901. The origin of stipules in *Liriodendron*. Bull. Torrey Bot. Club 28: 493–498.   Santamour, F. S. Jr. and F. G. Meyer. 1971. The two tuliptrees. Amer. Hort. Mag. 50(2): 87–89.

1. Liriodendron tulipifera Linnaeus, Sp. Pl. 1: 535. 1753

· Yellow-poplar, tulip-poplar, tuliptree, bois-jaune E F W1

Liriodendron procera Salisbury; *Tulipifera liriodendron* Miller

Trees, single-trunked, to 45 m. **Bark** light gray, thick, deeply furrowed. **Stipules** paired, light green, elliptic to oblanceolate, 20–45 mm; petiole 5–11.5 cm. **Leaf blade** commonly with 2 shallow upper lobes and 2 lateral lobes at broadest part, or sometimes squarrose and barely lobed, (6.5–)7.5–15(–23.5) × (8.5–)12.5–18.5(–25.5) cm; surfaces abaxially glaucous, adaxially bright green. **Flowers** campanulate; spathaceous bract 1, brownish, notched; tepals erect, adaxial orange blotch sometimes gummy, outermost tepals green to glaucous; stamens 20–50, 40–50 mm; filaments white; pistils 60–100. **Samaracetums** 4.5–8.5 cm, with numerous (1–)2-seeded, imbricate samaras 3–5.5 × 0.5–1 cm, falling separately at maturity; receptacles with basal pistil persistent. **Seeds** (1–)2. 2*n* = 38.

Flowering spring. Rich woodlands, bluffs, low mountains, and hills; 0–1500 m; Ont.; Ala., Ark., Conn., Del., D.C., Fla., Ga., Ill., Ind., Ky., La., Md., Mass., Mich., Miss., Mo., N.J., N.Y., N.C., Ohio, Pa., R.I., S.C., Tenn., Vt., Va., W.Va.

Leaf and flower color variation are widespread in this species, but the variation is continuous and without any discernible taxonomic significance.

Liriodendron tulipifera is widely cultivated; a few cultivars have been introduced to horticulture, and the hybrid *L. tulipifera* × *L. chinense* is known. *Liriodendron tulipifera* is reported to have escaped from cultivation in Texas, but I have seen no specimens. The specimens from Barry and Ozark counties, Missouri, may not be indigenous.

Liriodendron tulipifera is the state tree of both Indiana and Tennessee.

Native American tribes used *Liriodendron tulipifera* for making canoes. Cherokee and Rappahannock tribes used bark of the roots as a bitter tonic and heart stimulant, and it was considered useful in healing fevers, rheumatism, and digestive disorders (D. E. Moerman 1986).

The largest known tree of *Liriodendron tulipifera*, 44.5 m in height with a trunk diameter of 3.02 m, is recorded from Bedford, Virginia (American Forestry Association 1994).

2. ANNONACEAE Jussieu

· Custard-apple Family

Robert Kral

Trees, shrubs, rarely woody vines, deciduous or evergreen, with aromatic bark, leaves, and flowers. **Pith** septate to diaphragmed. **Leaves** alternate, simple, without stipules, petiolate. **Leaf blade** pinnately veined, unlobed, margins entire. **Inflorescences** axillary to leaf scars on old wood or to leaves on new shoots, solitary flowers or few-flowered fascicles, pedunculate; bracts or bracteoles present or absent. **Flowers** bisexual, rarely unisexual; receptacle becoming enlarged, elevated or flat; perianth hypogynous, segments valvate or imbricate; sepals persistent, (2–)3(–4), distinct or basally connate; petals either 6 in 2 unequal whorls of 3 with petals of outer whorl larger, petals of inner whorl fleshier than the outer, often with corrugate nectary zone, or petals 6–12(–15), nearly equal or unequal, veins impressed on inner face; stamens 10–20 or very numerous, hypogynous, spirally arranged, forming ball or flat-topped mass; filament short, stout; anther linear to oblong-linear, extrorse, longitudinally dehiscent; connective apically elongate, connivent; pistils 1–many, superior, 1-carpellate, 1-locular, distinct or connate to various degrees with at least stigmas distinct; placentation marginal, placenta 1; ovules 1–many per pistil; style short, thick; stigma terminal. **Fruits** berries, distinct, 1–8(–12) per flower, or coalescent, forming syncarps, 1 per flower. **Seeds** 1–many per pistil, arillate; endosperm ruminate, oily.

Genera ca. 128, species ca. 2300 (3 genera, 12 species in the flora): mostly circumtropical.

The family has particular importance in the tropics because of the edible syncarps of some species of *Annona;* in the eastern United States the fruit of *Asimina triloba* (pawpaw) was once much gathered and appreciated. Programs in breeding from selected stock of *Asimina* have been undertaken (G. A. Zimmerman 1941).

Currently, the Pawpaw Foundation is intensively researching means to develop commercially marketable fruits. Recent studies of the chemical properties of *Asimina* reveal its pesticidal possibilities, and its potential as an anticancer agent (E. M. Norman, pers. comm.). The warm-climate genera *Cananga, Rollinia,* and *Artabotrys* have been used as ornamentals.

SELECTED REFERENCES Fries, R. E. 1931. Revision der Arten einiger Annonaceen-Gattungen. Acta Horti Berg. 10: 1–341. Fries, R. E. 1934. Revision der Arten einiger Annonaceen-Gattungen. Acta Horti Berg. 12: 1–220. Fries, R. E. 1939. Revision der Arten einiger Annonaceen-Gattungen. Acta Horti Berg. 12: 289–577. Hutchinson, J. 1923. Contributions toward a phylogenetic classification of flowering plants. II. The genera of Annonaceae. Bull. Misc. Inform. Kew 1923: 241–261.   Kral, R. 1960. A revision of *Asimina* and *Deeringothamnus* (Annonaceae). Brittonia 12: 233–278.   Rusby, H. H. 1935. The custard-apple family in Florida. J. New York Bot. Gard. 36: 233–239.

1. Petals of both or all whorls nearly equal; receptacle apically flat or slightly convex; stamens
   10–20(–35) in flat-topped mass; pistils distinct; peduncular bracts absent . . . .   3. *Deeringothamnus*, p. 19
1. Petals of usually 2 whorls distinctly unequal in size and form; receptacle convex to ±
   globose or elongate; stamens very numerous, forming ball on elevated receptacle; pistils
   distinct or partially to completely connate; peduncular bracts or bracteoles present.
   2. Pistils 15 or more, variously syncarpous, remaining adnate to receptacle at maturity;
      ovules 1(–2) per pistil; fruits fleshy syncarps . . . . 1. *Annona*, p. 12
   2. Pistils 2–8(–12), distinct, falling independently from receptacle at maturity; ovules few
      to several per pistil; fruits pulpy, simple berries . . . . 2. *Asimina*, p. 14

1. ANNONA Linnaeus, Sp. Pl. 1: 536. 1753; Gen. Pl. ed. 5, 241. 1754 · Custard-apple, soursop, alligator-apple [native Hispaniolan *anon* or *hanon,* given to *A. muricata*]

Trees or shrubs, taprooted; trunks buttressed or not buttressed at base. **Bark** thin, mostly broadly and shallowly fissured, scaly, fissures anastomosing. **Shoots** slender, stiff, terete; lenticels raised; buds naked. **Leaves** persistent or deciduous to late deciduous. **Leaf blade** leathery or membranous, glabrous to pubescent. **Inflorescences** axillary or supra-axillary, occasionally from axillary buds on main stem or older stems, solitary flowers or fascicles; peduncle bracteolate. **Flowers:** receptacle convex to ± globose or elongate, elevated; sepals deciduous, 3(–4), smaller than outer petals, valvate in bud; petals 6(–8) in 2 whorls, usually fleshy, those of outer whorl larger, valvate in bud, those of inner whorl more ascending, distinctly smaller or reduced, rarely absent, valvate or imbricate in bud; nectaries present as darker-pigmented, usually corrugate zones adaxially near petal bases; stamens very numerous, packed into ball, club-shaped, curved; connective dilated, hooded or pointed beyond anther sac; pistils numerous, sessile, partially connate to various degrees with at least stigmas distinct; ovules 1(–2) per pistil; style and stigma club-shaped or narrowly conic. **Fruits** fleshy syncarps, 1 per flower, usually ovoid to nearly globose, surface variable depending on orientation, structure, and relative connation of pistil apices. **Seed** usually 1 per pistil, ovoid to ellipsoid, beanlike, coat tough, margins various, narrow. $x = 7$.

Species 110 (2 in the flora): mostly neotropical; North America; 10 in Africa.

SELECTED REFERENCES   Safford, W. E. 1914. Classification of the genus *Annona,* with descriptions of new and imperfectly known species. Contr. U.S. Natl. Herb. 18: i–ix, 1–68.   Sargent, C. S. 1922. Manual of the Trees of North America (Exclusive of Mexico), ed. 2. Boston and New York. Pp. 354–356. [Facsimile edition in 2 vols. 1961, reprinted 1965, New York.]   Wood, C. E. 1958. The genera of the woody Ranales in the southeastern United States. J. Arnold Arbor. 39: 296–346.

1. Sepals reniform-cordate; petals ovate, adaxially concave, those of inner whorl at least 2/3 length of
   outer whorl; syncarp smooth. . . . . 1. *Annona glabra*
1. Sepals deltate; outer petals oblong or lance-oblong, adaxially keeled, abaxially furrowed, those of
   inner whorl minute; syncarp muricate. . . . . 2. *Annona squamosa*

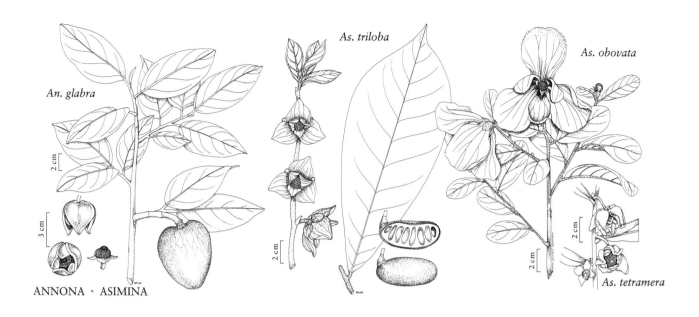

An. glabra

As. triloba

As. obovata

As. tetramera

ANNONA · ASIMINA

---

1. Annona glabra Linnaeus, Sp. Pl. 1: 537. 1753 · Pond-apple F

Annona palustris Linnaeus

Shrubs or trees, to ca. 15 m; trunks commonly buttressed at base. **Principal leaves** late deciduous; petiole 10–20 mm. **Leaf blade** ovate to elliptic, 5–15 × 6(–8) cm, base broadly cuneate to rounded, apex acute to short-acuminate; surfaces glabrous. **Inflorescences** from leaf axils on new shoots, solitary flowers; peduncle stout, linear, club-shaped, to 2 cm, becoming enlarged. **Flowers:** sepals reniform-cordate, 5–6 mm, glabrous; outer petals cream-white, ovate-cordate, adaxially concave, 2.5–3 cm, apex acute; inner petals cream-white, inside base deep purple, oblong-ovate, 2–2.5 cm, base cupped, incurved-cuneate, at least 2/3 length of outer petals, corrugate; stamens linear, 3–4 mm; connective thickened above anther tip; pistils conically massed, connate. **Syncarp** pendulous on thickened peduncle, dull yellow blotched with brown, ± ovoid, 5–12 cm, smooth with reticulate pattern formed by pistil boundaries. **Seed** ellipsoid to obovoid, 1–1.5 cm. $2n = 28$.

Flowering spring–early summer. Wet substrates, brackish to fresh, in pond borders, tidally influenced stream banks, banks of estuaries and lakes; 0–50 m; Fla.; West Indies; Mexico; Central America; South America; w Africa; South Asia (Sri Lanka).

Annona glabra (*Annona* sect. *Phelloxylon* Safford) has edible although scarcely desirable, yellow-fleshed fruits. The sectional name (Greek *phellos,* cork and

xylon, wood) is descriptive because small sections of the very light wood have been used as floats by fishermen. Forms with smaller leaves and fruit have been considered a distinct species, *A. palustris* (J. K. Small 1933); W. E. Safford (1914) considered them to be of no taxonomic consequence.

2. Annona squamosa Linnaeus, Sp. Pl. 1: 537. 1753 · Sweetsop, sugar-apple

Annona asiatica Linnaeus

Shrubs or trees, to ca. 8 m; trunks short, not buttressed at base. **Principal leaves** late deciduous; petiole 4–22 mm. **Leaf blade** narrowly elliptic to oblong or lanceolate, 5–17 × 2–5.5 cm, base broadly cuneate to rounded, apex acute to obtuse; surfaces glaucous, abaxially variably pubescent, adaxially glabrate. **Inflorescences** solitary flowers or fascicles; peduncle slender, to 2 cm, becoming enlarged in fruit. **Flowers:** sepals deltate, 1.5–2 mm, apex acute, surfaces abaxially pubescent or glabrous; outer petals pale green above purplish base, oblong or lance-oblong, 1.5–3 cm, base slightly concave, surfaces abaxially furrowed, pubescent, adaxially thickened, keeled; inner petals ovate, keeled, minute, nearly as long as stamens; stamens club-shaped, curved, 1–3 mm; connective dilated, flattened and truncate; pistils conically massed, separable at anthesis, later connate. **Syncarp** pendulous on thickened peduncle, greenish yellow, glaucous, mostly ± globose, 5–10 cm, muricate. **Seed** ellipsoid to obovoid, 1–1.4 cm.

Flowering spring–early summer. Dryish sandy substrates, dry hammocks; 0–50 m; introduced; Fla.; native to West Indies; naturalized or cultivated circumtropically.

The fruit of *Annona squamosa* (*Annona* sect. *Atta* C. Martius) has delicious whitish pulp, and it is popular in tropical markets.

2. ASIMINA Adanson, Fam. Pl. 2: 365. 1763 · Pawpaw [American Indian *assimin* through French *asiminier*]

Orchidocarpum Michaux; *Pityothamnus* Small; *Porcelia* Persoon 1807, not Ruiz & Pavón 1794; *Uvaria* Torrey & A. Gray

Shrubs or small trees, taprooted; trunks straight in arborescent forms. **Bark** thin, cracking longitudinally (shallowly furrowed in larger trees of *Asimina triloba*). **Shoots:** secondary branching sparse to copious, 2-ranked; primary shoots of shrubs usually several, erect to arching or decumbent; lenticels scattered, small, prominent; buds naked. **Leaves** deciduous. **Leaf blade** membranous or leathery, glabrous to variously hairy. **Inflorescences** axillary, on new shoots or from above leaf scars of previous seasons, fascicles; peduncle bracteate or bracteolate. **Flowers** 1–4 per leaf axil, nodding to nearly erect, becoming enlarged; receptacle convex to ± globose or elongate, elevated; sepals deciduous, 3(–4), nearly distinct, triangular to deltate-ovate, valvate in bud; petals in 2 unequal whorls, each of 3–4 equal members, with prominent veins, imbricate in bud, those of outer whorl larger, thinner than those of inner whorl; nectaries present, usually with differently pigmented, thickened-corrugate zones adaxially near petal bases; stamens very numerous, short-columnar, forming ball; filament short; connective blunt to nearly spheric or cuboid; pistils 2–8(–12), sessile to stipitate, distinct, glabrate; ovules several (sometimes few) in 2 staggered rows. **Fruits** simple berries, usually 3–5 per flower, spreading from swollen receptacles, unevenly oblong-cylindric, pulpy. **Seeds** 3–many per pistil, bean-shaped, slightly compressed laterally, coat tough, margins absent. $x = 9$.

Species 8 (8 in the flora): subtropical and temperate North America.

SELECTED REFERENCES Exell, A. W. 1927. William Bartram and the genus *Asimina* in North America. J. Bot. 65: 65–70. Nash, G. V. 1896. Revision of the genus *Asimina* in North America. Bull. Torrey Bot. Club 23: 234–242. Uphof, J. C. T. 1933. Die nordamerikanischen Arten der Gattung *Asimina*. Mitt. Deutsch. Dendrol. Ges. 45: 61–76. Wilbur, R. L. 1970. Taxonomic and nomenclatural observations on the eastern North American genus *Asimina* (Annonaceae). J. Elisha Mitchell Sci. Soc. 86: 88–96. Zimmerman, G. A. 1941. Hybrids of the American pawpaw. J. Heredity 32: 82–91.

1. Leaf blade commonly membranous, oblong-obovate to oblanceolate, apex acute to acuminate; flowers maroon, rarely yellow, fetid, from previous year's shoots before or during new leaf emergence; inner petals with saccate base.
   2. Flowers 2–4(–5) cm diam.; peduncle 1 cm or more at anthesis, hairs dark brown to red-brown; corrugate zone of inner petals distinct . . . . 1. *Asimina triloba*
   2. Flowers 1–1.7 cm diam.; peduncle less than 1 cm at anthesis, hairs red-brown to tan; corrugate zone of inner petals indistinct or absent. . . . . 2. *Asimina parviflora*
1. Leaf blade commonly leathery, linear to obovate, apex acute to broadly rounded or notched; flowers white or yellow-white to red or maroon, fetid or sweet-smelling, from previous year's growth or from new shoots.
   3. Flowers from nodes of previous year's growth, emerging before or with leaves; leaf blade elliptic or oblong, obovate or cuneate, never linear, lance-linear, or oblanceolate; flowers large, sweetly fragrant.
      4. Abaxial and adaxial surfaces of young leaf blade densely tomentose, hairs pale blond or tan; fully emerged leaf blade oblong to obovate, margins obscurely revolute, inner petals yellow-white with deep yellow corrugate nectary zone. . . . . 3. *Asimina incana*

4. Abaxial surface of young leaf blade densely orange-hairy, adaxial surface sparsely orange-hairy; fully emerged leaf blade narrowly obovate to oblong or elliptic, margins strongly to moderately revolute; inner petals white or yellowish white, rarely pink or cherry red, mostly with deep maroon to purple corrugate zone. . . . . 4. *Asimina reticulata*

3. Flowers from shoots of current year's growth, emerging after leaves are expanding or full-size, either axillary to new leaves or terminal on short shoots; leaf blade linear to narrowly obovate or spatulate; flowers large to small, fragrant or fetid.

    5. New shoots, petiole, abaxial surface of leaf blade along veins, and peduncle with dense, bright red hairs; new growth with flowers (buds) terminal . . . . 5. *Asimina obovata*

    5. New shoots, petiole, abaxial surface of leaf blade, and peduncle glabrous to sparsely hairy; new growth with flowers axillary to new leaves.

        6. Mature shrubs seldom more than 0.5 m, shoots decumbent to arching; perianth maroon to red, outer petals 1.5–3 cm. . . . . 6. *Asimina pygmaea*

        6. Mature shrubs seldom less than 1 m, shoots erect to nearly erect; perianth white to red or maroon, outer petals 2–8 cm.

            7. Leaf blade oblong-oblanceolate to broadly spatulate; expanded flowers reddish, fetid; outer petals 2–2.5 cm; perianth 3–4-merous; Florida. . . . . 7. *Asimina tetramera*

            7. Leaf blade linear, linear-elliptic, linear-oblanceolate, or narrowly spatulate; expanded flowers mostly with outer petals white or cream and inner petals red or white, sweet-smelling; outer petals 3–8 cm; perianth mostly 3-merous; n Florida, Georgia, Alabama . . . 8. *Asimina longifolia*

1. **Asimina triloba** (Linnaeus) Dunal, Monogr. Anonac., 83. 1817 · Pawpaw, dog-banana, Indian-banana, aciminier [E] [F] [W1]

Annona triloba Linnaeus, Sp. Pl. 1: 537. 1753; *A. pendula* Salisbury; *Orchidocarpum arietinum* Michaux; *Porcelia triloba* (Linnaeus) Persoon; *Uvaria triloba* (Linnaeus) Torrey & A. Gray

Shrubs or trees, 1.5–11(–14) m; trunks slender, to 20(–30) cm diam.; bark shallowly furrowed in larger trees. **Branches** spreading-ascending, slender; new shoots moderately to copiously brown-hairy apically, aging glabrous. **Leaves:** petiole 5–10 mm. **Leaf blade** oblong-obovate to oblanceolate, 15–30 cm, membranous, base narrowly cuneate, margins scarcely or not revolute, apex acute to acuminate; surfaces abaxially densely hairy, later sparsely so on veins, adaxially sparsely appressed-pubescent on veins, becoming glabrous. **Inflorescences** from previous year's shoots before or during new leaf emergence; peduncle nodding, (1–)1.5–2(–2.5) cm, densely hairy, hairs dark brown to red-brown; bracteoles 1–2, basal, usually ovate-triangular, rarely over 2–3 mm, hairy. **Flowers** maroon, fetid, 2–4(–5) cm diam.; sepals triangular-deltate, 8–12 mm, abaxially densely pilose; outer petals excurved, oblong-elliptic, 1.5–2.5 cm, abaxially puberulent on veins; inner petals elliptic, 1/3–1/2 length of outer petals, base saccate, apex recurved, surfaces abaxially glabrate, veins impressed adaxially, corrugate nectary zone distinct; pistils 3–7(–12). **Berries** yellow-green, 5–15 cm. **Seeds** brown to chestnut brown, 1.5–2.5 cm.

Flowering early spring. Mesic woods, often alluvial sites, fencerows; 0–1500 m; Ont.; Ala., Ark., Del., D.C., Fla., Ga., Ill., Ind., Iowa, Kans., Ky., La., Md., Mich., Miss., Mo., Nebr., N.J., N.Y., N.C., Ohio, Okla., Pa., S.C., Tenn., Tex., Va., W.Va.

2. **Asimina parviflora** (Michaux) Dunal, Monogr. Anonac., 82, plate 9. 1817 · Small-flowered pawpaw, small-fruited pawpaw, dwarf pawpaw [E]

Orchidocarpum parviflorum Michaux, Fl. Bor.-Amer. 1: 329. 1803; *Porcelia parviflora* (Michaux) Persoon; *Uvaria parviflora* (Michaux) Torrey & A. Gray

Shrubs or low trees, to 6 m; trunks slender, at most to 10 cm diam. **Branches** spreading-ascending, slender; new shoots red-brown, distally minutely reddish tomentose. **Leaves:** petiole 5–10 mm. **Leaf blade** obovate to oblanceolate, 6–15(–20) cm, membranous, rarely leathery, base narrowly to broadly cuneate, margins barely revolute, apex acute to acuminate; surfaces abaxially minutely rusty-tomentose, aging hairy on veins only, adaxially puberulent on veins, aging glabrous. **Inflorescences** from previous year's shoots before or during new leaf emergence; peduncle 0.3–0.8 cm, tomentose, hairs red-brown to tan; bracteoles 1(–2), basal, usually ovate-triangular, rarely over 2–3 mm, hairy. **Flowers** maroon, rarely yellow, faintly fetid, 1–1.7 cm diam.; sepals triangular-deltate, 4–7 mm, abaxially rusty-tomentose; outer petals oblong to ovate, 1–1.3 cm, apex excurved, lingulate, surfaces abaxially minutely rusty-tomentose; inner petals ovate, ca. 1/2 length of outer petals, base saccate, apex strongly recurved, veins adaxially incised, corrugate nectary zone indistinct or absent; pistils 5–7. **Berries** greenish yellow, 3–6(–7) cm. **Seeds** chestnut brown, 1–1.5(–2) cm. $2n = 18$.

Flowering early spring. Sands, sandy loams, or sandy alluvium of rich woods, alluvial terraces, and upland dry woods; 0–700 m; Ala., Ark., Fla., Ga., La., Miss., N.C., S.C., Tenn., Tex., Va.

Asimina parviflora is primarily a miniature version of *A. triloba* in flower, shoot, and leaf; it reaches tree size only in the karst country of Florida. Peduncles at anthesis are often so short that flowers appear sessile. Pubescence is of a lighter red or brown than that of *A. triloba*. Putative hybrids between the two have been observed in northern Alabama.

3. **Asimina incana** (W. Bartram) Exell, J. Bot. 65: 69. 1927 · Polecat-bush, flag pawpaw [E] [W1]

Annona incana W. Bartram, Travels Carolina, 171. 1791 (as *incarna*); *A. speciosa* Nash; *Pityothamnus incanus* (W. Bartram) Small

Shrubs, to 1.5 m. **Branches** copious, stiff; shoots red-brown or tan with dense, pale tomentum, maturing tomentose only distally. **Leaves:** petiole 2–6 mm. **Leaf blade** oblong to obovate, 5–8 cm, leathery, base round to abruptly broadly cuneate, margins obscurely revolute, apex obtuse to rounded, often notched; surfaces abaxially and adaxially densely tomentose with pale blond or tan hairs, becoming moderately or weakly so. **Inflorescences** from previous year's growth; peduncle 2–3.5 cm, pale-tomentous; bracteoles 1–2, basal, usually ovate-triangular, rarely more than 2–3 mm, hairy. **Flowers** 1–4 per node, fragrant, large; sepals triangular-deltate, 8–12 mm, abaxially pale-tomentose; outer petals 3(–4), white or cream, ovate to oblong or obovate, 3.5–7 cm, abaxially pale-puberulent, veins impressed adaxially; inner petals yellow-white, lance-hastate, 1/3–1/2 length of outer petals, base saccate, adaxially deep yellow corrugate zone; pistils 3–5(–11). **Berries** yellow-green, to 8 cm. **Seeds** pale to rich brown, dull, 1–2 cm. $2n = 18$.

Flowering spring–early summer. Sands or sandy loams, upland oak–pineland, pastures, disturbed sandy sites, mostly in longleaf pine–turkey oak systems; 0–150 m; Fla., Ga.

Asimina incana hybridizes with *A. longifolia* and *A. reticulata,* and possibly *A. pygmaea.*

4. **Asimina reticulata** Shuttleworth ex Chapman, Fl. South. U.S. ed. 2, 603. 1883 · Flag pawpaw [E] [W1]

Asimina cuneata Shuttleworth ex A. Gray; *Pityothamnus reticulatus* (Chapman) Small

Shrubs, to 1.5 m; crown much branched. **Shoots** red-brown to tan, distally red or pale-hairy, becoming gray-brown, distally glabrous or sparsely pale-hairy. **Leaves:** petiole 2–6 mm. **Leaf blade** oblong to elliptic or narrowly obovate, 5–8 cm, leathery, base abruptly and broadly cuneate or rounded, margins strongly to moderately revolute, apex acute to broadly rounded, occasionally notched; surfaces abaxially densely orange-hairy, becoming sparsely so on veins, adaxially sparsely orange-hairy, becoming glabrous and often glaucous. **Inflorescences** on previous year's growth; peduncle slender, 2–3.5 cm, tomentose; bracteoles 1–2, basal, usually ovate-triangular, rarely more than 2–3 mm, hairy. **Flowers** 1–3 per node, fragrant, large; sepals triangular, 8–10 mm, abaxially orange-puberulent; outer petals spreading, white or cream, narrowly oblong to obovate, 2.5–6 cm, abaxially puberulent on veins; inner petals incurved, white, yellowish white, rarely pink or cherry red, mostly with deep maroon to purple corrugate zone, lance-hastate, 1/3–1/2 length of outer petals, fleshier, base saccate, margins revolute; pistils 3–8. **Berries** yellow-green, 4–7 cm. **Seeds** dark to pale brown, lustrous, 1–2 cm. $2n = 18$.

Flowering winter–spring. Moist sands and sandy peat of pine-palmetto flats, savannas, low fields; 0–100 m; Fla.

Asimina reticulata hybridizes with *A. incana* and *A. pygmaea.* Hybrids with the latter frequently have cherry-red inner petals.

5. **Asimina obovata** (Willdenow) Nash, Bull. Torrey Bot. Club 23: 240. 1896 · Flag pawpaw [E] [F]

Annona obovata Willdenow, Sp. Pl. 2(2): 1269. 1799; *A. grandiflora* W. Bartram; *Asimina grandiflora* (W. Bartram) Dunal; *Orchidocarpum grandiflorum* (W. Bartram) Michaux; *Pityothamnus obovatus* (Willdenow) Small; *Porcelia grandiflora* (W. Bartram) Persoon; *Uvaria obovata* (Willdenow) Torrey & A. Gray

Shrubs or small trees, 2–3(–4.5) m, much branched. **New shoots** densely tomentose, hairs bright red, maturing glabrous. **Leaves:** petiole 2–6 mm, hairs bright red. **Leaf blade** obovate to oblong, oblanceolate or ovate, 4–10(–12) cm, leathery, base rounded to broadly cuneate, margins scarcely to prominently revolute, apex rounded to obtuse, often notched; surfaces abaxially densely hairy along veins, later sparsely so, hairs

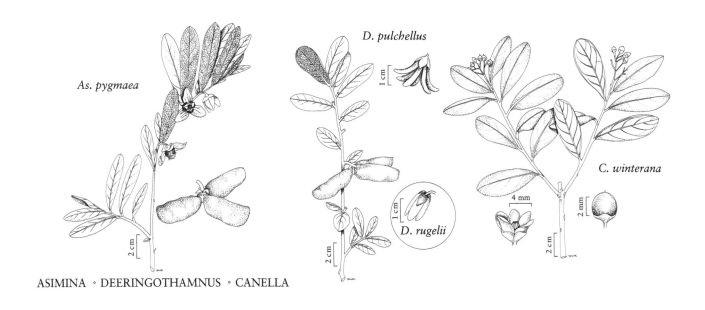

As. pygmaea

D. pulchellus

D. rugelii

C. winterana

ASIMINA · DEERINGOTHAMNUS · CANELLA

bright red, adaxially sparsely red-hairy, maturing glabrous, lustrous. **Inflorescences** terminal on new growth, sometimes directly from axils of previous season; peduncle usually erect, shootlike, to 0.5 cm, tomentose, hairs bright red; bracts leafy, often with 1–2 bracteoles basally. **Flowers** yellow-white with lemony fragrance, large; sepals elliptic to ovate, 5–15 mm, adaxially red-hairy; outer petals spreading, elliptic to obovate, 4–6 (–8) cm, veins abaxially puberulent, glabrescent, impressed adaxially; inner petals incurved, rarely pink or red, oval to oblong, rarely lanceolate, 1/5–1/2 length of outer petals, thick, base slightly saccate, corrugate zone adaxially purple; pistils 3–8(–11). **Berries** yellow-green, 5–9 cm. **Seeds** brown to chestnut brown, 1–2 cm. $2n = 18$.

Flowering late winter–early summer. White and yellow sand ridges, mostly in oak-pine woods, scrub, coastal dunes, and hammock edges; 0–150 m; Fla.

Taller strains of *Asimina obovata,* particularly those with pink or red inner petals, have great horticultural potential. W. Bartram figured this species adequately, but subsequent authors tended to confuse it and his somewhat similar *Annona incana,* with which it is sympatric in northern peninsular Florida.

6. **Asimina pygmaea** (W. Bartram) Dunal, Monogr. Anonac., 84, plate 10. 1817 · Gopher-berry, dwarf pawpaw E F

Annona pygmaea W. Bartram, Travels Carolina, 18, plate 1. 1791; *Asimina secundiflora* Shuttleworth ex Exell; *Orchidocarpum pygmaeum* (W. Bartram) Michaux; *Pityothamnus pygmaeus* (W. Bartram) Small; *Porcelia pygmaea* (W. Bartram) Persoon; *Uvaria pygmaea* (W. Bartram) Torrey & A. Gray

Shrubs, low, 2–3(–5) dm. **New shoots** 1–several, sparingly branched or unbranched, arching, red-brown with sparse, appressed, red hairs, glabrate. **Leaves:** petiole 3–10 mm. **Leaf blade** ascending along and above shoots, obovate or oblanceolate, rarely linear-elliptic, 4–7(–11) cm, leathery, base variously abruptly cuneate or narrowly rounded, margins revolute, apex rounded to obtuse or notched; surfaces abaxially and adaxially sparsely red-puberulent, glabrescent, abaxially pale, prominently reticulate. **Inflorescences** secund from axils of new shoot leaves, solitary flowers; peduncle slender, glabrate, 1.5–3(–4) cm; bracteoles 1–2, basal or 1 suprabasal, lance-oblong, less than 1 cm, hairy. **Flowers** maroon, fetid, large; sepals ovate, 5–10 mm, glabrate; outer petals maroon or pink with maroon streaks, oblong to ovate-lanceolate, 1.5–3 cm, fleshy, margins revolute, apex spreading; inner petals erect, deep maroon, ovate-acute to lance-ovate, 1/3–2/3 length of outer petals, fleshier, base saccate, apex excurved, corrugate zone adaxially deep purple; pistils 2–5. **Berries** yellow-green, 3–4(–5) cm. **Seeds** shiny brown, ca. 1 cm.

Flowering spring–early summer. Sandy peat of pine-palmetto flats, savannas, low sandy fields, and low sand ridges; 0–100 m; Fla., Ga.

Asimina pygmaea hybridizes with *A. longifolia* var. *longifolia* and *A. reticulata,* with backcrosses introducing various heights, larger and variously paler flowers, and degrees of fragrance grading from fetid in the maroon types to progressively more fragrant in the larger-flowered swarm components.

7. **Asimina tetramera** Small, Torreya 26: 56. 1926 · Four-petaled pawpaw C E F

Pityothamnus tetramerus (Small) Small

Shrubs, 1–3 m. **Branches** 1–several from base, nearly erect to arching; new shoots red-brown to tan, distally minutely red-tomentose, glabrate. **Leaves:** petiole 2–3 mm. **Leaf blade** spreading to upswept, oblong-oblanceolate to broadly spatulate, 5–10(–18) cm, leathery, base narrowly cuneate, margins slightly revolute, apex rounded to obtuse-angled; surfaces abaxially glabrate, adaxially glabrous. **Inflorescences** from axils of new shoots, solitary flowers; peduncle slender, 1–2 cm; bracteole 1, basal, lance-oblong, 5–7 mm, hairy. **Flowers** nodding to nearly erect, reddish, fetid; perianth (3–)4-merous; sepals elliptic or ovate, ca. 1 cm, abaxially sparsely rusty-hairy; outer petals ascending with spreading pale tips, lanceolate to oblong-lanceolate, 2–2.5 cm, margins slightly revolute, surfaces abaxially glabrate; inner petals ovate-acute or short-acuminate, ca. 1/2 length of outer petals, fleshier than outer petals, base large, saccate, corrugate zone adaxially deep purple; pistils 3–11. **Berries** yellow-green, 5–9 cm. **Seeds** chestnut brown, 1–2 cm. $2n = 18$.

Flowering spring–summer, all year if disturbed. Sands of ancient coastal dunes and adjacent open-wooded hammocks, flatwoods; of conservation concern; 0–50 m; Fla.

Morphologically, *Asimina tetramera* is much like a tall *A. pygmaea* in characters of flower and leaf.

8. **Asimina longifolia** Kral, Brittonia 12: 265. 1960 · Polecat-bush

Asimina angustifolia A. Gray 1886, not *A. angustifolia* Rafinesque 1840; *Pityothamnus angustifolius* (A. Gray) Small

Shrubs, 1–1.5(–1.75) m. **Shoots** stiffly erect to arching or spreading, red-brown to yellow-brown, glabrous or glabrate. **Leaves:** petiole 2–4 mm. **Leaf blade** ascending-spreading, upswept, linear-elliptic to narrowly spatulate or linear-oblanceolate, 5–15(–20) cm, leathery, base broadly to narrowly cuneate or attenuate, margins obscurely to very revolute, apex acute to obtuse, rounded, or rarely notched; surfaces abaxially glabrate, raised-reticulate, adaxially glabrate. **Inflorescences** from axils along new shoots, solitary flowers; peduncle slender, 2–4 cm, glabrate; bracteoles 1–2, often suprabasal, lance-linear, to 10 mm, hairy. **Flowers** nodding, fragrant; perianth 3(–6)-merous; sepals elliptic to ovate, 0.5–1 cm, glabrescent; outer petals spreading, white or cream, rarely pink or striped with red or pink, elliptic, oblong, or obovate, 3–8 cm, abaxially glabrate; inner petals erect, maroon to cherry red or white, oblong to lanceolate, to 1/2 length of outer petals, base broad, saccate, corrugate zone adaxially purple, apex spreading; pistils 2–7(–12). **Berries** yellow-green, 4–10 cm. **Seeds** dark brown, lustrous, 1–2 cm. $2n = 18$.

Varieties 2 (2 in the flora): subtropical North America.

1. Leaf blade linear-elliptic to linear-oblanceolate, apex acute to obtuse (rarely notched or rounded), broadest near middle, margins revolute; new growth mostly straight, ascending or nearly erect; foliage, flowers, and new shoots minutely tomentose when young; outer petals sparsely hairy abaxially; Florida, Georgia. . . . . 8a. *Asimina longifolia* var. *longifolia*
1. Leaf blade linear-spatulate to spatulate, apex obtuse, rounded, or notched, rarely broadly acute, broadest distally, margins slightly revolute; new growth arching to almost decumbent; foliage, flowers, and new shoots glabrous or sparsely short-hairy when young; Alabama, Florida, Georgia . . . . 8b. *Asimina longifolia* var. *spatulata*

8a. **Asimina longifolia** Kral var. **longifolia** E

Primary shoots mostly stiff, straight, and erect to ascending, at first distally minutely tomentose. **Leaf blade** linear, elliptic-linear, rarely linear-oblanceolate, mostly broadest toward middle, less than 1.5 cm wide, margins revolute, apex acute to obtuse, rarely notched or rounded; surfaces sparsely hairy abaxially when young. **Flowers:** outer petals white or cream-white, sparsely hairy abaxially; inner petals white, rarely pink or red.

Flowering spring–early summer, all year if disturbed. Sands and sandy peats or loams of yellow sandhills, sandy fields and pastures, open longleaf pine–turkey oak woods and savannas; 0–100 m; Fla., Ga.

Asimina longifolia var. *longifolia* hybridizes with *A. incana* (= *A.* ×*nashii* Kral) to produce spectacular, fragrant flowers with red or pink inner petals, and the hybrid is frequent over their nearly identical ranges.

Asimina longifolia var. *longifolia* crosses with *A. pygmaea,* particularly along the Suwannee-Okefenokee axis, ultimately to form swarms, the individuals varying in height, degree of arching of branches, flower sizes, pigments, and fragrances. Some of those hybrids were noted by W. Bartram.

8b. Asimina longifolia Kral var. **spatulata** Kral, Brittonia 12: 266, fig. 9a–b. 1960 [E]

Primary shoots ± lax, usually much arched to almost decumbent, glabrous or sparsely short-hairy when young. **Leaf blade** oblong-linear, oblong, or commonly oblanceolate or spatulate, margins slightly revolute, glabrous, apex obtuse, rounded or notched, rarely broadly acute, broadest distally. **Flowers:** outer petals abaxially glabrous or sparsely short-hairy when young; inner petals pink or red.

Flowering spring–early summer (to fall). Sands or sandy peats or loams of rises in pine savanna, pine flatwoods, or sandhills, sandy fields, and pastures; 0–100 m; Ala., Fla., Ga.

Asimina longifolia var. *spatulata* flowers as late as fall in response to disturbance. Because *A. longifolia* var. *spatulata* much resembles the hybrids between *A. longifolia* var. *longifolia* and *A. pygmaea,* it is not unlikely that it originated as such a hybrid and has extended west and north from the early swarms, these presumably along the Suwannee-Okefenokee drainage.

3. DEERINGOTHAMNUS Small, Bull. Torrey Bot. Club 51: 389. 1924 · [For Charles Deering, frequent sponsor of J. K. Small in his botanical explorations]

Shrubs, 2–3(–5) dm, from stout-linear or fusiform taproots. **Bark** thin, glabrate. **Shoots** arching to semidecumbent, simple or sparingly branched, glabrate; lenticels pale; buds naked. **Leaves** tardily deciduous. **Leaf blade** leathery, glabrate. **Inflorescences** from axils of new shoot leaves, solitary flowers; peduncle not bracteate. **Flowers** ascending to nodding; receptacle flat-surfaced; sepals persistent, (2–)3(–4), oblong to ovate, imbricate in bud; petals 6–9(–15) in 2 whorls, ± equal, fleshy, veins impressed adaxially, imbricate in bud; stamens 10–20(–25), erect in flat-topped mass; connective short-cylindric, extruding, blunt; pistils 1–5(–7), distinct, narrowly fusiform, glabrate; ovules 3–8 per pistil in 2 staggered rows. **Fruits** simple berries, 1–5 per flower, yellow-green, unevenly oblong-cylindric, pulpy, insipid. **Seeds** 3–8 per pistil, bean-shaped to ovoid, coat tough, margins absent.

Species 2 (2 in the flora): peninsular Florida.

SELECTED REFERENCES Kral, R. 1983c. *Deeringothamnus.* In: R. Kral, ed. 1983. A Report on Some Rare, Threatened, or Endangered Forest-related Plants of the South. Washington. Pp. 452–456. [U.S.D.A. Forest Service Techn. Publ. R8-TP 2.] Wilbur, R. L. 1970. Taxonomic and nomenclatural observations on the eastern North American genus *Asimina* (Annonaceae). J. Elisha Mitchell Sci. Soc. 86: 88–96.

1. Petals 6–12(–15), ascending to apically recurved, white to pink, linear or narrowly oblong, 2–3 cm × ca. 2 mm; peninsular Florida. . . . . 1. *Deeringothamnus pulchellus*
1. Petals prevalently 6, erect or slightly spreading, canary yellow, lance-oblong, ca. 1.5 cm × ca. 3–4 mm; e, c Florida. . . . . 2. *Deeringothamnus rugelii*

1. **Deeringothamnus pulchellus** Small, Bull. Torrey Bot. Club 51: 390. 1924 · Beautiful pawpaw [C] [E] [F]

Asimina pulchella (Small) Rehder & Dayton

Leaves: petiole 2–4 mm. **Leaf blade** spreading-ascending, mostly elevated above shoots, olong to oblong-ovate or spatulate, 4–7 cm, base cuneate to narrowly rounded, margins narrowly revolute, apex obtuse to broadly rounded, often notched. **Inflorescences:** peduncle slender, 1–3 cm. **Flowers** nodding to ascending, white to pale pink, sweetly fragrant; sepals 2–4, erect, ovate-triangular, ca. 5 mm; petals 6–12(–15), ascending to apically recurved, linear or narrowly oblong, 2–3 cm × ca. 2 mm; pistils (1–)5(–7). **Berries** 4–7 cm. **Seeds** 1–1.5 cm, slightly flattened laterally.

Flowering late winter–spring, all year on disturbance. Sandy peats of slash pine–palmetto flats, savannas; of conservation concern; 0–50 m; Fla.

Deeringothamnus pulchellus commonly associates with *Asimina reticulata* and overlaps with that species in flowering times; no hybrids between them have been observed. When protected from fire, the elongate and branching shoots of *D. pulchellus* and *D. rugelii* will persist, but such new growth rarely produces flowers. A conservation measure for such species must involve periodic burning.

SELECTED REFERENCES Rehder, A. J. and W. A. Dayton. 1944. A new combination in *Asimina*. J. Arnold Arbor. 25: 84. Small, J. K. 1926. *Deeringothamnus pulchellus*. Addisonia 11: 33–34.

2. **Deeringothamnus rugelii** (B. L. Robinson) Small, Addisonia 15: 17, plate 489. 1930 · Rugel's pawpaw [C] [E] [F]

Asimina rugelii B. L. Robinson in A. Gray et al., Syn. Fl. N. Amer. 1(1): 465. 1897

Leaves: petiole 1–2 mm. **Leaf blade** oblong to obovate or oblanceolate, 1–7 cm, margins revolute, apex acute to obtuse, broadly rounded, often notched. **Inflorescences:** peduncle slender, 1–2.5 cm. **Flowers** spreading-ascending to recurved, canary yellow, rarely purple, faintly fragrant; sepals 2–3, erect or slightly spreading, ovate to oblong, ca. 10 mm, apex acute; petals prevalently 6, ascending to slightly spreading, lance-oblong, ca. 1.5 cm × ca. 3–4 mm; pistils (2–)5(–7), fusiform. **Berries** 3–6 cm. **Seeds** 1–1.5 cm, ± compressed laterally.

Flowering spring, all year as disturbance reaction. Moist sandy peats of slash pine–palmetto flats, savannas; of conservation concern; 0–50 m; Fla.

Deeringothamnus rugelii commonly associates with both *Asimina pygmaea* and *A. reticulata,* and overlaps in flowering time with both. No hybrids between the two genera had been observed until a recent discovery of a putative hybrid between *D. rugelii* and *Asimina reticulata* by Dr. E. Norman (pers. comm. 1994).

SELECTED REFERENCE Small, J. K. 1930. *Deeringothamnus rugelii*. Addisonia 15: 17–18.

3. CANELLACEAE Martius

• Canella or Wild-cinnamon Family

Thomas K. Wilson

Trees [or prostrate shrubs], evergreen, aromatic. **Pith** homogeneous. **Leaves** alternate, simple, without stipules; petiole short. **Leaf blade** pinnately veined, unlobed, margins entire; pellucid dots (oil cells in tissue of leaf) conspicuous or inconspicuous. **Inflorescences** terminal and axillary, cymes [racemes or solitary flowers], pedunculate; bracts present or absent. **Flowers** bisexual; perianth hypogynous, segments imbricate; sepals persistent, 3; petals 5[–12] in 1[–4] whorl(s); stamens [7–]10[–12], hypogynous, monadelphous; filaments connate, forming tube around pistil; anthers extrorse, longitudinally dehiscent; pistil 1, superior, 2–6-carpellate; ovary 1-locular; placentation parietal, placentas 2[–6]; ovules 2–3 per placenta; style 1, generally short; stigma 1, usually 2–6-lobed. **Fruits** berries. **Seeds** 2 or more, not arillate; endosperm oily [ruminate].

Genera 6, species ca. 20 (1 genus, 1 species in the flora): primarily tropical, Western Hemisphere, Africa (including Madagascar).

The family was placed in Hypericales by J. K. Small (1933); the combination of oil cells in most of the tissue, pollen with a single distal furrow (occasionally with 3-radiate surficial furrow), very long vessels, and scalariform vessel perforation plates with many bars suggests Magnoliales or Annonales (A. Cronquist 1981; R. M. T. Dahlgren 1980; H. Melchoir and W. Schultze-Motel 1959; R. F. Thorne 1976; T. K. Wilson 1960; C. E. Wood Jr. 1958). Further support of this placement comes from leaf architecture studies by L. J. Hickey (1971) and P-type plastids in sieve elements (H.-D. Behnke 1988).

SELECTED REFERENCES Behnke, H.-D. 1988. Sieve-element plastids, phloem protein, and evolution of flowering plants. III. Magnoliidae. Taxon 37: 699–732. Correll, D. S. and H. B. Correll. 1982. Flora of the Bahama Archipelago. Vaduz. Dahlgren, R. M. T. 1980. A revised system of classification of the angiosperms. Bot. J. Linn. Soc. 80: 91–124. Gilg, E. 1925. Canellaceae. In: H. G. A. Engler et al., eds. 1924 +. Die natürlichen Pflanzenfamilien . . . , ed. 2. 26 + vols. Leipzig and Berlin. Vol. 21, pp. 323–328. Melchior, H. and W. Schultze-Motel. 1959. Canellaceae. (Supplement to Vol. 21.) In: H. G. A. Engler et al., eds. 1924 +. Die natürlichen Pflanzenfamilien . . . , ed. 2. 26 + vols. Leipzig and Berlin. Vol. 17a(2), pp. 221–224. Tomlinson, P. B. 1980. The Biology of Trees Native to Tropical Florida. Allston, Mass. Wilson, T. K. 1960. The comparative morphology of the Canellaceae. I. Synopsis of genera and wood anatomy. Trop. Woods 112: 1–27. Wood, C. E. Jr. 1958. The genera of the woody Ranales in the southeastern United States. J. Arnold Arbor. 39: 296–346.

1. CANELLA P. Browne, Civ. Nat. Hist. Jamaica, 275, plate 27, fig. 3. 1756 • Canella or wild-cinnamon [Latin *canella,* cinnamon, related to *cana,* cane or reed, and *-ella,* diminutive, because of the tightly rolled bark when dried]

Trees [or large, sprawling shrubs], (3–)8–10(–15) m. **Bark** whitish gray. **Stems** erect [to prostrate]. **Leaf blade** deep green, shiny, obovate to oblanceolate, thick, leathery, base acute, apex rounded, notched, or blunt; oil cells possibly evident as pellucid dots, emitting strong aromatic odor when broken, causing sharp burning sensation on tongue when bitten. **Inflorescences** of 5–40 flowers, crowded toward end of stem. **Flowers** bisexual, protogynous; sepals green, imbricate, thick; petals basally connate, dark red to violet, lighter at base, thick; stamens 10; filaments connate into tube surrounding pistil, tube protruding slightly beyond anthers and nearly equal to length of petals; anthers extrorsely dehiscent; pistil flask-shaped; ovary conic; style short; stigma 2-lobed. **Berry** changing from green through red to dark purple with age, globose, fleshy. **Seeds** shiny, hard. $x = 14$.

Species 1 (1 in the flora): tropical regions in North America, West Indies, and ne South America.

An early report of a second species in the Maracaibo region of Venezuela and reports of either species in Colombia appear unfounded.

SELECTED REFERENCES Wilson, T. K. 1964. Comparative morphology of the Canellaceae. III. Pollen. Bot. Gaz. 125: 192–197. Wilson, T. K. 1966. Comparative morphology of the Canellaceae. IV. Floral morphology and conclusions. Amer. J. Bot. 53: 336–343. Wilson, T. K. 1986. The natural history of *Canella alba* (Canellaceae). In: R. R. Smith, ed. 1986. Proceedings of the First Symposium on the Botany of the Bahamas. . . . San Salvador, Bahamas. Pp. 101–115.

1. Canella winterana (Linnaeus) Gaertner, Fruct. Sem. Pl. 1: 373, plate 77, fig. 2. 1788 · Canella, wild-cinnamon, cinnamon-bark F

Laurus winterana Linnaeus, Sp. Pl. 1: 371. 1753; *Canella alba* Murray

Leaf blade 5–15 × 2–5 cm, apex rounded or blunt; abaxial surface pellucid-dotted. **Flowers** ca. 7 mm diam., pedicel short; sepals green, 2–3 mm, fleshy; petals deep red to magenta, basally light red to yellow, 4.5–6 mm, thick and fleshy; anthers light red, becoming yellow at anthesis; stigma yellow. **Seeds** 1–5, black. $2n = 28$.

Flowering spring–summer (mid May–Jul). Coastal thickets, hammocks, commonly found on limestone or calcareous soils; 0–3 m; Fla.; West Indies; ne South America.

Although flowers of *Canella winterana* are bisexual and protogynous (with the gynoecium of each flower functionally mature before the androecium), they are functionally unisexual, because normally all flowers on a plant are at the same stage (either male or female) at any given time (T. K. Wilson 1986). *Canella winterana* is locally abundant in some areas, but with the clearing and development of the Florida Keys, it is becoming less common except in protected areas, such as Everglades National Park. Nurseries in extreme southern Florida occasionally market canella. It is an attractive, small- to medium-sized, very slow-growing tree.

4. CALYCANTHACEAE Lindley

• Strawberry-shrub Family

George P. Johnson

Shrubs, deciduous, rhizomatous, aromatic. **Leaves** opposite, without stipules, petiolate. **Leaf blade:** margins entire or, on fast-growing shoots, occasionally with a few teeth. **Inflorescences** terminal on short, leafy branches of current growth, solitary flowers. **Flowers** bisexual; hypanthium urn-shaped; tepals and stamens perigynous; inner tepals, stamens, and staminodes with succulent, white food bodies at tips; tepals 15–30, fleshy, petaloid, spirally inserted on outside of hypanthium; stamens 5–30, spirally inserted at top of hypanthium; staminodes 10–25, on inner surface of hypanthium; locules opening by longitudinal slits; pistils 5–35, 1-carpellate, spirally inserted within hypanthium; ovules 2 per locule, distal ovule abortive, submarginal; style elongate, filiform; stigma dry, decurrent. **Fruits** achenes, 1–35, dark reddish brown, cylindric, pubescent, enclosed within enlarged, fibrous persistent hypanthium (pseudocarp). **Seeds:** endosperm absent.

Genera 3, species 6 (1 genus, 2 species in the flora): temperate North America and China.

Calycanthus is often grown ornamentally. Also in cultivation is *Chimonanthus praecox* (Linnaeus) Link, an eastern Asiatic shrub. *Chimonanthus* differs from *Calycanthus* in its scaly buds and yellow flowers that appear long before the leaves.

Calycanthus contains calycanthine, an alkaloid similar to strychnine, and it is toxic to humans and livestock (W. H. Lewis and M. P. F. Elvin-Lewis 1977).

SELECTED REFERENCES Nicely, K. A. 1965. A monographic study of the Calycanthaceae. Castanea 30: 38–81. Wood, C. E. Jr. 1958. The genera of the woody Ranales in the southeastern United States. J. Arnold Arbor. 39: 296–346.

1. CALYCANTHUS Linnaeus, Syst. Nat. ed. 10, 2: 1066. 1759, name conserved

· Strawberry-shrub, sweetshrub, spicebush [Greek *kályx*, covering, cup, and *anthos*, flower]

Butneria Duhamel

Twigs quadrangular to nearly terete, pubescent to glabrous. **Buds** naked. **Leaves** 2-ranked. **Leaf blade** elliptic, ovate, or ovate-lanceolate; surfaces adaxially scabrous. **Flowers** maroon or reddish brown, rarely greenish or green-tipped, with strawberry or pineapple scent. $x = 11$.

Species 3 (2 in the flora): temperate North America and Asia (China).

1. Tepals oblong-elliptic to obovate-lanceolate, apex acute; stamens 10–20, oblong; mature hyphanthium cylindric, ellipsoid, pyriform, or globose; lateral bud partially hidden by petiole base; e North America. . . . . .1. *Calycanthus floridus*
1. Tepals linear to linear-spatulate or ovate-elliptic, apex rounded; stamens 10–15, linear to oblong-linear; mature hyphanthium campanulate; lateral bud exposed; California. . . . . .2. *Calycanthus occidentalis*

1. Calycanthus floridus Linnaeus, Syst. Nat. ed. 10, 2: 1066. 1759 · Carolina-allspice, sweetshrub $\boxed{E}$

Shrubs, to 3.5 m. **Lateral bud** partially hidden by petiole base. **Leaves:** petiole 3–10 mm, pubescent to glabrous. **Leaf blade** elliptic, broadly elliptic, oblong, or ovate, 5–15 × 2–6 cm, base acute to truncate, apex acute, acuminate, or blunt; surfaces abaxially green or glaucous, pubescent to glabrous. **Flowers:** hyphanthium cylindric, ellipsoid, pyriform, or globose at maturity, 2–6 × 1–3 cm; tepals oblong-elliptic to obovate-lanceolate, 2–4 cm × 3–8 mm, apex acute; stamens 10–20, oblong. 2*n* = 22, 33.

Varieties 2: temperate North America.

The Cherokee used *Calycanthus floridus* for medicinal purposes (D. E. Moerman 1986).

1. Twigs pubescent; petiole and abaxial surface of leaf blade pubescent . . . . .1a. *Calycanthus floridus* var. *floridus*
1. Twigs glabrous or with scattered trichomes; petiole and abaxial surface of leaf blade glabrous or with scattered trichomes . . . . .1b. *Calycanthus floridus* var. *glaucus*

1a. Calycanthus floridus Linnaeus var. floridus $\boxed{E}$ $\boxed{F}$

Calycanthus brockianus Ferry & Ferry f.; *C. mohrii* (Small) Pollard

Twigs pubescent. **Leaves:** petiole and abaxial surface of leaf blade pubescent.

Flowering late spring, fruiting mid fall. Deciduous or mixed woodlands, along streams and rivers, margins of woodlands; 0–1850 m; Ala., Fla., Ga., Md., Miss., N.C., Pa., S.C., Tenn., Va.

Calycanthus floridus var. *floridus* grades into var. *glaucus* in northeastern Alabama, northwestern Georgia, and southeastern Tennessee, where determination to variety is not always possible. *Calycanthus brockianus* was distinguished primarily by having green flowers instead of maroon, although flowers with greenish or green-tipped tepals are occasionally seen on maroon-flowered plants of both varieties of *C. floridus* and *C. occidentalis*. Viable seeds were produced after self-pollinations of plants identified as *C. brockianus*, but not after crosses with *C. floridus* var. *floridus* or when insects were excluded (R. J. Ferry Sr. and R. J. Ferry Jr. 1987). Agamospermy has been noted in both varieties of *C. floridus* (K. A. Nicely 1965), and pollination is required for the de-

velopment of the endosperm (C. E. Wood Jr. 1958). *Calycanthus brockianus* may therefore represent agamospermous triploids within diploid populations of *C. floridus*. Studies of cytology and of reproductive biology of plants identified as *C. brockianus* are necessary for determination of the status of that taxon.

1b. Calycanthus floridus Linnaeus var. glaucus (Willdenow) Torrey & A. Gray, Fl. N. Amer. 1(3): 475. 1840 $\boxed{E}$

Calycanthus glaucus Willdenow, Enum. Pl., 559. 1809; *Calycanthus fertilis* Walter; *C. floridus* var. *laevigatus* (Willdenow) Torrey & A. Gray; *C. floridus* var. *oblongifolius* (Nuttall) Boufford & Spongberg; *C. nanus* (Loiseleur-Deslongchamps) Small

Twigs glabrous or with scattered trichomes. **Leaves:** petiole and abaxial surface of leaf blade glabrous or with scattered trichomes.

Flowering late spring, fruiting mid fall. Deciduous or mixed woodlands, along streams and rivers, margins of woodlands; 0–1850 m; Ala., Fla., Ga., Ky., Miss., N.C., Ohio, Pa., S.C., Tenn., Va., W.Va.

The nomenclature for *Calycanthus floridus* var. *glaucus* is somewhat confusing. (For an in-depth study, see D. E. Boufford and S. A. Spongberg 1981 and J. W. Hardin 1984.) Variation in the amount of pubescence is common, and determination of variety is not always possible.

The Cherokee are reported to have used this plant as a urinary aid (D. E. Moerman 1986, listed under *C. fertilis*, a synonym for *C. floridus* var. *glaucus*).

2. Calycanthus occidentalis Hooker & Arnott, Bot. Beechey Voy., 340, plate 84. 1841 · California spicebush, sweetshrub $\boxed{E}$ $\boxed{F}$

Butneria occidentalis (Hooker & Arnott) Greene

Shrubs, to 4 m. **Lateral bud** exposed. **Leaves:** petiole 5–10 mm, pubescent to glabrous. **Leaf blade** ovate-lanceolate to oblong-lanceolate or ovate-elliptic, 5–15 × 2–8 cm, base rounded to nearly cordate, apex acute to obtuse; surfaces abaxially green, pubescent to glabrous. **Flowers:** hyphanthium campanulate or ovoid-campanulate at maturity, 2–4 × 1–2 cm; tepals linear to linear-spatulate or ovate-elliptic, 2–6 × 0.5–1 cm,

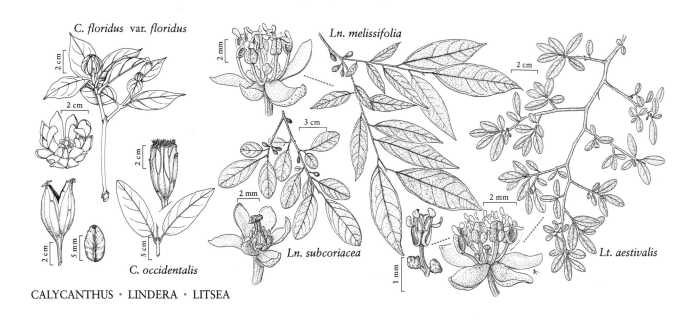

C. floridus var. *floridus*

Ln. melissifolia

Ln. subcoriacea

Lt. aestivalis

C. occidentalis

CALYCANTHUS · LINDERA · LITSEA

apex rounded; stamens 10–15, linear to oblong-linear. $2n = 22$.

Flowering late spring–early fall, fruiting mid fall. Along streams and on moist canyon slopes; 200–1600 m; Calif., Wash.

Calycanthus occidentalis grows in the northern Coast Range, the southern Cascades Range, and the western Sierra Nevada. It is ecologically similar to *C. floridus;* it consistently differs from that taxon in a number of vegetative and floral characteristics. Because of an apparent lack of hardiness, *C. occidentalis* is cultivated less often than *C. floridus.*

Some Native American Indians used scraped bark of *Calycanthus occidentalis* medicinally in treating severe colds (D. E. Moerman 1986).

5. LAURACEAE Jussieu

· Laurel Family

Henk van der Werff

Shrubs to tall trees, evergreen or rarely deciduous (*Cassytha* a parasitic vine with leaves reduced to scales), usually aromatic. **Leaves** alternate, rarely whorled or opposite, simple, without stipules, petiolate. **Leaf blade:** unlobed (unlobed or lobed in *Sassafras*), margins entire, occasionally with domatia (tufts of hair, or crevices or hollows serving as lodging for mites) in axils of main lateral veins (in *Cinnamomum*). **Inflorescences** in axils of leaves or deciduous bracts, panicles (rarely heads), racemes, compound cymes, or pseudoumbels (spikes in *Cassytha*), sometimes enclosed by decussate bracts. **Flowers** bisexual or unisexual, usually yellow to greenish or white, rarely reddish; hypanthium well developed, resembling calyx tube, tepals and stamens perigynous; tepals 6, in 2 whorls of 3, equal or rarely unequal, if unequal then usually outer 3 smaller than inner 3 (occasionally absent in *Litsea*); stamens (3–6)9(–12), in whorls of 3, but 1 or more whorls frequently staminodial or absent; stamens of 3d whorl with 2 glands near base; anthers 2- or 4-locular, locules opening by valves; pistil 1, 1-carpellate; ovary 1-locular; placentation basal; ovule 1; stigma subsessile, discoid or capitate. **Fruits** drupes, drupe borne on pedicel with or without persistent tepals at base, or seated in ± deeply cup-shaped receptacle (cupule), or enclosed in accrescent floral tube. **Seed** 1; endosperm absent.

Genera ca. 50, species 2000–3000 (9 genera, 13 species in the flora): pantropical, a few species also in subtropical and temperate regions; best represented in tropical America, Asia.

Cassytha is sometimes placed in its own family, Cassythaceae; it is here retained in Lauraceae.

SELECTED REFERENCE   Wood, C. E. Jr. 1958. The genera of the woody Ranales in the southeastern United States. J. Arnold Arbor. 39: 296–346.

Key to Genera Based on Flowering Material

1. Parasitic vines, leaves reduced to minute scales; stems pale green to yellow-green or orange, twining . . . . 9. *Cassytha*, p. 35
1. Shrubs or trees, leafy; stems various in color but not orange, not twining.

2. Plants deciduous; flowers appearing before or with new leaves.
   3. Flowers in racemes or panicles; leaf blade often lobed . . . . 3. *Sassafras*, p. 29
   3. Flowers in pseudoumbels; leaf blade always unlobed.
      4. Anthers 2-locular . . . . 1. *Lindera*, p. 27
      4. Anthers 4-locular . . . . 2. *Litsea*, p. 29
2. Plants evergreen; flowers appearing when leaves mature.
   5. Flowers in pseudoumbels . . . . 4. *Umbellularia*, p. 31
   5. Flowers in panicles or compound cymes.
      6. Stamens 3, anthers 2-locular . . . . .6. *Licaria*, p. 32
      6. Stamens 9, anthers 4-locular.
         7. Outer 3 tepals shorter than inner 3 . . . . 8. *Persea*, p. 34
         7. Tepals equal.
            8. Leaf blade pinnately veined, domatia absent; terminal bud not covered by
               imbricate scales . . . . 7. *Nectandra*, p. 33
            8. Leaf blade with (1–)3 primary veins, pubescent domatia in axils of main
               lateral veins; terminal bud covered by imbricate scales, young twigs with
               clusters of scars from fallen scales . . . . 5. *Cinnamomum*, p. 32

Key to Genera Based on Fruiting Material

1. Parasitic vines, leaves reduced to minute scales . . . . 9. *Cassytha*, p. 35
1. Shrubs or trees, leafy.
   2. Leaf blade usually lobed (often unlobed) . . . . 3. *Sassafras*, p. 29
   2. Leaf blade always unlobed.
      3. Tepals persistent at base of fruit; cupule absent . . . . 8. *Persea*, p. 34
      3. Tepals deciduous; small cupule present.
         4. Cupule usually double-rimmed . . . . .6. *Licaria*, p. 32
         4. Cupule single-rimmed.
            5. Fruit at maturity 2 cm or more in greatest dimension; California, Oregon . . .
            . . . . 4. *Umbellularia*, p. 31
            5. Fruit at maturity less than 2 cm in greatest dimension; e of Rocky Mountains.
               6. Infructescences umbellate or not branched, about 1 cm.
                  7. Leaf blade 4 × 1.5 cm or less . . . . 2. *Litsea*, p. 29
                  7. Leaf blade 4 × 2 cm or more . . . . 1. *Lindera*, p. 27
               6. Infructescences paniculate, more than 4 cm.
                  8. Leaf blade pinnately veined, domatia absent; terminal bud not covered
                     by imbricate scales . . . . 7. *Nectandra*, p. 33
                  8. Leaf blade with (1–)3 primary veins, pubescent domatia in axils of
                     main lateral veins; terminal bud covered by imbricate scales, young
                     twigs with clusters of scars from fallen scales . . . . 5. *Cinnamomum*, p. 32

1. LINDERA Thunberg, Nova Gen. Pl. 3: 64. 1783, name conserved · Spicebush [for John Linder, 1676–1723, Swedish botanist]

B. Eugene Wofford

Benzoin Boerhaave ex Schaeffer

Shrubs or small trees, deciduous. **Bark** grayish, becoming darker with age. **Leaves** alternate, aromatic when crushed (at least when young). **Leaf blade** pinnately veined, membranous to nearly leathery; surfaces glabrous to densely pubescent; domatia absent. **Inflorescences** appearing before leaves, axillary, clusters (pseudoumbels), clusters subsessile, nearly umbellate, each subtended by 2 pairs of decussate bracts. **Flowers** unisexual, staminate and pistillate on different plants, a few bisexual flowers on some plants; tepals deciduous, yellow, pellucid-dotted, equal, glabrous. **Staminate flowers:** stamens 9; anthers 2-locular, 2-valved, introrse. **Pistillate flowers:** staminodes variously developed; ovary globose. **Drupe** bright red, ellipsoid to nearly globose, borne on pedicel, with or without persistent tepals at base. $x = 12$.

Species ca. 100 (3 in the flora): North America, e Asia.

SELECTED REFERENCES   McCartney, R. D., K. Wurdack, and J. Moore. 1989. The genus *Lindera* in Florida. Palmetto 9: 3–8.   Steyermark, J. A. 1949. *Lindera melissaefolia*. Rhodora 51: 153–162.   Wofford, B. E. 1983. A new *Lindera* from North America. J. Arnold Arbor. 64: 325–331.

1. Leaf blade somewhat leathery, larger blades usually less than 8 × 4 cm, young leaves faintly aromatic when crushed, becoming essentially odorless with age . . . . .1. *Lindera subcoriacea*
1. Leaf blade membranous, larger blades usually more than 8 × 4 cm, crushed leaves strongly aromatic throughout growing season.
   2. Leaves horizontal to mostly ascending; blade obovate, base cuneate, apex acuminate on larger leaves; fruiting pedicels of previous season not persistent on stem, not conspicuously enlarged at apex; shrubs or small trees . . . . 2. *Lindera benzoin*
   2. Leaves drooping; blade elliptic to ovate, base rounded to widely cuneate, apex acute; fruiting pedicels of previous season persistent on stem, enlarged at apex; low shrubs rarely over 1.5 m . . . . 3. *Lindera melissifolia*

1. Lindera subcoriacea Wofford, J. Arnold Arbor. 64: 325. 1983 · Bog spicebush C E F

Shrubs, 2 m (to 4 m when shaded). **Young twigs** pubescent, glabrescent with age. **Leaves** horizontal to mostly ascending, faintly aromatic (piny lemon) when young, becoming essentially odorless with age; petiole 3–10 mm, pubescent. **Leaf blade** elliptic to oblanceolate, 4–8 × 2–4 cm, somewhat leathery, base cuneate, margins ciliate when young, apex obtuse to rounded; surfaces abaxially pubescent, adaxially pubescent when young, becoming glabrous with age. **Drupe** ellipsoid, ca. 10 mm; pedicels of previous season not persistent on stem, slender, to 4 mm, apex not conspicuously enlarged.

Flowering spring. Evergreen-shrub bogs, acidic swamps of blackwater swamp forests, acidic seepage bogs; Gulf and Atlantic coastal plains and adjacent Piedmont; of conservation concern; 0–200 m; Ala., Fla., Ga., La., Miss., N.C., S.C., Va.

Lindera subcoriacea was described originally from Mississippi and Louisiana. R. D. McCartney et al. (1989) reported it from the other sites.

2. Lindera benzoin (Linnaeus) Blume, Mus. Bot. 1: 324. 1851 · Spicebush, Benjamin bush E

Laurus benzoin Linnaeus, Sp. Pl. 1: 370. 1753; *Benzoin aestivale* (Linnaeus) Nees; *Lindera benzoin* var. *pubescens* (Palmer & Steyermark) Rehder

Shrubs or small trees, to 5 m. **Young twigs** glabrous or sparsely pubescent. **Leaves** horizontal to ascending, strongly aromatic (spicy) throughout growing season; petiole ca. 10 mm, glabrous or pubescent. **Leaf blade** obovate, smaller blades generally elliptic, (4–)6–15 × 2–6 cm, membranous, base cuneate, margins ciliate, apex rounded to acuminate on larger leaves; surfaces abaxially glabrous to densely pubescent, adaxially glabrous except for a few hairs along midrib. **Drupe** oblong, ca. 10 mm; fruiting pedicels of previous season not persistent on stem, slender, 3–5 mm, apex not conspicuously enlarged. $2n = 24$.

Flowering spring. Stream banks, low woods, margins of wetlands; uplands, especially with exposed limestone; 0–1200 m; Ont.; Ala., Ark., Conn., Del., D.C., Fla., Ga., Ill., Ind., Kans., Ky., La., Maine, Md., Mass., Mich., Miss., Mo., N.H., N.J., N.Y., N.C., Ohio, Okla., Pa., R.I., S.C., Tenn., Tex., Vt., Va., W.Va.

The flowers of *Lindera benzoin* have an unusually sweet fragrance.

Among the Cherokee, Creek, Iroquois, and Rappahannock tribes, *Lindera benzoin* was used for various medicinal purposes (D. E. Moerman 1986).

3. Lindera melissifolia (Walter) Blume, Mus. Bot. 1: 324. 1851 (as melissaefolia) C E F

Laurus melissifolia Walter, Fl. Carol., 134. 1788; *Benzoin melissifolia* (Walter) Nees (as *melissaefolia*)

Low shrubs, rarely over 1.5 m. **Young twigs** pubescent. **Leaves** drooping, strongly aromatic (similar to sassafras or root beer) throughout growing season; petiole ca. 10 mm, pubescent. **Leaf blade** elliptic to ovate, 8–16 × 3–6 cm, membranous, base rounded to widely cuneate, margins ciliate, apex acute; surfaces abaxially and adaxially pubescent. **Drupe** nearly globose, ca. 12 mm; fruiting pedicels of previous season persistent on stem, stout, 9–12 mm, apex enlarged.

Flowering spring. Low woods, depressions, pond and sink margins; Coastal Plain and Mississippi Embayment; of conservation concern; 0–100 m; Ark., Ga., Miss., Mo., N.C., S.C.

Lindera melissifolia has not been seen in Alabama, Florida, or Louisiana in over a century.

The orthographic variants *"melissaefolia"* and *"melisaefolium"* have sometimes been used.

2. **LITSEA** Lamarck in J. Lamarck et al., Encycl. 3: 574. 1792, name conserved · [Litsea, the Chinese name for the plant]

Henk van der Werff

Shrubs [or trees], deciduous [or evergreen]. **Leaves** alternate, often aromatic. **Leaf blade** pinnately veined, rarely with 3 primary veins, leathery; surfaces glabrous or variously pubescent; domatia absent. **Inflorescences** appearing with or before new leaves, axillary, pseudoumbels, subtended by decussate bracts. **Flowers** unisexual, staminate and pistillate on different plants; tepals deciduous, yellow, green, or white, equal, glabrous. **Staminate flowers:** stamens 9 (or 12); anthers 4-locular, 4-valved, introrse. **Pistillate flowers:** staminodes 9 (or 12); ovary globose. **Drupe** red, globose, seated in small, single-rimmed cupule.

Species ca. 400 (1 in the flora): North America, Mexico, Central America, mostly in Asia.

Litsea merits revision and, as accepted here, it is probably polyphyletic. It is very similar to *Lindera* and best recognized by its 4-locular anthers (2-locular in *Lindera*).

1. **Litsea aestivalis** (Linnaeus) Fernald, Rhodora 47: 140. 1945 · Pond-spice [E]

Laurus aestivalis Linnaeus, Sp. Pl. 1: 370. 1753; *Glabraria geniculata* (Walter) Britton; *Litsea geniculata* (Walter) G. Nicholson

Shrubs, to 3 m. **Branches** glabrous or sparsely pubescent, with typical zigzag shape. **Leaf blade** lanceolate or narrowly elliptic, 1.5–4 × 0.4–1.5 cm, mostly glabrous; surfaces abaxially frequently with spreading hairs along base of midrib. **Inflorescences** in axils of deciduous bracts, flowers in 4–5-flowered pseudoumbels. **Flowers:** ca. 6 mm diam., from exposed overwintering buds; tepals yellow. **Staminate flowers:** stamens 9; pistillode absent. **Pistillate flowers:** staminodes 9; style slender; stigma capitate. **Drupe** red, ca. 10 mm.

Flowering late winter–spring. Within basins of limesinks or other depressional ponds or Carolina bays; 10–200 m; Ala., Fla., Ga., La., Miss., N.C., S.C., Va.

Litsea aestivalis is mostly an outer coastal plain species, although we do have a substantial number of coastal plain records for the extreme southern Appalachians.

3. **SASSAFRAS** J. Presl in F. Berchtold and J. S. Presl., Prir. Rostlin 2: 30, 67. 1825 · [Spanish *sasafras*]

Henk van der Werff

Trees, deciduous. **Bark** red-brown, aromatic, thick, furrowed on old trunks. **Leaves** alternate, aromatic. **Leaf blade** pinnately veined, lobed or unlobed, basal veins more strongly developed than distal ones, papery; surfaces abaxially glaucous, glabrous or slightly pubescent; domatia absent. **Inflorescences** appearing before or with young leaves, situated in axils of bracts surrounding terminal buds, racemes or racemose panicles, flowers subtended by 1 linear bract.

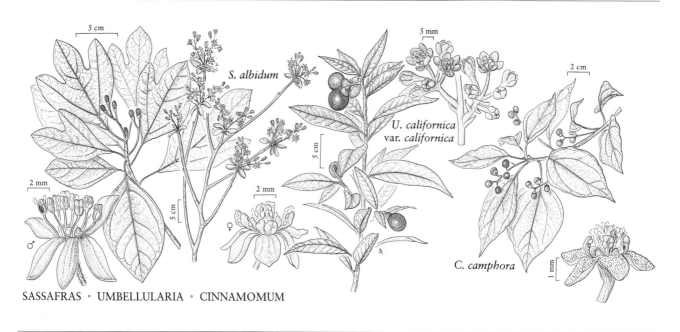

SASSAFRAS · UMBELLULARIA · CINNAMOMUM

Flowers unisexual, staminate and pistillate on different plants; tepals deciduous, yellowish, equal, pubescent. **Staminate flowers:** stamens 9, anthers 4-locular, 4-valved, introrse. **Pistillate flowers:** staminodes absent or present; ovary ovoid. **Drupe** blue, ovoid, seated in shallow, single-rimmed cupule.

Species 2 or 3 (1 in the flora): North America, Asia.

1. **Sassafras albidum** (Nuttall) Nees, Syst. Laur., 490. 1836 · Sassafras, white sassafras, filé, gombo filé E F WI

Laurus albida Nuttall, Gen. N. Amer. Pl. 1: 259. 1818; *L. sassafras* Linnaeus; *Sassafras albidum* var. *molle* (Rafinesque) Fernald; *S. officinalis* T. Nees ex C. H. Ebermaier; *S. sassafras* (Linnaeus) H. Karsten; *S. variifolium* (Salisbury) Kuntze

Trees, to 35 m. **Twigs** pale green with darker olive mottling, terete. **Leaf blade** ovate to elliptic, unlobed or 2–3-lobed (rarely more), 10–16 × 5–10 cm, apex obtuse to acute. **Inflorescences** to 5 cm, silky-pubescent; floral bract to 1 cm. **Flowers:** fragrant (sweet, lemony), glabrous; tepals greenish yellow. **Staminate flowers:** inner 3 stamens with 2 conspicuously stalked glands near base of filament, filament slender; pistillodes usually absent (sometimes present in terminal flower of inflorescence). **Pistillate flowers:** staminodes 6; style slender, 2–3 mm; stigma capitate. **Drupe** ca. 1 cm; pedicel reddish, club-shaped, ± fleshy. $2n = 48$.

Flowering spring (Apr–May). Habitat varied, forests, woodlands, fencerows, old fields (sometimes ag-gressively colonial), and disturbed areas; 0–1500 m; Ont.; Ala., Ark., Conn., Del., D.C., Fla., Ga., Ill., Ind., Iowa, Kans., Ky., La., Maine, Md., Mass., Mich., Miss., Mo., N.H., N.J., N.Y., N.C., Ohio, Okla., Pa., R.I., S.C., Tenn., Tex., Vt., Va., W.Va.

Infraspecific taxa have been based on amount of pubescence of leaves and color of young twigs; these taxa are not recognized here.

Traditionally, "sassafras tea" was prepared by steeping the bark of the roots (D. S. Correll and M. C. Johnston 1970). It was once considered a relatively pleasant drink. Several indigenous populations used sassafras twigs as chewing sticks, and sassafras root is used occasionally in commercial dental poultices. Sassafras root was one of the ingredients of root beer; this use has now been banned.

4. UMBELLULARIA (Nees) Nuttall, N. Amer. Sylv. 1: 87. 1842 · California bay [Latin *umbellula,* partial umbel]

Henk van der Werff

Oreodaphne Nees & Martius ex Nees subg. *Umbellularia* Nees, Syst. Laur., 464. 1836

Shrubs or trees, evergreen. **Bark** dark brown, thin. **Leaves** alternate, with pungent odor when crushed. **Leaf blade** pinnately veined, leathery; surfaces glabrous or sparsely pubescent, gland-dotted; domatia absent. **Inflorescences** appearing when mature leaves are present, axillary, pseudoumbels, stalked, young pseudoumbel enclosed by spirally arranged bracts. **Flowers** bisexual; tepals deciduous, yellowish, equal, glabrous; stamens 9, anthers 4-locular, 4-valved, outer anthers introrse, inner anthers extrorse; staminodes 3, small, stipitate; ovary superior, ovoid. **Drupe** greenish, dark purple when dried, spheric-ovoid, seated in flat, small, single-rimmed cupule.

Species 1: w North America.

1. **Umbellularia californica** (Hooker & Arnott) Nuttall, N. Amer. Sylv. 1: 87. 1842 · California bay, Oregon-myrtle, myrtle-wood, California laurel, pepperwood [F]

Tetranthera californica Hooker & Arnott, Bot. Beechey Voy., 159. 1833

Trees or shrubs, to 45 m; twigs terete, glabrous or sparsely appressed-pubescent, rarely minutely tomentose. **Leaf blade** deep yellow-green, shiny, narrowly oblong or narrowly elliptic, 3–10 × 1.5–3 cm, base acute or obtuse, apex acute; surfaces abaxially glabrous, sparsely appressed-pubescent or minutely tomentose, adaxially glabrous; domatia absent. **Inflorescences** pubescent. **Flowers** 5–10; tepals 6–8 mm. **Drupe** usually solitary, 2 cm or more diam. $2n = 24$.

Varieties 2 (2 in the flora): w coast, North America.

Native Americans used *Umbellularia californica* for medicinal purposes and occasionally as an insecticide (D. E. Moerman 1986).

1. Leaf blade abaxially glabrous or sparsely appressed-pubescent. . . . . .1a. *Umbellularia californica* var. *californica*
1. Leaf blade abaxially minutely tomentose. . . . . 1b. *Umbellularia californica* var. *fresnensis*

1a. **Umbellularia californica** (Hooker & Arnott) Nuttall var. **californica** [F]

Leaf blade abaxially glabrous or sparsely appressed-pubescent. **Inflorescence stalk** glabrous or pubescent, pubescence generally less dense and shorter than in var. *fresnensis.*

Flowering fall–early spring (Nov–Apr). Various habitats, mostly in canyons and valleys; 0–1500 m; Calif., Oreg.; Mexico (Baja California).

1b. **Umbellularia californica** (Hooker & Arnott) Nuttall var. **fresnensis** Eastwood, Leafl. W. Bot. 4: 166. 1945 [C] [E]

Leaf blade minutely tomentose abaxially. **Inflorescence stalk** with longer and denser pubescence than in var. *californica.*

Flowering fall–early spring (Nov–Apr). In woods; of conservation concern; 200 m; Calif.

Umbellularia californica var. *fresnensis* is known only from Fresno County, California. The type seems quite distinct from specimens of var. *californica,* but status and distribution of var. *fresnensis* should be further researched. Because I have not seen intermediate specimens, I am not willing to reduce var. *fresnensis* to synonymy.

5. CINNAMOMUM Schaeffer, Bot. Exped., 74. 1760, name conserved · [Greek *kinnamomon*, cinnamon]

Henk van der Werff

Trees or shrubs, evergreen. **Bark** gray [or brown], furrowed [or smooth]; bark and leaves often aromatic. **Leaves** alternate, infrequently opposite. **Leaf blade** with (1–)3 primary veins [or infrequently pinnately veined], papery to leathery; surfaces glabrous or variously pubescent; domatia frequently present. **Inflorescences** appearing when mature leaves are present, axillary, panicles. **Flowers** bisexual; tepals deciduous or persistent, white, green, or yellow, equal; stamens 9, anthers 4-locular, 4-valved (rarely with anthers of inner 3 stamens 2-locular), extrorse; staminodes 3, apex sagittate or cordate; ovary ovoid-ellipsoid. **Drupe** bluish black, nearly globose, seated in small cupule with entire single rim or tepals persistent.

Species 300 or more (1 in the flora): tropical and subtropical regions, North America, Central America, South America, Asia, Pacific Islands, Australia.

The neotropical species were formerly included in *Phoebe,* but they are better placed in *Cinnamomum.*

1. **Cinnamomum camphora** (Linnaeus) J. Presl in F. Berchtold and J. S. Presl, Prir. Rostlin 2: 47. 1825 · Camphor tree F

Laurus camphora Linnaeus, Sp. Pl. 1: 369. 1753; *Camphora camphora* (Linnaeus) H. Karsten

Trees, to 15 m. **Branches** terete, glabrous, terminal and axillary buds covered by imbricate bracts, young twigs with clusters of scars from fallen bracts. **Leaves** alternate; petiole to 3 cm. **Leaf blade** ovate to elliptic or elliptic-lanceolate, with (1–)3 primary veins, 7–12 × 3–5 cm, base rounded to cuneate, apex sharply acute; surfaces glabrous except for pubescent domatia in axils of main lateral veins. **Flowers:** tepals greenish white, 1–2 mm, glabrous abaxially, pubescent adaxially; stamens arranged in outer whorl of 6 (actually 2 whorls of 3) and inner whorl of 3. **Drupe** to 9 mm diam. $2n = 24$.

Flowering spring (Apr–May). Moist subtropical areas, including the Gulf Coast; 0–150 m; introduced; Ala., Miss., Fla., Ga., La., N.C., S.C., Tex.; native, e Asia.

Cinnamomum camphora is naturalized locally in the flora. Its crushed leaves have a strong smell of camphor. This species yields commercial camphor.

The name *Cinnamomum camphora* (Linnaeus) J. Presl has priority over *C. camphora* (Linnaeus) T. Nees & C. H. Ebermaier.

6. LICARIA Aublet, Hist. Pl. Guiane 1: 313. 1775 · Licari [local name in French Guiana]

Henk van der Werff

Acrodiclidium Nees & Martius; *Chanekia* Lundell; *Misanteca* Schlechtendal & Chamisso

Shrubs or trees, evergreen. **Bark** gray [pinkish, purplish, or maroon-brown], smooth with small, wartlike lenticels. **Leaves** alternate [rarely opposite or clustered at tips of branches]. **Leaf blade** pinnately veined, papery or leathery; surfaces glabrous or variously pubescent; domatia absent. **Inflorescences** appearing when mature leaves are present, axillary, panicles (rarely heads). **Flowers** bisexual; tepals deciduous, white or green, equal (rarely unequal), at anthesis mostly erect, glabrous (or pubescent); stamens 3, anthers 2-locular, 2-valvate, apical (or introrse); staminodes 9, 6, 3, or 0; ovary superior, enclosed by deep floral tube. **Drupe** red or purple, ellipsoid, seated in deep, usually double-rimmed cupule.

Species ca. 40 (1 in the flora): restricted to neotropics.

SELECTED REFERENCE   Kurz, H. 1983. Fortpflanzungsbiologie einiger Gattungen neotropischer Lauraceen und Revision der Gattung *Licaria* (Lauraceae). Ph.D. thesis. University of Hamburg.

1. Licaria triandra (Swartz) Kostermans, Recueil Trav. Bot. Néerl. 34: 588. 1937 · Gulf triandra [F]

Laurus triandra Swartz, Prodr., 65. 1788; *Acrodiclidium triandrum* (Swartz) Lundell; *Misanteca triandra* (Swartz) Mez

Trees, 8–12 m. **Leaf blade** narrowly elliptic, 5–15 × 2–7 cm, base acute, apex slightly acuminate; surfaces glabrous when mature. **Inflorescences** to 10 cm, flowers often densely grouped at tips of inflorescence branchlets. **Flowers** 2–3 mm; tepals glabrous; stamens longer than tepals; pistil glabrous. **Drupe** to 3 cm.

Flowering late winter–spring. In hammocks; 0–10 m; Fla.; West Indies; Central America; South America (Andes).

7. NECTANDRA Rolander ex Rottbøll, Acta Lit. Univ. Hafn. 1: 279. 1778, name conserved · [Latin *nectar,* from Greek *nektar,* and Greek *andro,* male]

Henk van der Werff

Trees or shrubs, evergreen. **Bark** dark reddish brown [brown, or gray], smooth with small wartlike lenticels. **Leaves** alternate. **Leaf blade** pinnately veined, papery or leathery; surfaces variously pubescent; domatia absent [or present]. **Inflorescences** appearing when mature leaves are present, axillary, panicles, usually many-flowered. **Flowers** bisexual, 5–17 mm diam.; tepals deciduous, white or greenish, equal, spreading at anthesis, with papillose hairs on adaxial surface; stamens 9, anthers 4-locular, anthers of outer 6 stamens introrse, locules arranged in arc, anthers of inner 3 stamens extrorse; staminodes 3, very small, sometimes absent. **Drupe** dark blue or black, ± elongate, seated in shallow [or cup-shaped], single-rimmed cupule.

Species ca. 120 (1 in the flora): nearly all neotropical.

SELECTED REFERENCE   Rohwer, J. G. 1993. Lauraceae: *Nectandra.* In: Organization for Flora Neotropica. 1968+. Flora Neotropica. 65+ vols. New York. Monogr. 60.

1. Nectandra coriacea (Swartz) Grisebach, Fl. Brit. W. I., 281. 1860 · Lancewood [F]

Laurus coriacea Swartz, Prodr., 65. 1788; *L. catesbyana* Michaux; *Nectandra catesbyana* (Michaux) Sargent; *Ocotea catesbyana* (Michaux) Sargent; *O. coriacea* (Swartz) Britton

Trees, to 10 m. **Branches** dense; terminal bud not covered by imbricate bracts. **Leaf blade** narrowly elliptic, 7–12 × 2–4 cm, base and apex acute; surfaces glabrous; domatia absent. **Inflorescences** to 8 cm, nearly glabrous or sparsely pilose. **Flowers:** tepals white, rotate; ovary superior, glabrous. **Drupe** ± ellipsoid, to 18 mm, seated in shallow, bowl-shaped cupule to 7 mm diam.

Flowering late winter–spring (Mar–Jun). In woods; 0–100 m; Fla.; Mexico (Yucatán); West Indies; Central America (Caribbean coasts of Belize, Guatemala, and Honduras).

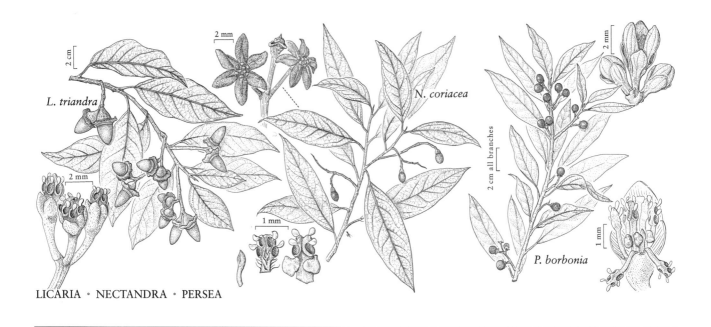

LICARIA · NECTANDRA · PERSEA

8. PERSEA Miller, Gard. Dict. Abr. ed. 4., vol. 3. 1754, name conserved · Red bay [used by Theophrastus for an oriental tree]

B. Eugene Wofford

Tamala Rafinesque

Shrubs to medium-sized trees, evergreen. **Bark** reddish brown, thin, fissured. **Leaves** alternate, aromatic. **Leaf blade** pinnately veined; surfaces pubescent, especially abaxially, becoming glabrescent with age; domatia absent. **Inflorescence** appearing when mature leaves are present, axillary, flowers in pedunculate, compound cymes. **Flowers** bisexual; tepals persistent, yellowish, pubescent, outer tepals slightly shorter than inner; stamens 9, anthers 4-locular, 4-valved, anthers of outer 6 stamens introrse, anthers of inner 3 latrorse; staminodes 3, sagittate; ovary nearly globose. **Drupe** dark blue to black, nearly globose, borne on pedicel with tepals persistent at base; cupule absent. $x = 12$.

Species ca. 150 (3 in the flora): tropical and subtropical regions, Western Hemisphere.

Persea americana Miller, the avocado of commerce, with large, fleshy fruits and deciduous tepals, is widely grown in California and Florida. It may persist after cultivation.

Pubescence type and density are the most reliable characteristics in identifying species and are best seen on young leaves and branches.

1. Pubescence of branches and leaves erect, crisped; peduncle longer than subtending leaf petiole . . . . 1. *Persea palustris*
1. Pubescence of branches and leaves appressed; peduncle equal to or shorter than subtending leaf petiole.
    2. Abaxial leaf surface moderately pubescent, glabrescent with age; leaf blade 6–16 cm . . . . 2. *Persea borbonia*
    2. Abaxial leaf surface densely pubescent, appearing sooty with age; leaf blade 5–8 cm . . . . 3. *Persea humilis*

1. **Persea palustris** (Rafinesque) Sargent, Bot. Gaz. 67: 229. 1919 · Swamp red bay

Tamala palustris Rafinesque, Sylva Tellur., 137. 1838; *Persea borbonia* (Linnaeus) Sprengel var. *pubescens* (Pursh) Little; *P. pubescens* (Pursh) Sargent; *Tamala pubescens* (Pursh) Small

Trees, to 15(–20) m. **Branches** pubescent, hairs erect, crisped. **Leaf blade** ovate to elliptic, 8–14 × 2–5 cm; surfaces abaxially pale, glaucous, densely pubescent when young with rusty brown, erect, crisped hairs, glabrescent with age, adaxially green, lustrous. **Inflorescences:** peduncle longer than subtending leaf petiole, pubescent. **Drupe** ca. 8 mm wide. $2n = 24$.

Flowering spring–early summer. Primarily in wetlands but not restricted to them; swamps, marshes, low pinewoods, savannas, maritime forests (sometimes mixed with *Persea borbonia*); Atlantic and Gulf coastal plains, less common in the Piedmont; 0–185 m; Ala., Del., Fla. Ga., La., Md., Miss., N.C., S.C., Tex., Va.; West Indies (Bahamas).

2. **Persea borbonia** (Linnaeus) Sprengel, Syst. Veg. 2: 268. 1825 · Red bay E F

Laurus borbonia Linnaeus, Sp. Pl. 1: 370. 1753; *Persea littoralis* Small; *Tamala borbonia* (Linnaeus) Rafinesque; *T. littoralis* (Small) Small

Trees, to 25 m. **Branches** appressed-pubescent. **Leaf blade** narrowly elliptic to widely ovate, 6–16 × 2–5 cm; surfaces abaxially pale, glaucous, moder-ately pubescent when young with rusty brown, appressed hairs, glabrescent with age, adaxially green, lustrous. **Inflorescences:** peduncle equal to or shorter than subtending leaf petiole, pubescent. **Drupe** (8–)10 mm diam., usually glaucous. $2n = 24$.

Flowering spring–early summer. In hammocks, mixed hardwoods, coastal dunes, maritime forests; outer Atlantic and Gulf coastal plains, rarely in the Piedmont; 0–100 m; Ala., Ark., Fla., Ga., La., Miss., N.C., S.C., Tex.

Some Native Americans used *Persea borbonia* for medicinal purposes (D. E. Moerman 1986).

3. **Persea humilis** Nash, Bull. Torrey Bot. Club 22: 157. 1895 · Silk bay E

Persea borbonia (Linnaeus) Sprengel var. *humilis* (Nash) L. E. Kopp; *Tamala humilis* (Nash) Small

Shrubs or small trees, to 10 m. **Branches** appressed-pubescent. **Leaf blade** elliptic to lance-elliptic, 5–8 × 1.5–2.5 cm; surfaces abaxially obscured by dense (silky), rusty brown, appressed hairs when young, appearing sooty with age, adaxially green, lustrous. **Inflorescences:** peduncle mostly shorter than subtending petiole, pubescent. **Drupe** 10–12 mm wide, glaucous.

Flowering spring–early summer. Sand pine-scrub; 0–100 m; Fla.

Persea humilis produces fewer flowers and fruits per peduncle than the preceding two species.

9. CASSYTHA Linnaeus, Sp. Pl. 1: 35. 1753; Gen. Pl. ed. 5, 22. 1754 (as "Cassyta") · [Greek *kasytas,* name for *Cuscuta*]

Henk van der Werff

Vines, parasitic, with threadlike stems. **Leaves** reduced to minute scales, spirally arranged, glabrous or pubescent. **Inflorescences** spikes [panicles or racemes], rarely reduced to single flower. **Flowers** bisexual, sessile or shortly pedicellate, subtended by bract and 2 bracteoles; tepals persistent at apex of accrescent floral tube that surrounds fruit, greenish white or whitish, outermost row similar to bracts, innermost row larger; stamens 9 (or 6), anthers 2-locular, anthers of outer 6 stamens introrse, of inner 3 extrorse; staminodes 3 (or 6); ovary globose. **Drupe** black, globose, enclosed in floral tube, remnants of perianth apical.

Species ca. 17 (1 in the flora): tropical and subtropical regions, North America, mostly Australia, a few in Africa.

SELECTED REFERENCE Weber, J. L. 1981. A taxonomic revision of *Cassytha* (Lauraceae) in Australia. J. Adelaide Bot. Gard. 3: 187–262.

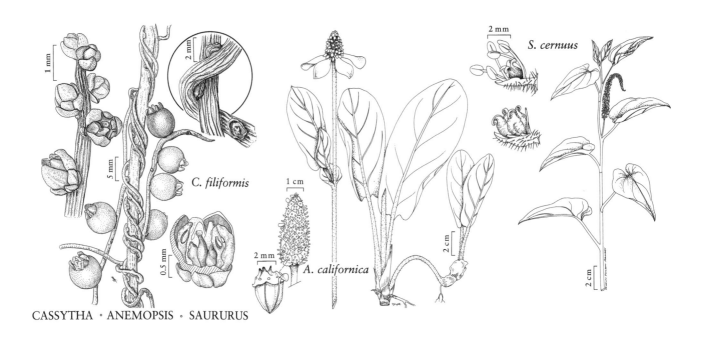

C. filiformis

A. californica

S. cernuus

CASSYTHA · ANEMOPSIS · SAURURUS

1. **Cassytha filiformis** Linnaeus, Sp. Pl. 1: 35. 1753 · Love-vine [F]

Stems twining, pale green to yellow-green to orange, filiform, glabrous or pubescent. **Leaves** alternate, ca. 1 mm. **Inflorescences** spikes, rarely reduced to single flower. **Flowers** bisexual, subtended by ciliate bract and bracteoles; outer 3 tepals 0.1–1 mm, similar to bracteoles, inner 3 tepals 1–1.8 mm, apex incurved; pistil 1.2 mm, glabrous. **Drupe** to 7 mm diam. $2n = 48$.

Flowering spring–summer (May–Jul). Coastal vegetation; 0–20 m; Fla.; Mexico; West Indies; Central America; South America; Asia; Africa (including Madagascar); Indian Ocean Islands; Pacific Islands; Australia.

Cassytha filiformis is a very distinctive plant that can be confused only with *Cuscuta*, a vining parasite of the Cuscutaceae.

6. SAURURACEAE E. Meyer

· Lizard's-tail Family

George F. Buddell II

John W. Thieret

Herbs, perennial, rhizomatous, aromatic, glabrous to pubescent, in wet places. **Stems** simple or branched; vascular strands in 1(–2) ring(s). **Leaves** basal and/or cauline, alternate, simple; stipules present, adnate to petiole; petioles usually present. **Leaf blade:** margins entire. **Inflorescences** terminal, compact, conic spikes or lax, spikelike racemes terminal and/or opposite leaves. **Flowers** bisexual; perianth absent, each flower subtended by nonpeltate bract; stamens (3–)6(–8), hypogynous or epigynous; anthers 2-locular; pistil 1, 3–5(–7)-carpellate; ovary 1- or 3–5(–7)-locular, superior or inferior; placentation parietal or marginal; ovules 2 or 18–40; styles and stigmas distinct. **Fruits** capsules or schizocarps. **Seeds** 1 or many (–40); endosperm scanty; perisperm abundant; embryo minute.

Genera 5, species 7 (2 genera, 2 species in the flora): North America, Central America, Asia.

SELECTED REFERENCES Engler, H. G. A. 1888. Saururaceae. In: H. G. A. Engler and K. Prantl, eds. 1887–1915. Die natürlichen Pflanzenfamilien. . . . 254 fasc. Leipzig. Fasc. 14[III,1], pp. 1–3. Raju, M. V. S. 1961. Morphology and anatomy of the Saururaceae. I. Floral anatomy and embryology. Ann. Missouri Bot. Gard. 48: 107–124. Wood, C. E. Jr. 1971b. The Saururaceae in the southeastern United States. J. Arnold Arbor. 52: 479–485.

1. Inflorescences compact, conic, terminal spikes subtended by petaloid bracts; leaves mostly basal; fruits capsules, seeds 18–40 . . . . 1. *Anemopsis*, p. 37
1. Inflorescences lax, spikelike racemes, opposite leaves and/or terminal, not subtended by petaloid bracts; leaves cauline; fruits schizocarps, mericarps (3–)4–5(–7), seed 1 per mericarp. . . . . 2. *Saururus*, p. 38

1. ANEMOPSIS Hooker & Arnott, Bot. Beechey Voy., 390. 1841 · Yerba mansa [Greek *anemone,* the windflower, and *opsis,* appearance]

Stems simple, nodes 1(–2). **Leaves** mostly basal. **Inflorescences** terminal, spikes, compact, conic, subtended by petaloid bracts. **Flowers** 75–150, coalescent; stamens epigynous, 6(–8); pistil 1, 3(–4)-carpellate; ovary inferior; placentae 3(–4), parietal; ovules 6–10 per placenta; styles and stigmas 3(–4), distinct. **Fruits** capsules, dehiscent. **Seeds** 18–40. $x = 11$.

Species 1: w North America.

SELECTED REFERENCES   Holm, T. 1905. *Anemiopsis* [sic] *californica* (Nutt.) H. and A. An anatomical study. Amer. J. Sci., ser. 4, 19: 76–82.   Kelso, L. 1932. A note on *Anemopsis californica*. Amer. Midl. Naturalist 13: 110–113.   Nozeran, R. 1955. Contribution à l'étude de quelques structures florales. (Essai de morphologie florale comparée). Ann. Sci. Nat., Bot., sér. 11, 16: 1–224.   Quibell, C. H. 1941. Floral anatomy and morphology of *Anemopsis californica*. Bot. Gaz. 102: 749–758.

1. Anemopsis californica Hooker & Arnott, Bot. Beechey Voy., 390. 1841   [F] [W2]

Anemopsis californica var. *subglabra* Kelso

Herbs, 8–80 cm, densely pubescent to nearly glabrous, producing stolons. **Basal leaves** 5–60 cm; petiole 2–40 cm; blade elliptic-oblong, 1–25 × 1–12 cm, base cordate to obtuse, apex rounded. **Cauline leaves** dimorphic; primary leaf 1(–2), usually bearing secondary leaves in axil; blade sessile, broadly to narrowly ovate, 1–9 × 1–4 cm, base clasping, apex rounded to acute. **Secondary leaves** 1–4, 2–20 cm; petiole 1–12 cm; blade elliptic-oblong, 2–10 × 1–5 cm, base cordate to rounded, apex rounded to acute. **Spikes** erect, fragrant, conic, 1–4 cm, subtended by bracts; bracts 4–9, white to reddish, petaloid, 5–35 × 5–15 mm. **Floral bracts** white, ± orbiculate, 3.5–6 mm (distinct portion), clawed, each adnate to an ovary. **Capsules** brown, 5–7 mm, coalescent but easily separable. **Seeds** brown, 1–1.5 × 0.8–1 mm, reticulate. $2n = 22$.

Flowering early spring–summer. Wet, alkaline, saline, and coastal marsh areas; 0–2000 m; Ariz., Calif., Colo., Kans., Nev., N.Mex., Okla., Oreg., Tex., Utah; n Mexico.

Some American Indians used *Anemopsis californica* for a variety of medicinal purposes (D. E. Moerman 1986).

2. SAURURUS Linnaeus, Sp. Pl. 1: 341. 1753; Gen. Pl. ed. 5, 159. 1754 · Lizard's-tail

[Greek *sauros*, lizard, and *oura*, tail]

Stems simple or branched, nodes more than 2. **Leaves** cauline. **Inflorescences** opposite leaves and/or terminal, racemes, lax, spikelike, not subtended by petaloid bracts. **Flowers** 175–350, not coalescent; stamens (3–)6(–8), hypogynous or epigynous; pistil 1, (3–)4–5(–7)-carpellate, connate only at base; ovary superior; placentation marginal; ovules 2 per placenta (only 1 developing into seed). **Fruits** schizocarps; mericarps dry, indehiscent. **Seed** 1 per mericarp. $x = 11$.

Species 2 (1 in the flora): e North America, e Asia.

SELECTED REFERENCES   Baldwin, J. T. and B. M. Speese. 1949. Cytogeography of *Saururus cernuus*. Bull. Torrey Bot. Club 76: 213–216.   Hall, T. F. 1940. The biology of *Saururus cernuus* L. Amer. Midl. Naturalist 24: 253–260.   Holm, T. 1926. *Saururus cernuus* L. A morphological study. Amer. J. Sci., ser. 5, 12: 162–168.   Johnson, D. S. 1900. On the development of *Saururus cernuus* L. Bull. Torrey Bot. Club 27: 365–372.   Tucker, S. C. 1975. Floral development in *Saururus cernuus* (Saururaceae). 1. Floral initiation and stamen development. Amer. J. Bot. 62: 993–1007.   Tucker, S. C. 1976. Floral development in *Saururus cernuus* (Saururaceae). 2. Carpel initiation and floral vasculature. Amer. J. Bot. 63: 289–301.   Tucker, S. C. 1979. Ontogeny of the inflorescence of *Saururus cernuus* (Saururaceae). Amer. J. Bot. 66: 227–236.

1. Saururus cernuus Linnaeus, Sp. Pl. 1: 341. 1753 · Lizard's-tail, saurure penché   [E] [F] [W2]

Herbs, 15–120 cm, mostly pubescent when young, often glabrate, having rhizomes, often with adventitious roots. **Leaves** 4–25 cm; petiole 1–10 cm; blade ovate, 2–17 × 1–10 cm, base cordate, apex acuminate. **Racemes** nodding to erect, fragrant, narrow, 5–35 cm. **Floral bracts** green, boat-shaped, 1.5–3 mm (distinct portion), adnate to pedicel. **Schizocarps** brown, 1.5–3 mm, rugose. **Seed** brown, 1–1.3 × 0.7–1 mm, smooth. $2n = 22$.

Flowering spring–summer, sometimes early fall. Wet soil, fresh or slightly brackish water to depth of 5 dm; 0–500 m; Ont., Que.; Ala., Ark., Conn., Del., Fla., Ga., Ill., Ind., Kans., Ky., La., Md., Mich., Miss., Mo., N.J., N.Y., N.C., Ohio, Okla., Pa., R.I., S.C., Tenn., Tex., Va., W.Va.

Some American Indians used *Saururus cernuus* for medicinal purposes (D. E. Moerman 1986).

7. PIPERACEAE C. Agardh

· Pepper Family

David E. Boufford

Small trees, shrubs, or perennial or annual herbs, often rhizomatous, sometimes aromatic, glabrous, pubescent, or glandular-dotted, terrestrial or epiphytic. **Stems** simple or branched; vascular bundles in more than 1 ring or scattered. **Leaves** basal and/or cauline, alternate, opposite, or whorled, simple; stipules present, adnate to petiole; petioles usually present. **Leaf blade:** margins entire. **Inflorescences** terminal, opposite leaves, or axillary, spikes. **Flowers** bisexual; perianth absent, each flower subtended by peltate bract; stamens 2–6, hypogynous; anthers 2-locular; pistil 1, 1- or 3–4-carpellate; ovary 1-locular, superior; placentation basal; ovule 1; stigmas usually 3–4. **Fruits** drupelike. **Seed** 1; endosperm scanty; perisperm abundant; embryo minute.

Genera 15, species 2000 (2 genera, 9 species in the flora): primarily tropical and subtropical regions worldwide.

Piper peltatum Linnaeus, a soft-wooded shrub to ca. 2 m, included by some authors in *Pothomorphe* or *Lepianthes,* has been collected as "growing wild" in Dade County, Florida (A. Herndon, pers. comm.). It differs from other species of *Piper* in the flora by its erect habit, by having axillary inflorescences, and by the spikes arranged in umbels.

SELECTED REFERENCES Candolle, C. de. 1869. Piperaceae. In: A. P. de Candolle and A. L. P. de Candolle, eds. 1823–1873. Prodromus Systematis Naturalis Regni Vegetabilis. . . . 17 vols. Paris etc. Vol. 16, part 1, pp. 235–471. Trelease, W. and T. G. Yuncker. 1950. The Piperaceae of Northern South America. 2 vols. Urbana.

1. PIPER Linnaeus, Sp. Pl. 1: 28. 1753; Gen. Pl. ed. 5, 18. 1754

Small trees, shrubs, subshrubs, or rarely herbs, erect or reclining, glabrous or pubescent. **Leaves** alternate, pubescent. **Leaf blade** conspicuously pinnately veined, lateral veins ascending-arching, connected by fainter, ladderlike, tertiary veins. **Spikes** opposite leaves, ascending-arching, densely flowered, distally drooping. **Flowers** sessile, borne on surface of rachis; floral bracts fringed with whitish hairs; stamens 2[–6]; stigmas [2–]3[–4]. **Fruits** sessile, oblong (inversely pyramidal–3-angled in *P. auritum*); beak minute.

Species 1000 (2 in the flora): primarily tropics and subtropics.

This genus includes *Piper nigrum* Linnaeus, the source of black pepper and white pepper.

Measurements for spike length in both descriptions include the peduncle.

1. Leaf blade obliquely rounded to obliquely cuneate at base; petiole not winged. . . . . 1. *Piper aduncum*
1. Leaf blade narrowly and deeply obliquely cordate at base; petiole winged at base . . . . 2. *Piper auritum*

1. Piper aduncum Linnaeus, Sp. Pl. 1: 29. 1753 [F]

Shrubs or small trees, erect, 2–7 m, sparsely pubescent. **Leaves:** petiole ca. 1/20 length of leaf blade, not winged. **Leaf blade** oblong, ovate, widely lanceolate to elliptic, 11–24 × 4–8 cm, base obliquely rounded to obliquely cuneate, apex acuminate; surfaces abaxially soft-pubescent, adaxially scabrous. **Spikes** 8–15 cm. **Fruits** oblong, 2 sides flattened longitudinally, both ends truncate, apex depressed, regularly pitted or reticulate; beak minute, ca. 0.1 mm.

Flowering all year. Thickets, woodland margins; 0–20 m; introduced, Fla.; West Indies; Central America; South America.

2. Piper auritum Kunth in A. von Humboldt et al., Nov. Gen. Sp. 1: 54. 1816

Shrubs or subshrubs (rarely herbaceous), reclining, 2–5 m, glabrous. **Leaves:** petiole 1/6–1/3 length of leaf blade, winged at base. **Leaf blade** ovate, 20–35 × 17–20 cm, base narrowly and deeply obliquely cordate, apex abruptly short-acuminate to acute; surfaces abaxially and adaxially minutely pubescent, most conspicuously so along veins. **Spikes** 12–25 cm. **Fruits** not seen.

Flowering all year. Thickets; 0–20 m; introduced, Fla.; Mexico; West Indies; Central America; n South America.

Piper auritum has been collected as "wild" in Broward County, Florida.

2. PEPEROMIA Ruiz & Pavón, Fl. Peruv. Prodr., 8. 1794 · Peperomia

Herbs, annual or perennial, erect, decumbent, or prostrate, terrestrial or epiphytic, glabrous or pubescent, sometimes glandular-dotted. **Leaves** alternate, opposite, or whorled, glabrous or pubescent, or glandular. **Leaf blade** conspicuously or inconspicuously veined, lateral veins ascending-arching, or inconspicuous, tertiary veins apparently absent or very faint. **Spikes** terminal, terminal and axillary, or opposite leaves, densely to loosely flowered. **Flowers** sessile, borne on surface or in pitlike depressions of rachis, floral bracts glabrous or glandular-dotted; stamens 2, attached at base of ovary; stigma 1, sometimes cleft. **Fruits** sessile or stipitate, globose, ovoid, oblong, or pyriform, surface warty, minutely reticulate, or faintly striate, ± viscid; beak mammiform or elongate, straight, bent, or hooked.

Species ca. 1000 (7 in the flora): mostly tropical and subtropical worldwide, especially tropical America and s Asia.

Many species of *Peperomia* are used as houseplants, greenhouse plants, and, in warm regions, garden plants.

In addition to the species below, *Peperomia simplex* Hamilton has been attributed to southern Florida, but no verifying specimens have been seen. A single specimen of *Peperomia emarginella* (Swartz ex Wikström) C. de Candolle, labeled "Alto, 7-16-1915, F. & S. 8725" (NY), may be from the United States.

SELECTED REFERENCE Boufford, D. E. 1982. Notes on *Peperomia* (Piperaceae) in the southeastern United States. J. Arnold Arbor. 63: 317–325.

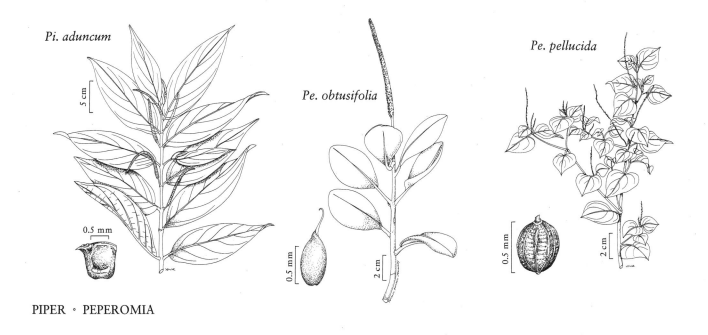

Pi. aduncum

Pe. obtusifolia

Pe. pellucida

PIPER · PEPEROMIA

1. Stems strigose; leaves opposite or whorled, sometimes proximal leaves alternate . . . . 1. *Peperomia humilis*
1. Stems glabrous or mostly so; leaves alternate.
  2. Plants with numerous black, glandular dots . . . . 2. *Peperomia glabella*
  2. Plants without black, glandular dots, or leaves occasionally with yellowish, resinous or pellucid
    dots.
    3. Leaf blade rounded, truncate, cordate, or auriculate at base.
      4. Leaf blade auriculate and ± clasping at base; petiole absent or to ca. 3 mm . . . . . . 6. *Peperomia amplexicaulis*
      4. Leaf blade rounded, truncate, or cordate at base, never clasping; petiole ca. 1/2 length of
        blade. . . . . 7. *Peperomia pellucida*
    3. Leaf blade narrowly to broadly cuneate, attenuate, or acuminate at base.
      5. Beak of fruit mammiform or conic, 0.2 mm or less; leaf blade elliptic, lanceolate, or oblan-
        ceolate; petiole slightly dilated and clasping at base, decurrent in lines or conspicuous in-
        terrnodal wings along stem . . . . 3. *Peperomia alata*
      5. Beak of fruit elongate, 0.5–1 mm; leaf blade oblanceolate, elliptic-obovate, ovate, or spatu-
        late; petiole not dilated at base, if appearing to be clasping, then not decurrent in lines or
        wings along stem.
        6. Peduncle with microscopic, spicule!ike hairs; beak of fruit filiform above conic base,
          abruptly hooked near apex . . . . .4. *Peperomia obtusifolia*
        6. Peduncle glabrous; beak of fruit tapering smoothly from broadened base to sharply
          acute apex, straight, bent, or gradually hooked from about middle.
          7. Leaves petiolate; leaf blade slightly decurrent along petiole, neither auriculate nor
            clasping. . . . . .5. *Peperomia magnoliifolia*
          7. Leaves sessile or nearly sessile; leaf blade auriculate at base, ± clasping. . . . . 6. *Peperomia amplexicaulis*

1. Peperomia humilis (Vahl) A. Dietrich, Sp. Pl. 1: 168. 1831

Piper humile Vahl, Enum. Pl. 1: 349. 1804; *Micropiper humile* (Vahl) Small; *M. leptostachyon* (Nuttall) Small; *Peperomia cumulicola* Small

Herbs, perennial, rhizomatous, erect or decumbent, simple or branched, 10–50 cm, strigose, without black, glandular dots. **Leaves** usually opposite or whorled, sometimes proximal leaves alternate; petiole ca. 1/4 length of blade, with ascending, incurved hairs. **Leaf blade** prominently to obscurely 3-veined from base, usually elliptic to obovate, sometimes proximal blades depressed-obovate, 1.2–6.5 × 0.4–3.5 cm, base cuneate to widely cuneate, apex acute or short-acuminate; surfaces abaxially and adaxially densely to sparsely strigose. **Spikes** axillary and terminal, 1–4, loosely flowered, 3–14 cm, mature fruiting spikes 1.5–2.5 mm diam. **Fruits** sessile, broadly ellipsoid to globose, 0.5–0.7 × 0.4–0.7 mm, warty, viscid; beak minute, knoblike, blunt, ca. 0.1 mm.

Flowering all year. Hammocks, often on limestone; 0–20 m; introduced, Fla.; native, West Indies (Jamaica).

2. Peperomia glabella (Swartz) A. Dietrich, Sp. Pl. 1: 156. 1831

Piper glabellum Swartz, Prodr., 16. 1788

Herbs, perennial, rhizomatous, erect, decumbent, or reclining, simple or branched, 8–45 cm, mostly glabrous, with numerous black, glandular dots. **Leaves** alternate; petiole 1/8–1/4 length of blade, pubescent in lines with upwardly curved hairs. **Leaf blade** 3–5-veined from base, ovate to narrowly or broadly elliptic, 2–6.5 × 1–4 cm, base nearly rounded to broadly cuneate, apex acute to short-acuminate; surfaces mostly glabrous. **Spikes** terminal or terminal and axillary, 1–4, densely flowered, 3–13 cm, mature fruiting spikes 2–3 mm diam. **Fruits** sessile, very broadly ovoid to globose, 0.7–0.8 × 0.6–0.7 mm, warty; beak obliquely conic, 0.1–0.2 mm.

Flowering all year. Epiphytic or terrestrial; 0–20 m; Fla.; West Indies; n South America.

3. Peperomia alata Ruiz & Pavón, Fl. Peruv. 1: 31. 1798

Herbs, perennial, rhizomatous, erect or reclining, simple or sparsely branched, 1–3 dm, glabrous, without black, glandular dots. **Leaves** alternate; petiole 1/10–1/5 length of blade, slightly dilated and clasping at base, decurrent in lines or conspicuous internodal wings. **Leaf blade** 3–5-veined from base, narrowly to broadly elliptic or elliptic-lanceolate or oblanceolate, 5–13 × 1.5–4 cm, base attenuate to acuminate, apex acute to short-acuminate; surfaces glabrous. **Spikes** terminal, 1–3, sometimes also with solitary spikes from proximal axils, densely flowered, 6–13 cm, mature fruiting spikes 2–3 mm diam. **Fruits** sessile, globose to very broadly ovoid, 0.5–0.8 × 0.6–0.8 mm, minutely warty; beak minute, mammiform or conic, 2 sides flattened, 0.1–0.2 mm.

Flowering all year. Swamps; 0–10 m; Fla.; West Indies; Central America; South America.

Peperomia alata is known from three collections (one without specific locality, the other two from Collier County, Florida). It is a common and widespread species of the West Indies, Central America, and South America. The characteristic wings on the stem are not so conspicuous as on many specimens from those regions.

4. Peperomia obtusifolia (Linnaeus) A. Dietrich, Sp. Pl. 1: 154. 1831 [F]

Piper obtusifolium Linnaeus, Sp. Pl. 1: 30. 1753; *Peperomia floridana* Small; *Rhynchophorum floridanum* Small; *R. obtusifolium* (Linnaeus) Small

Herbs, perennial, rhizomatous, erect, decumbent, reclining, or somewhat scandent, simple or sparsely branched, 8–40 cm, mostly glabrous, without black, glandular dots. **Leaves** alternate; petiole 1/5–1/2, mostly ca. 1/3, length of blade, glabrous. **Leaf blade** highly variable in size and shape, prominently to obscurely pinnately veined, elliptic, broadly elliptic, or spatulate to ovate, obovate, or very broadly ovate or broadly obovate, 2.2–13 × 1.5–5.5 cm, lateral veins arching-ascending, originating from base to about middle of blade, base attenuate, mostly narrowly to broadly cuneate, rarely slightly acuminate, apex obtuse, rounded, retuse, or notched; surfaces mostly glabrous. **Spikes** mostly terminal, 1–2, densely flowered, 5.5–23 cm; peduncle with microscopic, spiculelike hairs, mature fruiting spikes 3.5–5.5 mm diam. **Fruits** sessile, ellipsoid, both ends rounded, 0.8–1.1 × 0.4–0.5 mm, faintly longitudinally striate; beak elongate, 0.5–0.7 mm, filiform above narrowly to broadly conic base, straight or S-shaped, abruptly hooked near apex.

Flowering all year. Hammocks, epiphytic or terrestrial on rotten logs and humus; 0–20 m; Fla.; Mexico; West Indies; n South America.

5. Peperomia magnoliifolia (Jacquin) A. Dietrich, Sp. Pl. 1: 153. 1831 (as magnoliaefolia)

Piper magnoliifolium Jacquin, Collectanea 3: 210. 1791; *Peperomia spathulifolia* Small; *Rhynchophorum spathulifolium* (Small) Small

Herbs, perennial, rhizomatous, erect, decumbent, or reclining, simple or sparsely branched, 10–40 cm, glabrous, without black, glandular dots. **Leaves** alternate; petiole 1/6–1/4 length of blade, glabrous. **Leaf blade** prominently to obscurely pinnately veined, spatulate to broadly ovate or broadly elliptic, 3–14 × 1.5–6 cm, lateral veins arching-ascending, originating from base to near apex of blade, base attenuate or narrowly to broadly cuneate, not auriculate, not clasping, slightly decurrent along petiole, apex notched or retuse; surfaces resinous-dotted, especially abaxially. **Spikes** mostly terminal, 1–3, densely flowered, 6–17 cm; peduncle glabrous, mature fruiting spikes 2–3 mm diam. **Fruits** sessile, ellipsoid, base rounded or tapering, tapering gradually to beak, 0.7–0.9 × 0.5–0.6 mm, minutely warty; beak elongate, 0.5–1 mm, tapering smoothly from broadened base to sharply acute apex, without filiform apical portion, straight, bent, or gradually hooked from about middle.

Flowering all year. Hammocks, terrestrial or epiphytic; 0–20 m; Fla.; Mexico; West Indies; Bermuda; n South America.

6. **Peperomia amplexicaulis** (Swartz) A. Dietrich, Sp. Pl. 1: 144. 1831

Piper amplexicaule Swartz, Prodr., 16. 1788

Herbs, perennial, rhizomatous, erect, decumbent, or reclining, simple or sparsely branched, 10–60 cm, glabrous, without black, glandular dots. **Leaves** alternate; petiole absent or to ca. 3 mm, glabrous. **Leaf blade** prominently to obscurely pinnately veined, narrowly to broadly oblanceolate, 7–20 × 2–4.5 cm, lateral veins ascending, originating from base to near apex of blade, base decurrent, auriculate and ± clasping (amplexicaulous), apex acute or obtuse. **Spikes** mostly terminal, 1–3, densely flowered, 5–19 cm; peduncle glabrous, mature fruiting spikes 2–3 mm diam. **Fruits** sessile, ellipsoid, ca. 1.7 × 1 mm, minutely warty; beak elongate, 0.3–0.5 mm, tapering smoothly from broadened base to sharply acute apex, without filiform apical portion, straight, bent, or gradually hooked from about middle.

Flowering all year. Roadsides, woodlands; 0–20 m; introduced, Fla.; native, West Indies (apparently endemic to Jamaica).

7. **Peperomia pellucida** (Linnaeus) Kunth in A. von Humboldt et al., Nov. Gen. Sp. 1: 64. 1816 [F]

Piper pellucidum Linnaeus, Sp. Pl. 1: 30. 1753

Herbs, annual or short-lived perennial, erect or decumbent, freely branched, 10–50 cm, glabrous, without black, glandular dots. **Leaves** alternate; petiole ca. 1/2 length of blade, glabrous. **Leaf blade** palmately 5–7-veined, broadly ovate to deltate, 0.6–4 × 0.5–3 cm, base truncate, rounded, or cordate, apex acute to slightly acuminate; surfaces glabrous. **Spikes** axillary, terminal, and opposite leaves, solitary, rarely 2–more, loosely flowered, 2–6 cm, mature fruiting spikes 1–2 mm diam. **Fruits** sessile, very broadly ovoid to globose, 0.5–0.7 × 0.4–0.5 mm, longitudinally ribbed with ladderlike reticulations; beak minute, conic, ca. 0.1 mm.

Flowering all year. In shaded woods and around nurseries and greenhouses, along coastal plain; 0–20 m; Fla., Ga., La.; Mexico; West Indies; Central America; n South America.

Peperomia pellucida has shown antibacterial activity against *Staphylococcus aureus, Bacillus subtilis, Pseudomonas aeruginosa,* and *Escherichia coli;* it could have potential as a broad-spectrum antibiotic (A. C. Bojo et al. 1994).

8. ARISTOLOCHIACEAE Jussieu

• Dutchman's-pipe Family

Kerry Barringer

Alan T. Whittemore

Herbs or lianas [shrubs, rarely trees], deciduous or evergreen, often aromatic. **Wood** with broad medullary rays. **Leaves** alternate, simple, petiolate. **Leaf blade** unlobed, margins entire. **Inflorescences** terminal or axillary, racemes or solitary flowers, rarely fan-shaped cymes. **Flowers** bisexual; calyx enlarged, petaloid, usually tubular, [1-,] 3-, [6-, rarely 5-]merous, lobes valvate; corolla usually reduced to scales or absent; stamens 5, 6, or 12 [multiples of 3 or 5], free or adnate to styles and stigmas, forming gynostemium; anthers extrorse; pistil 1, 4–6-carpellate; ovary inferior, partly inferior, or superior; placentation axile (and ovaries 4–6-locular) or parietal; ovules many per locule, anatropous. **Fruits** capsules [follicles], regularly to irregularly loculicidal, rarely indehiscent [septicidal]. **Seeds** often flattened; endosperm copious.

Genera 5, species ca. 600 (3 genera, 28 species in the flora): primarily pantropical and subtropical.

SELECTED REFERENCES Duchartre, P. 1864. Aristolochiaceae. In: A. P. de Candolle and A. L. P. de Candolle, eds. 1823–1873. Prodromus Systematis Naturalis Regni Vegetabilis. . . . 17 vols. Paris etc. Vol. 15, pp. 421–498. Gregory, M. P. 1956. A phyletic rearrangement of the Aristolochiaceae. Amer. J. Bot. 43: 110–122. Schmidt, O. C. 1935. Aristolochiaceae. In: H. G. A. Engler et al., eds. 1924+. Die natürlichen Pflanzenfamilien . . . , ed. 2. 26+ vols. Leipzig and Berlin. Vol. 16b, pp. 202–242. Solereder, H. 1889. Aristolochiaceae. In: H. G. A. Engler and K. Prantl, eds. 1887–1915. Die natürlichen Pflanzenfamilien. . . . 254 fasc. Leipzig. Fasc. 35[III,1], pp. 264–273. Solereder, H. 1889b. Beiträge zur vergleichenden Anatomie der Aristolochiaceen. Bot. Jahrb. Syst. 10: 410–523, tables 12–14.

1. ARISTOLOCHIA Linnaeus, Sp. Pl. 2: 960. 1753; Gen. Pl. ed. 5, 410. 1754

· Dutchman's-pipe, aristoloche [Greek *aristolocheia*, birthwort, from *aristos*, best, and *lochia*, delivery, in reference to ancient use of herb as aid in childbirth]

Kerry Barringer

Endodeca Rafinesque; *Isotrema* Rafinesque; *Siphisia* Rafinesque

Herbs or lianas, perennial. **Stems** erect, twining, or procumbent. **Leaves** alternate, 2-ranked (evident on young growth, becoming obscure with age in some species); true stipules absent; pseudostipules absent [present]; petiole sometimes very short. **Leaf blade** membranous to leathery. **Inflorescences** on new growth or on older stems, axillary, racemes or solitary flowers; bracts present. **Flowers:** calyx usually mixture of purple, brown, green, or red, bilaterally symmetric, tubular, usually bent or curved, 1- or 3-lobed, not fleshy, base with utricle (basal, inflated portion of calyx surrounding or containing gynostemium); tube narrowed, sometimes extended proximally as cylindric syrinx (tubular or ringlike structure at juncture of tube and utricle, projecting into utricle cavity) and distally as annulus (circular flange at juncture of tube and limb) on limb; corolla absent; stamens 5–6, adnate to styles and stigmas, forming gynostemium; ovary inferior, 3-, 5-, or 6-locular; styles 3, 5, or 6, connate in column. **Capsule** dry, dehiscent. **Seeds** flattened or rounded, sometimes winged. $x = 6, 7, 8$.

Species ca. 300 (12 in the flora): nearly worldwide.

Most European and tropical species of *Aristolochia* are believed to be pollinated by small flies attracted to the flowers by the fetid odors and purple-brown color. Flies enter the flower when the stigmas are receptive and are trapped until after the anthers dehisce (H. Solereder 1889, 1889b). No formal studies of pollination of the North American species have been reported.

Many species of *Aristolochia* have been used in the treatment of snakebite; the treatment may or may not be effective. All species contain aristolochic acid, which is variously reported to be tumor-causing or tumor-inhibiting (J. A. Duke 1985).

The leaves of many species are eaten by pipe-vine swallowtail butterflies. The larvae eat leaves of these species and sequester aristolochic acid in their bodies, making them unpalatable to birds (W. H. Howe 1975).

SELECTED REFERENCES Pfeifer, H. W. 1966. Revision of the North and Central American hexandrous species of *Aristolochia* (Aristolochiaceae). Ann. Missouri Bot. Gard. 53: 1–114. Pfeifer, H. W. 1970. A Taxonomic Revision of the Pentandrous Species of *Aristolochia*. [Storrs.]

1. Leaf base attenuate, cuneate, or truncate (truncate to slightly cordate in *A. maxima*).
  2. Woody lianas; peduncle not bracteolate; capsule 4–7 cm wide, dehiscence acropetal; syrinx absent
    . . . . 11. *Aristolochia maxima*
  2. Herbs; peduncle bracteolate; capsule 1–3 cm wide, dehiscence basipetal; syrinx present.
    3. Inflorescences on new growth, axillary, solitary flowers; utricle ovoid to obconic; gynostemium 5-lobed; capsule ovoid. . . . . .6. *Aristolochia erecta*
    3. Inflorescences from base of stem, racemes; utricle pear-shaped to ovoid; gynostemium 3-lobed; capsule globose. . . . . 2. *Aristolochia serpentaria*
1. Leaf base cordate, auriculate, sagittate, or hastate.
  4. Lianas.
    5. Leaf venation palmate; calyx limb 1-lobed; young stem smooth; syrinx present; gynostemium 5-lobed. . . . . .9. *Aristolochia pentandra*
    5. Leaf venation pinnate or palmate-pinnate; calyx limb 3-lobed; young stem ribbed; syrinx absent; gynostemium 3- or 6-lobed.
      6. Leaf sinus 2–6 cm deep; capsule 1–3 cm wide; gynostemium 5–10 mm . . . . 4. *Aristolochia californica*
      6. Leaf sinus 1–4.5 cm deep; capsule 4–10 cm wide; gynostemium 1–4 mm.

7. Stem and leaf glabrous or puberulent; peduncle bracteolate; utricle globose to cylindric; annulus smooth . . . . 3. *Aristolochia macrophylla*
7. Stem tomentose; leaves abaxially tomentose; peduncle not bracteolate; utricle globose to cylindric; annulus rugulose . . . . 5. *Aristolochia tomentosa*
4. Herbs.
  8. Leaf venation palmate; leaf blade as wide as long or wider.
    9. Petiole 4–7 cm; stem erect; peduncle not bracteolate; gynostemium 3-lobed; young stem ribbed . . . . 12. *Aristolochia clematitis*
    9. Petiole 1–3 cm (to 4 cm in *A. coryi*); stem procumbent; peduncle bracteolate; gynostemium 5-lobed; young stem smooth or slightly striate.
      10. Stem and leaf velutinous with yellowish trichomes; leaf blade ovoid to hastate, (0–)3-lobed; calyx brown-purple, straight; gynostemium ovoid . . . . 10. *Aristolochia wrightii*
      10. Stem and leaf glabrous or puberulent; leaf blade deltate, not 3-lobed; calyx yellow-green or yellow-brown, curved; gynostemium crown-shaped. . . . . 7. *Aristolochia coryi*
  8. Leaf venation pinnate or palmate-pinnate; leaf blade longer than wide.
    11. Leaf apex obtuse or rounded; gynostemium 5–10 mm; syrinx indistinct or absent. . . . 1. *Aristolochia reticulata*
    11. Leaf apex acuminate or acute; gynostemium 1–4 mm; syrinx present.
      12. Calyx limb 1-lobed; inflorescences on new growth, axillary, solitary flowers; utricle ovoid to narrowly ellipsoid; gynostemium 5-lobed; capsule ovoid to obovoid; annulus absent on calyx limb. . . . . 8. *Aristolochia watsonii*
      12. Calyx limb 3-lobed; inflorescences from base of stem, rarely an additional flower in axil of stem leaf, racemes; utricle pear-shaped to ovoid; gynostemium 3-lobed; capsule globose; annulus present on calyx limb, smooth . . . . 2. *Aristolochia serpentaria*

1. Aristolochia reticulata Nuttall, Trans. Amer. Philos. Soc., n.s. 5: 162. 1835 · Texas Dutchman's-pipe [E][F]

Herbs, erect to sprawling, to 0.4 m. **Young stem** ribbed, hispid. **Leaves:** petiole 0.1–0.8 cm. **Leaf blade** ovate, 7–12 × 3–6 cm, base sagittate to auriculate, sinus depth 0.5–1.2 cm, apex obtuse or rounded; surfaces abaxially hispid; venation palmate-pinnate. **Inflorescences** from base of stem, racemes; peduncle bracteolate, 0.5–0.7 cm; bracteoles ovate-lanceolate, to 2 mm. **Flowers:** calyx brown-purple, bent; utricle pendent, pear-shaped to somewhat globose, 0.4–0.5 cm; syrinx indistinct or absent, oblique; tube horizontal, funnel-shaped, 5–7 × 1–3 cm; annulus absent; limb purplish brown, 3-lobed, lobes 0.4–0.6 × 0.5 cm, glabrous; gynostemium 3-lobed, globose, 5–10 mm; anthers 6; ovary 6-locular, 0.5–0.7 cm. **Capsule** globose, 1.2 × 1–3 cm, dehiscence basipetal; valves 6; septa entire, not attached to valves. **Seeds** rounded, ovate, 0.3 × 0.3 cm. $2n = 28$.

Flowering late spring–summer. Moist, sandy soils; 30–600 m; Ark., La., Okla., Tex.

The dried rhizome of *Aristolochia reticulata* is sometimes sold as serpentary. It is used as a tonic to calm the stomach, promote urination, and increase perspiration. The active ingredient is aristolochic acid, a potent gastric irritant that, in large doses, can cause respiratory paralysis. The leaves are eaten by larvae of the eastern pipe-vine swallowtail butterfly, *Battus philenor philenor* (Linnaeus) (W. H. Howe 1975).

2. Aristolochia serpentaria Linnaeus, Sp. Pl. 2: 961. 1753 · Virginia snakeroot [E]

Aristolochia convolvulacea Small; *A. hastata* Nuttall; *A. nashii* Kearney; *A. serpentaria* var. *hastata* (Nuttall) Duchartre

Herbs, erect to decumbent, to 0.6 m. **Young stem** ridged, glabrous to hispid. **Leaves:** petiole 0.5–3.5 cm. **Leaf blade** lanceolate to ovate, 5–15 × 1–5 cm, base truncate to cordate, sinus depth 0–1.5 cm, apex acute to acuminate; surfaces abaxially glabrous or hispid; venation pinnate. **Inflorescences** from base of stem, an additional flower in axil of stem leaf, racemes; peduncle bracteolate, to 1.5 cm; bracteoles lanceolate, to 3 mm. **Flowers:** calyx brown-purple, bent; utricle pendent, pear-shaped to ovoid, 0.5–5 cm; syrinx present, ringlike, 1 mm, oblique; tube bent, cylindric, 1 cm; annulus smooth; limb purplish brown, 3-lobed, lobes 0.5 × 0.5 cm, glabrous; gynostemium 3-lobed, globose to crown-shaped, 1.5 mm; anthers 6; ovary 3-locular, to 1.5 cm. **Capsule** globose, 0.8–2 × 1–2 cm, dehiscence basipetal; valves 6; septa absent. **Seeds** rounded, ovate, 0.5 × 0.4 cm. $2n = 28$.

Flowering late spring–summer. Mesic forests; 50–1300 m; Ala., Ark., Conn., Del., D.C., Fla., Ga., Ill., Ind., Iowa, Kans., Ky., La., Md., Mich., Miss., Mo., N.J., N.Y., N.C., Ohio, Okla., Pa., S.C., Tenn., Tex., Va., W.Va.

Inflorescences of *Aristolochia serpentaria* often bear closed flowers that appear to be cleistogamous. Leaf shape varies greatly among populations, especially

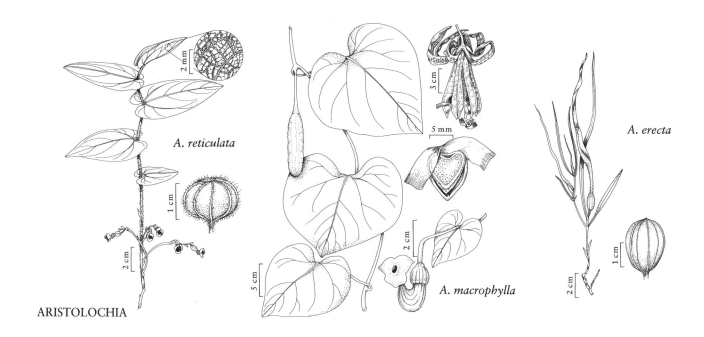

A. *reticulata*

A. *erecta*

A. *macrophylla*

ARISTOLOCHIA

with regard to leaf width and size of basal lobes. This variability is especially interesting because eastern pipe-vine swallowtail butterflies, *Battus philenor philenor* (Linnaeus), use leaf shape as a search image when looking for *Aristolochia* leaves on which to lay their eggs.

The dried rhizome, called Virginia snakeroot or serpentary, is a popular herbal tonic. In small doses, it is a gastric stimulant and diuretic. Large doses can cause violent gastric distress and respiratory paralysis (J. A. Duke 1985). The rhizome contains aristolochic acid and trimethylamine, both potential carcinogens.

Several Native American tribes used *Aristolochia serpentaria* for diverse medicinal purposes, including treatment of rheumatism, various pains, obstructions, worms, toothaches, sore throats, fever, sore noses, and colds, as a tonic, and mixed with saliva for snakebites (D. E. Moerman 1986).

3. **Aristolochia macrophylla** Lamarck in J. Lamarck et al., Encycl. 1: 255. 1783 · Dutchman's-pipe E F

Aristolochia sipho L'Héritier

Lianas, twining, to 20 m, woody. **Young stem** ribbed, glabrous. **Leaves:** petiole 4–6 cm. **Leaf blade** reniform, 7–34 × 10–35 cm, base cordate, sinus depth 1–4.5 cm, apex obtuse to acute or acuminate; surfaces abaxially glabrous to slightly puberulent; venation palmate-pinnate. **Inflorescences** on new growth, axillary, solitary flowers; peduncle brac-

teolate, 3–7 cm; bracteoles reniform, to 15 mm. **Flowers:** calyx yellow-brown marked with red-brown, strongly curved; utricle pendent, globose to cylindric, 0.5–1.5 × 0.8–1 cm; syrinx absent; tube curved or bent and angled upward, cylindric, 1–3 × 0.3–0.5 cm; annulus smooth; limb yellow to brown, 3-lobed, lobes 1.5–2 × 1.5–2 cm, glabrous; gynostemium 3-lobed, globose to crown-shaped, 4 mm; anthers 6; ovary 3-locular, 3–7 cm. **Capsule** ovoid to cylindric, 6–8 × 4–10 cm, dehiscence basipetal; valves 6; septa entire, not attached to valves. **Seeds*** flat, triangular, 1 × 1 cm. $2n = 28$.

Flowering late spring–summer. Forests, often on dissected uplands and rugged, rocky slopes; Cumberland and Blue Ridge mountains; 50–1300 m; Ont.; Ga., Ky., Md., N.C., Pa., S.C., Tenn., Va., W.Va.

Aristolochia macrophylla has possibly escaped from cultivation in Connecticut, Maryland, Massachusetts, New Jersey, New York, and Vermont; these are not documented.

---

* The unusual seeds of *Aristolochia macrophylla,* as released from the fruit, are tripartite structures. The seed proper (i.e., the embryo-containing portion) is flat, ± deltate, and ca. 8 × 7 mm. An appendage, resembling the seed proper in shape and size and attached to it by a thin thread, is said to develop through proliferation of tissue of the funiculus (E. J. H. Corner 1976). Finally, borne at the seed's broad end and centrally attached thereto is a translucent wing, which perhaps derives from the endocarp. Except for a few data on morphology of the seed in E. J. H. Corner (1976, vol. 1) and in H. Solereder (1889, 1889b), little seems known about these structures.—Ed.

The name *Aristolochia durior* Hill has been misapplied to this species.

The Cherokee applied decoctions made from the roots of *Aristolochia macrophylla* directly to feet and legs to alleviate swelling; they ingested a compound infusion of "stalk chips" for yellowish urine (D. E. Moerman 1986).

The leaves of *Aristolochia macrophylla* are eaten by larvae of the eastern pipe-vine swallowtail butterfly, *Battus philenor philenor* (Linnaeus) (W. H. Howe 1975).

4. Aristolochia californica Torrey, Pacif. Railr. Rep. 4(5): 128. 1857 · California snakeroot E

Lianas, twining, to 5 m, woody. **Young stem** ribbed, tomentose. **Leaves:** petiole 1–2.5 cm. **Leaf blade** ovate to reniform, 4–12 × 3–10 cm, base cordate, sinus depth 2–6 cm, apex obtuse to rounded; surfaces abaxially tomentose; venation palmate-pinnate. **Inflorescences** on new growth, axillary, solitary flowers; peduncle bracteolate, 1–4 cm; bracteoles reniform, to 8 mm. **Flowers:** calyx brown-purple, bent; utricle pendent, ellipsoid to narrowly ellipsoid, 0.5–5 × 1–1.5 cm; syrinx absent; tube bent, cylindric, 1–3 × 1–1.7 cm; annulus absent; limb purplish, 3-lobed, lobes 2 × 2 cm, glabrous; gynostemium 3-lobed, cylindric, 5–10 mm; anthers 6; ovary 6-locular, 1–4 cm. **Capsule** ovoid to cylindric, 5–6 × 1–3 cm, dehiscence basipetal; valves 6; septa entire, not attached to valves. **Seeds** flat, triangular, 1–1.2 × 1 cm. $2n = 28, 32$.

Flowering spring. Along streams, in forest thickets, chaparral; 50–700(–1000) m; Calif.

The leaves of *Aristolochia californica* are eaten by larvae of the western swallowtail butterfly, *Battus philenor hirsuta* (Skinner) (W. H. Howe 1975).

5. Aristolochia tomentosa Sims, Bot. Mag. 33: plate 1369. 1811 · Pipevine E

Isotrema tomentosa (Sims) H. Huber

Lianas, twining, to 25 m, woody. **Young stem** ribbed, tomentose. **Leaves:** petiole 1–5.5 cm. **Leaf blade** ovate to reniform, 9–20 × 8–15 cm, base cordate, sinus depth 1–2 cm, apex obtuse to acute; surfaces abaxially tomentose; venation palmate-pinnate. **Inflorescences** on new growth, axillary, solitary flowers; peduncle not bracteolate, 1–7 cm. **Flowers:** calyx yellow-green, sharply bent; utricle pendent, globose to cylindric, 0.5–1 × 0.5–0.8 cm;

syrinx absent; tube bent, cylindric, 1–3 × 0.5 cm; annulus rugulose; limb yellow, 3-lobed, lobes triangular, 2 × 2 cm, glabrous; gynostemium 3-lobed, globose, 3 mm; anthers 6; ovary 6-locular, 1–7 cm. **Capsule** ellipsoid to cylindric, 6–8 × 4–6 cm, dehiscence basipetal; valves 3; septa entire, not attached to valves. **Seeds** flat, triangular, 1 × 1 cm. $2n = 28$.

Flowering late spring–summer. Alluvial soils along rivers and streams; to 500 m; Ala., Ark., Fla., Ga., Ill., Ind., Kans., Ky., La., Miss., Mo., N.C., Okla., Tenn., Tex.

Aristolochia tomentosa has been reported from South Carolina. It has escaped from cultivation in various places, including Virginia; this is not documented.

6. Aristolochia erecta Linnaeus, Sp. Pl. ed. 2, 2: 1362. 1763 · Swanflower F

Aristolochia longiflora Engelmann & A. Gray

Herbs, erect to sprawling, to 0.25 m. **Young stem** ribbed, glabrous. **Leaves:** petiole 0.1–0.8 cm. **Leaf blade** linear-lanceolate, 10–15 × 0.5–1 cm, base cuneate to attenuate, apex acuminate; surfaces abaxially glabrous to puberulent; venation palmate-pinnate. **Inflorescences** on new growth, axillary, solitary flowers; peduncle bracteolate, 1–4 cm; bracteoles triangular to lanceolate, to 12 mm. **Flowers:** calyx brown-purple, curved; utricle angled upward, ovoid to obconic, 1.5 × 0.8 cm; syrinx present, broad and shallow, 1.5 mm; tube curved upward, cylindric, 1–3 × 0.4 cm; annulus absent; limb purple, 1-lobed, lobe strap-shaped, 8 × 1 cm, papillate; gynostemium 5-lobed, cylindric, 5–10 mm; anthers 5; ovary 5-locular, 1–4 cm. **Capsule** ovoid, 2.5 × 1–3 cm, dehiscence basipetal; valves 5; septa entire, not attached to valves. **Seeds** flat, triangular, 0.5 × 0.5 cm.

Flowering late spring–summer. Riverbanks, hillsides, chaparral; to 500 m; Tex.; Mexico.

7. Aristolochia coryi I. M. Johnston, J. Arnold Arbor. 21: 256. 1940 · Cory's Dutchman's-pipe

Herbs, procumbent, to 0.4 m. **Young stem** smooth, glabrous or puberulent. **Leaves:** petiole 1–3 cm. **Leaf blade** deltate, 3–5 × 3–5 cm, base cordate, sinus depth 1–2 cm, apex acute; surfaces abaxially puberulent; venation palmate. **Inflorescences** on new growth, axillary, solitary flowers; peduncle bracteolate, 1–4 cm; bracteoles lanceolate, to 5 mm. **Flowers:** calyx yellow-green or yellow-brown, curved; utricle horizontal, ovoid to obconic, 0.8 × 0.6 cm; syrinx present, funnel-shaped, 1–2 mm; tube curved, cylindric, 1–3 × 0.3 cm; annulus absent; limb brown with purple spots, 1-lobed, lobe strap-shaped, 1.5–2 × 1.5 cm, glabrous; gynostemium 5-lobed, crown-shaped, 1–4 mm; anthers 5; ovary 5-locular, 1–4 cm. **Capsule** globose, 1.2 × 1–3 cm, dehiscence basipetal; valves 5; septa entire, not attached to valves. **Seeds** flat, triangular, 0.5 × 0.6 cm.

Flowering late spring–summer. Rocky slopes, especially on limestone, occasionally river bottoms; to 600 m; Tex.; Mexico.

8. Aristolochia watsonii Wooton & Standley, Contr. U.S. Natl. Herb. 16: 117. 1913 · Indianroot

Aristolochia porphyrophylla Pfeifer

Herbs, procumbent, to 0.5 m. **Young stem** smooth, glabrous. **Leaves:** petiole 0.4–1 cm. **Leaf blade** lanceolate, 8 × 5 cm, base strongly sagittate, sinus depth 0.5–2 cm, apex acuminate to acute; surfaces abaxially tomentose to tomentulose; venation pinnate. **Inflorescences** on new growth, axillary, solitary flowers; peduncle bracteolate, to 1 cm; bracteoles lanceolate, to 6 mm. **Flowers:** calyx brown-purple, straight to curved; utricle angled upward, ovoid to narrowly ellipsoid, 0.6–0.7 × 0.5 cm; syrinx tubular, 1.5 mm; tube angled upward, cylindric, 1 × 0.3 cm; annulus absent; limb orangish with maroon spots, 1-lobed, lobe ovate, 1–6 × 0.6 cm, pilose; gynostemium 5-lobed, crown-shaped, 1–4 mm; anthers 5; ovary 5-locular, to 1 cm. **Capsule** ovoid to obovoid, 1.5 × 1–2 cm, dehiscence acropetal; valves 5; septa entire, not attached to valves. **Seeds** flat, triangular, 0.3–0.4 × 0.4 cm.

Flowering spring–summer. Rocky slopes; 50–1000 m; Ariz., N.Mex.; Mexico.

9. Aristolochia pentandra Jacquin, Enum. Syst. Pl., 30. 1760 · Marsh's Dutchman's-pipe [E]

Aristolochia marshii Standley; *A. racemosa* Brandegee

Lianas, twining, to 5 m, herbaceous. **Young stem** smooth, glabrous. **Leaves:** petiole 1–3 cm. **Leaf blade** ovate to hastate, 3-lobed, 3–8 × 2–7 cm, base cordate to sagittate, sinus depth to 1 cm, apex acute or acuminate; surfaces glabrous; venation palmate. **Inflorescences** on new growth, axillary, solitary flowers; peduncle bracteolate, 1–1.5 cm; bracteoles ovate to lanceolate, to 15 mm. **Flowers:** calyx brown-purple, straight; utricle

angled upward, ovoid, 0.5–5 × 0.5 cm; syrinx funnel-shaped, 2 mm; tube upright, cylindric, 1 × 0.25 cm; annulus absent; limb brownish red, 1-lobed, lobe narrowly cordate to narrowly ovate, 1–6 × 0.75 cm, glabrous; gynostemium 5-lobed, crown-shaped, 2–3 mm; anthers 5; ovary 5-locular, 1–1.5 cm. **Capsule** globose to ovoid, 1.5 × 1–3 cm, dehiscence basipetal; valves 5; septa entire, not attached to valves. **Seeds** flat, triangular, 0.5 × 0.5 cm.

Flowering winter–early spring. Wet Caribbean-type lowlands; 0–50 m; Fla., Tex.

10. Aristolochia wrightii Seemann, Bot. Voy. Herald, 331. 1856 · Wright's Dutchman's-pipe

Aristolochia brevipes Bentham var. *wrightii* (Seemann) Duchartre

Herbs, procumbent, to 0.4 m. **Young stem** slightly striate, velutinous with yellowish trichomes. **Leaves:** petiole 1–3 cm. **Leaf blade** ovate to hastate, (0–)3-lobed, 1–4 × 2–4 cm, base cordate to hastate, sinus 0.2–0.6 cm, apex obtuse to acute; surfaces velutinous with yellowish trichomes; venation palmate. **Inflorescences** on new growth, axillary, solitary flowers; peduncle bracteolate, 1–2 cm; bracteoles lanceolate, to 6 mm. **Flowers:** calyx brown-purple, straight; utricle angled upward, ovoid, to 0.7 × 0.4 cm; syrinx funnel-shaped, 2 mm; tube upright, cylindric, 1.5 × 0.3 cm; annulus absent; limb purplish, 1-lobed, lobe narrowly cordate to narrowly ovate, 1–6 × 0.9 cm, papillate; gynostemium 5-lobed, ovoid, 5–10 mm; anthers 5; ovary 5-locular, 1–2 cm. **Capsule** ± globose, 1.2 × 1–3 cm, dehiscence acropetal; valves 5; septa absent. **Seeds** flat, triangular, 0.4 × 0.5 cm.

Flowering all year. Rocky slopes; 500–2000 m; N.Mex., Tex.; Mexico.

11. Aristolochia maxima Jacquin, Enum. Syst. Pl., 30. 1760 · Florida Dutchman's-pipe

Lianas, twining, to 20 m, woody. **Young stem** ribbed, strigose. **Leaves:** petiole 0.1–0.8 cm. **Leaf blade** obovate, 7–15 × 3–7 cm, base truncate to slightly cordate, sinus depth 0–0.2 cm, apex acuminate; surfaces abaxially strigose; venation palmate-pinnate. **Inflorescences** on new growth, axillary, solitary flowers or rhipidiate from base of plant; peduncle not bracteolate, 2–2.5 cm. **Flowers:** calyx brown-purple, bent; utricle horizontal, ovoid, 2–3.5 × 1 cm; syrinx absent; tube upright, funnel-shaped, 1–3 cm; annulus absent; limb dark purple, 1-

lobed, lobe ovate, 1–6 × 1 cm, glabrous; gynostemium 6-lobed, crown-shaped, 5–10 mm; anthers 6; ovary 6-locular, 2–2.5 cm. **Capsule** ovoid, 10–15 × 4–7 cm, dehiscence acropetal; valves 6; septa persistent, clathrate. **Seeds** flat, triangular, 1–1.5 × 1.5 cm.

Flowering all year. Hammocks; 0–50 m; introduced; Fla.; native to Central America; South America.

12. Aristolochia clematitis Linnaeus, Sp. Pl. 2: 962. 1753 · Birthwort, aristoloche clématite, asarabacca

Herbs, erect, to 1 m. **Young stem** ribbed, glabrous to puberulent. **Leaves:** petiole 4–7 cm. **Leaf blade** deltate to reniform, 4–6 × 5–6.5 cm, base cordate, sinus depth 1–2 cm, apex obtuse to rounded; surfaces abaxially glabrous; venation palmate. **Inflorescences** on new growth, axillary, fascicles or solitary flowers; peduncle not bracteolate, 1–4 cm. **Flowers:** calyx yellow-brown, curved; utricle angled upward, ovoid to globose, 0.5 × 0.2–0.3 cm; syrinx absent; tube curved upward, cylindric, 1–3 × 0.2–0.3 cm; annulus absent; limb purple, 1-lobed, lobe funnel-shaped, 1–6 cm, pilose; gynostemium 3-lobed, globose, 1–4 mm; anthers 6; ovary 6-locular, 1–4 cm. **Capsule** ± globose, 3 × 2–3 cm, dehiscence basipetal; valves 6; septa entire, not attached to valves. **Seeds** flat, triangular, 1 × 1.2 cm.

Flowering summer. Waste places, roadsides, ballast; 0–100 m; introduced; Ont., Que.; Md., Mass., N.Y., Ohio, Pa.; Europe.

Aristolochia clematitis is occasionally cultivated, sometimes escapes, and probably does not persist.

2. ASARUM Linnaeus, Sp. Pl. 1: 442. 1753; Gen. Pl. ed. 5, 201. 1754 · Wild-ginger, asaret, gingembre sauvage [Ancient Greek *asaron,* name of an unknown plant]

Alan T. Whittemore

Michael R. Mesler

Karen L. Lu

Herbs, perennial, deciduous, rhizomatous, without aerial stems. **Leaves** alternate (sometimes appearing opposite because of crowding), 2-ranked; stipules absent; petiolate foliage leaves and sessile, triangular scale-leaves both present. **Leaf blade** membranous or leathery, pubescent at least abaxially and on margins. **Inflorescences** terminal on rhizome, flowers solitary; bracts absent. **Flowers:** sepals distinct, usually mixture of white, green, tan, red, or purple, proximally touching valvately and forming well-defined false tube, externally usually villous, inner surface strigose, smooth or with weak longitudinal ribs, never with network of low ridges; vestigial petals present or absent; stamens 12, distinct; filaments longer than pollen sacs; terminal appendage of anther well developed; ovary inferior, 6-locular; styles connate in column. **Capsule** fleshy, dehiscence irregular. **Seeds** ovoid, not winged, with fleshy appendage. $x = 13$.

Species ca. 10 (6 in the flora): North America, Eurasia.

The species seem amply distinct, but herbarium material can be difficult to key for several reasons. First, the diagnostic colors of some organs (especially of the connective and the inner hairs of the calyx) often darken on drying. Second, immature flowers and young fruit are superficially similar to mature flowers, but color and posture of floral organs may be different at those stages. For instance, posture of the distal portion of sepals at anthesis (whether erect, spreading, or reflexed) is diagnostic for the species, but sepals in all species are erect in bud and in fruit. Third, as in *Hexastylis,* distortion of the flower in pressing makes it difficult to interpret calyx structure. In particular, the distinction between proximal portions of the sepals, which meet valvately to form a well-defined false calyx tube, and distal portions, which do not, is obvious in fresh material but often unclear in the herbarium.

The flowers of *Asarum* are predominantly self-pollinated, but they are occasionally visited by mycotrophic flies (K. L. Lu 1982).

SELECTED REFERENCE   Mesler, M. R. and K. L. Lu. 1990. The status of *Asarum marmoratum* (Aristolochiaceae). Brittonia 42: 33–37.

1. Adaxial leaf surface almost always with white or silvery variegations; sterile tip of connective on inner stamens at least as long as pollen sacs; underground stems erect or ascending, deeply buried, internodes 0.2–1.5 cm.
   2. False calyx tube subglobose, inner surface dark red with purple hairs; distal part of sepal tan or greenish (inner surface rarely red proximally), erect or spreading at anthesis, 17–52 mm; marginal hairs of leaf ± perpendicular . . . . 6. *Asarum marmoratum*
   2. False calyx tube cylindric, inner surface white with brownish purple stripes and white hairs (turning brown with age); distal part of sepal reddish, spreading perpendicularly or reflexed at anthesis (but erect in bud and fruit), 12–27 mm; marginal hairs of leaf strongly curved toward apex. . . . . 5. *Asarum hartwegii*
1. Leaf surface never variegated; sterile tip of connective on inner stamens shorter than (rarely about as long as) pollen sacs; rhizomes horizontal, shallow (deeply buried in *A. wagneri*), internodes 0.5–6.5 cm.
   3. Flower descending; divergent part of sepal strongly reflexed at anthesis, 4–8 mm, acute to apiculate or short-acuminate. . . . . 4. *Asarum lemmonii*
   3. Flower horizontal to erect; divergent part of sepal spreading or reflexed at anthesis, 6–75 mm, apiculate to acuminate or filiform-attenuate.
      4. Flowers horizontal; divergent part of sepal (11–)30–75 mm; leaves cordate. . . . . 3. *Asarum caudatum*
      4. Flowers erect or ascending; divergent part of sepal 6–24 mm; leaves cordate-reniform to reniform.
         5. Adaxial surface of distal sepals purple; false calyx tube cylindric, outer surface usually tan or purplish; e North America. . . . . 1. *Asarum canadense*
         5. Adaxial surface of distal sepals white or light green (at least distally); false calyx tube subglobose to cylindric-urceolate or urceolate, outer surface light green; s Oregon . . . . 2. *Asarum wagneri*

1. Asarum canadense Linnaeus, Sp. Pl. 1: 442. 1753
· Asaret du canada, gingembre sauvage E F

Asarum acuminatum (Ashe) E. P. Bicknell; *A. canadense* var. *acuminatum* Ashe; *A. canadense* var. *ambiguum* (E. P. Bicknell) Farwell; *A. canadense* var. *reflexum* (E. P. Bicknell) B. L. Robinson; *A. reflexum* E. P. Bicknell; *A. rubrocinctum* Peattie

Rhizomes horizontal, shallow, internodes 1.0–3.5 cm. **Leaves:** petiole 6–20 cm, crisped-hirsute. **Leaf blade** not variegated, cordate-reniform to reniform, 4–8(–20) × 8–14(–21.5) cm, apex rounded or obtuse; surfaces abaxially appressed-hirsute, usually sparsely so, adaxially appressed-hirsute, at least along main veins, marginal hairs perpendicular to margin or curved toward apex. **Flowers** erect or ascending; peduncle 1.5–3 cm; false calyx tube cylindric, externally tan or purplish, hirsute (often densely), internally white or pale green, occasionally mottled with purple, with white or purple hairs; distal portion of sepal spreading or reflexed at anthesis, 6–24 mm, apex apiculate to acuminate or filiform-attenuate, abaxially green or purple, hirsute, adaxially purple, puberulent with crisped purple or pale hairs; pollen sacs 1–1.5 mm, sterile tip of connective on inner stamens purple, 0.5–1 mm, shorter than or about as long as pollen sacs. $2n = 26$.

Flowering spring–early summer (Mar–Jul). Understory of deciduous (rarely coniferous) forests; 0–1300 m; Man., N.B., Ont., Que; Ala., Ark., Conn., Del., D.C., Ga., Ill., Ind., Iowa, Kans., Ky., La., Maine, Md., Mass., Mich., Minn., Miss., Mo., Nebr., N.H., N.J., N.Y., N.C., N.Dak., Ohio, Okla., Pa., S.C., S.Dak., Tenn., Vt., Va., W.Va., Wis.

The rhizomes of *Asarum canadense* are occasionally used for seasoning. Handling the leaves is said to cause dermatitis for some people.

Native Americans used *Asarum canadense* medicinally to treat flux, poor digestion, swollen breasts, coughs and colds, typhus and scarlet fever, nerves, sore throats, cramps, heaves, earaches, headaches, convulsions, asthma, tuberculosis, urinary disorders, and venereal disease; as a stimulant, a seasoning, and a charm; and to strengthen other herbal concoctions and heighten appetite (D. E. Moerman 1986).

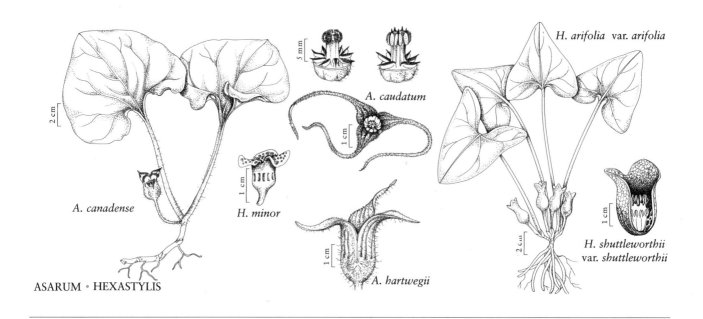

A. *canadense*

H. *minor*

A. *caudatum*

A. *hartwegii*

H. *arifolia* var. *arifolia*

H. *shuttleworthii*
var. *shuttleworthii*

ASARUM · HEXASTYLIS

2. Asarum wagneri K. L. Lu & Mesler, Brittonia 35: 331. 1983 [E]

Asarum caudatum Lindley var. *viridiflorum* M. Peck

Rhizomes horizontal, ± deeply buried, internodes 0.5–2.1 cm. **Leaves:** petiole 3–15 cm, sparsely crisped-hirsute. **Leaf blade** not variegated, broadly reniform to cordate-reniform, 3–8 × 4–11 cm, apex obtuse to rounded (broadly acute); surfaces abaxially sparsely hirsute, adaxially sparsely hirsute only along veins, marginal hairs mostly curved toward apex. **Flowers** erect or ascending; peduncle 0.8–3 cm; false calyx tube subglobose to cylindric-urceolate or urceolate, externally light green, sparsely to moderately hirsute, internally white or light green, bordered and occasionally striped with purple, with purple hairs; distal portion of sepal spreading perpendicularly from base at anthesis, bent abruptly upward at midpoint, 8–20 mm, apex filiform-acuminate, abaxially white to pale green, sparsely villous to villous, adaxially white or light green, at least distally, bordered with purple and occasionally with purple band across base, puberulent with crisped purple-tipped hairs; pollen sacs 1–2 mm, sterile tip of connective on inner stamens dark red, 0.25–1 mm, shorter than pollen sacs.

Flowering spring–summer (May–Jul). Understory of *Abies* forests and open boulder fields in *Tsuga* forests near timberline; 1500–3200 m; Oreg.

Asarum wagneri is endemic to the Cascade Range of southern Oregon (K. L. Lu and M. R. Mesler 1983).

3. Asarum caudatum Lindley, Edwards's Bot. Reg. 17: footnote after plate 1399. 1831 [E][F]

Rhizomes horizontal, shallow, internodes (0.5–)1.5–6.5 cm. **Leaves:** petiole 7.5–17 cm, sparsely crisped-hirsute. **Leaf blade** not variegated, cordate, 3–8.5 × 4.5–12 cm, apex usually obtuse, occasionally broadly acute; surfaces abaxially sparsely appressed-hirsute, at least proximally, adaxially glabrous or sparsely appressed-hirsute, marginal hairs perpendicular to margin or curved toward apex. **Flowers** horizontal; peduncle 1.5–5 cm; false calyx tube cylindric, externally brown-purple, rarely greenish, hirsute, internally white, usually with median purple stripe, with usually purple, rarely white hairs; distal portion of sepal spreading or weakly (rarely strongly) reflexed at anthesis, (11–)30–75 mm, apex filiform-attenuate, abaxially purple or greenish, sparsely hirsute, adaxially purple, puberulent with crisped purple hairs; pollen sacs 1.5–2 mm, sterile tip of connective on inner stamens purple, 0.5–1 mm, shorter than pollen sacs. $2n = 26$.

Flowering spring–summer (Apr–Jul). Understory of conifer forests, usually in mesic or wet places; 0–1200(–2200) m; B.C.; Calif., Idaho, Mont., Oreg., Wash.

In most populations of *Asarum caudatum*, the distal portion of the sepal is spreading or weakly reflexed and 30–75 mm. A single population south of Mt. Shasta, California, has the distal sepals strongly reflexed and unusually short, often as little as 1.1 cm. Flowers of these plants superficially resemble those of *A. lem-*

monii; they differ in being horizontal, not descending as in *A. lemmonii,* and in the filiform-attenuate sepals.

Native Americans used *Asarum caudatum* medicinally to treat headaches, intestinal pain, knee pain, indigestion, boils, tuberculosis, and colic, and as a general tonic (D. E. Moerman 1986).

4. Asarum lemmonii S. Watson, Proc. Amer. Acad. Arts 14: 294. 1879 E

Rhizomes horizontal, shallow, internodes (0.6–)1–4 cm. **Leaves:** petiole (3.5–)6–26 cm, sparsely crisped-hirsute. **Leaf blade** not variegated, cordate to almost reniform, 4–6.5 × 7–10.5 cm, apex broadly rounded-acute to rounded; surfaces abaxially sparsely appressed-hirsute, adaxially glabrous or sparsely hirsute along veins, marginal hairs perpendicular to margin or curved toward apex. **Flowers** descending; peduncle 2–3.5 cm; false calyx tube cylindric, externally reddish, at least in part, glabrous or sparsely hirsute, internally white, rarely with longitudinal red stripes, with white or purple hairs; distal portion of sepal strongly reflexed at anthesis, 4–8 mm, apex acute to apiculate or short-acuminate, abaxially purple, glabrous or sparsely hirsute, adaxially red, puberulent with crisped pale or purple hairs; pollen sacs 1–1.5 mm, sterile tip of connective on inner stamens purple or brown, 0.5–1 mm, shorter than pollen sacs. $2n = 26$.

Flowering spring–summer (May–Jul). Wet places, usually near creeks, in understory of conifer forests; 1100–1900 m; Calif.

Asarum lemmonii is endemic to the Sierra Nevada.

5. Asarum hartwegii S. Watson, Proc. Amer. Acad. Arts 10: 346. 1875 E F

Rhizomes erect or ascending, deeply buried, internodes 0.2–1.3 cm. **Leaves:** petiole 9–21 cm, sparsely crisped-hirsute. **Leaf blade** almost always variegated with white or silver along veins, cordate to cordate-reniform, 5.3–10 × 7–14 cm, apex rounded-acute to rounded; surfaces abaxially appressed-hirsute, usually sparsely so, adaxially glabrous or sparsely hirsute along veins, mar-

ginal hairs strongly curved toward apex. **Flowers** erect; peduncle 1–2.3 cm; false calyx tube cylindric, externally reddish, sometimes mottled red and green, hirsute, internally white with brownish purple stripes and white hairs (becoming brown with age); distal portion of sepal usually spreading at anthesis, sometimes reflexed or nearly erect, 12–27 mm, apex filiform-attenuate, abaxially reddish, hirsute, adaxially reddish, puberulent with crisped pale hairs; pollen sacs 2 mm, sterile tip of connective on inner stamens pale (sometimes dark in dried specimens), 3–5 mm, longer than pollen sacs. $2n = 26$.

Flowering spring–summer (Apr–Jul). Rocky slopes in dry conifer or oak forests; 150–2200 m; Calif.

Asarum hartwegii was confused with *A. marmoratum* until very recently; reports of *A. hartwegii* from southern Oregon are errors for *A. marmoratum* (M. R. Mesler and K. L. Lu 1990).

6. Asarum marmoratum Piper, Proc. Biol. Soc. Wash. 29: 99. 1916 E

Rhizomes erect or ascending, deeply buried, internodes 0.2–1.5 cm. **Leaves:** petiole 5–20 cm, sparsely crisped-hirsute. **Leaf blade** almost always variegated with white or silver along veins, cordate to cordate-reniform, 4–14 × 3–12 cm, apex acute to broadly acuminate, rarely obtuse; surfaces abaxially sparsely hirsute, adaxially glabrous or sparsely hirsute along veins, marginal hairs ± perpendicular to margin. **Flowers** erect or nearly so; peduncle 1.2–1.9 cm; false calyx tube subglobose, externally mottled red, sparsely to moderately hirsute, internally dark red, with purple hairs; distal portion of sepal erect or spreading at anthesis, 17–52 mm, apex filiform-attenuate, abaxially pale green, hirsute, adaxially tan or brownish green, rarely red proximally, puberulent with crisped purple hairs; pollen sacs 0.8–2.4 mm, sterile tip of connective on inner stamens dark red-brown, 1.2–3.8 mm, longer than pollen sacs.

Flowering late winter–spring (Mar–Jun). Understory of dry or mesic forests, or exposed rocky slopes or roadcuts; 200–1800 m; Calif., Oreg.

Asarum marmoratum is found only in the Cascades and the Siskiyou Mountains of southern Oregon and extreme northwestern California (M. R. Mesler and K. L. Lu 1990).

3. HEXASTYLIS Rafinesque, Neogenyton, 3. 1825 · Heartleaf [Greek *hexastylis,* with six styles]

Alan T. Whittemore

L. L. Gaddy

Herbs, perennial, evergreen, rhizomatous, without aerial stems. **Leaves** alternate, 2-ranked; stipules absent; petiolate foliage leaves and sessile, triangular scale-leaves both present. **Leaf blade** membranous or leathery, glabrous. **Inflorescences** terminal on rhizome, solitary flowers, subtended by triangular bract. **Flowers:** sepals connate for most of length, usually mixture of brown, purple, or yellow, externally glabrous, inner surface glabrous, with weak to strong network of ridges, calyx tube never forming differentiated utricle or syrinx; vestigial petals absent; stamens 12, distinct; filaments shorter than pollen sacs; terminal appendage of anther absent or rudimentary; ovary superior or partly inferior, 6-locular; styles 6, distinct (except sometimes at extreme base). **Capsule** fleshy, dehiscence irregular. **Seeds** ovoid, not winged, with fleshy appendage. $x = 13$.

Species 10 (all in the flora): North America.

Hexastylis is very similar to the Asiatic genera *Heterotropa* C. Morren & Decaisne and *Asiasarum* F. Maekawa; a strong case could be made for combining the three genera. Nevertheless, all three of these genera seem distinct from *Asarum* (in which they have been included by some authors; e.g., K. Barringer 1993) in their connate sepals, distinct styles, nonappendiculate anthers, and superior or partly inferior ovaries.

Herbarium specimens of *Hexastylis* are difficult to work with. The form of the calyx is very important taxonomically, but the calyx is fleshy and brittle and does not press well. Allowing flowers to wilt for several hours before pressing may help to reduce distortion, because the calyx becomes more flexible and less likely to split in the press. Features of the inner surface of the calyx are also important, but collectors seldom cut open flowers and press them with the inside visible. For this reason, herbarium specimens of *Hexastylis* are difficult to identify reliably, and meaningful work on the group requires field studies. L. L. Gaddy (1987) and H. L. Blomquist (1957) gave photographs and drawings of flowers of all species, and an extensive collection of liquid-preserved flowers is housed at the University of Tennessee; these are very helpful in identifying *Hexastylis* specimens.

SELECTED REFERENCES Blomquist, H. L. 1957. A revision of *Hexastylis* of North America. Brittonia 8: 255–281. Gaddy, L. L. 1987. A review of the taxonomy and biogeography of *Hexastylis* (Aristolochiaceae). Castanea 52: 186–196.

1. Sterile tip of style deeply 2-cleft, sinus reaching stigma; leaf blade triangular to ovate-sagittate or subhastate.
    2. Calyx not abruptly contracted near middle, tube urceolate-campanulate or ovoid, smooth internally, lobes erect or spreading. . . . . 1. *Hexastylis arifolia*
    2. Calyx abruptly contracted near middle, tube proximally narrowly cup-shaped, distally broadly cylindric, reticulately ridged internally, lobes spreading . . . . 2. *Hexastylis speciosa*
1. Sterile tip of style undivided or shallowly 2-cleft, sinus not reaching stigma; leaf blade cordate to orbiculate, triangular-cordate, or subreniform.
    3. Inner surface of calyx lobes pilose; rhizomes dimorphic, internodes of flowering rhizomes short, leaves crowded at rhizome apex, internodes of sterile rhizomes often long, leaves scattered along length of rhizome. . . . . .3. *Hexastylis lewisii*
    3. Inner surface of calyx lobes puberulent; rhizomes not dimorphic, internodes short, leaves crowded at rhizome apex.
        4. Calyx tube broadly urceolate-campanulate or rhombic-ovoid, conspicuously tapered above middle.
            5. Calyx tube urceolate-campanulate; calyx lobes 10–22 mm wide . . . . .10. *Hexastylis shuttleworthii*
            5. Calyx tube rhombic-ovoid; calyx lobes 3–8 mm wide

6. Internal reticulations of calyx tube well developed, ridges 1.5–2 mm high; ovary superior. . . . . 8. *Hexastylis rhombiformis*
6. Internal reticulations of calyx tube absent or poorly developed, ridges 0–1 mm high; ovary ca. 1/3-inferior . . . . .9. *Hexastylis contracta*
4. Calyx tube cylindric to narrowly cylindric-urceolate, not much tapered above middle.
  7. Calyx lobes erect or weakly spreading, 2–4 mm; calyx tube cylindric to narrowly cylindric-urceolate . . . . 4. *Hexastylis virginica*
  7. Calyx lobes moderately spreading to reflexed, 4–15 mm; calyx tube cylindric.
    8. Calyx tube wider than long, its opening wider than length of lobes. . . . . 7. *Hexastylis minor*
    8. Calyx tube at least as long as wide, its opening narrower than length of lobes.
      9. Calyx tube 4–7 mm wide, lobes 4–7 mm wide. . . . . 6. *Hexastylis naniflora*
      9. Calyx tube 7–14 mm wide, lobes 6–17 mm wide . . . . 5. *Hexastylis heterophylla*

1. Hexastylis arifolia (Michaux) Small in N. L. Britton, Man. Fl. N. States, 348. 1901   [E]

Asarum arifolium Michaux, Fl. Bor.-Amer. 1: 279–280. 1803

Rhizomes: internodes short, leaves crowded at rhizome apex (or internodes somewhat elongate, leaves scarcely crowded in var. *arifolia* and var. *callifolia* growing in wet places). **Leaf blade** commonly variegated, triangular-sagittate to subhastate, infrequently ovate-sagittate to deltate. **Flowers:** calyx tube narrowly to broadly urceolate-campanulate or ovoid, 12–30 × 6–12 mm, inner surface smooth; lobes erect or spreading, 2–8 × 2–9 mm, adaxially puberulent; stamen connective extending slightly beyond pollen sacs; ovary ca. 1/3-inferior; ovules ca. 6 per locule; styles 2-cleft to stigma.

Varieties 3 (all in the flora): North America.

Hexastylis arifolia is the most widespread species in the genus. Along the boundaries where the ranges of the varieties meet, intermediate specimens are occasionally found.

The Catawba tribe used *Hexastylis arifolia* (no varieties specified) medicinally for stomach pains, miscellaneous pains, heart trouble, and backaches; the Rappahannock, for treating whooping cough and asthma (D. E. Moerman 1986).

1. Calyx lobes erect, 2–4 × 2–4 mm. . . . . .1c. *Hexastylis arifolia* var. *ruthii*
1. Calyx lobes spreading, 2.5–8 × 3–9 mm.
  2. Calyx tube 13–18 × 6–10 mm. . . . . .1a. *Hexastylis arifolia* var. *arifolia*
  2. Calyx tube 20–25 × 10–12 mm. 1b. *Hexastylis arifolia* var. *callifolia*

1a. Hexastylis arifolia (Michaux) Small var. **arifolia**   [E] [F]

Flowers: calyx tube narrowly urceolate-campanulate, 13–18 × 6–10 mm; lobes spreading, 2.5–8 × 3–9 mm. $2n = 26$.

Flowering late winter–spring (Mar–May). Upland deciduous and mixed deciduous-conifer forests; 0–600 m; Ala., Fla., Ga., La., Miss., N.C., S.C., Tenn., Va.

1b. Hexastylis arifolia (Michaux) Small var. **callifolia** (Small) H. L. Blomquist, Brittonia 8: 268. 1957   [E]

Asarum callifolium Small, Bull. Torrey Bot. Club 24: 334. 1897; *A. arifolium* Michaux var. *callifolium* (Small) Barringer; *Hexastylis callifolia* (Small) Small

Flowers: calyx tube broadly urceolate-campanulate, 20–25 × 10–12 mm; lobes spreading, 3–6 × 4–9 mm.

Flowering late winter–spring (Mar–May). Upland and wetland deciduous, mixed and coniferous forests; often found along edges of wetlands; 0–100 m; Ala., Fla., La., Miss.

1c. Hexastylis arifolia (Michaux) Small var. **ruthii** (Ashe) H. L. Blomquist, Brittonia 8: 268. 1957   [E]

Asarum ruthii Ashe, J. Elisha Mitchell Sci. Soc. 14: 35–36. 1897; *A. arifolium* Michaux var. *ruthii* (Ashe) Barringer; *Hexastylis ruthii* (Ashe) Small

Flowers: calyx tube ovoid, 10–20 × 8–10 mm; lobes erect, 2–4 × 2–4 mm. $2n = 26$.

Flowering late winter–spring (Mar–Jun). Upland deciduous and mixed deciduous-conifer forests; often associated with shrubby Ericaceae in n part of range; 300–800 m; Ala., Ga., Ky., N.C., Tenn., Va.

2. **Hexastylis speciosa** R. M. Harper, Torreya 24: 79–80. 1924 [C] [E]

Asarum speciosum (R. M. Harper) Barringer

Rhizomes: internodes short, leaves crowded at rhizome apex. **Leaf blade** variegated or not, triangular to broadly triangular, subhastate. **Flowers:** calyx tube abruptly contracted near middle, proximally narrowly cup-shaped, distally broadly cylindric, 10–20 × 10–20 mm, inner surface with well-developed reticulations, lobes spreading, 4–8 × 6–8 mm, adaxially puberulent; stamen connective extending slightly beyond pollen sacs; ovary ca. 1/3-inferior; ovules ca. 6 per locule; style 2-cleft to stigma. $2n = 26$.

Flowering spring (April–May). Shaded sites, well-drained sandy loam in open pine-deciduous forests above acidic streams or bogs; of conservation concern; 70 m; Ala.

Hexastylis speciosa is endemic to a small area north of Montgomery, Alabama.

3. **Hexastylis lewisii** (Fernald) H. L. Blomquist & Oosting, Spring Fl. Piedmont, 50. 1948 [E]

Asarum lewisii Fernald, Rhodora 45: 398. 1943; *Hexastylis pilosiflora* H. L. Blomquist

Rhizomes dimorphic: internodes of flowering rhizomes short, leaves crowded at rhizome apex; internodes of sterile rhizomes often long, leaves scattered along length of rhizome. **Leaf blade** variegated, triangular-cordate, cordate, or orbiculate-cordate. **Flowers:** calyx tube cylindric-campanulate to urceolate-campanulate, sometimes with prominent transverse ridge just above middle, 14–20 × 16–22 mm, inner surface longitudinally ridged, without reticulations, lobes spreading, 8–15 × 10–15 mm, adaxially pilose; stamen connective extending beyond pollen sacs; ovary superior; ovules 10 per locule; style notched at apex. $2n = 26$.

Flowering spring (Apr–May). Upland and lowland (floodplain) forests; sometimes found along shores of Carolina bays; 0–200 m; N.C., Va.

Hexastylis lewisii resembles *H. shuttleworthii* in its calyx shape, the presence of elongate rhizomes, and the late flowering time.

4. **Hexastylis virginica** (Linnaeus) Small in N. L. Britton, Man. Fl. N. States, 348. 1901 [E]

Asarum virginicum Linnaeus, Sp. Pl. 1: 442. 1753; *A. memmingeri* Ashe; *Hexastylis memmingeri* (Ashe) Small

Rhizomes: internodes short, leaves crowded at rhizome apex. **Leaf blade** variegated or not, cordate, subcordate, or subreniform. **Flowers:** calyx tube cylindric to narrowly cylindric-urceolate, sometimes with prominent transverse ridge just below sinuses, 8–15 × 6–12 mm, inner surface with high reticulations, lobes erect or weakly spreading, 2–4 × 7–9 mm, adaxially puberulent; stamen connective not extending beyond pollen sacs; ovary ca. 1/3-inferior; ovules 8 per locule; styles notched at apex. $2n = 26$.

Flowering spring (Apr–Jun). Deciduous and mixed deciduous-conifer forests; 0–700 m; Ky., Md., N.C., Tenn., Va., W.Va.

Plants of *Hexastylis virginica* with small, cylindric-urceolate calyces have been treated as a distinct species, *H. memmingeri*. The two calyx types are often found in the same population, however, so *H. memmingeri* seems unworthy of taxonomic recognition at any rank.

Prior to the study by H. L. Blomquist (1957), many botanists interpreted *Hexastylis virginica* in a very broad sense, so old herbarium specimens of many other species of *Hexastylis* are often annotated as *H. virginica*.

5. **Hexastylis heterophylla** (Ashe) Small in N. L. Britton, Man. Fl. N. States, 347. 1901 [E]

Asarum heterophyllum Ashe, J. Elisha Mitchell Sci. Soc. 14: 33–34. 1897

Rhizomes: internodes short, leaves crowded at rhizome apex. **Leaf blade** variegated or not, cordate to orbiculate-cordate. **Flowers:** calyx tube cylindric, sometimes with prominent transverse ridge just above middle, 8–15 × 7–14 mm, inner surface longitudinally ridged with low reticulations between ridges, lobes spreading, 5–15 × 6–17 mm, adaxially puberulent; stamen connective extending slightly beyond pollen sacs; ovary superior; ovules 6–8 per locule; style notched at apex. $2n = 26$.

Flowering late winter–spring (Mar–Jun). Slopes and bluffs in deciduous and mixed deciduous-conifer forests, usually associated with *Kalmia latifolia*; 300–700 m; Ala., Ga., Ky., N.C., S.C., Tenn., Va., W.Va.

Hexastylis heterophylla is probably the most variable species in the genus. Flowers from the northern part of its range usually have a large calyx with a prominent transverse ridge just above the middle, while those from the southern part of its range usually have a small calyx and no ridge. Both flower types are sometimes found in the same population.

Living plants of *Hexistylis heterophylla, H. naniflora,* and *H. minor* are distinguishable, but herbarium specimens of the three cannot always be reliably determined.

6. Hexastylis naniflora H. L. Blomquist, Brittonia 8: 280. 1957 [C] [E]

Rhizomes: internodes short, leaves crowded at rhizome apex. **Leaf blade** variegated, cordate to orbiculate-cordate. **Flowers:** calyx tube cylindric to cylindric-campanulate, sometimes with prominent transverse ridge just above middle, 6–13 × 4–7 mm, inner surface longitudinally ridged with low reticulations between ridges, lobes spreading, 4–7 × 4–7 mm, adaxially puberulent; stamen connective not extending beyond pollen sacs; ovary 1/2-inferior; ovules 6 per locule; style notched at apex. 2*n* = 26.

Flowering late winter–spring (Mar–Jun). Acidic sandy loam on bluffs and in ravines in deciduous forests, often associated with *Kalmia latifolia;* of conservation concern; 500–700 m; N.C., S.C.

Asarum naniflorum (H. L. Blomquist) Pfeifer has never been validly published.

7. Hexastylis minor (Ashe) H. L. Blomquist, Brittonia 8: 275. 1957 [E] [F]

Asarum minus Ashe, J. Elisha Mitchell Sci. Soc. 14: 33. 1897

Rhizomes: internodes short, leaves crowded at rhizome apex. **Leaf blade** variegated, cordate to subreniform. **Flowers:** calyx tube cylindric to cylindric-campanulate, sometimes with prominent transverse ridge just below sinuses, 9–16 × 8–16 mm, inner surface longitudinally ridged with low reticulations between ridges, lobes spreading, 6–10 × 8–12 mm, adaxially puberulent; stamen connective not extending beyond pollen sacs; ovary ca. 1/3-inferior; ovules 6 per locule; style notched at apex. 2*n* = 26.

Flowering winter–spring (Feb–May). On slopes, bluffs, and flood plains of small streams, deciduous forests; 100–500 m; N.C., S.C., Va.

8. Hexastylis rhombiformis Gaddy, Brittonia 38: 82. 1986 [C] [E]

Rhizomes: internodes short, leaves crowded at rhizome apex. **Leaf blade** not variegated, cordate to subreniform. **Flowers:** calyx tube rhombic-ovoid, conspicuously tapered above middle, 16–24 × 12–19 mm, inner surface with high reticulations, ridges 1.5–2 mm high, lobes erect or weakly spreading, 3–6 × 3–6 mm, adaxially puberulent; stamen connective extending slightly beyond pollen sacs; ovary superior; ovules 8 per locule; style notched at apex.

Flowering spring (Apr–Jun). Sandy river bluffs or ravines, deciduous forests, with *Kalmia latifolia* and *Rhododendron maximum;* of conservation concern; 500–1000 m; N.C., S.C.

Hexastylis rhombiformis is endemic to a small area south of Asheville, North Carolina, extending south to the headwaters of the Saluda River, South Carolina. L. L. Gaddy (1986) illustrated the variation in flower shape in the species. Collections of this species made in the late 1800s were often identified as *Asarum memmingeri.*

9. Hexastylis contracta H. L. Blomquist, Brittonia 8: 279. 1957 [E]

Asarum contractum (H. L. Blomquist) Barringer

Rhizomes: internodes short, leaves crowded at rhizome apex. **Leaf blade** rarely variegated, orbiculate-cordate. **Flowers:** calyx tube rhombic-ovoid, conspicuously tapered above middle, with prominent constrictions just above base and just below sinuses, 15–27 × 12–17 mm, inner surface with reticulations absent or poorly developed, ridges 0–1 mm high, lobes erect to spreading, 4–5 × 7–8 mm, adaxially puberulent; stamen connective extending beyond pollen sacs; ovary ca. 1/3-inferior; ovules 6 per locule; styles 2-cleft halfway to stigma.

Flowering spring (May–Jun). Acid soils in deciduous forests, with *Kalmia* and *Rhododendron;* 300–1000 m; Ky., N.C., Tenn.

The distribution of *Hexastylis contracta* is unique; it is disjunct between the Cumberland Plateau of central Tennessee and Kentucky and the southeastern Blue Ridge Province of western North Carolina.

10. Hexastylis shuttleworthii (Britten & Baker f.) Small in N. L. Britton, Man. Fl. N. States, 347. 1901

Asarum shuttleworthii Britten & Baker f., J. Bot. 36: 98. 1898

Rhizomes: internodes short or long, leaves crowded at rhizome apex or scattered along its length. **Leaf blade** variegated or not, orbiculate to cordate. **Flowers:** calyx tube broadly urceolate-campanulate, conspicuously tapered above middle, 15–40 × 15–25 mm, inner surface with well-developed reticulations, lobes spreading or ascending, 6–13 × 10–22 mm, adaxially puberulent; stamen connective extending beyond pollen sacs; ovary superior; ovules 10–14 per locule; styles notched at apex.

Varieties 2 (both in the flora): North America.

1. Internodes of rhizome short; leaves confined to terminal portion of rhizome. . . . .
. . . . . 10a. *Hexastylis shuttleworthii* var. *shuttleworthii*
1. Internodes of rhizome long; leaves not confined to terminal portion of rhizome but scattered along its length. . . . .
. . . . .10b. *Hexastylis shuttleworthii* var. *harperi*

10a. Hexastylis shuttleworthii (Britten & Baker f.) Small var. **shuttleworthii** E F

Rhizomes: internodes short, leaves crowded at rhizome apex. **Leaf blade** cordate, 6–10 × 5–9 cm. $2n = 26$.

Flowering spring–early summer (May–Jul). Acidic soils, often along creeks beneath *Rhododendron maximum*, deciduous or mixed deciduous-conifer forests; 400–1300 m; Ala., Ga., N.C., S.C., Tenn., Va.

H. L. Blomquist (1957) reported *Hexastylis shuttleworthii* var. *shuttleworthii* from Morgantown in northern West Virginia; we have not seen this specimen.

10b. Hexastylis shuttleworthii (Britten and Baker f.) Small var. **harperi** Gaddy, Sida 12: 54. 1987 C E

Asarum shuttleworthii Britten & Baker f. var. *harperi* (Gaddy) Barringer

Rhizomes: internodes long, leaves scattered along length of rhizome. **Leaf blade** orbiculate to orbiculate-cordate, 3–7 × 3–7 cm.

Flowering spring (Apr–Jun). Forested bog edges and acidic hammocks in bogs; of conservation concern; 100–200 m; Ala., Ga., Miss.

9. ILLICIACEAE (de Candolle) A. C. Smith

· Star-anise Family

Michael A. Vincent

Shrubs or small trees, evergreen, glabrous or obscurely pubescent. **Leaves** alternate, simple, without stipules, petiolate. **Leaf blade** fragrant (especially when bruised or crushed), translucent-dotted, ovate to lanceolate, pinnately veined, thin to leathery, margins entire. **Flowers** bisexual, axillary or from main stem or older branches, solitary, pedunculate; peduncle bracteolate; perianth hypogynous; tepals 7–33, imbricate, distinct, in 2–3 series, outer tepals small, sometimes bractlike, innermost more petaloid; stamens 4–50, hypogynous, distinct, in 1–several series; filaments ligulate to terete; anthers basifixed, 4-locular, dehiscence longitudinal; pollen 3-aperturate; pistils simple, 5–21, distinct, closely laterally appressed in 1 series on convex axis, attached obliquely by broad base; placentation nearly basal; ovule 1 per locule; stigmatic surface adaxial. **Fruits** aggregates of radially arranged follicles; dehiscence adaxial. **Seed** 1 per follicle, ellipsoid to obovoid, somewhat flattened laterally; endosperm copious, oily; embryo minute.

Genus 1, species 42 (2 species in the flora): North America, West Indies, Central America, and ne South America; chiefly e Asia.

Aromatic oils obtained from some members of this genus are used for flavorings and as carminatives; oil derived from *Illicium anisatum* Linnaeus is poisonous. Chinese star-anise, used widely for flavoring wine and cooking, is obtained from *I. verum* Hooker f. (J. Hutchinson 1973; C. E. Wood Jr. 1958). The Chinese drug pa-chio-hui-hsiang, used to treat vomiting, epigastric pain, and abdominal colic, is derived from ripe fruits of *I. verum* (Xiao P. G. 1989).

Illiciaceae are considered closely allied to Schisandraceae. Anatomic details of wood of Illiciaceae and Schisandraceae are nearly indistinguishable, differing only in details linked to the climbing habit of members of the latter (I. W. Bailey and C. G. Nast 1948; S. Carlquist 1982; A. C. Smith 1947). Studies of fossil pollen led J. W. Walker and A. G. Walker (1984) to conclude that Illiciaceae and Schisandraceae are allies of Winteraceae. Based on analysis of nucleotide sequences from the plastid gene *rbc*L, however, M. W. Chase et al. (1993) and Qiu Y. L. et al. (1993) concluded that Illiciaceae and Schisandraceae are closely allied and closely related to Austrobaileyaceae but distant from Winteraceae.

SELECTED REFERENCES   Bailey, I. W. and C. G. Nast. 1948. Morphology and relationships of *Illicium, Schisandra* and *Kadsura.* J. Arnold Arbor. 29: 77–89.   Carlquist, S. 1982. Wood anatomy of *Illicium* (Illiciaceae): Phylogenetic, ecological, and functional interpretations. Amer. J. Bot. 69: 1587–1598.   Hutchinson, J. 1973. The Families of Flowering Plants, ed. 3. Oxford. Pp. 157–158.   Smith, A. C. 1947. The families Illiciaceae and Schisandraceae. Sargentia 7: 1–224.   Wood, C. E. Jr. 1958. The genera of the woody Ranales in the southeastern United States. J. Arnold Arbor. 39: 296–346.

1. ILLICIUM Linnaeus, Syst. Nat. ed. 10, 2: 1042, 1050, 1370. 1759 · Star-anise [Latin *illicere,* to allure]

Badianifera Kuntze

Leaves sometimes appearing whorled at ends of branches. **Flowers** erect to drooping; peduncle smooth to rugulose; perianth white, yellow, yellow-green, pink, or red to maroon, often glandular; outer tepals often bractlike and reduced, sometimes ciliolate, inner tepals often much larger and ligulate, sometimes only slightly larger and broadly obtuse, innermost often reduced, sometimes transitional to stamens; pistils orbicular to deltoid, styles narrow and acute, often recurved. **Fruits** dark brown, woody to leathery at maturity; peduncle sometimes thickened. **Seeds** glossy, dark brown to tawny or golden. $x = 13, 14$.

Species 42 (2 in the flora): North America, West Indies, Central America, and ne South America; chiefly e Asia.

SELECTED REFERENCES   Dirr, M. A. 1986. Hardy *Illicium* species display commendable attributes. Amer. Nurseryman 163: 92–100.   Jones, S. B. and N. C. Coile. 1988. The Distribution of the Vascular Flora of Georgia. Athens, Ga.   Roberts, M. L. and R. R. Haynes. 1983. Ballistic seed dispersal in *Illicium* (Illiciaceae). Pl. Syst. Evol. 143: 227–232.   Thien, L. B., D. A. White, and L. Y. Yatsu. 1983. The reproductive biology of a relict—*Illicium floridanum* Ellis. Amer. J. Bot. 70: 719–727.

1. Flowers 2.5–5 cm diam.; tepals 21–33, inner tepals ligulate, red-maroon, rarely white or pink; leaf apex acute to acuminate . . . . 1. *Illicium floridanum*
1. Flowers 0.8–1.2 cm diam.; tepals 11–16, inner tepals orbiculate-obovate, yellow-green; leaf apex obtuse to rounded . . . . 2. *Illicium parviflorum*

1. Illicium floridanum J. Ellis, Philos. Trans. 60: 524. 1770 · Florida-anise, anis étoilé F

Badianifera floridana (J. Ellis) Kuntze; *Illicium mexicanum* A. C. Smith

Leaves: petiole 6–26 mm. **Leaf blade** dark olive-green, elliptic to lanceolate, 5–21 × 1.5–6 cm, base cuneate, apex acute to acuminate. **Flowers** 2.5–5 cm diam.; peduncle 1–11 cm; bracteoles 3–6; tepals 21–33, red-maroon, rarely white or pink, inner tepals ligulate; stamens 25–50; pistils 11–21. **Fruit aggregates** collectively 2.5–4 cm diam., usually with 10–15 pistils at maturity. **Seeds** pale brown. $2n = 26$.

Flowering mid spring–early summer. Along streams, in marshy areas, moist woods; 0–500 m; Ala., Fla., Ga., La., Miss.; ne Mexico.

Illicium floridanum was placed in *Illicium* sect. *Badiana* by A. C. Smith (1947). The flowers of the species are pollinated by a variety of insects; fruit set is low (L. B. Thien et al. 1983). The seeds are dispersed by explosive dehiscence of the follicles (M. L. Roberts and R. R. Haynes 1983).

This species is cultivated in southeastern United States (M. A. Dirr 1986) and elsewhere. *Illicium mexicanum* A. C. Smith was considered a separate species by A. C. Smith (1947); expressions of all characters used to differentiate the two species overlap, however, and it seems best to consider them conspecific.

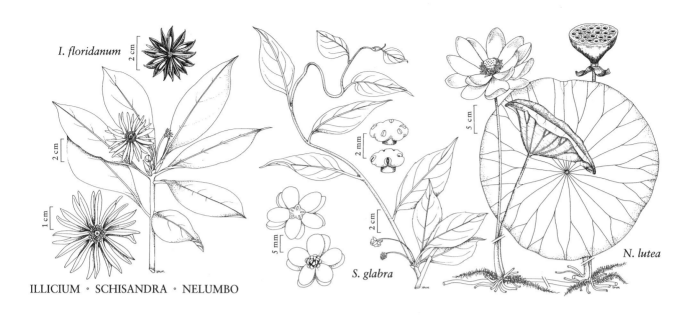

I. floridanum

2 cm

2 cm

1 cm

ILLICIUM · SCHISANDRA · NELUMBO

2 mm

5 mm

S. glabra

2 cm

5 cm

N. lutea

2. Illicium parviflorum Michaux ex Ventenat, Tabl. Règn. Vég. 3: 71. 1799 · Swamp star-anise, yellow-anise C E

Badianifera parviflora (Michaux ex Ventenat) Kuntze

Leaves: petiole 8–16 mm. **Leaf blade** dull green to olive, elliptic to obovate-elliptic, 5–13(–21) × 2–4(–6) cm, base acute, apex obtuse to rounded. **Flowers** 0.8–1.2 cm diam.; peduncle 0.7–2.4 cm; bracteoles 2–5; tepals 11–16, yellow-green, inner tepals orbiculate-obovate; stamens 6–7; pistils 11–14. **Fruit aggregates** 2–3.5 cm diam., usually with 10–13 pistils at maturity. **Seeds** hazel brown. $2n = 28$.

Flowering late spring. Moist woods, swamps; of conservation concern; 0–70 m; Fla., Ga.

A. C. Smith (1947) placed *Illicium parvifolium* in *Illicium* sect. *Cymbostemon* (Spach) A. C. Smith. *Illicium parviflorum* is endemic to Lake, Marion, Orange, Polk, Seminole, and Volusia counties in Florida, and S. B. Jones and N. C. Coile (1988) reported the species as an escape in Thomas County, Georgia. *Illicium parvifolium* is cultivated in the southeastern United States and elsewhere; it has been frequently misidentified as *I. anisatum* Linnaeus, a taxon that differs from *I. parviflorum* in having larger, white tepals (M. A. Dirr 1986). The species may also be confused with the infrequently cultivated *I. verum*, which has fewer pistils (7–9[–10]) and more stamens (11–20).

10. SCHISANDRACEAE Blume

• Star-vine Family

Michael A. Vincent

Shrubs, evergreen or deciduous, scandent or twining, glabrous. **Leaves** alternate, simple, without stipules, petiolate. **Leaf blade** often fragrant (especially when bruised or crushed), translucent-dotted, ovate to lanceolate, pinnately veined, thin to leathery, margins entire to remotely dentate. **Flowers** unisexual, staminate and pistillate on same [different] plants, axillary, solitary, pedunculate; peduncle bracteolate; perianth hypogynous; tepals 5–20, imbricate, distinct, in 2[–3] series, all similar but innermost more petaloid. **Staminate flowers:** stamens 4–80, hypogynous, distinct or connate partially or completely into fleshy, globose or discoid mass, in 1–several series; anthers basifixed, 4-locular, longitudinally dehiscent; pollen 3- or 6-aperturate; pistil absent. **Pistillate flowers:** pistils simple, 6–300, distinct, closely set in few to many series on globose to elongate axis, attached obliquely; placentation marginal to pendulous; ovules [1–]2–3[–10] per locule; stigmatic surface adaxial; stamens absent. **Fruits** aggregates of berries, produced on globose or elongate axis. **Seeds** [1–]2–3[–10] per berry, flattened; endosperm copious, oily; embryo small.

Genera 2, species 47 (1 genus, 1 species in the flora): North America, chiefly e Asia.

The Chinese drug wu-wei-zi and its substitutes are obtained from species of *Schisandra.* The drug is considered a potential source of expectorants, immune-response boosters, and anti-ulcer compounds. Several lignan compounds from species of the genus are used as a treatment for hepatitis and as central nervous system depressants (S. Foster 1989).

Schisandraceae were considered closely allied to Illiciaceae by A. C. Smith (1947). Embryologic studies (R. N. Kapil and S. Jalan 1964) were interpreted as showing that the former are considerably removed from the latter. In his monograph, A. C. Smith also postulated a close relationship between Schisandraceae and Magnoliaceae. J. Hutchinson (1973) stated that Schisandraceae were probably derived from Magnoliaceae. Studies of fossil pollen led J. W. Walker and A. G. Walker (1984) to conclude that Schisandraceae and Illiceaceae are closely allied with Winteraceae. Based on analysis of nucleotide sequences from the plastid gene *rbc*L, however, M. W. Chase et al. (1993) and Qiu Y. L. et al. (1993) concluded that Schisandraceae and Illiciaceae are closely allied and closely related to Austrobaileyaceae but distant from Winteraceae.

SELECTED REFERENCES Hutchinson, J. 1973. The Families of Flowering Plants, ed. 3. Oxford. Pp. 161–162. Smith, A. C. 1947. The families Illiciaceae and Schisandraceae. Sargentia 7: 1–224.

1. SCHISANDRA Michaux, Fl. Bor.-Amer. 2: 218. 1803, name conserved · Star-vine

[Greek *schisis,* splitting, and *andro,* male]

Shrubs, twining. **Leaves** remote on long shoots, close on short lateral shoots. **Flowers:** perianth of 9–12 tepals in 2 series. **Staminate flowers:** stamens 5, partially or completely connate. **Pistillate flowers:** ovules 2–3 per locule. **Berries** on elongate receptacle when mature. $x = 14$.

Species 25 (1 in the flora): North America, mainly e Asia.

The orthographic variant *Schizandra,* adopted by several authors for *Schisandra* Michaux, was first used by A. P. de Candolle ([1817]1818–1821, vol. 1, pp. 511, 544) and was perpetuated by subsequent authors. Convincing arguments for retention of the original orthography were presented by A. J. Rehder (1944). *Schisandra* is conserved over the earlier published name *Stellandria.*

SELECTED REFERENCES Kapil, R. N. and S. Jalan. 1964. *Schisandra* Michaux—its embryology and systematic position. Bot. Not. 117: 285–306. Rehder, A. J. 1944. *Schisandra* Michaux, nomen genericum conservandum. J. Arnold Arbor. 25: 129–131. Stone, D. E. 1968. Cytological and morphological notes on the southeastern endemic *Schisandra glabra* (Schisandraceae). J. Elisha Mitchell Sci. Soc. 84: 351–356.

1. **Schisandra glabra** (Brickell) Rehder, J. Arnold Arbor. 25: 131. 1944 · Star-vine E F

Stellandria glabra Brickell, Med. Repos. 6(3): 327. 1803; *Schisandra coccinea* Michaux

Plants sprawling or twining over shrubs and trees, to 20 m or more. **Bark** gray-brown, flaking; pith tan, becoming dark brown, continuous; young twigs pale brown and smooth, older stems to 4 cm diam. **Leaves:** petiole slender, 1–7 cm. **Leaf blade** ovate-elliptic to cordate, 2–13 × 1–8 cm (larger on vigorous shoots), thin, margins entire to remotely dentate, glabrous. **Flowers** unisexual, staminate and pistillate on same plant; peduncle slender, 2.5–5 cm; outer tepals greenish white, inner tepals rose to red, broadly ovate, 3–9 × 2–5 mm. **Staminate flowers:** anthers embedded in sessile 5-sided disc. **Pistillate flowers:** pistils 6–12, reddish. **Berries** red, globose to ellipsoid, 4–8 × 5–15 mm; receptacle 4–7 cm. **Seeds** 1–3, red-brown, reniform, 4–4.5 × 5–5.5 mm, smooth to slightly rugose. $2n = 28$.

Flowering late spring–early summer. Mesic wooded bluffs, ravines, stream banks; 0–500 m; Ala., Ark., Fla., Ga., Ky., La., Miss., N.C., S.C., Tenn.

Schisandra glabra was placed in *Schisandra* sect. *Schisandra* (as sect. *Euschisandra,* not valid) by A. C. Smith (1947); other species of the section are found in southern Japan, southern Korea, and eastern China.

11. NELUMBONACEAE Dumortier

· Lotus-lily Family

John H. Wiersema

Herbs, perennial, aquatic, rhizomatous; roots adventitious; air chambers conspicuous in vegetative portions of plant. **Rhizomes** branched, repent, slender, terminal portions becoming tuberous-thickened late in growing season. **Leaves** arising directly from rhizome, alternate, floating or emersed; petiole long. **Leaf blade** peltate, orbiculate, margins entire; laticifers present. **Inflorescences** axillary, solitary flowers. **Flowers** bisexual, protogynous, diurnal, borne above water surface; peduncle long; involucre absent; perianth not persistent in fruit (outermost tepals of *Nelumbo lutea* normally persistent), hypogynous; tepals numerous, distinct, outermost reduced, inner tepals becoming larger and more petaloid; stamens numerous; filaments slender; anthers dehiscing by longitudinal slits, connective appendage incurved; pistils numerous, 1-carpellate, separately embedded in flattened top of turbinate receptacle; ovary 1-locular; placentation apical; ovule 1; style short; stigma nearly sessile, capitate. **Fruits** nutlike, indehiscent, loose in cavities of strongly accrescent receptacle. **Seed** 1, aril absent; endosperm and perisperm absent; embryo completely filling seed; cotyledons 2, fleshy.

Genus 1, species 2 (2 in the flora): temperate and tropical regions.

Nelumbonaceae are pollinated by insects, often by beetles.

Formerly Nelumbonaceae frequently were included in Nymphaeaceae in the broad sense.

SELECTED REFERENCE   Wood, C. E. Jr. 1959. The genera of the Nymphaeaceae and Ceratophyllaceae in the southeastern United States. J. Arnold Arbor. 40: 94–112.

1. NELUMBO Adanson, Fam. Pl. 2: 76, 582. 1763 · Lotus [Ceylonese vernacular name]

Rhizomes forming swollen storage tubers late in growing season. **Leaf blade** peltate, orbiculate, margins entire. **Flowers** exceeding leaves, to 2.5 dm or more diam.; tepals caducous or outermost persistent, 14–30, white to pink-purple or yellow; stamens 100–200. **Fruits** somewhat globose or ovoid; receptacle gradually or abruptly narrowed from top to base, base tapered to rounded, lateral surface striate or rugose. *x* = 8.

Species 2 (2 in the flora): North America, West Indies, Central America, Asia, Australia.

A study by T. Borsch and W. Barthlott (1994[1996]) led them to conclude that the two species recognized here and universally elsewhere should be treated as subspecies under *Nelumbo nucifera*. Apart from the admitted difference in tepal color, however, most of the distinguishing characters I have employed were not mentioned or were not thoroughly studied.

Furthermore, the nearly identical *rbc*L sequence data reported for the two taxa by Les et al. (1991) and used to support this merger remains to be verified using wild-collected *N. nucifera*, as the possibility exists that the cultivated sample of *N. nucifera* used may have been of hybrid origin (D. H. Les, pers. comm.). Molecular studies by Jun Wen (pers. comm.) have revealed clear differences between authentic samples of the two taxa.

SELECTED REFERENCE Borsch, T. and W. Barthlott. 1994[1996]. Classification and distribution of the genus *Nelumbo* Adans. (Nelumbonaceae). Beitr. Biol. Pflanzen 68: 421–450.

1. Tepals pale yellow, outermost normally persistent; mature fruits mostly less than 1.25 times longer than wide. . . . . 1. *Nelumbo lutea*
1. Tepals pink or white, all normally caducous; mature fruits mostly more than 1.5 times longer than wide . . . . 2. *Nelumbo nucifera*

1. Nelumbo lutea Willdenow, Sp. Pl. 2: 1259. 1799 (as Nelumbium luteum) · American lotus, yellow lotus, water-chinquapin, volée [F] [W2]

Nelumbo nucifera Gaertner subsp. *lutea* (Willdenow) Borsch & Barthlott

Leaves: petiole to 2 m or more. **Leaf blade** to 6 dm or more. **Flowers:** tepals pale yellow, 1–13 cm, outermost 1–5 normally persistent; anthers 1–2 cm. **Fruits** somewhat globose, 10–16 × 8–13 mm, mostly less than 1.25 times longer than wide; receptacle to 1 dm diam. at maturity, abruptly narrowed ca. 1–2 cm below flattened top, base tapered, lateral surface usually distinctly striate. $2n = 16$.

Flowering late spring–summer. Mostly flood plains of major rivers in ponds, lakes, pools in marshes and swamps, and backwaters of reservoirs; 0–400 m; introduced at other sites; Ont.; Ala., Ark., Conn., Del., D.C., Fla., Ga., Ill., Ind., Iowa, Kans., Ky., La., Maine, Md., Mass., Mich., Minn., Miss., Mo., Nebr., N.J., N.Y., N.C., Ohio, Okla., Pa., R.I., S.C., Tenn., Tex., Va., W.Va., Wis.; Mexico; West Indies (Cuba, Jamaica, and Hispaniola); Central America (Honduras).

Nelumbo lutea is a species as magnificent as its Asian relative, *N. nucifera*, but it is less cultivated for ornament. It was probably originally confined to flood plains of major rivers and their tributaries in the east-central United States and carried northward and eastward by aborigines who used the seeds and tubers for food. The species is sometimes an aggressive, difficult-to-eradicate weed in ponds, lakes, and reservoirs.

Although *Nelumbo lutea* is often attributed to (Willdenow) Persoon, the spelling *Nelumbium luteum* used by Willdenow is an orthographic error for *Nelumbo lutea* that should be corrected (W. Greuter et al. 1994, Art. 61.4), and Persoon's later combination is superfluous.

The name *Nelumbo pentapetala* (Walter) Fernald, sometimes used for this taxon, was based on *Nymphaea pentapetala* Walter, a name of uncertain application that has been recently proposed for rejection (J. H. Wiersema and J. L. Reveal 1991).

SELECTED REFERENCES Hall, T. F. and W. T. Penfound. 1944. The biology of the American lotus *Nelumbo lutea* (Willd.) Pers. Amer. Midl. Naturalist 31: 744–758. Schneider, E. L. and J. D. Buchanan. 1980. Morphological studies of the Nymphaeaceae. XI. The floral biology of *Nelumbo pentapetala*. Amer. J. Bot. 67: 182–193.

2. Nelumbo nucifera Gaertner, Fruct. Sem. Pl. 1: 73. 1788 · Indian lotus, sacred lotus, oriental lotus, lotus

Leaves: petiole to 2 m or more. **Leaf blade** to 6 dm or more. **Flowers:** tepals normally all caducous, pink, pink-tinged, or fading to white, 1–13 cm; anthers 1–2 cm. **Fruits** ovoid, 10–20 × 7–13 mm, mostly more than 1.5 times longer than wide; receptacle to 1 dm diam. at maturity, gradually tapered or rounded from flattened top to base, base rounded or very slightly obtuse-tapered, lateral surface rugose or only weakly striate.

Flowering late spring–summer. Ponds and lakes; 0–400 m; introduced; Ala., Ark., Fla., Ga., La., Md., Mass., Miss., Mo., N.J., N.C., S.C., Tenn., Tex., W.Va.; naturalized, s Europe; Asia; n Australia.

Nelumbo nucifera is an ornamental species sporadically naturalized from cultivation over the area mapped in the southeastern United States. Although Virginia, Kentucky, and Delaware are within the range, I know of no collections from Virginia or Delaware and of no wild-growing specimens of *N. nucifera* in Kentucky. The species was listed in a flora of New York (R. S. Mitchell 1986); I have not seen a voucher specimen for the report.

The seeds of *Nelumbo nucifera* have been shown to remain viable for several hundred years under certain conditions (D. A. Priestley and M. A. Posthumus 1982).

12. NYMPHAEACEAE Salisbury

· Water-lily Family

John H. Wiersema

C. Barre Hellquist

Herbs, perennial, aquatic, rhizomatous; roots adventitious; air chambers conspicuous in vegetative portions of plant. **Rhizomes** branched or unbranched, erect or repent, tuberous-thickened, sometimes bearing stolons. **Leaves** arising directly from rhizome, alternate, floating, submersed, or emersed; stipules present or absent; petioles long. **Leaf blade** lanceolate to ovate or orbiculate, with basal sinus [peltate], margins entire to spinose-dentate. **Inflorescences** axillary or extra-axillary, flowers solitary. **Flowers** bisexual, protogynous [homogamous in some species of *Nymphaea*], diurnal or nocturnal, borne at or above water surface, occasionally submersed; peduncle long; involucre absent; perianth often persistent in fruit, hypogynous to perigynous [epigynous]; sepals usually 4–9(–12), distinct; petals numerous [rarely absent], often transitional to stamens; stamens numerous; filaments broad to slender; anthers dehiscing by longitudinal slits, with or without connective appendage; pistil 1, 3–35-carpellate and -locular; placentation laminar; ovules numerous per locule; stigma sessile, radiate on stigmatic disk. **Fruits** berrylike, indehiscent or irregularly dehiscent. **Seeds** several–numerous; aril present or absent; endosperm sparse; perisperm abundant; embryo minute; cotyledons 2, fleshy.

Genera 6 (including *Barclaya*), species ca. 50 (2 genera, 17 species in the flora): nearly worldwide.

Nymphaeaceae are insect-pollinated, often by beetles.

Formerly Nymphaeaceae often were treated to include Cabombaceae and Nelumbonaceae, but these are now generally segregated.

SELECTED REFERENCES Schneider, E. L. 1979. Pollination biology of the Nymphaeaceae. In: D. M. Caron, ed. 1979. Increasing Production of Agricultural Crops through Increased Insect Pollination: Proceedings of the IVth International Symposium on Pollination. College Park, Md. Pp. 419–429. [Maryland Agric. Exp. Sta. Special Misc. Publ. 1.] Wood, C. E. Jr. 1959. The genera of the Nymphaeaceae and Ceratophyllaceae in the southeastern United States. J. Arnold Arbor. 40: 94–112.

1. NUPHAR Smith in J. Sibthorp and J. E. Smith, Fl. Graec. Prodr. 1: 361. 1809, name conserved · Spatterdock, cow-lily, yellow pond-lily, nénuphar [ancient Arabic or Persian name]

Rhizomes branched, repent; stolons absent. **Leaves** floating, submersed, or emersed. **Leaf blade** orbiculate to linear, basal lobes divergent to overlapping, margins entire, apex of lobe narrowly obtuse to broadly rounded; primary venation mostly pinnate, basal section of midrib with several parallel veins. **Flowers** floating or held above surface of water, opening diurnally; perianth hypogynous, not spreading, nearly globose at anthesis; sepals 5–9(–12), outwardly yellow to green, becoming yellow within, often red-tinged, oblong or obovate to somewhat orbiculate; petals numerous, spirally arranged, inconspicuous, stamenlike; stamens yellow- or red-tinged, inserted below ovary, recurved at dehiscence, distal connective appendage absent; ovary longer than petals and stamens; stigmatic disk with margin entire to crenate or dentate, appendages absent. **Fruits** borne on straight peduncles. **Seeds** ovoid, to 6 mm; aril absent. $x = 17$.

Species 10–12, most intergrading and sometimes treated at subspecies rank (8 in the flora): north temperate, North America, Europe, and Asia.

The taxonomy of the genus is problematic. E. O. Beal (1956) departed dramatically from previous North American treatments in recognizing a single polymorphic species, *Nuphar lutea* (name of European origin), with several subspecies formerly treated as species. Subsequent research (C. E. DePoe and E. O. Beal 1969; E. O. Beal and R. M. Southall 1977) has supported Beal's treatment for some southeastern subspecies, but most other taxa have not been studied as extensively. Beal's treatment, for the most part, has not been adopted in the Northeast and elsewhere in North America or in Europe. Molecular studies of *Nuphar* currently in progress (D. J. Padgett, pers. comm.) have clearly shown the North American taxa to be distinct from the Eurasian *Nuphar lutea;* Beal's nomenclature under that taxon cannot be upheld. Continuing to treat those taxa at subspecific rank would require new combinations under *Nuphar sagittifolia* (Walter) Pursh, the oldest name that has hitherto been applied only in a geographically restricted sense. Until the molecular studies are completed, creating new names is premature. We therefore return to the previous treatment of the taxa as species.

Prior to conservation in its current sense, the name *Nymphaea* was frequently used for *Nuphar.* Although often treated as neuter, *Nuphar* was originally assigned the feminine gender (W. T. Stearn 1956; H. W. Rickett and F. A. Stafleu 1959).

The name *Nymphozanthus* Richard has been rejected.

SELECTED REFERENCES Beal, E. O. 1956. Taxonomic revision of the genus *Nuphar* Sm. of North America and Europe. J. Elisha Mitchell Sci. Soc. 72: 317–346. Miller, G. S. Jr. and P. C. Standley. 1912. The North American species of *Nymphaea.* Contr. U.S. Natl. Herb. 16: 63–108.

1. Sepals 5 (to 6 in *Nuphar rubrodisca*); fruit deeply constricted below stigmatic disk; stigmatic disk red.
  2. Anthers 1–3 mm; stigmatic disk with 6–10 deep crenations, stigmatic rays terminating 0–0.2 mm from margin of disk, constriction below disk 1.5–5 mm diam.; leaf sinus 2/3 or more length of midrib . . . . 1. *Nuphar microphylla*
  2. Anthers (2–)3–6 mm; stigmatic disk with 8–15 shallow crenations, stigmatic rays terminating 0–1.6 mm from margin of disk, constriction below disk 5–10 mm diam.; leaf sinus ca. 1/2 length of midrib. . . . . 2. *Nuphar rubrodisca*
1. Sepals 6–9(–12); fruit slightly or not constricted below stigmatic disk; stigmatic disk green, yellow, or occasionally reddened.
  3. Leaf blade more than 2 times as long as wide; sinus less than 1/3 length of midrib.
    4. Leaf blade lanceolate to ovate, ca. 2.5 times as long as wide; stigmatic rays ± elliptic. . . . . 5. *Nuphar ulvacea*
    4. Leaf blade linear to lanceolate, 3–5 times as long as wide; stigmatic rays linear. . . . . 6. *Nuphar sagittifolia*

3. Leaf blade to 2 times as long as wide; sinus 1/3 or more length of midrib.
   5. Leaf petiole adaxially flattened and winged along margins; fruit usually purplish; sepals red or maroon at base adaxially. . . . . 3. *Nuphar variegata*
   5. Leaf petiole terete or slightly flattened adaxially; fruit mostly greenish or yellowish, rarely reddened; sepals yellow or red at base adaxially.
      6. Sepals usually (6–)9(–12); stigmatic rays terminating within 1(–1.5) mm from margin of disk. . . . . 8. *Nuphar polysepala*
      6. Sepals usually 6; stigmatic rays terminating 1–3 mm from margin of disk.
         7. Leaves mostly floating; leaf blade orbiculate or nearly so, abaxially densely pubescent. . . 7. *Nuphar orbiculata*
         7. Leaves mostly emersed; leaf blade ovate to nearly orbiculate, abaxially glabrous to sparsely pubescent . . . . 4. *Nuphar advena*

1. Nuphar microphylla (Persoon) Fernald, Rhodora 19: 111. 1917 (as microphyllum) · Petit nénuphar jaune ⬚E⬚ ⬚F⬚

Nymphaea microphylla Persoon, Syn. Pl. 2: 63. 1807; *Nuphar kalmiana* (Michaux) W. T. Aiton; *N. minima* (Willdenow) Smith

Rhizomes 1–2 cm diam. **Leaves** mostly floating, occasionally submersed; petiole flattened to filiform. **Leaf blade** abaxially often purple, adaxially green to greenish purple, broadly elliptic to ovate, 3.5–10(–13) × 3.5–7.5(–8.5) cm, 1–1.5 times as long as wide, sinus 2/3 or more length of midrib, lobes divergent and forming V-shaped angle; surfaces abaxially glabrous to densely pubescent. **Flowers** 1–2 cm diam.; sepals 5(–10), abaxially green to adaxially yellow toward base; petals broadly spatulate and thin, or notched and thickened; anthers 1–3 mm, shorter than filaments. **Fruit** yellow, green, brown, or rarely purple, mostly globose-ovoid, occasionally flask-shaped, 1–2 cm, smooth basally, faintly ribbed toward apex, deeply constricted below stigmatic disk, constriction 1.5–5 mm diam.; stigmatic disk red, 2.5–7 mm diam., with 6–10 deep crenations; stigmatic rays 6–11, linear, terminating 0–0.2 mm from margin of disk. **Seeds** ca. 3 mm. $2n = 34$.

Flowering summer–early fall. Ponds, lakes, sluggish streams, sloughs, ditches, and occasionally tidal waters; 0–400 m; Man., N.B., N.S., Ont., Que.; Conn., Maine, Mass., Mich., Minn., N.H., N.J., N.Y., Pa., Vt., Wis.

Intermediates between *Nuphar microphylla* and *N. variegata*, probably of hybrid origin, are treated as *N. rubrodisca*. A form with ten sepals (*Nuphar microphyllum* forma *multisepalum* Lakela) occurs in northeastern Minnesota. Recent observations of the Eurasian *N. pumila* (Timm) de Candolle by C. B. Hellquist in Siberia suggest that Beal's lumping of *N. microphylla* under *N. lutea* subsp. *pumila* (Timm) E. O. Beal should be further studied.

2. Nuphar rubrodisca Morong, Bot. Gaz. 11: 167. 1886 (as rubrodiscum) · Nénuphar à disque rouge ⬚E⬚ ⬚F⬚

Nuphar lutea (Linnaeus) Smith subsp. *rubrodisca* (Morong) Hellquist & Wiersema; *Nymphaea hybrida* Peck; *N. rubrodisca* (Morong) Greene

Rhizomes 1–2.5(–4) cm diam. **Leaves** mostly floating, occasionally submersed; petiole flattened. **Leaf blade** often abaxially purple on new leaves, adaxially green to greenish purple, broadly ovate to oblong, 5–25 × 4.5–15 cm, 1.1–1.7 times as long as wide, sinus ca. 1/2 length of midrib, lobes overlapping to divergent and forming V-shaped angle; surfaces glabrous. **Flowers** 3 cm or more diam.; sepals 5–6, abaxially green to yellow, adaxially often red-tinged toward base; petals broadly spatulate and thin, or notched and thickened; anthers (2–)3–6 mm, shorter than filaments. **Fruit** purple, dark brown, or rarely green, globose-ovoid, occasionally flask-shaped, 1.5–2.5 cm, strongly ribbed, deeply constricted below stigmatic disk, constriction 5–10 mm diam.; stigmatic disk red, 8–14 mm diam., with 8–14 shallow crenations; stigmatic rays 8–15, linear, terminating 0–1.6 mm from margin of disk. **Seeds** 2.5–3.5 mm.

Flowering summer. Ponds, lakes, sluggish streams, sloughs, and occasionally tidal waters; 0–400 m; N.B., N.S., Ont., Que.; Conn., Maine, Mass., Mich., Minn., N.H., N.J., N.Y., Pa., Vt., Wis.

Nuphar rubrodisca is generally considered to be a hybrid between *N. microphylla* and *N. variegata* because it displays characteristics intermediate between the two taxa. It is reportedly sterile in some areas and completely fertile in others.

As this volume went to press, Hellquist and I discovered *Nuphar rubrodisca* in the southeastern corner of Manitoba.

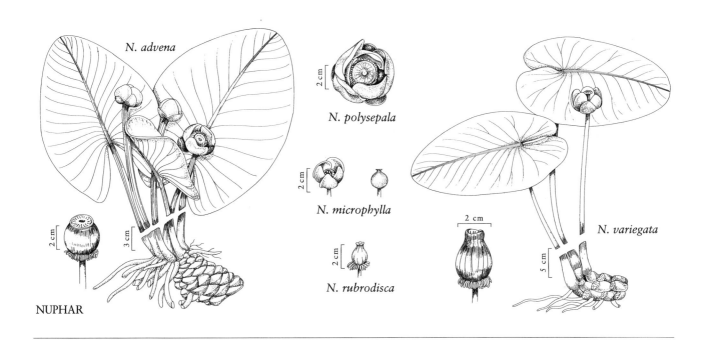

N. advena

N. polysepala

N. microphylla

N. rubrodisca

N. variegata

NUPHAR

3. **Nuphar variegata** Durand in G. W. Clinton, Rep. (Annual) Regents Univ. State New York State Cab. Nat. Hist. 19: 73. 1866 (as variegatum) · Grand nénuphar jaune E F

Nuphar fraterna (G. S. Miller & Standley) Standley; *N. lutea* (Linnaeus) Smith subsp. *variegata* (Durand) E. O. Beal; *Nymphaea americana* G. S. Miller & Standley; *N. fraterna* G. S. Miller & Standley

Rhizomes 2.5–7 cm diam. **Leaves** mostly floating, occasionally submersed; petiole adaxially flattened, with median ridge, winged along margins. **Leaf blade** abaxially and adaxially green, sometimes abaxially purple-tinged, broadly ovate to oblong, 7–35 × 5–25 cm, 1.2–1.6 times as long as wide, sinus 1/3–1/2 length of midrib, lobes approximate to overlapping; surfaces glabrous. **Flowers** 2.5–5 cm diam.; sepals mostly 6, abaxially green to yellow, adaxially usually with red or maroon toward base; petals oblong, thick; anthers 3–9 mm, longer than filaments. **Fruit** mostly purple-tinged, ovoid, 2–4.3 × 2–3.5 cm, strongly ribbed, slightly constricted below stigmatic disk; stigmatic disk green, rarely reddened, 8–20 mm diam., entire to deeply crenate; stigmatic rays 7–28, linear to narrowly lanceolate, terminating 0–1(–1.5) mm from margin of disk. **Seeds** 2.5–5 mm. 2*n* = 34.

Flowering late spring–summer. Ponds, lakes, sluggish streams, and ditches; 0–2000 m; St. Pierre and Miquelon; Alta., B.C., Man., N.B., Nfld., N.W.T., N.S., Ont., P.E.I., Que., Sask., Yukon; Conn., Del., Idaho, Ill., Ind., Iowa, Maine, Md., Mass., Mich.,

Minn., Mont., Nebr., N.H., N.J., N.Y., N.Dak., Ohio, Pa., R.I., S.Dak., Vt., Wis.

Nuphar variegata is distinct throughout most of its range. The leaves are characteristically floating, being emergent only under low-water conditions. Intermediates between *N. variegata* and *N. microphylla,* probably of hybrid origin, are treated as *N. rubrodisca.* Some intergrading of characteristics occurs where the range overlaps with *N. advena* (E. O. Beal 1956). This can be observed in the mid-Atlantic region. Intermediates between *N. variegata* and *N. polysepala* occur in eastern British Columbia where the two species are sympatric. Authorship and typification of this name were discussed by E. G. Voss (1965).

4. **Nuphar advena** (Aiton) W. T. Aiton in W. Aiton & W. T. Aiton, Hortus Kew. 3: 295. 1811 F W1

Nymphaea advena Aiton, Hort. Kew. 2: 226. 1789; *Nuphar fluviatilis* (R. M. Harper) Standley; *N. lutea* subsp. *advena* (Aiton) Kartesz & Gandhi; *N. lutea* subsp. *macrophylla* (Small) E. O. Beal; *N. lutea* subsp. *ozarkana* (G. S. Miller & Standley) E. O. Beal; *N. ovata* (G. S. Miller & Standley) Standley; *N. ozarkana* (G. S. Miller & Standley) Standley; *N. puteora* Fernald; *N.* ×*interfluitans* Fernald; *Nymphaea advena* subsp. *macrophylla* (Small) G. S. Miller & Standley; *N. chartacea* G. S. Miller & Standley; *N. fluviatilis* R. M. Harper; *N. ludoviciana* G. S. Miller & Standley; *N. macrophylla* Small; *N. microcarpa* G. S. Miller & Standley; *N. ovata* G. S. Miller & Standley; *N. ozarkana* G. S. Miller & Standley; *N. puberula* G. S. Miller & Standley

Rhizomes mostly 5–10 cm diam. **Leaves** mostly emersed, occasionally floating or submersed; petiole terete or adaxially slightly flattened. **Leaf blade** abaxially and adaxially green, broadly ovate to nearly orbiculate, 12–40 × 7–30 cm, 1–2 times as long as wide, sinus 1/3–1/2 length of midrib, lobes overlapping to divergent, often forming angle of 90° or greater; surfaces abaxially glabrous to sparsely pubescent. **Flowers** to 4 cm diam.; sepals mostly 6, abaxially green to adaxially yellow, rarely red-tinged toward base; petals oblong, thick; anthers 3–7 mm, longer than filaments. **Fruit** green, ovoid, 2–5 × 2–5 cm, moderately ribbed, slightly constricted below stigmatic disk; stigmatic disk green, occasionally reddened, 13–25 mm diam., entire to crenate; stigmatic rays 9–23, linear to lanceolate, terminating 1–3 mm from margin of disk. **Seeds** 3–6 mm.

Flowering mid spring–early fall, extended farther south. Ponds, lakes, sluggish streams and rivers, springs, marshes, ditches, canals, sloughs, and tidal waters; 0–450 m; Ont.; Ala., Ark., Conn., Del., D.C., Fla., Ga., Ill., Ind., Kans., Ky., La., Maine, Md., Mich., Miss., Mo., N.J., N.Y., N.C., Ohio, Okla., Pa., S.C., Tenn., Tex., Vt., Va., W.Va., Wis.; Mexico (Coahuila, Tamaulipas); West Indies (Cuba).

Nuphar advena is extremely variable and intergrades with *N. orbiculata, N. ulvacea,* and *N. sagittifolia* in areas of sympatry. Intergradation with *N. variegata* (E. O. Beal 1956) can be observed in the mid-Atlantic region, although most floristic treatments from the area of overlap treat the two taxa as distinct species. Local variation in the Ozark Mountains and in Texas, the basis for the names *Nymphaea ozarkana, N. ovata,* and *N. puberula,* is not considered sufficient to warrant recognition.

SELECTED REFERENCES Schneider, E. L. and L. A. Moore. 1977. Morphological studies of the Nymphaeaceae. VII. The floral biology of *Nuphar lutea* subsp. *macrophylla*. Brittonia 29: 88–99. Wiersema, J. H. and C. B. Hellquist. 1994. Nomenclatural notes in Nymphaeaceae for the North American flora. Rhodora 96: 170–178.

5. **Nuphar ulvacea** (G. S. Miller & Standley) Standley, Field Mus. Nat. Hist., Bot. Ser. 8(5): 311. 1931 (as ulvaceum) C E

Nymphaea ulvacea G. S. Miller & Standley, Contr. U.S. Natl. Herb. 16: 97. 1912; *Nuphar lutea* (Linnaeus) Smith subsp. *ulvacea* (G. S. Miller & Standley) E. O. Beal

Rhizomes 2–5 cm diam. **Leaves** floating or submersed; petiole terete. **Leaf blade** abaxially and adaxially green, lanceolate to ovate, 15–25 × 8–10 cm, ca. 2.5 times as long as wide, sinus less than 1/3 length of midrib, lobes often overlapping; surfaces glabrous. **Flowers** 2–3 cm diam.; sepals 6–9, abaxially green to adaxially yellow toward base; petals oblong, thick; anthers 2–5 mm, slightly longer than filaments. **Fruit** green, nearly globose, 1.5–2.5 × 1.5–2.5 cm, strongly ribbed, slightly constricted below stigmatic disk; stigmatic disk green, 11–18 mm diam., undulate; stigmatic rays 9–16, ± elliptic, terminating within 1 mm from margin of disk. **Seeds** 3.5–4 mm.

Flowering spring–early fall. Blackwater rivers and streams; of conservation concern; 0–100 m; Fla.

Nuphar ulvacea is probably better treated as a subspecies. Gulf coast plants intermediate between *N. ulvacea* and *N. advena,* which have been treated as *Nymphaea* [*Nuphar*] *chartacea,* are treated under *N. advena.*

6. **Nuphar sagittifolia** (Walter) Pursh, Fl. Amer. Sept. 2: 370. 1814 (as sagittaefolia) C E

Nymphaea sagittifolia Walter, Fl. Carol., 155. 1788; *Nuphar lutea* (Linnaeus) Smith subsp. *sagittifolia* (Walter) E. O. Beal

Rhizomes 2–2.5 cm diam. **Leaves** floating or submersed; petiole terete. **Leaf blade** abaxially and adaxially green, linear to lanceolate, 15–30(–50) × 5–10(–11.5) cm, 3–5 times as long as wide, sinus less than 1/3 length of midrib, lobes usually divergent and forming V-shaped angle; surfaces glabrous. **Flowers** 2–3 cm diam.; sepals 6, abaxially green to adaxially yellow toward base; petals oblong, thick; anthers 3–5 mm, barely or not at all longer than filaments. **Fruit** green, ovoid, 3–3.5 × 2–3 cm, smooth basally, strongly ribbed toward apex, slightly constricted below stigmatic disk; stigmatic disk green, 14–18 mm diam., nearly entire; stigmatic rays 10–14, linear, mostly terminating 1–2 mm from margin of disk. **Seeds** 4–5 mm.

Flowering mid spring–early fall. Freshwater streams, rivers, ponds, and lakes of coastal plain, extending to freshwater tidal areas; of conservation concern; 0–50 m; N.C., S.C., Va.

Nuphar sagittifolia is probably best treated as a subspecies. Plants intermediate between it and *N. advena* are treated under *N. advena.* The clinal variation pattern between the two taxa is apparently maintained via selection by vernalization (C. E. DePoe and E. O. Beal 1969; E. O. Beal and R. M. Southall 1977).

This taxon is the Cape Fear spatterdock of the aquarium trade.

7. **Nuphar orbiculata** (Small) Standley, Field Mus. Nat. Hist., Bot. Ser. 8(5): 311. 1931 (as orbiculatum)   E

Nymphaea orbiculata Small, Bull. Torrey Bot. Club 23: 128. 1896; *Nuphar lutea* (Linnaeus) Smith subsp. *orbiculata* (Small) E. O. Beal; *Nymphaea bombycina* G. S. Miller & Standley

Rhizomes ca. 7–8 cm diam. **Leaves** mostly floating, occasionally submersed; petiole terete. **Leaf blade** abaxially and adaxially green, often suffused with purple, orbiculate or nearly so, 20–45 × 20–45 cm, 1–1.2 times as long as wide, sinus ca. 1/2 length of midrib, lobes approximate to overlapping; surfaces abaxially densely pubescent. **Flowers** 4–8 cm diam.; sepals 6, abaxially green to adaxially yellow, never red-tinged toward base; petals oblong, thick; anthers 5–6 mm, longer than filaments. **Fruit** greenish or yellowish, cylindric to nearly globose, 3.5–5 cm, smooth basally, finely ribbed toward apex, slightly constricted below stigmatic disk; stigmatic disk green, yellow, or sometimes reddened, ca. 30–35 mm diam., undulate; stigmatic rays 12–28, linear or lanceolate, terminating 1–3 mm from margin of disk. **Seeds** 4–6 mm.

Flowering mid spring–early fall. Acidic ponds; 0–100 m; Ala., Fla., Ga.

Nuphar orbiculata is perhaps best treated as a subspecies. Plants intermediate between it and *N. advena* occur in southern Georgia and northern Florida.

8. **Nuphar polysepala** Engelmann, Trans. Acad. Sci. St. Louis 2: 282. 1865 (as polysepalum)   E F W1

Nuphar lutea (Linnaeus) Smith subsp. *polysepala* (Engelmann) E. O. Beal; *Nymphaea polysepala* (Engelmann) Greene

Rhizomes 3–8 cm diam. **Leaves** mostly floating, occasionally emersed or submersed; petiole terete. **Leaf blade** abaxially and adaxially green, widely ovate, 10–40(–45) × 7–30 cm, ca. 1.2–1.5 times as long as wide, sinus 1/3–2/3 length of midrib, lobes divergent to overlapping; surfaces glabrous. **Flowers** 5–10 cm diam.; sepals mostly (6–)9(–12), abaxially green to adaxially yellow, sometimes red-tinged toward base; petals oblong, thick; anthers 3.5–9 mm, slightly shorter than filaments. **Fruit** green to yellow, cylindric to ovoid, 4–6(–9) × 3.5–6 cm, strongly ribbed, slightly constricted below stigmatic disk; stigmatic disk green, 20–35 mm diam., entire to crenate; stigmatic rays 8–26(–36), linear to lanceolate, terminating within 1 (–1.5) mm from margin of disk. **Seeds** 3.5–5 mm. $2n = 34$.

Flowering spring (later in north) to summer. Ponds, lakes, and sluggish streams; 0–3700 m; B.C., N.W.T., Yukon; Alaska, Ariz., Calif., Colo., Idaho, Mont., Nev., N.Mex., Oreg., Utah, Wash., Wyo.

Plants intermediate between *Nuphar polysepala* and *N. variegata* occur in eastern British Columbia.

2. NYMPHAEA Linnaeus, Sp. Pl. 1: 510. 1753; Gen. Pl. ed. 5, 227. 1754, name conserved · Water-lily, nymphéa, lis d'eau, nénuphar blanc [Greek *nymphaia* and Latin *nymphaea*, water-lily, from Latin (*nympha*) or Greek (*nymphe*) mythology, goddess of mountains, waters, meadows, and forests]

John H. Wiersema

Castalia Salisbury

Rhizomes branched or unbranched, erect or repent; elongate stolons present or absent. **Leaves** mostly floating (vernal leaves submersed; blades sessile, broad). **Leaf blade** orbiculate to widely ovate or elliptic, basal lobes divergent to overlapping, margins entire to spinose-dentate, apex of lobe acute or acuminate to widely rounded; primary venation mostly palmate, midrib with 1 vein. **Flowers** floating or emersed, opening diurnally or nocturnally; perianth perigynous, spreading at anthesis; sepals 4, mostly greenish, ovate to elliptic; petals 8–many, spirally arranged or wholly or partially whorled, showy, white, pink, blue, or yellow, broadly lanceolate or ovate to obovate, grading into stamens; stamens yellow or cream-colored, inserted on lateral surface of ovary, spreading at anthesis, sometimes with distal connective appendage; ovary shorter than petals and stamens; stigmatic disk with prominent, distinct, upwardly incurved appendages around margin. **Fruits** borne on curved or coiled peduncles. **Seeds** nearly globose to ellipsoid, to 5 mm; aril present. $x = 14$.

Species 35–40 (9 in the flora): worldwide.

Nymphaea is an important genus of ornamental plants, with numerous cultivars or wild forms grown in water gardens. Some have become naturalized in some places, particularly in Florida, and two such taxa are included in this treatment. A third, *N.* ×*daubenyana* W. T. Baxter ex Daubeny (*N. micrantha* Guillemin & Perrottet × *N. caerulea* Savigny), with blue flowers and entire leaves and with a proliferous mound of fibrous tissue above insertion of petiole, may also be encountered in Florida.

Prior to conservation in its current sense, the name *Nymphaea* was frequently used for the genus now known as *Nuphar*.

SELECTED REFERENCES Conard, H. S. 1905. The waterlilies: A monograph of the genus *Nymphaea*. Publ. Carnegie Inst. Wash. 4: 1–279. Meeuse, B. J. D. and E. L. Schneider. 1980. *Nymphaea* revisited: A preliminary communication. Israel J. Bot. 28: 65–79. Moseley, M. F. Jr. 1961. Morphological studies of the Nymphaeaceae. II. The flower of *Nymphaea*. Bot. Gaz. 122: 233–259. Ward, D. B. 1977. Keys to the flora of Florida. 4. *Nymphaea* (Nymphaeaceae). Phytologia 37: 443–448. Wiersema, J. H. 1987. A monograph of *Nymphaea* subgenus *Hydrocallis* (Nymphaeaceae). Syst. Bot. Monogr. 16: 1–112. Wiersema, J. H. 1988. Reproductive biology of *Nymphaea* (Nymphaeaceae). Ann. Missouri Bot. Gard. 75: 795–804. Wiersema, J. H. 1996. *Nymphaea tetragona* and *Nymphaea leibergii* (Nymphaeaceae): Two species of diminutive waterlilies in North America. Brittonia 48: 520–531.

1. Sepals abaxially flecked with short dark lines.
  2. Leaf blade with dentate to spinose-dentate margins; petals white; connective appendage projecting to 3 mm or more beyond anther . . . . 1. *Nymphaea ampla*
  2. Leaf blade with entire to sinuate margins; petals pale violet to nearly white; connective appendage mostly projecting 1 mm or less beyond anther . . . . 2. *Nymphaea elegans*
1. Sepals abaxially uniformly greenish, reddish, or yellowish.
  3. Petals blue, lavender, or purple; connective appendage projecting to 4 mm or more beyond anther . . . . 3. *Nymphaea capensis*
  3. Petals white to pink, cream-colored, or yellow; connective appendage projecting less than 2 mm beyond anther.
    4. Leaf blade with spinose-dentate margins, abaxially slightly to densely puberulent . . . . 5. *Nymphaea lotus*
    4. Leaf blade with entire or sinuate margins, abaxially glabrous.
      5. Appendages at margin of stigmatic disk slightly club-shaped; flowers opening nocturnally; leaf blade with central web of cross veins between major veins . . . . 4. *Nymphaea jamesoniana*
      5. Appendages at margin of stigmatic disk tapered or boat-shaped; flowers opening diurnally; leaf blade with radiate venation, without central weblike pattern.
        6. Petals yellow; plants bearing stolons . . . . 6. *Nymphaea mexicana*
        6. Petals white; plants not bearing stolons.
          7. Petals 17–43; filament widest below middle; rhizomes repent . . . . 9. *Nymphaea odorata*
          7. Petals 8–17; filament widest above middle; rhizomes erect.
            8. Appendages at margin of stigmatic disk 0.6–1.5 mm; lines of insertion of sepals on receptacle not prominent . . . . 7. *Nymphaea leibergii*
            8. Appendages at margin of stigmatic disk mostly 3 mm or more; lines of insertion of sepals very prominent, forming tetragon on receptacle . . . . 8. *Nymphaea tetragona*

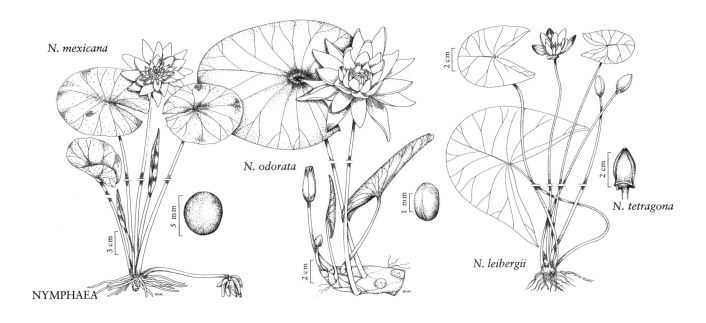

N. mexicana

N. odorata

2 cm

N. tetragona

N. leibergii

NYMPHAEA

1. Nymphaea ampla (Salisbury) de Candolle, Syst. Nat. 2: 54. 1821 [W1]

Castalia ampla Salisbury, Ann. Bot. (König & Sims) 2: 73. 1805

Rhizomes unbranched, erect, ovoid; stolons absent. **Leaves:** petiole glabrous. **Leaf blade** abaxially purple, often spotted, adaxially green, ovate to nearly orbiculate, 15–45 × 15–45 cm, margins dentate to spinose-dentate; venation radiate and prominent centrally, without weblike pattern, principal veins 13–29; surfaces glabrous. **Flowers** emersed, 7–18 cm diam., opening and closing diurnally, only sepals and outermost petals in distinct whorls of 4; sepals green, abaxially flecked with short dark streaks, faintly veined, lines of insertion on receptacle not prominent; petals 12–21, white; stamens 50–190, yellow, outer with connective appendage projecting 3–10 mm beyond anther; filaments widest at or below middle, mostly equal to or shorter than anthers; pistil 14–25-locular, appendages at margin of stigmatic disk short-triangular, to 3 mm. **Seeds** nearly globose to ellipsoid, 1.2–1.6 × 0.9–1.3 mm, 1.2–1.5 times as long as broad, with longitudinal rows of hairlike papillae 40–180 μm.

Flowering all year. Ditches, canals, ponds, and freshwater tidal margins; 0–350 m; Fla., Tex.; Mexico; West Indies; Central America; South America.

Reports of *Nymphaea ampla* in southern Texas by H. S. Conard (1905) are confirmed by specimens although no recent collections have been seen.

SELECTED REFERENCE Wunderlin, R. P. and D. H. Les. 1980. *Nymphaea ampla* (Nymphaeaceae), a waterlily new to Florida. Phytologia 45: 82–84.

2. Nymphaea elegans Hooker, Bot. Mag. 77: plate 4604. 1851 · Blue water-lily [W1]

Castalia elegans (Hooker) Greene

Rhizomes unbranched, erect, ovoid; stolons absent. **Leaves:** petiole glabrous. **Leaf blade** abaxially purple, adaxially green, ovate to nearly orbiculate, 8–20(–30) × 7–16(–25) cm, margins entire to sinuate; venation radiate and prominent centrally, without weblike pattern, compressed upon drying, principal veins 9–17; surfaces glabrous. **Flowers** emersed, 6–13 cm diam., opening and closing diurnally, only sepals and outermost petals in distinct whorls of 4; sepals green, abaxially flecked with short dark lines, faintly to obscurely veined, lines of insertion on receptacle not prominent; petals 8–27, pale violet to nearly white; stamens 55–145, yellow, connective appendage projecting to 1 mm, rarely more, beyond anther; filaments mostly widest above middle, shorter than anthers; pistil 12–20(–25)-locular, appendages at margin of stigmatic disk triangular, 1 mm or less. **Seeds** nearly globose, ca. 1.1–1.6 × 0.9–1.3 mm, 1.1–1.3 times as long as broad, with short appressed papillae absent or to 30 μm.

Flowering spring–fall. Freshwater ponds, lakes, sloughs, pools in swamps and marshes, ditches, and canals; 0–150 m; Fla., La., Tex.; ne, w Mexico; West Indies (Bahamas).

SELECTED REFERENCE   Schneider, E. L. 1982. Notes on the floral biology of *Nymphaea elegans* (Nymphaeaceae) in Texas. Aquatic Bot. 12: 197–200.

3. Nymphaea capensis Thunberg, Prodr. Pl. Cap. 2: 92. 1800 · Cape blue water-lily ⊡W1

Rhizomes unbranched, erect, ovoid; stolons absent. **Leaves:** petiole glabrous. **Leaf blade** abaxially suffused with maroon, adaxially green, ovate to orbiculate, to 35(–40) × 30 cm, margins sinuate to almost dentate; venation radiate and somewhat impressed centrally, without weblike pattern, principal veins mostly 20–25; surfaces glabrous. **Flowers** emersed, ca. 7–8 cm diam., opening and closing diurnally, only sepals and outermost petals in distinct whorls of 4; sepals green, shaded abaxially with reddish brown, margins purplish, evidently veined, lines of insertion on receptacle not prominent; petals 12–24, blue, lavender, or purple; stamens 100–200, colored as petals toward apex, outer with connective appendage projecting to 4–5 mm or more beyond anther; filaments widest above middle, shorter than anthers; pistil 15–31-locular, appendages at margin of stigmatic disk tapered, to 3 mm. **Seeds** ellipsoid, ca. 1.5 × 1 mm, ca. 1.4–1.8 times as long as broad, with longitudinal rows of hairlike papillae.

Flowering mid spring–summer. Sandy-bottomed ditches; 0–100 m; introduced; Fla.; e Africa.

The Florida introductions probably represent *Nymphaea capensis* var. *zanzibariensis* (Caspary) Conard, which B. Verdcourt (1989), in combining African and Asian plants into a single species with several varieties, has recently treated as *N. nouchali* N. L. Burman var. *zanzibariensis* (Caspary) Verdcourt. His interpretation of the type of *N. capensis,* differing from that of previous workers, should be further studied. The traditional view is retained here.

4. Nymphaea jamesoniana Planchon, Fl. Serres Jard. Eur. 8: 120. 1852

Rhizomes branched or unbranched, erect, ovoid to cylindric; stolons absent. **Leaves:** petiole glabrous. **Leaf blade** abaxially and adaxially green, sometimes with darker flecks, elliptic, 8–25 × 7–19 cm, margins entire; venation conspicuously weblike centrally with cross veins between major veins, principal veins 11–17; surfaces glabrous. **Flowers** floating, 7–15 cm diam., opening and closing nocturnally, sepals, petals, and outer stamens in distinct whorls of 4; sepals uniformly green, obscurely veined, lines of insertion on receptacle not prominent; petals 12, 16, or 20, outer greenish at least abaxially, inner creamy white; stamens 35–85, creamy white, connective appendage projecting to 1 mm, rarely more, beyond anther; filaments widest below middle, longer than anthers; pistil 19–33-locular, appendages at margin of stigmatic disk slightly club-shaped, 3–7(–9) mm. **Seeds** nearly globose, 0.5–0.8 × 0.45–0.7 mm, 1.1–1.2 times as long as broad, with scattered papillae 10–50 μm. 2*n* = 28.

Flowering late summer–fall. Shallow water of mostly temporary ponds or ditches; 0–100 m; native, Fla.; Mexico; Central America; South America.

All material of the predominantly neotropical *Nymphaea* subg. *Hydrocallis* in the flora is referable to this taxon. D. B. Ward (1977b) and others listed *N. blanda* G. Meyer [= *N. glandulifera* Rodschied], a Central and South American taxon, for Florida.

5. Nymphaea lotus Linnaeus, Sp. Pl. 1: 511. 1753 · Egyptian lotus (in part)

Rhizomes branched or unbranched, erect, ovoid; stolons slender. **Leaves:** petiole sparsely to densely puberulent. **Leaf blade** abaxially purplish, adaxially green, nearly orbiculate, to ca. 3 × 3 dm, margins spinose-dentate; venation radiate and prominent centrally, without weblike pattern, principal veins ca. 15; surfaces abaxially sparsely to densely puberulent. **Flowers** emersed, 12–25 cm diam., opening nocturnally, many flowers not closing until late morning, only sepals and outermost petals in distinct whorls of 4; sepals abaxially uniformly green, prominently veined, lines of insertion on receptacle not prominent; petals 16–20, white; stamens ca. 75, yellow, outer with connective appendage projecting less than 2 mm beyond anther; filaments widest below middle, slightly shorter to longer than anthers; pistil ca. 20(–30)-locular, appendages at margin of stigmatic disk linear, 6–12 mm. **Seeds** ellipsoid, 1.4–1.8 × 0.9–1.2 mm, ca. 1.5–1.6 times as long as broad, with longitudinal ridges bearing papillae 20–150 μm.

Flowering spring–summer. Ponds, ditches, and canals; 0–100 m; introduced; Fla., La.; Africa.

SELECTED REFERENCE   Wiersema, J. H. 1982. Distributional records for *Nymphaea lotus* (Nymphaeaceae) in the Western Hemisphere. Sida 9: 230–234.

6. Nymphaea mexicana Zuccarini, Abh. Math.-Phys. Cl. Königl. Bayer. Akad. Wiss. 1: 365. 1832 · Banana water-lily, yellow water-lily, herbe au coeur F W2

Castalia flava (Leitner) Greene; *Nymphaea flava* Leitner

Rhizomes unbranched, erect, cylindric; stolons elongate, spongy, developing clusters of curved, fleshy, overwintering roots resembling tiny bananas at terminal nodes. **Leaves:** petiole glabrous. **Leaf blade** abaxially purplish with dark flecks, adaxially green, often with brown mottling, ovate to elliptic or nearly orbiculate, 7–18(–27) × 7–14(–18) cm, margins entire or sinuate; venation radiate and impressed centrally, without web-like pattern, principal veins 11–22; surfaces glabrous. **Flowers** floating or emersed, 6–11 cm diam., opening and closing diurnally, only sepals and outermost petals in distinct whorls of 4; sepals uniformly yellowish green, often red-tinted, evidently veined, lines of insertion on receptacle often slightly prominent; petals 12–30, yellow; stamens ca. 50–60, yellow, connective appendage minute or absent; filaments widest below middle, longer than anthers; pistil 7–10-locular, appendages at margin of stigmatic disk oblong-tapered, to 4.5 mm. **Seeds** globose, ca. 5 × 5 mm, uniformly covered with hairlike papillae 100–220 μm. $2n = 56$.

Flowering spring–fall, mainly summer farther north. Outer coastal plain in alkaline lakes, ponds, warm springs, pools in marshes, sloughs, sluggish streams, ditches, and canals; 0–1100 m; Ala., Ariz., Calif., Fla., Ga., La., Miss., N.C., Okla., S.C., Tex.; ne, c Mexico.

Nymphaea mexicana is probably introduced in most inland sites and in California, where it is considered a problematic weed in waterways; it is not common in most states except Florida. The distribution of this species is similar to that of the winter distribution of canvasback ducks, for which the bananalike tubers are an important food (J. E. Cely 1979). This species forms natural hybrids with *N. odorata;* the hybrids have been named *N. ×thiona* D. B. Ward (D. B. Ward 1977). Except for stem characteristics, which resemble one or the other parent, and their added vigor, the hybrids are generally intermediate in morphology. They are completely sterile; however, hybrids with the stolon-bearing habit of *N. mexicana* can form extensive clones and, although somewhat larger in stature than *N. mexicana,* they closely resemble that less aggressive parent and could easily be mistaken for it. Some of the introductions, such as in southeastern Nevada and north-central Kentucky, are clearly this hybrid.

SELECTED REFERENCES Capperino, M. E. and E. L. Schneider. 1985. Floral biology of *Nymphaea mexicana* Zucc. (Nymphaeaceae). Aquatic Bot. 23: 83–93. Cely, J. E. 1979. The ecology and distribution of banana waterlily and its utilization by canvasback ducks. Proc. Annual Conf. SouthE. Assoc. Fish Wildlife Agencies 33: 43–47.

7. Nymphaea leibergii Morong, Bot. Gaz. 13: 124, in note. 1888 · Pygmy water-lily, small white water-lily E F

Castalia leibergii Morong; *Nymphaea tetragona* Georgi subsp. *leibergii* (Morong) A. E. Porsild

Rhizomes unbranched, erect, cylindric; stolons absent. **Leaves:** petiole glabrous. **Leaf blade** abaxially green to deep purple, adaxially green, ovate to elliptic, 3–19 × 2–15 cm, margins entire; venation radiate centrally, without weblike pattern, principal veins 7–13; surfaces glabrous. **Flowers** floating, 3–7.5 cm diam., opening and closing diurnally, only sepals and outermost (occasionally innermost) petals in distinct whorls of 4; sepals uniformly green, obscurely veined, lines of insertion on receptacle not prominent; petals 8–15, white; stamens 20–40, yellow, connective appendage projecting less than 0.2 mm beyond anther; filaments widest above middle, much longer than anthers; pistil 5–12-locular, appendages at margin of stigmatic disk tapered or slightly boat-shaped, 0.6–1.5 × 0.8–1.4 mm. **Seeds** ovoid, ca. 2–3 × 1.5–2 mm, ca. 1.3–1.5 times as long as broad, lacking papillae on surface.

Flowering summer. Ponds, lakes, and quiet streams; 0–1000 m; Alta., B.C., Man., Ont., Que., Sask.; Idaho, Maine, Mich., Minn., Mont.

Although widely distributed in northern North America, *Nymphaea leibergii* is apparently not common in much of its range. It is closely related to *N. tetragona* (see also), with which it has been confused; it differs in a number of floral and foliar characteristics. Although the two species are sympatric over central and western Canada, the distinctions are maintained. Coexistent populations are unknown at this time; such populations would provide useful study opportunities and should be sought in the area of overlap, especially in southeastern and central Manitoba and east-central British Columbia where nearby populations exist.

A presumed natural hybrid between *Nymphaea leibergii* and *N. odorata* has recently been found by Hellquist in northern Maine and by both of us in eastern Saskatchewan. It is intermediate in morphology between the two parental species and in Maine appears to be sterile. Further study of the Saskatchewan plants is needed as true *N. odorata* is absent from the area.

8. Nymphaea tetragona Georgi, Bemerk. Reise Russ. Reich 1: 220. 1775 · Pygmy water-lily, small white water-lily, nymphéa tétragonal [F]

Rhizomes unbranched, erect, cylindric; stolons absent. **Leaves:** petiole glabrous. **Leaf blade** abaxially green to dull purple, adaxially green, sometimes mottled with reddish brown or purple, especially on young leaves, ovate to elliptic, 3–13(–14) × 2–11 (–13) cm, margins entire; venation radiate centrally, without weblike pattern, principal veins 7–13; surfaces glabrous. **Flowers** floating, 3–7.5 cm diam., opening and closing diurnally, only sepals and outermost (occasionally innermost) petals in distinct whorls of 4; sepals uniformly green, obscurely veined, lines of insertion on receptacle very prominent, protuberant, forming tetragons; petals 10–17, white; stamens 30–70, yellow-orange, abaxially usually suffused with purple, connective appendage projecting less than 0.2 mm beyond anther; filaments widest above middle, longer than anthers; pistil 5–10-locular, appendages at margin of stigmatic disk boat-shaped, (2–)3–4 × 2–4 mm. **Seeds** ovoid, ca. 2–3 × 1.5–2 mm, ca. 1.3–1.5 times as long as broad, lacking papillae on surface. $2n = 112$.

Flowering summer. Ponds, lakes, and quiet streams; 0–1200 m; Alta., B.C., Man., N.W.T., Sask.; Alaska, Wash.; Eurasia.

Although broadly distributed in the northwest part of the flora, *Nymphaea tetragona* is apparently not common over the Canadian portion of its range. It was collected once in extreme northwestern Washington but is believed to be extirpated there. True *N. tetragona* is absent from northeastern North America and, now, from the conterminous United States, where this name has usually been applied to what is here segregated as *N. leibergii*. In size and shape of leaves and flowers the two taxa are very similar. They differ in the leaf mottling often present in developing leaves of *N. tetragona* but absent in *N. leibergii*; the distinctly tetragonal appearance of the receptacle in *N. tetragona*; and in the longer carpellary appendages, the presence usually of more stamens, and purple-colored stamens and pistils in *N. tetragona*. Only in living plants is it apparent that leaves of *N. leibergii* are thicker with impressed veins abaxially compared to the relatively thin leaves with raised veins in *N. tetragona*. Although distinctions in sepal and petal apices (often acute in *N. tetragona* and often rounded in *N. leibergii*) were the basis for the establishment of *Castalia leibergii,* the characters are variable in both taxa and thus of limited utility in distinguishing them.

9. Nymphaea odorata Aiton, Hort. Kew. 2: 227. 1789 · Fragrant water-lily, white water-lily, pond-lily [W2]

Rhizomes frequently branched, repent, cylindric; stolons absent. **Leaves:** petiole glabrous or pubescent. **Leaf blade** abaxially green to purple, adaxially green, ovate to nearly orbiculate, (5–)10–40 × (5–)10–40 cm, margins entire; venation radiate centrally, without weblike pattern, principal veins 6–27; surfaces glabrous. **Flowers** floating, 6–19 cm diam., opening and closing diurnally, only sepals and outermost petals in distinct whorls of 4; sepals uniformly green or reddened, obscurely to prominently veined, lines of insertion on receptacle not prominent; petals 17–43, white, rarely pink; stamens 35–120, yellow, connective appendage projecting less than 1(–2) mm beyond anther; filaments widest below middle, mostly longer than anthers; pistil 10–25-locular, appendages at margin of stigmatic disk linear-tapered, 3–8(–10) mm. **Seeds** ovoid, ca. 1.5–4.5 × 0.9–3 mm, 1.5–1.75 times as long as broad, lacking papillae on surface.

Subspecies 2 (2 in the flora): North America, Mexico, West Indies, Central America.

Nymphaea odorata is a polymorphic species, particularly in and around the Great Lakes region. Over much of that area the two entities treated here as subspecies are allopatric and can be readily distinguished; however, in areas of sympatry some populations are intermediate or contain some intermediate plants without any apparent loss of fertility. Although traditional treatments distinguish the two at specific rank, recent floristic works have accepted only one variable species with no infraspecific taxa. While calling attention to this taxonomic problem, field studies from within the Great Lakes region have not sufficiently accounted for the observed variation. Although evidence (P. H. Monson 1960; G. R. Williams 1970; I. L. Bayly and K. Jongejan 1982) suggests that some variability may be induced by environmental conditions, both extremes have been found growing together under seemingly identical conditions. Further study, especially involving artificial hybridization and/or molecular approaches, should be undertaken to clarify the relationship.

The geographic patterning of the overall variation and usefulness of retaining a separate status for those morphs previously classified as *Nymphaea tuberosa* justify the recognition of two subspecies at this time. The key, while useful in separating the two extremes in this morphologic continuum, is of limited use in identifying intermediate plants. Compounding the problem of identification is the fact that key characters are often poorly represented on herbarium material, thus some guidance should be taken from the distributional notes provided with each subspecies. Truly intermediate plants, known in Minnesota, Wisconsin, Michigan, New York, and Vermont, and in southern Ontario and

Quebec, may be treated as *N. odorata* without regard to subspecies. Sporadic populations, most probably introduced, on the Great Plains and farther west are difficult to place to subspecies and are best treated similarly although they are here included under subsp. *odorata*.

1. Petiole not striped, rarely faintly striped; leaf blade abaxially usually reddish purple, occasionally green; seeds 1.5–2.5 mm . . . . 9a. *Nymphaea odorata* subsp. *odorata*
1. Petiole with brown-purple stripes; leaf blade abaxially green or faintly purple; seeds mostly 2.8– 4.5 mm . . . . . . 9b. *Nymphaea odorata* subsp. *tuberosa*

9a. Nymphaea odorata Aiton subsp. **odorata** · Nymphéa odorant F

Castalia lekophylla Small; *C. minor* (Sims) de Candolle; *C. odorata* (Aiton) Woodville & Wood; *Nymphaea odorata* forma *rubra* (E. Guillon) Conard; *N. odorata* var. *gigantea* Tricker; *N. odorata* var. *godfreyi* D. B. Ward; *N. odorata* var. *minor* Sims; *N. odorata* var. *rosea* Pursh; *N. odorata* var. *stenopetala* Fernald

Rhizomes not constricted at branch joints, or only rarely so. **Leaves:** petiole uniformly greenish or more commonly reddish purple, rarely faintly striped, slender to stout. **Leaf blade** abaxially usually deeply reddish or purplish, occasionally greenish. **Flowers:** petals white, rarely pink, mostly lanceolate to elliptic, outer usually slightly or strongly tapering to apex, apex acute to rounded. **Seeds** 1.5–2.5 mm. $2n = 56, 84$.

Flowering spring–early fall, mainly summer farther north. Acidic or alkaline ponds, lakes, sluggish streams and rivers, pools in marshes, ditches, canals, or sloughs; 0–1700 m; B.C., Man., N.B., Nfld., N.S., Ont., P.E.I., Que., Sask.; Ala., Ariz., Ark., Calif., Colo., Conn., Del., D.C., Fla., Ga., Idaho, Ill., Kans., Ky., La., Maine, Md., Mass., Mich., Minn., Miss., Mo., Mont., Nebr., Nev., N.H., N.J., N.Mex., N.Y., N.C., Okla., Oreg., Pa., R.I., S.C., Tenn., Tex., Utah, Vt., Va., Wash., W.Va., Wis.; Mexico; West Indies (Bahamas, Cuba); Central America (Honduras, El Salvador, Nicaragua); South America (n Guyana, naturalized).

Nymphaea odorata subsp. *odorata* is introduced in British Columbia and in Arizona, California, Colorado, Idaho, Montana, Nevada, New Mexico, Oregon, Utah, and Washington. Intermediates to subsp. *tuberosa* (see previous comment for distribution) cannot be keyed satisfactorily to either subspecies. Plants of west-central Manitoba and east-central Saskatchewan, though included here, are intermediate in morphology to *N. leibergii* and are probably of hybrid origin. Unusually dwarfed plants that have been treated as *Nym-*

phaea odorata var. *minor* may be responses to highly acidic conditions. Very robust forms recognized by some as *N. odorata* var. *gigantea* occur sporadically along the coastal plain from New Jersey southward, perhaps in response to some unknown environmental factor. Further study should be undertaken. Occasional pink-flowered forms, treated as *N. odorata* var. *rosea* or forma *rubra,* are known from several states; all existing populations appear to be introductions for ornamental purposes. Natural hybrids with *N. leibergii* and *N. mexicana* are discussed under those species. *Nymphaea reniformis* Walter, an earlier name of uncertain application that has at times been applied to this taxon, has been recently proposed for rejection (J. H. Wiersema and J. L. Reveal 1991).

Flowering responses in the northern part of the range, where the flowers generally open slightly later in the morning and close much later in the afternoon, are much more variable than those farther south.

9b. Nymphaea odorata Aiton subsp. **tuberosa** (Paine) Wiersema & Hellquist, Rhodora 96: 170. 1994 · Nymphéa tubéreux E W1

Nymphaea tuberosa Paine, Cat. Pl. Oneida Co., 132. 1865

Rhizomes often constricted at branch joints to form detachable tubers. **Leaves:** petiole green with brown-purple stripes, stout. **Leaf blade** abaxially green or faintly purple. **Flowers:** petals white, rarely pink, elliptic to oblanceolate, outer usually with broadly rounded apex. **Seeds** mostly 2.8–4.5 mm.

Flowering late spring–summer. Mainly alkaline ponds, lakes, and sluggish streams and rivers, usually in very oozy sediments; 100–400 m; Man., Ont., Que.; Conn., Ill., Ind., Iowa, Kans., Maine, Mass., Mich., Minn., Mo., Nebr., N.H., N.Y., Ohio, Okla., Pa., Vt., Wis.

This taxon, which has been included within *Nymphaea odorata* by some recent workers, was formerly almost universally accepted as a distinct species. In the southern parts of the range of subsp. *tuberosa,* where subsp. *odorata* is absent, subsp. *tuberosa* is easily distinguished morphologically from subsp. *odorata.* Farther north, where their ranges overlap, the distinctions break down in some populations but are maintained in others. Some western populations are probably the result of introductions. A pink-flowered form seen in southeastern Ohio appears to be derived from this subspecies.

13. CABOMBACEAE A. Richard
· Water-shield Family

John H. Wiersema

Herbs, perennial, aquatic, rhizomatous; roots adventitious; air chambers conspicuous in vegetative portions of plant. **Rhizomes** branched, repent, slender. **Leaves** arising from erect stems with shortened internodes, submersed and floating; stipules absent. **Leaf blade:** laticifers commonly present. **Submersed leaves** (*Cabomba* only) opposite [whorled], short-petioled; leaf blade palmately dissected. **Floating leaves** alternate; petiole short to long; leaf blade peltate, linear to broadly elliptic, sometimes with basal sinus (some *Cabomba*), margins entire. **Inflorescences** axillary, solitary flowers. **Flowers** bisexual, protogynous, diurnal, borne at or above water surface (occasionally submersed); peduncle long; involucre absent; perianth persistent in fruit, hypogynous; sepals usually 3, distinct or nearly so; petals usually 3, alternate with sepals; stamens 3–36, rarely more; filaments slender; anthers dehiscing by longitudinal slits, without connective appendage; pistils [1–]2–18, 1-carpellate, inversely club-shaped or swollen on one side; ovary 1-locular; placentation laminar; ovules [1–]2–3[–5]; style short; stigma capitate or linear-decurrent. **Fruits** achenelike or folliclelike, leathery, indehiscent. **Seeds** 1–3, aril absent; endosperm sparse; perisperm abundant; embryo minute; cotyledons 2, fleshy.

Genera 2, species 6 (2 genera, 2 species in the flora): temperate and tropical.

Cabombaceae are pollinated by insects (*Cabomba*) or by wind (*Brasenia*).

Formerly Cabombaceae often have been included in Nymphaeaceae in the broad sense.

SELECTED REFERENCE Wood, C. E. Jr. 1959. The genera of the Nymphaeaceae and Ceratophyllaceae in the southeastern United States. J. Arnold Arbor. 40: 94–112.

1. Plants with opposite, dissected, submersed leaves and (when flowering) with inconspicuous, alternate, linear-elliptic floating leaves 0.6–3 cm; stamens 3–6; submersed parts barely coated with mucilage . . . . .1. *Cabomba,* p. 79
1. Plants with only alternate, entire, broadly elliptic floating leaves 3.5–13.5 cm; stamens 18–36(–51); submersed parts heavily coated with mucilage . . . . 2. *Brasenia,* p. 79

1. CABOMBA Aublet, Hist. Pl. Guiane, 321. 1775 · Fanwort [probably an aboriginal name]

Herbs, young vegetative parts often with rust-colored pubescence, barely mucilaginous. **Leaves** submersed and floating; petiole short to long. **Submersed leaves:** blade in 3–7 dichotomously [trichotomously] branched, linear segments. **Floating leaves:** blade terminal, linear-elliptic [broadly elliptic], margins entire or notched at base, appearing only during flowering, inconspicuous [more evident in *Cabomba aquatica*, a tropical species]. **Flowers:** sepals 3, petaloid, obovate; petals 3, oval, with proximal, yellow, nectar-bearing auricles, base clawed; stamens 3–6, opposite petals; pistils [1–]2–4, simple, 1-locular; ovules [1–]3[–5]; stigma capitate. **Fruits** elongate-pyriform, tapered to apex. **Seeds** ovoid [somewhat globose], tuberculate. $x = 13$.

Species 5 (1 in the flora): mostly tropical regions, Western Hemisphere.

SELECTED REFERENCES Fassett, N. C. 1953. A monograph of *Cabomba*. Castanea 13: 116–128. Ørgaard, M. 1991. The genus *Cabomba* (Cabombaceae)—A taxonomic study. Nordic J. Bot. 11: 179–203.

1. Cabomba caroliniana A. Gray, Ann. Lyceum Nat. Hist. New York 4: 47. 1837 · Fanwort F W1

Cabomba caroliniana var. *pulcherrima* R. M. Harper; *C. pulcherrima* (R. M. Harper) Fassett

Submersed leaves: petiole to 4 cm; leaf blade 1–3.5 × 1.5–5.5 cm, terminal segments 3–200, linear to slightly spatulate, to 1.8 mm wide. **Floating leaves:** blade 0.6–3 cm × 1–4 mm, margins entire or notched to sagittate at base. **Flowers** 6–15 mm diam.; sepals white to purplish [yellow] or with purple-tinged margins, 5–12 × 2–7 mm; petals colored as sepals but with proximal, yellow, nectar-bearing auricles, 4–12 × 2–5 mm, apex broadly obtuse or notched; stamens 3–6, mostly 6; pistils 2–4, mostly 3, divergent at maturity; ovules 3. **Fruits** 4–7 mm. **Seeds** 1–3, 1.5–3 × 1–1.5 mm, tubercles in 4 longitudinal rows. $2n =$ ca. 78, ca. 104.

Flowering late spring–early fall, earlier and later farther south. Acidic to alkaline ponds, lakes, pools in marshes, rivers, streams, ditches, canals, and reservoirs; 0–300 m; Ont.; Ala., Ark., Conn., D.C., Fla., Ga., Ill., Ind., Ky., La., Md., Mass., Mich., Miss., Mo., N.H., N.J., N.Y., N.C., Ohio, Okla., Oreg., Pa., R.I., S.C., Tenn., Tex., Va.; s South America.

Cabomba caroliniana, an important aquarium plant, is introduced in Oregon and probably in the northern part of its range where it is uncommon in several states. Formerly known from Kansas, it is thought to be extirpated there now. Although Delaware and West Virginia lie within the mapped area, I know of no collections from those states. In New England and parts of southeast United States, it is sometimes an aggressive weed. In parts of the southeastern United States, plants with purple-tinted flowers, possibly a response to some environmental factor, have been treated as *Cabomba caroliniana* var. *pulcherrima*. South American plants with yellow flowers have been called *C. caroliniana* var. *flavida* Ørgaard.

The submersed leaves of *Cabomba caroliniana* are similar in form to those of *Limnophila* (Scrophulariaceae; introduced in southeastern United States). The latter has whorled leaves in contrast to the opposite leaves of *Cabomba*.

SELECTED REFERENCE Schneider, E. L. and J. M. Jeter. 1982. Morphological studies of the Nymphaeaceae. XII. The floral biology of *Cabomba caroliniana*. Amer. J. Bot. 69: 1410–1419.

2. BRASENIA Schreber, Gen. Pl. 1: 372. 1789 · Water-shield [for Christoph Brasen, 1738–1774, Moravian missionary and plant collector in Greenland and Labrador]

Herbs, young vegetative parts heavily coated with mucilage. **Leaves** floating; petiole long. **Leaf blade** elliptic. **Flowers:** sepals 3, not petaloid, linear-oblong to narrowly ovate; petals 3, linear-oblong, lacking auricles, base not clawed; stamens 18–36(–51), opposite both sepals and petals; pistils 4–18, simple, 1-locular; ovules (1–)2; stigma linear-decurrent. **Fruits** slightly to strongly fusiform. **Seeds** ovoid, tubercles absent. $x = 40$.

Species 1 (1 in the flora): worldwide except Europe, mainly temperate areas and upland tropics.

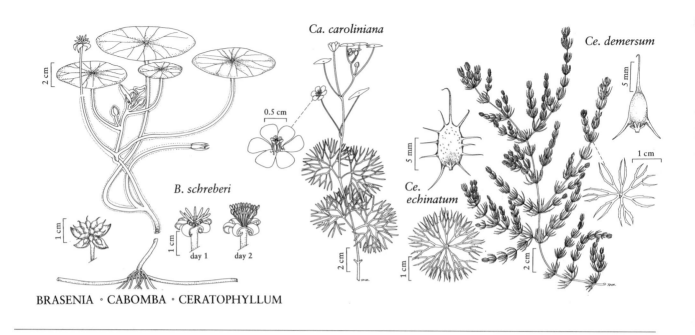

Ca. caroliniana

Ce. demersum

B. schreberi

Ce. echinatum

BRASENIA · CABOMBA · CERATOPHYLLUM

Brasenia is known from the fossil record in Europe although it is not known to grow there currently.

1. Brasenia schreberi J. F. Gmelin, Syst. Nat. 1: 853. 1791 · Water-shield, purple wen-dock, brasénie de Schreber [F] [W2]

Leaf blade 3.5–13.5 × 2–8 cm. **Flowers** ca. 2 cm diam.; perianth parts dull purple, 10–20 × 2–7 mm; petals slightly longer and narrower than sepals. **Fruits** 6–10 mm. **Seeds** 1–2, 2.5–4 × 2–3 mm. $2n = 80$.

Flowering late spring–summer. Oligotrophic or mesotrophic ponds, lakes, and sluggish streams; 0–2000 m; Alta., B.C., Man., N.B., Nfld., N.S., Ont., Que.; Ala., Alaska, Ark., Calif., Conn., Del., D.C., Fla., Ga., Idaho, Ill., Ind., Iowa, Kans., Ky., La., Maine, Md., Mass., Mich., Minn., Miss., Mo., Mont., N.H., N.J., N.Y., N.C., Ohio, Okla., Oreg., Pa., R.I., S.C., Tenn., Tex., Vt., Va., Wash., W.Va., Wis.; Mexico; West Indies (Cuba, Jamaica, Hispaniola); Central America (Guatemala and Belize); South America (se Venezuela); e Asia; Africa; e Australia.

SELECTED REFERENCE   Osborn, J. M. and E. L. Schneider. 1988. Morphological studies of the Nymphaeaceae *sensu lato*. XVI. The floral biology of *Brasenia schreberi*. Ann. Missouri Bot. Gard. 75: 778–794.

14. CERATOPHYLLACEAE Gray

• Hornwort Family

Donald H. Les

Herbs perennial, aquatic, without rhizomes; roots absent; air chambers conspicuous. **Leaves** arising from branched stems, sessile, whorled; stipules absent. **Leaf blade** undivided to finely dissected, margins serrulate, laticifers absent; 1st leaves of plumule simple or forked. **Inflorescences** axillary, solitary flowers. **Flowers** unisexual, staminate and pistillate on same plant, submersed, sessile or nearly so, subtended by involucre of 8–15 linear bracts; perianth absent; stamens 3–50; anthers nearly sessile, dehiscing by longitudinal slits, connective prolonged; pistil 1, simple; ovary 1-locular; placentation laminar; ovule 1; style elongate; stigma decurrent. **Fruits** achenes, indehiscent. **Seed** 1, aril absent; endosperm and perisperm absent; embryo well developed; cotyledons 2, fleshy.

Genus 1, species 6 (3 species in the flora): worldwide.

Ceratophyllaceae are water-pollinated.

The extant species of this aquatic family are obviously highly specialized; recent research indicates that Ceratophyllaceae are an ancient lineage. From comparative morphologic data, D. H. Les (1988) hypothesized that Ceratophyllaceae are vestiges of ancient angiosperms that diverged early from the line leading to most other modern taxa. This hypothesis is consistent with recent phylogentic analyses of DNA sequence data (D. H. Les et al. 1991; M. W. Chase et al. 1993; Qiu Y. L. et al. 1993) derived from the chloroplast gene *rbc*L. The great age of Ceratophyllaceae is indicated by fossil evidence documenting the existence of the family in the Cretaceous Aptian (D. L. Dilcher 1990). The hypothesis that Ceratophyllaceae represent the oldest living angiosperm lineage deserves further scrutiny; meanwhile recent morphologic, molecular, and paleobotanical evidence favors this interpretation over other proposed phylogenetic affinities of the family.

SELECTED REFERENCES Dilcher, D. L. 1990. The occurrence of fruits with affinities to Ceratophyllaceae in lower and mid-Cretaceous sediments. [Abstract.] Amer. J. Bot. 76: 162. Les, D. H. 1988. The origin and affinities of the Ceratophyllaceae. Taxon 37: 326–345. Les, D. H. 1993. Ceratophyllaceae. In: K. Kubitzki et al., eds. 1990+. The Families and Genera of Vascular Plants. 2+ vols. Berlin etc. Vol. 2, pp. 246–250. Wood, C. E. Jr. 1959. The genera of the Nymphaeaceae and Ceratophyllaceae in the southeastern United States. J. Arnold Arbor. 40: 94–112.

1. CERATOPHYLLUM Linnaeus, Sp. Pl. 2: 992. 1753; Gen. Pl. ed. 5, 428. 1754

· Hornwort, cornifle [Greek *ceratos*, horn, and *phyllon*, leaf]

Plants, submersed perennials. **Stems** suspended or anchored by branches buried in substrate; branches 0–3 per node. **Leaves** 3–11 per whorl, cauline; petiole inconspicuous. **Leaf blade** simple or forked into linear-filiform, denticulate segments. **Inflorescences** extra-axillary, alternating with leaves. **Flowers:** bracts basally connate, foliaceous; pedicel less than 1 mm or essentially absent. **Staminate flowers:** anther 2-locular, connective projecting as apical appendage. **Pistillate flowers:** ovary tapering to persistent, spinelike style; ovule pendulous. **Achene** ellipsoid, moderately compressed, surface smooth or tuberculate, basal or marginal spines (or both) present or absent, terminal spine present, marginal wing present or absent. $x = 12, 19, 20$.

Useful in identification of species of *Ceratophyllum* are leaf-forking characteristics. Leaves with no forking are "0-order"; they consist only of a primary segment. Those forking once are "1st-order"; their ultimate segments are secondary. Those in which at least one secondary segment forks are "2d-order"; their ultimate segments are tertiary. Those in which at least one tertiary segment forks are "3d-order"; their ultimate segments are quaternary. Those in which at least one quaternary segment forks are "4th order."

Plumule features used in the key below are observable through dissection of softened (soaked) seeds removed from achenes.

SELECTED REFERENCES Fassett, N. C. 1953. North American *Ceratophyllum*. Comun. Inst. Trop. Invest. Ci. Univ. El Salvador 2: 25–45.   Herendeen, P. S., D. H. Les, and D. L. Dilcher. 1990. Fossil *Ceratophyllum* (Ceratophyllaceae) from the Tertiary of North America. Amer. J. Bot. 77: 7–16.   Jones, E. N. 1931. The morphology and biology of *Ceratophyllum demersum*. Stud. Nat. Hist. Iowa Univ. 13: 11–55.   Les, D. H. 1985. The taxonomic significance of plumule morphology in *Ceratophyllum* (Ceratophyllaceae). Syst. Bot. 10: 338–346.   Les, D. H. 1986. The phytogeography of *Ceratophyllum demersum* and *C. echinatum* in glaciated North America. Canad. J. Bot. 64: 498–509.   Les, D. H. 1986b. The evolution of achene morphology in *Ceratophyllum* (Ceratophyllaceae), I. Fruit spine variation and relationships of *C. demersum, C. submersum,* and *C. apiculatum*. Syst. Bot. 11: 549–558.   Les, D. H. 1988b. The evolution of achene morphology in *Ceratophyllum* (Ceratophyllaceae), II. Fruit variation and systematics of the "spiny-margined" group. Syst. Bot. 13: 73–86.   Les, D. H. 1988c. The evolution of achene morphology in *Ceratophyllum* (Ceratophyllaceae), III. Relationships of the "facially-spined" group. Syst. Bot. 13: 509–518.   Les, D. H. 1989. The evolution of achene morphology in *Ceratophyllum* (Ceratophyllaceae), IV. Summary of proposed relationships and evolutionary trends. Syst. Bot. 14: 254–262.   Lowden, R. M. 1978. Studies in the submerged genus *Ceratophyllum* L. in the neotropics. Aquat. Bot. 4: 127–142.   Muenscher, W. C. 1940. Fruits and seedlings of *Ceratophyllum*. Amer. J. Bot. 27: 231–233.

1. Forking of largest leaves 1st or 2d order (rarely 3d order); leaves coarse-textured, marginal denticles usually strongly raised on broad base of green tissue; achene margin wingless, basal spines or tubercles 2 (rarely absent)..1. *Ceratophyllum demersum*
1. Forking of largest leaves 3d or 4th order; leaves fine-textured, marginal denticles not raised on broad base of green tissue, sometimes nearly absent; achene margin winged, spines (0–)2–20, 0.1–6.5 mm, basal spines 2 (rarely absent).
  2. First leaves of plumule simple; achene body (excluding terminal spine) 3–4.5 mm; Florida, Georgia, North Carolina...2. *Ceratophyllum muricatum*
  2. First leaves of plumule forked; achene body (excluding terminal spine) 4.5–6 mm; widespread in North America...3. *Ceratophyllum echinatum*

1. Ceratophyllum demersum Linnaeus, Sp. Pl. 2: 992. 1753 · Coontail, hornwort, cornifle nageante [F] [W2]

Ceratophyllum apiculatum Chamisso

Stems to 3 m; apical leaf whorls densely crowded. **Leaves** bright green, coarse-textured. **Leaf blade** simple or forked into 2–4(–5) ultimate segments (forking of largest leaves 1st or 2d order, rarely 3d order), segments not inflated, mature leaf whorls 1.5–6 cm diam., marginal denticles conspicuous, usually strongly raised on broad base of green tissue; 1st leaves of plumule simple. **Achene** dark green or reddish brown, body (excluding spines) 3.5–6 × 2–4 × 1–2.5 mm, basal spines or tubercles 2 (rarely absent), straight or curved, 0.1–12 mm, spine bases occasionally inconspicuously webbed, marginal spines absent, terminal spine straight, 0.5–14 mm, margins wingless. $2n = 24, 38, 40, 48$.

Flowering spring–late fall. Fresh to slightly brackish rivers, streams, ditches, lakes, ponds, pools, marshes, swamps; 0–1700 m; Alta., B.C., Man., N.B., N.W.T., N.S., Ont., P.E.I., Que., Sask., Yukon; Ala., Alaska, Ariz., Ark., Calif., Colo., Conn., Del., D.C., Fla., Ga., Idaho, Ill., Ind., Iowa, Kans., Ky., La., Maine, Md., Mass., Mich., Minn., Miss., Mo., Mont., Nebr., Nev., N.H., N.J., N.Mex., N.Y., N.C., N.Dak., Ohio, Okla., Oreg., Pa., R.I., S.C., S.Dak., Tenn., Tex., Utah, Vt., Va., Wash., W.Va., Wis., Wyo.; worldwide.

Specimens of *Ceratophyllum demersum* with short basal spines or tubercles have been misidentified as *C. submersum* Linnaeus, a species not known in the New World despite reports to the contrary. *Ceratophyllum demersum* is the most common species of *Ceratophyllum* in North America and also the least likely to be found with fruit, its reproduction being primarily asexual. Predominantly low leaf order is, therefore, the most reliable means of identifying this species.

Noted for its prolific growth, *Ceratophyllum demersum* occasionally has attained status as a serious weed.

2. Ceratophyllum muricatum Chamisso, Linnaea 4: 504. 1829 · Hornwort

Ceratophyllum submersum Linnaeus subsp. *muricatum* (Chamisso) Wilmot-Dear

Subspecies 3 (1 in the flora): nearly worldwide.

2a. Ceratophyllum muricatum Chamisso subsp. **australe** (Grisebach) Les, Syst. Bot. 13: 85. 1988

Ceratophyllum australe Grisebach, Symb. Fl. Argent., 14. 1879; *C. demersum* Linnaeus var. *cristatum* K. Schumann; *C. floridanum* Fassett; *C. llerenae* Fassett

Stems to 1 m; apical leaf whorls not densely crowded. **Leaves** light green to yellow-green, fine-textured. **Leaf blade** simple or forked into 2–10 ultimate segments (forking of largest leaves 3d or 4th order), segments sometimes conspicuously inflated, mature leaf whorls 2.5–6 cm diam., marginal denticles weak and inconspicuous, not raised on broad base of green tissue, sometimes nearly absent; 1st leaves of plumule simple. **Achene** green or dark brown, body (excluding spines) 3–4.5 × 2–3 × 1–2 mm, basal spines 2 (rarely absent), straight or curved, 0.3–4.5 mm, marginal spines (0–)2–20, 0.1–4 mm, terminal spine straight or curved, 1.5–7.5 mm, margins slightly winged.

Flowering spring. Fresh water of ponds and lakes; 0–50 m; Fla., Ga., N.C.; Mexico; West Indies; Central America; South America.

Principally of tropical distribution, *Ceratophyllum muricatum* is known in North America north of Mexico only from ephemeral habitats in coastal southeastern United States. The wide range of variation of spines on the fruits in *C. muricatum* has led to the describing of several variants (typically when spine lengths have been reduced) as different species. As in *C. demersum,* spineless phenotypes from North America have been called *C. submersum* Linnaeus, which does not occur in the New World. The affinity of *C. muricatum* for shallow, ephemeral habitats results in its sporadic and nonpersistent occurrence in present North American localities. Fruit variants of this species were once recognized as *C. floridanum,* which was considered to be an extinct Florida endemic. Efforts to protect *C. muricatum* should be given consideration; it is seldom collected in North America. Fossil records of *C. muricatum* from the lower and middle Eocene document its occurrence in more inland sites, presumably when the climate of the interior stations was more similar to tropical conditions. Of the three North American species of *Ceratophyllum,* this species is most likely to be collected (in season) with fruit present.

3. **Ceratophyllum echinatum** A. Gray, Ann. Lyceum Nat. Hist. New York 4: 49. 1837 · Hornwort, cornifle échinée E F

Ceratophyllum demersum Linnaeus var. *echinatum* (A. Gray) A. Gray; *C. submersum* Linnaeus var. *echinatum* (A. Gray) Wilmot-Dear

Stems to 1 m; apical leaf whorls not densely crowded. **Leaves** dark green or olive-green, fine-textured. **Leaf blade** simple or forked into 2–10 ultimate segments (forking of largest leaves 3d or 4th order), proximal segments often conspicuously inflated, mature leaf whorls 25–55 mm diam., marginal denticles weak and inconspicuous, weakly exserted, not raised on broad base of green tissue; 1st leaves of plumule forked. **Achene** dark green or brown, body (excluding spines) 4.5–6 × 2.5–4.5 × 1.5–3 mm, basal spines 2 (rarely absent), straight or curved, 1–5 mm, marginal spines 2–13, 0.5–6.5 mm, terminal spine straight, 1.5–7.5 mm, margins winged. $2n = 24$.

Flowering spring–summer. Fresh water of lakes, ponds, marshes, swamps; 0–500 m; B.C., Man., N.B., N.S., Ont., Que.; Ala., Ark., Conn., Del., D.C., Fla., Ga., Ill., Ind., Iowa, Kans., Ky., La., Maine, Md., Mass., Mich., Minn., Miss., Mo., N.H., N.J., N.Y., N.C., Ohio, Okla., Oreg., Pa., R.I., S.C., Tenn., Tex., Vt., Va., Wash., W.Va., Wis.

Principally an eastern North American species—and the only species of its genus endemic to North America—*Ceratophyllum echinatum* is disjunct in the Pacific Northwest as a result of repeated Pleistocene glaciation. The habitats of *C. echinatum* are typically more acidic (avg. pH 6.6) than those of *C. demersum* (avg. pH 7.4). The two species only rarely coexist. *Ceratophyllum echinatum* also thrives in cooler, clearer, and more oligotrophic water than *C. demersum* and often is found in more ephemeral sites, such as shrub swamps (e.g., with *Cephalanthus occidentalis*) and beaver ponds.

This species, relatively uncommon, is fast disappearing from much of its range because of habitat alteration or destruction and the introduction of nonindigenous species; steps should be taken to secure its conservation. Unlike *Ceratophyllum demersum*, *C. echinatum* does not attain status as a serious weed.

15. RANUNCULACEAE Jussieu

· Crowfoot Family

Alan T. Whittemore

Bruce D. Parfitt

Herbs, sometimes woody or herbaceous climbers or low shrubs, perennial or annual, often rhizomatous. **Stems** unarmed. **Leaves** usually basal and cauline, alternate or sometimes opposite, rarely whorled, simple or variously compound; stipules present or absent; petioles usually present, often proximally sheathing. **Leaf blade** undivided or more commonly divided or compound, base cordate, sometimes truncate or cuneate, margins entire, toothed, or incised; venation pinnate or palmate. **Inflorescences** terminal or axillary, racemes, cymes, umbels, panicles, or spikes, or flowers solitary, flowers pedicellate or sessile. **Flowers** bisexual, sometimes unisexual, inconspicuous or showy, radially or bilaterally symmetric; sepaloid bracteoles 0(–3); perianth hypogynous; sepals usually imbricate, 3–6(–20), distinct, often petaloid and colored, occasionally spurred; petals 0–26, distinct (connate in *Consolida*), plane, cup-shaped, funnel-shaped, or spurred, conspicuous or greatly reduced; nectary usually present, rarely absent; stamens 5–many, distinct; anthers dehiscing longitudinally; staminodes absent (except in *Aquilegia* and *Clematis*); pistils 1–many; styles present or absent, often persistent in fruit as beak. **Fruits** achenes, follicles, or rarely utricles, capsules, or berries, often aggregated into globose to cylindric heads. **Seeds** 1–many per ovary, never stalked, not arillate; endosperm abundant; embryo usually small.

Genera ca. 60, species 2500 (22 genera, 284 species in the flora): worldwide.

The flowers of many species of Ranunculaceae begin to open long before anthesis, while the floral organs are just partly expanded. Only mature flowers with open anthers should be used for determination of diagnostic characteristics (especially measurements).

The literature is inconsistent about the term for the whorl of organs between sepals and stamens; these may be conspicuous and petaloid, or reduced to stalked nectaries, or intermediate between the two states. They have been called petals, honey-leaves, or (when they are inconspicuous) staminodes or nectaries. We follow M. Tamura (1993) and treat as petals all organs between the sepals and stamens, except in *Clematis* where they usually bear rudimentary anthers and clearly represent staminodes.

SELECTED REFERENCES Brayshaw, T. C. 1989. Buttercups, Waterlilies, and Their Relatives (the Order Ranales) in British Columbia. Victoria. [Roy. Brit. Columbia Mus. Mem. 1.] Duncan, T. and C. S. Keener. 1991. A classification of the Ranunculaceae with special reference to the Western Hemisphere. Phytologia 70: 24–27. Tamura, M. 1963. Morphology, ecology and phylogeny of the Ranunculaceae I. Sci. Rep. Coll. Gen. Educ. Osaka Univ. 11: 115–126.   Tamura, M. 1993. Ranunculaceae. In: K. Kubitzki et al., eds. 1990+. The Families and Genera of Vascular Plants. 2+ vols. Berlin etc. Vol. 2, pp. 563–583.   Ziman, S. N. and C. S. Keener. 1989. A geographical analysis of the family Ranunculaceae. Ann. Missouri Bot. Gard. 76: 1012–1049.

1. Flowers bilaterally symmetric; sepals showy; petals smaller than sepals.
  2. Upper (adaxial) sepal (hood) saccate or helmet-shaped; petals completely hidden by sepals . . . . 15. *Aconitum*, p. 191
  2. Upper (adaxial) sepal spurred; petals at least partly exserted from calyx.
    3. Perennials; pistils 3(–5); petals 4, distinct . . . . 16. *Delphinium*, p. 196
    3. Annuals; pistil 1; petals 2, connate . . . . 17. *Consolida*, p. 240
1. Flowers radially symmetric; sepals showy or not; petals present or absent, smaller to larger than sepals.
  4. Fruits achenes or utricles; ovule 1 per pistil.
    5. Sepals spurred; leaves all basal, blade linear or narrowly oblanceolate . . . . 3. *Myosurus*, p. 135
    5. Sepals plane; leaves either not all basal, or blade not linear or narrowly oblanceolate.
      6. Leaves all cauline and opposite; stems ± woody, at least at base . . . . 6. *Clematis*, p. 158
      6. Leaves cauline and alternate (rarely opposite), or basal, or plants with basal leaves and opposite or whorled involucral bracts; stems herbaceous.
        7. Plants with 1 or more pairs (opposite) or whorls of involucral bracts, these leaflike or calyxlike.
          8. Achenes with conspicuous veins or ribs on lateral surfaces; style absent . . . . 22. *Thalictrum thalictroides*, p. 261
          8. Achenes without veins on lateral surfaces; style present . . . . 5. *Anemone*, p. 139
        7. Plants without involucral bracts (inconspicuous, linear-lanceolate involucral bracts in *Trautvetteria*), cauline leaves if present alternate (rarely a pair of opposite, unlobed leaves in *Ranunculus* sect. *Flammula*).
          9. Petals absent; inflorescences panicles, racemes, or corymbs (umbels in *Thalictrum thalictroides*); filaments filiform or dilated distally.
            10. Leaves simple, blade lobed; flowers bisexual; inflorescences corymbs . . . . 4. *Trautvetteria*, p. 138
            10. Leaves compound; flowers unisexual or bisexual; inflorescences panicles, racemes, corymbs, or umbels . . . . 22. *Thalictrum*, p. 258
          9. Petals present (rarely absent in *Ranunculus pedatifidus*); inflorescences simple or compound cymes or flowers solitary; filaments filiform.
            11. Petals without nectaries; sepals 5(–8) . . . . 12. *Adonis*, p. 184
            11. Petals with basal nectaries; sepals 3–5(–6) . . . . 2. *Ranunculus*, p. 88
  4. Fruits follicles, capsules, or berries; ovules 2 or more per pistil (1 of 2 aborting in *Xanthorhiza*, leaving 1 seed at maturity).
    12. Leaves dissected into linear, threadlike segments; pistils compound; fruits capsules . . . . 11. *Nigella*, p. 184
    12. Leaves not dissected, if parted or compound the segments not linear; pistils simple; fruits aggregates of follicles or solitary or aggregate berries.
      13. Shrubs; beak of fruit lateral, strongly incurved against abaxial surface of follicle . . . . 19. *Xanthorhiza*, p. 245
      13. Herbs; beak of fruit, if present, terminal or nearly so, straight or slightly curved, sometimes hooked at tip.
        14. Petals prominent, spurred . . . . 21. *Aquilegia*, p. 249
        14. Petals if present inconspicuous, plane or funnel-shaped.
          15. Flowers 12–50, in racemes or racemelike panicles.
            16. Pistils 1–8; fruits follicles, usually aggregate; petals 2-cleft or absent . . . . 8. *Cimicifuga*, p. 177
            16. Pistil 1; fruits berries; petals unlobed . . . . 9. *Actaea*, p. 181
          15. Flowers 1–10, in leafy cymes or solitary.
            17. Leaves simple, blade often lobed 1/2–3/4 its length, margins entire, crenate, or toothed; petals absent.

1. HYDRASTIS Linnaeus, Syst. Nat. ed. 10, 2: 1088. 1759 · Goldenseal, orangeroot, yellow-puccoon, sceau d'or [referring to superficial resemblance to some species of *Hydrophyllum*]

Bruce A. Ford

Herbs, perennial, from creeping, yellow, thick rhizomes. **Leaves** basal and cauline, simple, petiolate; cauline leaves alternate. **Leaf blade** palmately 3–9-lobed, broadly cordate-orbiculate, margins serrate. **Inflorescences** terminal, solitary flowers; bracts absent. **Flowers** bisexual, radially symmetric; sepals not persistent in fruit, 3, greenish white to creamy white, plane, ovate, oval, or elliptic, 3.5–7 mm; petals absent; stamens 50–75; filaments abruptly narrowed near apex; staminodes absent between stamens and pistils; pistils 5–15, simple; ovules 2 per ovary; style short. **Fruits** berries, aggregate, sessile, spheric, sides not veined; beak terminal, ± straight, 0.6–1 mm. **Seeds** black, ellipsoid, smooth, lustrous. $x = 13$.

Species 1 (1 in the flora): e North America.

Many classifications have included *Hydrastis* within Ranunculaceae (A. Cronquist 1981; W. C. Gregory 1941; O. F. Langlet 1932; M. Tamura 1963, 1968, 1993). Recent workers (H. Tobe and R. C. Keating 1985; A. L. Takhtajan 1987), however, have assigned *Hydrastis* to its own family, intermediate between Ranunculaceae and Berberidaceae. Phylogenetic studies by S. B. Hoot (1991) confirmed the isolated position of *Hydrastis* and suggested that characteristics such as the following justify its exclusion from Ranunculaceae: undifferentiated mesophyll, xylem straight (instead of V-shaped) in cross section, scalariform perforations in the vessels, micropyle defined by two integuments, pollen with a distinct striate-reticulate tectum, and a base chromosome number of $x = 13$ (as opposed to 6, 7, 8, or 9).

C. S. Keener (1993) challenged these conclusions, stating that features such as scalariform perforations in vessels, striate-reticulate pollen, and the micropyle defined by two integuments are found in other genera within Ranunculaceae. Only straight xylem in cross section and the base chromosome number are distinctive. Whether these features warrant segregation of *Hydrastis* into its own family is debatable. Decisions involving the circumscription of this genus await a molecular study involving Berberidaceae, Ranunculaceae, and related families.

SELECTED REFERENCES   Eichenberger, M. D. and G. R. Parker. 1976. Goldenseal (*Hydrastis canadensis* L.) distribution, phenology and biomass in an oak-hickory forest. Ohio J. Sci. 76: 204–210.   Keener, C. S. 1993. A review of the classification of the genus *Hydrastis* (Ranunculaceae). Aliso 13: 551–558.   Massey, J. R., D. K. S. Otte, T. A. Atkinson, and R. D. Whetstone. 1983. An Atlas and Illustrated Guide to Threatened and Endangered Vascular Plants of the Mountains of North Carolina and Virginia. Washington. [U.S.D.A. Forest Serv., Gen. Techn. Rep. SE-20.]   Tobe, H. and R. C. Keating. 1985. The morphology and anatomy of *Hydrastis* (Ranunculales): Systematic reevaluation of the genus. Bot. Mag. (Tokyo) 98: 291–316.

1. Hydrastis canadensis Linnaeus, Syst. Nat. ed. 10, 2: 1088. 1759   E F

Herbs, 15–50 cm. **Rhizomes** with tough fibrous roots. **Stems** erect, unbranched, pubescent. **Leaves:** basal leaf often quickly deciduous, 1; cauline leaves 2, similar to basal. **Leaf blade** 3–10 cm wide at anthesis, to 25 cm wide in fruit; lobes variously incised, margins singly or doubly serrate. **Flowers** 8–18 mm wide; peduncle 5–38 mm, ± closely subtended by distalmost cauline leaf; sepals not clawed, 3.5–7 mm, glabrous; stamens strongly exserted, white, showy, 4–8 mm; pistils 1-carpellate, distinct; stigma 2-lipped. **Berry aggregates** dark red, 10–15 × 8–15(–20) mm, each berry 5–8 × 1.5–5 mm. **Seeds** 1–2 per pistil, 2.5–4.5 mm.

Flowering spring. Mesic, deciduous forests, often on clay soil; 50–1200 m; Ont.; Ala., Ark., Conn., Del., Ga., Ill., Ind., Iowa, Ky., Md., Mass., Mich., Minn., Miss., Mo., N.J., N.Y., N.C., Ohio, Pa., Tenn., Vt., Va., W.Va., Wis.

A decrease in undisturbed, deciduous woodlands and commercial harvesting of the rhizomes for herbal medicine have contributed to a decline of this species. The species is considered very infrequent in Canada (G. W. Argus and K. M. Pryer 1990) and in some U.S. states (D. J. White and H. L. Dickson 1983). The raspberrylike fruit is considered inedible.

Native Americans used *Hydrastis canadensis* medicinally for treating cancer, whooping cough, diarrhea, liver trouble, earaches, sore eyes, fevers, pneumonia, heart trouble, tuberculosis, chapped or cut lips, and dyspepsia; to improve appetite; and as a tonic, and as a wash for inflammation (D. E. Moerman 1986).

2. **RANUNCULUS** Linnaeus, Sp. Pl. 1: 548. 1753; Gen. Pl. ed. 5, 243. 1754 · Buttercup, crowfoot, renoncule [Latin *rana,* frog, *unculus,* little, allusion to the wet habitats in which some species grow]

Alan T. Whittemore

Herbs, annual or perennial, from tuberous roots, caudices, rhizomes, stolons, or bulbous stem bases. **Leaves** basal, cauline, or both, simple, variously lobed or parted, or compound, all petiolate or distal leaves sessile; cauline leaves alternate (rarely a distal pair opposite in *Ranunculus* sect. *Flammula*). **Leaf blade** reniform to linear, margins entire, crenate, or toothed. **Inflorescences** terminal or axillary, 2–50-flowered cymes to 25 cm or solitary flowers; bracts present or absent, small or large and leaflike, not forming involucre. **Flowers** bisexual, radially symmetric; sepals sometimes persistent in fruit, 3–5(–6), green or sometimes purple, yellow, or white, plane (base saccate in *R. ficaria*), oblong to elliptic, ovate, or lanceolate, 1–15 mm; petals 0–22(–150), distinct, yellow, rarely white, red, or green, plane, linear to orbiculate, 1–26 mm, nectary on abaxial surface, usually covered by scale; stamens (5–)10–many; filaments filiform; staminodes absent between stamens and pistils; pistils 4–250, simple; ovule 1 per ovary; style present or absent. **Fruits** achenes, rarely utricles, aggregate, sessile, discoid, lenticular, globose, obovoid, or cylindric, sides sometimes veined; beak present or absent, terminal, straight or curved, 0–4.5 mm. $x = 7, 8$.

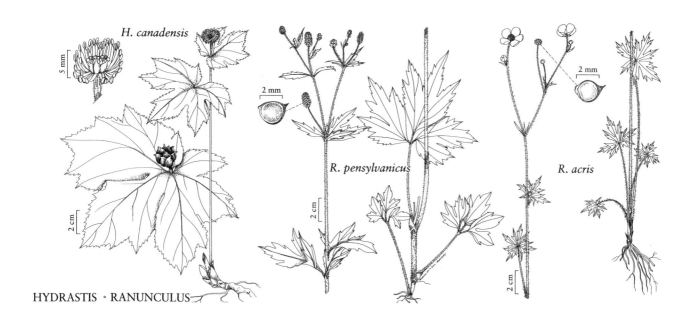

H. canadensis

5 mm

2 mm

2 cm

R. pensylvanicus

2 mm

2 cm

R. acris

2 cm

2 cm

HYDRASTIS · RANUNCULUS

Species about 300 (77 in the flora): worldwide except lowland tropics.

Most *Ranunculus* species are poisonous to stock; when abundant, they may be troublesome to ranchers. A few species with acrid juice were formerly used as vesicatories. The genus is badly in need of biosystematic work. Apomixis and interspecific hybridization occur in several Old World groups of buttercups; some of the taxonomic complexity of the New World species probably results from these processes.

Considerable disagreement exists among authors on the proper generic and infrageneric classification of *Ranunculus*. Most of the subgenera accepted here have been treated as separate genera at one time or another. All recent studies have been based on local or continental floras, however, and classifications proposed for one region may not work for the plants of other regions. Like most North American workers, I have followed the generic and infrageneric classification of L. D. Benson (1948), who gave by far the most thorough and best documented study of the problem. The genus and its subdivisions should be studied on a worldwide basis.

SELECTED REFERENCES   Benson, L. D. 1948. A treatise on the North American *Ranunculi*. Amer. Midl. Naturalist 40: 1–261.   Benson, L. D. 1954. Supplement to a treatise on the North American *Ranunculi*. Amer. Midl. Naturalist 52: 328–369.   Cook, C. D. K. 1966. A monographic study of *Ranunculus* subgenus *Batrachium* (DC) A. Gray. Mitt. Bot. Staatssamml. München 6: 47–237.   Duncan, T. 1980. A taxonomic study of the *Ranunculus hispidus* Michaux complex in the Western Hemisphere. Univ. Calif. Publ. Bot. 77: 1–125.   Nesom, G. L. 1993. *Ranunculus* (Ranunculaceae) in Nuevo León, with comments on the *R. petiolaris* group. Phytologia 75: 391–398.

Key to subgenera and sections of *Ranunculus*

1. All leaves simple and unlobed.
    2. Cauline leaves absent or scalelike; sepals deciduous or persistent.
       3. Petals 5; achene beak 0.1–0.2 mm; plants stoloniferous, without caudices . . . .
       . . . . 2b.2. *Ranunculus* (subg. *Cyrtorhycha*) sect. *Halodes* (*R. cymbalaria*), p. 126
       3. Petals 8–18; achene beak 0.8–1.4 mm; plants not stoloniferous, stems erect from
          short caudices.
          4. Leaf blades undivided, margins entire or serrulate; sepals persistent in fruit; petals 7–12 mm . . . . .2d. *Ranunculus* subg. *Oxygraphis* (*R. kamtschaticus*), p. 129

4. Leaf blades shallowly lobed, margins crenate; sepals deciduous; petals 2–4 mm
. . . . 2b.4.   *Ranunculus* (subg. *Cyrtorhyncha*) sect. *Pseudaphanostemma* (*R. hystriculus*), p. 127
2. Cauline leaves present, well developed; sepals deciduous.
  5. Tuberous roots present; flowers yellow; sepals 3 . . . . . 2h. *Ranunculus* subg. *Ficaria* (*R. ficaria*), p. 133
  5. Tuberous roots absent (roots thickened proximally in some 5-sepaled species of sect.
    Flammula); flowers yellow, white, or pink.
    6. Achenes 4.2–5.2 mm, achene body prolonged beyond seed as corky distal ap-
      pendage; sepals 3, 6–10 mm; petals white to pink . . . .
      . . . . 2f. *Ranunculus* subg. *Pallasiantha* (*R. pallasii*), p. 132
    6. Achenes 0.8–2.8 mm, achene body not prolonged beyond seed; sepals (3–)4–5,
      1.5–12 mm; petals yellow.
      7. Sepals covered with dense brown pubescence; distal leaves and bracts apically
        3-crenate or shallowly 3-lobed, otherwise undivided. . . . .
        . . . . 2a.3. *Ranunculus* (subg. *Ranunculus*) sect. *Epirotes* in part (*R. macauleyi*), p. 107
      7. Sepals glabrous or with colorless hairs; distal leaves simple and undivided.
        8. Achene wall papery, longitudinally ribbed; leaf apex broadly rounded to
          truncate, margins crenate. . . . .
          . . . . 2b.2. *Ranunculus* (subg. *Cyrtorhyncha*) sect. *Halodes* (*R. cymbalaria*), p. 126
        8. Achene wall thick, not ornamented, smooth (sometimes pubescent); leaf
          apex acuminate to rounded-obtuse, margins entire or finely toothed. . . . . .
          . . . . 2a.4. *Ranunculus* (subg. *Ranunculus*) sect. *Flammula*, p. 117
1. Some or all leaves simple and lobed, or compound.
  9. Leafy stems creeping and rooting at nodes or floating in water, then rootless.
    10. Leaves 3-foliolate. . . . . 2a.1. *Ranunculus* (subg. *Ranunculus*) sect. *Ranunculus*, p. 91
    10. Leaves simple, lobed to filiform-dissected or occasionally undivided.
      11. Achene body prolonged beyond seed as corky distal appendage; sepals 3, pet-
        als 5–11.
        12. Leaf blade as wide as long, ternately divided to base. . . . .
        . . . . 2g. *Ranunculus* subg. *Coptidium* (*R. lapponicus*), p. 133
        12. Leaf blade much longer than wide, unlobed or lobed. . . . .
        . . . . 2f. *Ranunculus* subg. *Pallasiantha* (*R. pallasii*), p. 132
      11. Achene body not prolonged beyond seed; either sepals 5 or sepals 3–4 and
        petals also 3–4.
        13. Petals white or white with yellow claws; achenes with strong coarse wrin-
          kles. . . . . .2e. *Ranunculus* subg. *Batrachium*, p. 130
        13. Petals yellow; achenes smooth (faintly wrinkled in *R. sceleratus* var. *scel-*
          eratus) . . . . 2a.5. *Ranunculus* (subg. *Ranunculus*) sect. *Hecatonia*, p. 122
  9. Leafy stems erect or if decumbent rooting only at base, never floating.
    14. Petals pure red, or white when immature; fruits winged achenes or utricles . . . . .
    . . . . .2c. *Ranunculus* subg. *Crymodes*, p. 128
    14. Petals yellow, rarely also with some red pigmentation abaxially, or greenish yellow;
    fruits achenes, rarely winged.
      15. Cauline leaves absent or scalelike, tuberous roots absent; leaves sometimes
      deeply parted or dissected, never compound.
        16. Plants villous; sepals 3–6 × 1–2 mm, persistent in fruit; fruit wall firm,
          smooth, beak much longer than achene body . . . .
          . . . . .2i. *Ranunculus* subg. *Ceratocephalus* (*R. testiculatus*), p. 134
        16. Plants glabrous; sepals 6–13 × 3–7 mm, deciduous in fruit; fruit wall
          thin, veined, beak much shorter than achene body.
          17. Leaves shallowly 5–7-lobed; petals inconspicuous, 2–4 mm . . . .
            2b.4. *Ranunculus* (subg. *Cyrtorhyncha*) sect. *Pseudaphanostemma* (*R. hystriculus*), p. 127
          17. Leaves 3–5-parted; petals showy, 8–12 mm . . . .
          . . . . .2b.3. *Ranunculus* (subg. *Cyrtorhyncha*) sect. *Arcteranthis*, (*R. cooleyae*), p. 126
      15. Cauline leaves present, simple, lobed or dissected, or compound (rarely re-
      duced to scales in *R. fascicularis* with tuberous roots and 3–5-foliolate leaves).

18. Style absent, stigma sessile; achene margins thick and corky; emergent aquatic, sometimes also found on very wet soil . . . .
. . . . 2a.5. *Ranunculus* (subg. *Ranunculus*) sect. *Hecatonia* in part (*R. sceleratus*), p. 122
18. Style present; achene margins not corky; in various habitats but rarely aquatic.
    19. Achene wall thin, longitudinally striate; scale of nectary reduced to low ridge, not covering nectary. . . . .
    . . .2b.1. *Ranunculus* (subg. *Cyrtorhyncha*) sect. *Cyrtorhyncha* (*R. ranunculinus*), p. 125
    19. Achene wall thick, smooth, papillose, or spiny; scale of nectary well-developed flap or pocket completely covering nectary.
        20. Achenes thick-lenticular or asymmetrically thick-lenticular to compressed-globose, 1.2–2 times as wide as thick; nectary scale joined with petal on 3 sides, forming pocket enclosing nectary (sometimes with apex free, forming flap shorter than pocket); basal leaves various, unlobed to deeply divided, margins entire to crenate but never at all serrate. . . . .
        . . . . 2a.3. *Ranunculus* (subg. *Ranunculus*) sect. *Epirotes*, p. 107
        20. Achenes strongly flattened, at least 3–15 times as wide as thick; nectary scale free from petal for at least 1/2 its length, thus forming free scale over nectary (scale sometimes free for less than 1/2 its length in *R. recurvatus*, with serrate to crenate-serrate leaf margins); basal leaves always deeply lobed or compound (except sometimes in *R. marginatus* and *R. orthorhynchus*), margins various.
            21. Achenes papillose or spiny (sometimes smooth in *R. sardous*); flowers small, petals 1–6 mm, scarcely longer than sepals, sometimes absent (larger and much longer than sepals in *R. sardous*). . . . .
            . . . . 2a.2. *Ranunculus* (subg. *Ranunculus*) sect. *Echinella*, p. 104
            21. Achenes smooth, glabrous or pubescent; flowers small to large, petals always present, 2–22 mm . . . .
            . . . . 2a.1. *Ranunculus* (subg. *Ranunculus*) sect. *Ranunculus*, p. 91

2a. RANUNCULUS Linnaeus subg. RANUNCULUS

2a.1. RANUNCULUS Linnaeus (subg. RANUNCULUS) sect. RANUNCULUS

Ranunculus sect. *Chrysanthe* (Spach) L. D. Benson

Plants usually ± hispid, sometimes glabrous. **Stems** erect to decumbent, sometimes stoloniferous, sometimes bulbous-based, without bulbils. **Roots** basal, sometimes also nodal, sometimes tuberous. **Leaves** basal and cauline, petiolate; basal deeply parted or compound (except sometimes in *R. orthorhynchus* var. *bloomeri*), blades with segments lobed or parted, margins toothed; cauline deeply parted or compound, similar to basal or with shorter petioles and/or blades with narrower segments (rarely poorly developed in *R. fascicularis*). **Inflorescences** 1–50-flowered cymes. **Flowers** pedicellate; sepals deciduous soon after anthesis, 5 (sometimes 6 in *R. occidentalis* var. *hexasepalus*); petals always present, yellow, sometimes abaxially reddish, rarely poorly developed; nectary scale attached basally, free from petal for at least 1/2 its length, forming flap covering nectary (or sometimes attached on 3 sides, forming pocket in

R. recurvatus), glabrous, free margin entire; style present. **Fruits** achenes, 1-locular; achene body discoid, strongly flattened, 3–15 times as wide as thick, not prolonged beyond seed; wall thick, not ornamented; margin low or high narrow ridge or wing; beak much shorter than achene body.

Species ca. 100 (19 in the flora): worldwide except lowland tropics.

1. Petals 2–6 mm, usually no longer than sepals (sometimes longer than sepals in *R. uncinatus*).
   2. Basal leaves simple and lobed or parted.
      3. Receptacle glabrous; base of stem not bulbous; w North America . . . . 1. *Ranunculus uncinatus*
      3. Receptacle hispid; base of stem bulbous, cormlike; e North America . . . . 2a. *Ranunculus recurvatus* var. *recurvatus*
   2. Basal leaves compound.
      4. Petals 2–4 × 1–2.5 mm; heads of achenes cylindric, 5–7 mm wide . . . . .3. *Ranunculus pensylvanicus*
      4. Petals 4–6 × 3.5–5 mm; heads of achenes globose to ovoid, 7–10 mm wide . . . . .4. *Ranunculus macounii*
1. Petals (6–)7–26 mm (rarely shorter in *R. occidentalis* and *R. canus,* in which petals are much longer than sepals).
   5. Sepals spreading, sometimes reflexed from base with age.
      6. Basal leaves deeply parted or dissected, ultimate segments linear to broadly linear, margins entire (occasionally a lobe reduced to large tooth); Rocky Mountains . . . 8a. *Ranunculus acriformis* var. *acriformis*
      6. Basal leaves variously parted or compound but not as above, segments seldom linear, margins toothed; widespread.
         7. Leaf blades simple, 3–5-parted or -divided.
            8. Basal leaf blades pentagonal in outline; beak of achene 0.2–1 mm; widespread . . . . .7. *Ranunculus acris*
            8. Basal leaf blades cordate to reniform in outline; beak of achene 1.6–2 mm; Mackenzie Delta to ne Alaska . . . . 6. *Ranunculus turneri*
         7. Leaf blades 3–5-foliolate.
            9. Tuberous roots present.
               10. Petals 10–22; c Texas. . . . . 15. *Ranunculus macranthus*
               10. Petals 5(–7); widespread, e North America. .16. *Ranunculus fascicularis*
            9. Tuberous roots absent.
               11. Beak of achene curved, 0.8–1.2 mm; stems decumbent to creeping. . . . . .18. *Ranunculus repens*
               11. Beak of achene straight or somewhat curved, 0.8–2.6 mm; stems erect to decumbent . . . . .17. *Ranunculus hispidus*
   5. Sepals reflexed along well-defined transverse fold 1–3 mm above base.
      12. Petals 8–22; Pacific Coast, Arizona, Texas.
         13. Tuberous roots absent; beak of achene lanceolate to deltate-apiculate, curved, 0.2–1.6 mm; California to British Columbia.
            14. Achenes 3.4–4.2 mm, beak deltate or lance-deltate; Transverse Ranges, s California. . . . . .11b. *Ranunculus canus* var. *ludovicianus*
            14. Achenes 1.8–3.2 mm, beak lanceolate; California to British Columbia.
               15. Beak of achene 1.2–1.6 mm; sepals 5–6; Queen Charlotte Islands, British Columbia . . . . 9f. *Ranunculus occidentalis* var. *hexasepalus*
               15. Beak of achene 0.2–0.8 mm; sepals 5; California, Oregon, Mexico . . . . 12. *Ranunculus californicus*
         13. Tuberous roots present; beak of achene subulate, straight, (1.8–)2–4 mm (but sometimes deciduous); Texas, Arizona.
            16. Petals 2–5 mm wide; achenes 2–2.4 mm wide, margins forming ribs or narrow wings 0.1–0.4 mm wide; sepals always reflexed; Arizona, trans-Pecos Texas . . . . 14. *Ranunculus fasciculatus*
            16. Petals 4–9 mm wide; achenes 2.8–3.4 mm wide, margins forming narrow wings 0.4–0.6 mm wide; sepals spreading or weakly reflexed; c Texas. . . . . 15. *Ranunculus macranthus*
      12. Petals 5–7; widespread.
         17. Receptacle glabrous; basal leaves 3-parted or -foliolate, segments or leaflets linear to cuneate.
            18. Beak of achene deltate or lance-deltate, curved; Sacramento Valley, California and adjacent foothills . . . . .11a. *Ranunculus canus* var. *canus*
            18. Beak of achene lanceolate to lance-subulate, straight or curved; widespread.

19. Basal leaves deeply 3-divided, segments again deeply parted, ultimate segments linear or nearly so; Rocky Mountains. . . . .
. . . . 8b. & 8c. *Ranunculus acriformis* var. *montanensis* and var. *aestivalis*
19. Basal leaves deeply 3-parted to 3-foliolate, margins of segments or leaflets toothed or shallowly lobed, ultimate lobe triangular or broadly lanceolate; Pacific Slope.
    20. Petals yellow on both surfaces; widespread. . . . . 9. *Ranunculus occidentalis*
    20. Petals reddish abaxially; sw Oregon. 10. *Ranunculus austro-oreganus*
17. Receptacle hispid; basal leaves usually pinnately 3–7-foliolate (sometimes merely 3-parted with orbiculate to ovate segments in *Ranunculus orthorhynchus* var. *bloomeri*).
    21. Beak of achene oblong or triangular, curved, 0.2–0.8 mm; introduced weeds.
        22. Base of stem bulbous cormlike; petals 9–13 × 8–11 mm. . . . . 19. *Ranunculus bulbosus*
        22. Base of stem not bulbous; petals 7–10 × 4–8 mm. . . . .
        . . . . 20. *Ranunculus sardous* (sect. *Echinella*)
    21. Beak of achene lanceolate or subulate, straight or somewhat curved (sometimes tip weakly hooked in *R. pacificus*), (0.8–)1–3.8(–4.8) mm; native.
        23. Stems decumbent, sometimes rooting at nodes; e North America . . . .
        . . . . 17c. *Ranunculus hispidus* var. *nitidus*
        23. Stems erect to decumbent, never rooting at nodes; w North America.
            24. Beak of achene 1–1.8 mm; leaflets lobed; Alaska panhandle. . . . .
            . . . . 5. *Ranunculus pacificus*
            24. Beak of achene 2–3.8(–4.8) mm (1.8–2.2 mm in var. *bloomeri,* with unlobed leaflets); throughout w North America. . . . .
            . . . . 13. *Ranunculus orthorhynchus*

---

1. Ranunculus uncinatus D. Don in G. Don, Gen. Hist. 1: 35. 1831 [E]

Ranunculus bongardii Greene; *R. bongardii* var. *tenellus* (A. Gray) Greene; *R. uncinatus* var. *earlei* (Greene) L. D. Benson; *R. uncinatus* var. *parviflorus* (Torrey) L. D. Benson

Stems erect, never rooting nodally, hispid or glabrous, base not bulbous. **Roots** never tuberous. **Basal leaf blades** cordate to reniform in outline, 3-parted or sometimes 3-foliolate, 1.8–5.6 × 2.8–8.3 cm, segments again lobed, ultimate segments elliptic to lanceolate, margins toothed or crenate-toothed, apex acute to rounded-obtuse. **Flowers:** receptacle glabrous; sepals reflexed or sometimes spreading, 2–3.5 × 1–2 mm, pubescent; petals 5, yellow, 2–4(–6) × 1–2(–3) mm. **Heads of achenes** globose or hemispheric, 4–7 × 4–7 mm; achenes 2–2.8 × 1.6–2 mm, glabrous or sparsely hispid, margin forming narrow rib 0.1–0.2 mm wide; beak persistent, lanceolate, curved, hooked, 1.2–2.5 mm. $2n = 28$.

Flowering spring–summer (Apr–Aug). Moist meadows or woods, often along streams; 0–3400 m; Alta., B.C.; Alaska, Ariz., Calif., Colo., Idaho, Mont., Nev., N.Mex., Oreg., Wash., Wyo.

Plants with hispid stems and achenes are often separated as *Ranunculus uncinatus* var. *parviflorus;* these two characters are poorly correlated, however, and sometimes vary between plants in a single collection.

Ranunculus uncinatus was reported from northeastern Alberta and adjacent Northwest Territories by H. J. Scoggan (1978–1979, part 3). The specimens have hairy receptacles and straight, broad achene beaks; they apparently represent small individuals of *R. macounii.*

Some Native Americans used *Ranunculus uncinatus* as an antirheumatic, a diaphoretic, a disinfectant, and an orthopedic aid, as well as in herbal steam baths intended to soothe sore muscles and rheumatism (D. E. Moerman 1986).

2. Ranunculus recurvatus Poiret in J. Lamarck et al., Encycl. 6: 125. 1804 · Renoncule recourbée

Varieties 2 (1 in the flora): e North America; West Indies.

2a. Ranunculus recurvatus Poiret var. **recurvatus** [E]

Ranunculus recurvatus var. *adpressipilis* Weatherby

Stems erect, never rooting nodally, hirsute, base bulbous, cormlike. **Roots** never tuberous. **Basal leaf blades** broadly cordate in outline, 3-cleft or -parted, 2–7.5 × 3–11.6 cm, segments undivided or shallowly 1–2-cleft, ultimate segments rhombic, margins crenate-toothed, apex acute or obtuse. **Flowers:** receptacle hispid; sepals reflexed from near base, 3–6 × 1.5–2.5 mm, pilose; petals 5, yellow, 3–5 × 1–2 mm. **Heads of achenes** globose, 5–6 mm wide; achenes 1.6–2.2 × 1.4–1.8 mm, glabrous, margin forming narrow rib 0.1–0.2 mm wide; beak persistent, lanceolate, curved, hooked, 1–1.4 mm. $2n = 32$.

Flowering spring (Apr–Jun). Mesic woods, stream banks, and swamps; 0–1200 m; N.B., Nfld., N.S., Ont., P.E.I., Que.; Ala., Ark., Conn., Del., D.C., Fla., Ga., Ill., Ind., Iowa, Kans., Ky., La., Maine, Md., Mass., Mich., Minn., Miss., Mo., Nebr., N.H., N.J., N.Y., N.C., N.Dak., Ohio, Okla., Pa., R.I., S.C., Tenn., Tex., Vt., Va., W.Va., Wis.

The Cherokee used *Ranunculus recurvatus* var. *recurvatus* as a sedative, a dermatological aid, and an oral aid; the Iroquois, as a laxative, a venereal aid, and a toothache remedy (D. E. Moerman 1986).

3. Ranunculus pensylvanicus Linnaeus f., Suppl. Pl., 272. 1782 · Renoncule de Pennsylvanie [E] [F]

Stems erect, never rooting nodally, hispid, base not bulbous. **Roots** never tuberous. **Basal leaf blades** broadly cordate in outline, 3-foliolate, 1.6–7 × 3–9 cm, leaflets cleft, usually deeply so, ultimate segments narrowly elliptic, margins toothed, apex acute. **Flowers:** receptacle hirsute; sepals reflexed ca. 1 mm above base, 3–5 × 1.5–2 mm, ± hispid; petals 5, yellow, 2–4 × 1–2.5 mm. **Heads of achenes** cylindric, 9–12 × 5–7 mm; achenes 1.8–2.8 × 1.6–2 mm, glabrous, margin forming narrow rib 0.1–0.2 mm wide; beak persistent, broadly lanceolate or nearly deltate, straight or nearly so, 0.6–0.8 mm. $2n = 16$.

Flowering late spring–summer (Jun–Aug). Stream banks, bogs, moist clearings, depressions in woodlands; 0–1700 m; Alta., B.C., Man., N.B., Nfld., N.W.T., N.S., Ont., P.E.I., Que., Sask.; Alaska, Ariz., Colo., Conn., Del., D.C., Idaho, Ill., Ind., Iowa, Maine, Md., Mass., Mich., Minn., Mont., Nebr., N.H., N.J., N.Mex., N.Y., N.Dak., Ohio, Pa., R.I., S.Dak., Vt., Wash., W.Va., Wis., Wyo.

Ojibwa tribes used *Ranunculus pensylvanicus* as a hunting medicine; the Potawatomi used it as an astringent for miscellaneous diseases (D. E. Moerman 1986).

4. Ranunculus macounii Britton, Trans. New York Acad. Sci. 12: 3. 1892 · Renoncule de Macoun [E]

Ranunculus macounii var. *oreganus* (A. Gray) K. C. Davis

Stems prostrate to nearly erect, often rooting nodally, hirsute or glabrous, base not bulbous. **Roots** never tuberous. **Basal leaf blades** cordate to reniform in outline, 3-foliolate, 3.7–7.5 × 4.5–9.5 cm, leaflets 3-lobed or -parted, ultimate segments elliptic or lance-elliptic, margins toothed or lobulate, apex acute to broadly acute. **Flowers:** receptacle hirsute; sepals spreading or reflexed ca. 1 mm above base, 4–6 × 1.5–3 mm, glabrous or hirsute; petals 5, yellow, 4–6 × 3.5–5 mm. **Heads of achenes** globose or ovoid, 7–11 × 7–10 mm; achenes 2.4–3 × 2–2.4 mm, glabrous, margin forming narrow rib 0.1–0.2 mm wide; beak persistent, lanceolate to broadly lanceolate, straight or nearly so, 1–1.2 mm. $2n = 32, 48$.

Flowering spring–summer (May–Sep). Meadows, depressions in woodlands, ditches, edges of streams and ponds, on wet soil or emergent from shallow water; 0–2900 m; Alta., B.C., Man., Nfld., N.W.T., Ont., Que., Sask., Yukon; Alaska, Ariz., Calif., Colo., Idaho, Mich., Minn., Mont., Nebr., Nev., N.Mex., N.Dak., Oreg., S.Dak., Utah, Wash., Wyo.

Through most of its range, *Ranunculus macounii* has conspicuously hispid herbage. Glabrous plants are found, however, in the lower Columbia River Valley (southwestern Washington and adjacent Oregon). This variant has been called *R. macounii* var. *oreganus*.

5. Ranunculus pacificus (Hultén) L. D. Benson, Amer. Midl. Naturalist 40: 79. 1948 [E]

Ranunculus septentrionalis Poiret subsp. *pacificus* Hultén, Acta Univ. Lund, n.s. 40: 769. 1944

Stems erect or reclining, never rooting nodally, hispid or glabrous, base not bulbous. **Roots** never tuberous. **Basal leaf blades** broadly triangular to cordate in outline, 3-foliolate, 6–13 × 8–16 cm, leaflets lobed, ultimate segments elliptic to lance-elliptic or oblong, apex acute or obtuse. **Flowers:** receptacle hispid; sepals reflexed 1–2 mm above base, 6–9 × 3–4 mm, sparsely hispid; petals 5, abaxially yellow or purplish, adaxially yellow, 9–11 × 6–8 mm. **Heads of achenes** ovoid to globose, 9–11 × 8–11 mm; achenes 3.2–3.8 × 2–3 mm, glabrous, margin forming narrow rib 0.1–0.2 mm wide; beak persistent, lanceolate or subulate from triangular base, straight or tip weakly hooked, 1–1.8 mm.

Flowering summer (Jul). Along streams and in meadows; 0 m; Alaska.

Ranunculus pacificus is endemic to the Alaska panhandle.

6. **Ranunculus turneri** Greene, Pittonia 2: 296. 1892 [C]

Ranunculus occidentalis Nuttall var. *turneri* (Greene) L. D. Benson

Stems erect, never rooting nodally, hirsute, base not bulbous. **Roots** never tuberous. **Basal leaf blades** cordate to reniform in outline, 3-parted, 1.3–3 × 1.8–3.8 cm, segments cleft, ultimate segments elliptic to lanceolate, margins toothed, apex acute. **Flowers:** receptacle glabrous; sepals spreading, 7–9 × 2–4 mm, hirsute; petals 5, yellow, 10–15 × 8–11 mm. **Heads of achenes** nearly globose, 7–10 mm wide; achenes 2.4 × 2.6–2.7 mm, glabrous, margin forming narrow rib 0.1–0.2 mm wide; beak persistent, lanceolate, strongly hooked or curved, 1.6–2 mm.

Flowering summer (Aug). Damp meadows; of conservation concern; 0 m; N.W.T., Yukon; Alaska; Asia.

Ranunculus turneri was considered an arctic race of *R. acris* by E. Hultén (1971). It occurs from the Mackenzie Delta to northeastern Alaska.

7. **Ranunculus acris** Linnaeus, Sp. Pl. 1: 554. 1753 · Renoncule âcre, bouton d'or [F] [W2]

Ranunculus acris var. *latisectus* Beck

Stems erect from short caudex or rhizome, never rooting nodally, hispid, strigose, or glabrous, base not bulbous. **Roots** never tuberous. **Basal leaf blades** pentagonal in outline, deeply 3–5-parted, 1.8–5.2 × 2.7–9.8 cm, segments 1–2×-lobed or -parted, ultimate segments narrowly elliptic or oblong to lanceolate, margins toothed or lobulate, apex acute to rounded. **Flowers:** receptacle glabrous; sepals spreading, 4–6(–9) × 2–5 mm, hispid; petals 5, yellow, 8–11(–17) × 7–13 mm. **Heads of achenes** globose, 5–7(–10) mm wide; achenes 2–3 × 1.8–2.4 mm, glabrous, margin forming narrow rib 0.1–0.2 mm wide; beak persistent, deltate, usually with tip short or long, straight or curved, subulate, 0.2–1 mm. 2*n* = 14.

Flowering spring–summer (May–Sep). Meadows, stream banks, roadsides, and old fields; 0–2300 m; largely introduced; Greenland; St. Pierre and Miquelon; Alta., B.C., Man., N.B., Nfld., N.W.T., N.S., Ont., P.E.I., Que., Sask.; Ala., Alaska, Ariz., Calif., Conn., Del., D.C., Ga., Idaho, Ill., Ind., Iowa, Kans., Ky., Maine, Md., Mass., Mich., Minn., Mo., Mont., Nebr.,

Nev., N.H., N.J., N.Y., N.C., N.Dak., Ohio, Oreg., Pa., R.I., S.C., S.Dak., Tenn., Utah, Vt., Va., Wash., W.Va., Wis., Wyo.; South America; Eurasia; Pacific Islands; Australia.

Ranunculus acris is variable in form and division of leaves, size of achene beak, and form of indument on the proximal stem. Most North American plants are weedy and have poorly differentiated caudices; these forms probably were introduced from Eurasia. Rhizomatous plants with large flowers (parenthetic measurements above) found in the Aleutian Islands of Alaska and in Greenland are probably native. Aleutian populations of this form have been called *R. acris* var. *frigidus* Regel or *R. grandis* Honda var. *austrokurilensis* (Tatewaki) H. Hara. Both names were originally applied to Asiatic plants, and their applicability to American specimens is open to question.

Some Native American tribes used *Ranunculus acris* as an analgesic, a dermatological or oral aid, an antidiarrheal, antihemorrhagic, and a sedative (D. E. Moerman 1986).

8. **Ranunculus acriformis** A. Gray, Proc. Amer. Acad. Arts 21: 374. 1886 [E]

Stems erect, not rooting nodally, hirsute or strigose, base not bulbous. **Roots** never tuberous, sometimes thick and ± fleshy proximally. **Basal leaf blades** broadly ovate to cordate or sometimes reniform in outline, deeply 3-divided or occasionally 3-foliolate, 2.2–6 × 2.5–7.7(–10) cm, divisions 1–2×-deeply parted or -dissected, ultimate segments linear to broadly linear, margins entire (occasionally a lobe reduced to large tooth), apex acute or rounded-acute. **Flowers:** receptacle glabrous; sepals spreading or variously reflexed, 4–6 × 2–4 mm, appressed-hirsute; petals 5(–10), yellow, 7–13 × 4–10 mm. **Heads of achenes** hemispheric or globose, 5–8 × 6–8(–10) mm; achenes 2.2–3.4 × 2–3 mm, glabrous, margin forming narrow rib 0.1–0.2 mm wide; beak persistent, lanceolate, strongly curved, 0.4–1.6 mm.

Varieties 3 (3 in the flora): w North America.

1. Sepals spreading, or reflexed from base, pubescence of appressed hairs; Wyoming, Colorado . . . . 8a. *Ranunculus acriformis* var. *acriformis*
1. Sepals reflexed 1–3 mm above base, pubescence of spreading hairs; Wyoming, Idaho, Montana, Utah.
  2. Beak of achene 1.2–1.6 mm; receptacle hemispheric to spheric; Utah to Idaho and Montana . . . .8b. *Ranunculus acriformis* var. *montanensis*
  2. Beak of achene 0.4–1 mm; receptacle obpyriform to cylindric; Utah . . . . 8c. *Ranunculus acriformis* var. *aestivalis*

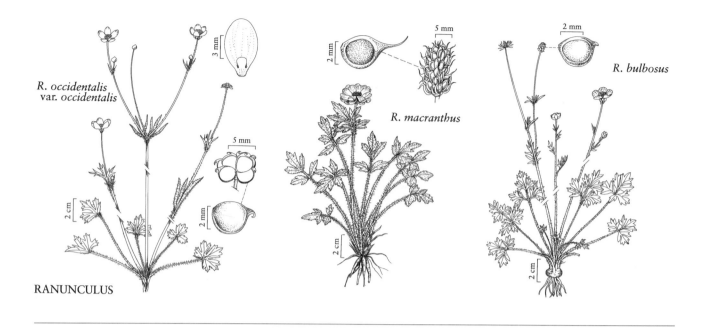

R. occidentalis
var. *occidentalis*

R. macranthus

R. bulbosus

RANUNCULUS

8a. Ranunculus acriformis A. Gray var. **acriformis** E

Stems strigose. **Flowers:** receptacle spheric or cylindric; sepals spreading or reflexed from base, pubescence of appressed hairs; petals 7–9 mm. **Achenes:** beak 1–1.2 mm. $2n = 28$.

Flowering spring–summer (May–Jul). Wet meadows; 1800–2800 m; Colo., Wyo.

8b. Ranunculus acriformis A. Gray var. **montanensis** (Rydberg) L. D. Benson, Amer. Midl. Naturalist 40: 43. 1948 E

Ranunculus montanensis Rydberg, Mem. New York Bot. Gard. 1: 166. 1900

Stems hirsute. **Flowers:** receptacle hemispheric to spheric; sepals reflexed 1–3 mm above base, pubescence of spreading hairs; petals 7–13 mm. **Achenes:** beak 1.2–1.6 mm.

Flowering spring–summer (May–Aug). Wet meadows; 1500–2400 m; Idaho, Mont., Utah, Wyo.

8c. Ranunculus acriformis A. Gray var. **aestivalis** L. D. Benson, Amer. Midl. Naturalist 40: 250. 1948 · Autumn buttercup E C

Ranunculus acris Linnaeus var. *aestivalis* (L. D. Benson) S. L. Welsh; *R. aestivalis* (L. D. Benson) R. Van Buren & K. T. Harper

Stems hirsute. **Flowers:** receptacle obpyriform to cylindric; sepals reflexed ca. 2 mm above base, pubescence of spreading hairs; petals 8–11 mm. **Achenes:** beak 0.4–1 mm.

Flowering summer (Jul–Sep). Meadows; 2100–2900 m; of conservation concern; Utah.

Ranunculus acriformis var. *aestivalis* is known from two sites in southern Utah, a saline meadow south of Panguitch in the Sevier River Valley and a riparian meadow in Boulger Canyon, Emery County. R. Van Buren et al. (1994) treated this variety as a distinct species, *R. aestivalis,* emphasizing the saline habitat of the Sevier Valley population. The morphologic differences between *R. acriformis* var. *aestivalis* and the other varieties of *R. acriformis* are minor, and the habitat of the single Emery County specimen is typical of the species. I prefer to retain this taxon in *R. acriformis.* The DNA data of Van Buren et al., based on a total of eight populations of the three varieties, are consistent with either interpretation, because the three varieties form a single clade with *R. acriformis* var. *aestivalis* as sister group to the other two.

SELECTED REFERENCE   Van Buren, R., K. T. Harper, W. R. Andersen, D. J. Stanton, S. Seyoum, and J. L. England. 1994. Evaluating the relationship of autumn buttercup (*Ranunculus acriformis* var. *aestivalis*) to some close congeners using random amplified polymorphic DNA. Amer. J. Bot. 81: 514–519.

9. Ranunculus occidentalis Nuttall in J. Torrey and A. Gray, Fl. N. Amer. 1: 22. 1838   E W1

Stems erect to reclining, not rooting nodally, hirsute or sometimes pilose or glabrous, base not bulbous. **Roots** never tuberous. **Basal leaf blades** broadly ovate to semicircular or reniform in outline, 3-parted or -foliolate, 1.5–5.3 × 2.2–8 cm, segments usually again 1(–2)×-lobed, ultimate segments oblong or elliptic to lanceolate or oblanceolate, margins dentate (sometimes dentate-lobulate or entire), apex acute to rounded-obtuse. **Flowers:** receptacle glabrous; sepals reflexed 2–3 mm above base, 4–7(–9) × 2–4 mm, hirsute; petals 5–14, yellow, 5–13 × 1.5–8 mm. **Heads of achenes** hemispheric, 3–7 × 5–9 mm; achenes 2.6–3.6(–4.8) × 1.8–3(–3.2) mm, glabrous, rarely hispid, margin forming narrow rib 0.1–0.2 mm wide; beak persistent, lanceolate to lance-subulate, straight or curved, 0.4–2.2 mm.

Varieties 7 (7 in the flora): w North America.

The seeds of *Ranunculus occidentalis* were eaten by some Californian Indians. Juice of the flowers of this taxon have been identified as an Aleut poison (D. E. Moerman 1986).

1. Petals 8–14; Queen Charlotte Islands, B.C. . . . .
. . . . 9f. *Ranunculus occidentalis* var. *hexasepalus*
1. Petals 5–6; widespread.
  2. Stem 4–8 mm thick; beak of achene 1.8–2.4 mm, curved; coastal Alaska. . . . .
. . . . 9g. *Ranunculus occidentalis* var. *nelsonii*
  2. Stem 1–3(–4) mm thick; beak of achene either 0.4–1.4 mm and curved, or 1.2–2.2 mm and straight; widespread.
    3. Beak of achene straight, 1.2–2.2 mm; Oregon and northernmost California.
      4. Ultimate segments of leaves lanceolate to oblanceolate . . . .
. . . . 9c. *Ranunculus occidentalis* var. *dissectus*
      4. Ultimate segments of leaves elliptic . . .
. . . . . 9d. *Ranunculus occidentalis* var. *howellii*
    3. Beak of achene curved, 0.4–1.4 mm; widespread.
      5. Petals 1.5–2.5 mm wide; beak of achene 0.4–1.2 mm; stems ± reclining; Sierra Nevada, above 1000 m . . . . . .
      9b. *Ranunculus occidentalis* var. *ultramontanus*
      5. Petals 3–8 mm wide; beak of achene (0.6–)1–1.4 mm; stems erect or reclining; widespread.
        6. Stems pilose or glabrous; Alaska to c British Columbia and Alberta. . . .
        9e. *Ranunculus occidentalis* var. *brevistylis*

        6. Stems hirsute, sometimes glabrous; California to sw British Columbia
        9a. *Ranunculus occidentalis* var. *occidentalis*

9a. Ranunculus occidentalis Nuttall var. **occidentalis**   E F

Ranunculus occidentalis var. *eisenii* (Kellogg) A. Gray; *R. occidentalis* var. *rattanii* A. Gray

Stems erect to reclining, 1–3(–4) mm thick, hirsute or sometimes glabrous. **Basal leaf blades** 3-parted or -foliolate, ultimate segments oblong or elliptic to lanceolate or oblanceolate, margins dentate or dentate-lobulate. **Flowers:** sepals 5, 4–7 mm; petals 5–6, 5–10 × 3–6 mm. **Achenes** 2.6–3.6(–4) × 1.8–3(–3.2) mm, glabrous or hispid; beak lanceolate, curved, (0.6–)1–1.4 mm.

Flowering late winter–spring (Mar–Jun). Grassy slopes of meadows or open woodlands; 0–1500 m; B.C.; Calif., Nev., Oreg., Wash.

L. D. Benson (1948) divided *Ranunculus occidentalis* var. *occidentalis* into three varieties. The name *R. occidentalis* var. *occidentalis* was applied only to plants from Oregon northward, in which leaves are rarely compound and never have lanceolate ultimate segments, and achenes are always glabrous and have beaks over 1 mm. California plants were treated as *R. occidentalis* var. *rattanii* (plants with small [5–8 mm] petals from the Coast Ranges) and *R. occidentalis* var. *eisenii* (plants with larger petals from the foothills surrounding the Central Valley). Most of those plants from California, however, cannot be distinguished from more northern plants, and forms with small petals are found throughout the range of the variety.

9b. Ranunculus occidentalis Nuttall var. **ultramontanus** Greene, Pittonia 3: 13. 1896   E

Stems ± reclining, 1–3 mm thick, hirsute or sometimes glabrous. **Basal leaf blades** 3-foliolate, rarely merely 3-parted, ultimate segments oblong or elliptic to lanceolate or oblanceolate, margins dentate or dentate-lobulate. **Flowers:** sepals 5, 4–6 mm; petals 5–6, 6–8 × 1.5–2.5 mm. **Achenes** 3–3.4 × 2–2.6 mm, glabrous, rarely hispid; beak lanceolate, curved, 0.4–1.2 mm.

Flowering spring–summer (May–Jul). Meadows; 1300–2100 m; Calif., Nev., Oreg.

Ranunculus occidentalis var. *ultramontanus* is found at middle elevations in the Sierra Nevada, southern Cascades, and northeastern Coast Ranges. It intergrades with var. *occidentalis* in the region of elevational overlap.

9c. Ranunculus occidentalis Nuttall var. dissectus L. F. Henderson, Rhodora 32: 25. 1930 [E]

Stems erect to spreading, 1–3 mm thick, hirsute or sometimes glabrous. **Basal leaf blades** 3-parted or sometimes 3-foliolate, ultimate segments lanceolate or oblanceolate, margins entire or sparsely dentate. **Flowers:** sepals 5, 4–6 mm; petals 5–6, 6–10 × 3–6 mm. **Achenes** 2.6–3.6 × 2–2.8 mm, glabrous; beak lance-subulate, straight, 1.2–2.2 mm.

Flowering spring–summer (May–Jul). Wet to dry meadows; 1000–1800 m; Calif., Oreg.

Ranunculus occidentalis var. *dissectus* is found in the Great Basin, Klamath region, and southern Cascade Mountains. As noted below, it is very similar to var. *howellii.*

9d. Ranunculus occidentalis Nuttall var. howellii Greene, Pittonia 3: 14. 1896 [E]

Stems erect to reclining, 1–3 mm thick, hirsute or sometimes glabrous. **Basal leaf blades** 3-parted, ultimate segments narrowly elliptic, margins entire or dentate. **Flowers:** sepals 5, 4–6 mm; petals 5–6, 6–10 × 3–5 mm. **Achenes** 3.4–4.8 × 2.6–3.2 mm, glabrous; beak lanceolate, straight, 1.6–2.2 mm.

Flowering spring–summer (Apr–Jul). Meadows; 900–1400 m; Calif., Oreg.

Ranunculus occidentalis var. *howellii, R. occidentalis* var. *dissectus,* and *R. austro-oreganus* form a distinctive group distinguishable in the *R. occidentalis* complex by the straight achene beak. They may be difficult to separate, however; further study of their relationships is needed.

9e. Ranunculus occidentalis Nuttall var. brevistylis Greene, Pittonia 3: 14. 1896 [E]

Ranunculus occidentalis subsp. *insularis* Hultén

Stems erect to reclining, 2–3 mm thick, pilose or glabrous. **Basal leaf blades** 3-parted, ultimate segments elliptic or narrowly elliptic, margins dentate. **Flowers:** sepals 5, 5–7 mm; petals 5–6, 8–12 × 3–8 mm. **Achenes** 3–3.4

× 2.6–3 mm, glabrous; beak lanceolate, curved, 1.2–1.4 mm. $2n = 28$.

Flowering spring–summer (May–Aug). Coastal and mountain meadows; 0–1000 m; Alta., B.C., Yukon; Alaska.

Ranunculus occidentalis var. *brevistylis* may be difficult to distinguish from var. *occidentalis;* the two have sometimes been combined. The pubescence character distinguishing them is well correlated with geography, however, so I am provisionally maintaining both of them.

9f. Ranunculus occidentalis Nuttall var. hexasepalus L. D. Benson, Bull. Torrey Bot. Club 68: 167. 1941 [E]

Ranunculus hexasepalus (L. D. Benson) L. D. Benson

Stems erect or spreading, 2–6 mm thick, hirsute or strigose, at least distally. **Basal leaf blades** 3-parted, ultimate segments elliptic or narrowly elliptic, margins dentate. **Flowers:** sepals 5–6, 4–5 mm; petals 8–14, 9–13 × 3–7 mm. **Achenes** 2.6–3 × 2.2–3 mm, glabrous; beak lanceolate, curved, at least distally, 1.2–1.6 mm.

Flowering summer. Open areas near coast; 0 m; B.C.

Ranunculus occidentalis var. *hexasepalus* is endemic to the Queen Charlotte Islands.

9g. Ranunculus occidentalis Nuttall var. nelsonii (de Candolle) L. D. Benson, Bull. Torrey Bot. Club 68: 166. 1941 [E]

Ranunculus recurvatus Poiret var. (*β*) *nelsonii* de Candolle, Syst. Nat. 1: 290. 1817; *R. occidentalis* subsp. *nelsonii* (de Candolle) Hultén

Stems erect or spreading, 4–8 mm thick, hirsute. **Basal leaf blades** 3-parted, rarely 3-foliolate, ultimate segments elliptic, margins dentate. **Flowers:** sepals 5, 4–9 mm; petals 5–6, 7–15 × 4–11 mm. **Achenes** 3–4 × 2.2–2.8 mm, glabrous; beak lanceolate, curved, 1.8–2.4 mm.

Flowering spring (May–Jun). Moist places near coast; 0 m; Alaska.

10. Ranunculus austro-oreganus L. D. Benson, Amer. Midl. Naturalist 52: 341. 1954 [C][E]

Stems erect or ascending, never rooting nodally, crisped-pilose, base not bulbous. **Roots** never tuberous. **Basal leaf blades** broadly rhombic to semicircular in outline, 3-parted, 2.8–4.3 × 3–5.5 cm, segments 3-lobed, ultimate segments lanceolate, margins entire or toothed, apex narrowly acute or acuminate. **Flowers:** receptacle glabrous; sepals reflexed 1 mm above base, 4–6 × 1.5–3 mm, densely pilose; petals 5, abaxially red, adaxially yellow, 10–12 × 4–6 mm. **Heads of achenes** hemispheric, 4–7 × 7–10 mm; achenes 3.4–4.2 × 2.8–3.2 mm, sometimes basally pilose, margin forming narrow rib 0.1–0.2 mm wide; beak persistent, lance-subulate, straight or somewhat curved distally, 1.6–2.6 mm.

Flowering spring (May). Grassy hillsides; of conservation concern; 500 m; Oreg.

Ranunculus austro-oreganus is doubtfully distinct from *R. occidentalis* var. *howellii.* L. D. Benson (1954) described the stem as bulbous-based and similar to that of *R. bulbosus,* but a differentiated base is not evident in material I have seen (some of which was cited by Benson).

11. Ranunculus canus Bentham, Pl. Hartw., 294. 1849 [E]

Stems erect to decumbent, never rooting nodally, hirsute, pilose, or glabrous, base not bulbous. **Roots** never tuberous. **Basal leaf blades** ovate to narrowly ovate in outline, 3-parted or -foliolate, 3.3–9.5 × 3.5–9.4 cm, leaflets or segments 1–3×-lobed, ultimate segments ovate or oblong-ovate to lanceolate, margins toothed, apex acute or obtuse. **Flowers:** receptacle glabrous; sepals reflexed 1–2 mm above base, 3–8 × 2–4 mm, hirsute; petals 5–17, yellow, 6–12 × 3–6 mm. **Heads of achenes** hemispheric to globose, 6–9 × 7–10 mm; achenes 3.4–4.4 × 2.4–3.6 mm, glabrous or rarely hispid, margin forming narrow rib 0.1–0.2 mm wide; beak persistent, deltate or lance-deltate, curved, 0.2–1.2 mm.

Varieties 2 (2 in the flora): California.

Ranunculus canus intergrades with *R. occidentalis* var. *occidentalis* and *R. californicus,* and some populations can be difficult to assign to species. The deltate or lance-deltate achene beak of *R. canus,* however, which is usually 0.8–1.2 mm wide at the base and less than 1.5 times as long as wide, contrasts with the narrower beaks of the other two species, which are usually less than 0.6 mm wide and at least twice as long as wide.

1. Petals 5–7; Sacramento Valley and adjacent foothills. . . . . .11a. *Ranunculus canus* var. *canus*
1. Petals 13–17; Transverse Ranges. . . . .
. . . . .11b. *Ranunculus canus* var. *ludovicianus*

11a. Ranunculus canus Bentham var. **canus** [E]

Ranunculus californicus Bentham var. *canus* (Bentham) W. H. Brewer & S. Watson; *R. canus* var. *laetus* (Greene) L. D. Benson

Leaf blades: ultimate segments ovate or oblong-ovate to lanceolate, margins toothed, apex acute or obtuse. **Flowers:** petals 5–7.

Flowering spring–summer (Mar–Jul). Grassland or very open oak woodland; 0–1200 m; Calif.

Ranunculus canus var. *canus* is endemic to the Sacramento Valley and adjacent foothills. As with *R. californicus,* the leaves vary in form. Plants from the low Sierra Nevada foothills often have simple, 3-parted leaves, while plants from the west side of the valley usually have compound leaves with leaflets parted into lanceolate segments. Many exceptions occur, however, and all possible leaf forms can be found in the eastern Sacramento Valley.

11b. Ranunculus canus Bentham var. **ludovicianus** (Greene) L. D. Benson, Bull. Torrey Bot. Club 68: 171. 1941 [E]

Ranunculus ludovicianus Greene, Bull. Calif. Acad. Sci. 2: 58. 1886; *R. californicus* Bentham var. *ludovicianus* (Greene) K. C. Davis

Leaf blades: ultimate segments lanceolate to oblong-lanceolate, margins entire or toothed, apex acute or rounded-acute. **Flowers:** petals 13–17.

Flowering spring–summer (Mar–Aug). Meadows; 1000–2300 m; Calif.

Ranunculus canus var. *ludovicianus* is endemic to the Transverse Ranges of California. Pending further study, I am reluctantly following L. D. Benson's (1948) placement of this taxon. The plants have the floral characters of *R. californicus* and have often been included in that species. Indeed, forms of *R. californicus* from the foothills west of the San Joaquin Valley may also have large achenes with deltate beaks. Those plants differ from *R. canus* var. *ludovicianus* only in their ovate leaf segments. Further study of the group is much needed.

12. Ranunculus californicus Bentham, Pl. Hartw., 295. 1849 [W2]

Stems erect to prostrate, never rooting nodally, hirsute, strigose, or glabrous, base not bulbous. **Roots** never tuberous. **Basal leaf blades** broadly ovate or cordate in outline, 3-lobed or -parted to 3-foliolate, 1.8–5.8 × 2.3–7.6 cm, leaflets or segments undivided or 1–2× -lobed or -parted, ultimate segments linear to orbiculate, margins toothed, crenate, or entire, apex acute to rounded. **Flowers:** receptacle glabrous or rarely hispid; sepals reflexed 2–3 mm above base, 4–8 × 2–4 mm, pilose; petals 9–17, yellow, (6–)7–14 × 2–6 mm. **Heads of achenes** globose or hemispheric, 3–7 × 4–9 mm; achenes 1.8–3.2 × 1.4–3.2 mm, glabrous, rarely hispid, margin forming narrow rib 0.1–0.2 mm wide; beak persistent, lanceolate, curved, 0.2–0.8 mm.

Varieties 2 (2 in the flora): Calif., Oreg.; Mexico (Baja California).

In addition to the range given, localized populations of *Ranunculus californicus* have been reported recently from a few islands in the vicinity of Victoria (British Columbia and Washington) (M. F. Denton 1978; T. C. Brayshaw 1989). Those populations are small and introgress freely with *R. occidentalis* wherever they come together. Denton referred her specimens to *R. californicus* var. *cuneatus*; Brayshaw reported both varieties from the same small populations, but his data are consistent with populations of *R. californicus* var. *cuneatus* that are introgressing extensively with *R. occidentalis*. Although both Denton and Brayshaw treat *R. californicus* as a native species in that region, several reasons support the belief that it is introduced there. No reports of *R. californicus* in the area occur prior to 1978, although the area is quite well collected (especially Victoria, B.C., and the San Juan Islands, Washington); a long history of extensive marine trade between Victoria and San Francisco has resulted in the introduction of a number of other California species to the area; and for scattered small populations of *R. californicus* to have persisted for long periods in the face of free introgression from *R. occidentalis* seems unlikely. Given the small population size and the introgression from *R. occidentalis,* it is questionable whether *R. californicus* can persist in the area.

SELECTED REFERENCES Brayshaw, T. C. 1989. Buttercups, Waterlilies, and Their Relatives (the Order Ranales) in British Columbia. Victoria. [Roy. Brit. Columbia Mus. Mem. 1.]   Denton, M. F. 1978. *Ranunculus californicus*, a new record for the state of Washington. Madroño 25: 132.

1. Stems erect to decumbent, hirsute or glabrous; ultimate segments of basal leaf blades acute to rounded-acute at apex . . . .
. . . . 12a. *Ranunculus californicus* var. *californicus*
1. Stems prostrate, strigose; ultimate segments of basal leaf blades rounded to obtuse at apex . . . .
. . . . 12b. *Ranunculus californicus* var. *cuneatus*

12a. Ranunculus californicus Bentham var. **californicus**

Ranunculus californicus var. *austromontanus* L. D. Benson; *R. californicus* var. *gratus* Jepson; *R. californicus* var. *rugulosus* (Greene) L. D. Benson

Stems erect or decumbent, hirsute or glabrous. **Basal leaf blades** 2.8–5.8 × 4–6 cm, leaflets or segments undivided or 1–2×-lobed or -parted, ultimate segments oblong-elliptic to lanceolate or linear, margins toothed or entire, apex acute to rounded-acute. **Achenes** 1.8–3.2 × 1.4–2.4 mm. $2n = 28$.

Flowering winter–summer (Mar–Aug). Grassland and open woodland; 0–2000 m; Calif., Oreg.; Mexico (Baja California).

Ranunculus californicus var. *californicus* is variable and has been divided into varieties based on indument, achene size, and especially leaf form. Those characters are poorly correlated, however, and most of the named varieties do not seem natural.

12b. Ranunculus californicus Bentham var. **cuneatus** Greene, Fl. Francisc. 2: 299. 1892 [E]

Stems prostrate, strigose. **Basal leaf blades** 4–5 × 4–6 cm, leaflets or segments undivided or 1×-lobed, ultimate segments elliptic to orbiculate, margins toothed or crenate, apex rounded to obtuse. **Achenes** 1.8–2.2 × 1.4–1.8 mm.

Flowering winter–spring (Jan–Apr). Bluffs and hillsides near coast; 0–200 m; Calif., Oreg.

13. Ranunculus orthorhynchus Hooker, Fl. Bor.-Amer. 1: 21, plate 9. 1829 [E] [W1]

Stems nearly erect or decumbent, never rooting nodally, hispid, strigose, or glabrous, base not bulbous. **Roots** sometimes fleshy and ± tuberous. **Basal leaf blades** narrowly ovate to oblong or semicircular in outline, simple to 3–5-lobed or -foliolate, 2.8–12.5 × 2.5–14 cm, leaflets or segments undivided or 1–2×-lobed or -parted, ultimate segments circular to linear, margins dentate, crenate, or entire, apex rounded to narrowly acute. **Flowers:** receptacle hispid; sepals reflexed 1–2 mm above base, 5–11 × 2–4 mm, hispid, hirsute, or glabrous; petals 5–6, abaxially yellow or red, adaxially yellow, 8–18 × 4–11 mm. **Heads of achenes** hemispheric to ovoid, 5–13 × 6–10 mm; achenes 2.8–4.5 × 1.8–3.2 mm, glabrous, margin forming narrow rib 0.1–0.2 mm wide; beak persistent, narrowly lanceolate to subulate, straight, 1.8–3.8(–4.8) mm.

Varieties 3 (3 in the flora): w North America.

The first two varieties (*Ranunculus orthorhynchus*

var. *orthorhynchus* and *R. orthorhynchus* var. *platyphyllus*) are rather weak, intergrading extensively in California and Oregon. By contrast, *R. orthorhynchus* var. *bloomeri* often grows with the others with little or no intergradation (although intermediate populations are found in some areas), and it has been treated as a distinct species, *R. bloomeri*, by many taxonomists.

1. Basal leaves simple or 3-foliolate, if compound then leaflets undivided, margins crenate; petals retuse.... 13c. *Ranunculus orthorhynchus* var. *bloomeri*
1. Basal leaves 3–5-foliolate, leaflets lobed or parted, margins entire or dentate; petals truncate or rounded.
    2. Heads of achenes hemispheric or sometimes globose, 5–7 mm; petals often abaxially red
        13a. *Ranunculus orthorhynchus* var. *orthorhynchus*
    2. Heads of achenes globose or ovoid, 8–13 mm; petals yellow...............
        ...13b. *Ranunculus orthorhynchus* var. *platyphyllus*

13a. Ranunculus orthorhynchus Hooker var. orthorhynchus [E]

Ranunculus orthorhynchus subsp. *alaschensis* (L. D. Benson) Hultén; *R. orthorhynchus* var. *alaschensis* L. D. Benson; *R. orthorhynchus* var. *hallii* Jepson

Basal leaf blades narrowly ovate to semicircular in outline, 3–5-foliolate, leaflets 1–2×-lobed or -parted, segments elliptic to linear, margins entire to dentate-lobulate, apex obtuse to narrowly acute. **Flowers:** petals abaxially red or yellow, 4–6 mm broad, rounded. **Heads of achenes** hemispheric or sometimes globose, 5–7 mm; beak 3–3.8 (–4.8) mm. $2n = 32$.

Flowering spring–summer (Apr–Jul). Meadows and marshy areas; 0–2100 m; B.C.; Alaska, Calif., Oreg., Wash.

13b. Ranunculus orthorhynchus Hooker var. platyphyllus A. Gray, Proc. Amer. Acad. Arts 21: 377. 1886 [E]

Basal leaf blades ovate to semicircular in outline, 3–5-foliolate, leaflets 1×-lobed or -parted, segments narrowly elliptic to linear, margins dentate or dentate-lobulate, apex obtuse to narrowly acute. **Flowers:** petals abaxially yellow, 7–11 mm broad, rounded. **Heads of achenes** globose or ovoid, 8–13 mm; beak 2–3 mm. $2n = 32$.

Flowering late winter–summer (Mar–Aug). Meadows and marshy areas; 0–2700 m; B.C.; Calif., Idaho, Mont., Nev., Oreg., Utah, Wash., Wyo.

13c. Ranunculus orthorhynchus Hooker var. bloomeri (S. Watson) L. D. Benson, Amer. Midl. Naturalist 40: 101. 1948 [E]

Ranunculus bloomeri S. Watson, Bot. California 2: 426. 1880

Basal leaf blades cordate to oblong or circular in outline, simple and undivided or 3-foliolate, if compound the leaflets undivided, margins crenate, apices rounded to obtuse or occasionally acute. **Flowers:** petals abaxially yellow, 6–9 mm broad, retuse. **Heads of achenes** hemispheric to ovoid, 7–9 mm; beak 1.8–2.2 mm. $2n = 32$.

Flowering late winter–spring (Mar–May). Meadows and marshy areas; 0–1500 m; Calif., Oreg.

Ranunculus orthorhynchus var. *bloomeri* is disjunct between the Coast Ranges of west-central California and south-central Oregon.

14. Ranunculus fasciculatus Sessé & Moçiño, Fl. Mexic. ed. 2, 134. 1894

Ranunculus macranthus Scheele var. *arsenei* L. D. Benson; *R. petiolaris* Humboldt, Bonpland, & Kunth ex de Candolle var. *arsenei* (L. D. Benson) T. Duncan

Stems erect or decumbent, not rooting nodally, hirsute to nearly glabrous, base not bulbous. **Roots** fleshy and somewhat tuberous. **Basal leaf blades** ovate to deltate in outline, 3(–5)-foliolate, 2.5–14.9 × 2.3–19.9 cm, leaflets 1–2×-lobed or -parted, ultimate segments narrowly oblong to elliptic or lanceolate, margins toothed, apex narrowly acute to rounded-acute. **Flowers:** receptacle hispid; sepals reflexed 1–2 mm above base, 5–10 × 3–5 mm, hispid; petals 11–16, yellow, 8–21 × 2–5 mm. **Heads of achenes** globose to ovoid, 6–13 × 7–9 mm; achenes 2.4–3.4 × 2–2.4 mm, glabrous, margin forming rib or narrow wing 0.1–0.4 mm wide; beak filiform from deltate base, straight, 1.8–2.5 mm, filiform tip often deciduous, leaving 1–1.2 mm deltate beak.

Flowering late spring–summer (Jun–Aug). Stream banks, lakeshores, and marshes; 1000–2200 m; Ariz., Tex.; Mexico.

I am following G. L. Nesom (1993) in treating *Ranunculus fasciculatus* as a distinct species. This taxon was considered a variety of *R. macranthus* by L. D. Benson (1948) and a variety of *R. petiolaris* by T. Duncan (1980). These disparate opinions result from different interpretations of Mexican members of the *R. petiolaris* group.

15. Ranunculus macranthus Scheele, Linnaea 21: 585. 1848 [E] [F]

Ranunculus fascicularis Muhlenberg ex J. M. Bigelow var. *cuneiformis* (Small) L. D. Benson

Stems erect or decumbent, never rooting nodally, hispid, base not bulbous. **Roots** tuberous. **Basal leaf blades** ovate in outline, 3–5-foliolate, 3.8–10 × 2.7–9 cm, leaflets 1×-lobed, ultimate segments narrowly elliptic to oblanceolate, margins entire or with few teeth, apex broadly acute to rounded-obtuse. **Flowers:** receptacle hispid; sepals spreading or weakly reflexed ca. 1 mm above base, 7–10 × 3–5 mm, hispid; petals 10–22, yellow, 12–22 × 4–9 mm. **Heads of achenes** globose to cylindric, 8–14 × 8–10 mm; achenes 2.2–4.2 × 2.8–3.4 mm, glabrous, margin forming narrow wing 0.4–0.6 mm wide; beak usually persistent, filiform from deltate base, straight, 2–4 mm. $2n = 32$.

Flowering late winter–spring (Mar–May). Riverbanks and wet meadows; 0–400 m; Tex.

16. Ranunculus fascicularis Muhlenberg ex J. M. Bigelow, Fl. Boston., 137. 1814 [E] [W2]

Ranunculus fascicularis var. *apricus* (Greene) Fernald

Stems erect or ascending, never rooting nodally, strigose or spreading-strigose, base not bulbous. **Roots** always both filiform and tuberous on same stem. **Basal leaf blades** ovate to broadly ovate in outline, 3–5-foliolate, 2.1–4.7 × 1.9–4.5 cm, leaflets undivided or 1×-lobed or -parted, ultimate segments oblanceolate or obovate, margins entire or with few teeth, apex rounded-acute to rounded-obtuse. **Flowers:** receptacle hispid or glabrous; sepals spreading or sometimes reflexed from base, 5–7 × 2–3 mm, hispid or glabrous; petals 5(–7), yellow, 8–14 × 3–6 mm. **Heads of achenes** globose or ovoid, 5–9 × 5–8 mm; achenes 2–2.8 × 1.8–2.2 mm, glabrous, margin forming narrow rib 0.1–0.2 mm wide; beak persistent, filiform, straight, 1.2–2.8 mm. $2n = 32$.

Flowering winter–spring (Jan–Jun). Grassland or deciduous forest; 0–300 m; Man., Ont; Ala., Ark., Conn., Ga., Ill., Ind., Iowa, Kans., Ky., La., Md., Mass., Mich., Minn., Miss., Mo., Nebr., N.J., N.Y., N.C., Ohio, Okla., Pa., R.I., S.C., Tenn., Tex., Vt., Va., Wis.

Ranunculus fascicularis is very similar to *R. hispidus* var. *hispidus,* and herbarium specimens without underground parts may be difficult to identify. *Ranunculus fascicularis* grows in drier habitats; segments of its leaves are commonly oblanceolate and blunt, with few or no marginal teeth; and its petals are widest at or below the middle. *Ranunculus hispidus* var. *hispidus* is usually larger in all its parts (leaves, flowers, heads of achenes); leaf segments are variable in shape but their apices are normally sharper and their marginal teeth more numerous, and petals are widest above the middle.

17. Ranunculus hispidus Michaux, Fl. Bor.-Amer. 1: 321. 1803 · Renoncule hispide [E]

Stems erect or decumbent, sometimes rooting nodally, hispid or strigose, base not bulbous. **Roots** never tuberous. **Basal leaf blades** ovate to deltate in outline, 3-foliolate or outer blades merely 3-parted, 2–13.4 × 2.4–16.8 cm, leaflets undivided to lobed or parted, ultimate segments narrowly elliptic or oblanceolate to circular, margins toothed, apex acuminate to rounded. **Flowers:** receptacle hispid; sepals spreading or reflexed, 4–10 × 2–5 mm, hispid; petals 5, yellow, 8–16 × 3–9 mm. **Heads of achenes** hemispheric to short-ovoid, 6–10 × 7–10 mm; achenes 2.2–5.2 × 2–3.8 mm, glabrous, margin forming narrow rib or broad wing 0.1–1.2 mm wide; beak persistent, lance-subulate, straight or somewhat curved, 0.8–2.6 mm.

Varieties 3 (3 in the flora): North America.

Until recently, the varieties of *Ranunculus hispidus* were usually treated as distinct species. Arguments for restoring species status to *R. hispidus* var. *nitidus* were given by G. L. Nesom (1993).

1. Stems erect or nearly erect, never rooting at nodes. . . . . 17a. *Ranunculus hispidus* var. *hispidus*
1. Stems decumbent, sometimes rooting at nodes.
  2. Sepals spreading or reflexed from base; achene margins 0.1–0.2 mm broad. . . . . 17b. *Ranunculus hispidus* var. *caricetorum*
  2. Sepals reflexed 1 mm above base; achene margins 0.4–1.2 mm broad . . . . 17c. *Ranunculus hispidus* var. *nitidus*

17a. Ranunculus hispidus Michaux var. **hispidus** [E]

Ranunculus hispidus var. *eurylobus* L. D. Benson; *R. hispidus* var. *falsus* Fernald; *R. hispidus* var. *greenmanii* L. D. Benson; *R. hispidus* var. *marilandicus* (Poiret) L. D. Benson

Stems erect or nearly erect, never rooting at nodes. **Flowers:** sepals spreading or reflexed from base. **Achene margin** 0.1–0.2 mm broad. $2n = 32$.

Flowering late winter–fall (Mar–Oct). Dry woods, grasslands, roadbanks; 0–1200 m; Ont.; Ala., Ark., Conn., Del., D.C., Ga., Ill., Ind., Kans., Ky., Md., Mass., Mich., Miss., Mo., N.J., N.Y., N.C., Ohio, Okla., Pa., S.C., Tenn., Vt., Va., W.Va.

17b. Ranunculus hispidus Michaux var. **caricetorum** (Greene) T. Duncan, Univ. Calif. Publ. Bot. 77: 40. 1980 [E]

Ranunculus caricetorum Greene, Pittonia 5: 194. 1903; *R. septentrionalis* Poiret var. *caricetorum* (Greene) Fernald

Stems decumbent, sometimes rooting at nodes. **Flowers:** sepals spreading or reflexed from base. **Achene margin** 0.1–0.2 mm broad. $2n = 64$.

Flowering spring–summer (Apr–Aug). Swampy woods, marshes, stream banks, ditches; 0–600 m; Man., N.B., Nfld., N.S., Ont., P.E.I., Que.; Conn., Ill., Ind., Iowa, Maine, Mass., Mich., Minn., Mo., N.H., N.J., N.Y., N.Dak., Ohio, Pa., R.I., S.Dak., Vt., Va., Wis.

The name *Ranunculus septentrionalis* Poiret has often been used for *R. hispidus* var. *caricetorum*. The type specimen, however, belongs to var. *nitidus* (T. Duncan 1980).

17c. Ranunculus hispidus Michaux var. **nitidus** (Chapman) T. Duncan, Univ. Calif. Publ. Bot. 77: 44. 1980 [E]

Ranunculus repens Linnaeus var. *nitidus* Chapman, Fl. South. U.S., 8. 1860; *R. carolinianus* de Candolle; *R. palmatus* Elliott; *R. septentrionalis* Poiret var. *pterocarpus* L. D. Benson

Stems decumbent, sometimes rooting at nodes. **Flowers:** sepals reflexed 1 mm above base. **Achene margin** 0.4–1.2 mm broad. $2n = 32$.

Flowering late winter–summer (Mar–Jul). Wet woods, swamps, ditches; 0–200 m; Ont.; Ala., Ark., D.C., Fla., Ga., Ill., Ind., Iowa, Kans., Ky, La., Md., Mich., Minn., Miss., Mo., Nebr., N.J., N.Y., N.C., Ohio, Okla., Pa., S.C., S.Dak., Tenn., Tex., Va., W.Va., Wis.

18. Ranunculus repens Linnaeus, Sp. Pl. 1: 554. 1753 · Renoncule rampante [W2]

Ranunculus repens var. *erectus* de Candolle; *R. repens* var. *glabratus* de Candolle; *R. repens* var. *linearilobus* de Candolle; *R. repens* var. *pleniflorus* Fernald; *R. repens* var. *villosus* Lamotte

Stems decumbent or creeping, rooting nodally, hispid to strigose or almost glabrous, base not bulbous. **Roots** never tuberous. **Basal leaf blades** ovate to reniform in outline, 3-foliolate, 1–8.5 × 1.5–10 cm, leaflets lobed, parted, or parted and again

lobed, ultimate segments obovate to elliptic or sometimes narrowly oblong, margins toothed, apex obtuse to acuminate. **Flowers:** receptacle hispid or rarely glabrous; sepals spreading or reflexed from base, 4–7 (–10) × 1.5–3(–4) mm, hispid or sometimes glabrous; petals 5(–150), yellow, 6–18 × 5–12 mm. **Heads of achenes** globose or ovoid, 5–10 × 5–8 mm; achenes 2.6–3.2 × 2–2.8 mm, glabrous, margin forming narrow rib 0.1–0.2 mm wide; beak persistent, lanceolate to lance-filiform, curved, 0.8–1.2 mm. $2n = 14, 32$.

Flowering late winter–summer (Mar–Aug). Meadows, borders of marshes, lawns, roadsides; 0–2500 m; introduced; Greenland; St. Pierre and Miquelon; Alta., B.C., N.B., Nfld., N.S., Ont., P.E.I., Que., Yukon; Ala., Alaska, Ark., Calif., Conn., Del., D.C., Idaho, Ill., Ind., Iowa, Ky., Maine, Md., Mass., Mich., Minn., Mo., Mont., Nebr., Nev., N.H., N.J., N.Y., N.C., Ohio, Oreg., Pa., R.I., S.C., S.Dak., Tenn., Tex., Utah, Vt., Va., Wash., W.Va., Wis., Wyo.; Central America; South America; native to Eurasia; Pacific Islands; Australia.

Ranunculus repens is widely naturalized in many parts of the world. Plants with sparse pubescence have been called *R. repens* var. *glabratus*. Horticultural forms with the outer stamens transformed into numerous extra petals occasionally become established and have been called *R. repens* var. *pleniflorus*. These variants have no taxonomic significance.

19. Ranunculus bulbosus Linnaeus, Sp. Pl. 1: 554. 1753 · Renoncule bulbeuse [F] [W2]

Ranunculus bulbosus var. *dissectus* Babey; *R. bulbosus* var. *valdepubens* (Jordan) Briquet

Stems erect, never rooting nodally, strigose or hirsute, base bulbous and cormlike. **Roots** never tuberous. **Basal leaf blades** ovate to cordate in outline, 3-foliolate, rarely merely deeply divided, 2–5.3 × 2.4–5.4 cm, leaflets 1–2×-lobed, ultimate segments oblong to obovate, margins toothed, apex rounded in outline. **Flowers:** receptacle pubescent; sepals reflexed 2–3 mm above base, 6–9 × 2–4 mm, pilose; petals 5, yellow, 9–13 × 8–11 mm. **Heads of achenes** ovoid, 6–9 × 5–7 mm; achenes 2.2–3.2 × 2.2–2.8 mm, glabrous, margin forming narrow rib 0.1–0.2 mm wide; beak persistent, lanceolate to deltate, 0.2–0.8 mm, slender tip hooked when present.

Flowering spring (Apr–Jun). Meadows; 0–700 m; introduced; B.C., Nfld., N.S., Ont., Que.; Ala., Ark., Calif., Conn., Del., D.C., Ga., Ill., Ind., Kans., Ky., La., Maine, Md., Mass., Mich., Mo., Nebr., N.H., N.J., N.Y., N.C., Ohio, Oreg., Pa., R.I., S.C., Tenn.,

Vt., Va., Wash., W.Va.; South America; native to Eurasia; Pacific Islands; Australia.

Ranunculus bulbosus is native to Europe and the Near East but has become naturalized in many other parts of the world. It is considered an introduced weed in the flora.

The Iroquois used *Ranunculus bulbosus* as a toothache remedy and as a treatment for venereal disease (D. E. Moerman 1986).

2a.2. RANUNCULUS Linnaeus (subg. RANUNCULUS) sect. ECHINELLA de Candolle in A. P. de Candolle and A. L. P. de Candolle, Prodr. 1: 41. 1824

Plants pubescent or glabrous. **Stems** erect or decumbent, not bulbous-based, without bulbils. **Roots** basal, never tuberous. **Leaves** basal and cauline or sometimes only cauline; basal and lower cauline leaves similar, petiolate, blades undivided, lobed, or compound, segments lobed or unlobed, margins entire or dentate. **Inflorescences** 2–16-flowered cymes (cymes sometimes sympodial, then flowers seemingly inserted opposite bracts). **Flowers** pedicellate (sessile in *R. platensis*); sepals deciduous soon after anthesis, 5 (3 in *R. platensis*); petals yellow (sometimes absent in *R. hebecarpus*); nectary scale attached basally, free from petal for at least 1/2 its length, forming flap covering nectary, glabrous, free margin entire; style present. **Fruits** achenes, 1-locular; achene body discoid, strongly flattened, 3–7 times as wide as thick, not prolonged beyond seed; wall thick, spiny, tuberculate, or papillose (sometimes smooth in *R. sardous*); margin low or high, broad or narrow ridge; beak much shorter than achene body.

Species ca. 15 (8 in the flora): worldwide except lowland tropics.

1. Petals 1–2 mm; achenes finely papillose, papillae terminating in slender hooked hairs or spines; native or introduced.
   2. Flowers sessile; sepals 3 . . . . 27. *Ranunculus platensis*
   2. Flowers pedicellate; sepals 5.
      3. Achenes ornamented on faces and margin, with lanceolate beaks; native of Pacific Slope . . . .
         . . . . 25. *Ranunculus hebecarpus*
      3. Achenes ornamented on faces but not margin, with deltate beaks; introduced weed, widespread in North America . . . . .26. *Ranunculus parviflorus*
1. Petals 4–10 mm; achenes coarsely papillose, tuberculate, or spinose, rarely smooth; introduced weeds.
   4. Sepals spreading; achenes 5–9 in single whorl, the faces and margin long-spinose . . . . .24. *Ranunculus arvensis*
   4. Sepals reflexed; achenes 10–60 in ovoid or globose heads, the faces papillose to spinose or rarely smooth, margin smooth.
      5. Basal leaves compound; achene beak 0.3–0.7 mm.
         6. Achene faces sparsely papillate or sometimes smooth; petals more than 5 mm . . . . .20. *Ranunculus sardous*
         6. Achene faces densely tuberculate; petals less than 5 mm. . . . . .21. *Ranunculus trilobus*
      5. Basal leaves simple; achene beak 1–2.5 mm.
         7. Achene faces but not margin covered with long spines; beaks 2–2.5 mm . . . . 22. *Ranunculus muricatus*
         7. Achene faces but not margin covered with high sharp tubercles or low spines; beaks ca. 1 mm . . . . 23. *Ranunculus marginatus*

20. Ranunculus sardous Crantz, Stirp. Austr. Fasc. 2: 84. 1763 F W2

Ranunculus parvulus Linnaeus

Stems nearly erect, hispid, base not bulbous. **Basal and lower cauline leaf blades** ovate to cordate, 3-foliolate, 2–6 × 2–6 cm, leaflets again parted, leaflet base truncate to acute, margins crenate-dentate to crenate-lobulate, apex rounded to obtuse. **Flowers** pedicellate; receptacle pilose; sepals 5, reflexed, 3–8 × 1.5–3 mm, pilose; petals 5, 7–10 × 4–8 mm. **Heads of achenes** globose or ovoid, 5–8 × 6–7 mm; achenes 15–35 per head, 2–3 × 2–3 mm, faces sparsely papillate or sometimes smooth, glabrous, margin smooth; beak oblong to deltate, curved, 0.4–0.7 mm.

Flowering late winter–summer (Mar–Aug). Roadsides, fields, open woods; 0–200 m; introduced; B.C.; Ala., Ark., Calif., Fla., Ga., Ill., Kans., Ky., La., Md., Miss., Mo., N.J., N.Y., N.C., Okla., Oreg., Pa., R.I., S.C., Tenn., Tex., Va.; native to Europe; Pacific Islands; Australia.

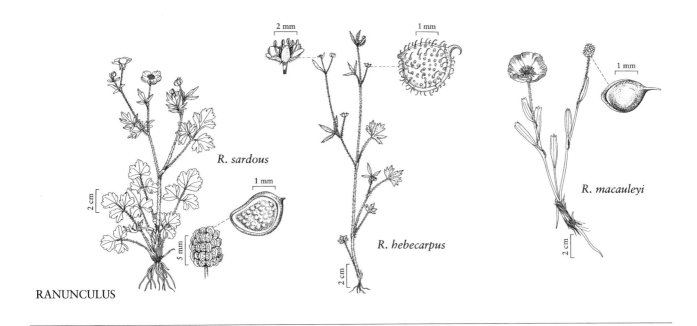

R. sardous

R. hebecarpus

R. macauleyi

RANUNCULUS

Ranunculus sardous was collected in New Brunswick and Ontario in the 1800s, but it apparently has not persisted in those provinces.

21. Ranunculus trilobus Desfontaines, Fl. Atlant. 1: 437. 1799

Stems erect or reclining, hispid or sometimes glabrous. **Basal and lower cauline leaf blades** cordate-ovate, 3-foliolate, 5–8 × 3–6(–7) cm, leaflets again cleft or parted, leaflet base obtuse to acute, margins dentate, apex rounded to acute. **Flowers** pedicellate; receptacle hispid; sepals 5, reflexed, 3–4 × 2 mm, sparsely hispid; petals 5, 4–5 × 2–3 mm. **Heads of achenes** ovoid, 8–9 × 6–7 mm; achenes 30–60 per head, 2 × 2 mm, faces densely tuberculate, glabrous, margin smooth; beak semicircular, hooked, 0.3 mm.

Flowering spring (Apr). Fields and roadsides; 0–100 m; introduced; Ala., Fla., La., Miss., S.C., Tex.; native to Europe.

22. Ranunculus muricatus Linnaeus, Sp. Pl. 1: 555. 1753 W2

Stems reclining or erect, glabrous or sparsely pilose. **Basal and lower cauline leaf blades** broadly cordate to reniform or semicircular, undivided or 3-lobed, 2–5 × 3–6.5 cm, base rounded to cordate, margins coarsely crenate, apex rounded. **Flowers** pedicellate; receptacle hispid; sepals 5, reflexed, 4–7 × 2–3 mm, sparsely bristly; petals 5, 4–8 × 2–4.5 mm. **Heads of achenes** globose, 13–16 × 13–16 mm; achenes 10–20 per head, 5–5.5 × 3–3.5 mm, faces covered with long spines, glabrous, margin smooth; beak lanceolate, curved, 2–2.5 mm.

Flowering spring (Mar–Jun). Fields and roadsides; 0–2000 m; introduced; Ala., Ark., Calif., Fla., Ga., La., Miss., N.C., Oreg., S.C., Tex., Wash.; South America; native to Eurasia; Africa; Pacific Islands; Australia.

23. Ranunculus marginatus d'Urville, Mém. Soc. Linn. Paris 1: 318. 1822

Stems erect, sparsely pilose. **Basal leaf blades** reniform to semicircular, undivided or shallowly 3-lobed, 2.5–5 × 3–7 cm, base truncate to nearly cordate, margins crenate or dentate, apex rounded or obtuse. **Flowers** pedicellate; receptacle hispid; sepals 5, reflexed 1 mm from base, 3–5 × 1–2 mm, hispid; petals 5, 4–5 × 2–3.5 mm. **Heads**

of achenes hemispheric to globose, 4 × 6 mm; achenes 15–20 per head, 2.5–3 × 2–2.5 mm, faces covered with high sharp tubercles or low spines, glabrous, margin smooth; beak deltate, straight, 0.8–1 mm.

Flowering spring (Mar–May). Roadsides; 0–100 m; introduced; Ala., La., Tex.; native to Europe.

A form of *Ranunculus marginatus,* with the fruits weakly tuberculate or almost smooth, similar to the smooth-fruited forms of *R. sardous,* is found in Europe; it is not known from North America.

24. Ranunculus arvensis Linnaeus, Sp. Pl. 1: 555. 1753 [W2]

Ranunculus arvensis var. *tuberculatus* de Candolle

Stems erect or ascending, sparsely pilose. **Basal and lower cauline leaf blades** obovate to rhombic in outline, 3-parted or 3(–5)-foliolate, 1.8–5.2 × 1.6–4.2 cm, leaflets oblanceolate or divided into oblanceolate or linear segments, leaflet base narrowly acuminate, margins entire or distally dentate, apex rounded or acuminate. **Flowers** pedicellate; receptacle sparsely hispid; sepals 5, spreading, 4–7 × 1–2 mm, strigose; petals 5, 5–8 × 2–4 mm. **Heads of achenes** discoid, 8–9 mm across; achenes 5–8 per head, 4–6.4 × 2.8–4.4 mm, faces and margin covered with long spines, glabrous; beak lance-subulate, straight, 1.6–3.8 mm.

Flowering spring (Mar–Jun). Grasslands, ephemeral pools, disturbed areas; 0–1200 m; introduced; Ark., Calif., D.C., Ga., Idaho, Ill., Kans., La., Miss., Mo., N.J., N.Y., N.C., Oreg., Pa., S.C., Tenn., Utah, Wash.; South America; native to Eurasia; Pacific Islands; Australia.

25. Ranunculus hebecarpus Hooker & Arnott, Bot. Beechey Voy., 316. 1838 [F]

Stems erect, pilose. **Basal and lower cauline leaf blades** cordate-reniform, deeply 3-parted, 0.6–2.3 × 1.2–3.5 cm, segments undivided or 2–4-lobed, base shallowly cordate, margins entire or 2–4-dentate, apex of ultimate segments acute. **Flowers** pedicellate; receptacle glabrous; sepals 5, spreading, 1.1–1.8 × 0.5–1 mm, pilose; petals 0–5, 1.3–2 × 0.3–0.7 mm. **Heads of achenes** discoid, 4–6 × 3 mm; achenes 4–9 per head, 1.7–2.3 × 1.7–2 mm, faces and margin papillose, each papilla crowned with hooked bristle, otherwise gla-

brous; beak lanceolate, hooked distally, 0.5–0.7 mm.

Flowering spring (Mar–May). Grasslands, open woodlands, and chaparral; 50–900 m; Calif., Idaho, Oreg., Wash.; Mexico (Baja California).

Ranunculus hebecarpus is native to western North America.

26. Ranunculus parviflorus Linnaeus, Sp. Pl. ed. 2, 1: 780. 1762 [W2]

Stems erect or nearly erect, hispid. **Basal and lower cauline leaf blades** semicircular or reniform, 3-parted or -divided, 1.5–3.2 × 1–2.4 cm, again lobed, base cordate, margins dentate, apex rounded. **Flowers** pedicellate; receptacle glabrous; sepals 5, reflexed, 1.5–2 × 0.8–1.2 mm, densely pubescent; petals 0–5, 1.1–1.8 × 0.2–0.7 mm. **Heads of achenes** globose, 3–5 × 3–5 mm; achenes 10–20 per head, 1.7–2 × 1.4–1.7 mm, faces papillose, each papilla crowned with hooked bristle, otherwise glabrous, margin smooth; beak deltate with slender recurved tip, 0.4–0.6 mm.

Flowering spring–summer (Mar–Jul). Roadsides, fields, and woods; 50–200 m; introduced; Ala., Ark., Calif., Fla., Ga., Ill., Ky., La., Md., Miss., Mo., N.J., N.Y., N.C., Okla., S.C., Tenn., Tex., Va.; native to Europe; Pacific Islands; Australia.

27. Ranunculus platensis A. Sprengel in K. Sprengel, Syst. Veg. 5: 586. 1828

Stems decumbent, sparsely pilose. **Basal and lower cauline leaf blades** reniform, 3-parted, 0.8–1.7 × 1.2–2 cm, segment base cordate, margins dentate or lobulate, apex rounded-obtuse. **Flowers** sessile; receptacle glabrous or pilose; sepals 3, spreading, 1–2 × 0.5–1 mm, hirsute; petals 3, 1–2 × 0.5–0.8 mm. **Heads of achenes** hemispheric, 2–2.5 × 3–3.5 mm; achenes 10–15 per head, 1.3–1.7 × 1–1.3 mm, faces and proximal margin finely papillate, each papilla crowned with hooked bristle, otherwise glabrous; beak semicircular to deltate, curved, 0.2–0.5 mm.

Flowering winter–spring (Feb–Apr). Weed in disturbed areas, usually near coast; 0–200 m; introduced; Ala., Fla., Ga., La., Miss., N.C., Tex.; native to South America (s Brazil, Uruguay, and Argentina).

2a.3. RANUNCULUS Linnaeus (subg. RANUNCULUS) sect. EPIROTES (Prantl) L. D. Benson, Amer. J. Bot. 23: 169. 1936

Ranunculus sect. *Marsypadenium* Prantl e. *Epirotes* Prantl, Bot. Jahrb. Syst. 9: 266. 1888

Plants glabrous or sometimes pilose or pubescent. **Stems** erect, not bulbous-based, without bulbils. **Roots** basal, sometimes tuberous. **Leaves** basal and cauline, simple or compound; basal leaves petiolate, blades variously divided, unlobed to deeply parted, compound, or filiform-dissected, segments undivided or again lobed or parted, margins entire or crenate, never serrate; cauline leaves sessile to nearly sessile or sometimes with much shorter petioles than basal leaves, blades lobed to compound or dissected (rarely unlobed in *R. macauleyi* and *R. glaberrimus*). **Inflorescences** 1–50-flowered cymes. **Flowers** pedicellate; sepals deciduous soon after anthesis, 5; petals yellow (rarely absent in *R. pedatifidus*); nectary scale joined with petal on 3 sides, forming pocket enclosing nectary (sometimes with apex free, forming flap shorter than pocket), glabrous or setose, free margin entire or fringed; style present. **Fruits** achenes, 1-locular; achene body thick-lenticular or asymmetrically thick-lenticular to compressed-globose, 1.2–2 times as wide as thick, not prolonged beyond seed; wall thick, smooth; margin low narrow ridge, often inconspicuous; beak much shorter than achene body.

Species ca. 70 (22 in the flora): worldwide except lowland tropics.

1. Abaxial surface of sepals with dense brown pubescence.
  2. Basal leaf blades narrowly elliptic to lanceolate or oblanceolate; s Rocky Mountains . . . . . . 30. *Ranunculus macauleyi*
  2. Basal leaf blades orbiculate to reniform; Canadian Rocky Mountains and Arctic.
    3. Receptacle brown-pilose; basal leaf blades usually shallowly lobed, or unlobed with crenate margins . . . . .29. *Ranunculus sulphureus*
    3. Receptacle glabrous; basal leaf blades 3-parted . . . . 28. *Ranunculus nivalis*
1. Abaxial surface of sepals glabrous or with colorless hairs.
  4. Basal leaves deciduous before anthesis; nectary scale ciliate, petals 2–3 times as long as wide . . .
    . . . . 39a. *Ranunculus arizonicus* var. *arizonicus*
  4. Basal leaves persistent; nectary scale glabrous, (glabrous or ± pilose in *R. cardiophyllus* and *R. glaberrimus*, which have petals 1–1.5 times as long as wide).
    5. Some or all basal leaf blades unlobed.
      6. All basal leafs undivided, margins entire or with 3 broad shallow rounded teeth; heads of achenes globose, 6–20 mm wide . . . . 49. *Ranunculus glaberrimus*
      6. Basal leaves not as above: either margins crenate to crenate-lobulate, with more than 5 rounded teeth, or innermost basal leaves lobed or divided; heads of achenes usually ovoid to cylindric (sometimes globose and 3–8 mm wide).
        7. Petals 1–3.5 mm.
          8. Stems villous, sometimes sparsely so; receptacle glabrous; base of some roots ± swollen and tuberous, usually 1–2 mm thick; leaf base usually obtuse or truncate . . . . .
          . . . . 44. *Ranunculus micranthus*
          8. Stems glabrous; receptacle pilose, sometimes sparsely so; base of roots never much swollen, 0.2–1.5 mm thick; leaf base ± cordate.
            9. Sepals hispid; achene beak 0.6–1 mm . . . . 46. *Ranunculus allegheniensis*
            9. Sepals glabrous; achene beak 0.1–0.2 mm . . . . 45. *Ranunculus abortivus*
        7. Petals 4–18 mm.
          10. Leaf blades wider than long; e of Great Plains, not in Great Lakes area.
            11. Pedicels glabrous (see also *R. eschscholtzii*, with lobed leaves); se United States . . . . 43a. *Ranunculus harveyi* var. *harveyi*
            11. Pedicels pubescent, sometimes sparsely so; e Canada . . . . 36. *Ranunculus allenii*
          10. Leaf blades at least as long as wide; Great Lakes area and Great Plains w to Great Basin.
            12. Sepals 5–8 × 3–7 mm; nectary scale ciliate, sometimes glabrous; leaf base cordate to broadly obtuse . . . . 40. *Ranunculus cardiophyllus*
            12. Sepals 3–6 × 1.5–3 mm; nectary scale glabrous; leaf base obtuse or acute to rounded.

13. Basal leaf blades ovate, obovate, or orbiculate; heads of achenes cylindric, 7–17 mm; achene beak 0.4–2 mm; Rocky Mountains, Great Basin, and Black Hills . . . . .37. *Ranunculus inamoenus*
13. Basal leaf blades ovate to rhombic; heads of achenes depressed-globose, 4–6 mm; achene beak 0.2–0.3 mm; n Great Plains and eastward. . . . 38. *Ranunculus rhomboideus*
5. All basal leaf blades lobed or parted.
  14. Basal leaf blades dissected into linear segments.
    15. Roots slender, 0.8–1.4 mm thick; achene beak 1.2–1.7 mm . . . . 32. *Ranunculus adoneus*
    15. Roots 2–3 mm thick; achene beak 0.8–1.5 mm . . . . 48. *Ranunculus triternatus*
  14. Basal leaves at most 1×-divided, segments not linear.
    16. Some roots clavate and tuberous, 2.5–5 mm thick; basal leaf blades deeply divided into 3 oblanceolate segments . . . . 47. *Ranunculus jovis*
    16. Roots not clavate or tuberous, 0.1–1.6 mm thick; leaves various but not as above.
      17. Petals 7–15 mm.
        18. Pedicels glabrous . . . . 31. *Ranunculus eschscholtzii*
        18. Pedicels pubescent.
          19. Heads of achenes globose to short-ovoid; basal leaves 3-parted with segments again lobed or parted, margins toothed; Greenland . . . . . . 42. *Ranunculus auricomus*
          19. Heads of achenes cylindric; basal leaves pedately (5–)7(–9)-parted or -divided, segments sometimes again lobed, margins never toothed; throughout n North America . . . . 41a. *Ranunculus pedatifidus* var. *affinis*
      17. Petals 1–8 mm.
        20. Flowering stems 0.6–3.5 cm (sometimes longer in fruit); petals 1–3.5 mm . . . . 34. *Ranunculus pygmaeus*
        20. Flowering stems (1–)4–15(–27) cm; petals 3–8 mm.
          21. Petals 6–16 mm; beak of achene straight, 0.6–1.8 mm; pedicels glabrous . . . . .31. *Ranunculus eschscholtzii*
          21. Petals 4–8 mm; beak of achene straight or curved, 0.3–0.7 mm; pedicels glabrous or pilose.
            22. Base of basal leaves obtuse; petals 5–8 mm; arctic Alaska, Canada, and Greenland . . . . 35. *Ranunculus sabinei*
            22. Base of basal leaves nearly cordate to truncate; petals 4–5 mm; Alaska, Yukon, and Rocky Mountains . . . . 33. *Ranunculus gelidus*

28. Ranunculus nivalis Linnaeus, Sp. Pl. 1: 553. 1753 · Renoncule nivale

Stems erect from short caudices, 4–22 cm, glabrous or sparsely pilose, each with 1 flower. **Roots** slender, 0.4–0.8 mm thick. **Basal leaves** persistent or deciduous, blades reniform, 3-parted, 0.6–2 × 1.3–3 cm, at least lateral segments again lobed or margins toothed, base truncate or cordate, apices of segments rounded-apiculate. **Flowers:** pedicels glabrous or brown-pilose; receptacle glabrous; sepals 6–8 × 3–5 mm, abaxially densely brown-hispid; petals 5(–6), 8–11 × 7–12 mm; nectary scale glabrous. **Heads of achenes** cylindric or ovoid-cylindric, 7–14 × 5–6 mm; achenes 1.5–2.2 × 1.2–1.6 mm, glabrous; beak slender, straight, 1–2 mm. $2n = 48$.

Flowering late spring–summer (Jun–Aug). Wet or dry alpine meadows, often around late snowbeds, cliffs, and streamsides; 0–1300 m; Greenland; Alta., B.C., Nfld., N.W.T., Que., Yukon; Alaska; Eurasia.

29. Ranunculus sulphureus Solander in C. J. Phipps, Voy. North Pole, 202. 1774 · Renoncule soufrée

Ranunculus sulphureus var. *intercedens* Hultén

Stems erect from short caudices, 3–20 cm, sparsely pilose to glabrous, each with 1–3 flowers. **Roots** slender, 0.4–1 mm thick. **Basal leaves** persistent, blades transversely elliptic to orbiculate, 1–3 × 1–3 cm, base obtuse to nearly truncate, margins crenate or else blades shallowly 3-lobed with crenate lateral lobes, apex rounded or rounded-apiculate. **Flowers:** pedicels sparsely brown-pilose; receptacle brown-pilose; sepals 6–8 × 3–6 mm, abaxially densely brown-hispid; petals 5(–6), 8–12 × 6–10 mm; nectary scale glabrous. **Heads of achenes** ovoid-cylindric or ovoid, 6–7(–9) × 5–6 mm; achenes 1.8–2.2 × 1.4–1.8 mm, glabrous or sparsely brown-hispid; beak slender, straight or curved, 0.8–1.4 mm. $2n = 42$, ca. 80, ca. 84, 96, ca. 98.

Flowering late spring–summer (Jun–Sep). Meadows and seepy slopes, often around late snowbeds, bogs, and streamsides; 0–1100 m; Greenland; B.C., Nfld., N.W.T., Que., Yukon; Alaska; Eurasia.

Material of *Ranunculus sulphureus* from the Aleutian Islands has 3-lobed leaves similar to those of *R. nivalis.* These plants are sometimes separated as *R. sulphureus* var. *intercedens.*

30. Ranunculus macauleyi A. Gray, Proc. Amer. Acad. Arts 15: 45. 1879 [E] [F]

Stems erect from short caudices, 6–15 cm, glabrous or sometimes pilose, each with 1–2 flowers. **Roots** slender, 0.7–1.3 mm thick. **Basal leaves** persistent, blades narrowly elliptic to lanceolate or oblanceolate, undivided, 1.5–4.5 × 0.5–1.1(–2.8) cm, base acute or long-attenuate, margins entire except for apex, apex truncate or rounded and 3(–5)-toothed. **Flowers:** pedicels glabrous or brown-pilose; receptacle glabrous; sepals 6–12 × 2.5–8 mm, abaxially densely brown-pilose; petals 5(–8), 10–19 × 6–17 mm; nectary scale glabrous. **Heads of achenes** ovoid or cylindric, 5–10 × 4–5.5 mm; achenes 1.5–1.7 × 1.2–1.3 mm, glabrous; beak slender, straight or recurved, 0.5–1.5(–2.2) mm.

Flowering late spring–summer (Jun–Aug). Sunny open soil of alpine meadows and slopes; 3300–3700 m; Colo., N.Mex.

The type specimen of *Ranunculus macauleyi* var. *brandegeei* L. D. Benson, from the Sangre de Cristo Mountains, Colorado, differs from typical *R. macauleyi* in its tall stem, broad, crenate-laciniate leaves, and sepals with pale or transparent hairs. These characteristics are suggestive of *R. inamoenus,* and the plant may be of hybrid ancestry.

31. Ranunculus eschscholtzii Schlechtendal, Animadv. Bot. Ranunc. Cand. 2: 16, plate 1. 1820

Stems erect or decumbent from short or long caudices, 4–27 cm, glabrous, each with 1–3 flowers. **Roots** slender, 0.4–1.6 mm thick. **Basal leaves** persistent, blades reniform or cordate to obovate or broadly oblong, lobed or 3-parted, 0.5–4.1 × 0.8–3.7 cm, segments again 1(–2)×-lobed, base obtuse to cordate, apices of segments rounded in outline. **Flowers:** pedicels glabrous; receptacle glabrous or sparsely pilose; sepals 4–8 × 2–6 mm, abaxially glabrous or pilose; petals 5–8, 6–16 × 4–16 mm; nectary scale glabrous. **Heads of achenes** cylindric or ovoid, 5–10 × 4–7 mm; achenes 1.4–2 × 1–1.6 mm, glabrous or sparsely pubescent; beak lanceolate or subulate, straight (sometimes curved when immature), 0.6–1.8 mm.

Varieties 6 (5 in the flora): North America; e Siberia.

1. Ultimate segments and sinuses of basal leaves acute or acuminate.
　2. Leaf blade reniform, base truncate or cordate . . . . . . 31b. *Ranunculus eschscholtzii* var. *suksdorfii*
　2. Leaf blade obovate to broadly oblong, base obtuse or rounded. . . . . .31c. *Ranunculus eschscholtzii* var. *eximius*
1. Ultimate segments and sinuses of basal leaves rounded or obtuse (sometimes broadly rounded-acute).
　3. Middle segments of many basal leaves lobed and toothed or 2×-lobed. . . . . 31d. *Ranunculus eschscholtzii* var. *trisectus*
　3. Middle segments of basal leaves unlobed or 1×-lobed.
　　4. Caudices with few or no persistent leaf bases; basal leaves always 3-parted; widespread in w North America. . . . . 31a. *Ranunculus eschscholtzii* var. *eschscholtzii*
　　4. Caudices densely clothed with persistent leaf bases; basal leaves sometimes parted but usually merely lobed; California and w Nevada . . . . 31e. *Ranunculus eschscholtzii* var. *oxynotus*

31a. Ranunculus eschscholtzii Schlechtendal var. **eschscholtzii** [E] [F]

Stems: caudex 1–3 cm, with few or no persistent leaf bases. **Basal leaf blades** reniform or cordate, always 3-parted, 1–2.3 × 1.5–3.7 cm, at least lateral segments again lobed, base truncate or cordate, middle segment unlobed or 1×-lobed, ultimate segments and sinuses rounded-obtuse or broadly rounded-acute. **Flowers:** petals 6–12 mm. 2*n* = 32, 48.

Flowering late spring–summer (Jun–Aug). Open rocky slopes and meadows, usually arctic or alpine; 0–3600 m; Alta., B.C., N.W.T., Yukon; Alaska, Ariz., Calif., Colo., Idaho, Mont., Nev., N.Mex., Oreg., Utah, Wash., Wyo.

31b. Ranunculus eschscholtzii Schlechtendal var. **suksdorfii** (A. Gray) L. D. Benson, Amer. J. Bot. 23: 170. 1936 [E]

Ranunculus suksdorfii A. Gray, Proc. Amer. Acad. Arts 21: 371. 1886

Stems: caudex 1–3 cm, with few or no persistent leaf bases. **Basal leaf blades** reniform, 3-parted, 0.8–2.1 × 1.5–3.5 cm, all segments again lobed, base truncate or cordate, middle segment always 1×-lobed, ultimate seg-

ments and sinuses acute or acuminate. **Flowers:** petals
7–11 mm.

Flowering spring–summer (May–Sep). Open rocky
slopes and meadows; 1200–3200 m; Alta., B.C.;
Calif., Idaho, Mont., Wash., Wyo.

31c. Ranunculus eschscholtzii Schlechtendal var.
eximius (Greene) L. D. Benson, Bull. Torrey Bot. Club
68: 654. 1941 [E]

Ranunculus eximius Greene, Erythea
3: 19. 1895

Stems: caudex 1.5–3.5 cm, with
few or no persistent leaf bases.
Basal leaf blades obovate to
broadly oblong, distally 5–9-
lobed or sometimes 3-parted
with lateral segments again
lobed, 1.4–4.1 × 1.4–3.3 cm,
base obtuse or rounded, middle segment unlobed, ultimate segments and sinuses acute or acuminate. **Flowers:** petals 9–16 mm.

Flowering late spring–summer (Jun–Aug). Alpine
meadows; 2000–3500 m; B.C.; Ariz., Idaho, Mont.,
Utah, Wyo.

31d. Ranunculus eschscholtzii Schlechtendal var.
trisectus (Eastwood ex B. L. Robinson) L. D. Benson,
Amer. J. Bot. 23: 170. 1936 [E]

Ranunculus trisectus Eastwood ex
B. L. Robinson, Proc. Amer. Acad.
Arts 45: 394. 1910

Stems: caudex 2–5 cm, sometimes sheathed with persistent
leaf bases and sometimes with
few or none. **Basal leaf blades**
cordate or semicircular, 3-
parted, 0.8–2.1 × 1.5–3.5 cm,
all segments again 1–2×-lobed or -parted, base truncate or cordate, at least some of middle segments
lobed and toothed or 2×-lobed, ultimate segments
and sinuses rounded to rounded-acute. **Flowers:** petals
7–11 mm.

Flowering summer (Jul–Aug). Open rocky slopes
and meadows; 900–3500 m; Idaho, Oreg., Wyo.

31e. Ranunculus eschscholtzii Schlechtendal var.
oxynotus (A. Gray) Jepson, Fl. Calif. 1: 537. 1922 [E]

Ranunculus oxynotus A. Gray, Proc.
Amer. Acad. Arts 10: 68. 1874

Stems: caudex 2.5–4.5 cm,
densely clothed with persistent
leaf bases. **Basal leaf blades** reniform, shallowly 5–9-lobed or
sometimes 3-parted with lateral
segments again lobed, 1.5–1.5
× 0.8–2 cm, base truncate, mid-

dle segment unlobed, ultimate segments and sinuses
rounded or obtuse. **Flowers:** petals 6–13 mm.

Flowering summer (Jul–Sep). Open rocky alpine
slopes and meadows; 2900–4000 m; Calif., Nev.

32. Ranunculus adoneus A. Gray, Proc. Acad. Nat. Sci.
Philadelphia 15: 56. 1863 [E]

Ranunculus adoneus var. *alpinus*
(S. Watson) L. D. Benson; *R.
eschscholtzii* Schlechtendahl var.
adoneus (A. Gray) C. L. Hitchcock;
R. eschscholtzii var. *alpinus*
(S. Watson) C. L. Hitchcock

Stems erect from large caudices,
9–25 cm, glabrous, each with 1–
3 flowers. **Roots** slender, 0.8–
1.4 mm thick. **Basal leaves** persistent, blades circular
to reniform in outline, 2–3×-dissected into linear segments, 0.9–2.5 × 1.1–2.8 cm, base obtuse, margins
entire, apices of segments narrowly rounded to acute.
Flowers: pedicels glabrous; receptacle glabrous; sepals
4–11 × 3–7 mm, abaxially sparsely pilose, hairs colorless; petals 5–10, 8–15 × 8–19 mm; nectary scale
glabrous. **Heads of achenes** ovoid, 6–12 × 5–9 mm;
achenes 1.8–2.4 × 1–1.4 mm, glabrous or nearly so;
beak subulate, straight, 1.2–1.7 mm. $2n = 16$.

Flowering spring–summer (May–Sep). Alpine and
subalpine meadows, usually around melting snowbanks; 2500–4000 m; Colo., Idaho, Mont., Nev.,
Utah, Wyo.

Most collections of *Ranunculus adoneus* from Colorado, including the type specimen, tend to be small,
with narrow leaf segments (only 0.5–1 mm wide) and
large flowers. The more widespread form, with leaf
segments 1–2 mm wide and more variable flowers, has
been called *R. adoneus* var. *alpinus*. The leaf and
flower characteristics are very poorly correlated, however, and specimens referable to var. *alpinus* vary
greatly in stature and flower size, so the two forms
scarcely merit formal recognition.

33. Ranunculus gelidus Karelin & Kirilov, Bull. Soc. Imp.
Naturalistes Moscou 15: 133. 1842

Ranunculus gelidus subsp. *grayi*
(Britton) Hultén; *R. grayi* Britton; *R.
verecundus* B. L. Robinson ex Piper

Stems erect or decumbent from
short caudices, 3–22 cm, glabrous, each with 1–5 flowers.
Roots slender, 0.5–1 mm thick.
Basal leaves persistent, blades
cordate or reniform, 3-parted,
0.5–1.8 × 0.8–3 cm, segments again lobed, base truncate or nearly cordate, apices of segments rounded.
Flowers: pedicels pubescent or glabrous; receptacle
glabrous or pubescent; sepals 3–5 × 1–4 mm, pubescent or glabrous; petals 5, 3–6 × 1–5 mm; nectary

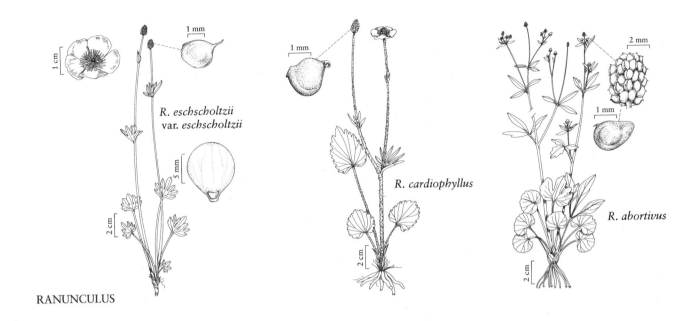

R. *eschscholtzii*
var. *eschscholtzii*

R. *cardiophyllus*

R. *abortivus*

RANUNCULUS

scale glabrous. **Heads of achenes** cylindric to ovoid-cylindric, 4–13 × 4–6 mm; achenes 1.2–2.4 × 0.8–2 mm, glabrous; beak subulate, curved or hooked, 0.4–0.8 mm. $2n = 16$.

Flowering late spring–summer (Jun–Aug). Open arctic and alpine slopes; 0–4000 m; Alta., B.C., N.W.T., Yukon; Alaska, Colo., Idaho, Mont., Oreg., Utah, Wash., Wyo.; Asia.

Plants with small achenes are often separated as *Ranunculus verecundus*. Achene size varies continuously over the range given, however, and it is not correlated with the minor shape difference mentioned by L. D. Benson (1948).

34. Ranunculus pygmaeus Wahlenberg, Fl. Lapp., 157. 1812

Ranunculus pygmaeus var. *langeana* Nathorst

Stems erect or ascending from short caudices, 0.6–3.5 cm (sometimes longer in fruit), each with 1–2 flowers. **Roots** slender, 0.1–0.6 mm thick. **Basal leaves** persistent, blades reniform to transversely elliptic or semicircular, 3-parted or -divided, 0.45–0.9 × 0.6–1.3 cm, at least lateral segments again lobed, base truncate or nearly cordate, margins entire, apex rounded to obtuse. **Flowers:** pedicels glabrous or pubescent; receptacle glabrous; sepals 2–4 × 1.2–1.6 mm, abaxially sparsely hairy, hairs colorless; petals 5, 1.2–3.5 × 1.1–2.8 mm; nectary scale glabrous. **Heads of achenes** nearly globose to cylindric, 2.5–7 × 2.5–5 mm;

achenes 1–1.2 × 0.8–1.1 mm, glabrous; beak subulate, straight or curved, 0.3–0.7 mm. $2n = 16$.

Flowering summer (Jul–Sep). Arctic and alpine meadows and slopes, usually around persistent snow patches; 0–4000 m; Greenland; Alta., B.C., Nfld., N.W.T., Que., Yukon; Alaska, Colo., Idaho, Mont., Utah, Wyo.; Europe (Spitsbergen).

Ranunculus pygmaeus var. *langeana* has been described as having deeply divided basal leaves and strongly elongate heads of achenes (at least 5 mm). These characteristics are not well correlated with one another, however, and the variety does not seem natural.

35. Ranunculus sabinei R. Brown, J. Voy. N.-W. Passage, Bot., 264. 1824 [E]

Ranunculus pygmaeus Wahlenberg subsp. *sabinei* (R. Brown) Hultén

Stems erect or decumbent, 1–12 cm, sparsely pilose, each with 1–3 flowers. **Roots** slender, 0.3–0.8 mm thick. **Basal leaves** persistent, blades broadly obovate to transversely elliptic, 3-lobed or -parted, 0.9–3 × 0.8–3.4 cm, segments undivided or again lobed, base obtuse, margins entire, apices of segments rounded to rounded-obtuse. **Flowers:** pedicels pilose; receptacle pilose; sepals 4–7 × 2–3 mm, abaxially pilose, hairs colorless; petals 5, 5–8 × 3–4 mm; nectary scale glabrous. **Heads of achenes** cylindric, 6–9 × 4 mm; achenes 1.2–1.4 × 0.8–1 mm, glabrous; beak lance-subulate, straight or curved, 0.4–0.6 mm. $2n = 64$.

Flowering summer (Jul–Aug). Slopes and hummocks in tundra, in sandy or gravelly soil; 0 m; Greenland; N.W.T., Yukon; Alaska.

An alternative interpretation of this taxon is given by E. Hultén (1971) who considered it to be the hybrid *Ranunculus nivalis* × *R. pygmaeus,* and considered all specimens referable here to be hybrids or members of stabilized populations of hybrid origin.

36. Ranunculus allenii B. L. Robinson, Rhodora 7: 220. 1905 · Renoncule d'Allen [E]

Stems erect or ascending, 5–19 cm, sparsely pilose, each with 1–4 flowers. **Roots** slender, 0.2–0.8 mm thick. **Basal leaves** persistent, blades reniform or semicircular, undivided or some 3-parted, 1.4–2.1 × 1.7–2.8 cm, base cordate or truncate, margins crenate with more than 5 crenae, apex rounded. **Flowers:** pedicels pilose, sometimes sparsely so; receptacle hispid; sepals 4–6 × 2–3 mm, abaxially pilose, hairs colorless; petals 5, 4–5 × 2–4 mm; nectary scale glabrous. **Heads of achenes** ovoid to cylindric, 4–7 × 4–6 mm; achenes 1.6–1.8 × 1.2–1.4 mm, glabrous; beak lance-subulate, curved, 0.4–0.6 mm. $2n = 32$.

Flowering summer (Jul–Aug). Wet alpine meadows, often around late snowbeds, shores of lakes and streams; 700–1300 m; Nfld., N.W.T., Quebec.

37. Ranunculus inamoenus Greene, Pittonia 3: 91. 1896 [E]

Stems erect, 5–33 cm, pilose or glabrous, each with 3–7 flowers. **Roots** slender, 0.6–1.2 mm thick. **Basal leaves** persistent, blades ovate, obovate or orbiculate, rarely reniform, undivided or innermost with 2 clefts or partings near apex, 1–3.7 × 1.1–3.5 cm, base acute to rounded, margins crenate, apex rounded. **Flowers:** pedicels appressed-pubescent; receptacle pilose or glabrous; sepals 3–5 × 2–3 mm, abaxially pilose, hairs colorless; petals 5, 4–9 × 2–5 mm; nectary scale glabrous. **Heads of achenes** cylindric, 7–17 × 5–8 mm; achenes 1.5–2 × 1.3–1.8 mm, canescent or glabrous; beak subulate, straight or hooked, 0.4–2 mm.

Varieties 2 (2 in the flora): w North America.

The Navajo-Ramah considered *Ranunculus inamoenus* to be an effective hunting medicine, used to protect hunters from their prey (D. E. Moerman 1986).

1. Sepals 3–5 mm; beaks of achenes 0.4–0.9 mm
     . . . . 37a. *Ranunculus inamoenus* var. *inamoenus*
1. Sepals 5–7 mm; beaks of achenes 1.4–2 mm . . .
     . . . . 37b. *Ranunculus inamoenus* var. *subaffinis*

37a. Ranunculus inamoenus Greene var. **inamoenus** [E]

Ranunculus inamoenus var. *alpeophilus* (A. Nelson) L. D. Benson

Flowers: sepals 3–5 mm. **Achene beaks** 0.4–0.9 mm. $2n = 32, 48$.

Flowering late winter–summer (Mar–Aug). Meadows, rocky slopes, occasionally open woods; 2000–3500 m; Alta., B.C., Sask.; Ariz., Colo., Idaho, Mont., Nebr., Nev., N.Mex., S.Dak., Utah, Wash., Wyo.

The type collection of *Ranunculus inamoenus* var. *alpeophilus* is a mixed collection, and some apparent "isotype" material is actually *R. eschscholtzii.*

37b. Ranunculus inamoenus Greene var. **subaffinis** (A. Gray) L. D. Benson, Amer. J. Bot. 27: 187. 1940 [C] [E]

Ranunculus arizonicus Lemmon ex A. Gray var. *subaffinis* A. Gray, Proc. Amer. Acad. Arts 21: 370. 1886

Flowers: sepals 5–7 mm. **Achene beaks** 1.4–2 mm.

Flowering late spring–summer (Jun–Jul). Moist ground near timberline; of conservation concern; 3000–3700 m; Ariz.

Ranunculus inamoenus var. *subaffinis* is endemic to the San Francisco Peaks, north-central Arizona.

38. Ranunculus rhomboideus Goldie, Edinburgh J. Sci. 6: 329. 1822 [E]

Stems erect, 5–22 cm, pilose or occasionally glabrous, each with 3–12 flowers. **Roots** slender, 0.8–1.8 mm thick. **Basal leaves** persistent, blades ovate to rhombic, undivided or rarely innermost 3-parted, 1.1–5.3 × 0.9–3.6 cm, base obtuse, margins crenate with 5 crenae, apex rounded. **Flowers:** pedicels pilose; receptacle pilose; sepals 4–6 × 1.5–3 mm, abaxially pilose, hairs colorless; petals 5, 6–8 × 2–4 mm; nectary scale glabrous. **Heads of achenes** depressed-globose, 4–6 × 5–7 mm; achenes 1.8–2.2 × 1.2–1.8 mm, glabrous; beak slender, curved, 0.2–0.3 mm. $2n = 16$.

Flowering spring (Apr–Jun). Prairies, or occasionally open woods or thickets; 0–900 m; Alta., B.C., Man., N.W.T., Ont., Sask.; Ill., Iowa, Mich., Minn., Mont., Nebr., N.Dak., S.Dak., Wis.

In addition to the range given above, L. D. Benson (1948) cited nineteenth-century specimens from Quebec, New York, Massachusetts, and Rhode Island. No modern specimens have been seen from those areas.

39. Ranunculus arizonicus Lemmon ex A. Gray, Proc. Amer. Acad. Arts 21: 370. 1886

Varieties 2 (1 in the flora): North America; Mexico.

39a. Ranunculus arizonicus Lemmon ex A. Gray var. **arizonicus**

Stems erect, 14–38 cm, sparsely pilose, each with 2–12 flowers. **Roots** cylindric, 1–2.5 mm thick. **Basal leaves** deciduous, all (except sometimes innermost) shed before anthesis, leaving dense brush of fibers, blades semicircular to reniform, outer leaves undivided, inner 5–7-parted, 3.2–4 × 4–6.2 cm, base obtuse to cordate, margins entire or toothed, apices of segments acute to rounded. **Flowers:** pedicels pubescent or glabrous; receptacle canescent; sepals 3–5 × 1–3 mm, abaxially glabrous or sparsely pilose, hairs colorless; petals 5(–11), 7–15 × 2–6 mm; nectary scale ciliate. **Heads of achenes** ovoid or cylindric, 4–10 × 4–6 mm; achenes 1.8–2.5 × 1.5–2 mm, finely and densely canescent; beak subulate, straight, 0.8–1.2 mm, brittle and often broken.

Flowering summer (Jul–Sep). Stream banks; 1500–2400 m; Ariz., N.Mex.; Mexico (Sonora, Chihuahua, and Durango).

40. Ranunculus cardiophyllus Hooker, Fl. Bor.-Amer. 1: 14, plate 15, fig. B. 1829 · Renoncule pédatifide [E] [F]

Ranunculus cardiophyllus var. *coloradensis* L. D. Benson; *R. cardiophyllus* var. *subsagittatus* (A. Gray) L. D. Benson; *R. pedatifidus* Smith var. *cardiophyllus* (Hooker) Britton

Stems erect, 11–53 cm, pilose or glabrous, each with 1–5 flowers. **Roots** cylindric, 1.3–2 mm thick. **Basal leaves** persistent, blades ovate or elliptic, undivided or innermost 3–5-parted, 2.2–6.9 × 1.8–4.5 cm, base cordate to broadly obtuse, margins crenate with more than 5 crenae, apex rounded to broadly acute. **Flowers:** pedicels pilose; receptacle canescent; sepals 5–8 × 3–7 mm, abaxially pilose, hairs colorless; petals (0–)5–10, 6–13 × 4–13 mm; nectary scale ciliate or sometimes glabrous. **Heads of achenes** ovoid or cylindric, 5–16 × 5–9 mm; achenes 1.8–2.2 × 1.5–2 mm, finely canescent; beak subulate, curved or straight, 0.6–1.2 mm. 2*n* = 32.

Flowering spring–summer (May–Sep). Wet or dry meadows; 600–3400 m; Alta., B.C., Sask.; Ariz., Colo., Mont., N.Mex., N.Dak., S.Dak., Utah, Wash., Wyo.

Ranunculus cardiophyllus is quite variable. Through most of its range, leaves always have rounded marginal crenae and cordate or truncate bases, stems are often densely pilose (but may be sparsely pilose or glabrous), and achene beaks are curved. In plants from Arizona and New Mexico, however, leaves may have obtuse marginal crenae or broadly obtuse bases, stems are never densely pilose, and achene beaks are sometimes straight. Forms showing some or all of these charactersistics are often separated as *R. cardiophyllus* var. *subsagittatus*. The characteristics are poorly correlated, however, and taxonomic recognition is not warranted.

Most specimens of *Ranunculus cardiophyllus* have all the basal leaves unlobed, but plants with the innermost basal leaf 3–5-lobed are common. A few specimens, mostly from the northern part of its range, have all the basal leaves 5-parted or -divided. Those plants approach *R. pedatifidus* in their morphology, and *R. cardiophyllus* has sometimes been considered a variety of that species.

41. Ranunculus pedatifidus Smith in A. Rees, Cycl. 29: Ranunculus no. 72. 1815

Varieties 2 (1 in the flora): North America; Asia.

41a. Ranunculus pedatifidus Smith var. **affinis** (R. Brown) L. D. Benson, Amer. Midl. Naturalist 52: 355. 1954 [E]

Ranunculus affinis R. Brown, J. Voy. N.-W. Passage, Bot., 265. 1824; *R. eastwoodianus* L. D. Benson; *R. pedatifidus* subsp. *affinis* (R. Brown) Hultén

Stems erect, 6–33(–46) cm, pilose or glabrous, each with 1–7 flowers. **Roots** slender, 0.4–1.2 mm thick. **Basal leaves** persistent, blades cordate or reniform in outline, pedately (5–)7(–9)-parted or -divided, 0.8–3.8 × 1–4.8 cm, segments undivided or again lobed or parted, base truncate to cordate, margins never toothed, apices of segments acute. **Flowers:** pedicels pilose; receptacle canescent; sepals 4–6 × 3–5 mm, abaxially pilose, hairs colorless; petals (0–)5–10, 7–10 × 5–9 mm; nectary scale glabrous. **Heads of achenes** cylindric, 7–15 × 5–8 mm; achenes 1.8–2.4 × 1.6–1.8 mm, finely canescent or glabrous; beak lanceolate or subulate, curved, 0.5–1 mm. 2*n* = 32, 48.

Flowering late spring–summer (Jun–Aug). Dry rocky places on open arctic and alpine slopes and shores, moist grassland depressions, and open aspen woods; 0–3700 m; Greenland; Alta., B.C., Man., Nfld., N.W.T., Ont., Que., Sask., Yukon; Alaska, Ariz., Colo., Mont., Utah, Wyo.

Specimens of *Ranunculus pedatifidus* from southern Siberia, China, and central Asia, including the type specimen, are relatively small and slender, with all the main divisions of the leaf parted into narrow (1.5–2.5 mm wide) segments. American material, including the type of *R. affinis*, is more robust; the main divisions of the leaf are mostly undivided, and the ultimate segments are broader (2–4 mm). L. D. Benson (1954) referred all material from Asia and islands in the Bering Sea to var. *pedatifidus*. Other authors (P. N. Ovchinnikov 1970; E. Hultén 1968; A. I. Tolmatchew 1971), on the other hand, have limited typical *R. pedatifidus* to material from temperate Asia and have included material from northern Siberia in *R. pedatifidus* subsp. *affinis*. T. G. Tutin and J. R. Akeroyd (1993) included both *R. pedatifidus* and *R. affinis* in *Flora Europaea*, but, as noted below (under *R. auricomus*), Tutin and Akeroyd's descriptions do not match the American plants. Further comparisons of American and Eurasian material are needed.

Glabrous-fruited forms have been called *Ranunculus pedatifidus* var. *leiocarpus* (Trautvetter) Fernald; the variety was first described from Asia.

42. Ranunculus auricomus Linnaeus, Sp. Pl. 1: 551. 1753

Ranunculus auricomus subsp. *boecheri* Fagerström & G. Kvist; *R. auricomus* subsp. *glabratus* (Lynge) Fagerström & G. Kvist; *R. auricomus* subsp. *hartzii* Fagerström & G. Kvist; *R. auricomus* var. *glabratus* Lynge

Stems erect or ascending, 12–30 cm, glabrous, each with 1–4 flowers. **Roots** filiform, 0.2–0.6 mm thick. **Basal leaves** persistent, blades reniform, 3-parted, 1.2–2.8 × 1.6–4.6 cm, segments again lobed or parted, base cordate, margins toothed, apices of segments rounded in outline. **Flowers:** pedicels appressed-pubescent; receptacle finely canescent; sepals 4–7 × 2.5–4 mm, abaxially sparsely pilose, hairs colorless; petals 5, 6–10(–15) × 5–9 mm; nectary scale glabrous. **Heads of achenes** globose to short-ovoid, 5–8 × 5–6 mm; achenes 2–2.2 × 1.8–2 mm, glabrous or sparsely and finely canescent distally; beak lanceolate, weakly to strongly curved, 1.2–2 mm. $2n = 16$.

Flowering summer (Jul–Aug). Moist arctic shrubland or herbland; 100–500 m; Greenland; Eurasia.

Ranunculus auricomus is predominantly apomictic, with irregular meiosis. Different European races may show a variety of characteristics not found in Greenland material (pilose stems, undivided, crenate-dentate leaves, densely canescent achenes, and glabrous receptacles). T. W. Böcher et al. (1968) recognized three distinct races from Greenland, distinguished by minor differences in stature and branching of the plants, width of cauline leaf segments, and indument of achenes. Those races, which were evidently based on very few collections, were named as subspecies by

L. Fagerström and G. Kvist (1983). More ample material collected during the 1980s does not support the Fagerström and Kvist classification. Except for stature and branching, which both depend on the general vigor of the plants, these characteristics are poorly correlated with one another and with geographic place of collection. Cauline leaf segments sometimes vary in width within a single collection, and achenes of the Ymer Island collection that they referred to as *R. auricomus* subsp. *glabratus* are sparsely canescent, not glabrous as stated. These observations, together with the uniformity of indument, leaf shape, and receptacle and achene morphology in the Greenland populations, suggest that they are best considered a single race within the variable apomictic complex comprising *R. auricomus* in the broad sense.

T. G. Tutin and J. R. Akeroyd (1993) treated *Ranunculus auricomus* var. *glabratus* as a synonym of *R. affinis* (= *R. pedatifidus* var. *affinis*; see above). The characteristics in that key and description (leaves palmately 3–5-lobed, achenes in globose heads) do not match American material of *R. pedatifidus* var. *affinis*.

SELECTED REFERENCE L. Fagerström and G. Kvist. 1983. Vier neue arktische und subarktische *Ranunculus auricomus*-Sippen. Ann. Bot. Fenn. 20: 237–243.

43. Ranunculus harveyi (A. Gray) Britton, Mem. Torrey Bot. Club 5: 159. 1894

Ranunculus abortivus Linnaeus var. *harveyi* A. Gray, Proc. Amer. Acad. Arts 21: 372. 1886

Varieties 2 (1 in the flora): North America; Mexico (Baja California).

43a. Ranunculus harveyi (A. Gray) Britton var. harveyi E

Stems erect, 12–40 cm, glabrous or sparsely pilose, each with 3–20 flowers. **Roots** dimorphic, some filiform, 0.2–0.5 mm thick and some with tuberous bases 1–2.5 mm thick. **Basal leaves** persistent, blades reniform, undivided or innermost 3-parted, 1.1–2.8 × 1.5–3.8 cm, base cordate or truncate, margins crenate with more than 5 rounded teeth, apex rounded or rounded-obtuse. **Flowers:** pedicels glabrous; receptacle glabrous or very sparsely pilose; sepals 2–4 × 1.5–3 mm, abaxially glabrous or sparsely pilose, hairs colorless; petals 5–7, 4–6.5 × 1.5–3 mm; nectary scale glabrous. **Heads of achenes** globose or ovoid, 3–6 × 3–5 mm; achenes 1.4–1.7 × 1–1.3 mm, glabrous; beak slender, straight, 0.2–0.6 mm.

Flowering spring (Mar–May). Woods and grasslands; 50–300 m; Ala., Ark., Ill., Mo., Okla., Tenn.

44. Ranunculus micranthus Nuttall in J. Torrey and A. Gray, Fl. N. Amer. 1: 18. 1838 [E] [W2]

Ranunculus micranthus var. *cymbalistes* (Greene) Fernald; *R. micranthus* var. *delitescens* (Greene) Fernald

Stems erect, 11–40 cm, villous, each with 8–35 flowers. **Roots** dimorphic, some filiform, 0.2–0.6 mm thick and some with tuberous bases 1–2 mm thick. **Basal leaves** persistent, blades ovate, orbiculate, or transversely elliptic, outer blades undivided, inner 3-parted or 3-foliolate, 1–3.3 × 1–3 cm, base truncate to broadly obtuse or sometimes weakly cordate, margins crenate, apex rounded-obtuse. **Flowers:** pedicels glabrous or villous; receptacle glabrous; sepals 2–4 × 1–1.5 mm, abaxially glabrous or pubescent, hairs colorless; petals 5, 1.5–3.5 × 0.5–1.5 mm; nectary scale glabrous. **Heads of achenes** globose to cylindric, 3–7 × 2–4 mm; achenes 1.1–1.5 × 1–1.3 mm, glabrous; beak subulate, straight or curved, 0.2–0.3 mm. 2*n* = 16.

Flowering spring (Mar–May). Woods, meadows, and clearings; 0–1000 m; Ala., Ark., Conn., D.C., Ill., Ind., Kans., Ky., Md., Mass., Miss., Mo., N.J., N.Y., N.C., Ohio, Okla., Pa., S.Dak., Tenn., Va., W.Va.

45. Ranunculus abortivus Linnaeus, Sp. Pl. 1: 551. 1753 [E] [F] [W2]

Ranunculus abortivus subsp. *acrolasius* (Fernald) B. M. Kapoor & Á. Löve; *R. abortivus* var. *acrolasius* Fernald; *R. abortivus* var. *eucyclus* Fernald; *R. abortivus* var. *indivisus* Fernald

Stems erect or nearly erect, 10–60 cm, glabrous, each with 3–50 flowers. **Roots** filiform, sometimes enlarged basally, 0.5–1.5 mm thick. **Basal leaves** persistent, blades reniform or orbiculate, undivided or sometimes innermost 3-parted or -foliate, 1.4–4.2 × 2–5.2 cm, base shallowly to deeply cordate, margins crenulate to crenate-lobulate, apex rounded to rounded-obtuse. **Flowers:** pedicels glabrous or nearly so; receptacle sparsely to very sparsely pilose; sepals 2.5–4 × 1–2 mm, abaxially glabrous; petals 5, 1.5–3.5 × 1–2 mm; nectary scale glabrous. **Heads of achenes** ovoid, 3–6 × 2.5–5 mm; achenes 1.4–1.6 × 1–1.5 mm, glabrous; beak subulate, curved, 0.1–0.2 mm. 2*n* = 16.

Flowering late winter–summer (Mar–Jul). Woods, meadows, fallow fields, and clearings; 0–3100 m; St. Pierre and Miquelon; Alta., B.C., Man., N.B., Nfld., N.W.T., N.S., Ont., P.E.I., Que., Sask., Yukon; Ala., Alaska, Ark., Colo., Conn., Del., D.C., Fla., Ga.,

Idaho, Ill., Ind., Iowa, Kans., Ky., La., Maine, Md., Mass., Mich., Minn., Miss., Mo., Mont., Nebr., N.H., N.J., N.Mex., N.Y., N.C., N.Dak., Ohio, Okla., Pa., R.I., S.C., S.Dak., Tenn., Tex., Vt., Va., Wash., W.Va., Wis., Wyo.

Three varieties of *Ranunculus abortivus* are sometimes recognized. Plants from New England and the northern Appalachians often have thick stems and orbiculate leaves with narrow, deep basal sinuses; this form has been called *R. abortivus* var. *eucyclus*. Plants from southeastern Virginia may have the upper bracts merely lobed rather than deeply divided as is usual in *R.* sect. *Epirotes;* those have been called *R. arbortivus* var. *indivisus*.

Native American tribes have used *Ranunculus abortivus* medicinally for a variety of purposes (D. E. Moerman 1986).

46. Ranunculus allegheniensis Britton, Bull. Torrey Bot. Club 22: 224. 1895 [E]

Stems erect or nearly erect, 10–50 cm, glabrous, each with 9–40 flowers. **Roots** slender, 0.2–0.8 mm thick. **Basal leaves** persistent, blades reniform, outer undivided, inner 3-lobed to 3-foliolate, 1–3.5 × 1.5–4.5 cm, base truncate or cordate, margins crenate, apex rounded or obtuse. **Flowers:** pedicels pubescent or glabrous; receptacle sparsely pilose; sepals 2–3 × 1–2 mm, abaxially sparsely hispid, hairs colorless; petals 5, 1–2 × 0.5–1 mm; nectary scale glabrous. **Heads of achenes** globose to ovoid, 3–7 × 3–5 mm; achenes 1.5–2 × 1.4–1.8 mm, glabrous; beak slender, strongly curved, 0.6–1 mm. 2*n* = 16.

Flowering spring–summer (Apr–Jul). Woods and pastures; 0–1100 m; Conn., Ky., Md., Mass., N.Y., N.C., Ohio, Pa., R.I., Tenn., Vt., Va., W.Va.

47. Ranunculus jovis A. Nelson, Bull. Torrey Bot. Club 27: 261. 1900 [E]

Stems erect, 2.5–7.5 cm, glabrous, each with 1–4 flowers. **Roots** tuberous, 2.5–5 mm thick. **Basal leaves** persistent, blades obdeltate in outline, 1–2.8 cm, segments 0.2–0.6 cm wide, deeply divided into 3 oblanceolate segments with lateral segments often again lobed or parted, base long-attenuate, margins entire, apex rounded. **Flowers:** pedicels glabrous; receptacle glabrous or sparsely pilose; sepals 3–7 × 1.5–3 mm, abaxially glabrous; petals 5, 6–12 × 2–5 mm; nectary

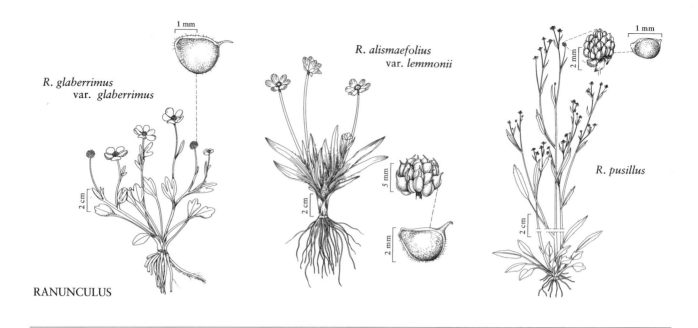

RANUNCULUS

scale glabrous. **Heads of achenes** globose to cylindric, 3.5–8 × 3–7 mm; achenes 1.1–1.4 × 0.8–1.1 mm, finely pubescent; beak subulate, straight, 0.2–0.8 mm.

Flowering spring–summer (Apr–Jul). Dry, open slopes, often around persistent snowbanks; 1700–3000 m; Colo., Idaho, Mont., Nev., Utah, Wyo.

48. Ranunculus triternatus A. Gray, Proc. Amer. Acad. Arts 21: 370. 1886   C E

Ranunculus glaberrimus Hooker var. *reconditus* L. D. Benson

Stems prostrate or ascending, 5–20 cm, glabrous, each with 1–6 flowers. **Roots** cylindric, 2–3 mm thick. **Basal leaves** persistent, blades rhombic to deltate or reniform in outline, 3–4×-dissected, 1.1–3.4 × 2–3.1 cm, segments linear, base obtuse, margins crenate, apices of segments narrowly rounded. **Flowers:** pedicels glabrous; receptacle short-pubescent; sepals 4–8 × 2–5 mm, abaxially glabrous or nearly so; petals 5, 6–15 × 4–10 mm; nectary scale glabrous. **Heads of achenes** depressed-globose, 4–6 × 5–8 mm; achenes 1.8–2.2 × 1.2–1.8 mm, finely pubescent; beak subulate, straight, 0.8–1.5 mm.

Flowering spring (Apr–May). Grassland or sagebrush; of conservation concern; 900–1700 m; Idaho, Nev., Oreg., Wash.

C. L. Hitchcock et al. (1955–1969, vol. 2) considered the name *Ranunculus triternatus* A. Gray to be an illegitimate homonym and used the illegitimate (superfluous) name *R. reconditus* A. Nelson & J. F. Macbride for this species. The name *Ranunculus triterna-tus* Poiret was not validly published (not accepted by Poiret) and does not invalidate *R. triternatus* A. Gray.

49. Ranunculus glaberrimus Hooker, Fl. Bor.-Amer. 1: 12, plate 5, fig. A. 1829   E

Stems prostrate or ascending, 4–15 cm, glabrous, each with 1–4 flowers. **Roots** cylindric, 1–3 mm thick. **Basal leaves** persistent, blades reniform or obovate to very narrowly elliptic, 0.7–5.2 × 1–2 cm, base truncate, obtuse or attenuate, margins entire or with 3 broad, apical crenae, apex rounded to acute. **Flowers:** pedicels glabrous or nearly so; receptacle glabrous; sepals 5–8 × 3–7 mm, abaxially glabrous or sparsely pilose, hairs colorless; petals 5–10, 8–13 × 5–12 mm; nectary scale glabrous or ciliate. **Heads of achenes** globose, 7–12 (–20) × 6–11(–20) mm; achenes 1.4–2.2 × 1.1–1.8 mm, usually finely pubescent; beak subulate or lance-subulate, straight or curved, 0.4–1 mm.

Varieties 2 (2 in the flora): North America.

Usually only a minority of the ovaries develop, and the fruiting receptacle is completely hidden by aborted ovaries. Populations growing at high elevations (*Ranunculus glaberrimus* var. *ellipticus*) and low elevations (var. *glaberrimus*) are usually well differentiated, but these varieties intergrade at intermediate elevations.

The Thompson Indians rubbed the flowers or the whole plant of *Ranunculus glaberrimus* on arrow points as a poison (D. E. Moerman 1986).

1. Basal leaf blades ovate to obovate, usually shallowly lobed; bracts 3-lobed, lobes equal in size; 400–2000 m. . . . .
. . . . 49a. *Ranunculus glaberrimus* var. *glaberrimus*

1. Basal leaf blades elliptic to oblanceolate, usually undivided; bracts 3-lobed, middle lobe much larger; 1200–3600 m . . . . 49b. *Ranunculus glaberrimus* var. *ellipticus*

49a. Ranunculus glaberrimus Hooker var. glaberrimus E F

Basal leaf blades elliptic to oblong or reniform, 0.7–1.9 × 1–1.7 cm, base obtuse to truncate, margins entire to deeply 3-crenate, apex rounded. **Inflorescences:** bracts with lobes equal. $2n = 128$.

Flowering late winter–spring (Mar–Jun). Meadows, open woods, and shrublands; 400–2000 m; B.C.; Calif., Idaho, Mont., Nev., Oreg., S.Dak., Utah, Wash., Wyo.

49b. Ranunculus glaberrimus Hooker var. ellipticus (Greene) Greene, Fl. Francisc. 1: 298. 1891 E

Ranunculus ellipticus Greene, Pittonia 2: 110. 1890; *R. oreogenes* Greene

Basal leaf blades ovate or elliptic to very narrowly elliptic, 1.5–5.2 × 0.7–2 cm, base obtuse to attenuate, margins entire or rarely with 3 broad, shallow, distal crenae, apex acute to rounded. **Inflorescences:** bracts with middle lobe much larger than lateral lobes. $2n = 80$.

Flowering late winter–summer (Mar–Jul). Moist seepy slopes and moist depressions in grassland; 500–3600 m; Alta., B.C., Sask.; Ariz., Calif., Colo., Idaho, Mont., Nebr., Nev., N.Mex., N.Dak., Oreg., S.Dak., Utah, Wash., Wyo.

2a.4. RANUNCULUS Linnaeus (subg. RANUNCULUS) sect. FLAMMULA (Webb) Rouy & Foucaud, Fl. France 1: 82. 1893

Ranunculus subg. *Flammula* Webb in E. Spach, Hist. Nat. Vég. 7: 208. 1838

Plants glabrous or sometimes pilose. **Stems** erect to decumbent or prostrate, not bulbous-based, without bulbils. **Roots** basal and sometimes also nodal, sometimes tuberous-thickened proximally. **Leaves** basal and cauline or all cauline; basal leaves when present similar to proximal cauline leaves; proximal cauline leaves petiolate, blades undivided, margins entire or finely toothed; distal cauline leaves simple, unlobed. **Inflorescences** 1–15(–40)-flowered cymes. **Flowers** pedicellate; sepals deciduous soon after anthesis, 3–5; petals yellow; nectary scale attached on 3 sides, forming pocket enclosing nectary, or sometimes projecting as free flap shorter than pocket, glabrous or ciliate, free margin entire or lobed; style present. **Fruits** achenes, 1-locular; achene body globose-lenticular to globose, 1–2 times as wide as thick, not prolonged beyond seed; achene wall thick, smooth, not ornamented (sometimes pubescent); margins low narrow ridge; beak much shorter than achene body.

Species ca. 25 (10 in the flora): widespread except lowland tropics.

The species of *Ranunculus* sect. *Flammula* are distinctive and easily recognized. Like many amphibious plants, however, they are very variable morphologically, and the species are difficult to define. The taxonomic status of the local endemic species recognized below should be reinvestigated. Depauperate plants of *R. glaberrimus* (*R.* sect. *Epirotes*) may have few or none of the leaves lobed. Such plants were formerly treated under the name *R. oreogenes* and referred to *R.* sect. *Flammula*. In the treatment of *R.* sect. *Flammula* below, they will key to *R. alismifolius* var. *alismellus,* from which they differ in their usually ciliate nectary scales, larger sepals and petals, and larger heads of achenes.

The name *Ranunculus lindheimeri* Engelmann has been applied to specimens belonging to this section. The name has never been validly published, and its correct application is unclear.

1. Petals 1–3, 1.5–2.5 mm.
　2. Bracts elliptic to ovate; achenes 1.4–1.8 mm; sepals 3 . . . . 59a. *Ranunculus bonariensis* var. *trisepalus*
　2. Bracts linear to lanceolate or oblanceolate; achenes 1–1.2 mm; sepals 4–5 . . . . 58. *Ranunculus pusillus*
1. Petals 4–12, 2–14 mm.

3. Bases of roots conspicuously fusiform-thickened.
   4. Stems prostrate, sometimes rooting nodally . . . . 52. *Ranunculus gormanii*
   4. Stems erect or ascending, not rooting nodally.
      5. Bases of basal leaf blades acute . . . . 50. *Ranunculus alismifolius*
      5. Bases of basal leaf blades broadly obtuse to cordate. . . . . 53. *Ranunculus populago*
3. Bases of roots not thickened.
   6. Achenes 0.8–1 mm; se United States . . . . 57. *Ranunculus laxicaulis*
   6. Achenes 1.2–2.8 mm; widespread.
      7. Stems erect or ascending, not rooting nodally.
         8. Roots canescent; petals 4–6 mm . . . . 51. *Ranunculus oresterus*
         8. Roots glabrous; petals 5–14 mm . . . . 50. *Ranunculus alismifolius*
      7. Stems erect or ascending to prostrate, rooting nodally.
         9. Blades of well-developed leaves 5.9–12.2 × 1.1–2.4 cm, acuminate at apex . . . . . 56. *Ranunculus ambigens*
         9. Blades of well-developed leaves not as above, either 0.8–1.9(–2.7) cm long or 0.04–1 cm wide; acute or rounded to filiform at apex.
           10. Leaf blades lanceolate to oblanceolate or filiform, base acute to filiform; beak of achene 0.1–0.6 mm; widespread . . . . 54. *Ranunculus flammula*
           10. Leaf blades ovate to broadly ovate, base rounded to weakly cordate; beak of achene 0.4–1 mm; desert southwest . . . . 55. *Ranunculus hydrocharoides*

50. Ranunculus alismifolius Geyer ex Bentham, Pl. Hartw., 295. 1849 (as **alismaefolius**)

Stems erect or ascending, not rooting nodally, glabrous or hirsute. **Roots** slender or fusiform-thickened basally, glabrous. **Leaves:** basal leaf blades with base acute; proximal cauline leaf blades lanceolate, ovate, or elliptic, 1.8–14.1 × 0.7–2.9 cm, base acuminate to rounded, margins entire or serrulate, apex obtuse to acuminate. **Inflorescences:** bracts lanceolate. **Flowers:** receptacle glabrous; sepals 5, spreading or reflexed from base, 2–6 × 1–4 mm, glabrous or hirsute; petals 5–12, 5–14 × 2–8 mm; nectary scales glabrous. **Heads of achenes** hemispheric to globose, 3–7 × 4–8 mm; achenes 1.6–2.8 × 1.2–2 mm, glabrous or rarely hispid; beak lance-subulate, straight or weakly curved, 0.4–1.2 mm.

Varieties 6 (6 in the flora): North America; Mexico (Baja California).

1. Petals 7–12 on most flowers, 2–3 mm broad . . .
    . . . . 50f. *Ranunculus alismifolius* var. *montanus*
1. Petals 4–6(–8), 2–8 mm broad.
   2. Stems 3.5–8 mm thick; leaf margins serrulate
     . . . . . 50a. *Ranunculus alismifolius* var. *alismifolius*
   2. Stems 1–3 mm thick; leaf margins entire.
     3. Leaf blades ovate or elliptic. . . . .
      . . . . . 50e. *Ranunculus alismifolius* var. *alismellus*
     3. Leaf blades broadly to narrowly lanceolate.
       4. Petals 10–14 mm; California and Nevada. . . . .
       . . . 50c. *Ranunculus alismifolius* var. *lemmonii*
       4. Petals 5–8 mm; widespread.
         5. Roots fusiform-thickened proximally; stems and petioles glabrous or pubescent . . . .
         . . 50d. *Ranunculus alismifolius* var. *davisii*

         5. Roots not or scarcely fusiform-thickened proximally; stems and petioles glabrous. . . . .
         50b. *Ranunculus alismifolius* var. *hartwegii*

50a. Ranunculus alismifolius Geyer ex Bentham var. alismifolius [E]

Stems 20–70 cm × 3.5–8 mm, glabrous. **Roots** not or scarcely fusiform-thickened proximally. **Leaves:** petiole glabrous. **Leaf blade** lanceolate, 5.8–14.1 × 1.2–2.9 cm, base acuminate or sometimes acute, margins serrulate. **Flowers:** petals 4–6, 7–11 × 4–8 mm.

Flowering spring (Apr–Jun). Wet meadows, bogs, shallow water of streams and ponds; 0–1700 m; B.C.; Calif., Idaho, Mont., Oreg., Wash.

50b. Ranunculus alismifolius Geyer ex Bentham var. hartwegii (Greene) Jepson, Fl. Calif. 1: 534. 1922 [E]

Ranunculus hartwegii Greene, Erythea 3: 45. 1895

Stems 13–30 cm × 1–3 mm, glabrous. **Roots** not or scarcely fusiform-thickened proximally. **Leaves:** petiole glabrous. **Leaf blade** lanceolate (usually narrowly so), 3.4–9.5 × 0.8–1.5 cm, base acute to acuminate (or sometimes obtuse), margins entire. **Flowers:** petals 5–6, 6–8 × 3–5 mm.

Flowering spring–summer (May–Aug). Meadows, open slopes and stream banks; 1400–2600 m; Calif., Idaho, Nev., Oreg., Wash., Wyo.

This variety is poorly defined and grades into several other varieties.

50c. Ranunculus alismifolius Geyer ex Bentham var. **lemmonii** (A. Gray) L. D. Benson, Amer. J. Bot. 23: 172. 1936 [E] [F]

Ranunculus lemmonii A. Gray, Proc. Amer. Acad. Arts 10: 68. 1875

Stems 7–35 cm × 1–3 mm, glabrous or pilose. **Roots** often distinctly fusiform-thickened proximally. **Leaves:** petiole glabrous or pilose. **Leaf blade** narrowly lanceolate, 2.7–10.8 × 0.6–1.3 cm, base acuminate, margins entire. **Flowers:** petals 5–6(–8), 10–14 × (3–)5–8 mm.

Flowering spring–summer (May–Jul). Meadows; 1400–2900 m; Calif., Nev.

Ranunculus alismifolius var. *lemmonii* is endemic to the Sierra Nevada.

50d. Ranunculus alismifolius Geyer ex Bentham var. **davisii** L. D. Benson, Amer. Midl. Naturalist 40: 179. 1948 [E]

Stems 13–32 cm × 1–2 mm, glabrous or hirsute. **Roots** fusiform-thickened proximally. **Leaves:** petiole glabrous or strigose. **Leaf blade** broadly to narrowly lanceolate, 3.5–7.2 × 0.5–1.6 cm, base broadly acute, margins entire. **Flowers:** petals 5–6, 5–7 × 3–6 mm.

Flowering late spring–summer (Jun–Aug). Meadows and bogs; 1300–2600 m; Idaho, Mont., Nev., Oreg., Wyo.

50e. Ranunculus alismifolius Geyer ex Bentham var. **alismellus** A. Gray, Proc. Amer. Acad. Arts 7: 327. 1868

Stems 6–21 cm × 1–3 mm, glabrous. **Roots** not or weakly fusiform-thickened proximally. **Leaves:** petiole glabrous. **Leaf blade** ovate or elliptic, 1.8–4.8 × 0.7–1.6 cm, base rounded to broadly acute, margins entire. **Flowers:** petals 5(–8), 5–8 × 2–6 mm.

Flowering spring–summer (May–Aug). Damp meadows, woods, streams; 1400–3600 m; Calif., Idaho, Mont., Nev., Oreg., Wash.; Mexico (Baja Calif.)

50f. Ranunculus alismifolius Geyer ex Bentham var. **montanus** S. Watson, List Pl. Nevada Utah, 7. 1871 [E]

Stems 15–40 cm × 2–7 mm, glabrous. **Roots** ± fusiform-thickened proximally. **Leaves:** petiole glabrous. **Leaf blade** lanceolate to narrowly ovate or elliptic, 2.4–9.7 × 0.9–2.2 cm, base acute to rounded-obtuse, margins entire, rarely serrulate. **Flowers:** petals (5–)7–12, 5–9 × 2–3 mm. $2n = 16$.

Flowering spring–summer (May–Aug). Wet meadows or stream banks; 1800–3500 m; Colo., Idaho, Nev., Utah, Wyo.

51. Ranunculus oresterus L. D. Benson, Amer. J. Bot. 26: 555. 1939 [E]

Stems erect or ascending, not rooting nodally, glabrous. **Roots** not thickened basally, canescent. **Proximal cauline leaf blades** narrowly elliptic to linear, 2.5–4.8 × 0.2–0.9 cm, base narrowly acuminate, margins entire, apex acute to acuminate. **Inflorescences:** bracts lanceolate to linear. **Flowers:** receptacle glabrous; sepals 5, spreading or reflexed from base, 3–5 × 1–2 mm, glabrous; petals 5, 4–6 × 1–2 mm; nectary scales glabrous. **Heads of achenes** hemispheric, 2–4 × 3–5 mm; achenes 1–1.4 × 0.8 mm, glabrous; beak subulate, straight, 0.2–0.4 mm.

Flowering spring (May). Moist meadows; 1200–1700 m; Idaho, Oreg.

52. Ranunculus gormanii Greene, Pittonia 3: 91. 1896 [E]

Stems prostrate, sometimes rooting nodally, glabrous. **Roots** thickened basally, glabrous. **Proximal cauline leaf blades** narrowly to broadly ovate, 1.2–4 × 0.7–2 cm, base rounded, truncate or sometimes obtuse, margins entire or denticulate, apex obtuse or acute. **Inflorescences:** bracts ovate or sometimes lanceolate. **Flowers:** receptacle glabrous; sepals 5, spreading or reflexed from near base, 2–4 × 1–3 mm, glabrous; petals 5–6, 4–6 × 2–4 mm; nectary scales glabrous. **Heads of achenes** hemispheric, 2–3 × 3–4 mm; achenes 1.2–2 × 1.2–1.4 mm, glabrous; beak lanceolate to subulate, straight or curved, 0.6–0.8 mm.

Flowering spring–summer (May–Jul). Damp soil

of meadows and stream banks; 900–3300 m; Calif., Oreg.

Ranunculus gormanii is restricted to middle elevations in the Klamath and southern Cascade Mountains.

53. Ranunculus populago Greene, Erythea 3: 19. 1895 E

Stems erect or ascending, never rooting nodally, glabrous. **Roots** thickened basally, glabrous. **Leaves:** basal leaf blades with base obtuse to cordate; proximal cauline leaf blades semicircular to cordate or ovate, 1.2–5.1 × 1.5–2.9 cm, base cordate to broadly obtuse, margins entire or crenulate, apex broadly acute to rounded. **Inflorescences:** bracts narrowly elliptic to ovate or lanceolate. **Flowers:** receptacle glabrous or hispidulous; sepals 4–5, spreading or reflexed from base, 3–5 × 2–4 mm, glabrous; petals 5–6, 4–9 × 2–5 mm; nectary scales glabrous. **Heads of achenes** hemispheric, 3 × 4–5 mm; achenes 1.6–1.8 × 1.2 mm, glabrous; beak lance-subulate, straight, 0.2–1 mm.

Flowering spring–summer (Apr–Aug). Wet ground and shallow water, in wet meadows, bogs, streams, lakes; 1300–2000 m; Calif., Idaho, Mont., Oreg., Wash.

54. Ranunculus flammula Linnaeus, Sp. Pl. 1: 548. 1753 · Spearwort

Stems erect to prostrate, usually rooting nodally, glabrous or sparsely strigose. **Roots** not thickened basally, glabrous. **Proximal cauline leaf blades** lanceolate to oblanceolate or filiform, 0.7–6.5 × 0.04–1 cm, base acute to filiform, margins entire or serrulate, apex acute to filiform. **Inflorescences:** bracts lanceolate to oblanceolate. **Flowers:** receptacle glabrous; sepals 5, spreading or weakly reflexed, 1.5–4 × 1–2 mm, glabrous or appressed-hispid; petals 5–6, 2.5–7 × 1–4 mm; nectary scales glabrous. **Heads of achenes** globose or hemispheric, 2–4 × 3–4 mm; achenes 1.2–1.6 × 1–1.4 mm, glabrous; beak lanceolate to linear, straight or curved, 0.1–0.6 mm.

Varieties 3 (3 in the flora): North America; Eurasia.

In Eurasia, this taxon is usually treated as two closely related species. *Ranunculus flammula* in the strict sense has relatively stout (0.8–3 mm thick) stems that are erect or ascending from prostrate bases, lanceolate to oblanceolate leaves 3–10 mm broad, sepals 3–4 mm, and petals 5–7 × 3–4 mm. *Ranunculus reptans* has slender (0.2–1 mm thick) stems that are usually prostrate except for the pedicels, leaves linear or filiform, to 2 mm broad, sepals 1–2 mm, and petals 3–5 × 1–2.5 mm. In North America, this distinction

holds up relatively well east of the Great Plains, where plants with the characteristics of *R. flammula* in the strict sense are found in eastern Canada (Newfoundland and northern Nova Scotia) while plants with the characteristics of *R. reptans* are widespread. In the western part of the continent, however, the situation is much less clear. Collections from the Great Plains and Rocky Mountains resemble *R. reptans* in most characters, but they often have broader leaves (up to 5 mm broad). Plants from farther west are very confusing; specimens showing the typical morphology of *R. flammula* in the strict sense and *R. reptans* are found over a wide area, but most specimens from this area combine the characteristics of the two taxa in various ways. For this reason, it is not possible to separate these taxa at the species level. Three varieties are usually recognized, but further study will probably alter the varietal classification (see comments below, under *R. flammula* var. *ovalis*).

1. Stems erect to prostrate; sepals 3–4 mm; petals 5–7 × 3–4 mm. . . . . 54a. *Ranunculus flammula* var. *flammula*
1. Stems prostrate or sometimes ascending; sepals 1–3 mm; petals 3–5 × 1–3 mm.
  2. Leaf blades 0.2–0.8 cm wide . . . . 54b. *Ranunculus flammula* var. *ovalis*
  2. Leaf blades 0.04–0.1 cm wide . . . . 54c. *Ranunculus flammula* var. *reptans*

54a. Ranunculus flammula Linnaeus var. flammula

Ranunculus flammula var. *angustifolius* Wallroth

Stems erect to prostrate, 0.8–3 mm thick. **Leaf blades** lanceolate to oblanceolate, 1.8–4.5 × 0.3–1 cm. **Flowers:** sepals 3–4 mm; petals 5–7 × 3–4 mm. $2n = 32$.

Flowering late spring–summer (Jun–Sep). Shallow water or muddy shores of ponds; 0–1000 m; St. Pierre and Miquelon; B.C., Nfld., N.S.; Calif., Idaho, Oreg., Wash.; Eurasia.

L. D. Benson (1948) reported *Ranunculus flammula* var. *flammula* only from eastern Canada and referred all material from the Pacific slope to *Ranunculus flammula* var. *ovalis*. Benson's treatment is not tenable, however, because some western collections are indistinguishable from the eastern plants.

54b. Ranunculus flammula Linnaeus var. **ovalis** (J. M. Bigelow) L. D. Benson, Bull. Torrey Bot. Club 69: 305. 1942 $\boxed{\text{E}}$

Ranunculus filiformis Michaux var. *ovalis* J. M. Bigelow, Fl. Boston. ed. 2, 224. 1824; *R. flammula* var. *samolifolius* (Greene) L. D. Benson; *R. reptans* Linnaeus var. *ovalis* (J. M. Bigelow) Torrey & A. Gray

Stems prostrate or sometimes ascending, 0.5–2 mm thick. **Leaf blades** lance-elliptic to lanceolate or linear, 0.8–3.3 × 0.2–0.8 cm. **Flowers:** sepals 2–3 mm; petals 3–5 × 2–3 mm. $2n = 32$.

Flowering spring–summer (May–Sep). Muddy ground or shallow water; 0–2900 m; Alta., B.C., Man., N.B., Nfld., N.W.T., N.S., Ont., P.E.I., Que., Sask.; Alaska, Ariz., Calif., Colo., Conn., Idaho, Maine, Mass., Mich., Minn., Mont., Nev., N.H., N.Mex., N.Y., N.Dak., Oreg., Pa., Utah, Vt., Wash., Wyo.

Ranunculus flammula var. *ovalis,* as currently understood, is heterogeneous. Many specimens from throughout the cited range scarcely differ from specimens of *R. flammula* var. *reptans* and perhaps should be included in the latter variety. Material from the Pacific slope, however, may be intermediate between *R. flammula* var. *reptans* and *R. flammula* var. *flammula* or may show various combinations of the distinguishing characteristics of the two. Biosystematic study of *R. flammula* as a whole will be needed for a meaningful treatment of these populations to be possible.

54c. Ranunculus flammula Linnaeus var. **reptans** (Linnaeus) E. Meyer, Pl. Labrador., 96. 1830 · Renoncule rampante

Ranunculus reptans Linnaeus, Sp. Pl. 1: 549. 1753; *R. flammula* var. *filiformis* (Michaux) Hooker

Stems prostrate, 0.2–1 mm thick. **Leaf blades** linear or filiform, 0.7–3 × 0.04–0.1 cm. **Flowers:** sepals 1–2 mm; petals 3–5 × 1–2.5 mm. $2n = 32$.

Flowering late spring–summer (Jun–Sep). Shallow water and wet ground, open shores of lakes and rivers and in temporary ponds; 0–2500 m; Greenland; St. Pierre and Miquelon; Alta., B.C., Man., N.B., Nfld., N.W.T., N.S., Ont., Que., Sask., Yukon; Alaska, Conn., Idaho, Maine, Mass., Mich., Minn., Mont., N.H., N.J., N.Y., Pa., Vt., Wash., Wis., Wyo.; Eurasia.

55. Ranunculus hydrocharoides A. Gray, Mem. Amer. Acad. Arts, ser. 2, 5: 306. 1855

Ranunculus hydrocharoides var. *stolonifer* (Hemsley) L. D. Benson

Stems erect to prostrate, usually rooting nodally, glabrous or strigose. **Roots** not thickened basally, glabrous. **Proximal cauline leaf blades** ovate to broadly ovate, 0.8–2.7 × 0.8–1.9 cm, base rounded to weakly cordate, margins entire or dentate, apex rounded or obtuse. **Inflorescences:** bracts lanceolate to oblanceolate or sometimes ovate. **Flowers:** receptacle glabrous; sepals 5, spreading or reflexed from base, 1.5–3 × 1–2 mm, glabrous; petals 5–6, 3–5 × 1–2 mm; nectary scales glabrous. **Heads of achenes** hemispheric or globose, 2–4 × 3–4 mm; achenes 1.2–1.4 × 1–1.2 mm, glabrous; beak lanceolate to lance-filiform, straight or curved, 0.4–1 mm.

Flowering late spring–summer (Jun–Aug). Wet soil or shallow water, in marshes and edges of streams and lakes; 2000–2900 m; Ariz., Calif., N.Mex.; Mexico; Central America (Guatemala).

56. Ranunculus ambigens S. Watson, Bibl. Index N. Amer. Bot. 1: 16. 1878 $\boxed{\text{E}}$

Stems erect or ascending, rooting at proximal nodes, glabrous or sparsely hirsute. **Roots** not thickened basally, glabrous or somewhat canescent proximally. **Proximal cauline leaf blades** lanceolate, 5.9–12.2 × 1.1–2.4 cm, base rounded-obtuse to acuminate, margins denticulate, apex acuminate. **Inflorescences:** bracts linear or lanceolate. **Flowers:** receptacle glabrous; sepals 5, spreading or sometimes reflexed from base, 3–5 × 2–3 mm, glabrous; petals 5, 5–8 × 2–3 mm; nectary scales glabrous. **Heads of achenes** short-ovoid to depressed-globose, 5–7 × 4–8 mm; achenes 1.8 × 1.2–1.4 mm, glabrous; beak lanceolate, straight, 0.6–1.2 mm.

Flowering late spring–summer (May–Aug). Creeks, ponds, ditches, marshes; 0–700 m; Ala., Conn., Del., D.C., Ga., Ill., Ind., Ky., La., Maine, Md., Mass., Mich., N.H., N.J., N.Y., N.C., Ohio, Pa., R.I., Tenn., Va., W.Va.

The name *Ranunculus obtusiusculus* Rafinesque has been mistakenly used for *R. ambigens.*

57. Ranunculus laxicaulis Darby, Man. Bot. 2: 4. 1841 E

Ranunculus mississippiensis Small; *R. subcordatus* E. O. Beal; *R. texensis* Engelmann ex Engelmann & A. Gray

Stems erect or ascending, often rooting at proximal nodes, glabrous or sparsely pilose. **Roots** not thickened basally, glabrous. **Proximal cauline leaf blades** ovate to lanceolate or narrowly elliptic, 1.5–5.7 × 0.4–2.4 cm, base cordate to acute, margins finely denticulate or entire, apex broadly rounded to acuminate. **Inflorescences:** bracts linear to lanceolate or oblanceolate. **Flowers:** receptacle glabrous; sepals 4–5, spreading or reflexed from base, 2–3 × 1.5–3 mm, glabrous or pubescent; petals 4–6, 2–6 × 1–2 mm; nectary scales glabrous. **Heads of achenes** hemispheric to ovoid, 2–4 × 2–3 mm; achenes 0.8–1 × 0.8 mm, glabrous; beak deciduous, leaving stump 0.1–0.2 mm.

Flowering late winter–summer (Mar–Jul). Around ponds and ditches, in meadows, roadsides, and open woods; 0–100 m; Ala., Ark., Del., Ga., Ill., Ind., Kans., Ky., La., Md., Miss., Mo., N.J., N.C., Okla., S.C., Tenn., Tex., Va.

58. Ranunculus pusillus Poiret in J. Lamarck et al., Encycl. 6: 99. 1804 E F

Ranunculus oblongifolius Elliott; *R. pusillus* var. *angustifolius* (Engelmann ex Engelmann & A. Gray) L. D. Benson; *R. tener* C. Mohr

Stems erect or ascending, rooting at most proximal nodes, glabrous. **Roots** not thickened basally, glabrous. **Proximal cauline leaf blades** ovate or lanceolate, 1.2–4.2 × 0.5–1.2 cm, base acute to truncate, margins entire or denticulate, apex acuminate to rounded. **Inflorescences:** bracts linear to lanceolate or oblanceolate. **Flowers:** receptacle glabrous; sepals 4–5, spreading or reflexed from base, 1.5–3 × 1–1.5 mm, glabrous or sparsely hirsute; petals 1–3, 1.5–2 × 0.5–1 mm; nectary scales glabrous. **Heads of achenes** hemispheric to cylindric, 2–8 × 2–3 mm; achenes 1–1.2 × 0.6–0.8 mm, ± tuberculate, glabrous; beak absent or nearly so, to 0.1 mm.

Flowering spring (Apr–Jun). Ditches, ponds, and swamps; 0–300 m; Ala., Ark., Calif., Del., D.C., Fla., Ga., Ill., Ind., Ky., La., Md., Miss., Mo., N.J., N.Y., N.C., Ohio, Okla., Pa., S.C., Tenn., Tex., Va., W.Va.

In most specimens of *Ranunculus pusillus,* the heads of achenes are hemispheric to short-ovate and only 2–3 mm. Occasional plants with cylindric heads of achenes 4–6 mm from the Gulf Coast states have been called *R. pusillus* var. *angustifolius.*

59. Ranunculus bonariensis Poiret in J. Lamarck et al., Encycl. 6: 102. 1804

Varieties 3 (1 in the flora): North America; temperate South America.

59a. Ranunculus bonariensis Poiret var. **trisepalus** (Gillies ex Hooker & Arnott) Lourteig, Darwiniana 9: 465. 1951

Ranunculus trisepalus Gillies ex Hooker & Arnott, Bot. Misc. 3: 133. 1834; *R. alveolatus* A. M. Carter

Stems erect or decumbent, rooting at proximal nodes, glabrous. **Roots** not thickened basally, glabrous. **Proximal cauline leaf blades** elliptic to ovate, 0.8–2.3 × 0.5–1.2 cm, base rounded or obtuse, margins entire to finely denticulate, apex broadly rounded-acute to rounded. **Inflorescences:** bracts elliptic to ovate. **Flowers:** receptacle glabrous; sepals 3, spreading or reflexed from base, 1.5–3 × 0.5–2 mm, glabrous; petals 1–3, 1.5–2.5 × 0.5–1 mm; nectary scales glabrous. **Heads of achenes** globose to ovoid, 2–5 × 2–4 mm; achenes 1.4–1.8 × 1–1.2 mm, smooth, glabrous; beak absent.

Flowering spring (Mar–May). Vernal pools and edges of streams; 30–1000 m; Calif.; South America.

2a.5. RANUNCULUS Linnaeus (subg. RANUNCULUS) sect. HECATONIA de Candolle in A. P. de Candolle and A. L. P. de Condolle, Prodr. 1: 30. 1824

Plants glabrous or sometimes hirsute. **Stems** creeping (erect in *Ranunculus sceleratus*), not bulbous-based, without bulbils. **Roots** basal and usually also nodal, never tuberous. **Leaves** all cauline or both basal and cauline; basal and lower cauline leaves similar, petiolate, blades lobed to divided or (submerged leaves) deeply dissected, segments undivided or lobed. **Inflorescences** 2–25-flowered cymes or axillary solitary flowers. **Flowers** pedicellate; sepals deciduous soon after anthesis, 3–5; petals 3–14, yellow; nectary scale variable, forming crescent-shaped ridge surrounding but not covering nectary, or funnel-shaped tube surrounding nectary, or distal margin of funnel expanded to form large flap, then nectary sometimes displaced onto flap, glabrous, free margin entire; style present or absent. **Fruits** achenes, 1-locular; achene body thick-lenticular or compressed-ellipsoid to discoid, 1.2–3 times as wide as thick, not prolonged beyond seed; achene wall thick, smooth or with weak transverse wrinkles, glabrous; margin a thick, low or high corky band or ridge; beak much shorter than achene body or sometimes absent.

Species ca. 15 (4 in the flora): widespread, in marshy or aquatic habitats.

1. Stems erect, rooting only at base (very rarely also at proximal nodes) . . . . 61. *Ranunculus sceleratus*
1. Stems prostrate and rooting at nodes, or floating and rootless.
  2. Leaf blades 0.3–1.2 cm, deeply 3-lobed or -parted, terminal segment entire or distally crenulate; nectary-scale low crescent-shaped ridge surrounding nectary; style 0.1–0.2 mm . . . . 60. *Ranunculus hyperboreus*
  2. Leaf blades 0.6–7.3 cm, 3-parted, terminal segment again lobed or dissected; nectary-scale a free flap, nectary on surface of flap; style 0.2–1.2 mm.
    3. Styles 0.2–0.4 mm in flower; achenes 1–1.6 mm, beaks 0.4–0.8 mm; petals 3–7 mm . . . . . 62. *Ranunculus gmelinii*
    3. Styles 0.8–1.2 mm in flower; achenes 1.8–2.2 mm, beaks 1–1.8 mm; petals 7–12 mm . . . 63. *Ranunculus flabellaris*

60. **Ranunculus hyperboreus** Rottbøll, Skr.
Kiøbenhavnske Selsk. Laerd. Elsk. 10: 458. 1770
· Renoncule hyperboréale [F]

Ranunculus hyperboreus subsp. *arnellii* Scheutz; *R. hyperboreus* subsp. *intertextus* (Greene) B. M. Kapoor & Á. Löve; *R. hyperboreus* var. *samojedorum* (Ruprecht) Perfiljev; *R. hyperboreus* var. *tricrenatus* Ruprecht; *R. hyperboreus* var. *turquetilianus* Polunin; *R. natans* C. A. Meyer var. *intertextus* (Greene) L. D. Benson

Stems prostrate, glabrous, rooting nodally. **Leaves:** basal leaves absent, cauline leaf blades reniform to broadly flabellate, deeply 3-lobed or 3-parted, 0.3–1.2 × 0.5–2.1 cm, base obtuse to cordate, lobes undivided or lateral lobes cleft, terminal segment entire or distally crenulate, apex rounded. **Flowers:** receptacle glabrous; sepals 3–4, spreading or reflexed from base, 2–4 × 1–3 mm, glabrous; petals 3–4, 2–4 × 1–3 mm; nectary on petal surface, scale poorly developed and forming crescent-shaped ridge surrounding but not covering nectary; style 0.1–0.2 mm. **Heads of achenes** globose or short-ovoid, 3–5 × 2–5 mm; achenes 1–1.4 × 0.8–1.2 mm, glabrous; beak linear, curved, 0.1–0.4 mm. $2n = 32$.

Flowering late spring–summer (Jun–Aug). Floating in shallow water or stranded on exposed mud at margins of streams and ponds and open wet soil and marshes, in tundra or boreal or subalpine forest; 0–3400 m; Greenland; Alta., B.C., Man., Nfld., N.W.T., Ont., Que., Sask., Yukon; Alaska, Colo., Idaho, Mont., Nev., Utah, Wyo.; Eurasia.

Specimens of *Ranunculus hyperboreus* from the central and southern Rocky Mountains have the leaves always cordate and the fruiting heads always 4–5 mm; they have been separated as *R. hyperboreus* subsp. *intertextus*. Although Arctic specimens are more variable, they often have shallowly cordate leaf bases and equally large heads of achenes, so segregation of the subspecies seems inappropriate.

61. **Ranunculus sceleratus** Linnaeus, Sp. Pl. 1: 551. 1753
· Cursed crowsfoot, renoncule scélérate [W2]

Hecatonia scelerata (Linnaeus) Fourreau

Stems erect, glabrous, rooting at base, only very rarely rooting at proximal nodes. **Leaves** basal and cauline, basal and proximal cauline leaf blades reniform to semicircular in outline, 3-lobed or -parted, 1–5 × 1.6–6.8 cm, base truncate to cordate, segments usually again lobed or parted, sometimes undivided, margins crenate or crenate-lobulate, apex rounded or occasionally obtuse. **Flowers:** receptacle pubescent or glabrous; sepals 3–5, reflexed at or near base, 2–5 × 1–3 mm, glabrous or sparsely hirsute; petals 3–5, 2–5 × 1–3 mm; nectary on petal surface, scale poorly developed and forming crescent-shaped or circular ridge surrounding but not covering nectary; style absent. **Heads of achenes** ellipsoid or cylindric, 5–13 × 3–7 mm; achenes 1–1.2 × 0.8–1 mm, glabrous; beak deltate, usually straight, 0.1 mm.

Varieties 2 (2 in the flora): North America; Eurasia.

Ranunculus sceleratus varieties were used by the Thompson Indians as a poison for their arrow points (D. E. Moerman 1986).

1. Faces of achene with fine transverse wrinkles; leaf blades lobed or parted, segments undivided or lobed, margins crenate . . . . 61a. *Ranunculus sceleratus* var. *sceleratus*
1. Faces of achene smooth; leaf blades always parted (often deeply so), segments lobed or parted, margins deeply crenate or lobulate . . . . 61b. *Ranunculus sceleratus* var. *multifidus*

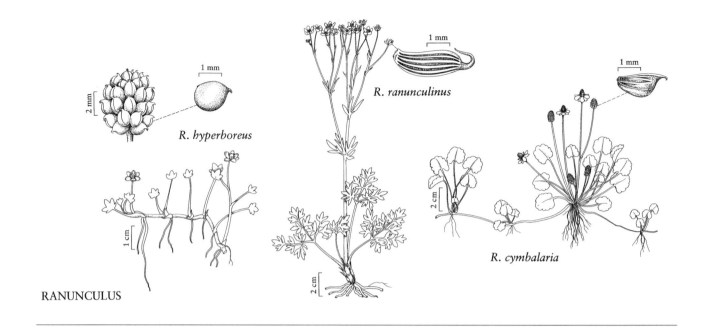

R. ranunculinus

R. hyperboreus

R. cymbalaria

RANUNCULUS

61a. Ranunculus sceleratus Linnaeus var. **sceleratus**

Leaf blades lobed or parted, segments undivided or lobed, margins crenate. **Achene** faces with fine transverse wrinkles. $2n = 32$.

Flowering late winter–summer (Mar–Sep). Ponds, ditches and riverbanks; 0–700 m; St. Pierre and Miquelon; Alta., B.C., Man., N.B., Nfld., N.S., Ont., Que., Sask.; Ala., Ark., Calif., Conn., Del., D.C., Fla., Ga., Ill., Ind., Iowa, Kans., Ky., La., Maine, Md., Mass., Mich., Minn., Miss., Mo., Nebr., N.H., N.J., N.Y., N.C., N.Dak., Ohio, Okla., Oreg., Pa., R.I., S.C., S.Dak., Tenn., Tex., Vt., Va., Wash., W.Va., Wis.; Europe.

Ranunculus sceleratus var. *sceleratus* is a serious weed of watercourses and marshy fields. It is a naturalized weed in western North America; it is not clear whether it is native in the eastern part of the continent or was introduced from Europe.

61b. Ranunculus sceleratus Linnaeus var. **multifidus** Nuttall in J. Torrey and A. Gray, Fl. N. Amer. 1: 19. 1838 [E]

Ranunculus sceleratus subsp. *multifidus* (Nuttall) Hultén

Leaf blades always parted (often deeply so), segments lobed or parted, margins deeply crenate or lobulate. **Achene** faces smooth. $2n = 64$.

Flowering spring–summer (Mar–Sep). Wet ground or shallow water, in bogs and ponds, on shores of lakes and rivers; 0–2600 m; Alta., B.C., Man., N.W.T., Ont., Que., Sask., Yukon; Alaska, Ariz., Calif., Colo., Idaho, Iowa, Minn., Mont., Nebr., Nev., N.Mex., N.Dak., Oreg., S.Dak., Utah, Wash., Wyo.

Reports of *Ranunculus sceleratus* var. *multifidus* for eastern Quebec and New Brunswick are errors for *R. sceleratus* var. *sceleratus*.

62. Ranunculus gmelinii de Candolle, Syst. Nat. 1: 303. 1817 · Renoncule de Gmelin

Ranunculus gmelinii subsp. *purshii* (Richardson) Hultén; *R. gmelinii* var. *hookeri* (D. Don) L. D. Benson; *R. gmelinii* var. *limosus* (Nuttall) H. Hara; *R. gmelinii* var. *prolificus* (Fernald) H. Hara; *R. purshii* Richardson

Stems prostrate or sometimes floating, glabrous or hirsute, rooting nodally. **Leaves:** basal leaves absent, cauline leaf blades reniform to circular, 3-parted, 0.6–6.5 × 1.1–9 cm, base cordate, segments again 1–3×-lobed to -dissected, margins entire or crenate, apex rounded to filiform. **Flowers:** receptacle sparsely hispid; sepals 4–5, spreading or reflexed from base, 2–5 × 2–4 mm, glabrous or sparsely pilose; petals 4–14, 3–7 × 2–5 mm; nectary scale variable, crescent-shaped, funnel-shaped, or flaplike; style 0.2–0.4 mm. **Heads of achenes** globose or ovoid, 3–8 × 3–7 mm; achenes 1–1.6 × 1–1.2 mm, glabrous; beak narrowly lanceolate or filiform, 0.4–0.8 mm. $2n = 16, 32, 64$.

Flowering spring–summer (May–Sep). Shallow water or drying mud, wet meadows, swamps, marshes, ponds, shores of rivers; 0–2800 m; Alta., B.C., Man., N.B., Nfld., N.W.T., N.S., Ont., P.E.I., Que., Sask., Yukon; Alaska, Colo., Idaho, Ill., Iowa, Maine, Mich.,

Minn., Mont., Nev., N.Mex., N.Dak., Oreg., Utah, Wash., Wis., Wyo.; Eurasia.

Ranunculus gmelinii has been divided into varieties on the basis of indument and flower size. These characters are variable and poorly correlated with one another, however, and these varieties scarcely seem natural.

63. Ranunculus flabellaris Rafinesque, Amer. Monthly Mag. & Crit. Rev. 2: 344. 1818 · Renoncule à évantails E W1

Ranunculus delphiniifolius Torrey

Stems floating or prostrate, glabrous, rooting at proximal nodes. **Leaves:** basal leaves seldom present, cauline leaf blades semicircular to reniform, 1–6× -lobed, parted, or dissected 1.2– 7.3 × 1.9–10.8 cm, base truncate or cordate, segment margins entire or crenate, apex rounded to filiform. **Flowers:** receptacle sparsely hispid; sepals 5, spreading or weakly reflexed, 5–7 × 3–6 mm, glabrous; petals 5– 6(–14), 7–12 × 5–9 mm; nectary scale variable, crescent-shaped, funnel-shaped, or flaplike; style 0.8– 1.2 mm. **Heads of achenes** ovoid, 8–10 × 7–8 mm; achenes 1.8–2.2 × 1.6–2.2 mm, glabrous; beak lanceolate, straight, 1–1.8 mm. $2n = 32$.

Flowering late spring–summer (May–Aug). Shallow water or drying mud; 0–1500 m; Alta., B.C., Man., N.B., Ont., Que.; Ala., Ark., Calif., Conn., Del., Idaho, Ill., Ind., Iowa, Kans., Ky., La., Maine, Md., Mass., Mich., Minn., Miss., Mo., Nebr., Nev., N.H., N.J., N.Y., N.C., N.Dak., Ohio, Okla., Oreg., Pa., R.I., S.Dak., Tenn., Tex., Utah, Vt., Va., Wash., Wis., Wyo.

Many specimens of *Ranunculus flabellaris* from North Dakota and Manitoba have small flowers and fruit, and they have been considered transitional to *R. gmelinii* by some authors (L. D. Benson 1948; B. Boivin 1967–79, part 2; G. E. Larson 1986). Boivin also refers the Alberta collection to *R. gmelinii*, but the measurements he gives are well within the normal range of variation of *R. flabellaris*.

The Fox tribes used *Ranunculus flabellaris* as a cold remedy and a respiratory aid (D. E. Moerman 1986).

2b. Ranunculus Linnaeus subg. Cyrtorhyncha (Nuttall) A. Gray in A. Gray et al., Syn. Fl. N. Amer. 1: 23. 1895

Cyrtorhyncha Nuttall in J. Torrey and A. Gray, Fl. N. Amer. 1: 26. 1838

2b.1. Ranunculus Linnaeus (subg. Cyrtorhyncha) sect. Cyrtorhyncha

Plants glabrous. **Stems** erect, not bulbous-based, without bulbils. **Roots** basal, never tuberous. **Leaves** basal and cauline; basal leaves petiolate, 2×-compound, leaflets again parted; cauline leaves sessile or nearly so, smaller, otherwise similar. **Inflorescences** 3–20-flowered, compound cymes. **Flowers** pedicellate; sepals deciduous soon after anthesis, 5; petals yellow, rarely absent; nectary scale rudimentary, attached basally, forming ridge below nectary, glabrous, free margin entire; style present. **Fruits** achenes, 1-locular; achene body cylindric, about as wide as thick, not prolonged beyond seed; achene wall thin, papery, not loose, longitudinally striate; margin an inconspicuous ridge similar to veins; beak much shorter than achene body.

Species 1 (1 in the flora): North America (Rocky Mountains).

64. Ranunculus ranunculinus (Nuttall) Rydberg, Bot. Surv. Nebraska 3: 23. 1894 E F

Cyrtorhyncha ranunculina Nuttall in J. Torrey and A. Gray, Fl. N. Amer. 1: 26. 1838

Stems erect from short caudices, not rooting nodally, glabrous, not bulbous-based. **Tuberous roots** absent. **Leaves:** basal leaf blades ovate to semicircular in outline, ternately or pinnately 2×-compound, 2.4–8.2 × 2–8.5 cm, leaflets parted and again lobed, ultimate segments elliptic to linear, margins entire (or occasionally a lobe reduced to tooth), apex acuminate to rounded. **Flowers:** receptacle glabrous; sepals spreading or reflexed from base, 3–6 × 1–3 mm, glabrous; petals (0–)5–6, yellow, 3–8 × 1–3 mm. **Heads of achenes** hemispheric to globose, 4–5 × 6–7 mm; achenes 2.2–3.6 × 1.2–1.8 mm, glabrous; beak filiform, strongly reflexed from base, 0.8– 1.5 mm, brittle, often broken. $2n = 32$.

Flowering spring–summer (Apr–Aug). Open grassy or brushy slopes; 1700–2600 m; Colo., N.Mex., Utah, Wyo.

2b.2. RANUNCULUS Linnaeus (subg. CYRTORHYNCHA) sect. HALODES (A. Gray) L. D. Benson, Amer. J. Bot. 27: 805. 1940

Ranunculus sect. *Halodes* A. Gray, Proc. Amer. Acad. Arts 21: 366. 1886

Plants glabrous or sparsely hirsute. **Stems** dimorphic, with erect flowering stems and prostrate stolons, not bulbous-based, without bulbils. **Roots** basal and nodal, never tuberous. **Leaves** basal and cauline; basal, proximal cauline, and stolon leaves similar, petiolate, base broadly rounded to truncate, margins crenate or crenate-serrate; distal cauline leaves much reduced, scalelike, simple and undivided. **Inflorescences** 1–5-flowered cymes. **Flowers** pedicellate; sepals deciduous soon after anthesis, 5; petals 5, yellow; nectary scale attached basally, forming flap over nectary, glabrous, free margin entire; style present. **Fruits** achenes, 1-locular; achene body discoid to flattened-lenticular, oblong to obovate in outline, 2–4 times as wide as thick, not prolonged beyond seed; achene wall papery, not loose, longitudinally ribbed, glabrous; margin prominent narrow ridge; beak much shorter than achene body.

Species ca. 10 (1 in the flora): North America, Mexico, South America, Eurasia.

65. Ranunculus cymbalaria Pursh, Fl. Amer. Sept. 2: 392. 1814 · Renoncule cymbalaire [F] [W1]

Halerpestes cymbalaria (Pursh) Greene; *Ranunculus cymbalaria* var. *alpinus* Hooker; *R. cymbalaria* var. *saximontanus* Fernald

Stems dimorphic, flowering stems erect or ascending, stolons prostrate, rooting nodally, glabrous or sparsely hirsute, not bulbous-based. **Tuberous roots** absent. **Basal leaves** simple and undivided, blades oblong to cordate or circular, 0.7–3.8 × 0.8–3.2 cm, base rounded to cordate, margins crenate or crenate-serrate, apex rounded. **Flowers:** receptacle hispid or glabrous; sepals spreading, 2.5–6 × 1.5–3 mm, glabrous; petals 5, yellow, 2–7 × 1–3 mm. **Heads of** achenes long-ovoid or cylindric, 6–12 × 4–5(–9) mm; achenes 1–1.4(–2.2) × 0.8–1.2 mm, glabrous; beak persistent, conic, straight, 0.1–0.2 mm. $2n = 16$.

Flowering late spring–summer (May–Sep). Bogs, marshes, ditches, stream banks, often saline, 0–3200 m; Greenland; St. Pierre and Miquelon; Alta., B.C., Man., N.B., Nfld., N.W.T., N.S., Ont., P.E.I., Que., Sask., Yukon; Alaska, Ariz., Calif., Colo., Conn., Idaho, Ill., Iowa, Kans., Maine, Mass., Mich., Minn., Mo., Mont., Nebr., Nev., N.H., N.J., N.Mex., N.Y., N.Dak., Okla., Oreg., R.I., S.Dak., Tex., Utah, Wash., Wis., Wyo.; Mexico; South America; Eurasia.

Various Navajo groups used *Ranunculus cymbalaria* as a treatment for venereal disease, an emetic, and a ceremonial medicine. The Kawaiisu used it as a dermatological aid (D. E. Moerman 1986).

2b.3. RANUNCULUS Linnaeus (subg. CYRTORHYNCHA) sect. ARCTERANTHIS (Greene) L. D. Benson, Amer. J. Bot. 23: 174. 1936

Arcteranthis Greene, Pittonia 3: 190. 1897

Plants glabrous. **Stems** erect, not bulbous-based, without bulbils. **Roots** basal, never tuberous. **Leaves** all basal or sometimes with single, scalelike, cauline leaf; basal leaves petiolate, blades 3–5-parted, segments again lobed and crenate; cauline leaf (when present) petiolate or sessile, blade small, palmately or pedately 3–9-lobed. **Inflorescences** terminal, solitary flowers. **Flowers:** sepals deciduous soon after anthesis, 5; petals yellow, showy; nectary scale a crescentic ridge surrounding nectary, glabrous, free margin entire; style present. **Fruits** achenes, 1-locular; achene body cylindric, 1.4–2 times as wide as thick, not prolonged beyond seed; wall thin, papery, loose, compressed, at maturity forming narrow ventral wing, longitudinally veined; margin an inconspicuous ridge similar to veins; beak 1–1.8 mm, much shorter than achene body.

Species 1 (1 in the flora): nw North America.

66. Ranunculus cooleyae Vasey & Rose ex Rose, Contr. U.S. Natl. Herb. 1: 289, plate 22. 1893 [E][F]

Arcteranthis cooleyae (Rose) Greene; *Kumlienia cooleyae* (Rose) Greene

Stems erect from short caudices, not rooting nodally, glabrous, not bulbous-based. **Tuberous roots** absent. **Basal leaf blades** circular to reniform in outline, 3–5-parted, 0.8–3.8 × 1.7–6.9 cm, segments again lobed, ultimate segments elliptic or oblong, margins crenate, apex rounded; cauline leaf 0–1, scalelike. **Flowers:** receptacle glabrous; sepals spreading, yellow, 7–11 × 4–7 mm, glabrous; petals 11–15, yellow, 8–12 × 3–5 mm. **Heads of achenes** hemispheric or spheric, 7–8 × 9–10 mm; achenes 2.4–4.6 × 1.2–2.2 mm, glabrous; beak persistent, filiform, hooked distally, 1–1.8 mm. $2n = 16$.

Flowering late spring–summer (Jun–Aug). Slopes near persistent snowbanks; 500–1800 m; B.C.; Alaska, Wash.

2b.4. RANUNCULUS Linnaeus (subg. CYRTORHYNCHA) sect. PSEUDAPHANOSTEMMA (A. Gray) L. D. Benson, Amer. J. Bot. 23: 174. 1936

Ranunculus sect. *Pseudaphanostemma* A. Gray, Proc. Amer. Acad. Arts 21: 365. 1886; *Kumlienia* Greene

Plants glabrous. **Stems** erect, stolons absent, not bulbous-based, without bulbils. **Roots** basal, never tuberous. **Leaves** all basal or sometimes with 1–2 cauline leaves; basal leaves petiolate, blades shallowly lobed; cauline leaves (when present) sessile or petiolate, blades small, scalelike, rarely shallowly 3-lobed, margins crenate. **Inflorescences** 1–2-flowered cymes. **Flowers** pedicellate; sepals deciduous soon after anthesis, 5; petals 8–12, yellow or greenish; nectary scale crescentic ridge surrounding nectary, glabrous, free margin entire; style present. **Fruits** achenes, 1-locular; achene body cylindric, about as wide as thick, not prolonged beyond seed; wall papery, not loose, longitudinally veined; margin an inconspicuous ridge similar to veins; beak much shorter than achene body.

Species 1 (1 in the flora): w North America (California).

67. Ranunculus hystriculus A. Gray, Proc. Amer. Acad. Arts 7: 328. 1867 [E][F]

Kumlienia hystriculus (A. Gray) Greene

Stems erect from short caudices, not rooting nodally, glabrous, not bulbous-based. **Tuberous roots** absent. **Leaves:** basal leaf blades semicircular or reniform in outline, shallowly 5–7-lobed, 1.2–4.6 × 1.8–6.6 cm, ultimate segments semicircular, margins crenate, apex rounded or weakly apiculate; cauline leaves 0–2, scalelike. **Flowers:** receptacle glabrous; sepals spreading, white or pale yellow, 6–13 × 3–6 mm, glabrous; petals 8–12, greenish, 2–4 × 0.6–1.6 mm. **Heads of achenes** ovoid, 6–7 × 6–8 mm; achenes 3.8–4.2 × 0.8–1 mm, canescent; beak persistent, filiform, 1.2–1.4 mm, hooked distally.

Flowering winter–summer (Feb–Jul). Wet places near streams, especially around waterfalls; 300–2300 m; Calif.

Ranunculus hystriculus is endemic to the west slope of the Sierra Nevada.

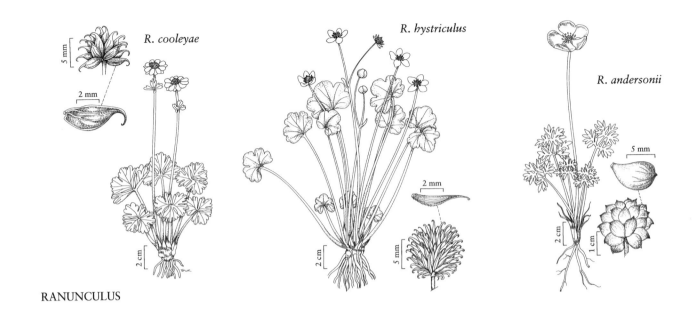

RANUNCULUS

2c. RANUNCULUS Linnaeus subg. CRYMODES (A. Gray) A. Gray in A. Gray et al., Syn. Fl. N. Amer. 1: 22. 1895

Ranunculus sect. *Crymodes* A. Gray, Proc. Amer. Acad. Arts 21: 365. 1886; *Beckwithia* Jepson

Plants distally glabrous or brown-pilose. **Stems** erect, not bulbous-based, without bulbils. **Roots** basal, never tuberous. **Leaves** all basal or basal and cauline, compound or simple and deeply parted; basal leaves petiolate, leaflets parted; cauline leaves (when present) sessile or nearly so, much smaller, otherwise similar to basal leaves. **Inflorescences** 1–3-flowered cymes. **Flowers** pedicellate; sepals persistent in fruit, 5; petals 5, red at maturity, sometimes white when young; nectary scale attached on 3 sides, forming pocket that may or may not cover nectary, glabrous, free margin entire; style present. **Fruits** utricles or winged achenes, 1-locular; fruit body obovoid to flattened-obovoid, 1–2.5 times as wide as thick, not prolonged beyond seed; fruit wall membranous, loose and sometimes inflated, smooth or veined; margin an inconspicuous vein or thin wing; beak much shorter than fruit body.

Species 4 (2 in the flora): North America, Eurasia.

1. Sepals brown-pilose; fruit wall smooth, loose but not inflated, winged, fruits thus winged achenes; arctic . . . . 68. *Ranunculus glacialis*
1. Sepals glabrous; fruit wall veined, inflated, not winged, fruits thus utricles; Great Basin . . . . . . 69. *Ranunculus andersonii*

68. Ranunculus glacialis Linnaeus, Sp. Pl. 1: 553. 1753

Beckwithia glacialis (Linnaeus) Á. Löve & D. Löve

Stems erect or ascending from caudices, not rooting nodally, distally glabrous or brown-pilose, not bulbous-based. **Tuberous roots** absent. **Basal leaf blades** reniform to broadly triangular in outline, 3-foliolate or deeply 3-parted, 1–4 × 1.6–4.4 cm, leaflets or segments 1–2×-parted, ultimate segments elliptic to oblanceolate or almost linear, margins entire or occasionally with 1–2 teeth, apex rounded to obtuse. **Flow**ers: receptacle glabrous or brown-pilose; sepals spreading, 7–12 × 4–9 mm, brown-pilose; petals initially white, usually becoming red with age, 9–15 × 7–14 (–18) mm. **Fruiting heads** hemispheric, 5–8 × 7–16 mm; fruit wall smooth, not veined, loose but not inflated, winged along suture, fruits winged achenes; achenes 2.6–3 × 1.4–2 mm, glabrous; beak persistent, lanceolate, 0.8–2.3 mm.

Varieties 2 (2 in the flora): North America; Eurasia.

1. Stems glabrous or with few hairs at nodes; caudices well-developed; basal leaves compound, ultimate segments elliptic to oblanceolate; receptacle glabrous; Greenland . . . .
. . . . 68a. *Ranunculus glacialis* var. *glacialis*
1. Stems distally brown-pilose; caudices short; basal leaves deeply parted or compound, ultimate segments oblanceolate to almost linear; receptacle brown-pilose; Alaska. . . . .
. . . . 68b. *Ranunculus glacialis* var. *camissonis*

68a. Ranunculus glacialis Linnaeus var. **glacialis**

Stems glabrous or with few hairs at nodes, from well-developed caudices. **Basal leaves** compound, leaflets parted and lobed, ultimate segments elliptic to oblanceolate. **Flowers:** receptacle glabrous. $2n = 16$.

Flowering summer (Jun–Aug). Open, often rocky, places, in meltwater streams and margins of rivers; 0–2100 m; Greenland; Europe.

68b. Ranunculus glacialis Linnaeus var. **camissonis** (Schlechtendal) L. D. Benson, Amer. Midl. Naturalist 40: 226. 1948

Ranunculus camissonis Schlechtendal, Animadv. Bot. Ranunc. Cand. 1: 12. 1819; *Beckwithia camissonis* (Schlechtendahl) Tolmatchew; *Ranunculus glacialis* subsp. *camissonis* (Schlechtendal) Hultén

Stems distally brown-pilose, from short caudices. **Basal leaf blades** deeply 3-parted or 3-foliolate, segments or leaflets 1–2×-parted, ultimate segments oblanceolate to almost linear. **Flowers:** receptacle brown-pilose.

Flowering summer (Jun–Aug). Wet meadows or marshlands; 0–1200 m; Alaska; Siberia.

The incorrect spelling "*chamissonis*" is frequently seen.

69. Ranunculus andersonii A. Gray, Proc. Amer. Acad. Arts 7: 327. 1867   E F

Beckwithia andersonii (A. Gray) Jepson; *Ranunculus andersonii* var. *juniperinus* (M. E. Jones) S. L. Welsh; *R. andersonii* var. *tenellus* S. Watson; *R. juniperinus* M. E. Jones

Stems erect from short caudices, not rooting nodally, glabrous, not bulbous-based. **Tuberous roots** absent. **Basal leaf blades** cordate in outline, ternately 1–2×-compound, 1.5–3.8 × 2.1–3.8 cm, leaflets 2–3×-parted, ultimate segments elliptic to linear, margins entire or with occasional teeth, apex obtuse to acuminate. **Flowers:** receptacle hispid; sepals spreading, 9–15 × 5–9 mm, glabrous; petals pinkish white, 12–18 × 9–13 mm. **Fruiting heads** globose or depressed-globose, 13–27 × 21–29 mm; fruit wall veined, inflated, not winged, fruits thus utricles; utricles 6–12 × 4–6 mm, glabrous; beak persistent, deltate or subulate from deltate base, 0.2–0.6 mm.

Flowering spring (Apr–May). Slopes in sagebrush or pinyon-juniper woodland; 900–2300 m; Ariz., Calif., Idaho, Nev., Oreg., Utah.

2d. RANUNCULUS Linnaeus subg. OXYGRAPHIS (Bunge) L. D. Benson, Amer. J. Bot. 27: 806. 1940

Oxygraphis Bunge, Verz. Altai Pfl., 46. 1836

Plants glabrous. **Stems** erect from short caudices or rhizomes, stolons absent, not bulbous-based, without bulbils. **Roots** from rhizome only, never tuberous. **Leaves** all basal, simple, petiolate; blades undivided, unlobed, margins entire or serrulate. **Inflorescences** terminal, solitary flowers. **Flowers:** sepals persistent in fruit, 5; petals 9–18, yellow, 7–12 mm; nectary scale attached basally, forming low ridge not covering nectary, glabrous, free margin entire; style present. **Fruits** achenes, 1-locular; achene body ellipsoid, about as wide as thick, not prolonged beyond seed; wall membranous, loose and ± inflated, with 1–2 longitudinal veins on each face; margin narrow ridge; beak 0.8–1.4 mm, shorter than achene body.

Species 5 (1 in the flora): North America, Asia.

70. Ranunculus kamtschaticus de Candolle, Syst. Nat. 1: 302. 1817 [F]

Oxygraphis glacialis (de Candolle) Bunge

Stems erect from short caudices or rhizomes, not rooting nodally, glabrous, not bulbous-based. **Tuberous roots** absent. **Basal leaf blades** circular to ovate or obovate, undivided, 0.8–2.6 × 0.6–1.4 cm, margins entire or serrulate, apex obtuse or rounded. **Flowers:** receptacle glabrous; sepals spreading, 4–8 × 2–4 mm, glabrous; petals yellow, 5–12 × 1–3 mm. **Heads of achenes** depressed-hemispheric or -conic, 3–4 × 5 mm; achenes 2–2.6 × 1–1.4 mm, glabrous; beak persistent, lanceolate, straight, 0.8–1.2 mm. $2n = 16$.

Flowering late spring–summer (May–Aug). Tundra on moist slopes; 0–300 m; Alaska; Asia.

2e. RANUNCULUS Linnaeus subg. BATRACHIUM (de Candolle) A. Gray, Proc. Amer. Acad. Arts 21: 363. 1886

Ranunculus sect. *Batrachium* de Candolle, Syst. Nat. 1: 233. 1817; *Batrachium* (de Candolle) Gray

Plants glabrous (or sheathing leaf bases sometimes pubescent). **Stems** creeping or floating, not bulbous-based, without bulbils. **Roots** nodal, never tuberous. **Leaves** cauline, either simple and laminate (with expanded blade), or completely dissected into filiform segments, or with proximal leaves filiform-dissected and distal leaves laminate; laminate leaves petiolate, undivided or 3-parted, segments entire to shallowly cleft and crenate; filiform-dissected leaves petiolate or sessile. **Inflorescences** axillary, solitary flowers. **Flowers** pedicellate; sepals deciduous soon after anthesis, 5; petals 5, white or white with yellow claws; nectary scale attached on 3 sides, forming ridge or shallow pocket not covering nectary, glabrous, free margin entire; style present. **Fruits** achenes, 1-locular; achene body ellipsoid or flattened-ellipsoid, 1–2 times as wide as thick, not prolonged beyond seed; wall thick, with coarse transverse ridges; margin low narrow ridge; beak much shorter than achene body.

Species ca. 15 (3 in the flora): widespread except lowland tropics.

Unlike most species of *Ranunculus,* members of this subgenus are not poisonous.

1. Leaves all dissected into filiform segments. . . . . 73b. *Ranunculus aquatilis* var. *diffusus*
1. Leaves not all filiform-dissected, all or some laminate and shallowly to deeply 3-lobed or -parted.
    2. Leaves all laminate, margins shallowly crenate-lobulate, segments rounded; e North America . . .
        . . . . .71. *Ranunculus hederaceus*
    2. Leaves of 2 types: laminate and filiform-dissected (filiform-dissected leaves sometimes few and inconspicuous in *R. lobbii*); laminate leaves 3-parted, segments elliptic, obovate, or fan-shaped; w North America.
        3. Receptacle glabrous; style 1–1.5 mm. . . . . 72. *Ranunculus lobbii*
        3. Receptacle hispid; style 0.1–0.8 mm . . . . .73a. *Ranunculus aquatilis* var. *aquatilis*

71. Ranunculus hederaceus Linnaeus, Sp. Pl. 1: 556. 1753

Stems glabrous. **Leaves** all laminate, filiform-dissected leaves absent. **Leaf blade** reniform, 0.4–1 × 0.6–1.7 cm, margins shallowly crenate-lobulate, segments rounded. **Flowers:** receptacle glabrous; sepals recurved, 1.5–2 × 1–1.5 mm, glabrous; petals 5, 2–4 × 1–1.5 mm; style 0.1 mm. **Fruiting pedicels** recurved. **Heads of achenes** globose or depressed-globose, 3–4 × 3–4 mm; achenes 1.2–1.6 × 1–1.2 mm, glabrous; beak deciduous, sometimes leaving stub to 0.1 mm. $2n = 16$ (Europe).

Flowering spring–summer (Apr–Aug). Edges of lakes and ponds on the coastal plain; 0–150 m; Nfld.; Md., N.C., Pa., S.C., Va.; Europe.

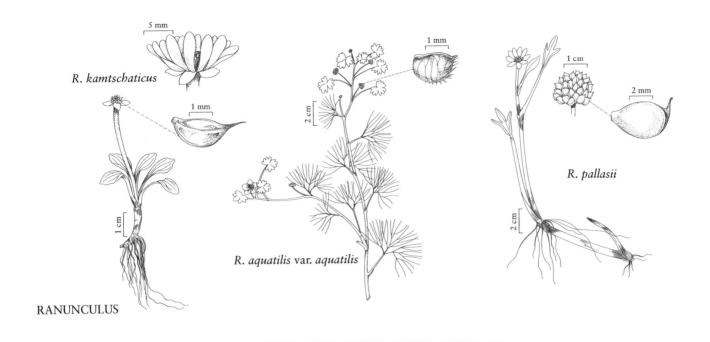

R. kamtschaticus

R. aquatilis var. aquatilis

R. pallasii

RANUNCULUS

72. Ranunculus lobbii (Hiern) A. Gray, Proc. Amer. Acad. Arts 21: 364. 1886 E

Ranunculus hydrocharis Spenner f. *lobbii* Hiern, J. Bot. 9: 66, plate 114. 1871

Stems glabrous. **Leaves** laminate and filiform-dissected; laminate leaf blades reniform in outline, deeply 3-parted, 0.5–0.8 × 0.9–1.5 cm, segments elliptic or obovate, margins sometimes notched; filiform-dissected leaves sometimes few and inconspicuous, stipules gradually tapering upward, connate for whole length, mostly petiolate. **Flowers:** receptacle glabrous; sepals spreading, 2–3 × 1–1.5 mm, glabrous; petals 5, 4–6 × 2–5 mm; style 1–1.5 mm. **Fruiting pedicels** recurved. **Heads of achenes** hemispheric, 3 × 4–5 mm; achenes 2–2.4 × 1.4–1.8 mm, glabrous; beak deciduous, sometimes leaving stub to 0.2 mm.

Flowering late winter–spring (Mar–May). Shallow ponds and vernal pools; 0–300 m; B.C.; Calif., Oreg.

73. Ranunculus aquatilis Linnaeus, Sp. Pl. 1: 556. 1753 · White water crowsfoot, renoncule aquatique

Batrachium aquatile (Linnaeus) Dumortier

Stems glabrous. **Leaves** laminate and filiform-dissected; laminate leaf blades reniform, 3-parted, 0.4–1.1 × 0.7–2.3 cm, segments obovate or fan-shaped, shallowly cleft, margins crenate; filiform-dissected leaves with stipules gradually tapering upward, connate for their whole length. **Flowers:** receptacle hispid, rarely glabrous; sepals spreading or reflexed, 2–4 × 1–2 mm, glabrous; petals 5, 4–7 × 1–5 mm; style 0.1–0.8 mm.

Fruiting pedicels usually recurved. **Heads of achenes** hemispheric to ovoid, 2–4 × 2–5 mm; achenes 1–2 × 0.8–1.4 mm, glabrous or hispid; beak persistent, filiform, 0.1–1.2 mm.

Varieties 2 (2 in the flora): North America; Mexico; South America; Eurasia; Australia.

Ranunculus aquatilis is very variable. In the past it has often been treated as three or four species and many varieties. These segregate taxa have been based on the size, petiolation and rigidity of the leaves (now known to be primarily under environmental control), petal size and curvature of the fruiting pedicel (both sometimes variable along a single stem), and number of achenes per head and length of achene beak (which vary continuously and are not correlated with one another). Unless reliable characters can be found, only two of the taxa recognized in older floras can be maintained.

1. Leaves of 2 types: laminate and filiform-dissected; w North America . . . . 73a. *Ranunculus aquatilis* var. *aquatilis*
1. Leaves all filiform-dissected; widespread . . . . 73b. *Ranunculus aquatilis* var. *diffusus*

73a. Ranunculus aquatilis Linnaeus var. **aquatilis** [F]

Ranunculus aquatilis var. *hispidulus* Drew; *R. trichophyllus* (Chaix) Bosch var. *hispidulus* (Drew) W. B. Drew

Leaves of 2 types: laminate and filiform-dissected. **Achenes** 1.6–2 × 1–1.4 mm; beak 0.1–0.4 mm. $2n = 48$ (Europe).

Flowering spring–summer (Apr–Sep). Marshes, ditches, streams, ponds, and lakes; 0–2500 m; B.C., N.W.T.; Alaska, Calif., Idaho, Mont., Nev., Oreg., Utah, Wash., Wyo.; South America; Eurasia.

Plants growing in deep water may flower without producing floating leaves. Such plants cannot be distinguished from specimens of *Ranunculus aquatilis* var. *diffusus* except by culture in shallow water.

73b. Ranunculus aquatilis Linnaeus var. **diffusus** Withering, Arr. Brit. Pl. ed. 3, 2: 507. 1796 [W1]

Batrachium circinatum (Sibthorp) Spach subsp. *subrigidum* (W. B. Drew) Á. Löve & D. Löve; *B. flaccidum* (Persoon) Ruprecht; *B. longirostris* (Godron) F. W. Schultz; *B. trichophyllum* (Chaix) Bosch; *B. trichophyllum* subsp. *lutulentum* (Perrier & Songeon) Janchen ex V. V. Petrovsky; *Ranunculus aquatilis* var. *calvescens* (W. B. Drew) L. D. Benson; *R. aquatilis* var. *capillaceus* (Thuillier) de Candolle; *R. aquatilis* var. *eradicatus* Laestadius; *R. aquatilis* var. *harrisii* L. D. Benson; *R. aquatilis* var. *longirostris* (Godron) G. Lawson; *R. aquatilis* var. *porteri* (Britton) L. D. Benson; *R. aquatilis* var. *subrigidus* (W. B. Drew) Breitung; *R. circinatus* Sibthorp var. *subrigidus* (W. B. Drew) L. D. Benson; *R. confervoides* (Fries) Fries; *R. longirostris* Godron; *R. subrigidus* W. B. Drew; *R. trichophyllus* Chaix; *R. trichophyllus* var. *calvescens* W. B. Drew; *R. trichophllus* var. *eradicatus* (Laestadius) W. B. Drew; *R. trichophyllus* subsp. *lutulentum* (Perrier & Songeon) Vierhapper

Leaves all filiform-dissected. **Achenes** 1–1.8 × 0.8–1.2 mm; beak 0.2–1.2 mm. $2n = 16, 32, 48$.

Flowering spring–summer (Apr–Sep). Ponds, lakes, streams, ditches, edges of rivers; 0–3200 m; Greenland; Alta., B.C., Man., N.B., Nfld., N.W.T., N.S., Ont., P.E.I., Que., Sask., Yukon; Ala., Alaska, Ariz., Ark., Calif., Colo., Conn., Del., Idaho, Ill., Ind., Iowa, Kans., Ky., Maine, Md., Mass., Mich., Minn., Mo., Mont., Nebr., Nev., N.H., N.Mex., N.J., N.Y., N.C., N.Dak., Ohio, Okla., Oreg., Pa., R.I., S.C., S.Dak., Tenn., Tex., Utah, Vt., Va., Wash., W.Va., Wis., Wyo.; Mexico; Eurasia; Australia.

Populations of *Ranunculus aquatilis* var. *diffusus* with long achene beaks are not known from the Old World. In North America, beak length varies continuously over the whole range given for the variety, and separation of plants with unusually long beaks as *R. longirostris* is not tenable. *Ranunculus aquatilis* var. *diffusus* shows geographic variation, and some regional forms have been recognized as separate varieties. Dwarf creeping arctic plants may be called *R. aquatilis* var. *eradicatum,* plants with sparsely pubescent or glabrous receptacle from eastern North America may be called *R. aquatilis* var. *calvescens,* plants with linear, noncapillary leaf segments from the northern Great Basin may be called *R. aquatilis* var. *porteri,* and very robust plants from Oregon and northernmost California may be called *R. aquatilis* var. *harrisii.* Extreme forms of these races are recognizable, but they intergrade and many specimens cannot be confidently assigned to one or another of them.

The Eurasian species *Ranunculus circinatus* Sibthorp has been reported from North America. These reports are based on specimens of *R. aquatilis* var. *diffusus.*

2f. Ranunculus Linnaeus subg. **Pallasiantha** L. D. Benson, Amer. J. Bot. 27: 807. 1940

Plants glabrous. **Stems** leafy, creeping or floating, not bulbous-based, without bulbils. **Roots** nodal, never tuberous. **Leaves** cauline, simple, petiolate; blade undivided or 3-lobed, much longer than wide, margins entire. **Inflorescences** axillary, solitary flowers. **Flowers** pedicellate (pedicels naked or leafy); sepals deciduous soon after anthesis, 3; petals 7–11, white or reddish; nectary scale attached on 3 sides for at least half its length, forming pocket (sometimes mouth of pocket prolonged as short flap), glabrous, free margin entire; style present. **Fruits** achenes, 1-locular, 4.2–5.2 mm; achene body oblong-lenticular, 1.3–1.7 times as wide as thick, prolonged beyond seed as corky distal appendage; wall thick, smooth; margin low corky band; beak much shorter than achene body.

Species 1 (1 in the flora): North America, Eurasia.

74. Ranunculus pallasii Schlechtendal, Animadv. Bot. Ranunc. Cand. 1: 15. 1819 · Renoncule de Pallas [F]

Stems creeping or floating, rooting nodally, glabrous, not bulbous-based. **Tuberous roots** absent. **Basal leaf blades** linear to obovate, undivided or 3-lobed, 1.5–3.6 × 0.3–2 cm, lobes lanceolate or elliptic, margins entire, apex rounded to acuminate. **Flowers:** receptacle glabrous; sepals spreading, 6–10 × 4–7 mm, glabrous; petals 7–11, white or pink, 8–13 × 3–6 mm. **Heads of achenes** globose or hemispheric, 5–12 × 9–15 mm; achenes 4.2–5.2 × 2.4–3.2 mm, glabrous; beak persistent, lanceolate, straight or curved, 1–1.2 mm. $2n = 32$.

Flowering summer (Jul–Aug). Shallow water of bogs and pools in muskeg and tundra; 0–700 m; Man., Nfld., N.W.T., Ont., Que., Yukon; Alaska; Eurasia.

2g. RANUNCULUS Linnaeus subg. **COPTIDIUM** (Nyman) L. D. Benson, Amer. J. Bot. 27: 807. 1940

Ranunculus V. *Coptidium* Nyman, Consp. Fl. Eur. 1: 13. 1878

Plants glabrous. **Stems** creeping, not bulbous-based, without bulbils. **Roots** nodal, never tuberous. **Leaves** cauline, simple, petiolate; blade deeply 3-parted to base, as wide as long, segments toothed or shallowly lobed. **Inflorescences** axillary, solitary flowers. **Flowers** pedicellate (pedicels naked, rarely leafy); sepals deciduous soon after anthesis, 3; petals 5–8, yellow; nectary scale attached on 3 sides, forming pocket, glabrous, free margin entire; style present. **Fruits** achenes, 1-locular; achene body oblong-lenticular, 1.9–2.6 times as wide as thick, prolonged beyond seed as corky distal appendage; wall thick, smooth; margin low corky band; beak much shorter than achene body.

Species 1 (1 in the flora): North America, Eurasia.

75. Ranunculus lapponicus Linnaeus, Sp. Pl. 1: 553. 1753 · Renoncule de Lapponie [F]

Stems prostrate, buried, rooting nodally, glabrous, not bulbous-based. **Tuberous roots** absent. **Basal leaf blades** reniform, deeply 3-parted, 1.1–2.6 × 1.6–4.3 cm, segments undivided or 1× cleft, margins crenate, apex rounded. **Flowers:** receptacle glabrous; sepals spreading or reflexed from base, 4–7 × 2–5 mm, glabrous; petals yellow, 5–6 × 2–3 mm. **Heads of achenes** hemispheric, 5–7 × 8–10 mm; achenes 3.8–4.2 × 2–2.2 mm, glabrous; beak persistent, lanceolate, curved, tip hooked, 1.6–2.4 mm. $2n = 16$.

Flowering late spring–summer (Jun–Jul). Boggy places and lakesides in tundra, muskeg, and boreal forest; 0–900 m; Greenland; Alta., B.C., Man., N.B., Nfld., N.W.T., Ont., Que., Sask., Yukon; Alaska, Maine, Mich., Minn.; Eurasia.

Starving individuals among western Eskimo groups ate the soaked plant of *Ranunculus lapponicus* as a dietary aid before consuming other food (D. E. Moerman 1986).

2h. RANUNCULUS Linnaeus subg. **FICARIA** (Schaeffer) L. D. Benson, Amer. J. Bot. 27: 807. 1940

Ficaria Schaeffer, Bot. Exped., 156–157. 1760

Plants glabrous. **Stems** erect or decumbent, not bulbous-based, sometimes producing axillary bulbils. **Roots** basal, some tuberous. **Leaves** basal and cauline, simple, petiolate; basal and cauline leaf blades similar, undivided, margins entire or crenulate. **Inflorescences** terminal, solitary flowers. **Flowers:** sepals deciduous soon after anthesis, 3; petals 7–10, yellow; nectary

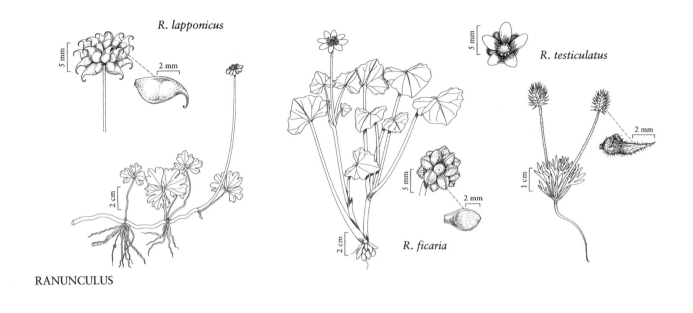

RANUNCULUS

scale attached on 3 sides, forming pocket, glabrous, free margin entire; style absent. **Fruits** achenes, 1-locular; achene body ellipsoid, about as wide as thick, not prolonged beyond seed; wall thick, smooth; margin low narrow ridge; beak absent.

Species 5 (1 in the flora): North America, Pacific Islands, Australia, Eurasia, and n Africa.

76. Ranunculus ficaria Linnaeus, Sp. Pl. 1: 550. 1753
· Lesser celandine, pilewort, ficaire, renoncule ficaire [F]

Ranunculus ficaria subsp. *bulbifer* Lambinon; *R. ficaria* subsp. *calthifolius* (Reichenbach) Arcangeli; *R. ficaria* var. *bulbifera* Albert

Stems erect to decumbent, not rooting nodally, glabrous, not bulbous-based, spheric or ellipsoid bulbils sometimes formed in leaf axils. **Tuberous roots** present. **Basal leaves** simple and undivided, blades cordate to deltate or semicircular, 1.8–3.7 × 2–4 cm, base cordate, margins entire or crenate, apex rounded or obtuse. **Flowers:** receptacle glabrous; sepals spreading, saccate at extreme base, 4–9 × 3–6 mm, glabrous; petals yellow, 10–15 × 3–7 mm. **Heads of**

achenes hemispheric, 4–5 × 6–8 mm; achenes 2.6–2.8 × 1.8–2 mm, pubescent; beak absent.

Flowering spring (Apr–May). Shaded stream banks and moist disturbed areas; 0–300 m; introduced; B.C., Nfld., Ont., Que.; Conn., D.C., Ill., Ky., Md., Mass., Mich., Mo., N.J., N.Y., Ohio, Oreg., Pa., R.I., Tenn., Va., Wash., W.Va.; native to Europe.

In North America, *Ranunculus ficaria* seems to be expanding its range rapidly in areas with cool mesic climates.

The species is extremely variable (especially in leaf size and stem posture), and many attempts have been made to divide it into varieties or subspecies (see P. D. Sell 1994). The different forms, however, intergrade extensively, and the varieties are often impossible to distinguish.

2i. RANUNCULUS Linnaeus subg. **CERATOCEPHALUS** (Moench) L. D. Benson, Amer. J. Bot. 27: 806. 1940

Ceratocephalus Moench, Methodus, 218. 1794

Plants villous. **Stems** erect, not bulbous-based, without bulbils. **Roots** basal, never tuberous. **Leaves** all basal, petiolate; blade 1–2×-dissected. **Inflorescences** terminal, flowers solitary. **Flowers:** sepals persistent in fruit, 5, 3–6 × 1–2 mm; petals 5, yellow; nectary scale attached basally, forming flap over nectary, glabrous, free margin entire; style present. **Fruits** achenes,

3-locular, lateral locules empty; achene body ellipsoid, about as wide as thick, not prolonged beyond seed; wall firm, smooth; margin low narrow ridge; beak much longer than achene body.

Species 10 (1 in the flora): North America, Pacific Islands, Australia, Europe, c, sw Asia, and n Africa.

Ranunculus subg. *Ceratocephalus* is very different in aspect and fruit structure from other species of the genus; it is often treated as a distinct genus.

SELECTED REFERENCE Klokov, M. V. 1978. The genus *Ceratocephala* Moench., on aspects of its overall biological differentiation. Novosti Sist. Vyssh. Rast. 15: 7–73. [In Russian.]

77. Ranunculus testiculatus Crantz, Stirp. Austr. Fasc. 2: 97. 1763 [F] [W2]

Ceratocephalus orthoceras de Candolle; *C. testiculatus* (Crantz) Roth

Stems erect or ascending, not rooting nodally, villous, not bulbous-based. **Tuberous roots** absent. **Basal leaf blades** broadly spatulate in outline, 1–2×-dissected, 0.9–3.8 × 0.5–1.5 cm, segments linear, margins entire, apex obtuse to acuminate. **Flowers:** receptacle glabrous; sepals spreading, 3–6 × 1–2 mm, villous; petals yellow, 3–5 × 1–3 mm. **Heads of achenes** cylindric, 9–16(–27) × 8–10 mm; achenes 1.6–2 × 1.8–2 mm, tomentose; beak persistent, lanceolate, 3.5–4.5 mm.

Flowering spring (Apr–May). Disturbed areas, especially in grassland; 400–2500 m; introduced; B.C., Sask.; Ariz., Calif., Colo., Idaho, Iowa, Kans., Minn., Mo., Mont., Nebr., Nev., N.Mex., N.Dak., Ohio, Oreg., S.Dak., Utah, Wash., Wyo.; native to Eurasia.

In North America, *Ranunculus testiculatus* seems to be expanding its range rapidly in arid and semiarid areas. A second species of this subgenus, *R. falcatus* Linnaeus [*Ceratocephala falcata* (Linnaeus) Persoon], has been reported from North America, but all reports seem to be based on misidentified material of *R. testiculatus*.

3. MYOSURUS Linnaeus, Sp. Pl. 1: 284. 1753; Gen. Pl. ed. 5, 137. 1754 · Mousetail [Greek *mus*, mouse, and *oura*, tail, from shape and texture of the fruiting head of *M. minimus*]

Alan T. Whittemore

Herbs, annual. **Leaves** basal, simple, tapering to filiform base. **Leaf blade** linear or very narrowly oblanceolate, margins entire. **Inflorescences** terminal, solitary flowers; bracts absent. **Flowers** bisexual, radially symmetric; sepals not persistent in fruit, (3–)5(–8), green or with scarious margins, spurred, oblong to elliptic, lanceolate, or oblanceolate, 1.5–4 mm; petals 0–5, distinct, white, plane, linear to very narrowly spatulate, long-clawed, 1–2.5 mm, nectary on abaxial surface; stamens 5–25; filaments filiform; staminodes absent between stamens and pistils; pistils 10–400, simple; ovule 1 per pistil; style persistent. **Fruits** achenes, aggregate, sessile, prismatic, exposed face forming plane outer surface, sides faceted or curved by compression against adjacent achenes; sides not veined; beak terminal, straight, 0.05–1.8 mm. $x = 8$.

Species 15 (5 in the flora): temperate regions worldwide.

Mature fruits are crucial to accurate identification of most North American species of *Myosurus*.

Flowers of *Myosurus* are unique in the family in that the receptacle continues to elongate and, in some species, to initiate new ovaries after the flowers open and pollen is shed (D. E. Stone 1959).

SELECTED REFERENCES Mason, H. L. and D. E. Stone. 1957. *Myosurus*. In: Mason, H. L. 1957. A Flora of the Marshes of California. Berkeley. Pp. 497–505 Stone, D. E. 1959. A unique balanced breeding system in the vernal pool mouse-tails. Evolution 13: 151–174.

1. Outer face of achene orbiculate to square or broadly rhombic, 0.8–1.3 times as high as wide.
    2. Outer face of achene bordered by prominent ridge . . . . .4. *Myosurus cupulatus*
    2. Outer face of achene not bordered . . . . .5. *Myosurus nitidus*
1. Outer face of achene narrowly rhombic to elliptic or oblong to linear, 1.5–5 times as high as wide.
    3. Beak of achene 0.05–0.4 mm, 0.05–0.3 times as long as body of achene, parallel to outer face of achene, heads of achenes thus appearing smooth . . . . .1. *Myosurus minimus*
    3. Beak of achene 0.6–1.8 mm, 0.4–1.1 times as long as body of achene, divergent, heads of achenes thus strongly roughened by projecting achene beaks.
        4. Scapes 0–0.2 cm; heads of achenes immersed in leaves . . . . 2. *Myosurus sessilis*
        4. Scapes 0.9–10.5 cm; heads of achenes projecting beyond leaves. .3. *Myosurus apetalus*

1. Myosurus minimus Linnaeus, Sp. Pl. 1: 284. 1753 F W2

Myosurus lepturus (A. Gray) Howell; *M. lepturus* var. *filiformis* (Greene) Greene ex Abrams; *M. minimus* subsp. *major* (Greene) G. R. Campbell; *M. minimus* var. *filiformis* Greene; *M. minimus* var. *major* (Greene) K. C. Davis

Herbs, 4–16.5 cm. **Leaf blades** narrowly oblanceolate or linear, 2.2–11.5 cm. **Inflorescences:** scape 1.8–12.8 cm. **Flowers:** sepals faintly or distinctly 3–5-veined, scarious margins narrow or absent; petal claw 1–2 times as long as blade. **Heads of achenes** 16–50 × 1–3 mm, exserted beyond leaves. **Achenes:** outer face narrowly rhombic to elliptic or oblong, 0.8–1.4 × 0.2–0.6 mm, 1.5–5 times as high as wide, not bordered; beak 0.05–0.4 mm, 0.05–0.3 as long as body of achene, parallel to outer face of achene, heads of achenes thus appearing smooth. 2*n* = 16.

Flowering spring (Mar–Jun). Wet fields, vernal pools, banks of streams and lakes; 0–3000 m; Alta., B.C., Man., Ont., Sask.; Ala., Ariz., Ark., Calif., Colo., Idaho, Ill., Ind., Iowa, Kans., Ky., La., Minn., Miss., Mo., Mont., Nebr., N.Mex., N.C., N.Dak., Okla., Oreg., Pa., S.C., S.Dak., Tenn., Tex., Utah, Va., Wash., Wyo.; Mexico (Baja California); Europe; sw Asia; n Africa.

Plants of *Myosurus minimus* from a few sites in coastal southern California, northern Baja California, and immediately west of Riley, Oregon, sometimes have short scapes, so that the heads of achenes are immersed in the leaves. These plants, which have been called *M. minimus* subsp. *apus* (Greene) G. R. Campbell, *M. minimus* var. *apus* Greene, or *M. clavicaulis* M. E. Peck, are indistinguishable from some recombinant lines found in *M. minimus* × *sessilis* hybrid swarms (see discussion under *M. sessilis*), but they occur outside the current range of *M. sessilis*. D. E. Stone (1959) has suggested that they resulted from past hybridization between the two species, perhaps at a time when *M. sessilis* had a wider range than it does now.

The Navajo-Ramah used *Myosurus minimus* medicinally to apply to ant bites (D. E. Moerman 1986).

2. Myosurus sessilis S. Watson, Proc. Amer. Acad. Arts 17: 362. 1882 C E F

Myosurus minimus Linnaeus var. *sessiliflorus* (Huth) G. R. Campbell

Herbs, 0.8–2.5 cm. **Leaf blades** linear or nearly so, 1.6–7.4 cm. **Inflorescences:** scapes 0–0.2 cm. **Flowers:** sepals indistinctly 3-veined, scarious margins broad; petal claw about as long as blade. **Heads of achenes** 8–25 × 2–4 mm, immersed among leaves. **Achenes:** outer face narrowly rhombic or occasionally narrowly oblong, 1.2–2 × 0.4–0.8 mm, 2–4.5 times as high as wide, not bordered; beak 0.8–1.8 mm, 0.6–1.1 times as long as body of achene, diverging from outer face of achene, heads of achenes thus strongly roughened by projecting achene beaks. 2*n* = 16.

Flowering spring (Mar–May). Vernal pools and alkalai flats; of conservation concern; 10–1600 m; Calif., Oreg.

Myosurus sessilis often grows with *M. minimus*, and hybrids between the two species are common. Because the plants mostly set seed by self-pollination, these hybrids often give rise to inbred lines showing the characteristics of the two parents in various combinations (D. E. Stone 1959). The names *M. alopecuroides* Greene and *M. sessilis* subsp. *alopecuroides* (Greene) D. E. Stone are based on the hybrid *M. minimus* × *sessilis* or on inbred lines descended from this hybrid.

3. Myosurus apetalus Gay, Fl. Chil. 1: 31, plate l, fig. 1. 1845

Herbs, 1.5–12.5 cm. **Leaf blades** linear or narrowly oblanceolate, 0.9–4 cm. **Inflorescences:** scapes 0.9–10.5 cm. **Flowers:** sepals 1–3-veined, scarious margins narrow or broad; petal claw 1–2 times as long as blade. **Heads of achenes** 4–26 × 1.5–2 mm, exserted beyond leaves. **Achenes:** outer face narrowly rhombic to elliptic

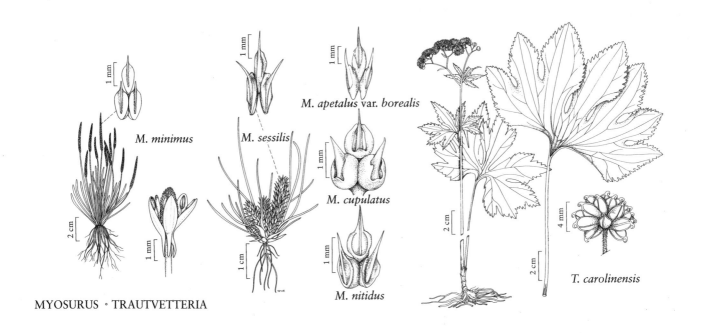

M. minimus · M. sessilis · M. apetalus var. borealis · M. cupulatus · M. nitidus · T. carolinensis

MYOSURUS · TRAUTVETTERIA

or oblong, 1–2.2 × 0.4–1 mm, 2–5 times as high as wide, not bordered; beak 0.6–1.4 mm, 0.4–1 times as long as body of achene, diverging from outer face of achene, heads of achenes thus strongly roughened by projecting achene beaks.

Varieties 3 (2 in the flora): North America; South America.

The illegitimate names *Myosurus aristatus* Bentham ex Hooker and *M. minimus* var. *aristatus* (Bentham ex Hooker) B. Boivin have been used for this species.

The Navajo-Ramah used *Myosurus apetalus* medicinally as a life medicine and as a protection against witches (D. E. Moerman 1986, citing *M. aristatus*).

1. Heads of achenes 4–9 mm; sepals 1-veined, scarious margins broad. . . . . .3a. *Myosurus apetalus* var. *borealis*
1. Heads of achenes 11–26 mm; sepals faintly 3-veined, scarious margins narrow. . . . . 3b. *Myosurus apetalus* var. *montanus*

3a. Myosurus apetalus Gay var. **borealis** Whittemore, Novon 4: 78. 1994 E F

Flowers: sepals 1-veined, scarious margins broad. **Heads of achenes** 4–9 mm; outer face of achene narrowly elliptic or rectangular.

Flowering spring (May). Flats and marshes, often in sagebrush; 600–1500 m; Alta., B.C., Sask.; Calif., Idaho, Mont., Nev., Oreg., Utah, Wash., Wyo.

3b. Myosurus apetalus Gay var. **montanus** (G. R. Campbell) Whittemore, Novon 4: 78. 1994 E

Myosurus minimus Linnaeus subsp. *montanus* G. R. Campbell, Aliso 2: 394. 1952

Flowers: sepals faintly 3-veined, scarious margins narrow. **Heads of achenes** 11–26 mm; outer face of achene narrowly rhombic. $2n = 16$.

Flowering late spring (May–Jun). Wet meadows, vernal pools and sloughs, bogs, muddy shores of lakes and streams; 1500–2900 m; Alta., Sask.; Ariz., Calif., Colo., Mont., Nev., N.Mex., N.Dak., Oreg., Utah, Wyo.

The illegitimate name *Myosurus aristatus* subsp. *montanus* (G. R. Campbell) H. Mason has been used for this taxon.

4. Myosurus cupulatus S. Watson, Proc. Amer. Acad. Arts 17: 362. 1882 F

Herbs, 3.3–16 cm. **Leaf blades** linear or very narrowly oblanceolate, 1.8–9.5 cm. **Inflorescences:** scapes 2.2–12 cm. **Flowers:** sepals faintly 3-veined, scarious margins narrow or absent; petal claw 1–3 times as long as blade. **Heads of achenes** 13–42 × 2–3 mm, long-exserted from leaves. **Achenes:** outer face orbiculate or sometimes square, 0.8–1.2 × 0.6–1 mm, 0.8–1.2 times as high as wide, bordered by prominent ridge; beak 0.6–

1.2 mm, 0.6–1.2 times as long as body of achene weakly divergent from outer face of achene, heads of achenes thus roughened by projecting achene beaks. $2n = 16$.

Flowering late winter–spring (Mar–May). Dry hillsides or canyon bottoms in shrubland; 350–1800 m; Ariz., Calif., Colo., Nev., N.Mex., Tex., Utah; Mexico (Baja California and Sonora).

The Navajo-Ramah used *Myosurus cupulatus* medicinally both externally and internally as an aid for ant bites or swallowing an ant (D. E. Moerman 1986).

5. Myosurus nitidus Eastwood, Bull. Torrey Bot. Club 32: 194. 1905 E F

Myosurus egglestonii Wooton & Standley

Herbs, 1–1.9 cm. **Leaf blades** linear or very narrowly oblanceolate, 0.6–1.2 cm. **Inflorescences:** scape 0.5–0.9 cm. **Flowers:** sepals sometimes strongly veined, scarious margins narrow or broad; petal claw 3 times as long as blade. **Heads of achenes** 8–17 × 2–2.5 mm, exserted from leaves. **Achenes:** outer face orbiculate to rhombic, 0.8–1 × 0.6–0.8 mm, 1–1.3 times as high as wide, not bordered; beak 0.8–1 mm, 0.8–1.2 as long as outer face of achene, diverging (sometimes weakly) from outer face of achene, heads of achenes thus ± roughened by projecting achene beaks.

Flowering spring. Under sagebrush; 1800–2100 m; Ariz., Colo., N.Mex.

4. TRAUTVETTERIA Fischer & C. A. Meyer in F. E. L. von Fischer et al., Index Sem. Hort. Petrop. 1: 22. 1835 · False bugbane, tassel-rue [for Ernst Rudolph von Trautvetter (1809–1889), Russian botanist]

Bruce D. Parfitt

Herbs, perennial, from short, slender rhizomes. **Leaves** basal and cauline, simple, proximal leaves petiolate, distal leaves sessile; cauline leaves alternate. **Leaf blade** palmately 5–11-lobed; segments broadly cuneate, margins lacerate to serrate. **Inflorescences** terminal, ± many-flowered corymbs, 2.5–43 cm; bracts inconspicuous, linear-lanceolate, not forming involucre. **Flowers** bisexual, radially symmetric; sepals not persistent in fruit, 3–5(–7), greenish white, concave-cupped, broadly ovate to obovate, clawed, 3–6 mm; petals absent; stamens ca. 50–100; outer filaments 15–20, spatulate, often distally wider than anthers, inner filaments not dilated; staminodes absent between stamens and pistils; pistils ca. 15, simple; ovule 1 per pistil; style present. **Fruits** utricles, aggregate, sessile, ellipsoid to obovoid, 4-angled in cross section, sides prominently veined; beak terminal, curved to hooked, 0.4–0.8 mm. $x = 8$.

Species 1 (1 in the flora): North America, e Asia.

1. Trautvetteria caroliniensis (Walter) Vail, Mem. Torrey Bot. Club 2: 42. 1890 · False bugbane, tassel-rue F

Hydrastis caroliniensis Walter, Fl. Carol., 156. 1788; *Trautvetteria caroliniensis* var. *borealis* (H. Hara) T. Shimizu; *T. caroliniensis* var. *occidentalis* (A. Gray) C. L. Hitchcock; *T. grandis* Nuttall; *T. palmata* (Michaux) Fischer & C. A. Meyer

Herbs, 0.5–1.5 m. **Rhizome** with fascicles of fibrous roots. **Stems** 1–several, erect, usually unbranched below inflorescence, 0.5–1.5 m, glabrous or glabrate. **Leaves:** basal leaves with petiole to 4.5 dm, blade 1–3(–4) dm wide, lobe apex acute; cauline leaves reduced toward apex of stem. **Inflorescences:** peduncle 1–8 dm; pedicel densely pubescent with minute, hooked trichomes. **Flowers:** stamens white, 5–10 mm. **Utricles** papery, veins prominent along angles and on 2 adaxial faces. $2n = 16$.

Flowering summer. Wooded seepage slopes, stream banks, bogs, rarely prairies or bluffs, western spruce-fir forests and subalpine meadows; 0–3800 m; B.C.; Ala., Ariz., Ark., Calif., Colo., Fla., Ga., Idaho, Ill., Ky., Md., Mo., Mont., N.Mex., N.C., Oreg., Pa., S.C., Tenn., Utah, Va., Wash., W.Va., Wyo.; Mexico; e Asia.

Trautvetteria caroliniensis apparently has been extirpated from Indiana.

The numerous white stamens make *Trautvetteria caroliniensis* an attractive ornamental, and it is reportedly easy to grow.

Populations of *Trautvetteria caroliniensis* in western North America have been distinguished from the east-ern typical material as *T. caroliniensis* var. *borealis* (H. Hara) T. Shimizu [synonym: *T. caroliniensis* var. *occidentalis* (A. Gray) C. L. Hitchcock]. Asian populations, long treated as the distinct species *T. japonica* Siebold & Zuccarini, were most recently regarded (T. Shimizu 1981; M. Tamura 1991) as conspecific with the North American populations [as *T. caroliniensis* var. *japonica* (Siebold & Zuccarini) T. Shimizu]. Aside from geography, varietal differences seem rather arbitrary.

The Bella Coola applied poultices made from the pounded roots of *Trautvetteria caroliniensis* to boils (on adults only) (D. E. Moerman 1986).

5. ANEMONE Linnaeus, Sp. Pl. 1: 538. 1753; Gen. Pl. ed. 5, 241. 1754 · Windflower, anémone [etymology not clear: probably Greek *anemos,* wind; possibly from *Naaman,* Semitic name for Adonis, whose blood, according to myth, produced *Anemone coronaria*]

Bryan E. Dutton
Carl S. Keener
Bruce A. Ford

Anemonastrum Holub; *Anemonidium* (Spach) Holub; *Anemonoides* Miller; *Hepatica* Miller; *Jurtsevia* Á. Löve & D. Löve; *Pulsatilla* Miller

Herbs, perennial, from rhizomes, caudices, or tubers. **Leaves** basal, simple or compound, petiolate. **Leaf blade** lobed or parted or undivided, reniform to obtriangular or lanceolate, margins entire or variously toothed. **Inflorescences** terminal, 2–9-flowered cymes or umbels, or flowers solitary, to 60 cm; involucres present, often with primary involucres subtending inflorescences, and secondary and tertiary involucres subtending inflorescence branches or single flowers (primary, secondary, and tertiary involucres appearing to be in tiers), involucral bracts 2–7(–9), leaflike or sepaloid, distant from or close to flowers. **Flowers** bisexual, radially symmetric; sepals not persistent in fruit, 4–20(–27), white, purple, blue, green, yellow, pink, or red, plane, linear to oblong or ovate to obovate, 3.5–40 mm; petals usually absent (present in *A. patens*), distinct, plane, obovate to elliptic, 1.5–2 mm, nectary on abaxial surface; stamens 10–200; filaments filiform or somewhat broadened at base; staminodes absent between stamens and pistils; pistils many, simple; ovule 1 per pistil; style present. **Fruits** achenes, aggregate, sessile or stalked, ovoid to obovoid, sides not veined; beak (persistent style) present, sometimes rudimentary, terminal, straight or curved, to 40(–50) mm, sometimes plumose. $x = 7$ or 8.

Species ca. 150 (25 in the flora): nearly worldwide, primarily in cooler temperate and arctic regions.

The taxonomy of *Anemone* continues to be problematic. *Anemone occidentalis* and *A. patens* var. *multifida* (the first two taxa in this treatment) are frequently placed in the genus *Pulsatilla* Miller on the basis of the long plumose achene beaks, and *A. acutiloba* and *A. americana* (the last two taxa in this treatment) in the genus *Hepatica* Miller, primarily on the basis of the involucre immediately subtending the flower and the lobed, persistent leaves. Recent phylogenetic analyses of *Anemone* in the broad sense, however, indicate that both *Pulsatilla* and *Hepatica* should be subsumed within *Anemone*. While traditional morphologic characters are useful in distinguishing between *Pulsatilla* and *Hepatica* species, respectively, many other morphologic and molecular attributes are shared with *Anemone,* strongly suggesting that these genera should be united (S. B. Hoot et al. 1994). In addition, a number of genera that have been recognized primarily on a cytotaxonomic basis (e.g., *Anemonastrum, Anemonidium, Anemonoides,* and *Jurtsevia*) are reduced to synonymy here. Some North American species of *Anemone* are closely related to plants in Europe, Asia, and South America and continue to be recognized at different ranks. For example, *Anemone patens* Linnaeus var. *multifida* (a species included in this treatment) was called *Pulsatilla multifida* (Pritzel) Juzepczuk for the former Soviet Union by S. V. Juzepczuk (1970) and *Pulsatilla patens* (Linnaeus) Miller var. *multifida* (Pritzel) S. H. Li & Y. H. Huang for China by Wang W.-T. (1980). Moreover, interspecific hybridization among some sympatric or nearly sympatric North American species also contributes to the confusion (see N. L. Britton 1891; C. L. Hitchcock et al. 1955–1969, vol. 2; R. S. Mitchell and J. K. Dean 1982). Additional analyses (e.g., G. Boraiah and M. Heimburger 1964; M. Heimburger 1959; C. Joseph and M. Heimburger 1966; and C. S. Keener et al. 1996) may prove to be helpful in resolving the taxonomy within this morphologically diverse genus.

Anemone nemorosa Linnaeus, *A. ranunculoides* Linnaeus, and *A. blanda* Schott & Kotschy, all native to Europe, are cultivated and may persist in the flora. Although apparently they rarely become naturalized, *A. nemorosa* is established at two sites in Newfoundland and Quebec, and *A. ranunculoides* in Quebec. Both are close relatives of *A. quinquefolia* and its allies.

Anemone ranunculoides is the only species in North America combining yellow sepals with rhizomes and 1–2-ternate leaves. *Anemone blanda* will key to *A. caroliniana* or *A. berlandieri* in this treatment. It can be distinguished by its short-pilose achenes, in contrast to the densely woolly achenes of *A. caroliniana* and *A. berlandieri. Anemone nemorosa* will key to *A. quinquefolia;* it differs in having 6–8 sepals and brown or black (never white) rhizomes with a 3–5 mm diameter in contrast to the 5 sepals and white or black rhizomes with 1–3 mm diameter of *A. quinquefolia.*

Protoanemonin, an irritating acrid oil, is an enzymatic breakdown product of the glycoside ranunculin and is found in many species of *Anemone.* While protoanemonin can cause severe topical and gastrointestinal irritation, it is unstable and changes into harmless anemonin when plants are dried (N. J. Turner and A. F. Szczawinski 1991).

A caudex, as the term is used here, is the "woody," perennating base of an aerial shoot (inflorescences and basal leaves). The word tuber refers to a swollen, more or less vertical underground stem. The aerial shoots arise from the apex of either of those persistent structures. Rhizome, as the term is used here, refers to an underground, usually horizontal stem (more or less vertical in *Anemone piperi*), that is nearly uniform in diameter (about 1–4 mm diameter, depending on the species) along its length. Aerial shoots arise directly from nodes at or near the apex of the rhizome.

Many species of *Anemone* have only one type of underground stem. Some species, however, have both rhizomes and caudices. In such cases the aerial shoots arise from the apex of a caudex attached to the rhizome. Some other species sometimes have both tubers and rhizomes. In those, one or more horizontal rhizomes arise near the apex of the tuber; the aerial shoots arise from the apex of the tuber.

Proportions given in the key for the middle lobes of basal leaves are calculated as follows: measure length of lobe from apex to a line connecting bases of sinuses; and measure total length of blade from leaf apex to summit of petiole.

SELECTED REFERENCES Fernald, M. L. 1928b. The North American species of *Anemone* § *Anemonanthea*. Rhodora 30: 180–188. Frodin, D. G. 1964. A Preliminary Revision of the Section *Anemonanthea* of *Anemone* in Eastern North America, with Special Reference to the Southern Appalachian Mountains. M.S. thesis. University of Tennessee. Hoot, S. B., A. A. Reznicek, and J. D. Palmer. 1994. Phylogenetic relationships in *Anemone* (Ranunculaceae) based on morphology and chloroplast DNA. Syst. Bot. 19: 169–200. Wang W.-T. 1980. *Anemone*. In: Academia Sinica. 1959+. Flora Reipublicae Popularis Sinicae. 80+ vols. Beijing. Vol. 28, pp. 1–56.

1. Achene beak 20 mm or more, plumose.
   2. Involucral bracts compound, ultimate segments of lateral leaflets 2–3 mm wide; leaflets of basal leaves pinnatifid to dissected, lateral leaflets 2×-parted; petals absent; sepals white, purple tinged (rarely abaxially blue proximally, white distally, and adaxially white)...................1. *Anemone occidentalis*
   2. Involucral bracts simple, each leaf deeply divided into 4–6 segments, segments 1–2(–3) mm wide; leaflets of basal leaves dichotomously dissected, lateral leaflets 3–4×-parted; petals present; sepals blue or purple (rarely nearly white)... 2. *Anemone patens*
1. Achene beak 6 mm or less, not plumose.
   3. Basal leaves simple (often deeply divided).
      4. Involucral bracts remotely subtending flowers, margins crenate or sharply and irregularly serrate, ± similar to basal leaves.
         5. Achenes winged, beak straight or indistinct; tiers of involucral bracts usually 2 .... 3. *Anemone canadensis*
         5. Achenes not winged, beak recurved; tier of involucral bracts 1 .... 4. *Anemone richardsonii*
      4. Involucral bracts closely subtending flowers, margins entire, not resembling basal leaves.
         6. Leaf lobes acute or acuminate, middle lobe 70–90% of total blade length; involucral bracts ± acute.. 24. *Anemone acutiloba*
         6. Leaf lobes rounded, middle lobe 50–70(–75%) of total blade length; involucral bracts obtuse... 25. *Anemone americana*
   3. Basal leaves compound.
      7. Involucral bracts of distalmost tier simple (sometimes deeply lobed and sessile, occasionally appearing compound in primary involucres of *A. tuberosa* and *A. okennonii*).
         8. Achene body winged, glabrous; inflorescences umbels or flowers solitary........... 5. *Anemone narcissiflora*
         8. Achene body not winged, hairy; inflorescences cymes or flowers solitary (sometimes umbelliform in *A. cylindrica*).
            9. Involucral bracts in 2 or more tiers.
               10. Involucral bracts ± similar to basal leaves............................6. *Anemone tuberosa*
               10. Involucral bracts dissimilar to basal leaves.
                  11. Basal leaves usually ternate; sepals (8–)10–20, 2–3 mm wide........... 7. *Anemone edwardsiana*
                  11. Basal leaves 2–3-ternate; sepals 7–11, 3–4.5 mm wide .... 8. *Anemone okennonii*
            9. Involucral bracts in 1 tier.
               12. Aerial shoots from tubers (tuber caudex-like in *A. tuberosa*), tubers bearing rhizomes in some species; stamens 70 or fewer; mostly s United States.
                  13. Sepals 8–10; aerial shoots from caudex-like tuber, rhizomes absent ....6. *Anemone tuberosa*
                  13. Sepals (7–)10–20(–30); aerial shoots from tubers, sometimes with rhizomes arising near apex of tubers (tubers rarely bearing rhizomes in *Anemone berlandieri*).
                     14. All basal leaf blades lobed or dissected differently from involucral bracts; involucres borne above middle of scape at anthesis .... 9. *Anemone berlandieri*
                     14. One or more basal leaf blades dissected similarly to involucral bracts; involucres borne below middle of scape at anthesis .... 10. *Anemone caroliniana*
               12. Aerial shoots from rhizomes (from short caudices on rhizomes in *A. parviflora*); stamens 70 or more; nw United States and Canada.
                  15. Leaflets of basal leaves (0.5–)0.7–1.8(–2.2) × 0.5–1.3 cm; sepals abaxially hairy; stamens 70–80; achenes ca. 1 mm wide, beak 1–2.5 mm......... 11. *Anemone parviflora*
                  15. Leaflets of basal leaves (2.5–)3–5(–6) × (2–)2.5–3.5 cm; sepals abaxially glabrous; stamens 100–120; achenes 2–3 mm wide, beak ca. 0.5 mm........15. *Anemone deltoidea*
      7. Involucral bracts of distalmost tier compound (petiole sometimes short and flat).
         16. Stamens 80 or more; involucral bracts 2-ternate; involucral bracts in 1 tier.
            17. Sepals white or adaxially white and abaxially bluish; stamens whitish; style white .... 12. *Anemone drummondii*
            17. Sepals uniformly dark blue to purple; stamens purple; style red .... 13. *Anemone multiceps*
         16. Stamens 80(–90) or fewer; involucral bracts 1(–2)-ternate; involucral bracts in 1 or more tiers.

18. Aerial shoots from caudices or from caudices on rhizomes; involucral bracts in 1–2 (–3) tiers.
  19. Ultimate lobes of leaflets of involucral bracts 1.5–3(–4.3) mm wide or less . . . . .14. *Anemone multifida*
  19. Ultimate lobes of leaflets of involucral bracts (4–)6 mm wide or more.
    20. Heads of achenes usually spheric or oblong-ellipsoid; achene beak 1–1.5 mm; involucral bracts of primary involucre 3(–5) . . . . .22. *Anemone virginiana*
    20. Heads of achenes cylindric; achene beak 0.5–1 mm; involucral bracts 3–7 (–9) . . . . .23. *Anemone cylindrica*
18. Aerial shoots always from rhizomes; involucral bracts in 1 tier.
  21. Rhizomes mostly vertical . . . . .18. *Anemone piperi*
  21. Rhizomes mostly horizontal.
    22. Lateral leaflets of basal leaves and involucral bracts mostly 1×-lobed or -parted . . . . .16a. *Anemone quinquefolia* var. *quinquefolia*
    22. Lateral leaflets of basal leaves and involucral bracts only occasionally lobed.
      23. Achene body 2.5–3 mm . . . . .16b. *Anemone quinquefolia* var. *minima*
      23. Achene body 3–5 mm.
        24. Achene beak (0.5–)1–1.5 mm; sepals 20 mm or less; stamens 75 or fewer.
          25. Leaflets of basal leaves 0.7–2.5(–3.5) cm wide; peduncle distally villous to pilose; w North America . . . . 19. *Anemone oregana*
          25. Leaflets of basal leaves 2.5–4(–6) cm wide; peduncle ± glabrous; e North America . . . . 17. *Anemone lancifolia*
        24. Achene beak 0.5–1 mm; sepals 15 mm or less; stamens 40 or fewer.
          26. Sepals 7–15 × 4–8 mm; pedicel (0.5–)3–10 cm in fruit; achene beak 0.6–1 mm; stamens 25–40 . . . . 21. *Anemone grayi*
          26. Sepals 3.5–8(–10) × 1.5–3(–3.5) mm; pedicel 1–3(–4) cm in fruit; achene beak ca. 0.5 mm; stamens 10–30(–35) . . . . 20. *Anemone lyallii*

1. Anemone occidentalis S. Watson, Proc. Amer. Acad. Arts 11: 121. 1876 · Western pasqueflower, mountain pasqueflower, pulsatille E

Anemone occidentalis var. *subpilosa* Hardin; *Pulsatilla occidentalis* (S. Watson) Freyn

Aerial shoots 10–60(–75) cm, from caudices, caudices ascending to vertical. **Basal leaves** (2–)3–6(–8), primarily 3-foliolate with each leaflet pinnatifid to dissected; petiole 6–8(–12) cm; terminal leaflet petiolulate, ovate in outline, (2.5–)3–6(–8) cm, base cuneate, margins pinnatifid to dissected throughout, apex narrowly acute, surfaces villous; lateral leaflets 2×-parted, pinnatifid; ultimate segments 2–3 mm wide. **Inflorescences** 1-flowered; peduncle woolly or densely villous, glabrate; involucral bracts 3, occasionally more, 1-tiered, ± similar to basal leaves, 3-foliolate, ovate in outline, bases distinct; terminal leaflet petiolulate, 2.5–7 cm (2.5 cm in flower, 7 cm or less in fruit), margins pinnatifid throughout, apex narrowly acute, surfaces villous; lateral leaflets 2×-parted, pinnatifid; ultimate segments 2–3 mm wide. **Flowers:** sepals 5–7, white, tinged purple (rarely abaxially blue proximally, white distally, and adaxially white), ovate to obovate, rarely elliptic, 15–30 × 10–17(–19) mm, abaxially hairy, adaxially glabrous; sta-

mens 150–200. **Heads of achenes** spheric, rarely cylindric; pedicels 15–20(–22) cm. **Achenes:** body ellipsoid, 3–4 × ca. 1.5 mm, not winged, villous; beak curved or recurved, reflexed with age, (18–)20–40(–50) mm, long-villous, plumose. 2*n* = 16.

Flowering spring–summer (May–Aug/Sep). Gravelly, rocky slopes, moist meadows; 500–3700 m; Alta., B.C.; Calif., Idaho, Mont., Oreg., Wash.

North American reports of the Old World species *Anemone alpina* Linnaeus from the early nineteenth century (e.g., W. J. Hooker 1829) are referable to *A. occidentalis*.

The Thompson Indians and the Okanagan used decoctions prepared from the roots of *Anemone occidentalis* to treat stomach and bowel troubles (D. E. Moerman 1986).

2. Anemone patens Linnaeus, Sp. Pl. 1: 538. 1753

Varieties ca. 4 (1 in the flora): North America, Eurasia.

2a. Anemone patens Linnaeus var. **multifida** Pritzel, Linnaea 15: 581. 1841 · Pasqueflower, prairie-smoke, prairie-crocus, pulsatille [F]

Anemone patens var. *nuttalliana* (de Candolle) A. Gray; *A. patens* var. *wolfgangiana* (Besser) Koch; *Pulsatilla patens* (Linnaeus) Miller subsp. *asiatica* Krylov & Sergievskaja; *P. patens* (Linnaeus) Miller subsp. *multifida* (Pritzel) Zämelis

Aerial shoots 5–40(–60) cm, from caudices, caudices ascending to vertical. **Basal leaves** (3–)5–8(–10), primarily 3-foliolate with each leaflet dichotomously dissected; petiole 5–10(–13) cm; terminal leaflet petiolulate to nearly sessile, obovate in outline, (2.5–)3–5 cm, base narrowly cuneate, margins dichotomously dissected throughout, apex acute to obtuse, surfaces villous, rarely glabrous; lateral leaflets 3–4×-parted (± dichotomously); ultimate segments 2–4 mm wide. **Inflorescences** 1-flowered; peduncle villous or glabrate; involucral bracts primarily 3, 1-tiered, simple, dissimilar to basal leaves, (2–)2.5–4 cm, bases clasping, connate, margins deeply laciniate throughout, surfaces villous, rarely glabrous to nearly glabrous; segments usually 4–6, filiform to linear, unlobed, 1–2(–3) mm wide. **Flowers:** sepals 5–8, blue, purple, to rarely nearly white, oblong to elliptic, (18–)20–40 × (8–)10–15 mm, abaxially villous, adaxially glabrous; petals present; stamens 150–200. **Heads of achenes** spheric to ovoid; pedicels 10–18(–22) cm. **Achenes:** body ellipsoid to obovoid, 3–4(–6) × ca. 1 mm, not winged, villous; beak curved, 20–40 mm, long-villous, plumose. $2n = 16$.

Flowering spring–summer (Apr–Aug). Prairies, open slopes, sometimes open woods or granite outcrops in woods; 100–3800 m; Alta., B.C., Man., N.W.T., Ont., Sask., Yukon; Alaska, Colo., Idaho, Ill., Iowa, Minn., Mont., Nebr., N.Mex., N.Dak., S.Dak., Utah, Wis., Wyo.; Eurasia.

Anemone patens var. *multifida* has frequently been recognized as a subspecies of *A. patens*. Although *A. patens* var. *wolfgangiana* has been used by some authors, *A. patens* var. *multifida* has priority.

Pasqueflower (as *Anemone patens* var. *wolfgangiana*) is the floral emblem of Manitoba and (as *Pulsatilla hirsutissima*) the state flower of South Dakota.

Native Americans used fresh leaves of *Anemone patens* var. *multifida* medicinally to treat rheumatism and neuralgia; crushed leaves for poultices; pulverized leaves to smell to alleviate headaches; and made decoctions from roots to treat lung problems (D. E. Moerman 1986).

The names *Pulsatilla hirsutissima* (Pursh) Britton and *P. ludoviciana* (Nuttall) A. Heller are illegitimate.

3. Anemone canadensis Linnaeus, Syst. Nat. ed. 12, 3: 231. 1768 · Canada anemone, anémone du Canada [E] [F]

Anemone dichotoma Linnaeus var. *canadensis* (Linnaeus) MacMillan

Aerial shoots (15–)20–80 cm, from caudices on rhizomes, caudices ascending, rhizomes ascending to horizontal. **Basal leaves** 1–5, simple, deeply divided; petiole 8–22(–37) cm; leaf blade orbiculate, 4–10 × 5–15(–20) cm, base sagittate to nearly truncate, margins serrate and incised on distal 1/3–1/2, apex acuminate, surfaces puberulous (more so abaxially); segments primarily 3, lanceolate to oblanceolate; lateral segments again 1×-lobed or -parted (proximal lobe occasionally lobed again); ultimate segments 10–30(–35) mm wide. **Inflorescences** 1(–3+)-flowered, rarely cymes; peduncle puberulous to villous, distally densely villous; involucral bracts 3 (secondary involucres with 2), remotely subtending flowers, (1–)2-tiered, simple, ± similar to basal leaves, broadly obtriangular, 3-cleft, 3–10 cm, bases broadly cuneate, connate, margins sharply, irregularly serrate and incised on distal 1/3–1/2, apex acuminate, surfaces puberulous, more so abaxially; segments 3, lanceolate to oblanceolate; lateral segments unlobed or 1×-lobed; ultimate lobes (8–)10–15(–20) mm wide. **Flowers:** sepals (4–)5(–6), white, obovate, (8–)10–20(–25) × 5–15 mm, hairy or glabrous; stamens 80–100. **Heads of achenes** spheric to ovoid; pedicels 7.5–11.5 cm. **Achenes:** body obovoid to ellipsoid, (2.5–)3–6 × 3.5–6 mm, winged, strigose or glabrate; beak straight, 2–6 mm, strigose, not plumose. $2n = 14$.

Flowering spring–summer (May–Aug). Damp thickets, meadows, wet prairies, lake shores, streamsides, clearings, occasionally swampy areas; 200–2800 m; Alta., B.C., Man., N.B., Nfld., N.W.T., N.S., Ont., P.E.I., Que., Sask.; Colo., Conn., Ill., Ind., Iowa, Kans., Maine, Md., Mass., Mich., Minn., Mo., Mont., Nebr., N.H., N.J., N.Mex., N.Y., N.Dak., Ohio, Pa., S.Dak., Vt., W.Va., Wis., Wyo.

Various parts of *Anemone canadensis* were used medicinally by Native Americans in the treatment of wounds, nasal hemorrhages, eye problems, and sore throats, to counteract witch medicines, and as a general panacea (D. E. Moerman 1986).

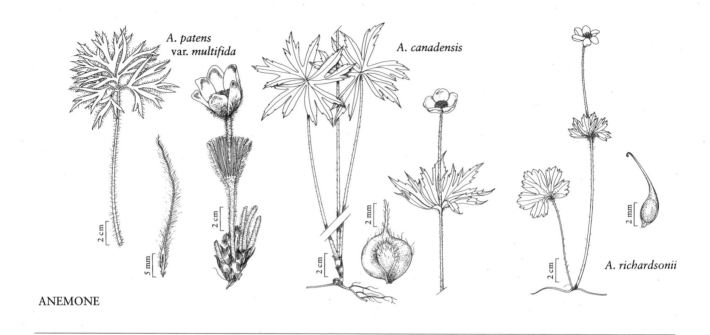

ANEMONE

---

4. Anemone richardsonii Hooker, Fl. Bor.-Amer. 1: 6. 1829 · Yellow anemone, anémone de Richardson F

Anemone vahlii Hornemann

Aerial shoots 5–30 cm, from rhizomes, rhizomes horizontal. Basal leaves 1, simple; petiole (0.8–)2–8(–12) cm; leaf blade reniform to ± orbiculate, (1–)1.5–3(–4) × 1.5–4 cm, base somewhat cordate, margins crenate to serrate ± throughout, apex acute to broadly acute, surfaces glabrous or ± villous, or abaxially glabrescent; segments primarily 3, rhombic; lateral segments unlobed or 1×-lobed; ultimate lobes 4–10(–15) mm wide. Inflorescences 1-flowered; peduncle puberulous; involucral bracts 2–3, 1-tiered, simple, ± similar to basal leaves, primarily rhombic, 1–3.5 cm, bases distinct, cuneate, margins crenate on distal 1/2, apex acute to narrowly acute, surfaces glabrous or ± villous, or abaxially glabrescent; segments 3, primarily rhombic; lateral segments unlobed, 2–10 mm wide. Flowers: sepals (4–)6(–8), yellow (rarely abaxially yellow, tinged blue, and adaxially yellow), elliptic, 8–15 × 4–10 mm, abaxially villous, especially proximally, adaxially glabrous; stamens 25–55. Heads of achenes loosely spheric; pedicels 3–20 cm. Achenes: body ovoid to oblong, 3–4 × ca. 1.5 mm, not winged, glabrous; beak recurved, 4–6 mm, glabrous. 2n = 14.

Flowering spring–summer (May–Aug). Thickets, moist woods, meadows, slopes; 20–2200 m; Greenland; Alta., B.C., Man., Nfld., N.W.T., Que., Sask., Yukon; Alaska; Asia.

5. Anemone narcissiflora Linnaeus, Sp. Pl. 1: 542. 1753 · Narcissus-flowered anemone

Aerial shoots 7–60 cm, from caudices, caudices ascending to vertical. Basal leaves 3–10, ternate; petiole (2–)4–20 cm; terminal leaflet ± sessile, obtriangular to oblanceolate, (2.5–)3–6(–9) × 2–10 cm, base narrowly cuneate to cuneate, margins incised (sometimes with few serrate teeth) on distal 1/3, apex acute to obtuse, surfaces glabrous or puberulous to villous or pilose; lateral leaflets 1–3×-parted and -lobed; ultimate lobes 3–10 mm wide. Inflorescences 2–8-flowered umbels or flowers solitary; peduncle puberulous to villous or pilose to nearly glabrous; involucral bracts (2–)3, 1-tiered, simple, greatly reduced, otherwise similar to basal leaves, obtriangular, distally 3-cleft and pinnatifid, (1–)1.5–5(–5.5) cm, bases clasping, ± connate, margins incised on distal 1/3, apex acuminate-acute to obtuse, surfaces glabrous or puberulous to villous or pilose; segments primarily 3, subulate or narrowly obtriangular; lateral segments unlobed or 2–3×-parted and -lobed; ultimate lobes 3–10 mm wide. Flowers: sepals 5–9, white or yellow, or abaxially white, tinged blue, white, or blue, and adaxially white, ovate to rhombic or obovate, 8–20 × 5–13(–15) mm, glabrous; stamens 40–80(–100). Heads of achenes spheric; pedicels (4.5–)5–14(–18.5) cm. Achenes: body ellipsoid to ovate, flat, 5–9 × (3–)4–6 mm, winged, glabrous; beak curved to recurved, 0.8–1.5 mm, glabrous. 2n = 14.

Varieties ca. 12 (3 in the flora): North America, Eurasia.

J. Jalas (1988), W. Greuter (1989), W. Greuter et al. (1989), J. Jalas and J. Suominen (1989), and T. G. Tutin et al. (1993+, vol. 1) have recently used the name *Anemone narcissifolia* Linnaeus because they considered *Anemone narcissiflora* an illegitimate name. B. E. Dutton et al. (1995) recently proposed to conserve the orthography of *Anemone narcissiflora,* and the authors of this treatment follow 14A.1 of the *Code,* which recommends following "existing usage as far as possible pending the General Committee's recommendation on the proposal" (W. Greuter et al. 1994).

The taxonomy of this highly variable, widespread species is extremely controversial. The conservative approach taken here most closely approximates S. L. Welsh's (1974) treatment for the Alaskan varieties. E. Hultén's discussion (1941–1950, vol. 4, pp. 735–736) of local races and the variation within this species, however, clearly illustrates the need for a thorough biosystematic investigation. Recognition of about 12 varieties is in light of S. V. Juzepczuk's (1970) work; however, he elevated local races to specific rank in his treatment.

The Aleuts used *Anemone narcissiflora* (no varieties specified) medicinally as an antihemorrhagic (D. E. Moerman 1986).

1. Lateral segments of involucral bracts lobed; petiole of basal leaves 15–20 cm; inflorescences umbels; stamens 60–80(–100) . . . . 5b. *Anemone narcissiflora* var. *villosissima*
1. Lateral segments of involucral bracts unlobed; petiole of basal leaves (2–)4–10 cm; inflorescences umbels or flowers solitary; stamens 40–60.
    2. Sepals (when fresh) white or abaxially white or blue and adaxially white (drying yellow); body of achenes 6–9 mm . . . . 5a. *Anemone narcissiflora* var. *monantha*
    2. Sepals (when fresh) yellow; body of achenes ca. 5 mm. . . . . 5c. *Anemone narcissiflora* var. *zephyra*

5a. Anemone narcissiflora Linnaeus var. **monantha** de Candolle, Syst. Nat. 1: 213. 1817

Anemone narcissiflora subsp. *alaskana* Hultén; *A. narcissiflora* subsp. *interior* Hultén; *A. narcissiflora* subsp. *sibirica* (Linnaeus) Hultén; *A. narcissiflora* var. *uniflora* Eastwood

Aerial shoots 10–50 cm. **Basal leaves** 3–5; petiole (2–)4–7 cm; leaflets puberulous to moderately villous and glabrate; lateral leaflets 1–2×-lobed; ultimate lobes 3–6 mm wide. **Inflorescences** 2–3(–5)-flowered umbels or flowers solitary; peduncle puberulous to villous to nearly glabrous; involucral bracts (2–)3, obtriangular, 3-cleft, (1–)1.5–2.5 cm, surfaces puberulous to moderately villous and glabrate; lateral segments unlobed. **Flowers:** sepals 5–8, white, or abaxially white or blue and adaxially white; stamens 40–60. **Pedicels** (4.5–)7–14

(–18.5) cm in fruit. **Achenes:** body 6–9 mm.

Flowering spring–summer (May–Jul). Moist slopes, grassy areas, thickets; 0–1700 m; B.C., N.W.T., Yukon; Alaska; Asia.

5b. Anemone narcissiflora Linnaeus var. **villosissima** de Candolle in A. P. de Candolle and A. L. P. de Candolle, Prodr. 1: 22. 1824 [F]

Anemone narcissiflora subsp. *villosissima* (de Candolle) Hultén

Aerial shoots (15–)25–60 cm. **Basal leaves** (3–)5–10; petiole (10–)15–20 cm; leaflets villous; lateral leaflets 2–3×-lobed; ultimate lobes 3–10 mm wide. **Inflorescences** (3–)5–8-flowered umbels; peduncle villous; involucral bracts obtriangular, distally pinnatifid, (2.5–)3–5(–5.5) cm, surfaces villous; lateral segments 2–3×-lobed. **Flowers:** sepals 5–6, white or often abaxially white, tinged blue, and adaxially white; stamens 60–80(–100). **Pedicels** 5–10(–13) cm in fruit. **Achenes:** body 6–9 mm.

Flowering spring–summer (May–Aug). Meadows, tundra; 0–900 m; Alaska; Asia.

5c. Anemone narcissiflora Linnaeus var. **zephyra** (A. Nelson) B. E. Dutton & Keener, Phytologia 77: 85. 1994 [E]

Anemone zephyra A. Nelson, Bot. Gaz. 42: 51. 1906; *A. narcissiflora* subsp. *zephyra* (A. Nelson) Á. Löve, D. Löve, & B. M. Kapoor

Aerial shoots 7–40 cm. **Basal leaves** 3–7; petiole (4–)5–10 (–15) cm; leaflets glabrous, sparsely pilose, or villous; lateral leaflets 2(–3)×-lobed; ultimate lobes 3–8 mm wide. **Inflorescences** 2–3(–4)-flowered umbels or flowers solitary; peduncle pilose or villous; involucral bracts obtriangular, 3-cleft, (1.5–)2.5–5 cm, surfaces glabrous, sparsely pilose, or villous; lateral segments unlobed. **Flowers:** sepals 5–9, yellow or yellowish; stamens 40–60. **Pedicels** 5–10 cm in fruit. **Achenes:** body ca. 5 mm.

Flowering summer (Jun–Aug). Forest margins, alpine meadows; 2700–4000 m; Colo., Wyo.

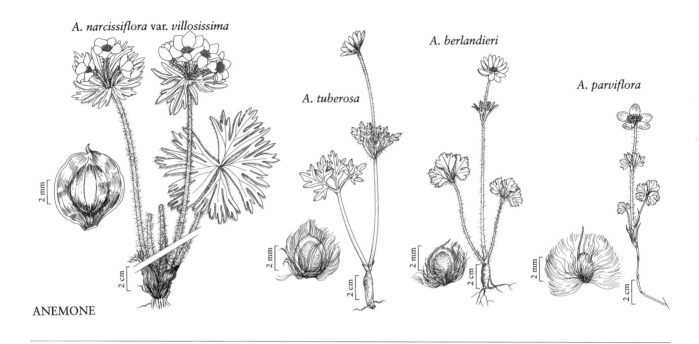

A. narcissiflora var. *villosissima*

A. tuberosa

A. berlandieri

A. parviflora

ANEMONE

6. Anemone tuberosa Rydberg, Bull. Torrey Bot. Club 29: 151. 1902 · Desert anemone F

Aerial shoots 10–30(–40) cm, from caudex-like tubers, tubers ascending to vertical. **Basal leaves** 1–3(–5), 1–2-ternate; petiole 5–7 cm; terminal leaflet sessile, rarely petiolulate, irregularly oblanceolate, (1.5–)2–3(–3.5) × 1–2(–2.5) cm, base narrowly cuneate, margins incised to dissected on distal 2/3, apex broadly acute, surfaces nearly glabrous; lateral leaflets 1–2×-parted and/or -lobed; ultimate lobes 4–8(–12) mm wide. **Inflorescences** 2–3(–5)-flowered cymes or flowers solitary; peduncle proximally nearly glabrous, distally villous; involucral bracts primarily 3, (1–)2-tiered, simple, ± similar to basal leaves, pinnatifid, 1.5–5.5 cm (primary involucral bracts 2–5.5 cm, secondary involucral bracts 1.5–3.5 cm), bases distinct, broadly to narrowly cuneate to clasping, margins irregularly serrulate and pinnatifid on ca. distal 1/2, apex narrowly acute to acuminate, surfaces thinly pilose; segments primarily 3, linear to pinnatifid; lateral segments 1–2×-parted and/or -lobed; ultimate lobes 1.5–2.5 mm wide. **Flowers:** sepals 8–10, pink to white, linear-oblong, 10–14(–20) × (2–)3–5(–6) mm, sparsely hairy; stamens 50–60. **Heads of achenes** fusiform; pedicels (5–)7–15(–22) cm. **Achenes:** body orbiculate, flat, 2.5–3.5 × 2–2.5 mm, not winged, densely villous; beak straight, ca. 1.5 mm, minutely puberulous, not plumose. $2n = 16$.

Flowering spring (Apr–May). Rocky slopes, stream-sides; 800–2500 m; Ariz., Calif., Nev., N.Mex., Tex., Utah; n Mexico.

The tuberous anemones in North America (*A. tuberosa*, *A. edwardsiana*, *A. okennonii*, *A. berlandieri*, and *A. caroliniana*) are closely related to each other and to the South American species *A. decapetala* Arduino, *A. triternata* Vahl, *A. cicutifolia* I. M. Johnston, and *A. sphenophylla* Poeppig. Particularly useful characters in identifying North American plants are number of tiers of involucral bracts, presence or absence of heterophylly between basal leaves and involucral bracts, and presence or absence of rhizomes. The current treatment of this group primarily follows C. S. Keener and B. E. Dutton (1994), who included a discussion of the relationships among its members. For a cytotaxonomic discussion of most members of this group, see C. Joseph and M. Heimburger (1966).

7. Anemone edwardsiana Tharp, Amer. Midl. Naturalist 33: 669, figs. 2,3. 1945 E

Aerial shoots (20–)30–50 cm, from tubers, tubers ± vertical. **Basal leaves** 3–6, 1–(nearly 2)-ternate; petiole 8–15 cm; terminal leaflet petiolulate, oblanceolate to obovate, 2–5 × 2–2.5 cm, base broadly cuneate, margins coarsely crenate on distal 1/2, apex obtuse to broadly acute, surfaces glabrous; lateral leaflets 1×-lobed; ultimate lobes 4–8(–12) mm wide. **Inflorescences** (1–)2–5-flowered cymes; peduncle proximally nearly glabrous, distally densely pubescent; involucral bracts primarily 3, 2-tiered, simple, dissimilar to basal leaves, 3-cleft to pinnatifid, 2–5 cm, bases clasping, ± connate, margins coarsely crenate throughout, apex narrowly acute to acuminate, surfaces glabrous; segments primarily 3, linear; lateral segments unlobed or occasionally 1×-lobed; ultimate lobes 3–5 mm wide. **Flowers:** sepals (8–)10–20, white to bluish, oblanceolate, 10–16 × 2–3 mm, sparsely hairy to nearly glabrous; stamens 50–60. **Heads of achenes** obconic to fusiform; pedicels 10–20 cm. **Achenes:** body orbiculate, flat, not winged, ± dull or varnished, woolly with long tuft of hairs at base, or glabrous; beak straight, rarely curved, 0.5–1 mm, glabrous or sparsely villous proximally, not plumose. 2*n* = 16.

Anemone edwardsiana is found only in Texas on the Edwards Plateau.

Varieties 2 (2 in the flora): North America.

1. Achenes woolly, ± dull . . . .
. . . . 7a. *Anemone edwardsiana* var. *edwardsiana*
1. Achenes glabrous, varnished. . . . .
. . . . 7b. *Anemone edwardsiana* var. *petraea*

7a. Anemone edwardsiana Tharp var. **edwardsiana** E

Achenes ± dull, woolly.

Flowering winter–spring (Feb–Apr). Moist, limestone bluffs and ledges; 200–500 m; Tex.

7b. Anemone edwardsiana Tharp var. **petraea** Correll, Madroño 19: 189. 1968 · Edge Falls anemone C E

Achenes varnished, glabrous.

Flowering winter–spring (Feb–Apr). Moist, rocky crevices; of conservation concern; 300–500 m; Tex.

8. Anemone okennonii Keener & B. E. Dutton, Sida 16: 198. 1994 · O'kennon's anemone E

Anemone tuberosa Rydberg var. *texana* Enquist & Crozier

Aerial shoots 20–30 cm, from tubers, tubers ± vertical. **Basal leaves** 4–10, 2–3-ternate or irregularly so; petiole 5–10 cm; terminal leaflet petiolulate, obovate, 1.2–1.8(–20) × ca. 1.5 (–2.5) cm, base narrowly cuneate, margins coarsely serrate to nearly incised on distal 1/2, ciliate or not, apex broadly acute, surfaces nearly glabrous; lateral leaflets 1–2×-lobed or -parted; ultimate lobes 2–5 mm wide. **Inflorescences** (1–)2–3-flowered cymes; peduncle proximally nearly glabrous,

distally downy; involucral bracts primarily 3, (1–)2-tiered, simple, dissimilar to basal leaves, obtriangular, 3-cleft to pinnatifid, 2–5 cm, bases clasping, ± connate, margins incised throughout, apex acuminate, surfaces appressed-pilose; segments primarily 3, linear; lateral segments unlobed or 1×-lobed; ultimate segments 1–2 mm wide. **Flowers:** sepals 7–11, greenish white to abaxially reddish, oblong, 6–12(–14) × (2–) 3–4.5 mm, abaxially hairy, adaxially glabrous; stamens 35–55. **Heads of achenes** oblong-ellipsoid; pedicels 10–18 cm. **Achenes:** body ovate, flat, not winged, densely white-villous (with long tuft of hairs at base); beak ± straight, ca. 1 mm, hidden in achene indument, white-tomentose, not plumose.

Flowering spring (Mar–Apr). Dry, open ledges and slopes; 500–1500 m; Tex.

9. Anemone berlandieri Pritzel, Linnaea 15: 628. 1841 · Tenpetal anemone F

Anemone heterophylla Nuttall ex A. W. Wood

Aerial shoots (10–)30–50 cm, from tubers, rarely also from rhizomes, tubers vertical, rarely ascending. **Basal leaves** (1–)3–6(–9), 1–(nearly 2)-ternate; petiole 3–20 cm; terminal leaflet sessile, rarely petiolulate, ovate to obovate, 1–3(–6) × (0.4–)0.7–3(–4) cm, base broadly cuneate, margins crenate to serrate on distal 1/2, rarely on distal 2/3, apex obtuse to broadly acute, surfaces pubescent; lateral leaflets unlobed or 1×-lobed or -parted; ultimate lobes (6–)8–12 mm wide. **Inflorescences** 1-flowered; peduncle villous; involucral bracts 3, borne above middle of scape at anthesis, 1-tiered, simple, dissimilar to basal leaves, obtriangular, 3-cleft, (1–)1.5–4(–5) cm, bases clasping, ± connate, margins incised throughout, apex narrowly acute to acuminate, surfaces pilose; segments primarily 3, broadly linear; lateral segments unlobed or occasionally 1×-lobed; ultimate lobes 1.5–2.5 mm wide. **Flowers:** sepals (7–)10–17(–25), abaxially blue to violet, adaxially white, linear-oblong, rarely narrowly elliptic, (10–)15–20 × 2–3(–4) mm, abaxially hairy, especially toward base, adaxially glabrous or hairy toward base; stamens 60–70. **Heads of achenes** cylindric; pedicels (7–)9–15(–28) cm. **Achenes:** body elliptic, flat, 2.7–3.5 × 2.2–2.5 mm, not winged, densely white-woolly; beak subulate, curved, 1–2.3 mm, hidden in achene indument, white tomentose, not plumose. 2*n* = 16.

Flowering winter–spring (Feb–Apr). Open grasslands, prairies, hillsides, often limy substrate, also woods over thin shale; 60–1100 m; Ala., Ark., Fla., Kans., La., Miss., N.C., Okla., S.C., Tex., Va.; n Mexico.

10. Anemone caroliniana Walter, Fl. Carol., 157. 1788 · Carolina anemone E

Aerial shoots 5–35(–60) cm, from tubers with rhizomes near apex, tubers ascending or vertical, rhizomes horizontal or ascending. **Basal leaves** 1–3(–5), 1–2-ternate; petiole 3–9(–10) cm; terminal leaflet petiolulate, rarely sessile, obtriangular, 2–3-cleft, 1–3 × 0.8–2(–2.5) cm, base narrowly attenuate to strongly and narrowly cuneate, margins serrate, rarely dissected on distal 2/3, apex obtuse to broadly acute, surfaces glabrous; lateral leaflets ± pinnately 1–2(–3)×-lobed or -parted; ultimate lobes 2–10 mm wide. **Inflorescences** 1-flowered; peduncle proximally glabrous, distally villous; involucral bracts 3, borne below middle of scape at anthesis, 1-tiered, simple, ± similar to basal leaves, obtriangular, 3-cleft, 1–2.5 cm, bases somewhat clasping, connate, margins incised on distal 1/2, apex acuminate, surfaces sparsely pilose; segments 3, broadly linear; lateral segments 1–2×-lobed or -parted; ultimate lobes 1–2.5 mm wide. **Flowers:** sepals 10–20(–30), white to rose or blue to purple, linear to oblong, (6–)10–20(–22) × 2–5(–7) mm, abaxially sparsely hairy, adaxially glabrous or nearly so; stamens 50–60. **Heads of achenes** fusiform; pedicels (5–)10–20 cm. **Achenes:** body ovoid, 1.5–2.5 (–3) × ca. 2 mm, not winged, densely woolly; beak straight, 1.3–2 mm, projecting beyond achene indument, proximally tomentose, not plumose. $2n = 16$.

Flowering winter–spring (Feb–Apr). Dry prairies, barrens, pastures, meadows, rocky, open woods; 60–700 m; Ala., Ark., Ga., Ill., Iowa, Kans., La., Minn., Miss., Mo., Nebr., Okla., S.C., S.Dak., Tenn., Tex., Wis.

Anemone caroliniana was recorded from Indiana (Vigo County); it has since been extirpated.

11. Anemone parviflora Michaux, Fl. Bor.-Amer. 1: 319. 1803 · Northern anemone, small-flowered anemone F

Anemone borealis Richardson; *A. parviflora* Michaux var. *grandiflora* Ulbrich

Aerial shoots 5–30(–35) cm, from short caudices on rhizomes, rhizomes primarily horizontal. **Basal leaves** 1–5(–7), ternate; petiole 1–10 cm; terminal leaflet sessile, obtriangular, (0.5–)0.7–1.8(–2.2) × 0.5–1.3 cm, base cuneate, margins crenate to broadly serrate on distal 1/3, apex obtuse to truncate, surfaces villous to nearly glabrous; lateral leaflets usually 1×-lobed or -parted; ultimate lobes 4–15 mm wide. **Inflorescences** 1-flowered; peduncle villous; involucral bracts 2–3, 1-tiered, simple, ± similar to terminal leaflets of basal leaves, obtriangular, 3-cleft, 0.5–2.5 cm, bases distinct, cuneate, margins crenate to broadly serrate, surfaces villous to nearly glabrous; segments 3, oblanceolate to obovate; lateral segments unlobed, 2–8 mm wide. **Flowers:** sepals 4–7, white or tinged blue or abaxially white, proximally blue, and adaxially white, broadly elliptic to ovate, (7–)8–20 × 4–9 mm, abaxially hairy, adaxially glabrous; stamens 70–80. **Heads of achenes** spheric; pedicels 4–18 cm. **Achenes:** body obovoid, 2–2.5 × ca. 1 mm, not winged, densely woolly; beak straight, 1–2.5 mm, glabrous. $2n = 16$.

Flowering spring–summer (May–Aug). Streamsides, meadows, rocky slopes; 0–3800 m; Alta., B.C., Man., N.B., Nfld., N.W.T., Ont., Que., Sask., Yukon; Alaska, Colo., Idaho, Mont., Oreg., Utah, Wash., Wyo.; Asia.

12. Anemone drummondii S. Watson, Bot. California 2: 424. 1880 · Drummond's anemone

Aerial shoots (7–)10–25(–30) cm, from caudices, caudices ascending to primarily vertical. **Basal leaves** 5–15, 2-ternate, occasionally irregularly so; petiole 2–10 cm; terminal leaflet sessile or basally attenuate and appearing petiolulate, obovate to obtriangular, 0.5–3 × 0.5–2 cm, base narrowly cuneate to cuneate, margins incised to dissected on distal 1/3–1/2, apex broadly acute to obtuse, surfaces villous; lateral leaflets 2×-parted, division frequently irregular; ultimate segments 1–2.6 mm wide. **Inflorescences** 1(–2)-flowered; peduncle villous; involucral bracts 3(–4), 1-tiered, ± similar to basal leaves, highly reduced, 2-ternate or irregularly so, bases distinct; terminal leaflet sessile or basally attenuate and appearing petiolulate, obovate to pinnatifid, 1–3.5 × 0.5–2 cm, bases narrowly cuneate to cuneate, margins incised to dissected on distal 1/3–1/2, apex broadly acute to obtuse, surfaces villous; lateral leaflets 2×-parted, division frequently irregular; ultimate segments 1–2.5 mm wide. **Flowers:** sepals (5–)6–9, white, or abaxially white, tinged blue, and adaxially white, ovate, rarely oblong or narrowly obovate, 8–20 × 6–10 mm, abaxially hairy, rarely glabrous, adaxially glabrous; stamens 80–100, whitish; styles white. **Heads of achenes** spheric, rarely cylindric; pedicels (2–)3–10 cm. **Achenes:** body ovoid, 2–4 × 1–1.5 mm, not winged, woolly; beak straight, 2–4(–6) mm, glabrous.

Varieties 2 (2 in the flora): North America, Asia.

Anemone drummondii is an extremely variable species whose circumscription is controversial. Some plants appear intermediate between this species and *A. multifida*; cytologically the two are quite distinct (G. Boraiah and M. Heimburger 1964; C. L. Hitchcock et al. 1955–1969, vol. 2).

1. Ultimate segments of lateral leaflets 1–1.5(–2) mm wide, leaflets villous . . . . 12a. *Anemone drummondii* var. *drummondii*

1. Ultimate segments of lateral leaflets 1.5–2.6 mm
wide, leaflets nearly glabrous or pilose . . . .
. . . . .12b. *Anemone drummondii* var. *lithophila*

12a. Anemone drummondii S. Watson var. drummondii

Anemone cairnesiana Greene; *A. californica* Eastwood

Aerial shoots (7–)10–25(–30) cm. **Basal leaves** 5–15; petiole 2–10 cm; leaflets villous; ultimate segments of lateral leaflets 1–1.5(–2) mm wide. **Inflorescences:** peduncle villous; involucral bracts villous; ultimate segments of lateral leaflets 1–1.5(–2) mm wide. **Flowers:** sepals white, tinged blue, or abaxially white, tinged blue, and adaxially white, abaxially hairy, rarely glabrous. **Heads of achenes** spheric, rarely cylindric. $2n = 32$.

Flowering summer (Jun–Aug). Gravelly slopes, dry rocky ledges; 1200–3300 m; B.C.; Calif., Idaho, Oreg., Wash.; Asia.

12b. Anemone drummondii S. Watson var. lithophila (Rydberg) C. L. Hitchcock in C. L. Hitchcock et al., Vasc. Pl. Pacif. N.W. 2: 325. 1964 Ⓔ

Anemone lithophila Rydberg, Bull. Torrey Bot. Club 29: 152. 1902; *A. globosa* var. *lithophila* (Rydberg) M. Peck

Aerial shoots (7–)10–25 cm. **Basal leaves** 5–12(–15); petiole 5–9 cm; leaflets nearly glabrous or pilose; ultimate segments of lateral leaflets 1.5–2.6 mm wide. **Inflorescences:** peduncle pilose; involucral bracts nearly glabrous or pilose; ultimate segments of lateral leaflets 1.5–2.6 mm wide. **Flowers:** sepals white, tinged blue, rarely abaxially white and blue, adaxially white, abaxially hairy; filaments yellow. **Heads of achenes** spheric. $2n = 48$.

Flowering summer (Jun–Aug). Slopes, ridges; 2100–3300 m; Alta., B.C., N.W.T., Yukon; Alaska, Idaho, Mont., Wyo.

13. Anemone multiceps (Greene) Standley, Field Mus. Nat. Hist., Bot. Ser. 8(5): 310. 1931 (as multifida) · Purple anemone Ⓔ

Pulsatilla multiceps Greene, Erythea 1: 4. 1893

Aerial shoots (3–)5–15 cm, from caudices, caudices ascending to primarily vertical. **Basal leaves** 5–20(–30+), 2-ternate, occasionally irregularly so; petiole 2–5(–10) cm; terminal leaflet sessile or basally attenuate and ap-pearing petiolulate, obovate to obtriangular, 0.5–2(–3) × 0.5–2 cm, bases narrowly cuneate to cuneate, margins incised to dissected on distal 1/3–1/2, apex broadly acute to obtuse, surfaces sparsely villous, often nearly glabrous; lateral leaflets 2×-lobed, lobing frequently irregular; ultimate lobes 1–2.6 mm wide. **Inflorescences** 1-flowered; peduncle villous; involucral bracts 3(–4), 1-tiered, ± similar to basal leaves, highly reduced, 2-ternate or irregularly so, bases distinct; terminal leaflet sessile or basally attenuate and appearing petiolulate, obovate to pinnatifid, 1–2(–2.5) × 0.5–2 cm, base narrowly cuneate to cuneate, margins incised to dissected on distal 1/3–1/2, apex broadly acute to obtuse, surfaces villous; lateral leaflets 2×-parted, division frequently irregular; ultimate segments 1–2.5 mm wide. **Flowers:** sepals 5–6, blue to purple, oblong or narrowly obovate, 8–20 × 6–10 mm, abaxially villous; stamens 80–100, purple; styles red. **Heads of achenes** spheric, rarely cylindric; pedicels 2–6(–18) cm. **Achenes:** body ovoid, 2–4 × 1–1.5 mm, not winged, villous; beak straight, 2–4(–6) mm, glabrous. $2n = 16$.

Flowering spring–summer (May–Jul). Tundra, slopes, ridges, limestone outcrops, and screes; 100–2100 m; Yukon; Alaska.

Anemone multiceps has uniformly purple sepals, purple stamens, and red styles, whereas *A. drummondii* (morphologically most similar to *A. multiceps*) has white or adaxially white and abaxially bluish sepals, whitish stamens, and white styles. Future biosystematic analysis of these entities is needed to determine whether *A. multiceps* is indeed distinct from *A. drummondii*.

14. Anemone multifida Poiret in J. Lamarck et al., Encycl., suppl. 1: 364. 1810 · Cut-leaved anemone, anémone multifide

Aerial shoots 10–70 cm, from caudices, caudices ascending to vertical. **Basal leaves** 3–6(–10), 1–2-ternate; petiole (2–)4–10(–14) cm; terminal leaflet petiolulate to ± sessile, broadly and irregularly rhombic to obovate, (1.5–)2.5–4.5(–5.5) × (1–)3–10 cm, base narrowly cuneate, margins incised on distal 1/3, apex broadly acute to nearly obtuse; surfaces abaxially villous-silky, hispid to villous, or sparsely long-pilose, adaxially glabrous, nearly glabrous, villous-silky, or hispid to villous; lateral leaflets (2–)3×-parted; ultimate segments (1.5–)2–3.5(–5) mm wide. **Inflorescences** 2–7-flowered cymes or flowers solitary; peduncle villous, pilose, or hispid to villous; involucral bracts usually 3–5, occasionally 2 in secondary involucres, 1–2-tiered, ternate, occasionally incompletely ternate, ± similar to basal leaves, greatly reduced, bases distinct; terminal leaflet petiolulate to ± sessile, broadly and irregularly rhombic to obovate, (1.5–)3–4(–5) × 0.5–1(–2) cm, base narrowly cuneate, margins incised on distal 1/3, apex broadly acute to nearly obtuse, surfaces abaxially hispid to villous, villous-silky, or sparsely long-pilose, adaxially glabrous, nearly glabrous, hispid to villous, or villous-silky; lateral leaflets (2–)3×-parted or -lobed, lobes frequently unequal; ultimate lobes 1.5–3(–4.3) mm wide. **Flowers:** sepals 5–9, green to yellow, blue, purple, red, or occasionally white, or abaxially blue, red, yellow and red, or purple, or tinged purple, adaxially white, yellow, yellow and red, blue, or tinged purple, ovate to oblong, 5–17 × (3.5–)5–7(–9) mm, abaxially hairy, adaxially glabrous; stamens 50–80. **Heads of achenes** spheric; pedicels 6–15(–23) cm. **Achenes:** body irregularly ellipsoid or elliptic, flat, 3–4 × 1.5–2 mm, not winged, tomentose, woolly, or villous; beak ± straight, distally recurved or strongly hooked, 1–6 mm, glabrous.

Varieties 4 (4 in the flora): North America, South America.

G. Boraiah and M. Heimburger (1964) conducted an extensive cytotaxonomic analysis of this wide-ranging and extremely variable species and its relatives. The present treatment takes a broader view of the species (and its variation) and recognizes fewer entities. In addition, *Anemone tetonensis* and *A. stylosa*, plants treated as closely related species by G. Boraiah and M. Heimburger, are treated here as varieties of *A. multifida*.

Early-season plants of *Anemone multifida* var. *multifida* have solitary flowers and will key to var. *saxicola*. *Anemone multifida* var. *tetonensis* and especially var. *saxicola* might be based on characteristics that are influenced primarily by environment; further study is warranted.

Native Americans used *Anemone multifida* (no varieties specified) medicinally as an antirheumatic, cold remedy, nosebleed cure, and general panacea, as well as a means of killing lice and fleas (D. E. Moerman 1986).

1. Flowers in cymes; involucral bracts 2-tiered.
  2. Beak straight.
    3. Aerial shoots (30–)40–70 cm; flowers (2–) 5–7; bracts silky; sepals ovate or oblong ....14a. *Anemone multifida* var. *multifida*
    3. Aerial shoots (10–)20–40 cm; flowers (1–) 2–3; bracts villous; sepals elliptic....... .... 14b. *Anemone multifida* var. *saxicola*
  2. Beak recurved or hooked.
    4. Beak recurved .... 14c. *Anemone multifida* var. *tetonensis*
    4. Beak hooked .... 14d. *Anemone multifida* var. *stylosa*
1. Flowers solitary; involucral bracts 1-tiered.
  5. Abaxial color of sepals different from adaxial color.
    6. Aerial shoots 20–40 cm; achene beak ± straight ... 14b. *Anemone multifida* var. *saxicola*

6. Aerial shoots 10–20(–25) cm; achene beak recurved .... 14c. *Anemone multifida* var. *tetonensis*
  5. Abaxial color of sepals same as adaxial color.
    7. Sepals blue, purple, or sometimes white; achene beak recurved .... 14c. *Anemone multifida* var. *tetonensis*
    7. Sepals purple to red, or green to red; achene beak strongly hooked distally.... .... 14d. *Anemone multifida* var. *stylosa*

14a. Anemone multifida Poiret var. **multifida** [F]

Anemone globosa Nuttall ex A. Nelson; *A. multifida* var. *hudsoniana* de Candolle; *A. multifida* var. *nowasadii* B. Boivin; *A. multifida* var. *richardsiana* Fernald; *A. multifida* var. *sansonii* B. Boivin

Aerial shoots (30–)40–70 cm. **Basal leaves** silky. **Inflorescences** (2–)5–7-flowered cymes; peduncle villous; involucral bracts (1–)2-tiered, silky. **Flowers:** sepals 5(–9), green to yellow, blue, purple, or red (rarely abaxially blue or red and adaxially white or yellow), ovate or oblong, 6–17 mm. **Achenes:** beak ± straight. 2*n* = 32.

Flowering spring–summer (Apr–Jul). Open forests, grassy slopes, often rocky areas, also moist prairie depressions, coulees, sandhill shrubland; 0–3200 m; Alta., B.C., Man., N.B., Nfld., N.W.T., Ont., Que., Sask., Yukon; Alaska, Calif., Colo., Idaho, Maine, Mich., Minn., Mont., Nebr., Nev., N.Mex., N.Y., N.Dak., Oreg., S.Dak., Utah, Vt., Wash., Wis., Wyo.; South America (Chile and Argentina).

According to D. E. Moerman (1986), the Blackfeet burned *Anemone multifida* var. *multifida* on hot coals to alleviate headaches.

14b. Anemone multifida Poiret var. **saxicola** B. Boivin, Canad. Field-Naturalist 65: 2. 1951 [E]

Anemone multifida var. *hirsuta* C. L. Hitchcock

Aerial shoots (10–)20–40 cm. **Basal leaves** villous. **Inflorescences** (1–)2–3-flowered cymes; peduncle villous; involucral bracts 1–2-tiered, villous. **Flowers:** sepals 5, abaxially red, purple, or blue, adaxially white or yellow, elliptic, 6–10 mm. **Achenes:** beak ± straight. 2*n* = 32.

Flowering summer (Jun–Aug). Rocky slopes, meadows; 1300–4200 m; Alta., B.C., N.W.T.; Alaska, Colo., Idaho, Mont., Nev., Oreg., Utah, Wash., Wyo.

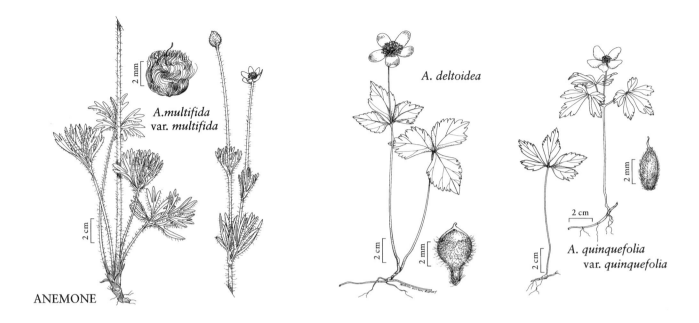

ANEMONE

A.*multifida*
var. *multifida*

A. deltoidea

A. quinquefolia
var. *quinquefolia*

14c. Anemone multifida Poiret var. **tetonensis** (Porter ex Britton) C. L. Hitchcock in C. L. Hitchcock et al., Vasc. Pl. Pacif. N.W. 2: 327. 1964 [E]

Anemone tetonensis Porter ex Britton, Ann. New York Acad. Sci. 6: 224. 1891

Aerial shoots 10–20(–25) cm. **Basal leaves** hispid to villous. **Inflorescences** (2–)3-flowered cymes or flowers solitary; peduncle hispid to villous; involucral bracts 1(–2)-tiered, hispid to villous. **Flowers:** sepals 5–7(–9), usually blue or purple, sometimes white, or abaxially yellow and red, blue, or tinged purple, and adaxially yellow, red, blue, or tinged purple, oblong to elliptic or ovate, 5–15 mm. **Achenes:** beak recurved. $2n = 32$.

Flowering spring–summer (Mar–Aug). Talus slopes, canyons; 2200–3700 m; Idaho, Mont., Nev., Oreg., Utah, Wyo.

14d. Anemone multifida Poiret var. **stylosa** (A. Nelson) B. E. Dutton & Keener, Phytologia 77: 84. 1994 [C][E]

Anemone stylosa A. Nelson, Bot. Gaz. 42: 52. 1906

Aerial shoots 10–15 cm. **Basal leaves** abaxially sparsely long-pilose, adaxially glabrous or nearly glabrous. **Inflorescences:** flowers solitary; peduncle sparsely long-pilose; involucral bracts 1-tiered, abaxially sparsely long-pilose, adaxially glabrous or nearly gla-

brous. **Flowers:** sepals 6(–8), purple to red or green to red, ovate to oblong, 8–10 mm. **Achenes:** beak strongly hooked distally. $2n = 32$.

Flowering summer (Jul–Aug). Rocky slopes; of conservation concern; 2700–3700 m; Ariz., Utah.

15. Anemone deltoidea Hooker, Fl. Bor.-Amer. 1: 6, plate 3, fig. A. 1829 · Columbia windflower, western white anemone [E][F]

Aerial shoots (7.5–)10–30 cm, from rhizomes, rhizomes horizontal. **Basal leaves** 0–2, ternate; petiole 10–15 cm; terminal leaflet sessile, ovate to rhombic, (2.5–)3–5(–6) × (2–)2.5–3.5 cm, base slightly oblique to broadly cuneate, margins crenate to serrate on distal 2/3, apex acuminate, surfaces abaxially glabrous, rarely puberulous on veins, adaxially usually glabrous; lateral leaflets unlobed, rarely 1×-lobed; ultimate lobes (8–)10–20 mm wide. **Inflorescences** 1-flowered; peduncle sparsely hispid or glabrous; involucral bracts 3, 1-tiered, simple, ± similar to terminal leaflets of basal leaves, ovate, rarely rhombic, (2.5–)4–8 cm, bases distinct, broadly cuneate, margins crenate to serrate, or dentate on distal 2/3, apex acuminate, surfaces usually glabrous; lateral leaves unlobed or occasionally 1×-lobed; ultimate lobes 8–15 mm wide. **Flowers:** sepals usually 5, white, ovate to obovate, rarely oblanceolate, (12–)15–25 × (8–)10–15(–20) mm, glabrous; stamens 100–120. **Heads of achenes** spheric; pedicels (5–)7–12 cm.

Achenes: body ovoid, swollen, 2.5–4 × 2–3 mm, not winged, proximally hispid, distally glabrous; beak straight to subulate, 0.5 mm or less, glabrous. $2n = 14$.

Flowering spring–summer (Apr–Jul). Open to deep woods; 200–2000 m; Calif., Oreg., Wash.

Reports of *Anemone deltoidea* in British Columbia are erroneous (H. J. Scoggan 1978–1979, vol. 3).

Rhizomes 7 m long and about 2 mm in diameter have been observed in *Anemone deltoidea.*

16. Anemone quinquefolia Linnaeus, Sp. Pl. 1: 541. 1753
· Wood anemone, anémone à cinq folioles

Aerial shoots 5–30 cm, from rhizomes, rhizomes primarily horizontal. **Basal leaves** 0–1, ternate; petiole 4–25 cm; terminal leaflet ± sessile, rhombic to oblanceolate or obliquely oblanceolate, 1–4.5 × 1.4–3.7 cm, base narrowly cuneate, margins crenate-serrate, serrate, occasionally dentate or incised on distal 1/3–1/2, apex acuminate to obtuse, surfaces glabrous or puberulous to pilose; lateral leaflets unlobed or 1×-lobed or -parted; ultimate lobes 4–15 mm wide. **Inflorescences** 1-flowered; peduncle puberulous, villous, or glabrous; involucral bracts 3, 1-tiered, ternate, ± similar to basal leaves, bases distinct; terminal leaflet petiolulate or sessile, rhombic to oblanceolate or obliquely oblanceolate, 1–5(–5.5) × 0.5–3 cm, base narrowly cuneate, margins crenate to serrate, serrate to dentate, occasionally incised on distal 1/3–1/2, apex acuminate to acute, surfaces glabrous or puberulous to pilose; lateral leaflets unlobed or 1×-lobed or -parted, ultimate lobes 5–20 mm wide. **Flowers:** sepals (4–)5(–6), white, rarely pink, or abaxially white or tinged pink or blue and adaxially white, oblong to ovate, 6–25 × 4–8 mm, glabrous; stamens 30–60. **Heads of achenes** spheric or nearly so; pedicels 1–6(–8) cm. **Achenes:** body ellipsoid or ovoid, 2.5–4.5 × 1–1.5 mm, not winged, puberulous or villous; beak straight to curved, rarely recurved, 0.5–2 mm, ± glabrous to sparsely puberulous, not plumose.

Varieties 2 (2 in the flora): North America.

Anemone quinquefolia and its relatives (*A. grayi, A. lancifolia, A. lyallii, A. oregana,* and *A. piperi*) are remarkably similar morphologically. Consideration of the sizes of various plant parts, degree of lobing of lateral leaflets, and orientation of the rhizome are particularly critical in distinguishing these species. This species complex also shares close morphologic affinities with both European (e.g., *A. nemorosa* Linnaeus) and Asian (e.g., *A. altaica* Fischer ex Ledebour) plants.

1. Achene body 3–4.5 mm; lateral leaflets of basal leaves frequently 1×-lobed or -parted. . . . .
. . . . .16a. *Anemone quinquefolia* var. *quinquefolia*
1. Achene body 2.5–3 mm; lateral leaflets of basal leaves unlobed or occasionally 1×-lobed. . . . . .
. . . . 16b. *Anemone quinquefolia* var. *minima*

16a. Anemone quinquefolia Linnaeus var. **quinquefolia** [E] [F]

Anemone nemorosa Linnaeus var. *bifolia* (Farwell) B. Boivin; *A. quinquefolia* var. *bifolia* Farwell

Aerial shoots 5–30 cm. **Leaves:** petiole 5–25 cm; terminal leaflet of basal leaves 1.5–4.5 × 1.4–3.7 cm, abaxially puberulous or glabrous, adaxially puberulous; lateral leaflets frequently 1×-lobed or -parted or unlobed; ultimate lobes 5–15 mm wide. **Inflorescences:** involucral bracts with terminal leaflet 1–5(–5.5) × 1–3 cm, base cuneate or ± oblique, margins crenate to serrate, sometimes incised, apex acuminate to acute, surfaces glabrous or puberulous; lateral leaflets frequently 1×-lobed or -parted, sometimes unlobed; ultimate lobes 5–20 mm wide. **Flowers:** sepals 6–25 × 4–8 mm; stamens 30–60. **Pedicels** 2–6(–8) cm in fruit. **Achenes:** body 3–4.5 mm. $2n = 32$.

Flowering spring–early summer (Mar–Jun). Moist, open woods, thickets, clearings, streamsides, occasionally swampy areas; 30–1900 m; Alta., Man., N.B., N.S., Ont., Que., Sask.; Ala., Conn., Del., D.C., Ga., Ill., Ind., Iowa, Ky., Maine, Md., Mass., Mich., Minn., Mo., N.H., N.J., N.Y., N.C., N.Dak., Ohio, Pa., R.I., S.C., S.Dak., Tenn., Vt., Va., W.Va., Wis.

Anemone quinquefolia var. *interior* Fernald is an often used, although illegitimate, name for *A. quinquefolia* var. *bifolia.* This taxon, characterized by its villous flowering stems, is scarcely worthy of varietal recognition inasmuch as local populations may have plants referable to both var. *bifolia* and var. *quinquefolia* (C. S. Keener 1975b).

Intergradation of characteristics between *Anemone quinquifolia* var. *minima* and var. *quinquefolia,* and between var. *quinquefolia* and *A. lancifolia* frequently makes identification extremely difficult in the area where the taxa overlap.

16b. Anemone quinquefolia Linnaeus var. **minima** (de Candolle) Frodin, Phytologia 77: 86. 1994 [E]

Anemone minima de Candolle, Syst. Nat. 1: 206. 1817

Aerial shoots 5–15 cm. **Leaves:** petiole 4–12 cm; terminal leaflet of basal leaves 1–2.5 × 1–2 cm, pilose; lateral leaflets unlobed or occasionally 1×-lobed; ultimate lobes 4–8 mm wide. **Inflorescences:** involucral bracts with terminal leaflet of 1–2 × 0.5–1 cm wide, base ± narrowly cuneate, margins crenate or serrate to dentate, apex acuminate to narrowly acute, surfaces pilose; lateral leaflets unlobed, rarely slightly 1×-lobed). **Flow-**

ers: sepals (7–)8(–11) × ca. 4 mm; stamens 30–40. **Pedicel** 1–4 cm in fruit. **Achenes:** body 2.5–3 mm. $2n = 32$.

Flowering spring (Apr–May). Damp, frequently acidic, wooded hillsides; 600–1000 m; N.C., Tenn., Va., W.Va.

17. Anemone lancifolia Pursh, Fl. Amer. Sept. 2: 386. 1814 · Lance-leaved anemone E

Anemone quinquefolia Linnaeus var. *lancifolia* (Pursh) Fosberg

Aerial shoots 10–40 cm, from rhizomes, rhizomes horizontal. **Basal leaves** 1, ternate; petiole 10–25 cm; terminal leaflet sessile, oblanceolate to ovate, (3.5–)4–7(–8) × 2.5–4(–6) cm, base broadly cuneate, margins coarsely serrate on distal 1/2–2/3, apex acuminate, surfaces ± glabrous; lateral leaflets unlobed or occasionally 1×-lobed; ultimate lobes 8–25 mm wide. **Inflorescences** 1-flowered; peduncle ± glabrous; involucral bracts 3, 1-tiered, ternate, ± similar to basal leaves, bases distinct; terminal leaflet sessile, oblanceolate to ovate, (2–)3–8.7 × 0.8–3 cm, bases narrowly cuneate to cuneate, margins coarsely serrate on distal 1/2–2/3, apex acuminate, surfaces ± glabrous; lateral leaflets unlobed or occasionally 1×-lobed; ultimate lobes 15–30 mm wide. **Flowers:** sepals (4–)5(–7), white, oblong to elliptic, (13–)15–20(–25) × 5–10 mm, glabrous; stamens 50–70. **Heads of achenes** nearly spheric; pedicels (3–)4–8(–10) cm. **Achenes:** body elliptic, flat, 3.5–5 × 1–1.5 mm, not winged, puberulous; beak straight or slightly curved, 1–1.5 mm, ± glabrous, not plumose.

Flowering spring–summer (Apr–Jun). Damp, rich woods; 800–1500 m; N.C., S.C., Va., W.Va.

18. Anemone piperi Britton ex Rydberg, Bull. Torrey Bot. Club 29: 153. 1902 · Piper's anemone E

Aerial shoots 10–35 cm, from rhizomes, rhizomes primarily vertical, occasionally strongly ascending. **Basal leaves** (0–)1–2, ternate; petiole 10–20 cm; terminal leaflet sessile, rhombic, lanceolate, or oblanceolate, (1.5–)2.5–6 × (1–)2–4 cm, base narrowly cuneate, margins coarsely serrate to coarsely dentate on distal 1/2–2/3, apex acuminate to narrowly acute, surfaces pilose or glabrous; lateral leaflets unlobed or sometimes 1×-lobed; ultimate lobes 10–19 mm wide. **Inflorescences** 1-flowered; peduncle coarsely pilose distally; involucral bracts 3, 1-tiered, ternate, ± similar to basal leaves, bases dis-

tinct; terminal leaflet sessile, rhombic, lanceolate, or oblanceolate, (1.5–)2–5.5(–7) × (0.6–)1–2.5 cm, base narrowly cuneate, margins coarsely serrate to coarsely dentate on distal 1/2–2/3, apex acuminate to narrowly acute, surfaces pilose or glabrous; lateral leaflets unlobed or sometimes 1×-lobed; ultimate lobes (5–)8–18 mm wide. **Flowers:** sepals 5–7, white, rarely pinkish, elliptic-obovate to ovate, (6–)8–20 × 6–8 mm, glabrous; stamens 35–55(–90). **Heads of achenes** nearly spheric; pedicels (1.5–)2–5 cm. **Achenes:** body ellipsoid to obliquely ovoid, 3–4 × 1.5–2 mm, not winged, villous; beak straight or slightly curved, 0.5–1 mm, glabrous or proximally minutely puberulous, not plumose.

Flowering spring–summer (Apr–Aug). Shaded, moist woods; 400–3000 m; B.C.; Idaho, Mont., Oreg., Utah, Wash.

Plants of *Anemone piperi* from southeastern Washington and northeastern Oregon (i.e., the westernmost limits of the species) are sometimes intermediate between *A. piperi* and *A. oregana*. Although they possess vertical rhizomes characteristic of *A. piperi*, they have the bluish or pinkish sepals of *A. oregana*. These plants are best referred to *A. piperi*, pending detailed biosystematic analysis.

19. Anemone oregana A. Gray, Proc. Amer. Acad. Arts 22: 308. 1887 · Oregon anemone, western wood anemone E

Aerial shoots 5–30(–35) cm, from rhizomes, rhizomes horizontal. **Basal leaves** 0–1, ternate; petiole 4–20 cm; terminal leaflet sessile to petiolulate, oblanceolate to rhombic, oblong, or ovate, 1–5(–6) × 0.7–2.5(–3.5) cm, base narrowly cuneate, margins sharply serrate on distal 1/2(–2/3), apex acuminate to acute, surfaces strigose to nearly glabrous; lateral leaflets unlobed or 1×-lobed; ultimate lobes 0.4–10 mm wide. **Inflorescences** 1-flowered; peduncle proximally glabrous, distally villous to pilose; involucral bracts 3, 1-tiered, ternate, ± similar to basal leaves, bases distinct; terminal leaflet sessile to petiolulate, oblanceolate to rhombic, oblong, or ovate, 1–8 × 0.8–3(–3.5) cm, bases narrowly cuneate to cuneate, margins crenate to serrate on distal 1/2(–2/3), apex acuminate to acute, surfaces abaxially glabrous or strigose, adaxially nearly glabrous to strigose; lateral leaflets unlobed or 1×-lobed; ultimate lobes 0.5–10 mm wide. **Flowers:** sepals 5–7(–8), blue to purple, reddish, or purple to pink (rarely nearly white or abaxially reddish, violet, or marginally purple, adaxially white), ovate, oblong, or elliptic, 10–20 × 5–8(–10) mm, glabrous; stamens 30–75. **Heads of achenes** nearly spheric; pedicels (1.5–)2–5(–7) cm. **Achenes:** body oblong to ellipsoid, 4–5 × 1.5–2 mm, not winged, puberulous to pilose, rarely glabrous; beak ± straight, (0.5–)1–1.5 mm, glabrous.

Varieties 2 (2 in the flora): North America.

1. Stamens 30–60; abaxial color of sepals same as adaxial color; margins of involucral bracts similar to those of basal leaves . . . . 19a. *Anemone oregana* var. *oregana*
1. Stamens 60–75; abaxial color of sepals different from adaxial color; margins of involucral bracts more coarsely serrate than those of basal leaves . . . . 19b. *Anemone oregana* var. *felix*

19a. Anemone oregana A. Gray var. oregana E F

Anemone adamsiana Eastwood; *A. quinquefolia* Linnaeus var. *oregana* (A. Gray) B. L. Robinson

Aerial shoots 8–30(–35) cm. **Basal leaves:** lateral leaflets 1×-lobed; ultimate lobes 0.4–2 mm wide. **Inflorescences:** involucral bracts with terminal leaflet 2–8 × 1–3(–3.5) cm, base cuneate, margins crenate to deeply serrate, apex acuminate to acute; lateral leaflets unlobed or 1×-lobed; ultimate lobes 0.5–2 mm wide. **Flowers:** sepals blue to purple, reddish, or purple to pink, rarely nearly white; stamens 30–60, occasionally more. **Achenes:** body puberulous to pilose, rarely glabrous. 2*n* = 16.

Flowering spring (Mar–Jun). Shaded, moist woods, open hillsides; 100–1800 m; Calif., Oreg., Wash.

19b. Anemone oregana A. Gray var. felix (M. Peck) C. L. Hitchcock in C. L. Hitchcock et al., Vasc. Pl. Pacif. N.W. 2: 329. 1964. C E

Anemone felix M. Peck, Torreya 32: 149. 1932

Aerial shoots 5–20(–30) cm. **Basal leaves:** lateral leaflets usually 1×-lobed; ultimate lobes 6–10 mm wide. **Inflorescences:** involucral bracts with terminal leaflet 1–4 × 0.8–2.5 cm, base narrowly cuneate, margins serrate, more sharply serrate than in basal leaves, apex acuminate to narrowly acute; lateral leaflets usually 1×-lobed; ultimate lobes 4–10 mm wide. **Flowers:** sepals abaxially reddish, violet, or marginally purple, adaxially white; stamens 60–75. **Achenes:** body puberulous to pilose.

Flowering spring (Apr–Jun). Marshes, sphagnum bogs; of conservation concern; 300–1000 m; Oreg., Wash.

20. Anemone lyallii Britton, Ann. New York Acad. Sci. 6: 227. 1891 · Little mountain anemone E

Anemone oligantha Eastwood; *A. quinquefolia* Linnaeus var. *lyallii* (Britton) B. L. Robinson

Aerial shoots 5–30(–40) cm, from rhizomes, rhizomes horizontal. **Basal leaves** 0–1, ternate; petiole 5–8 cm; terminal leaflet sessile to petiolulate, ovate to oblanceolate, (1–)1.5–3(–4) × (0.4–)0.7–1.5(–2) cm, base narrowly cuneate, margins serrate to crenate on distal 1/2–2/3, apex acuminate to narrowly acute, rarely obtuse, surfaces glabrous, rarely puberulous; lateral leaflets unlobed or occasionally nearly lobed. **Inflorescences** 1-flowered; peduncle distally puberulous or nearly glabrous throughout; involucral bracts 3, 1-tiered, ternate, ± similar to basal leaves, bases distinct; terminal leaflet sessile to petiolulate, ovate to oblanceolate, (1–)1.5–3(–4) × (0.4–)0.7–1.5(–2) cm, bases narrowly cuneate, margins serrate to crenate on distal 1/2–2/3, apex acuminate to narrowly acute, rarely obtuse, surfaces glabrous, rarely puberulous; lateral leaflets unlobed or occasionally nearly lobed. **Flowers:** sepals 5(–7), white, pink, or sometimes blue-tinged, oblong, rarely narrowly ovate, 3.5–8(–10) × 1.5–3(–3.5) mm, glabrous; stamens 10–30(–35). **Heads of achenes** nearly spheric; pedicels 1–3(–4) cm in fruit. **Achenes:** body elliptic, flat, 3–4 × 1–1.5 mm, not winged, white-puberulous; beak subulate, sometimes slightly curved, ca. 0.5 mm, glabrous. 2*n* = 16.

Flowering spring–summer (Mar–Jul). Shaded woods, subalpine ridges; 200–1900 m; B.C.; Calif., Oreg., Wash.

Anemone lyallii may occasionally intergrade with *A. oregana* west of the Cascades in northern Oregon (C. L. Hitchcock et al. 1955–1969, vol. 2). The area of probable intergradation should be extended to the southern limits of both species where they are sympatric.

21. Anemone grayi Behr & Kellogg, Bull. Calif. Acad. Sci. 1: 5. 1884 · Western windflower E

Anemone quinquefolia Linnaeus var. *grayi* (Behr & Kellogg) Jepson; *A. quinquefolia* var. *minor* (Eastwood) Munz

Aerial shoots (3–)10–30(–40) cm, from rhizomes, rhizomes horizontal, occasionally ascending. **Basal leaves** usually absent, occasionally 1, ternate; petiole (1.5–)2.5–20(–25) cm; terminal leaflet sessile or petiolulate, ovate to rhombic or oblanceolate, (1–)1.5–4.5(–5) × (0.6–)1–2.5(–3.5) cm, base cuneate, mar-

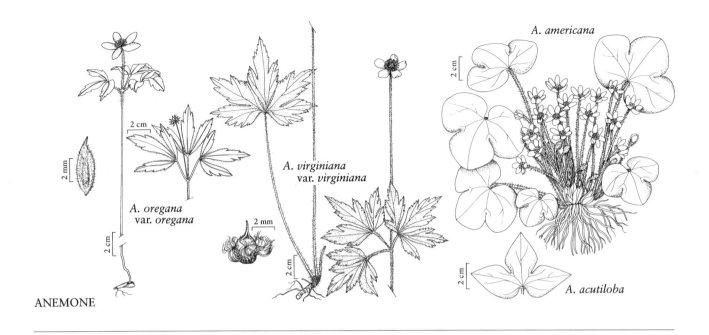

A. americana

A. virginiana
var. virginiana

A. oregana
var. oregana

A. acutiloba

ANEMONE

gins crenate, sometimes coarsely serrate, on distal 1/2–2/3, apex narrowly acute to acute, surfaces white-puberulous to pilose to nearly glabrous; lateral leaflets unlobed, rarely scarcely 1×-lobed; ultimate lobes 4–12 mm wide. **Inflorescences** 1-flowered; peduncle proximally glabrous, distally villous or pilose; involucral bracts 3, 1-tiered, ternate, ± similar to basal leaves, bases distinct; terminal leaflet sessile to petiolulate, ovate to rhombic or oblanceolate, (1–)1.5–4.5 (–5) × (0.6–)1–2.5(–3.5) cm, bases cuneate, margins crenate, sometimes coarsely serrate, on distal 1/2–2/3, apex narrowly acute to acute, surfaces white-puberulous to pilose to nearly glabrous; lateral leaflets unlobed, rarely scarcely 1×-lobed; ultimate lobes 4–12 mm wide. **Flowers:** sepals 5–6, white or blue, elliptic to obovate, rarely ovate, 7–15 × 4–8 mm, glabrous; stamens 25–40. **Heads of achenes** nearly spheric; pedicels (0.5–)3–10 cm in fruit. **Achenes:** body elliptic, flat, 3–4 × 1.5–2 mm, not winged, puberulous to pilose; beak curved or recurved, 0.6–1 mm, glabrous. 2*n* = 16.

Flowering winter–early summer (Feb–Jun). Shaded woods, wooded slopes; 100–900 m; Calif., Oreg.

The rhizomes of *Anemone grayi* tend to be knobby rather than filiform as in *A. lyallii* and *A. oregana*. Investigation of the populations in which these three species overlap and appear to intergrade in southern Oregon and northern California is needed to clarify the taxonomic relationships among them.

22. **Anemone virginiana** Linnaeus, Sp. Pl. 1: 540. 1753
· Thimbleweed, tall anemone, anémone de Virginie E

Aerial shoots 30–100(–110) cm, from caudices, rarely with ascending rhizomes, caudices ascending to vertical. **Basal leaves** 1–5, ternate; petiole 5–35 cm; terminal leaflet sessile or nearly so, oblanceolate to obovate, 2–9(–12) × 2–5(–7) cm, base cuneate to broadly cuneate, margins coarsely serrate and incised on distal 1/2, apex acuminate to narrowly acute, surfaces pilose, more so abaxially; lateral leaflets usually 1–2×-lobed or -parted, occasionally unlobed; ultimate lobes 10–30(–40) mm wide. **Inflorescences** (1–)3–9-flowered cymes; peduncle villous; primary involucral bracts 3 (–5), secondary involucral bracts 2(–3), (1–)2-tiered, ternate, ± similar to basal leaves, bases distinct; terminal leaflet ± sessile, elliptic to oblanceolate, 2–10(–12) cm (2 cm in secondary involucre) × 2–5(–7) cm, bases cuneate, margins coarsely serrate and incised on distal 1/2, apex acuminate to narrowly acute, surfaces pilose, more so abaxially; lateral leaflets unlobed or 1×-lobed or -parted; ultimate lobes 8–25(–35) mm wide. **Flowers:** sepals usually 5, green, yellow, or red (rarely white or abaxially green to green-yellow and adaxially green or yellow and tinged red), oblong, ovate, or obovate, 6.5–20 × 2.5–10 mm, abaxially hairy, adaxially glabrous or nearly so; stamens 50–70. **Heads of achenes** oblong-ellipsoid, rarely obconic; pedicels 13–25(–30) cm. **Achenes:** body obovoid, 2–3.7 × 1.5–2 mm, not winged, densely woolly; beak curved, 1–1.5 mm, puberulous, not plumose.

Varieties 3 (3 in the flora): North America.

See C. S. Keener et al. (1996) for an analysis of infraspecific variation within *Anemone virginiana* from which the current treatment has been adapted.

Varieties of *Anemone virginiana* used medicinally by native Americans were not specified; the species was used as an antidiarrheal, an aid for whooping cough, a stimulant, an emetic, a love potion, a remedy for tuberculosis, and a protection against witchcraft medicine (D. E. Moerman 1986).

1. Sepals usually 5–10 mm, abaxially densely tomentose; anthers 0.7–1(–1.2) mm; primarily Canadian, in dry woods, sandy ridges, and grasslands. . . . . . 22b. *Anemone virginiana* var. *cylindroidea*
1. Sepals (6–)10–21 mm, abaxially usually thinly pubescent; anthers (0.8–)1–1.7 mm; widely distributed, in moist habitats.
 2. Base of involucral bracts cordate or reniform, rarely subtruncate, terminal leaflets light green, margins proximally mostly straight- to convex-sided, variously lobed or serrate, variously pubescent; anthers typically greater than 1.1 mm; heads of achenes ovoid to ovoid-cylindric, (9–)11–14 mm diam.; widely distributed . . . .
 . . . . 22a. *Anemone virginiana* var. *virginiana*
 2. Base of involucral bracts usually truncate to subtruncate, sometimes reniform or cordate, terminal leaflets deep green, margins proximally concave- to straight-sided, distally incised, thinly pubescent; anthers typically less than 1.2 mm; heads of achenes ± ovoid-cylindric, 8–10(–11) mm diam.; distributed primarily in New England, Great Lakes area, and adjacent Canada. . . . .
 . . . . 22c. *Anemone virginiana* var. *alba*

22a. Anemone virginiana Linnaeus var. **virginiana** E F

Inflorescences: involucral bracts with base cordate or reniform, rarely subtruncate; terminal leaflets light green, margins proximally mostly straight to convex-sided, variously lobed or serrate, variously pubescent. **Flowers:** sepals (6–)10–20 mm, abaxially usually thinly pubescent; anthers (0.9–)1–1.7 mm. **Heads of achenes** ovoid to ovoid-cylindric, (9–)11–14 mm diam. $2n = 16$.

Flowering summer (Jun–Aug). Dry, rocky, open woods, thickets, river banks; 0–2000 m; Ont., Que.; Ala., Ark., Conn., Del., D.C., Ga., Ill., Ind., Iowa, Kans., Ky., La., Maine, Md., Mass., Mich., Minn., Miss., Mo., Nebr., N.H., N.J., N.Y., N.C., N.Dak., Ohio, Okla., Pa., R.I., S.C., S.Dak., Tenn., Vt., Va., W.Va., Wis., Wyo.

22b. Anemone virginiana Linnaeus var. **cylindroidea** B. Boivin, Phytologia 18: 281. 1969 E

Inflorescences: involucral bracts ± light green. **Flowers:** sepals 5–10 mm, abaxially densely tomentose; anthers 0.7–1(–1.2) mm. **Heads of achenes** ovoid to ovoid-cylindric, 7–11 mm diam.

Flowering summer (Jun–Aug). Dry woods, sandy ridges, grasslands; 400–3000 m; Alta., B.C., Man., N.B., Ont., Que., Sask.; Minn., N.Y.

22c. Anemone virginiana Linnaeus var. **alba** (Oakes) A. W. Wood, Class-book Bot. ed. 2(a), 140. 1847 E

Anemone cylindrica A. Gray var. *alba* Oakes, Mag. Hort. Bot. 7: 182. 1841; *A. riparia* Fernald; *A. virginiana* var. *riparia* (Fernald) B. Boivin

Inflorescences: involucral bracts with base often truncate to subtruncate, sometimes reniform or cordate; terminal leaflets dark green, margins proximally concave to straight-sided, distally incised, abaxially thinly pubescent. **Flowers:** sepals (9–)10–21 mm, abaxially thinly pubescent; anthers 0.8–1.2(–1.7) mm. **Heads of achenes** ± ovoid-cylindric, 8–10(–11) mm diam. $2n = 16$.

Flowering spring–summer (May–Jul). Riparian areas, moist cliffs and ledges; 10–1000 m; N.B., Nfld., N.S., Ont., Que.; Conn., Maine, Mass., Mich., Minn., N.H., N.J., N.Y., Vt., Wis.

23. Anemone cylindrica A. Gray, Ann. Lyceum Nat. Hist. New York 3: 221. 1835 · Long-headed anemone, thimbleweed, candle anemone, anémone cylindrique E

Aerial shoots (20–)30–70(–80) cm, from caudices, rarely with very short ascending rhizomes, caudices ascending to vertical. **Basal leaves** (2–)5–10(–13), ternate; petiole 9–21 cm; terminal leaflet sessile, broadly rhombic to oblanceolate, (2.5–)3–5(–6) × (3–)4–10(–14) cm, base narrowly cuneate, margins crenate, or serrate and deeply incised on distal 1/2, apex narrowly acute, surfaces strigose, more so abaxially; lateral leaflets 1–2×-parted and -lobed; ultimate lobes 4–10(–13) mm wide. **Inflorescences** (1–)2–8-flowered cymes, sometimes appearing umbellike; peduncle villous to densely villous; involucral bracts 3–7(–9), 2(–3)-tiered (can appear 1-tiered), ternate, ± similar to basal leaves, bases distinct; terminal leaflet sessile, rhombic to oblanceolate, 2.5–6.5 ×

(1–)1.5–2(–2.5) cm, bases narrowly cuneate, margins serrate and incised on distal 1/3–1/2, apex narrowly acute, surfaces puberulous, more so abaxially; lateral leaflets 1(–2)×-parted or -lobed; ultimate lobes (4–)6–10(–15) mm wide. **Flowers:** pedicels usually appearing bractless; sepals 4–5(–6), green to whitish, oblong to elliptic or ovate, 5–12(–15) × 3–6 mm, abaxially silky, adaxially glabrous; stamens 50–75. **Heads of achenes** cylindric; pedicels 10–30 cm. **Achenes:** body ovoid, (1.8–)2–3 × 1.5–2 mm, not winged, woolly; beak usually recurved, (0.3–)0.5–1 mm, hidden by achene indument, glabrous. 2*n* = 16.

Flowering summer (Jun–Jul). Prairies, dry, open woods, pastures, roadsides; 300–3000 m; Alta., B.C., Man., Ont., Que., Sask.; Ariz., Colo., Conn., Idaho, Ill., Ind., Iowa, Kans., Maine, Mass., Mich., Minn., Mo., Mont., Nebr., N.H., N.J., N.Mex., N.Y., N.Dak., Ohio, Pa., R.I., S.Dak., Vt., Wis., Wyo.

The cymes of *Anemone cylindrica* may appear 1-tiered because the second tier of involucres is closely nestled among the leaves of the first tier. The cymes then resemble umbels with unusually leafy involucral bracts; they might be misinterpreted as such.

Anemone cylindrica was used medicinally by Native Americans for headaches, sore eyes, and bad burns, as a psychological aid, and as a relief from tuberculosis (D. E. Moerman 1986).

24. **Anemone acutiloba** (de Candolle) G. Lawson, Proc. & Trans. Roy. Soc. Canada 2(4): 30. 1884 · Sharp-lobed hepatica, anémone à lobes aigus, hépatique à lobes aigus E F

Hepatica acutiloba de Candolle, Prodr. 1: 22. 1824; *H. acuta* (Pursh) Britton; *H. nobilis* Miller var. *acuta* (Pursh) Steyermark; *H. triloba* Chaix var. *acuta* Pursh; *H. triloba* var. *acutiloba* (de Candolle) R. Warner

Aerial shoots 5–19 cm, from rhizomes, rhizomes ascending to horizontal. **Basal leaves** 3–15, often purplish abaxially, simple, deeply divided; petiole 3–19 cm; leaf blade widely orbiculate, 1.3–6.5 × 1.8–11.5 cm, base cordate, margins entire, apex acute or acuminate, surfaces strongly villous to glabrescent; lobes 3(–5), deltate, 0.7–4 cm wide; middle lobe 70–90% of total blade length. **Inflorescences** 1-flowered, villous to pilose; involucral bracts 3, 1-tiered, simple, dissimilar to basal leaves, lanceolate to ovate, 0.53–1.8 × 0.27–0.95 cm, sessile, calyxlike, closely subtending flowers, bases distinct, cuneate, margins entire, apex acute, strongly villous to glabrescent. **Flowers:** sepals 5–12, white to pink or bluish, ovate to obovate, 6–14.6 × 2.2–5.8 mm, glabrous; petals absent; stamens 10–30. **Heads of achenes** spheric; pedicels 0.1–0.4 cm. **Achenes:** body narrowly ovoid, 3.5–

4.7 × 1.3–1.9 mm, slightly winged, hispid, gradually tapering; beak indistinct. 2*n* = 14.

Flowering spring. Deciduous woods, often in calcareous soils; 0–1200 m; Ont., Que.; Ala., Ark., Conn., Ga., Ill., Ind., Iowa, Ky., Maine, Md., Mass., Mich., Minn., Mo., N.H., N.Y., N.C., Ohio, Pa., S.C., Tenn., Vt., Va., W.Va., Wis.

In North America, *Anemone acutiloba* and *A. americana* are sufficiently well differentiated to enable the distinction of the two species. Some intermediates do occur but it is uncertain as to whether these are true intermediates or hybrids. The fact that the two species are highly sympatric and still maintain their differences implies that they should still be recognized as distinctive species (see G. L. Stebbins 1993).

The two North American species formerly placed in *Hepatica* are closely allied to the Eurasian *Anemone hepatica* Linnaeus [= *Hepatica nobilis* Miller, *Hepatica hepatica* (Linnaeus) Karst]. Among European collections, plants approach either *A. acutiloba* or *A. americana* in leaf morphology, but some intermediates are found (J. A. Steyermark and C. S. Steyermark 1960). North American plants differ from *A. hepatica* in having narrower sepals, larger involucral bracts, and shorter and less pubescent scapes. Further research, including a comparative study of breeding systems, is needed to clarify the relationship between *Anemone hepatica*, *A. acutiloba*, and *A. americana*. Pending such work, the eastern North American hepaticas are here recognized as distinct species.

D. E. Moerman (1986) lists *Hepatica acutiloba* as one of the plants used medicinally by Native Americans in the treatment of abdominal pains, poor digestion, and constipation, as a wash for "twisted mouth or crossed eyes," and as a gynecological aid.

25. **Anemone americana** (de Candolle) H. Hara, J. Jap. Bot. 33: 271. 1958 · Round-lobed hepatica, anémone d'Amérique, hépatique d'Amérique E F

Hepatica triloba Chaix var. *americana* de Candolle, Syst. Nat. 1: 216. 1817; *H. americana* (de Candolle) Ker Gawler; *H. nobilis* Miller var. *obtusa* (Pursh) Steyermark; *H. triloba* var. *obtusa* Pursh

Aerial shoots 5–18 cm, from rhizomes, rhizomes ascending to horizontal. **Basal leaves** 3–15, often purplish abaxially, simple, deeply divided; petiole 5–20 cm; leaf blade widely orbiculate, 1.5–5 × 2–10 cm, base cordate, margins entire, apex rounded, surfaces strongly villous to glabrescent; lobes 3(–4), widely ovate, 1–4 cm wide; middle lobe 50–70(–75)% of total blade length. **Inflorescences** 1-flowered, villous to pilose; involucral bracts 3, 1-tiered, simple, dissimilar to basal leaves, sessile, calyxlike, closely subtending flower, ovate or elliptic, 0.65–1.8 × 0.5–1.2 cm, bases distinct, cuneate, margins entire, apex obtuse, strongly villous to glabrescent. **Flowers:** sepals 5–12, white to pink or bluish, ovate to obovate 7.5–14.5 × 3.5–7.7 mm, glabrous; petals absent; stamens 10–30. **Heads of achenes** spheric; pedicel 0.1–0.4 cm. **Achenes:** body narrowly ovoid, 3.5–5 × 1.2–1.6 mm, slightly winged, hispid, gradually tapering; beak indistinct. $2n = 14$.

Flowering spring. Mixed woods, often in association with both conifers and deciduous trees; 0–1200 m; Man., N.B., N.S., Ont., Que.; Ala., Ark., Conn., Del., D.C., Fla., Ga., Ill., Ind., Iowa, Ky., Maine, Md., Mass., Mich., Minn., Miss., Mo., N.H., N.J., N.Y., N.C., Ohio, Pa., R.I., S.C., Tenn., Vt., Va., W.Va., Wis.

Anemone americana is found in habitats similar to those of *A. acutiloba* but usually in drier sites with more acid soils.

6. CLEMATIS Linnaeus, Sp. Pl. 1: 543. 1753; Gen. Pl. ed. 5, 242. 1754 · Clematis, clématite [Greek *clema,* plant shoot, ancient name of a vine]

James S. Pringle

Vines, ± woody, sometimes only at base, climbing by means of tendril-like petioles and leaf rachises, or erect, herbaceous perennials, from elongate rhizomes. **Leaves** cauline, opposite, simple or compound, sessile or petiolate. **Leaf blade** undivided or 1–3-pinnately or -ternately compound; leaf or leaflets cordate to orbiculate, oblong, lanceolate, or oblanceolate, lobed or unlobed, margins entire or toothed. **Inflorescences** axillary and/or terminal, 1–many-flowered cymes or panicles or flowers solitary or in fascicles, to 15 cm; bracts present and leaflike or ± scalelike or absent, not forming involucre. **Flowers** bisexual or unisexual, radially symmetric; sepals not persistent in fruit, 4, white, blue, violet, red, yellow, or greenish, plane, ovate to obovate or linear, 6–60 mm; petals absent; sometimes anther-bearing staminodes between sepals and stamens; stamens many; filaments filiform to flattened; pistils 5–150, simple; ovule 1 per pistil; beak present. **Fruits** achenes, aggregate, sessile, lenticular, nearly terete, or flattened-ellipsoid, sides not prominently veined; beak terminal, straight or curved, 12–110 mm. $x = 8$.

Species ca. 300 (32 in the flora): worldwide, mostly temperate, a few subarctic, subalpine, or tropical.

Clematis is highly diverse in vegetative and floral aspects and has been divided into three or more genera by some authors, the groups segregated in some literature being *Clematis* subg. *Atragene* as the genus *Atragene* and *Clematis* subg. *Viticella* as the genus *Viticella*. Species in *Clematis* subg. *Viorna* have been crossed with highly dissimilar species in *Clematis* subg. *Clematis* and *Clematis* subg. *Viticella,* and species in *Clematis* subg. *Clematis* have been crossed with species in *Clematis* subg. *Viticella*. Chromosome morphology is strikingly similar in all subgenera.

The circumscription of subgenera in this work follows C. S. Keener and W. M. Dennis (1982). Major realignments have been proposed by F. B. Essig (1992) on the basis of seedling morphology, including the transfer of *Clematis recta* and *C. terniflora* to *Clematis* subg. *Viorna.*

Many species are valued as ornamentals; some have escaped from cultivation and have become established in the flora.

SELECTED REFERENCES Essig, F. B. 1992. Seedling morphology in *Clematis* (Ranunculaceae) and its taxonomic implications. Sida 15: 377–390. James, J. F. 1883. Revision of the genus *Clematis* of the United States. J. Cincinnati Soc. Nat. Hist. 6: 118–135. Keener, C. S. 1975. Studies in the Ranunculaceae of the southeastern United States. III. *Clematis* L. Sida 6: 33–47. Keener, C. S. and W. M. Dennis. 1982. The subgeneric classification of *Clematis* (Ranunculaceae) in temperate North

America north of Mexico. Taxon 31: 37–44.   Kuntze, O. 1885. Monographie der Gattung *Clematis*. Verh. Bot. Vereins Prov. Brandenburg 26: 6–202.   Torrey, J. and A. Gray. 1838–1843. A Flora of North America. 2 vols. in 7 parts. New York, London, and Paris. Vol. 1, pp. 7–11.

1. Sepals ± thick, leathery, connivent proximally and usually much of length; perianth bell- to urn-shaped, blue, violet, or yellowish white . . . . 6d. *Clematis* subg. *Viorna*, p. 167
1. Sepals thin, spreading, not connivent; perianth widely bell-shaped to rotate, or if narrowly bell-shaped, bright yellow.
    2. Staminate flowers with petaloid staminodes between stamens and sepals; perianth widely bell-shaped or tardily rotate . . . . 6c. *Clematis* subg. *Atragene*, p. 165
    2. Staminate flowers without staminodes between stamens and sepals; perianth rotate, sepals wide-spreading, or sepals recurved at least toward tip.
        3. Flowers 1–many (if flowers solitary, either unisexual or with yellow sepals), gener- ally in cymes or panicles, unisexual or bisexual; sepals white or yellow, linear- oblong, elliptic, lanceolate, ovate, oblanceolate, or obovate . . . . 6a. *Clematis* subg. *Clematis*, p. 159
        3. Flowers 1–3 (if 1, sepals not yellow), in axillary clusters, bisexual; sepals blue to violet, rarely white, broadly obovate to elliptic-rhombic . . . . 6b. *Clematis* subg. *Viticella*, p.164

6a. CLEMATIS Linnaeus subg. CLEMATIS

Nancy P. Moreno

Frederick Essig

Woody vines (erect, herbaceous perennials in *C. recta*). **Leaf blade** 1–2-pinnate; leaflets lobed or unlobed, margins entire or toothed. **Inflorescences** terminal and/or axillary on current year's stems, cymes or panicles or flowers solitary or paired, bracteate. **Flowers** bisexual, or unisex- ual with staminate and pistillate on different plants, not nodding, or ± nodding in yellow- flowered species; perianth rotate; sepals spreading, not connivent, linear, oblong, elliptic, lan- ceolate, ovate, oblanceolate, or obovate, thin and white or somewhat thickened and yellow; filaments filiform, slender, glabrous or pubescent; staminodes absent from staminate flowers, usually present in pistillate flowers; pistils rudimentary or absent in staminate flowers. **Achenes** flattened or nearly terete; beak more than 1.5 cm, plumose.

Species 50–100 (11 in the flora): worldwide.

The Asian (Korean) species *Clematis serratifolia* Rehder, with light yellow sepals and purple stamens, may escape from cultivation and spread locally.

SELECTED REFERENCES   Essig, F. B. 1990. The *Clematis virginiana* (Ranunculaceae) complex in the southeastern United States. Sida 14: 49–68.   Grey-Wilson, C. 1989. *Clematis orientalis* (Ranunculaceae) and its allies. Kew Bull. 44: 33–60.   Hara, H. 1975. The identity of *Clematis terniflora* DC. J. Jap. Bot. 50: 155–158.

Key to the species of *Clematis* subg. *Clematis*
1. Sepals greenish yellow to bright yellow, ascending or wide-spreading and recurved.
    2. Leaflet margins entire or coarsely few-toothed; sepals greenish yellow, pubescent or abaxially gla- brous . . . . 1. *Clematis orientalis*
    2. Leaflet margins serrate; sepals bright yellow, adaxially glabrous . . . . 2a. *Clematis tangutica* var. *tangutica*
1. Sepals white to cream, wide-spreading, not recurved.
    3. Flowers bisexual.
        4. Stems herbaceous, not viny . . . . 5. *Clematis recta*
        4. Stems ± woody, climbing with tendril-like petioles and rachises of compound leaves.
            5. Pistils 10 or fewer per flower; achenes flattened, with conspicuous rims . . . . 3. *Clematis terniflora*
            5. Pistils 20 or more per flower; achenes nearly terete, without conspicuous rims . . . . 4. *Clematis vitalba*
    3. Flowers unisexual.
        6. Leaflets deltate to ovate, strongly 3-parted to 3-sect, segments ovate, deltate, or linear; achene with beak 4–9 cm; sw United States, Mexico . . . . 6. *Clematis drummondii*

6. Leaflets ovate to lanceolate, variously lobed or toothed, but without narrow segments; achene with beak to 5.5 cm.
   7. Flowers solitary (rarely 3 in simple cymes); pedicel (or peduncle and pedicel combined for solitary flowers) stout, 3.5 cm or more; pistils 75–100 per flower . . . . 8. *Clematis lasiantha*
   7. Flowers (1–)3 or more, in simple or compound cymes or in panicles, occasionally solitary in *C. pauciflora;* pedicel slender, less than 3.5 cm; pistils fewer than 70 per flower.
      8. Achene body broadly ovate to nearly orbiculate, glabrous; sepals hairy only abaxially; leaflets to 3.5 cm . . . . 7. *Clematis pauciflora*
      8. Achene body ovate, pubescent; sepals hairy on both surfaces; leaflets usually longer than 3 cm.
         9. Leaf blade 3-foliolate . . . . 9. *Clematis virginiana*
         9. Leaf blade pinnately 5-foliolate to 2-pinnate.
            10. Inflorescences compound cymes, often distinctly corymbiform, flowers crowded; pistils 25–65 per flower; leaf blade somewhat succulent, ultimate venation obscure; w North America . . . . 10. *Clematis ligusticifolia*
            10. Inflorescences panicles, not usually corymbiform or with flowers crowded; pistils 18–35 per flower; leaf blade membranous, ultimate venation conspicuous; primarily e North America . . . . 11. *Clematis catesbyana*

1. Clematis orientalis Linnaeus, Sp. Pl. 1: 543. 1753 [F]

Viticella orientalis (Linnaeus) W. A. Weber

Stems climbing, 2–8 m. **Leaf blade** pinnately 5–7-foliolate, proximal leaflets sometimes 3-foliolate; leaflets lanceolate to elliptic or ovate, usually 2–3-lobed proximally, 1–5.5 × 0.5–3.5 cm, margins entire or coarsely few-toothed; surfaces at least abaxially pubescent, glaucous. **Inflorescences** axillary, sometimes terminal, 3–many-flowered cymes or flowers solitary. **Flowers** bisexual; pedicel (0.5–)1–11 cm; sepals wide-spreading and recurved, greenish yellow, ovate-lanceolate to elliptic, 0.8–2 cm, length ca. 2.5 times width, margins densely pubescent, abaxially and adaxially pubescent or abaxially glabrous; stamens 20–40; filaments pilose proximally; staminodes absent; pistils 75–150. **Achenes** turgid, not conspicuously rimmed, pilose; beak 2–5 cm.

Flowering summer–fall. Roadsides, other secondary habitats, open woods; 0–2600 m; introduced; Ont.; Colo., Nev., N.Mex., Utah; native to Eurasia.

Clematis orientalis has been reported from Idaho; it probably can be expected elsewhere.

This species has been divided by C. Grey-Wilson (1989) into five varieties, partly correlated with their distribution in Asia. Naturalized plants in North America seem best referred to *C. orientalis* var. *robusta* Grey-Wilson, native to Afghanistan.

Although *Clematis orientalis* has been naturalized in the Rocky Mountains since the late nineteenth century, it has spread especially rapidly since ca. 1975, becoming weedy and, in some localities, constituting a threat to young trees and native shrubby and herbaceous species.

2. Clematis tangutica (Maximowicz) Korshinsky, Izv. Imp. Akad. Nauk, ser. 5, 9: 399. 1898

Clematis orientalis Linnaeus var. *tangutica* Maximowicz, Fl. Tangut., 3. 1889

Varieties 2 (1 in the flora): North America, Asia (China, India, Mongolia).

2a. Clematis tangutica (Maximowicz) Korshinsky var. tangutica [F]

Stems climbing, 2–4 m. **Leaf blade** pinnately 5–7-foliolate, proximal leaflets rarely 3-foliolate; leaflets lanceolate to oblong or elliptic, unlobed or lobed proximally, 1.8–5.7 × 0.4–1.6 cm, margins serrate; surfaces usually abaxially pubescent, not glaucous. **Inflorescences** terminal and axillary, 1(–3)-flowered. **Flowers** bisexual; pedicel 0.6–3 cm; sepals ascending, scarcely spreading except near tip (forming narrowly bell-shaped perianth), bright yellow, lanceolate to elliptic, 1.8–3.4 cm, length ca. 2.5 times width, margins silky-pubescent, abaxially silky-pubescent, adaxially glabrous; stamens 20–50; filaments pilose proximally; staminodes absent; pistils 80–150. **Achenes** turgid, not conspicuously rimmed, pilose; beak 2–4.5 cm. $2n = 16$.

Flowering summer–early fall. Roadsides, thickets, other ± open, disturbed sites; 700–1000 m; introduced; Alta., B.C., Man., Sask., Yukon; native to Asia (China, India).

Clematis tangutica var. *tangutica,* the most commonly cultivated yellow-flowered species of *Clematis,* should probably be expected elsewhere. It is closely related to, and sometimes treated as a variety of, *C. orientalis.*

3. Clematis terniflora de Candolle, Syst. Nat. 1: 137. 1817 · Sweet autumn clematis, yam-leaved clematis

Clematis dioscoreifolia H. Léveillé & Vaniot; *C. dioscoreifolia* var. *robusta* (Carriér) Rehder; *C. maximowicziana* Franchet & Savatier; *C. paniculata* Thunberg (1794), not J. F. Gmelin (1791)

Stems climbing with tendril-like petioles and leaf rachises, 3–6 m. **Leaf blade** pinnately 3- or 5-foliolate; leaflets ovate or broadly lanceolate to narrowly deltate, to 6.5 × 3.5 cm, margins entire; surfaces abaxially glabrous or very sparingly appressed-strigose on major veins. **Inflorescences** axillary, 3–12-flowered cymes or compound cymes or paniculate with cymose subunits. **Flowers** bisexual, often some unisexual (staminate) in same inflorescence; pedicel 1–3.5 cm, slender; sepals wide-spreading, not recurved, white, linear or elliptic to lanceolate or narrowly obovate, 0.9–2.2 cm, length ca. 2–3 times width, abaxially tomentose along margins, adaxially glabrous; stamens ca. 50; filaments glabrous; staminodes absent; pistils 5–10. **Achenes** broad, flat, conspicuously rimmed, minutely appressed-silky, sometimes sparsely so; beak 2–6 cm.

Flowering summer (Jul–Sep). Roadsides, thickets, and other secondary sites, edges of woods near creeks; 0–1000 m; introduced; Ont.; Ala., Ark., Conn., Fla., Ga., Ill., Ind., Kans., Ky., La., Md., Mass., Miss., Mo., Nebr., N.H., N.J., N.Y., N.C., Ohio, Okla., Pa., S.C., Tenn., Tex., Va., W.Va.; native to Asia (China, Korea, Japan).

Clematis terniflora is commonly cultivated as an ornamental. It is widely naturalized in the eastern United States. The illegitimate homonym *C. paniculata* Thunberg is frequently applied to this species.

Some authors have recognized two or more varieties in this species, correlated with their distribution in Asia, but in the study by H. Hara (1975), all of the varietal names were reduced to synonymy.

4. Clematis vitalba Linnaeus, Sp. Pl. 1: 544. 1753 · Traveler's joy, old man's beard

Stems climbing with tendril-like petioles and leaf-rachises, to 12 m. **Leaf blade** pinnately 5-foliolate; leaflets cordiform, 8 × (2–)3–5(–6) cm, margins entire to regularly crenate or dentate; surfaces abaxially minutely pubescent on veins, adaxially glabrous. **Inflorescences** axillary and terminal, (3–)5–22-flowered cymes. **Flowers** bisexual; pedicel 1–1.5 cm, slender; sepals wide-spreading, not recurved, white to cream, elliptic or oblanceolate to obovate, ca. 1 cm, length ca. 2 times width, abaxially and adaxially tomentose; stamens ca. 50; filaments glabrous; staminodes absent; pistils 20 or more. **Achenes** nearly terete, not conspicuously rimmed, densely pubescent; beak ca. 3.5 cm.

Flowering summer (Jun–Aug). Roadsides, waste ground, secondary growth; 0–100 m; introduced; B.C., Ont.; Maine, Oreg., Wash.; native to Europe, n Africa.

Clematis vitalba is naturalized in only a few sites in eastern North America and northwestern Oregon to Puget Sound.

5. Clematis recta Linnaeus, Sp. Pl. 1: 544. 1753

Stems herbaceous, ascending to erect, not climbing, 0.6–1.5 m. **Leaf blade** pinnately 5–9-foliolate; leaflets lanceolate to ovate, 3–9 × 0.8–4 cm, margins entire; surfaces glabrous. **Inflorescences** axillary and terminal, many-flowered cymes and panicles. **Flowers** bisexual; pedicel 8–20 mm, slender; sepals wide-spreading, not recurved, white, oblanceolate to oblong, 8–20 mm, length ca. 4 times width, margins tomentose, otherwise glabrous; stamens 20–50; filaments glabrous; staminodes absent; pistils 8–25. **Achenes** broad, flat, conspicuously rimmed, glabrous; beak 1.2–2 cm.

Flowering summer (Jun–Jul). Old fields and thickets; 0–100 m; introduced; Ont.; N.Y.; native of Eurasia.

Clematis recta should probably be expected elsewhere.

6. Clematis drummondii Torrey & A. Gray, Fl. N. Amer. 1: 7. 1838 · Barbas de chivato

Clematis nervata Bentham

Stems scrambling to climbing with tendril-like petioles and leaf-rachises, 4–5 m or more. **Leaf blade** odd-pinnate, usually 5-foliolate; leaflets deltate to ovate, strongly 3-parted to 3-cleft, proximal leaflets sometimes 3-cleft, 1.5–5.5 × 0.5–4.5 cm, membranous to leathery; segments ovate, deltate, or linear, margins dentate; surfaces pilose, abaxially more densely so. **Inflorescences** usually axillary, 3–12-flowered simple cymes or compound with central axis or flowers solitary or paired. **Flowers** unisexual; pedicel slender, (1.1–)1.5–7 cm; sepals wide-spreading, not recurved, white to cream, oblong or elliptic to obovate or oblanceolate, (7–)9–13(–15) mm, abaxially and adaxially pubescent; stamens 40–90; filments glabrous; staminodes 17–35 when present; pistils 35–90.

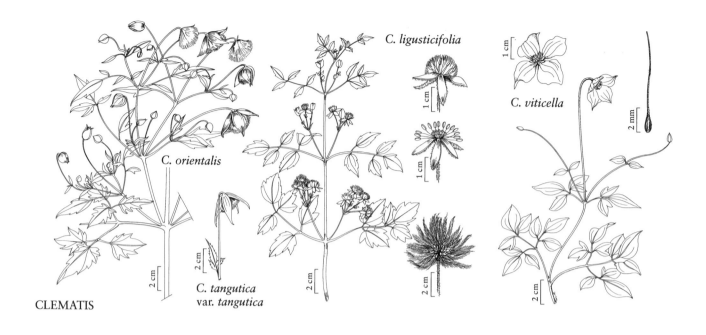

C. orientalis

C. ligusticifolia

C. viticella

C. tangutica
var. tangutica

CLEMATIS

Achenes elliptic to ovate, 3–5 × 1.5–2.5 mm, rimmed, short-silky; beak 4–9 cm.

Flowering spring–fall (Mar–Oct). Chaparral, xeric scrub, oak scrub, and grasslands; pastures, fencerows, and other secondary sites; often along streams or on slopes; 0–2200 m; Ariz., Calif., Colo., N.Mex., Okla., Tex.; n Mexico.

As with many other members of the subgenus, the leaves of *Clematis drummondii* are reputedly used in a poultice to treat irritations of the skin in humans and other animals.

Clematis coahuilensis D. J. Keil is found in central and north-central Mexico in habitats similar to those of *C. drummondii*; it is distinguished by entire or few-toothed, to sometimes more or less shallowly 3-lobed leaflets and by 2–3.5(–4) cm beaks.

7. Clematis pauciflora Nuttall in J. Torrey and A. Gray, Fl. N. Amer. 1: 9, 657. 1838 · Ropevine

Stems scrambling to climbing, 2–3 m. **Leaf blade** 1–2-pinnate, usually 5-foliolate, sometimes proximal and lateral leaflets also 3-foliolate; leaflets ovate to lanceolate, usually 3-lobed, 1–3.5 × 0.7–3.2 cm, membranous to leathery, margins each usually with 1–3 acute to rounded teeth, rarely entire; surfaces glabrous or very sparsely silky. **Inflorescences** axillary, 3(–12)-flowered cymes or flowers solitary or paired. **Flowers** unisexual; pedicel slender, 1–3.5 cm; sepals wide-spreading, not recurved,

white to cream, obovate to oblanceolate, 7–12 mm, abaxially pilose to silky, adaxially glabrous; stamens 30–50; staminodes absent or fewer than stamens; filaments glabrous; pistils 20–50. **Achenes** broadly ovate to nearly orbiculate, 4–4.5 × 2.5–3.5 mm, conspicuously rimmed, glabrous; beak 2.5–4 cm.

Flowering winter–fall (Jan–Oct; usually Mar–Apr). Dry chaparral, edges of meadows and cultivated fields; 0–2200 m; Calif.; Mexico (Baja California).

Clematis pauciflora is a distinctive species easily identified by the small, 3-lobed leaflets, glabrous or almost glabrous leaves and stems, and large, suborbicular achenes. Intermediates between *C. pauciflora* and *C. lasiantha* are present where the two species occur together.

8. Clematis lasiantha Nuttall in J. Torrey and A. Gray, Fl. N. Amer. 1: 9. 1838 · Pipestem

Stems scrambling to climbing, 3–4 m. **Leaf blade** 3-foliolate; leaflets ovate, largest leaflets usually 3-lobed, 1.5–6 × 1.5–5 cm; terminal leaflet occasionally 3-cleft, margins usually toothed; surfaces glabrous or sparsely silky. **Inflorescences** axillary, flowers solitary, rarely 3-flowered cymes. **Flowers** unisexual; pedicel (including peduncle) stout, 3.5–11 cm; sepals wide-spreading, not recurved, white to cream, ovate or elliptic to obovate or oblanceolate, 10–21 mm, abaxially and adaxially pilose; stamens 50–100; filaments glabrous; staminodes absent or 50–100; pistils 75–100. **Achenes** asymmetric-ovate, not broadly orbiculate, 3–4 × 1.5–2 mm, not conspicuously rimmed, glabrous; beak 3.5–5.5 cm. $2n = 16$.

Flowering winter–spring (Jan–Jun). Chaparral, open woodlands; 0–2000 m; Calif.; Mexico (Baja California).

Clematis lasiantha is common in the Coast Ranges and the foothills of the Sierra Nevada of California.

The Shasta used pounded stems or chewed or burned roots of *Clematis lasiantha* medicinally in the treatment of colds (D. E. Moerman 1986).

9. **Clematis virginiana** Linnaeus, Cent. Pl. I, 15. 1755 · Virgin's-bower, clématite de Virginie E

Clematis canadensis Miller; *C. holosericea* Pursh; *C. missouriensis* Rydberg; *C. virginiana* var. *missouriensis* (Rydberg) E. J. Palmer & Steyermark

Stems climbing, 2–7 m. **Leaf blade** 3-foliolate; leaflets ovate to lanceolate, 3.5–9 × 1.5–7.5 cm, margins coarsely toothed to entire; surfaces abaxially sparsely to densely pilose, adaxially glabrate. **Inflorescences** axillary, 3–many-flowered simple or compound cymes. **Flowers** unisexual; pedicel slender, 1–2 cm; sepals wide-spreading, not recurved, white to cream, elliptic or nearly oblong to oblanceolate, 6–14 mm, abaxially densely white-hairy, adaxially sparsely white-hairy; stamens ca. 30–50+; filaments glabrous; staminodes absent or fewer than stamens; pistils 40–70; beak nearly equaling sepals. **Achenes** ovate, 2.5–3.5 × 1.5 mm, conspicuously rimmed, sparsely short-hairy; beak 2.5–5 cm. $2n = 16$.

Flowering summer (Jun–Sep). Streamsides, wet roadsides, fencerows, and other moist, disturbed, wooded or open sites, locally abundant; 0–1500 m; Man., N.B., N.S., Ont., P.E.I., Que.; Ala., Ark., Conn., Del., Fla., Ga., Ill., Ind., Iowa, Kans., Ky., La., Maine, Md., Mass., Mich., Minn., Miss., Mo., Nebr., N.H., N.J., N.Y., N.C., N.Dak., Ohio, Okla., Pa., R.I., S.C., S.Dak., Tenn., Tex., Vt., Va., W.Va., Wis.

Clematis virginiana is the most frequent and widespread virgin's-bower in eastern North America. It is easily distinguished from *C. catesbyana* by the presence of three ovate leaflets.

Native Americans used infusions prepared from the roots of *Clematis virginiana* medicinally to treat kidney ailments, and mixed them with milkweed to heal backaches and venereal sores. Decoctions of stems were ingested to induce strange dreams. In addition, the plant was used as an ingredient in green corn medicine (D. E. Moerman 1986).

10. **Clematis ligusticifolia** Nuttall in J. Torrey and A. Gray, Fl. N. Amer. 1: 9. 1838 · Virgin's-bower, old man's beard, hierba de chivo F

Clematis ligusticifolia var. *brevifolia* Nuttall; *C. ligusticifolia* var. *californica* S. Watson; *C. neomexicana* Wooton & Standley; *C. suksdorfii* B. L. Robinson

Stems clambering or climbing, to 6(–20) m. **Leaf blade** pinnately 5-foliolate or 2-pinnate and 9–15-foliolate, somewhat succulent; leaflets lanceolate to broadly ovate, lobed or unlobed, (1–)3–9 × 0.9–7.2 cm, margins entire or variously toothed; surfaces abaxially glabrous or sparsely pilose or silky, especially on veins; ultimate venation obscure. **Inflorescences** axillary, usually 7–20(–65)-flowered compound cymes, often distinctly corymbiform, flowers crowded. **Flowers** unisexual; pedicel slender, 0.5–3 cm; sepals wide-spreading, not recurved, white to cream, obovate to oblanceolate, 6–10 mm, abaxially and adaxially pilose; stamens 25–50; staminodes absent or fewer than stamens; pistils 25–65. **Achenes** elliptic, 3–3.5 × ca. 1.5 mm, prominently rimmed, silky; beak 3–3.5 cm. $2n = 16$.

Flowering summer (Jun–Sep). Forest edges, woods, riparian deciduous woodlands, moist wooded draws, scrub, secondary sites derived from these, or clearings and pastures, usually near streams or on moist slopes; 0–2600 m; Alta., B.C., Man., Sask.; Ariz., Calif., Colo., Idaho, Mont., Nebr., Nev., N.Mex., N.Dak., Oreg., S.Dak., Utah, Wash., Wyo.; nw Mexico.

Clematis ligusticifolia is the common virgin's-bower in the western United States and Canada. In California it might hybridize with *C. lasiantha*. In Mexico it is found only in the extreme north; it is probably related to the Mexican and Central American species, *C. grossa* Bentham.

The name *Clematis neomexicana* has been applied to the populations with crenate leaflets in New Mexico and northern Chihuahua. This distinction is tenuous, however, given the wide variation present in leaflet margins within this and other species in *Clematis* subg. *Clematis*.

Two varieties of *Clematis ligusticifolia* have been weakly distinguished based on the presence or absence of 2-pinnate leaves.

Infusions prepared from the plants of *Clematis ligusticifolia* were used medicinally by Native Americans as a wash for skin eruptions, a lotion for backaches or swollen limbs, and a lotion to protect one against witches; stems and leaves were chewed to treat colds and sore throats; decoctions of leaves were also used as a wash and for stomach aches and cramps; and lathers of leaves were used to treat boils on humans and on animals (D. E. Moerman 1986).

11. Clematis catesbyana Pursh, Fl. Amer. Sept. 2: 736. 1814 · Virgin's-bower [E]

Clematis cordata Pursh; *C. micrantha* Small

Stems climbing, 3–6 m. **Leaf blade** pinnate or 2-ternate, 5–9-foliolate, membranous; leaflets ovate to lanceolate, largest 4–9 × 2.5–9 cm, surfaces abaxially sparsely to densely pilose, adaxially glabrate; proximal and lateral leaflets typically 3-cleft, unlobed or few-lobed, margins coarsely toothed or entire, surfaces sparsely pubescent or nearly glabrous; ultimate venation conspicuous. **Inflorescences** axillary, 3–many-flowered simple to compound panicles with central axis. **Flowers** unisexual; pedicel slender, 11–13 mm, to 2 cm in fruit; sepals wide-spreading, not recurved, white to cream, oblong or obovate to oblanceolate, 6–14 × 2–5 mm, abaxially densely white-hairy, adaxially sparsely white-hairy; stamens ca. 30–50; staminodes absent or fewer than stamens; pistils 18–35. **Achenes** ovate, ca. 3.5 × 1.5 mm, conspicuously rimmed, sparsely short hairy; beak 2.5–3.5 cm.

Flowering spring–summer (May–Jul). Disturbed or open, well-drained sites, frequently on limestone outcrops, coastal sands; 0–1200 m; Ala., Ark., Fla., Ga., Kans., Ky., La., Miss., Mo., N.C., Okla., S.C., Tenn., Va., W.Va.

6b. CLEMATIS Linnaeus subg. VITICELLA (Moench) Keener & W. M. Dennis, Taxon 31: 42. 1982

James S. Pringle

Viticella Moench, Methodus, 296. 1794

Woody vines. Leaf blade [simple] 1–2-pinnate, lobed or few-lobed, margins entire. **Inflorescences** axillary, on suppressed shoots from previous year's stems, 2–3-flowered bracteate fascicles or cymes, or flowers solitary, peduncle bracteate. **Flowers** bisexual, slightly or not nodding; sepals usually wide-spreading, not connivent, blue, violet, or rose violet [rarely white], broadly obovate to elliptic-rhombic, thin; filaments flattened but slender, glabrous or sparsely pubescent; staminodes absent. **Achenes** flattened; beak usually less than 1.5 cm, glabrous or silky.

Species 6 (1 in the flora): temperate regions, North America, Eurasia.

Clematis subg. *Viticella* has sometimes been included in *Clematis* subg. *Viorna*, as suggested by F. B. Essig (1992), although it is very different from subg. *Viorna* in the strict sense in the aspect of its flowers and was treated as a distinct genus by earlier authors.

Several other large-flowered European and Asiatic species and hybrids, including *Clematis campaniflora* Brotero, with broadly campanulate perianth, and *C. florida* Thunberg, *C. lanuginosa* Lindley, and *C.* ×*jackmanii* T. Moore (= *C. lanuginosa* × *C. viticella*), with rotate perianth often with 6–8 sepals, occasionally persist after cultivation; in some cases they have become very locally naturalized.

12. Clematis viticella Linnaeus, Sp. Pl. 1: 543. 1753 [F]

Viticella viticella (Linnaeus) Small

Vines 2–4(–6) m. **Leaf blade:** leaflets 3–7, proximal leaflets sometimes 3-foliolate, lanceolate to broadly ovate or elliptic, unlobed or 1–3-lobed, 1.5–7 cm, somewhat leathery, margins entire. **Flowers:** sepals 4, blue to violet or rose-violet, 1.5–4 cm, length ca. 1.2–2 times width, abaxially pubescent; stamens green; beak glabrous.

Flowering summer–fall. Roadsides, thickets and other secondary habitats; 200 m; introduced; Ont.; native to Europe.

Clematis viticella has also been reported from Quebec, New York, and Tennessee, but the reports have not been verified. It probably should be expected elsewhere.

6c. CLEMATIS Linnaeus subg. ATRAGENE (Linnaeus) Torrey & A. Gray, Fl. N. Amer. 1: 10. 1838

James S. Pringle

Atragene Linnaeus, Sp. Pl. 1: 542. 1753; Gen. Pl. ed. 5, 241. 1754

Woody vines or rhizomatous herbs with short, tufted stems. **Leaf blade** 1–3-ternate, lobed or unlobed, margins entire or coarsely serrate. **Inflorescences** terminal on short shoots or rarely terminal on long shoots, 1[–2]-flowered with bractless peduncles subtended by 1 or 2 pairs of leaves. **Flowers** bisexual, ± nodding; perianth widely bell-shaped to rotate; sepals ascending or tardily spreading, not connivent, usually violet-blue, sometimes reddish violet, or white, ovate to oblong, thin, margins densely pubescent, abaxially sparsely pubescent; staminodes present between stamens and sepals, flattened, petaloid, bearing reduced, sterile anthers; filaments flattened, pubescent at least on margins. **Achenes** flattened; beak over 2 cm, plumose.

Species ca. 5 (2 in the flora): temperate to subarctic and subalpine North America and Eurasia.

F. B. Essig (1992) suggested that *Clematis* subg. *Atragene* might be included in *Clematis* subg. *Clematis*. Because of its distinctive inflorescence and floral morphology, however, and because it has not been successfully crossed with species in any other subgenus, its subgeneric status is retained here.

The two North American species have been known to hybridize in Montana.

SELECTED REFERENCE Pringle, J. S. 1971. Taxonomy and distribution of *Clematis*, sect. *Atragene* (Ranunculaceae), in North America. Brittonia 23: 361–393.

Key to the species of *Clematis* subg. *Atragene*

1. Leaf blade consistently 2–3-ternate . . . . 13. *Clematis columbiana*
1. Leaf blade 1-ternate (or terminal leaflet sometimes ternate in var. *dissecta*) . . . . 14. *Clematis occidentalis*

13. Clematis columbiana (Nuttall) Torrey & A. Gray, Fl. N. Amer. 1: 11. 1838 ☐E

Atragene columbiana Nuttall, J. Acad. Nat. Sci. Philadelphia 7: 7. 1834

Stems viny, climbing or trailing (mainly rhizomatous, not viny in var. *tenuiloba*). **Leaf blade** consistently 2–3-ternate; leaflets diverse in shape, thin or ± succulent, usually deeply lobed, margins serrate. **Flowers:** sepals violet-blue (rarely white in var. *columbiana*), lance-ovate to ovate.

Varieties 2 (2 in the flora): North America.

The name *Clematis columbiana* has been and still is widely misapplied to *C. occidentalis* var. *grosseserrata*; it is often associated with that taxon in some horticultural and popular publications. In such works, true *C. columbiana* is usually called *C. pseudoalpina*.

The two varieties of *Clematis columbiana,* although strikingly dissimilar in their extremes, intergrade extensively. The phenotype of *C. columbiana* var. *tenuiloba* may be at least in part a response to habitat; in some areas it grows on exposed summits while var. *columbiana* occurs nearby at lower elevations. In other areas, however, such as the Killdeer Mountains of North Dakota and the Black Hills of South Dakota and Wyoming, only the *tenuiloba* extreme is present.

1. Aerial stems elongating, viny, 0.5–1.5(–3.5) m; ultimate divisions of leaves often more than 5 mm wide, thin. . . . . 13a. *Clematis columbiana* var. *columbiana*
1. Aerial stems tufted, not viny, usually less than 0.1 m; ultimate divisions of leaves mostly 1.5–5 mm wide, ± succulent. . . . . 13b. *Clematis columbiana* var. *tenuiloba*

13a. Clematis columbiana (Nuttall) Torrey & A. Gray var. **columbiana** ☐E ☐F

Clematis pseudoalpina (Kuntze) A. Nelson

Aerial stems viny, climbing or trailing, 0.5–1.5(–3.5) m. **Leaf blade** 2–3-ternate; leaflets mostly lanceolate to ovate, few-lobed, lobes often over 5 mm wide, thin, margins serrate. **Flowers:** sepals violet-blue, rarely white, 2.5–6 cm.

Flowering spring–early summer(–fall). Rocky, open woods and thickets; 1700–3200 m; Ariz., Colo., Idaho, Mont., N.Mex., Tex., Utah, Wyo.

Clematis columbiana var. *columbiana* flowers in

early summer at high elevations and occasionally flowers in fall on new growth. D. S. Correll and M. C. Johnston (1970) included the taxon in *C. alpina* Linnaeus, a broader circumscription than that accepted here.

13b. Clematis columbiana (Nuttall) Torrey & A. Gray var. **tenuiloba** (A. Gray) J. S. Pringle, Brittonia 23: 382. 1971 [E] [F]

Clematis alpina (Linnaeus) Miller var. *occidentalis* (Hornemann) A. Gray subvar. *tenuiloba* A. Gray in H. Newton and W. P. Jenney, Rep. Geol. Resources Black Hills, 531. 1880; *C. tenuiloba* (A. Gray) C. L. Hitchcock

Stems mainly subterranean, rhizomatous, aerial stems not viny, mostly less than 0.1 m (to 1.5 m in forms transitional to var. *columbiana*), tufted. **Leaf blade** mostly 3-ternate, ± succulent; leaflets or lobes mostly 1.5–5 mm wide. **Flowers:** sepals violet-blue, 1.5–5 cm. $2n = 16$.

Flowering late spring–early summer. Cliffs, rocky summits, usually in open sites or open pine forest; 1000–3000 m; Colo., Mont., N.Dak., S.Dak., Utah, Wyo.

14. Clematis occidentalis (Hornemann) de Candolle in A. P. de Candolle and A. L. P. de Candolle, Prodr. 1: 10. 1824 · Purple clematis, clématite occidentale [E]

Atragene occidentalis Hornemann, Hort. Bot. Hafn. 2: 520. 1815

Stems viny, climbing or trailing (plants scarcely viny perennials in var. *dissecta*). **Leaf blade** 1-ternate (or terminal leaflet sometimes ternate in var. *dissecta*), ± firm but not succulent; leaflets lance-ovate to triangular or suborbiculate, lobed or unlobed, margins entire or toothed. **Flowers:** sepals violet-blue, reddish violet, or white, lanceolate to ovate or elliptic-oblong.

Varieties 3 (3 in the flora): North America.

1. Leaflets lobed or unlobed, margins entire or crenate-serrate (or terminal leaflet sometimes ternate); stems tufted or, if viny, up to 0.5(–1.5) m . . . . 14c. *Clematis occidentalis* var. *dissecta*
1. Leaflets unlobed or some 1–3-lobed, margins entire or shallowly serrate; stems ± viny, climbing or trailing, 0.25–3.5 m.
    2. Sepals reddish violet, rounded-mucronate to nearly acuminate . . . . 14a. *Clematis occidentalis* var. *occidentalis*
    2. Sepals violet-blue to pale blue, rarely white, usually distinctly acuminate . . . . 14b. *Clematis occidentalis* var. *grosseserrata*

14a. Clematis occidentalis (Hornemann) de Candolle var. **occidentalis** [E]

Atragene americana Sims; *Clematis verticillaris* de Candolle; *C. verticillaris* var. *cacuminis* Fernald; *C. verticillaris* var. *grandiflora* B. Boivin

Stems ± viny, climbing or trailing, 0.25–3.5 m. **Leaves:** leaflets unlobed or occasionally 1–3-lobed, (2–)3–6(–10) cm, margins entire or shallowly crenate-serrate. **Flowers:** sepals remaining moderately divergent, reddish violet, ovate to oblong-elliptic, 2.5–6 cm, margins not fluted, tips rounded-mucronate to nearly acuminate. $2n = 16$.

Flowering spring. Calcareous cliffs, rock ledges, talus slopes, gravelly embankments, rocky woods, and clearings; 0–1300 m; N.B., Ont., Que.; Conn., Del., Ill., Iowa, Maine, Md., Mass., Mich., Minn., N.H., N.J., N.Y., N.C., Pa., R.I., Vt., Va., W.Va., Wis.

Clematis occidentalis var. *occidentalis* formerly occurred in Ohio. Plants in the western part of the range of this variety tend to have larger, more abruptly tapering sepals; they have been segregated as var. *grandiflora* B. Boivin, but they do not appear to constitute a distinct taxon.

14b. Clematis occidentalis (Hornemann) de Candolle var. **grosseserrata** (Rydberg) J. S. Pringle, Brittonia 23: 370. 1971 [E]

Atragene grosseserrata Rydberg, Bull. Torrey Bot. Club 29: 156. 1902; *Clematis occidentalis* subsp. *grosseserrata* (Rydberg) Roy L. Taylor & MacBryde

Stems ± viny, climbing or trailing, 0.25–2.5 m. **Leaves:** leaflets unlobed or occasionally 1–3-lobed, 2–11 cm, margins entire or less often shallowly crenate-serrate. **Flowers:** sepals often eventually wide-spreading, violet-blue to pale blue or rarely white, lance-ovate, 3–6 cm, margins often ± fluted, tips acuminate. $2n = 16$.

Flowering spring–early summer. Often deep, fine soils in shady forest, also cliffs and other rocky sites in open woods and thickets; 400–2800 m; Alta., B.C., Sask., Yukon; Colo., Idaho, Mont., Oreg., Utah, Wash., Wyo.

The names *Clematis columbiana* (Nuttall) Torrey & A. Gray and *C. verticillaris* var. *columbiana* (Nuttall) A. Gray have long and frequently been misapplied to this taxon. This erroneous usage continues in some horticultural references.

14c. Clematis occidentalis (Hornemann) de Candolle var. **dissecta** (C. L. Hitchcock) J. S. Pringle, Brittonia 23: 371. 1971 E

Clematis columbiana (Nuttall) Torrey & A. Gray var. *dissecta* C. L. Hitchcock, Univ. Wash. Publ. Biol. 17(2): 341. 1964

Stems short and tufted or, if viny, up to 0.5(–1.5) m and trailing. **Leaves:** leaflets lobed or unlobed, 1.5–6.5(–9) cm, margins crenate-serrate; terminal leaflet, at least, usually deeply lobed, or sometimes ternate. **Flowers:** sepals moderately divergent, reddish violet or less often violet-blue, lance-ovate, 2.5–4.5(–6) cm, margins not fluted, tips acute to acuminate.

Flowering spring. Cliffs and other rocky sites in open woods and thickets; 700–1900 m; Wash.

Clematis occidentalis var. *dissecta* occurs only in the Wenatchee and adjacent ranges of the Cascade Mountains.

6d. CLEMATIS Linnaeus subg. VIORNA A. Gray in A. Gray et al., Syn. Fl. N. Amer. 1(1): 5. 1895

James S. Pringle

Viorna Spach 1839, not *Viorna* (Persoon) Reichenbach 1837

Woody vines or erect, ± herbaceous perennials, clumped (or patch-forming from rhizomes in *C. socialis*). **Leaves** simple. **Leaf blade** 1- or 2-pinnate, ternate, or finely dissected; ultimate divisions lobed or unlobed, margins entire or few-toothed. **Inflorescences** terminal and/or axillary, on current year's stems; 3–7-flowered bracteate cymes or flowers solitary or paired, peduncles bracteate [or several–many-flowered panicles]. **Flowers** bisexual, usually nodding (± erect in some spp., esp. *C. morefieldii*); sepals ascending, connivent at least proximally and usually much of length, variously colored, lanceolate or oblong to broadly ovate, thick, usually leathery, abaxially glabrous to silky, hirsute, or tomentose; filaments slender, usually pubescent (except *C. pitcheri* var. *dictyota*), connectives often ± prolonged (especially in *C. pitcheri*); staminodes absent. **Achenes** flattened; beak variable in length, plumose to nearly glabrous.

Species ca. 25 (18 in the flora): temperate, mostly North America, a few in Eurasia.

Clematis integrifolia Linnaeus, with relatively wide-spreading, blue sepals, is locally naturalized in Ontario.

SELECTED REFERENCES Dennis, W. M. 1976. A Biosystematic Study of *Clematis* Section *Viorna* Subsection *Viornae*. Ph.D. dissertation. University of Tennessee. Erickson, R. O. 1943. Taxonomy of *Clematis* section *Viorna*. Ann. Missouri Bot. Gard. 30: 1–62, plate 1. Fernald, M. L. 1943. Morphological differentiation of *Clematis ochroleuca* and its allies. Rhodora 45: 401–412, figs. 776–782. Keener, C. S. 1967. A biosystematic study of *Clematis* subsection *Integrifoliae* (Ranunculaceae). J. Elisha Mitchell Sci. Soc. 83: 1–41.

Key to the species of *Clematis* subg. *Viorna*

1. Leaves all simple, blade rarely so deeply lobed that proximal 2 lobes appear as distinct linear leaflets; plants herbaceous or ± woody at base, erect, not viny.
   2. Sepals glabrous or sparsely villous, tips acuminate; achene body cobwebby-tomentose distally, or long-pubescent.
      3. Leaf blade 3.5–11 cm wide, prominently reticulate adaxially; beak not plumose, proximally silky-tomentose, sparsely appressed-pubescent to nearly glabrous distally; Kansas, Nebraska, Missouri. . . . . 29. *Clematis fremontii*
      3. Leaf blade 0.2–2(–3.5) cm wide, not prominently reticulate adaxially; beak distinctly plumose; Florida. . . . . 30. *Clematis baldwinii*
   2. Sepals (except margins) minutely puberulent, silky, woolly, or nearly glabrous, tips obtuse to acute; achene body pilose or short-pilose.
      4. Leaf blade abaxially moderately to densely soft-pubescent, rarely nearly glabrous; secondary and tertiary veins forming prominent reticulum on adaxial surface.
         5. Stems and abaxial surface of leaf blades moderately silky-pilose with spreading hairs, rarely nearly glabrous; beak yellowish brown to reddish brown; hairs of achene rim appressed-ascending. . . . . 25. *Clematis ochroleuca*

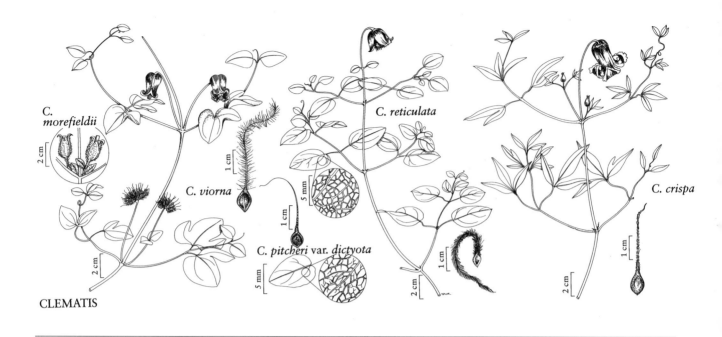

CLEMATIS

5. Stems and abaxial surface of leaf blades densely silky-tomentose with appressed hairs; beak white to pale yellow; hairs of achene rim spreading. . . . . .26. *Clematis coactilis*
4. Leaf blade abaxially glabrous or nearly so, sparsely or rarely densely villous on veins; secondary and tertiary veins not forming prominent reticulum on adaxial surface.
  6. Sepals abaxially silky to woolly; beak white to pale yellow. . . . . .27. *Clematis albicoma*
  6. Sepals abaxially glabrous to minutely puberulent; beak coppery brown.
    7. Leaf blade abaxially glaucous and glabrous; sepals abaxially glabrous . . . . 19. *Clematis addisonii*
    7. Leaf blade abaxially not glaucous, sparsely to densely villous on veins; sepals abaxially nearly glabrous or minutely puberulent . . . . 28. *Clematis viticaulis*
1. At least some distal leaves of main stems distinctly compound or deeply much-dissected; plants erect or viny.
  8. Plants erect or sprawling, not viny; stems to 0.65 m.
    9. Larger leaf blades (1–)2–3-pinnate or -ternate or deeply dissected.
      10. Leaflets usually less than 1.5 cm wide, mostly more than 2.5 times as long as wide, mostly unlobed, if lobed then with lateral lobes 1 or 2, usually small, distinctly narrower than central portion; blade abaxially sparsely to densely hirsute; beak plumose . . . . 32. *Clematis hirsutissima*
      10. Leaflets usually more than 1.5 cm wide and/or less than 2.5 times as long as wide, mostly lobed, lateral lobes often nearly as wide as central portion; blade glabrous or nearly so (rachis and petiolules may be ± hirsute); beak glabrous or inconspicuously appressed-pubescent. . . . . 23. *Clematis bigelovii*
    9. Leaf blade 1-pinnate or simple.
      11. Terminal leaflets usually tendril-like; blades of simple leaves and lateral leaflets of compound leaves usually more than 2 cm wide; sepals purple or reddish purple, whitish toward tips . . . . 19. *Clematis addisonii*
      11. Terminal leaflets with expanded blade, not tendril-like; blades of simple leaves and lateral leaflets usually less than 2 cm wide; sepals uniformly violet-blue.
        12. Plants strongly rhizomatous, forming patches; sepals 2–2.5(–3) cm; beak 1.5–2.5 cm . . . . 31. *Clematis socialis*
        12. Plants not rhizomatous; sepals 2.5–5.5 cm; beak 6–10 cm . . . . .30. *Clematis baldwinii*
  8. Plants viny, petioles and/or rachises of leaves often functioning as tendrils; stems usually 1–5 m.
    13. Largest leaf blades 1–2-pinnate, leaflets mostly deeply lobed; beak 1–3 cm, inconspicuously appressed-pubescent to nearly glabrous. . . . . 22. *Clematis pitcheri*
    13. Largest leaves simple, blades 1-pinnate, or if some 2-pinnate, leaflets of 2-pinnate leaves usually unlobed, rarely 2–5-lobed; beak 2–7 cm, plumose (appressed-puberulent in *C. crispa* and *C. pitcheri*).

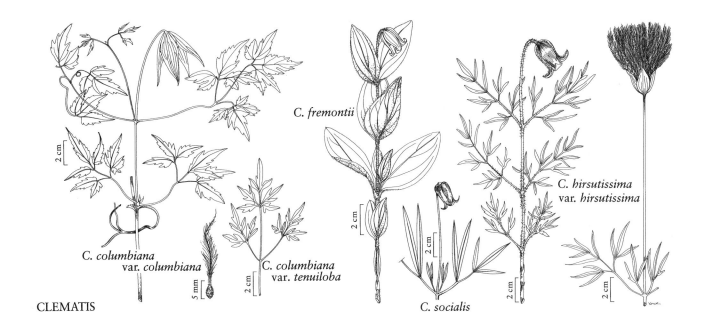

14. Leaflets abaxially glaucous and glabrous, rarely with a few scattered hairs.
    15. Leaves mostly simple, only distalmost compound . . . . 19. *Clematis addisonii*
    15. Leaves all or mostly compound, simple leaves, if any, only on small branches and
        near base of main stem.
        16. Leaf blade ± thin, secondary and tertiary veins not forming prominent reticu-
            lum on adaxial surface . . . . .18. *Clematis glaucophylla*
        16. Leaf blade leathery, secondary and tertiary veins forming prominent reticulum
            on adaxial surface.
            17. Sepals rose-red to scarlet abaxially and at tip adaxially; tips recurved. . . . . 20. *Clematis texensis*
            17. Sepals pale lavender or blue-violet to reddish purple abaxially, often paler
                or greenish adaxially, tips slightly spreading . . . . .17. *Clematis versicolor*
14. Leaflets not glaucous, usually abaxially pubescent, sometimes glabrous.
    18. Sepals with thin, crispate margins to 6 mm wide distally . . . . 24. *Clematis crispa*
    18. Sepals without expanded, thin, crispate margins or with margins less than 2.5 mm
        wide.
        19. Leaf blade leathery, secondary and tertiary veins forming prominent reticulum
            on adaxial surface.
            20. Beak plumose, with spreading hairs; leaf blade finely reticulate (ultimate
                closed areoles mostly less than 2 mm in longer dimension), even quaternary
                veins distinctly raised on adaxial surface . . . . .21. *Clematis reticulata*
            20. Beak sparsely pubescent to silky, with ascending to appressed hairs; leaf
                blade less finely reticulate (ultimate closed areoles mostly more than 2 mm
                in longer dimension), tertiary and quaternary veins scarcely or not raised
                on adaxial surface . . . . 22. *Clematis pitcheri*
        19. Leaf blade thin, secondary and tertiary veins not forming prominent reticulum
            on adaxial surface.
            21. Stems generally cobwebby as well as villous; bracts near base of peduncle/
                pedicel; sepals densely silky-pubescent abaxially, pinkish suffused with
                green, tips spreading to short-reflexed . . . . 16. *Clematis morefieldii*
            21. Stems without cobwebby pubescence; bracts well above base of peduncle/
                pedicel; sepals sparsely pubescent abaxially, pale lavender to reddish purple,
                tips recurved . . . . .15. *Clematis viorna*

15. **Clematis viorna** Linnaeus, Sp. Pl. 1: 543. 1753 · Leather-flower E F

Clematis beadlei (Small) R. O. Erickson; *C. viorna* var. *flaccida* (Small ex Rydberg) R. O. Erickson; *Viorna beadlei* Small; *V. flaccida* (Small ex Rydberg) Small; *V. gattingeri* (Small) Small; *V. viorna* (Linnaeus) Small

Stems viny, to 4 m, without cobwebby pubescence, nearly glabrous, or moderately pilose-pubescent proximal to nodes. **Leaf blade** mostly 1-pinnate, some simple; leaflets usually 4–8 plus additional tendril-like terminal leaflet, lanceolate to ovate, unlobed or 2–3-lobed, or most proximal 3-foliolate, 2–12 × 1–5(–6) cm, thin, not conspicuously reticulate; surfaces abaxially sparsely to densely pilose, not glaucous. **Inflorescences** axillary, 1–7-flowered; bracts well above base of peduncle/pedicel. **Flowers** broadly urn-shaped to bell-shaped; sepals pale lavender to reddish purple, grading to cream-yellow toward tip, ovate-lanceolate, 1.5–3 cm, margins not expanded, very thick, not crispate, tomentose, tips acuminate, recurved, abaxially sparsely to densely pubescent. **Achenes:** bodies silky-pubescent; beak 2.5–6 cm, plumose. $2n = 16$.

Flowering spring–summer. Wooded cliffs and stream banks; 0–1400 m; Ala., Ark., Del., D.C., Ga., Ill., Ind., Ky., Md., Miss., Mo., N.C., Ohio, Pa., S.C., Tenn., Va., W.Va.

Formerly *Clematis viorna* was locally naturalized near Guelph, Ontario; no recent reports are known. D. S. Correll and M. C. Johnston (1970) mention "a report of a specimen" from Texas; neither the specimen nor further details have been located.

The Fox Indians prepared a drink from the roots of *Clematis viorna* to use medicinally as a panacea (D. E. Moerman 1986).

16. **Clematis morefieldii** Kral, Ann. Missouri Bot. Gard. 74: 665, figs. 1, 2. 1987 · Morefield's clematis, Morefield's leather-flower C E F

Stems viny, to 5 m, cobwebby-tomentose and pilose. **Leaf blade** 1-pinnate; leaflets 4–10 plus additional tendril-like terminal leaflet, narrowly to broadly ovate, unlobed or 2–3-lobed, 3.5–10 × 2–6.5 cm, thin, reticulate; surfaces abaxially densely silky-pilose, not glaucous. **Inflorescences** axillary, 1–5-flowered; bracts at or near base of peduncle/pedicel. **Flowers** urn-shaped; sepals pinkish, suffused with green, oblong-lanceolate, 2–2.5 cm, margins not expanded, thick, not crispate, tomentose, tips acuminate, slightly spreading to short-reflexed, ab-axially densely silky-pubescent. **Achenes:** bodies silky-pubescent; beak 3–3.5 cm, plumose.

Flowering spring–early summer. Open woods among limestone boulders; of conservation concern; 200–300 m; Ala.

Clematis morefieldii is known only from limestone uplands east of Huntsville, Madison County, Alabama.

From all variants of the closely related *Clematis viorna*, *C. morefieldii* differs in the cobwebby-tomentose as well as villous pubescence of its stems, and in having bracts at or very near the base of the peduncle rather than well above the base.

17. **Clematis versicolor** Small ex Rydberg in N. L. Britton, Man. Fl. N. States, 421. 1901 · Pale clematis E

Viorna versicolor (Small ex Rydberg) Small

Stems viny, to 5 m, glabrous or sparsely pubescent. **Leaf blade** 1-pinnate; leaflets 8–10 plus additional tendril-like terminal leaflet, elliptic to ovate, usually unlobed, occasionally 2–3-lobed, 2–8 × 1.5–6.5 cm, leathery, abaxially and adaxially prominently reticulate; surfaces abaxially glabrous and glaucous. **Inflorescences** axillary, 1–7-flowered. **Flowers** broadly urn-shaped; sepals abaxially pale lavender to reddish purple, grading to pale green toward tip, narrowly ovate, 1.5–3 cm, margins not expanded, thick, not crispate, tomentose, tips acute, slightly spreading, glabrous. **Achenes:** bodies appressed-pubescent; beak (3–)5–6 cm, plumose. $2n = 16$.

Flowering spring–summer (Jun–Aug). Calcareous outcrops, sandy soils, dry woods and thickets, barrens, and roadsides; 30–100 m; Ala., Ark., Ky., Mo., Okla., Tenn., Tex.

18. **Clematis glaucophylla** Small, Bull. Torrey Bot. Club 24: 337. 1897 · Glaucous clematis E

Viorna glaucophylla (Small) Small

Stems viny, to 5 m, glabrous. **Leaf blade** 1-pinnate; leaflets 4–10 plus additional tendril-like terminal leaflet, proximal leaflets usually 3-lobed or 3-foliolate, distal leaflets usually unlobed, ovate, 3–10 × 2–7.5 cm, ± thin, not prominently reticulate adaxially; surfaces abaxially glabrous and glaucous. **Inflorescences** axillary, 1–3-flowered; bracts about 1/3 distance from base of peduncle. **Flowers** ovoid; sepals deep rose-red to purplish red, ovate-lanceolate, 2–2.5 cm, margins not expanded, thick, not crispate, tomentose, tips long-acuminate, ± recurved, abaxially glabrous. **Achenes:** bodies appressed-pubescent; beak 5–6 cm, plumose. $2n = 16$.

Flowering spring–summer. Stream banks in rich, neutral to slightly acid soils; 0–100 m; Ala., Fla., Ga., Miss., Okla.

Reports of *Clematis glaucophylla* from other southeastern states have been based on misidentified specimens (W. M. Dennis 1976). Recent reports of the species in Arkansas, Kentucky, Louisiana, and Virginia have not been confirmed.

19. **Clematis addisonii** Britton, Mem. Torrey Bot. Club 2: 28, plate 3. 1890 · Addison's virgin's-bower, Addison Brown's clematis, Addison's leather-flower, Addison Brown's leather-flower [C] [E]

Viorna addisonii (Britton) Small

Stems usually ascending to erect, occasionally somewhat viny, 0.6–1 m, glabrous. **Leaves** all simple, blade often 1-pinnate on distal and middle leaves on vigorous plants 4–13 × 2–9.5 cm; leaflets 2–6 plus additional tendril-like terminal leaflet, ovate, unlobed, 1.5–6 × 1–4.5 cm, not prominently reticulate; surfaces abaxially glabrous and glaucous. **Inflorescences** terminal and axillary, flowers solitary. **Flowers** ovoid to broadly urn-shaped; sepals purple or reddish purple, whitish toward tips, ovate-lanceolate, 1.2–2.5 cm, margins not expanded, thick, not crispate, tomentose, tips acute, spreading, abaxially glabrous. **Achenes:** bodies puberulent; beak 2.5–3.5 cm, plumose. $2n = 16$.

Flowering spring–early summer. Calcareous, dry woods, glades, rock outcrops; of conservation concern; 200–600 m; Va.

Clematis addisonii is known only from Botetourt, Montgomery, Roanoke, and Rockbridge counties in western Virginia. Reports of this infrequent species from other southeastern states have been based on misidentified specimens (W. M. Dennis 1976).

20. **Clematis texensis** Buckley, Proc. Acad. Nat. Sci. Philadelphia 13: 448. 1862 · Scarlet clematis, crimson clematis [E]

Viorna coccinea (A. Gray) Small

Stems viny, to 3 m, glabrous or sometimes ± hirsute near nodes. **Leaf blade** 1-pinnate; leaflets 6–10 plus additional tendril-like terminal leaflet, ovate to nearly round, unlobed, 2–3-lobed, or most proximal occasionally 3-foliolate, 1–9 × 1–6 cm, leathery, ± prominently reticulate adaxially; surfaces abaxially usually glabrous, occasionally sparsely pubescent, glaucous. **Inflorescences** axillary, 1–7-flowered. **Flowers** ovoid to urn-shaped; sepals rose-red to scarlet abaxially and at tip adaxially, ovate-lanceolate, 1.5–3 cm, margins not expanded, thick, not crispate, tomentose, tips acute to acuminate, recurved, abaxially glabrous. **Achenes:** bodies appressed-pubescent; beak 4–7 cm, plumose. $2n = 16$.

Flowering spring–summer (Mar–Jun). Woodlands, calcareous cliffs, and stream banks; 80–700 m; Tex.

Although widely cultivated because it is the only species of *Clematis* with truly red flowers, *C. texensis* is native only to the southeastern part of the Edwards Plateau, Texas.

21. **Clematis reticulata** Walter, Fl. Carol., 156. 1788 [E] [F]

Viorna reticulata (Walter) Small; *V. subreticulata* Harbison ex Small

Stems viny, to 4 m, glabrous or sparsely pilose-pubescent, sometimes more densely pubescent near nodes. **Leaf blade** 1-pinnate; leaflets 6–8 plus additional tendril-like terminal leaflet, elliptic to ovate, unlobed, 1–3-lobed, or proximal 3-foliolate, 1–9 × 0.5–5(–7.5) cm, leathery, prominently and finely reticulate abaxially and adaxially; surfaces abaxially silky-pubescent, not glaucous. **Inflorescences** axillary, 1–3-flowered; bracts about 1/3 distance from base of peduncle. **Flowers** urn-shaped; sepals pale lavender to purple, greenish toward tip, ovate-lanceolate, 1.2–3 cm, margins not expanded, ± thick, not crispate, densely tomentose, tips acute, recurved, abaxially usually ± densely yellowish pubescent, occasionally nearly glabrous. **Achenes:** bodies appressed-pubescent; beak 4–6 cm, plumose. $2n = 16$.

Flowering spring–summer (May–Jun). Dry woods and thickets in sandy soils; 0–150 m; Ala., Ark., Fla., Ga., La., Miss., Okla., S.C., Tex.

In immature fruit, especially, the vesture of the beaks of *Clematis reticulata* might not consistently suffice to distinguish it from *C. pitcheri*, which has appressed-pubescent beaks. *Clematis reticulata* is distinguished from *C. pitcheri* by the very fine reticulation of the leaves, with the smallest areoles completely enclosed by veinlets generally less than 1 mm long and even the quaternary veins prominently raised on the adaxial surface.

22. **Clematis pitcheri** Torrey & A. Gray, Fl. N. Amer. 1: 10. 1838 · Bellflower clematis, Pitcher's clematis

Viorna pitcheri (Torrey & A. Gray) Britton

Stems viny, to 4 m, very sparsely short-pilose, sometimes nearly glabrous. **Leaf blade** mostly 1–2 pinnate, many leaves simple; primary leaflets 2–8 plus additional tendril-like terminal leaflet, deeply 2–5-lobed or unlobed or 3-foliolate, leaflets or major lobes lanceolate to broadly ovate, 1–11 × 1–6 cm, leathery (thin in var. *pitcheri*), ± prominently reticulate adaxially; surfaces abaxially nearly glabrous to densely pubescent, not glaucous. **Inflorescences** axillary, 1–7-flowered. **Flowers** ovoid to urn-shaped; sepals pale to dark bluish or reddish purple, sometimes whitish toward tip, ovate-lanceolate, 1.2–3(–4) cm (larger sepals mostly in w part of range), margins narrowly expanded distally to about 1 mm wide, thin, crispate toward tip, tomentose, tips acuminate, recurved, abaxially sparsely to densely appressed-puberulent. **Achenes:** bodies appressed-pubescent; beak 1–3 cm, nearly glabrous to ± appressed-pubescent or silky.

Varieties 2 (2 in the flora): North America, Mexico.

Clematis pitcheri is highly variable, notably in the size and thickness of the leaflets, the external sepal color and internal color of the recurved tips, and the amount of pubescence of the beaks. Additional varieties might be recognized, as some authors have done in the past, but the extent of intergradation and the lack of correlation among varying traits tend to make recognition of additional varieties impractical (W. M. Dennis 1976). The two varieties recognized here show very extensive intergradation in the western part of the range of the species.

Although otherwise similar to *Clemitis reticulata, C. pitcheri* differs distinctly in its more coarsely reticulate leaves, with the smallest closed areoles mostly over 2 mm long, and its scarcely raised tertiary and quaternary veins.

1. Leaflets mostly 4–11 cm, thin; stamens with filaments and extended connectives usually pubescent. . . . . 22a. *Clematis pitcheri* var. *pitcheri*
1. Leaflets mostly 3–4 cm, somewhat leathery; stamens with filaments and extended connectives usually glabrous or nearly so . . . . 22b. *Clematis pitcheri* var. *dictyota*

22a. Clematis pitcheri Torrey & A. Gray var. **pitcheri**

Clematis filifera Bentham; *C. pitcheri* var. *filifera* (Bentham) R. O. Erickson

Leaf blade usually 1-pinnate, distalmost leaf occasionally simple; leaflets usually unlobed, occasionally some 1–2-lobed, mostly 4–11 × 1.5–6 cm (occasionally smaller in w part of range), thin. **Flowers:** sepals mostly 1–3(–3.5) cm; stamens with filaments and extended connectives pubescent, occasionally glabrous. $2n = 16$.

Flowering spring–fall (Mar–Oct). Limestone out-

crops in dry to moist woods and thickets, disturbed sites; 0–2200 m; Ark., Ill., Ind., Iowa, Kans., Ky., Mo., Nebr., N.Mex., Okla., Tenn., Tex.; Mexico.

22b. Clematis pitcheri Torrey & A. Gray var. **dictyota**
(Greene) W. M. Dennis, Sida 8: 194. 1979 [F]

Clematis dictyota Greene, Pittonia 5: 133. 1903

Leaf blade mostly 2-pinnate or 2-ternate; leaflets often deeply lobed, mostly 3–4 cm, somewhat leathery. **Flowers:** sepals mostly 2–4 × 1–3 cm; stamens with filaments and extended connectives usually glabrous, occasionally with a few hairs. $2n = 16$.

Flowering spring–fall (Apr–Sep). Rocky sites; 1200–2300 m; N.Mex., Tex.; Mexico (Coahuila).

In typical *Clematis pitcheri* var. *dictyota*, the stamen filaments and the extended anther connectives are glabrous or nearly so, whereas in typical *C. pitcheri* var. *pitcheri* they are pubescent. In some parts of the range of the species, however, these character states are not well correlated with leaflet size and thickness.

The names *Clematis filifera* Bentham and *C. pitcheri* var. *filifera* (Bentham) R. O. Erickson have generally been applied to this variety or to a taxon consisting in large part of this variety, but the type specimen is referable to *C. pitcheri* var. *pitcheri* (W. M. Dennis 1979).

23. Clematis bigelovii Torrey, Pacif. Railr. Rep. 4(5): 61. 1857 ["1856"] · Bigelow's clematis [E]

Stems erect or sprawling, 0.1–0.6 m, short pubescent-pilose, sometimes sparsely so. **Leaf blade** 1–2(–3)-pinnate; primary leaflets 7–11, mostly deeply 2–several-lobed, leaflets and larger lobes mostly ovate, 0.8–3.5 × 0.5–1.5 cm (lateral lobes nearly as wide as central portion), thin, not prominently reticulate; surfaces glabrous, somewhat glaucous. **Inflorescences** terminal, 1-flowered. **Flowers** broadly urn- to bell-shaped; sepals purple, lanceolate, 1.5–3 cm, margins narrowly expanded distally to ca. 1 mm wide, thin, crispate, tomentose, tips acuminate, spreading, abaxially sparsely pubescent. **Achenes:** bodies appressed–long-pubescent; beak 2–3 cm, glabrous or inconspicuously appressed-pubescent, sparsely so toward tip.

Flowering spring–fall. Mountain slopes, moist sites in canyons; 1700–2400 m; Ariz., N.Mex.

Clematis bigelovii is locally common and is restricted to New Mexico and a few sites in Arizona. Although usually grouped with *C. hirsutissima* because of its scarcely viny habit, this species appears to represent the extreme expression of clinal variation including *C. pitcheri* var. *dictyota* and, at the other extreme, eastern populations of *C. pitcheri* var. *pitcheri*; it might well be treated as *C. pitcheri* var. *bigelovii* (Torrey) B. L. Robinson.

Clematis palmeri Rose is represented by a small number of specimens from New Mexico and one from eastern Arizona, and its distinctness as a species has been questioned. The few published descriptions scarcely suffice to indicate its distinguishing features and are sometimes at variance with each other and with specimens so identified by R. O. Erickson (1943) or others. Almost all specimens were found where *C. bigelovii* was reported nearby. Pending further studies, it seems likely that *C. palmeri* comprises somewhat aberrant specimens of *C. bigelovii*, *C. hirsutissima* var. *scottii*, and/or *C. pitcheri* var. *dictyota*, perhaps also some herbarium sheets of flowering and fruiting material inadvertently collected from different species, and/ or hybrids involving the species named above in one or more combinations.

24. **Clematis crispa** Linnaeus, Sp. Pl. 1: 543. 1753 · Marsh clematis, curly clematis, blue-jasmine E F

Viorna crispa (Linnaeus) Small; *V. obliqua* Small

Stems viny, to 3 m, glabrous or sparsely to moderately pilose-pubescent, denser at nodes. **Leaf blade** 1–2-pinnate or rarely a few simple or 3-foliolate; leaflets 4–10 plus additional ± tendril-like terminal leaflet, usually lanceolate to ovate, occasionally linear, unlobed or proximally 3–5-lobed, (1.5–)3–10 × (0.1–)0.4–4(–5) cm, thin, not conspicuously reticulate; surfaces glabrous, not glaucous. **Inflorescences** terminal, 1-flowered; bracts absent. **Flowers** bell-shaped; sepals distally strongly spreading to recurved, violet-blue, lanceolate, 2.5–5 cm, margins proximally thick and tomentose, distally broadly expanded, 2–6 mm wide, thin, crispate, less conspicuously tomentose than proximal portion, or glabrate, tips acuminate, abaxially glabrous. **Achenes:** bodies appressed-puberulent; beak 2–3.5 cm, appressed-puberulent. $2n = 16$.

Flowering spring–summer. Low woods, bottomlands, swamps; 0–200 m; Ala., Ark., Fla., Ga., Ill., Ky., La., Miss., Mo., N.C., Okla., S.C., Tenn., Tex., Va.

Clematis crispa is highly variable in leaflet width, and conspicuous variation may occur on a single plant (R. O. Erickson 1943); no discontinuity or geographic correlation exists that would permit the recognition of varieties. The dilated, petaloid sepal tips and thin,

crispate, broadly expanded sepal margins are diagnostic for this species.

25. **Clematis ochroleuca** Aiton, Hort. Kew. 2: 260. 1789 · Erect silky leather-flower, curly-heads E

Viorna ochroleuca (Aiton) Small

Stems erect to ± sprawling, not viny, 2–7 dm, sparsely to ± densely pilose. **Leaves** simple. **Leaf blade** narrowly to broadly ovate, unlobed or rarely few-lobed, 3–14 × (1.5–)2.5–8(–9.5) cm, ± leathery, reticulate adaxially; surfaces abaxially moderately silky-pilose with spreading hairs or rarely nearly glabrous, not glaucous. **Inflorescences** terminal, flowers solitary; bracts absent. **Flowers** narrowly urn-shaped; sepals pale yellow to pale purple, lanceolate, 1–3.5 cm, margins not expanded, thin, not crispate, tomentose, tip obtuse, spreading to recurved, abaxially silky-pubescent. **Achenes:** bodies pilose, hairs appressed-ascending; beak yellowish brown to reddish brown, 3–6 cm, plumose. $2n = 16$.

Flowering spring. Dry to moist woods, thickets, roadsides, and other shady to open, ± disturbed sites, mostly on mafic substrates; 0–500 m; D.C., Ga., Md., N.J., N.Y., N.C., S.C., Va.

In New York, *Clematis ochroleucra* is known only from Staten Island and, formerly, from western Long Island (Brooklyn).

26. **Clematis coactilis** (Fernald) Keener, J. Elisha Mitchell Sci. Soc. 83: 36. 1967 C E

Clematis albicoma Wherry var. *coactilis* Fernald, Rhodora 45: 407, plate 780. 1943

Stems erect, not viny, 2–4.5 dm, densely silky, hirsute, or ± tomentose with appressed hairs. **Leaves** simple. **Leaf blade** rarely 1-pinnate, narrowly to broadly ovate, unlobed or sometimes few-lobed, 5–12 × 3–9.5 cm, leathery, ± prominently reticulate adaxially; surfaces abaxially densely silky-tomentose with appressed hairs, not glaucous. **Inflorescences** terminal, flowers solitary; bracts absent. **Flowers** broadly urn-shaped; sepals pale yellow to rarely purple-tinged, lanceolate, 1.9–3.4 cm, margins not expanded or narrowly expanded to 1.7 mm wide, thin, not crispate, tomentose, tips obtuse, spreading, abaxially finely tomentose. **Achenes:** body pilose, hairs of rim spreading; beak (2.5–)3–4.5(–5.5) cm, plumose. $2n = 16$.

Flowering spring–early summer. Shale barrens, rarely on sandstone, dolomite, or limestone outcrops; of conservation concern; 300–600 m; Va.

Clematis coactilis is known only from western Virginia. C. S. Keener (1967, 1975) suggested that this species may be a stabilized derivative of past hybridization between *C. albicoma* and *C. ochroleuca*.

In fruit, *Clematis coactilis* is distinguishable from *C. ochroleuca* by its combination of spreading to reflexed hairs on the achene rims and whitish to pale yellow (rarely tawny) hairs on the beaks, contrasting with the strongly ascending hairs on the achene rims and tawny (rarely yellowish white) hairs on the beaks of *C. ochroleuca*. This species and *C. ochroleuca* lack stomates on the adaxial surface of the leaves, whereas the closely related species *C. albicoma*, *C. fremontii*, and *C. viticaulis* have stomates on both leaf surfaces (C. S. Keener 1967).

27. Clematis albicoma Wherry, J. Wash. Acad. Sci. 21: 198, fig. 1. 1931 · Erect mountain clematis, white-haired leather-flower E

Stems erect, not viny, 2–4(–6) dm, pubescent or pilose to ± tomentose or hirsute. Leaves simple. Leaf blade elliptic-lanceolate to ovate, unlobed, 3.5–8(–10) × 1.5–5(–6.5) cm, thin, not conspicuously reticulate; surfaces abaxially glabrous to sparsely (rarely more densely) villous on veins, not glaucous. Inflorescences terminal, flowers solitary; bracts absent. Flowers narrowly urn-shaped; sepals purplish, yellowish toward tips, oblong-lanceolate, (1.1–)1.4–3 cm, margins not expanded or less than 1 mm wide, thin, not crispate, tomentose, tips obtuse, spreading to recurved, abaxially silky- to woolly-pubescent. Achenes: bodies pilose; beak white to pale yellow, (1.5–)2–4(–4.5) cm, plumose. 2n = 16.

Flowering spring–early summer. Shale barrens; 300–800 m; Va., W.Va.

Clematis albicoma is known only from shale barrens predominantly developed from the Upper Devonian Brallier Formation in nine counties of western Virginia and adjacent West Virginia.

28. Clematis viticaulis Steele, Contr. U.S. Natl. Herb. 13: 364. 1911 · Millboro leather-flower, grape clematis, grape leather-flower C E

Stems erect, 2–5 dm, finely and densely hirtellous. Leaves simple. Leaf blade elliptic-lanceolate to narrowly ovate, unlobed, (2–)4–8 × 1.5–3.5(–4.5) cm, thin, not conspicuously reticulate; surfaces abaxially sparsely (rarely more densely) villous on veins, not glaucous. Inflorescences termi-

nal, flowers solitary; bracts absent. Flowers urn-shaped; sepals pale purple, often suffused with green abaxially, lanceolate, 1.4–2.5 cm, margins not expanded, thin, not crispate, puberulent, tips obtuse to acute, spreading to recurved, abaxially nearly glabrous to minutely puberulent. Achenes: bodies short-pilose; beak coppery brown, 2–3.5(–4) cm, plumose. 2n = 16.

Flowering spring–early summer. Shale barrens; of conservation concern; 400–500 m; Va.

Clematis viticaulis is known only from shale barrens developed from the Upper Devonian Brallier Formation in Bath and Rockbridge counties of western Virginia.

The coppery brown hairs on the mature beaks are useful for distinguishing this species (C. S. Keener 1967).

29. Clematis fremontii S. Watson, Proc. Amer. Acad. Arts 10: 339. 1875 · Fremont's clematis E F

Clematis fremontii var. *riehlii* R. O. Erickson; *Viorna fremontii* (S. Watson) A. Heller

Stems erect, 1.5–4(–7) dm, moderately to densely villous, sometimes sparsely so near nodes. Leaves simple. Leaf blade ovate-elliptic to broadly ovate, unlobed, 5–14 × 3.5–11 cm, leathery, prominently reticulate adaxially; surfaces abaxially glabrous or sparsely villous-tomentose on veins, not glaucous. Inflorescences terminal, flowers solitary; bracts absent. Flowers narrowly urn-shaped; sepals pale blue-violet to purple, pale green toward tips, lanceolate, 2–4 cm, margins not expanded or 1–3 mm wide, thin or ± thick, not crispate or slightly crispate, tomentose, tips acuminate, spreading to recurved, abaxially glabrous to sparsely villous. Achenes: bodies proximally pubescent with silky or sparse short hairs, distally cobwebby-tomentose; beak 1.5–3(–3.5) cm, proximally silky-tomentose, sparsely appressed-pubescent to nearly glabrous toward tip. 2n = 16.

Flowering spring. Calcareous prairies and glades; 100–700 m; Kans., Mo., Nebr.

R. O. Erickson (1943) separated the Missouri plants as *Clematis fremontii* var. *riehlii* R. O. Erickson on the basis of their supposedly greater height and more widely spaced, proportionately narrower leaves. As noted by C. S. Keener (1967), however, so much overlap occurs in the ranges of variation of the Missouri and the Kansas-Nebraska populations that recognition of these varieties is not appropriate.

30. Clematis baldwinii Torrey & A. Gray, Fl. N. Amer. 1: 8. 1838 · Pine-woods clematis, pine-hyacinth [E]

Clematis baldwinii var. *latiuscula* R. W. Long; *Viorna baldwinii* (Torrey & A. Gray) Small

Stems erect, 2–6 dm, nearly glabrous to moderately pilose. **Leaves** usually simple. **Leaf blade** unlobed or 2–3-lobed, occasionally divided into 3–5 leaflets; leaflet blades or lobes linear to narrowly elliptic-lanceolate, or unlobed leaf blades elliptic to ovate, 1.5–10 × 0.2–2.5(–3.5) cm, thin, not prominently reticulate; surfaces glabrous, not glaucous. **Inflorescences** terminal, flowers solitary; bracts absent or sometimes distal pair of leaves smaller, bractlike. **Flowers** narrowly bell-shaped; sepals uniformly violet-blue, oblong-lanceolate, 2–5.5 cm, margins narrowly expanded distally to ca. 1 mm wide, thin, crispate, proximally tomentose, glabrous where expanded, distally ± tomentose, tips acuminate, spreading to recurved, abaxially glabrous. **Achenes:** bodies long-pubescent; beak 6–10 cm, plumose. 2*n* = 16.

Flowering all year. Sandy, flat pine woods; 0–50 m; Fla.

The long peduncles (10–30 cm) elevating the flowers well above the leaves are unique to *Clematis baldwinii* among the simple-leaved species of *Clematis* subg. *Viorna* in the flora. Broad-leaved, large-flowered plants have been segregated as *C. baldwinii* var. *latiuscula*, but many intermediates connect the extremes, and flower size is not well correlated with leaf shape. As noted by C. S. Keener (1975), leaf shape appears to be uncorrelated with distribution; collections from a single population often include broad- and narrow-leaved plants.

31. Clematis socialis Kral, Rhodora 84: 287, fig. 1, 2. 1982 · Alabama leather-flower [C][E][F]

Stems erect, not viny, 0.2–0.3 (–0.5) m, glabrous or slightly pubescent, arising from horizontal, branching rhizomes and forming patches. **Leaves:** proximal simple, blades unlobed or 2–3-lobed, distal blades 1-pinnate; leaflets and unlobed blades linear-elliptic to narrowly lanceolate or oblanceolate, (3–)4–12(–15) × (0.3–)0.5–1(–1.5) cm, thin, not prominently reticulate; surfaces abaxially nearly glabrous to sparsely villous on veins, not glaucous. **Inflorescences** terminal, flowers solitary; bracts absent. **Flowers** narrowly urn-shaped; sepals uniformly violet-blue, oblong-lanceolate, 2–2.5(–3) cm,

margins narrowly expanded distally to about 1 mm wide, thin, crispate, proximally tomentose, tips spreading to recurved, acute to acuminate, abaxially sparsely puberulent. **Achenes:** bodies appressed-puberulent; beak 1.5–2.5 cm, appressed-puberulent.

Flowering spring–fall. Openings in wet bottomland woods; of conservation concern; 200 m; Ala.

Clematis socialis, the only species of *Clematis* subg. *Viorna* in the flora with horizontal, patch-forming rhizomes, is known only from three small populations in St. Clair and Cherokee counties south of Ashville, in northeastern Alabama.

32. Clematis hirsutissima Pursh, Fl. Amer. Sept. 2: 385. 1814 [E]

Coriflora hirsutissima (Pursh) W. A. Weber

Stems erect, not viny, 1.5–6.5 dm, hirsute (sometimes sparsely so in var. *hirsutissima*) or densely short, soft-pubescent to nearly glabrous. **Leaf blade** 2–3-pinnate; leaflets often deeply 2–several-lobed, if lobed than lateral lobes usually small and distinctly narrower than central portion, leaflets or lobes linear to lanceolate, 1–6 × 0.05–1.5 cm, thin, not prominently reticulate; surfaces sparsely to densely silky-hirsute, not glaucous. **Inflorescences** terminal, flowers solitary. **Flowers** broadly cylindric to urn-shaped; sepals very dark violet-blue or rarely pink or white, oblong-lanceolate, 2.5–4.5 cm, margins narrowly expanded distally, 0.5–2 mm wide, thin, distally ± crisped, tomentose, tips obtuse to acute, slightly spreading, abaxially usually densely hirsute, occasionally moderately so. **Achenes:** bodies densely long-pubescent; beak 4–9 cm, plumose.

Varieties 2 (2 in the flora): North America.

The varieties of *Clematis hirsutissima,* although highly dissimilar in their extreme forms, intergrade extensively in Wyoming, Colorado, and Utah.

1. Leaflets and lobes linear to narrowly lanceolate, 0.5–6(–10) mm wide . . . . 32a. *Clematis hirsutissima* var. *hirsutissima*
1. Leaflets and lobes narrowly to broadly lanceolate or ovate, 5–15 mm wide . . . . 32b. *Clematis hirsutissima* var. *scottii*

32a. Clematis hirsutissima Pursh var. **hirsutissima** · Hairy clematis, sugar-bowls [E][F]

Clematis hirsutissima var. *arizonica* (A. Heller) R. O. Erickson; *Viorna arizonica* (A. Heller) A. Heller; *V. bakeri* (Greene) Rydberg; *V. eriophora* (Rydberg) Rydberg; *V. jonesii* (Kuntze) Rydberg; *V. wyethii* (Nuttall) Rydberg

Stems generally simple, erect. **Leaf blade:** primary leaflets 7–13 or not distinctly differentiated; leaflets and larger lobes narrowly linear to narrowly lanceolate, 1–6 cm × 0.5–6(–10) mm; surfaces nearly glabrous to densely silky-hirsute. $2n = 16$.

Flowering spring–summer. Moist mountain meadows, prairies, and open woods and thickets; 700–3300 m; Ariz., Colo., Idaho, Mont., N.Mex., Oreg., Utah, Wash., Wyo.

Plants from the vicinity of Flagstaff, Coconino County, Arizona (and in post-1943 identifications, some from New Mexico), with the lobes of the leaflets ca. 1 mm wide, were recognized by R. O. Erickson (1943) as *C. hirsutissima* var. *arizonica,* but these scarcely appear to constitute a distinct taxon; some plants from Washington, Oregon, Colorado, and elsewhere have leaflets quite as narrowly lobed, and other plants in the Flagstaff area have more widely lobed leaflets. The widely spreading leaves allegedly characteristic of *C. hirsutissima* var. *arizonica* likewise occur elsewhere in the range of the species. *Clematis hirsutissima* var. *hirsutissima,* as circumscribed here, is highly variable in the density of leaf pubescence throughout most of its range.

32b. Clematis hirsutissima Pursh var. **scottii** (Porter) R. O. Erickson, Ann. Missouri Bot. Gard. 30: 47. 1943 · Scott's clematis E

Clematis scottii Porter in T. C. Porter and J. M. Coulter, Syn. Fl. Colorado, 1. 1874; *Viorna scottii* (Porter) Rydberg

Stems simple or branched, erect or ± sprawling. **Leaf blade:** primary leaflets 7–13; leaflets and larger lobes narrowly to broadly lanceolate or ovate, 1–6 cm × 5–15 mm, surfaces nearly glabrous to sparsely hirsute.

Flowering spring–summer. Dry to moist mountain meadows, thickets, and rocky slopes; 1500–3200 m; Colo., Nebr., N.Mex., Okla., S.Dak., Wyo.

The leaflets of *Clematis hirsutissima* var. *scottii* are usually more widely spaced than those of *C. hirsutissima* var. *hirsutissima.*

7. HELLEBORUS Linnaeus, Sp. Pl. 1: 557. 1753; Gen. Pl. ed. 5, 244. 1754 · Hellebore, hellébore [Greek *helleborus,* ancient name for this plant]

Bruce A. Ford

Herbs [subshrubs], perennial, from tough, short rhizomes [rhizomes absent]. **Leaves** basal and cauline, basal leaf much larger [all leaves cauline], petiolate; cauline leaves alternate. **Leaf blade** pedately or palmately compound or deeply parted [undivided], lobes narrowly elliptic to oblanceolate or lanceolate, margins sharply toothed [entire]. **Inflorescences** terminal, 3–4-flowered cymes, to 25 cm or flowers solitary or paired; bracts ± leaflike, divided, not forming involucre. **Flowers** bisexual, radially symmetric; sepals persistent in fruit [not persistent], 5, yellowish green [white, pink, or purple], plane, ovate to elliptic, 19–30(–50) mm; petals 5–15, distinct, green or brown, funnel-shaped, ± 2-lipped, clawed, 4–8 mm, nectary in center of "funnel"; stamens 30–60; filaments filiform; staminodes absent between stamens and pistils; pistils [2–]3–6[–10], simple, proximally connate [distinct or completely connate]; ovules several per pistil; style present. **Fruits** follicles [capsules], aggregate, sessile, oblong, sides with prominent transverse veins; beak terminal, straight, 5–15 mm. **Seeds** usually ± carinate. $x = 8$.

Species ca. 25 (1 in the flora): North America, Europe, Asia (in Asia Minor and Tibet).

Although other species of *Helleborus* are grown as ornamentals, only the green-flowered *H. viridis* appears to persist after cultivation. *Helleborus niger* Linnaeus (Christmas-rose) is a more popular ornamental because of its showy, white to pinkish flowers. It does not appear to persist away from cultivation; it was reported as an escape in 1880 at Sennet, New York, and in 1919 in Washtenaw County, Michigan (R. S. Mitchell and J. K. Dean 1982; E. G. Voss 1972+, vol. 2). *Helleborus niger* can be distinguished from *H. viridis* by its flower color and its simple, distal cauline leaves with entire margins.

Both living and dried plants of all species of *Helleborus* are extremely poisonous. Plants

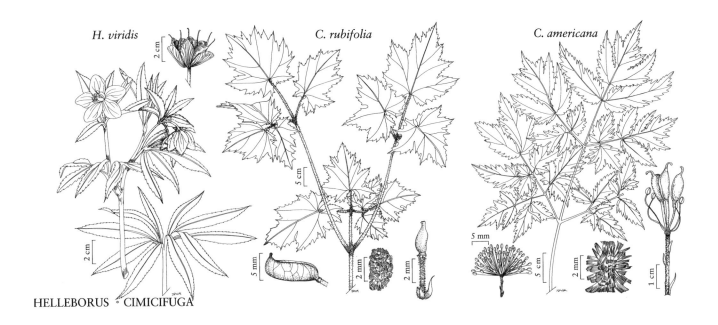

HELLEBORUS · CIMICIFUGA

contain a cardiac glycoside (helleborin), which acts directly on the heart muscle, causing convulsions, delirium, and sometimes death. Poisoning from contaminated hay has been known to cause livestock fatalities in some areas (R. S. Mitchell and J. K. Dean 1982).

1. Helleborus viridis Linnaeus, Sp. Pl. 1: 558. 1753 · Green hellebore F

Herbs, 1.2–3.5 dm. Stems fluted and ridged. Leaves 2–5 or more; basal leaves with petioles to 30 cm; blades to 40 cm wide, lobes 6–15, 2-cleft or incised, 6–21 × 1.5–4.2 cm, margins sharply serrate; cauline leaves similar to basal but smaller, sessile or short-petioled. Inflorescences: peduncles 2–5 cm. Flowers pendent, 35–60 mm diam.; sepals scarcely imbricate, 9–20 mm wide; petals upwardly curved, cornucopia-like with involute margins. Follicles 3–6, connate at base, swollen; body 14–25 mm; beak persistent.

Flowering winter–early spring (Dec–Mar). Waste places, abandoned gardens, shaded roadsides, and calcareous woodlands; 0–400 m; introduced; Ill., Md., Mich., N.J., N.Y., Ohio, Pa., W.Va.; Europe.

This species is not as commonly planted as it once was, and most records are old.

8. CIMICIFUGA Wernischeck, Gen. Pl., 298. 1763 · Bugbane, snakeroot, cohosh [Latin *cimex*, bug, and *fugare*, to drive away]

Gwynn W. Ramsey

Herbs, perennial, from hard, knotted, long-lived rhizomes. Leaves basal and cauline, compound, petiolate with basal wings clasping stem; cauline leaves alternate. Leaf blade 1–3-ternately compound; leaflets ovate-lanceolate to broadly obovate or orbiculate, 2–5-lobed, lobe margins toothed or shallow to deeply incised. Inflorescences terminal, many-flowered panicles of racemelike branches [spikes in Asian spp.], 7–60 cm; bracts 1 or 3, alternate, subtending pedicel (pedicels bracteolate in *C. americana*), not forming involucre. Flowers bisexual [unisexual], radially symmetric; sepals not persistent in fruit, (2–)4–5(–6), greenish white or cream to greenish yellow, sometimes pinkish or tinged with red, plane or ± concave, ovate to obovate, 3–6 mm; petals 0–8, distinct, white or yellowish, plane, apex 2-cleft [entire], sometimes clawed, 3–6 mm, nectariferous area sometimes present; stamens 20–110; filaments filiform [flattened]; staminodes absent between stamens and pistils; pistils 1–8, simple; ovules 4–15 per pistil; style present. **Fruits** follicles, usually aggregate, sessile or stipitate, ovoid to obovoid, weakly to strongly compressed, sides not prominently veined; beak terminal, straight or hooked at tip, 0.5–2.5 mm. **Seeds** pale brown to reddish or purplish brown, angled or laterally compressed, hemispheric, lenticular, or cylindric, smooth, slightly ridged, verrucose, or densely scaly. $x = 8$.

Species 12 (6 in the flora): North America and Eurasia.

Cimicifuga may be divided into two natural groups: those with seeds scaly and those with seeds lacking scales or nearly so. *Cimicifuga racemosa* and *C. elata* of North America, with scaleless seeds, are most closely related to *C. biternata* (Siebold & Zuccarini) Miquel and *C. japonica* (Thunberg) Sprengel of Asia.

Four or five species of *Cimicifuga* are cultivated as ornamentals, and at least five named cultivars have been developed.

SELECTED REFERENCES Compton, J. 1992. *Cimicifuga* L.: Ranunculaceae. Plantsman 14: 99–115. Ramsey, G. W. 1965. A Biosystematic Study of the Genus *Cimicifuga* (Ranunculaceae). Ph.D. thesis. University of Tennessee. Ramsey, G. W. 1987. Morphological considerations in North American *Cimicifuga*. Castanea 52: 129–141.

1. Seeds without scales or scales very short; stigma 0.5 mm wide.
  2. Follicles ovoid; seeds hemispheric, smooth or ± rough-ridged, brown; pedicel subtended by 1 bract; base of terminal leaflet with 3 prominent veins; petals (1–)4(–8), oblong; e North America . . . . 1. *Cimicifuga racemosa*
  2. Follicles oblong; seeds lenticular, usually verrucose, rarely with very short scales, reddish to purplish brown; pedicel subtended by 3 bracts; base of terminal leaflet with 5–7 prominent veins; petals absent; w North America . . . . 2. *Cimicifuga elata*
1. Seeds scaly; stigma minute, 0.2–0.3 mm wide.
  3. Pistils sessile, 1–2(–4 in *C. arizonica*); seeds lenticular, covering of scales giving cylindric appearance.
    4. Petals absent; filaments 3–5 mm; pistils 1(–2), sparsely glandular; style short, straight or slightly recurved; terminal leaflet deeply cordate with 5–9 prominent veins at base; e North America . . . . 3. *Cimicifuga rubifolia*
    4. Petals (0–)2; filaments 5–8 mm; pistils 1–3(–4), glandular; style long, hooked; terminal leaflet somewhat cordate with 3 prominent veins at base; w North America . . . . 4. *Cimicifuga arizonica*
  3. Pistils short-stipitate, (2–)3–8; seeds lenticular, scales not causing cylindric appearance.
    5. Pistils glabrous; pedicel bracteolate, granular; petiole glabrous; seeds ca. 3.5 mm, covered with broad, lacerate scales; stamens 40–70; filaments 6–10 mm; e North America . . . . 5. *Cimicifuga americana*
    5. Pistils densely pubescent; pedicel not bracteolate; petiole densely pubescent; seeds ca. 2.5–3 mm, loosely covered with narrow, lacerate scales; stamens 20–25; filaments 4–5 mm; w North America . . . . 6. *Cimicifuga laciniata*

1. **Cimicifuga racemosa** (Linnaeus) Nuttall, Gen. N. Amer. Pl. 2: 15. 1818 · Black snakeroot, black cohosh [E] [W1]

Actaea racemosa Linnaeus, Sp. Pl. 1: 504. 1753; *A. monogyna* Walter

Stems 75–250 cm, glabrous. **Leaves:** petiole angled or ± terete, 15–60 cm, usually not grooved abaxially, glabrous. **Leaf blade** 2–3-ternately compound; leaflets 20–70; terminal leaflet of central segment ovate to obovate, 3-lobed, 6–15 × 6–16.5 cm, with 3 prominent veins arising basally, base somewhat cuneate to somewhat cordate, margins dentate to deeply dentate-serrate or incised, apex acute to acuminate, surfaces glabrous, abaxially rarely pubescent on veins; other leaflets 4–12 × 3–8 cm. **Inflorescences** erect panicles of 4–9 wandlike, racemelike branches, 10–60 cm, distally pubescent; bracts 1, subtending pedicel, subulate, 3–4 mm; pedicel 4–10 mm, pubescent, bracteoles absent. **Flowers:** sepals 4, greenish white; petals

(1–)4(–8), white, oblong, ca. 3 mm, clawed; nectary basal; stamens 55–110; filaments 5–10 mm; pistils 1 (–2), sessile, ± pubescent; style short, thick; stigma 0.5 mm wide. **Follicles** 1, sessile, ovoid, ± laterally compressed, 5–10 mm, thick walled. **Seeds** brown, hemispheric, 3 mm, smooth or ± rough-ridged, without scales. $2n = 16$.

Flowering summer (Jun–Sep). Moist, mixed deciduous forests, wooded slopes, ravines, creek margins, thickets, moist meadowlands, forest margins, and especially mountainous terrain; 0–1500 m; Ont.; Ala., Ark., Conn., Del., D.C., Ga., Ill., Ind., Ky., Md., Mass., Mo., N.J., N.Y., N.C., Ohio, Pa., S.C., Tenn., Va., W.Va.

Maine and Vermont specimens probably were planted originally.

Several varieties (A. Gray et al. 1878–1897, vol. 1(1,1), pp. 53–55) or forms (M. L. Fernald 1950) have been named. Specimens with extremely dissected leaves from Connecticut to Delaware and Virginia have been called *Cimicifuga racemosa* var. *dissecta* A. Gray, or *C. racemosa* forma *dissecta* (A. Gray) Fernald. Of the approximately 2500 specimens of *C. racemosa* examined, only twelve represent var. *dissecta,* and only two of those have flowers or fruits. Because of the limited knowledge concerning the dissected-leaf form, and because plants similar to those referred to by Gray and Fernald have not been collected in this century, the form is of uncertain taxonomic significance. Further study is needed.

Native Americans used infusions of plants of *Cimicifuga racemosa* medicinally to stimulate menstruation, to treat rheumatic pains, coughs and colds, constipation, and kidney trouble, to make babies sleep, and to promote milk flow in women (D. E. Moerman 1986).

2. **Cimicifuga elata** Nuttall in J. Torrey and A. Gray, Fl. N. Amer. 1: 316. 1838 · Tall bugbane [C] [E]

Stems 120–180 cm, sparsely puberulent, glandular, or lanate. **Leaves:** petiole angled, to 40 cm, deeply and broadly grooved adaxially, glabrous or densely pubescent in groove. **Leaf blade** 2-ternately compound; leaflets 9–27; terminal leaflet of central segment ovate to orbiculate, often 2–3-lobed, 8–18 × 9–23 cm, with 5–7 prominent veins arising basally, base deeply cordate, margins coarsely dentate to serrate, teeth gland-tipped, apex acute to acuminate, surfaces abaxially pubescent, adaxially glabrous; other leaflets 5–15 × 7–20 cm. **Inflorescences** erect panicles of 4–14 racemelike branches, 7–17 cm, glandular to lanate; bracts 3, subtending pedicel, central bract largest, lance-subulate, lateral bracts ovate-deltate; pedicel 1–8 mm, densely pubescent, bracteoles absent. **Flowers:** sepals 5, white or pinkish; petals absent; stamens 20–30; filaments 5–6 mm; pistils 1–3, sessile, glandular-pubescent; style short; stigma 0.5 mm wide. **Follicles** usually 1(–3 in proximal flowers), sessile or nearly sessile (stipe 0–2 mm), oblong, ± laterally compressed, 8–12 mm, thin walled. **Seeds** reddish to purplish brown, lenticular, 2 mm, usually verrucose, rarely with very short scales. $2n = 16$.

Flowering summer (Jun–early Aug). Moist, wooded slopes, damp forest margins and roadsides, along shaded streams, rather open to closed woods, mountain hemlock habitats; of conservation concern; 60–900 m; B.C.; Oreg., Wash.

Cimicifuga elata has the most extensive range of the three western North American species. It is very likely to be threatened by human activities. Even though a number of historic records occur for this species and its preferred habitat is fairly extensive (albeit not undisturbed), the number of colonies actually known to exist is not great. In addition, few of these populations are of sufficient size and extent to be viable over the long term.

3. **Cimicifuga rubifolia** Kearney, Bull. Torrey Bot. Club 24: 561. 1897 · Appalachian bugbane [E] [F]

Stems 30–140 cm, glabrous. **Leaves:** petiole angled, 20–50 cm, proximally deeply grooved adaxially, densely pubescent in groove, otherwise glabrous or sparsely pubescent. **Leaf blade** 1–2-ternately compound; leaflets 3–9(–17); terminal leaflet of central segments broadly obovate, deeply 3–5-lobed, 9–30 × 9–25 cm, with 5–9 prominent veins arising basally, base deeply cordate, margins coarsely and irregularly dentate, apex sharply acuminate, surfaces abaxially with long, appressed, delicate hairs, adaxially glabrous; other leaflets 8–24 × 6–22 cm. **Inflorescences** erect panicles of 2–6 racemelike branches, terminal raceme 15–30 cm, puberulent to short-pubescent; bracts 3, subtending pedicel, central bract larger, lanceolate, 2 mm, lateral bracts ovate-deltate; pedicel to 5 mm, short-pubescent, bracteoles absent. **Flowers:** sepals 5, white; petals absent; stamens 35–65; filaments 3–5 mm; pistils 1(–2), sessile, sparsely glandular; style short, straight or slightly recurved; stigma minute, 0.2–0.3 mm wide. **Follicles** 1(–2), sessile, oblong, strongly compressed laterally, 8–20 mm, thin walled. **Seeds** reddish brown, lenticular, covering of scales giving cylindric appearance, ca. 3 mm, covered with reddish brown, membranous scales. $2n = 16$.

Flowering summer–fall (Aug–Oct). North-facing, limestone talus slopes and river bluffs, ravines, and coves, and along rivers and creeks; 300–900 m; Ala., Ill., Ind., Ky., Tenn., Va.

In addition to being found in the Ridge and Valley Province of Tennessee and Virginia, *Cimicifuga rubifolia* has disjunct populations in Alabama, the lower Ohio River Valley, and in the middle Tennessee counties of Davidson, Montgomery, and Stewart.

The name *Cimicifuga racemosa* var. *cordifolia* has been misapplied to *C. rubifolia*. See discussion of *C. cordifolia* under *C. americana*.

4. **Cimicifuga arizonica** S. Watson, Proc. Amer. Acad. Arts 20: 352. 1885 · Arizona bugbane [C] [E]

Stems 70–150 cm, glabrous. **Leaves:** petiole rounded abaxially, 10–35 cm, usually deeply and broadly grooved adaxially, glabrous. **Leaf blade** 2(–3)-ternately compound; leaflets (9–)21–45; terminal leaflet of central segment ovate or oblong-ovate, 3-lobed, 8–17.5 × 6–12.5 cm, with 3 prominent veins arising basally, base somewhat cordate, margins serrate to dentate-serrate, teeth gland-tipped, apex acuminate, rarely nearly truncate, surfaces nearly glabrous; other leaflets 7–12 × 4–7 cm. **Inflorescences** erect to lax panicles of 2–8 racemelike branches, 10–45 cm, short-pubescent distally; bracts 3, subtending pedicel, central bract larger and elongate-subulate, 3–4 mm, lateral bracts ovate-deltate; pedicel to 8 mm, pubescent, bracteoles absent. **Flowers:** sepals 5, outer 2 greenish yellow, inner creamy white; petals (0–)2, white, ovate to obovate, 3 mm, long-clawed; nectary absent; stamens 40–75; filaments 5–8 mm; pistils 1–3(–4), sessile, glandular; style long, hooked; stigma minute, 0.2 mm wide. **Follicles** (1–)2(–3), sessile, oblong, ± compressed laterally, 10–18 mm, membranous walled. **Seeds** pale brown, lenticular, covering of scales giving cylindric appearance, 2.5–3 mm, loosely covered with pale brown, membranous scales.

Flowering summer (Jul–Aug). Rich soil and deep shade in moist wooded ravines; of conservation concern; 1600–2100 m; Ariz.

Cimicifuga arizonica is known from only a few localities in central Arizona. It grows in moist, loamy soil of the ecotone between the coniferous forest and the riparian habitat and does not spread into the forest. It appears to be adapted to deep shade.

5. **Cimicifuga americana** Michaux, Fl. Bor.-Amer. 1: 316. 1803 · American bugbane [E] [F]

Cimicifuga cordifolia Pursh; *C. racemosa* (Linnaeus) Nuttall var. *cordifolia* (Pursh) A. Gray

Stems 60–250 cm, usually glabrous, rarely sparsely pubescent. **Leaves:** petiole rounded abaxially, 8–16 cm, broadly and shallowly grooved adaxially, glabrous, rarely with a few lax hairs in groove. **Leaf blade** (2–)3-ternately compound; leaflets 32–100; terminal leaflet of central segment ovate to oblong, incisely 3-lobed, 6–15 × 4–14 cm, with 3 veins arising basally, base cuneate to somewhat cordate, margins dentate to serrate, apex acute to acuminate, surfaces glabrous or glabrate; other leaflets 3–15 × 3–10 cm. **Inflorescences** ± lax panicles of 3–10 racemelike branches, 10–50 cm, densely short-pubescent and granular; bract 1, subtending pedicel, broadly triangular, acute, 1–3 mm; pedicel to 20 mm, granular, bracteoles present along most of pedicel, narrowly triangular, acute. **Flowers:** sepals 5, 3 outer white tinged red, 2 inner yellowish green; petals usually 2, rarely more, sessile, body yellowish with white lobes, ovate, 3–6 mm; nectary basal; stamens 40–70; filaments 6–10 mm; pistils 3–8, short-stipitate, body glabrous, stipe granular; style subulate, often curved; stigma minute, 0.2 mm wide. **Follicles** 2–4, stipitate, obovate, laterally compressed, 8–17 mm, membranous walled, glabrous. **Seeds** pale brown, lenticular, ca. 3.5 mm, covered with whitish, broad, lacerate scales but not appearing cylindric.

Flowering summer–fall (Aug–Oct). Moist, rich, rocky and boulder-strewn, wooded slopes and coves; 300–2000 m; Ga., Ill., Ky., Md., N.C., Pa., S.C., Tenn., Va., W.Va.

Much confusion has developed over the name *Cimicifuga cordifolia* Pursh. Although F. Pursh (1814, vol. 2, pp. 372–373) considered it synonymous with *C. americana*, several other authors have variously misapplied the name *C. cordifolia* to *C. racemosa* and *C. rubifolia*.

6. **Cimicifuga laciniata** S. Watson, Proc. Amer. Acad. Arts 20: 352. 1885 · Mount Hood bugbane, cut-leaved bugbane [E]

Stems 110–130 cm, glabrous or sparsely pubescent. **Leaves:** petiole rounded abaxially or sometimes laterally compressed, 10–45 cm, adaxially narrowly and deeply grooved, densely pubescent. **Leaf blade** 2–3-ternately compound; leaflets 20–35; terminal leaflet of central segment ovate to ovate-lanceolate, deeply cleft into 3 primary lobes, 7–12 × 7–12 cm, with 3 veins arising basally, base somewhat cordate, margins incised and laciniate, apex acuminate, surfaces abaxially tomentose on larger veins, adaxially glabrous; other leaflets 5–10 × 5–11 cm. **Inflorescences** erect panicles of 4–8 racemelike branches, 10–35 cm, tomentose; bract 1, subtending pedicel, lance-subulate, 3–4 mm; pedicel 3–13 mm, tomentose, bracteoles absent. **Flowers:** sepals 4 or 5, white; petals 1–5, body yellowish, lobes white, ovate, ca. 3–4 mm, clawed; nectary basal; stamens 20–30; filaments 4–5 mm; pistils (2–)3–5, short-stipitate, densely pubescent; style subulate; stigma minute, 0.3 mm wide. **Follicles** (2–)3–4(–5), slender-stipitate, oblong-obovate, laterally compressed, 7–13 mm, thin walled. **Seeds** pale brown, lenticular, 2.5–3 mm, loosely covered with elongate, narrow, white, lacerate scales but not appearing cylindric.

Flowering summer (Aug–Sep). Moist, open woods, boggy flats, thickets near heads of drainages, streamsides, meadow margins, and lakesides; 500–1800 m; Oreg., Wash.

Historically *Cimicifuga laciniata* had been collected only at Lost Lake on Mount Hood, Oregon. The discovery of many new sites in recent years has led to the removal of this species from state and federal lists of protected plants.

9. ACTAEA Linnaeus, Sp. Pl. 1: 504. 1753; Gen. Pl. ed. 5, 222. 1754 · Baneberry, necklaceweed, cohosh, acteé [Greek *aktea,* ancient name for the elder, probably for leaf similarity]

Bruce A. Ford

Herbs, perennial, from caudices ca. 1 cm thick. **Leaves** cauline, alternate, petiolate. **Leaf blade** 1–3-ternately or -pinnately compound; leaflets ovate to narrowly elliptic, unlobed to 3-lobed, margins sharply cleft, irregularly dentate. **Inflorescences** terminal or axillary, 25(–more)-flowered racemes, 2–17 cm; bracts leaflike, sometimes present between leaves and inflorescence, bracteoles 1–2, at base of each pedicel, not forming involucre. **Flowers** bisexual, radially symmetric; sepals not persistent in fruit, 3–5, whitish green, plane, orbiculate, 2–4.5 mm; petals 4–10, distinct, cream colored, plane, spatulate to obovate, clawed, 2–4.5 mm, nectary absent; stamens 15–50; filaments filiform; staminodes absent between stamens and pistils; pistil 1, simple; ovules many per pistil; style very short or absent. **Fruits** berries, solitary, sessile, broadly ellipsoid to nearly globose, sides smooth; beak a wart, terminal, to 1 mm. **Seeds** dark brown to reddish brown, obconic to wedge-shaped, rugulose. $x = 8$.

Species ca. 8 (2 in the flora): temperate to cool forests throughout Northern Hemisphere.

The two species in North America are similar to each other vegetatively and differ primarily in flower and fruit characteristics.

SELECTED REFERENCES Hultén, E. 1971. The circumpolar plants. II. Dicotyledons. Kongl. Svenska Vetenskapsakad. Handl., ser. 4, 13: 1–463. Keener, C. S. 1977. Studies in the Ranunculaceae of the southeastern United States. VI. Miscellaneous genera. Sida 7: 1–12. Pellmyr, O. 1985. The pollination biology of *Actaea pachypoda* and *A. rubra* (including *A. erythrocarpa*) in northern Michigan and Finland. Bull. Torrey Bot. Club 112: 265–273.

1. Pedicel in fruit bright red, stout, (0.7–)0.9–2.2(–3) mm diam., ± as thick as axis of raceme; stigma at anthesis as broad as or broader than ovary, 1.5–2.8 mm diam. in flower and fruit; petals truncate or cleft, often antherlike at apex; berries white, very rarely red. . . . . 1. *Actaea pachypoda*
1. Pedicel in fruit dull green or brown, slender, 0.3–0.7 mm diam., thinner than axis of raceme; stigma at anthesis narrower than ovary, 0.7–1.2 mm diam. in flower and fruit; petals acute or obtuse, not cleft or antherlike at apex; berries red or white. . . . . 2. *Actaea rubra*

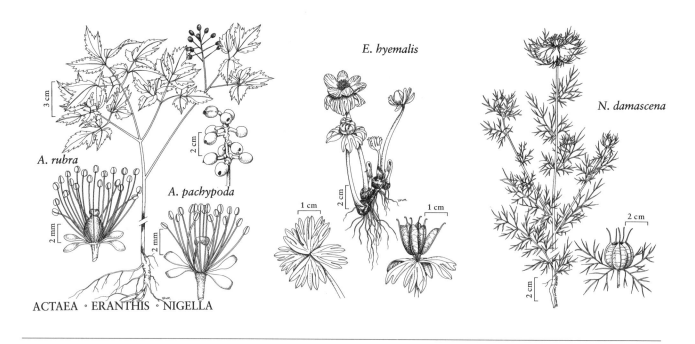

E. *hyemalis*

N. *damascena*

A. *rubra*

A. *pachypoda*

ACTAEA · ERANTHIS · NIGELLA

1. Actaea pachypoda Elliott, Sketch. Bot. S. Carolina 2: 15. 1821 · Doll's-eyes, actée à gros pédicelles [E] [F]

Leaf blade: leaflets abaxially ± glabrous. **Inflorescences** at anthesis often longer than wide, short-cylindric. **Flowers:** petals truncate or cleft, often antherlike at apex; stigma sessile, 1.5–2.8 mm diam. during anthesis, as broad as or broader than ovary. **Berries** white, very rarely red, widely ellipsoid to nearly globose, 6.5–9 mm; pedicel bright red, stout, (0.7–)0.9–2.2(–3) mm diam., ± as thick as axis of raceme. **Seeds** 3.4–4.5 mm. $2n = 16$.

Flowering spring–early summer. Deciduous forests, less often with pines, junipers, or other conifers; 0–1200 m; N.B., N.S., Ont., P.E.I., Que.; Ala., Ark., Conn., Del., Fla., Ga., Ill., Ind., Iowa, Kans., Ky., La., Maine, Md., Mass., Mich., Minn., Miss., Mo., Nebr., N.H., N.J., N.Y., N.C., Ohio, Okla., Pa., R.I., S.C., Tenn., Vt., Va., W.Va., Wis.

The "eye" formed by the persistent stigma in *Actaea pachypoda* is larger than that in *A. rubra*.

Red- and pink-berried plants have been called *Actaea pachypoda* forma *rubrocarpa* (Killip) Fernald or *A. ludovicii* Boivin. Some of these plants are intermediate in morphology between *A. pachypoda* and *A. rubra;* they may be of hybrid origin. The sterility of fruits in many such plants lends support to this theory (R. S. Mitchell and J. K. Dean 1982).

Actaea pachypoda has been called *A. alba* (Linnaeus) Miller in some manuals (e.g., H. A. Gleason and A. Cronquist 1991; S. M. Walters et al. 1984 +, vol. 3;

Great Plains Flora Association 1986). Other authors (e.g., M. L. Fernald 1940; C. S. Keener 1977) have argued that the name *A. alba* is based on an illustration that is conspecific with the type of the European *A. spicata* Linnaeus and does not apply to plants here called *A. pachypoda.*

Native Americans prepared infusions from *Actaea pachypoda* to use medicinally as a gargle or throat aid (D. E. Moerman 1986).

SELECTED REFERENCE Fernald, M. L. 1940. What is *Actaea alba?* Rhodora 42: 260–265.

2. Actaea rubra (Aiton) Willdenow, Enum. Pl. 1: 561. 1809 · Actée rouge [E] [F] [W2]

Actaea spicata Linnaeus var. *rubra* Aiton, Hort. Kew. 2: 221. 1789; *A. arguta* Nuttall; *A. eburnea* Rydberg; *A. rubra* subsp. *arguta* (Nuttall) Hultén; *A. neglecta* Gillman; *A. rubra* var. *dissecta* Britton; *A. viridiflora* Greene

Leaf blade: leaflets abaxially glabrous or pubescent. **Inflorescences** at anthesis often as long as wide, pyramidal. **Flowers:** petals acute to obtuse at apex; stigma nearly sessile, 0.7–1.2 mm diam. during anthesis, much narrower than ovary. **Berries** red or white, widely ellipsoid, 5–11 mm; pedicel dull green or brown, slender, 0.3–0.7 mm diam., thinner than axis of raceme. **Seeds** 2.9–3.6 mm. $2n = 16$.

Flowering spring–early summer. Mostly deciduous forests, also mixed coniferous forests, open pine or spruce woodlands, swales, stream banks, and swamps; 0–3500 m; Alta., B.C., Man., N.B., Nfld., N.W.T.,

N.S., Ont., P.E.I., Que., Sask., Yukon; Alaska, Ariz., Calif., Colo., Conn., Idaho, Ill., Ind., Iowa, Maine, Mass., Mich., Minn., Mont., Nebr., Nev., N.H., N.J., N.Mex., N.Y., N.Dak., Ohio, Oreg., Pa., R.I., S.Dak., Utah, Vt., Wash., Wis., Wyo.

The "eye" formed by the persistent stigma in *Actaea rubra* is smaller than that in *A. pachypoda*.

Actaea rubra is part of a circumboreal complex and is very similar to the black-fruited European species *A. spicata* Linnaeus, with which it is sometimes considered conspecific. The western North American plants of *A. rubra* have been called *A. arguta* and were distinguished on the basis of their smaller berries, more pubescent leaves, and narrow, more dissected leaflets. Those distinctions, however, are weak; specimens from the West often have fruits and leaves similar to those of plants from the East. A thorough study of *A. spicata* in the broad sense, on a worldwide scale, is needed to resolve the delimitation of taxa within this complex.

Plants with white fruit, sometimes distinguished as *Actaea rubra* forma *neglecta* (Gillman) H. Robinson, are frequent and are more common than the red-fruited form in many localities.

Native Americans used various preparations made from the roots of *Actaea rubra* medicinally to treat coughs and colds, sores, hemorrhages, stomach aches, syphilis, and emaciations; preparations from the entire plant as a purgative; and infusions from the stems to increase milk flow. It was also used in various ceremonies (D. E. Moerman 1986).

10. ERANTHIS Salisbury, Trans. Linn. Soc. London 8: 303. 1807, name conserved · Winter-aconite [Greek *er*, spring, and *anthos*, flower]

Bruce D. Parfitt

Herbs, perennial, from ca. 1-cm-thick rhizomes formed of several subglobose segments (tubers). **Leaves** basal, petiolate. **Leaf blade** deeply, palmately divided or 3–5-foliolate; leaflets ovate, unlobed to 3-lobed, margins sharply cleft, irregularly dentate. **Inflorescences** terminal, flowers solitary; involucres present, involucral bracts 3, leaflike, closely subtending flower. **Flowers** bisexual, radially symmetric; sepals not persistent in fruit, 5–8, yellow, plane, narrowly obovate or elliptic, ca. 15–18 mm; petals 5–8, distinct, yellow [white], cupped around nectary, 2-lipped, ± clawed, ca. 4 mm, nectary abaxial, covered by scale; stamens 30–36; filaments filiform; staminodes absent between stamens and pistils; pistils 3–9(–11), 1-carpellate; ovules 6–9 per pistil; style present. **Fruit** follicles, aggregate, stipitate, flat, [linear] oblong-lanceolate, sides with veins conspicuous upon drying; beak terminal, straight or distally curved, 2–3 mm. **Seeds** olive-brown, ovoid to ellipsoid, somewhat flattened, smooth. $x = 8$.

Species 8 (1 in the flora): North America, Europe, Asia.

The seven Asian species are sometimes segregated as *Shibateranthis* Nakai. Of the species that are cultivated in North America, only one is known to escape and infrequently become established; more often it is found persisting after cultivation.

SELECTED REFERENCE McIntosh, F. 1989. *Eranthis*. In: S. M. Walters et al., eds. 1984+. The European Garden Flora. 4+ vols. Cambridge, etc. Vol. 3, p. 331.

1. Eranthis hyemalis (Linnaeus) Salisbury, Trans. Linn. Soc. London 8: 304. 1807 · Winter-aconite [F]

Helleborus hyemalis Linnaeus, Sp. Pl. 1: 557. 1753; *Cammarum hyemale* (Linnaeus) Greene; *Eranthis cilicicus* Schott & Kotschy; *E.* ×*tubergenii* Hoog

Leaves basal, 3–10 cm. **Leaf blade** 3–5 cm diam.; lateral leaflets 2-parted, segments 2–3-cleft, terminal leaflet 3-cleft. **Inflorescences:** involucral bracts 2 cm. **Flowers** cup-shaped, 20–45 mm diam.; sepals 15–22 × 5–11 mm; petals shorter than stamens, abaxial lip ca. 2–2.5 times length of adaxial lip. **Follicles:** body 8–14 mm; stipe ca. 3 mm. **Seeds** ca. 2 mm.

Flowering late winter–early spring. Mostly moist places; introduced; Ont.; D.C., Ill., Ohio, N.J., Pa.; native to Eurasia.

Eranthis hyemalis is a garden plant that sometimes persists after cultivation and occasionally becomes established locally. It has been reported for Newfoundland but it is probably not established there.

Eranthis cilicicus from southeastern Turkey, usually with more leaflets and smaller flowers, may be distinguished from *E. hyemalis* in gardens but not in the wild. Hybrids between the two species are referred to as *E.* ×*tubergenii.*

11. NIGELLA Linnaeus, Sp. Pl. 1: 534. 1753; Gen. Pl. ed. 5, 238. 1754 · Love-in-a-mist [Latin *niger,* black, and *ella,* diminutive; pertaining to seeds]

Bruce A. Ford

Herbs, annual, from taproots. **Leaves** basal and cauline, petiolate or distal leaves sessile; cauline leaves alternate. **Leaf blade** 2–3-pinnately dissected, segments linear, threadlike [ovate or oblong or sometimes undivided, short], margins entire. **Inflorescences** terminal or axillary, flowers solitary; involucres present [absent], involucral bracts 5–6, finely pinnately dissected, closely subtending flower. **Flowers** bisexual, radially symmetric; sepals persistent in fruit, 5–25, blue to white or pink [yellowish white or green], plane, ovate, clawed [not clawed], 8–25 mm, apex acuminate; petals (0–)5–10, distinct, lead-colored, hooded, obovate, 2-labiate, 2–5 mm, nectary apical; stamens 15–75; filaments filiform; staminodes absent between stamens and pistils; pistil compound [carpels connate proximally], [2–]5–10-carpellate; ovules 25–100; style present. **Fruits** capsular [partially connate follicles], sessile, inflated-spheric [not inflated], sides not prominently veined; beak terminal, straight, 13–20 mm. **Seeds** black, broadly obovate, reticulate, with raised ridges. $x = 6$.

Species ca. 20 (1 in the flora): North America, s Europe, sw, c Asia, n Africa.

SELECTED REFERENCES Mitchell, R. S. and J. K. Dean. 1982. Ranunculaceae (Crowfoot Family) of New York State. Bull. New York State Mus. Sci. Serv. 446. Zohary, M. 1983. The genus *Nigella* (Ranunculaceae)—A taxonomic revision. Pl. Syst. Evol. 142: 71–107.

1. Nigella damascena Linnaeus, Sp. Pl. 1: 534. 1753 ⅁

Stems erect, slender, 10–75 cm, glabrous. **Leaves** 2–16 cm; basal leaves petiolate, segments wider than ± sessile cauline leaves. **Inflorescences:** involucral bracts whorled, similar to cauline leaves, curving up to surround flower. **Flowers** 10–50(–60) mm diam.; sepals blue, sometimes pink or white, short-clawed, 8–25 × 3–15 mm, apex entire to irregularly incised or lobed, occasionally lacerate; petals clawed, abaxial lip distally 2-lobed, bearing 2–3 nectar glands or apex expanded, adaxial lip scalelike. **Capsules** smooth, 8–35 mm; locules 5–10; beak persistent, slender.

Flowering late spring–early fall. Dump sites and waste places; 0–400 m; introduced; B.C., Ont., Que.; Ill., Kans., Md., Mich., Mo., N.Y., Ohio, Oreg., Pa., Tenn., W.Va.; native to Eurasia.

Nigella damascena is frequently cultivated as an ornamental and for dried-flower arrangements. It occasionally escapes cultivation and may become established. Populations in Ontario and Quebec, and probably elsewhere, are short-lived.

Most North American populations of *Nigella damascena* are represented by a mixture of single- and double-flowered (having supernumerary flower parts) individuals. Sepals tend to be larger and more variable in color than in Eurasian plants. Single-flowered plants usually have petals; petals appear to be absent in double-flowered individuals.

12. ADONIS Linnaeus, Sp. Pl. 1: 547. 1753; Gen. Pl. ed. 5, 242. 1754 · Pheasant's-eye, adonis [Greek mythology: sprouted from blood of Adonis, lover of Aphrodite, based on the blood red flowers]

Bruce D. Parfitt

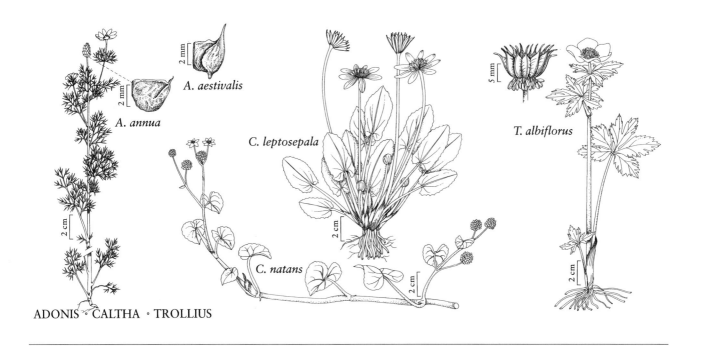

ADONIS · CALTHA · TROLLIUS

Herbs, annual, or perennial from stout rhizomes. **Leaves** basal and cauline (cauline often absent at flowering time), proximal leaves petiolate, distal leaves sessile; cauline leaves alternate. **Leaf blade** 1–3-pinnately dissected, segments narrowly linear, margins entire or with occasional tooth. **Inflorescences** terminal, flowers solitary; bracts absent. **Flowers** bisexual, radially symmetric; sepals not persistent in fruit, 5(–8), nearly colorless or green, plane, obovate, 6–22 mm, apex ± erose; petals 3–20, distinct, yellow to red [white], often striped or basally darkened with black, purple, or blue, plane, oblanceolate, 8–35 mm, nectary absent; stamens 15–80; filaments filiform; staminodes absent between stamens and pistils; pistils ca. 20–50, simple; ovule 1 per pistil; style present. **Fruits** achenes, aggregate, sessile, nearly globose, sides veined or rugose; beak terminal, straight or strongly curved, 0.5–1 mm. $x = 8$.

Species ca. 35 (3 in the flora): North America, Eurasia, n Africa.

Several species of *Adonis* are frequently cultivated in North America for their showy flowers. In eastern North America the species sometimes escape from cultivation; although they may become briefly naturalized, most populations evidently do not persist. In the western United States *A. aestivalis* apparently was introduced as a contaminant in agricultural seed and might have escaped from cultivation. It has become naturalized, and populations tend to persist.

Reliance on herbarium specimens, rather than population studies, of annual *Adonis* in North America has resulted in confused species concepts. Flower color is variable within the species and changes with drying, and diagnostic features of the fruit are reliable only when achenes are fully mature (C. C. Heyn and B. Pazy 1989).

While most Ranunculaceae attract pollinators to often highly specialized floral nectaries, *Adonis* lacks nectaries, instead offering pollen as a reward. The fruits are harvested by ants.

Leaves and roots are said to be poisonous to humans and livestock.

SELECTED REFERENCES Heyn, C. C. and B. Pazy. 1989. The annual species of *Adonis* (Ranunculaceae)—A polyploid complex. Pl. Syst. Evol. 168: 181–193. McIntash, F. 1989b. *Adonis*. In: S. M. Walters, et al., eds. 1984 +. The European Garden Flora. 4 + vols. Cambridge etc. Vol. 3.

1. Herbs perennial; flowers 4–8 cm diam., anthers yellow; achenes pubescent, beak strongly recurved. . . . 3. *Adonis vernalis*
1. Herbs annual; flowers 1.5–3 cm diam., anthers purple-black (olive green with age); achenes glabrous,
   beak straight or nearly so, erect.
  2. Body of achene without transverse flange, abaxial keel not toothed . . . . 1. *Adonis annua*
  2. Body of achene with transverse flange (usually most pronounced abaxially), abaxial keel with
     small tooth at junction with flange . . . . 2. *Adonis aestivalis*

1. Adonis annua Linnaeus, Sp. Pl. 1: 547. 1753 · Autumn
adonis, bird's-eye [F] [W1]

Adonis autumnalis Linnaeus

Herbs, annual, taprooted. **Stem** 1, 10–40 cm. **Leaves:** basal leaves 2–5 cm, similar to cauline, petiolate; cauline leaves sessile or subsessile. **Leaf blade** 2–3-pinnatifid. **Flowers** 1.5–2.5 cm diam.; sepals appressed to petals, broadly obovate, ± erose in distal 1/3, minutely ciliate, otherwise glabrous; petals 6–10, spreading, dark red [orange to red-purple], usually with dark basal blotch, ± plane, 8–15 mm, ca. 1.5 times length of calyx, apex slightly erose; stamens 15–20; anthers purple-black (olive green with age); pistils 20–30. **Heads of achenes** short-cylindric, 13–20 × 6–10 mm; pedicels conspicuously exserted well beyond leaves. **Achenes** 3–5 mm, glabrous, adaxial margin nearly straight, abaxial keel gibbous, not toothed; beak straight, erect, 0.5–1 mm.

Flowering spring–summer (Apr–Jun). Prairies, fields, river banks, and waste places; 100–200 m; introduced; Ala., Ark., La., Mo., Tenn., Tex.; native to Eurasia.

Adonis annua also has been reported for the lower Mississippi Valley (W. C. Muenscher 1980), Ontario and Manitoba (H. J. Scoggan 1978–1979, part 3), some of the western states, and New England. I have not found more recent records to confirm abundance of the species in those regions. Muenscher's report was presumably based on J. K. Small (1933), who cited the species as abundant along newly built levees of the Mississippi River. Although the species may escape frequently in some areas, the populations are often tenuously established and tend not to persist. Records of *A. annua* from western states were based on misidentified *A. aestivalis*.

Herbarium specimens without mature achenes may be impossible to distinguish from the closely related *Adonis aestivalis*, which differs in achene morphology and, reportedly, a slight difference in orientation of sepals and petals.

Several infraspecific taxonomies have been proposed for *Adonis annua*; none are well supported by observations in the field (C. C. Heyn and B. Pazy 1989).

Petal color has been variously reported; in the flora only dark red with a dark basal blotch is confirmed by herbarium specimens.

2. Adonis aestivalis Linnaeus, Sp. Pl. ed. 2, 1: 771. 1762
· Summer adonis [F]

Herbs, annual, taprooted. **Stem** 1, 20–50(–100) cm. **Leaves:** basal leaves 3–5 cm, similar to cauline, petiolate; cauline leaves sessile or subsessile. **Leaf blade** 2–3-pinnatifid. **Flowers** 1.5–3.5 cm diam.; sepals appressed to petals, broadly obovate, ± erose in distal 1/3, minutely ciliate, otherwise glabrous; petals 6–8, ± erect, orange [yellow, scarlet, or bright red-purple], usually with dark purple basal blotch, ± plane, 10–17 mm, ca. 1.5 times length of calyx, apex slightly erose; stamens ca. 30; anthers purple-black (olive green with age); pistils 30–40. **Heads of achenes** cylindric, 20–30 × 7–10 mm; pedicels conspicuously exserted well beyond leaves; achenes 4–6 mm, glabrous, adaxial margin with low, rounded tooth, transverse flange around middle (most pronounced abaxially), abaxial keel bearing small obtuse tooth at junction with flange; beak straight, erect, 1.5–2 mm.

Flowering spring–summer (May–Jul). Roadsides, fields, sagebrush scrub, and open pine or aspen forests in valleys and foothills; 1200–2400 m; introduced; Calif., Idaho, Mont., Oreg., Utah; native to Eurasia.

I have not seen specimens from Washington and Wyoming; *Adonis aestivalis* is documented from Idaho immediately across the border from those states.

Except for details of the achenes and the subtle differences in orientation of sepals and petals, *Adonis aestivalis* is much like *A. annua*. *Adonis aestivalis* has been divided into three variable subspecies, although only two [*Adonis aestivalis* subsp. *aestivalis* and subsp. *parviflora* (de Candolle) N. Busch] seem to be well delimited morphologically and geographically in Eurasia (C. C. Heyn and B. Pazy 1989). North American material is not readily assigned to either subspecies.

Petal color has been variously reported; in the flora only orange with a dark basal blotch is confirmed.

3. Adonis vernalis Linnaeus, Sp. Pl. 1: 547. 1753 · Spring adonis, ox-eye [W1]

Herbs, perennial, stout-rhizomatous. **Stems** 1–4, 5–35 cm at anthesis (10–40 cm at fruiting time). **Leaves:** basal leaves scalelike, 1 cm, sessile; proximal cauline leaves with sheathlike petiole base, distal leaves short-petiolate to sessile; leaf blade 1–2-pinnatisect. **Flowers** 4–8 cm diam.; sepals appressed to petals, ovate-elliptic to obovate, margins and abaxial surface villous, apex slightly crenulate to ± erose; petals 10–20, spreading, yellow, without basal blotch, plane, 25–35

mm, 1.5 times length of calyx, apex erose or irregularly crenulate; stamens ca. 80; anthers yellow; pistils 40–50. **Heads of achenes** globose to ovoid, 15–20 × 12–15 mm; pedicels hidden among dense leaves; achenes pubescent, 3.5–5.5 mm, transverse flange absent, basal tooth absent, adaxial margin very short (1 mm), straight, abaxial margin strongly gibbous; beak strongly recurved against abaxial surface, 0.5–1 mm.

Flowering spring (Apr–May). Roadsides and sites adjacent to gardens; introduced; N.Y.; native to Eurasia.

Adonis vernalis occasionally escapes from cultivation as an ornamental. It is much less frequently encountered in the flora than the other two species of *Adonis*.

13. CALTHA Linnaeus, Sp. Pl. 1: 558. 1753; Gen. Pl. ed. 5, 244. 1754 · Marsh-marigold, populage [Greek name for some yellow-flowered plants]

Bruce A. Ford

Herbs, perennial, from thick caudices 0.5–2 cm or slender stolons. **Leaves** basal and cauline, simple; proximal leaves petiolate, distal leaves sessile or nearly so; cauline leaves alternate. **Leaf blade** unlobed, oblong-ovate to orbiculate-reniform or cordate, margins entire, dentate, or crenate. **Inflorescences** terminal or axillary, 1–6-flowered cymes or flowers solitary, to 22 cm; bracts leaflike, not forming involucre. **Flowers** bisexual, radially symmetric; sepals not persistent in fruit, 5–12, white, pinkish, yellow, or orange, plane, oval-orbiculate to narrowly obovate, 4–23 mm; petals absent; stamens 10–40; filaments filiform; staminodes absent between stamens and pistils; pistils 5–55, simple; ovules 15–35 per pistil; style 0.1–2 mm. **Fruits** follicles, aggregate, sessile or stipitate, linear-oblong to ellipsoid, sides prominently veined or not; beak terminal, straight or weakly curved, 0.2–2 mm. **Seeds** brown, elliptic to broadly elliptic, rugulose. $x = 8$.

Species 10 (3 in the flora): primarily temperate wetlands, worldwide.

SELECTED REFERENCE Smit, P. G. 1973. A revision of *Caltha* (Ranunculaceae). Blumea 21: 119–150.

1. Stems leafless, or with 1 leaf; sepals white to yellow, not orange or pinkish. . . . . 1. *Caltha leptosepala*
1. Stems leafy; sepals white, yellow, or orange, or pinkish.
  2. Stems creeping or floating, rooting at nodes; sepals white or pinkish, 4–7(–8) mm; follicles 20–55, bodies 3.2–6.5 mm; seeds 0.5–0.8 mm. . . . . 2. *Caltha natans*
  2. Stems erect, or sprawling with age and then producing shoots and roots at nodes; sepals yellow or orange, (6–)10–25 mm; follicles 5–15(–25), bodies 8–15 mm; seeds 1.5–2.5 mm. . . . . 3. *Caltha palustris*

1. Caltha leptosepala de Candolle, Syst. Nat. 1: 310. 1817 [E][F]

Caltha biflora de Candolle; *C. biflora* subsp. *howellii* (Huth) Abrams; *C. biflora* var. *rotundifolia* (Huth) C. L. Hitchcock; *C. howellii* (Huth) Greene; *C. leptosepala* var. *rotundifolia* Huth; *C. leptosepala* var. *sulfurea* C. L. Hitchcock; *C. uniflora* Rydberg; *Psychropila leptosepala* (de Candolle) W. A. Weber

Stems leafless or with 1 leaf,

erect. **Basal leaves:** blade oblong-ovate to orbiculate-reniform, largest 1.5–11.5(–15) × 1–13 cm, margins entire or crenate to dentate. **Inflorescences** 1–2(–4)-flowered. **Flowers** 15–40 mm diam.; sepals white to yellow (abaxially bluish), 8.5–23 mm. **Follicles** 4–15, spreading, short-stipitate or sessile, linear-oblong; bodies 10–20 × 3–4.5 mm; style and stigma straight or curved, 0.5–1.8 mm. **Seeds** elliptic, 1.9–2.5 mm. $2n = 48, 96$.

Flowering late spring–summer (Jun–Aug). Open, wet,

subalpine and alpine marshes, wet seepages; 750–3900 m; Alta., B.C., Yukon; Alaska, Ariz., Calif., Colo., Idaho, Mont., Nev., N.Mex., Oreg., Utah, Wash., Wyo.

Caltha leptosepala is morphologically complex, and a number of segregate taxa have been described. Plants are most commonly assigned to two species, however. *Caltha leptosepala* in strict sense is found in the Rocky Mountains of Arizona and New Mexico north to Alaska and is characterized by longer-than-broad leaves with small, nonoverlapping basal lobes, solitary-flowered inflorescences, and sessile follicles. Plants in the Coast Ranges of central California north to the coastal islands of southern Alaska, distinguished by broader-than-long leaves with large, overlapping basal lobes, 2-flowered inflorescences, and stipitate follicles, have been called *C. biflora*. My comparison of specimens from the Rocky Mountains and the Coast Ranges indicated that no clear distinction could be made (table 15.1). While plants are often distinctive in the southern part of their range, a continuous intergradation between the two extremes exists over much of their range.

SELECTED REFERENCES Morris, M. I. 1973. A biosystematic analysis of the *Caltha leptosepala* (Ranunculaceae) complex in the Rocky Mountains. III. Variability in seed and gross morphological characters. Canad. J. Bot. 51: 2259–2268. Smit, P. G. and W. Punt. 1969. Taxonomy and pollen morphology of the *Caltha leptosepala* complex. Proc. Kon. Ned. Akad. Wetensch. C. 72: 16–27.

2. **Caltha natans** Pallas, Reise Russ. Reich. 3: 284. 1776 [F]

Stems leafy, floating or creeping, rooting at nodes. **Basal leaves:** blade ovate-reniform or cordate, largest 1–2.5 × 1–5 cm, margins nearly entire. **Inflorescences** 2–6-flowered. **Flowers** 8–13 mm diam.; sepals white or pinkish, 4–7(–8) mm. **Follicles** 20–55, widely spreading, sessile, oblong; bodies 3.2–6.5 × 1–2.5 mm; style and stigma curved, 0.1–0.4 mm. **Seeds** broadly elliptic, 0.5–0.8 mm. 2*n* = 16, 32.

TABLE 15.1. Morphologic comparison of *Caltha leptosepala* from the Rocky Mountains and Coast Ranges.

	Rocky Mountains[*] C. leptosepala *in strict sense*	Coast Ranges[**] C. bicolor *in strict sense*
Leaf (L:W ratio)	0.8–2.2(–3.1)	0.4–1.5
Flower number	2–4	1–4
Stipe (mm)	0–3.2	0–2.7

[*]Including Alaska, Arizona, Colorado, Idaho, and Montana.

[**]Including British Columbia, California, Oregon, and Washington.

Flowering late spring–summer (Jun–Aug). Floating or on moist mud, ponds, lakes, slow-moving rivers and streams; 25–1500 m; Alta., B.C., Man., N.W.T., Ont., Sask., Yukon; Alaska, Minn., Wis.; Eurasia.

Unlike the other species of *Caltha* in North America, *C. natans* is relatively invariable morphologically and has not been divided into segregate taxa.

3. **Caltha palustris** Linnaeus, Sp. Pl. 1: 558. 1753 · Cowslip, cowflock, kingcup, buttercup, populage des marais, soucis d'eau

Caltha arctica R. Brown; *C. asarifolia* de Candolle; *C. palustris* subsp. *arctica* (R. Brown) Hultén; *C. palustris* subsp. *asarifolia* (de Candolle) Hultén; *C. palustris* var. *arctica* (R. Brown) Huth; *C. palustris* var. *asarifolia* (de Candolle) Huth; *C. palustris* var. *flabellifolia* (Pursh) Torrey & A. Gray

Stems leafy, permanently erect, or sprawling with age and producing roots and shoots at nodes. **Basal leaves:** blade rounded to ovate, reniform, or cordate, largest (0.5–)2–13 × (1–)2–19 cm, margins entire or crenate to dentate. **Inflorescences** 1–7-flowered. **Flowers** 10–45 mm diam.; sepals yellow or orange, (6–)10–25 mm. **Follicles** 5–15(–25), spreading, sessile, ellipsoid; bodies 8–15 × 3–4.5 mm; style and stigma straight or curved, 0.5–2 mm. **Seeds** elliptic, 1.5–2.5 mm. 2*n* = 32, 56, 60.

Flowering spring–summer (Apr–Jul). Marshes, fens, ditches, wet woods, and swamps, thriving best in open or only partly shaded sites; 0–1500 m; Alta., B.C., Man., N.B., Nfld., N.W.T., N.S., Ont., P.E.I., Que., Sask., Yukon; Alaska, Conn., Del., Ill., Ind., Iowa, Ky., Maine, Md., Mass., Mich., Minn., Mo., Nebr., N.H., N.J., N.Y., N.C., N.Dak., Ohio, Oreg., Pa., R.I., S.Dak., Tenn., Vt., Va., Wash., W.Va., Wis.; Eurasia.

Caltha palustris has been divided into different taxa, although plants have been most commonly assigned to two varieties in North America. Typical *C. palustris* var. *palustris* is characterized by permanently erect, stout stems that do not produce roots and shoots at the nodes after anthesis. The basal leaves are broadly cordate to reniform with coarsely crenate-dentate margins and overlapping basal lobes. Generally more than three flowers occur on a stem. In contrast, *C. palustris* var. *flabellifolia* [= var. *arctica*, var. *radicans* (T. F. Forster) Beck] is characterized by stems that sprawl with age and produce roots and shoots at the nodes after anthesis. The basal leaves are ± reniform with denticulate margins, and the basal lobes are widely divergent and do not overlap. Often fewer than three flowers occur on a stem. *Caltha palustris* var. *flabellifolia* is distributed locally throughout the range of *C. palustris* var. *palustris;* it often grows in places with more extreme environmental conditions, such as shorelines, tidal areas, swiftly running streams and rivers, and areas with an arctic climate. Many arctic specimens can be assigned to this variety.

While *Caltha palustris* var. *palustris* and var. *flabellifolia* are distinctive in their extremes, they appear to represent elements along a morphologic continuum rather than recognizable taxonomic entities. For example, P. G. Smit (1973) found plants from Point Barrow, Alaska, to be dwarfed, few flowered, and prostrate, while specimens from southern Alaska were robust, many flowered, and erect. Between these two extremes a complete series of intermediates occurs. Based on that evidence, and considering the phenotypic plasticity known to exist in this species, the various specific and infraspecific segregates of *C. palustris* in North America are not recognized.

Native Americans used various preparations of the roots of *Caltha palustris* medicinally to treat colds and sores, as an aid in childbirth, and to induce vomiting, and as a protection against love charms; infusions of leaves were taken for constipation (D. E. Moerman 1986).

SELECTED REFERENCES Smit, P. G. 1967. Taxonomical and ecological studies in *Caltha palustris* L. (preliminary report). Proc. Kon. Ned. Akad. Wetensch. C. 70: 500–510. Smit, P. G. 1968. Taxonomical and ecological studies in *Caltha palustris* L. II. Proc. Kon. Ned. Akad. Wetensch. C. 71: 280–292. Woodell, S. R. J. and M. Kootin-Sanwu. 1971. Intraspecific variation in *Caltha palustris*. New Phytol. 70: 173–186.

14. TROLLIUS Linnaeus, Sp. Pl. 1: 556. 1753; Gen. Pl. ed. 5, 243. 1754 · Globe-flower [German *Trollblume,* globe-flower]

Bruce D. Parfitt

Herbs, perennial, from short caudices. **Leaves** basal and cauline, proximal leaves petiolate, distal leaves sessile or nearly so; cauline leaves alternate. **Leaf blade** deeply palmately divided into (3–)5–7 segments, segments obovate, ± 3-lobed, margins coarsely toothed, often incised. **Inflorescences** terminal, 1–3[–7]-flowered open cymes or flowers solitary; peduncle 2–30 cm; bracts leaflike, not forming involucre. **Flowers** bisexual, radially symmetric; sepals not persistent in fruit, (4–)5–9[–30], white to orange-yellow [orange-red or purplish], ± plane [strongly concave and incurved], elliptic, orbiculate, or obovate, sometimes short-clawed, 10–30 mm; petals 5–25, distinct, yellow or orange, plane with cupped base of blade, linear-oblong [ovate], ± clawed, 2–10[–40] mm, nectary within pocketlike base of blade; stamens 20–75; filaments filiform; staminodes absent between stamens and pistils; pistils 5–28[–50], simple; ovules 4–5(–9) per pistil; style present. **Fruits** follicles, aggregate, sessile, oblong, sides transversely veined; beak terminal, straight or somewhat curved, 2–4 mm. **Seeds** black or dark brown, faceted to angular, dull or lustrous. $x = 8$.

Species ca. 30 (3 in the flora): north temperate and arctic regions, North America, Europe, Asia.

As many as 10 Eurasian species of *Trollius* have been cultivated in North America as ornamentals. Of these, only *T. europaeus* Linnaeus has been reported to escape. The species infrequently persists near old dwellings in New Brunswick (B. Boivin 1966; H. Hinds, pers. comm.). *Trollius europaeus* may be distinguished from all North American species by its globose flowers with strongly incurved sepals (in North American species flowers are shallowly bowl-shaped with sepals ± spreading).

SELECTED REFERENCE Doroszewska, A. 1974. The genus *Trollius* L., a taxonomical study. Monogr. Bot. 41: 1–167, plates 1–16.

1. Petals 5–7, ± length of stamens when pollen shed, 8–12 mm; Alaska . . . . 1. *Trollius riederianus*
1. Petals 10–25, 1/2–2/3 length of stamens when pollen shed, 3–6 mm; Washington to Rocky Mountains, and e United States.
   2. Sepals white when fresh; plants of various moist habitats (often ± acidic), w North America. . . . 2. *Trollius albiflorus*
   2. Sepals yellow when fresh; plants restricted to alkaline meadows and swamps (fens), e United States . . . . 3. *Trollius laxus*

1. Trollius riederianus Fischer & C. A. Meyer in F. E. L. von Fischer et al., Index Sem. Hort. Petrop. 4: 48. 1838

Stems 1.1–4.5(–6) dm, base with persistent thatch of petioles from previous years. **Leaves:** basal leaves with petioles 5–20(–40) cm; cauline leaves 2–5, with broad, clasping, membranous sheaths. **Flowers** 3–6 cm diam.; sepals 5–7(–12), spreading, somewhat incurved, orange-yellow when fresh, obovate to ovate-orbiculate, 15–27 mm; petals 5–7, orange, ± length of stamens when pollen shed, 8–12 mm. **Follicles** 5–10, 7–10 mm including beak; beak straight.

Flowering summer. Moist meadows; 0–200 m; Alaska; Asia.

Trollius riederianus has been found in the flora only on the south side of Kiska Island in the western Aleutian Islands and at Cold Bay at the western end of the Alaska Peninsula; it is apparently native. The species is more widespread in eastern Asia, ranging from northern Japan and the coastal plain of China to the Russian Far East.

2. Trollius albiflorus (A. Gray) Rydberg, Mem. New York Bot. Gard. 1: 152. 1900 · White globe-flower [E] [F]

Trollius laxus Salisbury var. *albiflorus* A. Gray, Amer. J. Sci. Arts, ser. 2, 33: 241. 1862

Stems 0.7–5.5 dm (to 8 dm in fruit), base with few petioles persistent from previous year. **Leaves:** basal leaves with petioles 4–25 cm, some leaves reduced to sessile, ovate membranous scales; cauline leaves 1–3(–5), with broad, clasping, membranous sheaths. **Flowers** 2.5–5 cm diam.; sepals 5–9, spreading, white when fresh (pale yellow to greenish white before anthesis), ovate to obovate or nearly orbiculate, 10–20 mm; petals 15–25, yellow, 1/2–2/3 length of stamens when pollen shed, 3–6 mm. **Follicles** usually 11–14, 8–16 mm including beak; beak often somewhat recurved, sometimes straight. 2*n* = 16.

Flowering summer. Open, wet places, ± acidic, montane to alpine; 1200–3800 m; Alta., B.C.; Colo., Idaho, Mont., Utah, Wash., Wyo.

The diploid *Trollius albiflorus* is isolated from the tetraploid *T. laxus* ecologically, geographically, and reproductively, although it often has been treated as a variety of the latter.

Identities of specimens of *Trollius albiflorus* and the superficially similar *Anemone narcissiflora* subsp. *zephyra* in Colorado and Wyoming are sometimes confused. Close examination reveals a number of differences. The anemone has sepals yellow (not white), leaf blades and flowering stems pilose to villous (not glabrous), achenes (not follicles), and leaflike bracts subtending the pedicels whorled (leaves alternate in *Trollius*).

3. Trollius laxus Salisbury, Trans. Linn. Soc. London 8: 303. 1807 · Spreading globe-flower, American globe-flower [E]

Trollius americanus de Candolle

Stems to 0.5–5.2 dm, base with few petioles (occasionally thatchlike) persisting from previous year. **Leaves:** basal leaves with petioles to 5–30 cm; cauline leaves 2–5, base scarcely membranous or clasping. **Flowers** 2.5–5 cm diam.; sepals 5–7 (–9), spreading, bright to pale yellow when fresh, ovate to obovate or nearly orbiculate, 10–20 mm; petals 10–15(–25), yellow, 1/2–2/3 length of stamens when pollen shed, 3–6 mm. **Follicles** usually 5–12, 8–12 mm including beak; beak usually straight or slightly incurved (rarely slightly recurved). 2*n* = 32.

Flowering spring–summer. Calcareous soils (rarely not calcareous) in alkaline meadows and open swamps; 100–500 m; Conn., Del., N.J., N.Y., Pa., Ohio.

Trollius laxus is closely related to the widespread, relatively common, western species *T. albiflorus*. Morphologically, the two are separated only by sepal color and the tendency for *T. albiflorus* to have slightly recurved follicular beaks and slightly smaller seeds.

Trollius laxus has been extirpated from central Pennsylvania. The species has been reported from Michigan; E. G. Voss (1972+, vol. 2) cast doubt on its occurrence there because no specimens to substantiate the reports have been found.

The Cherokee used infusions prepared from the leaves and stems of *Trollius laxus* to treat "thrash" (D. E. Moerman 1986).

15. ACONITUM Linnaeus, Sp. Pl. 1: 532. 1753; Gen. Pl. ed. 5, 236. 1754 · Monkshood, aconite, wolfsbane, aconit [Greek *akoniton*, leopard's-bane, wolfsbane; possibly from the name Aconis, an ancient city of Bithynia in Asia Minor]

D. E. Brink

J. A. Woods

Herbs, perennial, from tubers or elongate, fascicled roots. **Leaves** basal and cauline, proximal leaves petiolate, distal leaves sessile or nearly so; cauline leaves alternate. **Leaf blade** palmately divided into 3–7 segments, ultimate segments narrowly elliptic or lanceolate to linear, margins incised and toothed. **Inflorescences** terminal, sometimes also axillary, 1–32(–more) racemes or panicles, to 28 cm; bracts leaflike, not forming involucre. **Flowers** bisexual, bilaterally symmetric; sepals not persistent in fruit; lower sepals (pendents) 2, plane, 6–20 mm; lateral sepals 2, round-reniform; upper sepal (hood) 1, saccate, arched, crescent-shaped or hemispheric to rounded-conic or tall and cylindric, usually beaked, 10–50 mm; petals 2, distinct, bearing near apex a capitate to coiled spur, concealed in hood, long-clawed, nectary on spur; stamens 25–50; filaments with base expanded; staminodes absent between stamens and pistils; pistils 3(–5), simple; ovules 10–20 per pistil; style present. **Fruits** follicles, aggregate, sessile, oblong, sides prominently transversely veined; beak terminal, straight, 2–3 mm. **Seeds** deltoid, usually with small, transverse, membranous lamellae. $x = 8$.

Species ca. 100 (5 in the flora): circumboreal, southward into n Mexico and n Africa.

The greatest concentration of species of *Aconitum* is in Asia, with a smaller group in Europe.

Aconitum is phylogenetically most closely related to *Delphinium* Linnaeus as evidenced by similarities in karyotype, production of diterpene alkaloids, and in floral morphology. Distinctive and unique floral morphology clearly distinguishes *Aconitum* from all other genera.

The aconites have been of interest since ancient times because they contain diterpene alkaloids that range from relatively nontoxic to deadly poisonous. In various parts of the world they have been used medicinally and as a source of poisons throughout history (D. E. Brink 1982). Use of *Aconitum* alkaloids in modern medicine was largely discontinued by the late 1930s and early 1940s (E. E. Swanson et al. 1938; H. C. Wood and A. Osol 1943; A. Osol et al. 1960).

Aconitum is a circumboreal arctic and alpine genus that extends into lower latitudes where there is suitable mesic habitat at high elevations along the north-south chains of mountains in eastern and western North America, and also in outlying, scattered, mesic, interglacial refugia, occasionally at low elevations.

The genus *Aconitum* worldwide is notorious for complex patterns of morphologic intergradation that blur the lines between taxa. Aconites from different regions may be morphologically distinct but connected by a series of intermediate races. *Aconitum columbianum* exemplifies this in North America, and *A. delphiniifolium* may extend this complex of variation into Asia. Intergradation between *A. columbianum* and *A. delphiniifolium* should be more fully investigated.

Cultivated aconites with origins outside North America sometimes persist in old gardens or are encountered as garden escapes, especially in eastern Canada (New Brunswick, Newfoundland, Nova Scotia, Ontario, and Quebec). These may include *Aconitum lycoctonum* Linnaeus, *A. napellus* Linnaeus, *A. variegatum* Linnaeus, and *A. bicolor* Schultes. *Aconitum lycoctonum* is similar to *A. reclinatum* of the southeastern United States in having the tall, conic-cylindric hood that is characteristic of species in *Aconitum* sect. *Lycoctonum* de Candolle. *Aconitum reclinatum* has white flowers whereas *A. lycoctonum* has lilac-purple flowers.

Aconitum napellus and *A. variegatum* are European introductions with leaves divided to the base as in *A. delphiniifolium,* which is native to Canada and Alaska. The introduced species have taller hoods and relatively short-petiolate leaves compared to *A. delphiniifolium;* in *A. delphiniifolium* petioles are longer, i.e., mostly as long as blades. *Aconitum napellus* has smooth seeds that lack the undulating membranous lamellae present in *A. delphiniifolium* and *A. variegatum.*

Aconitum bicolor is a reputed hybrid between *A. napellus* and *A. variegatum,* having leaves like the former and flowers similar to the latter. It is always sterile; seeds are not viable. Flowers of *A. bicolor* are frequently white with purple margins.

A more complete treatment of the cultivated aconites likely to be encountered in North America can be found in H. J. Scoggan (1978–1979, part 3, pp. 718–720) and P. A. Munz (1945).

SELECTED REFERENCES Brink, D. E. 1982. Tuberous *Aconitum* (Ranunculaceae) of the continental United States: Morphological variation, taxonomy and disjunction. Bull. Torrey Bot. Club 109: 13–23. Brink, D. E., J. A. Woods, and K. R. Stern. 1994. Bulbiferous *Aconitum* (Ranunculaceae) of the western United States. Sida 16: 9–15. Hardin, J. W. 1964. Variation in *Aconitum* of eastern United States. Brittonia 16: 80–94. Kadota, Y. 1987. A Revision of *Aconitum* Subgenus *Aconitum* (Ranunculaceae) of East Asia. Utsunomiya. Munz, P. A. 1945. The cultivated aconites. Gentes Herb. 6: 462–505. Shteinberg, E. I. 1970. *Aconitum* L. In: V. L. Komarov et al., eds. 1963+. Flora of the U.S.S.R. (Flora SSSR). Translated from Russian. 22+ vols. Jerusalem. Vol. 7, pp. 143–184.

1. Aerial stem arising from elongate, fascicled roots; hoods very tall, conic or nearly cylindric, flowers white to cream colored; e United States . . . . 1. *Aconitum reclinatum*
1. Aerial stem arising from tuber; hoods low, conic-hemispheric to crescent-shaped, flowers commonly blue to purple, occasionally white or yellowish; w Canada, e, w United States (including Alaska).
  2. Cauline leaf blades deeply divided to very base or with at most 2(–4) mm tissue between base and deepest sinus; hood low, crescent-shaped to conic-hemispheric; w Canada, Alaska. . . . . . . 2. *Aconitum delphiniifolium*
  2. Cauline leaf blades usually not divided to the very base, usually more than 2 mm between base and deepest sinus; hood conic-hemispheric, occasionally crescent-shaped; widespread.
    3. Tuber distally bulblike; terminal inflorescence often contracted, capitate; Alaska (Alaska Peninsula, and Aleutian Islands) . . . . 3. *Aconitum maximum*
    3. Tuber distally not obviously bulblike; inflorescence elongate, not capitate; sw Canada, United States except Alaska.
      4. Parent and daughter tubers separated by connecting rhizome 5–30 mm; e United States excluding Iowa, New York, and Wisconsin. . . . . 4. *Aconitum uncinatum*
      4. Parent and daughter tubers contiguous or nearly so; sw Canada, w United States, and Iowa, New York, Ohio, and Wisconsin . . . . 5. *Aconitum columbianum*

1. Aconitum reclinatum A. Gray, Amer. J. Sci. Arts 42: 34. 1842 · Trailing wolfsbane [E] [F]

Roots slender, elongate, fascicled. **Stems** erect, reclining or climbing, 6–25 dm. **Cauline leaves:** blade 3–7-divided with more than 4 mm leaf tissue between deepest sinus and base of blade, 12–20 cm wide, segment margins cleft and toothed. **Inflorescences** open racemes or panicles. **Flowers** white to cream colored, 18–30 mm from tips of pendent sepals to top of hood; pendent sepals 7–10 mm; hood conic to nearly cylindric, 15–23 mm from receptacle to top of hood, 4–12 mm wide from receptacle to beak apex.

Flowering late spring–summer (Jun–Sep). Shaded ravines of woods in mountains and upper piedmont; to 1700 m; N.C., Pa., Va., W.Va.

The only American species of *Aconitum* sect. *Lycoctonum* de Candolle, *A. reclinatum* exhibits elongate, fasciculate roots and the tall, conic-cylindric hood characteristic of that section.

2. Aconitum delphiniifolium de Candolle, Syst. Nat. 1: 380. 1817 [F]

Aconitum delphiniifolium subsp. *chamissonianum* (Reichenbach) Hultén; *A. delphiniifolium* subsp. *paradoxum* (Reichenbach) Hultén; *A. delphiniifolium* var. *paradoxum* (Reichenbach) S. L. Welsh

Roots tuberous, tuber ca. 10–15 × 5 mm, with 1 (rarely more) contiguous daughter tuber. **Stem** erect, slender, 1–10 dm. **Cauline leaves:** blade 3-divided to base, (1–)7(–13) cm wide, occasionally with up to 2(–4) mm of leaf tissue between deepest sinus and base of blade, 2 lateral segments each deeply 2-parted; margins cleft into narrow, linear-oblong lobes. **Inflorescences** open racemes or panicles. **Flowers** blue, purple, occasionally greenish purple, yellowish, or white, 20–40 mm from tips of pendent sepals to top of hood; pendent sepals 9–20 mm; hood low, crescent-shaped to hemispheric to conic-hemispheric, 10–24 mm high receptacle to top of hood, 13–30 mm broad from receptacle to beak apex. $2n = 16$.

Flowering summer (mid Jun–Sep). Meadows, along creeks, thickets, woods, rocky slopes, and alpine tundra; 0–1700 m; Alta., B.C., N.W.T., Yukon; Alaska; Asia (Russia).

Aconitum delphiniifolium is variable and was divided into three intergrading subspecies by E. Hultén (1941–1950, vol. 4) on basis of plant size, flower size, number of flowers, petal morphology, and narrow versus somewhat broader leaf lobing. Because of the "paucity of definitive criteria," J. A. Calder and R. L. Taylor (1968, vol. 1) and S. L. Welsh (1974) chose not to recognize infraspecific taxa within *A. delphiniifolium*. Similarly, we have deferred formal recognition of infraspecific taxa within this species pending population studies. The most distinctive group within *A. delphiniifolium* has one to several large flowers with shallow nectaries borne on diminutive plants. These are sometimes distinguished as subsp. *paradoxum,* and they exhibit a different pattern of variation from that found in *A. columbianum,* in which large flower size is associated almost invariably with large, robust plants and deep nectaries.

More typical forms of *Aconitum delphiniifolium* are similar to *A. columbianum* except that leaves tend to be more deeply divided and the hood tends to be lower, i.e., more crescent-shaped and less conic. Some plants and populations of *A. columbianum* also have leaves divided almost to the base and low hoods. Intergradation between the two species should be investigated more fully.

The Salishan used *Aconitum delphiniifolium* for unspecified medicinal purposes (D. E. Moerman 1986).

3. Aconitum maximum Pallas ex de Candolle, Syst. Nat. 1: 380. 1817

Subspecies 2 (1 in the flora): maritime and subalpine areas of w North America, and Asia (Russian Far East and Japan).

3a. Aconitum maximum Pallas ex de Candolle subsp. **maximum** [F]

Roots tuberous, tubers distally bulbous, 20–50 × 5–20 mm, parent and daughter tubers separated by connecting rhizome usually 10–15 mm. **Stem** erect, ca. 3–20 dm. **Cauline leaves:** blade 3-divided with more than 2 mm leaf tissue between deepest sinus and base of blade, 5–20 cm wide, segment margins deeply cleft and toothed. **Inflorescences** racemes or panicles, terminal portion often contracted and capitate. **Flowers** blue, purple, or occasionally bluish white, 25–50 mm from tips of pendent sepals to top of hood; pendent sepals ca. 15 mm; hood conic-hemispheric, ca. 17 mm from receptacle to top of hood, ca. 20 mm wide from receptacle to beak apex.

Flowering summer (mid Jul–Sep). Meadows, thickets, and forests; 0–300 m; Alaska; Asia (Russian Far East and Japan).

Aconitum maximum subsp. *maximum* is a variable taxon. The only other species of *Aconitum* occurring with it on the Alaska Peninsula and the Aleutian Islands is *A. delphiniifolium. Aconitum maximum* subsp. *maximum* is reputedly extremely poisonous and was used as a source of arrow poison by the Aleuts (D. E. Moerman 1986; D. E. Brink 1982).

4. Aconitum uncinatum Linnaeus, Sp. Pl. ed. 2, 1: 750. 1762 · Wild monkshood [E][F]

Aconitum uncinatum subsp. *muticum* (de Candolle) Hardin; *A. uncinatum* var. *acutidens* Fernald

Roots tuberous, tubers distally not obviously bulblike, 10–30 × 5–15 mm, parent tuber producing several (ca. 5) daughter tubers separated from parent by connecting rhizomes 5–30 mm. **Stems** erect, reclining or climbing, 3–25 dm. **Cauline leaves:** blade 3–5-divided, usually with more than 2 mm leaf tissue between deepest sinus and base of blade, 4–10 cm wide, segment margins cleft and toothed. **Inflorescences** open racemes or panicles. **Flowers** commonly blue, 2.5–5 cm from tips of pendent sepals to top of hood; pendent sepals 10–18 mm; hood conic-hemispheric, 15–27 mm from receptacle to

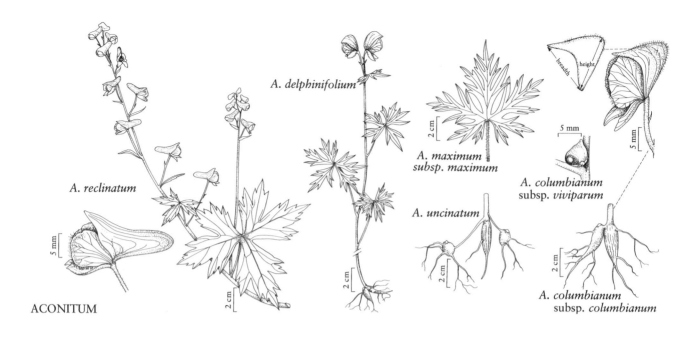

ACONITUM

top of hood, 13–24 mm wide from receptacle to beak apex.

Flowering late summer (mid Aug–late Sep). Wet areas along streams and in springs, also less mesic locations in woods and clearings; 200–2000 m; Ga., Ind., Ky., Md., N.C., Ohio, Pa., S.C., Tenn., Va., W.Va.

Aconitum uncinatum grows primarily in the Appalachian Mountains, on the Piedmont, and on the upper Atlantic Coastal Plain. It is a relatively homogeneous group divided into two intergrading subspecies by J. W. Hardin (1964).

Daughter tubers and connecting rhizomes are seldom present on herbarium specimens because they are easily dislodged during collection.

Available information suggests that *Aconitum uncinatum* is probably not one of the extremely toxic aconites (D. E. Brink 1982).

5. Aconitum columbianum Nuttall in J. Torrey and A. Gray, Fl. N. Amer. 1: 34. 1838 [W2]

Roots tuberous, tuber distally not obviously bulblike, to 60 × 15 mm, parent tuber producing 1 (rarely 2) daughter tubers with connecting rhizome very short, i.e., tubers ± contiguous. **Stems** erect and stout to twining and reclining, 2–30 dm. **Cauline leaves:** blade deeply 3–5(–7)-divided, usually with more than 2 mm leaf tissue between deepest sinus and base of blade, 5–15 cm wide, segment margins variously cleft and toothed. **Inflorescences** open racemes or panicles. **Flowers** commonly blue, sometimes white, cream colored, or blue tinged at sepal margins, 18–50 mm from tips of pendent sepals to top of hood; pendent sepals 6–16

mm; hood conic-hemispheric, hemispheric, or crescent-shaped, 11–34 mm from receptacle to top of hood, 6–26 mm wide from receptacle to beak apex.

Subspecies 2 (2 in the flora): moist areas, primarily in w North America, sporadic in e U.S.

Available information suggests that *Aconitum columbianum* is probably not one of the extremely toxic aconites (D. E. Brink 1982; J. D. Olsen et al. 1990).

SELECTED REFERENCES Brink, D. E. 1980. Reproduction and variation in *Aconitum columbianum* (Ranunculaceae), with emphasis on California populations. Amer. J. Bot. 67: 263–273. Brink, D. E. and J. M. J. de Wet. 1980. Interpopulation variation in nectar production in *Aconitum columbianum* (Ranunculaceae). Oecologia 47: 160–163.

1. Leaf axils and inflorescence without bulbils. . . .
. . . .5a. *Aconitum columbianum* subsp. *columbianum*
1. Leaf axils and/or inflorescence with conspicuous bulbils. . . 5b. *Aconitum columbianum* subsp. *viviparum*

5a. Aconitum columbianum Nuttall subsp. **columbianum** [F]

Aconitum columbianum var. *ochroleucum* A. Nelson; *A. columbianum* subsp. *pallidum* Piper; *A. geranioides* Greene; *A. infectum* Greene; *A. leibergii* Greene; *A. mogollonicum* Greene; *A. noveboracense* A. Gray; *A. uncinatum* subsp. *noveboracense* (A. Gray) Hardin

Bulbils absent in leaf axils and inflorescences. $2n = 16, 18$.

Flowering summer (Jul–Sep). Spring-fed bogs, seep areas, meadows, along streams, and in other wet areas; 300–3500 m; B.C.; Ariz., Calif., Colo., Idaho, Iowa, Mont., Nev., N.Mex., N.Y., Ohio, Oreg., S.Dak., Utah, Wash., Wis., Wyo.; Mexico.

Disjunct, outlying populations of *Aconitum columbianum* subsp. *columbianum* in Iowa, Wisconsin, Ohio, and New York occur at relatively low elevations (as low as 300 m), sometimes in frigid air drainages from caves, or in other microhabitats that simulate conditions of higher elevations. They are probably relict populations that have persisted locally since the last glacial period. These northern wild monkshoods have been treated as a species (*Aconitum noveboracense*, which has had U.S. federal conservation status), or as a subspecies of *A. uncinatum*. We find, however, that they are part of the *A. columbianum* complex. They have a single daughter tuber that is separated from the parent tuber by a connecting rhizome no more than 5 mm long. This is like *A. columbianum*, and unlike *A. uncinatum*, which has several daughter tubers separated from the parent by elongate connectives. Leaf morphology is also typical of *A. columbianum*, and unlike *A. uncinatum*. Floral morphology is similar to that found in diminutive races of *A. columbianum* in California, Wyoming, and South Dakota. Several populations in Iowa and Wisconsin are at the diminutive extreme of the range of variation in *A. columbianum* floral characters such as nectary depth and hood height. Data for Iowa and Wisconsin populations can be found in D. E. Brink (1982, also 1980). Plants in an Ohio population were too stressed and depauperate for data collection. Data collected in New York populations by Brink in 1982 are not published.

Aconitum columbianum subsp. *columbianum* is exceedingly variable. Plants often occur in dense, highly localized populations; they are very similar morphologically within populations and within regional groups of populations. Extreme differences occur between the geographic races. Specimens of the most diminutive races rarely exceed 1 m in height, whereas plants of the largest races may exceed 3 m, with correlated differences in size and number of plant parts. A complete range of variation exists between the extremes if many regional groups of populations are considered. Geographic patterns of morphologic varia-

tion have been considered too complex to accord formal taxonomic rank to the variants, so the group has been treated as one large, intergrading species complex, with bulbil-bearing and nonbulbil-bearing subspecies. White-flowered variants occur within populations, but white-flowered populations and groups of populations also occur. In each case, these seem to be sporadic variants within larger, regional patterns of morphologic variation. Consequently, white-flowered morphs are not accorded formal taxonomic rank.

5b. Aconitum columbianum Nuttall subsp. **viviparum** (Greene) Brink, Sida 16: 13. 1994 [E] [F]

Aconitum viviparum Greene, Repert. Spec. Nov. Regni Veg. 7: 2. 1909; *A. bulbiferum* Howell 1897, not *A. bulbiferum* Reichenbach 1819; *A. columbianum* var. *howellii* (A. Nelson & J. F. Macbride) C. L. Hitchcock; *A. hansenii* Greene

Bulbils conspicuous in leaf axils and sometimes in inflorescence in place of flowers. $2n = 18, 19, 20$.

Flowering summer (Jul–Sep). Spring-fed bogs, seep areas, meadows, along streams, and other moist areas in mountains; 900–2500 m; Calif., Oreg.

One group of populations occurs in the Sierra Nevada, south of Lake Tahoe, in California. Another group of bulbiferous populations begins ca. 350 miles north in the Klamath Mountains of California and extends to northern Oregon in the Cascade Range. These bulbiferous populations form a geographically and morphologically distinct group. We have not noted intergradation between bulbiferous and nonbulbiferous subspecies of *A. columbianum*.

Bulbils are an effective means of vegetative reproduction. They fall to the ground late in the season and sprout vigorously, giving rise to new plants. Bulbil production should not be confused with the production of one to several small daughter tubers at the first few nodes above the parent tuber, usually below ground, which can occur on a small percentage of the plants in bulbiferous and nonbulbiferous populations. In other respects, these bulbiferous populations are similar to adjoining races of *Aconitum columbianum* subsp. *columbianum*.

16. DELPHINIUM Linnaeus, Sp. Pl. 1: 530. 1753; Gen. Pl. ed 5, 236. 1754 · Larkspur, delphinium [Greek *delphinion,* derived from *delphin,* possibly for fancied resemblance of flowers of some species to classical sculptures of dolphins]

Michael J. Warnock

Herbs, perennial, from fasciculate roots or rhizomes. **Leaves** basal and/or cauline, petiolate, petioles gradually to abruptly shorter on distal leaves; basal leaves usually larger than cauline; cauline leaves alternate. **Leaf blade** deeply palmately divided, round to pentagonal or reniform, margins entire or lobes apically crenate or lacerate, lobes of basal blades wider and fewer than those of cauline blades. **Inflorescences** terminal, 2–100(–more)-flowered racemes (occasionally branched, thus technically panicles), 5–40 cm or more; bracts subtending inflorescence branches; pedicels present; bracteoles (on pedicels) subopposite-subalternate, not forming involucre. **Flowers** bisexual, bilaterally symmetric; sepals not persistent in fruit, 5; upper sepal 1, spurred, 8–24 mm; lateral sepals 2, ± ovate to elliptic, 8–18 mm; lower sepals 2, similar to lateral sepals; upper petals 2, spurred, enclosed in upper sepal, nectary inside tip of spur; lower petals 2, plane, ± ovate, ± 2-lobed, clawed, 2–12 mm, nectary absent; stamens 25–40; filaments with base expanded; staminodes absent between stamens and pistils; pistils 3(–5), simple; ovules 8–20 per pistil; style present. **Fruits** follicles, aggregate, sessile, ± curved-cylindric, sides prominently veined or not; beak terminal, straight, 1–3 mm. **Seeds** dark brown to black (often appearing white because of air in seed coat cells), rectangular to pyramidal, often ± rough surfaced. $x = 8$.

Species ca. 300 (61 in the flora): n temperate and arctic; subtropical and, in Eastern Hemisphere, tropical mountains (s of equator in Africa).

Three Eurasian species of *Delphinium*—*D. elatum* Linnaeus, *D. grandiflorum* Linnaeus, and *D. tatsienense* Franchet—have been commonly cultivated in North America. Of the nonnative taxa, only *D. elatum* may be sporadically naturalized, as far as is known. Isolating mechanisms in *Delphinium* appear to be primarily ecological, geographic, and/or temporal. Where these distinctions are disrupted, introgression often exists. Hybridization occurs regularly between certain taxa, particularly in areas of disturbance (e.g., roadcuts, drainage ditches, clearcuts). The more common and easily recognized hybrids are included in the key.

Many names have been misapplied in *Delphinium.* The few misapplied names mentioned in discussions below refer to relatively widespread problems.

Unless otherwise noted, the key and descriptions refer to fresh material. Some features may be significantly altered by pressing; they can, however, usually be determined with a certain amount of effort and experience. In the descriptions, "base of cleft" refers to the point where the cleft or sinus reaches most deeply into the petal blade. All species are apparently toxic.

SELECTED REFERENCES Ewan, J. 1945. A synopsis of the North American species of *Delphinium.* Univ. Colorado Stud., Ser. D, Phys. Sci. 2: 55–244.   Lewis, H. and C. Epling. 1954. A taxonomic study of Californian delphiniums. Brittonia 8: 1–22. Olsen, J. D., G. D. Manners, and S. W. Pelletier. 1990. Poisonous properties of larkspur (*Delphinium* spp.). Collect. Bot. (Barcelona) 19: 141–151.   Taylor, R. J. 1960. The genus *Delphinium* in Wyoming. Univ. Wyoming Publ. 24: 9–21.   Warnock, M. J. 1993. *Delphinium.* In: J. C. Hickman, ed. 1993. The Jepson Manual. Higher Plants of California. Berkeley, Los Angeles, and London. Pp. 916–922.   Warnock, M. J. 1995. A taxonomic conspectus of North American *Delphinium.* Phytologia 78: 73–101.

Key to the sections of *Delphinium*

1. Lower petal blades less than 1/5 length of lateral sepals; sepals never red or yellow. . . . .
   . . . . 16a. *Delphinium* sect. *Elatopsis,* p. 197
1. Lower petal blades more than 1/5 length of lateral sepals (may be less if sepals red or
   yellow); sepals blue, purple, white, red, or yellow . . . . 16b. *Delphinium* sect. *Diedropetala,* p. 197

16a. DELPHINIUM Linnaeus sect. ELATOPSIS Huth, Bull. Herb. Boissier 1: 330. 1893

Roots 20–60 cm, fibrous, dry; buds more than 3 mm. **Stems** 2–8 per root; base firmly attached to root. **Leaves** cauline, gradually reduced into bracts. **Flowers:** sepals blue, purple, or white; lower petal blades less than 1/5 length of lateral sepals, darker colored than lateral sepals.

16a.1. DELPHINIUM Linnaeus subsect. ELATA W. T. Wang, Acta Bot. Sin. 10: 81. 1962

Roots 20–50 cm, twisted fibrous, dry, 5–11-branched; buds more than 3 mm, usually present throughout dormant season. **Stems** 2–8 per root, usually unbranched, elongation commencing within 2 weeks of leaf initiation; base not narrowed, firmly attached to root. **Leaves** largest at or slightly below middle of stem; petiole ascending; blade shape and lobing similar throughout. **Inflorescences** ± dense, usually 3–8 flowers per 5 cm, narrowly pyramidal to cylindric; pedicel spreading-ascending, usually less than 3 cm; bracts ± similar to leaves but smaller. **Fruits** erect. **Seeds** rectangular to crescent-shaped, 1.5–3.5 × 1.2–2.5 mm, not ringed at proximal end, wing-margined; seed coats with irregular, small wavy ridges, cells elongate, margins straight.

Species ca. 30 (1 in the flora): North America, Eurasia.

1. Delphinium elatum Linnaeus, Sp. Pl. 1: 531. 1753

Stems 40–200 cm; base green, pubescent or glabrous. **Leaves** cauline, 7–26 at anthesis; petiole 1–18 cm. **Leaf blade** round to pentagonal, 3–15 × 6–22 cm, ± puberulent; ultimate lobes 3–9, width 8–30 mm. **Inflorescences** 25–100(–more)-flowered; pedicels 1–3(–5) cm, glabrous to pubescent; bracteoles 2–5(–9) mm from flowers, green, linear, 5–9 mm, ± puberulent. **Flowers:** sepals blue, white, or purple, ± puberulent, lateral sepals spreading, 12–23 × 4–12 mm, spurs straight, ascending ca. 45° above horizontal, 15–22 mm; lower petal blades elevated, exposing stamens, 3–5 mm, clefts 0.2–1 mm; hairs sparse or dense, mostly near center of blade, yellow or white. **Fruits** 13–20 mm, 3.5–4.5 times longer than wide, ± puberulent. **Seeds** winged; seed coats ± with irregular, small wavy ridges, cells elongate, surface roughened.

Flowering summer, more than 8 weeks after snowmelt. Old homesites; 50–3000 m; introduced; B.C., Man., Sask., and probably elsewhere; native to Europe and w Asia.

Delphinium elatum is cultivated as a garden plant or for cut flowers. It is not known to be naturalized extensively in North America; it may persist long after cultivation in cooler parts of the region.

16b. DELPHINIUM Linnaeus sect. DIEDROPETALA Huth, Bot. Jahrb. Syst. 20: 420. 1895

Roots 2–80 cm, tuberlike or fibrous, dry or fleshy; buds often less than 3 mm. **Stems** 1–8 (–19) per root; base firmly attached to root or not. **Leaves** cauline and/or in basal rosette, gradually or abruptly reduced into bracts. **Flowers:** sepals blue, purple, white, red, or yellow; lower petal blades often same color as lateral sepals, usually greater than 1/5 length of lateral sepals (exceptions in red- and yellow-flowered species).

Key to subsections of *Delphinium* sect. *Diedropetala*

1. Roots very easily separable from stems; if not extracted with digging tool, roots typically breaking cleanly from stem and completely absent from herbarium specimens.
    2. Primary root segments ± succulent, usually brittle; roots with single primary segment (cormlike), or branched from within 1 cm of stem attachment . . . .
    . . . . 16b.9. *Delphinium* subsect. *Grumosa*, p. 231
    2. Primary root segments dry, braided, tough; major root branches at least 1 cm from stem attachment.

3. Fruits spreading; seeds ringed at proximal end; inflorescence seldom more than 3 times longer than wide. . . . . 16b.8. *Delphinium* subsect. *Bicoloria*, p. 227

3. Fruits erect; seeds not ringed at proximal end; inflorescence seldom less than 4 times longer than wide . . . . 16b.4. *Delphinium* subsect. *Subscaposa*, p. 210

1. Roots not easily separable from stems; if not extracted with digging tool, roots typically breaking raggedly, leaving portion attached to lower stem. Specimens completely lacking roots should be keyed here.

    4. Large buds (more than 3 mm) present at anthesis on rootcrowns (1 or more typically found at base of specimens pulled from ground); roots usually more than 20 cm (often more than 50 cm); stems often 2 or more per root, more than 100 cm (exceptional plants less than 50 cm); basal leaves and lower cauline leaves absent at anthesis.

        5. Proximal internodes similar in length to those of midstem; basal rosettes absent; leaves monomorphic; largest leaves found near midstem, gradually reduced upward; bracts (if present) similar to and gradually smaller than leaves . . . .
. . . . .16b.1. *Delphinium* subsect. *Exaltata*, p. 200

        5. Proximal internodes much shorter than those of midstem; basal rosettes (often absent at anthesis) formed 3–28 weeks before stem elongation; leaves ± dimorphic (rosette leaves with fewer and wider lobes than cauline leaves); largest leaves found near base of stem (sometimes absent at anthesis), often abruptly reduced upward; bracts (if present) markedly smaller and fewer lobed than leaves.

           6. Inflorescences ± pyramidal; pedicel more than 3 cm . . . .
. . . . 16b.2. *Delphinium* subsect. *Wislizenana*, p. 206

           6. Inflorescences cylindric; pedicel less than 3 cm.

               7. Seed coat cells more than 3 times longer than wide; pedicel ascending . . . .
. . . . .16b.3. *Delphinium* subsect. *Multiplex*, p. 208

               7. Seed coat cells less than 3 times longer than wide; pedicel spreading to ascending . . . . 16b.2. *Delphinium* subsect. *Wislizenana*, p. 206

    4. Large buds absent at anthesis from rootcrowns; roots less than 20 cm (or completely absent from specimen), substantial portions of root often present on herbarium specimens; stems 1 per root, less than 100 cm (exceptional plants more than 200 cm); basal leaves and/or lower cauline leaves often present at anthesis.

        8. Seeds with transverse wavy ridges visible without magnification; pedicel appressed-ascending. . . . . 16b.6. *Delphinium* subsect. *Virescens*, p. 220

        8. Seeds lacking transverse wavy ridges visible without magnification (possibly other protrusions from surface); pedicel rarely appressed-ascending.

           9. Rachis to midpedicel angle less than 30°; proximal 2/3 spur length often intersecting rachis.

              10. Stems less than 60(–80) cm; pedicel closely ascending rachis (straight); spurs often intersecting rachis . . . . 16b.5. *Delphinium* subsect. *Depauperata*, p. 218

              10. Stems rarely less than 60 cm; if shorter, then pedicel remotely ascending (sigmoid); spurs sometimes intersecting rachis. . . .16b.3. *Delphinium* subsect. *Multiplex*, p. 208

           9. Rachis to midpedicel angle more than 30°; spurs rarely intersecting rachis (except sometimes in distal 1/3 spur length).

              11. Proximal internodes similar in length to those of midstem; basal rosettes absent; leaves monomorphic, largest found near midstem, gradually reduced upward; bracts (if present) similar to and gradually smaller than leaves . . .
. . . . .16b.1. *Delphinium* subsect. *Exaltata*, p. 200

              11. Proximal internodes much shorter than those of midstem; basal rosettes (often absent at anthesis) formed 3–28 weeks before stem elongation; leaves ± dimorphic (rosette leaves with fewer and wider lobes than cauline leaves), largest found on lower stem (sometimes absent at anthesis), often abruptly reduced upward; bracts (if present) markedly smaller and fewer lobed than leaves.

                12. Roots completely absent from specimens containing stem base.

                  13. Fruits erect; seeds not ringed at proximal end; inflorescence usually

at least 4 times longer than wide. . . . . 16b.4. *Delphinium* subsect. *Subscaposa*, p. 210

13. Fruits spreading; seeds ringed at proximal end; inflorescence seldom more than 4 times longer than wide.
    14. Sepals red or yellow. . . . . 16b.8. *Delphinium* subsect. *Bicoloria*, p. 227
    14. Sepals blue, purple, white, or pink.
        15. Most proximal pedicel less than 1/4 inflorecence length (except sometimes in *D. basalticum*) . . . .
        . . . . 16b.9. *Delphinium* subsect. *Grumosa*, p. 231
        15. Most proximal pedicel at least 1/4 inflorescence length.
            16. Seed coat cell surfaces roughened . . . .
            . . . . 16b.9. *Delphinium* subsect. *Grumosa*, p. 231
            16. Seed coat cell surfaces smooth.
                17. Inflorescences at least 3 times longer than wide
                . . . . 16b.9. *Delphinium* subsect. *Grumosa*, p. 231
                17. Inflorescences less than 3 times longer than wide.
                    18. Lateral sepals usually spreading (reflexed only in *D. antoninum*, with succulent leaves). . . . . 16b.8. *Delphinium* subsect. *Bicoloria*, p. 227
                    18. Lateral sepals reflexed (succulent leaves never present). . . . . 16b.9. *Delphinium* subsect. *Grumosa*, p. 231

12. Roots present (at least vestigially) on specimens containing stem base or stem base not included in specimen.
    19. Fruits spreading; seeds ringed at proximal end.
        20. Primary root segments ± succulent, usually brittle; roots with single primary segment (cormlike), or branched from within 1 cm of stem attachment. . . . . 16b.9. *Delphinium* subsect. *Grumosa*, p. 231
        20. Primary root segments usually dry, braided, tough; major root branches usually at least 1 cm from stem attachment. . . . . . .
        . . . . 16b.8. *Delphinium* subsect. *Bicoloria*, p. 227
    19. Fruits erect; seeds not ringed at proximal end.
        21. Roots ± succulent, not braided, usually less than 10 cm, thin threadlike segments restricted to major segment termini. . . . .
        . . . . 16b.7. *Delphinium* subsect. *Echinata*, p. 222
        21. Roots dry, braided, usually more than 10 cm, thin threadlike segments apparent nearly entire length (or specimens lacking roots).
            22. Inflorescences ± pyramidal; lowermost pedicel more than 3 cm.
                23. Sepals red. . . . . 16b.2. *Delphinium* subsect. *Wislizenana*, p. 206
                23. Sepals blue, white, or yellow . . . .
                . . . . 16b.4. *Delphinium* subsect. *Subscaposa*, p. 210
            22. Inflorescences cylindric; lowermost pedicel less than 3 cm.
                24. Seed coat cells more than 3 times longer than wide
                .16b.3. *Delphinium* subsect. *Multiplex*, p. 208
                24. Seed coat cells less than 3 times longer than wide.
                    25. Proximal portions of stems pubescent with straight hairs; glandular hairs absent; flowering after 1 July (except *D. geyeri*). . . . .
                    . . . . 16b.2. *Delphinium* subsect. *Wislizenana*, p. 206
                    25. Proximal portions of stems glabrous or pubescent with arched hairs; glandular hairs present or absent on pedicel; flowering completed by 1 July
                    . . . . 16b.4. *Delphinium* subsect. *Subscaposa*, p. 210

16b.1. DELPHINIUM Linnaeus subsect. EXALTATA N. I. Malyutin, Bot. Zhurn. (Moscow & Leningrad) 72: 688. 1987

Roots 4–15-branched, 10–80 cm, twisted fibrous, dry; buds more than 3 mm, usually present throughout dormant season. **Stems** (1–)3–8(–19) per root, usually unbranched, elongation commencing within 2 weeks of leaf initiation; base not narrowed, firmly attached to root; proximal internodes similar in length to those of midstem. **Leaves** cauline, largest at or slightly below middle of stem, others gradually reduced into bracts; petiole ± ascending; blade shape and lobing similar throughout. **Inflorescences** usually 5–10 flowers per 5 cm, ± dense, cylindric (greatly shortened in some), spurs rarely intersecting rachis; pedicel ± spreading, usually less than 2 cm; rachis to midpedicel angle more than 30°; bracts ± similar to leaves but smaller. **Fruits** erect. **Seeds** rectangular to crescent-shaped, 1.5–3.5 × 1.2–2.5 mm, not ringed at proximal end, wing-margined or not; seed coats ± with small irregular wavy ridges or ripples, cells elongate, cell margins straight.

Species 13 or more (12 in the flora): North America, Mexico, at least 1 species and probably more in Asia.

Members of *Delphinium* subsect. *Exaltata* are the typical tall larkspurs of poisonous-plant literature. Their abundance on some ranges, combined with their large size and toxicity, make them significant sources of livestock poisoning. Several of the species in this subsection (*Delphinium andesicola, D. californicum, D. glaucum, D. novomexicanum, D. robustum, D. sapellonis,* and the Mexican *D. valens* Standley) form a tightly knit group in which the degree of difference between members of the group and patterns of variation within the members appear largely determined by the degree and length of isolation in the various mountain ranges where the plants are found. *Delphinium andesicola, D. novomexicanum, D. robustum, D. sapellonis,* and *D. valens* form the southern Cordilleran complex.

Key to the species of *Delphinium* subsect. *Exaltata*
1. Leaves present on proximal 1/5 of stem at anthesis.
  2. Lateral sepals acute at apex; Alaska, Yukon. . . . . .3. *Delphinium brachycentrum*
  2. Lateral sepals rounded at apex; Colorado, New Mexico.
    3. Stems less than 30 cm . . . . .13. *Delphinium alpestre*
    3. Stems more than (45–)70 cm . . . . .12. *Delphinium ramosum*
1. Leaves absent from proximal 1/5 of stem at anthesis.
  4. Sepals brownish, yellowish, or purple (either permanently or with age).
    5. Sepals (in bud) yellowish or brownish purple. . . . . 9. *Delphinium sapellonis*
    5. Sepals (in bud) purple to lavender. . . . . 8. *Delphinium novomexicanum*
  4. Sepals in bud blue or purple, rarely white or pink, not brownish or yellowish with age (although sometimes in press).
    6. Hairs in inflorescence gland-based.
      7. Inflorescence more than 3 times longer than wide . . . .*Delphinium barbeyi* × *D. glaucum*
      7. Inflorescence less than 3 times longer than wide . . . . .2. *Delphinium barbeyi*
    6. Hairs in inflorescence (if present) not gland-based.
      8. Lobes of midcauline leaves less than 3 times longer than wide.
        9. Sepals whitish to pale lavender or purple; e of Great Plains . . . . 4. *Delphinium exaltatum*
        9. Sepals lavender to greenish white; coastal California . . . . 5. *Delphinium californicum*
      8. Lobes of midcauline leaves more than 3 times longer than wide (neither e of Great Plains nor along coastal California).
        10. Stems finely, evenly puberulent throughout . . . . .12. *Delphinium ramosum*
        10. Stems glabrous, or if pubescent, then only in inflorescences.
          11. Leaf blades laciniate, lobe tips gradually tapered to mucronate apex.
            12. Spurs blunt tipped, purple . . . . 11. *Delphinium andesicola*
            12. Spurs pointed, dark blue . . . . .10. *Delphinium robustum*

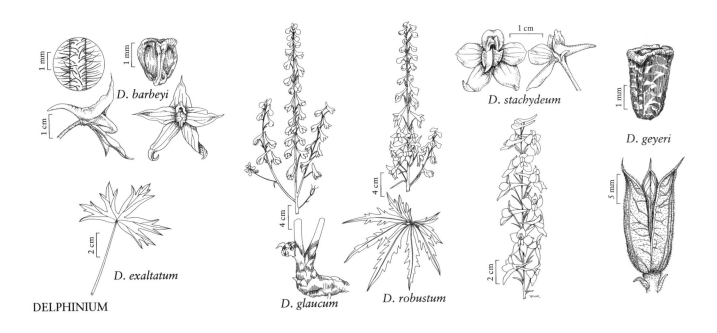

DELPHINIUM

D. barbeyi

D. exaltatum

D. glaucum

D. robustum

D. stachydeum

D. geyeri

11. Leaf blades seldom laciniate, lobe tips abruptly tapered to mucronate apex.
    13. Rarely more than 25 flowers per stem; sepals dark blue . . . . 6. *Delphinium glaucescens*
    13. Rarely fewer than 25 flowers per stem; sepals bluish purple to lavender . . . 7. *Delphinium glaucum*

2. Delphinium barbeyi (Huth) Huth, Bull. Herb. Boissier 1: 335. 1893 · Subalpine larkspur, tall larkspur, Barbey's larkspur E F W2

Delphinium exaltatum Aiton var. (ε) *barbeyi* Huth, Helios 10: 35. 1892; *D. occidentale* (S. Watson) S. Watson var. *barbeyi* (Huth) S. L. Welsh

Stems 50–150 cm; base green, glabrous. **Leaves** cauline, 8–24, absent from proximal 1/5 of stem at anthesis; petiole 1–14 cm. **Leaf blade** round to reniform, 4–8 × 7–15 cm, glabrous; ultimate lobes 5–9, width 8–50 mm. **Inflorescences** 10–50-flowered; pedicels 0.5–6 cm, glandular-puberulent; bracteoles 1–4(– 8) mm from flowers, blue to green, awl-shaped, 5–14 mm, puberulent. **Flowers:** sepals dark bluish purple, apex acute, sparsely puberulent, lateral sepals forward pointing, 13–23 × 5–8 mm, spurs ascending ca. 45°, downcurved apically, 10–18 mm; lower petal blades ± covering stamens, 4–7 mm, clefts 2–3 mm; hairs centered near base of cleft, sparse elsewhere, white or yellow. **Fruits** 17–22 mm, 2.5–3 times longer than wide, puberulent. **Seeds** wing-margined; seed coat cells narrow but short, surfaces pustulate. $2n = 16$.

Flowering summer. Subalpine and alpine sites in wet soils; 2500–4100 m; Ariz., Colo., N.Mex., Utah, Wyo.

Delphinium barbeyi hybridizes extensively with *D.*

glaucum in western Colorado and eastern Utah, where plants appearing to be hybrid [*D.* ×*occidentale* (S. Watson) S. Watson] are often far more common than plants of either putative parent. Several other names have been used for these plants, including *D. elatum* var. *occidentale* S. Watson, *D. abietorum* Tidestrom, and *D. scopulorum* subsp. *occidentale* (S. Watson) Abrams. *Delphinium barbeyi* is also known to hybridize with *D. ramosum* and *D. sapellonis*.

3. Delphinium brachycentrum Ledebour, Fl. Ross. 1: 60. 1841 · Arctic larkspur

Delphinium chamissonis Pritzel ex Walpers

Stems 20–50(–80) cm; base often reddish, pubescent. **Leaves** cauline, 5–14, on proximal 1/5 of stem at anthesis; petiole 0.4– 9 cm. **Leaf blade** round to pentagonal, 1.5–6 × 3–11 cm, puberulent; ultimate lobes 5–15, width 1–7(–15) mm. **Inflorescences** 5–18-flowered; pedicels 1–4.5 cm, pubescent; bracteoles 2–7 mm from flowers, blue, linear, 5–8 mm, puberulent. **Flowers:** sepals dark blue, apex acute, puberulent, lateral sepals forward pointing, 17–24 × 7–9 mm, spurs straight to slightly decurved, ascending 30–60° above horizontal, 15–17 mm; lower petal blades ± covering stamens,

5–7 mm, clefts 0.5–1.5 mm; hairs sparse, local, centered above base of cleft, white. **Fruits** 14–17 mm, 3.5–4 times longer than wide, pubescent. **Seeds** wing-margined; seed coat cells ± elongate, surfaces roughened.

Flowering summer. Well-drained tundra slopes; 0–1600 m; Yukon; Alaska; e Asia.

Hybrids with *Delphinium glaucum* are known as *D.* × *nutans* A. Nelson.

4. Delphinium exaltatum Aiton, Hort. Kew. 2: 244. 1789 · Tall larkspur E F

Stems 70–200 cm; base reddish, nearly glabrous. **Leaves** cauline, 7–24, absent from proximal 1/5 of stem at anthesis; petiole 1–15 cm. **Leaf blade** pentagonal, 2–7 × 3–9 cm, puberulent; ultimate lobes 3–7, width 5–25 mm; mid-cauline leaf lobes less than 3 times longer than wide. **Inflorescences** 8–30-flowered; pedicels 0.5–2 cm, puberulent; bracteoles 2–4 mm from flowers, green, linear, 2–4 mm, puberulent. **Flowers:** sepals whitish to pale lavender or purple, puberulent, lateral sepals forward pointing, 9–11 × 4–6 mm, spurs straight, as much as 45° above or below horizontal, 9–12 mm; lower petal blades ± covering stamens, 3–5 mm, clefts 1–2 mm; hairs centered mostly near base of cleft, white. **Fruits** 7–12 mm, 2–2.5 times longer than wide, ± puberulent. **Seeds** ± wing-margined; seed coat cells elongate, surfaces smooth.

Flowering summer. Rocky slopes in open deciduous woods and barrens, mainly on calcareous substrates, also shale and mafic and ultramafic rocks; 150–2000 m; Ky., Md., Mo., N.C., Ohio, Pa., Tenn., Va., W.Va.

5. Delphinium californicum Torrey & A. Gray, Fl. N. Amer. 1: 31. 1838 E

Stems (60–)100–160(–230) cm; base reddish, puberulent-glabrous. **Leaves** cauline, 6–20, absent from proximal 1/5 of stem at anthesis; petiole 0.5–7 cm. **Leaf blade** pentagonal to round, 3–9 × 3–12 cm, puberulent; ultimate lobes 3–15, width 5–60 mm; mid-cauline leaf lobes less than 3 times longer than wide. **Inflorescences** (16–)40–90(–162)-flowered; pedicels 0.5–2.5(–6.5) cm, puberulent; bracteoles 2–5 mm from flowers, reddish, linear, (3–)7–12(–15) mm, puberulent. **Flowers:** sepals lavender to greenish white, abaxially puberulent, lateral sepals forward pointing, 6–11 × 5–7 mm, spurs straight, slightly ascending, apically downcurved, 7–11(–14) mm; lower petal blades elevated, ± exposing stamens, 3–5 mm, clefts 1.5–2.5 mm; hairs covering adaxial surface, white or light yellow. **Fruits** 11–16 mm, 2.5–3.5 times longer than wide,

glabrous to sparsely puberulent. **Seeds** unwinged; seed coat cells elongate, surfaces ± roughened.

Subspecies 2 (2 in the flora): California.

Delphinium californicum can be confused only with *D. glaucum* or *D. exaltatum.* The three species may be separated by the abundant pubescence and early flowering date of *D. californicum* as compared to the other two taxa, low elevation sites of *D. californicum* compared to *D. glaucum*, and western geographic distribution of *D. californicum* compared to *D. exaltatum.*

1. Sepals lavender; inflorescences puberulent; upper petals hairy. . . . .
. . . . . 5a. *Delphinium californicum* subsp. *californicum*
1. Sepals greenish white; inflorescences nearly glabrous; upper petals glabrous. . . . .
. . . . 5b. *Delphinium californicum* subsp. *interius*

5a. Delphinium californicum Torrey & A. Gray subsp. **californicum** · Coast larkspur E

Stems puberulent most of length. **Leaf blades** abaxially and marginally puberulent. **Inflorescences** puberulent. **Flowers:** sepals forward pointing, lavender, densely puberulent, lateral sepals 7–11 mm, spurs 7–14 mm; upper petals with simple, nearly apical hairs. $2n = 16$.

Flowering late spring. Generally coastward slopes in dense coastal chaparral, often within sight of Pacific Ocean; 0–1000 m; Calif.

Delphinium californicum subsp. *californicum* is locally common, but populations tend to be scattered and often are in areas subject to human encroachment. A general trend of increased density and length of pubescence, correlated with increased frequency and density of coastal fog, is apparent. Specimens called *D. californicum* forma *longipilis* Ewan are so named for their abundance of pubescence; they grow in areas of abundant fog. This subspecies has formed garden hybrids with *D. cardinale.*

5b. Delphinium californicum Torrey & A. Gray subsp. **interius** (Eastwood) Ewan, Univ. Colorado Stud., Ser. D, Phys. Sci. 2: 146. 1945 · Hospital Canyon larkspur C E

Delphinium californicum Torrey & A. Gray var. *interius* Eastwood, Leafl. W. Bot. 2: 137. 1938

Stems proximally sparsely puberulent, remainder ± glabrous. **Leaf blades** glabrous. **Inflorescences** nearly glabrous. **Flowers:** sepals spreading to forward pointing, greenish white, hairs usually near apex only, lateral sepals 6–8 mm, spurs 7–10 mm; upper petals glabrous.

Flowering late spring. Usually on inland facing slopes in open woods; of conservation concern; 300–800 m; Calif.

Delphinium californicum subsp. *interius* is known from fewer than a dozen localities. It apparently hybridizes with *D. hesperium* subsp. *pallescens.*

6. Delphinium glaucescens Rydberg, Mem. New York Bot. Gard. 1: 155. 1900 · Glaucous larkspur E

Stems 35–70(–100) cm; base ± reddish, glabrous, glaucous. **Leaves** cauline, 6–22, absent from proximal 1/5 of stem at anthesis; petiole 1–10 cm. **Leaf blade** pentagonal to round, 2–9 × 3–14 cm, margins seldom laciniate, nearly glabrous; ultimate lobes 5–17, width 2–8 mm, tips abruptly tapered to mucronate apex; midcauline leaf lobes more than 3 times longer than wide. **Inflorescences** 10–25-flowered; pedicels 0.5–1.5 cm, puberulent to glabrous; bracteoles 2–20 mm from flowers, green, linear, 3–8 mm, nearly glabrous. **Flowers:** sepals dark blue, nearly glabrous, lateral sepals spreading to reflexed, 8–12 × 3–4 mm, spurs straight, ascending 40–90°, 9–13 mm; lower petal blades ± covering stamens, 3–5 mm, clefts 1–2 mm; hairs centered near base of cleft, white. **Fruits** 7–13 mm, 3.5–4 times longer than wide, puberulent. **Seeds** ± wing-margined; seed coat cells narrow but short, surfaces smooth.

Flowering summer. Rocky slopes in subalpine coniferous woods; 2000–4000 m; Idaho, Mont., Wyo.

Hybrids with *Delphinium glaucum* are known.

7. Delphinium glaucum S. Watson, Bot. California 2: 427–428. 1880 · Mountain larkspur, Brown's larkspur, Hooker's larkspur, pale-flowered Brown's larkspur, duncecap larkspur, giant larkspur, tall larkspur, western larkspur E F W2

Delphinium scopulorum A. Gray var. *glaucum* (S. Watson) A. Gray; *D. splendens* G. N. Jones

Stems (60–)100–200(–300) cm; base usually green, glabrous, glaucous. **Leaves** cauline, 15–20, absent from proximal 1/5 of stem at anthesis; petiole 1–14 cm. **Leaf blade** round to pentagonal, 2–11 × 3–18 cm, margins seldom laciniate, glabrous; ultimate lobes 5–9(–15), width 5–24(–35) mm, tips abruptly tapered to mucronate apex; midcauline leaf lobes more than 3 times longer than wide. **Inflorescences** (13–)40–90(–140)-flowered; pedicels 1–3(–5) cm, puberulent or glabrous; bracteoles 2–6(–10) mm from flowers, green to blue, linear, 2–7 mm, puberulent or glabrous. **Flowers:** sepals bluish purple to lavender, puberulent, lateral sepals forward pointing to spreading, 8–14(–21) × 3–6 mm, spurs straight, ascending to ca. 45°, 10–15(–19) mm; lower petal blades ± covering stamens, 4–6 mm, clefts 1–3 mm; hairs centered near base of cleft, white. **Fruits** 9–20 mm, 3.5–4.5 times longer than wide, glabrous to puberulent. **Seeds** wing-margined; seed coat cells elongate but short, surfaces smooth or roughened. $2n = 16$.

Flowering summer. Meadows, wet thickets, bogs, streamsides, open coniferous woods; 0–3200 m; Alta., B.C., Man., N.W.T., Sask., Yukon; Alaska, Calif., Colo., Idaho, Mont., Nev., Oreg., Utah, Wash., Wyo.

At the sites in Manitoba and Saskatchewan, *Delphinium glaucum* is reportedly naturalized, not native.

Delphinium glaucum hybridizes extensively with *D. barbeyi* in Utah and Colorado to the extent that hybrids [*D.* ×*occidentale* (S. Watson) S. Watson] are more common in many areas than individuals of either parental stock. It occasionally hybridizes with *D. distichum, D. polycladon, D. ramosum,* and *D. stachydeum.* Hybrids with *D. brachycentrum* are called *D.* ×*nutans* A. Nelson.

Tremendous variation is apparent in what is here recognized as *Delphinium glaucum.* This is the northern expression of the complex described in the discussion under *Delphinium* subsect. *Exaltata.* Although some geographic patterns are apparent in the variation within *D. glaucum,* infraspecific entities are not here recognized. Apparently because of rather recent and/or incomplete genetic isolation, the degree of differentiation between these units is not such that they can be consistently recognized.

Specimens named *Delphinium splendens* represent plants grown in high-moisture, low-light conditions and may occur as sporadic individuals anywhere from California to Alaska. Type specimens of *D. brownii* Rydberg, *D. canmorense* Rydberg, and *D. hookeri* A. Nelson represent plants grown on relatively dry sites at high latitudes. Plants from dry sites at low latitudes are represented by *D. bakerianum* Bornmüller and *D. occidentale* var. *reticulatum* A. Nelson. Plants with lavender to white flowers are represented by type specimens of *D. brownii* forma *pallidiflorum* B. Boivin and *D. cucullatum* A. Nelson. Type specimens of *D. alatum* A. Nelson and *D. glaucum* var. *alpinum* F. L. Wynd (an invalid name) represent plants growing above or near treeline.

Delphinium glaucum may be confused with *D. californicum, D. exaltatum, D. polycladon,* or *D. stachydeum.* For distinctions from *D. californicum,* see discussion under that species. Absence of basal or proximal cauline leaves, generally much larger plants (greater than 1.5 m), more flowers in the inflorescence, and shorter petioles on the leaves of *D. glaucum* are features that serve to distinguish this species from *D. polycladon.* In the latter, the leaves are primarily on the proximal stem, plants often less than 1.5 m, flowers more scattered, and petioles more than twice the length of leaf blades. Features of the sepals may be used to distinguish *D. glaucum* (dark lavender to blue purple, usually only minutely puberulent) from *D. stachydeum* (bright blue, densely puberulent). Vegetative parts of *D. stachydeum* are also densely puberulent, while those of *D. glaucum* typically are glabrous.

8. Delphinium novomexicanum Wooton, Bull. Torrey Bot. Club 37: 37. 1910 · New Mexico larkspur, White Mountain larkspur [E]

Delphinium sierrae-blancae Wooton

Stems 90–180(–250) cm; base usually green, glabrous. **Leaves** cauline, 12–20, absent from proximal 1/5 of stem at anthesis; petiole 3–13 cm. **Leaf blade** round to pentagonal, 5–10 × 8–18 cm, nearly glabrous; ultimate lobes 5–21, width 4–15 mm. **Inflorescences** (20–)30–70(–140)-flowered; pedicels 0.5–1.5 cm, puberulent; bracteoles 1–3 mm from flowers, green, linear, 5–8 mm, puberulent. **Flowers:** sepals (in bud) purple to lavender, fading brownish, puberulent, lateral sepals ± forward pointing, 7–11 × 4–5 mm, spurs straight to gently decurved, ascending 30–45° above horizontal, 7–11 mm; lower petal blades ± covering stamens, 3.5–6 mm, clefts 1–2 mm; hairs centered between base of cleft and junction of blade and claw, white or yellow. **Fruits** 12–16 mm, 3–4 times longer than wide, puberulent. **Seeds** wing-margined; seed coat cells elongate, surfaces ± roughened. $2n = 16$.

Flowering summer to early autumn. Meadows in coniferous forest; 2200–3900 m; N.Mex.

Delphinium novomexicanum represents the southern Cordilleran complex in the Sacramento and White mountains.

9. Delphinium sapellonis Tidestrom, Bot. Gaz. 34: 453. 1902 · Sapello Canyon larkspur [E]

Stems (50–)100–180(–220) cm; base sometimes reddish, glabrous sometimes glaucous. **Leaves** cauline, 10–20, absent from proximal 1/5 of stem at anthesis; petiole 5–12 cm. **Leaf blade** round to pentagonal, 6–10 × 8–16 cm, nearly glabrous; ultimate lobes 5–15, width 5–25 mm. **Inflorescences** (12–)30–80(–120)-flowered; pedicels 0.5–2 cm, glandular-puberulent; bracteoles 3–5 mm from flowers, green to purple, linear, 5–8 mm, glandular-puberulent.

Flowers: sepals (in bud) yellowish or brownish purple, becoming browner or yellower with age, glandular-puberulent, lateral sepals forward pointing, 8–12 × 3–5 mm, spurs straight, ascending 20–45° above horizontal, 8–11 mm; lower petal blades slightly elevated, ± exposing stamens, 2.5–5 mm, clefts 1–2 mm; hairs centered above base of cleft, yellow. **Fruits** 12–18 mm, 3–4 times longer than wide, puberulent. **Seeds** wing-margined; seed coat cells elongate, surfaces smooth.

Flowering summer. Subalpine meadows and open coniferous forest; 2600–3500 m; N.Mex.

Delphinium sapellonis hybridizes with *D. barbeyi* and *D. robustum*. It replaces *D. robustum* and represents the southern Cordilleran complex at higher elevations of the southern Sangre de Cristo Mountains east of Santa Fe. It is not known elsewhere.

10. Delphinium robustum Rydberg, Bull. Torrey Bot. Club 28: 276. 1901 · Robust larkspur [C][E][F]

Stems 100–200(–250) cm; base sometimes reddish, glabrous, glaucous. **Leaves** cauline, 12–22, absent from proximal 1/5 of stem at anthesis; petiole 5–13 cm. **Leaf blade** round to pentagonal, 7–12 × 10–20 cm, nearly glabrous; ultimate lobes 5–15, width 6–30 mm, tips gradually tapered to mucronate apex; midcauline leaf lobes more than 3 times longer than wide. **Inflorescences** 40–90 (–180)-flowered; pedicel 0.5–2 cm, puberulent; bracteoles 4–6 mm from flowers, green to purple, linear, 5–8 mm, puberulent. **Flowers:** sepals bluish purple to pale lavender, nearly glabrous, lateral sepals ± forward pointing, 9–14 × 4–6 mm, spurs slightly decurved, 30–45° above horizontal, dark blue, 10–13 mm; lower petal blades ± covering stamens, 5–7 mm, clefts 2–3 mm; hairs centered on inner lobes near base of cleft, yellow to white. **Fruits** 13–18 mm, 3–4 times longer than wide, puberulent. **Seeds** wing-margined; seed coat cells elongate, surfaces smooth.

Flowering summer. Riparian woodlands, subalpine meadows; of conservation concern; 2200–3000 m; Colo., N.Mex.

Delphinium robustum is the representative of the southern Cordilleran complex from the Jemez, San Antonio, San Juan, San Pedro, and Sangre de Cristo mountains. Hybrids are known with *D. sapellonis*.

11. Delphinium andesicola Ewan, J. Wash. Acad. Sci. 29: 476. 1939 E

Delphinium andesicola Ewan subsp. *amplum* (Ewan) Ewan

Stems 60–200+ cm; base reddish or not, glabrous, glaucous. **Leaves** cauline, 10–30, absent from proximal 1/5 of stem at anthesis; petiole 1–15 cm. **Leaf blade** cordate to semicircular, 5–8 × 5–12 cm, nearly glabrous; ultimate lobes 5–16, width 3–20 mm, tips gradually tapered to mucronate apex; midcauline leaf lobes more than 3 times longer than wide. **Inflorescences** 20–80-flowered; pedicels 1–2(–3) cm, puberulent; bracteoles 1–3 mm from flowers, green to brown, linear-lanceolate, 3–6 mm, puberulent. **Flowers:** sepals purple, puberulent, lateral sepals spreading, 9–12 × 5–7 mm, spurs ascending ca. 45°, curved downward apically, purple, 10–13 mm, blunt tipped; lower petal blades ± covering stamens, 4–6 mm, clefts 1.5–2.5 mm; hairs centered on inner lobes near base of cleft, white. **Fruits** 12–15 mm, 3.5–4 times longer than wide, sparsely puberulent. **Seeds** unwinged; seed coat cells elongate, surfaces pustulate. $2n = 16$.

Flowering summer–early fall. Meadows and coniferous woods; 2200–3200 m; Ariz.

Delphinium andesicola, the westernmost representative of the southern Cordilleran complex, is found in the Chiricahua, Huachuca, Graham, and White mountains. Hybrids with *Delphinium scopulorum* are known.

12. Delphinium ramosum Rydberg, Bull. Torrey Bot. Club 28: 276. 1901 E

Stems (45–)70–100 cm; base sometimes reddish, puberulent. **Leaves** cauline, 8–24, absent from or on proximal 1/5 of stem at anthesis; petiole 1–12 cm. **Leaf blade** round to pentagonal, 2–8 × 4–14 cm, nearly glabrous; ultimate lobes 5–21, width 1–5 mm; midcauline leaf lobes more than 3 times longer than wide. **Inflorescences** (10–)15–40(–120)-flowered; pedicels 1–2.5(–4) cm, puberulent; bracteoles 1–3 mm from flowers, green, sometimes margins white, linear or lanceolate, 3–5(–8) mm, puberulent. **Flowers:** sepals bright dark blue, apex rounded, puberulent, lateral sepals forward pointing to spreading, 11–13 × 4–6 mm, spurs straight, ascending ca. 30° above horizontal, 9–13 mm; lower petal blades ± covering stamens, 5–7 mm, clefts 2–3 mm; hairs centered above base of cleft, white. **Fruits** 11–17 mm, 3–4 times longer than wide, puberulent. **Seeds** wing-margined; seed coat cells narrow, short, surfaces roughened. $2n = 16$.

Flowering summer. Meadows, aspen woodlands, *Artemisia* scrub; 2100–3200(–3400) m; Colo., N.Mex.

Delphinium ramosum hybridizes with *D. barbeyi* and *D. glaucum.*

13. Delphinium alpestre Rydberg, Bull. Torrey Bot. Club 29: 146. 1902 · Alpine larkspur C E

Delphinium ramosum Rydberg var. *alpestre* (Rydberg) W. A. Weber

Stems 5–25 cm; base green, puberulent. **Leaves** cauline, 5–20, on proximal 1/5 of stem at anthesis; petiole 1–10 cm. **Leaf blade** round to pentagonal, 1.5–5 × 2–5 cm, puberulent; ultimate lobes 3–15, width 2–11 mm. **Inflorescences** 2–8-flowered; pedicels 1–4 cm, puberulent; bracteoles 1–3 mm from flowers, green, linear-lanceolate, 6–10 mm, puberulent. **Flowers:** sepals dark blue, apex rounded, puberulent, lateral sepals spreading to forward pointing, 11–14 × 5–7 mm, spurs straight except usually slightly downcurved at apex, varying from 20° above to 20° below horizontal, 8–12 mm; lower petal blades ± covering stamens, 4–6 mm, clefts 2–4 mm; hairs sparse, mostly near base of cleft, centered on inner lobes, white. **Fruits** 7–12 mm, 3.5–4 times longer than wide, puberulent. **Seeds** unwinged; seed coat cells elongate, surface roughened.

Flowering mid–late summer. Exposed talus slopes on high peaks; of conservation concern; (3400–)3800 m and above; Colo., N.Mex.

Delphinium alpestre is very similar to *D. ramosum,* possibly divergent from that taxon only since the most recent glaciation of North America, during which ancestors of *D. alpestre* might have survived on peaks above the ice, while ancestors of *D. ramosum* survived in valleys below the ice. Since glaciation, *D. ramosum* apparently has migrated upslope, near but not adjoining populations of *D. alpestre.*

16b.2. DELPHINIUM Linnaeus subsect. WISLIZENANA M. J. Warnock, Madroño 31: 243. 1984

Roots 3–8-branched, (5–)15–24(–80) cm, twisted fibrous, dry to fleshy, thin threadlike segments apparent nearly entire length; buds sometimes more than 3 mm, usually absent during dormant season. **Stems** 1(–4) per root, usually unbranched, elongation delayed 2–10 weeks after leaf initiation; base usually not narrowed, firmly attached to root; proximal internodes much shorter than those of midstem. **Leaves** basal and cauline, largest near base of stem, others usually abruptly smaller on distal portion of stem; basal petioles spreading, cauline petioles ascending; basal leaves more rounded and with fewer, wider lobes than cauline leaves. **Inflorescences** usually 2–9 flowers per 5 cm, open to dense, cylindric–narrow-pyramidal, spurs rarely intersecting rachis; pedicels spreading to ascending, often more than 2 cm, rachis to midpedicel angle more than 30°; bracts usually smaller and fewer lobed than leaves. **Fruits** erect. **Seeds** rectangular to crescent-shaped, 2–3.5 × 1.3–2.5 mm, not ringed on proximal end, wing-margined or not; seed coats ± covered with small irregular wavy ridges, cells short (less than 3 times longer than wide), cell margins usually straight.

Species 12 (5 in the flora): North America, Mexico.

Key to the species of *Delphinium* subsect. *Wislizenana*

1. Sepals red . . . . 18. *Delphinium cardinale*
1. Sepals bluish or purplish, rarely white, extremely rarely pink, not red.
   2. Flowers ± bright blue, sepals with lighter median line adaxially; leaf blade light green, veins obscure; leaf lobe apex gradually tapering to point.
      3. Leaf blade sparsely puberulent; stems more than (40–)70 cm . . . . 15. *Delphinium stachydeum*
      3. Leaf blade densely puberulent; stems less than 60(–80) cm . . . . 16. *Delphinium geyeri*
   2. Flowers purple to blue (if bright blue, then lower stems glabrous), sepal color ± uniform; leaf blade dark green, at least adaxially, veins prominent; leaf lobe apex abruptly tapered, usually mucronate.
      4. Midstems and leaf blades pubescent; sepals dark blue to purple . . . . 14. *Delphinium geraniifolium*
      4. Midstems and leaf blades glabrous to subglabrous; sepals bright dark blue . . . . 17. *Delphinium scopulorum*

14. Delphinium geraniifolium Rydberg, Bull. Torrey Bot. Club 26: 583. 1899 · Mogollon larkspur [E]

Delphinium tenuisectum Greene subsp. *amplibracteatum* (Wooton) Ewan

Stems 60–100 cm; base reddish, puberulent, midstems pubescent. **Leaves** mostly on proximal 1/3 of stem; basal leaves (0–)2–7 at anthesis; cauline leaves 12–20 at anthesis; petiole 1–14 cm. **Leaf blade** dark green, at least adaxially, fan-shaped to reniform, 2–5 × 3–7 cm, pubescent, especially abaxially; ultimate lobes 5–15, width 4–15 mm (basal lobes 5–15 mm), apex abruptly tapered, usually mucronate; veins prominent. **Inflorescences** 20–90-flowered, dense, cylindric; pedicels ascending to spreading, 1–2 cm, puberulent; bracteoles 1–2 mm from flowers, green, linear, 4–6 mm, puberulent. **Flowers:** sepals dark blue to purple, puberulent, lateral sepals spreading to slightly forward pointing, 10–14 × 3–5 mm, spurs ascending 20–70°, truncate or downcurved apically, 12–15 mm; lower petal blades slightly elevated, ± exposing sta-

mens, 4–6 mm, clefts 0.5–2 mm; hairs sparse, local below junction of blade and claw, scattered on margins, white. **Fruits** 13–18 mm, 3–3.5 times longer than wide, puberulent. **Seeds** unwinged; seed coat cells with margins straight, surfaces ± roughened. 2*n* = 16.

Flowering summer. Heavy clay soil, dry meadows in coniferous woods; 1800–3400 m; Ariz., N.Mex.

Delphinium geraniifolium is a more pubescent analog of the closely related *D. scopulorum,* the former occurring in heavier soils at higher elvation.

15. Delphinium stachydeum (A. Gray) Tidestrom, Proc. Biol. Soc. Wash. 27: 61. 1914 · Umatilla larkspur [E][F]

Delphinium scopulorum A. Gray var. *stachydeum* A. Gray, Bot. Gaz. 12: 52. 1887; *D. umatillense* Ewan

Stems (40–)70–150(–200) cm; base reddish, puberulent. **Leaves** mostly on proximal 1/2 stem; basal leaves absent at anthesis; cauline leaves 8–16 at anthesis; petiole 0.5–17 cm. **Leaf blade** light green, ± round, 2–8 × 3.5–11 cm, sparsely puberulent; ultimate lobes 7–19, width 1–8 mm, apex tapering to point; veins obscure. **Inflorescences** (14–)30–60(–102)-flowered, dense, cylindric; pedicels spreading, 0.8–2(–3) cm, puberulent; bracteoles 1–4 mm from flowers, green, linear, 2–7(–10) mm, puberulent. **Flowers:** sepals bright blue, puberulent, lateral sepals spreading, 9–13 × 4–7 mm, spurs straight, within 30° above or below horizontal, 11–17 mm; lower petal blades ± covering stamens, 4–8 mm, clefts 0.5–2 mm; hairs sparse, centered mostly near junction of blade and claw above base of cleft, white. **Fruits** 10–15 mm, 3.5–4.5 times longer than wide, puberulent. **Seeds** wing-margined; seed coat cells with margins straight, surfaces ± roughened.

Flowering summer. Swales in *Artemisia* scrub; 1300–3000 m; Calif, Idaho, Nev., Oreg., Wash.

Populations of *Delphinium stachydeum* are widely scattered in isolated mountain ranges surrounded by desert or grassland. The species has been reported (visual sightings) from northwestern Utah; no specimens have been seen from there. Hybrids between *D. stachydeum* and *D. glaucum* have been observed. Although *D. stachydeum* has been seen flowering within 30 m of flowering *D. depauperatum*, no hybrids have been observed.

Delphinium stachydeum may possibly be confused with *D. geyeri,* from which it may be distinguished by its usually greater plant size, less pubescent foliage, and later flowering date. *Delphinium stachydeum* also may be confused with *D. glaucum;* see discussion under that species.

16. **Delphinium geyeri** Greene, Erythea 2: 189. 1894 · Geyer's larkspur, poisonweed [E] [F] [W2]

Stems (15–)30–60(–80) cm; base usually reddish, puberulent. **Leaves** distribution variable; basal leaves 0–4 at anthesis; cauline leaves 4–22 at anthesis; petiole 0.4–10 cm. **Leaf blade** light green, ± round, 1–5 × 1–6 cm, densely puberulent; ultimate lobes 7–20, width 2–5 mm (basal), 2–4 mm (cauline), apex gradually tapering to point; veins obscure. **Inflorescences** 6–30(–60)-flowered, ± open, cylindric; pedicels ascending to spreading, 1–3(–4) cm, puberulent; bracteoles 1–3 mm from flowers, green, lanceolate, 3–6 mm, puberulent. **Flowers:** sepals bright blue, puberulent, lateral sepals spreading, 10–18 × 4–8 mm, spurs straight to slightly downcurved, ascending 0–30°, 11–16 mm; lower petal blades slightly elevated, ± exposing stamens, 4–8 mm, clefts 0.5–2 mm; hairs centered on inner lobes near base of cleft, white to light yellow. **Fruits** 11–15 mm, 3–3.5 times longer than wide, sparsely puberulent.

Seeds unwinged; seed coat cells with margins straight, surfaces ± roughened. $2n = 16$.

Flowering late spring–early summer. Grasslands or *Artemisia*-*Cercocarpus* scrub; 1400–3000 m; Colo., Utah, Wyo.

Apparently closely related to *Delphinium stachydeum, D. geyeri* is generally smaller, earlier flowering, with more finely dissected leaves, and a more eastern geographic distribution.

17. **Delphinium scopulorum** A. Gray, Smithsonian Contr. Knowl. 5(6): 9. 1853 · Rocky Mountain larkspur [F]

Delphinium macrophyllum Wooton

Stems 50–120 cm; base often reddish, puberulent, midstems glabrous to subglabrous. **Leaves** mostly on proximal 1/3 of stem; basal leaves (0–)3–7 at anthesis; cauline leaves 6–15 at anthesis; petiole 0.5–15 cm. **Leaf blade** round to pentagonal, 1.5–10 × 2–16 cm, nearly glabrous; ultimate lobes 5–19, width 5–30 mm (basal), 1–10 mm (cauline). **Inflorescences** 10–30-flowered, open, cylindric; pedicels ascending to spreading, 1–3(–4.5) cm, puberulent; bracteoles 2–7 mm from flowers, green, linear, 2–4 mm, puberulent. **Flowers:** sepals bright dark blue, nearly glabrous, lateral sepals forward pointing, 12–15 × 4–6 mm, spurs gently decurved, slightly ascending, 15–20 mm; lower petal blades ± covering stamens, 5–8 mm, clefts 1–3 mm; hairs sparse, centered below junction of blade and claw, white. **Fruits** 16–20 mm, 4–4.5 times longer than wide, nearly glabrous. **Seeds** wing-margined; seed coat cells with margins straight, cell surfaces ± roughened. $2n = 16$.

Flowering late summer–early autumn. Riparian forests and open woodlands; 1700–2600 m; Ariz., N.Mex; Mexico (Sonora).

Hybrids are known with *Delphinium andesicola.*

18. **Delphinium cardinale** Hooker, Bot. Mag. 81: plate 4887. 1855 · Scarlet larkspur

Stems (33–)50–150(–280) cm; base reddish, ± puberulent. **Leaves** mostly cauline; basal leaves absent at anthesis except in small plants; cauline leaves 5–18 at anthesis; petiole 1–12 cm. **Leaf blade** round to reniform, 3–7 × 5–10 cm, nearly glabrous; ultimate lobes 0–27, width 5–40 mm (basal), 0.5–6 mm (cauline). **Inflorescences** 10–40(–80)-flowered, open, narrowly pyramidal; pedicels spreading, (1–)2–5 cm, ± puberulent; bracteoles (2–)7–15(–25) mm from flowers, green, linear, 3–7 mm,

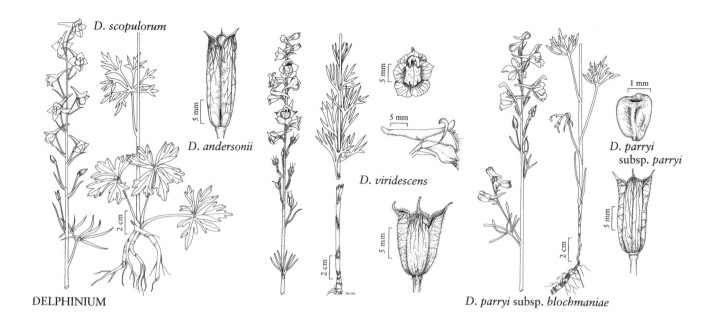

D. scopulorum

D. andersonii

D. viridescens

D. parryi subsp. parryi

DELPHINIUM

D. parryi subsp. blochmaniae

glabrous to puberulent. **Flowers:** sepals red, glabrous, lateral sepals forward pointing, 11–15 × 5–8 mm, spurs straight, stout, slightly ascending, 15–24 mm; lower petal blades nearly coplanar with claw, exposing stamens, 2–5 mm, clefts 0.5–1.5 mm; hairs centered at base of cleft, short, sparse, yellow. **Fruits** erect, 9–18 mm, 2.5–4 times longer than wide, glabrous. **Seeds** unwinged; seed coat cells with margins undulate, surfaces roughened. $2n = 16$.

Flowering spring–early summer. Slopes (often unstable) in chaparral; 50–1500 m; Calif.; Mexico (Baja California, Baja California Sur).

Hybrids between *Delphinium cardinale* and *D. parryi* have been named *D. ×inflexum* Davidson. Because of horticultural interest in red-flowered delphiniums, garden hybrids have been made with *D. elatum, D. hesperium, D. hutchinsoniae, D. nudicaule, D. parishii, D. penardii, D. scopulorum, D. tatsienense* Franchet, *D. uliginosum,* and *D. zalil* Aitchison & Hemsley, al-though *D. cardinale* does not grow with any of these in the wild.

Plants of *Delphinium cardinale* are quite variable in size, leaf distribution, and pubescence, resulting in considerable differences between, and sometimes within, populations. No patterns could be seen, however, to justify recognition of separate taxa within *D. cardinale.* Populations farther south (in Baja California, Mexico) may represent a distinct entity; they require further study.

The only possible confusion between *Delphinium cardinale* (seeds not ringed, fruits erect, grows in relatively dry sites) and another taxon might occur with *D. nudicaule* (seeds ringed, fruits spreading, grows in moist habitats). The two are separated geographically and phenologically (although *D. cardinale* may begin flowering in southern California before *D. nudicaule* has finished in northern California).

16b.3. DELPHINIUM Linnaeus subsect. MULTIPLEX M. J. Warnock, Phytologia 78: 81. 1995

Roots 2–6-branched, (5–)10–40(–50) cm, twisted fibrous, dry to fleshy, thin threadlike segments apparent nearly entire length; buds more than 3 mm, sometimes present during dormant season. **Stems** 1–4(–8) per root, usually unbranched, rarely less than 60 cm, elongation delayed 1–3 weeks after leaf initiation; base usually not narrowed, firmly attached to root; proximal internodes shorter than those of midstem. **Leaves** basal and cauline, largest near base of stem, others usually abruptly smaller on distal portion of stem; basal petioles ± spreading, cauline petioles ascending; basal leaves more rounded and with fewer, wider lobes than cauline leaves. **Inflorescences** usually 3–12 flowers per 5 cm, dense to open, cylindric, spurs sometimes intersecting rachis; pedicels ascending, usually less than 2 cm; if stems less than 60 cm, then pedicels remotely ascending (sigmoid); bracts usually markedly smaller and fewer lobed than leaves. **Fruits** erect. **Seeds** rectangular to crescent-shaped, 1.8–3 × 1.3–2.5 mm, not ringed at proximal end, wing-margined or not; seed coats with small irregular wavy ridges or ripples, cells elongate (more than 3 times longer than wide), cell margins straight.

Species 4 (4 in the flora): w United States.

Key to the species of *Delphinium* subsect. *Multiplex*

1. Sepals yellowish green . . . . 22. *Delphinium viridescens*
1. Sepals bluish or purplish (sometimes white, or lavender), not yellow or yellowish.
  2. Leaves usually absent from proximal 1/5 of stem at anthesis . . . . 21. *Delphinium multiplex*
  2. Leaves present on proximal 1/5 of stem at anthesis.
    3. Bracteoles less than 4 mm; spurs 9–12 mm; pedicels 0.3–1.5(–2.5) cm; sepals white to light blue . . . . 20. *Delphinium inopinum*
    3. Bracteoles more than 4 mm; spurs 11–22 mm; pedicels 1–4(–15) cm; sepals bluish purple . . . . 19. *Delphinium polycladon*

19. Delphinium polycladon Eastwood, Bull. Torrey Bot. Club 28: 669. 1901 · High mountain larkspur [E]

Delphinium scopulorum A. Gray var. *luporum* (Greene) Jepson

Stems (20–)60–100(–160) cm; base reddish or not, glabrous. **Leaves** mostly on proximal 1/3 of stem, on proximal 1/5 at anthesis; basal leaves 0–3 at anthesis; cauline leaves 4–7 at anthesis; petiole 1.5–17 cm. **Leaf blade** round to pentagonal, 1.5–7 × 2–14 cm, glabrous; ultimate lobes 3–12, width 4–30(–45) mm (basal), 3–30 mm (cauline). **Inflorescences** 3–15(–35)-flowered, open, often ± secund; pedicels 1–4(–15) cm, glabrous to puberulent; bracteoles 2–8(–37) mm from flowers, green, linear, 4–7(–11) mm, nearly glabrous. **Flowers:** sepals bluish purple, nearly glabrous, lateral sepals spreading, (10–)12–18 × 7–10 mm, spurs usually downcurved, ca. 30° below horizontal, 11–22 mm; lower petal blades slightly elevated, ± exposing stamens, 4–6 mm, clefts 1–2 mm; hairs mostly near base of cleft on inner lobes, yellow, sometimes white. **Fruits** 13–20 mm, 3.5–4 times longer than wide, puberulent. **Seeds** unwinged; seed coat cells with surfaces roughened. $2n = 16$.

Flowering summer–early autumn. Wet sites near springs, streamsides, bogs, and wet talus; 2200–3600 m; Calif., Nev.

Delphinium polycladon hybridizes with *D. depauperatum* and *D. glaucum*. Plants of *D. polycladon* are extremely variable. Individuals from very rocky, thin-soiled, sunny sites at higher elevations tend to be quite compact; they show the features of the species in a dwarfed state. Proximal internodes are especially shortened. Plants from areas of deeper soil (high or low elevations), especially those growing among shrubs, usually are much taller, with elongate proximal internodes, and other vegetative parts proportionally larger. Shorter plants may be confused with *D. depauperatum* or *D. nuttallianum;* see discussion under those species for distinguishing features. Taller plants may be confused with *D. glaucum;* they can be distinguished by their leaves predominately on proximal part of stem, sigmoid pedicels, and fewer flowers.

20. Delphinium inopinum (Jepson) H. F. Lewis & Epling, Brittonia 8: 11. 1954 · Unexpected larkspur [E]

Delphinium parishii A. Gray var. *inopinum* Jepson, Fl. Calif. 1: 526. 1915

Stems 70–110(–150) cm; base reddish or not, glabrous, often glaucous. **Leaves** mostly on proximal 1/3 of stem, on proximal 1/5 at anthesis; basal leaves 0–4 at anthesis; cauline leaves 6–12 at anthesis; petiole 1–18 cm. **Leaf blade** ± pentagonal, 1–5 × 1.5–7 cm, glabrous; ultimate lobes 3–9, width 5–28 mm (basal), 3–18 mm (cauline). **Inflorescences** 20–35(–51)-flowered, usually dense; pedicels 0.3–1.5(–2.5) cm, glabrous; bracteoles 2–4 mm from flowers, green, linear, 1–2(–4) mm, nearly glabrous. **Flowers:** sepals white to light blue, glabrous, lateral sepals spreading to forward pointing, 8–12 × 3–5 mm, spurs straight to gently upcurved, ascending 30–60° above horizontal, 9–12 mm; lower petal blades slightly elevated, ± exposing stamens, 3–5 mm, clefts 1–2 mm; hairs centered near base of cleft, white. **Fruits** 12–20 mm, 2.6–4 times longer than wide, glabrous. **Seeds** wing-margined; seed coat cells surfaces smooth.

Flowering summer. Rock outcrops in open coniferous woods; 2200–2800 m; Calif.

Delphinium inopinum is apparently endemic to a white metamorphic rock substrate in the Piute Mountains and southern Sierra Nevada. It is not known to hybridize with any other species, although *D. patens* subsp. *montanum* has been collected (when both were flowering) within 1 km of *D. inopinum* and probably occurs much closer. *Delphinium inopinum* is often confused with *D. parishii* subsp. *pallidum* and superficially resembles some white-flowered individuals of *D. hansenii*, as well as *D. gypsophilum* and *D. hesperium* subsp. *pallescens*. The massive roots with prominent buds readily distinguish *D. inopinum* from all of these. In addition, the pubescence found on *D. hansenii* will separate it from the glabrous *D. inopinum*. Leaves are rarely seen at anthesis near the base of the stem in *D. hesperium* subsp. *pallescens* or *D. gypsophilum*; they are present in *D. inopinum*.

21. Delphinium multiplex (Ewan) C. L. Hitchcock in C. L. Hitchcock et al., Vasc. Pl. Pacif. N.W. 2: 357. 1964 [E]

Delphinium cyanoreios Piper forma *multiplex* Ewan, Univ. Colorado Stud., Ser. D, Phys. Sci. 2: 163. 1945

Stems (40–)80–130(–180) cm; base usually green, glabrous. **Leaves** mostly on proximal 1/3 of stem, usually absent from proximal 1/5 stem at anthesis; basal leaves 0–4 at anthesis; cauline leaves 10–20 at anthesis; petiole 0.5–15 cm. **Leaf blade** round on proximal stem, ± cuneate on distal stem, 1.5–9 × 2–14 cm, glabrous; ultimate lobes 3–15, width 15–30 mm (basal), 1–20 mm (cauline). **Inflorescences** 25–50-flowered, ± dense; pedicels 0.5–2 cm, glandular-puberulent; bracteoles 4–5 mm from flowers, green to blue, lanceolate-linear, 3–7 mm, glandular-puberulent. **Flowers:** sepals dark blue, puberulent, lateral sepals forward pointing to spreading, 12–15 × 5–7 mm, spurs straight, within 20° of

horizontal, 12–16 mm; lower petal blades slightly elevated, exposing stamens, 6–8 mm, clefts 2–3 mm; hairs centered and local near base of cleft, white. **Fruits** 10–15 mm, 3–4 times longer than wide, glandular-puberulent. **Seeds** narrowly wing-margined; seed coat cells with surfaces somewhat roughened. $2n = 16$.

Flowering late spring–summer. Rocky streambeds; 1500–1800 m; Wash.

Delphinium multiplex hybridizes with *D. glaucum* and *D. distichum*.

22. Delphinium viridescens Leiberg, Proc. Biol. Soc. Wash. 11: 39. 1897 · Wenatchee larkspur [C][E][F]

Stems 90–150 cm; base usually green, glabrous. **Leaves** cauline, 17–30 at anthesis; petiole 0.2–8 cm. **Leaf blade** cuneate to semicircular, 2–5 × 3–12 cm, nearly glabrous; ultimate lobes 3–21, width 1–8 mm. **Inflorescences** 25–80-flowered, dense; pedicels 0.5–2 cm, glandular-pubescent; bracteoles 1–4 mm from flowers, green, lanceolate, 3.5–6 mm, glandular-pubescent. **Flowers:** sepals yellowish green, nearly glabrous, lateral sepals forward pointing, 7–9 × 3–4 mm, spurs decurved, 30–45° below horizontal, often hooked apically, 8–11 mm; lower petal blades ± covering stamens, 4–6 mm, clefts 0.5–1.5 mm; hairs centered near junction of blade and claw, yellow. **Fruits** 8–11 mm, 2.5–3 times longer than wide, puberulent. **Seeds** ± wing-margined; seed coat cells with surfaces ± roughened.

Flowering summer. Wet meadows and streamsides in coniferous forest, heavy clay soils; of conservation concern; 500–1000 m; Wash.

Delphinium viridescens is local in mountains southwest of Wenatchee, Washington.

16b.4. DELPHINIUM Linnaeus subsect. SUBSCAPOSA Ewan, Bull. Torrey Bot. Club 63: 330. 1936

Roots 3–8(–12)-branched at least 1 cm from stem attachment, (4–)10–30(–40) cm, fibrous, twisted, dry, thin threadlike segments apparent nearly entire length; buds minute. **Stems** 1(–3) per root, unbranched, elongation delayed 4–10 weeks after leaf initiation; base often narrowed, firmly attached to root; proximal internodes much shorter than those of midstem. **Leaves** basal and cauline, largest near base of stem, others often abruptly smaller on distal stem; basal petioles spreading, cauline petioles ascending; basal leaves more rounded and with fewer, wider lobes than cauline leaves. **Inflorescences** usually 4–8 flowers per 5 cm, ± open, cylindric to narrowly pyramidal, seldom less than 4 times longer than wide, spurs rarely intersecting rachis; pedicels spreading to ascending, usually less than 2 cm, rachis to midpedicel angle more than 30°; bracts markedly smaller, fewer lobed than leaves. **Fruits** erect. **Seeds** rectangular to crescent-shaped, 1.3–3.3 × 1.1–2.3 mm, not ringed on proximal end, ± wing-margined; seed coats usually lacking wavy ridges (present only in *D. purpusii*), cells brick-shaped (less than 3 times longer than wide), cell margins undulate or straight.

Species 10+ (10 in the flora): North America, possibly others in Asia.

Key to the species of *Delphinium* subsect. *Subscaposa*

1. Yellowish gland-based hairs present (at least apically) on pedicel; sepals reflexed.
    2. Sepals light blue to lavender . . . . 32. *Delphinium lineapetalum*
    2. Sepals rose to pinkish or yellow
        3. Sepals magenta to rose . . . . 30. *Delphinium purpusii*
        3. Sepals yellow . . . . 31. *Delphinium xantholeucum*
1. Yellowish gland-based hairs absent from pedicel; sepals spreading or reflexed.
    4. Cells of seed coat with undulate margins visible at 10×; mature fruits usually 3 or fewer times longer than wide; sepals usually light blue to white.
        5. Sepals rarely reflexed; stems more than (50–)60 cm . . . . 29. *Delphinium gypsophilum*
        5. Sepals (especially laterals) usually reflexed; stems less than 60(–100) cm.
            6. Lower petals white (contrasted with blue sepals) . . . . 26. *Delphinium recurvatum*
            6. Lower petals blue or white to pink (same color as sepals) . . . . 25. *Delphinium parishii*
    4. Cells of seed coat with smooth margins visible at 10× (roughened in *D. parryi*); mature fruits usually more than 3 times longer than wide; sepals usually blue to dark blue.
        7. Sepals strongly reflexed . . . . 32. *Delphinium lineapetalum*
        7. Sepals spreading or slightly reflexed, not strongly reflexed.
            8. Green leaves usually absent on proximal 1/5 of stem at anthesis; if present, then proximal portion of stem and/or petioles covered with short, arched hairs.
                9. Lobes of proximal leaves less than 5 mm wide, or plants from less than 10 km inland or less than 400 m elevation . . . . 23. *Delphinium parryi*
                9. Lobes of proximal leaves 5 mm or more wide, and plants from more than 10 km inland and greater than 400 m elevation . . . . *Delphinium umbraculorum* × *D. parryi*
            8. Green leaves usually present on proximal 1/5 of stem at anthesis; proximal portion of stem and petioles glabrous to nearly glabrous (straight hairs).
                10. Leaves almost entirely on proximal 1/5 of stem at anthesis . . . . 27. *Delphinium scaposum*
                10. Leaves primarily cauline at anthesis.
                    11. Lobes of proximal leaves less than 4 mm wide; lateral sepals spreading to reflexed
                    . . . . 24. *Delphinium andersonii*
                    11. Lobes of proximal leaves at least 4 mm wide; lateral sepals spreading . . 28. *Delphinium umbraculorum*

23. Delphinium parryi A. Gray, Bot. Gaz. 12: 53. 1887

Stems (10–)40–80(–110) cm; base reddish, puberulent. **Leaves** variably distributed; green leaves usually absent on proximal 1/5 of stem at anthesis; basal leaves 0–9 at anthesis; cauline leaves 2–7 at anthesis; petiole 1–13 cm. **Leaf blade** pentagonal, 1–7 × 2–10 cm, ± puberulent; ultimate lobes 3–27, width 1–20 mm (basal), 0.5–5 mm (cauline). **Inflorescences** (2–)8–24(–48)-flowered, cylindric; pedicels ± spreading, (0.5–)1–3 (–6.8) cm, usually puberulent; bracteoles 2–7(–16) mm from flowers, green to blue, lance-linear, 2–6(–10) mm, puberulent. **Flowers:** sepals dark blue to bluish purple, puberulent, lateral sepals spreading or reflexed, (7–) 10–20(–25) × 4–9 mm, spurs straight, ascending 0–30° above horizontal, 9–17(–21) mm; lower petal blades slightly elevated, ± exposing stamens, 3–10 mm, clefts 2–6 mm; hairs mostly near base of cleft, centered or on inner lobes, white. **Fruits** 10–19 mm, 2.8–4 times longer than wide, puberulent or glabrous.

Seeds: seed coat cells ± brick-shaped, cell margins undulate, surfaces ± roughened.

Subspecies 5 (5 in the flora): North America (California); Mexico.

A number of local phases are found in *Delphinium parryi*. Five of these appear consistently distinct and are recognized here. Other phases may be locally distinct but grade into other nearby phases. *Delphinium parryi* hybridizes with *D. cardinale* (*D.* ×*inflexum* Davidson).

The Kawaiisu used the ground root of *Delphinium parryi* medicinally as a salve for swollen limbs (D. E. Moerman 1986, no subspecies specified).

1. Basal leaves usually absent at anthesis.
    2. Lateral sepals 16–25 mm. . . . .
        . . . . . . .23d. *Delphinium parryi* subsp. *blochmaniae*
    2. Lateral sepals 9–15 mm. . . . .
        . . . . .23a. *Delphinium parryi* subsp. *parryi*
1. Basal leaves usually present at anthesis.

3. Lateral sepals 7–11 mm; above 700 m elevation . . . . . 23c. *Delphinium parryi* subsp. *purpureum*
3. Lateral sepals (9–)12–20 mm; below 700 m elevation.
    4. Sepals usually reflexed . . . .
       . . . . 23e. *Delphinium parryi* subsp. *eastwoodiae*
    4. Sepals usually spreading . . . .
       . . . . . .23b. *Delphinium parryi* subsp. *maritimum*

23a. Delphinium parryi A. Gray subsp. **parryi** · Parry's larkspur F

Delphinium hesperium A. Gray var. *seditiosum* Jepson; *D. parryi* subsp. *seditiosum* (Jepson) Ewan

Roots 5–20 cm. **Stems** (35–)60–90(–110) cm. **Leaves:** basal leaves usually absent at anthesis; cauline leaves with ultimate lobes 7–27, width 1–6 mm. **Inflorescences:** bracteoles 3–7 mm. **Flowers:** sepals usually spreading, lateral sepals 9–15 mm, spurs 8–15 mm; lower petal blades 3–8 mm. $2n = 16$.

Flowering spring. Locally abundant in oak woodland, chaparral; 200–1700 m; Calif., Mexico (Baja California).

Two morphotypes may be recognized in *Delphinium parryi* subsp. *parryi*. That corresponding to the type specimen of subsp. *parryi* has larger flowers (especially lower petal blades), less abundant pubescence, and somewhat more coarsely dissected leaves; it is usually found in woodlands or relatively moist chaparral. The second morphotype, in its extreme represented by the type specimen of *D. parryi* var. *seditiosum,* has smaller flowers, more pubescence, and more finely dissected leaves and is usually found in chaparral and, less often, in dry woodlands. It may occur sporadically throughout the range of *D. parryi* subsp. *parryi*, although it is most common north of the Transverse Ranges. Hybrids with *D. cardinale* have been named *D.* ×*inflexum*. Hybrids are also known with *D. gypsophilum* subsp. *parviflorum, D. hesperium* subsp. *pallescens, D. umbraculorum,* and *D. variegatum.*

Delphinium parryi subsp. *parryi* may be confused with the blue-flowered phases of *D. hesperium;* see discussion under that species for distinguishing features.

23b. Delphinium parryi A. Gray subsp. **maritimum** (Davidson) M. J. Warnock, Phytologia 68: 2. 1990 · Maritime larkspur

Delphinium parryi var. *maritimum* Davidson, Muhlenbergia 4: 35. 1908

Roots less than 10 cm. **Stems** 15–40 cm. **Leaves** well distributed; basal and cauline leaves present at anthesis; ultimate lobes 5–10, width usually more than 6 mm. **Inflorescences:** bracteoles 4–9 mm. **Flowers:** sepals usually spreading, lateral sepals 9–20 mm, spurs 8–21 mm; lower petal blades 4–11 mm. $2n = 16$.

Flowering late winter–spring. Coastal chaparral; 0–300 m; Calif.; Mexico (Baja California).

Populations of *Delphinium parryi* subsp. *maritimum* are local, very near the coast. The species also occurs on islands off southern California and northern Baja California, Mexico. Occasional individuals with white to grayish blue sepals occur (mostly in Ventura and western Los Angeles counties, where entire populations may consist of such individuals). Collections of subsp. *maritimum* made before 1940 are numerous; recent collections are much less common. Population reductions have probably resulted from urbanization of its preferred habitat.

Confused with *Delphinium variegatum, D. parryi* subsp. *maritimum* lacks the long hairs of that species. Some plants of subsp. *maritimum* (unringed seeds, erect fruits, and arched hairs) from very near the coast appear superficially like some plants of *D. nuttallianum* (ringed seeds, spreading fruits, and no arched hairs).

23c. Delphinium parryi A. Gray subsp. **purpureum** (F. H. Lewis & Epling) M. J. Warnock, Phytologia 68: 2. 1990 · Mount Pinos larkspur C E

Delphinium parishii A. Gray subsp. *purpureum* F. H. Lewis & Epling, Brittonia 8: 15. 1954

Roots usually more than 10 cm. **Stems** 30–90 cm. **Leaves** mostly on proximal 1/3 of stem; basal leaves usually present at anthesis; blade with ultimate lobes 3–20, width less than 4 mm. **Inflorescences:** bracteoles 3–6 mm. **Flowers:** sepals usually reflexed, lateral sepals 7–11 mm, spurs 10–13 mm; lower petal blades 3–5 mm. $2n = 16$.

Flowering late spring–early summer. Dry chaparral, sage scrub, lower montane woods; of conservation concern; 1000–1600(–2400) m; Calif.

Delphinium parryi subsp. *purpureum* may be imperiled by human encroachment.

This taxon may be confused with *Delphinium par-*

ishii, but its abundant arched hairs and lack of wider-lobed basal leaves will distinguish *D. parryi* subsp. *purpureum*. It hybridizes with *D. parishii* subsp. *pallidum*.

23d. Delphinium parryi A. Gray subsp. blochmaniae
(Greene) F. H. Lewis & Epling, Brittonia 8: 19. 1954 · Dune larkspur, Blochman's larkspur [C] [E] [F]

Delphinium blochmaniae Greene, Erythea 1: 247. 1893 (as *blochmanae*)

Roots less than 10 cm. **Stems** (19–)30–50(–65) cm. **Leaves:** basal leaves usually absent at anthesis; blade with ultimate lobes 7–25, width less than 3 mm. **Inflorescences:** bracteoles 7–10 mm. **Flowers:** sepals reflexed, lateral sepals 16–25 mm, spurs 11–16 mm; lower petal blades 7–10 mm, colored lighter than sepals (more so when dry). 2*n* = 16.

Flowering early spring. Coastal chaparral, deep sand of dunes; of conservation concern; 0–200 m; Calif.

Delphinium parryi subsp. *blochmaniae* is very local; no other *Delphinium* is normally found within its limited range. It is easily recognized by its very large sepals and lower petals. The only species with which it might be confused is the large-flowered phase of *D. variegatum;* the latter has long hairs on proximal petioles vs. long hairs absent in *D. parryi* subsp. *blochmaniae.*

23e. Delphinium parryi A. Gray subsp. eastwoodiae
Ewan, Univ. Colorado Stud., Ser. D, Phys. Sci. 2: 182. 1945 · Eastwood's larkspur [C] [E]

Roots less than 10 cm. **Stems** 15–40 cm. **Leaves** mostly on proximal 1/3 of stem; basal leaves usually present at anthesis; blade with ultimate lobes 5–15, width less than 7 mm. **Inflorescences:** bracteoles 7–10 mm. **Flowers:** sepals usually reflexed, lateral sepals 11–20 mm, spurs 11–17 mm; lower petal blades 6–9 mm.

Flowering spring. Serpentine endemic in grasslands surrounded by coastal chaparral; of conservation concern; 50–500 m; Calif.

Delphinium parryi subsp. *eastwoodiae* is usually very local, although in a few localities it is abundant. It is likely to be confused only with *D. variegatum; D. parryi* subsp. *eastwoodiae* does not have long hairs as are present on proximal petioles of *D. variegatum.*

24. Delphinium andersonii A. Gray, Bot. Gaz. 12: 53.
1887 · Anderson's larkspur [E] [F]

Delphinium andersonii subsp. *cognatum* (Greene) Ewan

Stems (20–)30–60(–90) cm; base reddish, glabrous. **Leaves** mostly on proximal 1/2 of stem; green leaves usually present on proximal 1/5 of stem at anthesis; basal leaves 0–8 at anthesis; cauline leaves (0–)3–8 at anthesis; petiole 0.5–8 cm. **Leaf blade** round, 1.5–4 × 2–6 cm, nearly glabrous; ultimate lobes 5–30, width 2–8 mm (basal), 1–4 mm (cauline); lobe width of proximal leaves less than 4 mm. **Inflorescences** 10–25-flowered, cylindric; pedicels sigmoid (proximally spreading, distally ascending), 1–4(–6.8) cm, glabrous to puberulent; bracteoles 2–6(–8) mm from flowers, green, linear, 4–6(–11) mm, ± puberulent. **Flowers:** sepals dark blue, nearly glabrous, lateral sepals spreading to reflexed, 9–16 × 3–7 mm, spurs horizontal to slightly ascending, often decurved apically, 12–18 mm; lower petal blades elevated, ± exposing stamens, 4–8 mm, clefts 1–4 mm; hairs centered and local between claw and base of cleft, white. **Fruits** 17–32 mm, 4–5.5 times longer than wide, glabrous. **Seeds:** seed coat cells ± brick-shaped, cell margins ± undulate, surfaces smooth. 2*n* = 16.

Flowering late spring–early summer. Talus, cold desert scrub, often growing up through shrubs, low places where snow collects; 1300–2000 m; Calif., Idaho, Mont., Nev., Oreg., Utah.

In much of its range *Delphinium andersonii* hybridizes occasionally with members of the *D. nuttallianum* complex and apparently with *D. parishii* in at least one site in California. These three taxa, with *D. scaposum*, form an interesting group in that they appear to be ecological replacements for one another, with *D. parishii* occupying arid, hot deserts to the south and southwest, *D. andersonii* growing in cooler, higher latitude and altitude deserts farther north, *D. scaposum* in cool deserts farther east, and *D. nuttallianum* at higher elevations in much of the geographic range of the other three species. *Delphinium andersonii* is often mistaken for *D. nuttallianum*. Most individuals of *D. andersonii* (roots much larger and more fibrous; stems solidly attached to roots; fruits long, narrow, erect; inflorescences usually longer and narrower at base; and pedicels sigmoid) can easily be distinguished from *D. nuttallianum* (roots smaller and not fibrous; stems tenuously attached to roots; fruits shorter, proportionally thicker, spreading; inflorescences relatively shorter and wider at base; and pedicels nearly straight).

Although roots of *Delphinium andersonii* are quite

similar to those of *D. antoninum,* the two taxa may be readily distinguished by most features that separate *D. nuttallianum* from *D. andersonii.* The name *Delphinium menziesii* was misapplied to *D. andersonii* by S. Watson.

25. Delphinium parishii A. Gray, Bot. Gaz. 12: 53. 1887

Stems (17–)30–60(–100) cm; base reddish or not, glabrous to puberulent. **Leaves** scattered or mostly on proximal 1/3 of stem; basal leaves 0–11 at anthesis; cauline leaves 3–7 at anthesis; petiole 1–13 cm. **Leaf blade** pentagonal, 0.7–5 × 1–8 cm, glabrous to puberulent; ultimate lobes 3–18, width 2–18 mm (basal), 0.5–8 mm (cauline). **Inflorescences** (6–)10–40(–74)-flowered, cylindric; pedicel ascending-spreading, (0.3–)1–2.5(–4.8) cm, glabrous to puberulent; bracteoles 2–5(–10) mm from flowers, green, lance-linear, 2–6(–16) mm, glabrous to puberulent. **Flowers:** sepals dark blue to white to pink, often puberulent, lateral sepals reflexed or spreading, (7–)9–13 × 2–7 mm, spurs ± decurved, ascending 20–45° above horizontal, 7–15 mm; lower petal blades ± elevated, exposing stamens, blue or white to pink (concolorous with sepals), 3–6 mm, clefts 1–3 mm; hairs mostly near base of cleft, centered or on inner lobes, white. **Fruits** 9–21 mm, 2–4 times longer than wide, glabrous to puberulent. **Seeds:** seed coat cells ± brick-shaped, cell margins undulate, surfaces roughened.

Subspecies 3 (3 in the flora): sw United States, nw Mexico.

1. Lateral sepals reflexed; sepals bright, ± sky blue . . . . .25a. *Delphinium parishii* subsp. *parishii*
1. Lateral sepals not reflexed; sepals dark blue or white to pink.
  2. Sepals dark blue; flowers rarely present after 20 May . . . . 25b. *Delphinium parishii* subsp. *subglobosum*
  2. Sepals white, pinkish, or purplish; flowers rarely present before 20 May .25c. *Delphinium parishii* subsp. *pallidum*

25a. Delphinium parishii A. Gray subsp. **parishii**
· Parish's larkspur, Apache larkspur, Clary's larkspur, Mohave larkspur

Delphinium amabile Tidestrom; *D. amabile* subsp. *apachense* (Eastwood) Ewan

Stems (17–)30–60(–100) cm. **Leaves** basal and cauline, basal often absent at anthesis; basal leaves with ultimate lobes 3–5; cauline leaves usually much smaller, ultimate lobes 3–15, narrower than those of basal leaves. **Inflorescences:** pedicels 10–48 mm, 8–25 mm apart. **Flowers:** sepals bright, ± sky blue, reflexed, lateral sepals 8–12 × 3–

6 mm, spurs 8–15 mm; lower petal blades 3–6 mm. **Fruits** 9–21 mm. 2*n* = 16.

Flowering spring. Desert scrub and juniper woods; 200–3900 m; Ariz., Calif., Nev., Utah; Mexico (Baja California, Sonora).

The typical phase of *Delphinium parishii* subsp. *parishii* is found on the floor of desert canyons just east of the peninsular ranges. These plants have sky blue flowers and, often, weak stems; blades of proximal leaves are rarely present at anthesis. The phase represented by the type specimen of *D. amabile* grows in low elevation desert in most of the range of *D. parishii* subsp. *parishii* in California, Nevada, Utah, and western Arizona. These plants also have sky blue sepals but stout stems; they often retain blades of proximal leaves at anthesis. The phase named *D. amabile* subsp. *clarianum* is found primarily at higher elevations of desert mountains within the range of *D. parishii* subsp. *parishii* and is most easily recognized by its darker blue sepals. The type specimen of *D. apachense* represents a phase that grows under relatively high moisture conditions, grows taller, and retains more proximal leaves at anthesis; its sepals may be sky blue or dark blue.

Delphinium parishii subsp. *parishii* hybridizes with *D. andersonii,* *D. cardinale,* *D. hansenii* subsp. *kernense,* and *D. nudicaule* (in gardens). The subspecies is likely to be confused only with *D. andersonii.* See discussion under that species for distinguishing features and ecological relationships of the two taxa.

25b. Delphinium parishii A. Gray subsp. **subglobosum** (Wiggins) F. H. Lewis & Epling, Brittonia 8: 15. 1954

Delphinium subglobosum Wiggins, Contr. Dudley Herb. 1: 99. 1929

Stems (19–)30–50(–78) cm. **Leaves** basal and cauline; ultimate lobes 7–12, lobes of basal leaves nearly as narrow as lobes of cauline. **Inflorescences:** pedicels 3–20 mm, 8–17 mm apart. **Flowers:** sepals dark blue, usually spreading, lateral sepals 9–13 × 5–7 mm, spurs 12–14 mm; lower petal blades 4–6 mm. **Fruits** 8–11 mm.

Flowering late winter–mid spring. Dry chaparral and desert scrub; 600–1300 m; Calif.; Mexico (Baja California).

Delphinium parishii subsp. *subglobosum* occurs only on the eastern side of peninsular ranges.

Delphinium parishii subsp. *subglobosum* hybridizes with *D. parryi* subsp. *parryi.* Likely to be confused only with *D. parryi, D. parishii* subsp. *subglobosum* may be differentiated from that species by its lack of arched hairs on stems.

25c. Delphinium parishii A. Gray subsp. **pallidum**
(Munz) M. J. Warnock, Phytologia 68: 2. 1990 · Pale-
flowered Parish's larkspur E

Delphinium parishii var. *pallidum*
Munz, Bull. S. Calif. Acad. Sci. 31:
61. 1932

Stems (27–)40–60(–95) cm.
Leaves primarily on proximal
1/3 of stem; ultimate lobes 3–7,
abruptly smaller with narrower
lobes on distal portion of stem.
Inflorescences: pedicels usually
less than 15 mm, 4–17 mm apart. **Flowers:** sepals
white to pink or blue, spreading to erect, lateral sepals
6–11 × 2–4 mm, spurs 7–13 mm; lower petal blades
3–4 mm. **Fruits** 11–14 mm. $2n = 16$.

Flowering spring–early summer. *Artemisia* scrub,
open pine woods, chaparral; 900–1900 m; Calif.

Nowhere common, *Delphinium parishii* subsp. *pallidum* may be in danger of significant population reductions because of human encroachment.

Delphinium parishii subsp. *pallidum* hybridizes
with *D. parryi* subsp. *purpureum*. It is frequently confused with *D. inopinum*. Most of the data attributed
to *D. inopinum* in the paper by H. Lewis and
C. Epling (1954) pertain to *D. parishii* subsp. *pallidum*. Refer to discussion under *D. inopinum* for features used to distinguish these two taxa. Plants of *D.
parishii* subsp. *pallidum* might be confused with *D.
gypsophilum* subsp. *parviflorum*. Distinguishing characteristics are found in discussion of that taxon.

26. Delphinium recurvatum Greene, Pittonia 1: 285. 1889
· Valley larkspur, recurved larkspur C E

Delphinium hesperium A. Gray var.
recurvatum (Greene) K. C. Davis

Stems (18–)30–50(–85) cm; base
reddish, glabrous. **Leaves:** basal
leaves 0–2 at anthesis; cauline
leaves 3–7 at anthesis; petiole 1–
8 cm. **Leaf blade** round to pentagonal, 1–4 × 1.5–6 cm,
nearly glabrous; ultimate lobes
3–11, width 3–15 mm (basal), 1–10 mm (cauline). **Inflorescences** (8–)10–25(–47)-flowered, narrowly pyramidal; pedicels ± spreading, (0.5–)1.5–4(–6) cm,
nearly glabrous; bracteoles 3–8(–18) mm from flowers, green, sometimes margins white, lanceolate to linear, 3–5(–8) mm, nearly glabrous. **Flowers:** sepals
light to sky blue (becoming bluer upon drying), puberulent, lateral sepals reflexed, 11–16 × 5–7(–9) mm,
spurs straight to gently upcurved, ascending 0–30°
above horizontal, 10–15(–18) mm; lower petal blades
elevated, ± exposing stamens, 5–8 mm, clefts 0.5–2.5
mm; hairs centered on inner lobes near base of cleft,
white. **Fruits** 8–21 mm, 2.2–3 times longer than wide,

puberulent. **Seeds:** seed coat cells brick-shaped, cell
margins undulate, surfaces roughened. $2n = 16$.

Flowering spring. Grassland, *Atriplex* scrub; of
conservation concern; 30–600 m; Calif.

Delphinium recurvatum has a very restricted distribution in the Central (especially San Joaquin) Valley.
This species was probably much more common in the
past; most of its habitat has been converted into irri-
gated croplands. *Delphinium recurvatum* grows in
poorly drained, alkaline soils on valley floors.

Hybrids are known between *Delphinium recurvatum* and *D. gypsophilum*, *D. hesperium*, *D. parryi*,
and *D. variegatum*. *Delphinium recurvatum* is most
likely to be confused with *D. gypsophilum* or *D.
hesperium* subsp. *pallescens*. Distinguishing features
are found in discussions of those taxa.

27. Delphinium scaposum Greene, Bot. Gaz. 6: 156.
1881, not **D. scaposum** W. T. Wang 1957 E F

Delphinium andersonii A. Gray var.
scaposum (Greene) S. L. Welsh

Stems 25–50(–65) cm; base usually reddish, glabrous, glaucous.
Leaves mostly basal, 4–10 at anthesis; cauline leaves 1–4 at anthesis; green leaves usually present on proximal 1/5 of stem at
anthesis; petiole 1–12 cm, gla-
brous. **Leaf blade** ± round, 0.5–4 × 0.5–6 cm, puberulent to glabrous; ultimate lobes 3–9, width 2–15
mm (basal), 0.5–3 mm (cauline). **Inflorescences** 10–
25(–40)-flowered, cylindric; pedicels ascending, 0.6–
2.5 cm, glabrous; bracteoles 2–5 mm from flowers,
green to blue, linear-lanceolate, 2–4 mm, glabrous.
Flowers: sepals bright dark blue, glabrous, lateral sepals spreading, 11–14 × 4–6 mm, spurs straight,
sometimes decurved, ascending 30–45° above horizontal, 13–18 mm; lower petal blades elevated,
exposing stamens, 5–8 mm, cleft 2–4 mm; hairs centered on inner lobes near junction of blade and claw,
white. **Fruits** 12–16 mm, 2.5–3 times longer than
wide, glabrous. **Seeds:** seed coat cells ± brick-shaped,
cell margins straight, surfaces smooth. $2n = 16$.

Flowering spring. Juniper woods, grassland; 1200–
2700 m; Ariz., Colo., N.Mex., Utah.

Delphinium scaposum is reportedly used in Navajo
and Hopi religious ceremonies, as well as for a wash
following childbirth (D. E. Moerman 1986).

28. Delphinium umbraculorum F. H. Lewis & Epling, Brittonia 8: 19. 1954 · Umbrella larkspur [E]

Stems 40–70(–90) cm; base often reddish, glabrous or puberulent. Leaves usually present on proximal 1/5 of stem at anthesis; basal leaves 0–3 at anthesis; cauline leaves 3–7 at anthesis; petiole 0.8–12 cm. Leaf blade round to pentagonal, 1.5–4 × 2–6 cm, nearly glabrous; ultimate lobes 3–13, width 3–20 mm (basal), 1–8 mm (cauline). Inflorescences (5–)10–25(–45)-flowered, open, narrowly pyramidal; pedicels 0.5–3(–7) cm, glabrous to puberulent; bracteoles 3–7 mm from flowers, green, linear, 3–6 mm, puberulent. Flowers: sepals dark blue, puberulent, lateral sepals spreading, 9–16 × 4–7 mm, spurs gently upcurved, ascending 30–45° above horizontal, 8–14 mm; lower petal blades elevated, exposing stamens, 3.5–6 mm, clefts 0.5–1.5 mm; hairs densest near junction of blade and claw above base of cleft, centered or on inner lobes, white. Fruits 9–16(–19) mm, 2.5–3(–4) times longer than wide, puberulent. Seeds: seed coat cells brick-shaped, cell margins straight, surfaces smooth. 2n = 16.

Flowering late spring–early summer. Slopes in oak forests; 400–1600 m; Calif.

Delphinium umbraculorum is most often confused with *D. patens* subsp. *hepaticoideum;* refer to discussion of that taxon for distinguishing features. Hybrids occur with *D. parryi* and *D. patens* subsp. *montanum.*

29. Delphinium gypsophilum Ewan, Univ. Colorado Stud., Ser. D, Phys. Sci. 2: 189, fig. 54. 1945 [E]

Stems (30–)60–100(–150) cm; base usually reddish, glabrous, glaucous. Leaves mostly cauline; basal leaves absent at anthesis; cauline leaves 3–7 at anthesis; petiole 2–10 cm. Leaf blade round to pentagonal, 1.5–6 × 2–12 cm, nearly glabrous; ultimate lobes 3–12, width 3–24 mm (basal), 1–8 mm (cauline). Inflorescences 15–30(–64)-flowered, cylindric; pedicels spreading, (0.5–)1.5–3.5 cm, glabrous; bracteoles 2–6 mm from flowers, green, linear, 2–5 mm, glabrous. Flowers: sepals rarely reflexed, white to pink, nearly glabrous, lateral sepals spreading, 7–19 × 3–10 mm, spurs straight to upcurved, ascending 30–45° above horizontal, 7–15 mm; lower petal blades elevated, exposing stamens, 3–8 mm, clefts 1–4 mm; hairs centered near base of cleft, ± evenly distributed, white. Fruits 9–18 mm, 2.5–3.2 times longer than wide, puberulent. Seeds: seed coat cells brick-shaped, cell margins undulate, surfaces roughened.

Subspecies 2 (2 in the flora): California.

Delphinium gypsophilum is sometimes confused with *D. hesperium* subsp. *pallescens, D. recurvatum,* and the white-flowered phases of *D. hansenii* subsp. *kernense.* The echinate seeds and long-haired petioles of *D. hansenii* immediately distinguish it from *D. gypsophilum,* which has neither.

Delphinium gypsophilum is related, and similar in many respects, to *D. recurvatum.* The two may be distinguished morphologically by their sepals. *Delphinium recurvatum* has reflexed, blue sepals; those of *D. gypsophilum* are spreading and white, although they may change to light blue when dry. Plants of *D. recurvatum* normally are less than 60 cm; those of *D. gypsophilum* are usually more than 60 cm. Ecologically, *D. recurvatum* occupies level ground among shrubs, typically in alkaline valley bottoms; *D. gypsophilum* is found on well-drained hillsides among grasses and in chaparral and oak woodland.

From *Delphinium hesperium* subsp. *pallescens,* specimens of *D. gypsophilum* may be separated by their much more finely dissected leaves, with less surface area, stem base usually reddish, stems frequently glaucous proximally, undulate margins of seed coat cells, and absence of striations in stem base of dried specimens. In contrast, *D. hesperium* subsp. *pallescens* has leaves less dissected, with greater surface area, stem base rarely reddish, stems not glaucous proximally, seed coat cells with straight margins, and striations present on the proximal stem of dried specimens.

1. Lower petals 5–8 mm; lateral sepals 10 mm or more; pedicels usually more than 1 cm apart . . .
. . . . 29a. *Delphinium gypsophilum* subsp. *gypsophilum*
1. Lower petals 3–5 mm; lateral sepals 10 mm or less; pedicels usually less than 1 cm apart . . . . .
. . . . 29b. *Delphinium gypsophilum* subsp. *parviflorum*

29a. Delphinium gypsophilum Ewan subsp. **gypsophilum** · Gypsum-loving larkspur [E]

Stems (50–)70–100(–150) cm. Inflorescences open, with 1–5 flowers per 5 cm. Flowers: sepals white, spreading, lateral sepals 10–15(–19) × 5–9 mm, spurs 10–15 mm; lower petal blades white, 5–8 mm. 2n = 16, 32.

Flowering spring. Slopes in grassland and open oak woods; 150–1200 m; Calif.

Tetraploid individuals of *Delphinium gypsophilum* subsp. *gypsophilum* occur intermingled with diploid individuals and are normally indistinguishable morphologically (H. Lewis et al. 1951).

Hybridization may occur with *Delphinium recurvatum* in the San Joaquin Valley, with *D. parryi* in the southern Coast Ranges, with *D. hansenii* in the foothills of the southern Sierra Nevada and the Tehachapi Mountains, and probably with *D. hesperium* subsp.

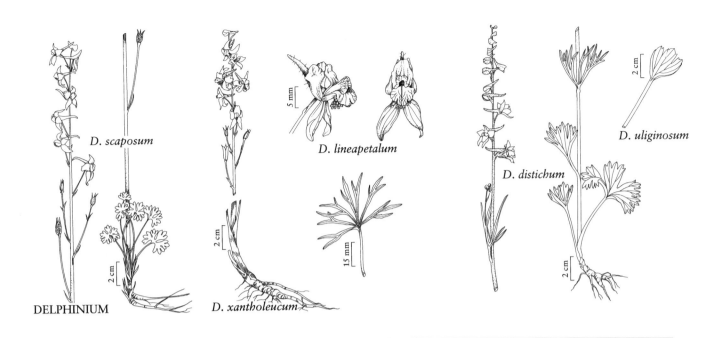

D. scaposum

D. lineapetalum

D. uliginosum

D. distichum

DELPHINIUM

D. xantholeucum

pallescens in Coast Ranges. For a summary on the possible hybrid origin of *D. gypsophilum,* see H. Lewis and C. Epling (1959).

29b. Delphinium gypsophilum Ewan subsp. parviflorum F. H. Lewis & Epling, Brittonia 8: 5. 1954 · Small-flowered gypsum-loving larkspur C E

Stems (30–)60–90(–140) cm. **Inflorescences** ± dense, with 5–13 flowers per 5 cm. **Flowers:** sepals white to pink, erect, lateral sepals 7–10 × 3–7 mm, spurs 7–11 mm; lower petal blades white or yellow, 3–5 mm. $2n = 16$.

Flowering spring. Open oak woodlands, chaparral, and grassland; of conservation concern; 200–600 m; Calif.

Delphinium gypsophilum subsp. *parviflorum* is more infrequent than *D. gypsophilum* subsp. *gypsophilum.*

This subspecies is sometimes confused with *Delphinium parishii* subsp. *pallidum.* Plants of *D. gypsophilum* subsp. *parviflorum* are usually taller and grow in grasslands and woodlands; *D. parishii* subsp. *pallidum* plants are usually shorter and grow in chaparral and shrubland. Hybrids occur with *Delphinium parryi.*

30. Delphinium purpusii Brandegee, Bot. Gaz. 27: 444. 1899 · Purpus's larkspur, rose-colored larkspur C E

Stems (30–)50–80(–120) cm; base reddish or not, nearly glabrous. **Leaves** mostly on proximal 1/2 of stem; basal leaves 3–5 at anthesis; cauline leaves 4–12 at anthesis; petiole 3–17 cm. **Leaf blade** round, 1.5–6 × 2–10 cm, ± puberulent; ultimate lobes 0–5, width 5–30(–50) mm (basal), 5–20(–40) mm (cauline). **Inflorescences** (8–)12–20(–32)-flowered, ± cylindric; pedicels ± ascending, (0.5–)1–4(–5) cm, glandular-puberulent; bracteoles 1–6 mm from flowers, green to magenta, linear, 2–4 mm, glandular-puberulent. **Flowers:** sepals magenta to rose, nearly glabrous, lateral sepals reflexed, 10–16 × 3–7 mm, spurs straight, 30–45° above horizontal, (10–)14–19 mm; lower petal blades nearly coplanar with claws, exposing stamens, 3–4 mm, clefts 0.5–1.5 mm; hairs sparse, scattered, white. **Fruits** (11–)18–29 mm, 4–4.5 times longer than wide, glabrous. **Seeds:** seed coats with small wavy ridges, cells brick-shaped, cell margins undulate, surfaces smooth. $2n = 16$.

Flowering spring. Talus, cliffs, on and near large boulders; of conservation concern; 300–1300 m; Calif.

Delphinium purpusii is not likely to be confused with any other *Delphinium* in North America. Hybrids with *Delphinium hansenii* subsp. *kernense* are known to occur.

31. Delphinium xantholeucum Piper, Contr. U.S. Natl. Herb. 11: 280. 1906 C E F

Stems 40–60(–100) cm; base often reddish, glabrous, ± glaucous. **Leaves** basal and cauline; basal leaves usually absent at anthesis; cauline leaves 4–6 at anthesis; petiole 0.8–16 cm. **Leaf blade** round, 2–6 × 4–10 cm, nearly glabrous; ultimate lobes 3–15, width 3–8 mm (basal), 1–5 mm (cauline). **Inflorescences** 10–20(–60)-flowered, narrowly pyramidal; pedicels spreading, 1.5–3 cm, ± yellowish, glandular-puberulent; bracteoles 6–12 mm from flowers, green to light brown, linear to lanceolate, 4–7 mm, nearly glabrous. **Flowers:** sepals yellow, glabrous, lateral sepals reflexed, 9–12 × 3–5 mm, spurs straight, ascending ca. 45° above horizontal, 11–15 mm; lower petal blades elevated, exposing stamens, 3–5 mm, clefts 1–2 mm; hairs centered on inner lobes near base of cleft, white. **Fruits** 15–22 mm, 3–4 times longer than wide, glabrous to glandular-puberulent. **Seeds:** seed coat cells narrow, short, cell margins straight, surfaces smooth. $2n = 16$.

Flowering spring. Slopes in open yellow pine forests, grasslands, sage scrub; of conservation concern; 150–600 m; Wash.

Delphinium xantholeucum is very local; much of the habitat of this species has been converted to orchards.

32. Delphinium lineapetalum Ewan, Univ. Colorado Stud., Ser. D, Phys. Sci. 2: 126. 1945 C E F

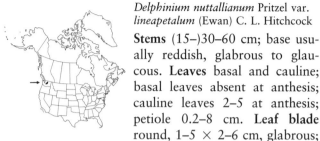

Delphinium nuttallianum Pritzel var. *lineapetalum* (Ewan) C. L. Hitchcock

Stems (15–)30–60 cm; base usually reddish, glabrous to glaucous. **Leaves** basal and cauline; basal leaves absent at anthesis; cauline leaves 2–5 at anthesis; petiole 0.2–8 cm. **Leaf blade** round, 1–5 × 2–6 cm, glabrous; ultimate lobes 5–16, width 1–4 mm (basal), 0.5–3 mm (cauline). **Inflorescences** (3–)9–24(–40)-flowered, pyramidal; pedicels spreading, 1–2.5 cm, glabrous or glandular-puberulent; bracteoles 4–7 mm from flowers, blue or green, linear, 1–3 mm, glabrous or glandular-puberulent. **Flowers:** sepals light blue to lavender, nearly glabrous, lateral sepals reflexed, 10–13 × 3–5 mm, spurs straight to slightly decurved, nearly horizontal to ascending ca. 30°, 11–17 mm; lower petal blades elevated, exposing stamens, 3–5 mm, clefts 0.5–2 mm; hairs centered on inner lobes near base of cleft, white. **Fruits** 13–21 mm, 4–4.5 times longer than wide, puberulent to glabrous. **Seeds:** seed coat cells narrow, short, cell margins straight, surfaces smooth.

Flowering spring. Open pine woods, dry meadows; of conservation concern; 400–1800 m; Wash.

16b.5. DELPHINIUM Linnaeus subsect. DEPAUPERATA M. J. Warnock, Phytologia 78: 84. 1995

Roots 2–8-branched, 2–10(–15) cm, twisted-fibrous, dry to fleshy; buds minute. **Stems** 1(–2) per root, usually unbranched, less than 60(–80) cm, elongation delayed 2–8 weeks after leaf initiation; base usually not narrowed, firmly attached to root. **Leaves** basal and cauline, largest near base of stem, others usually abruptly smaller on distal portion of stem; basal petioles ± spreading, cauline petioles ascending; basal leaves more rounded and with fewer, wider lobes than cauline leaves. **Inflorescences** usually 1–10 flowers per 5 cm, dense to open, cylindric, spurs often intersecting rachis; pedicel closely ascending rachis (straight), usually less than 2 cm, rachis to midpedicel angle less than 30°; bracts markedly smaller and fewer lobed than leaves. **Fruits** erect. **Seeds** rectangular to crescent-shaped, 1.3–2.3 × 0.8–1.4 mm, not ringed on proximal end, ± wing-margined; seed coats without wavy ridges, cells elongate, cell margins straight.

Species 3 or more (3 in the flora): North America, possibly others in Asia.

Key to the species of *Delphinium* subsect. *Depauperata*

1. Margins of basal leaf, measured less than 1 cm from blade base, demarcating less than 90° of arc when leaf laid flat; sepal spurs at 30–45° angle to inflorescence axis . . . . 35. *Delphinium uliginosum*
1. Margins of basal leaf, measured less than 1 cm from blade base, demarcating considerably more than 90° of arc when leaf laid flat; sepal spurs at near 90° angle to inflorescence axis.

2. Cauline leaf blades (at least most of them) exceeding internodes; basal leaves absent at anthesis; cauline leaves 9 or more at anthesis. . . . . .34. *Delphinium distichum*

2. Cauline leaf blades much shorter than internodes; basal leaves usually present at anthesis; cauline leaves 3 or fewer at anthesis. . . . . 33. *Delphinium depauperatum*

33. Delphinium depauperatum Nuttall in J. Torrey and A. Gray, Fl. N. Amer. 1: 33. 1838 · Dwarf larkspur, Blue Mountain larkspur, mountain larkspur E

Delphinium cyanoreios Piper; *D. diversifolium* Greene; *D. diversifolium* subsp. *harneyense* Ewan

Stems unbranched, 10–40(–70) cm; base usually not reddish, nearly glabrous. **Leaves** mostly on proximal 1/3 of stem; basal leaves 3–4 at anthesis; cauline leaves 0–3 at anthesis; petiole 0.4–8 cm. **Leaf blade** ± round, 1–5 × 1.5–9 cm, nearly glabrous; ultimate lobes 3–10, width 2–12 mm (basal), 0.5–3 mm (cauline); margins of basal leaf, measured less than 1 cm from blade base, demarcating considerably more than 90° of arc when leaf laid flat; cauline leaf blades much shorter than internodes. **Inflorescences** 3–15(–25)-flowered, open, ± secund; pedicels 0.5–3(–7) cm, ± glandular-puberulent; bracteoles 2–6(–12) mm from flowers, green to blue, linear or lanceolate, 4–5(–8) mm, ± glandular-puberulent. **Flowers:** sepals dark blue to bluish purple, puberulent, lateral sepals spreading, 10–14 × 4–7 mm, spurs straight, horizontal or nearly so, 12–16 mm; lower petal blades somewhat elevated, ± exposing stamens, 4–7 mm, clefts 2–4 mm; hairs local near base of cleft, white or light yellow. **Fruits** 9–16 mm, 3.5–4.5 times longer than wide, ± puberulent. **Seeds:** seed coat cells with surfaces roughened. $2n = 16$.

Flowering summer. Moist meadows; 1800–2600 m; Calif., Idaho, Mont., Nev., Oreg., Wash.

Delphinium depauperatum and *D. nuttallianum* are often found in the same meadows, with *D. depauperatum* occupying wetter sites, often very near streams, while *D. nuttallianum* is found in drier, better-drained sites. In typical years, the substrate will be dry around *D. nuttallianum* plants, while the substrate is damp near *D. depauperatum* plants as they flower. In addition, within a meadow, *D. depauperatum* flowers later than *D. nuttallianum,* so there is normally little overlap in flowering phenology of the two taxa. Although hybridization between *D. depauperatum* and *D. nuttallianum* is uncommon, hybrids do occur; they have been named *D.* × *burkei* Greene. Burke's specimens at Kew represent a good series of permutations of this cross and successive backcrosses.

Specimens labeled *Delphinium depauperatum* subsp. *harneyense* represent the phase with more abundant yellow-glandular trichomes in the inflorescence and slightly larger flowers. Considerable variation in these features may be found within populations. Presence of yellow-glandular hairs is generally greater in more northern populations. Type specimens of *Delphinium diversifolium* are intermediate in amount of glandular pubescence.

Often confused with *Delphinium nuttallianum*, *D. depauperatum* may be distinguished by its cylindric inflorescences, less dissected leaves, winged seeds, and erect fruits. These character states contrast with the pyramidal inflorescences, more dissected leaves, ringed seeds, and spreading fruits of *D. nuttallianum.*

Dwarfed phases of *Delphinium polycladon* may be confused with *D. depauperatum;* they can be distinguished on the basis of bluish purple flowers, sigmoid pedicels, and prominent buds in the former, and dark blue flowers, straight pedicels, and absence of prominent buds in the latter.

34. Delphinium distichum Geyer ex Hooker, London J. Bot. 6: 68. 1847 · Strict larkspur, meadow larkspur E F

Delphinium strictum A. Nelson var. *distichiflorum* (Hooker) H. St. John

Stems (25–)30–60(–80) cm; base sometimes reddish, puberulent. **Leaves** basal and cauline; basal leaves usually absent at anthesis; cauline leaves 9–18(–28) at anthesis; petiole 0–9 cm. **Leaf blade** cuneate to semicircular, 1–5 × 1.5–7 cm, puberulent; ultimate lobes 5–19, width 2–8(–15) mm (basal), 0.5–3(–5) mm (cauline); margins of basal leaf, measured less than 1 cm from blade base, demarcating considerably more than 90° of arc when leaf laid flat; most cauline leaf blades exceeding internodes. **Inflorescences** 8–30(–40)-flowered, usually dense; pedicels 0.5–1.5 cm, puberulent; bracteoles 0–3 mm from flowers, green to blue, linear, 4–8 mm, puberulent. **Flowers:** sepals dark blue to bluish purple, puberulent, lateral sepals ± erect, 8–12 × 3.5–5 mm, spurs straight, horizontal or nearly so, 9–15 mm; lower petal blades ± covering stamens, 4.5–6.5 mm, clefts 2–3 mm; hairs centered near base of cleft, white. **Fruits** 7–13 mm, 3.5–4 times longer than wide, ± puberulent. **Seeds:** seed coat cells with surfaces roughened. $2n = 16$.

Flowering late spring–early summer. Wet meadows; 100–2400 m; B.C.; Idaho, Mont., Oreg., Wash., Wyo.

Delphinium distichum hybridizes with *D. glaucum*, *D. multiplex* and *D. nuttallianum* (*D.* × *diversicolor* Rydberg). The name *D. burkei* has often been misapplied to *D. distichum*.

35. **Delphinium uliginosum** Curran, Proc. Calif. Acad. Sci. 1: 151. 1885 · Swamp larkspur E F

Stems 10–30(–70) cm; base reddish or not, nearly glabrous. **Leaves** mostly on proximal 1/4 of stem; basal leaves 6–8 at anthesis; cauline leaves 0–5 at anthesis; petiole 0.3–7 cm. **Leaf blade** obdeltoid, apically several parted, 1–8 × 1–7 cm, ± fleshy, glabrous; ultimate lobes 0–3, width 3–20 mm (cauline only); margins of basal leaf, measured less than 1 cm from blade base, demarcating less than 90° of arc when leaf laid flat. **Inflorescences** 5–20(–48)-flowered, ± open; pedicels 0.3–3(–10) cm, glabrous to puberulent; bracteoles 2–3(–5) mm from flowers, green to blue, lanceolate-linear, 3–4(–7) mm, puberulent. **Flowers:** sepals dark blue, nearly glabrous, lateral sepals spreading, 9–15 × 5–8 mm, spurs usually upcurved, ascending 30–45° above horizontal, 10–14 mm; lower petal blades slightly elevated, ± exposing stamens, 4–5 mm, clefts 2–3 mm; hairs centered, densest on inner lobe above base of cleft, also on margins, white. **Fruits** 10–18 mm, 4.1–4.5 times longer than wide, puberulent. **Seeds:** seed coat cells with surfaces bumpy or wavy. $2n = 16$.

Flowering late spring–early summer. Serpentine streamsides, chaparral, grassland; 400–600 m; Calif.

Although some populations are large, *Delphinium uliginosum* is very local. Hybrids with *D. hesperium* subsp. *pallescens* have been seen.

Delphinium uliginosum is a very distinctive species, not likely to be confused with any other. The fan-shaped, slightly dissected leaves are apparently unique in the genus.

16b.6. DELPHINIUM Linnaeus subsect. VIRESCENS M. J. Warnock, Phytologia 78: 85. 1995

Roots 2–8-branched, fusiform, fascicled, 3–8(–15) cm, ± fleshy; buds minute. **Stems** 1(–2) per root, usually unbranched, elongation delayed 4–16 weeks after leaf initiation; base not narrowed, firmly attached to root. **Leaves** basal and cauline, largest near base of stem, others often abruptly smaller on distal portion of stem; basal petioles spreading, cauline petioles ascending; basal leaves more rounded and with fewer, wider lobes than cauline leaves. **Inflorescences** usually with 3–8 flowers per 5 cm, ± dense, cylindric; pedicel appressed-ascending, usually less than 2 cm; bracts markedly smaller and fewer lobed than leaves. **Fruits** erect. **Seeds** crescent-shaped to obpyramidal, 1.5–2.5 × 0.8–1.8 mm, not ringed, not wing-margined; seed coats with prominent multicellular, transverse, wavy ridges, cells ± brick-shaped, cell margins straight.

Species 3 (3 in the flora): North America, n Mexico.

SELECTED REFERENCE   Warnock, M. J. 1981. Biosystematics of the *Delphinium carolinianum* complex (Ranunculaceae). Syst. Bot. 6: 38–54.

Key to the species of *Delphinium* subsect. *Virescens*

1. Sepals white to lavender, reflexed; leaves grayish pubescent with short hairs . . . . 37. *Delphinium wootonii*
1. Sepals white to blue; if sepals lavender, then not reflexed and leaves not grayish pubescent.
  2. Plants with combination of midcauline leaf blades round with at least 7 lobes more than 3/10 length of longest lobe; sepals dark to light blue; distalmost petiole immediately below inflorescence at least 1 cm long . . . . .38. *Delphinium madrense*
  2. Plants deviating in at least one feature from combination above; leaf blades round or pentagonal with more than, fewer than, or equal to 7 lobes; sepals purple to blue to white; distalmost petiole immediately below inflorescence more or less than 1 cm long. . . . . 36. *Delphinium carolinianum*

36. **Delphinium carolinianum** Walter, Fl. Carol., 155. 1788

Stems (20–)40–90(–150) cm; base reddish or not, ± pubescent. **Leaves** basal and cauline; basal leaves 0–10 at anthesis; cauline leaves 4–12 at anthesis; petiole 0.1– 14 cm. **Leaf blade** round to pentagonal, 1–8 × 2–12 cm, pubescence variable; ultimate lobes 3–29, width 2–10 mm (basal), 1–7 mm (cauline). **Inflorescences** (3–)8–27(–94)-flowered; pedicels (0.4–)0.7–1.8(–5.7) cm, nearly glabrous to glandular; bracteoles 1–3.5(–6) mm from flowers, green or blue, linear, 2–7 mm, pubescence nearly glabrous to glandular. **Flowers:** sepals purple to blue to white, nearly glabrous, lateral sepals spreading, (7–)9–14(–17) × (3–)3.5–6(–8) mm, spurs ± upcurved, ascending 20–90° above vertical, (9–)11–17(–19) mm; lower petal blades ± covering stamens, 5–7 mm, cleft 2–4 mm; hairs centered, densest near base of cleft, white, sometimes blue or yellow. **Fruits** (10–)12.5–18.5(–27) mm, 4–4.5 times longer than wide, glabrous to puberulent. **Seeds:** seed coat cells with surfaces pustulate or smooth.

Subspecies 4 (4 in the flora): c, se North America, Mexico.

1. Basal leaves absent at anthesis, cauline leaves divided into many narrow (less than 2 mm wide) segments, blade not distinctly 3-parted; distalmost petiole less than 5 mm; sepals usually blue or purple (rarely white) . . . .
 . . . 36a. *Delphinium carolinianum* subsp. *carolinianum*
1. Basal leaves usually present at anthesis, and/or cauline leaf lobes usually wider than 2 mm, blade often distinctly 3-parted or more; distalmost petiole more than 5 mm; sepals blue or white.
  2. Leaf blade distinctly 3-parted with few additional divisions; sepals blue to white; roots usually ± vertical, often without major branches. . . . .
   . . . 36c. *Delphinium carolinianum* subsp. *vimineum*
  2. Leaf blade with 3–5 or more major divisions, each further divided into segments; sepals white; roots ± horizontal with several major branches.
   3. Stems usually less than 45 cm; in thin soils over limestone in clearings of deciduous woods; leaf blade with 3 major divisions; e of Mississippi River. . . . .
    36b. *Delphinium carolinianum* subsp. *calciphilum*
   3. Stems usually more than 45 cm; in deeper soils in grasslands; leaf blade with 5 or more major divisions; w of Mississippi River. . . . .
    . . 36d. *Delphinium carolinianum* subsp. *virescens*

36a. Delphinium carolinianum Walter subsp. **carolinianum** · Blue larkspur, Carolina larkspur

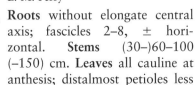

Delphinium carolinianum var. *crispum* L. M. Perry; *D. carolinianum* var. *nortonianum* (Mackenzie & Bush) L. M. Perry

Roots without elongate central axis; fascicles 2–8, ± horizontal. **Stems** (30–)60–100 (–150) cm. **Leaves** all cauline at anthesis; distalmost petioles less than 5 mm. **Leaf blade** not distinctly 3-parted; ultimate lobes of midcauline leaf blades 12–25, width

0.5–1.5 mm. **Flowers:** sepals various shades of blue or purple, rarely white. $2n = 16, 32$.

Flowering late winter–early summer. Prairies and forest openings; 50–700 m; Ala., Ark., Fla., Ga., Ill., Iowa, Kans., Miss., Mo., N.C., Okla., S.C.

Delphinium carolinianum subsp. *carolinianum* occasionally hybridizes with *D. treleasei*.

36b. Delphinium carolinianum Walter subsp. **calciphilum** M. J. Warnock, Phytologia 68: 4. 1990

Roots without elongate central axis; fascicles 2–8, ± horizontal. **Stems** 20–40(–70) cm. **Leaves** usually basal and cauline at anthesis; distalmost petioles more than 5 mm. **Leaf blade** with 3 major divisions; ultimate lobes of midcauline leaf blades 5–12, width 2–4 mm. **Flowers:** sepals white.

Flowering spring. Limestone glades; 150–600 m; Ala., Ga., Ky., Tenn.

Very similar to *Delphinium carolinianum* subsp. *virescens* and often identified as that subspecies, subsp. *calciphilum* usually has shorter stems, fewer leaf lobes, and narrower leaf lobes.

36c. Delphinium carolinianum Walter subsp. **vimineum** (D. Don) M. J. Warnock, Syst. Bot. 6: 49. 1981 · Pine woods larkspur, blue larkspur, Gulf Coast larkspur

Delphinium vimineum D. Don in R. Sweet, Brit. Fl. Gard. 7: plate 374. 1837

Roots usually with elongate single central axis, axis ± vertical, terminal branches (if present) sometimes approaching horizontal. **Stems** 20–60(–100) cm. **Leaves** usually basal and cauline at anthesis; distalmost petioles more than 5 mm. **Leaf blade** distinctly 3-parted; ultimate lobes of midcauline leaf blades 3–9, width 3–10 mm. **Flowers:** sepals various shades of blue to white. $2n = 16, 32$.

Flowering late winter–early summer. Sandy or clay soils, coastal prairies, pine woods, *Prosopis* or *Quercus* scrub; 0–300 m; Ark., La., Okla., Tex.; Mexico (Coahuila, Nuevo León, and Tamaulipas).

Delphinium carolinianum subsp. *vimineum* hybridizes with *D. madrense*, especially on the southern edge of Edwards Plateau in Texas.

36d. Delphinium carolinianum Walter subsp. **virescens** (Nuttall) R. E. Brooks, Phytologia 60: 8. 1982 · Plains larkspur, white larkspur, Penard's larkspur E W2

Delphinium virescens Nuttall, Gen. N. Amer. Pl. 2: 14. 1818; *D. virescens* var. *macroseratilis* (Rydberg) Cory; *D. virescens* var. *penardii* (Huth) L. M. Perry

Roots without elongate central axis; fascicles 2–8, ± horizontal. **Stems** 50–80 cm. **Leaves** mostly cauline at anthesis; distalmost petioles more than 5 mm. **Leaf blade** with 5 or more major divisions; ultimate lobes of midcauline leaf blades 5–15, width 2–6 mm. **Flowers:** sepals white. $2n = 16$.

Flowering late winter–early summer. Prairies, sandy, loamy, or clay soils; 250–2000 m; Man.; Colo., Iowa, Kans., Minn., Mo., Nebr., N.Dak., Okla., S.Dak., Tex., Wis.

Delphinium carolinianum subsp. *virescens* is known to hybridize with *D. madrense* and with *D. wootonii.*

37. Delphinium wootonii Rydberg, Bull. Torrey Bot. Club 26: 587. 1900 · Wooton's larkspur F

Delphinium virescens Nuttall subsp. *wootonii* (Rydberg) Ewan

Stems (15–)30–50(–60) cm; base sometimes reddish, pubescent. **Leaves** mostly on proximal 1/4 of stem at anthesis; basal leaves 0–10 at anthesis; cauline leaves 3–10 at anthesis; petiole 0.5–8 (–14) cm. **Leaf blade** reniform to fan-shaped, 1.5–3 × 2.5–4 cm, puberulent; ultimate lobes 5–24, width 2–6 mm (basal), 1–3 mm (cauline). **Inflorescences** 15–30(–49)-flowered; pedicels 1–2(–3.6) cm, puberulent; bracteoles 1–5 mm from flowers, green, linear, 1–4(–12) mm, puberulent. **Flowers:** sepals white to lavender, nearly glabrous, lateral sepals usually reflexed, (6–)8–13 × 3–7 mm, spur straight, ascending 30–80° above horizontal, (8–)13–20(–25)

mm; lower petal blades ± elevated, 6–9 mm, cleft 2.5–4.5 mm; hairs centered, densest near base of cleft, white. **Fruits** (10–)11–20(–24) mm, 3–3.5 times longer than wide, glabrous. **Seeds:** seed coat cells with surfaces pustulate. $2n = 16$.

Flowering spring. Oak woods, grasslands, desert scrub; 700–1800 m; Ariz., Colo., N.Mex., Tex.; Mexico (Sonora, Chihuahua, and Coahuila).

Delphinium wootonii hybridizes with *D. madrense* in the eastern Big Bend area of Texas, and with *D. carolinianum* subsp. *virescens.*

38. Delphinium madrense S. Watson, Proc. Amer. Acad. Arts 17: 141. 1890 · Sierra Madre larkspur, Edwards Plateau larkspur F

Stems 30–80(–100) cm; base often reddish, puberulent. **Leaves** cauline and basal; basal leaves 0–8 at anthesis; cauline leaves 3–11 at anthesis; petiole 1–15(–25) cm. **Leaf blade** semicircular to cordate, 2–8 × 2–10 cm, nearly glabrous; ultimate lobes 3–12, width 3–10 mm (basal), 2–6 mm (cauline). **Inflorescences** 5–75(–98)-flowered; pedicels 1–2.5(–5) cm, puberulent; bracteoles 2–4(–8) mm from flowers, green, lanceolate-linear, 3–5 mm, puberulent. **Flowers:** sepals dark blue to light blue, puberulent, lateral sepals spreading, 9–15 × 5–7 mm, spurs straight, ascending ca. 45(–90)°, 10–15(–19) mm; lower petal blades elevated, exposing stamens, 4–7 mm, clefts 2–4 mm; hairs centered, densest on inner lobes near base of cleft, white. **Fruits** 15–21 mm, 3.5–4.5 times longer than wide, nearly glabrous. **Seeds:** seed coat cells with surfaces pustulate.

Flowering spring–early summer. Calcareous slopes, oak woods or desert scrub; 300–2100 m; Tex.; Mexico (Coahuila, Nuevo León, and Tamaulipas).

Delphinium madrense hybridizes with *D. carolinianum* subspp. *vimineum* and *virescens.*

16b.7. DELPHINIUM Linnaeus subsect. **ECHINATA** (Ewan) N. I. Malyutin, Bot. Zhurn. (Moscow & Leningrad) 72: 689. 1987

Delphinium series *Echinatae* Ewan, Bull. Torrey Bot. Club 69: 139. 1942

Roots 3–9(–20)-branched, 3–8(–30) cm, fascicled, dry to fleshy, ± succulent, thin threadlike segments restricted to termini of major segments; buds minute. **Stems** 1(–2) per root, unbranched, elongation delayed 2–10 weeks after leaf initiation; base usually not narrowed, firmly attached to root; proximal internodes much shorter than those of midstem. **Leaves** basal and cauline, largest near base of stem, others often abruptly smaller on distal portion of stem; basal petioles spreading, cauline petioles ascending; blades of basal leaves more rounded and

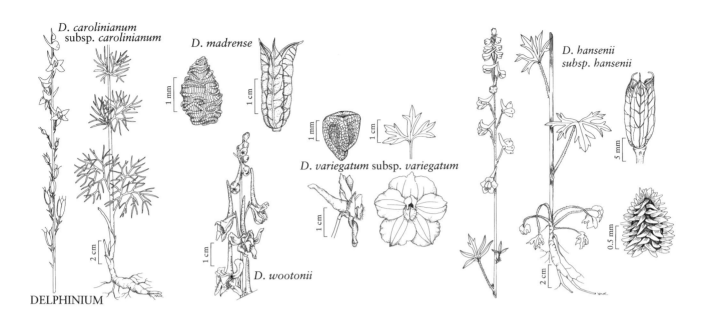

D. carolinianum subsp. *carolinianum*

D. madrense

D. variegatum subsp. *variegatum*

D. hansenii subsp. *hansenii*

D. wootonii

DELPHINIUM

with fewer, wider lobes than cauline leaves. **Inflorescences** usually with 2–11 flowers per 5 cm, dense to open, cylindric, spurs rarely intersecting rachis; pedicel ascending, usually less than 2 cm, rachis to midpedicel angle more than 30°; bracts markedly smaller and fewer lobed than leaves. **Fruits** erect. **Seeds** ± rectangular, 1.5–2.5 × 0.8–1.5 mm, not ringed at proximal end, ± wing-margined; seed coats without wavy ridges (sometimes with prism-shaped structures in *D. hansenii*), cells ± brick-shaped, cell margins straight or undulate.

Species 4 (4 in the flora): California.

Key to the species of *Delphinium* subsect. *Echinata*

1. Proximal petioles glabrous to puberulent with short (less than 0.5 mm) appressed and/or arched hairs.
  2. Sepal color and/or stem pubescence different from that of other individuals with which they occur; stem base puberulent with short arched hairs . . . . *Delphinium hesperium* subsp. *pallescens* × *D. parryi*
  2. Sepal color and stem pubescence similar to that of other individuals with which they occur; stem base glabrous to pubescent with short straight hairs . . . . 39. *Delphinium hesperium*
1. Proximal petioles with long (more than 0.5 mm), straight spreading hairs.
  3. Seeds echinate, appearing fuzzy to naked eye; lateral sepals 13 mm or shorter; flowers usually more than 12 per main inflorescence branch . . . . 42. *Delphinium hansenii*
  3. Seeds not echinate, appearing ± smooth to naked eye; lateral sepals 10 mm or longer; flowers usually fewer than 12 per main inflorescence branch.
    4. Stem base longitudinally ridged . . . . 39. *Delphinium hesperium*
    4. Stem base not longitudinally ridged.
      5. Margins of lower petals ciliate; bracteoles usually 7 mm or less from sepals; sepal spur straight . . . . 40. *Delphinium variegatum*
      5. Margins of lower petals glabrous; bracteoles usually 7 mm or more from sepals; sepal spur often downcurved for more than 3 mm at apex. . . . . 41. *Delphinium hutchinsoniae*

39. Delphinium hesperium A. Gray, Bot. Gaz. 12: 53. 1887 [E]

Stems (11–)40–80(–120) cm; base usually reddish, longitudinally ridged, puberulent. **Leaves** mostly cauline; basal leaves 0–3 at anthesis; cauline leaves 3–8 at anthesis; petiole 0.5–7 cm, petioles of proximal leaves glabrous to puberulent. **Leaf blade** round to pentagonal, 1–4 × 2–6 cm, usually puberulent, especially abaxially; ultimate lobes 3–14, width 3–8 mm (basal), 2–5 mm (cauline). **Inflorescences** (5–)15–30(–100)-flowered, moderately open; pedicels (0.5–)1–2.5(–7.5) cm, puberulent; bracteoles 2–6(–12) mm from flowers, green to blue, margins often white, linear-lanceolate, 3–7(–12) mm, puberulent. **Flowers:** sepals dark blue to white, puberulent, lateral sepals spreading, 7–13(–16) × 3–10 mm, spurs straight to upcurved, ascending 30–45° above horizontal, 9–18 mm; lower petal blades slightly elevated, ± exposing stamens, 3–8 mm, clefts 1–5 mm; hairs centered, denser on inner lobes near base of cleft, white. **Fruits** 8–18 mm, 2.2–3(–4.2) times longer than wide, sparse puberulent. **Seeds** not echinate, appearing ± smooth to naked eye; seed coat cells with margins straight, surfaces smooth or roughened.

Subspecies 3 (3 in the flora): California.

Delphinium hesperium is often confused with *D. hansenii*. See discussion under that species for distinguishing features.

1. Lateral sepals 4(–5) mm wide or less; inflorescence dense (individual flowers often touching in pressed specimens). . . . . 39b. *Delphinium hesperium* subsp. *cuyamacae*
1. Lateral sepals more than 4 mm wide; inflorescence open (individual flowers seldom touching in pressed specimens).
　　2. Sepals dark blue-purple. .39a. *Delphinium hesperium* subsp. *hesperium*
　　2. Sepals white to pinkish, rarely light blue. . . . . .39c. *Delphinium hesperium* subsp. *pallescens*

39a. Delphinium hesperium A. Gray subsp. hesperium
· Western larkspur E

Stems glabrous to simple-puberulent, long hairs rare. **Inflorescences** usually fewer than 30-flowered, open; pedicel ascending. **Flowers:** sepals dark blue-purple, spreading, lateral sepals 8–16 mm, more than 4 mm wide, spurs 9–16 mm; lower petal blades 5–8 mm. **Fruits** 3.5–4.2 times longer than wide. $2n = 16$.

Flowering mid–late spring. Open oak woods, grasslands, openings in coastal chaparral; 10–1100 m; Calif.

Populations of *Delphinium hesperium* subsp. *hesperium* are found only on western slopes of the Coast Ranges. Hybrids with *D. parryi* and *D. variegatum* are known.

Although *Delphinium hesperium* subsp. *hesperium* has been confused with *D. parryi*, it may be recognized by its usually darker blue sepals, absence of arched hairs on proximal portion of stems, and pronounced venation abaxially on leaves. *Delphinium parryi* has brighter blue sepals, arched hairs present on proximal portion of stems, and leaves not prominently veined abaxially.

39b. Delphinium hesperium A. Gray subsp. cuyamacae
(Abrams) H. F. Lewis & Epling, Brittonia 8: 11. 1954 · Cuyamaca larkspur C E

Delphinium cuyamacae Abrams, Bull. Torrey Bot. Club 32: 538. 1905

Stems proximally puberulent with arched hairs. **Inflorescences** generally more than 25-flowered, dense; pedicels ascending. **Flowers:** sepals dark blue-purple, spreading to erect, lateral sepals 7–10 mm, less than 4(–5) mm wide, spurs 8–12 mm; lower petal blades 3–5 mm. **Fruits** 2.2–2.5 times longer than wide. $2n = 16$.

Flowering early summer. Grassland, open pine woods; of conservation concern; 1100–1500 m; Calif.

Delphinium hesperium subsp. *cuyamacae* is local near Cuyamaca Lake. Although it has been reported also from Mt. Palomar, specimens have not been seen.

No other *Delphinium* with similar features grows in the region where *D. hesperium* subsp. *cuyamacae* grows. Superficially, specimens of this taxon resemble those of some *D. hansenii* subsp. *hansenii*. Seeds are quite different, as are pubescence patterns and venation on abaxial surface of leaves. *Delphinium parryi* occurs near *D. hesperium* subsp. *cuyamacae*; flowers of *D. parryi* in that area are much larger and more widely spaced on the inflorescence than in *D. hesperium* subsp. *cuyamacae*.

39c. Delphinium hesperium A. Gray subsp. pallescens
(Ewan) F. H. Lewis & Epling, Brittonia 8: 10. 1954 · Pale-flowered western larkspur E

Delphinium hesperium A. Gray forma *pallescens* Ewan, Univ. Colorado Stud., Ser. D, Phys. Sci. 2: 179. 1945

Stems usually glabrous. **Inflorescences** usually fewer than 25-flowered, open; pedicels ± spreading. **Flowers:** sepals white to pinkish, rarely light blue, spreading, lateral sepals 10–15 mm, more than 4 mm wide, spurs 10–17 mm; lower petal blades 4–7 mm. **Fruits** 3–3.5 times longer than wide. $2n = 16$.

Flowering spring. Open oak woods and grassland; 20–1000 m; Calif.

Delphinium hesperium subsp. *pallescens* grows only east of the Coast Range divide. It hybridizes with *D. gypsophilum*, *D. parryi*, *D. recurvatum*, *D. uliginosum*, and *D. variegatum*. Light blue sepals probably result from introgression involving *D. variegatum*, *D. parryi*, or *D. hesperium* subsp. *hesperium*. Intermediates are occasionally found between *D. hesperium* subsp. *pallescens* and subsp. *hesperium* (particularly in Napa County), but the two taxa remain, for the most part, geographically isolated.

Delphinium hesperium subsp. *pallescens* has sometimes been confused with *D. gypsophilum* or with white-flowered plants of *D. hansenii;* see discussion under those species.

40. **Delphinium variegatum** Torrey & A. Gray, Fl. N. Amer. 1: 32. 1838; not **D. variegatum** Baillon 1883 [E]

Stems (10–)20–50(–80) cm; base reddish or not, not longitudinally ridged, long-pubescent. **Leaves** mostly on proximal 1/3 of stem; basal leaves 0–6 at anthesis; cauline leaves 1–6 at anthesis; petiole 0.5–9 cm, petioles of proximal leaves long-pubescent. **Leaf blade** round to pentagonal, 1–4.5 × 2–6 cm, pubescent; ultimate lobes 3–15, width 3–8 mm (basal), 2–5 mm (cauline). **Inflorescences** (4–)6–17(–20)-flowered, open; pedicels 0.5–3(–7) cm, usually puberulent; bracteoles 2–7(–23) mm from flowers, green, linear-lanceolate, 3–8 mm, puberulent. **Flowers:** sepals usually deep royal blue, sometimes bright blue to light blue or white, ± finely pubescent, lateral sepals spreading, 10–18(–25) × 6–10 mm, spurs straight, ascending ca. 30° above horizontal, 10–19 mm; lower petal blades slightly elevated, ± exposing stamens, 4–11 mm, clefts 0.5–3.5 mm; hairs sparse, more on inner lobes than outer lobe or centered, scattered on margins, white or yellow. **Fruits** 9–19 mm, 2.2–3.8 times longer than wide, ± puberulent. **Seeds** not echinate, appearing ± smooth to naked eye; seed coat cells with margins ± undulate, surfaces roughened.

Subspecies 3 (3 in the flora): California.

1. Long hairs dense on base of stem; sepals usually deep royal blue; mainland . . . .
    . . . . . . 40a. *Delphinium variegatum* subsp. *variegatum*
1. Long hairs sparse on base of stem; sepals white to bright blue; islands.
   2. Sepals bright blue to light blue. . . . .
       . . . . . . 40c. *Delphinium variegatum* subsp. *thornei*
   2. Sepals white to light blue. . . . .
       . . . . 40b. *Delphinium variegatum* subsp. *kinkiense*

40a. **Delphinium variegatum** Torrey & A. Gray subsp. **variegatum** · Royal larkspur [E] [F]

Delphinium emiliae Greene; *D. variegatum* Torrey & A. Gray subsp. *apiculatum* (Greene) Ewan; *D. variegatum* var. *apiculatum* (Greene) Greene

Stem: base densely long-pubescent. **Inflorescences** usually with fewer than 10 flowers per branch. **Flowers:** sepals deep

royal blue, rarely white or lavender, lateral sepals 10–25 mm; lower petal blades 4–11 mm. $2n = 16, 32$.

Flowering spring. Grassland, open oak woods; 20–800 m; Calif.

Some populations of *Delphinium variegatum* subsp. *variegatum* may be found on serpentine; they are not well marked morphologically and are not recognized as distinct taxa. This taxon is one of the most commonly encountered plants in California, at times coloring foothills of the Coast Ranges and Sierra Nevada. Hybrids with *D. hansenii, D. hesperium, D. parryi,* and *D. recurvatum* are known.

Delphinium variegatum has sometimes been confused with *D. hansenii* or with *D. hesperium.* Distinguishing features may be found in discussion under those taxa. Plants recognized under the name *variegatum* exhibit considerable morphologic variation. This variation could not be correlated, however, in such a way as to make defensible taxonomic segregates within *D. variegatum* subsp. *variegatum.* Further study may indicate some means of consistently recognizing some phases. For instance, the type specimen of *D. variegatum* forma *superbum* Ewan represents a phase with huge flowers. Plants with this feature may be found scattered almost throughout the range of *D. variegatum* subsp. *variegatum* (either as isolated plants or as populations made up largely of this morphotype); they are most common in San Francisco Bay area. The type specimen of *D. subnudum* Eastwood represents plants with few cauline leaves, a feature that also appears apparently at random throughout the range of subsp. *variegatum.* Type specimens (isotypes) of *D. emiliae* appear to be plants introgressed with *D. hesperium.* The type specimen of *D. apiculatum* represents the most common and widespread phase of subsp. *variegatum,* with flowers intermediate in size between those of the type specimen of *D. variegatum* and those of *D. variegatum* forma *superbum.*

40b. **Delphinium variegatum** Torrey & A. Gray subsp. **kinkiense** (Munz) M. J. Warnock, Phytologia 68: 2. 1990 · San Clemente Island larkspur [C] [E]

Delphinium kinkiense Munz, Aliso 7: 69. 1969

Stem: base sparsely long-pubescent. **Inflorescences** usually with fewer than 12 flowers per branch. **Flowers:** sepals white or light blue, lateral sepals 11–18 mm; lower petal blades 4–9 mm.

Flowering early spring. Chaparral, oak woods, grassland; of conservation concern; 0–500 m; Calif.

Delphinium variegatum subsp. *kinkiense* has an extremely restricted distribution, is very poorly known, and is imperiled by grazing. *Delphinium parryi* and *D. variegatum* subsp. *thornei* are the only other taxa of *Delphinium* naturally found on San Clemente Island.

40c. Delphinium variegatum Torrey & A. Gray subsp. thornei Munz, Aliso 7: 70. 1969 · Thorne's larkspur, Thorne's royal larkspur C E

Stem: base sparsely long-pubescent. **Inflorescences** usually with fewer than 16 flowers per branch. **Flowers:** sepals bright blue to light blue, lateral sepals 17–21 mm; lower petal blades 6–11 mm.

Flowering spring. Grassland, oak woods; of conservation concern; 0–500 m; Calif.

Delphinium variegatum subsp. *thornei* has an extremely restricted distribution, is very poorly known, and is imperiled by grazing. *Delphinium parryi* and *D. variegatum* subsp. *kinkiense* are the only other taxa of *Delphinium* naturally found on San Clemente Island.

41. Delphinium hutchinsoniae Ewan, Bull. Torrey Bot. Club 78: 379. 1951 · Hutchinson's larkspur, Hutchinson's delphinium C E

Stems (25–)50–80(–100) cm; base reddish, not longitudinally ridged, variably puberulent. **Leaves** mostly cauline at anthesis; basal leaves 0–3 at anthesis; cauline leaves 2–13 at anthesis; petiole 1–19 cm, petioles of proximal leaves ± long-pubescent. **Leaf blade** round to pentagonal, 1–6 × 1.5–10 cm, puberulent; ultimate lobes 3–17, width 4–16(–25) mm (basal), 1–8(–19) mm (cauline). **Inflorescences** (2–)7–20(–31)-flowered, open; pedicels 1–4(–6) cm, puberulent; bracteoles (2–)8–12 mm from flowers, green, linear, 3–6(–9) mm, puberulent. **Flowers:** sepals dark bluish purple, puberulent, lateral sepals spreading, (12–)14–19(–24) × 7–12(–15) mm, spurs ascending, decurved apically, 11–19 mm; lower petal blades slightly elevated, mostly covering stamens, 5–10 mm, cleft 2–3 mm; hairs sparse, mostly on inner lobes, absent on margins, white. **Fruits** 9–21 mm, 2.5–4.2 times longer than wide, sparsely puberulent. **Seeds** not echinate, ± smooth to naked eye; seed coat cells with margins ± undulate, surfaces smooth.

Flowering spring. Coastal chaparral, clearings in coniferous woods; of conservation concern; 0–400 m; Calif.

Delphinium hutchinsoniae is known from only a few populations near Monterey and south to the Big Sur region. Hybrids have been produced between *D. hutchinsoniae* and *D. cardinale* grown in a common garden. Hybrids also occur with *D. parryi* subsp. *maritimum*.

Delphinium hutchinsoniae is similar, and probably closely related, to *D. variegatum*. The two may be distinguished by the decurved spur of *D. hutchinsoniae*; the spur of *D. variegatum* is normally straight (or decurved nearer apex). *Delphinium hutchinsoniae* lacks marginal hairs on lower petals; such hairs are present in *D. variegatum*. The two species are also geographically separated.

42. Delphinium hansenii (Greene) Greene, Pittonia 3: 94. 1896 E

Delphinium hesperium A. Gray var. *hansenii* Greene, Fl. Francisc. 3: 304. 1892

Stems (25–)40–80(–180) cm; base usually reddish, pubescent. **Leaves** cauline and basal; basal leaves 0–5 at anthesis; cauline leaves 2–8 at anthesis; petiole 0.5–8 cm, petioles of proximal leaves long-pubescent. **Leaf blade** pentagonal, 1.5–5 × 2.5–8 cm, long-pubescent, especially abaxially; ultimate lobes 0–18, width 4–20 mm (basal), 2–9 mm (cauline). **Inflorescences** (9–)15–40(–160)-flowered, dense to open; pedicel 0.3–2.5(–6) cm, puberulent; bracteoles 1–5(–8) mm from flowers, green, sometimes white-margined, linear-lanceolate, 2–6(–8) mm, puberulent. **Flowers:** sepals violet to white, ± puberulent, lateral sepals spreading to forward pointing, 7–10(–13) × 3–6(–8) mm, spurs gently up-curved, ascending 0–30° above horizontal, (6–)9–13(–16) mm; lower petal blades elevated, ± exposing stamens, 3–7 mm, cleft 1–2(–4) mm; hairs densest on inner lobes near base of cleft, white. **Fruits** 8–20 mm, 2.2–4 times longer than wide, glabrous. **Seeds** echinate, appearing fuzzy to naked eye; seed coat cells with margins straight, surfaces sparsely pustulate.

Subspecies 3 (3 in the flora): California.

Although *Delphinium hansenii* has often been confused with *D. hesperium*, seeds will instantly allow identification. Seeds of *Delphinium hansenii* are, as far as known, unique, bearing numerous, elongate, prismlike raised structures (extensions of single cells or small groups of cells) over the entire seed coat. If seeds are absent, larger flowers, more open inflorescences (except in *D. hesperium* subsp. *cuyamacae*), and general absence of pubescence of long hairs in *D. hesperium* are apparent upon comparison of the two species. Separating *D. hansenii* from *D. variegatum* may also be difficult. Again, seeds leave no doubt. In addition, smaller flowers and greater number of flowers per plant of *D. hansenii* should serve to distinguish *D. hansenii* from *D. variegatum*. White-flowered *D. hansenii* has been confused with *D. gypsophilum* and with *D. hesperium* subsp. *pallescens.* Other than seeds, pubescence of long hairs and smaller flowers present in *D. hansenii* and absent in the others will distinguish them.

1. Sepals reddish purple to dark maroon . . . .
   . . . . 42c. *Delphinium hansenii* subsp. *ewanianum*
1. Sepals dark blue-purple to white.
   2. Leaves primarily basal (although leaves may be dry at anthesis and thus lost in herbarium specimens, petiole base will be present); cauline leaves usually fewer than 3; sepals bright blue to white . . . .
      . . . . 42b. *Delphinium hansenii* subsp. *kernense*
   2. Leaves primarily cauline, basal leaves usually absent at anthesis; cauline leaves 3 or more; sepals dark blue-purple to white or pink . . . .
      . . . . 42a. *Delphinium hansenii* subsp. *hansenii*

42a. Delphinium hansenii (Greene) Greene subsp. **hansenii** · Hansen's larkspur E F

Delphinium hesperium A. Gray var. *hansenii* Greene; *D. hansenii* subsp. *arcuatum* (Greene) Ewan

Stems 40–80(–180) cm, base simple- and/or long-puberulent. **Leaves** mainly cauline; basal leaves usually absent at anthesis; cauline leaves 3 or more, gradually smaller than basal leaves. **Inflorescences** (2–)6–11 flowers per 5 cm, dense or less commonly open. **Flowers:** sepals dark blue-purple to white or pink, lateral sepals 7–10(–13) mm, spurs 8–13 mm. $2n = 16, 32$.

Flowering spring(–early summer). Open oak woods, grasslands; 100–3000 m; Calif.

Delphinium hansenii subsp. *hansenii* produces natural hybrids with *D. gypsophilum, D. hesperium,* and *D. variegatum.*

42b. Delphinium hansenii (Greene) Greene subsp. **kernense** (Davidson) Ewan, Univ. Colorado Stud., Ser. D, Phys. Sci. 2: 193. 1945 · Kern County larkspur E

Delphinium hansenii var. *kernense* Davidson, Muhlenbergia 4: 37. 1908

Stems (34–)50–80(–120) cm, base simple-puberulent. **Leaves** nearly all basal, often present but dried at anthesis, rarely on herbarium specimens; cauline leaves usually fewer than 3, abruptly smaller than basal leaves. **Inflorescences** 1–8 flowers per 5 cm, open. **Flowers:** sepals bright blue to white, lateral sepals 7–13 mm, spurs 8–16 mm. $2n = 16$.

Flowering spring. Open oak woods, s Sierra chaparral; 800–1900 m; Calif.

Delphinium hansenii subsp. *kernense* hybridizes with *D. gypsophilum, D. parishii,* and *D. purpusii.*

42c. Delphinium hansenii (Greene) Greene subsp. **ewanianum** M. J. Warnock, Phytologia 68: 3. 1990 · Ewan's larkspur C E

Stems (25–)60–100(–130) cm, base usually simple-puberulent. **Leaves** mostly cauline; cauline leaves 3 or more, ± gradually smaller than basal leaves. **Inflorescences** 2–5(–8) flowers per 5 cm, ± open. **Flowers:** sepals reddish purple to dark maroon, lateral sepals 8–12 mm, spurs (6–)9–12(–16) mm. $2n = 32$.

Flowering early–mid spring. Rock outcrops in open oak woods, grasslands; of conservation concern; 60–600 m; Calif.

Delphinium hansenii subsp. *ewanianum* is extremely local and imperiled by human encroachment.

16b.8. DELPHINIUM Linnaeus subsect. **BICOLORIA** (Rydberg) N. I. Malyutin, Bot. Zhurn. (Moscow & Leningrad) 72(5): 687. 1987

Delphinium sect. *Bicoloria* Rydberg, Fl. Rocky Mts., 309. 1917

Roots 2–9-branched at least 1 cm from stem attachment, (5–)10–30(–40) cm, diffuse-fibrous, ± braided, dry; buds minute. **Stems** 1(–2) per root, usually unbranched, elongation delayed 2–6(–10) weeks after leaf initiation; base ± narrowed, ± tenuously attached to root; proximal internodes much shorter than those of midstem. **Leaves** basal and cauline, largest leaves near base of stem, others often abruptly smaller on distal portion of stem; basal petioles spreading, cauline petioles ascending; blade shape and lobing similar throughout. **Inflorescences** usually with 2–6 flowers per 5 cm, open, ± pyramidal, seldom more than 3 times longer than wide, spurs rarely intersecting rachis; pedicels spreading, usually more than 1.5

cm, rachis to midpedicel angle more than 30°; bracts markedly smaller and fewer lobed than leaves. **Fruits** spreading. **Seeds** obpyramidal, 1.6–2.9 × 0.7–1.9 mm, ± ringed at proximal end, usually wing-margined; seed coats lacking wavy ridges, cells elongate, cell margins straight.

Species 6 (6 in the flora): w North America.

Key to the species of *Delphinium* subsect. *Bicoloria*

1. Sepals red or yellow.
  2. Sepals red to reddish orange, not waxy . . . . 47. *Delphinium nudicaule*
  2. Sepals bright yellow, appearing waxy . . . . 48. *Delphinium luteum*
1. Sepals usually bluish, not red or yellow (sometimes maroon, white, or pink).
  3. Sepals maroon; plants distinctly different from other individuals within populations.
    4. Leaf segments more than 5, 5 mm or less wide . . . . *Delphinium nudicaule* × *D. depauperatum*
    4. Leaf segments 5 or fewer, more than 5 mm wide.
      5. Proximal portion of stem hairy . . . . *Delphinium nudicaule* × *D. decorum*
      5. Proximal portion of stem glabrous . . . . *Delphinium nudicaule* × *D. patens*
  3. Sepals blue or pink; plants usually similar to other individuals within populations.
    6. Leaves mostly above proximal 1/3 of stem; flowers usually more than 15 per main inflorescence axis . . . . *Delphinium trolliifolium* × *D. nudicaule*
    6. Leaves mostly on proximal 1/3 of stem; flowers usually fewer than 20 per main inflorescence axis.
      7. Sepals pinkish; plants distinctly different from other individuals within populations.
        8. Plants in populations with red flowers . . . . *Delphinium nudicaule* × *D. trolliifolium*
        8. Plants in populations with blue flowers . . . . *Delphinium antoninum* × *D. nudicaule*
      7. Sepals dark blue to white; plants usually similar to other individuals within populations.
        9. Leaves ± succulent, clustered on proximal portion of stem.
          10. Lateral sepals dark blue to white, spreading to reflexed . . . . 46. *Delphinium antoninum*
          10. Lateral sepals dark blue, spreading . . . . 44. *Delphinium glareosum*
        9. Leaves not succulent, clustered on proximal portion of stem or not.
          11. Lower petal blade cleft usually less than 1/3 blade length . . . . 43. *Delphinium bicolor*
          11. Lower petal blade cleft more than 1/3 blade length . . . . 45. *Delphinium basalticum*

43. **Delphinium bicolor** Nuttall, J. Acad. Nat. Sci. Philadelphia 7: 10. 1834 · Flathead larkspur, low larkspur, little larkspur [E] [W2]

Stems 10–40(–70) cm; base often reddish, glabrous to puberulent. **Leaves** mostly on proximal 1/3 of stem; basal leaves (0–)2–7 at anthesis; cauline leaves 3–6 at anthesis; petiole 0.3–8 cm. **Leaf blade** round, 1–4 × 1.5–7 cm, not succulent, glabrous to puberulent; ultimate lobes 3–19, width 1–8 mm (basal), 1–3 mm (cauline). **Inflorescences** 3–12(–22)-flowered; pedicels 1–4 (–8) cm, ± puberulent; bracteoles 2–7(–17) mm from flowers, green, sometimes white-margined, lanceolate, 4–6(–8) mm, puberulent. **Flowers:** sepals dark blue, puberulent, lateral sepals usually spreading, 16–21 × 6–12 mm, spurs straight to gently decurved, ascending 0–40° above horizontal, 13–23 mm; lower petal blades covering stamens, 7–12 mm, clefts 0.1–3 mm; hairs sparse, short, mostly on inner lobes below junction of blade and claw, white or yellow. **Fruits** (12–)16–22 mm, 4–4.5 times longer than wide, usually puberulent. **Seeds** often winged; seed coat cells with surfaces ± smooth.

Subspecies 2 (2 in the flora): w North America.

Delphinium bicolor is closely related to *D. glareosum*; it differs in its wider-lobed cauline leaves, shallower petal clefts, and narrower fruits.

1. Sepals (especially in fresh material) dark blue to purple; cleft in lower petals 2 mm or less; soils not derived from limestone . . . . 43a. *Delphinium bicolor* subsp. *bicolor*
1. Sepals (especially in fresh material) bright dark blue; cleft in lower petals at least 2 mm; soils derived from limestone . . . . 43b. *Delphinium bicolor* subsp. *calcicola*

43a. **Delphinium bicolor** Nuttall subsp. **bicolor** [E] [F]

Delphinium nuttallianum Pritzel var. *pilosum* C. L. Hitchcock

Flowers: sepals dark blue to purple, lateral sepals 16–21 × 6–11 mm, spurs 13–18 mm; lower petal blades with cleft less than 2 mm; hairs usually white.

Flowering late spring–early summer. Dry meadow edges, sage scrub, open woodlands and edges, not on soils derived from limestone; 600–3100 m; Alta., B.C., Sask.; Idaho, Mont., Nebr., N.Dak., S.Dak., Wyo.

Rydberg's *Delphinium bicolor* var. *montanense* tends to have more pubescence and larger flowers but is otherwise typical and apparently fully intergrades with *D. bicolor* subsp. *bicolor*. Often referred to as one of the low larkspurs in poisonous-plant literature, the plant is abundant on some ranges and is the cause of some livestock poisonings.

43b. Delphinium bicolor Nuttall subsp. **calcicola** M. J. Warnock & Vanderhorst, Phytologia 78: 90. 1995 $\boxed{\text{E}}$

Flowers: sepals bright dark blue, lateral sepals 16–21 × 9–12 mm, spurs 15–23 mm; lower petal blades with cleft 2 mm or more; hairs usually yellow.

Flowering late spring–early summer. Rocky soils in short-grass prairie and sagebrush communities on limestone outcrops; 1300–2100 m; Mont.

In the field, *Delphinium bicolor* subsp. *calcicola* is readily separated from *D. bicolor* subsp. *bicolor* on the basis of its brighter colored, slightly larger flowers and its edaphic preference. *Delphinium bicolor* subsp. *calcicola* usually flowers somewhat later where the two are sympatric. These differences are often not apparent from herbarium specimens. Specimens of this taxon misidentified as *D. geyeri* account for most reports of that species from Montana.

44. Delphinium glareosum Greene, Pittonia 3: 257. 1898 · Olympic Mountain larkspur $\boxed{\text{E}}$ $\boxed{\text{F}}$

Delphinium caprorum Ewan

Stems 20–40 cm; base reddish or not, glabrous, glaucous. **Leaves:** distribution variable; basal leaves 0–4 at anthesis; cauline leaves 3–6 at anthesis; petiole 1–17 cm. **Leaf blade** round, 2–4 × 3–7 cm, ± succulent, glabrous; ultimate lobes 7–15, width 3–10 mm (basal), 2–7 mm (cauline). **Inflorescences** 5–12-flowered; pedicels 1–5(–10) cm, nearly glabrous to glandular-puberulent; bracteoles 2–10 (–20) mm from flowers, green, linear, 3–8 mm, puberulent. **Flowers:** sepals dark blue, nearly glabrous, lateral sepals spreading, 14–22 × 6–10 mm, spurs ± straight, orientation varies from 30° above to 30° below horizontal, 16–20 mm; lower petal blades slightly elevated, ± covering stamens, 6–9 mm, clefts 2–4 mm; hairs local, densest on inner lobes near base of cleft, white to light yellow. **Fruits** 12–17 mm, 2.5–3 times longer than wide, glabrous. **Seeds** ± wing-margined; seed coat cells with surfaces smooth.

Flowering summer. Steep rocky subalpine to alpine slopes; 1500–2800 m; B.C.; Oreg., Wash.

See discussion under *Delphinium bicolor.*

45. Delphinium basalticum M. J. Warnock, Phytologia 78: 91. 1995 · Columbia Gorge larkspur $\boxed{\text{C}}$ $\boxed{\text{E}}$

Stems 20–50(–65) cm; base often reddish, puberulent. **Leaves** basal and cauline; basal leaves 1–4 at anthesis; cauline leaves 2–5 at anthesis; petiole 0–15 cm. **Leaf blade** round, 2–6 × 5–9 cm, not succulent, nearly glabrous; ultimate lobes 5–19, width 3–15 mm (basal), 1–12 mm (cauline).

Inflorescences (2–)6–16(–26)-flowered; pedicels 2–7 cm, nearly glabrous; bracteoles 4–12 mm from flowers, green, linear, 3–7 mm, nearly glabrous. **Flowers:** sepals dark blue, nearly glabrous, lateral sepals spreading, 15–21 × 7–10 mm, spur straight to decurved, ascending 30–45° above horizontal, 14–18 mm; lower petal blades slightly elevated, ± exposing stamens, 7–9 mm, cleft 4–5 mm; hairs centered, mostly on inner lobes above base of cleft, yellow to white. **Fruits** 12–17 mm, 3.5–4 times longer than wide, glabrous. **Seeds** ± wing-margined; seed coat cells with surfaces smooth.

Flowering spring(–early summer). Basaltic cliff faces, n and e slopes at base of cliffs; of conservation concern; 200–500 m; Oreg., Wash.

Hybrids between *Delphinium basalticum* and *D. trolliifolium* are known.

46. Delphinium antoninum Eastwood, Leafl. W. Bot. 3: 126. 1942 · Anthony Peak larkspur $\boxed{\text{C}}$ $\boxed{\text{E}}$

Stems (7–)15–30(–60) cm; base reddish, glabrous to puberulent. **Leaves** mostly on proximal 1/3 of stem; basal leaves 3–12 at anthesis; cauline leaves 3–5 at anthesis; petiole 2–12 cm. **Leaf blade** round, 0.8–4 × 1.2–8 cm, ± succulent, nearly glabrous; ultimate lobes 3–15, width 1–8 mm (basal), 1–4 mm (cauline). **Inflorescences** 3–25-flowered; pedicels (0.6–)1.2–3.2 cm, usually puberulent; bracteoles 4–6 mm from flowers, green, linear, 3–5 mm, puberulent. **Flowers:** sepals dark blue to white, nearly glabrous, lateral sepals spreading to reflexed, 11–13 × 4–6 mm, spurs straight to gently decurved, within 30° above or below horizontal, 12–16 mm; lower petal blades elevated, exposing stamens, 3–7 mm, clefts 1.5–3 mm; hairs centered, mostly on inner lobes above base of cleft, white. **Fruits** 14–22 mm, 3.5–4 times longer than wide, puberulent. **Seeds** unwinged; seed coat cells with surfaces smooth.

Flowering late spring–early summer. Uncommon on moist talus slopes; of conservation concern; 1100–2700 m; Calif.

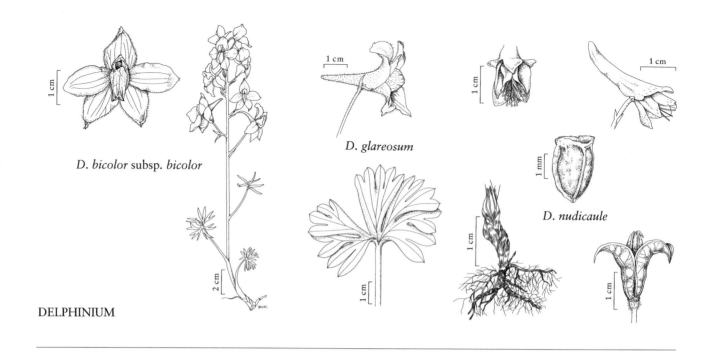

D. bicolor subsp. *bicolor*

D. glareosum

D. nudicaule

DELPHINIUM

Delphinium antoninum hybridizes with *D. decorum* subsp. *tracyi* and with *D. nudicaule*. Most often confused with *D. decorum* subsp. *tracyi*, it is separable on the basis of its longer root, usually reflexed sepals, and typically colorfast flowers. It also may be confused with *D. andersonii*; see discussion under that species.

47. Delphinium nudicaule Torrey & A. Gray, Fl. N. Amer. 1: 33. 1838 · Red larkspur, scarlet larkspur E F

Delphinium armeniacum A. Heller 1940, not *D. armeniacum* Stapf ex Huth 1895

Stems (15–)20–50(–125) cm; base reddish, glabrous. **Leaves** mostly on proximal 1/3 of stem; basal leaves 2–7 at anthesis; cauline leaves 3–4 at anthesis; petiole 0.5–14 cm. **Leaf blade** round to pentagonal, 2–6 × 3–10 cm; ultimate lobes 3–12, width 5–40 mm (basal), 2–20 mm (cauline). **Inflorescences** 5–20(–69)-flowered; pedicels (1.5–)2–6(–8) cm, glabrous to glandular-pubescent; bracteoles 14–20(–30) mm from flowers, green to red, linear, 2–4(–9) mm, glabrous to puberulent. **Flowers:** sepals scarlet to reddish orange, rarely dull yellow, glabrous, lateral sepals forward-pointing to form pseudotube, (6–)8–13(–16) × 3–6 mm, spurs straight, slightly ascending, (12–)18–27(–34) mm; lower petal blades elevated, exposing stamens, 2–3 mm, clefts 0.5–1 mm; hairs sparse, evenly dispersed, yellow. **Fruits** 13–26 mm, 3.5–4.5 times longer than wide, glabrous. **Seeds** unwinged or sometimes slightly wing-margined; seed coat cells with surfaces smooth. $2n = 16$.

Flowering late winter–early summer. Moist talus, cliff faces; 0–2600 m; Calif., Oreg.

Delphinium nudicaule hybridizes with most other taxa of *Delphinium* that it encounters. Apparent hybrids involving *D. nudicaule*, and seen by the author (either afield or as specimens), include *D. andersonii*, *D. antoninum*, *D. decorum*, *D. luteum*, *D. nuttallianum*, *D. patens*, and *D. trolliifolium*. In addition, garden-grown plants have been hybridized with *D. cardinale*, *D. elatum*, *D. menziesii*, *D. parishii*, *D. penardii*, *D. tatsienense* Franchet, *D. triste* Fischer ex de Candolle, and *D. uliginosum*; *D. nudicaule* does not naturally occur with these species. *Delphinium nudicaule* is one of the earliest larkspurs to flower in any given locality. Douglas's type collection of *D. nudicaule* represents plants (synonyms *D. sarcophyllum* Hooker & Arnott and *D. peltatum* Hooker, an invalid name) grown under very moist conditions, probably quite near the ocean. The type specimen of *D. armeniacum* A. Heller represents plants grown under unusually dry conditions.

The Mendocino Indians consider *Delphinium nudicaule* a narcotic (D. E. Moerman 1986).

48. **Delphinium luteum** A. Heller, Bull. S. Calif. Acad. Sci. 6: 68–69. 1903 · Yellow larkspur C E

Stems 20–40(–55) cm; base often reddish, nearly glabrous. **Leaves** mostly on proximal 1/4 of stem; basal leaves 4–9 at anthesis; cauline leaves 2–4 at anthesis; petiole 0.5–6 cm. **Leaf blade** round to pentagonal, 1–5 × 2–10 cm, nearly glabrous; ultimate lobes 3–5, width 8–30 mm (basal), 5–15 mm (cauline). **Inflorescences** 5–25(–37)-flowered; pedicels (1–)3–5(–7) cm, puberulent; bracteoles 6–10 (–17) mm from flowers, green, linear-lanceolate, 6–7 mm, nearly glabrous. **Flowers:** sepals bright yellow, puberulent, appearing waxy, lateral sepals ± forward pointing, (11–)14–16 × (6–)9–13 mm, spur straight, ca. 30° below horizontal, 11–20 mm; lower petal blades elevated, exposing stamens, 3–4 mm, clefts 0.5–1.5 mm; hairs sparse or absent, ± evenly distributed if present, white to yellow. **Fruits** 11–14 mm, 3.5–4.5 times longer than wide, glabrous. **Seeds** unwinged; seed coat cell surfaces smooth.

Flowering late winter–mid spring. Wet cliffs, coastal grassland or chaparral; of conservation concern; 0–50 m; Calif.

Delphinium luteum is presently known from only three populations. It is known to hybridize with *D. decorum* and with *D. nudicaule*. Populations of *D. hesperium* subsp. *hesperium* also occur at the type locality; *D. luteum* flowers earlier and hybrids are not known.

Delphinium luteum is not likely to be mistaken for any other species of *Delphinium*. It has been treated as a variety of *D. nudicaule* and is closely related to that species. Sepals of the infrequent yellow-flowered phase of *D. nudicaule*, however, have a much drabber appearance compared with the bright shining yellow of the sepals in *D. luteum*.

16b.9. **Delphinium** Linnaeus subsect. **Grumosa** (N. I. Malyutin) M. J. Warnock, Phytologia 78: 92. 1995

Delphinium sect. *Grumosa* N. I. Malyutin, Bot. Zhurn. (Leningrad & Moscow) 72: 689. 1987

Roots 1–5(–9)-branched from within 1 cm of stem attachment, 2–7(–16) cm, cormlike to fascicled or fibrous, ± fleshy, ± succulent, usually brittle; buds minute. **Stems** 1(–2) per root, usually unbranched, elongation delayed 2–10 weeks after leaf initiation; base narrowed, tenuously attached to root; proximal internodes much shorter than those of midstem. **Leaves** basal and cauline, largest leaves near base of stem, others usually gradually smaller on distal portion of stem; basal petioles spreading, cauline petioles ascending; blade shape and lobing similar throughout. **Inflorescences** usually with 2–6 flowers per 5 cm, open, ± pyramidal, seldom more than 3 times longer than wide, spurs rarely intersecting rachis; pedicel spreading, usually more than 1.5 cm, rachis to midpedicel angle more than 30°; bracts markedly smaller and fewer lobed than leaves. **Fruits** ± spreading. **Seeds** obpyramidal, 1.5–2.7 × 0.7–2 mm, ringed at proximal end, wing-margined or not; seed coats usually lacking wavy ridges, cells usually elongate (short and narrow in *D. alabamicum, D. newtonianum, D. nuttallianum,* and *D. treleasei*), smooth or roughened, cell margins straight.

Species 13 or more (13 in the flora): North America, possibly others in Asia.

Delphinium subsect. *Grumosa* is an extremely difficult complex, with many variations in a number of morphologic traits. Two divergent views of the complex were represented in the work of D. M. Sutherland (1967), in which the species recognized were large agglomerations, and that of P. A. Rydberg (1917), in which species rank was used for some edaphic variants. The complex has been and continues to be a major source of confusion for identification of *Delphinium* in North America. Most of the low larkspurs in poisonous-plant literature are members of this subsection.

Key to the species of *Delphinium* subsect. *Grumosa*

1. Sepals yellowish, white, or pink; if white then white in most members of the population.
  2. Flowers with no trace of pink or lavender.
    3. Spurs less than 12 mm; sepals spreading to erect . . . . 54. *Delphinium nuttallii*

3. Spurs more than 11 mm; sepals widely spreading.
    4. Stems less than 60 cm; widest leaf lobe less than 1 cm wide . . . . .52. *Delphinium menziesii*
    4. Stems more than 60 cm; widest leaf lobe more than 1 cm wide . . . .
    . . . .*Delphinium menziesii* × *D. trolliifolium* (= *D.* ×*pavonaceum*)
2. Flowers (especially sepal spur) with some element of pink or lavender.
    5. Stems less than 50(–70) cm; plants with sepal color similar to that of most other individuals
       in the population.
       6. Seeds with elongate blunt hairs on surface . . . . .55. *Delphinium tricorne*
       6. Seeds lacking elongate blunt hairs . . . . 61. *Delphinium nuttallianum*
    5. Stems more than (40–)50 cm; plants with sepal color dissimilar from that of most other indi-
       viduals in the population.
       7. Plants in populations with red flowers . . . .*Delphinium nudicaule* × *D. trolliifolium*
       7. Plants in populations with blue flowers . . . .*Delphinium trolliifolium* × *D. nudicaule*
1. Sepals bluish, maroon, or purple, not yellowish; if white then only as sporadic individuals in a population.
    8. Sepals maroon; plants usually distinctly different individuals within populations.
       9. Stems puberulent proximally . . . . *Delphinium nudicaule* × *D. decorum*
       9. Stems glabrous proximally. . . . .*Delphinium nudicaule* × *D. patens*
    8. Sepals bluish or purple; plants usually similar to other individuals in populations.
      10. Leaves mostly borne on distal 2/3 of stems at anthesis; flowers usually more than 15 per
         main inflorescence axis.
         11. Inflorescences as wide as long or nearly so . . . . .56. *Delphinium newtonianum*
         11. Inflorescences at least 2 times longer than wide.
            12. Sepals light blue. . . . .*Delphinium trolliifolium* × *D. nudicaule*
            12. Sepals dark blue to bluish purple.
               13. Lobes of midcauline leaves more than 6 mm wide; stems more than (40–)60 cm.
                  14. Leaf margins crenate . . . . .50. *Delphinium bakeri*
                  14. Leaf margins ± incised . . . . .49. *Delphinium trolliifolium*
               13. Lobes of midcauline leaves less than 6 mm wide; stems less than 60(–90) cm.
                  15. Sepals bluish purple (± drab) and often partly fading upon drying (espe-
                     cially veins, giving sepals of dried specimens mottled appearance) . . . . .54. *Delphinium nuttallii*
                  15. Sepals dark deep blue and retaining color upon drying.
                     16. Lower petals white, yellowish, or tan; lower petal blade clefts no more
                       than 1/3 blade length. . . . . 53. *Delphinium sutherlandii*
                   16. Lower petals blue to purple; lower petal blade clefts at least 1/3 blade
                     length. . . . . 61. *Delphinium nuttallianum*
      10. Leaves mostly borne on proximal 1/3 of stems; flowers often fewer than 20 per main inflo-
        rescence axis.
        17. Sepals dark blue-purple (± drab), often partly fading upon drying (especially veins, giv-
          ing sepals of dried specimens mottled appearance), distinctly puberulent externally, usu-
          ally not reflexed; lower stem pubescent.
          18. Seeds with swollen, blunt, hairlike protuberances; at least 2/3 of leaves in lower 1/4
             of stems; stems more than (45–)60 cm . . . . .58. *Delphinium alabamicum*
          18. Seeds without swollen, blunt, hairlike protuberances; at least 1/3 of leaves above
             lower 1/4 of stems; stems less than 60(–85) cm.
             19. Lower petal blade clefts at least 1/3 blade length. . . . . 51. *Delphinium decorum*
             19. Lower petal blade clefts no more than 1/4 blade length . . . . .52. *Delphinium menziesii*
        17. Sepals bright blue or purple (not drab blue-purple), usually retaining color upon drying,
          usually glabrous, often reflexed; if sepals drab blue-purple, puberulent, and not reflexed,
          then proximal portion of stems nearly glabrous to glabrous.
          20. Lateral sepals strongly reflexed; leaf blade with usually 5 or fewer lobes extending
             3/5 distance to petiole (if more than 5, then pedicel puberulent), lobes often more
             than 7 mm wide.
             21. Pedicels spreading from rachis at nearly 90°; terminal leaf lobes distinctly
                wedge-shaped, widest in distal 1/2. . . . . .59. *Delphinium gracilentum*
             21. Pedicels spreading from rachis at usually less than 70°; terminal leaf lobes sel-
                dom wedge-shaped, widest near midpoint. . . . . .60. *Delphinium patens*
          20. Lateral sepals not reflexed or only weakly so; leaf blade with more than 5 lobes
             extending more than 3/5 distance to petiole, lobes less than 7 mm wide.
             22. Cauline leaves 2 or fewer, less than 1/2 size of basal leaves; stems glaucous
               . . . . .57. *Delphinium treleasei*

22. Cauline leaves 3 or more, similar in size to basal leaves; stems not glaucous.
    23. Inflorescences at least 3 times longer than wide; lower petals tan or yellow-ish, at least 8 mm .... 53. *Delphinium sutherlandii*
    23. Inflorescences less than 3 times longer than wide; lower petals blue (except sometimes in white-flowered plants), 3–8(–11) mm.
        24. Seeds with swollen, blunt, hairlike protuberances on surface ...... 55. *Delphinium tricorne*
        24. Seeds without swollen, blunt, hairlike protuberances .... 61. *Delphinium nuttallianum*

49. Delphinium trolliifolium A. Gray, Proc. Amer. Acad. Arts 8: 375. 1872 · Cow-poison, poison larkspur E

Stems (40–)60–120(–180) cm; base usually reddish, glabrous to puberulent. **Leaves** mostly cauline at anthesis; basal leaves 0 (–3) at anthesis; cauline leaves 5–12 at anthesis; petiole 0.5–17 cm. **Leaf blade** ± pentagonal, 4–8 × 7–16 cm, margins ± incised, nearly glabrous; ultimate lobes 0–9, width 15–30 mm (basal), 5–20 mm (cauline). **Inflorescences** (5–)14–40(–75)-flowered, ± open, pedicels 1–4(–9) cm, puberulent to glabrous; bracteoles (2–)6–12 mm from flowers, green, linear, 5–9(–14) mm, puberulent. **Flowers:** sepals dark blue, glabrous, lateral sepals spreading, (8–)14–21 × 5–9 mm, spurs straight or downcurved at apex, within 20° of horizontal, (10–)16–23 mm; lower petal blades covering stamens, 5–10 mm, clefts 1.5–3 mm; hairs sparse, mostly near junction of blade and claw, centered or on inner lobes, well dispersed, yellow. **Fruits** (15–)23–34 mm, 3.8–5.5 times longer than wide, glabrous. **Seeds** unwinged; seed coats smooth. $2n = 16$.

Flowering spring. Oak woods, coastal chaparral, wet woodlands; 30–1100 m; Calif., Oreg.

Delphinium trolliifolium occurs in the northern Coast Range of California, the Columbia River Valley to just east of Mt. Hood, and the Willamette Valley of Oregon upstream to Lane County. California plants differ somewhat from Oregon plants in pubescence patterns and habitat preferences. Further study may show that two entities are involved.

Hybrids between *Delphinium trolliifolium* and *D. decorum, D. menziesii* subsp. *pallidum* (*D.* ×*pavonaceum* Ewan, Peacock larkspur), *D. menziesii* subsp. *menziesii, D. nudicaule, D. nuttallianum,* and *D. nuttallii* are known. *Delphinium trolliifolium* is likely to be confused only with *D. bakeri.* Refer to discussion under that species for differences.

50. Delphinium bakeri Ewan, Bull. Torrey Bot. Club 69: 144. 1942 · Baker's larkspur C E

Stems (45–)60–85(–100) cm; base reddish, glabrous. **Leaves** usually all cauline at anthesis; basal leaves 0–2 at anthesis; cauline leaves 3–8 at anthesis; petiole 0.4–18 cm. **Leaf blade** pentagonal to round, 1–6 × 1.5–8 cm, margins crenate, glabrous; ultimate lobes 3–5, width 15–30 mm (basal), 5–30 mm (cauline). **Inflorescences** 8–23-flowered, pedicels 1–6(–9) cm, glabrous to glandular-puberulent; bracteoles 4–6 mm from flowers, green to blue, lance-linear, 5–8(–13) mm, glabrous to glandular-puberulent. **Flowers:** sepals dark bluish purple, nearly glabrous, lateral sepals spreading, 9–11 × 4–5 mm, spur apex decurved, ± horizontal, 9–13 mm; lower petal blades ± covering stamens, 5–7 mm, clefts 2–3 mm; hairs sparse, mostly near base of cleft, centered or on inner lobes, white. **Fruits** 18–20 mm, 3.5–4 times longer than wide, glabrous. **Seeds** unwinged; seed coats smooth.

Flowering spring. Brushlands and coastal chaparral; of conservation concern; 100–300 m; Calif.

Delphinium bakeri is possibly extinct in the wild because of cultivation and sheep grazing in the small area where it grows. It is known from only two localities and has not been collected since 1960. Plants have been grown at Strybing Arboretum, Golden Gate Park, San Francisco. Although their geographic ranges are distinct, *D. bakeri* is most similar to, and probably closely related to, *D. trolliifolium.* The former has more rounded incisions on the leaves than the latter, and the pedicel of *D. bakeri* are consistently glandular. Glandular pedicels appear only occasionally in *D. trolliifolium.*

51. Delphinium decorum Fischer & C. A. Meyer in F. E. L. von Fischer et al., Ind. Sem. Hort. Petrop. 3: 650. 1837 E

Stems 5–25(–42) cm; base reddish or not, at least two proximal internodes long-pubescent. **Leaves** mostly on proximal 1/3 of stem; basal leaves 1–6 at anthesis; cauline leaves 2–4 at anthesis; petiole 0.5–7 cm. **Leaf blade** deltate to pentagonal, 1–2.5 × 1.5–4.5 cm, ± puberulent; ultimate lobes 3–15, width 3–17 mm (basal), 1–8 mm (cauline). **Inflorescences** 2–10(–20)-flowered; pedicels 1–4(–6) cm, puberulent to glabrous; bracteoles (3–) 7–11(–21) mm from flowers, green to blue, linear-lanceolate, 6–9(–12) mm, puberulent. **Flowers:** sepals bluish purple (somewhat faded on drying), puberulent, lateral sepals spreading, (11–)15–20(–24) × 6–11(–15) mm, spur straight, 0–30° above horizontal, 13–20 mm; lower petal blades ± covering stamens, 6–11 mm, cleft 2–4 mm; hairs on entire surface, densest on inner lobes, yellow or white. **Fruits** 9–20 mm, 3–4.5 times longer than wide, glabrous to puberulent. **Seeds** wing-margined; seed coat cells ± aggregate in small wavy ridges, cell surfaces ± roughened, without swollen, blunt hairs.

Subspecies 2 (2 in the flora): California.

Delphinium decorum and *D. menziesii* have been confused. *Delphinium menziesii* usually has darker, colorfast (nonfading) sepals with less lavender, and a more northern distribution than *D. decorum*. It also has more finely dissected leaves than *D. decorum* subsp. *decorum*.

Often confused with *Delphinium patens*, *D. decorum* is sometimes circumscribed to include that species. The spreading, fading, blue-purple sepals, pubescent proximal portion of stems, and large lower petal blade of *D. decorum*, compared to the reflexed, colorfast, bluer sepals, proximally glabrous stems, and smaller lower petal blades of *D. patens*, adequately distinguish the two taxa.

1. Leaf blade usually with 3–5 lobes longer than 1/2 leaf radius; 0–100 m . . . .
. . . . .51a. *Delphinium decorum* subsp. *decorum*
1. Leaf blade usually with more than 5 lobes longer than 1/2 leaf radius; 700–2300 m . . . .
. . . . .51b. *Delphinium decorum* subsp. *tracyi*

51a. Delphinium decorum Fischer & C. A. Meyer subsp. **decorum** · Coast larkspur [E]

Stems (8–)15–25(–35) cm. **Leaves:** ultimate lobes 3–5, longer than 1/2 leaf radius, merely cleft near apices. **Inflorescences** puberulent, hairs glandular or not. **Flowers:** lateral sepals 12–24 mm, spurs 13–19 mm; lower petal hairs yellow.

Flowering early–mid spring. Grasslands, open coastal chaparral; 0–100 m; Calif.

The names *Delphinium decorum* var. *racemosum* Eastwood and *D. decorum* var. *sonomensis* Eastwood have been used for various hybrids between *D. decorum* subsp. *decorum* and *D. patens*. The type specimen of *D. decorum* var. *sonomensis* appears to represent a nearly direct intermediate between *D. decorum* and *D. patens*. These hybrids are quite common in the San Francisco Bay region where habitats have been dis-turbed. Normally a woodland plant, *D. patens* contrasts with *D. decorum*, which occurs in grassland and brushland. *Delphinium decorum* subsp. *decorum* also hybridizes with *D. luteum*, *D. nudicaule*, and *D. trollii-folium*.

51b. Delphinium decorum Fischer & C. A. Meyer subsp. **tracyi** Ewan, Univ. Colorado Stud., Ser. D, Phys. Sci. 2: 100. 1945 · Tracy's larkspur [E]

Stems 7–20(–45) cm. **Leaves:** ultimate lobes 5–15, usually more than 5 lobes longer than 1/2 leaf radius, apical lobes pronounced. **Inflorescences** glabrous or nearly so. **Flowers:** lateral sepals 11–18 mm, spurs 13–20 mm; lower petal hairs white. $2n = 16$.

Flowering mid–late spring. Meadows in montane forests; 700–2300 m; Calif.

Although *Delphinium decorum* subsp. *tracyi* probably occurs in southern Oregon, no specimens have been seen from that state. Habitat appears to be the main isolating mechanism between *D. decorum* subsp. *tracyi* and *D. antoninum*; hybrids do occur. The subspecies also hybridizes with *D. nudicaule* and *D. trolliifolium*.

Delphinium decorum subsp. *tracyi* may be confused with *D. antoninum* or *D. nuttallianum*. For distinctions from the former, see discussion under that species. Distinctions between *D. decorum* subsp. *tracyi* and *D. nuttallianum* may be made on basis of spreading, fading, bluish purple sepals of the former as opposed to reflexed, colorfast, dark blue to white sepals of the latter. Pubescence of the proximal portion of stems and larger lower petal blades of subsp. *tracyi* are also useful to distinguish from usually glabrous stems and smaller lower petal blades of *D. nuttallianum*.

52. Delphinium menziesii de Candolle, Syst. Nat. 1: 355. 1817 · Menzies' larkspur [E] [W2]

Stems (10–)35–70(–85) cm; base often reddish, puberulent. **Leaves** basal and cauline; basal leaves 0–4 at anthesis; cauline leaves 3–7(–10) at anthesis; petiole 0.5–11 cm. **Leaf blade** round, 1.5–5 × 3–9 cm, puberulent; ultimate lobes 5–18, width 2–15 mm (basal), 1–10 mm (cauline). **Inflorescences** 3–15(–43)-flowered; pedicels 1.5–4(–7) cm, (glandular) puberulent; bracteoles 8–10(–24) mm from flowers, green to blue, linear, 4–6 (–9) mm, puberulent. **Flowers:** sepals bluish purple or yellowish, often partly fading upon drying, puberulent, lateral sepals spreading, (11–)13–20 × 5–11 mm, spurs straight, ascending less than 30° above horizontal, 11–17 mm; lower petal blades ± covering stamens, 8–12 mm, clefts 0.2–2.5 mm; hairs sparse, centered, mostly near junction of blade and claw above

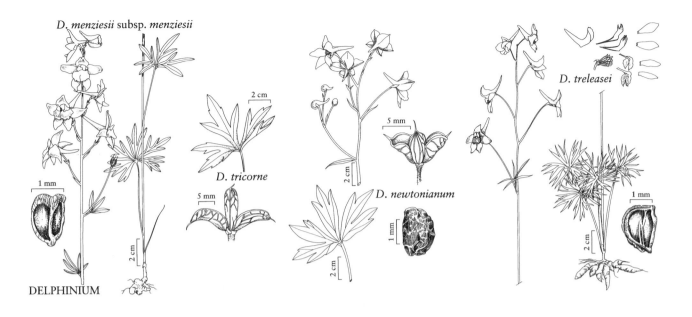

D. menziesii subsp. *menziesii*

D. tricorne

D. newtonianum

D. treleasei

DELPHINIUM

base of cleft, white or blue. **Fruits** 11–17 mm, 3.5–4 times longer than wide, puberulent. **Seeds** wing-margined; seed coat cell surfaces smooth, without swollen blunt hairs.

Subspecies 2 (2 in the flora): w North America.

Although *Delphinium menziesii* has often been confused with *D. nuttallii,* it may be distinguished by its consistently larger flowers and usually fewer flowers per plant. Interestingly, each species produces both blue-purple and yellowish flower colors in separate populations.

The Chehalis consider *Delphinium menziesii* poisonous, but they also apply it to sores. The women of the Thompson Indians use it as a love charm (D. E. Moerman 1986, subspecies not indicated).

1. Sepals blue to purple . . . .
. . . . 52a. *Delphinium menziesii* subsp. *menziesii*
1. Sepals yellowish to white . . . .
. . . . 52b. *Delphinium menziesii* subsp. *pallidum*

52a. Delphinium menziesii de Candolle subsp. **menziesii**
· Menzies' larkspur E F

Delphinium menziesii var. *pyramidalis* (Ewan) C. L. Hitchcock; *D. oreganum* Howell

Stems 35–70 cm. **Flowers:** sepals blue to purple, 13–20 × 5–11 mm; spurs 11–17 mm. $2n = 16$.

Flowering spring. Meadows, open woodlands; 0–1000 m; B.C.; Oreg., Wash.

Delphinium menziesii subsp. *menziesii* hybridizes with *D. trolliifolium* and *D. nuttallii.*

52b. Delphinium menziesii de Candolle subsp. **pallidum** M. J. Warnock, Phytologia 78: 96. 1995 · White-flowered Menzies' larkspur E

Stems 50–70 cm. **Flowers:** sepals white to yellowish, 15–20 × 7–11 mm; spurs 11–15 mm. $2n = 16.$

Flowering spring. Meadows, open woodlands; 50–100 m; Oreg.

Hybrids between *Delphinium menziesii* subsp. *pallidum* and *D. trolliifolium* have been named *D.* ×*pavonaceum* Ewan.

53. Delphinium sutherlandii M. J. Warnock, Phytologia 78: 97. 1995 · Sutherland's larkspur E

Stems (15–)30–70 cm; base often reddish, glabrous. **Leaves** basal and cauline; basal leaves 0–2 at anthesis; cauline leaves 3–7 at anthesis; petiole 0.4–14 cm. **Leaf blade** round, 1.5–10 × 3–18 cm, nearly glabrous; ultimate lobes 5–21, 5 or more extending more than 3/5 distance to petiole, width 3–9(–15) mm (basal), 0.5–6(–10) mm (cauline). **Inflorescences** (2–)11–26(–37)-flowered; pedicels 1–3 cm, puberulent; bracteoles 2–7 mm from flowers, green to blue, linear-lanceolate, 3–5 mm, puberulent. **Flowers:** sepals dark blue, retaining color upon drying, puberulent, lateral sepals spreading, 14–20 × 5–10 mm, spurs gently decurved, ascending 0–20° above horizontal, 14–18 mm; lower petal blades slightly elevated, ± covering stamens, white, yellowish, or tan, 8–12 mm, clefts 2–4 mm; hairs short, sparse, mostly below junction of blade and claw, slightly offset to inner lobes, white. **Fruits** (13–)18–25 mm, 4.5–5.2 times longer than wide, puberulent. **Seeds** winged on 1 margin; seed coat cell surfaces roughened. $2n = 16$.

Flowering spring. Dry meadows, rock outcrops, open conifer woods; 400–800 m; B.C.; Idaho, Mont., Wash.

54. Delphinium nuttallii A. Gray, Bot. Gaz. 12: 54. 1887 E

Stems 20–60(–90) cm; base usually reddish, pubescent. **Leaves** mostly on proximal 1/2 of stem; basal leaves 0–2 at anthesis; cauline leaves 3–10 at anthesis; petiole 0.3–19 cm. **Leaf blade** round to reniform, 2–8 × 3–14 cm, pubescent; ultimate lobes 5–18, width 4–7 mm (basal), 1–5 mm (cauline). **Inflorescences** 5–25(–40)-flowered; pedicels 1.5–4(–9) cm, puberulent; bracteoles 2–3 mm from flowers, green, linear, 4–6 mm, puberulent. **Flowers:** sepals bluish purple (± drab), to yellowish, often partly fading upon drying, puberulent, lateral sepals ± spreading, 8–11 × 3–6 mm, spurs straight, slightly ascending, 9–13 mm; lower petal blades ± covering stamens, 4–6 mm, clefts 0.5–2 mm; hairs well dispersed, mostly near margins and base of cleft, white to yellow or blue. **Fruits** 10–14(–18) mm, 3.5–4 times longer than wide, pubescent. **Seeds** wing-margined; seed coat cell surfaces smooth.

Subspecies 2 (2 in the flora): Oreg., Wash.

Delphinium nuttallii has often been confused with *D. menziesii*; it may be distinguished by consistently smaller flowers and usually more flowers per plant than in the latter. Interestingly, each species produces both blue-purple and yellowish flower colors, in separate populations.

1. Sepals blue or bluish purple; spur 10 mm or more . . . . 54a. *Delphinium nuttallii* subsp. *nuttallii*
1. Sepals white or light yellow; spur 11 mm or less . . . . 54b. *Delphinium nuttallii* subsp. *ochroleucum*

54a. Delphinium nuttallii A. Gray subsp. nuttallii
· Nuttall's larkspur E

Stems 40–70 cm. **Flowers:** sepals blue or bluish purple, spurs 10–13 mm; lower petal blades 4.5–6.5 mm. $2n = 16$.

Flowering late spring–early summer. Rock outcrops, rocky meadows; 20–300 m; Oreg., Wash.

54b. Delphinium nuttallii A. Gray subsp. ochroleucum
(Nuttall) M. J. Warnock, Phytologia 78: 98. 1995 · Light-yellow-flowered larkspur C E

Delphinium menziesii de Candolle var. (β) *ochroleucum* Nuttall in J. Torrey and A. Gray, Fl. N. Amer. 1: 31. 1838; *D. leucophaeum* Greene

Stems 30–60 cm. **Flowers:** sepals white or light yellow, spurs 9–11 mm; lower petal blades 4–6 mm. $2n = 16$.

Flowering late spring. Rock outcrops, rocky meadows; of conservation concern; 50–100 m; Oreg.

The range of morphologic features of *Delphinium nuttallii* subsp. *ochroleucum* (*D. leucophaeum*) is almost completely encompassed within that of *D. nuttallii* subsp. *nuttallii*. Sepal color is the only feature consistently separating the two subspecies. Were it not for the fact that any given population typically has plants of only one flower color, a rank of forma would be more appropriate.

55. Delphinium tricorne Michaux, Fl. Bor.-Amer., 314–315. 1803 · Dwarf larkspur E F

Stems 20–60 cm; base often reddish, nearly glabrous. **Leaves** mostly on proximal 1/3 of stem; basal leaves 0–4 at anthesis; cauline leaves 2–8 at anthesis; petiole 4–12 cm. **Leaf blade** round, 2–8 × 4–12 cm, nearly glabrous; ultimate lobes 3–18, 5 or more extending more than 3/5 distance to petiole, width 2–10 mm (basal), 4–10 mm (cauline). **Inflorescences** 5–15(–30)-flowered, less than 3 times longer than wide; pedicel 1–2.5 cm, puberulent; bracteoles 1–4(–6) mm from flowers, green, linear, 3–5 mm, puberulent. **Flowers:** sepals deep bluish purple to pink or white, puberulent, lateral sepals spreading, 11–19 × 4–7 mm, spurs straight, within 30° of horizontal, 13–16 mm; lower petal blades ± covering stamens, blue, except sometimes in white-flowered plants, 6–10 mm, clefts 0.5–2 mm; hairs sparse, mostly centered near junction of blade and claw, white. **Fruits** 14–22 mm, 4–4.5 times longer than wide, nearly glabrous. **Seeds** unwinged; surface of each seed coat cell with 1–5 small, swollen, elongate, blunt, hairlike structures, barely visible at 20×, otherwise smooth. $2n = 16$.

Flowering spring. Slopes in deciduous forests, thicket edges, moist prairies; 10–1500 m; Ala., Ark., D.C., Ga., Ill., Ind., Iowa, Kans., Ky., Md., Miss., Mo., Nebr., N.C., Ohio, Okla., Pa., S.C., Tenn., Va., W.Va.

Delphinium tricorne is the most commonly encountered larkspur east of the Great Plains.

The Cherokee prepared infusions of *Delphinium tricorne* to ingest for heart problems, although they believed the roots of the plant made cows drunk and killed them (D. E. Moerman 1986).

56. Delphinium newtonianum Dw. Moore, Rhodora 41: 196–197, plates 548, 549. 1939 · Ozark larkspur E F

Stems 40–90 cm; base often reddish, puberulent. **Leaves** mostly cauline; basal leaves usually absent at anthesis; cauline leaves 2–8 at anthesis; petiole 0.5–10 cm. **Leaf blade** round to pentagonal, 4–7 × 5–15 cm, nearly glabrous; ultimate lobes 3–7, width 8–20 mm (basal), 5–15 mm (cauline). **Inflorescences** 8–20(–40)-flowered, as wide as long or nearly so; pedicels 1–4(–6) cm, pubescent; bracteoles 6–15 mm from flowers, green, linear, 1.5–5 mm, puberulent. **Flowers:** sepals dark to light blue, rarely white, glabrous, lateral sepals spreading, 12–14 × 6–7 mm, spurs straight to decurved, within 30° of horizontal, 10–15 mm; lower petal blades slightly elevated, ± exposing stamens, 4–5 mm, clefts 2–3 mm; hairs mostly centered near base of cleft, yellow. **Fruits** 8–12 mm, 3–3.5 times longer than wide, nearly glabrous. **Seeds** unwinged; surface of each seed coat cell with swollen, blunt, hairlike structures, barely visible at 20×, otherwise smooth.

Flowering early summer. Slopes in deciduous forest; 500–700 m; Ark.

No cases of hybridization are known. *Delphinium newtonianum* often occurs in mixed populations with *D. tricorne*. It normally does not begin flowering until 4–6 weeks after *D. tricorne* has finished.

57. Delphinium treleasei Bush ex K. C. Davis, Minnesota Bot. Stud. 2: 444. 1900 · Trelease's larkspur E F

Stems (40–)60–80(–110) cm; base reddish, glabrous, glaucous. **Leaves** mostly basal; basal leaves 3–7 at anthesis; cauline leaves 1–3 at anthesis; petiole 0.5–19 cm. **Leaf blade** round to reniform, 0.5–8 × 2–20 cm, glabrous; ultimate lobes 3–35(–56), 5 or more extending more than 3/5 distance to petiole, width 2–6 mm (basal), 0.5–2 mm (cauline). **Inflorescences** 5–30(–56)-flowered; pedicels (1–)2.5–7 (–11) cm, glabrous; bracteoles 1–7(–46) mm from flowers, blue or green, awl-shaped, 1.5–4(–12) mm, glabrous. **Flowers:** sepals dark bright blue, usually retaining color upon drying, nearly glabrous, lateral sepals spreading, (10–)12–17(–20) × 4–8(–10) mm,

spurs straight, ascending ca. 45°, 15–17 mm; lower petal blades ± covering stamens, 5–8 mm, clefts (1–) 3–5 mm; hairs centered, densest above base of cleft, yellow. **Fruits** 12–18(–22) mm, 3–4 times longer than wide, glabrous. **Seeds** unwinged; seed coat cell surfaces smooth. $2n = 16$.

Flowering late spring. Open juniper glades on calcareous substrate; 250–450 m; Ark., Mo.

Delphinium treleasei is locally abundant but extremely limited by microhabitat within its distributional range.

Hybrids between *Delphinium treleasei* and *D. carolinianum* subsp. *carolinianum* are known.

58. Delphinium alabamicum Kral, Sida 6: 250. 1976 · Alabama larkspur C E

Stems (45–)60–90(–130) cm; base reddish, pubescent. **Leaves** mostly basal; basal leaves 2–6 at anthesis; cauline leaves 1–4 at anthesis; petiole 0.5–22 cm. **Leaf blade** reniform to semicircular, 2–11 × 3–19 cm, nearly glabrous; ultimate lobes 3–35, width 3–8 mm (basal), 0.5–3 mm (cauline). **Inflorescences** 5–27-flowered; pedicels 1–5 cm, puberulent; bracteoles 3–12 mm from flowers, green, linear, 2–8 mm, puberulent. **Flowers:** sepals royal blue, ± drab, often partly fading upon drying, puberulent, lateral sepals spreading, 12–20 × 5–10 mm, spurs straight, horizontal to slightly ascending, (13–)15–16.5(–19) mm; lower petal blades ± covering reproductive parts, 4–10 mm, clefts 3–5 mm; hairs centered between claw and base of cleft, covering most of adaxial surface, white. **Fruits** 11–18 mm, 2.5–3 times longer than wide, sparsely puberulent. **Seeds** unwinged; surface of each seed coat cell with swollen, blunt, hairlike structures, barely visible at 20×, otherwise smooth. $2n = 16$.

Flowering mid–late spring. Very local; thin soils in and on edges of *Juniperus* glades on limestone substrates; of conservation concern; 100–300 m; Ala., Ga.

In addition to the Alabama sites, *Delphinium alabamicum* is known from one population in Georgia; it might have been transplanted there. *Delphinium alabamicum* has not been seen in central Alabama since 1950; populations there might have been extirpated.

59. Delphinium gracilentum Greene, Pittonia 3: 15. 1896 · Greene's larkspur E

Delphinium patens Bentham subsp. *greenei* (Eastwood) Ewan; *D. pratense* Eastwood

Stems (15–)30–50(–80) cm; base reddish, nearly glabrous. **Leaves** mostly cauline; basal leaves 0–3 at anthesis; cauline leaves 2–5 at anthesis; petiole 3–15 cm. **Leaf blade** round to pentagonal, 1.5–4 × 3–7 cm, nearly glabrous; ultimate lobes 3–7, distinctly wedge-shaped, usually 5 or fewer extending 3/5 distance to petiole, width 5–20 mm (basal), 1–15 mm (cauline), widest in distal 1/2. **Inflorescences** 5–20 (–38)-flowered; pedicels spreading from rachis at nearly 90°, 1–3(–4) cm, glabrous or glandular-pubescent; bracteoles (7–)11–19 mm from flowers, blue or green, linear, 2–5 mm, puberulent to glabrous. **Flowers:** sepals dark bluish purple to pink or white, usually retaining color upon drying, glabrous, lateral sepals reflexed, 6–10(–13) × 3–6 mm, spurs often curved upward, within 30° above or below horizontal, 8–12(–14) mm; lower petal blades elevated, exposing stamens, 3–5 mm, clefts 1–3 mm; hairs almost exclusively near base of cleft, centered or mostly on inner lobes, usually yellow. **Fruits** 8–16 mm, 3–3.5 times longer than wide, glabrous to glandular-puberulent. **Seeds** unwinged; seed coats ± pitted, cell surfaces roughened.

Flowering spring–early summer. Open coniferous forest; 150–2700 m; Calif.

Delphinium gracilentum hybridizes with *D. patens* subsp. *patens* in the northern Sierra Nevada foothills and is very similar to that species, making hybrids difficult to discern. While *D. gracilentum* and *D. patens* are easily distinguished in most of their ranges, morphologic distinctions between the two taxa are blurred in the northern Sierra Nevada foothills region, particularly in Butte County, California. Coniferous woods are preferred by *D. gracilentum*; *D. patens* subspp. *patens* and *hepaticoideum* are more often found in broadleaf woods. The former species has more widely spreading pedicels than the latter, and *D. gracilentum* usually has wider leaf lobes than *D. patens* subsp. *patens*. In the southern Sierra Nevada, *D. gracilentum* may come in contact with *D. patens* subsp. *montanum*. Though hybrids are not common, some gene flow has apparently occurred.

Sepal color phases are not stable and considerable variation occurs within populations. The type specimen of *Delphinium gracilentum* represents the northern, lower-elevation, nonglandular, dark-flowered phase. The type specimen of *D. gracilentum* forma *versicolor* Ewan differs only by its pink or white

flowers. A limited range of intermediate colors occurs, and populations may be made up of plants of a single color or several different colors. The type specimen of *D. greenei* Eastwood represents the southern, higher elevation, glandular (at least on pedicels) expression. The type specimen includes representatives of dark- and light-flowered individuals of this phase. The type specimen of *D. gracilentum* forma *versicolor* (not seen by the author) is the "albino" phase referred to by Greene in his description of *D. gracilentum*. Several of the paratypes cited by Ewan have been seen, as have a number of individuals in natural populations.

Delphinium gracilentum has been confused with *D. patens* or *D. nuttallianum*. *Delphinium gracilentum* may be distinguished from *D. nuttallianum* by its wider leaf lobes, smaller fruits, and more elongate inflorescences, and from *D. patens* by its wider leaf lobes, more open inflorescences, and usually shorter fruits.

60. Delphinium patens Bentham, Pl. Hartw., 296. 1849 E

Stems (10–)20–50(–90) cm; base reddish, glabrous to puberulent. **Leaves** mostly on proximal 1/3 of stem; basal leaves 1–3 at anthesis; cauline leaves 2–4 at anthesis; petiole 1.5–12 cm. **Leaf blade** round to pentagonal, 1–5.5 × 2–7.5 cm, nearly glabrous; ultimate lobes 3–9, less often wedge-shaped, 5 or more extending more than 3/5 distance to petiole, width 5–30(–50) mm (basal), 1–15 mm (cauline), widest at middle or in proximal 1/2. **Inflorescences** 4–25(–36)-flowered; pedicels spreading from rachis at usually less than 70°, 1–4(–8) cm, glabrous to glandular; bracteoles 3–12(–23) mm from flowers, blue to green, linear, 3–8(–16) mm, glabrous to glandular. **Flowers:** sepals dark blue, usually retaining color upon drying, glabrous, lateral sepals reflexed, (6–)10–15(–20) × 4–8 mm, spurs straight, ascending ca. 30° above horizontal, 8–18 mm; lower petal blades elevated, exposing stamens, 3–6(–8) mm, clefts 1–3 mm; hairs centered on base of cleft or on inner lobes, scattered, white, rarely yellow. **Fruits** 12–23 mm, 3.3–3.6 times longer than wide, glabrous or puberulent. **Seeds** unwinged; seed coats ± pitted, cell surfaces roughened.

Subspecies 3 (3 in the flora): Calif.

1. Lobes of proximal cauline leaves usually more than 15 mm wide; basal and proximal cauline leaves rarely cleft more than 4/5 radius of blade . . . . 60b. *Delphinium patens* subsp. *hepaticoideum*
1. Lobes of proximal cauline leaves usually less than 15 mm wide; basal and proximal cauline leaves usually cleft more than 4/5 radius of blade
  2. Pedicels puberulent, usually glandular; most

leaf blades with more than 7 ultimate lobes
. . . . 60c. *Delphinium patens* subsp. *montanum*
2. Pedicels usually glabrous; most leaf blades
with fewer than 7 ultimate lobes . . . .
. . . . 60a. *Delphinium patens* subsp. *patens*

60a. Delphinium patens Bentham subsp. **patens** 〔E〕

Delphinium decorum Fischer & C. A. Meyer var. *patens* (Bentham) A. Gray

Leaves: basal leaves usually present at anthesis; basal and proximal cauline leaf blades usually cleft more than 4/5 radius of blade. **Leaf blade:** ultimate lobes 3–5, width usually less than 15 mm. **Inflorescences:** pedicel usually glabrous. **Flowers:** lateral sepals 9–20 mm, spurs 10–15 mm; lower petal blades 4–6 mm; hairs asymmetrically distributed. $2n = 16$.

Flowering spring. Grasslands, open woods; 80–1100 m; Calif.

Delphinium patens subsp. *patens* hybridizes with *D. decorum* (see discussion under *D. decorum*), *D. nudicaule*, and *D. variegatum*. Plants of *D. patens* subsp. *patens* have been confused with *D. decorum, D. gracilentum,* and *D. nuttallianum.* See discussions under those species for distinguishing features.

60b. Delphinium patens Bentham subsp. **hepaticoideum** Ewan, Univ. Colorado Stud., Ser. D, Phys. Sci. 2: 103. 1945 〔E〕

Leaves: basal leaves usually present at anthesis; basal and proximal cauline leaves rarely cleft more than 4/5 radius of blade. **Leaf blade:** ultimate lobes 3–5, width more than 15 mm. **Inflorescences:** pedicels usually glabrous. **Flowers:** lateral sepals 11–17 mm, spur 10–18 mm; lower petal blades 5–8 mm; hairs symmetrically distributed. $2n = 16$.

Flowering spring. Wooded ravines, near streams; 300–1300 m; Calif.

Although *Delphinium patens* has not been reported outside California, subspp. *hepaticoideum* and/or *montanum* may grow in mountains of northern Baja California, Mexico.

Although it hybridizes with *Delphinium parryi* and *D. umbraculorum, D. patens* subsp. *hepaticoideum* usually flowers early enough not to overlap in any given site with flowering time of those species.

Likely to be confused only with *Delphinium umbraculorum, D. patens* subsp. *hepaticoideum* may be distinguished from that species by its ringed, unwinged

seeds, recurved fruits, and lack of arched hairs. Contrasting features of *D. umbraculorum* are presence of unringed, winged seeds, erect fruits, and arched hairs.

60c. Delphinium patens Bentham subsp. **montanum** (Munz) Ewan, Bull. Torrey Bot. Club 69: 147. 1942 〔E〕

Delphinium parryi A. Gray var. *montanum* Munz, Bull. S. California Acad. Sci. 31: 61. 1932

Leaves: basal leaves usually absent at anthesis; basal and proximal cauline leaves usually cleft more than 4/5 radius of blade. **Leaf blade:** ultimate lobes 5–10, width less than 10 mm. **Inflorescences:** pedicels generally glandular-puberulent. **Flowers:** lateral sepals 7–11 mm, spurs 8–14 mm; lower petal blades 3–6 mm; hairs symmetrically distributed.

Flowering late spring–early summer. Open coniferous forest; 1500–2800 m; Calif.

Delphinium patens subsp. *montanum* is usually found on drier, leeward sides of mountain ranges; subsp. *hepaticoideum* is found on wetter, windward sides of many of the same mountain ranges.

Delphinium patens subsp. *montanum* is likely to be confused with *D. gracilentum* or *D. nuttallianum;* see discussion under those species for distinguishing features.

Hybrids with *D. umbraculorum* are known.

61. Delphinium nuttallianum Pritzel in W. G. Walpers, Repert. Bot. Syst. 1: 744. 1842 〔E〕〔W2〕

Delphinium pauciflorum Nuttall in J. Torrey and A. Gray, Fl. N. Amer. 1: 33. 1838, not *D. pauciflorum* D. Don 1825, not *D. pauciflorum* Reichenbach ex Chamisso 1831; *D. nuttallianum* var. *fulvum* C. L. Hitchcock; *D. nuttallianum* var. *levicaule* C. L. Hitchcock; *D. sonnei* Greene

Stems unbranched, 10–40(–70) cm; base reddish, pubescence variable. **Leaves** mostly on proximal 1/4 of stem; basal leaves 2–6 at anthesis; cauline leaves 2–10 at anthesis; petiole 0.4–12 cm. **Leaf blade** round, 1–6 × 2–12 cm, nearly glabrous; ultimate lobes 5–21, 5 or more extending more than 3/5 distance to petiole, width 1–7(–14) mm (basal), 0.5–6 mm (cauline). **Inflorescences** 4–18(–48)-flowered; pedicels 0.8–6 cm, pubescence variable; bracteoles 3–8(–18) mm from flowers, green to blue, linear, 3–7 mm, pubescence variable. **Flowers:** sepals usually bluish purple, rarely white to pink, puberulent, lateral sepals reflexed or spreading, 8–21 × 3–10 mm, spurs decurved to straight, ascending 20–60° above horizontal, 8–23 mm; lower petal blades elevated, exposing stamens, blue to purple, except sometimes in white-flowered plants, 4–11 mm, clefts 2–5 mm; hairs mostly on inner lobes below junction of blade and claw, white, rarely yellow. **Fruits** 7–22 mm, 3.5–5 times longer than wide, glabrous to puberulent. **Seeds** winged or not; seed coat cell surfaces smooth or roughened, blunt hairs absent. $2n = 16$.

Flowering spring(–early summer). Open coniferous woods, grassy sage scrub, meadow edges and well-drained streamsides (generally not in very wet sites); 300–3500 m; B.C.; Ariz., Calif., Colo., Idaho, Mont., Nev., N.Mex., Oreg., Utah, Wash., Wyo.

Delphinium nuttallianum represents an extremely difficult complex, with many variations in a number of morphologic traits. The complex has been and continues to be a major source of confusion for identification of *Delphinium* in North America. Type specimens of *D. nuttallianum* represent plants growing under dry conditions in open areas. These are typically found at 1200–2000 m in sage scrub or lower montane forest.

Delphinium nuttallianum may be confused with *D. andersonii*, *D. antoninum*, *D. depauperatum*, *D. gracilentum*, and two subspecies of *D. patens* (subsp. *patens* and subsp. *montanum*). Features that may be used to separate *D. nuttallianum* from the first four, are enumerated under the respective species discussions. From *D. patens* subsp. *patens*, *D. nuttallianum* may be distinguished by its narrower leaf lobes, larger fruits, and more compact inflorescence. The frequent presence of glandular hairs in the inflorescence of *D. patens* subsp. *montanum*, contrasted with their absence in *D. nuttallianum*, will separate these taxa. Dwarfed plants of *D. polycladon* may be confused with *D. nuttallianum*. The latter, however, may be distinguished by its ringed seeds, and it does not have prominent buds or sigmoid pedicel.

Hybrids have been seen between *Delphinium nuttallianum* and *D. andersonii*, *D. depauperatum* (*D. ×burkei* Greene), *D. distichum* (*D. ×diversicolor* Rydberg), *D. nudicaule*, and *D. polycladon*.

17. CONSOLIDA (de Candolle) Gray, Nat. Arr. Brit. Pl. 2: 711. 1821 · Larkspur [Latin *consolidatus,* to become solid or firm, from reputed ability to heal wounds]

Michael J. Warnock

Delphinium Linnaeus sect. *Consolida* de Candolle, Syst. Nat. 1: 341. 1817

Herbs, annual, from slender taproots. **Leaves** all cauline at anthesis, alternate, proximal leaves petiolate, distal leaves nearly sessile. **Leaf blade** palmately finely dissected, ± semicircular, lobes less than 2 mm wide (except sometimes on leaves near base of stem), margins entire. **Inflorescences** terminal, 3–45(–75)-flowered racemes, 5–30 cm; bracts leaflike, not forming involucre. **Flowers** bisexual, bilaterally symmetric; sepals not persistent in fruit, 5; upper sepal 1, spurred, 5–18 mm; lateral sepals 2, blue to purple, pink, or white, elliptic or lanceolate, sometimes clawed, 5–18 mm; lower sepals 2, similar to lateral sepals, often narrower; petals 2, connate, spurred, spur enclosed in upper sepal, 1–3-lobed, nectary on spur; stamens many; filaments with base expanded; staminodes absent between stamens and pistils; pistil 1, simple; ovules several per pistil; style present. **Fruits** follicles, solitary, sessile, ± cylindric, sides smooth; beak terminal, straight, 1–2 mm. **Seeds** dark brown to black, obpyramidal to 3-angled. $x = 8$.

Species ca. 40 (5 in the flora): temperate and subtropical regions, North America, Europe, Africa, Asia; introduced in South America, Australia.

All species of *Consolida* included here are cultivated in North America. *Consolida regalis* has been reported to self-sow for many years. Inclusion of the other four species is based on herbarium specimens with label data indicating that the specimens were collected from sites not under cultivation at the time.

SELECTED REFERENCES Chater, A. O. 1993. *Consolida.* In: T. G. Tutin et al., eds. 1993+. Flora Europaea, ed. 2. 1+ vols. Cambridge and New York. Vol. 1, pp. 260–262. Keener, C. S. 1976. Studies in the Ranunculaceae of the southeastern United

States. IV. Genera with zygomorphic flowers. Castanea 41: 12–20.   Munz, P. A. 1967. A synopsis of African species of *Delphinium* and *Consolida*. J. Arnold Arbor. 48: 30–55.   Munz, P. A. 1967b. A synopsis of the Asian species of *Consolida* (Ranunculaceae). J. Arnold Arbor. 48: 159–202.

1. Proximal bracts simple, sometimes lowermost 3-lobed.
    2. Spur less than 12 mm . . . . 5. *Consolida tenuissima*
    2. Spur 12 mm or more.
        3. Terminal lobe of petal more than 2 mm wide, sinus more than 0.5 mm deep; stems ± puberulent or not . . . . 4. *Consolida regalis*
        3. Terminal lobe of petal less than 2 mm wide, sinus less than 0.5 mm deep; stems densely pubescent . . . . 3. *Consolida pubescens*
1. Proximal bracts (at least 2 lowermost) dissected into 3 or more lobes.
    4. Bracteoles touching sepals, 1–5 mm from flower; spur 12 mm or less . . . . 2. *Consolida orientalis*
    4. Bracteoles (at least on proximal flowers) not touching sepals, 4–20 mm from flower; spur 12 mm or more.
        5. Follicles less than 1/2 length of proximal pedicels; inflorescence with 3 or more branches; stems pubescent . . . . 3. *Consolida pubescens*
        5. Follicles 1/2 or more length of proximal pedicels; inflorescence usually with 3 or fewer branches; stems glabrous to sparsely puberulent . . . . 1. *Consolida ajacis*

1. Consolida ajacis (Linnaeus) Schur, Verh. Mitth. Siebenbürg. Vereins Naturwiss. Hermannstadt 4: 47. 1853   F

Delphinium ajacis Linnaeus, Sp. Pl. 1: 531. 1753

Stems 3–8(–10) dm, glabrous to sparsely puberulent. **Leaves** 5–20 or more. **Leaf blade** semicircular, 12–60-lobed or more, 1–5 cm wide, glabrous to puberulent, lobes less than 1.5 mm wide. **Inflorescences** 6–30(–75)-flowered, simple or with 3 or fewer branches; bracts (at least lowermost 2) with 3 or more lobes; pedicels ascending-spreading, 1–3(–5) cm, ± puberulent; bracteoles not touching sepals, 4–20 mm from flower, ± linear, 1–3 mm, ± puberulent. **Flowers:** sepals blue to purple, rarely pink or white, nearly glabrous, lower sepals 8–18 × 4–8 mm, lateral sepals 8–18 × 6–14 mm, spur 12–20 mm; petals of same color as sepals or whiter, lateral lobes 3–6 mm, terminal lobes 5–8 × 2–4 mm, cleft 0.2–1 mm. **Follicles** 12–25 mm, puberulent.

Flowering summer. Waste places, old homesites, drainage ditches, roadsides, and railroads; 0–2000 m; introduced; B.C., Man., Ont.; Ala., Ariz., Ark., Calif., Conn., Del., D.C., Fla., Ga., Ill., Ind., Iowa, Kans., Ky., La., Md., Mass., Mich., Minn., Miss., Mo., Mont., Nebr., N.C., N.Dak., N.J., N.Y., Ohio, Okla., Pa., R.I., S.C., Tenn., Tex., Vt., Va., W.Va., Wis.; native to Europe and Africa; introduced in South America, Asia, and Australia.

In many floras the names *Consolida ambigua* (Linnaeus) Ball & Heywood and *Delphinium ambiguum* Linnaeus have been misapplied to this taxon.

Consolida ajacis has escaped and become more or less naturalized in many temperate and subtropical parts of the world. It is by far the most commonly encountered species of *Consolida* in North America.

The Cherokee used *Consolida ajacis* medicinally in infusions to treat heart problems (D. E. Moerman 1986, as *Delphinium ajacis*).

2. Consolida orientalis (Gay) Schrödinger, Abh. K. K. Zool.-Bot. Ges. Wien 4: 25. 1909   F

Delphinium orientale Gay in C. R. A. Des Moulins, Cat. Rais. Pl. Dordogne, 12. 1840

Stems 3–6(–10) dm, puberulent. **Leaves** 5–14. **Leaf blade** semicircular, 11–32-lobed, 2–4 cm wide, nearly glabrous, lobes less than 1.5 mm wide. **Inflorescences** 3–20-flowered, simple or with 3(–5) branches; at least lower 2 bracts with 5 or more lobes; pedicels spreading-ascending, 0.5–3(–4) cm, puberulent; bracteoles touching sepals, 1–5 mm from flower, lance-linear, 2–4 mm, puberulent. **Flowers:** sepals purple, rose, or white, ± pubescent, lower sepals 10–13 × 4–6 mm, lateral sepals 9–13 × 5–9 mm, spur 8–12 mm; petals of ± same color as sepals, lateral lobes 3–5 mm, terminal lobes 5–7 × 2–3 mm, cleft 0.5–1.5 mm. **Follicles** 14–21 mm, pubescent.

Flowering summer. Waste places, old homesites, roadsides, and railroads; 0–600 m; introduced; Kans., Mo.; native to s Europe, nw Africa, sc, sw Asia.

Consolida orientalis is cultivated and more or less naturalized in many temperate parts of the world.

3. Consolida pubescens (de Candolle) Soó, Oesterr. Bot. Z. 71: 241. 1922

Delphinium pubescens de Candolle in J. Lamarck and A. P. de Candolle, Fl. Franç. ed. 3, 5: 641. 1815

Stems 1–5 dm, pubescent. **Leaves** 3–18. **Leaf blade** semicircular, 9–32-lobed, 1–3 cm wide, pubescent, lobes less than 1.5 mm wide. **Inflorescences** 8–35-flowered or more, with 3 or more branches; proximal bracts simple or multifid; pedicels ascending, 1–3 cm, pubescent; bracteoles not touching sepals, 6–17 mm from flower, linear, 2.5–5 mm, pubescent. **Flowers:** sepals pale blue or pale purple to white, pubescent, lower sepals 8–12 × 3–5 mm, lateral sepals 9–15 × 5–10 mm, spur 12–20 mm; petals whitish to bluish, lateral lobes 3–5 mm, terminal lobe 3–4 × 1–2 mm, cleft 0.1–0.5 mm. **Follicles** 10–15 mm, pubescent.

Flowering summer. Waste places, railroads; 70–1000 m; introduced; Ill., Mo., N.Dak., Tenn., Wis.; native to sw Europe, nw Africa.

Consolida pubescens is not so commonly cultivated as *C. ajacis* or *C. orientalis*.

4. Consolida regalis Gray, Nat. Arr. Brit. Pl. 2: 711. 1821 · Rocket larkspur ⬚F⬚

Delphinium consolida Linnaeus, Sp. Pl. 1: 530. 1753

Stems 2–6(–11) dm, ± puberulent. **Leaves** 5–28. **Leaf blade** semicircular, 9–27-lobed, 1–5 cm wide, ± puberulent, lobes less than 1.5 mm wide. **Inflorescences** 6–41-flowered, often more than 3 branches; bracts simple (except lowermost sometimes 3-cleft); pedicels ascending, 1.5–5 cm, ± puberulent; bracteoles 4–15

mm from flower, lance-linear, 1.5–4 mm, ± puberulent. **Flowers:** sepals dark blue, rarely pink or white, puberulent, lower sepals 9–15 × 4–6 mm, lateral sepals 9–16 × 5–8 mm, spur 12–22 mm; petals blue to yellow, lateral lobes 4–5 mm, terminal lobes 3.5–5 × 2–3 mm, cleft 0.5–1.5 mm. **Follicles** 8–17 mm, glabrous to pubescent.

Flowering summer. Wheat fields, roadsides, waste places, old homesites; 0–800 m; introduced; Ont.; Ala., D.C., Mo., N.Y., N.C., Pa., Wis.; native to Europe, sw Asia.

Consolida regalis is widely and commonly cultivated. In some areas it was apparently introduced as a contaminant with wheat seed (N. G. Miller 1995) and subsequently became established.

5. Consolida tenuissima (Sibthorp & Smith) Soó, Oesterr. Bot. Z. 71: 241. 1922

Delphinium tenuissimum Sibthorp & Smith, Fl. Graec. Prodr. 1: 370. 1806

Stems 1–5 dm, pubescent. **Leaves** 5–26. **Leaf blade** semicircular, 9–28-lobed, 1–3 cm wide, ± pubescent, lobes less than 2 mm wide. **Inflorescences** 7–45-flowered or more, with 3 or more branches; bracts simple; pedicels ascending, 1–4 cm, pubescent; bracteoles 5–22 mm from flower, lanceolate to linear, 2.5–6 mm, pubescent. **Flowers:** sepals violet to dark blue, ± puberulent, lower sepals 5–7 × 2–3.5 mm, lateral sepals 5–8 × 3–5 mm, spur 7–10 mm; petals blue to white, lateral lobes 3–5 mm, terminal lobes 4–6 × 2–3 mm, cleft 0.2–0.6 mm. **Follicles** 5–8 mm, nearly glabrous.

Flowering summer. Waste places, roadsides, and railroads; 80–500 m; introduced; Mo.; native to Europe (Greece).

Consolida tenuissima is uncommon in cultivation.

18. COPTIS Salisbury, Trans. Linn. Soc. London 8: 305. 1807 · Goldthread, coptide [Greek *kopto,* to cut, referring to dissected leaves]

Bruce A. Ford

Herbs, perennial, from orange, yellow, or pale brown, slender rhizomes 0.5–2 mm thick. **Leaves** basal, petiolate. **Leaf blade** 1–2-ternately compound, 1–2-pinnately compound, or deeply divided; leaflets ovate or triangular, lobed or parted, margins sharply serrate to denticulate. **Inflorescences** scapose, 1–4-flowered cymes, to 3 cm (9 cm in fruit); bracts absent. **Flowers** bisexual and staminate (all bisexual in *C. trifoliata*), radially symmetric; sepals not persistent in fruit, 5–7, green to white, plane, linear-lanceolate, oblanceolate to obovate or elliptic, occasionally clawed, 4–11 mm; petals 5–7, distinct, green, plane or concave distally, either

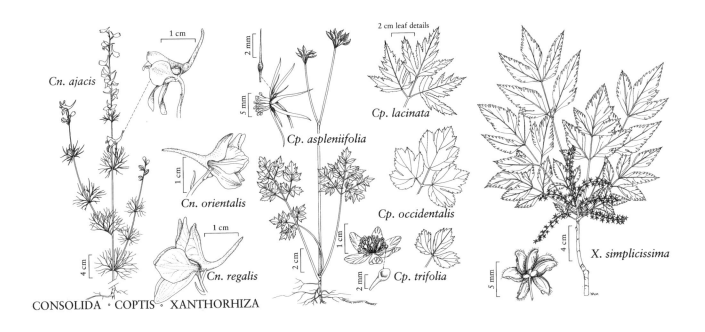

CONSOLIDA · COPTIS · XANTHORHIZA

clavate with adaxial nectary at apex or linear with adaxial nectary near base, clawed, 2–7 mm; stamens 10–60; filaments filiform; staminodes absent between stamens and pistils; pistils 4–15, simple; ovules 4–10 per pistil; style present. **Fruits** follicles, aggregate, stipitate, forming umbel-like clusters, oblong to ellipsoid, sides not veined; beak present or absent, terminal, straight or apically hooked, 0–4 mm. **Seeds** dark brown to tan, ellipsoid, shiny, often appearing wrinkled. $x = 9$.

Species ca. 10 (4 in the flora): temperate and boreal regions of North America and Asia.

SELECTED REFERENCES Fernald, M. L. 1929. *Coptis trifolia* and its eastern American representative. Rhodora 31: 136–142. Hultén, E. 1937. Flora of the Aleutian Islands and Westernmost Alaska Pensinsula with Notes on the Flora of Commander Islands. . . . Stockholm.

1. Leaflets unlobed or slightly lobed, ± sessile; sepals oblanceolate to obovate or elliptic; nectary near apex of petal; follicle body 3.9–7 mm; beak straight to ascending in fruit, 2–4 mm . . . . 4. *Coptis trifolia*
1. Leaflets deeply lobed or pinnatifid, short- to long-petiolulate; sepals linear-lanceolate; nectary near base of petal; follicle body 7–13.5 mm; beak recurved in fruit, less than 1 mm.
   2. Leaf blade 2-pinnate with pinnatifid leaflets, to 3-pinnate or occasionally 2-ternate; coastal British Columbia to Alaska . . . . 1. *Coptis aspleniifolia*
   2. Leaf blade (1–)2-ternate; sw Washington s to California, e to Montana.
      3. Flowers nodding; leaf blade 2-ternate or ternate with leaflets lobed almost to base; coastal Washington to California . . . . 2. *Coptis laciniata*
      3. Flowers erect; leaf blade ternate with leaflets lobed ca. 1/2 length to base; Rocky Mountains of Washington, Idaho, and Montana . . . . 3. *Coptis occidentalis*

1. Coptis aspleniifolia Salisbury, Trans. Linn. Soc. London 8: 306. 1807 [E] [F]

Rhizomes pale brown. **Leaves:** blade 2-pinnate with pinnatifid leaflets to 3-pinnate, occasionally 2-ternate; leaflets short- to long-petiolulate, blade ovate, deeply lobed or incised, margins sharply serrate. **Inflorescences** 2–3-flowered, often longer than leaves at anthesis, 8–12 cm, elongating to 35 cm in fruit. **Flowers** nodding; sepals reflexed and ascending, linear-lanceolate, 6–11(–15) × 0.3–1 mm; petals linear-lanceolate, nectary nearly basal, blade flattened, narrowly ligulate at apex; stamens 9–15. **Follicles** 6–10; stipe equal to or slightly longer than body; body oblong, 7–10 mm; beak recurved, less than 1 mm. **Seeds** 1.8–2.2 mm. $2n = 18$.

Flowering spring. Moist, coniferous forests, seeps, and bogs; 0–1500 m; B.C.; Alaska, Wash.

This species is widespread in coastal areas from southern British Columbia to southeastern Alaska. The Washington State Heritage Program tracks this species as "state-rare" in Snohomish County, Washington; I have not seen any specimens to confirm its presence in the state.

Coptis aspleniifolia, C. laciniata, and *C. occidentalis* form a group of morphologically similar, allopatric species that are probably recently derived. The species may have originated in response to the opening of the western Cordilleran landscape after Pleistocene glaciation and could be considered localized variants of a single species. Although most individuals can be readily distinguished, some can be difficult to place.

A putative hybrid between *Coptis aspleniifolia* and *C. trifolia* has been found along the Kennedy River of Vancouver Island, British Columbia (T. C. Brayshaw, pers. comm.). It has 3–5 deeply dissected leaflets per leaf and no complete flowers.

2. Coptis laciniata A. Gray, Bot. Gaz. 12: 297. 1887 [E] [F]

Rhizomes pale brown. **Leaves:** blade 2-ternate or ternate with leaflets deeply 3-lobed almost to base; leaflets short- to long-petiolulate, blade ovate to triangular, incised-lobate nearly to midvein, margins sharply serrulate-denticulate. **Inflorescences** 2–4-flowered, often shorter than leaves at anthesis, 5–10 cm, elongating to 25 cm in fruit. **Flowers** nodding; sepals reflexed and ascending, linear-lanceolate, 6–11 × 0.3–1 mm; petals linear-lanceolate, nectary nearly basal, blade flattened,

narrowly ligulate at apex; stamens 10–25. **Follicles** 5–11; stipe equal to or slightly shorter than body; body oblong, 8–13.5 mm; beak recurved, less than 1 mm. **Seeds** 2–2.5 mm.

Flowering spring–summer. Moist woods, stream banks, seeps, and wet cliffs in the coastal mountains; 500–2000 m; Calif., Oreg., Wash.

3. Coptis occidentalis (Nuttall) Torrey & A. Gray, Fl. N. Amer. 1: 28. 1838 [E] [F]

Chrysocoptis occidentalis Nuttall, J. Acad. Nat. Sci. Philadelphia 7: 8. 1834

Rhizomes pale brown. **Leaves:** blade ternate; leaflets long-petiolulate, blade widely ovate, incised with 2–3 lobes divided ca. 1/2 length to base, margins sharply serrate-denticulate. **Inflorescences** 2–3(–5)-flowered, often shorter than leaves at anthesis, 10–25 cm, elongating to 32 cm in fruit. **Flowers** erect; sepals spreading to reflexed, linear-lanceolate, 7–11 × 0.4–1 mm; petals linear-lanceolate, nectary nearly basal, blade flattened, narrowly ligulate at apex; stamens 10–35. **Follicles** 5–15; stipe slightly shorter than body; body oblong, 7.5–14 mm; beak recurved, less than 1 mm. **Seeds** 2.2–2.6 mm.

Flowering spring. Moist, coniferous woods; 500–2000 m; Idaho, Mont., Wash.

4. Coptis trifolia (Linnaeus) Salisbury, Trans. Linn. Soc. London 8: 305. 1807 · Goldthread, goldenroot, yellow snakeroot, savoyana, coptide trifoliolée [F] [W1]

Helleborus trifolius Linnaeus, Sp. Pl. 1: 558. 1753; *Coptis groenlandica* (Oeder) Fernald; *C. trifolia* subsp. *groenlandica* (Oeder) Hultén

Rhizomes bright yellow to orange. **Leaves:** blade ternate; leaflets sessile or short-petiolulate, blade cuneate-obovate, margins serrate, slightly lobed distally. **Inflorescences** 1-flowered, usually equal to or longer than leaves at anthesis, 3–17 cm, not elongating in fruit. **Flowers** erect; sepals spreading, oblanceolate to obovate or elliptic, 4–11 × 1–4 mm; petals clavate, nectary apical, blade absent; stamens 30–60. **Follicles** 4–7; stipe equal to or longer than body; body elliptic, 3.9–7 mm; beak straight to ascending, 2–4 mm. **Seeds** 1–1.5 mm. $2n = 18$.

Flowering late spring–summer (May–Aug). Wet to mesic, coniferous and mixed forests, bogs, willow scrub, and tundra, often associated with mosses; 0–1500 m; Greenland; St. Pierre and Miquelon; Alta., B.C., Man., N.B., Nfld., N.W.T., N.S., Ont., P.E.I.,

Que., Sask.; Alaska, Conn., Ind., Maine, Md., Mass., Mich., Minn., N.H., N.J., N.Y., Ohio, Pa., R.I., Vt., W.Va., Wis.; e Eurasia.

M. L. Fernald (1929) treated plants from Greenland and eastern North America as *Coptis groenlandica* and those from Alaska and eastern Asia as *C. trifolia*. He did concede, however, that intermediates do occur. E. Hultén (1937) treated the two taxa as subspecies but concluded that *C. trifolia* subsp. *trifolia* was restricted to the Aleutian Islands in North America. Plants from southeastern Alaska were comparable to material from Greenland and were called *C. trifolia* subsp. *groenlandica*. Hultén (1944) later reversed this decision, concluding that the differences between the two subspecies were too slight to warrant their continued recognition. Finally, R. L. Taylor and G. A. Mulligan (1968), in their study of North American *Coptis,* found that the two subspecies could be distinguished on the basis of sepals and seed shape: subsp. *trifolia* having clawed petals and quadrate seeds; subsp. *groenlandica* possessing sepals that gradually narrow toward the base and round seeds.

Petiolule length, sepal width, the length-to-width ratio of the nectary, follicle body and beak length, and sepal and seed shape have been used most commonly to distinguish these two taxa. My comparison of herbarium specimens from the Aleutian Islands, the rest

TABLE 15.2. Morphologic comparison of *Coptis trifolia* from the Aleutian Islands, the rest of North America, and Asia.

	Aleutians	Other N. Amer.	Asia
Petiolules (mm)	0.3–0.8	0.5–3.5	0.5–3
Sepals (mm wide)	1.7–2.9	1–3.6	1.8–4.5
Nectaries (L:W ratio)	0.9–1.3	0.8–1.3	0.9–1.6
Follicle bodies (mm)	3.9–5.3	4–7	4.8–5.5
Follicle beaks (mm)	3–3.5	2.3–3.8	2.2–3

of North America, and eastern Asia indicated that no clear distinction could be made (table 1). Some plants from eastern Asia and the Aleutian Islands had more distinctly clawed sepals than those from eastern North America but this was not evident on all individuals. Seeds from eastern North American plants were found to be variable in shape, with all seeds having at least one angle in cross section.

Native Americans used various preparations made from the roots of *Coptis trifolia* medicinally to treat stomach cramps, jaundice, sore mouth and throat, gum problems, and worms, to stop vomiting, especially for children, as eyedrops, for teething, and as an astringent (D. E. Moerman 1986).

19. XANTHORHIZA Marshall, Arbust. Amer., 167. 1785 · Yellowroot [Greek *xanthos*, yellow, and *rhiza*, root]

Bruce D. Parfitt

Shrubs, deciduous, from long, slender, yellow rhizomes. **Leaves** cauline, petiolate; cauline leaves alternate. **Leaf blade** 1–2-pinnately compound; leaflets lance-ovate to broadly ovate or rhombic, margins variously sharply cleft, incised, or serrate. **Inflorescences** axillary, many-flowered panicles, 5–21 cm; bracts scalelike, not forming involucre. **Flowers** bisexual, radially symmetric; sepals persistent in fruit, 5, dark brown-purple to greenish yellow, plane, lanceolate to oblanceolate, usually clawed, ca. 2.5–5 mm; petals 5, distinct, brown-purple, plane, peltate, 0.5–0.9 mm, nectary terminal, transversely oblong, 2-lobed; stamens 5 (or 10); filaments with base expanded; staminodes absent between stamens and pistils; pistils (2–)5–10, simple; ovules 2, basal ovule aborting; style present. **Fruits** follicles, aggregate, sessile, compressed, obliquely oblong or somewhat sickle-shaped, sides not veined; beak becoming abaxial, curved, 0.4–0.8 mm. **Seeds** reddish brown, ovoid, smooth. $x = 18$.

Species 1 (1 in the flora): North America.

1. Xanthorhiza simplicissima Marshall, Arbust. Amer., 167. 1785 · Yellowroot, shrub yellowroot, brook-feather E F

Xanthorhiza apiifolia L'Héritier

Stems 20–70 cm, 3–6 mm diam.; bark smooth, ringed with leaf scars, inner bark yellow. **Leaves** clustered near stem apex, to 18 cm; leaflets 3–5, 2.5–10 × 2–8 cm, sessile to short-petiolulate. **Inflorescences** broad-paniculate, arising from cluster of leaves, 6–21 cm, short-pilose; pedicels 2–5 mm. **Flowers:** sepals spreading, acuminate; petals with nectary transversely oblong, 2-lobed. **Follicles** yellowish brown, glossy, somewhat inflated, 3–4 mm, distally ciliate. $2n = 36$.

Flowering spring–summer (Apr–May). Shaded stream banks, moist woods, thickets, and rocky ledges; 0–1200 m; Ala., Fla., Ga., Ky., La., Maine, Md., Mass., Miss., N.H., N.Y., N.C., Ohio, S.C., Tenn., Tex., Va., W.Va.

Xanthorhiza simplicissima is cultivated as an effective ground cover in moist soils. In the northeastern and midwestern United States it occasionally escapes from cultivation, and either persists or becomes established. In southeastern Texas and western Louisiana the species is apparently native, though infrequent. The leaves bear a striking resemblance to those of celery, hence *apiifolia*, the epithet in synonymy.

The yellow inner bark and roots contain a bitter principle; the roots have been used medicinally and in making yellow dye. Bundles of yellowroot are widely sold in Alabama and Georgia for treatment of gastrointestinal disorders and fever blisters (R. D. Whetstone, pers. comm.).

Native Americans used *Xanthorhiza simplicissima* medicinally to treat ulcerated stomachs, colds, jaundice, piles, sore mouth, sore throat, cancer, and cramps, and as a blood tonic (D. E. Moerman 1986).

20. ENEMION Rafinesque, J. Phys. Chim. Hist. Nat. Arts 91: 70. 1820 · False rue-anemone

Bruce A. Ford

Herbs, perennial, from stout woody rhizomes or tuberous roots. **Leaves** basal and cauline, proximal leaves petiolate, distal leaves sessile or nearly so; cauline leaves alternate. **Leaf blade** 2-ternately compound; leaflets broadly ovate to cuneate-obovate, margins entire to deeply 3-notched. **Inflorescences** terminal or axillary, 2–10-flowered cymes or racemes or flowers solitary, to 4 cm; bracts absent. **Flowers** bisexual, radially symmetric; sepals not persistent in fruit, 5(–9), white, occasionally tinged pinkish, plane, ovate to obovate, 3.5–15 mm; petals absent; stamens 9–75; filaments filiform to clavate or narrowly triangular; staminodes absent between stamens and pistils; pistils 2–10, simple; ovules 2–6 per pistil; style present. **Fruits** follicles, aggregate, sessile or stipitate, oblong or elliptic to obovate, laterally compressed, sides with a few prominent veins and transverse veinlets; beak terminal, straight or curved, 0.5–3 mm. **Seeds** reddish brown, ovoid, smooth, rugulose, or minutely pubescent. $x = 7$.

Species 6 (5 in the flora): North America, Asia.

The delimitation of taxa within tribe Isopyreae Schrödinger has been open to considerable debate. North American taxonomists tend to retain the North American species in *Isopyrum* Linnaeus whereas taxonomists elsewhere recognize *Enemion* along with a number of other segregate genera [e.g., *Dichocarpum* (Tamura & Kosuge) W. T. Wang and Fu D.-Z., *Isopyrum* in the strict sense, *Leptopyrum* Reichenbach, and *Paraquilegia* Drummond & Hutchinson] (J. R. Drummond and J. Hutchinson 1920; Fu D.-Z. 1990; M. Tamura 1984, 1993; M. Tamura and L. A. Lauener 1968).

Enemion has no petals and is regarded as the most primitive member in tribe Isopyreae (Fu D.-Z. 1990; M. Tamura 1984; M. Tamura and L. A. Lauener 1968). The other closely related genera form a transition series: *Dichocarpum* has peltate petals similar to the stamens; *Isopyrum* and *Leptopyrum* possess peltate petals with larger, tubular limbs and shorter claws;

and *Paraquilegia* has nonpeltate petals that are flat and ± concave or swollen near the base and nearly sessile.

While some authors have argued that too much emphasis has been placed on petals as characters for segregating genera (e.g., J. A. Calder and R. L. Taylor 1963), *Enemion* is a well-defined taxon, easily distinguished from other members of tribe Isopyreae. Final decisions involving the circumscription of *Enemion* await molecular study involving all members of subfamily Isopyroideae Tamura.

All species of *Enemion* in the flora, with the exception of *E. biternatum,* are localized endemics of western North America from British Columbia to California.

SELECTED REFERENCES Drummond, J. R. and J. Hutchinson. 1920. A revision of *Isopyrum* (Ranunculaceae) and its nearer allies. Bull. Misc. Inform. Kew 1920: 145–169. Tamura, M. and L. A. Lauener. 1968. A revision of *Isopyrum, Dichocarpum* and their allies. Notes Roy. Bot. Gard. Edinburgh 28: 267–273.

1. Inflorescences terminal or axillary, well-defined 3–10-flowered cymes with small scalelike bracts; leaflets abaxially pubescent . . . . .2. *Enemion hallii*
1. Inflorescences axillary, flowers solitary or loosely grouped in 2–4-flowered leafy cymes or racemes, bracts similar to distal stem leaves; leaflets abaxially glabrous.
    2. Follicles stipitate; peduncle strongly clavate; stamens fewer than 15, filaments flat, narrowly triangular . . . . 5. *Enemion stipitatum*
    2. Follicles sessile; peduncle not clavate; stamens more than 20, filaments filiform to club-shaped.
        3. Lobes of leaflets with shallow glandular notches at apices; sepals (10–)12.6–15(–16.8) mm; follicle bodies 11–15 mm; coastal British Columbia . . . . 4. *Enemion savilei*
        3. Lobes of leaflets glandular-apiculate; sepals 5.5–13.5 mm; follicle bodies 3.5–11.5 mm; California, c, e North America.
            4. Follicle body 3.5–6.5 mm, gradually contracted into style beak; beak 1.7–3 mm; seeds 2.1–2.7 mm, minutely pubescent; roots fibrous; c, e North America . . . . 1. *Enemion biternatum*
            4. Follicle body (7.7–)8.5–11.5 mm, abruptly contracted into style beak; beak 0.8–1.7 mm; seeds 1.5–2 mm, glabrous; roots tuberous; c, s California. . . . . 3. *Enemion occidentale*

1. **Enemion biternatum** Rafinesque, J. Phys. Chim. Hist. Nat. Arts 91: 70. 1820 [E][F]

Isopyrum biternatum (Rafinesque) Torrey & A. Gray

Stems 10–40 cm, weakly rhizomatous; roots fibrous. **Leaves:** leaflets irregularly 2–3-lobed, lobes sometimes with 1–3 secondary lobes, apex rounded, glandular-apiculate; surfaces abaxially glabrous. **Inflorescences** axillary, flowers solitary or loosely grouped in 2–4-flowered leafy racemes; peduncle not strongly clavate. **Flowers:** sepals 5.5–13.5 × 3.5–8.5 mm; stamens 25–50; filaments filiform to club-shaped, 1.8–5.8 mm. **Follicles** sessile, upright to widely divergent; body widely elliptic to widely obovate, 3.5–6.5 mm, gradually contracted into style beak; beak 1.7–3 mm. **Seeds** 2.1–2.7 mm, minutely pubescent. 2*n* = 14.

Flowering spring. Moist deciduous woods of valleys, flood plains, and ravine bottoms, occasionally in open pastures, often on limey soils; 25–1000 m; Ont.; Ala., Ark., Fla., Ill., Ind., Iowa, Kans., Ky., Mich., Minn., Miss., Mo., N.Y., N.C., Ohio, Okla., S.C., S.Dak., Tenn., Va., W.Va., Wis.

Enemion biternatum has been mistaken for the superficially similar *Thalictrum thalictroides* because of its white flowers and compound *Thalictrum*-like leaves. *Enemion biternatum* is easily distinguished, however, by its few-seeded follicles and deeply lobed leaves with glandular-apiculate apices. *Thalictrum thalictroides,* on the other hand, is characterized by having achenes and somewhat crenate leaves with notched apices.

2. **Enemion hallii** (A. Gray) J. R. Drummond & Hutchinson, Bull. Misc. Inform. Kew 1920: 161. 1920 · Willamette rue-anemone [E][F]

Isopyrum hallii A. Gray, Proc. Amer. Acad. Arts 8: 374. 1872

Stems 35–85 cm, with short, stout, woody rhizome; roots fibrous. **Leaves:** leaflets variously lobed and sharply dentate, apex acute, glandular-apiculate; surfaces abaxially pubescent. **Inflorescences** terminal or axillary, well-defined 3–10-flowered cymes with small scalelike bracts; peduncle not strongly clavate. **Flowers:** sepals 5–10.5 × 2.5–6.5 mm; stamens 50–75; filaments filiform to club-shaped, 4.5–8.2 mm. **Follicles** sessile, upright to widely divergent; body widely elliptic to

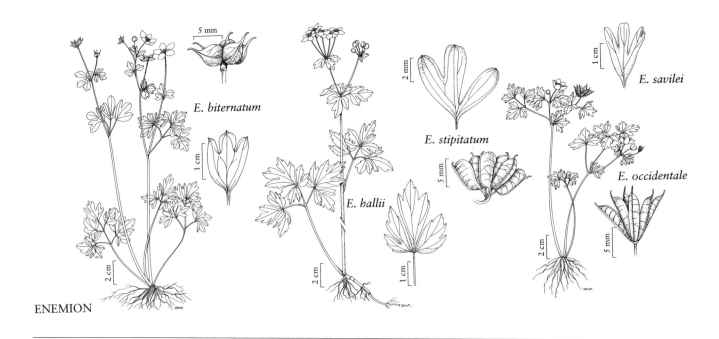

ENEMION

E. biternatum

E. hallii

E. stipitatum

E. savilei

E. occidentale

widely obovate, 3.8–7 mm, gradually contracted into style beak; beak 1.1–2.5 mm. **Seeds** 1.8–2.2 mm, glabrous.

Flowering late spring–early summer. Moist woods and streambanks; 100–1500 m; Oreg., Wash.

Enemion hallii differs from all other North American members of the genus in having well-defined cymose inflorescences. Its closest ally is thought to be the east-Asian species *E. raddeanum* Regel, from which it differs in having long-petiolate leaves and cymose inflorescences with bracteolate subumbels. *Enemion raddeanum* is characterized by sessile or short-petiolate leaves and simple, umbellate inflorescences.

3. **Enemion occidentale** (Hooker & Arnott) J. R. Drummond & Hutchinson, Bull. Misc. Inform. Kew 1920: 160. 1920 · Western rue-anemone [E] [F]

Isopyrum occidentale Hooker & Arnott, Bot. Beechey Voy., 316. 1841

Stems 10–40 cm, cespitose, not rhizomatous; roots tuberous. **Leaves** abaxially glabrous; leaflets irregularly 2–3-lobed, lobes often with 1–3 secondary lobes; apex rounded, glandular-apiculate. **Inflorescences** axillary, flowers solitary or loosely grouped in 2–3-flowered leafy cymes; peduncle not strongly clavate. **Flowers:** sepals (5–)7–11.5 × 2.8–7 mm; stamens 20–40; filaments filiform to club-shaped, 2.2–6 mm. **Follicles** sessile, upright to widely divergent; body oblong, (7.7–)8.5–11.5 mm, abruptly contracted into style beak; beak 0.8–1.7 mm. **Seeds** 1.5–2 mm, glabrous.

Flowering early spring. Shaded places, oak woodland, chaparral, and coniferous and deciduous woods; 200–1500 m; Calif.

Enemion occidentale is endemic to California where it is infrequent and local (P. A. Munz 1959).

4. **Enemion savilei** (Calder & Roy L. Taylor) Keener, Castanea 45: 278. 1981 [E] [F]

Isopyrum savilei Calder & Roy L. Taylor, Madroño 17: 70. 1963

Stem 10–35 cm, strongly rhizomatous; roots fibrous. **Leaves** abaxially glabrous; leaflets irregularly 2–3-lobed, lobes often with 1–3 secondary lobes; apex rounded, with shallow glandular notches. **Inflorescences** axillary, flowers solitary or occasionally in loose 2-flowered leafy cymes; peduncle not strongly clavate. **Flowers:** sepals (10–)12.6–15(–16.8) × (6.9–)8.2–10.2(–11.2) mm; stamens 40–60; filaments filiform to club-shaped, 5–8 mm. **Follicles** sessile, upright to widely divergent; body oblong, 11–15 mm, abruptly contracted into style beak; beak 0.8–1.7 mm. **Seeds** 2–2.5 mm, glabrous. $2n = 14$.

Flowering spring. Moist, shady, rocky crevices and talus slopes; 100–1000 m; B.C.

This distinctive species was discovered in 1957 on the Queen Charlotte Islands and was thought to be endemic to those islands. Subsequently it has been found on the Brooks Peninsula on the west coast of Vancouver Island and on Porcher Island, south of Prince Rupert (T. C. Brayshaw 1989).

5. Enemion stipitatum (A. Gray) J. R. Drummond & Hutchinson, Bull. Misc. Inform. Kew 1920: 160. 1920 · Siskiyou rue-anemone [E] [F]

Isopyrum stipitatum A. Gray, Proc. Amer. Acad. Arts 12: 54. 1876

Stems 3–12 cm, cespitose, not rhizomatous; roots tuberous. Leaves abaxially glabrous; leaflets entire or deeply 2–3-lobed, apex rounded, glandular-apiculate. Inflorescences axillary, flowers solitary; peduncle strongly clavate. Flowers: sepals 3.5–6 × 1.4–2.5 mm; stamens 8–15; filaments flat, narrowly triangular, 1.9–3.2 mm. Follicles stipitate, upright and appressed; body oblong, 4–8(–9) mm, abruptly contracted into style beak; beak 0.5–1.5 mm. Seeds 0.9–1.4(–1.7) mm, glabrous.

Flowering late winter–early spring. Shaded shrubby areas, oak woodlands, and moist deciduous or mixed evergreen forests, occasionally in open pastures; 200–1500 m; Calif., Oreg.

21. AQUILEGIA Linnaeus, Sp. Pl. 1: 533. 1753; Gen. Pl. ed. 5, 237. 1754 · Columbine, ancolie [derivation disputed; possibly Latin *aqua,* water, and *legere,* to draw or collect, because of the wet habitat of some species or quantity of liquid nectar borne in spurs, or Latin *aquila,* eagle, because of similarity in shape of curved spurs of some European species to an eagle's talons]

Alan T. Whittemore

Herbs, perennial, from slender woody rhizomes. Leaves basal and cauline, proximal leaves petiolate, distal leaves sessile; cauline leaves alternate. Leaf blade 1–3×-ternately compound, leaflets lobed or parted, margins crenate. Inflorescences terminal, 1–10-flowered cymes or solitary flowers, to 30 cm; bracts leaflike, not forming involucre. Flowers bisexual, radially symmetric; sepals not persistent in fruit, 5, white to blue, yellow, or red, plane, narrowly ovate to oblong-lanceolate, short-clawed, 7–51 mm; petals 5, distinct, white to blue, yellow, or red, blade oblong to rounded or spatulate, 0–30 mm, base backward-pointing tubular spur, apex plane, nectary in ± enlarged tip of spur; stamens many; filaments filiform; scalelike staminodes usually present between stamens and pistils; pistils 5–10, simple; ovules many per pistil; beak present. Fruits follicles, aggregate, sessile, cylindric, sides prominently veined; beak terminal, straight, 3–26 mm. Seeds black, obovoid, smooth. $x = 7$.

Species ca. 70 (21 in the flora): circumboreal.

Species of *Aquilegia* are polymorphic and difficult to define adequately. Some of the variability is because of introgressive hybridization. Even distantly related species of columbine are often freely interfertile, and many cases of natural hybridization and introgression are known from North America. Only the most important are mentioned below. In arid areas *Aquilegia* species tend to form small populations often completely isolated from one another. This leads to local fixation of genes and therefore increased variability in species such as *A. micrantha* and *A. desertorum.* In addition, populations with spurless petals are occasionally found in many species.

SELECTED REFERENCES Munz, P. A. 1946. The cultivated and wild columbines. Gentes Herb. 7: 1–150. Payson, E. B. 1918. The North American species of *Aquilegia.* Contr. U.S. Natl. Herb. 20: 133–157.

1. Spurs hooked, 3–22 mm; sepals white or blue.
    2. Spurs 14–22 mm; introduced species, at low elevations (0–1500 m) . . . . 4. *Aquilegia vulgaris*
    2. Spurs 3–10 mm; native, at high elevations or high latitudes.
        3. Basal leaves much shorter than stems . . . . 3. *Aquilegia brevistyla*
        3. Basal leaves about as long as stems.

      4. Sepals and spurs white or nearly so; Wyoming . . . . 2. *Aquilegia laramiensis*
      4. Sepals and spurs blue; Colorado . . . . 1. *Aquilegia saximontana*
1. Spurs straight or nearly so (sometimes tips incurved in *A. flavescens*), 8–180 mm; sepals blue, white,
   cream, yellow, pink, or red (*A. flavescens* with yellow or pink sepals).
    5. Sepals and spurs blue, white, cream, reddish purple, or occasionally pink (if pink then with no
       trace of yellow); flowers usually erect (sometimes nodding in *A. micrantha*); spurs slender (stout
       at least proximally in *A. jonesii*), evenly tapered from base.
      6. Leaflets viscid . . . . 13. *Aquilegia micrantha*
      6. Leaflets not viscid.
        7. Spurs 8–15 mm . . . . 5. *Aquilegia jonesii*
        7. Spurs 25–70(–72) mm.
          8. Leaflets glaucous on both sides, 5–14 mm, crowded, primary petiolules 3–15 mm; spurs
            25–40 mm. . . . . .6. *Aquilegia scopulorum*
          8. Leaflets green adaxially, 13–42(–61) mm, not crowded, primary petiolules (10–)20–70
            mm; spurs 28–72 mm. . . . . .7. *Aquilegia coerulea*
    5. Sepals and spurs yellow, pink and yellow, or red; flowers erect or nodding; spur shape various.
      9. Sepals red (at least proximally); spurs red (red proximally, then pink in *A. shockleyi*), stout (at
       least proximally), abruptly narrowed near middle, 12–32 mm; flowers nodding or pendent.
       10. Sepals perpendicular to floral axis; petal blades 0–6 mm.
         11. Mouth of spur cut obliquely backward; stamens 17–30 mm. . . . . 18. *Aquilegia eximia*
         11. Mouth of spur truncate or with short blade; stamens 12–17 mm.
           12. Leaflets glaucous on both sides; petal blades 2–5 mm . . . . 17. *Aquilegia shockleyi*
           12. Leaflets green adaxially; petal blades 0–6 mm. . . . . 16. *Aquilegia formosa*
       10. Sepals parallel to or divergent from floral axis; petal blades 4–12 mm.
         13. Sepals red proximally, yellow-green distally, not much longer than petal blades; sta-
           mens 8–14 mm . . . . .19. *Aquilegia elegantula*
         13. Sepals red or apex green or yellow-green, about 2 times length of petal blades; sta-
           mens 14–23 mm.
           14. Blades of petals pale yellow or yellow-green; basal leaves 2×-ternately com-
             pound, leaflets to 17–52 mm; e North America, w to c Texas . . . . 21. *Aquilegia canadensis*
           14. Blades of petals yellow or red and yellow; basal leaves 2–3×-ternately com-
             pound, leaflets to 9–26(–32) mm; Arizona, New Mexico, Utah . . . . 20. *Aquilegia desertorum*
      9. Sepals and spurs yellow or pink; spurs slender (except for *A. flavescens* and *A. barnebyi*),
       evenly tapered from base (sometimes abruptly narrowed near middle in *A. flavescens* and *A.
       micrantha*), 10–180 mm; flowers usually erect, sometimes nodding.
       15. Spurs 42–180 mm.
         16. Spurs 72–180 mm; petal blades spatulate . . . . 12. *Aquilegia longissima*
         16. Spurs 42–70 mm; petal blades oblong, not much broadened distally.
           17. Sepals 14–18 mm wide. . . . . 11. *Aquilegia hinckleyana*
           17. Sepals 5–10 mm wide. . . . . 10. *Aquilegia chrysantha*
       15. Spurs 10–40 mm.
         18. Spurs yellow, stout, ± incurved, 10–18 mm; flowers nodding . . . . 15. *Aquilegia flavescens*
         18. Spurs yellow to pink or cream, slender, straight, 15–40 mm; flowers erect to nodding.
           19. Beak 15–18 mm; sepals 9–19 mm, yellow; se New Mexico, w Texas . . . . 9. *Aquilegia chaplinei*
           19. Beak 8–12 mm; sepals not as above: either cream or pink or if yellow, then
             (15–)20–25 mm; Colorado, Arizona to California.
             20. Sepals (15–)20–25 mm, petal blades 8–17 mm, spurs 25–40 mm; flowers
               erect; California. . . . . .8. *Aquilegia pubescens*
             20. Sepals 8–20 mm, petal blades 6–10 mm, spurs 14–30 mm; flowers nodding
               or erect; Colorado, Arizona, Utah.
               21. Leaflets viscid, green adaxially . . . . 13. *Aquilegia micrantha*
               21. Leaflets not viscid, glaucous on both surfaces . . . . 14. *Aquilegia barnebyi*

1. Aquilegia saximontana Rydberg in A. Gray et al., Syn. Fl. N. Amer. 1: 43. 1895 [C][E][F]

Stems 5–25 cm. **Basal leaves** 1–2×-ternately compound, 5–25 cm, no longer than stems; leaflets green adaxially, 7–22 mm, not viscid; primary petiolules 7–40 mm (leaflets usually crowded), glabrous. **Flowers** nodding; sepals divergent from floral axis, blue, oblong-ovate, 9–18 × 5–8 mm, apex obtuse to acute; petals: spurs blue, hooked, 3–9 mm, stout, evenly tapered from base, blades yellowish, oblong or spatulate, 7–10 × 3–5 mm; stamens 5–8 mm. **Follicles** 7–10 mm; beak 3–5 mm. 2*n* = 14.

Flowering summer (Jul–Aug). Cliffs and rocky slopes, subalpine and alpine; of conservation concern; 3300–4000 m; Colo.

2. Aquilegia laramiensis A. Nelson, Wyoming Agric. Exp. Sta. Bull. 28: 78–79. 1896 · Laramie columbine [C][E]

Stems 5–25 cm. **Basal leaves** 1–2×-ternately compound, 5–25 cm, about as long as stems; leaflets green adaxially, to 9–27 mm, not viscid; primary petiolules 17–35 mm (leaflets not crowded), glabrous. **Flowers** nodding; sepals divergent from floral axis, greenish white, linear or lanceolate, 7–15 × 1–4 mm, apex acute to rounded; petals: spurs white, hooked, 5–8 mm, stout, evenly tapered from base, blades cream colored, oblong to elliptic, 5–12 × 3–7 mm; stamens 10–11 mm. **Follicles** 10–14 mm; beak 3–5 mm. 2*n* = 14.

Flowering summer (Jun–Jul). Rock crevices; of conservation concern; 2000–2500 m; Wyo.

Aquilegia laramiensis is endemic to the Laramie Mountains.

3. Aquilegia brevistyla Hooker, Fl. Bor.-Amer. 1: 24. 1829 [E][F]

Stems 20–80 cm. **Basal leaves** 2×-ternately compound, 5–30 cm, much shorter than stems; leaflets green adaxially, to 12–44 mm, not viscid; primary petiolules 10–55 mm (leaflets not crowded), distally pilose or occasionally glabrous. **Flowers** nodding; sepals divergent from floral axis, blue, lanceolate or narrowly elliptic, 13–16 × 3–6 mm, apex acuminate or acute; petals: spurs blue, hooked, 5–10 mm, stout, evenly tapered from base, blades white or pale yellow, oblong, 7–10 × 3–6 mm;

stamens 7–11 mm. **Follicles** 15–25 mm; beak 3–4 mm. 2*n* = 14, 16.

Flowering summer (Jun–Aug). Open woods, meadows, shores, and rock outcrops; 800–3500 m; Alta., B.C., Man., N.W.T., Ont., Sask., Yukon; Alaska, Mont., S.Dak., Wyo.

Aquilegia brevistyla has been reported from Minnesota. All Minnesota material seen, however, has been misidentified.

The chromosome number in this species needs to be reinvestigated.

4. Aquilegia vulgaris Linnaeus, Sp. Pl. 1: 533. 1753 · Ancolie vulgaire [W1]

Stems 30–72 cm. **Basal leaves** 2×-ternately compound, 10–30 cm, much shorter than stems; leaflets green adaxially, to 15–47 mm, not viscid; primary petiolules 22–60 mm (leaflets not crowded), pilose or rarely glabrous. **Flowers** nodding; sepals divergent from or perpendicular to floral axis, mostly blue or purple, lance-ovate, (10–)15–25 × 8–12 mm, apex broadly acute or obtuse; petals: spurs mostly blue or purple, hooked, 14–22 mm, stout, evenly tapered from base, blades mostly blue or purple, oblong, 10–13 × 6–10 mm; stamens 9–13 mm. **Follicles** 15–25 mm; beak 7–15 mm. 2*n* = 14 (Europe).

Flowering spring–summer (May–Jul). Disturbed habitats; 0–1500 m; introduced; B.C., Man., N.B., Nfld., N.S., Ont., P.E.I., Que.; Conn., Ill., Iowa, Maine, Mass., Mich., Minn., N.H., N.J., N.Y., N.C., Ohio, Pa., R.I., Vt., Wash., W.Va.; native to Europe.

Aquilegia vulgaris is cultivated as an ornamental and occasionally escapes into disturbed habitats. Most plants have blue or purple flowers (the wild type), but horticultural races with white or reddish flowers sometimes become established. Many cultivated columbines are derived from hybrids between *A. vulgaris* and related species. Some of our escaped plants are probably descended from such hybrids.

5. Aquilegia jonesii Parry, Amer. Naturalist 8: 211. 1874 [E][F]

Aquilegia jonesii var. *elatior* Boothman

Stems 3.5–12 cm. **Basal leaves** 1–2×-ternately compound, 2.5–10 cm, not much shorter than stems; leaflets to 3–12 mm, not viscid, glaucous on both sides; primary petiolules 1–10 mm (leaflets very crowded), pilose.

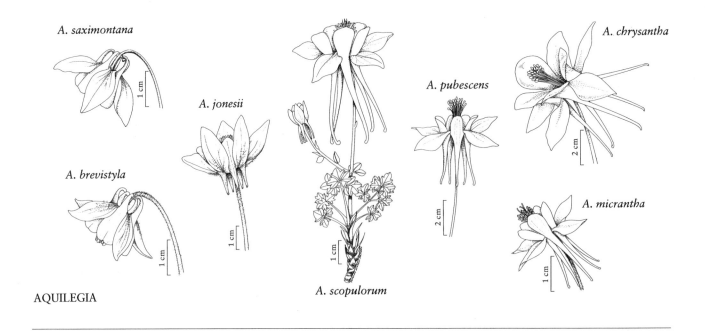

AQUILEGIA

Flowers usually erect; sepals divergent from floral axis, blue or purple, oblong-elliptic, 15–22 × 6–10 mm, apex acute to obtuse; petals: spurs blue, straight, ± parallel, 8–15 mm, stout (at least proximally), evenly tapered from base, blades blue, oblong to spatulate, 8–13 × 5–6 mm; stamens 9–14 mm. **Follicles** 14–22 mm; beak 8–12 mm. $2n = 14$.

Flowering summer (Jun–Jul). Rocky places in limestone areas, usually subalpine; 1800–3400 m; Alta.; Mont., Wyo.

6. Aquilegia scopulorum Tidestrom, Amer. Midl. Naturalist 1: 167, plate 11. 1910 [E][F]

Aquilegia scopulorum var. *calcarea* (M. E. Jones) Munz

Stems 5–30 cm. **Basal leaves** 2×-ternately compound, 3–12 cm, much shorter than stems; leaflets to 5–14 mm, not viscid, glaucous on both sides; primary petiolules 3–15 mm (leaflets densely crowded), glabrous or pilose. **Flowers** erect; sepals perpendicular to floral axis, blue to white or sometimes reddish purple, ovate-oblong, 13–22 × 4–10 mm, apex obtuse or broadly acute; petals: spurs blue to white or reddish purple, straight, ± parallel, 25–40 mm, slender, evenly tapered from base, blades white, yellow, blue, or reddish purple, oblong or spatulate, 8–14 × 4–7 mm; stamens 11–15 mm. **Follicles** 10–18 mm; beak 8–11 mm.

Flowering summer (Jun–Aug). Rocky slopes, woodlands, or meadows; 2000–3500 m; Nev., Utah.

Populations intermediate between *Aquilegia scopulorum* and *A. coerulea* are known from Utah. *Aquilegia scopulorum* has also been reported from southern Wyoming; the record is doubtful.

7. Aquilegia coerulea E. James, Account Exped. Pittsburgh 2: 15. 1823 [E]

Stems 15–80 cm. **Basal leaves** (1–)2(–3)×-ternately compound, 9–37 cm, much shorter than stems; leaflets green adaxially, to 13–42(–61) mm, not viscid; primary petiolules (10–)20–70 mm (leaflets not crowded), glabrous or occasionally pilose. **Flowers** erect; sepals perpendicular to floral axis, white, blue, or sometimes pink, elliptic-ovate to lance-ovate, 26–51 × 8–23 mm, apex obtuse to acute or acuminate; petals: spurs white, blue, or sometimes pink, straight, ± parallel or divergent, 28–72 mm, slender, evenly tapered from base, blades white, oblong or spatulate, 13–28 × 5–14 mm; stamens 13–24 mm. **Follicles** 20–30 mm; beak 8–12 mm.

Varieties 4 (4 in the flora): w North America.

Aquilegia coerulea shows considerable geographic variation in flower color and in size of different floral organs, reflecting adaptation to different pollinators in different parts of its range (R. B. Miller 1981). Four weakly differentiated varieties are recognized.

Aquilegia coerulea var. *coerulea* and *A. coerulea* var. *ochroleuca* intergrade to some extent; northwestern populations of var. *coerulea* often contain individuals with pale flowers, and eastern populations of var. *ochroleuca* often contain blue-flowered plants.

The Gosiute tribe chewed the seeds of *Aquilegia coerulea* or used an infusion made from the roots to treat abdominal pains or as a panacea (D. E. Moerman 1986).

Most authors have spelled the epithet "caerulea"; "cocrulca" is the original spelling.

Columbine (as *Aquilegia caerulea*) is the state flower of Colorado.

SELECTED REFERENCE Miller, R. B. 1981. Hawkmoths and the geographic patterns of floral variation in *Aquilegia caerulea*. Evolution 35: 763–774.

1. Petal blades 13–17 mm . . . .
. . . . 7c. *Aquilegia coerulea* var. *alpina*
1. Petal blades 19–28 mm.
  2. Sepals medium to deep blue. . . . .
  . . . . 7a. *Aquilegia coerulea* var. *coerulea*
  2. Sepals white, pale blue, or pink.
    3. Spurs 36–54 mm (means of populations 40–48 mm); stamens 13–18 mm; Utah to Nevada, Montana . . . .
    . . . . 7b. *Aquilegia coerulea* var. *ochroleuca*
    3. Spurs 45–72 mm (means of populations 50–58 mm); stamens 18–24 mm; Utah, Arizona . . . 7d. *Aquilegia coerulea* var. *pinetorum*

7a. Aquilegia coerulea E. James var. **coerulea** E

Aquilegia coerulea var. *daileyae* Eastwood

Leaves 2×-ternately compound. **Flowers:** sepals medium or deep blue, 28–43 mm; petals: spurs 34–48 mm, blades (17–)20–24 mm; stamens 13–19 mm. $2n = 14$.

Flowering summer (mid Jun–early Sep). Rocky slopes or near streams in open woodland or herbland; 2100–3600 m; Colo., N.Mex., Wyo.

7b. Aquilegia coerulea E. James var. **ochroleuca** Hooker, Bot. Mag., 90: subplate 5477. 1864 E

Leaves 2×-ternately compound. **Flowers:** sepals white or pale blue, 26–40 mm; petals: spurs 36–54 mm, blades 19–25 mm; stamens 13–18 mm. $2n = 14$.

Flowering summer (late Jun–Aug). Rocky slopes or near streams, in open woodland or herbland; 2000–3600 m; Idaho, Mont., Nev., Utah, Wyo.

7c. Aquilegia coerulea E. James var. **alpina** A. Nelson, Wyoming Agric. Exp. Sta. Bull. 28: 78. 1896 C E

Leaves 1–2×-ternately compound. **Flowers:** sepals pale blue, 26–36 mm; petals: spurs 28–36 mm, blades 13–17 mm; stamens 13–15 mm.

Flowering summer (Jun–Aug). Open rocky slopes; of conservation concern; 2100–3500 m; Wyo.

7d. Aquilegia coerulea E. James var. **pinetorum** (Tidestrom) Payson ex Kearney & Peebles, Fl. Pl. Ferns Ariz., 318. 1942 E

Aquilegia pinetorum Tidestrom, Amer. Midl. Naturalist 1: 166–167. 1910

Leaves 2–3×-ternately compound. **Flowers:** sepals white, pale blue, or pink, (22–)29–51 mm; petals: spurs 43–72 mm, blades 20–28 mm; stamens 17–24 mm.

Flowering spring–summer (May–Sep). Rocky slopes or near streams, in open woodland or herbland; 1800–3400 m; Ariz., Utah.

Aquilegia coerulea var. *pinetorum* is very close to *A. coerulea* var. *ochroleuca*; leaf differences given by P. A. Munz (1946) do not hold up. The varieties may be reliably distinguished if data on variation in the whole population are available (mean spur length 40–48 mm in populations of *A. coerulea* var. *ochroleuca*, 50–58 mm in populations of *A. coerulea* var. *pinetorum*), but the range for individual flowers overlaps somewhat.

8. Aquilegia pubescens Coville, Contr. U.S. Natl. Herb. 4: 56–57, plate 1. 1893 E F

Stems 20–50 cm. **Basal leaves** 1–2×-ternately compound, 8–25 cm, much shorter than stems; leaflets green adaxially, to 10–25 mm, not viscid; primary petiolules 12–48 mm (leaflets not crowded), glabrous or sometimes pilose. **Flowers** erect; sepals perpendicular to floral axis, cream to yellow or pink, lance-oblong to ovate, (15–)20–25 × 5–9 mm, apex obtuse to acuminate; petals: spurs cream to yellow or pink, straight, ± parallel or divergent, 25–40 mm, slender, evenly tapered from base, blades cream to yellow, oblong, 8–17 × 5–8 mm; stamens 13–16 mm. **Follicles** 20–25 mm; beak 10–12 mm. $2n = 14$.

Flowering summer (Jun–Aug). Open rocky places, alpine or subalpine; 3000–4000 m; Calif.

Aquilegia pubescens is endemic to the southern Sierra Nevada. It sometimes forms extensive hybrid swarms with *A. formosa*.

9. Aquilegia chaplinei Standley ex Payson, Contr. U.S. Natl. Herb. 20: 156–157. 1918 C E

Aquilegia chrysantha A. Gray var. *chaplinei* (Payson) E. J. Lott

Stems 20–50 cm. **Basal leaves** 2–3×-ternately compound, 7–25 cm, much shorter than stems; leaflets to 9–19 mm, not viscid, sometimes glaucous adaxially; primary petiolules to 10–95 mm (leaflets not crowded), glabrous or sparsely pilose. **Flowers** suberect to inclined; sepals perpendicular to floral axis, pale yellow, broadly lanceolate, 9–19 × 4–6 mm, apex obtuse to acuminate; petals: spurs yellow, straight, ± parallel or divergent, 30–40 mm, slender, evenly tapered from base, blades pale yellow, oblong, 7–14 × 5–6 mm; stamens 10–19 mm. **Follicles** 18–22 mm; beak 15–18 mm.

Flowering summer (Jul–Aug). Rocky places in canyons, mostly along streams; of conservation concern; 1900 m; N.Mex., Tex.

Aquilegia chaplinei is endemic to the Guadalupe Mountains.

10. Aquilegia chrysantha A. Gray, Proc. Amer. Acad. Arts 8: 621. 1873 F

Aquilegia chrysantha var. *rydbergii* Munz

Stems 30–120 cm. **Basal leaves** 2–3×-ternately compound, 9–45 cm, much shorter than stems; leaflets green adaxially, to 11–55 mm, not viscid; primary petiolules 20–50 mm (leaflets not crowded), glabrous or distally pilose. **Flowers** erect; sepals perpendicular to floral axis, yellow, lanceolate to ovate-lanceolate, 20–36 × 5–10 mm, apex narrowly acute or acuminate; petals: spurs yellow, straight, ± parallel or divergent, 42–65 mm, slender, evenly tapered from base, blades yellow, oblong, 13–23 × 6–15 mm; stamens 12–25 mm. **Follicles** 18–30 mm; beak 10–18 mm.

Flowering spring–summer (Apr–Sep). Damp places in canyons; 1000–3500 m; Ariz., Colo., N.Mex., Tex., Utah; nw Mexico.

Colorado populations supposedly having spurs only 35–40 mm have been called *Aquilegia chrysantha* var. *rydbergii*. Material seen from this area falls within the normal range of variation of the species. Populations

intermediate between *A. chrysantha* and *A. coerulea* var. *pinetorum* occur in northern Arizona (M. Butterwick et al. 1991).

11. Aquilegia hinckleyana Munz, Gentes Herb. 7: 141, fig. 38. 1946 · Hinckley's columbine C E

Aquilegia chrysantha A. Gray var. *hinckleyana* (Munz) E. J. Lott

Stems 50–70 cm. **Basal leaves** 2×-ternately compound, 30–40 cm, much shorter than stems; leaflets green adaxially, to 20–40 mm, not viscid; primary petiolules 25–50 mm (leaflets not crowded), pilose. **Flowers** suberect; sepals perpendicular to floral axis, yellow, ovate, 25–34 × 14–18 mm, apex obtuse; petals: spurs yellow, straight, ± parallel, 40–56 mm, slender, evenly tapered from base, blades yellow, oblong, 19–23 × 13–17 mm; stamens 17–20 mm. **Follicles** 20–25 mm; beak 20 mm.

Flowering spring (Mar–Apr). Dripping cliffs; of conservation concern; 1000 m; Tex.

Aquilegia hinckleyana is endemic to Capote Falls, Presidio County, Texas. The key and description above are based on specimens seen. E. J. Lott (1979) gave the range of sepal width in *A. hinckleyana* as 9–17 mm, thus overlapping the range in *A. chrysantha*. She considered sepal shape to be the most reliable key character for distinguishing these species, with sepals less than 2.5 times as long as wide in *A. hinckleyana* and more than 2.5 times as long as wide in *A. chrysantha*. Perhaps because of the overlap in characters, she later reduced *A. hinckleyana* to a variety of *A. chrysantha* (E. J. Lott 1985). Until her data are published, I prefer to follow the established taxonomy.

12. Aquilegia longissima A. Gray ex S. Watson, Proc. Amer. Acad. Arts 17: 317–318. 1882 · Long-spur columbine

Stems 25–90 cm. **Basal leaves** 3×-ternately compound, 20–45 cm, usually shorter than stems; leaflets green adaxially, to 20–40 mm, not viscid; primary petiolules 28–82 mm (leaflets not crowded), glabrous or sometimes pilose. **Flowers** erect; sepals perpendicular to floral axis, pale yellow, lanceolate, 25–40 × 6–11 mm, apex narrowly acute or acuminate; petals: spurs pale yellow, straight, ± parallel, 72–180 mm, very slender, evenly tapered from base, blades pale yellow, spatulate, 15–30 × 7–11 mm; stamens 20–33 mm. **Follicles** 24–31 mm; beak 16–26 mm.

Flowering summer (Jul–Sep). Near streams or in damp rocky places in canyons; 1370–1520 m; Ariz., Tex.; ne Mexico.

R. B. Miller (1985) suggested that Arizona reports of *Aquilegia longissima* are based on "unusually long-spurred individuals of *A. chrysantha,*" but it is not clear on what characters he based his interpretation. Specimens from Arizona's Baboquivari Mountains have spurs 8–10 cm long, far outside the range of *A. chrysantha,* and seem correctly identified as *A. longissima.*

13. Aquilegia micrantha Eastwood, Proc. Calif. Acad. Sci., ser. 2, 4: 559–560, plate 19. 1895 [E] [F]

Aquilegia flavescens S. Watson var. *rubicunda* (Tidestrom) S. L. Welsh; *A. micrantha* var. *mancosana* Eastwood

Stems 30–60 cm. **Basal leaves** 2–3×-ternately compound, 10–35 cm, much shorter than stems; leaflets green adaxially, to 13–32 mm, viscid; primary petiolules 21–64 mm (leaflets not crowded), glandular-pubescent or glandular. **Flowers** erect or nodding; sepals perpendicular to floral axis, white, cream, blue, or pink, oblong-lanceolate, 8–20 × 3–6 mm, apex acuminate to obtuse; petals: spurs white or colored like sepals, straight, ± parallel or divergent, 15–30 mm, slender, evenly tapered from base or occasionally ± abruptly narrowed near middle, blades white or cream, oblong, 6–10 × 3–7 mm; stamens 9–14 mm. **Follicles** 10–20 mm; beak 8–10 mm.

Flowering spring–summer (Apr–Sep). Seepy rock walls of canyons; 1000–2500 m; Ariz., Colo., Utah.

14. Aquilegia barnebyi Munz, Leafl. W. Bot. 5: 177–178. 1949 [E]

Stems 30–80 cm. **Basal leaves** 2–3×-ternately compound, 5–30 cm, much shorter than stems; leaflets to 8–20 mm, glaucous on both sides, not viscid; primary petiolules 17–34 mm (leaflets not crowded), glabrous. **Flowers** erect or nodding; sepals perpendicular to floral axis, pink, ovate-lanceolate to elliptic, 10–18 × 5–7 mm, apex acute or acuminate; petals: spurs pink, straight, ± parallel, 14–27 mm, stout proximally, slender distally, evenly tapered from base, blades yellow or cream, oblong, 6–10 × 4–6 mm; stamens 11–16 mm. **Follicles** 18–25 mm; beak 8–12 mm.

Flowering summer (Jun–Jul). Cliff walls and talus slopes, usually on shale; 1500–2600 m; Colo., Utah.

Aquilegia barnebyi is endemic to the Green River drainage. It is very similar to *A. micrantha,* and intermediate plants are found in Colorado.

15. Aquilegia flavescens S. Watson, Botany (Fortieth Parallel), 10. 1871 [E] [F]

Aquilegia flavescens var. *miniana* J. F. Macbride & Payson

Stems 20–70 cm. **Basal leaves** 2×-ternately compound, 8–30 cm, much shorter than stems; leaflets green adaxially, to 14–42 mm, not viscid; primary petiolules to 13–67 mm (leaflets not crowded), glabrous or pilose. **Flowers** nodding; sepals perpendicular to floral axis, yellow or tinged with pink, elliptic-lanceolate to oblong, 12–22 × 4–10 mm, apex obtuse to acute or sometimes acuminate; petals: spurs yellow, tips incurved, 10–18 mm, stout, evenly tapered from base or more abruptly narrowed near middle, blades cream colored, oblong, 7–10 × 4–8 mm; stamens 12–17 mm. **Follicles** 18–27 mm; beak 8–10 mm.

Flowering summer (Jun–Aug). Moist mountain meadows and alpine slopes; 1300–3500 m; Alta., B.C.; Idaho, Mont., Oreg., Utah, Wash., Wyo.

Aquilegia flavescens sometimes forms hybrid swarms with *A. formosa* var. *formosa,* which grows at lower elevations through much of its range. Intermediate specimens having pinkish red flowers and petal blades 5–6 mm are occasionally found where these species grow together. The name *A. flavescens* var. *miniana* has sometimes been mistakenly applied to these intermediates, but the type of var. *miniana* is a typical, pink-sepaled plant of *A. flavescens.*

16. Aquilegia formosa Fischer ex de Candolle in A. L. de Candolle and A. L. P. de Candolle, Prodr. 1: 50. 1824

Stems (15–)30–100 cm. **Basal leaves** 2×-ternately compound, 10–40 cm, much shorter than stems; leaflets green adaxially, to 14–68 mm, not viscid (petioles and petiolules viscid in var. *hypolasia*); primary petiolules 16–95 mm (leaflets not crowded), glabrous or pilose. **Flowers** nodding to pendent; sepals perpendicular to floral axis, red, elliptic to lanceolate, 10–26 × 4–9 mm, apex broadly acute to acuminate; petals: spurs red, straight, ± parallel, 13–21 mm, stout, abruptly narrowed near middle, blades yellow, oblong, rounded, or obsolete, 0–6 × 4–6 mm; stamens 12–17 mm. **Follicles** 15–25(–29) mm; beak 9–12 mm.

Varieties 3 (3 in the flora): North America, Mexico.

The type specimen of *Aquilegia formosa* var. *fosteri* S. L. Welsh, described from southwestern Utah, has viscid leaves with short petiolules and crowded leaflets, erect, dark red flowers with slender, evenly tapering spurs, and long (19 mm) stamens. Aside from the red sepals and spurs, it has little in common with *A. formosa* and its relatives. The crowded leaflets, erect flowers, and evenly tapering spurs are reminiscent of

A. scopulorum, and it could be a hybrid involving that species, but it is not clear what the other parent might be. The taxonomy of *Aquilegia* in southwestern Utah is complex, with six species known and several hybrid combinations apparently formed; satisfactory disposition of this name will require further work in the area.

Native Americans used *Aquilegia formosa* for various purposes: as a charm to gain the affections of men or to retain wealth and possessions; medicinally, seeds were chewed to alleviate stomach aches, and leaves were chewed or used in infusions to treat coughs, colds, and sore throats (D. E. Moerman 1986, varieties not indicated).

1. Blades of petals 3–6(–7) mm; sepals 14–26 mm, 0.9–1.3 times as long as spurs . . . .
. . . . .16a. *Aquilegia formosa* var. *formosa*
1. Blades of petals 0–3 mm; sepals 10–20 mm, 0.7–1.1 times as long as spurs.
    2. Stems and petioles glabrous or sparingly pilose, not at all viscid . . . .
    . . . . 16b. *Aquilegia formosa* var. *truncata*
    2. Stems and petioles densely pubescent, somewhat viscid . . . 16c. *Aquilegia formosa* var. *hypolasia*

16a. Aquilegia formosa Fischer ex de Candolle var. **formosa** E

Aquilegia formosa var. *communis* B. Boivin; *A. formosa* var. *megalantha* B. Boivin; *A. formosa* var. *wawawensis* (Payson) H. St. John

Stems and petioles glabrous or pilose, not viscid. **Flowers:** sepals 14–26 mm, 0.9–1.3 times as long as spurs; petal blades 3–6(–7) mm. $2n = 14$.

Flowering spring–summer (Apr–Aug). Mesic woods; 0–2500 m; Alta., B.C., Yukon; Alaska, Calif., Idaho, Mont., Nev., Oreg., Utah, Wash., Wyo.

16b. Aquilegia formosa Fischer ex de Candolle var. **truncata** (Fischer & C. A. Meyer) Baker, Gard. Chron., ser. 2, 10: 111. 1878 E F

Aquilegia truncata Fischer & C. A. Meyer in F. E. L. von Fischer et al., Index Sem. Hort. Petrop. 9, suppl.: 8. 1844; *A. formosa* var. *pauciflora* (Greene) Munz

Stems and petioles glabrous or sparsely pilose, not viscid. **Flowers:** sepals 10–20 mm, 0.7–1.1 times as long as spurs; petal blades 0–3 mm. $2n = 14$.

Flowering spring–summer (Apr–Aug). Mesic woods or shrublands; 0–3500 m; Calif., Nev., Oreg.

Aquilegia formosa var. *truncata* replaces *A. for-*

mosa var. *formosa* from the Sierra Nevada westward. The two varieties intergrade where they come together.

The common form of *Aquilegia formosa* var. *truncata* is 50–100 cm, with well-developed stem leaves. Montane forms with short stems and very small stem leaves are often separated as *A. formosa* var. *pauciflora.* Similar dwarf montane races with the floral characters of *A. formosa* var. *formosa* occur in the Pacific Northwest; these have never been separated taxonomically.

16c. Aquilegia formosa Fischer ex de Candolle var. **hypolasia** (Greene) Munz, Gentes Herb. 7: 106. 1946

Aquilegia hypolasia Greene, Leafl. Bot. Observ. Crit. 2: 141–142. 1911

Stems and petioles densely pilose, somewhat viscid. **Flowers:** sepals 13–20 mm, 0.7–1 times as long as spurs; petal blades 0–3 mm.

Flowering summer (Jun–Aug). Mesic woods or shrublands; 0–2000 m; Calif.; Mexico (Baja California).

Aquilegia formosa var. *hypolasia* is endemic to the peninsular ranges of California and adjacent Baja California. It is very close to *A. formosa* var. *truncata;* occasional plants of var. *hypoplasia* from the Peninsular Ranges may be almost glabrous, and occasional plants of var. *truncata* may be almost as pubescent as var. *hypoplasia.*

17. Aquilegia shockleyi Eastwood, Bull. Torrey Bot. Club 32: 193–194. 1905 E

Aquilegia formosa Fischer ex de Candolle var. *caelifax* (Payson) Munz; *A. mohavensis* Munz

Stems 40–100 cm. **Basal leaves** 2–3×-ternately compound, 9–45 cm, much shorter than stems; leaflets to 11–38 mm, sometimes viscid, glaucous on both surfaces; primary petiolules 23–80 mm (leaflets not crowded), glabrous or pilose. **Flowers** nodding; sepals perpendicular to floral axis, red (or sometimes partly yellow or green), lanceolate to elliptic, 10–20 × 4–8 mm, apex acuminate or acute; petals: spurs red or pink, straight, ± parallel, 12–25(–30) mm, rather stout, usually abruptly narrowed near middle, blades yellow, oblong or rounded, 2–5 × 4–7 mm; stamens 12–16 mm. **Follicles** 14–25 mm; beak 9–12 mm.

Flowering spring–summer (Apr–Aug). Moist places in dry woodlands and shrublands; 1200–2700 m; Calif., Nev.

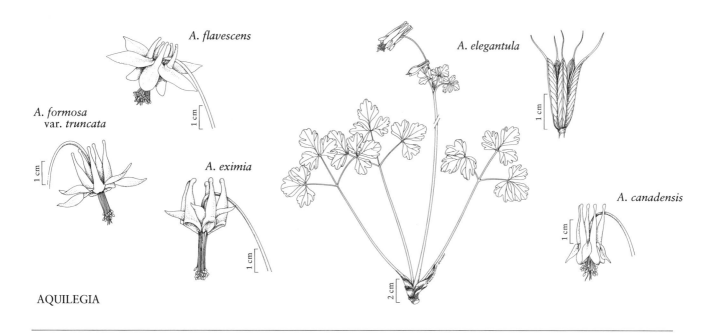

A. flavescens

A. formosa
var. *truncata*

A. eximia

A. elegantula

A. canadensis

AQUILEGIA

18. Aquilegia eximia Van Houtte ex Planchon, J. Gén. Hort. 12: 13, plate 1188. 1857 [E] [F]

Aquilegia fontinalis J. T. Howell

Stems 50–100 cm. **Basal leaves** 2–3×-ternately compound, 15–35 cm, much shorter than stems; leaflets green adaxially, to 10–48 mm, somewhat viscid; primary petiolules 21–90 mm (leaflets not crowded), glandular-pubescent. **Flowers** nodding; sepals perpendicular to floral axis, red, lance-ovate, 15–28 × 5–10 mm, apex narrowly acute or acuminate; petals: spurs red, straight, ± parallel, 18–32 mm, stout, abruptly narrowed near middle, mouth yellow, cut obliquely backward from insertion, blade thus absent; stamens 17–30 mm. **Follicles** 15–25 mm; beak 12–20 mm.

Flowering spring–summer (May–Aug). Damp rocky places; 0–2000 m; Calif.

19. Aquilegia elegantula Greene, Pittonia 4: 14–15. 1899 [F]

Stems 10–60 cm. **Basal leaves** 2×-ternately compound, 15–30 cm, usually shorter than stems; leaflets green adaxially, to 11–33 mm, not viscid; primary petiolules 17–58 mm (leaflets not crowded), glabrous or pilose. **Flowers** pendent; sepals erect, red proximally, yellow-green distally, elliptic-ovate, 7–11 × 4–5 mm, apex rounded to acute; petals: spurs red, straight, ± parallel, 16–23 mm, stout (at least proximally), abruptly narrowed near middle, blades yellow-green, oblong or rounded, 6–8 × 3–4 mm; stamens 8–14 mm. **Follicles** 13–20 mm; beak 13–15 mm.

Flowering spring–summer (May–Jul). Moist coniferous forests, especially along streams; 1500–3500 m; Ariz., Colo., N.Mex., Utah; n Mexico.

20. Aquilegia desertorum (M. E. Jones) A. Heller, Muhlenbergia 1: 27. 1901 [E]

Aquilegia formosa Fischer ex de Candolle var. *desertorum* M. E. Jones, Contr. W. Bot. 8: 2. 1898; *A. triternata* Payson

Stems 15–60 cm. **Basal leaves** 2–3×-ternately compound, 7–30 cm, much shorter than stems; leaflets to 9–26(–32) mm, not viscid, either green adaxially or glaucous on both sides; primary petiolules 15–57 mm (leaflets not crowded), glabrous or pilose, sometimes somewhat viscid. **Flowers** nodding; sepals divergent from floral axis, red or apex yellow-green, elliptic-ovate to ovate or lanceolate, 7–20 × 3–8 mm, apex obtuse to acuminate; petals: spurs red, straight, ± parallel, 16–32 mm, stout proximally, slender distally, abruptly narrowed near middle, blades yellow or red and yellow, oblong or rounded, 4–12 × 3–8 mm; stamens 14–19 mm. **Follicles** 15–30 mm; beak 8–12 mm.

Flowering spring–fall (May–Oct). Open rocky places; 2000–2500 m; Ariz., N.Mex., Utah.

Aquilegia desertorum is very similar to *A. cana-*

densis. Further research is needed to determine whether they are really distinct at the species level. The leaflets may be weakly viscid abaxially in plants from the Zion National Park area, Utah.

Plants from the eastern and southern parts of the range of *Aquilegia desertorum* have usually been considered a distinct species, *A. triternata,* mainly because of their longer sepals and petal blades (sepals narrowly ovate to lanceolate, 12–20 × 4–8 mm, apex acuminate, petal blades 6–12 mm in *A. triternata* versus sepals ovate or elliptic-ovate, 7–13 × 3–6 mm, apex obtuse or acute, petal blades 3–5 mm in *A. desertorum* in the strict sense). These sepal and petal types intergrade over much of central Arizona, however, and other characteristics supposedly diagnostic for *A. triternata* (leaves not glaucous, 3×-ternately compound) are scarcely correlated with the flower characteristics or with one another.

The Navajo-Kayenta used *Aquilegia desertorum* in ceremonies (D. E. Moerman 1986).

21. Aquilegia canadensis Linnaeus, Sp. Pl. 1: 533–534. 1753 · Canadian columbine, ancolie du Canada E F

Aquilegia australis Small; *A. canadensis* var. *australis* (Small) Munz; *A. canadensis* var. *coccinea* (Small) Munz; *A. canadensis* var. *eminens* (Greene) B. Boivin; *A. canadensis* var. *latiuscula* (Greene) Munz; *A. coccinea* Small

Stems 15–90 cm. **Basal leaves** 2×-ternately compound, 7–30 cm, much shorter than stems; leaflets green adaxially, 17–52 mm, not viscid; primary petiolules 17–93 mm (leaflets not crowded), glabrous or pilose, sometimes somewhat viscid. **Flowers** pendent; sepals divergent from floral axis, red or apex green, lance-ovate to oblong-ovate, 8–18 × 3–8 mm, apex broadly acute to acuminate; petals: spurs red, straight, ± parallel to divergent, 13–25 mm, stout (at least proximally), abruptly narrowed near middle, blades pale yellow or yellow-green, oblong to rounded, 5–9 × 4–8 mm; stamens 15–23 mm. **Follicles** 15–31 mm; beak 10–18 mm. $2n = 14$.

Flowering spring–summer (Mar–Jun). Shaded or open woods, often around cliffs, rock outcrops, and forest edges; 0–1600 m; Man., Ont., Que., Sask.; Ala., Ark., Conn., Del., Fla., Ga., Ill., Ind., Iowa, Kans., Ky., Maine, Md., Mass., Mich., Minn., Mo., Nebr., N.H., N.J., N.Y., N.C., N.Dak., Ohio, Okla., Pa., R.I., S.C., S.Dak., Tenn., Tex., Vt., Va., W.Va., Wis.

P. A. Munz divided this species into five varieties, based on size of the plants, sepals, and leaflets and whether the leaves are 2–3×-ternately compound. The variation in size of these organs is not discontinuous or even bimodal, however, and I have not seen any material with 3×-ternately compound leaves. For this reason, no varieties are recognized here. The name *Aquilegia canadensis* var. *hybrida* Hooker has been misapplied to this species; the type specimen actually belongs to *A. brevistyla* (B. Boivin 1953).

Aquilegia canadensis has also been reported from New Brunswick, but the specimen has been destroyed and the species has never been recollected in the province.

Native Americans prepared infusions from various parts of plants of *Aquilegia canadensis* to treat heart trouble, kidney problems, headaches, bladder problems, and fever, and as a wash for poison ivy; pulverized seeds were used as love charms; and a compound was used to detect bewitchment (D. E. Moerman 1986).

22. THALICTRUM Linnaeus, Sp. Pl. 1: 545. 1753; Gen. Pl. ed. 5, 242. 1754

· Meadow-rue, pigamon [*Thaliktron,* an ancient name used by Dioscorides]

Marilyn M. Park
Dennis Festerling Jr.

Herbs, perennial, from woody rhizomes, caudices, or tuberous roots. **Leaves** basal and cauline, proximal leaves petiolate, distal leaves sessile; cauline leaves alternate. **Leaf blade** 1–4×-ternately or -pinnately compound; leaflets cordate-reniform, obovate, lanceolate, or linear, sometimes 3-lobed or more, margins entire or crenate. **Inflorescences** terminal, sometimes also axillary, (1–)2–200-flowered panicles, racemes, corymbs, umbels, to 41 cm, or flowers solitary; involucres absent or present, involucral bracts 2–3 (these compound, often resembling whorl of 6–9 simple bracts), leaflike, not closely subtending flowers. **Flowers** all bisexual, bisexual and unisexual on same plant, or all unisexual with sexes on same or different plants, radially symmetric; sepals not persistent in fruit, 4–10, whitish to greenish yellow or purplish, plane, lanceolate to reniform or spatulate, 1–18 mm; petals absent; stamens 7–30; filaments filiform to clavate or distally dilated; staminodes absent between stamens and pistils; pistils 1–16, simple; ovule 1 per pistil; style present or absent. **Fruits** achenes, usually aggregate, sessile or stipitate, ovoid to obovoid, falcate, or discoid, sides prominently veined or ribbed; beak present or absent, terminal, straight to coiled, 0–4 mm. $x = 7$.

Species 120–200 (22 in the flora): nearly worldwide, mostly temperate.

Thalictrum is a taxonomically difficult genus that should be carefully researched through additional population-based field studies. Past treatments of *Thalictrum* have often emphasized leaf characters that are highly variable in most species; they are therefore of poor diagnostic value and not indicative of true relationships. Because of the paucity of field studies and a continuing emphasis on highly variable characters, the literature is replete with names that do not represent distinct entities. Often mixes of character states can be found within a single population; many of the character states used in past studies were neither ecologically nor geographically distinct.

Some species of *Thalictrum* have been divided into varieties by previous authors. In the absence of carefully collected, supporting evidence from field studies, we are unwilling to perpetuate the use of any infraspecific names.

Characters useful in identifying species of *Thalictrum* include leaflet shape, degree of dilation of filaments, anther length, shape of anther apex, achene shape and venation patterns, and vesture (glands and/or hairs) of leaves and achenes. Leaflets described in this treatment are the central, distalmost of a midstem leaf; proximal and distal leaves are more variable and often not representative of the species. Stigma and filament colors refer to fresh material in the following descriptions.

In *Thalictrum* species, the stigma extends down the side of the style, so length of style in fruit (beak) includes the stigma.

For many species no reliable characteristics for the identification of staminate material are known. Extensive field work and careful analysis are required to determine if such characteristics exist.

In a narrow strip from southeastern Ontario to Ohio to Louisiana, some individuals of some species in *Thalictrum* section *Leucocoma* may lack their normal vesture. In the absence of glands or pubescence, the differences among species are difficult to describe. The remaining characteristics overlap considerably. The species involved may be identified in the final couplets of the key as follows: if the plant in hand falls into the area of overlap for the first character of the couplet, go on to the next character, and so forth, until a distinguishing character is found. One or more of the characters offered should distinguish the infrequent, problematic individual.

Several species of *Thalictrum* are used as ornamentals. At least one species, *T. aquilegiifolium* Linnaeus, occasionally escapes cultivation in Ontario and Quebec and possibly elsewhere. The plant is tall (40–100 cm); flowers bisexual, mauve to pink; and achenes few, filiform, 3-winged, stipitate, very small, and hidden at anthesis among the bases of long, rigid stamens.

Numerous alkaloids have been identified from plants of the genus, some with pharmacologic potential. Some exhibit antimicrobial activity; others inhibit growth of tumors or lower blood pressure in mammals.

SELECTED REFERENCES Boivin, B. 1944. American *Thalictra* and their Old World allies. Rhodora 46: 337–377, 391–445, 453–487. Boivin, B. 1948. Key to Canadian species of *Thalictra*. Canad. Field-Naturalist 62: 169–170. Lecoyer, J. C. 1885. Monographie du genre *Thalictrum*. Bull. Soc. Roy. Bot. Belgique 24: 78–324. Park, M. M. 1992. A Biosystematic Study of *Thalictrum* Section *Leucocoma* (Ranunculaceae). Ph.D. dissertation. Pennsylvania State University. Tamura, M. 1968. Morphology, ecology, and phylogeny of the Ranunculaceae. VIII. Sci. Rep. Coll. Gen. Educ. Osaka Univ. 17: 41–56. Tamura, M. 1992. A new classification of the family Ranunculaceae. Acta Phytotax. Geobot. 43: 53–58.

1. Inflorescences umbels or flowers solitary (sect. *Anemonella*)............................1. *Thalictrum thalictroides*
1. Inflorescences panicles, racemes, or corymbs.
  2. Flowers bisexual; sepals 5 (often 4 in *T. alpinum*).
    3. Achenes sessile to nearly sessile; filaments filiform (sect. *Thalictrum*).
      4. Stems 15–150 cm; sepals 3–4 mm; achenes 3–15.........................2. *Thalictrum minus*
      4. Stems (3–)5–20(–30) cm; sepals 1–2.3(–2.7) mm; achenes 2–6 ....3. *Thalictrum alpinum*
    3. Achenes stipitate; filaments ± dilated distally.
      5. Filaments weakly dilated; achenes short-stipitate, stipe less than 1.5 mm, body (4–)5–6 mm;
        Canada, w United States (sect. *Omalophysa*)..........................6. *Thalictrum sparsiflorum*
      5. Filaments strongly clavate; achenes long-stipitate, stipe 1–3.5(–4) mm, body 2.5–5 mm; se
        United States (sect. *Physocarpum*).
        6. Adaxial margin of achene concave, ca. 2 times length of stipe; filaments 2.5–4 mm . . .
          ....4. *Thalictrum clavatum*
        6. Adaxial margin of achene straight, ± equaling length of stipe; filaments 2–3 mm ....5. *Thalictrum mirabile*
  2. Flowers unisexual, or unisexual and bisexual, rarely only bisexual; sepals 4(–6).
    7. Leaflets apically 3–12-lobed, lobe margins crenate (rarely entire in *T. debile*); filaments vari-
      ously colored, rarely white, filiform (sect. *Heterogamia*).
      8. Lateral veins of achene anastomosing-reticulate.........................17. *Thalictrum polycarpum*
      8. Lateral veins of achene not reticulate, veins parallel, converging, or rarely branched.
        9. Achenes laterally compressed.
          10. Leaf blade membranous, green; leaflets (5–)10–20 × (6–)8–12(–18) mm; stems
            (20–) 30–60(–150) cm; achenes 7–11(–14) per flower ....15. *Thalictrum fendleri*
          10. Leaf blade leathery, glaucous; leaflets 5–8 × 4–5 mm; stems 14–50 cm; achenes 4–
            5(–6) per flower.......................................16. *Thalictrum heliophilum*
        9. Achenes not laterally compressed, or very slightly so.
          11. Achenes stipitate; stipe 0.7–2.5 mm.
            12. Achenes erect; beak 1.5–3 mm ....8. *Thalictrum coriaceum*
            12. Achenes spreading to reflexed; beak 3–4.5(–6) mm.................14. *Thalictrum occidentale*
          11. Achenes nearly sessile; stipe 0–0.3 mm.
            13. Achenes incurved.
              14. Beak (2–)2.5–4(–5) mm; adaxial surface of achene 4–6 mm ....12. *Thalictrum confine*
             14. Beak 1.5–2.5(–3) mm; adaxial surface of achene 3–4(–6) mm ......13. *Thalictrum venulosum*
            13. Achenes straight.
              15. Roots fibrous; stems erect, 30–80 cm; largest leaflets more than 15 mm
                wide ....7. *Thalictrum dioicum*
              15. Roots tuberous; stems reclining to erect, usually less than 30(–45) cm;
                largest leaflets less than 15 mm wide.
                16. Beak 0.5–1 mm; achenes ovoid; stems erect; roots black when dry
                  ....11. *Thalictrum texanum*
                16. Beak 1.3–2 mm; achenes oblong to elliptic-lanceolate; stems reclining
                  or decumbent; roots brown.
                  17. Achenes 0.7–1.2 mm wide, veins 6–8, prominent; beak 1.3–3 mm
                    ....9. *Thalictrum debile*
                  17. Achenes 1.5–2 mm wide, veins 10–12; beak (1.3–)2.3–3 mm . . .
                    ....10. *Thalictrum arkansanum*
    7. Leaflets undivided or 3-lobed apically, lobe margins entire (some leaflet margins on some indi-
      viduals rarely crenate); filaments usually white, rarely lavender, filiform to clavate (sect. *Leuco-
      coma*).
      18. Achenes, peduncles, abaxial surfaces of leaflets, and/or petioles and rachises with stipitate
        glands ....19. *Thalictrum revolutum*
      18. Achenes, peduncles, abaxial surfaces of leaflets, and/or petioles and rachises without stipi-
        tate glands.
        19. Achenes, peduncles, abaxial surfaces of leaflets, and/or petioles and rachises with mi-
          nute papillae (i.e., sessile glands), may also be pubescent.
          20. Leaflet length 0.9–5.25 times width; nonglandular trichomes absent; filaments
            2.5–7.8 mm; anthers (0.7–)1.2–2.7(–3) mm; stipe 0.2–1.7 mm; e North America,
            rare w of Missouri...19. *Thalictrum revolutum*
          20. Leaflet length 0.9–2.6 times width; nonglandular trichomes present or absent;
            filaments 2–6.5 mm; anthers 1–3.6(–4) mm; stipe 0–1.1 mm; c North America,
            very rare e of Ohio ....20. *Thalictrum dasycarpum*

19. Achenes, peduncles, abaxial surfaces of leaflets, petioles, and rachises without papillae, may be pubescent or glabrous.

    21. Achenes, peduncles, abaxial surfaces of leaflets, and/or petioles and rachises pubescent.

        22. Anthers less than 1.5 mm, apex blunt or slightly apiculate; filaments rigid, ascending, prominently clavate; beak straight or coiled distally, ca. 1/2 as long as achene body. . . . . 18. *Thalictrum pubescens*

        22. Anthers usually 1–3.6(–4) mm, apex usually strongly apiculate; filaments flexible, drooping, filiform, scarcely dilated distally; beak ± straight, filiform, about as long as achene body . . . . 20. *Thalictrum dasycarpum*

    21. Achenes, peduncles, abaxial surfaces of leaflets, and/or petioles and rachises glabrous.

        23. Leaflets linear to narrowly lanceolate or oblanceolate, (2.6–)4–26 times longer than wide . . . . 22. *Thalictrum cooleyi*

        23. Leaflets nearly orbiculate to ovate, or lanceolate to obovate, usually less than 4 times longer than wide.

            24. Leaflets undivided or apically 2–3-lobed, largest usually less than 22 mm wide; filaments 1.8–4 mm; se United States . . . . 21. *Thalictrum macrostylum*

            24. Leaflets apically 3-lobed, seldom undivided, largest usually 15–60 mm or more wide; filaments 1.5–7.8 mm; Ontario to Ohio to Louisiana.

                25. Anthers 0.5–1.5(–2.1) mm; stigma straight or distally coiled; flowers often bisexual. . . . . .18. *Thalictrum pubescens*

                25. Anthers (0.7–)1–3.6 mm; stigma straight, ± filiform; flowers rarely bisexual (included here are very infrequent forms of *T. dasycarpum* and *T. revolutum*).

                    26. Leaflet length 0.9–5.25 times width; filaments 2.5–7.8 mm; anthers (0.7–)1.2–2.7(–3) mm; stipe 0.2–1.7 mm; e North America, infrequent w of Missouri. . . . . 19. *Thalictrum revolutum*

                    26. Leaflet length 0.9–2.6 times width; filaments 2–6.5 mm; anthers 1–3.6(–4) mm; stipe 0–1.1 mm; w, c North America, very infrequent e of Ohio . . . . 20. *Thalictrum dasycarpum*

22a. THALICTRUM Linnaeus sect. ANEMONELLA (Spach) Tamura, Acta Phytotax. Geobot. 43: 57. 1992

Anemonella Spach, Hist. Nat. Vég. 7: 239. 1838; *Syndesmon* Hoffmannsegg

Inflorescences umbels or flowers solitary. **Flowers** bisexual; sepals 5(–10); filaments narrowly clavate. **Achenes** sessile to nearly sessile.

Species 1 (1 in the flora): North America.

1. **Thalictrum thalictroides** (Linnaeus) A. J. Eames & B. Boivin, Bull. Soc. Roy. Bot. Belgique 89: 319. 1957 · Rue-anemone, windflower E F

Anemone thalictroides Linnaeus, Sp. Pl. 1: 542. 1753; *Anemonella thalictroides* (Linnaeus) Spach; *Syndesmon thalictroides* (Linnaeus) Hoffmannsegg; *Thalictrum anemonoides* Michaux

Roots black, tuberous. **Stems** erect, scapose, 10–30 cm, glabrous. **Leaves** basal; petiole 10–30 cm. **Leaf blade** 2×-ternately compound; leaflets widely ovate or obovate to nearly rotund, apically 3-lobed, 8–30 mm wide, surfaces glabrous. **Inflorescences** umbels or flowers solitary, (1–)3–6-flowered; involucral bracts usually 3-foliolate, petiolate and opposite, or sessile with leaflets appearing to be whorls of 6 petiolate leaves, otherwise similar to basal leaves. **Flowers:** sepals not caducous, white to pinkish, showy, elliptic to obovate, 5–18 mm, longer than stamens; filaments narrowly clavate, 3–4 mm; anthers 0.4–0.7 mm. **Achenes** (4–)8–12(–15), short-stipitate; stipe 0.1–0.4 mm; body ovoid to fusiform, 3–4.5 mm, prominently 8–10-veined.

Flowering spring (Mar–Jun). Deciduous woods, banks, and thickets; 0–300 m; Ont.; Ala., Ark., Conn., Del., D.C., Fla., Ga., Ill., Ind., Iowa, Kans., Ky., La., Maine, Mass., Mich., Minn., Miss., Mo., N.H., N.J., N.Y., N.C., Ohio, Okla., Pa., R.I., S.C., Tenn., Vt., Va., W.Va., Wis.

In *Thalictrum*, *T. thalictroides* is unique in having umbelliform inflorescences and is therefore easy to

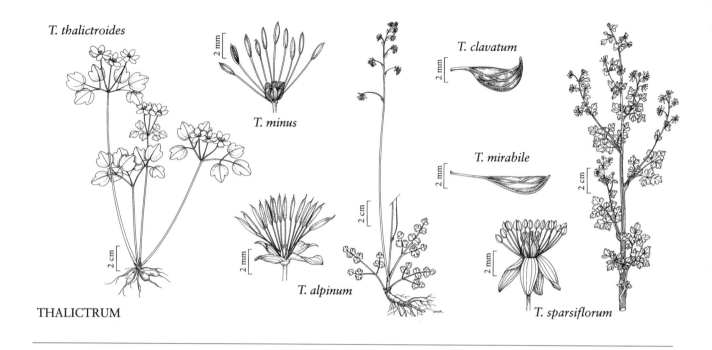

T. thalictroides

T. minus

T. clavatum

T. mirabile

T. alpinum

T. sparsiflorum

THALICTRUM

identify. Based on this one distinction, many botanists still place it in the genus *Anemonella*. The leaflets, flowers, and fruits, however, are not unlike those of *Thalictrum*.

The Cherokee used infusions prepared from the roots of *Thalictrum thalictroides* to treat diarrhea and vomiting (D. E. Moerman 1986).

22b. THALICTRUM Linnaeus sect. THALICTRUM

Thalictrum sect. *Homothalictrum* B. Boivin

Inflorescences panicles or racemes. **Flowers** bisexual; sepals 5 (often 4 in *T. alpinum*); filaments filiform. **Achenes** sessile to nearly sessile.

Species 33 (2 in the flora): temperate North America and Eurasia.

2. Thalictrum minus Linnaeus, Sp. Pl. 1: 546. 1753   [F]

Thalictrum minus subsp. *kemense* (Fries) Hultén; *T. minus* var. *stipellatum* (C. A. Meyer) Tamura

Stems erect, nearly cespitose or rhizomatous, 15–150 cm, glabrous or somewhat glandular. **Leaves** basal, 7–30 cm. **Leaf blade** 3–4-ternate; leaflets nearly orbiculate or broadly ovate, ir-regularly 2–3-lobed or margins dentate in distal 1/2, 15–30 mm, surfaces glabrous to glandular. **Inflorescences** panicles with long branches, many flowered. **Flowers:** sepals yellowish green, ovate, 3–4 mm; stamens 10–15; anthers yellowish, 2–3 mm. **Fruiting pedicels** not recurved. Achenes 3–15, sessile; body broadly ovoid to narrowly oblong-ovoid, 2.5–4 mm, ± weakly veined.

Flowering late spring–summer (Jun–Jul). Steppe meadows, shrub thickets, forest margins, and forest meadows; 0–300 m; Alaska; Eurasia.

Thalictrum minus has been reported from mainland Alaska (E. Hultén 1968); we have been able to confirm its occurrence only in the Aleutian Islands.

Initially pendent, the flowers become erect. The beak is 0.75–1 mm, much shorter than the achene, and not fimbriate.

3. **Thalictrum alpinum** Linnaeus, Sp. Pl. 1: 545. 1753
· Arctic meadow-rue, dwarf meadow-rue, pigamon
alpin F

Thalictrum alpinum var. *hebetum*
B. Boivin

Stems erect, scapose, or nearly scapose with very slender rhizomes, (3–)5–20(–30) cm, glabrous. **Leaves** all basal or single cauline leaf near base, 2–10 cm. **Leaf blade** 2×-pinnately compound, proximal primary divisions ternate; leaflets cuneate-obovate to orbiculate, apically 3–5-lobed, 2–10 mm, surfaces glabrous. In-florescences racemes, usually elongate, few flowered. **Flowers:** sepals early deciduous, purplish tinged, ovate or elliptic, 1–2.3(–2.7) mm; stamens 8–15; filaments purple; anthers bright yellow, 1.5–3 mm; stigmas purple. **Fruiting pedicels** recurved. Achenes 2–6, nearly sessile; body lance-obovoid, 2–3.5 mm, with thick veins. $2n = 14, 21$.

Flowering late spring–summer (Jun–Aug). Wet meadows, damp rocky ledges and slopes, and cold (often calcareous) bogs in willow-sedge, lodgepole pine, and spruce-fir; 0–3800 m; Greenland; B.C., Nfld., N.W.T., Que., Yukon; Alaska, Calif., Colo., Idaho, Mont., Nev., N.Mex., Oreg., Utah, Wyo.; n Eurasia.

22c. THALICTRUM Linnaeus sect. PHYSOCARPUM de Candolle, Syst. Nat. 1: 171. 1817

Physocarpum (de Candolle) Berchtold & J. Presl, not *Physocarpon* Necker ex Rafinesque 1840, not *Physocarpus* Maximowicz; *Sumnera* Nieuwland

Inflorescenes panicles to nearly corymbs. **Flowers** bisexual; sepals 5; filaments strongly clavate, petaloid. **Achenes** long-stipitate.

Species 45 (2 in the flora): temperate regions, North America, e Asia.

4. **Thalictrum clavatum** de Candolle, Syst. Nat. 1: 171. 1817 · Mountain meadow-rue E F

Thalictrum filipes Torrey & A. Gray; *T. nudicaule* Schweinitz ex Torrey & A. Gray

Roots few, blackish, filiform or somewhat tuberous. **Stems** erect, not scapose, 15–50(–60) cm, glabrous. **Leaves** basal and cauline; proximal cauline leaves petiolate, 2–3×-ternately compound; distal cauline leaves sessile or short-petiolate, 2×-ternately compound or simple. **Leaf blade:** leaflets reniform to obovate, apically 4–7-lobed, 10–30 mm wide, lobe margins crenate, surfaces abaxially glaucous. **Inflorescences** panicles or nearly corymbs, few flowered; pedicels very slender, elongate. **Flowers:** sepals white, obovate-spatulate, 2.5–4 mm; filaments white, 2.5–4 mm; anthers 0.3–0.5 mm. **Achenes** 3–8, spreading; stipe 1–3(–4) mm, usually ± 1/2 as long (sometimes nearly as long) as achene body; body flat, falcate, 3–5 mm, abaxially convex, adaxially concave, conspicuously 3-veined on each face; beak minute.

Flowering spring–summer (May–Jul). Rich moist woods, cliffs, seepage slopes, and mountain streams in mountains and piedmont; 500 m; Ga., Ky., N.C., S.C., Tenn., Va., W.Va.

5. **Thalictrum mirabile** Small, Bull. Torrey Bot. Club 27: 277. 1900 C E F

Stems weakly erect to reclining, 10–30 cm, glabrous. **Leaves** basal and cauline; basal petioles 2 cm, distal petioles shorter. **Leaf blade:** basal usually ternately compound, distal gradually less divided; leaflets nearly orbiculate to orbiculate-reniform, apically 4–7-lobed, 20–30 mm wide, lobe margins crenate, surfaces abaxially glaucescent. **Inflorescences** panicles, few flowered. **Flowers:** sepals white, spatulate to rhombic-spatulate, 1.5 mm; filaments white, 2–3 mm; anthers 0.3–0.5 mm. **Achenes** 3–8, spreading; stipe 2.5–3.5 mm, nearly as long as achene body; body flat, 2.5–4 mm, abaxial margin convex, adaxial margin straight, conspicuously 3-veined on each face; beak minute.

Flowering late spring–early summer (Jun). Moist sandstone bluffs, sinks, and rocky crevices; of conservation concern; 200–1500 m; Ala., Ga., Ky., N.C., Tenn.

Thalictrum mirabile is very similar to *T. clavatum*. Field studies are in progress to clarify the status of this species.

22d. THALICTRUM Linnaeus sect. OMALOPHYSA Turczaninow ex Fischer & C. A. Meyer in F. E. L. Fischer et al., Index Sem. Hort. Petrop. 1: 40. 1835

Inflorescences panicles or racemes. **Flowers** bisexual; sepals 5; filaments ± dilated. **Achenes** short-stipitate.

Species 24 (1 in the flora): North America, e Asia.

6. Thalictrum sparsiflorum Turczaninow ex Fischer & C. A. Meyer in F. E. L. von Fischer et al., Index Sem. Hort. Petrop. 1: 40. 1835 · Mountain meadow-rue, few-flowered meadow-rue F

Thalictrum sparsiflorum subsp. richardsonii (A. Gray) Cody; T. sparsiflorum var. nevadense B. Boivin; T. sparsiflorum var. richardsonii (A. Gray) B. Boivin; T. sparsiflorum var. saximontanum B. Boivin

Stems erect, leafy, slender, (20–)30–100(–120) cm, glabrous. **Leaves** mostly cauline, proximal leaves petiolate, distalmost sessile. **Leaf blade** (2–)3-ternate; leaflets obovate to orbiculate or cordate, usually 3-cleft and divisions 3-lobed, thin, 10–20 mm, surfaces abaxially often glandular-puberulent. **Inflorescences** axillary, 1–few flowers, diffuse, leafy; bracts leaflike, large. **Flowers:** sepals whitish or greenish, often purplish tinged, elliptic, 2–3.5(–4) mm; stamens 12–20, whitish; filaments 3–4.5 mm; anthers 0.5–0.8 mm. **Fruiting pedicels** abruptly recurved. **Achenes** (4–)6–12; stipe 0.3–1.5 mm; body obliquely obovate to half-rhombic, strongly compressed, (4–)5–6 × 3–4 mm, abaxial margin straight, glabrous or glandular-puberulent, faintly 3–4(–5)-veined; beak 1–1.5 mm.

Flowering late spring–summer (Jun–Aug). Meadows, damp thickets, bogs, and coniferous, deciduous, and riparian woods; 0–3000 m; Alta., B.C., Man., N.W.T., Ont., Sask., Yukon; Alaska, Calif., Colo., Idaho, Mont., Nev., Oreg., Utah, Wyo.; ne Asia.

The Cheyenne used the flowers and ground plants of *Thalictrum sparsiflorum* medicinally to make their horses "spirited, long-winded, and enduring" (D. E. Moerman 1986).

22e. THALICTRUM Linnaeus sect. HETEROGAMIA (de Candolle) B. Boivin, Rhodora 46: 432. 1944

Thalictrum [unranked] *Heterogamia* de Candolle, Syst. Nat. 1: 172. 1817 (as *Heterogama*)

Inflorescences panicles, racemes, or corymbs. **Flowers** unisexual (rarely also bisexual in *T. debile* and *T. fendleri*); sepals 4(–6); filaments filiform (to slightly clavate in *T. heliophilum*). **Achenes** sessile to nearly sessile.

Species 25 (11 in the flora): North America, Mexico.

7. Thalictrum dioicum Linnaeus, Sp. Pl. 1: 545. 1753 · Early meadow-rue, quicksilver-weed, pigamon dioïque E F

Roots yellow to light brown, fibrous, from stout caudex. **Stems** erect, 30–80 cm, glabrous or glandular. **Leaves** basal and cauline, petiolate. **Leaf blade** 1–4×-ternately compound; leaflets reniform or cordate to obovate or orbiculate, apically 3–12-lobed, 10–45 mm wide, lobe margins often crenate, surfaces abaxially glabrous or glandular. **Inflorescences** terminal and axillary, panicles to corymbs, many flowered. **Flowers:** sepals greenish to purple, ovate or obovate to oval, 1.8–4 mm; filaments yellow to greenish yellow, 3.5–5.5 mm; anthers 2–4 mm, mucronate to acuminate; stigma purple. **Achenes** (3–)7–13, not reflexed, sessile or nearly so; stipe terete, 0–0.2 mm; body ovoid to ellipsoid, not laterally compressed, 3.5–5 mm, glabrous, very strongly veined, veins not anastomosing-reticulate; beak 1.5–3 mm.

Flowering spring (Apr–Jun). Rocky woods, ravines, and alluvial terraces, mountains and piedmont; 10–1000 m; Man., Ont., Que.; Ala., Conn., D.C., Ga., Ill., Ind., Iowa, Kans., Ky., Maine, Md., Mass., Mich., Minn., Mo., Nebr., N.H., N.J., N.Y., N.C., N.Dak., Ohio, Pa., R.I., S.C., S.Dak., Tenn., Vt., Va., W.Va., Wis.

Glandular plants of *Thalictrum dioicum* have often been misidentified as *T. revolutum* despite important differences, especially the leaflets having crenate versus entire lobe margins, respectively. The stamens in both *T. dioicum* and *T. revolutum* are pendulous.

Native Americans used roots of *Thalictrum dioicum* in various preparations to treat diarrhea and vomiting and for heart palpitations (D. E. Moerman 1986).

8. Thalictrum coriaceum (Britton) Small, Mem. Torrey Bot. Club 4: 98. 1893 ☐E

Thalictrum dioicum Linnaeus var. *coriaceum* Britton, Bull. Torrey Bot. Club 18: 363. 1891; *T. caulophylloides* Small; *T. steeleanum* B. Boivin

Roots bright yellow, tuberous. **Stems** erect, coarse, 65–150 cm, glabrous. **Leaves** cauline; petioles of proximal leaves well developed, clasping stem, distal leaves sessile or nearly so. **Leaf blade** 1–4×-ternately compound; leaflets reniform or obovate to orbiculate, apically 3–9-lobed or toothed, 10–75 mm wide, lobe margins crenate, surfaces abaxially glabrous or glandular. **Inflorescences** panicles, pyramidal, loosely branched, many flowered. **Flowers:** sepals white to purplish, lanceolate-ovate, 1.5–5.5 mm; filaments maroon, 4–4.5 mm; anthers 2–5.5 mm, apiculate, subulate-tipped; stigma maroon. **Achenes** 3–15, erect, not reflexed, stipitate; stipe ± wing-angled, 0.7–2.5 mm; body obliquely ovoid to ellipsoid, not laterally compressed, 3–6.5 mm, strongly veined or ribbed, veins not anastomosing; beak 1.5–3 mm.

Flowering mid–late spring (late May–Jun). Rocky or mesic, open, deciduous woods, thickets, and moist alluvium, chiefly in mountains and piedmont; 3–1100 m; D.C., Ky., Md., N.C., Pa., Tenn., Va., W.Va.

Glandular plants of *Thalictrum coriaceum* have often been misidentified as *T. revolutum* despite important differences in the leaflets, the latter having entire rather than crenate lobe margins.

Studies by M. Park and L. Morse (unpubl.) for The Nature Conservancy confirmed that *Thalictrum steeleanum* is highly variable in all allegedly diagnostic characters and is not distinct from *T. coriaceum*.

9. Thalictrum debile Buckley, Amer. J. Sci. Arts 45: 175. 1843 ☐C☐E

Roots brownish, fusiform-tuberous with dried ribs. **Stems** reclining, branched and flexible proximally, 10–40 cm, glabrous. **Leaves:** proximal cauline leaves petiolate, 1–3×-ternately compound; distal cauline leaves petiolate or sessile, 1–2×-ternately compound or simple. **Leaf blade:** leaflets ovate or obovate to reniform or orbiculate, apically shallowly to deeply 3–7-lobed, rarely undivided, 4–15 mm wide, surfaces glabrous. **Inflorescences** termi-

nal and axillary, panicles, elongate, few flowered. **Flowers:** sepals whitish, lanceolate to obovate, 1.5–2.7 mm; filaments colored, not white, 1.5–2 mm; anthers 1.7–2.5 mm, mucronate; stigma color unknown. **Achenes** 1–6, not reflexed, nearly sessile; stipe 0.1–0.3 mm; body oblong to elliptic-lanceolate, not compressed, 3–3.7 × 0.7–1.2 mm, glabrous, prominently 6–8-veined, veins not anastomosing; beak 1.3–2 mm.

Flowering in early spring (Mar–Apr). Rich, rocky, limestone woods, often in wet, alluvial soil; of conservation concern; 50–300 m; Ala., Ga., Miss.

Thalictrum debile is closely related to *T. arkansanum* and *T. texanum*. The distinctions among the three species should be further studied.

10. Thalictrum arkansanum B. Boivin, Rhodora 46: 433. 1944 ☐C☐E

Roots brown, thick, tuberous. **Stems** decumbent, 20–40 cm, glabrous. **Leaves:** distal cauline long-petiolate. **Leaf blade** 1–3×-ternately compound; leaflets ovate to obovate to reniform or orbiculate, 3–(or more)-lobed, largest leaflets less than 15 mm wide, lobe margins crenate, surfaces glabrous. **Inflorescences** terminal, racemes, few flowered. **Flowers:** sepals whitish, ovate or elliptic, 0.9–2.8(–3) mm; filaments colored, 2–3 mm; anthers 1.8–2.3 mm, mucronate, tip 0.1–0.4 mm; stigma color unknown. **Achenes** few, not reflexed, nearly sessile; stipe 0.1–0.3 mm; body ellipsoid to oblong, 3.5–4.5 × 1.5–2 mm, glabrous, veins 10–12, not anastomosing-reticulate; beak (1.3–)2.3–3 mm.

Flowering early spring (Mar–Apr). Wet bottomland forest, sometimes upland woods; of conservation concern; 20–150 m; Ark., Okla., Tex.

Poorly known, *Thalictrum arkansanum* is closely related to *T. texanum* and *T. debile;* it possibly should be considered as a variety of the latter.

11. Thalictrum texanum (A. Gray) Small, Fl. S.E. U.S., 446. 1903 ☐C☐E

Thalictrum debile Buckley var. *texanum* A. Gray in A. Gray et al., Syn. Fl. N. Amer. 1: 18. 1895

Roots becoming black when dry, tuberous, not ribbed, irregular. **Stems** erect, 10–45 cm, rigid, glabrous. **Leaves:** distal cauline long-petiolate. **Leaf blade** 2×-ternately compound; leaflets cuneate to reniform, undivided, cleft, or lobed, 2–7 mm wide, margins entire or sometimes weakly crenate, surfaces glabrous, somewhat glaucous. **Inflorescences**

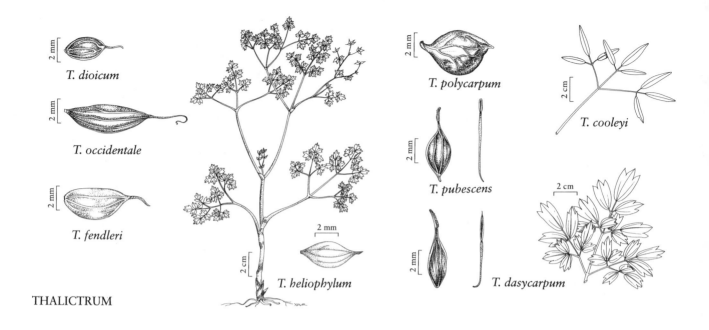

THALICTRUM

terminal, racemes, several flowered. **Flowers:** sepals lanceolate to obovate, in staminate flowers 1.7–3 mm, in pistillate flowers 0.7–1.5 mm; filaments colored, not white, ca. 1.5 mm; anthers 1.4–2 mm; stigma color unknown. **Achenes** few, not reflexed, nearly sessile; stipe 0.1–0.3 mm; body ovoid, not laterally compressed, adaxial surface 2.7–3.7 × 1.4–1.6 mm, glabrous, prominently 6–8-veined, veins not anastomosing-reticulate; beak 0.5–1 mm.

Flowering early spring (Mar–Apr). Margins or openings of mesic woodlands or forests; of conservation concern; 10–100 m; Tex.

Somewhat difficult to locate in the field, *Thalictrum texanum* is currently known from only two populations. It is closely related to *T. arkansanum* and *T. debile* and sometimes treated as a variety of the latter.

Thalictrum confine is quite similar to *T. occidentale* and *T. venulosum*; it has been treated as a variety or synonym of the latter (R. S. Mitchell 1988).

12. **Thalictrum confine** Fernald, Rhodora 2: 232. 1900 · Northern meadow-rue, pigamon des fronterières [E]

Thalictrum turneri B. Boivin; *T. venulosum* Trelease var. *confine* (Fernald) B. Boivin

Stems erect, to 100 cm, from rhizomes. **Leaves:** cauline leaves 3–4, those proximal to inflorescence petiolate, those subtending panicle branches sessile. **Leaf blade:** leaflets reniform-cordate, apically 3–5-lobed, 15–50 mm wide, lobe margins cre-

nate, surfaces glabrous to glandular. **Inflorescences** terminal, panicles, narrow with ascending branches, many flowered. **Flowers:** sepals yellowish to purple, oblong, 1.5–5 mm; filaments colored, not white; anthers 3–4.5 mm, mucronate; stigma purple. **Achenes** 4–13, erect, incurved, not reflexed, sessile; body fusiform to ovoid, not laterally compressed, adaxial surface 4–6 mm, glandular, veins prominent, not anastomosing-reticulate; beak (2–)2.5–4(–5) mm.

Flowering early–mid summer (Jun–Jul). Alluvial or shingly calcareous shores and talus; 0–200 m; N.B., Ont., P.E.I., Que.; N.Y., Vt.

The relationship between *Thalictrum confine* and *T. venulosum* is unclear and requires additional field study.

13. **Thalictrum venulosum** Trelease, Proc. Boston Soc. Nat. Hist. 23: 302. 1886 · Veiny meadow-rue, early meadow-rue [E]

Stems erect, 20–50 cm, glabrous, from rhizomes. **Leaves** basal and cauline; cauline 1–3, those proximal to inflorescence petiolate, those subtending panicle branches sessile. **Leaf blade** 3–4×-ternately compound; leaflets obovate to orbiculate, apically 3–5-lobed, 5–20 mm, lobe margins crenate, surfaces abaxially glabrous or glandular-puberulent. **Inflorescences** terminal, panicles, narrow and dense, many flowered. **Flowers:** sepals greenish white, lanceolate or broadly ovate to elliptic or ob-ovate, 2–4 mm; filaments colored, not white, (1.8–)3–5.5 mm; anthers 2–3.5 mm, blunt to mucronate; stigma commonly yellowish. **Achenes** 5–17, erect to spreading, not reflexed, nearly sessile; stipe 0.1–0.3 mm; body often distinctly curved, elliptic-oblong, nearly terete to slightly flattened, adaxial surface 3–4 (–6) mm, glabrous to glandular, veins distinct, not anastomosing-reticulate; beak 1.5–2.5(–3) mm.

Flowering early–mid summer (Jun–Aug). Prairies, riparian woods, and coniferous, deciduous, and mixed forests; 600–3700 m; Alta., B.C., Man., N.W.T., Ont., Que., Sask., Yukon; Colo., Idaho, Minn., Mont., Nebr., Nev., N.Mex., N.Dak., Oreg., S.Dak., Utah, Wash., Wis., Wyo.

Thalictrum venulosum is similar to *T. confine* and *T. occidentale.* Careful field studies are needed to clarify the relationships among these taxa.

14. Thalictrum occidentale A. Gray, Proc. Amer. Acad. Arts 8: 372. 1873 · Western meadow-rue E F

Thalictrum occidentale var. *macounii* B. Boivin; *T. occidentale* var. *palousense* H. St. John

Roots yellow to medium brown or black, thin, fibrous. **Stems** erect, 30–100(–120) dm, glabrous, often from well-developed rhizomes. **Leaves** mostly cauline. **Leaf blade** 3–4×-ternately compound; leaflets orbiculate to obovate-cuneate or cordate, apically 3-lobed, 10–30 mm wide, lobe margins coarsely crenate, surfaces glabrous to glandular. **Inflorescences** terminal (some flowers in axils of distal leaves), panicles, rather open, many flowered. **Flowers:** sepals whitish or greenish or purplish tinged, ovate, 3.5–4.4 mm in staminate flowers, 1.5–2 mm in pistillate flowers; filaments purplish, 4–10 mm; anthers 1.5–4 mm, long-apiculate; stigma often purplish. **Achenes** 6–9, spreading to reflexed, short-stipitate; stipe 0.4–1.2 mm; body fusiform, not laterally compressed, (4–)6–9(–10) mm, tapering at both ends, glandular, strongly 3-veined on each side, veins not anastomosing; beak 3–4.5(–6) mm.

Flowering early–mid summer (Jun–Aug). Open woods, meadows, and copses; 200–3400 m; Alta., B.C., Sask., Yukon; Alaska, Calif., Colo., Idaho, Mont., Nev., N.Dak., Oreg., Utah, Wash., Wyo.

Thalictrum occidentale is similar to *T. confine* and *T. venulosum;* thorough field studies are needed to determine whether or not they should be maintained as separate species. *Thalictrum occidentale* can usually be distinguished by its reflexed achenes.

Plants of northern British Columbia, sometimes called *Thalictrum occidentale* var. *breitungii* (B. Boivin) Brayshaw, appear to be intermediate between *T. occidentale* and *T. venulosum* (T. C. Brayshaw, pers. comm.); achenes are ascending, ± compressed, and beaks rather short (2–4 mm) (T. C. Brayshaw 1989).

Some of the Native Americans used *Thalictrum occidentale* medicinally for headaches, eye trouble, and sore legs, to loosen phlegm, and to improve blood circulation (D. E. Moerman 1986).

15. Thalictrum fendleri Engelmann ex A. Gray, Mem. Amer. Acad. Arts, ser. 2, 4: 5. 1849 · Fendler's meadow-rue F

Thalictrum fendleri var. *platycarpum* Trelease; *T. fendleri* var. *wrightii* (A. Gray) Trelease

Roots dark brown to ± black (when dry), fibrous. **Stems** mostly erect, sometimes reclining, (20–)30–60(–150) cm, glabrous, from rhizomes or branched caudices. **Leaves** mainly cauline, mostly short-petiolate. **Leaf blade** green, (2–)3–4×-ternately compound, membranous; leaflets obliquely orbiculate or nearly cordate, apically 3-lobed, (5–)10–20 × (6–)8–12(–18) mm wide, lobe margins crenate, surfaces abaxially often glandular. **Inflorescences** terminal and axillary, panicles, open and leafy, many flowered. **Flowers:** sepals whitish or greenish, in staminate flowers ovate to elliptic, 3–5 mm; in pistillate flowers ovate to rhombic or broadly lanceolate, 1.5–2 mm; filaments deep yellow or purplish, 4–7.5 mm; anthers 2.2–3.4 mm, apiculate with tip to 0.8 mm; stigma purplish. **Achenes** 7–11(–14), not reflexed, sessile to short-stipitate; stipe 0–2 mm; body oblanceolate to obliquely obovate-elliptic, strongly laterally compressed, (5–)9(–11) mm, glandular or glabrous, 3–4(–5)-veined on each side, veins ± parallel, converging toward ends (rarely branched or sinuous), not anastomosing-reticulate; beak 1.5–4 mm.

Flowering early–mid summer (Jun–Aug). Willow, birch, mountain brush, sagebrush-snowberry, boxelder-cottonwood, alder, ponderosa pine, lodgepole pine, aspen–tall forb, and spruce-fir communities; 1100–3300 m; Ariz., Calif., Colo., Idaho, Mont., Nev., N.Mex., Oreg., S.Dak., Tex., Utah, Wyo.; n Mexico.

The stems and achenes of *Thalictrum fendleri* are often purplish.

Decoctions prepared from the roots of *Thalictrum fendleri* were used medicinally by Native Americans to cure colds and gonorrhea, and in ceremonies (D. E. Moerman 1986).

16. Thalictrum heliophilum Wilken & DeMott, Brittonia 35: 156. 1983 [C] [E] [F]

Roots fibrous. **Stems** 14–50 cm, arising singly or in dense clusters of 2–3 from short, horizontal, fibrous-rooted rhizomes. **Leaves** basal and cauline, petiolate. **Leaf blade** ternately compound, cauline blades gradually reduced upward, distalmost 2-ternate; leaflets broadly obovate, apically 3-toothed, otherwise undivided, 5–8 × 4–5 mm, leathery, surfaces glabrous, glaucous. **Inflorescences** terminal, panicles, many flowered. **Flowers:** sepals 4, color unknown, lanceolate to ovate, 2–3 mm; filaments brownish, 2–3 mm; anthers 2–3 mm, apiculate; stigma color unknown. **Achenes** 4–5(–6), not reflexed, nearly sessile; stipe 0.1–0.2 mm; body oblique-obovate, strongly laterally compressed, 4–5 mm, glabrous, glaucous, prominently 3-veined on each side, veins converging near apex, rarely branched or sinuous, not anastomosing-reticulate; beak ca. 1.5 mm.

Flowering summer (Jun–Aug). Decomposing shale of Green River Formation; of conservation concern; 2500 m; Colo.

In a genus of primarily mesophytic plants, *Thalictrum heliophilum* is notable for its relatively xeric habitat. Known only from Garfield and Rio Blanco counties, northwestern Colorado, it is similar to the widespread *T. fendleri*; it may be distinguished by its smaller, leathery, glaucous leaflets and fewer achenes.

17. Thalictrum polycarpum (Torrey) S. Watson, Proc. Amer. Acad. Arts 14: 288. 1879 · Tall western meadow-rue [F]

Thalictrum fendleri Engelmann ex A. Gray var. *polycarpum* Torrey, Pacif. Railr. Rep. 4(5): 61. 1857

Roots fibrous. **Stems** erect, 6–18(–20) dm, glabrous. **Leaves** mostly cauline, petiolate. **Leaf blade** 3–4×-ternately compound; leaflets orbiculate to obovate, apically 3-cleft or 3-parted, divisions undivided or shallowly 3-lobed, 15–40 mm wide, lobes rounded or somewhat acute, surfaces glabrous or glandular. **Inflorescences** terminal, panicles, many flowered. **Flowers:** sepals whitish to purplish, elliptic to ovate or lanceolate, 2–4(–5) mm; filaments whitish to pinkish, 3–6 mm; anthers (1.4–) 2–4 mm, distinctly apiculate. **Achenes** 10–15, spreading in globose heads, not reflexed, sessile or nearly so; stipe 0–0.6 mm; body nearly globose to obovoid to obliquely obovate, laterally compressed, somewhat inflated and papery, 4–7(–8) mm, glabrous to glandular, often with 1 or 2 primary veins on each side, veins sinuous, branched, anastomosing-reticulate; beak 2–4 mm.

Flowering mid–late spring (Apr–Jun). Streamsides and other moist places, forests, and open woodlands; 600–3100 m; Calif., Nev., Oreg., Utah; Mexico (Baja California).

Thalictrum polycarpum is the only species in sect. *Heterogamia* with anastomosing-reticulate veins on the achene.

The stems and roots of *Thalictrum polycarpum* are considered poisonous when ingested by humans or cattle; Native Americans used this species medicinally as a wash for headaches, as an application for sprains, and as a universal charm and panacea (D. E. Moerman 1986).

22f. THALICTRUM Linnaeus sect. LEUCOCOMA (Greene) B. Boivin, Rhodora 46: 469. 1944

Thalictrum [unranked] *Leucocoma* Greene, Leafl. W. Bot. 2: 54–55. 1910; *Leucocoma* (Greene) Nieuwland 1914, not *Leucocoma* Rydberg 1917

Inflorescences panicles, racemes, or corymbs. **Flowers** bisexual and/or unisexual; sepals 4–6; filaments filiform to clavate. **Achenes** sessile to stipitate.

Species 5 (5 in the flora): North America.

18. **Thalictrum pubescens** Pursh, Fl. Amer. Sept. 2: 388.
1814, not **T. pubescens** Schleicher ex de Candolle 1817
· Tall meadow-rue, late meadow-rue, meadow-weed,
muskrat-weed, king-of-the-meadow, pigamon
pubescent E F W1

Thalictrum carolinianum Bosc ex de
Candolle [var.] *subpubescens* de
Candolle; *T. polygamum* Muhlenberg
ex Sprengel; *T. polygamum* var.
hebecarpum Fernald; *T. polygamum*
var. *intermedium* B. Boivin; *T.
polygamum* var. *pubescens* (Pursh)
K. C. Davis; *T. pubescens* var.
hebecarpum (Fernald) B. Boivin

Stems erect, coarse, 50–300 cm.
Leaves basal and cauline; basal and proximal cauline
leaves petiolate, distal cauline sessile; petioles and rachises frequently pubescent. **Leaf blade** ternately and
pinnately decompound; leaflets light to dark green,
cordate or nearly orbiculate to ovate or obovate, apically undivided to 2–3(–5)-lobed or -toothed, 11–68
× 5–70 mm, length 0.8–2.6 times width, membranous to firm, margins scarcely revolute, lobe margins
entire, surfaces abaxially pubescent to glabrous. **Inflorescences** racemes or panicles to corymbs, apically ±
rounded, many flowered; peduncles and pedicels often
pubescent. **Flowers** unisexual or bisexual (sometimes
bisexual with very few stamens); sepals 4(–6), white to
purplish, elliptic-rounded, 2–3.5 mm; filaments ascending, white to purplish, filiform to distinctly clavate, 1.5–7 mm, usually rigid; anthers 0.5–1.5(–2.1)
mm, usually blunt or only slightly apiculate. **Achenes**
numerous, sessile to stipitate; stipe 0.5–1.5(–2.4) mm;
body ellipsoid, 3–5 mm, prominently veined, usually
pubescent; beak usually persistent, straight or coiled
distally, 0.6–2.5 mm, about 1/2 length of achene body.
$2n = 126$.

Flowering late spring–summer (mid Jun–early Aug).
Full sun to deep shade, rich woods, low thickets,
swamps, wet meadows, and stream banks; 15–1500
m; St. Pierre and Miquelon; N.B., Nfld., N.S., Ont.,
P.E.I., Que.; Ala., Ark., Conn., Del., D.C., Ga., Ill.,
Ind., Ky., Maine, Md., Mass., Mich., Miss., N.H.,
N.J., N.Y., N.C., Ohio, Pa., R.I., S.C., Tenn., Vt., Va.,
W.Va.

The ovaries change from white to purplish, becoming light green, then darker green, and finally brown
as fruits mature.

Because of the polymorphic nature of *Thalictrum
pubescens,* a proliferation of names for minor morphologic variants has resulted. Field studies (M. Park
1992) have shown that too much morphologic variation occurs within populations to support the recognition of previously described taxa. Plants in New England and northeastern Canada often have a corymbose
inflorescence and longer filaments and achene beaks.

This species is often incorrectly treated in floras as
T. polygamum Sprengel, an invalid name.

19. **Thalictrum revolutum** de Candolle, Syst. Nat. 1: 173.
1817, not **T. revolutum** Le Lièvre 1873 · Skunk meadow-
rue, wax-leaved meadow-rue, purple meadow-rue E

Thalictrum amphibolum Greene; *T.
hepaticum* Greene; *T. moseleyi*
Greene; *T. revolutum* var.
glandulosior B. Boivin

Stems erect, coarse, 50–150 cm.
Leaves cauline, proximal leaves
petiolate, distal sessile; petioles
and rachises stipitate-glandular
to glabrous. **Leaf blade** 1–4×-
ternately compound; leaflets grayish or brownish
green or dark to bright green, lanceolate, elliptic,
ovate, reniform to obovate, apically undivided or 2–
3(–5)-lobed, 9–60 × 5–50 mm, length 0.9–2.7(–5.25)
times width, usually leathery, margins often revolute,
lobe margins entire; surfaces abaxially with sessile to
stalked glands or muriculate to whitish papillose. **Inflorescences** racemes to panicles, elongate, many
flowered; peduncles and pedicels sometimes stipitate-
glandular. **Flowers** usually unisexual, staminate and
pistillate on different plants; sepals 4(–6), whitish,
ovate to oblanceolate, (2–)3–4 mm; filaments white,
slightly clavate, 2.5–7.8 mm, ± flexible; anthers (0.7–)
1.2–2.7(–3) mm, blunt to apiculate. **Achenes** 8–16, sessile or slightly stipitate; stipe 0.2–1.7 mm; body lanceolate to ellipsoid, 3.5–5 mm, prominently veined, usually stipitate-glandular; beak ± persistent, linear-
filiform, (1–)1.5–3.3(–5) mm, ± equal to length of
achene body. $2n = 140$.

Flowering spring–summer (Mar–Jul). Dry open
woods, brushy banks, thickets, barrens, and prairies;
30–2000 m; Man., Ont., Que.; Ala., Ariz., Ark., Colo.,
Conn., Del., D.C., Fla., Ga., Ill., Ind., Iowa, Ky., La.,
Md., Mass., Mich., Minn., Miss., Mo., Nev., N.J.,
N.Mex., N.Y., N.C., Ohio, Okla., Pa., R.I., S.C.,
S.Dak., Tenn., Tex., Vt., Va., W.Va., Wis., Wyo.

Glandular individuals of *Thalictrum revolutum* have
been called var. *glandulosior.* They are seen throughout
the range of the species and do not represent a distinct
lineage. Occasional glandular plants with unusually
short anthers are often misidentified as *T. pubescens.*

Material of this species from the western United
States has been incorrectly assumed by previous authors to be *T. dasycarpum,* because *T. revolutum* is not
included in floras of that region.

20. Thalictrum dasycarpum Fischer & Avé-Lallemant in F. E. L. von Fischer et al., Index Sem. Hort. Petrop. 8: 72. 1842 · Purple meadow-rue [E] [F]

Thalictrum dasycarpum var. *hypoglaucum* (Rydberg) B. Boivin

Stems erect, stout, 40–150(–200) cm. **Leaves** chiefly cauline; basal and proximal cauline petiolate, distal cauline leaves sessile or nearly so; petioles and rachises glabrous or occasionally pubescent and/or stipitate-glandular. **Leaf blade:** basal and proximal cauline 3–5×-ternately compound; leaflets brownish green to dark green or bright green, ovate to cuneate-obovate, apically undivided or 2–3(–5)-lobed, 15–60 × 8–45 mm, length 0.9–2.6 times width, usually leathery with veins prominent abaxially, margins often revolute, lobe margins entire, surfaces abaxially usually pubescent and/or papillose (i.e., with very minute sessile glands). **Inflorescences** panicles, apically ± acutely pyramidal, many flowered; peduncles and pedicels usually glabrous, rarely pubescent or stipitate-glandular. **Flowers** usually unisexual, staminate and pistillate on different plants; sepals 4(–6), whitish, lanceolate, 3–5 mm; filaments white to purplish, filiform, scarcely dilated distally, 2–6.5 mm, flexible; anthers 1–3.6(–4) mm, usually strongly apiculate. **Achenes** numerous, sessile or nearly sessile; stipe 0–1.1 mm; body ovoid to fusiform, 2–4.6 mm, prominently veined, usually pubescent and/or glandular; beak often dehiscent as fruit matures, ± straight, filiform, 1.5–4.7(–6) mm, about as long as achene body.

Flowering late spring–summer (May–late Jul). Deciduous, riparian woods, damp thickets, swamps, wet meadows, and prairies; 80–2500 m; Alta., B.C., Man., Ont., Que., Sask., Yukon; Ala., Ariz., Ark., Colo., Idaho, Ill., Ind., Iowa, Kans., Ky., La., Mich., Minn., Miss., Mo., Mont., Nebr., N.Mex., N.Y., N.Dak., Ohio, Okla., Pa., S.Dak., Tenn., Tex., Utah, Wis., Wyo.

Thalictrum dasycarpum is a variable species similar to, and possibly intergrading with, *T. pubescens*. Glabrous variants of *T. dasycarpum* have been treated as *T. dasycarpum* var. *hypoglaucum*. Glabrous and glandular (stipitate and papillate) forms are found throughout the range of the species and occur together in some populations.

Native Americans used *Thalictrum dasycarpum* medicinally to reduce fever, cure cramps, as a stimulant for horses, and as a love charm (D. E. Moerman 1986).

21. Thalictrum macrostylum Small & A. Heller, Mem. Torrey Bot. Club 3: 8. 1892   [E]

Thalictrum subrotundum B. Boivin

Stems erect to ± reclining, slender, 50–200 cm, glabrous. **Leaves** basal and cauline; basal and proximal cauline petiolate, distal cauline sessile; petioles and rachises glabrous, neither pubescent nor glandular. **Leaf blade** ternately and pinnately decompound; leaflets grayish green to brownish to bright green, nearly orbiculate to ovate or obovate, apically undivided or shallowly 2–3-lobed, 5–16(–22) × 3–18 mm, length 1–3.3 times width, leathery and prominently reticulate abaxially, or sometimes quite membranous, margins sometimes revolute, lobe margins entire; surfaces abaxially glabrous. **Inflorescences** racemes or panicles, elongate, few flowered; peduncles and pedicels neither pubescent nor glandular. **Flowers** either unisexual with staminate and pistillate on different plants, or bisexual and unisexual with staminate and bisexual on some plants, pistillate and bisexual on others; sepals 4(–6), greenish to white, nearly orbiculate, 1–2 mm; filaments white, filiform or sometimes clavate, 1.8–4 mm, rigid to flexible; anthers 0.5–1.2 mm. **Achenes** numerous, slightly stipitate; stipe 0.3–0.7 mm; body ovoid, 3–4.5 mm, prominently veined, glabrous; beak 0.7–1.7 mm. $2n = 56$.

Flowering late spring–summer (early Jun–mid Jul). Low woods, rich wooded slopes, cliffs, swampy forests, meadows, and limestone sinks; 500–800 m; Ala., Fla., Ga., Miss., N.C., S.C., Va.

Much variation in *Thalictrum macrostylum* seems to be associated with habitat differences, especially the amount of sunlight received. The name *T. subrotundum* merely represents plants of *T. macrostylum* growing in deep shade. Common garden studies and cluster analyses do not support recognition of two species (M. Park 1992).

22. Thalictrum cooleyi H. E. Ahles, Brittonia 11: 68. 1959 · Cooley's meadow-rue [C] [E] [F]

Stems erect to reclining, slender, 60–200 cm. **Leaves:** proximal cauline petiolate, distal cauline sessile to nearly sessile; petioles and rachises glabrous, neither pubescent nor glandular. **Leaf blade:** proximal cauline mostly 2×-ternately compound, distal cauline usually ternately compound; leaflets linear to narrowly lanceolate or oblanceolate, apically occasionally 2–3-lobed, 12–68 × 1–12 mm, length (2.6–)4–26 times width, membranous to leathery, margins sometimes revolute, lobe margins entire; surfaces abaxially glabrous. **Inflorescences** racemes to panicles, elongate, few flowered; peduncles and pedicels neither pubescent nor glandular. **Flowers** usually unisexual, staminate and pistillate on different plants; sepals 4–5, white to yellowish in staminate flowers, greenish in pistillate flowers, obovate, 1.5 mm; filaments white to purple, 2.5–6 mm; anthers 0.9–2.5 mm. **Achenes** 5–6, sessile or nearly sessile; stipe 0–0.4 mm; body ellipsoid, 4.5–6 mm, prominently veined, glabrous; beak 1.3–2.4 mm. $2n = 210$.

Flowering summer (mid Jun–mid Jul). Boggy, savannalike borders of low woodlands, and disturbed areas such as roadside ditches, clearings, and edges of fre-quently burned savannas; of conservation concern; Fla., N.C.

Thalictrum cooleyi occurs commonly on Grifton soil and is associated with some sort of disturbance, including clearings, edges of frequently burned savannas, roadsides, and powerline rights-of-way that are maintained by fire or mowing. Silvicultural and agricultural practices and their associated suppression of fire have seriously affected populations of *T. cooleyi*. Furthermore, fruit production appears to be quite low in the species (S. W. Leonard 1987). Leaves of *Thalictrum cooleyi* have fewer leaflets than other species of *Thalictrum* sect. *Leucocoma*.

16. BERBERIDACEAE Jussieu

• Barberry Family

R. David Whetstone

T. A. Atkinson

Daniel D. Spaulding

Herbs or shrubs [trees], perennial, evergreen or deciduous, sometimes rhizomatous. **Stems** with or without spines. **Leaves** alternate, opposite, or fascicled, simple, 2–3-foliolate, or 1–3×-pinnately or 2–3(–4)×-ternately compound; stipules present or absent; venation pinnate or palmate. **Inflorescences** terminal or axillary, racemes, cymes, umbels (or umbel-like), spikes, or panicles, or flowers solitary or in pairs, flowers pedicellate or sessile. **Flowers** bisexual, inconspicuous or showy, radially symmetric; stipitate glands absent (except in *Vancouveria*); sepaloid bracteoles 0–9; perianth sometimes absent (*Achlys*), more frequently present, 2- or 3-merous, or sepals and petals intergrading (*Nandina*); sepals 6, distinct, often petaloid and colored, not spurred; petals 6–9, distinct, plane or hooded; nectary present; stamens 6; anthers dehiscing by valves or longitudinal slits; ovary superior, apparently 1-carpellate; placentation marginal or appearing basal; style present or absent, sometimes persistent in fruit as beak. **Fruits** follicles, berries, or utricles. **Seeds** 1–50, sometimes arillate; endosperm abundant; embryo large or small; mature seeds elevated on elongating stalk in *Caulophyllum*.

Genera 15, species ca. 650 (8 genera, 33 species in the flora): widespread, well represented in the north temperate zone.

Berberidaceae presents several interesting biogeographic features. *Achlys* is disjunct from western North America to eastern Asia with few morphologic differences between taxa. *Diphylleia*, *Jeffersonia*, and *Podophyllum*, each with a single eastern North American species, exhibit wide disjunctions to eastern Asia. *Caulophyllum* has three species, one in eastern Asia and two in the flora. *Vancouveria* is endemic to northwestern United States with nearest relations to *Epimedium* Linnaeus (H. Loconte and J. R. Estes 1989b; W. T. Stearn 1938), an exclusively Eastern Hemisphere genus.

Nandina, Berberis, Epimedium, and *Podophyllum* are commonly cultivated.

The perianth of Berberidaceae is commonly composed of three distinct types of organs, but terminology for the organs varies from author to author. In our treatment, we refer to the small, outer parts as bracteoles (collectively forming a calyculus); the large, middle parts as sepals; and the innermost parts, which are commonly nectariferous, as petals. Some authors have referred to the bracteoles as outer sepals and to the petals as staminodes.

SELECTED REFERENCES Ernst, W. R. 1964. The genera of Berberidaceae, Lardizabalaceae, and Menispermaceae in the southeastern United States. J. Arnold Arbor. 45: 1–35. Loconte, H. 1993. Berberidaceae. In: K. Kubitzki et al., eds. 1990+. The Families and Genera of Vascular Plants. 2+ vols. Berlin etc. Vol. 2, pp. 147–152. Loconte, H. and J. R. Estes. 1989b. Phylogenetic systematics of Berberidaceae and Ranunculales (Magnoliidae). Syst. Bot. 14: 565–579. Meacham, C. A. 1980. Phylogeny of the Berberidaceae with an evaluation of classifications. Syst. Bot. 5: 149–172. Ohwi, J. 1965. Flora of Japan (in English). . . . Washington. Nowicke, J. W. and J. J. Skvarla. 1981. Pollen morphology and phylogenetic relationships of the Berberidaceae. Smithsonian Contr. Bot. 50: 1–83. Terebayashi, S. 1985. The comparative floral anatomy and systematics of the Berberidaceae. I. Morphology. Mem. Fac. Sci. Kyoto Univ., Ser. Biol. 10: 73–90. Terebayashi, S. 1985b. The comparative floral anatomy and systematics of the Berberidaceae. II. Systematic considerations. Acta Phytotax. Geobot. 36: 1–13.

1. Shrubs (aerial stems sometimes very short in *Berberis repens* and *B. pumila*).
   2. Leaves 2–3×-pinnately compound; margins of leaflets entire; inflorescences panicles . . . 1. *Nandina,* p. 273
   2. Leaves simple or 1×-pinnately compound; if compound, margins of leaflets toothed; inflorescences racemes or umbels, or flowers solitary. . . . . 3. *Berberis,* p. 276
1. Perennial herbs.
   3. Leaves simple, 2–7(–9)-parted; usually both basal and cauline leaves present.
      4. Flowers 2 or more in cymes or umbels; leaves 2-parted, each segment 5–9-lobed; anthers dehiscent by 2 apically hinged valves; berries dark blue . . . . 4. *Diphylleia,* p. 287
      4. Flowers solitary; leaves 5–7(–9)-parted, segments lobed or not; anthers dehiscent by longitudinal slits; berries yellow, orange, red, or purplish red . . . . 5. *Podophyllum,* p. 287
   3. Leaves compound; leaves all basal or (in *Caulophyllum*) all cauline.
      5. Leaves 2–3-foliolate; flowers solitary or in dense spikes.
         6. Leaves 3-foliolate; perianth absent; flowers in dense spikes; British Columbia to California. . . . . 6. *Achlys,* p. 288
         6. Leaves 2-foliolate; perianth present; flowers solitary; e North America (w to Minnesota and Iowa). . . . . 8. *Jeffersonia,* p. 291
      5. Leaves 2–3×-ternately compound; flowers in racemes, panicles, or compound cymes.
         7. Petals well developed, 4–6 mm; ovary wall persisting and enclosing seeds in fruit; leaves all basal; California to Washington . . . . 7. *Vancouveria,* p. 290
         7. Petals inconspicuous, 1–3 mm; ovary wall rupturing, seeds naked at maturity; leaves cauline; e North America (w to Manitoba, Nebraska, and Arkansas). . . . . 2. *Caulophyllum,* p. 274

1. NANDINA Thunberg, Nov. Gen. Pl. 1: 14. 1781 · [Chinese name meaning "plant from the south"]

R. David Whetstone
T. A. Atkinson
Daniel D. Spaulding

Shrubs, evergreen, to ca. 2 m, glabrous. **Rhizomes** absent. **Aerial stems** monomorphic, mostly unbranched, with leaves densely clustered mostly along distal 1/3 of plant. **Leaves** persistent, alternate, 2–3×-pinnately compound; petiole attached at base of blade, petioles and petiolules swollen at base. **Leaf blade** broadly ovate in overall outline, 30–50 cm; leaflet blades elliptic to ovate to lanceolate, margins entire; venation pinnate. **Inflorescences** terminal or axillary panicles of dozens to hundreds of flowers. **Flowers** 3-merous, 5–7 mm; bracteoles present; all perianth parts caducous, cream to white; sepals and petals intergrading, 27–36; nectariferous petals absent; stamens 6; anthers dehiscing by longitudinal slits; pollen exine punctate; ovary

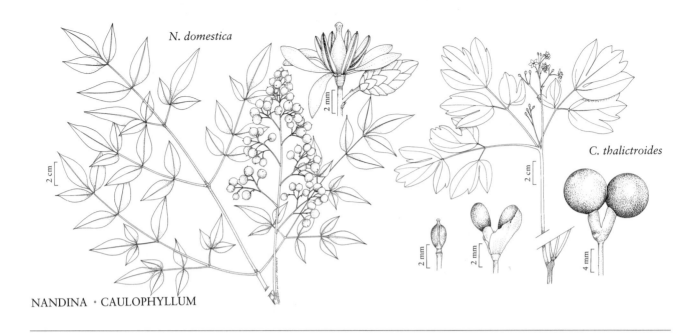

N. domestica

C. thalictroides

NANDINA · CAULOPHYLLUM

club-shaped; placentation submarginal; style central. **Fruits** berries, red to purplish, orbicular. **Seeds** 1–3, grayish or brownish; aril absent.

Species 1 (1 in the flora): North America, Asia.

Nandina is treated as a separate family, Nandinaceae, by A. Takhtajan (1986).

SELECTED REFERENCES   Nakai, T. 1915–1936. Flora Sylvatica Koreana. . . . 22 parts. Seoul and Keijyo. Part 21, pp. 61–62.   Shen Y. 1954. Phylogeny and wood anatomy of *Nandina*. Taiwania 5: 89–91.

1. **Nandina domestica** Thunberg, Nov. Gen. Pl. 1: 14. 1781 · Heavenly-bamboo [F]

Stems erect. **Wood and pith** bright yellow. **Leaves** frequently reddish tinged, 5–10 dm; petioles basally enlarged and clasping. **Leaflets** 9–81, nearly sessile, leaflet blades 4–11 × 1.5–3 cm, base cuneate, apex acuminate. **Inflorescences** with hundreds of flowers, 1–2 dm. **Flowers** fragrant, pedicellate; perianth segments imbricate, weakly 2–4-seriate. **Berries** 6–9 mm. **Seeds** mostly 2. $2n = 20$.

Flowering spring–summer (May–early Jul); fruiting summer–fall (Sep–Dec). Old home sites, woodlands, mesic flood plains, hammocks; 0–600 m; introduced; Ala., Ark., Fla., Ga., La., N.C., S.C., Tex.; native, Asia (Japan, China, India).

Nandina domestica is commonly cultivated as an ornamental. China and Japan have been considered the possible sources of cultivated material. In the flora, seedlings are frequent in the vicinity of plantings, and mature plants have been found far from areas of current cultivation in the southeastern United States.

2. **CAULOPHYLLUM** Michaux, Fl. Bor.-Amer. 1: 204. 1803 · Blue cohosh, caulophylle [Greek *caulos*, stem, and *phyllos*, leaf]

Henry Loconte

Herbs, perennial, deciduous, to 2–9 dm, glabrous. **Rhizomes** nodose, producing 2 leaves per year. **Aerial stems** present. **Leaves** caducous, cauline, 2-ranked, 1st leaf larger than 2d leaf, 2–4×-ternately compound; petioles short or absent. **Leaf blade** broadly obovate in overall outline; rachis pulvinate; leaflet blades broadly obovate, entire or lobed, margins not toothed; venation pinnate to palmate. **Inflorescences** terminal, compound cymes. **Flowers** 3-merous, 6–20 mm; bracteoles caducous, 3–4, sepaloid; sepals 6, yellow, purple, red, brown, or green, petaloid; petals 6, fan-shaped, bearing nectar; stamens 6; anthers dehiscing by 2 apically hinged flaps; pollen exine reticulate; ovary bladderlike; placentation appearing basal; styles eccentric. **Fruits** not developed, ovary wall soon rupturing. **Seeds** 2, elevated on elongating stalk, naked at maturity; seed coat blue, fleshy, glaucous; aril absent. $x = 8$.

Species 3 (2 in the flora): North America, Asia.

Caulophyllum species are understory herbs of mesophytic forests, alluvial flats, rich mesic slopes, and limestone slopes. The seeds of *Caulophyllum* are buoyant and showy and may be dispersed by water as well as other means; seed germination is hypogeal, the cotyledons remaining underground. *Caulophyllum* is occasionally cultivated in woodland gardens.

SELECTED REFERENCES Brett, J. F. 1981. The Morphology and Taxonomy of *Caulophyllum thalictroides* (L.) Michx. (Berberidaceae) in North America. M.S. thesis. University of Guelph.   Loconte, H. and W. H. Blackwell. 1985. Intrageneric taxonomy of *Caulophyllum* (Berberidaceae). Rhodora 87: 463–469.   Loconte, H. and J. R. Estes. 1989. Generic relationships within Leonticeae (Berberidaceae). Canad. J. Bot. 67: 2310–2316.   Moore, R. J. 1963. Karyotype evolution in *Caulophyllum*. Canad. J. Genet. Cytol. 5: 384–388.

1. Pistil 3–5 mm; style 1–2 mm; stamen filaments 1.5–2.5 mm; sepals 6–9 mm; inflorescences with 4–18 flowers; 1st leaf (2–)3-ternate; leaflets 5–10 cm. . . . . .1. *Caulophyllum giganteum*
1. Pistil 1–3 mm; style 0.25–1 mm; stamen filaments 0.5–1.5 mm; sepals 3–6 mm; inflorescences with 5–70 flowers; 1st leaf 3(–4)-ternate; leaflets 3–8 cm. . . . . .2. *Caulophyllum thalictroides*

1. Caulophyllum giganteum (Farwell) Loconte & W. H. Blackwell, Phytologia 49: 483. 1981 [E]

Caulophyllum thalictroides (Linnaeus) Michaux var. *giganteum* Farwell, Rep. (Annual) Michigan Acad. Sci. 20: 178. 1918

Stems 2–7 dm. **Leaves:** 1st leaf (2–)3×-ternate; 2d leaf (1–)2×-ternate; leaflets 5–10 × 3–8 cm. **Inflorescences** with 4–18 flowers. **Flowers:** bracteoles 2–4 mm; sepals purple, red, brown, yellow, 6–9 × 1–4 mm, laterally revolute; petals 2–3 mm; stamen filaments 1.5–2.5 mm; pistil 3–5 mm; style 1–2 mm. $2n = 16$.

Flowering early spring. Mesophytic forests; 0–700 m; Ont., Que.; Ky., Md., Mass., Mich., N.H., N.Y., N.C., Ohio, Pa., Tenn., Vt., Va., W.Va.

Caulophyllum giganteum has a relatively northern distribution, and it flowers precociously; *C. thalictroides* has a broader distribution, extending farther south, flowers and fruits later, and is less precocious. *Caulophyllum giganteum* is treated as part of *C. thalictroides* by most authors.

2. Caulophyllum thalictroides (Linnaeus) Michaux, Fl. Bor.-Amer. 1: 204. 1803 · Blue cohosh, squaw-root, papoose-root, caulophylle faux-pigamon [E] [F]

Leontice thalictroides Linnaeus, Sp. Pl. 1: 312. 1753

Stems 2–9 dm. **Leaves:** 1st leaf (3–)4×-ternate; 2d leaf (2–)3×-ternate; leaflets 3–8 × 2–10 cm. **Inflorescences** with 5–70 flowers. **Flowers:** bracteoles 1–3 mm; sepals yellow, purple, green, 3–6 × 2–3 mm, apically revolute; petals 1–2.5 mm; stamen filaments 0.5–1.5 mm; pistil 1–3 mm; style 0.25–1 mm. $2n = 16$.

Flowering late spring. Mesophytic forests; 0–1200 m; Man., N.B., Ont., Que.; Ala., Ark., Conn., Del., D.C., Ga., Ill., Ind., Iowa, Kans., Ky., Maine, Md., Mass., Mich., Minn., Mo., Nebr., N.H., N.J., N.Y., N.C., N.Dak., Ohio, Pa., S.C., S.Dak., Tenn., Vt., Va., W.Va., Wis.

Native Americans used various preparations of the root of *Caulophyllum thalictroides* medicinally to treat rheumatism, toothaches, profuse menstruation, indigestion and stomach cramps, fits and hysterics, genito-urinary dysfunction, gallstones, and fever, as an aid in childbirth, and as a general tonic (D. E. Moerman 1986).

3. BERBERIS Linnaeus, Sp. Pl. 1: 330. 1753; Gen. Pl. ed. 5, 153. 1754 · Barberry, Oregon-grape, berbéris, algerita [Mediaeval Latin *barbaris*]

Alan T. Whittemore

Mahonia Nuttall, name conserved; *Odostemon* Rafinesque

Shrubs or subshrubs, evergreen or deciduous, 0.1–4.5(–8) m, glabrous or with tomentose stems. **Rhizomes** present or absent, short or long, not nodose. **Stems** branched or unbranched, monomorphic or dimorphic, i.e., all elongate or with elongate primary stems and short axillary spur shoots. **Leaves** alternate, sometimes leaves of elongate shoots reduced to spines and foliage leaves borne only on short shoots; foliage leaves simple or 1×-odd-pinnately compound; petioles usually present. **Simple leaves:** blade narrowly elliptic, oblanceolate, or obovate, 1.2–7.5 cm. **Compound leaves:** rachis, when present, with or without swollen articulations; leaflet blades lanceolate to orbiculate, margins entire, toothed, spinose, or spinose-lobed; venation pinnate or leaflets 3–6-veined from base. **Inflorescences** terminal, usually racemes, rarely umbels or flowers solitary. **Flowers** 3-merous, 3–8 mm; bracteoles caducous, 3, scalelike; sepals falling immediately after anthesis, 6, yellow; petals 6, yellow, nectariferous; stamens 6; anthers dehiscing by valves; pollen exine punctate; ovary symmetrically club-shaped; placentation subbasal; style central. **Fruits** berries, spheric to cylindric-ovoid or ellipsoid, usually juicy, sometimes dry, at maturity. **Seeds** 1–10, tan to red-brown or black; aril absent. $x = 14$.

Species ca. 500 (22 in the flora): almost worldwide.

Many species of *Berberis* are grown as ornamental shrubs. Some species harbor the black stem-rust of wheat (*Puccinia graminis* Persoon); the sale or transport of susceptible or untested species is illegal in the United States and Canada. Data on susceptibility of *Berberis* spp. to infection by *Puccinia graminis* was supplied by Dr. D. L. Long, U.S. Department of Agriculture (pers. comm.).

The berries of many species are edible and frequently are used for jam and jelly.

The genus *Berberis* as recognized below is divided into two genera, *Berberis* and *Mahonia,* by some authors (e.g., L. Abrams 1934). Species 1–5 below represent *Berberis* in the narrow sense (characterized by dimorphic stems, with elongate primary stems and short axillary shoots; leaves of primary stems modified as spines; foliage leaves simple; and inflorescences usually rather lax, with acuminate bracts and 1–20 flowers; most species susceptible to *Puccinia*). Species 13–22 represent the segregate genus *Mahonia* (with stems never regularly dimorphic; stem spines absent; leaves pinnately compound; and inflorescences dense, with rounded or obtuse [rarely acute] bracts and 25–70 flowers; never susceptible to *Puccinia*). Species 6–12, traditionally included in *Mahonia* when that genus is recognized (L. Abrams 1934), are actually intermediate, resembling *Berberis* proper in their dimorphic stems, inflorescence structure, and susceptibility to *Puccinia,* and *Mahonia* in their spineless stems and compound leaves. Species showing different combinations of the characteristics of the two groups are found in other parts of the world (J. W. McCain and J. F. Hennen 1982; R. V. Moran 1982), so these segregate genera do not seem to be natural. *Mahonia* is often recognized in horticultural works, but it is seldom recognized by botanists.

SELECTED REFERENCES Abrams, L. 1934. The mahonias of the Pacific states. Phytologia 1: 89–94. McCain, J. W. and J. F. Hennen. 1982. Is the taxonomy of *Berberis* and *Mahonia* (Berberidaceae) supported by their rust pathogens *Cumminsiella santa* sp. nov. and other *Cumminsiella* species (Uredinales)? Syst. Bot. 7: 48–59. Moran, R. V. 1982. *Berberis claireae,* a new species from Baja California; and why not *Mahonia*. Phytologia 52: 221–226.

1. Stems spiny; leaves simple; plants deciduous or evergreen.
   2. Plants evergreen; leaf blades thick and rigid, each margin with 2–4 teeth or shallow lobes, each tooth or lobe 1–3 mm, tipped with spine 1.2–1.6 × 0.2–0.3 mm; stems tomentose. . . . . 5. *Berberis darwinii*
   2. Plants deciduous; leaf blades thin and flexible, margins entire or each with 3–30 teeth, each tooth 0–1 mm, tipped with bristle 0.2–1.4 × 0.1–0.2 mm; stems glabrous.
      3. Inflorescences of solitary flowers or umbellate; margins of leaf blade entire . . . . 4. *Berberis thunbergii*
      3. Inflorescences racemose; margins of leaf blade entire or toothed.
         4. Bark of 2d-year branches gray; each margin of leaf blade with (8–)16–30 teeth; racemes 10–20-flowered. . . . . 3. *Berberis vulgaris*
         4. Bark of 2d-year branches brown, purple, or reddish; leaf blade entire or each margin with 3–12 teeth; racemes 3–15-flowered.
            5. Leaf blade oblanceolate or sometimes narrowly elliptic, apex rounded or rounded-obtuse; surfaces adaxially ± glaucous . . . . 1. *Berberis canadensis*
            5. Leaf blade narrowly elliptic, apex acute to obtuse or rounded; surfaces adaxially not glaucous, often shiny . . . . 2. *Berberis fendleri*
1. Stems not spiny; leaves compound; plants evergreen.
   6. Racemes loose (rather dense in *B. harrisoniana*), 1–11-flowered; bracts acuminate.
      7. All leaves 3-foliolate; terminal leaflet sessile.
         8. Terminal leaflet blade 0.9–2 cm wide; berries red. . . . . 6. *Berberis trifoliolata*
         8. Terminal leaflet blade 2.2–3.2 cm wide; berries blue-black . . . . 7. *Berberis harrisoniana*
      7. Leaves 5–11-foliolate (sometimes a minority of leaves 3-foliolate); terminal leaflet stalked on most or all leaves.
         9. Marginal spines of leaflet blade 0.4–1.2 × 0.1–0.15 mm.
            10. Bracts (at least proximal ones) leathery, spine-tipped; berries white or red, somewhat glaucous, 9–16 mm, usually hollow; c Texas. . . . . 11. *Berberis swaseyi*
            10. Bracts usually membranous, seldom spine-tipped; berries yellowish red to red, not glaucous, 5–6 mm, solid; s California. . . . . 12. *Berberis nevinii*
         9. Marginal spines of leaflet blade 0.8–3 × 0.2–0.3 mm.
            11. Berries dry, inflated, 12–18 mm. . . . . 8. *Berberis fremontii*
            11. Berries juicy, solid, 5–8 mm.
               12. Blade of terminal leaflet mostly 2–5 times as long as wide; berries purplish red. . . . . 9. *Berberis haematocarpa*
               12. Blade of terminal leaflet mostly 1–2.5 times as long as wide; berries yellowish red . . . . 10. *Berberis higginsiae*
   6. Racemes dense, 25–70-flowered; bracts obtuse or acute.
      13. Bud scales persistent, 11–44 mm; leaflet blades 4–6-veined from base; filaments unappendaged.
         14. Shrubs 0.1–0.8(–2) m; teeth 6–13 per blade margin, 1–2(–3) mm, spines 0.1–0.2 mm thick; native, Pacific Coast states, B.C., and Idaho . . . . 21. *Berberis nervosa*
         14. Shrubs 1–2 m; teeth 2–7 per blade margin, 3–8 mm, spines 0.3–0.6 mm thick; locally naturalized, se United States. . . . . 22. *Berberis bealei*
      13. Bud scales 2–8(–14) mm, deciduous; leaflet blades 1–3-veined from base (sometimes 1–5-veined in *B. amplectens*); distal end of each filament with pair of recurved teeth (status of this character in *B. amplectens* unknown).
         15. Leaflet blades abaxially smooth and somewhat shiny (outer surface of cells of abaxial epidermis of leaf plane).
            16. Blade of terminal leaflet 1.3–1.9 times as long as wide; lateral leaflet blades elliptic to ovate or broadly lanceolate . . . . 20. *Berberis pinnata*
            16. Blade of terminal leaflet 1.7–2.5 times as long as wide; lateral leaflet blades lance-ovate or lance-elliptic . . . . 19. *Berberis aquifolium*
         15. Leaflet blades abaxially papillose and very dull (outer surface of cells of abaxial epidermis of leaf strongly bulging).
            17. Leaflet blades thin and flexible; teeth 6–24 per blade margin, 0.1–0.25 mm thick; plants 0.02–0.2(–0.6) m. . . . . 18. *Berberis repens*
            17. Leaflet blades thick and rigid; teeth 2–15 per blade margin, 0.2–0.6 mm thick; plants 0.3–2 m (0.1–0.4 m in *B. pumila*).
               18. Leaflet blades adaxially glossy.
                  19. Teeth 6–12 per blade margin; n California and Oregon . . . . 17. *Berberis piperiana*
                  19. Teeth 3–5 per blade margin; Arizona and New Mexico . . . . 16. *Berberis wilcoxii*

18. Leaflet blades adaxially dull, ± glaucous.

     20. Blade margins strongly crispate, each margin with 3–8 teeth . . . . 13. *Berberis dictyota*

     20. Blade margins plane to undulate or, if crispate, each margin with 9–15 teeth.

         21. Plants 0.2–1.2 m; each blade margin with 9–15 teeth . . . . 14. *Berberis amplectens*

         21. Plants 0.1–0.4 m; each blade margin with 2–10 teeth . . . . 15. *Berberis pumila*

1. Berberis canadensis Miller, Gard. Dict. ed. 8, Berberis no. 2. 1768   [E]

Shrubs, deciduous, 0.4–2 m. **Stems** dimorphic, with long primary shoots and short axillary shoots. **Bark** of 2d-year stems purple or brown, glabrous. **Bud scales** 1–1.5 mm, deciduous. **Spines** present, simple or 3(–7)-fid. **Leaves** simple; petioles 0.2–0.8(–1.3) cm. **Leaf blade** oblanceolate or sometimes narrowly elliptic, 1-veined from base, 1.8–7.5 × 0.8–3.3 cm, thin and flexible, base long-attenuate, margins plane, toothed, each with 3–12 teeth 0–1 mm high tipped with bristles to 0.2–1.2 × 0.1–0.15 mm, apex rounded or rounded-obtuse; surfaces abaxially dull, smooth, adaxially dull, ± glaucous. **Inflorescences** racemose, lax, 3–12-flowered, 2–5.5 cm; bracts membranous, apex acuminate. **Flowers:** filaments without distal pair of recurved lateral teeth. **Berries** red, oblong-ellipsoid, 10 mm, juicy, solid. $2n = 28.$

Flowering spring (Apr–May). In woods or glades, on rocky slopes and near rivers; 100–700 m; Ala., Ga., Ill., Ind., Ky., Md., Mo., N.C., Pa., Tenn., Va., W.Va.

Berberis canadensis is susceptible to infection by *Puccinia graminis.*

The Cherokee Indians used scraped bark of *Berberis canadensis* in infusions to treat diarrhea (D. E. Moerman 1986).

2. Berberis fendleri A. Gray, Mem. Amer. Acad. Arts, ser. 2, 4: 5. 1849   [E] [F]

Shrubs, deciduous, 1–2 m. **Stems** dimorphic, with elongate primary and short axillary shoots. **Bark** of 2d-year stems purple, glabrous. **Bud scales** 1–2 mm, deciduous. **Spines** present, simple or 1–2-pinnately branched. **Leaves** simple; petioles 0.2–0.7 cm. **Leaf blade** narrowly elliptic, 1-veined from base, 1.7–4.6 × 0.6–1.7 cm, thin and flexible, base long-attenuate, margins plane, entire or toothed, each with 3–12 teeth 0–1 mm high tipped with bristles to 0.4–1.4 × 0.1–0.2 mm, apex acute to obtuse or rounded; surfaces abaxially dull or glossy and smooth, adaxially dull or glossy and not glaucous.

Inflorescences racemose, lax, 4–15-flowered, 1.5–4.5 cm; bracts membranous, apex acuminate. **Flowers:** filaments without distal pair of recurved lateral teeth. **Berries** red, not glaucous, oblong-ellipsoid, 6–8 mm, juicy, solid.

Flowering spring–summer (May–Aug). Slopes and canyon bottoms; 1300–2700 m; Colo., N.Mex., Utah.

Berberis fendleri is susceptible to infection by *Puccinia graminis.*

3. Berberis vulgaris Linnaeus, Sp. Pl. 1: 330. 1753 · Common barberry, épine-vinette, berbéris vulgaire   [W2]

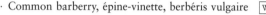

Shrubs, deciduous, 1–3 m. **Stems** dimorphic, with elongate primary and short axillary shoots. **Bark** of 2d-year stems gray, glabrous. **Bud scales** 2–3 mm, deciduous. **Spines** present, simple or 3-fid. **Leaves** simple; petioles 0.2–0.8 cm. **Leaf blade** obovate to oblanceolate or almost elliptic, 1-veined from base, 2–6(–8) × 0.9–2.8 cm, thin and flexible, base short- to long-attenuate, margins plane, finely serrate, each with (8–)16–30 teeth 0–1 mm high tipped with spines or bristles to 0.6–1.4 × 0.1 mm, apex rounded or obtuse; surfaces abaxially dull, smooth, adaxially dull, ± glaucous. **Inflorescences** racemose, lax, 10–20-flowered, 2–6 cm; bracts membranous, apex acute. **Flowers:** filaments without distal pair of recurved lateral teeth. **Berries** red or purple, ellipsoid, 10–11 mm, juicy, solid.

Flowering spring (May–Jun). Roadsides, woods, old fields; 0–1800 m; introduced; B.C., Man., N.B., N.S., Ont., P.E.I., Que.; Colo., Conn., Del., Idaho, Ill., Ind., Iowa, Kans., Maine, Md., Mass., Mich., Minn., Mo., Mont., Nebr., N.H., N.J., N.Y., N.Dak., Ohio, Pa., R.I., S.Dak., Vt., Va., Wash., W.Va., Wis.; native, Europe.

During the eighteenth and nineteenth centuries, *Berberis vulgaris* was very commonly cultivated in North America for thorn hedges and as a source of jam and yellow dye. It frequently escaped from cultivation and became naturalized over a wide area of eastern North America. It is susceptible to infection by *Puccinia graminis.* As the most important alternate host of this fungus, it has been the subject of vigorous eradication programs, and it is now infrequent or absent in many areas where it was once frequent (A. P. Roelfs 1982).

4. Berberis thunbergii de Candolle, Syst. Nat. 2: 19. 1821 · Japanese barberry W2

Shrubs, deciduous, 0.3–3 m. **Stems** dimorphic, with short axillary shoots. **Bark** of 2d-year stems purple or brown, glabrous. **Bud scales** 1–2 mm, deciduous. **Spines** present, simple or 3-fid. **Leaves** simple; petioles 0–0.8 cm. **Leaf blade** obovate to spatulate, 1-veined from base, (0.5–)1.2–2.4 × 0.3–1(–1.8) cm, thin and flexible, base long-attenuate, margins plane, entire, apex rounded or obtuse; surfaces abaxially dull, smooth, adaxially dull, scarcely glaucous. **Inflorescences** umbellate, 1–5-flowered, 1–1.5 cm; bracts membranous, apex acute. **Flowers:** filaments without distal pair of recurved lateral teeth. **Berries** red, ellipsoid or spheric, (7–)9–10 mm, juicy, solid.

Flowering late winter–spring (Mar–May). Woods, old fields, roadsides; 0–1300 m; introduced; N.B., N.S., Ont., P.E.I.; Conn., Del., Ga., Ill., Ind., Iowa, Kans., Ky., Maine, Md., Mass., Mich., Minn., Mo., Nebr., N.H., N.J., N.Y., N.C., Ohio, Pa., S.Dak., Vt., Va., W.Va., Wis., Wyo.; native, Asia (Japan).

The U.S. Department of Agriculture lists *Berberis thunbergii* as resistant to infection by *Puccinia graminis,* and the species is widely grown as an ornamental in the United States. Preliminary tests carried out by Agriculture Canada, however, suggest that some strains may be susceptible to *Puccinia graminis* infection, and cultivation of *B. thunbergii* is illegal in Canada.

5. Berberis darwinii Hooker, Icon. Pl. 7: 672. 1844

Shrubs, evergreen, 1–3 m. **Stems** dimorphic, with elongate primary and short axillary shoots. **Bark** of 2d-year stems brown, densely tomentose. **Bud scales** 2–4 mm, deciduous. **Spines** present, pedately 5–9-fid. **Leaves** simple; petioles 0.1–0.3 cm. **Leaf blade** obovate, 1-veined from base, 1.7–3 × 0.9–1.2 cm, thick and rigid, base acute or acuminate, margins reflexed, undulate, toothed or shallowly lobed, each with 2–4 teeth or lobes 1–3 mm high tipped with spines to 1.2–1.6 × 0.2–0.3 mm, apex obtuse or rounded; surfaces abaxially glossy, smooth, adaxially glossy, green. **Inflorescences** racemose, rather dense, 10–20-flowered, 3–4 cm; bracts membranous, apex acuminate. **Flowers:** filaments without distal pair of recurved lateral teeth. **Berries** dark purple, spheric, 6–7 mm, juicy, solid.

Flowering winter (Feb). Humid areas near coast; 0–

20 m; introduced; Calif., Oreg.; native, s South America.

Berberis darwinii only rarely escapes from cultivation. It is resistant to infection by *Puccinia graminis.*

6. Berberis trifoliolata Moricand, Pl. Nouv. Amér., 113. 1841 · Algerita, agarito, currant-of-Texas, agritos F

Mahonia trifoliolata (Moricand) Fedde

Shrubs, evergreen, 1–3.5 m. **Stems** ± dimorphic, with elongate primary and short axillary shoots. **Bark** of 2d-year stems gray or grayish purple, glabrous. **Bud scales** 2–3 mm, deciduous. **Spines** absent. **Leaves** 3-foliolate; petioles 0.8–5.4 cm. **Leaflet blades** thick and rigid; surfaces abaxially dull, papillose, adaxially dull, ± glaucous; terminal leaflet sessile, blade 2.3–5.8 × 0.9–2 cm, 1.6–3.1 times as long as wide; lateral leaflet blades narrowly lanceolate or narrowly elliptic, 1-veined from base, base acute or acuminate, rarely rounded-acute, margins plane, toothed or lobed, with 1–3 teeth or lobes 3–7 mm high tipped with spines to 1–2 × 0.2–0.3 mm, apex narrowly acute or acuminate. **Inflorescences** racemose, lax, 1–8-flowered, 0.5–3 cm; bracts membranous, apex acuminate. **Flowers:** filaments without distal pair of recurved lateral teeth. **Berries** red, sometimes glaucous, spheric, 6–11 mm, juicy, solid.

Flowering winter–spring (Feb–Apr). Slopes and flats in grassland, shrubland, and sometimes open woodland; 0–2000 m; Ariz., N.Mex., Tex.; n Mexico.

The illegitimate name *Berberis trifoliolata* Moricand var. *glauca* (I. M. Johnston) M. C. Johnston has been used for plants with very strongly glaucous leaves. Weakly and strongly glaucous plants are often found in the same population, however, indicating that they are not distinct varieties.

Berberis trifoliolata is susceptible to infection by *Puccinia graminis.*

7. Berberis harrisoniana Kearney & Peebles, J. Wash. Acad. Sci. 29: 477. 1939 C E

Shrubs, evergreen, 0.5–1.5 m. **Stems** often ± dimorphic, with elongate primary and somewhat elongate axillary shoots. **Bark** of 2d-year stems brown or gray, glabrous. **Bud scales** 1.5–3 mm, deciduous. **Spines** absent. **Leaves** 3-foliolate; petioles 1.5–5 cm. **Leaflet blades** thick and rigid; surfaces abaxially ± dull, papillose, adaxially dull, rarely glossy, somewhat glaucous; terminal leaflet ses-

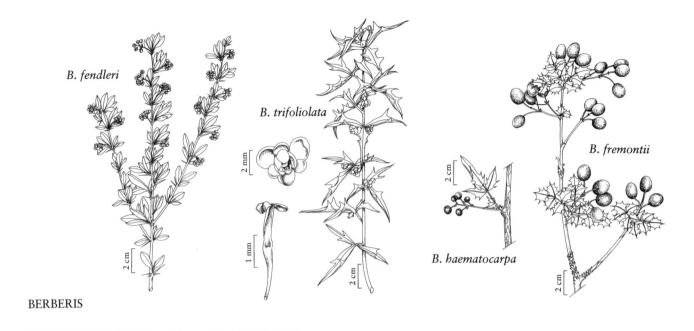

BERBERIS

sile, blade 2.9–5.4 × 2.2–3.2 cm, 1.3–2.4 times as long as wide; lateral leaflet blades ovate or rhombic to lanceolate, 1–3-veined from base, base acute to rounded-obtuse, margins plane or undulate, lobed, with 1–2 teeth 5–13 mm high tipped with spines to 2–3.4 × 0.3–0.4 mm, apex acuminate. **Inflorescences** racemose, rather dense, 6–11-flowered, 1.5–2.5 cm; bracts membranous, apex acute or obtuse. **Flowers:** filaments with distal pair of recurved lateral teeth. **Berries** blue-black, glaucous, spheric to short-ovoid, 5–6 mm, juicy, solid.

Flowering winter (Jan–Mar). Shady spots in rocky canyons; of conservation concern; 800–1100 m; Ariz.

Berberis harrisoniana is endemic to the Kofa and Ajo mountains. It has not been tested for resistance to infection by *Puccinia graminis*.

8. **Berberis fremontii** Torrey in W. H. Emory, Rep. U.S. Mex. Bound. 2(1): 30. 1859 ⎡E⎤ ⎡F⎤

Mahonia fremontii (Torrey) Fedde

Shrubs, evergreen, 1–4.5 m. **Stems** ± dimorphic, with elongate primary and short or somewhat elongate axillary shoots. **Bark** of 2d-year stems light brown or grayish purple, glabrous. **Bud scales** 2–4 mm, deciduous. **Spines** absent. **Leaves** 5–9(–11)-foliolate; petioles 0.2–0.8(–3) cm. **Leaflet blades** thick and rigid; surfaces abaxially dull, papillose, adaxially dull, glaucous; terminal leaflet stalked in most or all leaves, blade 1–2.6(–4) × 0.7–1.8(–2.5)

cm, 1–2.5 times as long as wide; lateral leaflet blades elliptic to ovate or orbiculate, 1–3-veined from base, base obtuse or truncate, margins strongly crispate, toothed or lobed, with 2–5 teeth 2–6 mm high tipped with spines to 0.8–2.2 × 0.2–0.3 mm, apex obtuse to acuminate. **Inflorescences** racemose, lax, 3–6-flowered, 2.5–6.5 cm; bracts membranous, apex acuminate. **Flowers:** filaments with distal pair of recurved lateral teeth. **Berries** yellow or red to brown, ± glaucous, spheric, 12–18 mm, dry, inflated.

Flowering spring (Apr–Jun). Slopes and flats in desert grassland and pinyon-juniper woodland; 1100–2400(–3400) m; Ariz., Calif., Colo., Nev., N.Mex., Utah.

Berberis fremontii is susceptible to infection by *Puccinia graminis*.

The Apache Indians used *Berberis fremontii* for ceremonial purposes; the Hopi used it medicinally to heal gums (D. E. Moerman 1986).

9. **Berberis haematocarpa** Wooton, Bull. Torrey Bot. Club 25: 304. 1898 · Algerita ⎡F⎤

Berberis nevinii A. Gray var. *haematocarpa* (Wooton) L. D. Benson; *Mahonia haematocarpa* (Wooton) Fedde

Shrubs, evergreen, 1–4 m. **Stems** ± dimorphic, with elongate primary and short or somewhat elongate axillary shoots. **Bark** of 2d-year stems grayish purple, glabrous. **Bud scales** 2–4 mm, deciduous. **Spines** absent. **Leaves** 3–9-foliolate; petioles 0.1–0.5 cm. **Leaflet blades** thick and rigid; surfaces abaxially dull, papillose, adaxially dull, glaucous; terminal leaflet stalked in most leaves, blade 1.5–3.8 × 0.5–1.1 cm, 2–5 times as long as wide; lateral leaflet blades oblong-ovate to ovate or lanceolate, 1(–3)-veined from base, base acute to obtuse, rarely subtruncate, margins undulate or crispate, toothed or lobed, with 2–4 teeth 1–4 mm high tipped with spines to 1.2–2 × 0.2–0.3 mm, apex narrowly acute or acuminate. **Inflorescences** racemose, lax, 3–7-flowered, 1.5–4.5 cm; bracts membranous, apex acuminate. **Flowers:** filaments without distal pair of recurved lateral teeth. **Berries** purplish red, glaucous, spheric or short-ellipsoid, 5–8 mm, juicy, solid.

Flowering winter–spring (Feb–Jun). Slopes and flats in desert shrubland, desert grassland, and dry oak woodland; 900–2300 m; Ariz., Calif., Colo., Nev., N.Mex, Tex.; Mexico (Sonora).

Typical populations of *Berberis haematocarpa* (with narrowly ovate or lanceolate leaflets and small, juicy, deep red berries) and *B. fremontii* (with ovate or orbiculate leaflets and large, dry, inflated, yellowish or brownish berries) are easily distinguished. These characteristics are not always well correlated, however, and intermediate populations, showing different combinations of leaflet shape and berry size, color, and inflation, are known.

Berberis haematocarpa is susceptible to infection by *Puccinia graminis*.

10. **Berberis higginsiae** Munz, Aliso 4: 91. 1958 [C]

Mahonia higginsiae (Munz) Ahrendt

Shrubs, evergreen, 1–3 m. **Stems** ± dimorphic, with elongate primary and short or somewhat elongate axillary shoots. **Bark** of 2d-year stems brown or purple, glabrous. **Bud scales** 2–3 mm, deciduous. **Spines** absent. **Leaves** 5–7-foliolate (or 3 by abortion of basal pair, leaving prominent articulation on petiole); petioles 0.1–0.4 cm. **Leaflet blades** thick and rigid; surfaces abaxially dull, papillose, adaxially dull, glaucous; terminal leaflet stalked (sessile in a few leaves), blade 1.4–3.4 × 1.1–2.4 cm, 1–2.5 times as long as wide; lateral leaflet blades oblong to ovate or elliptic, 1–3-veined from base, base obtuse or truncate, margins undulate or crispate, toothed, each with 2–5 teeth 1–4 mm high tipped with spines to 1.2–3 × 0.2–0.3 mm, apex rounded to acute. **Inflorescences** racemose, lax, 5–8-flowered, 2.5–8 cm; bracts membranous, apex acuminate. **Berries** yellowish red, slightly glaucous, spheric, 6–8 mm, juicy, solid.

Flowering spring (Apr–Jun). Chaparral and pinyon-juniper woodland; of conservation concern; 800–1200 m; Calif.; Mexico (Baja California).

Berberis higginsiae is endemic to the region immediately south and east of San Diego, California. The leaflet description above fits the few known California collections; specimens with narrower leaflets (terminal leaflets to 4.5 times as long as wide) have been collected just south of the Mexican border, where leaflet shape may be variable on a single specimen. *Berberis higginsiae* is intermediate between *B. fremontii* and *B. haematocarpa* in its variable leaflet shape and berries that are small and juicy but yellowish red. Further study may show that it is conspecific with one of these species (R. V. Moran 1982).

Berberis higginsiae is susceptible to infection by *Puccinia graminis*.

11. **Berberis swaseyi** Buckley ex M. J. Young, Famil. Lessons Bot., 152. 1873 [E]

Mahonia swaseyi (Buckley ex M. J. Young) Fedde

Shrubs, evergreen, 1–2 m. **Stems** ± dimorphic, with elongate primary and short or somewhat elongate axillary shoots. **Bark** of 2d-year stems purple, glabrous. **Bud scales** 1.5–4 mm, deciduous. **Spines** absent. **Leaves** 5–9-foliolate (basal pair of leaflets sometimes reduced to bristles); petioles 0.1–0.5 cm. **Leaflet blades** thin or thick and rigid; surfaces abaxially dull, papillose, adaxially dull, somewhat glaucous; terminal leaflet stalked (sessile in a few leaves), blades 1.8–3.5 × 0.7–1.7 cm, 1.3–4.7 times as long as wide; lateral leaflets oblong to elliptic or lanceolate, 1-veined from base, base truncate to obtuse, rarely acute, margins plane or undulate, toothed, each with 3–8 teeth 0.5–2 mm high tipped with spines to 0.6–1.2 × 0.1–0.2 mm, apex rounded to acuminate. **Inflorescences** racemose, lax, 2–6-flowered, 4–6 cm; bracts leathery, apex spinose-acuminate, sometimes with proximal bracts as described, distal membranous and acuminate. **Flowers:** filaments with distal pair of recurved lateral teeth. **Berries** white or red and somewhat glaucous, spheric, 9–16 mm, dry or juicy, hollow.

Flowering winter–spring (Feb–Apr). Limestone ridges and canyons; 150–600 m; Tex.

Berberis swaseyi is endemic to the Edwards Plateau. According to M. C. Johnston (pers. comm.), *B. swaseyi* and *B. trifoliolata* hybridize in central Texas.

Berberis swaseyi is susceptible to infection by *Puccinia graminis*.

12. Berberis nevinii A. Gray in A. Gray et al., Syn. Fl. N. Amer. 1: 69. 1895 [C] [E]

Mahonia nevinii (A. Gray) Fedde

Shrubs, evergreen, 1–4 m. **Stems** ± dimorphic, with elongate primary and short or somewhat elongate axillary shoots. **Bark of** 2d-year stems grayish or brownish purple, glabrous. **Bud scales** 2–3 mm, deciduous. **Spines** absent. **Leaves** 3–5(–7)-foliolate; petioles 0.2–0.7 cm. **Leaflet blades** thin but rigid; surfaces abaxially dull, papillose, adaxially dull, glaucous; terminal leaflet stalked in most or all leaves, blade 2.1–4.1 × 0.7–1.1 cm, 3–6 times as long as wide; lateral leaflet blades lance-ovate or lance-elliptic to lanceolate, 1-veined from base, base obtuse or rounded, margins plane or undulate, toothed, each with 4–11 teeth 0–1 mm high tipped with spines to 0.4–2 × 0.1–0.2 mm, apex acuminate. **Inflorescences** racemose, lax, 3–8-flowered, 2–5 cm; bracts membranous, apex acuminate, sometimes with proximal bracts leathery, spinose-acuminate. **Flowers:** filaments with distal pair of recurved lateral teeth. **Berries** yellowish red to red, not glaucous, spheric, 5–6 mm, juicy, solid.

Flowering winter–spring (Feb–May). Sandy slopes and washes in chaparral, coastal scrub, and riparian scrub; of conservation concern; 0–600 m; Calif.

Berberis nevinii is known from scattered populations from San Francisquito Canyon, north of Valencia, south to Dripping Springs, near Aguanga. It is susceptible to infection by *Puccinia graminis*.

13. Berberis dictyota Jepson, Bull. Torrey Bot. Club 18: 319. 1891 [E] [F]

Berberis aquifolium Pursh var. *dictyota* (Jepson) Jepson; *B. californica* Jepson; *Mahonia dictyota* (Jepson) Fedde

Shrubs, evergreen, (0.3–)0.5–2 m. **Stems** usually monomorphic, seldom with short axillary shoots. **Bark of** 2d-year stems brown or purple, glabrous. **Bud scales** 3–5 mm, deciduous. **Spines** absent. **Leaves** 5–7-foliolate; petioles 0.5–3 cm. **Leaflet blades** thick and rigid; surfaces abaxially dull, papillose, adaxially dull, glaucous; terminal leaflet stalked, rarely sessile in a few leaves, blades 2.2–8.8 × 1.8–6 cm, 1.2–2 times as long as wide; lateral leaflet blades elliptic to oblong or oblong-ovate, 1–3-veined from base, base obtuse or truncate, margins strongly crispate, toothed, each with 3–8 teeth 2–8 mm tipped with spines to 2–4.8 × 0.4–

0.6 mm, apex rounded or obtuse. **Inflorescences** racemose, dense, 25–50-flowered, 3–7 cm; bracts membranous, apex rounded or obtuse. **Flowers:** filaments with distal pair of recurved lateral teeth. **Berries** dark blue, at least sometimes glaucous, oblong-ovoid, 6–7 mm, juicy, solid. $2n = 28$.

Flowering winter–spring (Mar–Apr). Dry rocky places in chaparral and open woodland; 600–1800 m; Calif.

Berberis dictyota, *B. amplectens*, *B. pumila*, and *B. wilcoxii* are very similar, and the characters that separate them (height, glossiness and crispation of leaflets, and size and number of marginal teeth) are rather variable within the species. *Berberis piperiana* also belongs to this group, although it is usually more distinct because of its thinner leaflets with more slender, more numerous marginal spines.

Berberis dictyota is resistant to infection by *Puccinia graminis*.

Medicinally, the Kawaiisu used a decoction of the root of *Berberis dictyota* to treat gonorrhea (D. E. Moerman 1986).

14. Berberis amplectens (Eastwood) L. C. Wheeler, Rhodora 39: 376. 1937 [C] [E]

Mahonia amplectens Eastwood, Proc. Calif. Acad. Sci., ser. 4, 20: 145. 1931

Shrubs, evergreen, 0.2–1.2 m. **Stems** monomorphic, without short axillary shoots. **Bark of** 2d-year stems purple, glabrous. **Bud scales** 3–6 mm, deciduous. **Spines** absent. **Leaves** 5–7-foliolate; petioles 1.5–5 cm. **Leaflet blades** thick and rigid; surfaces abaxially dull, papillose, adaxially dull, ± glaucous; terminal leaflet stalked, blade 4.4–5.5 × 3.1–4.6 cm, 1.1–1.4 times as long as wide; lateral leaflet blades oblong or circular, 1–5-veined from base, base truncate or cordate, margins undulate or crispate, toothed, each with 9–15 teeth 1–3 mm tipped with spines to 1.4–2.4 × 0.2–0.4 mm, apex truncate or broadly rounded. **Inflorescences** racemose, dense, 25–35-flowered, 3–6 cm; bracts membranous, apex obtuse or rounded. **Flowers:** filaments distally with pair of recurved teeth: author had no data available. **Berries** dark blue, glaucous, ovoid to elliptic, 7–9 mm, juicy, solid. $2n = 28$.

Flowering spring (Apr–May). Rocky slopes in chaparral and open forest; of conservation concern; 900–1900 m; Calif.

Berberis amplectens is endemic to the Peninsular Ranges of southern California. It is resistant to infection by *Puccinia graminis*.

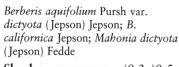

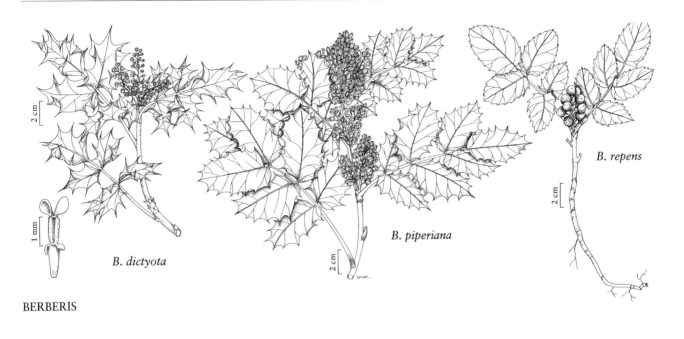

B. dictyota

B. piperiana

B. repens

BERBERIS

15. Berberis pumila Greene, Pittonia 2: 161. 1891 [E]

Mahonia pumila (Greene) Fedde

Shrubs, evergreen, 0.1–0.4 m. **Stems** monomorphic, without short axillary shoots. **Bark** of 2d-year stems gray-brown or purplish, glabrous. **Bud scales** 3–6 mm, deciduous. **Spines** absent. **Leaves** 3–9-foliolate; petioles 0.5–4 cm. **Leaflet blades** thick and rigid; surfaces abaxially dull, papillose, adaxially dull, glaucous; terminal leaflet stalked, at least on most leaves, blade 4–8 × 2–5 cm, 1.3–1.9 times as long as wide; lateral leaflet blades ovate to oblong-ovate or elliptic, 1(–3)-veined from base, base obtuse, rarely truncate, margins plane or undulate, toothed, with 2–10 teeth 1–3 mm tipped with spines to 1.6–3 × 0.3–0.4(–0.5) mm, apex obtuse or rounded, rarely broadly acuminate. **Inflorescences** racemose, dense, 30–45-flowered, 2–4 cm; bracts membranous, apex rounded or obtuse, sometimes apiculate. **Flowers:** filaments with distal pair of recurved lateral teeth. **Berries** dark blue, glaucous, oblong-ovoid to spheric, 5–8 mm, juicy, solid. $2n = 28$.

Flowering winter–spring (Mar–May). Open woods and rocky areas; 300–1200 m; Calif., Oreg.

Berberis pumila is resistant to infection by *Puccinia graminis*.

16. Berberis wilcoxii Kearney, Trans. New York Acad. Sci. 14: 29. 1894

Shrubs, evergreen, 0.3–2 m. **Stems** ± monomorphic, seldom with short axillary shoots. **Bark** of 2d-year stems purple or brown, glabrous. **Bud scales** 2–6 mm, deciduous. **Spines** absent. **Leaves** 5–9-foliolate; petioles 1–5 cm. **Leaflet blades** thick and rigid; surfaces abaxially dull, papillose, adaxially glossy, green; terminal leaflet stalked, blade 2.6–6.6 × 1.7–4.4 cm, 1–2.5 times as long as wide; lateral leaflet blades oblong to ovate or elliptic, 1–3-veined from base, base obtuse to rounded or truncate, margins plane to crispate, toothed, each with 3–5 teeth 1–5 mm tipped with spines to 1.2–3.8 × 0.2–0.6 mm, apex acute to rounded. **Inflorescences** racemose, dense, 30–50-flowered, 2–7 cm; bracts membranous, apex rounded or obtuse. **Flowers:** filaments with distal pair of recurved lateral teeth. **Berries** blue, glaucous, oblong-ovoid, 6–11 mm, juicy, solid.

Flowering spring (Apr–May). Slopes and canyons; 1500–2500 m; Ariz., N.Mex.; Mexico (Sonora).

Berberis wilcoxii has not been tested for resistance to infection by *Puccinia graminis*.

17. Berberis piperiana (Abrams) McMinn, Man. Calif. Shrubs, 125. 1939   E F

Mahonia piperiana Abrams, Phytologia 1: 91. 1934

Shrubs, evergreen, 0.3–0.8 m. **Stems** monomorphic, without short axillary shoots. **Bark** of 2d-year stems brown or purple, glabrous. **Bud scales** 3–8 mm, deciduous. **Spines** absent. **Leaves** 5–9-foliolate; petioles 1–6 cm. **Leaflet blades** rather thick and ± rigid; surfaces abaxially dull, papillose, adaxially glossy, green; terminal leaflet stalked, blade 3.6–9.5 × 2–5.2 cm, 1.3–2.1 times as long as wide; lateral leaflet blades lance-elliptic to elliptic-ovate or narrowly oblong, 1(–3)-veined from base, base rounded to obtuse or truncate, margins undulate, toothed, each with 6–12 teeth 1–4 mm, tipped with spines to 1.6–2.8 × 0.2–0.3 mm, apex acute to rounded-obtuse. **Inflorescences** racemose, dense, 25–60-flowered, 3–10 cm; bracts membranous, apex obtuse or rounded. **Flowers:** filaments with distal pair of recurved lateral teeth. **Berries** dark blue and glaucous, oblong-ovoid, 7–10 mm, juicy, solid. 2*n* = 28.

Flowering winter–spring (Mar–Jun). Open wooded and shrubby slopes; 900–1700 m; Calif., Oreg.

Berberis piperiana is resistant to infection by *Puccinia graminis.*

18. Berberis repens Lindley, Bot. Reg. 14: plate 1176. 1828   E F

Berberis aquifolium Pursh var. *repens* (Lindley) Scoggan; *B. sonnei* (Abrams) McMinn; *Mahonia repens* (Lindley) G. Don; *M. sonnei* Abrams

Shrubs, evergreen, 0.02–0.2 (–0.6) m. **Stems** monomorphic, usually without short axillary shoots. **Bark** of 2d-year stems grayish or purplish brown, glabrous. **Bud scales** 3–8 mm, deciduous. **Spines** absent. **Leaves** (3–)5–7-foliolate; petioles (1–)3–9 cm. **Leaflet blades** thin and flexible; surfaces abaxially dull, papillose, adaxially dull, rarely glossy, somewhat glaucous; terminal leaflet stalked, blade 3.2–9.5 × 2.3–6 cm, 1.2–2.2(–2.5) times as long as wide; lateral leaflets ovate or elliptic, 1(–3)-veined from base, base rounded to obtuse or truncate, margins plane, toothed, with 6–24 teeth 0.5–3 mm tipped with spines to 0.6–2.8 × 0.1–0.25 mm, apex rounded, rarely obtuse or even broadly acute. **Inflorescences** racemose, dense, 25–50-flowered, 3–10 cm; bracts membranous, apex rounded to obtuse or broadly acute. **Flowers:** filaments with

distal pair of recurved lateral teeth. **Berries** blue, glaucous, oblong-ovoid, 6–10 mm, juicy, solid. 2*n* = 28.

Flowering spring (Apr–Jun). Open forest, shrubland, and grassland; 200–3000 m; Alta., B.C.; Ariz., Calif., Colo., Idaho, Minn., Mont., Nebr., Nev., N.Mex., N.Dak., Oreg., S.Dak., Tex., Utah, Wash., Wyo.

Berberis sonnei was described based on plants with relatively narrow, rather shiny leaflets collected by C. F. Sonne in Truckee, California. Subsequent collections from this population show the morphology typical of *B. repens;* Sonne's collections evidently are an aberrant form of this species.

Berberis repens is resistant to infection by *Puccinia graminis.*

Various Native American tribes used preparations of the roots of *Berberis repens* to treat stomach troubles, to prevent bloody dysentary, and as a blood purifier; mixed with whiskey, it was used for bladder problems, venereal diseases, general aches, and kidney problems; and preparations made from the entire plant served as a cure-all and as a lotion for scorpion bites (D. E. Moerman 1986).

19. Berberis aquifolium Pursh, Fl. Amer. Sept., 219, plate 4. 1814 · Oregon-grape   E F

Mahonia aquifolium (Pursh) Nuttall

Shrubs, evergreen, 0.3–3(–4.5) m. **Stems** usually monomorphic, seldom with short axillary shoots. **Bark** of 2d-year stems gray-brown or purplish, glabrous. **Bud scales** 4–8(–14) mm, deciduous. **Spines** absent. **Leaves** 5–9-foliolate; petioles 1–6 cm. **Leaflet blades** thin and flexible or rather rigid; surfaces abaxially glossy, smooth, adaxially glossy, green; terminal leaflet stalked, blade 5.1–8.7(–14.5) × 2.4–4.5(–5.5) cm, 1.7–2.5 times as long as wide; lateral leaflet blades lance-ovate to lance-elliptic, 1(–3)-veined from base, base obtuse or truncate, rarely weakly cordate, margins plane or undulate, toothed, each with 5–21 teeth 0–2 mm tipped with spines to 0.8–2.2 × 0.2–0.3 mm, apex acute or sometimes obtuse or rounded. **Inflorescences** racemose, dense, 30–60-flowered, 3–9(–11) cm; bracts membranous, apex rounded or obtuse, sometimes apiculate. **Flowers:** filaments with distal pair of recurved lateral teeth. **Berries** blue, glaucous, oblong-ovoid, 6–10 mm, juicy, solid. 2*n* = 28, 56.

Flowering winter–spring (Mar–Jun). Open woods and shrublands; 0–2100 m; B.C.; Calif., Idaho, Mont., Oreg., Wash.

Berberis aquifolium is the state flower of Oregon. It

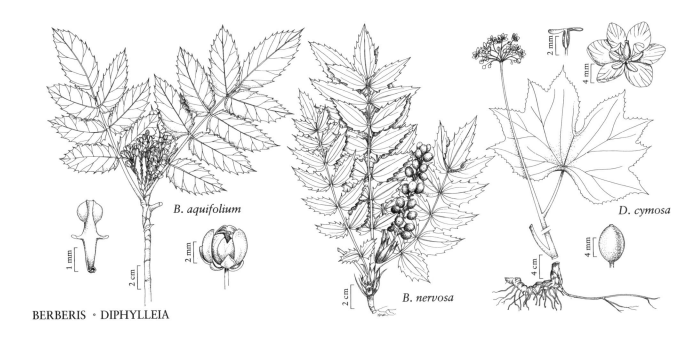

is widely used as an ornamental and has been reported as an escape from cultivation in scattered localities across the continent (Ontario, Quebec, central California, Michigan, and Nevada).

Berberis aquifolium is resistant to infection by *Puccinia graminis*.

Medicinally, various root preparations of *Berberis aquifolium* were used by Native Americans for stomach trouble, hemorrhages, and tuberculosis; as a panacea, a tonic, a gargle, and an eye wash; and to purify blood. Leaves and roots were used in steam baths to treat yellow fever; the tips of stems were used to treat stomach aches; and the entire plant was considered poisonous (D. E. Moerman 1986).

20. Berberis pinnata Lagasca, Elench. Pl., 14. 1816

Mahonia pinnata (Lagasca) Fedde

Shrubs, evergreen, 0.3–1.6(–7) m. **Stems** usually monomorphic, seldom with short axillary shoots. **Bark** of 2d-year stems grayish brown, glabrous. **Bud scales** 3–7 mm, deciduous. **Spines** absent. **Leaves** (3–)5–13-foliolate; petioles 0.5–4.5(–7.5) cm. **Leaflet blades** thin and ± rigid or flexible; surfaces abaxially glossy, smooth, adaxially glossy, green; terminal leaflet stalked, blade 2.6–6.2 × 2–4.5 cm, 1.3–1.9 times as long as wide; lateral leaflet blades elliptic to ovate or broadly lanceolate, 1(–3)-veined from base, base broadly obtuse, truncate, or weakly cordate, margins

plane to crispate, toothed, each with 5–22 teeth 0–2 mm tipped with spines to 1–3 × 0.1–0.3 mm, apex acute to rounded-obtuse. **Inflorescences** racemose, dense, 25–50-flowered, 2–9 cm; bracts membranous, apex rounded to broadly acute, sometimes apiculate. **Flowers:** filaments with distal pair of recurved lateral teeth. **Berries** blue, glaucous, oblong-ovoid to subspheric, 6–7 mm, juicy, solid.

Subspecies 2 (2 in the flora): Calif., Oreg.; Mexico (Baja California).

Berberis pinnata is very similar to *B. aquifolium,* and the two are sometimes difficult to separate. Some authors have used the spacing of the lateral leaflets (said to be contiguous or imbricate in *B. pinnata* and remote in *B. aquifolium*) to separate them, but the leaflets are often remote in both species and may be contiguous in *B. aquifolium.*

Berberis pinnata is resistant to infection by *Puccinia graminis.*

1. Margins of leaf blade undulate or crispate, marginal spines to 1.2–3 × 0.2–0.3 mm; shrubs 0.3–1.6 m, self-supporting . . . . .20a. *Berberis pinnata* subsp. *pinnata*
1. Margins of leaf blade plane or weakly undulate, marginal spines to 1–1.6 × 0.1–0.2 mm; shrubs 1–7 m, self-supporting or clambering over surrounding vegetation. . . . . 20b. *Berberis pinnata* subsp. *insularis*

20a. Berberis pinnata Lagasca subsp. **pinnata**

Shrubs, self-supporting, 0.3–1.6 m. **Leaflets** thin but ± rigid; lateral leaflet blades ovate to broadly lanceolate, margins undulate or crispate, marginal teeth tipped with spines to 1.2–3 × 0.2–0.3 mm. **Inflorescences** 2–5 cm.

Flowering winter–spring (Feb–May). Exposed rocky openings in woods and shrubland; 0–1200 m; Calif., Oreg.; Mexico (n Baja California).

20b. Berberis pinnata Lagasca subsp. **insularis** Munz, Aliso 2: 230. 1950 C E

Shrubs, self-supporting or clambering, 1–7 m. **Leaflets** thin and flexible; lateral leaflet blades lance-oblong or elliptic, margins plane or weakly undulate, marginal teeth tipped with spines to 1–1.6 × 0.1–0.2 mm. **Inflorescences** 6–9 cm.

Flowering winter (Mar). In shade beneath trees, pine and pine-oak forest; of conservation concern; 100–450 m; Calif.

Berberis pinnata subsp. *insularis* is endemic to the Channel Islands. Specimens of *B. pinnata* from the coast north of Santa Barbara often show some characteristics of subsp. *insularis*.

21. Berberis nervosa Pursh, Fl. Amer. Sept., 219, plate 5. 1814 E F

Berberis nervosa Pursh var. *mendocinensis* Roof; *Mahonia nervosa* (Pursh) Nuttall; *M. nervosa* var. *mendocinensis* (Roof) Roof

Shrubs, evergreen, 0.1–0.8(–2) m. **Stems** monomorphic, without short axillary shoots. **Bark** of 2d-year stems brown or yellow-brown, glabrous. **Bud scales** (13–)20–44 mm, persistent. **Spines** absent. **Leaves** 9–21-foliolate; petioles 2–11 cm. **Leaflet blades** thin and ± flexible; surfaces abaxially rather dull, smooth, adaxially dull, somewhat glaucous; terminal leaflet stalked, blade 2.9–8.4 × 1.2–4.8 cm, 1.8–3.2 times as long as wide; lateral leaflet blades lance-ovate to ovate, 4–6-veined from base, base rounded to cordate, margins plane, toothed, each with 6–13 teeth 1–2(–3) mm tipped with spines to 1–2.4 × 0.1–0.2 mm, apex acute or broadly acuminate. **Inflorescences** racemose, dense, 30–70-flowered, 6–17 cm; bracts membranous, apex acute, obtuse, or rounded. **Flowers:** filaments without distal pair of recurved lateral teeth. **Berries** blue, glaucous, oblong-ovoid or globose, 8–11 mm, juicy, solid. $2n = 56$.

Flowering winter–spring (Mar–Jun). Open or shaded woods, often in rocky areas; 0–1800 m; B.C.; Calif., Idaho, Oreg., Wash.

Plants of *Berberis nervosa* are usually very low (commonly 0.1–0.3 m), but occasional plants may be considerably taller (to 2 m). One such population from north of Westport, California, has been separated as *B. nervosa* var. *mendocinensis*. Similar populations occur sporadically throughout the range of *B. nervosa*, so the form should not be recognized taxonomically.

Berberis nervosa is resistant to infection by *Puccinia graminis*.

The Skagit tribe used *Berberis nervosa* medicinally in a root preparation to treat venereal disease (D. E. Moerman 1986).

22. Berberis bealei Fortune, Gard. Chron. 1850: 212. 1850

Mahonia bealei (Fortune) Carrière

Shrubs, evergreen, 1–2 m. **Stems** monomorphic, without short axillary shoots. **Bark** of 2d-year stems tan, glabrous. **Bud scales** 11–13 mm, persistent. **Spines** absent. **Leaves** 5–9-foliolate; petioles 2–8 cm. **Leaflet blades** thick and rigid; surfaces abaxially smooth, shiny, adaxially dull, gray-green; terminal leaflet stalked, blade 6.5–9.3 × 4–7 cm, 1.3–2.3 times as long as wide; lateral leaflet blades ovate or lance-ovate, 4–6-veined from base, base truncate or weakly cordate, margins plane, toothed, with 2–7 teeth 3–8 mm tipped with spines to 1.4–4 × 0.3–0.6 mm, apex acuminate. **Inflorescences** racemose, dense, 70–150-flowered, 5–17 cm; bracts ± corky, apex rounded to acute. **Berries** dark blue, glaucous, oblong-ovoid, 9–12 mm, juicy, solid.

Flowering fall–winter (Dec–Mar). Open woodlands and shrublands; 100–500 m; introduced; Ala., Ga., N.C., Va.; native, Asia (China).

Berberis bealei is commonly cultivated; although it rarely escapes, it is locally naturalized in the southeastern United States. It is resistant to infection by *Puccinia graminis*.

4. DIPHYLLEIA Michaux, Fl. Bor.-Amer. 1: 203, plates 19, 20. 1803 · [Greek *dis*, twice, and *phyllon*, leaf]

Lisa O'Rourke George

Herbs, perennial, deciduous, to 12 dm, glabrous or pubescent. **Rhizomes** formed of distinct annual increments, producing 1 leaf or flowering shoot per year. **Aerial stems** present. **Leaves** simple, 2-parted. **Leaf blade** 5–47 cm, parts lobed [or not], margins prominently dentate; venation palmate. **Leaves of nonflowering shoot** 1, basal; petiole centrally attached, erect, stemlike; blade orbiculate, peltate. **Leaves of flowering shoots** 2, cauline, alternate; petiole attached to blade near margin; blade reniform-orbiculate, peltate. **Inflorescences** terminal, cymose or umbelliform. **Flowers** 3-merous, 8–20 mm; bracteoles caducous, 2, scalelike; sepals falling early, 6, white or pale green; petals 6, white; stamens 6; anthers dehiscing by 2 apically hinged flaps; pollen exine spinose; ovaries ellipsoid; placentation appearing basal; style central. **Fruits** berries, dark blue, ellipsoid, glaucous. **Seeds** 2–11, red; aril absent. $x = 6$.

Species 3 (1 in the flora): widely disjunct temperate areas in e North America and e Asia.

SELECTED REFERENCE Ying T. S., S. Terabayashi, and D. E. Boufford. 1984. A monograph of *Diphylleia* (Berberidaceae). J. Arnold Arbor. 65: 57–94.

1. Diphylleia cymosa Michaux, Fl. Bor.-Amer. 1: 203, plates 19, 20. 1803 · Umbrella-leaf [E] [F]

Leaves of nonflowering shoots 2.5–9 dm. **Leaf blades** 13–53 cm diam. **Flowering shoots** 6–12 dm; leaves alternate, unequal in size; proximal leaf petioles 10–18 cm, blades 17–46 × 23–56 cm; distal leaf petioles 3–15 cm, blades 10–38 × 15–51 cm. **Leaf blade** divided at apex and base into 2 parts, each part 5–9-lobed, lobes broadly acuminate, abaxially sparsely pubescent with unicellular hairs. **Inflorescences** cymose, glabrous; peduncle 0.2–4 cm; pedicel 0.7–3.5 cm; peduncle and pedicel turning red at fruit maturity. **Flowers** 7–70 (or more); outer sepals 1.7–4.5 × 0.4 mm; inner sepals 2.5–6 × 3.5–4 mm; outer petals narrowly obovate, 9–11 × 4.5–6 mm; inner petals elliptic to obovate, 10–13 × 6–7 mm; stamens 3–4 mm; filaments 1–2 mm; anthers 2 × 1 mm; ovaries ellipsoid, 3–5 × 1.5–2.5 mm; stigma 0.5–1 mm. **Berries** 6–13 × 4–11 mm. **Seeds** 2–4, 4–7 × 2–5 mm, abaxially rounded, adaxially flattened to concave. $2n = 12$.

Flowering late spring, fruiting summer. Forming dense colonies on moist slopes in mixed deciduous forests, in seepages, or along streams; 800–1700 m; Ga., N.C., S.C., Tenn., Va.

Diphylleia cymosa is endemic to the Blue Ridge Mountains of the southern Appalachians. It is occasionally grown in woodland gardens.

Cherokee Indians are reported to have used *D. cymosa* to treat a variety of ailments and as a disinfectant (D. E. Moerman 1986).

5. PODOPHYLLUM Linnaeus, Sp. Pl. 1: 505. 1753; Gen. Pl. ed. 5, 22. 1754 · [Greek *podos*, foot, and *phyllon* leaf]

Lisa O'Rourke George

Herbs, perennial, deciduous, 2–6 dm, glabrous to sparsely pubescent. **Rhizomes** short to elongate, formed of distinct annual increments, producing 1 leaf or flowering shoot per year. **Aerial stems** present. **Leaves** simple, variously parted. **Leaves of nonflowering shoot** 1, basal; petiole centrally attached, erect, stemlike, blade orbiculate, peltate. **Leaves of flowering shoots** (0–)2(–3), cauline, alternate or nearly opposite; petiole attached near margin, blade reniform-orbiculate, peltate. **Leaf blades** 10–38 cm, parts entire or lobed, margins entire or serrate; venation palmate. **Inflorescences** terminal, flowers solitary. **Flowers** 3-merous, 30–55 mm; bracteoles absent; sepals caducous, 6, white or pale green; petals 6–9, white or pink; stamens equal to or 2 times number of petals; anthers dehiscing longitudinally; pollen exine finely reticulate to verrucate; ovaries ellipsoid; placentation marginal; style central. **Fruits** berries, yellow, orange, red, or maroon, ellipsoid. **Seeds** 20–50, yellow, orange, red, or maroon; aril yellow, rarely maroon, fleshy, enclosing seed. $x = 6$.

Species 2 (1 in the flora): widely disjunct temperate areas in e North America and e Asia.

Podophyllum hexandrum Royle, which is native to eastern Asia, is recognized in its own genus, *Sinopodophyllum* T. S. Ying, by some authors. Also occurring in east Asia is the segregate genus *Dysosoma* Woodson, with about seven species.

SELECTED REFERENCE Dewick, P. M. 1983. Tumour inhibitors from plants. In: G. E. Trease and W. C. Evans. 1983. Pharmacognosy, ed. 12. London. Pp. 629–647.

1. **Podophyllum peltatum** Linnaeus, Sp. Pl. 1: 505. 1753 · May-apple, Indian-apple, wild-mandrake, pomme de mai, podophylle pelté E F

Rhizomes: annual elongation increments (2–)6–20 cm. **Leaves of nonflowering shoots** 2–5 dm; blade 18–38 × 18–38 cm. **Flowering shoots** 3–6 dm; leaves nearly opposite, slightly unequal in size; petioles 5–15 cm; proximal blades 10–35 × 14–40 cm, distal blades 6–25 × 10–33 cm. **Leaf blades** 5–7(–9)-parted, parts lobed or not (frequently 2-lobed), margins entire or coarsely dentate, teeth apiculate; surfaces abaxially sparsely pubescent to glabrous. **Flowers** solitary, nodding, fragrant; peduncle arising from angle between petioles, 1.5–6 cm; sepals orbiculate, 10–18 × 10–18 mm; petals white, rarely pink, obovate, 15–35 × 10–25 mm; stamens 2 times number of petals, 8–13 mm; filaments 3–5 mm; anthers 5–8 × 1–1.5 mm; ovaries 6–12 × 4–8 mm; style 1–2 mm; stigmas 3–6 mm. **Berries** yellow, rarely orange or maroon, 3.5–5.5 × 2.0–4 cm. **Seeds** 30–50, ovoid, 6–8 × 4–6 mm. $2n = 12$.

Flowering spring, fruiting late spring–summer; summer deciduous. Mixed deciduous forest, fields, moist road banks, river banks; 50–800 m; Ont., Que.; Ala., Ark., Conn., Del., D.C., Fla., Ga., Ill., Ind., Iowa, Kans., Ky., La., Md., Mass., Mich., Minn., Miss., Mo., Nebr., N.H., N.J., N.Y., N.C., Ohio, Okla., Pa., R.I., S.C., Tenn., Tex., Vt., Va., W.Va., Wis.

The following forms have been described: *Podophyllum peltatum* forma *aphyllum* Plitt—fertile shoots with no foliage leaves; *Podophyllum peltatum* forma *biltmoreanum* Steyermark—fruits orange; *Podophyllum peltatum* forma *deamii* Raymond—fruits and seeds maroon, and flowers, placentae, and plant axes pink-tinged; *Podophyllum peltatum* forma *polycarpum* (Clute) Plitt—flowers with multiple, free carpels.

The ripe fruit of *Podophyllum peltatum* is considered edible; all other parts of the plant are toxic. Several lignans and their glycosides, present in the resin extracted from rhizomes and roots, exhibit antitumor activity. Etoposide, a semisynthetic derivative of one of the lignans, is currently used in the treatment of small-cell lung cancer and testicular cancer (P. M. Dewick 1983). Native Americans used *Podophyllum* for a wide variety of medicinal purposes and as an insecticide (D. E. Moerman 1986).

Podophyllum peltatum is sometimes cultivated in woodland gardens, and some populations on the periphery of its geographical range may be escapes from cultivation.

6. ACHLYS de Candolle, Syst. Nat. 2: 35. 1821 · [Greek *Achlus,* a god of night]

R. David Whetstone
T. A. Atkinson
Daniel D. Spaulding

Herbs, perennial, deciduous, 2.5–5 dm, glabrous. **Rhizomes** extensive, branching, producing 1–few foliage leaves or flowering shoots per year. **Aerial stems** absent. **Leaves** basal, alternate, 3-foliolate; petiole long, slender. **Leaf blade** orbiculate in gross outline, 20–40 cm; leaflet blades fan-shaped, entire or lobed, lateral leaflet blades strongly asymmetric, margins entire to coarsely sinuate; venation palmate. **Inflorescences** terminal, dense scapose-pedunculate spikes of inconspicuous flowers. **Flowers** 3-merous, white to cream, 6 mm or less; bracteoles

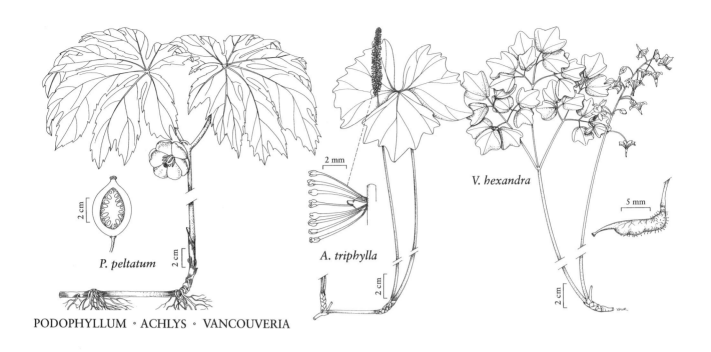

P. peltatum

A. triphylla

V. hexandra

PODOPHYLLUM · ACHLYS · VANCOUVERIA

absent; sepals absent; petals absent; stamens 8–10(–13); anthers dehiscing by 2 apically hinged flaps; pollen exine striate; ovaries asymmetrically ellipsoid; placentation marginal, placenta developed only near base of ovary. **Fruits** follicles with transverse dehiscence, purplish red or brown, curved, furrowed. **Seeds** 1, brown; aril absent. $x = 6$.

Species 3 (2 in the flora): North America, Asia (Japan).

Achlys is of particular interest because of its amphi-Pacific disjunction. Despite the 8000 km or more disjunction, the taxa are remarkably similar in morphology, ecology, and karyology. Japanese populations are diploid; American populations are diploid and tetraploid.

Two species are recognized in this treatment; some researchers prefer to treat them as varieties. In the Californian portion of the range, some field botanists believe the two taxa are sufficiently morphologically distinct to be called species; farther north these differences are reportedly less apparent.

SELECTED REFERENCES Fukuda, I. 1967. The biosystematics of *Achlys*. Taxon 16: 308–316. Fukuda, I. and H. G. Baker. 1970. *Achlys californica*—(Berberidaceae)—a new species. Taxon 19: 341–344.

1. Central leaflet blade 4–11 cm, distal margins (1–)3–4(–8)-lobed; stamens 3–4 mm; follicles red-purple . . . . .1. *Achlys triphylla*
1. Central leaflet blade ca. 7–16 cm, distal margins (3–)6–9(–12)-lobed; stamens 3.8–6 mm; follicles brown . . . . 2. *Achlys californica*

1. Achlys triphylla (Smith) de Candolle, Syst. Nat. 2: 35. 1821 · Deer-foot, vanilla-leaf E F

Leontice triphylla Smith in A. Rees, Cycl. 20: [5]. 1812

Plants, 2–4 dm. **Leaves:** petiole 1–3 cm. **Central leaflet blade** 4–11 × 4–8 cm, proximal margins entire, distal margins (1–)3–4 (–8)-lobed. **Inflorescences** 2.5–5 cm excluding peduncle. **Flowers:** stamens 3–4 mm; ovaries 1–1.5 mm. **Follicles** red-purple, 3–4.5 mm. $2n = 12$.

Flowering spring–summer (Apr–Jul). Mountain regions in Cascade Range and Coast Range of California in coniferous forests; 0–1500 m; B.C.; Calif., Oreg., Wash.

Medicinally, Native Americans used preparations of the leaves of *Achlys triphylla* to treat tuberculosis, for a hair wash, and as an emetic (D. E. Moerman 1986).

2. Achlys californica Fukuda & H. G. Baker, Taxon 19: 341. 1970 [E]

Plants, 3–5 dm. Leaves: petiole 1–3 cm. Central leaflet blade ca. 7–16 × 8–17 cm, proximal margins entire, distal margins (3–)6–9(–12)-lobed. Inflorescences 2.8–5.8 cm excluding peduncle. Flowers: stamens 3.8–6 mm; ovaries 1.5–2 mm. Follicles brown, 3.5–5 mm. $2n = 24$.

Flowering spring–summer (Apr–Jul). Coastal and mountainous regions, *Sequoia sempervirens* and *Pseudotsuga menziesii* forests; 0–1200 m; B.C.; Calif., Oreg., Wash.

Triploid plants have been reported from one locality in central Washington and from a site in northwestern California.

7. VANCOUVERIA C. Morren & Decaisne, Ann. Sci. Nat., Bot., sér. 2, 2: 351. 1834

· Inside-out flower [for George Vancouver (1757–1798), English navigator and explorer]

David Whetstone
Daniel D. Spaulding
T. A. Atkinson

Herbs, perennial, evergreen or deciduous, 1–5 dm, glabrous, glandular-pubescent, or sparsely hairy. Rhizomes extensive, creeping, nodose, producing 3 or more foliage leaves and flowering shoots per year. Aerial stems absent. Leaves basal, alternate, 2–3×-ternately compound (sometimes 3–5-foliolate in *V. chrysantha*); petiole long, slender. Leaf blade deltate in overall outline; rachis without pulvinae; leaflet blades rhomboid or rounded pentagonal to ovate to oblong, shallowly 3-lobed, margins entire to sinuate; venation palmate. Inflorescences terminal, racemes or panicles, open. Flowers 3-merous, 6–14 mm; bracteoles 6–9, sepaloid; sepals 6, white to yellow; petals 6, white to yellow, hooded with tip reflexed or flat, bearing nectar; stamens 6; anthers dehiscing by 2 apically hinged flaps; pollen exine striate; ovaries ellipsoid; placentation marginal; style lateral. Fruits follicles, brown, asymmetric, generally elliptic, dehiscing by 2 valves. Seeds 4–7, black to reddish brown; aril whitish, covering ca. 1/2–2/3 of seed. $x = 6$.

Species 3 (3 in the flora): w United States.

The fruits of *Vancouveria* are thin-walled follicles that are green or greenish brown at the time of dehiscence. The follicles dehisce by means of two valves that begin below the style and open to the base. The two valves recurve, exposing the seeds downward. In *V. hexandra* the follicle opens before the seeds are mature. The green seeds continue to grow and ripen in the open follicle. The appendage or aril on *Vancouveria* seeds has been shown to be a true elaiosome (R. Y. Berg 1972). Ants carry the seeds to their nests and harvest the appendage as a food source.

SELECTED REFERENCES Berg, R. Y. 1972. Dispersal ecology of *Vancouveria* (Berberidaceae). Amer. J. Bot. 59: 109–122. Stearn, W. T. 1938. *Epimedium* and *Vancouveria*, a monograph. J. Linn. Soc., Bot. 51: 409–535.

1. Leaflet margins not conspicuously thickened; leaves falling when fruits maturing; pedicels lacking glands . . . . 1. *Vancouveria hexandra*
1. Leaflet margins conspicuously thickened; leaves persistent; pedicels stipitate-glandular.
   2. Petals yellow, petal apex reflexed; follicles densely stipitate-glandular . . . . 2. *Vancouveria chrysantha*
   2. Petals white, sometimes lavender-tinged, petal apex not reflexed; follicles lacking glands . . . . . .
      . . . . 3. *Vancouveria planipetala*

1. **Vancouveria hexandra** (Hooker) C. Morren & Decaisne, Ann. Sci. Nat., Bot., sér. 2, 2: 351. 1834 · Northern inside-out flower [E] [F]

Epimedium hexandrum Hooker, Fl. Bor.-Amer. 1: 30. 1829

Leaves falling when fruits maturing, 2–3×-ternately compound, 8–30 cm; petiole 3–25 cm, pilose at base. **Leaflet blades** narrowly to broadly ovate to rhomboid or rounded pentagonal, often 3-lobed, base cordate, margins entire to slightly sinuate and not conspicuously thickened, apex rounded to notched; surfaces abaxially sparsely hairy, adaxially glabrous. **Inflorescences:** peduncle 2–3 dm; pedicels 1–3 cm, glands absent. **Flowers** 5–30; bracteoles 6–9, white, yellowish when dried, dotted with glandular trichomes; sepals 6, white, 5–12 mm; petals 6, white, yellowish when dried, 4–6 mm, margins entire, petal apex strongly reflexed, with nectar-bearing pocket, nectaries golden; filaments stipitate-glandular. **Follicles** greenish to light brown, 10–15 mm including beak, beak 2–3 mm, stipitate-glandular. **Seeds** 1–6, black, lunate to reniform, 3 mm. $2n = 12$.

Flowering and fruiting spring–summer (May–Jul). Redwood and Douglas-fir forests, deep shade; 100–1700 m; Calif., Oreg., Wash.

2. **Vancouveria chrysantha** Greene, Bull. Calif. Acad. Sci. 1: 66. 1885 · Siskiyou inside-out flower, golden inside-out flower [E]

Leaves persistent, 2×-ternately compound or 3–5-foliolate, 10–18 cm; petiole 3–12 cm, sparsely hairy. **Leaflet blades** ovate to oblong, slightly 3-lobed, leathery, base cordate, margins thickened, crisped, apex notched; surfaces abaxially pubescent, glaucous, adaxially glabrous to rarely pubescent. **Inflorescences:** peduncle 2–3 dm; pedicels 1–4 cm, stipitate-glandular. **Flowers** 4–15; bracteoles 6–9, tan to brown, 1–4 mm, caducous, stipitate-glandular; sepals 6, yellow, spatulate, 6–10 mm, stipitate-glandular; petals 6, yellow, 4–6 mm, margins entire, apex strongly reflexed, apical nectary darker yellow; filaments stipitate-glandular. **Follicles** greenish brown, 8–15 mm including beak, beak 3–4 mm, densely stipitate-glandular. **Seeds** 3–10, reddish brown, reniform to oblong, 3–4 mm. $2n = 12$.

Flowering spring (May–Jun); fruiting spring–summer (Jun–Jul). Open, mixed evergreen forests and thickets on serpentine substrates; 100–1500 m; Calif., Oreg.

3. **Vancouveria planipetala** Calloni, Malpighia 1: 266, plate 6. 1887 · Redwood-ivy, redwood inside-out flower [E]

Vancouveria parviflora Greene

Leaves persistent, 2(–3)×-ternately compound, 10–30 cm; petiole 1–15 cm, sparsely hairy, becoming glabrous. **Leaflet blades** rounded-deltate to rounded-pentagonal, often broader than long, obscurely 3-lobed, base cordate, margins conspicuously thickened, crisped, apex minutely notched; surfaces abaxially glabrous or sparsely hairy, adaxially glabrous. **Inflorescences:** peduncle 1–2 dm; pedicels 1–3 cm, stipitate-glandular. **Flowers** 20–50; bracteoles 6–9, white to yellow, glands absent; sepals 6, white to yellow, oblanceolate, 4–5 mm, glands absent; petals 6, white, sometimes lavender-tinged, 3–4 mm, margins entire, apex notched, not reflexed, lateral lobes bearing nectaries, nectaries golden; filaments without glands. **Follicles** greenish brown, 4–7 mm including beak, beak 2 mm, glands absent. **Seeds** 1–2, black, lunate, 3–4 mm. $2n = 24$.

Flowering spring (May–Jun); fruiting spring–summer (Jun–Jul). Redwood forests, shaded areas; 50–1700 m; Calif., Oreg.

---

8. **JEFFERSONIA** Barton, Trans. Amer. Philos. Soc. 3: 342, plate 1. 1793 · [For Thomas Jefferson (1743–1826), third president of the United States]

Lisa O'Rourke George

Herbs, perennial, deciduous, 1–4 dm, glabrous. **Rhizomes** short, producing 4–8 or more leaves per year. **Aerial stems** absent. **Leaves** alternate, 2-foliolate; petiole long, slender. **Leaf blade** reniform-orbiculate in overall outline, 1.5–18 cm, leaflet margins entire to shallowly lobed; venation palmate. **Inflorescences** terminal, solitary flowers borne on scapes. **Flowers** 4-merous, 15–30 mm; bracteoles absent; sepals caducous, (3–)4(–5), petaloid; petals 8, white [lavender];

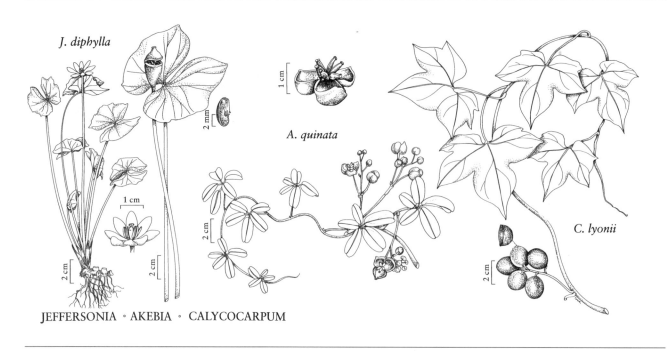

JEFFERSONIA · AKEBIA · CALYCOCARPUM

stamens 8; anthers dehiscing by 2 apically hinged flaps; pollen exine striate; ovaries obovoid; placentation marginal; style central. **Fruits** capsulelike, brownish, obovoid, with transverse or oblique dehiscence, top forming reflexed lid. **Seeds** 10–25, red; aril white, not enclosing seed. $x = 6$.

Species 2 (1 in the flora): e North America and e Asia.

1. **Jeffersonia diphylla** (Linnaeus) Persoon, Syn. Pl. 1: 418. 1805 · Twinleaf, rheumatism-root E F

Podophyllum diphyllum Linnaeus, Sp. Pl. 1: 505. 1753

Leaves: petiole slender, erect, 9–25 cm at anthesis, maturing to 18–43 cm. **Leaf blade** 2-foliolate, often with minute apiculation between leaflets; leaflets 1.2–4 × 0.6–2.5 cm at anthesis, maturing to 6–13 × 3–7 cm, lobes rounded to acute. **Scapes** 9–33 cm, frequently taller than petioles at anthesis. **Flowers:** sepals elliptic to obovate, 7–15 × 4–6 mm; petals white, elliptic to obovate, 11–22 × 9–12 mm; stamens 6–12 mm; filaments 2–3 mm; anthers 4–9 × 1–1.5 mm; ovaries 6–11 × 3–7 mm. **Fruits** 18–38 × 8–17 mm, leathery, opening transversely, apical quarter resembling lid, becoming reflexed. **Seeds** oblong, 4–7 × 2 mm; aril laciniate, attached at adaxial side of hilum. $2n = 12$.

Flowering early spring; fruiting spring. Rich moist woods to semiopen rocky slopes and outcrops, usually over limestone or other calcareous rocks; 100–800 m; Ont.; Ala., Ga., Ill., Ind., Iowa, Ky., Md., Mich., Minn., N.Y., N.C., Ohio, Pa., Tenn., Va., W.Va., Wis.

Plants of *Jeffersonia diphylla* were used medicinally by Native Americans for treatment of dropsy, "gravel" and urinary ailments, and for gall and diarrhea, and in poultices for sores and ulcers (D. E. Moerman 1986).

17. LARDIZABALACEAE Decaisne

• Lardizabala Family

John W. Thieret

John T. Kartesz

Vines [small trees or shrubs], woody, evergreen or deciduous. **Stems** without spines. **Leaves** alternate; stipules absent [except in *Lardizabala*]; petioles present. **Leaf blade** palmately compound [pinnately compound or 1–3×-ternate]. **Inflorescences** axillary, racemes, occasionally solitary flowers, flowers pedicellate. **Flowers** unisexual, staminate and pistillate on same plants in same or different inflorescences [staminate and pistillate on different plants or flowers bisexual, staminate, and pistillate], showy, 3-merous; sepaloid bracteoles absent; sepals 3–6, distinct, usually petaloid, not spurred; petals absent [6; nectary present]. **Staminate flowers:** stamens 6(–8), distinct [proximally connate]; anthers with 2 pollen sacs, extrorse, dehiscing longitudinally; pistillodes sometimes present. **Pistillate flowers:** staminodes sometimes present; pistils 3–15 in 1–5 whorls of 3 each, 1-carpellate, 1-locular; placentation submarginal or laminar; ovules 100–several hundred [1–20]; stigma 1, sessile to nearly sessile. **Fruits** aggregates of berries or of fleshy follicles (sometimes only 1 pistil maturing). **Seeds** [1–]100–several hundred, never stalked; aril absent; endosperm abundant; embryo small, straight.

Genera ca. 8, species ca. 30 (1 genus, 1 species in the flora): North America, South America, Asia.

Lardizabalaceae have a disjunct distribution. Two of the genera—*Decaisnea* and *Lardizabala*—are found in Chile; the others are Asiatic (Himalayas to Vietnam, Japan, and Korea).

SELECTED REFERENCES   Ernst, W. R. 1964. The genera of Berberidaceae, Lardizabalaceae, and Menispermaceae in the southeastern United States. J. Arnold Arbor. 45: 1–35.   Matthews, V. A. 1989. Lardizabalaceae. In: S. M. Walters et al., eds. 1984+. The European Garden Flora. 4+ vols. Cambridge etc. Vol. 3, pp. 396–398.   Prantl, K. 1888. Lardizabalaceae. In: H. G. A. Engler and K. Prantl, eds. 1887–1915. Die natürlichen Pflanzenfamilien. . . . 254 fasc. Leipzig. Fasc. 19[III,2], pp. 67–70.   Spongberg, S. A. and I. H. Burch. 1979. Lardizabalaceae hardy in temperate North America. J. Arnold Arbor. 60: 302–315.

1. AKEBIA Decaisne, Arch. Mus. Hist. Nat. 1: 195. 1839 · [Japanese *akebi*, name for *Akebia quinata*]

Vines, twining. **Leaves** palmately compound; leaflets 3–5, articulate at base of blade and at base of petiolule. **Inflorescences** racemose, pistillate flowers proximal to staminate flowers in each raceme. **Flowers** dimorphic: pistillate flowers larger and longer pediceled than staminate flowers; sepals mostly brownish to purplish. **Staminate flowers:** pistillodes present. **Pistillate flowers:** pistils (3–)–8(–15); placentation laminar; staminodes present. **Fruits** follicles, fleshy, dehiscent along adaxial suture. **Seeds** 100–several hundred. $x = 16$.

Species 4 (1 in the flora): North America, Asia.

Pistils in *Akebia* are incompletely closed distally (W. W. Payne and J. L. Seago 1968).

SELECTED REFERENCES Anderson, E. 1934. The genus *Akebia*. Bull. Popular Inform. Arnold Arbor., ser. 4, 2: 17–20. Payne, W. W. and J. L. Seago. 1968. The open conduplicate carpel of *Akebia quinata* (Berberidales: Lardizabalaceae). Amer. J. Bot. 55: 575–581. Sargent, C. S. 1891. The fruit of *Akebia quinata*. Gard. & Forest 4: 136–137.

1. **Akebia quinata** (Houttuyn) Decaisne, Arch. Mus. Hist. Nat. 1: 195, fig. 1a–c. 1839 · Chocolate-vine, five-leaf [F]

Rajania quinata Houttuyn, Nat. Hist. 11: 366, plate 75, fig. 1. 1779

Plants, deciduous to semi-evergreen, climbing to 12 m, glabrous. **Leaves:** petiole 1.6–12.5 cm; leaflets mostly 5, petiolules 0.2–2.2 cm, blades oblong to ovate-elliptic, 0.7–8.2 × 0.4–4.2 cm, base rounded, margins entire, apex retuse. **Inflorescences** pendent, 4.5–12 cm; pedicel with basal bracts. **Flowers** fragrant. **Staminate flowers** 4–15 per inflorescence, 1.2–1.6 cm diam.; sepals oblong to ovate or elliptic, 5–9 mm; stamens 4–5 mm. **Pistillate flowers** (0–)1–5 per inflorescence, 2–3 cm diam.; sepals elliptic to ovate or nearly orbiculate, 10–16 mm; pistils 3–7, 1 or more maturing. **Follicles** glaucous, violet to dark purple, oblong, 5–15 cm. **Seeds** black, ovoid, embedded in whitish pulp.

Flowering spring, fruiting fall (Sep–Oct). Waste places, open woodlands; 0–400 m; introduced; Conn., Ga., Ind., Ky., Md., Mass., Mich., N.J., N.Y., N.C., Ohio, Pa., Va., W.Va.; native, Asia.

No specimens are known from Rhode Island.

A fast-growing, invasive vine whose aggressiveness may at times approach that of *Lonicera japonica*, *Akebia quinata* is occasionally planted as an ornamental; it is of more botanical than horticultural interest. A greenish- to whitish-flowered variant, known from Asia, is cultivated in North America. The edible, though allegedly insipid, fruits are apparently uncommon in cultivation; cross-pollination appears to be necessary for their development (C. S. Sargent 1891).

18. MENISPERMACEAE Jussieu

• Moonseed Family

Donald G. Rhodes

Vines and lianas [shrubs or trees], deciduous, woody at least at base, twining or clambering. **Stems** striate, without spines. **Leaves** alternate, simple; stipules absent; petioles present. **Leaf blade** palmately veined, often palmately lobed. **Inflorescences** axillary or terminal, fascicles, cymes, racemes, or panicles, flowers pedicillate. **Flowers** unisexual, staminate and pistillate on different plants, never showy; sepaloid bracteoles absent; perianth hypogynous, segments distinct or fused, not showy, greenish white to white or cream, imbricate or valvate. **Staminate flowers:** sepals usually 6, not spurred; petals usually 6, sometimes absent, distinct or connate, ± concave, frequently minute; nectaries absent; stamens either opposite petals and equal in number, or numerous; filaments distinct or united; anthers dehiscing longitudinally [or transversely]; pistillodes sometimes present. **Pistillate flowers:** sepals (4–)6, sometimes reduced to 1; petals often 6 or reduced to 1, ± concave, usually minute; nectaries absent; staminodes frequently present; pistils 1–6; ovules 2, aborting to 1, amphitropous; style often recurved; stigma entire or lobed. **Fruits** drupes, straight or horseshoe-shaped; exocarp membranous; mesocarp ± pulpy; endocarp (stone) bony, often warty, ribbed. **Seeds** never stalked; endosperm present or absent; embryo usually curved.

Genera ca. 78, species ca. 525 (4 genera, 5 species in the flora): widespread in temperate and tropical regions.

SELECTED REFERENCES Diels, F. L. E. 1910. Menispermaceae. In: H. G. A. Engler, ed. 1900–1953. Das Pflanzenreich. . . . 107 vols. Berlin. Vol. 46[IV,94], pp. 334–345. Ernst, W. R. 1964. The genera of Berberidaceae, Lardizabalaceae, and Menispermaceae in the southeastern United States. J. Arnold Arbor. 45: 1–35.

1. Inflorescences with bracts; pistillate flowers: sepal 1, petal 1, pistil 1; staminate flowers (not known in flora): sepals 4, petals 4, connate, stamen filaments fused, anthers 1-locular; leaf blade usually reniform; drupes pubescent. . . . . 4. *Cissampelos*, p. 299
1. Inflorescences with minute bracts or bracts absent; pistillate and staminate flowers: sepals 6–9, petals distinct when present, stamen filaments distinct, anthers 2–4-locular, pistils 2–6; leaf blade various shapes, rarely reniform; drupes glabrous.
   2. Petals absent or vestigial; anthers 2-locular; ovary ellipsoid to fusiform; stigma

multicleft; leaf blade not mucronate at apex, surfaces sparsely bristly or glabrous; endocarp smooth, excavate on 1 side, margins erose. . . . . .1. *Calycocarpum*, p. 296

2. Petals well developed; anthers 4-locular; ovary slightly asymmetrically pouched; stigma
entire or slightly lobed; leaf blade mucronate at apex, surfaces soft-pubescent or glabrous; endocarp warty and ribbed, depressed but not excavate.

    3. Inflorescences usually racemes or racemose panicles; petals 6, to 2 mm, with auricu-
late basal lobes; stamens 6; pistils 6; leaf blades not peltate . . . . 2. *Cocculus*, p. 296

    3. Inflorescences usually of well-developed panicles; petals (4–)6(–12), to 4 mm, with-
out auriculate basal lobes; stamens 12–36; pistils (2–)3(–4); leaf blade usually pel-
tate, petiole inserted near margin . . . . .3. *Menispermum* p. 297

1. CALYCOCARPUM Nuttall ex Spach, Hist. Nat. Vég. 8: 7. 1838 · Cupseed [Greek, *calyx,* cup, and *carpos,* fruit]

Vines, twining or clambering. **Stems** often bluish green, apically tomentose grading to glabrate. **Leaves** not peltate. **Leaf blade** generally pentagonal, palmately 3–5-lobed, base cordate, mar-
gins of lobes entire or coarsely dentate, apex caudate-acuminate, not mucronate; surfaces sparsely bristly or glabrous. **Inflorescences** axillary, racemes or panicles; bracts absent. **Flowers** 3-merous; sepals 6–9, elliptic to obovate, glabrous. **Staminate flowers:** petals absent; stamens 6–12; filaments distinct; anthers 2-locular; pistillodes absent. **Pistillate flowers:** petals absent or vestigial; staminodes 6–91, poorly developed; pistils 3; ovary ellipsoid to fusiform, glabrous; stigma multicleft. **Drupes** globose to ellipsoid, glabrous; endocarp smooth, cup-shaped with erose margins, glabrous.

    Species 1 (1 in the flora): c, s United States.

1. Calycocarpum lyonii (Pursh) A. Gray, Gen. Amer. Bor. 1: 76. 1848 · Cupseed E F

Menispermum lyonii Pursh, Fl. Amer. Sept., 371. 1814

Vines, climbing to tops of trees; rhizomes slender, branched. **Leaves:** petioles to 28 cm. **Leaf blade** 10–25 × 10–29 cm, membranous; venation 5 or 7. **Inflorescences** to 32 cm; rachis puberulent, glandular. **Flowers:** sepals 1.6–4.6 × 0.6–2 mm, glabrous. **Staminate**
flowers: stamens to 3 mm. **Pistillate flowers:** ovary to 2 mm. **Drupes** green at maturity, darkening to deeper green or black upon drying, 15–25 × 10–20 mm.

    Flowering spring–summer. Primarily along rivers or smaller streams in deciduous forests; 0–350 m; Ala., Ark., Fla., Ga., Ill., Ind., Kans., Ky., La., Miss., Mo., Okla., S.C., Tenn.

    Calycocarpum lyonii has been reported as oc-
curring in Texas, but no Texas specimens have been seen.

2. COCCULUS de Candolle, Syst. Nat. 1: 515. 1818 · Coral beads [diminutive of Latin *coccus,* berry]

Vines, twining or clambering. **Stems** green, apically tomentose grading to pilose or glabrate on older portions. **Leaves** not peltate. **Leaf blade** generally ovate to hastate or oblong, base cordate, truncate, or rounded, margins entire, apex mucronate; surfaces soft-pubescent or glabrous. **Inflorescences** axillary or terminal, racemes or racemose panicles; bracts minute (bracte-
oles). **Flowers** 3-merous; sepals 6–9, ovate to elliptic or obovate, outer sepals glabrous or pilose to sparsely pilose abaxially, inner sepals glabrous; petals 6, free. **Staminate flowers:** petals to 2 mm, auriculate lobes at base inflexed over 6 stamens; filaments distinct; anthers 4-
locular; pistillodes 3–6 or absent, glandular. **Pistillate flowers:** perianth similar to staminate;

staminodes 6, poorly developed; pistils 6; ovary slightly asymmetrically pouched, glabrous; stigma entire. **Drupes** globose, glabrous; endocarp bony, depressed but not excavate, warty, ribbed. $x = 13$.

Species 11 (2 in the flora): temperate regions, North America, Mexico, Asia, Africa.

1. Stems with spreading pubescence; leaf blade typically ovate, occasionally sagittate or hastate, abaxially pubescent; sepals of outer series pubescent; drupes red . . . . .1. *Cocculus carolinus*
1. Stems with appressed pubescence; leaf blade typically oblong, sometimes basally lobed, abaxially glabrous; sepals of outer series glabrous; drupes black. . . . . .2. *Cocculus diversifolius*

1. Cocculus carolinus (Linnaeus) de Candolle, Syst. Nat. 1: 524. 1818 · Red-berried moonseed, coral vine, margil, hierba de ojo F

Menispermum carolinum Linnaeus, Sp. Pl. 1: 340. 1753; *Epibaterium carolinum* (Linnaeus) Britton

Vines, to 5 m or more; rhizomes to 1.4 cm diam. **Stems** with spreading pubescence. **Leaves:** petiole to 10 cm. **Leaf blade** generally ovate or deltate, sometimes sagittate or hastate, to 17 × 14 cm, membranous to leathery, base sometimes with 3–5 lobes, margins usually entire, apex acuminate to rounded, then often retuse, mucronate; surfaces abaxially slightly pale, rarely glaucous, sparsely to densely pilose, adaxially glabrous to sparsely pilose; venation 5. **Inflorescences** to 22 cm; bracteoles often present; rachis glabrous or tomentose, not glaucous. **Flowers:** perianth parts not glaucous; sepals in 3 series, outer sepals 0–3, ovate, 0.3–1.4 × 0.2–0.8 mm, pilose, middle 3 sepals ovate to elliptic or obovate, 1–3 × 0.6–2 mm, glabrous to pilose, inner 3 sepals elliptic to nearly orbiculate or obovate, 0.8–3 × 0.8–2 mm, glabrous to sparsely pilose; petals (5–)6, yellow, elliptic, deltate, rhombic, obovate, or nearly orbiculate, 0.6–2 × 0.4–1.4 mm, glabrous. **Staminate flowers:** stamens (5–)6, to 2.2 mm; pistillodes to 0.5 mm. **Pistillate flowers:** staminodes to 0.8 mm; pistils to 2 mm. **Drupes** red, 5–8 mm diam. $2n = 78$.

Flowering late spring–early fall. Woodland and shrub borders, along streams, fencerows, waste places; 0–500 m; Ala., Ark., Del., Fla., Ga., Ill., Ind., Kans., Ky., La., Miss., Mo., N.C., Okla., S.C., Tenn., Tex., Va.; Mexico.

Cocculus carolinus was used by some Native American tribes medicinally to treat blood ailments (D. E. Moerman 1986).

2. Cocculus diversifolius de Candolle, Syst. Nat. 1: 523. 1818 · Sarsaparilla, correhuela

Vines, to 3 m or more; rhizomes to 1.4 cm diam. **Stems** with appressed pubescence. **Leaves:** petiole to 1.8 cm. **Leaf blade** generally linear, lanceolate-oblong, oblong, or ovate-oblong, to 8.5 × 6 cm, ± leathery, base sometimes lobed, apex obtuse to rounded and often retuse, mucronate; surfaces glabrous, abaxially slightly pale; venation 3–5. **Inflorescences** to 7 cm; bracteoles and rachis glabrous or pubescent with short-appressed hairs, sometimes glaucous. **Flowers:** perianth parts glabrous, often glaucous; sepals in 3 series, outer sepals 0–3, ovate, 0.4–0.8 × 0.3–0.4 mm, middle 3 sepals ovate to elliptic or obovate, 0.6–1.6 × 0.4–1 mm, inner 3 sepals elliptic to nearly orbiculate, 1.2–2.2 × 1.2–1.8 mm; petals 6, usually yellowish, elliptic to obovate, 0.8–1.6 × 0.6–1 mm, glabrous. **Staminate flowers:** stamens 6, to 1.8 mm; pistillodes absent. **Pistillate flowers:** staminodes to 0.4 mm; pistils to 1.4 mm. **Drupes** black or bluish black, 4–6 mm diam., often glaucous.

Flowering spring–fall. Brushlands, prairies, palm groves, fencerows, along resecas and canyons; 0–1400 m; Ariz., Tex.; Mexico.

3. MENISPERMUM Linnaeus, Sp. Pl. 1: 340. 1753; Gen. Pl. ed. 5, 158. 1754

· Moonseed [Greek *mene,* moon, and *sperma,* seed]

Vines, twining, or lianas. **Stems** green, apically tomentose grading to glabrate on older portions. **Leaves** peltate, petiole inserted near margin, rarely with petiole attached basally. **Leaf blade** broadly ovate, generally (3–)5–7-lobed or angled, base cordate, truncate, or rounded, margins entire, apex acuminate to rounded, mucronate, rarely notched or obcordate; surfaces

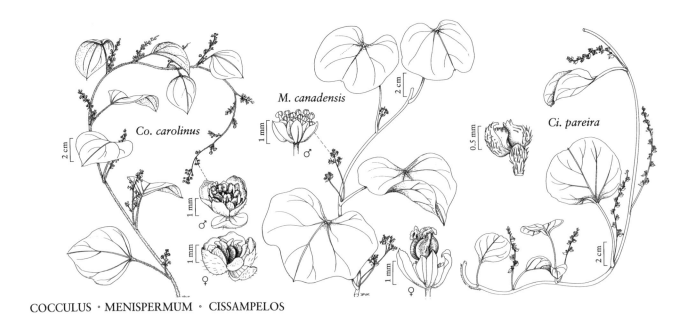

M. canadensis

Co. carolinus

Ci. pareira

COCCULUS · MENISPERMUM · CISSAMPELOS

abaxially very pale, sometimes silvery, glabrous to pilose, adaxially glabrous to sparsely pilose. **Inflorescences** axillary, panicles well developed; bracts minute (bracteoles). **Flowers** 3-merous; sepals usually 6, ovate to elliptic or obovate, glabrous; petals (4–)6(–12), free, to 4 mm, entire, auriculate basal lobes absent. **Staminate flowers:** petals entire; stamens 12–36; filaments distinct; anthers 4-locular; pistillodes absent. **Pistillate flowers:** staminodes 6–9, well developed; pistils (2–)3(–4); ovary slightly asymmetrically pouched, glabrous; stigma slightly lobed. **Drupes** globose, glabrous; endocarp bony, warty, ribbed.

Species 3 (1 in the flora): North America, Mexico, and Asia.

1. **Menispermum canadense** Linnaeus, Sp. Pl. 1: 340. 1753 · Moonseed, Canada moonseed, ménisperme du Canada, raison de couleuvre [E] [F]

Vines or lianas, twining, to ca. 5 m; rhizomes to 1 cm diam. **Leaves** peltate with petiole inserted to 11 mm from margin, rarely not peltate; petiole to 20 cm. **Leaf blade** ovate or nearly orbiculate, rarely reniform, to 23 × 24 cm, membranous; venation 7–12. **Inflorescences** to 18 cm; rachis glabrous or sparsely pilose. **Flowers:** sepals (4–)5–8, ovate, elliptic, or obovate, 1–4 × 0.4–1.8 mm, glabrous or sparsely pilose; petals 4–12, elliptic to nearly orbiculate or obovate, 0.6–2 × 0.6–2 mm, margins slightly involute, glabrous. **Staminate flowers:** stamens to 4 mm. **Pistillate flowers:** staminodes to 0.8 (–1.5) mm; pistils 2–4, to 1.4 mm. **Drupes** black or bluish black, 8–13 mm diam., often glaucous. $2n = 52$.

Flowering spring–summer. Deciduous woods and thickets, along streams, bluffs and rocky hillsides, fencerows, shade tolerant; 0–700 m; Man., Ont., Que.; Ala., Ark., Conn., Del., D.C., Fla., Ga., Ill., Ind., Iowa, Kans., Ky., Md., Mass., Mich., Minn., Miss., Mo., Nebr., N.H., N.J., N.Y., N.C., N.Dak., Ohio, Okla., Pa., R.I., S.C., S.Dak., Tenn., Tex., Vt., Va., W.Va., Wis.

The fruit of *Menispermum canadense* is thought to be poisonous. This species is sometimes grown as an ornamental.

Some Native American tribes used *Menispermum canadense* medicinally as dermatological, gastrointestinal, gynecological, and venereal aids, and as remedies for various other complaints (D. E. Moerman 1986).

4. CISSAMPELOS Linnaeus, Sp. Pl. 2: 1031. 1753; Gen. Pl. ed. 5, 455. 1754 · [Greek pertaining to ivy or vine)

Vines, twining or clambering. **Stems** green, glabrous to densely pilose. **Leaves** peltate or petiole inserted at base of blade. **Leaf blade** usually reniform, sometimes nearly orbiculate, base cordate to truncate, margins entire, apex acute to rounded, often retuse, mucronate; surfaces abaxially pale, pilose to densely pilose, adaxially glabrous to pilose. **Inflorescences:** pistillate inflorescences fascicles from axils of normal leaves, individual flowers in axils of reduced leaves or bracts upon secondary axillary branches. **Pistillate flowers:** sepal 1, obovate, abaxially pubescent; petal 1; staminodes absent; pistil 1; ovary asymmetrically pouched, pilose [or glabrous]; stigma 3–5-lobed. [**Staminate inflorescences and flowers, and mature fruits** have not been found in the flora; descriptions given here are for extraterritorial species. **Staminate inflorescences** axillary, multiflowered cymes from normal leaves, reduced leaves, or bracts. **Staminate flowers** 2-merous; sepals 4; petals 4, connate, corolla cup-shaped; stamens 4; anthers on column formed by fused filaments, 1-locular, dehiscing transversely. **Drupes** nearly globose, pubescent; endocarp bony, warty, ribbed.]

Species 19 (1 in the flora): tropical and subtropical regions, North America, South America, Africa, and Asia.

SELECTED REFERENCE Rhodes, D. G. 1975. A revision of the genus *Cissampelos*. Phytologia 30: 415–484.

1. **Cissampelos pareira** Linnaeus, Sp. Pl. 2: 1031. 1753 [F]

Vines, from thickened root. **Leaves** with petiole inserted to 2 mm from margin; petiole to 7.4 cm. **Leaf blade** 6–12 × 7–14 cm, membranous to nearly leathery; venation 7. **Staminate inflorescences** not seen. **Pistillate inflorescences** to 13 cm; bracts sessile or shortly petiolate, usually reniform, to 5 × 6 mm, grading to minute, membranous, glabrous to densely pilose. **Pistillate flowers:** sepals obovate, 1.2–1.4 × 0.8–1 mm; petals 4-angulate to deltate, 0.4–0.8 × 1–1.6 mm; ovary to 0.8 mm. **Drupes** not seen.

Flowering summer–winter. Hammocks and residential areas; 0–10 m; Fla.; West Indies; Central America; South America; Asia; Africa; Pacific Islands.

Only seven collections of *Cissampelos pareira* have been seen for North America, all from Dade County, Florida. It is odd that all of them had pistillate flowers. No fruiting material was observed. If staminate plants are present, they are yet to be collected.

19. PAPAVERACEAE Jussieu

· Poppy Family

Robert W. Kiger

Herbs, subshrubs, shrubs, or small trees, annual, biennial, or perennial, scapose or caulescent, usually from taproots, sometimes from rhizomes; sap clear, white, or colored, often sticky. **Stems** leafy or naked, erect, spreading, or decumbent, simple or branching. **Leaves** basal and/or cauline, alternate to opposite or whorled, simple, without stipules, petiolate or sessile; blade unlobed or with 1–3 odd-pinnate, subpalmate, or palmate orders of lobes. **Inflorescences** axillary or terminal, unifloral or else multifloral and cymiform, racemose, umbelliform, corymbiform, or paniculate, pedunculate or subsessile; bracts usually present. **Flowers** radially symmetric, pedicellate or sessile; receptacle sometimes expanded and forming cup or ring beneath calyx (only in *Dendromecon, Eschscholzia, Meconella,* and *Platystemon*); perianth and androecium sometimes perigynous; sepals caducous, 2 or 3, distinct or connate, usually obovate; petals distinct, usually obovate, mostly 2 times number of sepals, sometimes more or absent; stamens many or 4–15 (only in *Meconella* and *Canbya*); anthers 2-locular; pistil 1, 2–18 [–22]-carpellate; ovary 1–2-locular or incompletely to completely multilocular by placental intrusion; placentas 2 or more, parietal; style 1 or absent; stigmas or stigma lobes 2–many. **Fruits** capsular, dehiscence valvate, poricidal, or transverse, or carpels dissociating and breaking transversely into 1-seeded segments (only in *Platystemon*). **Seeds** usually many, small, sometimes arillate or carunculate.

Genera 25–30 (17 genera, 63 species in the flora): worldwide, mainly Northern Hemisphere.

According to W. R. Ernst (1962b), Papaveraceae "may be divided conveniently into four subfamilies." His scheme is followed here, but with the subfamilies taken up in alphabetic order; they seem to be natural groups, but their phylogenetic interrelationships are not yet clear. Similarly, the evolutionary relationships within the subfamilies remain ambiguous, and the genera in each are listed alphabetically. Subfamily Chelidonioideae Ernst includes genera 1–5; subf. Eschscholzioideae Ernst, genera 6–7; subf. Papaveroideae Ernst, genera 8–14; and subf. Platystemonoideae Ernst, genera 15–17.

Hunnemannia fumariifolia Sweet, native to the highlands of Mexico, is occasionally found in California as a garden escape. A glabrous perennial with glaucous, blue-gray stem and leaves, and glossy, yellow petals, it bears an overall resemblance to *Eschscholzia* but has distinct sepals, and a peltate stigma. Below, it would key out as *Arctomecon*.

SELECTED REFERENCES Ernst, W. R. 1962. A Comparative Morphology of the Papaveraceae. Ph.D. dissertation. Stanford University. Ernst, W. R. 1962b. The genera of Papaveraceae and Fumariaceae in the southeastern United States. J. Arnold Arbor. 43: 315–343. Ernst, W. R. 1967. Floral morphology and systematics of *Platystemon* and its allies *Hesperomecon* and *Meconella* (Papaveraceae: Platystemonoideae). Univ. Kansas Sci. Bull. 47: 25–70. Fedde, F. 1909. Papaveraceae–Hypecoideae et Papaveraceae–Papaveroideae. In: H. G. A. Engler, ed. 1900–1953. Das Pflanzenreich. 107 vols. Berlin. Vol. 40[IV,104], pp. 1–430. Fedde, F. 1936. Papaveraceae. In: H. G. A. Engler et al., eds. 1924 +. Die natürlichen Pflanzenfamilien, ed. 2. 26 + vols. Leipzig and Berlin. Vol. 17b, pp. 5–145. Grey-Wilson, C. 1993. Poppies: A Guide to the Poppy Family in the Wild and in Cultivation. Portland. Gunn, C. R. 1980. Seeds and fruits of Papaveraceae and Fumariaceae. Seed Sci. Techn. 8: 3–58. Gunn, C. R. and M. J. Seldin. 1976. Seeds and Fruits of North American Papaveraceae. Washington. [U.S.D.A. Agric. Res. Serv., Techn. Bull. 1517.] Harms, H. 1936. Reihe Rhoeadales. In: H. G. A. Engler et al., eds. 1924 +. Die natürlichen Pflanzenfamilien, ed. 2. 26 + vols. Leipzig and Berlin. Vol. 17b, pp. 1–4. Hutchinson, J. 1925. Contributions towards a phylogenetic classification of flowering plants: V. The genera of Papaveraceae. Bull. Misc. Inform. Kew 1925: 161–168. Kadereit, J. W. 1993. Papaveraceae. In: K. Kubitzki et al., eds. 1990 +. The Families and Genera of Vascular Plants. 2 + vols. Berlin etc. Vol. 2, pp. 494–506. Stermitz, F. R. 1968. Alkaloid chemistry and the systematics of *Papaver* and *Argemone*. Recent Advances Phytochem. 1: 161–183.

1. Leaves opposite or whorled.
 2. Plants glabrous or glabrate; basal leaves winged-petiolate; stamens 4–12; capsules linear . . . . 16. *Meconella,* p. 337
 2. Plants usually distinctly pubescent; basal leaves sessile or absent; stamens 12 or more; capsules not linear.
  3. Stems branching distally; stigmas and carpels 6 or more, carpels dissociating and breaking transversely into 1-seeded segments . . . . 17. *Platystemon,* p. 338
  3. Stems branching from base; stigmas and carpels 3, carpels not dissociating, not breaking into transverse segments, capsules valvate. . . . . 15. *Hesperomecon,* p. 336
1. Leaves alternate or subopposite (sometimes only 1 in *Sanguinaria*).
 4. Plants subshrubby or shrubby.
  5. Leaf blades and capsules harshly prickly. . . . . .9. *Argemone* (in part), p. 314
  5. Leaf blades and capsules not harshly prickly, sometimes bristly.
   6. Leaf blades lobed; petals white . . . . 13. *Romneya,* p. 334
   6. Leaf blades unlobed; petals yellow. . . . . 6. *Dendromecon,* p. 306
 4. Plants herbaceous.
  7. Sepals connate, calyptrate; evident receptacular cup beneath calyx . . . . 7. *Eschscholzia,* p. 308
  7. Sepals distinct, not calyptrate; receptacular cup absent or obscure.
   8. Inflorescences paniculate; petals absent . . . . 3. *Macleaya,* p. 304
   8. Inflorescences not paniculate; petals present.
    9. Inflorescences umbelliform.
     10. Stigma lobes 2; capsule 2-valved, dehiscing from base, glabrous. . . . 1. *Chelidonium,* p. 302
     10. Stigma lobes 3–4(–5); capsule (3–)4-valved, dehiscing from apex, pubescent. . . . . 5. *Stylophorum,* p. 305
    9. Inflorescences not umbelliform.
     11. Leaf blades and capsules harshly prickly . . . . 9. *Argemone* (in part), p. 314
     11. Leaf blades and capsules not harshly prickly.
      12. Stigmas radiating on sessile disc . . . . 11. *Papaver,* p. 323
      12. Stigmatic disc absent.
       13. Style absent or indistinct.
        14. Sap clear; leaf blades unlobed or lobed only distally; petals 5–7.
         15. Leaf blades glabrous; inflorescences 1-flowered; stamens 6–15 . . . . 10. *Canbya,* p. 323
         15. Leaf blades long-pilose; inflorescences 3–20-flowered; stamens many. 8. *Arctomecon* (in part), p. 312
        14. Sap yellow; leaf blades lobed throughout; petals 4.

1. CHELIDONIUM Linnaeus, Sp. Pl. 1: 505. 1753; Gen. Pl. ed. 5, 224. 1754

· Celandine, greater celandine, rock-poppy, swallowwort, chélidoine [Greek *cheilidon*, swallow (bird), perhaps from lore reported by Aristotle and others that mother swallows bathe eyes of their young with the sap]

Robert W. Kiger

Herbs, biennial or perennial, caulescent, from stout rhizomes or taproots; sap yellow to orange. **Stems** leafy. **Leaves** petiolate; basal rosulate, cauline alternate; blade 1–2×-pinnately lobed. **Inflorescences** axillary or terminal, umbelliform, few flowered; bracts present. **Flowers:** sepals 2, distinct; petals 4; stamens ca. 12–many; pistil 2-carpellate; ovary 1-locular; style ± distinct; stigma 2-lobed. **Capsules** erect, 2-valved, dehiscing from base. **Seeds** few to many, arillate. $x = 6$.

Species 1: North America, Eurasia.

1. Chelidonium majus Linnaeus, Sp. Pl. 1: 505. 1753 [F]
[W2]

Plants to 10 dm. **Stems** branching, ribbed. **Leaves** to 35 cm; petiole 2–10 cm; blade deeply 5–9-lobed; margins irregularly dentate or crenate, rarely laciniate. **Inflorescences:** peduncle 2–10 cm. **Flowers:** pedicels 5–35 mm; sepals to 1 cm; petals bright yellow, obovate to oblong, to 2 cm wide; style ca. 1 mm. **Capsules** linear to narrowly oblong, 2–5 cm, glabrous. **Seeds** black, reticulate-pitted.

Flowering spring–summer. Moist to dry woods, thickets, fields, hedgerows and fences, roadsides, railroads, and waste ground; 0–1000 m; introduced; B.C., N.B., N.S., Ont., P.E.I., Que.; Conn., Del., Ill., Ind., Iowa, Maine, Md., Mass., Mich., Minn., Mo., Mont., N.H., N.J., N.Y., N.C., Ohio, Pa., R.I., Utah, Vt., Va., Wash., W.Va., Wis.; Eurasia.

The irritating sap of *Chelidonium* has been used to treat warts. In the vegetative state, this weedy introduction from Eurasia is difficult to distinguish from the native *Stylophorum diphyllum*.

2. GLAUCIUM Miller, Gard. Dict. Abr. ed. 4, vol. 2. 1754 · Horned-poppy, sea-poppy

[Greek *glaukos*, gray-green, in reference to color of waxy bloom on all parts]

Robert W. Kiger

Herbs, annual, biennial, or perennial, caulescent, glaucous, from taproots; sap yellow. **Stems** leafy, sometimes becoming woody at base. **Leaves:** basal rosulate, petiolate; cauline alternate, sessile; blades 1–2×-pinnately lobed [unlobed]. **Inflorescences** axillary or terminal, 1-flowered;

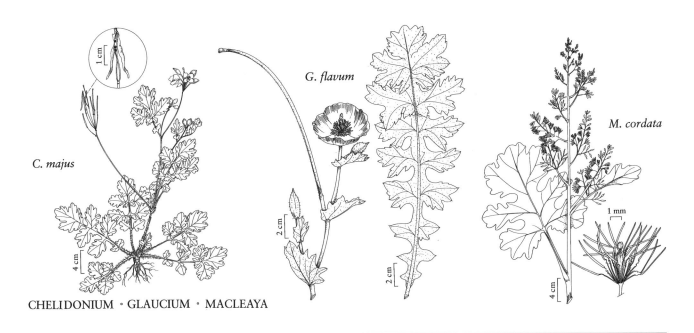

C. *majus*

G. *flavum*

M. *cordata*

CHELIDONIUM · GLAUCIUM · MACLEAYA

bracts present. **Flowers:** sepals 2, distinct; petals 4; stamens many; pistil 2-carpellate; ovary 2-locular; style absent or indistinct; stigma 2-lobed. **Capsules** erect, 2-valved, dehiscing from apex [base]. **Seeds** many, dark brown, reticulate-pitted, embedded in pithy septum, aril absent. $x = 6$.

Species 20–25 (2 in the flora): North America, Europe, c, sw Asia.

1. Basal leaves few, glabrate to moderately pubescent; blades of distal leaves not distinctly clasping stem; petals orange to reddish orange, usually with blackish basal spot; capsules straight to slightly curved, pubescent or glabrate . . . . 1. *Glaucium corniculatum*
1. Basal leaves numerous, densely pubescent; blades of distal leaves distinctly clasping stem; petals yellow or orange-yellow, sometimes with reddish to violet basal spot; capsules mostly distinctly curved, glabrous, tuberculate, or scabrous . . . . 2. *Glaucium flavum*

1. Glaucium corniculatum (Linnaeus) Rudolph, Fl. Jen., 13. 1781 [W2]

Chelidonium corniculatum Linnaeus, Sp. Pl. 1: 506. 1753

Plants annual or occasionally biennial, to 5 dm. **Stems** branching. **Leaves** to 25 cm; basal few, blade glabrate to moderately pubescent; basal and proximal cauline with blade lyrate, 7–9-lobed; distal with blade ovate, not distinctly clasping stem (sometimes slightly cordate-clasping); margins deeply dentate. **Flowers:** pedicels stout, to 5 cm; sepals 15–30 mm; petals orange to reddish orange, usually with blackish basal spot, obovate, to 40 mm. **Capsules** sublinear, straight or slightly curved, to 25 cm, appressed- to ascending-pubescent or glabrate.

Flowering late spring–summer. Open shores, fields, pastures, and canyon slopes; 0–1600 m; introduced; Kans., Mont., Nev., N.Y., Oreg., Pa., Tex.; Europe; sw Asia.

Glaucium corniculatum has been widely introduced outside its native Eurasian range as a crop weed and ballast waif. It can persist in a fairly broad range of climates and probably is established in North America more widely than existing herbarium records indicate.

2. Glaucium flavum Crantz, Stirp. Austr. Fasc. 2: 133. 1763 · Yellow horned-poppy [F]

Chelidonium glaucium Linnaeus, Sp. Pl. 1: 506. 1753; *Glaucium luteum* Scopoli

Plants biennial or perennial, to 8 dm. **Stems** branching. **Leaves** to 30 cm; basal numerous, blade densely pubescent; basal and proximal cauline with blade lyrate, 7–9-lobed; distal with blade ovate, cordate, distinctly clasping stem; margins deeply dentate. **Flowers:** pedicels stout, to 4 cm; sepals 20–30 mm; petals yellow, sometimes orangish, sometimes with reddish to violet basal spot, obovate, 25–40 mm. **Capsules** sublinear, mostly distinctly curved, sometimes straight, to 30 cm, glabrous, tuberculate, or scabrous. $2n = 12$.

Flowering summer. Open sandy shores and flats, waste places, and on ballast; 0–200 m; introduced; Ont.; Colo., Conn., Md., Mass., Mich., N.J., N.Y., Okla., Oreg., Pa., R.I., Va.; Europe; sw Asia.

Native from the Black Sea and Transcaucasus to coastal southern and western Europe, and also well established as a ruderal in central Europe, *Glaucium flavum* has spread far beyond that range as a ballast waif and occasional garden escape. It should be expected elsewhere in the flora.

3. **M A C L E A Y A** R. Brown in D. Denham and H. Clapperton, Narr. Travels Africa, app., 218. 1826 · Plume-poppy, tree-celandine [for Alexander Macleay, 1767–1848, Scottish botanist, entomologist, and Secretary to the Colony of New South Wales]

Robert W. Kiger

Herbs, perennial, caulescent, usually glaucous, from rhizomes; sap yellow. **Stems** hollow, leafy. **Leaves** alternate, petiolate; blade 1–2×-subpalmately or pinnately lobed. **Inflorescences** terminal, paniculate, many flowered; bracts present. **Flowers:** sepals 2, distinct; petals absent; stamens 25–30 [8–12]; pistil 2-carpellate; ovary substipitate, 1-locular; style short; stigma 2-lobed. **Capsules** nodding, substipitate, 2-valved, dehiscing from apex. **Seeds** [1–]4–6, arillate [not arillate]. $x = 10$.

Species 2 (1 in the flora): North America, Asia (China and Japan).

Native to temperate eastern Asia, *Macleaya* has sometimes been merged with neotropical *Bocconia*, which differs in having perennial stems, long-stipitate ovaries, fleshy, single-seeded capsules dehiscing from the base, and much larger seeds. The sap of *Macleaya* has been used in traditional Chinese medicine as an antiseptic for wounds (C. Grey-Wilson 1993).

SELECTED REFERENCE   Hutchinson, J. 1920. *Bocconia* and *Macleaya*. Bull. Misc. Inform. Kew 1920: 275–282.

1. **Macleaya cordata** (Willdenow) R. Brown in D. Denham and H. Clapperton, Narr. Travels Africa, app., 218. 1826 F

Bocconia cordata Willdenow, Sp. Pl. 2(2): 841. 1799

Plants to 25 dm. **Stems** simple or branching distally. **Leaves** 10–30 cm long and wide; petiole 2–15(–20) cm; blade broadly ovate-cordate, 7–9-lobed; margins irregularly and coarsely dentate; abaxial surface usually whitish and densely short-pubescent, adaxial glabrous. **Inflorescences** to 30 cm or more. **Flowers:** pedicels 4–10 mm; sepals white to cream colored, spatulate, 5–10 mm; style to ca. 1 mm. **Capsules** oblanceoloid, strongly compressed, 15–20 mm, glabrous. **Seeds** dark brown, reticulate. $2n = 20$.

Flowering summer. Deciduous woods, thickets, old fields, ditches, roadsides, pond margins, and along watercourses; 0–800 m; introduced; Ont., Que.; Ala., Conn., Ill., Ind., Maine, Mass., Mich., N.J., N.Y., N.C., Ohio, Pa., Va., W.Va.; e Asia.

A garden perennial esteemed for its vegetative features and large, plumelike inflorescence, *Macleaya cordata* occasionally escapes from cultivation, persisting and reproducing fairly successfully. It might be found almost anywhere in temperate North America east of the Mississippi River at elevations below 1000 m.

4. SANGUINARIA Linnaeus, Sp. Pl. 1: 505. 1753; Gen. Pl. ed. 5, 223. 1754

· Bloodroot, puccoon, sanguinaire du Canada, sang-dragon [Latin *sanguis*, blood, in reference to color of sap]

Robert W. Kiger

Herbs, perennial, scapose, from thick rhizomes; sap orange to red. **Leaves** 1, or few and rosulate, sheathing-petiolate; blade 1×-palmately lobed. **Inflorescences** terminal, 1(–3)-flowered; bracts absent. **Flowers:** sepals 2, distinct; petals 6–12, unequal; stamens many; pistil 2-carpellate; ovary 1-locular; style ± distinct; stigma 2-lobed. **Capsules** erect, 2-valved, dehiscing from base. **Seeds** few to many, arillate. $x = 9$.

Species 1: North America.

Sanguinaria is similar, and probably most closely related, to *Eomecon* Hance of eastern Asia, which is monotypic also.

SELECTED REFERENCES Greene, E. L. 1905. Suggestions regarding *Sanguinaria*. Pittonia 5: 306–308. Harshberger, J. W. 1903. Juvenile and adult forms of bloodroot. Pl. World 6: 106–108. Nieuwland, J. A. 1910. Notes on the seedlings of bloodroot. Amer. Midl. Naturalist 1: 199–203.

1. Sanguinaria canadensis Linnaeus, Sp. Pl. 1: 505. 1753 [E] [F]

Sanguinaria australis Greene; *S. canadensis* var. *rotundifolia* (Greene) Fedde; *S. dilleniana* Greene

Plants to 4(–6) dm, glabrous; rhizomes branching. **Leaves:** petiole to 15 cm; blade orbiculate-reniform to cordate-sagittate, mostly palmately 5–7-lobed, to 25 cm wide; margins crenate; adaxial surface glaucous. **Inflorescences:** scape to 15 cm. **Flowers:** sepals ca. 1 cm; petals white or pinkish, oblong to oblanceolate, 15–30 mm; style to 3 mm. **Capsules** fusiform, 35–60 mm, glabrous. **Seeds** black to red-orange, obscurely reticulate. $2n = 18$ (cult.).

Flowering earliest spring. Moist to dry woods and thickets, often on flood plains and shores or near streams on slopes, less frequently in clearings and meadows or on dunes, rarely in disturbed sites; 0–1300 m; Man., N.B., N.S., Ont., Que.; Ala., Ark., Conn., Del., D.C., Fla., Ga., Ill., Ind., Iowa, Kans., Ky., La., Maine, Md., Mass., Mich., Minn., Mo., Nebr., N.H., N.J., N.Y., N.C., N.Dak., Ohio, Okla., Pa., R.I., S.C., S.Dak., Tenn., Tex., Vt., Va., W.Va., Wis.

Sanguinaria canadensis has been reported from Mississippi, but no specimens are known.

The leaves of *Sanguinaria canadensis* are quite variable in shape and size, and the scape and petals vary considerably in length. In some plants the petals are clearly differentiated into sets of two different sizes, but in others the differentiation is barely perceptible. Extremes of variation in these characters have been the bases for recognizing several forms, varieties, and even distinct species, but intermediates of all degrees are found and the variation is only loosely correlated with geography or habitat. Thus, it seems best to limit formal recognition to a single, quite variable species.

Although bloodroot is an ingredient of some compound cough remedies, it contains the poisonous alkaloid sanguinarine, and the U.S. Food and Drug Administration has characterized *Sanguinaria canadensis* as an unsafe herb (J. A. Duke 1985).

5. STYLOPHORUM Nuttall, Gen. N. Amer. Pl. 2: 7. 1818 · [Greek *stylos*, style, and *phoros*, bearing, in reference to the conspicuous style, unusual in the family]

Robert W. Kiger

Herbs, perennial, caulescent, from stout rhizomes; sap yellow to orange. **Stems** naked above base except for 2–3 leaves subtending inflorescence. **Leaves:** basal rosulate, petiolate; cauline subopposite, sessile; blades 1–2×-pinnately lobed. **Inflorescences** terminal, umbelliform, few flowered; bracts present. **Flowers:** sepals 2, distinct; petals 4; stamens many; pistil [2–](3–)4-carpellate; ovary 1-locular; style distinct; stigma shallowly 3–4(–5)-lobed. **Capsules** nodding, [2–](3–)4-valved, dehiscing from apex. **Seeds** few to many, arillate. $x = 10$.

Species 3 (1 in the flora): North America, China.

Stylophorum is most closely related to *Hylomecon* Maximowicz, a monotypic genus from eastern Asia, and to *Chelidonium*.

1. **Stylophorum diphyllum** (Michaux) Nuttall, Gen. N. Amer. Pl. 2: 7. 1818 · Celandine-poppy, mock poppy, yellow-poppy, wood-poppy E F

Chelidonium diphyllum Michaux, Fl. Bor.-Amer. 1: 309. 1803; *Stylophorum ohiense* Sprengel

Plants to 5 dm, downy. **Stems** simple or branching. **Leaves** to 5 dm; petiole to 2 dm; blade pale abaxially, deeply 5–7-lobed; margins irregularly dentate or crenate. **Flowers:** pedicels 25–80 mm; sepals ca. 15 mm; petals yellow, obovate, 2–3 cm wide; style 3–6 mm. **Capsules** ellipsoid, 20–35 mm, pubescent. **Seeds** pale brown, reticulate-pitted. $2n = 20$ (cult.).

Flowering spring. Moist deciduous woods, thickets, and cedar barrens, often on slopes, occasionally in fields or on shaded dunes, in loam or sand; 100–600 m; Ont.; Ala., Ark., Ill., Ind., Ky., Mich., Mo., Ohio, Tenn., Va., W.Va.

Vegetatively, this native species closely resembles the more frequent and widespread *Chelidonium majus,* introduced from Eurasia. Various authors have reported *Stylophorum diphyllum* from western Pennsylvania, but W. E. Buker and S. A. Thompson (1986) could not confirm its past or current native presence there.

6. DENDROMECON Bentham, Trans. Hort. Soc. London, ser. 2, 1: 407. 1835 · [Greek *dendron,* tree, and *mekon,* poppy]

Curtis Clark

Shrubs or small trees, evergreen, glabrous; sap colorless. **Stems** leafy. **Leaves** alternate, short-petiolate; blade unlobed, leathery; margins entire or minutely denticulate. **Inflorescences** terminal on short branchlets, 1-flowered. **Flowers:** receptacle expanded, forming cup beneath calyx; sepals 2, distinct; petals 4, yellow, obovate or obcuneate, with satin sheen from microscopic linear grooves; stamens many; pistil 1, 2-carpellate; ovary cylindric to long-conic, 1-locular; style absent; stigmas 2, obcompressed. **Fruits** capsular, cylindric, 2-valved, dehiscing from base along placentas. **Seeds** many, brown or black, obovoid, smooth, pale carunculate, aril absent. $x = 28$.

Species 2 (2 in the flora): w North America.

1. Leaf blades lanceolate, 3–8 times longer than wide, margins minutely denticulate, apex acute to acuminate; mainland . . . . 1. *Dendromecon rigida*
1. Leaf blades elliptic, ovate, or oblong, 1.5–3 times longer than wide, margins entire, apex usually rounded; California Channel Islands . . . . 2. *Dendromecon harfordii*

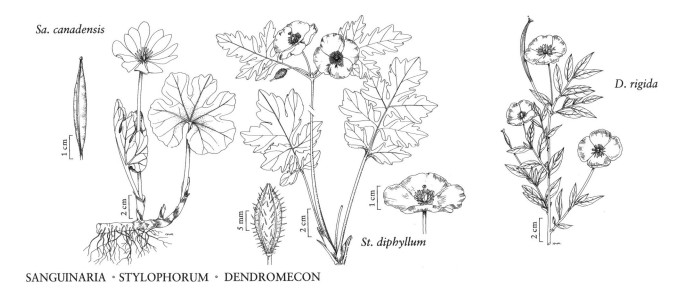

Sa. canadensis

D. rigida

St. diphyllum

SANGUINARIA · STYLOPHORUM · DENDROMECON

1. **Dendromecon rigida** Bentham, Trans. Hort. Soc. London, ser. 2, 1: 407. 1835 · Bush-poppy [F]

Shrubs, 1–3 m. **Leaves** 2.5–10 × 0.7–2.5 cm; blade lanceolate; margins minutely denticulate; apex acute to acuminate. **Flowers:** petals 2–3 cm. **Capsules** curved, 5–10 cm. **Seeds** 2–2.5 mm. $2n = 56$.

Flowering spring–early summer (Apr–Jul). Dry slopes, washes, especially in areas of recent burns; to 1800 m; Calif.; Mexico (Baja California).

2. **Dendromecon harfordii** Kellogg, Proc. Calif. Acad. Sci. 5: 102. 1873 · Island tree-poppy [E]

Dendromecon rigida Bentham subsp. *harfordii* (Kellogg) P. H. Raven

Trees or shrubs, 2–6 m. **Leaves** 3–8 × 1.5–4.5 cm; blade elliptic, ovate, or oblong; margins entire; apex usually rounded. **Flowers:** petals 2–3 cm. **Capsules** curved, 7–10 cm. **Seeds** 2.5–3 mm.

Flowering spring–early summer (Apr–Jul). Brushy slopes; less than 600 m; Calif.

Dendromecon harfordii grows only on the Channel Islands. Plants with paler leaves from the southern islands (extant on Santa Catalina Island, once present on San Clemente Island) have been called *D. rigida* var. *rhamnoides* (Greene) Munz [= *D. rigida* Bentham subsp. *rhamnoides* (Greene) Thorne] and are considered to be of conservation concern.

7. ESCHSCHOLZIA Chamisso in C. G. D. Nees, Horae Phys. Berol., 73. 1820 · [For Johann F. G. von Eschscholtz, 1793–1831, Estonian physician and biologist who traveled with Chamisso on the Romanzoff (or Kotzebue) Expedition to the Pacific Coast]

Curtis Clark

Herbs, annual or perennial, scapose or caulescent, from taproots; sap colorless or clear orange. **Stems** leafy. **Leaves** alternate, basal and sometimes cauline, petiolate; blade 1–4×-pinnately deeply lobed, lobes of each order usually 3; ultimate lobes narrow. **Inflorescences** terminal, cymose with bracts present, or 1-flowered. **Flowers:** receptacle expanded, forming cup beneath calyx, sometimes with free rim; perianth and androecium perigynous; sepals 2, connate, calyptrate, deciduous as unit; petals 4, rarely more (doubled flowers), obovate to obcuneate, with satin sheen from microscopic linear grooves; stamens 12–many; pistil 2-carpellate; ovary 1-locular; style absent; stigmas 4–8, spreading, linear. **Fruits** capsular, cylindric, 2-valved, dehiscing from base along placentas, often explosively. **Seeds** many, tan, brown, or black, globose to ovoid, reticulate, ridged and burlike, or pitted, aril absent. $x = 6, 7$.

Species 12 (10 in the flora): w North America (United States), nw Mexico.

Eschscholzia species are introduced from cultivation elsewhere in warm-temperate regions worldwide.

SELECTED REFERENCES Clark, C. and J. A. Jernstedt. 1978. Systematic studies of *Eschscholzia* (Papaveraceae). II. Seed coat microsculpturing. Syst. Bot. 3: 386–402. Ernst, W. R. 1964b. The genus *Eschscholzia* in the south Coast Ranges of California. Madroño 17: 281–294. Greene, E. L. 1905b. Revision of *Eschscholtzia*. Pittonia 5: 205–308. Lewis, H. and R. Snow. 1951. A cytotaxonomic approach to *Eschscholzia*. Madroño 11: 141–143.

1. Receptacular cup with spreading free rim . . . . 2. *Eschscholzia californica*
1. Receptacular cup without spreading free rim.
  2. Plants scapose; ultimate leaf lobes acute; petals yellow; calyx glabrous.
    3. Seeds burlike with raised ridges; petals 12 mm or shorter; California (Great Central Valley, Sierra Nevada foothills) . . . . 6. *Eschscholzia lobbii*
    3. Seeds minutely pitted, not burlike; petals usually 12 mm or longer; California, Arizona, Nevada, and Utah (Mojave and Sonoran deserts) . . . . 3. *Eschscholzia glyptosperma*
  2. Plants caulescent, with flowers borne on leafy stems (not readily apparent in young plants); ultimate leaf lobes acute or obtuse; petals yellow or orange; calyx glabrous or pubescent.
    4. Calyx pubescent; buds nodding; leaf blades sparsely pubescent.
      5. Petals orange or deep yellow throughout, 15–40 mm; receptacles broader than 1.5 mm . . . . 5. *Eschscholzia lemmonii*
      5. Petals yellow, sometimes with orange spot at base, 10–20 mm; receptacles narrower than 1.5 mm . . . . 4. *Eschscholzia hypecoides*
    4. Calyx glabrous; buds nodding or erect; leaf blades essentially glabrous, never consistently pubescent.
      6. Older buds nodding.
        7. Leaf blades bright green or yellow-green, terminal lobes slender, acute . . . . 8. *Eschscholzia parishii*
        7. Leaf blades grayish or bluish green, terminal lobes broadened at apex, usually obtuse . . . . 7. *Eschscholzia minutiflora*
      6. Older buds erect.
        8. Receptacle somewhat swollen and translucent distally, usually broader than 2 mm.
          9. Petals (15–)20 mm or more, orange or deep yellow; California (Kern County) . . . 5. *Eschscholzia lemmonii*
          9. Petals 15 mm or less, yellow; inland California Coast Ranges . . . . 10. *Eschscholzia rhombipetala*
        8. Receptacle strictly obconic, not translucent distally, narrower than 2.5 mm.
          10. Flower buds blunt or rounded short-acuminate, tip less than 1/4 length of bud; ultimate leaf lobes elongate, giving diffuse appearance; California Channel Islands . . . . 9. *Eschscholzia ramosa*
          10. Flower buds apiculate-acuminate, tip usually more than 1/4 length of bud; ultimate leaf lobes short, giving compact appearance; California and Oregon mainland foothills . . . . 1. *Eschscholzia caespitosa*

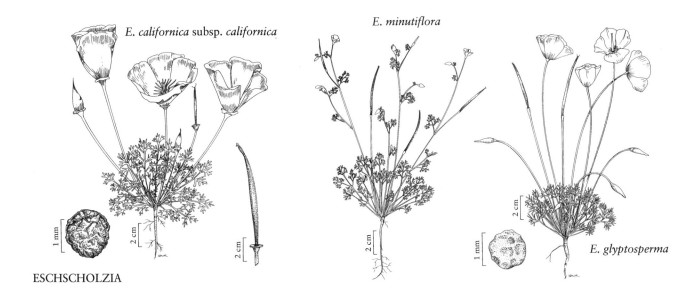

E. *californica* subsp. *californica*

E. *minutiflora*

E. *glyptosperma*

ESCHSCHOLZIA

1. **Eschscholzia caespitosa** Bentham, Trans. Hort. Soc. London, ser. 2, 1: 408. 1835

Plants annual, caulescent, erect, tufted, 5–30 cm, glabrous, sometimes slightly glaucous. **Leaves** basal and cauline; blade with ultimate lobes short (giving compact appearance), obtuse or acute. **Inflorescences** cymose or 1-flowered; buds erect, apiculate-acuminate, tip usually more than 1/4 length of bud. **Flowers:** receptacle obconic, less than 2.5 mm broad, cup without spreading free rim; calyx apiculate, glabrous; petals yellow, sometimes with orange spot at base, 10–25 mm. **Capsules** 4–8 cm. **Seeds** brown to black, ellipsoid to obovoid, 1.5–2.4 mm, reticulate. $2n = 12$.

Flowering spring (Mar–May). Open chaparral; 0–1500 m; Calif., Oreg.; Mexico (Baja California).

Eschscholzia caespitosa grows in the mainland foothills.

2. **Eschscholzia californica** Chamisso in C. G. D. Nees, Horae Phys. Berol., 73, plate 15. 1820 · California-poppy [W2]

Plants perennial or annual, caulescent, erect or spreading, 5–60 cm, glabrous, sometimes glaucous; taproot heavy in perennial forms. **Leaves** basal and cauline; blade with ultimate lobes obtuse or acute. **Inflorescences** cymose or 1-flowered; buds erect. **Flowers:** receptacle obconic, cup with spreading free rim; calyx acute to acuminate, glabrous, sometimes glaucous; petals yellow to orange, usually with orange spot at base, 20–60 mm. **Capsules** 3–9 cm. **Seeds** brown to black, globose to ellipsoid, 1.5–1.8 mm, reticulate.

Subspecies 2 (2 in the flora): North America (w United States), nw Mexico.

Eschscholzia californica is the state flower of California. Although it is toxic to humans, its roots are relished by gophers. Widely planted in North America and elsewhere as an ornamental, roadside, and reclamation plant, with many color forms in the horticultural trade, it often escapes but usually does not persist.

This species is highly variable (more than 90 infraspecific taxa have been described), not only among different plants and locations but also within individual plants over the course of the growing season, especially in petal size and color (see W. L. Jepson 1909–1943, vol. 1, part 7, pp. 564–569).

SELECTED REFERENCE Clark, C. 1978. Systematic studies of *Eschscholzia* (Papaveraceae). I. The origin and affinities of *E. mexicana*. Syst. Bot. 3: 374–385.

1. Spreading rim of receptacular cup prominent; cotyledons usually 2-lobed; inland valleys, California . . . 2a. *Eschscholzia californica* subsp. *californica*
1. Spreading rim of receptacular cup often inconspicuous; cotyledons unlobed; southwestern deserts . . . . . . 2b. *Eschscholzia californica* subsp. *mexicana*

2a. Eschscholzia californica Chamisso subsp. **californica** [F]

Eschscholzia californica var. *crocea* (Bentham) Jepson; *E. californica* var. *peninsularis* (Greene) Munz; *E. peninsularis* Greene; *E. procera* Greene

Plants usually perennial, sometimes annual; cotyledons usually 2-lobed. **Flowers:** receptacular cup with spreading rim prominent. $2n = 12$.

Flowering mid winter–late summer (Feb–Sep). Grassy, open areas; to 2000 m; Calif., Nev., Oreg., Wash.; Mexico (Baja California).

2b. Eschscholzia californica Chamisso subsp. **mexicana** (Greene) C. Clark, Syst. Bot. 3: 382. 1978

Eschscholzia mexicana Greene, Bull. Calif. Acad. Sci. 1: 69. 1885

Plants annual; cotyledons unlobed. **Flowers:** receptacular cup with spreading rim often inconspicuous. $2n = 12$.

Flowering mid winter–late summer (Feb–Sep). Deserts; to 2000 m; Ariz., Calif., N.Mex., Nev., Tex., Utah; Mexico (Sonora).

C. T. Mason Jr. (pers. comm.) has noted several specimens from Arizona as far east as Cochise County that appear to be perennial.

3. Eschscholzia glyptosperma Greene, Bull. Calif. Acad. Sci. 1: 70. 1885 [E] [F]

Plants annual, scapose, erect, 5–25 cm, glabrous, sometimes glaucous. **Leaves** basal; blade with ultimate lobes acute. **Inflorescences** 1-flowered; buds nodding or sometimes erect. **Flowers:** receptacle conic, cup without spreading free rim; calyx acuminate, glabrous, sometimes glaucous; petals yellow, (10–)12–25 mm. **Capsules** 4–7 cm. **Seeds** tan to brown, globose, 1.2–1.8 mm, minutely pitted. $2n = 14$.

Flowering spring (Mar–May). Desert washes, flats, slopes; 50–1500 m; Ariz., Calif., Nev., Utah.

Eschscholzia glyptosperma is known from the Mojave and Sonoran deserts.

4. Eschscholzia hypecoides Bentham, Trans. Hort. Soc. London, ser. 2, 1: 408. 1835 [E]

Eschscholzia caespitosa Bentham var. *hypecoides* (Bentham) A. Gray

Plants annual, caulescent, erect, 5–30 cm, sparsely pubescent. **Leaves** basal and cauline; blade with ultimate lobes usually obtuse. **Inflorescences** cymose or 1-flowered; buds nodding. **Flowers:** receptacle turbinate, less than 1.5 mm broad, cup without spreading free rim; calyx acute, pubescent; petals yellow, sometimes with orange spot at base, 10–20 mm. **Capsules** 3–7 cm. **Seeds** brown, globose to ellipsoid, 1–1.3 mm, reticulate. $2n = 12$.

Flowering spring (Mar–May). Grassy areas in woodland, chaparral; 500–1500 m; Calif.

5. Eschscholzia lemmonii Greene, W. Amer. Sci. 3: 157. 1887 [E]

Plants annual, caulescent, erect or spreading, 5–30 cm, glabrous to sparsely pubescent. **Leaves** basal and cauline; blade with ultimate lobes generally obtuse. **Inflorescences** cymose or 1-flowered; buds erect or nodding. **Flowers:** receptacle turbinate, distally translucent and somewhat swollen, more than 1.5 mm broad, cup without spreading free rim; calyx acute, glabrous or pubescent; petals orange or deep yellow, 15–40 mm. **Capsules** 3–7 cm. **Seeds** brown, 1.3–1.8 mm, reticulate. $2n = 12$.

Subspecies 2 (2 in the flora): w North America (California).

1. Plants spreading or erect; flower buds nodding; calyx generally pubescent. . . . .
. . . . 5a. *Eschscholzia lemmonii* subsp. *lemmonii*
1. Plants erect; flower buds generally erect; calyx glabrous. . . . 5b. *Eschscholzia lemmonii* subsp. *kernensis*

5a. Eschscholzia lemmonii Greene subsp. **lemmonii** [E] [F]

Plants spreading or erect. **Inflorescences:** buds nodding. **Flowers:** calyx generally pubescent; petals 15–40 mm. **Seeds** ellipsoid.

Flowering spring (Mar–May). Open grasslands; 200–1000 m; Calif.

This subspecies grows in the southern Sierra Nevada foothills, the western Tehachapi Mountains, the eastern Outer South Coast Ranges, and the Inner South Coast Ranges.

5b. Eschscholzia lemmonii Greene subsp. **kernensis** (Munz) C. Clark, Madroño 33: 225. 1986 [C] [E]

Eschscholzia caespitosa Bentham subsp. *kernensis* Munz, Aliso 4: 90. 1958

Plants erect. **Inflorescences:** buds generally erect. **Flowers:** calyx glabrous; petals 20–40 mm. **Seeds** globose to ellipsoid.

Flowering spring (Mar–May). Open grasslands; of conservation concern; 500–800 m; Calif.

This subspecies grows in the southwestern Tehachapi Mountains and in the northern part of the western Transverse Ranges.

6. Eschscholzia lobbii Greene, Pittonia 5: 290. 1905 · Frying-pans [E] [F]

Plants annual, scapose, erect, 5–15 cm, glabrous. **Leaves** basal; blade glabrous, ultimate lobes acute. **Inflorescences** 1-flowered; buds erect. **Flowers:** receptacle obconic, cup without spreading free rim; calyx acute, glabrous; petals yellow, 7–12 mm. **Capsules** 4–7 cm. **Seeds** brown, globose to ellipsoid, 1.4–2 mm, burlike with raised ridges. $2n = 12$.

Flowering late winter–spring (Mar–May). Open fields, grasslands; 0–600 m; Calif.

Eschscholzia lobbii grows in the Great Central Valley and in the foothills of the Sierra Nevada.

7. Eschscholzia minutiflora S. Watson, Proc. Amer. Acad. Arts 11: 122. 1876 [F]

Eschscholzia covillei Greene; *E. minutiflora* subsp. *covillei* (Greene) C. Clark; *E. minutiflora* subsp. *twisselmannii* C. Clark & M. Faull; *E. minutiflora* var. *darwinensis* M. E. Jones

Plants annual, caulescent, erect or spreading, 5–35 cm. **Leaves** basal and cauline; blade grayish or bluish green, glabrous, glaucous; ultimate lobes usually obtuse, terminal broadened at apex. **Inflorescences** cymose or 1-flowered; buds nodding. **Flowers:** receptacle obconic, cup without spreading free rim; calyx acuminate, glabrous, sometimes glaucous; petals yellow, sometimes with orange spot at base, 3–26 mm. **Capsules** 3–6 cm. **Seeds** brown to black, ellipsoid, 1–1.4 mm, reticulate. $2n = 12, 24, 36$.

Flowering late winter–spring (Mar–May). Desert washes, flats, and slopes; 0–2000 m; Ariz., Calif., Nev., Utah; nw Mexico.

Eschscholzia minutiflora is highly variable in flower size. Typically, plants are hexaploid ($2n = 36$) with petals 3–10 mm. Tetraploid plants ($2n = 24$) with petals 6–18 mm, from the northern and central Mojave Desert of California, have been distinguished as subsp. *covillei*. Diploid plants ($2n = 12$) with petals 10–26 mm, restricted to the El Paso and Rand mountains of the western Mojave Desert, have been distinguished as subsp. *twisselmannii* and are considered to be of conservation concern; previously they were misattributed to *E. parishii*.

SELECTED REFERENCES Clark, C. and M. Faull. 1991. A new subspecies and a combination in *Eschscholzia minutiflora* (Papaveraceae). Madroño 38: 73–79. Mosquin, T. 1961. *Eschscholzia covillei* Greene, a tetraploid species from the Mojave Desert. Madroño 16: 91–96.

8. Eschscholzia parishii Greene, Bull. Calif. Acad. Sci. 1: 183. 1885

Plants annual, caulescent, erect, 5–30 cm. **Leaves** basal and cauline; blade bright green or yellow-green, glabrous; ultimate lobes obtuse except terminal one slender, acute. **Inflorescences** cymose or 1-flowered; buds nodding. **Flowers:** receptacle obconic, cup without spreading free rim; calyx apiculate, glabrous, sometimes glaucous; petals yellow, 15–30 mm. **Capsules** 5–7 cm. **Seeds** tan to brown, globose to ellipsoid, 1–1.4 mm, reticulate. $2n = 12$.

Flowering spring (Mar–May). Desert slopes, hillsides; 0–1200 m; Calif.; Mexico (Baja California and Sonora).

Plants of the El Paso and Rand mountains in the western Mojave Desert that have been referred to this species are *Eschscholzia minutiflora*.

9. Eschscholzia ramosa (Greene) Greene, Bull. Torrey Bot. Club 13: 217. 1886 · Island-poppy

Eschscholzia elegans Greene var. *ramosa* Greene, Bull. Calif. Acad. Sci. 1: 182. 1885

Plants annual, caulescent, erect, 5–30 cm, glabrous, sometimes glaucous. **Leaves** basal and cauline; blade glabrous; ultimate lobes elongate, obtuse. **Inflorescences** cymose or 1-flowered; buds erect, blunt or rounded short-acuminate, tip less than 1/4 length of bud. **Flowers:** receptacle obconic, less than 2.5 cm broad, cup without spreading free rim; calyx acuminate, glabrous; petals yellow, sometimes with orange spot at base, 5–20 mm. **Capsules** 4–7 cm. **Seeds** brown, ellipsoid, 1.4–1.6 mm, reticulate. $2n = 24$.

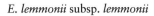

E. lemmonii subsp. *lemmonii*

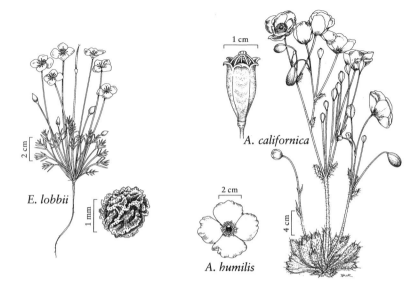

A. californica

E. lobbii

A. humilis

ESCHSCHOLZIA · ARCTOMECON

Flowering spring (Apr–Jun). Open places, especially in chaparral; 0–300 m; Calif.; Mexico.

Within the flora area, *Eschscholzia ramosa* is known only from the California Channel Islands.

10. Eschscholzia rhombipetala Greene, Bull. Calif. Acad. Sci. 1: 71. 1885 [C][E]

Plants annual, caulescent, erect, 5–30 cm, glabrous. **Leaves** basal and cauline; blade glabrous; ultimate lobes generally obtuse. **Inflorescences** cymose or 1-flowered; buds erect or nodding. **Flowers:** receptacle turbinate to obconic, distally translucent and somewhat swollen, more than 2

mm broad, cup without spreading free rim; calyx acute or acuminate, glabrous, sometimes glaucous; petals yellow, 3–15 mm. **Capsules** 4–7 cm. **Seeds** black, globose, 1.3–1.8 mm, reticulate.

Flowering spring (Mar–May). Fallow fields, open places; of conservation concern; 0–500 m; Calif.

Originally known from seven sites in California, *Eschscholzia rhombipetala* had not been seen since the early 1970s before it was rediscovered in 1993 on the Carrizo Plain in San Luis Obispo County.

8. ARCTOMECON Torrey & Frémont in J. C. Frémont, Rep. Exped. Rocky Mts., 312. 1845 · Desert bearclaw-poppy [Greek *arktos,* bear, alluding to the long-pilose pubescence, and *mekon,* poppy]

Susan E. Meyer

Herbs, perennial, evergreen, cespitose, scapose or very short-caulescent and subscapose, from taproots. **Stems** (caudices) when present leafy, erect, branching, covered with marcescent leaf bases. **Leaves** all or mostly basal and rosulate, petiolate; blade cuneate to fan-shaped, 1×-lobed distally, glaucous, long-pilose, hairs flexuous, barbed; margins entire; lobes acute, long-bristled. **Inflorescences** terminal, cymiform, simple or branching; bracts present; buds nodding. **Flowers:** sepals 2 or 3, distinct; petals caducous or persistent, 4 or 6; stamens many; pistil 3–6-

carpellate; ovary 1-locular; style 1 and short, or absent; stigmas 1 per carpel, connate, capitate, cordately 2-lobed. **Capsules** erect, 3–6-valved, dehiscing from apex leaving persistent placental ribs unseparated at apex, valve tips recurving. **Seeds** several to many, dark brown, shining, ovoid, 1.5–3 mm, aril present. $x = 12$.

Species 3 (3 in the flora): North America (Mojave Desert).

SELECTED REFERENCES Janish, J. R. 1977. Nevada's vanishing bear-poppies. Mentzelia 3: 2–5. Mozingo, H. N. and Margaret Williams. 1980. Threatened and Endangered Plants of Nevada: An Illustrated Manual. [Washington.] Nelson, D. R. and K. T. Harper. 1991. Site characteristics and habitat requirements of the endangered dwarf bear-claw poppy (*Arctomecon humilis* Coville, Papaveraceae). Great Basin Naturalist 51: 167–175. Nelson, D. R. and S. L. Welsh. 1993. Taxonomic revision of *Arctomecon* Torr. & Frem. Rhodora 95: 197–213. Raynie, D. E., D. R. Nelson, and K. T. Harper. 1991. Alkaloidal relationships in the genus *Arctomecon* (Papaveraceae) and herbivory in *A. humilis*. Great Basin Naturalist 51: 397–403. Welsh, S. L. and K. L. Thorne. 1979. Illustrated Manual of Proposed Endangered and Threatened Plants of Utah. [Washington.]

1. Petals 4, 1.2–2.5 cm, marcescent; leaf blades minutely hirsute but only sparsely long-pilose . . . . . . 1. *Arctomecon humilis*
1. Petals 6, 2.5–4 cm, caducous; leaf blades copiously long-pilose.
   2. Inflorescences 1(–6)-flowered; flower buds long-pilose; petals white . . . . 2. *Arctomecon merriamii*
   2. Inflorescences 3–20-flowered; flower buds glabrous; petals deep yellow. . . . . 3. *Arctomecon californica*

1. Arctomecon humilis Coville, Proc. Biol. Soc. Wash. 7: 67. 1892 (as **humile**) · Dwarf bearclaw-poppy [C] [E] [F]

Plants subscapose, to 2.5 dm. Leaves 1–8 cm; blade shallowly to deeply 3–5-lobed distally, to 1.6 cm wide distally, minutely hirsute, sparsely long-pilose. Inflorescences 2–7-flowered, branching, glabrous or nearly so; buds glabrous. Flowers: sepals glabrous; petals marcescent, 4, white, 1.2–2.5 cm; filaments somewhat dilated distally; style 1 mm. Capsules ovoid, 0.8–1.5 cm, dehiscing 1/3–1/2 length, deciduous soon after. $2n = 24$.

Flowering late spring; fruiting mid summer. Barren, heavily gypsiferous shales, creosote bush zone; of conservation concern; 700–1100 m; Utah.

Extant populations of *Arctomecon humilis* are known only from the area within 20 km of St. George in extreme southwestern Utah.

2. Arctomecon merriamii Coville, Proc. Biol. Soc. Wash. 7: 66. 1892 · Great bearclaw-poppy [E]

Plants scapose, to 4 dm. Leaves 4–12 cm; blade nearly unlobed to shallowly 3–7-lobed distally, to 3.5 cm wide distally, copiously long-pilose. Inflorescences 1(–6)-flowered, simple, glabrous; buds long-pilose. Flowers: sepals copiously long-pilose; petals caducous, 6, white, 2.5–4 cm; filaments dilated distally; style 1–1.5 mm. Capsules persistent, oblanceoloid, 2.5–4 cm, dehiscing not more than 1/4 length. $2n = 24$.

Flowering mid spring; fruiting early summer. Bar-

ren, calcareous, scree slopes and weakly gypsiferous shales interbedded with marine limestones, creosote bush and blackbrush zones; 600–1700 m; Calif., Nev.

Arctomecon merriamii is found in small populations at scattered stations from Death Valley in southeastern California to the Meadow Valley Wash of southeastern Nevada.

3. Arctomecon californica Torrey & Frémont in J. C. Frémont, Rep. Exped. Rocky Mts., 312, plate 2. 1845 · Golden bearclaw-poppy [C] [E] [F]

Plants subscapose, to 6 dm. Leaves 3–20 cm; blade shallowly to deeply 3–7-lobed distally, 5 cm wide distally, copiously long-pilose, sometimes also minutely hirsute. Inflorescences 3–20-flowered, branching, glabrous throughout or long-pilose proximally; buds glabrous. Flowers: sepals glabrous; petals caducous, 6, deep yellow, 2.5–4 cm; filaments not dilated distally; style absent. Capsules persistent, obconic, 1–2.5 cm, dehiscing not more than 1/4 length. $2n = 24$.

Flowering mid spring; fruiting early summer. Barren shales that are heavily gypsiferous or otherwise chemically unusual (borate-bearing, lithium-bearing), creosote bush zone; of conservation concern; 500–1000 m; Ariz., Nev., Utah.

Arctomecon californica is found in mostly small populations ranging from the vicinity of Las Vegas, Nevada, eastward to extreme northwestern Arizona near Lake Mead. It is known otherwise only from a single collection from an artificially established population in Washington County, Utah.

9. ARGEMONE Linnaeus, Sp. Pl. 1: 508. 1753; Gen. Pl. ed. 5, 225. 1754

· Prickly-poppy [a poppylike herb mentioned by Pliny]

Gerald B. Ownbey

Herbs or subshrubs, annual or perennial, caulescent, glaucous, from transitory or persistent taproots; sap white to orange. **Stems** leafy, branching. **Leaves** sessile; basal rosulate, cauline alternate; blade unlobed or commonly shallowly to deeply 1×-pinnately lobed; margins dentate, each tooth terminated by prickle; surfaces glaucous, often mottled over veins, unarmed or prickly, glabrous or hispid. **Inflorescences** terminal, cymose; bracts present. **Flowers** conspicuous, sometimes subtended by foliaceous bracts; sepals 2 or 3, unarmed or prickly, each with erect, subterminal, hollow horn tipped with prickle; petals 6, in 2 whorls of 3; stamens 20–250 or more; pistil 3–5(–7)-carpellate; ovary 1-locular; style short, to 3 mm in fruit; stigma 3–5(–7)-lobed. **Capsules** erect, 3–5(–7)-valved, grooved over sutures, prickly, rarely unarmed, dehiscing from apex ca. 1/3 length, valves separating from framework of vascular elements, to which persistent style and stigma remain attached. **Seeds** numerous, subglobose, minutely pitted, 1–3 mm, aril present. $x = 14$.

Species 32 (15 in the flora): North America, South America, Hawaii; introduced in other tropical and temperate regions of the world.

None of the North American species occurs in South America except for the pantropical weeds *Argemone mexicana* and, probably, *A. ochroleuca*. *Argemone glauca* is endemic to Hawaii. Three suffrutescent, perennial species are known from Mexico in Coahuila (*A. fruticosa*) and Chihuahua (*A. turnerae, A. ownbeyana*).

Most herbaceous species can be hybridized, but the F_1 plants are sterile when the parents differ in ploidy level. The F_2 generation, when achieved, consists mainly of plants of low vigor.

The alkaloids of *Argemone* have been studied extensively. F. S. Stermitz (1968) has suggested that the species fall into four groups (one with two subgroups) according to their alkaloidal properties, and that these groups coincide to a considerable degree with the informal species alliances suggested by G. B. Ownbey (1958). A full evaluation of the importance of alkaloidal content to *Argemone* taxonomy and evolution has not been published, but there is little doubt that it is highly significant.

The name *Argemone intermedia* Sweet (Hort. Brit. ed. 2, 1830) is encountered in several regional and local floras but is of uncertain application.

SELECTED REFERENCES Ownbey, G. B. 1958. Monograph of the genus *Argemone* for North America and the West Indies. Mem. Torrey Bot. Club 21(1): 1–159. Ownbey, G. B. 1961. The genus *Argemone* in South America and Hawaii. Brittonia 13: 91–109. Prain, D. 1895. An account of the genus *Argemone*. J. Bot. 33: 129–135, 176–178, 207–209, 307–312, 325–333, 363–371. Stermitz, F. R., D. E. Nicodem, Wei C. C., and K. D. McMurtrey. 1969. Alkaloids of *Argemone polyanthemos, A. corymbosa, A. chisosensis, A. sanguinea, A. aurantiaca* and general *Argemone* systematics. Phytochemistry 8: 615–620.

1. Petals yellow, golden, or bronze.
   2. Stamens 150 or more, filaments red or purplish; flowers 7–12 cm broad; petals bright yellow to golden or bronze . . . . 3. *Argemone aenea*
   2. Stamens 20–75, filaments yellow; flowers 3–7 cm broad; petals pale lemon yellow to bright yellow.
      3. Flower buds subglobose, body length ± equal to breadth; petals mostly bright yellow . . . . 1. *Argemone mexicana*
      3. Flower buds oblong, body length 1.5–2 times breadth; petals mostly pale lemon yellow . . .2. *Argemone ochroleuca*
1. Petals white or pale lavender.
   4. Leaf surfaces prickly on veins, often also minutely prickly or hispid between veins; stems usually closely prickly.
      5. Longest capsular prickles simple, 5–8(–10) mm.
         6. Distal leaf blades not clasping; leaf surfaces densely crisped-hispid between main veins. . . . 6. *Argemone hispida*
         6. Distal leaf blades usually definitely · clasping; leaf surfaces variously prickly but not crisped-hispid . . . . 7. *Argemone munita*

5. Longest capsular prickles branched, usually 8–35 mm.
    7. Longest capsular prickles 15–35 mm . . . . 4. *Argemone aurantiaca*
    7. Longest capsular prickles about 8–15 mm.
        8. Apices of leaf lobes angular, marginal teeth 3 mm or more; flowers usually closely sub-
            tended by 1–2 foliaceous bracts. . . . . 5. *Argemone squarrosa*
        8. Apices of leaf lobes usually definitely rounded, marginal teeth usually less than 1 mm;
            flowers usually not closely subtended by foliaceous bracts . . . . 7. *Argemone munita*
4. Leaf surfaces rarely prickly except on veins; stems with widely spaced prickles or almost unarmed
(copiously long-prickly in *A. arizonica*).
    9. Longest capsular prickles branched, usually 8–35 mm.
      10. Longest capsular prickles 15–35 mm . . . . 4. *Argemone aurantiaca*
      10. Longest capsular prickles 8–15 mm . . . . 5. *Argemone squarrosa*
    9. Longest capsular prickles simple, 4–10(–12) mm.
      11. Prickles (when present) of sepals and sepal horns ascending, or sepals unarmed; blades of
        proximal leaves often lobed less than 1/2 distance to midrib.
        12. Leaf blades thick and leathery; stamens 100–120 . . . . 12. *Argemone corymbosa*
        12. Leaf blades often succulent, but not leathery; stamens 150 or more.
          13. Flower buds obovoid or oblong; sepal horns usually flattened adaxially.
            14. Flower buds obovoid; surface of sepal horns prickly; blades of proximal
               leaves lobed ca. 3/4 distance to midrib, apices of lobes angular, marginal
               teeth prominent. . . . . 9. *Argemone pleiacantha*
            14. Flower buds oblong; surface of sepal horns unarmed; blades of proximal
               leaves lobed ca. 1/2 distance to midrib, apices of lobes rounded, marginal
               teeth very short . . . . 7. *Argemone munita*
          13. Flower buds usually subglobose, rarely ellipsoid or ellipsoid-oblong; sepal horns
            terete.
            15. Distal leaves not evidently clasping; stems with scattered prickles, the longest
              7–10 mm . . . . 13. *Argemone gracilenta*
            15. Distal leaves clasping; stems with shorter prickles.
              16. Sepal horns 6–10(–15) mm; flower buds ellipsoid-oblong; capsules nar-
               rowly to broadly ellipsoid . . . . 14. *Argemone polyanthemos*
              16. Sepal horns 3–6(–10) mm; flower buds subglobose to broadly ellipsoid;
               capsules oblong to oblong-ellipsoid or narrowly ellipsoid . . . . 15. *Argemone albiflora*
      11. Prickles (when present) of sepals and sepal horns patent; blades of proximal leaves mostly
        lobed more than 1/2 distance to midrib.
        17. Blades of distal leaves usually definitely clasping; blades of proximal leaves lobed ca.
          1/2–3/4 distance to midrib.
          18. Blades of proximal leaves lobed ca. 3/4 distance to midrib, apices of lobes an-
            gular. . . . . 9. *Argemone pleiacantha*
          18. Blades of proximal leaves lobed 1/2 or less distance to midrib, apices of lobes
            usually rounded . . . . 7. *Argemone munita*
        17. Blades of distal leaves not definitely clasping; blades of proximal leaves lobed 4/5
          distance to midrib or more.
          19. Flower buds usually subglobose; sepal horns terete, usually slender, unarmed
            . . . . 13. *Argemone gracilenta*
          19. Flower buds oblong to obovoid or ellipsoid; sepal horns terete or flattened or
            angular, usually prickly.
            20. Stems copiously long-prickly, often decumbent, diffusely branched; blades of
             basal and proximal leaves lobed nearly to midrib, lobe length often 5 times
             width; sepal horns 12–15 mm. . . . . 8. *Argemone arizonica*
           20. Stems sparsely to copiously prickly, erect, not diffusely branched; blades of
             basal and proximal leaves lobed 4/5 distance to midrib or less, lobes propor-
             tionately broader; sepal horns 5–12 mm.
             21. Flower buds usually oblong-obovoid, rarely subglobose; sepal horns
               usually flattened adaxially, rarely terete . . . . 9. *Argemone pleiacantha*
             21. Flower buds broadly ellipsoid to ellipsoid-oblong; sepal horns terete or
              angular.
              22. Stems often closely prickly; capsules narrowly ellipsoid-ovoid; seeds
               1.8–2.2 mm . . . . 11. *Argemone chisosensis*
              22. Stems not closely prickly; capsules narrowly to broadly ellipsoid;
               seeds about 1.5 mm. . . . . 10. *Argemone sanguinea*

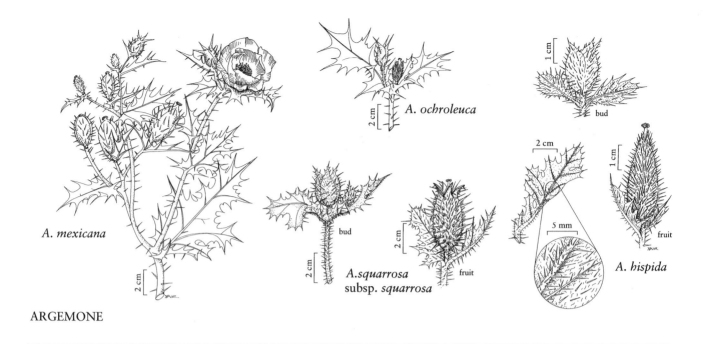

A. mexicana

A. ochroleuca

A. squarrosa subsp. squarrosa

A. hispida

ARGEMONE

---

1. Argemone mexicana Linnaeus, Sp. Pl. 1: 508. 1753

Argemone leiocarpa Greene [F] [W2]

Plants annual. **Stems** often branching from base, 2.5–8 dm, unarmed or sparingly prickly. **Leaf blades:** surfaces unarmed or sparingly prickly on veins; proximal lobed 1/2 or more distance to midrib; distal more shallowly lobed, mostly clasping. **Inflorescences:** buds subglobose, body 10–15 × 9–13 mm, unarmed or sparingly prickly; sepal horns terete, 5–10 mm, unarmed. **Flowers** 4–7 cm broad, subtended by 1–2 foliaceous bracts; petals bright yellow or rarely pale lemon yellow; stamens 30–50; filaments yellow; pistil 4–6-carpellate. **Capsules** oblong to broadly ellipsoid, 25–45 × 12–20 mm (including stigma and excluding prickles when present), unarmed or prickly, longest prickles 6–10 mm. **Seeds** 1.6–2 mm. $2n = 28$.

Flowering and fruiting spring–fall, or throughout year in tropics. Waste places, often a weed of roadsides, dooryards, fallow fields; 0–1500 m; Ont.; Ala., Conn., Fla., Ga., Ill., Ind., Kans., La., Md., Mass., Mich., Mo., Nebr., N.J., N.Y., N.C., Pa., S.C., Tenn., Tex., Va.; Mexico; West Indies; Central America.

Argemone mexicana is probably native to southern Florida as well as the Caribbean islands and has been introduced along the coast of the United States from New England to Texas and, more infrequently, inland. Although it has been reported from Mississippi, no specimens are known. It is widespread in temperate and tropical regions around the world by introduction.

2. Argemone ochroleuca Sweet, Brit. Fl. Gard. 3: plate 242. 1828 [F]

Argemone mexicana Linnaeus var. *ochroleuca* (Sweet) Lindley

Plants annual or short-lived perennial. **Stems** branching distally, 3–10 dm, sparingly prickly. **Leaf blades:** abaxial surface sparingly prickly on veins, adaxial surface usually unarmed; proximal deeply lobed nearly to midrib; distal more shallowly lobed, usually clasping. **Inflorescences:** buds oblong, body 8–18 × 4–11 mm, with 4–10 prickles per sepal; sepal horns terete, 5–12 mm, unarmed. **Flowers** 3–7 cm broad, closely subtended by 1–2 foliaceous bracts; petals lemon yellow or rarely darker yellow; stamens 20–75; filaments yellow; pistil 3–6-carpellate. **Capsules** ovoid-ellipsoid to oblong, 20–50 × 10–18 mm (including stigma and excluding prickles), longest prickles 8–12 mm. **Seeds** 1.5–2 mm. $2n = 56$.

Flowering and fruiting spring–fall. Disturbed soil, often a weed of fields, roadways, embankments, intermountain plains; 0–2250 m; introduced; Ariz.; Mexico.

Argemone ochroleuca is widespread in temperate and tropical regions of the world as an introduced weed.

3. Argemone aenea G. B. Ownbey, Mem. Torrey Bot. Club 21(1): 50. 1958

Plants annual or short-lived perennial. **Stems** 3–8 dm, prickly. **Leaf blades:** abaxial surface prickly on veins, adaxial surface unarmed or sparingly prickly on veins; proximal lobed 3/4–5/6 distance to midrib. **Inflorescences:** buds ellipsoid-oblong, body 15–20 × 13–16 mm, prickly; sepal horns terete, 7–12 mm, unarmed or with a few basal prickles. **Flowers** 7–12 cm broad, often subtended by 1–2 foliaceous bracts; petals bright yellow to golden or bronze; stamens ca. 150; filaments red or purplish; pistil 4–5-carpellate. **Capsules** narrowly ellipsoid-oblong, 25–35 × 12–16 mm (including stigma and excluding prickles), coarsely prickly, surface clearly visible, prickles very unequal, longest ca. 8 mm. **Seeds** 1.5–1.7 mm. $2n = 28$.

Flowering and fruiting early spring–summer. Dry plains and low hills, road and field margins; 0–1500 m; Tex.; Mexico (Coahuila, Nuevo León, and Tamaulipas).

4. Argemone aurantiaca G. B. Ownbey, Mem. Torrey Bot. Club 21(1): 53. 1958 [E]

Plants annual or biennial. **Stems** widely branching, to ca. 8 dm, prickly throughout. **Leaf blades:** abaxial surface sparingly to copiously prickly on and sometimes between veins, adaxial surface less so; proximal lobed 5/6 distance to midrib; distal often clasping. **Inflorescences:** buds oblong, body 18–25 × 14–18 mm, prickly, longest prickles sometimes branched; sepal horns angular in cross section or flattened, 8–12 mm, prickly, apical prickle indurate. **Flowers** 8–l2 cm broad, closely subtended by 1–2 foliaceous bracts; petals white; stamens 150 or more; filaments pale yellow; pistil 5–6-carpellate. **Capsules** ovoid, 40–50 × 15–25 mm (including stigma and excluding prickles); prickles herbaceous, erect or reflexed, often branched, very unequal, longest 15–35 mm. **Seeds** 2.5–3 mm. $2n = 84$.

Flowering spring–summer; fruiting summer–fall. Fields, pastures, hills; 150–500 m (transition zone between lowlands and plateau); Tex.

5. Argemone squarrosa Greene, Pittonia 4: 68. 1899 [E] [W2]

Plants perennial. **Stems** widely branching, 4–8 dm, moderately to copiously prickly throughout. **Leaf blades:** abaxial surface prickly on main veins, adaxial surface prickly or unarmed on main veins, densely prickly-hispid to glabrous between veins; basal lobed 2/3–4/5 distance to midrib, lobe apices angular, marginal teeth 3 mm or more; distal clasping. **Inflorescences:** buds subglobose to oblong, body 16–25 × 15–20 mm, prickly; sepal horns angular in cross section, 8–14(–18) mm, apical prickle indurate. **Flowers** 8–11 cm broad, closely subtended by 1–2 foliaceous bracts; petals white; stamens 150 or more; filaments pale yellow; pistil 4–5-carpellate. **Capsules** ellipsoid, oblong or lance-ovoid, 25–50 × 10–18 mm (including stigma and excluding prickles); prickles basally herbaceous, spreading or reflexed, unequal, longest branched, 8–15 mm. **Seeds** 2–2.5 mm.

Subspecies 2 (2 in the flora): w North America.

1. Capsules densely short-prickly as well as long-prickly. . . . . . 5a. *Argemone squarrosa* subsp. *squarrosa*
1. Capsules moderately short-prickly as well as long-prickly . . . 5b. *Argemone squarrosa* subsp. *glabrata*

5a. Argemone squarrosa Greene subsp. squarrosa [E] [F]

Stems with 80–200 prickles per cm². **Leaf blades:** surfaces prickly on veins and prickly-hispid, often densely so, between veins. **Capsules** densely short-prickly as well as long-prickly, longest prickles 10–15 mm. $2n = $ ca. 112.

Flowering spring–summer; fruiting summer–fall. Prairies and foothills; 1000–1800 m; Colo., Kans., N.Mex., Okla.

5b. Argemone squarrosa Greene subsp. glabrata G. B. Ownbey, Mem. Torrey Bot. Club 21(1): 62. 1958 [E]

Argemone squarrosa var. *glabrata* (G. B. Ownbey) Shinners

Stems with 40–100 prickles per cm². **Leaf blades:** surfaces essentially unarmed between veins, abaxial surface sparingly prickly on main veins, adaxial surface unarmed or very sparingly prickly on veins. **Capsules** moderately short-prickly as well as long-prickly, longest prickles 8–12 mm.

Flowering spring–summer; fruiting summer–fall. Arid slopes and valleys; 600–1500 m; N.Mex., Tex.

6. Argemone hispida A. Gray, Mem. Amer. Acad. Arts 4: 5. 1849 [E] [F]

Argemone bipinnatifida Greene; *A. platyceras* Link & Otto var. *hispida* (A. Gray) Prain

Plants perennial. **Stems** 3–6 dm, densely prickly and crisped-hispid. **Leaf blades:** surfaces sparsely to densely crisped-hispid between veins, abaxial surface mostly densely prickly on midrib and main veins, adaxial surface less so; proximal lobed 4/5 distance to midrib; distal not clasping. **Inflorescences:** buds oblong, body 16–20 × 14–18 mm, prickly and hispid; sepal horns 4–7 mm, prickly, apical prickle flattened, indurate. **Flowers** 7–10 cm broad; petals white; stamens 150 or more; filaments pale yellow; pistil 3–4-carpellate. **Capsules** ovoid, 30–40 × 12–18 mm (including stigma and excluding prickles), densely prickly, surface obscured, longest prickles straight or incurved, ca. 5 mm. **Seeds** ca. 2.5 mm.

Flowering spring–summer; fruiting summer–fall. Prairies, slopes, and eastern foothills of the Laramie and Rocky mountains; 1400–2100 m; Colo., N.Mex., Wyo.

7. Argemone munita Durand & Hilgard, J. Acad. Nat. Sci. Philadelphia, ser. 2, 3: 37. 1854

Plants annual or perennial. **Stems** 4–16 dm, densely to sparingly prickly. **Leaf blades:** surfaces copiously prickly on veins and intervein surfaces to sparingly prickly on main veins only; basal lobed ca. 1/2 distance to midrib, lobe apices usually distinctly rounded, marginal teeth usually less than 1 mm; distal usually definitely clasping. **Inflorescences:** buds oblong to ellipsoid to obovoid, body 12–22 × 10–16 mm, prickly; sepal horns terete, flattened or angular in cross section, (4–)6–8(–10) mm, unarmed to densely prickly. **Flowers** 5–10(–13) cm broad, not closely subtended by foliaceous bracts; petals white; stamens 150–250; filaments pale yellow; pistil 3–5-carpellate. **Capsules** ovoid, ellipsoid, or lanceoloid, 35–55 × 9–15(–18) mm (including stigma and excluding prickles), prickly, longest prickles to 10 mm, widely spaced, or shorter, more numerous, and interspersed with still shorter prickles, surface then partially obscured. **Seeds** 1.8–2.6 mm.

Subspecies 4 (4 in the flora): w North America.

1. Prickles 0–30 per cm^2 on main stem below capsule; leaf surfaces unarmed or sparingly prickly on veins and intervein areas.
    2. Prickles 0–10 per cm^2; abaxial leaf surface sparingly prickly on main veins, adaxial sur-

face usually unarmed. . . . .
       . . . . .7a. *Argemone munita* subsp. *robusta*
    2. Prickles 10–30 per cm^2; both leaf surfaces sparingly to moderately prickly on veins and often on intervein areas. . . . .
       . . . . .7b. *Argemone munita* subsp. *munita*
1. Prickles 100–500 per cm^2 on main stem below capsule; leaf surfaces mostly closely prickly on veins and intervein areas.
    3. Longest prickles of stem 4–6 mm; stems usually purplish; bud prickles sometimes branched. . . . 7c. *Argemone munita* subsp. *rotundata*
    3. Longest prickles of stem 7–10 mm; stems greenish white; bud prickles simple. . . . . . .
       . . . . .7d. *Argemone munita* subsp. *argentea*

7a. Argemone munita Durand & Hilgard subsp. **robusta** G. B. Ownbey, Mem. Torrey Bot. Club 21(1): 73. 1958 [C] [E]

Argemone munita var. *robusta* (G. B. Ownbey) Shinners

Stems 5–16 dm, with 0–10 prickles per cm^2 on main stem below capsule. **Leaf blades:** abaxial surface sparingly prickly on main veins, adaxial surface usually unarmed. **Capsules** scattered-prickly, surface not obscured, prickles nearly equal, longest 6–8 mm. $2n = 28$.

Flowering spring–summer; fruiting summer–fall. Dry, rocky soil, full sun; of conservation concern; 1500–1700 m; Calif.

Argemone munita subsp. *robusta* is endemic to the Santa Ana Mountains.

7b. Argemone munita Durand & Hilgard subsp. **munita**

Stems 6–15 dm, with 10–30 prickles per cm^2 on main stem below capsule. **Leaf blades:** surfaces sparingly to moderately prickly on veins and intervein areas. **Capsules** moderately prickly, surface not obscured, prickles uneven, longest 5–8 mm. $2n = 28$.

Flowering spring–summer; fruiting summer–fall. Dry foothills and slopes; 100–1500 m; Calif.; Mexico (Baja California).

Argemone munita subsp. *munita* grows in the Coast Ranges. Intermediates between it and subsp. *rotundata* are frequent.

7c. Argemone munita Durand & Hilgard subsp.
rotundata (Rydberg) G. B. Ownbey, Mem. Torrey Bot.
Club 21(1): 77. 1958 [E] [F]

Argemone rotundata Rydberg, Bull.
Torrey Bot. Club 29: 160. 1902; *A.
munita* var. *rotundata* (Rydberg)
Shinners

Stems usually purplish, 4–10
dm, with 120–500 prickles per
cm² on main stem below capsule, longest prickles 4–6 mm.
Leaf blades: surfaces usually
densely prickly on veins and intervein areas. **Inflorescences:** bud prickles sometimes branched. **Capsules**
usually densely prickly, surface partially obscured,
prickles unequal, longest to 8 mm. 2*n* = 28.

Flowering spring–fall; fruiting summer–fall. Open
slopes and foothills, full sun; 1200–2500 m; Ariz.,
Calif., Idaho, Nev., Oreg., Utah.

7d. Argemone munita Durand & Hilgard subsp. **argentea**
G. B. Ownbey, Mem. Torrey Bot. Club 21: 83. 1958 [E]

Argemone munita var. *argentea* (G. B.
Ownbey) Shinners

Stems greenish white, 4–8 dm,
with 100–300 prickles per cm²
on main stem below capsule,
longest prickles 7–10 mm. **Leaf
blades:** surfaces usually densely
prickly on veins and intervein
areas. **Inflorescences:** bud prick-
les simple. **Capsules** densely prickly, surface partially
obscured, prickles unequal, longest 7–10 mm.
2*n* = 28.

Flowering spring; fruiting spring–summer. Dry desert ranges; 150–850 m; Ariz., Calif., Nev.

8. Argemone arizonica G. B. Ownbey, Mem. Torrey Bot.
Club 21(1): 91. 1958 [C] [E]

Plants perennial. **Stems** 5–8 dm,
diffusely branched, often decumbent, copiously long-prickly.
Leaf blades: surfaces sparingly
long-prickly on veins; basal and
proximal lobed nearly to midrib,
lobe length often to 5 times
width; distal not clasping. **Inflorescences:** buds ellipsoid-oblong,
body 15–18 × 12–15 mm, prickly; sepal horns slender, terete, 12–15 mm, prickly at base. **Flowers** 7–10
cm broad; petals white; stamens 100 or more; filaments
pale yellow; pistil 3-carpellate. **Capsules** narrowly
ellipsoid-oblong, 35–45 × 10–14 mm (including
stigma and excluding prickles), closely prickly, longest
prickles 8–10 mm. **Seeds** ca. 2 mm.

Flowering spring; fruiting summer. Precipitous
slopes; of conservation concern; 1000–2000 m; Ariz.

Argemone arizonica grows well in Grand Canyon
National Park, especially along the Kaibab and Bright
Angel trails.

9. Argemone pleiacantha Greene, Repert. Spec. Nov.
Regni Veg. 6: 161. 1908

Plants annual or perennial. **Stems** branched, 5–12 dm,
sparingly to closely prickly. **Leaf blades:** abaxial surface
sparingly prickly on veins, adaxial surface unarmed or
sparingly prickly on veins, apices of lobes angular,
marginal teeth prominent; proximal lobed ca. 3/4 distance to midrib; distal sometimes clasping. **Inflorescences:** buds oblong-obovoid or subglobose, body 14–
20 × 11–18 mm, prickly, prickles simple or branched;
sepal horns terete or adaxially flattened, 4–12 mm.
Flowers 8–12(–16) cm broad; petals white; stamens ca.
150; filaments pale yellow to red; pistil 3–4-carpellate.
Capsules ovoid to ellipsoid-lanceoloid, 25–45 × 10–
16 mm (including stigma and excluding prickles),
closely to sparingly prickly, longest prickles 4–6(–8)
mm. **Seeds** 2–2.5 mm.

Subspecies 3 (3 in the flora): western North America
(United States and Mexico).

1. Capsules closely prickly; bud prickles often
   branched. . . . .
   . . . . 9a. *Argemone pleiacantha* subsp. *pleiacantha*
1. Capsules sparingly prickly; bud prickles simple.
   2. Blades of proximal leaves lobed 4/5 distance
      to midrib, distal not clasping . . . .
      . . . . . 9b. *Argemone pleiacantha* subsp. *pinnatisecta*
   2. Blades of proximal leaves lobed 1/2–2/3 dis-
      tance to midrib, distal often clasping. . . . . . .
      . . . . 9c. *Argemone pleiacantha* subsp. *ambigua*

9a. Argemone pleiacantha Greene subsp.
pleiacantha [F]

Stems 5–12 dm. **Leaf blades:**
proximal mostly lobed 4/5 distance to midrib; distal not clasping. **Inflorescences:** bud prickles
often branched. **Capsules** mostly
closely prickly, surface partially
obscured, prickles straight or incurved. 2*n* = 28.

Flowering spring–summer;
fruiting summer–fall. Foothills and adjacent high
plains; 800–2200 m; Ariz., N.Mex.; Mexico (Chihuahua and Sonora).

9b. Argemone pleiacantha Greene subsp. **pinnatisecta** G. B. Ownbey, Mem. Torrey Bot. Club 21(1): 99. 1958 [C][E][F]

Argemone pleiacantha var. *pinnatisecta* (G. B. Ownbey) Shinners

Stems 6–10 dm. **Leaf blades:** proximal lobed 4/5 distance to midrib; distal not clasping. **Inflorescences:** bud prickles simple. **Capsules** sparingly prickly.

Flowering spring–summer; fruiting summer–fall. Dry hills, pinyon-juniper zone; of conservation concern; 1900–2200 m; N.Mex.

This subspecies is restricted to the west slope of the Sacramento Mountains in New Mexico.

9c. Argemone pleiacantha Greene subsp. **ambigua** G. B. Ownbey, Mem. Torrey Bot. Club 21(1): 101. 1958 [E]

Argemone pleiacantha var. *ambigua* (G. B. Ownbey) Shinners

Stems 5–10 dm. **Leaf blades:** proximal lobed 1/2–2/3 distance to midrib; distal often clasping. **Inflorescences:** bud prickles simple. **Capsules** sparingly prickly.

Flowering spring–summer; fruiting summer–fall. Dry slopes and embankments, foothills; 1000–1800 m; Ariz.

10. Argemone sanguinea Greene, Pittonia 4: 68. 1899

Argemone platyceras Link & Otto var. *rosea* J. M. Coulter

Plants annual to short-lived perennial. **Stems** 4–8(–12) dm, sparingly prickly. **Leaf blades:** abaxial surface moderately prickly on veins, adaxial surface unarmed to sparingly prickly on main veins; basal and proximal lobed to 4/5 distance to midrib. **Inflorescences:** buds broadly ellipsoid-oblong, body 12–20 × 10–15 mm, somewhat prickly; sepal horns often angular in cross section, 5–10 mm, slightly prickly basally. **Flowers** 6–9 cm broad, usually closely subtended by 1–2 foliaceous bracts; petals white or lavender; stamens 150 or more; filaments lemon yellow to red; pistil 3–6-carpellate. **Capsules** narrowly to broadly ellipsoid, 25–40(–50) × 8–15(–18) mm (including stigma and excluding prickles), prickly, surface clearly visible; prickles unequal, longest 5–7(–10) mm. **Seeds** ca. 1.5 mm. $2n = 28, 56$.

Flowering early spring–summer; fruiting spring–summer. Coastal plains and westward; 10–1500 m; Tex.; Mexico (Coahuila, Durango, and Nuevo León).

11. Argemone chisosensis G. B. Ownbey, Mem. Torrey Bot. Club 21(1): 114. 1958

Plants biennial or perennial. **Stems** 4–8 dm, sparingly to copiously prickly, prickles long, patent. **Leaf blades:** abaxial surface prickly on main veins, adaxial surface less prickly to almost unarmed on veins. **Inflorescences:** buds broadly ellipsoid, body 15–20 × 11–15 mm, prickly, prickles patent; sepal horns terete or angular in cross section, 7–12 mm, basally prickly. **Flowers** 7–10 cm broad, usually closely subtended by 1–2 foliaceous bracts; petals white or pale lavender; stamens 150 or more; filaments pale lemon yellow or red; pistil 3–4-carpellate. **Capsules** narrowly ellipsoid-lanceoloid to ellipsoid-ovoid, 30–45 × 8–13 mm (including stigma and excluding prickles), prickly, surface clearly visible; longest prickles scattered, to 10 mm, interspersed with a few shorter ones. **Seeds** 1.8–2.2 mm.

Flowering and fruiting spring–fall. Arid plains and mountains; 900–2000 m; Tex.; Mexico (Chihuahua and Coahuila).

12. Argemone corymbosa Greene, Bull. Calif. Acad. Sci. 2: 59. 1886 [E]

Argemone intermedia Sweet var. *corymbosa* (Greene) Eastwood

Plants perennial. **Stems** 2–8 dm, scattered-prickly. **Leaf blades** shallowly to deeply lobed, thick, leathery, surfaces unarmed to sparingly prickly, especially abaxial surface on veins; distal clasping. **Inflorescences:** buds subglobose to ellipsoid-ovoid, body 10–18 × 8–14 mm, sparingly prickly; sepal horns terete or flattened in cross section, 5–7 mm. **Flowers** 4–9 cm broad; petals white; stamens 100–120; filaments pale yellow; pistil 3–5-carpellate. **Capsules** ovoid to lanceoloid, 25–35 × 10–16 mm (including stigma and excluding prickles), prickly, surface clearly visible, prickles simple, well spaced, subequal, longest 5–7 mm. **Seeds** 1.5–2 mm.

Subspecies 2 (2 in the flora): w North America.

1. Leaf blades essentially unlobed to lobed ca. 1/4 distance to midrib . . . .
. . . . 12a. *Argemone corymbosa* subsp. *corymbosa*
1. Leaf blades lobed 1/2–3/4 distance to midrib . . .
. . . . 12b. *Argemone corymbosa* subsp. *arenicola*

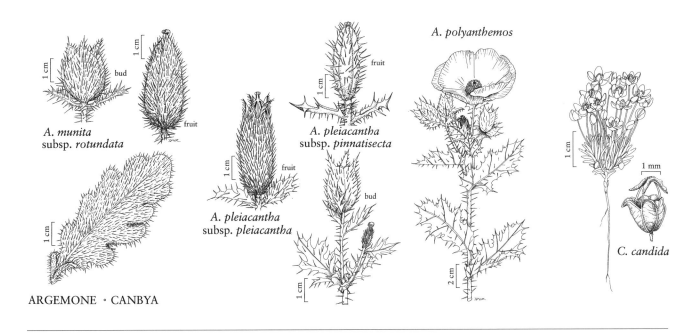

A. munita subsp. rotundata

A. pleiacantha subsp. pleiacantha

A. pleiacantha subsp. pinnatisecta

A. polyanthemos

C. candida

ARGEMONE · CANBYA

12a. Argemone corymbosa Greene subsp. **corymbosa** E

Leaf blades essentially unlobed to lobed ca. 1/4 distance to midrib.

Flowering and fruiting spring. Dry flats and slopes, stabilized dunes, washes, sandy or granitic soil; 400–900 m; Ariz., Calif.

Argemone corymbosa subsp. *corymbosa* grows in the Mojave Desert–Panamint Valley area.

12b. Argemone corymbosa Greene subsp. **arenicola** G. B. Ownbey, Mem. Torrey Bot. Club 21(1): 118. 1958 E

Argemone corymbosa var. *arenicola* (G. B. Ownbey) Shinners

Leaf blades lobed 1/2–3/4 distance to midrib. $2n = 56$.

Flowering spring; fruiting spring–summer. Dry valleys and plains, especially around sand dune areas; 900–1560 m; Ariz., Utah.

13. Argemone gracilenta Greene, Pittonia 3: 346. 1898

Plants perennial. **Stems** 4–12 (–18) dm, moderately prickly throughout, prickles patent, longest 7–10 mm. **Leaf blades:** abaxial surface prickly on main veins, adaxial surface unarmed or sparingly prickly on midrib; basal shallowly to often deeply lobed; distal not clasping. **Inflorescences:** buds subglobose to oblong, body 15–18 × 12–15 mm, sparingly prickly, prickles patent; sepal horns usually slender, terete, (5–)8–12(–14) mm, unarmed or with 1–2 prickles near base. **Flowers** 6–9 cm broad; petals white; stamens about 150; filaments pale yellow; pistil 3–4-carpellate. **Capsules** narrowly ellipsoid to narrowly ellipsoid-ovoid, 30–45 × 8–14 mm (including stigma and excluding prickles), scattered-prickly, surface clearly visible, longest prickles 6–8 (–10) mm. **Seeds** to 2 mm. $2n = 28$.

Flowering spring; fruiting spring–summer. Washes and outwash plains, deserts; 0–900 m (s to n); Ariz.; Mexico (Sonora and Baja California).

14. Argemone polyanthemos (Fedde) G. B. Ownbey, Mem. Torrey Bot. Club 21(1): 128. 1958 [E][F][W2]

Argemone intermedia Sweet var. *polyanthemos* Fedde in H. G. A. Engler, Pflanzenr. 40[IV,104]: 283. 1909

Plants annual or biennial. **Stems** 4–8(–12) dm, sparingly prickly. **Leaf blades:** abaxial surface scattered-prickly on main veins, adaxial surface unarmed; proximal lobed 2/3 distance to midrib; distal clasping. **Inflorescences:** buds ellipsoid-oblong, body 15–22 × 10–15 mm, sparingly prickly; sepal horns terete, 6–10(–15) mm, usually unarmed. **Flowers** 7–10 cm broad, usually closely subtended by 1–2 foliaceous bracts; petals white, very rarely lavender; stamens 150 or more; filaments lemon yellow; pistil 3–4-carpellate. **Capsules** narrowly to broadly ellipsoid, 35–50 × 10–17 mm (including stigma and excluding prickles), prickly, surface clearly visible, prickles widely spaced, longest 4–10(–12) mm, interspersed with a few shorter ones. **Seeds** ca. 2 mm. 2*n* = 28.

Flowering spring–summer; fruiting late spring–summer. Prairies, foothills and mesas; 300–2300 m; Colo., Kans., Mont., Nebr., N.Mex., N.Dak., Okla., S.Dak., Tex., Utah, Wyo.

Argemone polyanthemos is introduced in Utah.

15. Argemone albiflora Hornemann, Hort. Bot. Hafn. 2: 489. 1815 [E][W2]

Argemone alba F. Lestiboudois

Plants annual or biennial. **Stems** 4–10(–15) dm, sparingly to moderately prickly. **Leaf blades:** abaxial surface sparingly prickly on main veins, adaxial surface unarmed or sparingly prickly on main veins; basal and proximal lobed 1/2–4/5 distance to midrib; distal clasping. **Inflorescences:** buds broadly ellipsoid to subglobose, body 12–18 × 10–16 mm, sparingly to rather closely prickly; sepal horns terete, 3–6(–10) mm, usually unarmed. **Flowers** 5–10 cm broad, closely to more distantly subtended by 1–2 foliaceous bracts; petals white; stamens 150 or more; filaments lemon yellow; pistil (3–)4–5(–7)-carpellate. **Capsules** oblong, oblong-ellipsoid, or narrowly ellipsoid, 20–40(–45) × 10–15 (–25) mm (including stigma and excluding prickles),

prickly, surface clearly visible to partially obscured, longest prickles widely spaced, 6–10(–12) mm, sometimes interspersed with smaller ones. **Seeds** 1.6–2 mm.

Subspecies 2 (2 in the flora): North America.

1. Capsules oblong to oblong-ellipsoid, prickles equal, widely spaced . . . . 15a. *Argemone albiflora* subsp. *albiflora*
1. Capsules usually narrowly ellipsoid, prickles unequal, longer ones widely spaced, interspersed with shorter, unequal ones. . . . . 15b. *Argemone albiflora* subsp. *texana*

15a. Argemone albiflora Hornemann subsp. **albiflora** [E]

Stems sparsely prickly. **Leaf blades:** basal and proximal lobed 1/2–2/3 distance to midrib. **Inflorescences:** sepal horns 3–5 mm, unarmed. **Capsules** oblong to oblong-ellipsoid, surface clearly visible, prickles widely spaced, equal, 6–8(–10) mm. 2*n* = 28.

Flowering early spring–summer; fruiting late spring–summer. Waste places, roadsides, fields, beaches, dunes, coastal plains; 0–300 m; Ala., Conn., Fla., Ga., Ill., La., Mass., Mich., Miss., Mo., N.Y., N.C., S.C.

15b. Argemone albiflora Hornemann subsp. **texana** G. B. Ownbey, Mem. Torrey Bot. Club 21(1): 141. 1958 [E]

Argemone albiflora var. *texana* (G. B. Ownbey) Shinners

Stems sparsely to moderately prickly. **Leaf blades:** basal and proximal lobed nearly to midrib. **Inflorescences:** sepal horns 4–6 (–10) mm, unarmed or sparingly prickly near base. **Capsules** usually narrowly ellipsoid, surface partially obscured, longer prickles widely spaced, interspersed with numerous, shorter, unequal ones. 2*n* = 28.

Flowering early spring–summer; fruiting spring–fall. Pastures and waste ground; 0–300 m; Ark., La., Mo., Tex.

Argemone albiflora subsp. *texana* is introduced in Arkansas and Missouri.

10. CANBYA Parry ex A. Gray, Proc. Amer. Acad. Arts 12: 51, plate 1. 1876 · Pygmy-poppy [for William M. Canby, 1831–1904, Delaware botanist]

Curtis Clark

Robert W. Kiger

Herbs, annual, semiacaulescent, very small, glabrous, from taproots; sap clear. **Stems** branching at and just above ground level, leafy, branches short. **Leaves** alternate, congested in tufts, sessile; blade unlobed, linear-oblong, fleshy, margins entire. **Inflorescences** axillary, 1-flowered, much exceeding leaves; peduncle slender. **Flowers:** sepals 3, distinct; petals 5–7; stamens 6–15; pistil 3–4-carpellate; ovary ovoid, 1-locular; style absent; stigmas 3(–4), linear, radiate-recurved, appressed or adherent to ovary. **Capsules** erect, ovoid to oblong-ovoid, 3(–4)-valved, dehiscing from apex. **Seeds** many, brown, oblong-obovoid to ellipsoid, slightly arcuate, 0.6–0.8 mm, glossy, aril absent. $x = 8$.

Species 2 (2 in the flora): w United States.

These diminutive (but handsome) desert endemics are among the smallest plants in the family. Little is known about their life histories, reproduction, ecological interactions, or evolutionary histories, and they merit more field and laboratory study.

1. Petals white, marcescent; filaments shorter than anthers. . . . . .1. *Canbya candida*
1. Petals bright yellow, deciduous; filaments longer than anthers. . . . . 2. *Canbya aurea*

1. Canbya candida Parry ex A. Gray, Proc. Amer. Acad. Arts 12: 51, plate 1. 1876 · White pygmy-poppy C E F

Plants 1–3 cm. **Leaves** 5–9 mm. **Inflorescences:** peduncle 1–2 cm. **Flowers:** sepals 2 mm; petals marcescent, 5–7, white, elliptic, 3–4 mm; stamens 6–9; filaments shorter than anthers. **Capsules** 1.5–2.5 mm. $2n = 16$.

Flowering Apr–Jun. Dry, sandy ground; 600–1200 m; of conservation concern; Calif.

Canbya candida has been found only in the western Mojave Desert.

2. Canbya aurea S. Watson, Proc. Amer. Acad. Arts 21: 445. 1886 · Yellow pygmy-poppy E

Plants 1–2 cm. **Leaves** 3–10 mm. **Inflorescences:** peduncle 1–6 cm. **Flowers:** sepals 2 mm; petals deciduous, 6, bright yellow, broadly ovate, 3–4 mm; stamens 6–15; filaments longer than anthers. **Capsules** 2–4 mm. $2n = 16$.

Flowering Apr–Jul. Dry, sandy ground, usually with sagebrush; 900–1700 m; Oreg., Nev.

11. PAPAVER Linnaeus, Sp. Pl. 1: 506. 1753; Gen. Pl. ed. 5, 224. 1754 · Poppy, pavot [classical Latin name for poppy; perhaps from Greek *papa*, pap, alluding to the thick, sometimes milky sap]

Robert W. Kiger

David F. Murray

Herbs, annual, biennial, or perennial, scapose or caulescent, from taproots; sap white, orange, or red. **Stems** when present leafy. **Leaves:** basal rosulate, petiolate; cauline alternate, proximal petiolate, distal subsessile or sessile, sometimes clasping (in *P. somniferum*); blade unlobed or 1–3×-pinnately lobed or parted; margins entire or dentate, crenate, or incised. **Inflorescences**

cymiform, with flowers disposed in 1s, 2s, or 3s on long scapes or peduncles; bracts present; buds nodding [erect]. **Flowers:** sepals 2(–3), distinct; petals 4(–6); stamens many; pistil 3–18 [–22]-carpellate; ovary 1-locular, sometimes incompletely multilocular by placental intrusion; style absent; stigmas 3–18[–22], radiating on sessile, ± lobed disc, velvety. **Capsules** erect, 3–18[–22]-pored or -short valved immediately beneath persistent or sometimes deciduous (in *P. hybridum*) stigmatic disc. **Seeds** many, minutely pitted, aril absent. $x = 7$.

Species 70–100 (16 in the flora): temperate and arctic North America, Eurasia, n, s Africa, Australia.

Papaver is rich in alkaloids, notably opiates. The genus is quite complex cytologically; in addition to diploids, there are numerous polyploid species and some that apparently are aneuploid. Most commonly, $n = 7$ or a multiple, and $2n$ ranges from 14 to over 100. There are published chromosome counts for almost every taxon in the flora, but for the introduced species none has been made from wild-collected North American material.

The scapose poppies in the flora are native; the caulescent ones, except *Papaver californicum*, are introduced Eurasian ornamentals, crop weeds, and ballast waifs. All the scapose species are confined to arctic and alpine habitats. Plants of the introduced caulescent species, especially *P. rhoeas, P. dubium,* and *P. somniferum,* vary greatly in size, and surprisingly diminutive mature individuals are sometimes found, especially northward.

SELECTED REFERENCES Kadereit, J. W. 1988. Sectional affinities and geographical distribution in the genus *Papaver* L. (Papaveraceae). Beitr. Biol. Pflanzen 63: 139–156. Kadereit, J. W. 1990. Some suggestions on the geographical origin of the central, west and north European synanthropic species of *Papaver* L. Bot. J. Linn. Soc. 103: 221–231. Kiger, R. W. 1973. Sectional nomenclature in *Papaver* L. Taxon 22: 579–582. Kiger, R. W. 1975. *Papaver* in North America north of Mexico. Rhodora 77: 410–422. Kiger, R. W. 1985. Revised sectional nomenclature in *Papaver* L. Taxon 34: 150–152. Novák, J. and V. Preininger. 1987. Chemotaxonomic review of the genus *Papaver.* Preslia 59: 1–13.

1. Plants caulescent (sometimes subscapose), at least a few cauline leaves present.
　　2. Blades of distal leaves clasping stem (sect. *Papaver*)...................................14. *Papaver somniferum*
　　2. Blades of distal leaves not clasping stem.
　　　　3. Capsules setose (sect. *Argemonidium*).
　　　　　　4. Capsules obovoid-ellipsoid to subglobose, densely and firmly setose .... 2. *Papaver hybridum*
　　　　　　4. Capsules oblong to clavate, sparsely and weakly setose .... 1. *Papaver argemone*
　　　　3. Capsules glabrous.
　　　　　　5. Plants perennial; flowers 10 cm or more broad (sect. *Macrantha*) .... 4. *Papaver orientale*
　　　　　　5. Plants annual; flowers less than 10 cm broad.
　　　　　　　　6. Peduncles glabrous or sparsely pilose; petals with greenish basal spot; stigmatic disc conic, usually umbonate; capsules distinctly short-valvate (sect. *Californicum*). ..... 3. *Papaver californicum*
　　　　　　　　6. Peduncles hispid; petals unspotted or with dark basal spot; stigmatic disc ± flat; capsules poricidal (sect. *Rhoeadium*).
　　　　　　　　　　7. Peduncles markedly spreading-hispid distally; capsules less than 2 times longer than broad.. 16. *Papaver rhoeas*
　　　　　　　　　　7. Peduncles strongly appressed-hispid distally; capsules 2 times or more longer than broad.. 15. *Papaver dubium*
1. Plants strictly scapose, leaves all basal (sect. *Meconella*).
　　8. Leaf blades mostly with 3 primary lobes .... 12. *Papaver walpolei*
　　8. Leaf blades with 5–many primary lobes.
　　　　9. Capsules more than 4 times longer than broad............................. 6. *Papaver macounii*
　　　　9. Capsules 1–2.5 times longer than broad.
　　　　　　10. Trichomes on capsules ivory colored.
　　　　　　　　11. Plants tall, seldom less than 2 dm .... 5. *Papaver nudicaule*
　　　　　　　　11. Plants short, seldom more than 1.5 dm.
　　　　　　　　　　12. Leaf blades setose .... 10. *Papaver alboroseum*
　　　　　　　　　　12. Leaf blades glabrous or sparsely hirsute. .... 11. *Papaver pygmaeum*
　　　　　　10. Trichomes on capsules light to dark brown or black.
　　　　　　　　13. Leaf blades mostly with 5 primary lobes, lobes mostly simple.
　　　　　　　　　　14. Primary leaf lobes oblanceolate to strap-shaped.................... 13. *Papaver gorodkovii*
　　　　　　　　　　14. Primary leaf lobes broadly lanceolate to ovate .... 8. *Papaver radicatum*

13. Leaf blades with more than 5 primary lobes, lobes mostly divided.
    15. Scapes straight, erect, generally longer than 20 cm; capsules oblong-ellipsoid. . .7. *Papaver lapponicum*
    15. Scapes curved, erect or decumbent, less than 15 cm; capsules obconic to subglobose.
        16. Leaf blades generally green, not glaucous, primary lobes lanceolate, their apices acute to obtuse . . . . .8. *Papaver radicatum*
        16. Leaf blades generally gray- and blue-green, glaucous, primary lobes obovate to strap-shaped, their apices rounded. . . . . .9. *Papaver mcconnellii*

11a. PAPAVER Linnaeus sect. ARGEMONIDIUM Spach, Hist. Nat. Vég. 7: 19. 1839

Plants annual, caulescent. **Leaf blades** deeply 1–2×-lobed, distal not clasping stem. **Flowers:** filaments purple, clavate. **Capsules** poricidal, setose.

SELECTED REFERENCE Kadereit, J. W. 1986. A revision of *Papaver* section *Argemonidium*. Notes Roy. Bot. Gard. Edinburgh 44: 25–43.

1. Papaver argemone Linnaeus, Sp. Pl. 1: 506. 1753 [W2]

Plants to 5 dm, hispid. **Stems** simple or branching. **Leaves** to 12[–20] cm. **Inflorescences:** peduncle appressed-hispid. **Flowers:** petals dark red, sometimes with dark basal spot, to 25 mm; anthers pale blue; stigmas 4–6, disc convex and radially vaulted. **Capsules** sessile, oblong to clavate, distinctly ribbed, to 2 cm, sparsely and weakly setose.

Flowering spring–summer. Fields and disturbed sites; 0–300 m; introduced; Idaho, Oreg., Pa., Utah; Europe; sw Asia.

In its native range, *Papaver argemone* is a complex of five diploid, tetraploid, and hexaploid subspecies (J. W. Kadereit 1986, 1990). Apparently two or more of these have been represented among the crop weeds and ballast waifs introduced in North America, where plants are difficult to assign to particular subspecies. The species should be expected elsewhere in the flora. Collections attributed to Maryland, Ohio, and Virginia are known also, but they lack more specific citations of locality.

2. Papaver hybridum Linnaeus, Sp. Pl. 1: 506. 1753
· Rough poppy

Plants to 5 dm, hispid. **Stems** branching. **Leaves** to 10 cm. **Inflorescences:** peduncle appressed-hispid. **Flowers:** petals early caducous, red to purplish red, with dark basal spot, to 25 mm; anthers pale blue; ovaries setose; stigmas 4–8, disc convex and radially vaulted. **Capsules** sessile, obovoid-ellipsoid to subglobose, obscurely to distinctly ribbed, to 1.5 cm, densely and firmly setose, stigmatic disc often deciduous.

Flowering spring. Fields, vineyards, and disturbed sites; 0–700 m; introduced; Calif., N.C., Pa., S.C.; Eurasia; n Africa.

Papaver hybridum should be expected elsewhere in the flora. Some California collections of this crop weed have been misidentified as *Papaver apulum* Tenore var. *micranthum* (Boreau) Fedde, which is not known to occur in the flora.

11b. PAPAVER Linnaeus sect. CALIFORNICUM Kadereit, Rhodora 90: 11. 1988

Plants annual, caulescent. **Leaf blades** 1–2×-lobed, distal not clasping stem. **Flowers:** filaments greenish yellow, filiform. **Capsules** distinctly valvate, glabrous.

SELECTED REFERENCE Kadereit, J. W. 1988b. *Papaver* L. sect. *Californicum* Kadereit, a new section of the genus. Rhodora 90: 7–13.

3. Papaver californicum A. Gray, Proc. Amer. Acad. Arts 22: 313. 1887 · Western poppy [E]

Papaver lemmonii Greene

Plants to 6.5 dm, glabrate or sparsely pilose. **Stems** simple or branching. **Leaves** to 15 cm. **Inflorescences:** peduncle glabrous or sparsely pilose. **Flowers:** petals light orange or orange-red, with pink-edged, greenish basal spot, to 2.5 cm; anthers yellow; stigmas 4–8(–11), disc conic, usually umbonate. **Capsules** sessile, ellipsoid to obovoid-turbinate, distinctly ribbed, to 1.8 cm. $2n = 28$.

Flowering spring. Chaparral and oak woodlands, especially in grassy areas, clearings, burns and other disturbed sites; 0–900 m; Calif.

Papaver californicum grows in central western and southwestern California in the Coast, Transverse, and Peninsular ranges. This is the only caulescent poppy, and the only annual one, native to the flora. In the past it has been included in *Papaver* sect. *Rhoeadium*, together with the other annuals that have glabrous capsules and distal leaves not clasping, which are native to Eurasia. Recently, based on differences in filament color, stigmatic disc shape, and capsule dehiscence, J. W. Kadereit (1988b) assigned *P. californicum* to a new monotypic section and suggested that it originated from the same stock as the perennial, scapose, arctic-alpine poppies (*Papaver* sect. *Meconella*).

11c. PAPAVER Linnaeus sect. **MACRANTHA** Elkan, Tent. Monogr. Papaver, 19. 1839

Papaver sect. *Oxytona* (Bernhardi) Pfeiffer

Plants perennial, caulescent. **Leaf blades** unlobed or 1×-lobed, distal not clasping stem. **Flowers:** filaments purple, clavate. **Capsules** poricidal, glabrous.

SELECTED REFERENCE   Goldblatt, P. 1974. Biosystematic studies in *Papaver* section *Oxytona*. Ann. Missouri Bot. Gard. 61: 264–296.

4. Papaver orientale Linnaeus, Sp. Pl. 1: 508. 1753 · Oriental poppy

Papaver pseudo-orientale (Fedde) Medwedew

Plants to 10 dm, hispid. **Stems** simple or rarely branching. **Leaves** to 35 cm. **Inflorescences:** peduncle moderately to densely appressed pale hispid; bracts sometimes 1–4 just beneath flower. **Flowers:** petals light orange to orange-red, usually with pale basal spot, sometimes dark-spotted or unspotted, to 6 cm; anthers violet; stigmas (8–)11–15, disc flat or shallowly convex. **Capsules** sessile, subglobose, obscurely ribbed, to 2.5 cm, glaucous.

Flowering spring–summer. Fields, clearings, roadsides, and disturbed sites; introduced; Ont.; Colo., Iowa, Mich., N.J., Pa., Utah, Va., Wis.; sw Asia.

An alpine species in its native range, *Papaver orientale* is widely grown for ornament and sometimes persists after spreading from cultivation. It should be expected elsewhere in the flora.

Papaver bracteatum Lindley, which some authors have included in *P. orientale*, is similar but more robust, with buds erect, sepals subtended by 3–5 sepaloid and 2 foliaceous bracts, flowers to 20 cm broad, deep red petals with dark basal spot, and capsules to 4 cm. It is widely cultivated and may occasionally escape but apparently does not become naturalized.

11d. PAPAVER Linnaeus sect. **MECONELLA** Spach, Hist. Nat. Vég. 7: 19. 1839

Papaver sect. *Scapiflora* Elkan; *P.* sect. *Lasiotrachyphylla* (Bernhardi) Pfeiffer

Plants perennial, scapose. **Leaf blades** unlobed or 1–2×-lobed. **Flowers:** filaments white or yellow, filiform. **Capsules** obscurely valvate, ribbed, variously pubescent.

More study is needed for a comprehensive assessment of the North American scapose poppies in circumpolar context. The circumscriptions and interrelationships of the Alaskan and Asiatic members of this section are especially problematic.

Typification of *Papaver radicatum* Rottbøll also is problematic, and interpretation of its identity has significant consequences in the complex nomenclature of the arctic-alpine poppies. For details and discussion, see G. Knaben (1958); Á. Löve (1962b, 1962c); and G. Knaben and N. Hylander (1970).

SELECTED REFERENCES Knaben, G. 1958. *Papaver*-studier, med et forsvar for *P. radicatum* Rottb. som en Islandsk-Skandinavisk art. Blyttia 16: 61–80. Knaben, G. 1959. On the evolution of the *radicatum*-group of the *Scapiflora* Papavers as studied in 70 and 56 chromosome species. Part A. Cytotaxonomical aspects. Part B. Experimental studies. Opera Bot. 2(3), 3(3). Knaben, G. and N. Hylander. 1970. On the typification of *Papaver radicatum* Rottb. and its nomenclatural consequences. Bot. Not. 123: 338–345. Löve, Á. 1962b. Typification of *Papaver radicatum*—A nomenclatural detective story. Bot. Not. 115: 113–136. Löve, Á. 1962c. Nomenclature of North Atlantic Papavers. Taxon 11: 132–138. Löve, D. 1969. *Papaver* at high altitudes in the Rocky Mountains. Brittonia 21: 1–10. Rändel, U. 1974. Beitrage zur Kenntnis der Sippenstruktur der Gattung *Papaver* Sectio *Scapiflora* Reichenb. (Papaveraceae). Feddes Repert. 84: 655–732. Rändel, U. 1975. Die Beziehungen von *Papaver pygmaeum* Rydb. aus den Rocky Mountains zum nordamerikanischen *P. kluanense* D. Löve sowie zu einigen nordostasiatischen Vertretern der Sektion *Scapiflorae* Reichenb. im Vergleich mit *P. alpinum* L. (Papaveraceae). Feddes Repert. 86: 19–37. Rändel, U. 1977. Über Sippen des subarktisch-arktischen Nordamerikas, des Beringia-Gebietes und Nordost-Asiens der Sektion *Lasiotrachyphylla* Bernh. (Papaveraceae) und deren Beziehungen zu einander und zu Sippen anderer Arealteile der Sektion. Feddes Repert. 88: 421–450. Rändel, U. 1977b. Über die grönländischen Vertreter der Sektion *Lasiotrachyphylla* Bernh. (Papaveraceae) und deren Beziehungen zu Vertretern anderer arktischer Arealteile der Sektion. Feddes Repert. 88: 451–464.

5. Papaver nudicaule Linnaeus, Sp. Pl. 1: 507. 1753

Subspecies numerous (1 in the flora): boreal North America, Europe, Asia.

5a. Papaver nudicaule Linnaeus subsp. **americanum** Rändel ex D. F. Murray, Novon 5: 294. 1995 [E]

Plants loosely cespitose, to 4 dm. **Leaves** to 20 cm; petiole to 2/3 length of leaf; blade gray-green abaxially, green adaxially, lanceolate, deeply 1–2×-lobed with 3–4 pairs of primary lateral lobes, glabrate or setose adaxially with long white trichomes; primary lobes lanceolate or strap-shaped. **Inflorescences:** scape erect, hispid. **Flowers** to 6 cm broad; petals yellow or white; anthers yellow; stigmas 5–6, disc flat to convex-umbonate. **Capsules** subglobose to ellipsoid, conspicuously ribbed, to 2.5 times longer than broad, to 1.5 cm, strigose with light (ivory) trichomes having abruptly expanded bases. $2n = 28$.

Flowering Jun–Aug. Dry, exposed, rocky openings such as on steep slopes, screes, and outcrops; 300–1000 m; Yukon; Alaska.

Papaver nudicaule grows in the boreal forest of the continental interior and disjunct westward along the Yukon and Kuskokwim rivers. It includes numerous red, white, and yellow cultivars of the so-called Iceland poppy encountered along the road system in Alaska, Canada, southwestern Greenland, and no doubt elsewhere. The species was originally described from Siberian material, to which our subsp. *americanum* has obvious affinities (U. Rändel 1977).

6. Papaver macounii Greene, Pittonia 3: 247. 1897

Plants loosely cespitose, with persistent leaf bases. **Leaves:** petiole to 3/4 length of leaf; blade lanceolate, 1–2×-lobed; primary lobes generally lanceolate, sometimes strap-shaped, apex acute or obtuse. **Inflorescences:** scape erect. **Capsules** more than 4 times longer than broad.

Subspecies 2 (2 in the flora): North America, Russia.

1. Plants robust; persistent leaf bases numerous, firm, acuminate; leaf blades green on both surfaces, hispid . . . . .6a. *Papaver macounii* subsp. *macounii*
1. Plants slender; persistent leaf bases few, flexible, lanceolate; leaf blades light green or glaucous abaxially, dark green adaxially, glabrate to densely pilose . . . . 6b. *Papaver macounii* subsp. *discolor*

6a. Papaver macounii Greene subsp. **macounii** [E]

Papaver alaskanum Hultén var. *macranthum* Hultén

Plants robust, to 3 dm, persistent leaf bases numerous, firm, acuminate, ciliate, glabrous to densely hispid. **Leaves** to 10 cm; blade green, with 3–4 pairs of primary lateral lobes; surfaces hispid, trichomes white. **Inflorescences:** scape hispid, often densely so, with appressed and spreading trichomes. **Flowers** to 6 cm broad; petals yellow, or white and tinged yellow basally; anthers yellow; stigmas 5–6, disc convex to conic-umbonate. **Capsules** obovoid-obconic, to 2 cm, strigose with light brown trichomes. $2n = 28$.

Flowering Jun–Aug. Dry rocky and gravelly soils; 0–200 m; Alaska.

This subspecies has been collected repeatedly on St. Paul Island, the type locality, and is known also from St. George Island and several disjunct localities on the coastal mainland of western Alaska as far north as the Seward Peninsula. It can be distinguished from *Papaver macounii* subsp. *discolor* by its more uniformly green and generally hispid leaves, and from *P. radicatum* subsp. *alaskanum* by its longer and narrower capsules. No counterpart has been found in the Russian

Far East. A report of $2n = 42$, erroneously attributed to K. Horn (1938) by G. Knaben (1959), is a misprint.

6b. Papaver macounii Greene subsp. **discolor** (Hultén) Rändel ex D. F. Murray, Novon 5: 294. 1995 [F]

Papaver macounii var. *discolor* Hultén, Fl. Alaska Yukon 5: 803. 1945; *P. keelei* Porsild; *P. scammanianum* D. Löve

Plants slender, to 4 dm, persistent leaf bases few, flexible, lanceolate, ciliate, glabrous to hispid. **Leaves** to 12 cm; blade green or light green abaxially, dark green adaxially, with 2–3 pairs of primary lateral lobes; surfaces glabrate to densely pilose, trichomes light brown. **Inflorescences:** scape glabrate to hispid with appressed and spreading trichomes. **Flowers** to 5 cm broad; petals yellow or very rarely pink; anthers yellow; stigmas 4–6, disc conic-umbonate. **Capsules** clavate, to 2.5 cm, sparsely to sometimes densely strigose. $2n = 28, 70$ (as *P. radicatum* Rottb. *s.l.*).

Flowering Jun–Aug. Mesic arctic and alpine tundra and meadows; 0–1200 m; N.W.T., Yukon; Alaska; Russian Far East.

If this taxon is recognized at species rather than subspecies rank, the correct name is *Papaver keelei*. Specimens of *P. paucistaminum* Tolmatchew & V. V. Petrovsky from Chukotka in the Russian Far East fall within the range of variation described here.

7. Papaver lapponicum (Tolmatchew) Nordhagen, Bergens Mus. Årbok 2: 45. 1931 [F]

Papaver radicatum Rottbøll subsp. *lapponicum* Tolmatchew, Bot. Mater. Gerb. Glavn. Bot. Sada RSFSR 4: 86. 1923; *P. hultenii* Knaben; *P. hultenii* var. *salmonicolor* Hultén

Plants loosely cespitose, to 3.5 (seldom less than 2) dm. **Leaves** to 12 cm; petiole 1/2–3/4 length of leaf; blade green to gray-green on both surfaces, lanceolate, 1–2 ×-lobed with 2–3 pairs of primary lateral lobes; surfaces hirsute, sometimes densely so, with long white trichomes; primary lobes lanceolate, mostly divided, apex obtuse or acute to acuminate, frequently bristle-tipped. **Inflorescences:** scape erect, straight, generally longer than 20 cm, glabrate to hispid. **Flowers** to 3.5 cm broad; petals yellow, sometimes distally tinged with pink; anthers yellow; stigmas 5–7, disc convex. **Capsules** oblong-ellipsoid, to 2 cm, 1–2.5 times longer than broad, strigose with brown trichomes. $2n = 42$ (as *P. hultenii*), 56.

Flowering Jun–Aug. Mesic tundra and in sand and gravel of floodplain terraces and shorelines; 0–1000

m; B.C., Nfld., N.W.T., Que., Yukon; Alaska; Eurasia (northernmost Norway and Russia).

We recognize *Papaver lapponicum* in a much narrower sense than did G. Knaben (1959). Much further study is needed to assess the relationships of North American populations with several taxa from the Russian Far East. Plants with rose-colored petals have been distinguished as *A. lapponicum* var. *salmonicolor* (*P. alboroseum* of some authors, not Hultén). Such specimens from arctic Alaska appear to be the same as *P. shamurinii* Petrovsky from Russia. Knowledge of *P. lapponicum* from Greenland, where evidently it also occurs, is inadequate to permit an accurate account of its distribution there.

8. Papaver radicatum Rottbøll, Skr. Kiøbenhavnske Selsk. Laerd. Elsk. 10: 455. 1770 · Arctic poppy

Plants loosely to densely cespitose, to 1.5 dm. **Leaves** to 12 cm; petiole 2/3 length of leaf; blade green on both surfaces, not glaucous, lanceolate, 1–2 ×-lobed with 2–3(–4) pairs of primary lateral lobes; primary lobes broadly lanceolate or strap-shaped, apex obtuse to acute. **Inflorescences:** scape erect or bowed and decumbent, less than 15 cm, sparsely to densely hispid. **Flowers** to 6.5 cm broad; petals yellow or white, rarely pink tinged, or brick red; anthers yellow; stigmas 4–7, disc convex. **Capsules** obovoid to subglobose, 1–2.5 times longer than broad, strigose, trichomes light to dark brown or black.

Subspecies numerous (4 in the flora): arctic and alpine North America, Europe, Asia.

Many infraspecific taxa have been named from throughout the extensive range of this extremely variable species. Within North America, the following broadly circumscribed subspecies are generally, but not always, distinguishable.

1. Scapes densely hispid with dark trichomes; capsules with dark brown trichomes. . . . . 8c. *Papaver radicatum* subsp. *polare*
1. Scapes hispid with light-colored trichomes (sometimes dark brown in subsp. *alaskanum*); capsules with light to dark brown trichomes.
  2. Flowers 2 cm broad or less; capsules ellipsoid-subglobose to oblong-obconic; Rocky Mountain system from Alaska and Yukon southward . . . . 8d. *Papaver radicatum* subsp. *kluanensis*
  2. Flowers mostly greater than 2 cm broad; capsules broadly obovoid to ellipsoid; Aleutian Islands n, e across arctic Alaska and Canada to Greenland.
    3. Capsule trichomes generally with abruptly thickened bases; Aleutian Islands, islands of the Bering Strait and w coast of Alaska . . . . 8b. *Papaver radicatum* subsp. *alaskanum*
    3. Capsule trichomes generally without thickened bases; w, n Alaska e to Canada

and Greenland....................
.... 8a. *Papaver radicatum* subsp. *radicatum*

8a. Papaver radicatum Rottbøll subsp. **radicatum** [F]

Papaver lapponicum (Tolmatchew) Nordhagen subsp. *labradoricum* (Fedde) Knaben; *P. lapponicum* subsp. *occidentale* (Lundström) Knaben; *P. lapponicum* subsp. *porsildii* Knaben; *P. nigroflavum* D. Löve; *P. nudicaule* Linnaeus var. *labradoricum* Fedde; *P. radicatum* subsp. *labradoricum* (Fedde) Fedde; *P. radicatum* var. *labradoricum* (Fedde) J. Rousseau & Raymond; *P. radicatum* subsp. *occidentale* Lundström; *P. radicatum* subsp. *porsildii* (Knaben) Á. Löve

Plants loosely cespitose, to 1.7 dm, with persistent, broadly lanceolate leaf bases. **Leaves** to 12 cm; petiole to 2/3 length of leaf; blade green on both surfaces, lanceolate, with 2–4 pairs of primary lateral lobes, hirsute; primary lobes lanceolate to narrowly strapshaped, apex acute to rounded. **Inflorescences:** scapes erect and curved, hispid, trichomes light colored. **Flowers** to 5 cm broad; petals yellow. **Capsules** broadly obovoid to ellipsoid, to 1.5 cm, strigose, trichomes brown, generally without thickened bases. $2n = 56, 70$.

Flowering Jun–Aug. Dry, sandy soils, rocky tundra communities, fellfield, blockfields, and screes at highest elevations and northernmost latitudes; 0–2300 m; Greenland; Nfld., N.W.T., Que., Yukon; Alaska; circumpolar.

8b. Papaver radicatum Rottbøll subsp. **alaskanum**
(Hultén) J. P. Anderson, Fl. Alaska, 244, fig. 517. 1959

Papaver alaskanum Hultén, Fl. Aleut. Isl., 190, plate 10. 1937; *P. alaskanum* var. *grandiflorum* Hultén; *P. alaskanum* var. *latilobum* Hultén; *P. microcarpum* de Candolle subsp. *alaskanum* (Hultén) Tolmatchew

Plants loosely to densely cespitose, to 1.8 dm, with abundant long-lanceolate, acuminate, dull to shiny, persistent leaf bases. **Leaves** to 10 cm; petiole to 3/4 length of leaf; blade green on both surfaces, lanceolate, with 2–4 pairs of primary lateral lobes; primary lobes lanceolate to obovate, apex acute to rounded. **Inflorescences:** scape curved or straight, sparingly to densely hispid, trichomes appressed-ascending to spreading, light to dark brown. **Flowers** to 6.5 cm broad; petals yellow, occasionally pink tinged (w Aleutians). **Capsules** subglobose to obovoid-obconic, to 1.5 cm, strigose, trichomes light to dark brown, base abruptly expanded. The chromosome number ($2n = 42$) attributed to G. Knaben (1959) by U. Rändel (1977) cannot be confirmed.

Flowering Jun–Aug. Sands and gravels and on rocky tundra; 0–100 m; Alaska; Asia (Russian Far East, Kamchatka).

In the flora area, *Papaver radicatum* subsp. *alaskanum* is known from Kodiak Island, the Kenai Peninsula, the islands of the Bering Sea, the Aleutian Islands, the coastal Alaska Peninsula and Bristol Bay region, and disjunct northward in Alaska to the Chukchi Sea.

We distinguish these coastal plants from the ones at high elevations in the interior of Alaska and Yukon that also have numerous persistent leaf bases and have been named *Papaver alaskanum* by others (cf. E. Hultén 1968; S. L. Welsh 1974) but otherwise have the features of *P. radicatum* subsp. *radicatum* or subsp. *kluanensis*. Specimens of *P. radicatum* subsp. *alaskanum* with short, broad capsules and light-colored trichomes approach, in these respects, the Asiatic *P. microcarpum,* a similarity that Tolmatchew has already indicated with his combination *P. microcarpum* subsp. *alaskanum.*

8c. Papaver radicatum Rottbøll subsp. **polare**
Tolmatchew, Bot. Mater. Gerb. Glavn. Bot. Sada RSFSR 4: 87. 1923

Papaver cornwallisensis D. Löve; *P. nudicaule* Linnaeus var. *albiflora* Lange; *P. polare* (Tolmatchew) Perfiljev; *P. radicatum* var. *albiflorum* (Lange) Porsild

Plants loosely to densely cespitose, sometimes even pulvinate, to 1 dm, with numerous short, dull, persistent leaf bases. **Leaves** to 4 cm; petiole to 1/2 length of leaf; blade green and blue-green on both surfaces, lanceolate, with 2(–3) pairs of primary lateral lobes; primary lobes broadly lanceolate to ovate, apex obtuse to rounded. **Inflorescences:** scape curved, decumbent, often densely hispid, trichomes spreading, dark colored. **Flowers** to 3 cm broad; petals yellow or white. **Capsules** subglobose to obovoid, to 1.5 cm, strigose, trichomes dark brown. The report of $2n = 84$ (as *P. cornwallisensis*) attributed to G. Knaben (1959) by Á. Löve (1962b) cannot be confirmed.

Flowering Jun–Jul. Largely unvegetated, rocky tundra and clayey soils of the northernmost arctic islands; 0–1000 m; N.W.T.; circumpolar.

Similar but poorly known plants from the highest elevations of the St. Elias and Richardson mountains may belong here. *Papaver uschakovii* Tolmatchew & V. V. Petrovsky from arctic Russia also appears to fall within the range of variation described here. Knowledge of this subspecies in Greenland, where evidently it also occurs, is inadequate to permit an accurate account of its distribution there. The names *P. dahlianum* Nordhagen and *P. radicatum* subsp. *dahlianum* (Nordhagen) Rändel, which apply to plants from

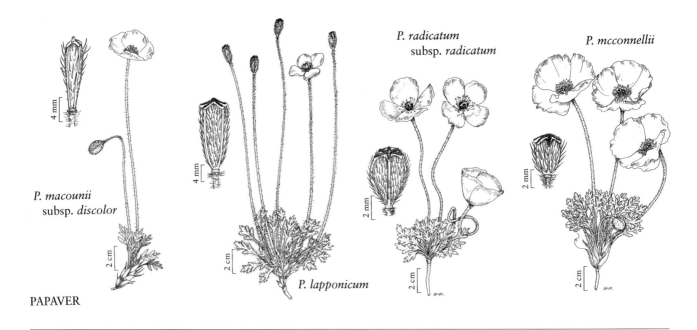

P. radicatum
subsp. *radicatum*

P. mcconnellii

P. macounii
subsp. *discolor*

P. lapponicum

PAPAVER

northernmost Norway and Svalbard, have been applied also to North American specimens of *P. radicatum* subsp. *polare*. It may be that these entities are in fact the same, or that *P. radicatum* subsp. *polare* occurs also in northern Scandinavia along with distinct *P. dahlianum,* but present knowledge is insufficient for a definite determination.

If this taxon, as here circumscribed, is recognized instead at species rank, the correct name is *Papaver cornwallisensis.*

8d. Papaver radicatum Rottbøll subsp. **kluanensis** (D. Löve) D. F. Murray, Novon 5: 294. 1995 E

Papaver kluanensis D. Löve, Bot. Not. 109: 178. 1956; *P. freedmanianum* D. Löve

Plants loosely cespitose, to 1.7 dm, mostly shorter, with short, soft, dull, persistent leaf bases. **Leaves** to 7 cm; petiole to 2/3 length of leaf; blade gray- to blue-green on both surfaces, lanceolate, with 2–3(–4) pairs of primary lateral lobes, distal pairs closely spaced, proximal often somewhat distant; primary lobes lanceolate, apex acute to obtuse. **Inflorescences:** scape erect, white-hispid, often densely so. **Flowers** to 2 cm broad; petals yellow, rarely pink tinged or brick red. **Capsules** ellipsoid-subglobose to oblong-obconic, to 1.2 cm, strigose, trichomes light brown. $2n = 42, 56.$

Flowering Jun–Aug. Dry, rocky alpine ridges; 1500–4000 m; Alta., B.C., Yukon; Alaska, Colo., Idaho, N.Mex., Utah, Wyo.

Plants from the northern part of the range (including the type locality) are taller, with leaf blades $1\times$-lobed, larger flowers, and ellipsoid-subglobose capsules, whereas those from the southern part are consistently shorter, with leaf blades $2\times$-lobed (more deeply so than in *Papaver radicatum* subsp. *radicatum* or subsp. *polare*), smaller flowers, and oblong-obconic capsules. Because of these differences, D. Löve (1969) maintained *P. radicatum* subsp. *kluanensis* as distinct from the Utah (and presumably other southern) populations of the *Papaver radicatum* complex. Nevertheless, we agree with U. Rändel (1975) that the northern and southern populations are linked (in Colorado and Wyoming) by intermediate forms, and hence all are here included within *P. radicatum* subsp. *kluanensis.*

9. Papaver mcconnellii Hultén, Fl. Alaska Yukon 5: 803, fig. 1. 1945 E F

Papaver denalii Gjaerevoll

Plants loosely cespitose, to 1.5 dm. **Leaves** to 10 cm; petiole to 1/2 length of leaf; blade gray- and blue-green on both surfaces, glaucous, lanceolate, $1–2\times$-lobed, commonly with 3 pairs of primary lateral lobes, hirsute; primary lobes obovate to strap-shaped, mostly divided, short, apex acute or more commonly obtuse or rounded. **Inflorescences:** scape erect or decumbent, generally curved, less than 15 cm, hispid. **Flowers** to 6 cm broad; petals yellow, or white with yellow basal spot; anthers yellow; stigmas 5–7, disc conic-umbonate. **Capsules** subglobose to obconic, to 1.5 cm, 1–2.5 times longer than broad, strigose, trichomes dark brown to black. $2n = 28$.

Flowering Jun–Aug. Calcareous mountain summits, ridges, and screes; 1000–1500 m; N.W.T., Yukon; Alaska.

In view of intermediate forms in the Alaska Range, we have broadened the concept of *Papaver mcconnellii* to include *P. denalii*.

10. Papaver alboroseum Hultén, Fl. Kamtchatka 2: 141, plate 3, fig. c. 1928

Plants cespitose, to 1.5 dm. **Leaves** to 4 cm; petiole 1/2 length of leaf or less; blade gray-green on both surfaces, broadly lanceolate, 1–2×-lobed with 1 or 2 pairs of primary lateral lobes, white- to brown-setose; primary lobes obovate to strap-shaped, margins sometimes toothed, apex obtuse-rounded to acute, bristle-tipped. **Inflorescences:** scape often decumbent, bowed, spreading-hispid. **Flowers** to 2.5 cm broad; petals white to rose with yellow basal spot; anthers yellow; stigmas 5–7, disc convex. **Capsules** subglobose to ellipsoid, to 1.3 cm, 1–2 times longer than broad, strigose, trichomes light (ivory). $2n = 28$.

Flowering Jun–Aug. Rocky tundra of ridges and mountain summits, ash and cinder slopes, and in sand and gravel of glacial outwash and river flood plains; 0–2000 m; B.C., Yukon; Alaska; Asia (Russian Far East, Kamchatka).

Papaver alboroseum is infrequent at scattered localities on high mountains within the area mapped. It is locally and unusually abundant in gravels below the terminus of the Portage Glacier, near Anchorage, Alaska. Reports of its presence in arctic Alaska are based on misidentifications of *P. lapponicum*.

11. Papaver pygmaeum Rydberg, Bull. Torrey Bot. Club 29: 159. 1902 [E]

Papaver radicatum Rottbøll var. *pygmaeum* (Rydberg) S. L. Welsh

Plants loosely cespitose, to 1.2 dm. **Leaves** to 5 cm; petiole to 2/3 length of leaf; blade blue-green on both surfaces, broadly ovate, 2×-lobed with 2 pairs of primary lateral lobes, glabrous or sparsely hirsute; primary lobes lanceolate to obovate, apex obtuse or rounded, sometimes bristle-tipped. **Inflorescences:** scape erect or curved, sparsely hispid, trichomes spreading. **Flowers** to 2 cm broad; petals yellow, or orange with yellow

basal spot, or orange-pink; anthers yellow; stigmas 4–5, disc convex. **Capsules** obovoid to obconic, 2–2.5 times longer than broad, to 1.5 cm, strigose, trichomes light (ivory). $2n = 14$.

Flowering Jul–Aug. Mountain summits, ridges, and screes; to 2900 m; Alta., B.C.; Mont.

This well-marked species has its closest relatives within the *Papaver alpinum* complex in the mountains of central and southern Europe, e.g., *P. pyrenaicum* (Linnaeus) Willdenow.

The report of $2n = 42$, attributed by D. Löve and N. J. Freedman (1956) to A. C. Fabergé (1944), is an error; it is based on material from Pikes Peak, Colorado, where *P. pygmaeum* has not been found (D. F. Murray 1995).

12. Papaver walpolei A. E. Porsild, Rhodora 41: 231, plate 552, figs. 4–10. 1939

Papaver walpolei var. *sulphureomaculata* Hultén

Plants densely cespitose, to 1(–2) dm. **Leaves** to 4 cm; petiole to 3/4 length of leaf; blade light green abaxially, dark green adaxially, sometimes glaucous, short-lanceolate, unlobed or 1×-lobed with 1(–2) pair(s) of lateral lobes, glabrous adaxially; terminal lobe rarely with small secondary lobes, apex rounded. **Inflorescences:** scape erect, glabrate to hispid. **Flowers** to 3 cm broad; petals yellow, or white with yellow basal spot; anthers yellow; stigmas 4–5, disc convex. **Capsules** turbinate to ellipsoid-obovoid, to 1 cm, sparsely to densely hirsute, trichomes light brown to black. $2n = 14$.

Flowering late May–Aug. Exposed tundra uplands, especially calcareous fellfield and river gravels; 0–900 m; Yukon; Alaska; Asia (Russian Far East, Chukotka).

13. Papaver gorodkovii Tolmatchew & V. V. Petrovsky, Bot. Zhurn. (Moscow & Leningrad) 58: 1128, fig. 1. 1973

Plants cespitose, sometimes densly so, to 1.5 dm. **Leaves** to 5 cm; petiole 1/2–3/4 length of leaf; blade light green abaxially, dark green adaxially, lanceolate, 1×-lobed with 1, occasionally 2, pairs of lateral lobes, hirsute; terminal lobe occasionally with small secondary lobes, apex obtuse, rounded. **Inflorescences:** scape erect, sparsely to densely hispid. **Flowers** to 3.5 cm broad; petals yellow or white; anthers yellow; stigmas 5–6, disc flat. **Capsules** subglobose to obconic, to 1.2 cm, 1–2.5 times longer than broad, densely hirsute, trichomes dark brown or black.

Flowering Jul–early Aug. Well-drained gravels of floodplain terraces and coastal arctic screes; 0–100 m; Alaska; Asia (Russian Far East).

According to A. I. Tolmatchew and V. V. Petrovsky (1975), this species is known in Alaska also from the Seward Peninsula, presumably based on a specimen at LE (St. Petersburg), which we have not seen.

11e. PAPAVER Linnaeus sect. PAPAVER

Plants annual, caulescent. **Leaf blades** unlobed or 1×-lobed, distal clasping stem. **Flowers:** filaments white, clavate. **Capsules** poricidal, glabrous.

SELECTED REFERENCE   Kadereit, J. W. 1986b. A revision of *Papaver* L. sect. *Papaver* (Papaveraceae). Bot. Jahrb. Syst. 108: 1–16.

14. Papaver somniferum Linnaeus, Sp. Pl. 1: 508. 1753

· Opium poppy, common poppy, pavot (commun)   [F] [W2]

Plants to 15 dm, glabrate, glaucous. **Stems** simple or branching. **Leaves** to 30 cm; blade sometimes sparsely setose abaxially on midrib; margins usually shallowly to deeply dentate. **Inflorescences:** peduncle often sparsely setose. **Flowers:** petals white, pink, red, or purple, often with dark or pale basal spot, to 6 cm; anthers pale yellow; stigmas 5–18, disc ± flat. **Capsules** stipitate, subglobose, not ribbed, to 9 cm, glaucous.

Flowering spring–summer. Fields, clearings, stream banks, railroads, roadsides, and other disturbed sites; 0–1300 m; introduced; Greenland; Alta., B.C., Man., N.B., Nfld., N.S., Ont., Que., Sask.; Ariz., Calif., Conn., Ill., Ky., Maine, Mass., Mich., Minn., Mo., N.H., N.J., N.Mex., N.Y., N.C., N.Dak., Ohio, Oreg., Pa., Tex., Utah, Vt., Va., Wis.; Europe; Asia.

Unknown in the wild, *Papaver somniferum* probably came originally from southeastern Europe and/or southwestern Asia. It has been cultivated for centuries as the source of opium (and its modern derivatives heroin, morphine, and codeine), and also for edible seeds and oil. Various color forms with laciniate and/or doubled petals are grown for ornament. Widely introduced from cultivation and also as a crop weed, it should be expected elsewhere in the flora.

SELECTED REFERENCE   Danert, S. 1958. Zur Systematik von *Papaver somniferum* L. Kulturpflanze 6: 61–88.

11f. PAPAVER Linnaeus sect. RHOEADIUM Spach, Hist. Nat. Vég. 7: 16. 1839

Plants annual, caulescent, sometimes subscapose. **Leaf blades** 1–2×-lobed, distal not clasping stem. **Flowers:** filaments purple, filiform. **Capsules** poricidal, glabrous.

SELECTED REFERENCE   Kadereit, J. W. 1989. A revision of *Papaver* L. section *Rhoeadium* Spach. Notes Roy. Bot. Gard. Edinburgh 45: 225–286.

15. Papaver dubium Linnaeus, Sp. Pl. 2: 1196. 1753

· Long-headed poppy   [F] [W1]

Plants to 7 dm, hirsute to hispid. **Stems** simple or branching. **Leaves** to 20 cm. **Inflorescences:** peduncle proximally spreading-hispid, distally appressed-hispid. **Flowers:** petals orange to red, rarely with dark basal spot, to 3 cm; anthers violet; stigmas 7–9, disc ± flat. **Capsules** sessile or substipitate, narrowly obovoid, usually distinctly ribbed, to 2 cm, 2 times or more longer than broad.

Flowering spring–summer. Fields, glades, dunes, stream banks, marshy areas, railroads, roadsides, and other disturbed sites; 0–900 m; introduced; Greenland; N.B., Ont., Que.; Ark., Conn., Del., D.C., Ill., Kans., Ky., Md., Mass., Mo., N.J., N.Y., N.C., Ohio, Okla., Oreg., Pa., R.I., S.C., Va., W.Va.; Europe; sw Asia.

In its native range, *Papaver dubium* is a tetraploid complex of five subspecies whose morphologies and distributions intersect to a considerable degree (J. W. Kadereit 1989, 1990). Probably several, if not all, of these entities have been introduced in North America, but it is fruitless to try to distinguish them here, where the species has arrived as a crop weed and the subspecies have no geographic integrity.

Papaver dubium sometimes seems to intergrade with

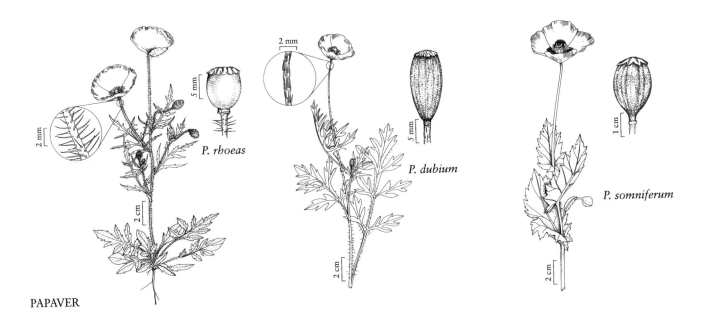

P. rhoeas

P. dubium

P. somniferum

PAPAVER

P. rhoeas, at least in North America. The most readily evident character for distinguishing them reliably is the nature of the distal pubescence on the peduncles—whether spreading or appressed.

16. Papaver rhoeas Linnaeus, Sp. Pl. 1: 507. 1753 · Common poppy, corn poppy, field poppy, Flanders poppy, coquelicot, amapola [F] [W1]

Plants to 8 dm, hispid to setulose. **Stems** simple or usually branching. **Leaves** to 15 cm; distal often somewhat clustered. **Inflorescences:** peduncle sparsely to moderately spreading-hispid throughout. **Flowers:** petals white, pink, orange, or red, often with dark basal spot, to 3.5 cm; anthers bluish; stigmas 5–18, disc ± flat. **Capsules** sessile or substipitate, turbinate to subglobose, obscurely ribbed, to 2 cm, less than 2 times longer than broad.

Flowering spring–summer. Fields, pastures, stream banks, railroads, roadsides, and other disturbed sites; 0–2000 m; introduced; Man., N.B., N.S., Ont., Que., Sask.; Alaska, Calif., Conn., D.C., Idaho, Ill., Iowa, Ky., La., Maine, Md., Mass., Mich., Minn., Mo., Mont., N.H., N.J., N.Mex., N.Y., N.C., Ohio, Oreg., Pa., R.I., Tex., Utah, Vt., Va., Wash., W.Va.; Europe; sw Asia; n Africa.

J. W. Kadereit (1990) suggested that *Papaver rhoeas* originated on the east coast of the Mediterranean, probably derived from one or more of the other species of the section that are native in that region, and only after (and because) "suitable habitats in sufficient extent were provided by man." Various forms with pale pink or white, unspotted, sometimes doubled petals are grown for ornament, notably the Shirley poppies. In North America, the species escapes from cultivation fairly readily and has been introduced also as a crop weed.

Excluded species:

Papaver dahlianum Nordhagen, Bergens Mus. Årbok 2: 46. 1931

Papaver radicatum Rottbøll subsp. *dahlianum* (Nordhagen) Rändel

We regard this species as being restricted to arctic Europe, a narrower circumscription than U. Rändel's (1977).

Papaver microcarpum de Candolle, Syst. Nat. 2: 71. 1821

We are so far unable to substantiate D. Löve's (1969) report of this essentially Asiatic species "from Seward and Kenai peninsulas in Alaska, the Aleutian Islands."

12. ROEMERIA Medikus, Ann. Bot. (Usteri) 1(3): 15. 1792 · [For Johann Jakob Roemer, 1763–1819, Swiss physician and naturalist at Zürich]

Robert W. Kiger

Herbs, annual, caulescent; sap yellow. **Stems** leafy. **Leaves:** basal usually few, rosulate, indistinctly petiolate; cauline usually few and remote, alternate, proximal indistinctly petiolate, distal sessile; blade 2–3×-pinnately lobed. **Inflorescences** axillary and terminal, 1-flowered; bracts present. **Flowers:** sepals 2, distinct; petals 4; stamens many; pistil 2–4(–6)-carpellate; ovary 1-locular; style absent or obscure; stigmas 2–4(–6), radiate. **Capsules** slenderly cylindric or terete, 2–4(–6)-valved, dehiscing from apex almost to base; peduncle straight or often recurved. **Seeds** many, aril absent.

Species 3–5 (1 in the flora): North America, s Europe, sw, sc Asia.

1. Roemeria refracta de Candolle, Syst. Nat. 2: 93. 1821 F W1

Plants to 6 dm, glabrescent to sparsely setose. **Stems** simple or usually branching. **Leaves** to 10 cm; petiole to 2 cm; blade with margins entire; ultimate lobes lanceolate to linear, usually tipped with short, delicate bristles. **Inflorescences:** peduncle to 15 cm. **Flowers:** petals to 4 cm, bright red with black basal spot sometimes edged white; filaments dark violet to blackish. **Capsules** to 10 cm, straight to slightly curved, glabrous to sparsely hirsute or setose, valves usually tipped with distinct awns extending beyond stigmas. **Seeds** gray, reticulate-pitted.

Flowering late spring–early summer. Cultivated fields; 1200–1500 m; introduced; Utah; s Europe; sw, sc Asia.

Roemeria refracta has been introduced in Cache and Box Elder counties in Utah, and perhaps elsewhere in temperate and subtropical North America. It is closely related to, and not always clearly distinct from, *R. hybrida* (Linnaeus) de Candolle, which also ranges from southern Europe to south-central Asia and could be introduced into North America. The most obvious difference between the two species is in petal coloration—bright red with a blackish basal spot sometimes edged with white in *R. refracta,* and violet with or without a blackish basal spot in *R. hybrida*—a distinction that apparently is consistent throughout their generally overlapping native ranges. However, some Asian and southeast European plants, and some of the North American introductions, are intermediate in one or both of the other characters that have been used to separate the two species: the nature of any capsule pubescence, and the presence of awns on the tips of the capsule valves. Typical *R. refracta* has capsules that are either glabrous or appressed-bristly, with valves bearing distinct terminal awns that clearly extend beyond the level of the stigmas, whereas typical *R. hybrida* has capsules that are either glabrous or spreading-bristly, with valves not awn-tipped. In some specimens of *R. refracta,* though, capsule pubescence is spreading or even patent rather than appressed, and in some the valve awns are reduced or absent. As well, some plants otherwise attributable to *R. hybrida* have capsule pubescence that is more appressed than spreading.

13. ROMNEYA Harvey, London J. Bot. 4: 74. 1845 · Matilija-poppy [for Rev. T. Romney Robinson, 1792–1882, Irish astronomer at Armagh and friend of Thomas Coulter, botanist at Dublin]

Curtis Clark

Subshrubs or shrubs, large, glaucous, from creeping rhizomes; sap clear. **Stems** herbaceous above base, branching, leafy, sparsely pilose or glabrate. **Leaves** alternate, petiolate; blade gray-green, 1–2×-pinnately deeply lobed; primary lobes 3–5, lanceolate or ovate. **Inflorescences** terminal, 1-flowered; bracts present. **Flowers:** sepals 3, distinct; petals 6, white, obovate, crinkled; stamens many; anthers yellow; pistil 7–12-carpellate; ovary oblong to ovoid,

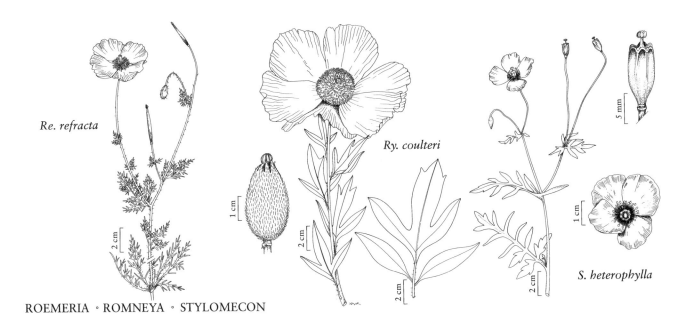

Re. refracta

Ry. coulteri

S. heterophylla

ROEMERIA · ROMNEYA · STYLOMECON

1-locular or incompletely to completely multilocular by placental intrusion; style absent; stigmas 7–12, flattened, coherent proximally, radiate-ascending, velvety. **Fruits** capsular, oblong to ovoid, bristly, dehiscing from apex. **Seeds** many, ovoid, 1.3–1.5 mm, aril absent. $x = 19$.

Species 2 (2 in the flora): California and nw Mexico (Baja California).

Both species of *Romneya*, closely related and sometimes merged, are cultivated widely in warm regions for their showy, white flowers, which are the largest in the family.

1. Calyx glabrous; seeds papillose; petals 6–10 cm; leaves 5–20 cm . . . . 1. *Romneya coulteri*
1. Calyx appressed-pubescent; seeds smooth; petals 4–8 cm; leaves 3–10 cm . . . . 2. *Romneya trichocalyx*

1. Romneya coulteri Harvey, London J. Bot. 4: 75, plate 3. 1845 [E] [F]

Plants 10–25 dm. Leaves 5–20 cm. **Flowers:** calyx distinctly beaked at bud apex, glabrous; petals 6–10 cm. **Capsules** 3–4 cm. **Seeds** dark brown, papillose. $2n = 38$.

Flowering spring–summer (Apr–Aug). Dry washes, canyons; less than 1200 m; Calif.

Romneya coulteri shares with *Hibiscus lasiocarpos* the distinction of having the largest flowers of any plant native to California.

2. Romneya trichocalyx Eastwood, Proc. Calif. Acad. Sci., ser. 3, 1: 133, plate 11. 1898

Romneya coulteri var. *trichocalyx* (Eastwood) Jepson

Plants 10–25 dm. Leaves 3–10 cm. **Flowers:** calyx not or only indistinctly beaked at bud apex, appressed-pubescent; petals 4–8 cm. **Capsules** 2.5–3.5 cm. **Seeds** light to medium brown, smooth. $2n = 38$.

Flowering spring–summer (Apr–Aug). Dry washes, canyons; less than 1100 m; Calif.; Mexico (Baja California).

14. STYLOMECON G. Taylor, J. Bot. 68: 140. 1930 · Wind-poppy [Greek *stylos,* style, and *mekon,* poppy]

Curtis Clark

Herbs, annual, caulescent; sap yellow, clear. **Stems** simple or branching, leafy. **Leaves** alternate, usually petiolate; blade 1–2× pinnately deeply lobed. **Inflorescences** axillary, 1-flowered; peduncle slender; bud drooping. **Flowers:** sepals 2, distinct; petals caducous, 4, obcuneate; stamens many; pistil 4–11-carpellate; ovary turbinate, 1-locular; style persistent; stigma 4–11-lobed. **Fruits** capsular, obovoid-ellipsoid to turbinate, ribbed, dehiscing by flaps beneath apex. **Seeds** many, brown or black, reniform, reticulate-rugose, aril absent. $x = 28$.

Species 1 (1 in the flora): California, nw Mexico.

1. Stylomecon heterophylla (Bentham) G. Taylor, J. Bot. 68: 140. 1930 [F]

Meconopsis heterophylla Bentham, Trans. Hort. Soc. London, ser. 2, 1: 408. 1835; *M. crassifolia* Bentham; *Papaver crassifolium* (Bentham) Greene; *P. heterophyllum* (Bentham) Greene; *P. heterophyllum* var. *crassifolium* (Bentham) Jepson

Plants 30–60 cm, glabrous or sometimes sparsely pilose proximally. **Leaves** 2–12 cm; blade with primary lobes 5–13, remote. **Inflorescences:** peduncle 5–20 cm. **Flowers:** sepals 4–10 mm; petals orange-red, spotted purple at base, 1–2 cm. **Capsules** 1–2 cm. **Seeds** 0.4 mm. $2n = 56$.

Flowering spring (Apr–May). Grassy areas, openings in chaparral; less than 1200 m; Calif.; Mexico (Baja California).

15. HESPEROMECON Greene, Pittonia 5: 146. 1903 · [Greek *hesperos,* evening or western, and *mecon,* poppy]

Gary L. Hannan

Herbs annual, short-caulescent, subscapose, distinctly pubescent, from fibrous roots; sap clear. **Stems** leafy at and near base, erect, branching from base. **Leaves** opposite, sessile; blade unlobed. **Inflorescences** axillary or terminal, 1-flowered; bracts absent; bud nodding. **Flowers:** sepals 3, with overlapping, loosely connivent flaps; petals 6, occasionally more on robust specimens; stamens 12 or more, usually in several series; filaments dilated distally; pistil 3-carpellate; ovary 1-locular; stigmas 3. **Capsules** erect, 3-valved, dehiscing from apex. **Seeds** many, aril absent. $x = 7$.

Species 1 (1 in the flora): California.

1. Hesperomecon linearis (Bentham) Greene, Pittonia 5: 146. 1903 [E] [F]

Platystigma lineare Bentham, Trans. Hort. Soc. London, ser. 2, 1: 407. 1834; *Meconella linearis* (Bentham) A. Nelson & J. F. Macbride; *M. linearis* var. *pulchellum* (Greene) Jepson; *Platystemon linearis* (Bentham) Curran

Plants 0.3–4 dm, sparsely pilose to densely hirsute. **Leaves** 0.5–8.5 cm; blade linear; margins entire. **Inflorescences:** peduncle 2.5–38 cm; bud globose to ovoid-cylindric. **Flowers:** petals white to cream colored, sometimes with yellow tips, sometimes tinged red in age, narrowly ovate to obovate, 0.3–2 × 0.2–1 cm, apex acute to obtuse; ovary ovoid-ellipsoid; stigmas deltoid, margins revolute. **Capsules** ovoid, to 1.5 cm. **Seeds** dark, shining, smooth. $2n = 14$.

Flowering early–late spring. Open, grassy areas in grasslands, sand dunes, oak and pine woodlands; 0–1000 m; Calif.

Depauperate plants of this species may be confused with *Meconella,* but *Hesperomecon* can be distinguished by its linear, pilose leaves and ovoid capsules. Robust plants may be confused with *Platystemon,* but the strictly 3-carpellate, strongly syncarpous gynoecium and valvate capsules of *Hesperomecon* are always diagnostic.

16. MECONELLA Nuttall in J. Torrey and A. Gray, Fl. N. Amer. 1: 64. 1838 · [Greek *mekon,* poppy, and *ella,* diminutive]

Gary L. Hannan

Herbs, annual, caulescent, glabrous or glabrate, from fibrous roots. **Stems** leafy, mostly at base, erect to decumbent, simple or branching. **Leaves:** basal rosulate, winged-petiolate; cauline, subsessile or sessile, proximal whorled, distal whorled or opposite; blades unlobed. **Inflorescences** axillary or terminal, 1-flowered; bracts absent; bud globose, nodding. **Flowers:** receptacle sometimes expanded into small ring beneath calyx; sepals 3, with overlapping, loosely connivent flaps; petals 6; stamens 4–6 in 1 series or ca. 12 in 2 series; filaments usually dilated distally; pistil 3-carpellate; ovary 1-locular, linear-oblong; stigmas 3. **Capsules** erect, linear, greatly elongate at maturity and often spirally twisted, dehiscing apically by separation of valvelike carpels. **Seeds** few, black, shiny, aril absent. $x = 8$.

Species 3 (3 in the flora): far w North America.

1. Receptacle ± as long as broad, not expanded into ring beneath calyx; anthers linear-oblong, usually at least 1/2 length of filaments. . . . . .2. *Meconella denticulata*
1. Receptacle shorter than broad, expanded into ring beneath calyx; anthers ovoid to globose, much shorter than filaments.
  2. Stamens ca. 12, sometimes fewer in depauperate specimens, in 2 series, outer filaments shorter than inner. . . . . 1. *Meconella californica*
  2. Stamens 4–6, in 1 series, filaments all ± equal. . . . . .3. *Meconella oregana*

1. Meconella californica Torrey & Frémont in J. C. Frémont, Rep. Exped. Rocky Mts., 312. 1845   E F

Meconella oregana Nuttall var. *californica* (Torrey & Frémont) Jepson

Plants 0.3–1.8 dm. **Stems** erect to ascending. **Leaves** 3–25 mm; basal with petiole to 17 mm; proximal cauline with blade linear-spatulate; distal with blade broadly linear; margins entire or rarely denticulate. **Inflorescences:** peduncle 3–8 cm. **Flowers:** receptacle shorter than broad, expanded into small ring beneath calyx; petals white to cream colored or outer sometimes yellow, alternately obovate and narrowly ovate, 2–7 × 1–5 mm, apex rounded; stamens ca. (6–)12, in 2 series; outer filaments shorter than inner; anthers ovoid to globose, much shorter than filaments, usually broader. **Capsules** to 50 × 1.5 mm. $2n = 16$.

Flowering late winter–late spring. Moist, sunny slopes in open grassy areas in grasslands, oak or pine woodlands; 0–1000 m; Calif.

The yellow color that occurs in the outer petals of some plants fades at night and returns in daylight.

2. Meconella denticulata Greene, Bull. Calif. Acad. Sci. 2: 59. 1886   E

Meconella oregana Nuttall var. *denticulata* (Greene) Jepson

Plants 0.3–2.1 dm. **Stems** erect to decumbent. **Leaves** 2–30(–40) mm; basal with petiole to 3 cm; proximal cauline with blade linear-spatulate; distal with blade broadly linear to ovate; margins entire or occasionally denticulate. **Inflorescences:** peduncle 1.5–5 cm. **Flowers:** receptacle ± as long as broad, not expanded into ring beneath calyx; petals white, occasionally with obscure yellowish green patch near base, narrowly ovate, nearly equal, 2–6 × 1–2 mm, apex rounded; stamens 6 in 1 series; filaments ± equal, nearly as broad as anthers, tapering from base; anthers linear-oblong, usually at least 1/2 length of filaments. **Capsules** to 30 × 2.5 mm. $2n = 16$.

Flowering early–late spring. Moist, partly shaded slopes in chaparral or oak-pine woodlands; 100–1000 m; Calif.

Plants from the southern end of the range tend to have longer anthers and shorter filaments.

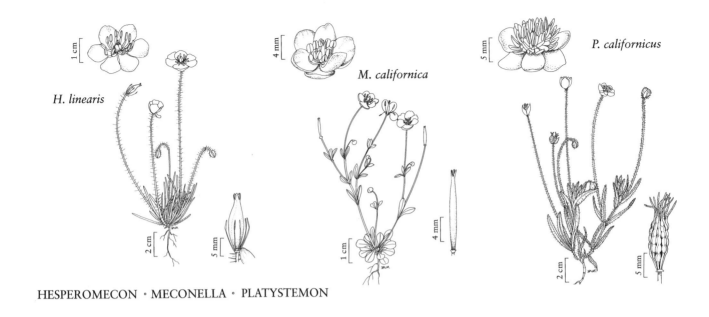

HESPEROMECON · MECONELLA · PLATYSTEMON

3. Meconella oregana Nuttall in J. Torrey and A. Gray, Fl. N. Amer. 1: 64. 1838 [C] [E]

Plants 0.2–1.6 dm. **Stems** erect to ascending. **Leaves** 3–18 mm; basal with petiole to 10 mm; proximal cauline with blade linear-spatulate; distal with blade broadly linear; margins entire. **Inflorescences:** peduncle 2–8 cm. **Flowers:** receptacle shorter than broad, expanded into small ring beneath calyx; petals white, alternately obovate and oblanceolate, 1–5 × 1–3 mm, apex rounded; stamens 4–6, in 1 series; filaments ± equal, usually as broad as anthers; anthers ovoid, minute, much shorter than filaments. **Capsules** to 25 × 1.5 mm. $2n = 16$.

Flowering early–late spring. Sandy bluffs, meadows and partly sunny, moist banks; of conservation concern; 0–300 m; B.C.; Calif., Oreg., Wash.

Flowers of *Meconella oregana* often display irregularities such as fusion, loss, or addition of parts (W. R. Ernst 1962). Some specimens from central California (Alameda and Contra Costa counties) are difficult to assign with certainty to either *M. oregana* or *M. californica*. Depauperate plants of the latter species sometimes can be distinguished from *M. oregana* only by their unequal and more numerous stamens (W. R. Ernst 1967).

17. PLATYSTEMON Bentham, Trans. Hort. Soc. London, ser. 2, 1: 405. 1834 · Creamcups [Greek *platus*, broad, and *stemon*, stamen]

Gary L. Hannan

Herbs, annual, caulescent, usually distinctly pubescent, from fibrous roots. **Stems** leafy, erect to decumbent, branching distally. **Leaves** mostly or all cauline, opposite or whorled, sessile; blade unlobed. **Inflorescences** axillary or terminal, 1-flowered; bracts absent; bud nodding. **Flowers:** receptacle slightly expanded beneath calyx; sepals 3, with overlapping, loosely connivent flaps; petals 6, occasionally more on robust specimens; stamens many in several series; filaments dilated distally; pistil 6–many-carpellate; ovary with each carpel forming almost closed locule with 4–24 ovules; stigmas 1 per carpel. **Capsules** erect or sometimes nodding, carpels dissociating, each torulose and breaking transversely into 1-seeded segments, sometimes also releasing free seeds. **Seeds** usually many, aril absent. $x = 6$.

Species 1 (1 in the flora): w United States; Mexico (Baja California).

SELECTED REFERENCE   Hannan, G. L. 1982. Correlation of morphological variation in *Platystemon californicus* (Papaveraceae) with flower color and geography. Syst. Bot. 7: 35–47.

1. **Platystemon californicus** Bentham, Trans. Hort. Soc. London, ser. 2, 1: 405. 1834 (as **californicum**) [F]

Platystemon arizonicus Greene; *P. australis* Greene; *P. californicus* var. *ciliatus* Dunkle; *P. californicus* var. *crinitus* Greene; *P. californicus* var. *horridulus* (Greene) Jepson; *P. californicus* var. *nutans* K. Brandegee; *P. californicus* var. *ornithopus* (Greene) Munz; *P. mohavensis* Greene

Plants 0.3–3 dm, pilose or hirsute, sometimes glabrate. **Leaves** 10–90 × 1.7–8.1 mm; blade broadly linear; margins entire; apex rounded to long-acute. **Inflorescences:** peduncle 3.4–25.8 cm; bud globose to ovoid-cylindric. **Flowers:** petals white to cream colored, sometimes with yellow tip and/or base, rarely gold overall, sometimes tinged red in age, narrowly ovate to obovate, 6–19 × 3.5–16 mm, apex acute to rounded; ovary cylindric to oblong-ellipsoid; stigmas linear, margins revolute. **Capsules** ellipsoid, to 1.6 cm. **Seeds** black, shining, smooth. $2n = 12$ (plus occasional supernumerary chromosomes).

Flowering early–late spring. Open, grassy areas with loose or disturbed soil or following burns; 0–1000(–2000) m; Ariz., Calif., Nev., Oreg., Utah; Mexico (Baja California).

This highly variable, wind-pollinated taxon has been split into as many as 57 species on the basis of characteristics showing little cohesiveness. Ecotypic variation has produced morphologic extremes ranging from semisucculent, nearly glabrous coastal forms to very robust, moderately pubescent plants of interior grassland to compact, densely pubescent plants of semidesert habitats (G. L. Hannan 1979, 1982). Several varieties are recognized in some currently used floras: *Platystemon californicus* var. *ciliatus* Dunkle, from Santa Barbara Island; *P. californicus* var. *nutans* K. Brandegee, from coastal San Diego County and Santa Rosa and Santa Cruz islands; and *P. californicus* var. *ornithopus* (Greene) Munz, from San Miguel, San Nicholas, and Santa Rosa islands. These geographically restricted morphotypes appear to result from the same sort of ecotypic variation found in many other parts of the range. Rather than naming each ecotype, it seems best to treat *Platystemon* as a single, highly variable species with many locally adapted, intergrading populations.

20. FUMARIACEAE Linnaeus

• Fumitory Family

Kingsley R. Stern

Herbs, annual or perennial, scapose or caulescent, from taproots, bulblets, tubers, or rhizomes; sap clear. **Stems** when present leafy, erect to prostrate or climbing, simple or branching. **Leaves** basal and/or cauline, alternate, mostly compound, sometimes simple, without stipules, petiolate; blade with 2–6 odd-pinnate orders of leaflets and/or lobes. **Inflorescences** terminal, axillary, extra-axillary, or leaf-opposed, unifloral or else multifloral and thyrsoid, paniculate, racemose, or corymbose; peduncles present; bracts present. **Flowers** bilaterally symmetric about 1 plane or each of 2 perpendicular planes; pedicel present; sepals caducous or persistent, 2, thin; petals 4, distinct or coherent basally to almost completely connate, in 2 whorls of 2; outer petals alike or dissimilar, 1 or both sometimes swollen or spurred basally; inner petals alike, apically connate, clawed, with somewhat hollow, membranous, wrinkled, abaxial median crests; stamens 6, in 2 bundles of 3 each, opposite outer petals; filaments of each bundle partially to completely connate, sometimes basally adnate to petals, with basal nectariferous tissue often in form of spur; anthers connivent, adhering to stigma, median anthers 2-locular, lateral anthers 1-locular; pistil 1, 2-carpellate; ovary 1-locular; placentae parietal; style threadlike, rigid; stigma 1, compressed, with 2 lobes or apical horns, and/or 2–8 papillar stigmatic surfaces. **Fruits** capsular, indehiscent or dehiscent and valvate. **Seeds** 1–many, small, elaiosome (oil-bearing appendage) often present.

Genera 19, species ca. 450 (4 genera, 23 species in the flora): North America, Eurasia.

The genera of Fumariaceae are distributed mostly in the Old World and primarily in temperate Eurasia. One acaulescent species of *Dicentra* occurs in Siberia, Kamchatka, and Japan; a caulescent species is found in western China and northern Burma; and nine climbing species are distributed throughout the Himalayan area and Burma. More than 400 taxa of *Corydalis* and 50 of *Fumaria,* distributed primarily throughout temperate, often montane, regions of Eurasia and Africa, have been described. *Adlumia* comprises only two species, which are quite similar morphologically, one from North America and the other from East Asia.

Most European and some American systematists treat Fumariaceae as a subfamily of Papaveraceae. However, although a few taxa are morphologically intermediate, the members of

340

Fumariaceae generally are quite distinct from those of Papaveraceae in several respects, including floral symmetry, sap character, and stamen number and fusion.

SELECTED REFERENCES Ernst, W. R. 1962. The genera of Papaveraceae and Fumariaceae in the southeastern United States. J. Arnold Arbor. 43: 315–343. Fedde, F. 1936. Papaveraceae. In: H. G. A. Engler et al., eds. 1924+. Die natürlichen Pflanzenfamilien, ed. 2. 26+ vols. Leipzig and Berlin. Vol. 17b, pp. 5–145. Gunn, C. R. 1980. Seeds and fruits of Papaveraceae and Fumariaceae. Seed Sci. Technol. 8: 3–58. Hutchinson, J. 1921. The genera of the Fumariaceae and their distribution. Bull. Misc. Inform. Kew 1921: 97–115. Lidén, M. 1986. Synopsis of Fumarioideae (Papaveraceae) with a monograph of the tribe Fumarieae. Opera Bot. 88: 1–133. Rachelle, L. D. 1974. Pollen morphology of the Papaveraceae of the northeastern U.S. and Canada. Bull. Torr. Bot. Club. 101: 152–159. Ryberg, M. 1960. Studies in the Morphology and Taxonomy of the Fumariaceae. Uppsala. Ryberg, M. 1960b. A morphological study of the Fumariaceae and the taxonomic significance of the characters examined. Acta Hort. Berg. 19: 121–248.

1. Petals almost completely connate, spongy; plants climbing, petiolules and reduced leaflets twining and tendril-like . . . . .2. *Adlumia*, p. 347
1. Petals coherent or connate only basally, not spongy; plants not climbing.
  2. Both outer petals swollen or spurred basally . . . . .1. *Dicentra*, p. 341
  2. Only 1 outer petal swollen or spurred basally.
    3. Fruit an elongate, dehiscent capsule; seeds more than 1, with elaiosome . . . . .3. *Corydalis*, p. 348
    3. Fruit a ± globose, indehiscent capsule; seeds 1, without elaiosome. . . . . .4. *Fumaria*, p. 356

1. DICENTRA Bernhardi, Linnaea 8: 457, 468. 1833, name conserved · Bleeding-heart, dicentre [Greek *dis*, twice, and *kentron*, spur]

Kingsley R. Stern

Bikukulla Adanson; *Capnorchis* Miller

Herbs, annual or perennial, scapose or caulescent, from taproots, bulblets, tubers, or rhizomes. **Stems** when present erect, simple or branching, hollow at maturity. **Leaves** basal or cauline, compound; blade with 2–4 orders of leaflets and lobes; margins entire, crenate, or serrate; surfaces glabrous, sometimes glaucous. **Inflorescences** axillary, extra-axillary, leaf-opposed, or terminal, unifloral or else multifloral and thyrsoid, paniculate, racemose, or corymbose. **Flowers** bilaterally symmetric about each of 2 perpendicular planes; sepals caducous; corolla cordate to oblong in outline; petals coherent or connate only basally, not spongy; outer petals both swollen or spurred basally, usually keeled apically; inner petals with blade fiddle-, spoon-, or arrowhead-shaped, claw linear-oblong to oblanceolate; stamens with nectariferous tissue borne on median filament in each bundle and sometimes forming spur or loop that projects into swollen base of adjacent outer petal; ovary broadly ovoid or obovoid to narrowly cylindric; stigma persistent, with 2 lobes or apical horns, sometimes also with 2 lateral papillae. **Capsules** indehiscent or dehiscent and 2-valved. **Seeds** few–many, elaiosome usually present. $x = 8$.

Species 20 (9 in the flora): temperate North America and eastern Asia.

About 35 isoquinoline alkaloids have been isolated from Fumariaceae, and such compounds are present in the tissues of all species. Some of these alkaloids have been used medicinally, mostly in the past. The drug complex corydalis, which contains several alkaloids extracted from the bulblets of *Dicentra canadensis* and *D. cucullaria*, has been used as a healing agent in chronic skin diseases, as a tonic and diuretic, and in the treatment of syphilis. The alkaloid bulbocapnine, obtained from all parts of *D. canadensis*, has been used in the treatment of Ménière's disease and muscular tremors, and as a pre-anaesthetic. Cattle find *D. cucullaria* and *D. canadensis* distasteful and usually do not ingest the plants unless suitable forage is unavailable; when they do, however, the toxic alkaloid cucullarine brings about local anaesthesia, narcosis, convulsions, and death. A decoction from the rhizome of *D. formosa* has been used in the Pacific Northwest to expel intestinal worms (D. E. Moerman 1986).

Dicentra spectabilis (Linnaeus) Lemaire is cultivated through much of the flora area. It was introduced in Europe only in the middle of the 19th century, but it has been cultivated for centuries in temperate China and Japan, where it is now so widespread that the limits of its natural distribution are obscure. It does not appear to be truly naturalized in North America, but it may be encountered as a transitory garden relic or escape. It differs from *D. ochroleuca* and *D. chrysantha* in having rose-purple to pink or sometimes white outer petals, pendent flowers, and reticulate seeds with elaiosomes.

SELECTED REFERENCES Berg, R. Y. 1969. Adaptation and evolution in *Dicentra* (Fumariaceae), with special reference to seed, fruit, and dispersal mechanism. Nytt Mag. Bot. 16: 49–75. Fahselt, D. 1970. The anthocyanins of *Dicentra* (Fumariaceae). Canad. J. Bot. 48: 49–53. Fahselt, D. and M. Ownbey. 1968. Chromatographic comparison of *Dicentra* species and hybrids. Amer. J. Bot. 55: 334–345. Stern, K. R. 1961. Revision of *Dicentra* (Fumariaceae). Brittonia 13: 1–57. Stern, K. R. 1962. The use of pollen morphology in the taxonomy of *Dicentra*. Amer. J. Bot. 49: 362–368. Stern, K. R. 1970. Pollen aperture variation and phylogeny in *Dicentra*. Madroño 20: 354–359. Stern, K. R. and M. Ownbey. 1971. Hybridization and cytotaxonomy of *Dicentra*. Amer. J. Bot. 58: 861–866.

1. Plants with evident stems, not scapose.
  2. Petals dull to yellowish white, 15–30 mm; inflorescences subglobose; seeds ca. 1.3 mm diam. (rarely to 2 mm diam.). . . . . 1. *Dicentra ochroleuca*
  2. Petals golden yellow, 10–22 mm; inflorescences elongate; seeds 1.5–2.2 mm diam . . . . 2. *Dicentra chrysantha*
1. Plants without evident stems, scapose.
  3. Plants with elongate rhizomes; bulblets or tubers absent.
    4. Reflexed portions of outer petals 4–8 mm; e United States . . . . 4. *Dicentra eximia*
    4. Reflexed portions of outer petals 2–5 mm; w North America.
      5. Stamen bundles with median filament essentially connate with others . . . . 3. *Dicentra formosa*
      5. Stamen bundles with median filament forming distinct angular loop extending outward above base. . . . . 5. *Dicentra nevadensis*
  3. Plants without elongate rhizomes (occasionally with rhizomes in *D. pauciflora*); short, bulblet- or tuber-bearing rootstock, or cluster of spindle-shaped tubers, or combination of tubers and bulblets present.
    6. Plants with well-developed bulblets; nectariferous tissue forming spur.
      7. Bulblets globose, yellow; nectariferous spur 1 mm or less . . . . 8. *Dicentra canadensis*
      7. Bulblets teardrop-shaped, pink to white; nectariferous spur 1–3(–4.5) mm. . . . . 9. *Dicentra cucullaria*
    6. Bulblets absent or minute (less than 1 mm diam.); nectariferous tissue not forming spur.
      8. Flowers 1–3 per inflorescence; non-reflexed portions of outer petals usually 12 mm or more; blades of inner petals narrowly spoon-shaped, never triangular or lanceolate . . . . 6. *Dicentra pauciflora*
      8. Flowers solitary; non-reflexed portions of outer petals usually less than 7 mm; blades of inner petals triangular to lanceolate or spoon-shaped . . . . 7. *Dicentra uniflora*

1. Dicentra ochroleuca Engelmann, Bot. Gaz. 6: 223. 1881 · White eardrops E

Plants perennial, caulescent, glaucous, from stout taproots. Stems 1 or more, rigidly erect, 10–20(–40) dm, 2–3 cm or more diam. at base. Leaves (12–)25–35(–50) × (6–)10–20(–35) cm; blade with 4 orders of leaflets and lobes; ultimate leaflets (10–)20–30(–60) × (5–)8–15(–30) mm, ultimate lobes linear-lanceolate. Inflorescences terminal, paniculate, 5–many-flowered, subglobose; bracts nearly round to ovate, 4–8 × 3–7 mm, margins entire. Flowers erect; pedicel rigid, 2–10 mm; sepals nearly round to ovate, (5–)7–10(–12) × 3–7(–9) mm; petals dull to yellowish white; outer petals purple-tipped, (15–)22–26(–30) mm, reflexed portion 5–12 mm; inner petals (15–)20–22(–24) mm, claw 6–9 mm and 2/3 width of blade, crest 3–5 mm diam., exceeding apex by 2–4 mm; filaments of each bundle connate from base to shortly below anthers, rarely distinct from near base; nectariferous tissue borne at base of median filament, not projecting into outer petal; stigma shallowly 2-horned, ca. 2 times broader than long. Capsules ovoid, attenuate at both ends, (10–)20–30(–35)

× 5–8 mm. **Seeds** slightly reniform, ca. 1.3 (rarely to 2) mm diam., densely covered with tiny protuberances, elaiosome absent. 2*n* = 32.

Flowering early spring–late summer. Dry gravelly hillsides, gullies, and disturbed areas, often invading after fire; 15–2200 m; Calif.

The seeds of *Dicentra ochroleuca* usually do not germinate unless desiccated or seared by fire.

2. **Dicentra chrysantha** (Hooker & Arnott) Walpers, Repert. Bot. Syst. 1: 118. 1842 · Golden eardrops

Dielytra chrysantha Hooker & Arnott, Bot. Beechey Voy., 320, plate 73. 1838

Plants perennial, caulescent, glaucous, from stout taproots. **Stems** 1 or more, rigidly erect, 5–15 dm, 1–2.5 cm diam. at base. **Leaves** (10–)15–30(–45) × (3–)5–9(–16) cm; blade with 3, rarely 4 orders of leaflets and lobes; ultimate leaflets (10–)20–30(–60) × (5–)10–20(–25) mm, ultimate lobes oblong, apex acute. **Inflorescences** paniculate, usually at least 5 times longer than broad; bracts lanceolate to ovate, 3–5 × 1–2 mm, margins entire; bracteoles absent, rarely present. **Flowers** erect, with slightly pungent odor; pedicels rigid, 2–10 mm; sepals ovate to cordate or nearly round, 3–7 × 2–5 mm; petals golden yellow; outer petals (10–)12–16(–22) × 2–4 mm, reflexed portion 4–7 mm; inner petals (8–)15–17(–18) mm, blade 3–5 mm, claw 6–9 mm, confluent with and nearly as wide as blade, crest usually about 1–3 mm diam., scarcely exceeding apex; filaments of each bundle connate from base to shortly below anthers, rarely distinct from near base; nectariferous tissue borne at base of median filament, not projecting into outer petal; stigma shortly 2-horned, ca. 2 times broader than long. **Capsules** ovoid, attenuate at both ends or sometimes rounded at base, (5–)15–25(–32) × 5–8 mm. **Seeds** slightly reniform, 1.5–2.2 mm diam., densely covered with tiny protuberances, elaiosome absent. 2*n* = 32. Artificially produced hybrids between *Dicentra ochroleuca* and *D. chrysantha* also had 2*n* = 32.

Flowering early spring–late summer. Dry gravelly hillsides, gullies, and disturbed areas, often invading after fire; 100–2200 m; Calif.; Mexico (Baja California).

The seeds of *Dicentra chrysantha* usually do not germinate unless desiccated or seared by fire.

3. **Dicentra formosa** (Haworth) Walpers, Repert. Bot. Syst. 1: 118. 1842 · Pacific bleeding-heart [E]

Fumaria formosa Haworth, Bot. Repos. 6: plate 393. 1800; *Dicentra saccata* (Nuttall ex Torrey & A. Gray) Walpers

Plants perennial, scapose, from elongate, stout rhizomes. **Leaves** (15–)25–40(–55) × (8–)12–20(–35) cm; blade with 3–5 orders of leaflets and lobes; abaxial surface and sometimes adaxial surface glaucous; penultimate lobes oblong, distal ones usually coarsely 3-toothed at apex, (4–)10–20(–50) × (1.5–)3–4(–8) mm. **Inflorescences** paniculate, 2–30-flowered, usually exceeding leaves; bracts linear-lanceolate, 4–7(–12) × 1–2 mm, apex acuminate. **Flowers** pendent; sepals lanceolate to ovate or nearly round, 2–7 × 2–3 mm; petals rose-purple, pink, cream, or pale yellow, rarely white; outer petals (12–)16–19(–24) × 3–6 mm, reflexed portion 2–5 mm; inner petals (12–)15–18(–22) mm, blade 2–4 mm wide, claw linear-elliptic to linear-lanceolate, 7–10(–12) × 1–2 mm, crest 1–2 mm diam., exceeding apex by 1–2 mm; filaments of each bundle connate from base to shortly below anthers except for a 2–3 mm portion of median filament just above base; nectariferous tissue borne along distinct portion of median filament; style 3–9 mm; stigma rhomboid, 2-horned. **Capsules** oblong, 4–5 mm diam. **Seeds** reniform, ca. 2 mm diam., finely reticulate, elaiosome present.

Subspecies 2 (2 in the flora): w North America.

Andrews has been cited almost universally as the author of *Fumaria formosa*. However, Haworth's authorship of the sixth volume of Andrews' *Botanists' Repository* (in which this species was originally described) generally has been overlooked, and it was actually Haworth who first delineated *F. formosa* (W. T. Stearn 1944).

Early attempts to cross *Dicentra formosa* with *D. eximia* (2*n* = 16) failed, possibly because the *D. formosa* parents were tetraploids. Several later hybrids between the two species received plant patents and have become widely marketed throughout the flora area and elsewhere (K. R. Stern 1961, 1968; K. R. Stern and M. Ownbey 1971).

Both subspecies, as well as hybrids between them and *Dicentra eximia,* are widely cultivated.

1. Petals rose-purple to pink, rarely white; leaf blades adaxially not glaucous (rarely glaucescent) . . . . 3a. *Dicentra formosa* subsp. *formosa*
1. Petals cream-colored or rarely pale yellow, rose-tipped; leaf blades adaxially distinctly glaucous . . . . 3b. *Dicentra formosa* subsp. *oregana*

3a. Dicentra formosa (Haworth) Walpers subsp.
formosa [E] [F]

Dicentra formosa (Haworth) Walpers
var. *breviflora* L. F. Henderson

Leaf blades abaxially glaucous,
adaxially not glaucous (rarely
glaucescent). **Flowers:** petals
rose-purple to pink, rarely
white. $2n = 16, 32$.

Flowering early spring–early
fall. Loam or gravel soils in
moist woods and clearings, and along banks of
streams; 0–2250 m; B.C.; Calif., Oreg., Wash.

This subspecies occurs in two chromosomal races:
tetraploids ($2n = 32$), distributed from the Cascade
Mountains of Oregon southward throughout the
Coast Ranges to central California, and diploids ($n =
16$), distributed from Vancouver Island and British
Columbia southward through the Cascades and Coast
Ranges, and along the western slopes of the Sierra Nevada to Monterey and Tulare counties, California. The
flowers of both races vary appreciably in color, shape,
and size.

3b. Dicentra formosa (Haworth) Walpers subsp. **oregana**
(Eastwood) Munz, Aliso 4: 91. 1958 [E]

Dicentra oregana Eastwood, Proc.
Calif. Acad. Sci., ser. 4, 20: 144. 1931

Leaf blades distinctly glaucous
on both surfaces. **Flowers:** petals
cream-colored, rarely pale yellow, rose-tipped. $2n = 32$.

Flowering early spring–early
fall. Open to shaded moist
woods and clearings, in loam or
gravelly soils; 500–2000 m; Calif. (Del Norte and
Siskiyou counties), Oreg. (Curry and Josephine counties).

4. Dicentra eximia (Ker Gawler) Torrey, Fl. New York 1:
46. 1843 · Eastern bleeding-heart [E]

Fumaria eximia Ker Gawler, Bot. Reg.
1: plate 50. 1815; *Bicuculla eximia*
(Ker Gawler) Millspaugh

Plants perennial, scapose, from
elongate, stout, scaly rhizomes.
Leaves (10–)20–35(–55) × (5–)
10–15(–30) cm; blade with 4
orders of leaflets and lobes; abaxial surface glaucous; penulti-
mate lobes lanceolate to oblong or ovate, (6–)10–20
(–35) × 2–5 mm. **Inflorescences** paniculate, 5–many-
flowered, usually exceeding leaves, (20–)30–45(–65)
cm; bracts lanceolate, 3–6(–11) × 1–2 mm, apex acu-

minate. **Flowers** pendent; sepals reniform, 2–5(–8) ×
1.5–4 mm, apex acuminate; petals rose-purple to pink,
rarely white; outer petals (15–)20–25(–30) × 2–5
mm, reflexed portion 4–8 mm; inner petals (15–)18–
22(–25) mm, blade 2–4 mm, claw linear-lanceolate, 5–
10(–14) × 1–2.5 mm, crest 1–3 mm diam., exceeding
apex by 2–3 mm; filaments of each bundle connate at
base and near apex, distinct in between, distinct portion of median filament forming loop that lies within
base of outer petal; nectariferous tissue borne toward
base of median filament; style 7–14 mm; stigma 2-
horned. **Capsules** oblong to ovoid, (15–)18–22(–27)
× ca. 4 mm. **Seeds** slightly reniform, ca. 2 mm diam.,
finely reticulate, elaiosome present. $2n = 16$.

Flowering mid spring–early fall. Dry to moist,
rocky, mountain woods, often in rock crevices at cliff
bases; 100–1700 m; Md., N.J., N.C., Pa., Tenn., Va.,
W.Va.

The natural range of *Dicentra eximia* extends along
the Appalachians from North Carolina and Tennessee
to Maryland and Pennsylvania. It is frequently cultivated and sometimes escapes outside that area, but it
evidently has not become truly naturalized beyond it.
Such garden escapes, perhaps including misidentified
plants of *D. formosa,* also widely cultivated, are almost surely the basis for reports of *D. eximia* from Illinois, Michigan, Ohio, New York, Connecticut, and
Vermont.

Several patented hybrids between *Dicentra eximia*
and *D. formosa* are sold in nurseries.

5. Dicentra nevadensis Eastwood, Proc. Calif. Acad. Sci.,
ser. 4, 20: 143. 1931 [C] [E]

Dicentra formosa (Haworth) Walpers
subsp. *nevadensis* (Eastwood) Munz

Plants perennial, scapose, variably glaucous, from elongate,
stout rhizomes. **Leaves** (10–)15–
25(–30) × (5–)8–12(–18) cm;
blade with 3–4 orders of leaflets
and lobes; penultimate lobes oblong, distal usually coarsely 3-
toothed at apex, (3–)6–12(–20) × 1.5–4 mm. **Inflorescences** paniculate, 2–20-flowered, shorter than to
exceeding leaves; bracts linear-lanceolate, 4–7(–10) ×
1–1.5 mm, apex acuminate. **Flowers** pendent; sepals
ovate to acuminate-lanceolate, (3–)6–7(–12) × 1–3
mm; petals white to pale yellow or rose-tinted; outer
petals 12–18 × 2–4 mm, reflexed portion 3–5 mm;
inner petals 11–17 mm, blade 2.5–3.5 mm, claw
linear-elliptic to linear-lanceolate, 6–9 × 1–2 mm,
crest 1–2 mm diam., exceeding apex by 1–2 mm; filaments of each bundle connate from base to shortly
below anthers except for a 2–6 mm portion of median

filament just above base, distinct portion of median filament forming angular loop that projects into base of outer petal; nectariferous tissue borne along loop; style 4–7 mm; stigma rhomboid, 2-horned. **Capsules** oblong, (10–)13–16(–20) × 4–5 mm. **Seeds** reniform, ca. 2 mm diam., finely reticulate, elaiosome present. $2n = 16$.

Flowering early–late summer. High meadows, in gravelly soils; of conservation concern; 2100–3300 m; Calif. (Tulare County).

In *Dicentra nevadensis* the median filament of each stamen bundle bends out in an angular loop between the base and midpoint; in *D. formosa* the median filaments lack such loops. Also, the flowers of *D. nevadensis* are smaller and narrower than those of *D. formosa*. Pressed flowers of *D. nevadensis* often turn black, suggesting possible chemical differences, other than in alkaloids, from *D. formosa*.

6. **Dicentra pauciflora** S. Watson, Bot. California 2: 429. 1880 E

Plants perennial, scapose, from rhizomes or clusters of spindle-shaped tubers, bulblets often present at proximal ends of tubers or along rhizomes. **Leaves** (7–)9–13(–16) × 3–7(–10) cm; petiole (2–)4–7(–10) cm; blade with 3–4 orders of leaflets and lobes; ultimate lobes linear-lanceolate, (2–)7–13(–18) × 1.5–3 mm, occasionally irregular, minutely apiculate. **Inflorescences** racemose, 1–3-flowered, barely exceeding leaves; bracts ovate, 4–5 × 2–3 mm. **Flowers** erect to nodding; pedicels 5–25 mm; sepals ovate to lanceolate, 5–8 × 2–4 mm; outer petals white to pink, (15–)18–22(–25) × 3–6 mm, reflexed portion (5–)7–8(–11) mm; inner petals purple, (15–)18–22(–24) mm, blade spoon-shaped, 2–3 mm, claw obovate-elliptic, ca. 10 × 3–4 mm, crest absent; filaments of each bundle connate at base and near apex, distinct in between, distinct portion of median filament forming loop that almost doubles back to its proximal end; nectariferous tissue borne at lowermost point of loop and often extending to base of median filament; style 7–11 mm; stigma 2-lobed, much reduced, ca. 2 times broader than style. **Capsules** spindle-shaped to ovoid, 10–15 × 4–6 mm. **Seeds** reniform, ca. 2 mm diam., smooth, elaiosome present. $2n = 16$.

Flowering late spring–late summer. Openings in coniferous forests, in volcanic and granitic soils; 1200–2700 m; Calif., Oreg.

7. **Dicentra uniflora** Kellogg, Proc. Calif. Acad. Sci. 4: 141. 1871 · Steer's-head E F

Plants perennial, scapose, from clusters of club- to spindle-shaped tubers, small bulblets often present at proximal ends of tubers. **Leaves** 4–7(–10) × (1–)3–4 cm; petiole 2–5(–8) cm; blade with 3–4 orders of leaflets and lobes; surfaces pubescent near base, rarely sparsely pubescent throughout; ultimate lobes variable in shape, often oblong to spoon-shaped, (2–)6–12(–18) × 1.5–6 mm, minutely apiculate, occasionally retuse. **Inflorescences** 1-flowered, barely exceeding to shorter than leaves; bracts 1 (rarely 2) per scape, lanceolate, 4–6 × 2–3 mm. **Flowers** nodding to erect; pedicels 2–7 mm; petals pink to white, suffused light brown; outer petals (7–)12–16(–20) mm, reflexed portion (4–)6–8(–10) mm; inner petals purple-tipped, (11–)13–15(–17) mm, blade triangular to lanceolate or spoon-shaped, 4–8 mm, claw linear-oblong, 4–6 mm, crest absent; filaments of each bundle usually distinct above base, occasionally connate just below anthers, median filament forming loop with slight indentation near base; nectariferous tissue borne along basal part of loop; style 2–5 mm; stigma shallowly 2-lobed, much reduced, slightly broader than style. **Capsules** ovoid, attenuate to style, 9–14 × 5–9 mm. **Seeds** reniform, ca. 1.5 mm diam., finely and obscurely reticulate, elaiosome absent. $2n = 16$.

Flowering very early spring–late summer. Rocky slopes and hillsides in gravelly soils, often flowering near edges of melting snowbanks; 1500–2200(–3300) m; B.C.; Calif., Idaho, Mont., Nev., Oreg., Utah, Wash., Wyo.

8. **Dicentra canadensis** (Goldie) Walpers, Repert. Bot. Syst. 1: 118. 1842 · Squirrel-corn, dicentre du Canada E

Corydalis canadensis Goldie, Edinburgh Philos. J. 6: 329. 1822; *Bicuculla canadensis* (Goldie) Millspaugh

Plants perennial, scapose, from short rootstocks bearing yellow, globose bulblets. **Leaves** (10–)14–24(–30) × (4–)6–14(–18) cm; petiole (5–)8–16(–22) cm; blade with 4 orders of leaflets and lobes; abaxial surface glaucous; ultimate lobes linear to linear-elliptic or linear-obovate, (2–)5–15(–23) × (0.4–)2–4 mm, usually minutely apiculate. **Inflorescences** racemose, 3–12-flowered, usually exceeding leaves, (10–)15–27(–33) cm; bracts ovate, 2–5 × 1–3 mm. **Flowers** pendent, very fragrant; pedicels (2–)3–7(–14) mm; sepals trian-

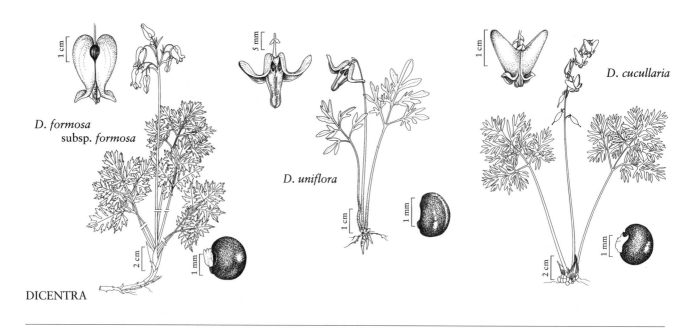

DICENTRA

gular to ovate, 2–4 × 1–2 mm, apex acuminate; petals white; outer petals (10–)12–16(–20) × (2–)4–5(–8) mm, reflexed portion 3–5 mm; inner petals (10–)12–15(–18) mm, blade 2–4 mm, claw linear-elliptic, 5–9 mm, crest prominent, ca. 2 mm diam., exceeding apex by ca. 2 mm; filaments of each bundle distinct nearly to base; nectariferous tissue forming 0.5–1 mm spur oriented vertically; style 4–7 mm; stigma shallowly 2-horned with 2 lateral papillae. **Capsules** ovoid, attenuate at both ends, (5–)9–13(–17) × 3–6 mm. **Seeds** slightly reniform, very obscurely reticulate, elaiosome present.

Flowering early–late spring. Deciduous woods, often among rock outcrops, in rich loam soils; 0–1500 m; Ont., Que.; Conn., D.C., Ill., Ind., Iowa, Ky., Maine, Md., Mass., Mich., Minn., Mo., N.H., N.J., N.Y., N.C., Ohio, Pa., Tenn., Vt., Va., W.Va., Wis.

See discussion under the following species.

9. **Dicentra cucullaria** (Linnaeus) Bernhardi, Linnaea 8: 457, 468. 1833 · Dutchman's-breeches, dicentre à capuchon [E] [F]

Fumaria cucullaria Linnaeus, Sp. Pl. 2: 699. 1753; *Bicuculla cucullaria* (Linnaeus) Millspaugh; *B. occidentalis* Rydberg; *Dicentra cucullaria* (Linnaeus) Bernhardi var. *occidentalis* (Rydberg) M. Peck

Plants perennial, scapose, from short rootstocks bearing pink to white, teardrop-shaped bulblets. **Leaves** (10–)14–16(–36) × (4–)6–14(–18) cm; petiole (5–)8–16(–24) cm; blade with 4 orders of leaflets and lobes; abaxial surface glaucous; ultimate lobes linear to linear-elliptic or linear-obovate, (2–)5–15(–23) × (0.4–)2–3(–4.2) mm, usually minutely apiculate. **Inflorescences** racemose, 3–14-flowered, usually exceeding leaves; bracts minute. **Flowers** pendent; pedicels (2–)4–7(–12) mm; sepals broadly ovate, 1.8–5 × 1.3–4 mm; petals white, frequently suffused pink, apex yellow to orange-yellow; outer petals (10–)12–16(–20) × (3–)6–10(–13) mm, reflexed portion 2–5 mm; inner petals (7.5–)9–12(–14) mm, blade 1.8–4 mm, claw linear, 4–8 × less than 1 mm, crest prominent, ca. 2 mm diam.; filaments of each bundle connate from base to shortly below anthers; nectariferous tissue forming 1–3(–4.5) mm spur diverging at angle from base of bundle; style 2–4 mm; stigma 2-horned with 2 lateral papillae. **Capsules** ovoid, attenuate at both ends, (7–)9–13(–16) × 3–5 mm. **Seeds** reniform, ca. 2 mm diam., very obscurely reticulate, elaiosome present. $2n = 32$.

Flowering early–late spring. Deciduous woods and clearings, in rich loam soils; 0–1500 m; N.B., N.S., Ont., P.E.I., Que.; Ala., Ark., Conn., Del., D.C., Ga., Idaho, Ill., Ind., Iowa, Kans., Ky., Maine, Md., Mass., Mich., Minn., Mo., Nebr., N.H., N.J., N.Y., N.C., N.Dak., Ohio, Okla., Oreg., Pa., R.I., S.C., S.Dak., Tenn., Vt., Va., Wash., W.Va., Wis.

Dicentra cucullaria is occasionally confused with *D. canadensis,* with which it is sympatric. It is distinguished from that species by its basally pointed (versus rounded) outer petal spurs, by its flowers lacking a fragrance, by flowering 7–10 days earlier, and by its pink to white, teardrop-shaped (versus yellow, pea-shaped) bulblets.

After fruit set, the bulblets of both *Dicentra cucul-*

laria and *D. canadensis* remain dormant until fall, when stored starch is converted to sugar. At this time also, flower buds and leaf primordia are produced below ground; these then remain dormant until spring (P. G. Risser and G. Cottam 1968; B. J. Kieckhefer 1964; K. R. Stern 1961). Pollination of both species is effected by bumblebees (*Bombus* spp.) and other long-tongued insects (L. W. Macior 1970, 1978; K. R. Stern 1961).

Flavonoid components indicate that *Dicentra canadensis* and *D. cucullaria* are more closely related to each other than to any other member of the genus (D. Fahselt 1971). Even so, species purported to be hybrids between them probably are not. There is considerable variation in floral morphology within *D. cucullaria*, which can have flowers superficially resem-

bling those of *D. canadensis*. However, when all characters of the plants are examined, these putative hybrids almost always are clearly assignable to one species or the other.

The western populations of *Dicentra cucullaria* appear to have been separated from the eastern ones for at least a thousand years. The western plants are generally somewhat coarser, which apparently led Rydberg to designate the western populations as a separate species. Plants from the Blue Ridge Mountains of Virginia, however, are virtually indistinguishable from those of the West, and much of the variation (which is considerable) within the species probably involves phenotypic response to the environment, or represents ecotypes within the species.

2. ADLUMIA Rafinesque ex de Candolle, Syst. Nat. 2: 111. 1821, name conserved · [For John Adlum, 1759–1836, a horticulturist born in York, Pa., died in Georgetown, D.C.]

David E. Boufford

Bicuculla Borkhausen

Vines or vinelike caulescent herbs, biennial, perhaps also annual, from taproots. **Stems** developing in 2d year, climbing, usually simple. **Leaves** cauline, compound, petiolate; blade with 3–5 orders of leaflets and lobes; margins entire; surfaces glabrous; distal petiolules and reduced leaflets twining and tendril-like. **Inflorescences** axillary, cymose-paniculate, multifloral. **Flowers** bilaterally symmetric about each of 2 perpendicular planes; sepals caducous, peltate with attachment near base; corolla persistent, compressed-urceolate, becoming spongy; outer petals connate except at apex, base saccate, apex with erect or reflexed, ovate to deltate lobe; inner petals similar but with apical lobes connate over stigma; filaments basally connate and adnate to petals; ovary linear or narrowly oblong; style persistent; stigma ± 2-lobed. **Capsules** dehiscent, 2-valved. **Seeds** ca. 6, elaiosome absent.

Species 2 (1 in the flora): e North America, Asia (Korea).

The Asian member of the genus usually is considered to be a distinct species, *Adlumia asiatica* Ohwi.

1. **Adlumia fungosa** (Aiton) Greene ex Britton, Sterns, & Poggenberg, Prelim. Cat., 3. 1888 · Climbing fumitory, Allegheny-vine, mountain-fringe E F

Fumaria fungosa Aiton, Hort. Kew. 3: 1. 1789

Plants 0.5–4 m. **Leaves** 2–13 × 1–8 cm. **Flowers:** corolla white to pale pink or purplish, 10–17 × 3–7 mm. **Capsules** compressed-cylindric, ca. 10 mm. **Seeds** compressed-globose, lustrous. $2n = 32$.

Flowering summer–early fall. Moist coves, rocky woods, ledges, alluvial slopes, and thickets; 0–1500 m; B.C., Man., N.B., N.S., Ont., Que.; Conn., Del., Ill., Ind., Ky., Maine, Md., Mass., Mich., Minn., N.H., N.J., N.Y., N.C., Ohio, Pa., R.I., Tenn., Vt., Va., W.Va., Wis.

Adlumia fungosa is apparently naturalized in southwestern British Columbia. It was reported to be "freely escaping" from a garden in Alberta (H. Groh 1949), and it is a casual, but usually not persisting, escape elsewhere.

3. CORYDALIS de Candolle in J. Lamarck and A. P. de Candolle, Fl. Franç. ed. 3, 4: 637. 1805, name conserved · [Greek *korydallis,* crested lark]

Kingsley R. Stern

Capnoides Adanson

Herbs, annual, biennial, or perennial, caulescent, from taproots, tubers, or rhizomes. **Stems** erect to prostrate, simple or branching. **Leaves** basal and/or cauline, simple or usually compound; blade with 2–6 orders of leaflets and/or lobes. **Inflorescences** axillary or terminal, paniculate or racemose, multifloral, sometimes cleistogamous-flowered (in *Corydalis flavula* and *C. micrantha*). **Flowers** bilaterally symmetric about 1 plane; sepals caducous or persistent; petals distinct or somewhat coherent basally, not spongy; outer petals dissimilar, each with median adaxial keel or crest, sometimes with distal marginal wing, 1 basally spurred, the other sometimes gibbous but not spurred; inner petals connate apically; stamens with nectariferous spur projecting from near base of median filament in bundle opposite spurred petal and adhering to inner surface of petal spur; ovary broadly ovoid to obovoid; stigma persistent, with or without 2 lobes or apical horns, or 4–8 papillar stigmatic surfaces. **Capsules** dehiscent, 2-valved. **Seeds** few–many, reniform to subglobose, elaiosome usually present. $x = 8$.

Species ca. 100 (10 in the flora): temperate North America, Eurasia, and Africa.

Several native species of *Corydalis* have been grown as ornamentals, particularly *C. scouleri, C. aurea, C. sempervirens,* and *C. caseana;* they may be found as garden escapes in areas of the continent outside their natural ranges. Two Eurasian species that are widely cultivated in the flora area also escape sometimes, but evidently they are not truly naturalized here. *Corydalis lutea* (Linneaus) de Candolle, reported from New York and Oregon, can be distinguished from the native species that also are rhizomatous perennials by its yellow petals and axillary racemes. *Corydalis solida* (Linneaus) Swartz (sometimes identified as *C. bulbosa* Persoon), reported from Connecticut, Massachusetts, Vermont, and southern Ontario, can be distinguished from the native species that also are tuberous perennials by its lack of sepals, pedicels usually longer than 10 mm, and sometimes yellow petals.

SELECTED REFERENCES Ownbey, G. B. 1947. Monograph of the North American species of *Corydalis*. Ann. Missouri Bot. Gard. 34: 187–260. Ownbey, G. B. 1951. On the cytotaxonomy of the genus *Corydalis*, section *Eucorydalis*. Amer. Midl. Naturalist 45: 164–186. Ryberg, M. 1955. A taxonomical survey of the genus *Corydalis* Ventenat, with reference to cultivated species. Acta Hort. Berg. 17: 155–175.

1. Plants perennial; petals not yellow; stigma as long as or longer than broad; capsules oblong to obovoid, reflexed.
    2. Roots small, tuberous, usually forked; petals blue, often purple-tinged; stigma rhomboid, narrower toward base; n British Columbia, w Northwest Territories, Yukon, and Alaska . . . . 4. *Corydalis pauciflora*
    2. Rhizome or roots large and fleshy; petals pink or white; stigma rectangular or triangular, not narrower toward base; w United States and s British Columbia.
        3. Primary inflorescence axis with ca. 15–35 flowers; inner petals not tipped deep red or purple; stigma ± triangular; capsules obovoid; seeds ca. 3.5 mm diam . . . . 1. *Corydalis scouleri*
        3. Primary inflorescence axis typically with 50 or more flowers; inner petals tipped red or purple; stigma roughly rectangular; capsules usually ellipsoid; seeds 2–2.5 mm diam.
            4. Blades of proximal cauline leaves with ultimate lobes mostly more than 15 mm; spurred petal with marginal wing conspicuous; petals with crest absent or beaklike . . . . 2. *Corydalis caseana*
            4. Blades of proximal cauline leaves with ultimate lobes 5–15 mm; spurred petal with marginal wing inconspicuous or absent; petals with crest present, wrinkled . . . . 3. *Corydalis aqua-gelidae*
1. Plants annual or biennial; petals yellow, or pink, tipped yellow; stigma broader than long; capsules narrowly to broadly linear, usually not reflexed.
    5. Petals pink, tipped yellow, crests absent, claws of inner petals distinctly longer than blades; seeds ca. 1 mm diam . . . . 5. *Corydalis sempervirens*

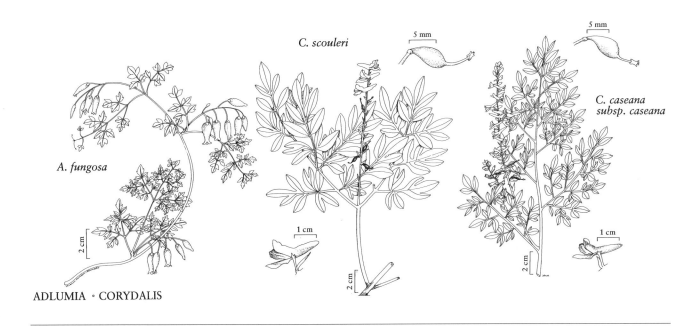

C. scouleri

A. fungosa

C. caseana subsp. caseana

ADLUMIA · CORYDALIS

5. Petals pale to bright yellow throughout, crests present, claws of inner petals equal to or shorter than blades; seeds 1.5–2 mm diam.
   6. Spurred petal 7–9 mm, spur incurved, ca. 2 mm; capsules pendent, pedicels long . . . . 6. *Corydalis flavula*
   6. Spurred petal typically 10–22 mm, spur not appreciably incurved, ca. 4–8 mm; capsules erect (or pendent at maturity in *Corydalis aurea*), pedicels short.
      7. Outer petals with crest conspicuous, marginal wing very broad; capsules usually with basally pustulate whitish hairs . . . . 7. *Corydalis crystallina*
      7. Outer petals with crest inconspicuous, marginal wing medium to narrow; capsules essentially glabrous.
         8. Plants frequently bearing inconspicuous cleistogamous flowers; seeds ca. 1.5 mm diam.
         . . . . 8. *Corydalis micrantha*
         8. Plants rarely bearing cleistogamous flowers; seeds ca. 2 mm diam.
            9. Seeds appearing distinctly rough under magnification .9. *Corydalis curvisiliqua*
            9. Seeds appearing essentially smooth under magnification . . . . 10. *Corydalis aurea*

1. Corydalis scouleri Hooker, Fl. Bor.-Amer. 1: 36, plate 14. 1829 E F

Plants perennial, from large, fleshy rhizomes. **Stems** 1 or more, mostly 5–10 dm. **Leaves** ca. 3, compound, 10 cm or more; blade with 3 orders of leaflets and lobes; ultimate lobes broadly elliptic or less commonly ovate to obovate with rounded apex, sometimes narrowly elliptic with acute apex, 1–8 × 0.5–4 cm, minutely apiculate. **Inflorescences** axillary or terminal, racemose or paniculate, 15–35-flowered on primary axis; bracts inconspicuous, proximal bracts narrowly elliptic, distal linear and much reduced. **Flowers** erect; pedicel 2–5 mm; sepals caducous, ovate to broadly lanceolate, 1–2 mm, margins lacerate or dentate; petals light to deep pink; spurred petal usually somewhat curved, 20–25 mm, spur lanceoloid, 14–20 mm, crest well developed, usually exceeding petal apices, marginal wing absent; unspurred outer petal boat-shaped, 12–15 mm; inner petals not tipped deep red or purple, usually 9–11 mm, blade much wider at apex, claw slender, equaling blade in length; nectariferous spur 1/2–2/3 length of petal spur, bent or hooked at apex; style ca. 3 mm; stigma roughly triangular, with 2 apical and 2 lateral papillae. **Capsules** reflexed, obovoid, 10–15 × 4–5 mm. **Seeds** ca. 3.5 mm diam., with numerous small protuberances. $2n$ = ca. 130.

Flowering mid spring–early summer. Moist, shady woods, particularly along streams; 0–1100 m; B.C.; Oreg., Wash.

Corydalis scouleri is restricted to cool, wet habitats from northwestern Oregon northward to Vancouver Island. It is most easily distinguished from *Corydalis caseana* by the usually highly developed crests and absence of wings on its outer petals. The stigma is essentially triangular (versus rectangular in *C. caseana*), and the capsule shape (typically obovoid) is rarely approached in *C. caseana*.

2. Corydalis caseana A. Gray, Proc. Amer. Acad. Arts 10: 69. 1874 · Fitweed [E] [W2]

Plants perennial, glaucous, from large, fleshy roots. **Stems** 1–several, 10–15 dm or more. **Leaves** ca. 5, compound, to 10 dm; blade with 2–4 orders of leaflets and lobes; ultimate lobes of proximal cauline leaves narrowly to broadly elliptic, 1–5 cm, apiculate. **Inflorescences** paniculate, 50 or more flowers on primary axis; bracts inconspicuous, linear or narrowly elliptic, rarely wider, proximal bracts ca. 10 mm, distal greatly reduced. **Flowers** ascending or spreading; pedicel 3–5 mm; sepals persistent, round and irregularly dentate to ovate, lunate, or attenuate-lanceolate with broad, sometimes lobed base, 2–3 mm; petals light pink to white, inner ones tipped reddish purple; spurred petal often incurved, 16–25 mm, spur tapered or not, 9–16 mm, apex obtuse, crest absent, inconspicuous, or extending into beak beyond petal apices, marginal wing narrow to broad, conspicuous; unspurred outer petal 10–15 mm; inner petals 7–12 mm, claw 3–5 mm; nectariferous spur 1/2–3/4 length of petal spur; style 3–4 mm; stigma roughly rectangular, with 8 papillae. **Capsules** reflexed, ellipsoid, 10–15 × ca. 3–5 mm. **Seeds** ca. 2.5 mm diam., with numerous minute protuberances.

Subspecies 5 (5 in the flora): w United States.

Significant livestock losses have been caused by ingestion of *Corydalis caseana*, which is palatable to both cattle and sheep, despite the toxicity.

1. Outer petals with marginal wing narrow or absent, apex of unspurred outer petal acute; California . . . . 2a. *Corydalis caseana* subsp. *caseana*
1. Outer petals with marginal wing moderately to highly developed, apex of unspurred outer petal not acute; other than California.
  2. Outer petals with marginal wing scarcely revolute, apex rounded, sometimes minutely apiculate or notched.
    3. Outer petals minutely apiculate; stems mostly 10–15 dm; Colorado and New Mexico. . . 2b.*Corydalis caseana* subsp. *brandegei*
    3. Outer petals not minutely apiculate, occasionally minutely notched; stems mostly 4–10 dm; Utah. . . . . 2c. *Corydalis caseana* subsp. *brachycarpa*
  2. Outer petals with marginal wing distinctly revolute, apex distinctly notched.
    4. Inflorescences profusely branching; outer petals with marginal wing moderately developed, minutely eroded; n Idaho. . . . . 2d. *Corydalis caseana* subsp. *hastata*
    4. Inflorescences not profusely branching;

outer petals with marginal wing highly developed, not minutely eroded; ne Oregon and s Idaho . . . . 2e. *Corydalis caseana* subsp. *cusickii*

2a. Corydalis caseana A. Gray subsp. **caseana** [E] [F]

Stems ca. 10 dm. **Flowers:** outer petals with marginal wing narrow or absent, apex of unspurred outer petal acute.

Flowering early–late summer. Very moist, often shady, forested areas, in springs, in and along streams, and on gravel bars; 1200–1850 m; Calif.

2b. Corydalis caseana A. Gray subsp. **brandegei** (S. Watson) G. B. Ownbey, Ann. Missouri Bot. Gard. 34: 203. 1947 [C] [E]

Corydalis brandegei S. Watson, Bot. California 2: 430. 1880

Stems mostly 10–15 dm. **Flowers:** outer petals with marginal wing highly developed, scarcely revolute, apex rounded, minutely apiculate.

Flowering early–late summer. Very moist, subalpine areas, particularly along streams; of conservation concern; 2400–3500 m; Colo., N.Mex.

Pressing sometimes distorts the features that distinguish this subspecies.

2c. Corydalis caseana A. Gray subsp. **brachycarpa** (Rydberg) G. B. Ownbey, Ann. Missouri Bot. Gard. 34: 204. 1947 [C] [E]

Capnoides brachycarpum Rydberg, Bull. Torrey Bot. Club 34: 426. 1907

Stems mostly 4–10 dm. **Flowers:** outer petals with marginal wing highly developed, scarcely revolute, apex rounded, occasionally minutely notched.

Flowering mid summer. On gravel bars along streams; of conservation concern; 2550–3050 m; Utah.

The epithet *brachycarpa*, apparently intended to denote a difference between the capsules of this taxon and those of its close relatives, is a misnomer since their fruits are all similar. It may have resulted from a misinterpretation of the swollen capsules commonly found on plants of this subspecies. Such fruits always contain insect larvae that stimulate production of abnormal tissue, which increases the fruit diameter significantly.

2d. Corydalis caseana A. Gray subsp. **hastata** (Rydberg) G. B. Ownbey, Ann. Missouri Bot. Gard. 34: 206. 1947 [E]

Capnoides hastatum Rydberg, Bull. Torrey Bot. Club 34: 426. 1907; *Corydalis caseana* A. Gray var. *hastata* (Rydberg) C. L. Hitchcock

Stems 10–18 dm. **Inflorescences** profusely branching. **Flowers:** outer petals with marginal wing moderately developed, distinctly revolute, minutely eroded, apex rounded, distinctly notched.

Flowering early–late summer. In and along streams; 900–1220 m; Idaho.

This subspecies appears to be confined to very wet habitats at intermediate elevations in a few Idaho counties.

2e. Corydalis caseana A. Gray subsp. **cusickii** (S. Watson) G. B. Ownbey, Ann. Missouri Bot. Gard. 34: 205. 1947 [E]

Corydalis cusickii S. Watson, Bot. California 2: 430. 1880; *Capnoides cusickii* (S. Watson) A. Heller; *Corydalis caseana* A. Gray var. *cusickii* (S. Watson) C. L. Hitchcock

Stems 8–15 dm. **Inflorescences** not profusely branching. **Flowers:** outer petals with marginal wing strongly developed, distinctly revolute, not eroded, apex rounded, distinctly notched.

Flowering early–late summer. In and along streams; 1500–2300 m; Idaho, Oreg.

3. Corydalis aqua-gelidae M. Peck & W. C. Wilson, Leafl. W. Bot. 8: 39. 1956 [E]

Plants perennial, subaquatic to aquatic, from relatively deep roots. **Stems** 1–several, hollow, 3–11 dm, succulent. **Leaves** several, compound, to 6 dm; blade with 4–6 orders of leaflets and lobes; abaxial surface glaucous; ultimate lobes of proximal cauline leaves elliptic to ovate, 5–12(–15) × 1.5–6 mm. **Inflorescences** racemose or paniculate, 20–70-flowered; bracts inconspicuous, proximal to 10 mm, distal reduced. **Flowers** ascending or spreading; sepals rounded and irregularly dentate to attenuate-lanceolate with wide base; petals pale to deep pink or lavender, red or purple at tips of inner petals; spurred petal (10–)12–20 mm, spur 9–11 mm, crest conspicuous, wrinkled, marginal wing absent or inconspicuous, unspurred outer petal 10–15 mm; inner petals 7–12 mm, claw 3–5 mm; nectariferous spur 1/2–

3/4 length of petal spur; style ca. 3 mm; stigma roughly rectangular, with 8 papillae. **Capsules** reflexed, narrowly ellipsoid, 8–15 × ca. 3–5 mm. **Seeds** ca. 2 mm diam.

Flowering early–late summer. In and along streams and springs, and in moist shady woods; 30–1100 m; Oreg., Wash.

Because of morphologic intergradation with *Corydalis caseana* subsp. *cusickii*, G. B. Ownbey (pers. comm.) does not consider *Corydalis aqua-gelidae* a distinct species; he believes that it probably should be treated as a subspecies of *C. caseana*.

4. Corydalis pauciflora (Stephan ex Willdenow) Persoon, Syn. Pl. 2: 269. 1807

Fumaria pauciflora Stephan ex Willdenow, Sp. Pl. 3(2): 861. 1803; *Corydalis arctica* Popov; *C. pauciflora* (Stephan ex Willdenow) Persoon var. *albiflora* A. E. Porsild

Plants perennial, from small, tuberous, usually forked taproots with fibrous rootlets at base, sometimes with 1–several accessory buds at summit. **Stems** usually 1–3, erect, mostly 0.8–2 dm, often with 1–2 basal scales. **Leaves** 2–5, simple; blade with 2 orders of lobes; ultimate lobes elliptic. **Inflorescences** terminal, racemose, 3–5-flowered, flowers crowded at summit of stout peduncle; bracts inconspicuous, ovate to obovate, 4–10 × 3–5 mm, proximal bract largest. **Flowers** erect; pedicel stout, 4–10 mm; sepals caducous, 1–2 × 1–2 mm, margins variously dentate; petals blue, often tinged purple; spurred petal 17–20 mm, spur abruptly incurved near apex, 7–10 mm, crest low, extending to apex, marginal wing narrow, unspurred outer petal 10–12 mm, margins revolute; inner petals 8–10 mm, blade obovate, claw slender, ca. 4–5 mm; nectariferous spur clavate, 2/3–3/4 length of petal spur; style ca. 2 mm; stigma 4-lobed, rhomboid, narrower toward base. **Capsules** reflexed, ellipsoid to obovoid, ca. 12 × 5 mm. **Seeds** essentially smooth. $2n = 16$.

Flowering early–mid summer. Tundra; 0–1100 m; B.C., N.W.T., Yukon; Alaska; Asia.

This distinctive species is the only blue-flowered member of the genus in the flora area. It is an essentially Asiatic species whose distribution extends across the Bering Strait into North America.

5. Corydalis sempervirens (Linnaeus) Persoon, Syn. Pl. 2: 269. 1807 · Pink and yellow corydalis, pale corydalis, harlequin-flower, corydale toujours verte [E] [F] [W1]

Fumaria sempervirens Linnaeus, Sp. Pl. 2: 700. 1753; *Capnoides sempervirens* (Linnaeus) Borkhausen

Plants biennial, from somewhat succulent roots. **Stems** usually 1, erect, 0.5–8 dm, very glaucous. **Leaves** compound; blade with 3–4 orders of leaflets and lobes; ultimate lobes oblong-elliptic, obtuse, apiculate. **Inflorescences** terminal, racemose or paniculate, 1–8-flowered on each axis; bracts inconspicuous, narrowly elliptic, 2–5 × 0.5–1 mm. **Flowers** erect; pedicel slender, 5–20 mm; sepals ovate, short-attenuate, to 3 mm; petals pink, tipped yellow; spurred petal 10–15 mm, spur blunt, 3–4 mm, crest absent, marginal wing relatively broad, revolute, unspurred outer petal 10–13 mm; inner petals 9–12 mm, blade broadly obovate, with high, angular keel, claw slender, 6–8 mm; nectariferous spur 1/3 length of petal spur, blunt; style ca. 4 mm; stigma triangular, with 4 papillae. **Capsules** erect, linear, straight, (25–) 30–35(–50) mm. **Seeds** ca. 1 mm diam., minutely decorated. $2n = 16$.

Flowering early summer–early fall. Rock crevices, talus, forest clearings, open woods, and on burned or otherwise disturbed areas in shallow, often dry soil; 10–1550 m; Alta., B.C., Man., N.B., Nfld., N.W.T., N.S., Ont., P.E.I., Que., Sask., Yukon; Alaska, Conn., Ga., Ill., Ind., Ky., Maine, Md., Mass., Mich., Minn., Mont., N.H., N.J., N.Y., N.C., Ohio, Pa., R.I., S.C., Tenn., Vt., Va., W.Va., Wis.

6. Corydalis flavula (Rafinesque) de Candolle in A. P. de Candolle and A. L. P. de Candolle, Prodr. 1: 129. 1824 [E]

Fumaria flavula Rafinesque, J. Bot. (Desvaux) 1: 224. 1808; *Capnoides flavulum* (Rafinesque) Kuntze

Plants annual, from somewhat succulent roots. **Stems** 1–several, initially erect, often becoming prostrate or ascending, usually 1.5–3 dm. **Leaves** compound; blade with 2 orders of leaflets and lobes; lobes elliptic, variable in size, margins incised, apex subapiculate. **Inflorescences** racemose, commonly 6–10-flowered, equaling or barely exceeding leaves, sometimes poorly developed, cleistogamous-flowered racemes present, inconspicuous, 1–5-flowered; bracts elliptic, 6–12 × 3–7 mm, proximal bracts often leaflike or variously incised, distal reduced and entire. **Flowers** erect; pedicel slender, 6–15 mm or more; sepals lanceolate, ca. 1 mm; petals pale yellow, spurred petal 7–9 mm, spur incurved, ca.

2 mm, crest high, marginal wing well developed, both crest and wing wrinkled or dentate, unspurred outer petal similar to spurred petal, 6–8 mm; inner petals 5–7 mm, blade apex ca. 2 times wider than distinctly lobed base, claw 2–3 mm; nectariferous spur less than 1/2 length of petal spur; style 1.5–2 mm; stigma broader than long, with 4 terminal papillae. **Capsules** pendent, linear, straight or sometimes reflexed, (14–) 18–20 (–22) mm. **Seeds** ca. 2 mm diam., minutely decorated on narrow marginal ring.

Flowering early–late spring. Wooded slopes, bottomlands, and rock outcrops, in moist, loose soil; 0–650 m; Ont.; Ala., Ark., Del., D.C., Fla., Ill., Ind., Iowa, Kans., Ky., La., Md., Mich., Miss., Mo., Nebr., N.J., N.Y., N.C., Ohio, Okla., Pa., S.C., Tenn., Va., W.Va.

7. Corydalis crystallina (Torrey & A. Gray) A. Gray, Manual ed. 5, 62. 1867; Bot. Gaz. 11: 189. 1886 [E]

Corydalis aurea Willdenow var. (β) *crystallina* Torrey & A. Gray, Fl. N. Amer. 1: 665. 1840

Plants winter annual, glaucous, from somewhat succulent roots. **Stems** 1–several, erect or ascending, 2–4 dm. **Leaves** compound; blade with 2 orders of leaflets and lobes; ultimate lobes lanceolate to linear-lanceolate, margins incised, apex subapiculate. **Inflorescences** racemose, (8–)12–15(–18)-flowered, primary racemes exceeding leaves, later secondary racemes fewer flowered, exceeded by leaves; bracts ovate to ovate-acuminate, 5–12 × 3–6 mm, distal usually much reduced. **Flowers** erect; pedicel stout, ca. 1 mm; sepals broadly ovate to cordate, somewhat attenuate, to 2 mm, margins sometimes incised; petals bright yellow; spurred petal 16–22 mm, spur straight, 6–8 mm, apex distinctly globose, crest very high, wrinkled or toothed, marginal wing very broad, revolute, unspurred outer petal 12–14 mm, marginal wing broad, not revolute, enclosing margins of spurred petal in bud; inner petals oblanceolate, 9–11 mm, blade apex ca. 2 times wider than base, basal lobes small, claw narrow, 4–5 mm; nectariferous spur clavate, 3.5–5 mm, curved near apex; style slender, ca. 4 mm; stigma broader than long, with 8 papillae. **Capsules** erect, linear, stout, straight to slightly incurved, 14–18 mm, with numerous basally pustulate, whitish hairs or rarely glabrate. **Seeds** ca. 2 mm diam., minutely decorated, marginal ring absent.

Flowering mid–late spring. Prairies, fields, open woods, and wasteland; 10–500 m; Ark., Kans., Mo., Okla., Tex.

The peculiar pubescence of the capsules distinguishes *Corydalis crystallina* from all others in the ge-

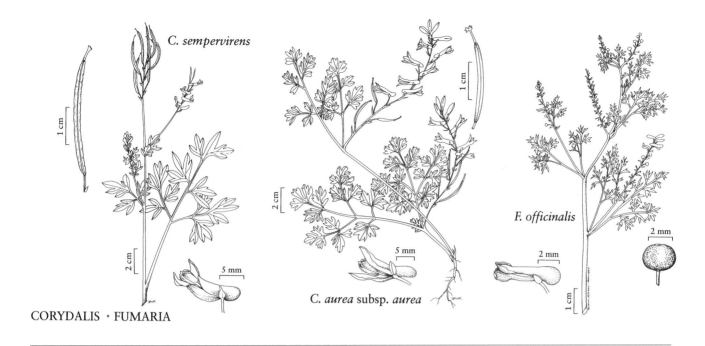

C. *sempervirens*

C. *aurea* subsp. *aurea*

F. *officinalis*

CORYDALIS · FUMARIA

nus. Also, the crest and marginal wing of the spurred petal are more highly developed than in any other yellow-flowered species.

8. Corydalis micrantha (Engelmann ex A. Gray) A. Gray, Bot. Gaz. 11: 189. 1886 [E]

Corydalis aurea Willdenow var. *micrantha* Engelmann ex A. Gray, Manual ed. 5, 62. 1867; *Capnoides micranthum* (Engelmann ex A. Gray) Britton

Plants winter-annual, glaucous to nearly green, from somewhat succulent roots. **Stems** 1–several, erect to prostrate-ascending, (1.5–)2–4(–6) dm. **Leaves** crowded, compound; blade with 2 orders of leaflets and lobes; lobes ovate, oblong-elliptic, or obovate, margins incised, apex subapiculate. **Inflorescences** racemose, (6–)10–16(–20)-flowered, primary racemes slightly to conspicuously exceeding leaves, secondary racemes fewer flowered, exceeded by leaves, cleistogamous-flowered racemes frequently present, 1–5-flowered, inconspicuous; bracts elliptic to attenuate-ovate, 5–8 × 2–4 mm, margins denticulate, distal bracts usually much reduced, those of cleistogamous racemes minute. **Flowers** erect or spreading; pedicel 2–6 mm; sepals ovate, to 1.5 mm, margins often sinuate or dentate; petals pale to medium yellow; spurred petal slightly to strongly curved, (11–)12–14(–15) mm, spur straight, 4–7 mm, apex obtuse or ± globose, crest low, wrinkled, rarely obsolescent, marginal wing well developed, sometimes revolute, unspurred outer petal slightly bent, 9–11 mm, crest low; inner petals oblanceolate, 7–10 mm, blade apex 2 times or more wider than base, basal lobes obscure, claw 3–4 mm; nectarif-

erous spur straight or curved, sometimes clavate, ca. 3/5 length of petal spur; style ca. 4 mm; stigma rectangular, 2-lobed, 1/2 as long as broad, with 8 papillae. **Capsules** erect, linear, slender, straight to slightly incurved, 10–35 mm, essentially glabrous, usually shorter in cleistogamous-flowered racemes. **Seeds** ca. 1.5 mm diam., concentrically and minutely decorated, marginal ring absent.

Subspecies 3 (3 in the flora): c, se United States.

Corydalis micrantha can be distinguished readily from other yellow-flowered North American species by its very small seeds. Cleistogamy is encountered regularly in *C. micrantha.* A single plant from any part of the range may have only cleistogamous flowers, only chasmogamous flowers, or both types. Plants having only cleistogamous flowers usually are much more profusely and delicately branched. In *C. micrantha,* at least, shade and age appear to play roles in the initiation of cleistogamy.

1. Racemes of chasmogamous flowers not greatly exceeding leaves, often short; petal spur ± globose at apex; capsules often stout, commonly 10–15 mm. . . 8a. *Corydalis micrantha* subsp. *micrantha*
1. Racemes of chasmogamous flowers often greatly exceeding leaves, elongate; petal spur blunt, not globose at apex; capsules slender, commonly 15–30 mm.
  2. Stems usually weak; capsules commonly 15–20 mm . . . . 8b. *Corydalis micrantha* subsp. *australis*
  2. Stems usually stout; capsules commonly 21–35 mm . . . . . 8c. *Corydalis micrantha* subsp. *texensis*

8a. Corydalis micrantha (Engelmann ex A. Gray) A. Gray subsp. micrantha E

Inflorescences: racemes of chasmogamous flowers not greatly exceeding leaves, often short; petal spur ± globose at apex. **Capsules** often stout, commonly 10–15 mm. $2n = 16$.

Flowering mid–late spring. Bluffs, rocky hills, open woods, and river banks, often in disturbed soil; 0–600 m; Ark., Ill., Iowa, Kans., Minn., Mo., Nebr., N.C., Okla., S.Dak., Tenn., Tex., Wis.

8b. Corydalis micrantha (Engelmann ex A. Gray) A. Gray subsp. australis (Chapman) G. B. Ownbey, Ann. Missouri Bot. Gard. 34: 222–224. 1947 E

Corydalis aurea Willdenow var. *australis* Chapman, Fl. South. U.S. ed. 2, 604. 1883; *Capnoides halei* Small; *Corydalis campestris* (Britton) J. T. Buchholz & E. J. Palmer; *C. halei* (Small) Fernald & Schubert; *C. micrantha* (Engelmann ex A. Gray) A. Gray var. *australis* (Chapman) Shinners

Stems usually weak. **Inflorescences:** racemes of chasmogamous flowers often greatly exceeding leaves, elongate; petal spur blunt, not globose at apex. **Capsules** slender, commonly 15–20 mm.

Flowering very early–late spring. Abandoned fields and waste areas, roadsides, and open woods, in disturbed, often sandy soil; 10–650 m; Ala., Ark., Fla., Ga., Ill., Kans., La., Miss., Mo., N.C., Okla., S.C., Tex.

8c. Corydalis micrantha (Engelmann ex A. Gray) A. Gray subsp. texensis G. B. Ownbey, Ann. Missouri Bot. Gard. 34: 225–226, map 9. 1947 E

Corydalis micrantha var. *texensis* (G. B. Ownbey) Shinners

Stems usually stout. **Inflorescences:** racemes of chasmogamous flowers often greatly exceeding leaves, elongate; petal spur blunt, not globose at apex. **Capsules** slender, commonly 21–35 mm.

Flowering very early–mid spring. Open alluvial plains and uplands, in moist, often sandy soil; 0–200 m; Tex.

9. Corydalis curvisiliqua (A. Gray) A. Gray, Manual ed. 5, 62. 1867 E

Corydalis aurea Willdenow var. *curvisiliqua* A. Gray, Proc. Acad. Nat. Sci. Philadelphia 15: 75. 1863

Plants winter-annual or perhaps biennial, glaucous, from somewhat succulent roots. **Stems** 1–several, erect to ascending, 1–4 dm. **Leaves** compound; blade with 2–3 orders of leaflets and lobes; ultimate lobes oblong, elliptic, or obovate, margins sometimes incised, apex obtuse or rounded. **Inflorescences** racemose, 6–18-flowered, primary racemes usually exceeding leaves, more numerous than secondary, fewer-flowered racemes; bracts ovate, to 15 × 6 mm, proximal bracts sometimes leaflike, distal usually slightly to greatly reduced. **Flowers** spreading, often strongly curved; pedicel stout, 1–3 mm; sepals ovate to attenuate-ovate, ca. 1 mm, margins often sinuate or dentate; petals bright yellow; spurred petal 15–18 mm, spur not appreciably incurved, 7–9 mm, apex blunt, somewhat globose, crest well developed to absent, wrinkled or dentate, marginal wing well developed, unspurred outer petal bent, 12–15 mm, crest similar to that of spurred petal; inner petals oblanceolate, 9–12 mm, claw slender, nearly 1/2 petal length; nectariferous spur clavate, 4–6 mm, bent near apex; style ca. 5 mm; stigma rectangular, 2-lobed, 1/2 as long as broad, with 8 papillae. **Capsules** erect, linear, slender, straight to strongly incurved, 20–35 mm. **Seeds** ca. 2 mm diam., appearing distinctly to faintly reticulate under magnification, marginal ring present or essentially absent.

Subspecies 2 (2 in the flora): c United States.

1. Capsules mostly 25–35 mm, usually abruptly acute at apex; petals with no or only moderately developed crest; seeds appearing distinctly reticulate under magnification. . . . . 9a. *Corydalis curvisiliqua* subsp. *curvisiliqua*
1. Capsules mostly 20–25 mm, attenuate to apex; petals with well-developed crest; seeds appearing only faintly reticulate under magnification . . . . 9b. *Corydalis curvisiliqua* subsp. *grandibracteata*

9a. Corydalis curvisiliqua (A. Gray) A. Gray subsp. curvisiliqua E

Flowers: petals with no or only moderately developed crest. **Capsules** usually abruptly acute at apex, mostly 25–35 mm. **Seeds** appearing distinctly reticulate under magnification.

Flowering early–late spring. Sandy bottoms, abandoned fields, open woods, hillsides, and valleys, in disturbed soils; 0–1500 m; Tex.

9b. Corydalis curvisiliqua (A. Gray) A. Gray subsp. **grandibracteata** (Fedde) G. B. Ownbey, Ann. Missouri Bot. Gard. 34: 227. 1947 E

Corydalis curvisiliqua var. *grandibracteata* Fedde, Repert. Spec. Nov. Regni Veg. 11: 291. 1912

Flowers: petals with well-developed crest. **Capsules** attenuate to apex, mostly 20–25 mm. **Seeds** appearing only faintly reticulate under magnification.

Flowering early–late spring. Disturbed areas; 100–500 m; Ill., Iowa, Kans., Okla., Tex.

10. Corydalis aurea Willdenow, Enum. Pl. 2: 740. 1809 · Golden corydalis, corydalis dorée W2

Capnoides aureum (Willdenow) Kuntze

Plants winter-annual or biennial, glaucous, from ± branched caudices. **Stems** 10–50, prostrate-ascending, 2–3.5 dm. **Leaves** compound; blade with 3 orders of leaflets and lobes; ultimate lobes elliptic, 1.5 times or more longer than wide, margins incised, apex subapiculate. **Inflorescences** racemose, 10–20(–30)-flowered, primary racemes shorter than to slightly exceeding leaves, secondary racemes fewer flowered; bracts elliptic to linear, 4–10 × 1–2 mm, rarely larger, margins often denticulate toward apex, distal bracts usually much reduced. **Flowers** at first erect, later reflexed; pedicel 5–10 mm; sepals ovate to attenuate-ovate, to 1–3 mm, margins often sinuate or dentate; petals pale to bright yellow; spurred petal 13–16 mm, spur straight or slightly incurved, 4–5 mm, apex subglobose, crest low and incised or absent, marginal wing moderately to well developed, unspurred outer petal 9–11 mm, crest same as that of spurred petal; inner petals oblanceolate, 8–10 mm, blade wider than claw and more prominently winged toward apex, claw 3.5–4.5 mm; nectariferous spur 2–3 mm; style ca. 3 mm; stigma 2-lobed, 1/2 as long as broad, with 8 papillae. **Capsules** erect to pendent at maturity, linear, often torulose, slender to somewhat stout, straight to moderately incurved, 12–24(–30) mm. **Seeds** nearly 2 mm diam., appearing essentially smooth under magnification, narrow marginal ring present or absent.

Subspecies 2 (2 in the flora): North America, Mexico.

1. Capsules slender, pendent or spreading at maturity, usually 18–24 mm; seeds without marginal ring; leaves generally exceeding racemes . . . . 10a. *Corydalis aurea* subsp. *aurea*
1. Capsules stout, erect at maturity, 12–20 mm; seeds with narrow marginal ring; racemes generally exceeding leaves . . . . 10b. *Corydalis aurea* subsp. *occidentalis*

10a. Corydalis aurea Willdenow subsp. **aurea** F

Inflorescences: racemes weak, generally exceeded by leaves. **Flowers:** petals sometimes crested. **Capsules** pendent or spreading at maturity, slender, 18–24(–30) mm. **Seeds** without marginal ring. $2n = 16$.

Flowering spring–late summer. Talus slopes, ledges, rocky hillsides, forest clearings, open shores, creek bottoms, gravel pits, road cuts, and burned-over areas, in loose, often gravelly soil; 100–3400 m; Alta., B.C., Man., N.W.T., Ont., Que., Sask., Yukon; Alaska, Ariz., Calif., Colo., Idaho, Ill., Mich., Minn., Mo., Mont., Nebr., Nev., N.H., N.Mex., N.Y., N.Dak., Ohio, Oreg., Pa., S.Dak., Tex., Utah, Vt., Wash., Wis., Wyo.; n Mexico.

Corydalis aurea subsp. *aurea* intergrades at times with *C. aurea* subsp. *occidentalis*, but usually the two can be distinguished readily when fruiting.

10b. Corydalis aurea Willdenow subsp. **occidentalis** (Engelmann ex A. Gray) G. B. Ownbey, Ann. Missouri Bot. Gard. 34: 234. 1947

Corydalis aurea var. *occidentalis* Engelmann ex A. Gray, Manual ed. 5, 62. 1867; *C. curvisiliqua* Engelmann subsp. *occidentalis* (Engelmann ex A. Gray) W. A. Weber; *C. montana* Engelmann ex A. Gray

Inflorescences: racemes robust, generally exceeding leaves. **Flowers:** petals crested. **Capsules** erect at maturity, stout, 12–20 mm. **Seeds** with narrow marginal ring.

Flowering early spring–late summer. Bottomlands, prairies, plains, foothills, mesas, ditches, railroad embankments, and washes, in loose, often sandy, dry soil; 300–2800 m; Ariz., Colo., Kans., Mo., Nebr., Nev., N.Mex., Okla., S.Dak., Tex., Utah, Wyo.; Mexico (Chihuahua, Durango, Sinaloa, and Sonora).

4. FUMARIA Linnaeus, Sp. Pl. 2: 699. 1753; Gen. Pl. ed. 5, 314. 1754 · Fumitory, fumeterre [Latin *fumus,* smoke, presumably alluding to odor of roots]

David E. Boufford

Herbs, annual, caulescent, from taproots. **Stems** erect to reclining, branching. **Leaves** cauline, sometimes also basal, compound; blade with 3–4 orders of leaflets and lobes, margins entire, surfaces glabrous. **Inflorescences** terminal on main stem and leaf-opposed branches, racemose. **Flowers** bilaterally symmetric about 1 plane; sepals peltate with attachment near base, ovate, base rounded, margins ± lacerate, apex acute to acuminate; outer petals inconspicuously crested, one basally spurred; stamens with filaments of each bundle completely connate, adhering basally to petals; ovary ovoid; style promptly deciduous after anthesis, elongate; stigma ± 2-lobed. **Capsules** indehiscent. **Seed** 1, elaiosome absent.

Species ca. 50 (3 in the flora): Eurasia, Africa, and Atlantic islands, with greatest diversity in w Mediterranean region.

Fumaria parviflora Lamarck was reported as adventive from Europe in central Texas by D. S. Correll and M. C. Johnston (1970) and in central and southern coastal California by J. C. Hickman (1993). In the former treatment *F. parviflora* is said to differ from *F. officinalis* by having leaf segments with channeled lobes, and capsules obtuse to apiculate or beaked, while in the latter treatment it is reported to differ from that species by having shorter (3–4 mm), cream-colored petals, the inner ones tipped purple, and ± crested capsules. Evidently, the species is found in North America only as a waif and is not naturalized here. *Fumaria martinii* Clavaud, a synonym of *F. reuteri* Boissier according to M. Lidén (1986), was reported from southwestern British Columbia by B. Boivin (1966), but I have not seen specimens. *Fumaria bastardii* Boreau also has been reported from British Columbia, as an infrequent garden escape on roadsides, in waste places, and at forest edges in the southern part of the province (G. W. Douglas et al. 1989). It differs from *F. officinalis* in that the corolla is dark pink and 10–12 mm long.

1. Pedicels rigidly arcuate-recurved in fruit; capsules smooth or nearly so; corolla 9–14 mm . . . . 1. *Fumaria capreolata*
1. Pedicels straight and spreading to ascending in fruit; capsules slightly wrinkled, warty, pebbled, or shallowly pitted; corolla 5–9.5 mm.
   2. Corolla 6–9.5 mm, spur ca. 2.5 mm . . . . 2. *Fumaria officinalis*
   2. Corolla 5–6 mm, spur 1–1.5 mm . . . . .3. *Fumaria vaillantii*

1. Fumaria capreolata Linnaeus, Sp. Pl. 2: 701. 1753

Plants 1–8 dm. **Inflorescences,** excluding peduncle, 2–3.5 cm; bracteoles equaling or shorter than pedicels. **Flowers:** pedicel rigidly arcuate-recurved in fruit, ca. 3 mm; corolla 9–14 mm, spur 2–3 mm; outer petals white; inner petals white near base, deep red or dark purple apically. **Capsules** globose, slightly compressed, 2–2.5 mm diam., smooth or nearly so. $2n = 64$.

Flowering spring. Waste places, ditches, cultivated fields; 0–50 m; introduced; Fla.; sw Europe; n Africa.

2. Fumaria officinalis Linnaeus, Sp. Pl. 2: 700. 1753
· Fumitory, fumeterre officinal F W2

Plants 1–7 dm. **Inflorescences,** excluding peduncle, 3–7 cm; bracteoles 1/2 to nearly as long as pedicels. **Flowers:** pedicel straight and ascending in fruit, ca. 3 mm; corolla 6–9.5 mm, spur ca. 2.5 mm; petals purplish pink or white near base, deep reddish purple to maroon apically. **Capsules** subglobose, sometimes slightly depressed, 1.5–2 mm diam., ± warty or pebbled. $2n = 32, 48$.

Flowering spring. Waste places, cultivated or fallow fields, thin woods, ditches, roadsides; 0–2100 m; introduced, scattered localities; St. Pierre and Miquelon;

Alta., B.C., Man., N.B., Nfld., N.S., Ont., P.E.I., Que., Sask; Ala., Calif., Colo., Conn., D.C., Fla., Ga., Idaho, Ill., Ind., Kans., La., Maine, Mass., Mich., Minn., Mo., Mont., N.H., N.J., N.Y., N.C., Ohio, Oreg., Pa., S.C., Tex., Utah, Vt., Va., W.Va., Wyo.; Europe; n Africa.

Some plants in North America have conspicuously warty capsules with persistent styles. Similar plants in Europe were treated by M. Lidén (1986) as *Fumaria officinalis* subsp. *wirtgenii* (Koch) P. D. Sell. In all other characters, those plants are not significantly different from other members of the species, and they are not distinguished formally in this treatment.

Weaker, somewhat scandent plants with smaller, perhaps cleistogamous, white flowers and smaller fruits seem to be correlated with shaded situations. They are otherwise indistinguishable from other members of the species.

3. **Fumaria vaillantii** Loiseleur-Deslongchamps, J. Bot. (Desvaux) 2: 358. 1809

Plants 5–50 cm. **Inflorescences,** excluding peduncle, 1–4.5 cm; bracteoles 1/2–3/4 as long as pedicels. **Flowers:** pedicel straight and spreading to ascending in fruit, 1–3 mm; corolla 5–6 mm, spur 1–1.5 mm; petals pink or deep pink near base, deep pink or purple apically. **Capsules** subglobose, frequently longer than broad, 1.7–2.5 mm diam., slightly wrinkled or shallowly pitted. $2n = 32$.

Flowering spring. Open, disturbed places; elevation unknown; introduced; N.Dak., S.Dak.; Europe; w Asia; n Africa.

The description here is taken in part from *Flora of the Great Plains* (Great Plains Flora Association 1986), where the specific epithet is misspelled "vaillentii," supplemented by measurements from specimens of European origin in the Harvard University Herbaria.

21. PLATANACEAE T. Lestiboudois ex Dumortier
• Plane-tree Family

Robert B. Kaul

Trees, deciduous, to 50 m. **Trunks** 1–several, erect to prostrate. **Bark** smooth at first, exfoliating in thin plates, exposing conspicuous mosaic of chalky white to buff or greenish new bark, becoming dark, thick, and fissured with age. **Axillary buds:** each hidden by swollen base of petiole. **Leaves** alternate, simple; stipules sometimes persisting, green, prominent, sheathing stem, flaring, margins entire to serrate. **Leaf blade** palmately (0–)3–7-lobed, base cordate, truncate, or cuneate; surfaces tomentose or glabrescent. **Inflorescences** axillary, solitary, appearing with leaves; staminate inflorescences with heads soon falling, 1–5, green, sessile, globose; pistillate with heads 1–7, terminal (and in some species lateral), sessile or pedunculate, globose, the whole much elongate and pendulous in fruit. **Flowers** unisexual, staminate and pistillate on same plants, very crowded, 3–4(–8)-merous, inconspicuous; perianth hypogynous, minute. **Fruits:** achenes maturing in fall, often persisting until spring, tan, club-shaped, quadrangular, with terminal stylar beak, surrounded by numerous hairs; hairs basally attached, thin, unbranched, 2/3 to nearly equal to length of achene.

Genus 1, species ca. 8 (1 genus, 3 species in the flora): temperate regions, North America, Europe, and Asia.

The family is well known in the North American fossil record and was once more widespread in the flora (S. R. Manchester 1986).

Native species of the flora area are cultivated for shade and ornament in and beyond their native ranges, as are a European species and a hybrid with many cultivars (F. S. Santamour Jr. 1986). The sycamores are known for their great size, imposing stature, smooth and light-colored bark, and tolerance of pruning. The trichomes falling from the new growth reportedly cause irritation to mucous membranes. The wood is difficult to work and therefore has limited commercial value; its resistance to splitting makes it useful for butcher blocks and buttons, hence the old vernacular name buttonwood.

Phytochemical, morphologic, and anatomic characters were summarized for the family by R. N. Schwarzwalder Jr. and D. L. Dilcher (1991).

358

SELECTED REFERENCES Boothroyd, L. E. 1930. The morphology and anatomy of the inflorescence and flower of the Platanaceae. Amer. J. Bot. 17: 678–693. Manchester, S. R. 1986. Vegetative and reproductive morphology of an extinct plane tree (Platanaceae) from the Eocene of western North America. Bot. Gaz. 147: 200–226. Santamour, F. S. Jr. 1986. Checklist of cultivated *Platanus* (planetree). J. Arboric. 12: 78–83. Schwarzwalder, R. N. Jr. and D. L. Dilcher. 1991. Systematic placement of the Platanaceae in the Hamamelidae. Ann. Missouri Bot. Gard. 78: 962–969.

1. PLATANUS Linnaeus, Sp. Pl. 2: 999. 1753; Gen. Pl. ed. 5, 433. 1754 · Sycamore, plane, platane, sycomore [Greek *platanos,* perhaps from *platys,* broad, for the wide leaves]

Leaves, twigs, and inflorescences densely invested with dendritic trichomes, at least when young; fragrance sometimes reminiscent of balsam poplars. **Staminate flowers:** sepals and petals 3–6 +, petals nearly obsolete; pistillodes sometimes present; stamens as many as and opposite sepals, much exceeding tiny perianth; anthers subsessile, 4-sporangiate; dehiscence latrorse, the connective distally expanded into terminal, pubescent, peltate appendage. **Pistillate flowers:** sepals 3–4, petals absent; staminodes 3–4, prominent, club-shaped, tomentose; pistils 3–9, free; ovules 1(–2), pendulous, orthotropous; stigma 1, dark red, much elongate, conspicuous. **Achenes** falling with ring of subtending hairs attached.

In North America *Platanus* is usually called sycamore, a name apparently borrowed from the European sycamore maple, *Acer pseudoplatanus* Linnaeus, which has similar leaves. That name in turn comes from the Middle Eastern sycomore fig, *Ficus sycomorus* Linnaeus, its specific epithet from the Greek *sykomoros,* mulberry.

SELECTED REFERENCES Benson, L. D. 1943. Revisions of status of southwestern trees and shrubs. Amer. J. Bot. 30: 230–240. Santamour, F. S. Jr. 1972b. Interspecific hybridization in *Platanus.* Forest Sci. 18: 236–239.

1. Leaf sinuses usually broad and gently concave, depth of distal sinuses mostly less than 1/2 distance from sinus to base of blade; terminal lobe mostly wider than long, margins entire to coarsely serrate; fruiting heads 1(–2); e North America, ne Mexico. . . . . 1. *Platanus occidentalis*
1. Leaf sinuses narrow to broad, deeply concave, depth of distal sinuses often more than 1/2 distance from sinus to base of blade; terminal lobe longer than wide, margins entire to serrate; fruiting heads (1–)2–7 on rachis; sw United States, nw Mexico.
  2. Terminal leaf lobe 1/3–2/3 length of blade; leaf blade abaxially persistently tomentose, adaxially glabrescent; fruiting heads (1–)2–7 on rachis, lateral ones sessile; California, Mexico (Baja California) . . . . 2. *Platanus racemosa*
  2. Terminal leaf lobe 2/3 or more length of blade; leaf blade abaxially and adaxially glabrescent; fruiting heads (1–)2–4 on rachis, lateral ones sessile or often pedunculate; Arizona, New Mexico; Mexico (Chihuahua, Sinaloa, and Sonora) . . . . 3. *Platanus wrightii*

1. Platanus occidentalis Linnaeus, Sp. Pl. 2: 999. 1753
· Sycamore, American plane-tree [E] [F]

Trees, to 50 + m, becoming massive; trunks straight and unbranched to great heights or low-branching or multitrunked, to 4 + m diam. **Leaves:** stipules entire to coarsely serrate. **Leaf blade** light green, usually shallowly 3–5(–7)-lobed, occasionally unlobed, 6–20 + × 6–25 + cm (to 30 × 40 cm on sucker shoots), not especially thick; lobes of blade mostly wider than long, basal lobes usually smaller, often strongly reflexed, sinuses broad and gently concave, depth of distal sinuses mostly less than 1/2 distance from sinus to base of blade, terminal leaf lobe 1/2–2/3 length of blade; margins entire to coarsely serrate, teeth sometimes short-awned, apex usually acuminate; surfaces glabrate, abaxially often persistently tomentose along veins. **Pistillate inflorescences:** heads 1(–2); fruiting heads 25–30 mm diam.; peduncle to 15 cm. **Achenes** 7–10 mm, basal hairs nearly as long. $2n = 42$.

Flowering spring; fruiting late fall. Often abundant on alluvial soils near streams and lakes and in moist ravines, sometimes on uplands, sometimes on limestone soils, cultivated in parks and gardens and as a street tree; 0–950 m; Ont.; Ala., Ark., Conn., Del., D.C., Fla., Ga., Ill., Ind., Iowa, Kans., Ky., La., Maine, Md., Mass., Mich., Miss., Mo., Nebr., N.H., N.J.,

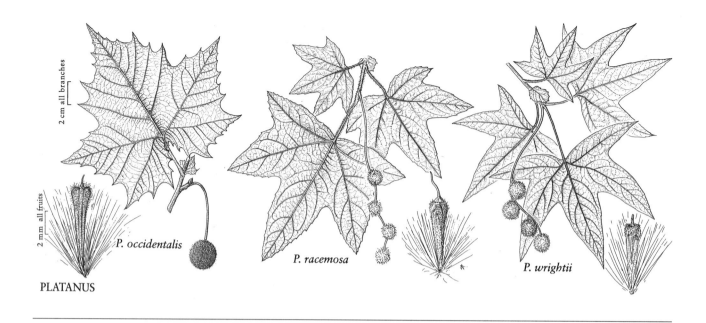

P. occidentalis

P. racemosa

P. wrightii

PLATANUS

N.Y., N.C., Ohio, Okla., Pa., R.I., S.C., Tenn., Tex., Vt., Va., W.Va., Wis.; Mexico (Coahuila, Nuevo León, San Luis Potosí, and Tamaulipas).

Of the angiospermous trees of North America, *Platanus occidentalis* is one of the tallest (to 50 + m) and reaches the greatest trunk diameter (to 4 + m). Trees with smaller and broader-than-long leaf blades, with lobes mostly entire, have been called *P. occidentalis* var. *glabrata* (Fernald) Sargent, especially in the western range of the species from Iowa to Mexico; the range of var. *glabrata* overlaps that of *P. rzedowskii* Nixon & Peale in Tamaulipas. Trees with the blade more deeply lobed, and the base long-cuneate and decurrent on the petiole, are occasional over much of the range of the species. They have been called *P. occidentalis* var. *attenuata* Sargent.

The cultivated London plane-tree [*Platanus* ×*acerifolia* (Aiton) Willdenow, *Platanus hybrida* Broterius] will key here. It is distinguished by the lobes of its larger leaves being somewhat longer and narrower (often longer than wide), the fruiting heads one or two on each rachis, and the bark often somewhat greener. Many cultivars are available, some with deeper lobed or variegated leaves or with upright habit (F. S. Santamour Jr. 1986). It is often planted in cities because it is exceptionally well adapted as a street tree. Apparently it has not escaped in North America, where it is mostly seed-propagated. It is only occasionally reported as naturalized in Europe; there it is clonally propagated and is variously reported to be fertile or sterile. Reputedly, it is a hybrid of *P. occidentalis* with the Eurasian *P. orientalis* Linnaeus. Such a hybrid has been synthesized (F. S. Santamour Jr. 1972b).

Native Americans used *Platanus occidentalis* for a variety of medicinal purposes, including cold and cough remedies, as well as dietary, dermatological, gynecological, respiratory, and gastrointestinal aids (D. E. Moerman 1986).

2. **Platanus racemosa** Nuttall, N. Amer. Sylv. 1: 47. 1842 · California sycamore F

Trees, to 15(–25) m, rather massive; trunks sometimes straight and erect, more commonly irregular, reclining, or prostrate with branches irregular and sometimes near ground, to 2 + m diam. **Leaves:** stipules entire to somewhat dentate. **Leaf blade** dark green, 3–5-lobed, 10–25 × 10–25 cm, rather thick; lobes of blade mostly longer than wide, basal lobes usually somewhat smaller and spreading, not reflexed, sinuses broad and deeply concave, depth of distal sinuses about 1/3–2/3 distance from sinus to base of blade, terminal leaf lobe ca. 1/3–2/3 length of blade; margins entire to remotely serrulate, apex acute to acuminate, sometimes rounded; surfaces abaxially persistently tomentose, adaxially glabrescent. **Pistillate inflorescences:** heads (1–)2–7; fruiting heads 20–25(–30) mm diam., lateral ones sessile; fruiting rachis to 25 cm. **Achenes** 7–10 mm, basal hairs about 2/3 length of achenes. $2n = 42$.

Flowering spring; fruiting late fall. Streamsides and moist, rocky canyons, often abundant; 0–1500 m; Calif.; Mexico (Baja California).

The Eurasian oriental plane, *Platanus orientalis* Linnaeus, is sometimes cultivated and would key to this or the next species. It has deeply lobed, serrate leaves resembling *Acer saccharinum* and *A. macrophyllum,* and the fruiting rachis bears three to six heads.

Native Americans used infusions prepared from the plants of *Platanus racemosa* medicinally as a panacea, and from the bark for indisposition (D. E. Moerman 1986).

3. **Platanus wrightii** S. Watson, Proc. Amer. Acad. Arts 10: 349. 1875 · Arizona sycamore [F]

Trees, to 25 m; trunks straight and erect to inclined or basally reclining or prostrate, to 1.2(–2) m diam., lower branches becoming thick, contorted. **Leaves:** stipules entire to somewhat dentate. **Leaf blade** dark green, deeply 3–5(–7)-lobed, 9–25 × 9–30 cm, rather thick; lobes of blade much longer than wide, basal lobes usually smaller and spreading, not reflexed, sinuses broad and very deeply concave, depth of distal sinuses equal to or greater than distance from sinus to base of blade, terminal leaf lobe about 2/3 or more length of blade; margins entire to serrulate, apex acute to acuminate; surfaces abaxially and adaxially glabrescent. **Pistillate inflorescences:** heads (1–)2–4; fruiting heads to 20 mm diam., lateral heads sessile or pedunculate, peduncle often eventually obscured by maturing achenes; fruiting rachis to 25 cm. **Achenes** 5–8 mm, basal hairs about 2/3 or equal to length of achenes. $2n = 42$.

Flowering spring; fruiting late fall. Often abundant in riparian forests, especially in montane canyons, extending into deserts along streams and near springs, and cultivated; 600–2000+ m; Ariz., N.Mex.; Mexico (Chihuahua, Sinaloa, and Sonora).

L. D. Benson (1943) reported intermediates between *Platanus wrightii* and *P. racemosa* in southern California. He reduced *P. wrightii* to *P. racemosa* var. *wrightii* (S. Watson) L. Benson. Most authors have retained that taxon at the specific level because of its more deeply lobed, more glabrate leaves and its often pedunculate fruiting heads. Very low yields of germinable seeds were obtained from crosses of *P. wrightii* and *P. racemosa* with *P. occidentalis* (F. S. Santamour Jr. 1972b).

22. HAMAMELIDACEAE R. Brown

· Witch-hazel Family

Frederick G. Meyer

Trees or shrubs, deciduous; pubescence stellate or sometimes simple. **Dormant buds** naked or scaly. **Leaves** alternate, simple, petiolate; stipules early deciduous. **Leaf blade** unlobed or deeply (3–)5–7-lobed, pinnately or palmately veined. **Inflorescences** axillary or terminal, capitate or spicate to ± racemose, bracteate. **Flowers** bisexual or unisexual; perianth epigynous, 1–2-seriate, 4–5-merous, often reduced or absent, hypanthium present; calyx persistent, 4–5 (–7)-lobed, or absent; petals distinct or absent; stamens 4–34; anthers basifixed, 2-locular, longitudinally dehiscent by simple slit or by 1–2 vertical valves; pistil 1, 2-carpellate; ovary half-inferior, 2-locular; placentation axile; ovules 1(–2) in each locule and apical, or many, anatropous, pendent; styles and stigmas persistent, 2 each, erect or contorted and recurved. **Fruits** capsular, with leathery exocarp and bony endocarp, loculicidal and 2-seeded or septicidal with most of many seeds per locule aborted; 1–2 viable seeds per locule (*Liquidambar*). **Seeds** fusiform, bony, lustrous, hilum apical, light colored, or seeds winged, hilum lateral (*Liquidambar*).

Genera ca. 31, species ca. 100 (3 genera, 5 species in the flora): temperate to tropical regions, e North America, Mexico, Central America, e Asia, Africa (including Madagascar), Pacific Islands, and Australia.

SELECTED REFERENCES Bogle, A. L. 1970. Floral morphology and vascular anatomy of the Hamamelidaceae: The apetalous genera of Hamamelidoideae. J. Arnold Arbor. 51: 310–366. Britton, N. L. 1905. Hamamelidaceae. In: N. L. Britton et al., eds. 1905+. North American Flora. . . . 47+ vols. New York. Vol. 22, pp. 185–187. Endress, P. K. 1987. Aspects of evolutionary differentiation of the Hamamelidaceae and the lower Hamamelidae. Pl. Syst. Evol. 162: 193–211. Endress, P. K. 1989. A suprageneric taxonomic classification of the Hamamelidaceae. Taxon 38: 371–376. Endress, P. K. 1989b. Phylogenetic relationships in the Hamamelidoideae. In: P. R. Crane and S. Blackmore, eds. 1989. Evolution, Systematics, and Fossil History of the Hamamelidae. 2 vols. Oxford. Vol. 1, pp. 227–248. [Syst. Assoc. Special Vol. 40A, B.] Ernst, W. R. 1963. The genera of Hamamelidaceae and Platanaceae in the southeastern United States. J. Arnold Arbor. 44: 193–210. Godfrey, R. K. 1988. Trees, Shrubs, and Woody Vines of Northern Florida and Adjacent Georgia and Alabama. Athens, Ga. Goldblatt, P. and P. K. Endress. 1977. Cytology and evolution in Hamamelidaceae. J. Arnold Arbor. 58: 67–71.

1. Trees with simple hairs; leaf blade palmately (3–)5(–7)-lobed and -veined; flowers unisexual, terminal, the staminate in ± racemose heads, each head a clustered mass of numerous stamens, pistillate inflorescences many flowered; capsules septicidal, fused at base into

long-pedunculate, globose, echinate heads; viable seeds winged, most seeds aborted. . . . .
. . . . .3. *Liquidambar*, p. 366
1. Shrubs or small trees with stellate pubescence; leaf blade unlobed, pinnately veined; flowers bisexual, in axillary, few-flowered clusters or in many-flowered, elongate spikes; capsules loculicidal, not fused into heads; viable seeds not winged.
  2. Inflorescences axillary, (1–)3(–5)-flowered clusters; calyx lobes 4, reflexed; petals 4, liguliform, yellow or reddish, deep red to orange; stamens 4, very short; staminodes 4; capsules solitary or 2–3 together. . . . . .1. *Hamamelis*, p. 363
  2. Inflorescences terminal, elongate, many-flowered spikes; calyx lobes 5–7(–9), erect; petals absent; stamens 12–34; filaments white, 4–17 mm; staminodes absent; capsules in groups of more than 3. . . . . 2. *Fothergilla*, p. 365

1. HAMAMELIS Linnaeus, Sp. Pl. 1: 124. 1753; Gen. Pl. ed. 5, 59. 1754 · Witch-hazel

[Greek name used by Hippocrates for medlar, *Mespilus germanica* Linnaeus]

Trilopus Adanson

Shrubs or small trees, suckering or bearing stolons, not aromatic and resinous; twigs, young leaves, and flower buds stellate-pubescent. **Bark** gray to gray-brown, smooth or slightly roughened. **Dormant buds** naked, stellate-pubescent; terminal bud and 1 of each pair of lateral buds stalked, with 2 subtending scales. **Leaves** short-petiolate. **Leaf blade** broadly elliptic to obovate, pinnately veined, base oblique, cuneate, margins repand to sinuate, apex rounded to acute or short-acuminate. **Inflorescences** axillary, (1–)3(–5)-flowered, stalked clusters. **Flowers** bisexual, appearing before or with leaves; calyx lobes 4, reflexed, adnate to ovary; petals 4, yellow or orange to deep red, liguliform, circinnate in bud, notched or truncate, sometimes pointed; stamens 4, very short within cup; anthers introrse, dehiscing by 2 valves hinged adaxially on connective; staminodes 4, opposite petals, bearing nectar; styles 2, subulate, spreading to recurved. **Capsules** solitary or 2–3 together, fused with persistent tubular calyx (stylar beaks very short), loculicidally 2-valved, woody, appressed stellate-pubescent, explosively dehiscent. **Seeds** 2 per capsule, black, glossy, bony, not winged. $x = 12$.

Species 4 (2 in the flora): temperate regions, e North America, e Asia.

In *Hamamelis,* the explosively dehiscent capsules may eject the seeds to 10 m. The Japanese species *H. japonica* Siebold & Zuccarini, with reddish to yellow flowers, suggests an affinity with *H. vernalis.* Both Asian species, *H. japonica* and *H. mollis* Oliver of China, and the hybrid *H.* ×*intermedia* Rehder (= *H. japonica* × *H. mollis*), with a number of cultivars, are widely cultivated.

SELECTED REFERENCES Bradford, J. L. and D. L. Marsh. 1977. Comparative studies of the witch hazels *Hamamelis virginiana* and *H. vernalis.* Proc. Arkansas Acad. Sci. 31: 29–31. De Steven, D. 1983. Floral ecology of witch-hazel (*Hamamelis virginiana*). Michigan Bot. 22: 163–171. Fulling, E. H. 1953. American witch-hazel—History, nomenclature, and modern utilization. Econ. Bot. 7: 359–389. Jenne, G. E. 1966. A Study of Variation in North American *Hamamelis* L. (Hamamelidaceae). M.S. thesis. Vanderbilt University. Sargent, C. S. 1890–1902. The Silva of North America. . . . 14 vols. Boston and New York. Vol. 5, pp. 3–5. Sargent, C. S. [1902–]1905–1913. Trees and Shrubs. . . . 2 vols. Boston and New York. Vol. 2, pp. 137–138. Shoemaker, D. N. 1905. On the development of *Hamamelis virginiana.* Bot. Gaz. 39: 248–266. Steyermark, J. A. 1934. *Hamamelis virginiana* in Missouri. Rhodora 36: 97–100.

1. Flowers appearing in autumn, faintly fragrant; petals pale to deep yellow, rarely reddish, 10–20 mm; staminodes conspicuously dilated; leaves not persistent in winter, blade broad-elliptic to nearly rounded or obovate, base strongly oblique and rounded, sometimes somewhat cuneate, surfaces abaxially pale green, not glaucous; plants suckering . . . . 1. *Hamamelis virginiana*
1. Flowers appearing in winter, distinctly fragrant; petals reddish or deep red to orange, occasionally yellow, 7–10 mm; staminodes not dilated or slightly so; leaves often persistent in winter, blade mostly obovate, base narrowed to somewhat cuneate, rarely rounded, weakly oblique, surfaces often abaxially glaucous; plants stoloniferous . . . . 2. *Hamamelis vernalis*

1. Hamamelis virginiana Linnaeus, Sp. Pl. 1: 124. 1753 · Witch-hazel, café du diable, hamémelis [E] [F]

Hamamelis androgyna Walter; *H. corylifolia* Moench; *H. dioica* Walter; *H. macrophylla* Pursh; *H. virginiana* var. *angustifolia* Nieuwland; *H. virginiana* var. *orbiculata* Nieuwland; *H. virginica* Linnaeus var. *macrophylla* (Pursh) Nuttall; *H. virginica* var. *parvifolia* Nuttall; *Trilopus dentata* Rafinesque; *T. estivalis* Rafinesque; *T. nigra* Rafinesque; *T. nigra* var. *catesbiana* Rafinesque; *T. parvifolia* (Nuttall) Rafinesque; *T. rotundifolia* Rafinesque; *T. virginica* (Linnaeus) Rafinesque

Shrubs or small trees, to 6(–10.6) m, suckering, forming dense clumps. **Leaves** not persistent in winter; petioles 6–15(–20) mm. **Leaf blade** broad-elliptic to nearly rounded or obovate, 3.7–16.7 × 2.5–13 cm, base strongly oblique and rounded, sometimes somewhat cuneate and weakly oblique, apex acute to short-acuminate or broadly rounded; surfaces abaxially pale green, not glaucous. **Flowers** appearing in autumn, faintly fragrant; calyx adaxially yellow-green to yellow; petals pale to deep yellow, rarely reddish, 10–20 mm; staminodes conspicuously dilated. **Capsules** 10–14 mm. **Seeds** 5–9 mm. $2n = 24$.

Flowering fall (Oct–Nov [Dec]). Dry woodland slopes, moist woods, bluffs, and high hammocks; 0–1500 m; N.B., N.S., Ont., Que.; Ala., Ark., Conn., Del., D.C., Fla., Ga., Ill., Ind., Iowa, Ky., La., Maine, Md., Mass., Mich., Minn., Miss., Mo., N.H., N.J., N.Y., N.C., Ohio, Okla., Pa., R.I., S.C., Tenn., Tex., Vt., Va., W.Va., Wis.

Hamamelis virginiana exhibits a complex range of variation, not easily reconciled taxonomically, especially in the leaves and flowers. In the northern part of the range, the leaves are larger, averaging 9 × 2.6 cm, the petals are bright yellow, and the plants are normally shrubby. In South Carolina, Georgia, and Florida, the leaves are usually smaller, averaging 6.2 × 4.1 cm, the petals are distinctly pale yellow, and the plants sometimes attain small tree proportions, to 30 cm in trunk diameter. Such plants have been referred to as *H. virginiana* var. *macrophylla*. On the Ozark Plateau, *H. virginiana* and *H. vernalis* are sympatric. There the petals of *H. virginiana* are often reddish at the base, indicating the role of hybridization in that part of the range. Infraspecific taxa are not recognized for *H. virginiana* because no consistently defined pattern of variation or geographic correlation can be identified with this plant.

Hamamelis virginiana was well known as a medicinal plant by Native Americans. Cherokee, Chippewa, Iroquois, Menominee, Mohegan, and Potowatomi tribes used it as a cold remedy, dermatological aid, febrifuge, gynecological aid, eye medicine, kidney aid, and in other ways (D. E. Moerman 1986).

Witch-hazel was subsequently used by the early European settlers in similar ways. A tea of the leaves was employed for a variety of medicinal purposes. The twigs were used as divining rods (water-witching), thus giving the vernacular name to the plant. Modern uses employ both the bark and leaves, and a good demand still exists for the pleasant-smelling water of witch-hazel, derived from the leaves and bark. The products are used in skin cosmetics, shaving lotions, mouth washes, eye lotion, ointments, and soaps.

Hamamelis virginiana is sometimes cultivated, largely for its autumn flowering.

The largest known tree of *Hamamelis virginiana*, 10.6 m in height with a trunk diameter of 0.4 m, is recorded from Bedford, Virginia (American Forestry Association 1994).

2. Hamamelis vernalis Sargent, Trees & Shrubs 2: 137. 1908 · Ozark witch-hazel [E]

Hamamelis vernalis var. *tomentella* (Rehder) E. J. Palmer

Shrubs, 2–4 m, stoloniferous, with widespreading roots. **Leaves** often persistent in winter; petioles 7–15(–18) mm. **Leaf blade** mostly obovate, 7–13 × 6.7–13 cm, base narrowed to somewhat cuneate, rarely rounded and weakly oblique, apex acute or rounded; surfaces often abaxially glaucous. **Flowers** appearing in winter on naked branches, distinctly fragrant; calyx often adaxially deep purple; petals reddish or deep red to orange, occasionally yellow, 7–10 mm; staminodes not dilated or slightly so. **Capsules** 10–15 mm. **Seeds** 7–9 mm. $2n = 24$.

Flowering winter (Dec–Mar). Gravel bars and rocky stream banks, developing thickets, rarely on wooded hillsides; 100–400 m; Ark., Mo., Okla.

In the original description of *Hamamelis vernalis*, C. S. Sargent erroneously figured the flowers of *H. virginiana* (plate 156) and failed to mention either the fragrance or the distinctive color range of the flowers in *H. vernalis*. He stated only that the petals were light yellow.

Hamamelis vernalis is restricted to the Ozark Plateau of Missouri, Oklahoma, and Arkansas, often in close proximity with the more widespread *H. virginiana*. It is difficult to explain the restricted occurrence of *H. vernalis* in the Ozark area, although ancient geology of the area, with predominently Paleozoic rocks, makes it a well-known refugium. *Hamamelis vernalis* and *H. virginiana* are sympatric, sometimes growing

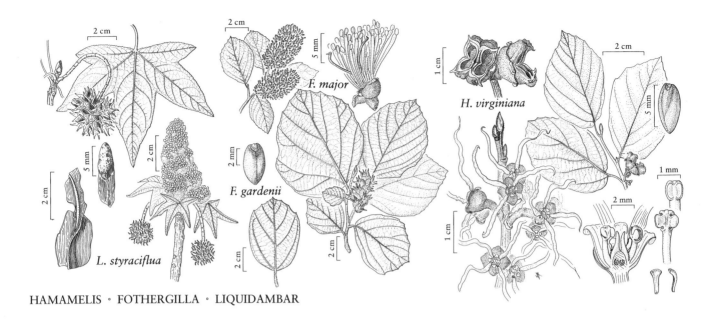

HAMAMELIS ◦ FOTHERGILLA ◦ LIQUIDAMBAR

within 30 m of each other, yet their identity is maintained, and the two species are easily distinguished through a composite of diagnostic characters (J. L. Bradford and D. L. Marsh 1977). *Hamamelis vernalis* shows an unusual color range of the flowers. Plants growing side by side commonly differ in flower color, varying from orange to deep red or, occasionally, deep

yellow. Sometimes flower color varies on the same plant; e.g., petals that are initially deep red can later fade to yellow.

Hamamelis vernalis is not well known in cultivation and is infrequently planted. It is desirable for the fragrance and color variation of the flowers.

2. FOTHERGILLA Linnaeus, Syst. Veg. ed. 13, 418. 1774 · Witch-alder [for Dr. John Fothergill, 1712–1780, London physician and patron of early American botany]

Shrubs, usually multitrunked, from short stolons, often forming dense clumps, not aromatic or resinous; twigs, leaves, and flower buds stellate-pubescent. **Bark** light gray on mature branches, smooth. **Dormant buds** naked, densely stellate-pubescent; terminal bud short-stalked or sessile, with 2–4 subtending scales. **Leaves** short-petiolate. **Leaf blade** elliptic to obovate to oblong, ovate, to nearly orbiculate, unlobed, pinnately veined, base oblique, cuneate, truncate to rounded, margins transparent, crenate or serrate-dentate, occasionally undulate to entire, apex rounded to acute. **Inflorescences** terminal, elongate, many-flowered, dense spikes, sessile or short-pedunculate. **Flowers** bisexual, fragrant, basal flowers functionally staminate, appearing before leaves; calyx lobes 5–7(–9), erect, minute, irregular, forming shallowly campanulate hypanthium with adnate androecium; sepals connate; petals absent; stamens 12–32, conspicuous, in centrifugal sequence on rim of hypanthium; filaments white, 4–17 mm, conspicuous, the longest somewhat club-shaped distally; anthers dehiscing by 2 flaps; staminodes absent; ovary adnate to hypanthium, ca. 1/3 its length; styles 2, becoming hornlike, with recurved tip. **Capsules** in groups of more than 3, stylar beaks prominent, appressed stellate-pubescent mixed with longer, straight hairs on calyx rim and distally, loculicidal. **Seeds** 2 per capsule, reddish brown, glossy, bony, not winged. $x = 12$.

Species 2 (2 in the flora): temperate regions, e North America.

The species of *Fothergilla* are spring-flowering shrubs with inflorescences of apetalous flowers and heads of showy white stamens. Both species are in cultivation. The relationship of *Fothergilla* to other genera of the Hamamelidaceae is unclear. The genus is restricted to the southeastern United States.

SELECTED REFERENCE Weaver, R. E. 1969. Studies in the North American genus *Fothergilla* (Hamamelidaceae). J. Arnold Arbor. 50: 599–619.

1. Shrubs, 3–10(–20) dm, branches slender; inflorescences 1.5–4.2 × 1.5–3.5 cm; stipules 1.5–4(–6.1) mm; leaf blade elliptic-oblong, obovate, or somewhat orbiculate, base symmetric, 1.9–6 × 1.3–4.5 cm, distal margins dentate and mucronate-toothed or entire; veins 4–5 pairs . . . . 1. *Fothergilla gardenii*
1. Shrubs, 7–65 dm, branches robust; inflorescences 3–6 × 2–3 cm; stipules 2.8–7(–10.2) mm; leaf blade broadly elliptic or somewhat orbiculate to obovate, base asymmetric, 2.5–13.5 × 4.2–12.5 cm, distal margins sinuate to repand and mucronate-toothed, rarely entire; veins (4–)5–6(–7) pairs . . . 2. *Fothergilla major*

1. Fothergilla gardenii Linnaeus, Syst. Veg. ed. 13, 418. 1774 E F

Fothergilla carolina (Linnaeus) Britton; *F. parvifolia* Kearney

Shrubs, 3–10(–20) dm; branches slender. **Leaves:** stipules 1.5–4 (–6.1) mm; petioles 3–8(–12) mm. **Leaf blade** elliptic-oblong, obovate, or somewhat orbiculate, symmetric, 1.9–6 × 1.3–4.5 cm, base rounded to truncate, rarely oblique, proximal margins entire, distal margins dentate and mucronate-toothed or entire, apex acute to obtuse or rounded; surfaces abaxially glaucous or green, adaxially green, both surfaces stellate-pubescent; veins 4–5 pairs. **Inflorescences** sessile or short-pedunculate, 1.5–4.2 × 1.5–3.5 cm. **Flowers:** calyx lobes obsolete in fruit; stamens 12–24; filaments 4–12 mm. **Fruiting spikes** 2.3–4.3 × 1.5–2 cm. **Capsules** 6–12 mm. **Seeds** 4–6 mm, apex blunt. 2*n* = 48.

Flowering spring (Mar–May). Sandy pine woods, sphagnum swamps and bogs, Atlantic and Gulf coastal plains; 0–185 m; Ala., Fla., Ga., N.C., S.C.

Nomenclaturally *Fothergilla alnifolia* Linnaeus f. is illegitimate.

2. Fothergilla major Loddiges, Bot. Cab. 16: plate 1520. 1829 E F

Fothergilla monticola Ashe

Shrubs, 7–65 dm; branches robust. **Leaves:** stipules 2.8–7 (–10.2) mm; petioles 3–10(–18) mm. **Leaf blade** broadly elliptic or somewhat orbiculate to obovate, asymmetric, 2.5–13.5 × 4.2–12.5 cm, base rounded to truncate, rarely cuneate, often oblique, proximal margins entire, distal margins coarsely sinuate to repand, rarely entire, apex short-acuminate to rounded and mucronate; surfaces abaxially glaucous or green, adaxially green, both surfaces stellate-pubescent or nearly glabrous; veins (4–)5–6 (–7) pairs. **Inflorescences** nearly sessile to short-pedunculate, 3–6 × 2–3 cm. **Flowers:** calyx lobes persistent in fruit; stamens (10–)22–34; filaments 6–17 mm. **Fruiting spikes** 3.5–7 × 1.5–2.5 cm. **Capsules** 5.5–13 mm. **Seeds** 5–6(–8) mm, apex pointed. 2*n* = 72.

Flowering spring (Apr–May). Bluffs, dry rocky woodlands, talus slopes, riverbanks, upper piedmont to mountains; 150–1300 m; Ala., Ark., Ga., N.C., S.C., Tenn.

The disjunct occurrence of *Fothergilla major* in Arkansas is a recent discovery.

3. LIQUIDAMBAR Linnaeus, Sp. Pl. 2: 999. 1753; Gen. Pl. ed. 5, 434. 1754

· Sweetgum [Latin *liquidus*, fluid, liquid, and Arabic *ambar*, amber]

Trees, aromatic and resinous, glabrous or with simple hairs. **Bark** gray-brown, deeply furrowed; twigs and branches sometimes corky-winged. **Dormant buds** scaly, pointed, shiny, resinous, sessile. **Leaves** long-petiolate. **Leaf blade** fragrant when crushed, (3–)5(–7)-lobed, palmately veined, base deeply cordate to truncate, margins glandular-serrate, apex of each lobe long-acuminate. **Inflorescences** terminal, many-flowered heads; staminate heads in pedunculate racemes, each head a cluster of many stamens; pistillate heads pendent, long-pedunculate, the flowers ± coalesced. **Flowers** unisexual, staminate and pistillate on same plant, appearing with leaves; calyx and corolla absent. **Staminate flowers:** anthers dehiscing longitudinally; staminodes absent. **Pistillate flowers** pale green to greenish yellow; staminodes 5–8; styles indurate and spiny in fruit, incurved. **Capsules** many, fused at base into long-pedunculate, spheric, echinate heads, 2-beaked, glabrous, septicidal. **Seeds** numerous, mostly aborting, 1–2 viable in each capsule, winged. $x = 16$.

Species 3(–4) (1 in the flora): North America, e Asia, and Turkey.

SELECTED REFERENCES Bogle, A. L. 1986. The floral morphology and vascular anatomy of the Hamamelidaceae: Subfamily Liquidambaroideae. Ann. Missouri Bot. Gard. 73: 325–347. Duncan, W. H. 1959. Leaf variation in *Liquidambar styraciflua* L. Castanea 24: 99–111. Holm, T. 1930. Leaf-variation in *Liquidambar styraciflua* L. Rhodora 32: 95–105. Sargent, C. S. 1890–1902. The Silva of North America. . . . 14 vols. Boston and New York. Vol. 5, pp. 7–12. Schery, R. W. 1952. Plants for Man. New York. Schmitt, D. 1966. Pistillate inflorescence of sweetgum (*Liquidambar styraciflua* L.). Silvae Genet. 15: 33–35. Wilson, P. 1905. Altingiaceae: *Liquidambar*. In: N. L. Britton et al., eds. 1905+. North American Flora. . . . 47+ vols. New York. Vol. 22, p. 189.

1. **Liquidambar styraciflua** Linnaeus, Sp. Pl. 2: 999. 1753 · Sweetgum, redgum F W1

Liquidambar barbata Stokes; *L. gummifera* Salisbury; *L. macrophylla* Oersted; *L. styraciflua* var. *mexicana* Oersted

Trees, to 41 m. **Leaves:** stipules linear-lanceolate, 3–4 mm, early deciduous, leaving 2 stipular scars adaxially near base of petiole; petioles (44–)60–100(–150) mm. **Leaf blade** palmately lobed, main lobes sometimes again dentate-lobed, 7–19(–25) × 4.4–16 cm; surfaces glabrous, except young leaves hairy on veins and main vein-axils at base with persistent reddish brown simple hairs. **Staminate flowers** in pedunculate clusters, 3–6 cm; perianth absent; stamens 4–8(–10) per flower, 150–176(–300) per cluster, falling after anthesis. **Pistillate flowers** without perianth; hypanthium disclike, with 5–8 staminodes around cycle of disc lobes; ovary (1–)2-locular; styles 2; stigmas introrsely curved. **Capsular heads** brown at maturity, globose, 2.5–4 cm diam. (including indurate styles). **Seeds** apically winged, 8–10 mm, marked with resin ducts; aborted seeds brownish, 1–2 mm, unwinged, irregular, resembling sawdust. $2n = 32$.

Flowering spring (Mar–May). Fields, woodlands, flood plains, low hammocks, swamps, riverbanks; 0–800 m; Ala., Ark., Conn., Del., D.C., Fla., Ga., Ill., Ind., Ky., La., Md., Miss., Mo., N.J., N.Y., N.C., Ohio, Okla., Pa., S.C., Tenn., Tex., Va., W.Va.; Mexico; Central America (Belize and Honduras to Nicaragua).

The leaves of *Liquidambar styraciflua*, fragrant when bruised, turn deep red to crimson in autumn. Although leaf variation is common in *L. styraciflua*, this deviation is randomly distributed and without any definable geographic correlation. *Liquidambar styraciflua* is often cultivated; a number of cultivars have been introduced in cultivation.

Liquidambar styraciflua was well known as a medicinal plant by Native Americans. Cherokee, Choctaw, Houma, Koasati, and Rappahannock tribes used it in various ways, especially the gum, bark, and root, as an antidiarrheal, dermatological aid, gynecological aid, sedative, febrifuge, and for related uses (D. E. Moerman 1986).

Liquidambar styraciflua produces a balsamic oleoresin called American styrax or storax, a thick, clear, brownish yellow, semisolid or solid with a pronounced aromatic odor. It is chewed as a sweet, natural gum. The balsam is collected from the inner bark of the tree after wounding or deliberate gashing. It is used in soaps and cosmetics, as a fixative in perfumes, adhesives, lacquers, and incense, and as a flavoring in tobacco. The wood is used for cabinet making, furniture, veneer, interior finish, barrels, and wooden dishes. Medicinally the gum has been used for catarrh, coughs, dysentery, sores, and wounds of both humans and domestic animals.

The largest known tree of *Liquidambar styraciflua*, 41.4 m in height with a trunk diameter of 2.25 m, is recorded from Craven County, North Carolina (American Forestry Association 1994).

23. ULMACEAE Mirbel

• Elm Family

Susan L. Sherman-Broyles

William T. Barker

Leila M. Schulz

Trees or shrubs, deciduous (sometimes tardily deciduous in *Ulmus*). **Bark** smooth to deeply fissured or scaly and flaky; sap watery. **Leaves** alternate [opposite], distichous [or not], simple; stipules present; petiole present. **Leaf blade:** base often oblique, margins entire or serrate, crenate, or toothed; venation pinnate to palmate-pinnate. **Inflorescences** axillary, cymes, racemes, fascicles, or flowers solitary, arising from branchlets of previous season (e.g., *Ulmus*) or of current season (e.g., *Celtis*). **Flowers** bisexual or unisexual, staminate and pistillate on same [different] plants; sepals persistent, (1–)5(–9), connate [distinct], imbricate or valvate; petals absent; stamens usually as many as calyx lobes, hypogynous, opposite calyx lobes, erect in bud; filaments free or arising from calyx tube, distinct, curved or sigmoid in bud; anthers 2-locular, dehiscence longitudinal; pistils 1, 2(–3)-carpellate; ovary 1(–2)-locular; ovules 1 per locule, pendulous from apex of locule, anatropous or amphitropous; styles (1–)2, distinct, receptive stigmatic area decurrent on distal inner margin of style branch. **Fruits** fleshy drupes, samaras, or nutlike. **Seeds** 1; arils absent; endosperm absent to scanty, consisting of 1 layer of thick-walled cells; embryo straight or curved.

Genera ca. 18, species ca. 150 (4 genera, 19 species in the flora): tropical and north temperate regions.

Plants of this family are wind pollinated (anemophilous).

Ulmaceae are frequently divided into two subfamilies, Ulmoideae and Celtoideae; they are sometimes separated into two families, Ulmaceae and Celtidaceae (I. A. Grudzinskaya 1965). These subfamilial or familial distinctions are supported by flavonoid chemistry (D. E. Giannasi and K. J. Niklas 1977; D. E. Giannasi 1978), pollen morphology (M. Zavada 1983), and some anatomic structures (E. M. Sweitzer 1971). Typically the Ulmoideae have flavonols, strictly pinnately veined leaves, and dry fruits; the Celtoideae have glycoflavones, pinnipalmately veined leaves, and drupaceous fruits. Some genera (e.g., *Zelkova*, with pinnately veined leaves and drupaceous fruits) are intermediate, and various authors place them in different subfamilies.

In this treatment *Ulmus* and *Planera* are considered part of the subfamily Ulmoideae; *Celtis* and *Trema* are in subfamily Celtoideae. *Zelkova serrata* is widely cultivated as an ornamental tree in North America, but it is not known to be naturalized in the flora. Chemical similarities between subfamilies include the presence of proanthocyanins with some tannins and scattered mucilaginous cells or canals. Additionally, members of the family share a strong tendency toward mineralization of the cell walls with calcium carbonate or silica and possess solitary or clustered crystals of calcium oxalate.

Ulmaceae include trees and shrubs of horticultural importance.

SELECTED REFERENCES Barker, W. T. 1986. Ulmaceae. In: Great Plains Flora Association. 1986. Flora of the Great Plains. Lawrence, Kans. Pp. 119–123. Elias, T. S. 1970. The genera of Ulmaceae in the southeastern United States. J. Arnold Arbor. 51: 18–40. Giannasi, D. E. 1978. Generic relationships in the Ulmaceae based on flavonoid chemistry. Taxon 27: 331–344. Giannasi, D. E. and K. J. Niklas. 1977. Flavonoids and other constituents of fossil Miocene *Celtis* and *Ulmus* (Succor Creek Flora). Science 197: 765–767. Grudzinskaya, I. A. 1965. The Ulmaceae and reasons for distinguishing the Celtidoideae as a separate family Celtidaceae Link. Bot. Zhurn. (Moscow & Leningrad) 52: 1723–1749. Sweitzer, E. M. 1971. The comparative anatomy of Ulmaceae. J. Arnold Arbor. 52: 523–585. Zavada, M. 1983. Pollen morphology of Ulmaceae. Grana 22: 23–30.

1. Leaf blade pinnately veined; fruits dry, nutlike or samaras.
   2. Flowers bisexual; fruits samaras . . . . 1. *Ulmus*, p. 369
   2. Flowers normally unisexual, inflorescences usually with a few bisexual flowers; fruits nutlike . . . . 2. *Planera*, p. 376
1. Leaf blade palmately veined at base, pinnately veined over remainder of blade; fruits drupes.
   3. Leaf blade entire or serrate to ca. 3/4 length; flowers solitary or in few-flowered clusters; drupes 1 . . . . 3. *Celtis*, p. 376
   3. Leaf blade crenate to serrate for entire length; flowers 12–20, in cymes . . . . 4. *Trema*, p. 379

1. ULMUS Linnaeus, Sp. Pl. 1: 225. 1753; Gen. Pl. ed. 5, 106. 1754 · Elm, orme [Latin *ulmus*, elm]

Susan L. Sherman-Broyles

Trees, less often shrubs, to 35 m; crowns variable. **Bark** gray, brown, or olive to reddish, tan, or orange, deeply furrowed, sometimes with plates (smooth when young in *Ulmus glabra*). **Branches** unarmed, slender to stout, some with corky wings; twigs glabrous to pubescent. **Leaves** sometimes tardily deciduous; stipules falling early. **Leaf blade** ovate to obovate or elliptic, base usually oblique, sometimes cordate or rounded to cuneate, margins serrate to doubly serrate; venation pinnate. **Inflorescences** fascicles, racemes, or cymes, pedunculate or subsessile, subtended by 2 bracts. **Flowers** on branches of previous season, appearing in spring before leaves or in fall, bisexual, pedicellate or sessile; calyx 3–9-lobed; stamens 3–9; styles persistent, deeply 2-lobed. **Fruits** samaras, usually flattened, membranously winged. $x = 14$.

Species 20–40 (10 in the flora): temperate regions, Northern Hemisphere, most in Eurasia.

A recent chloroplast DNA study (S. J. Wiegrefe et al. 1994) has led to the proposal of a new subgeneric and sectional classification of elms. The chloroplast DNA data are supported by morphologic, chemical, and nuclear ribosomal DNA evidence and indicate that the "rock" or hard elms (*Ulmus serotina, U. thomasii, U. crassifolia,* and *U. alata*) may belong in two sections of the same subgenus and are more closely related than indicated by previous subgeneric treatments (C. K. Schneider 1916; I. A. Grudzinskaya 1980).

Most identification manuals include the introduced species, *Ulmus glabra, U. procera,* and *U. parvifolia,* and indicate that they are frequently naturalized. That may well be true. Available herbarium specimens are often inadequately labeled or do not reflect current occurrences.

Ease of naturalization can be neither corroborated nor disproved. I include the three species in this treatment because they are known to persist and sometimes naturalize locally where the species have been planted. Extensive field work and collection of *U. glabra* and *U. procera* are needed to document their naturalized distributions. *Ulmus parvifolia* reportedly has been widely planted in groves and hedgerows in the Midwest and might well be expected to have become naturalized in more rural settings.

Street and field elms throughout much of North America have been killed by Dutch elm disease. The pathogen responsible for the disease is *Ceratocystis ulmi,* a fungus native to Europe that was first discovered in North America in Colorado in the 1930s. Since the rapid spread of the disease in the 1960s, much research has been devoted to development of disease-resistant elms (R. J. Stipes and R. J. Campana 1981). Various hybridization projects, including cloning of disease-resistant elms by the Elm Research Institute, have been started across the country. *Ulmus parvifolia* and *U. pumila* have varying degrees of disease resistance and are utilized as shade trees or in breeding programs (see *U. pumila* below). Apparently Dutch elm disease also affects *U. parviflora, U. glabra,* and *U. procera;* certainly the latter two species are more common as seedlings than as trees.

SELECTED REFERENCES Green, P. S. 1964. Registration of cultivar names in *Ulmus*. Arnoldia (Jamaica Plain) 24: 41–80. Sherman, S. L. 1987. Flavonoid Systematics of *Ulmus* L. in the United States. M.S. thesis. University of Georgia. Sherman, S. L. and D. E. Giannasi. 1988. Foliar flavonoids of *Ulmus* in eastern North America. Biochem. Syst. & Ecol. 16: 51–56. Stipes, R. J. and R. J. Campana, eds. 1981. Compendium of Elm Diseases. St. Paul. Stockmarr, J. 1974. SEM studies on pollen grains of North European *Ulmus* species. Grana 14: 103–107. Wheeler, E., C. A. LaPasha, and Regis B. Miller. 1988. Wood anatomy of elm (*Ulmus*) and hackberry (*Celtis*) species native to the United States. I. A. W. A. Bull., N.S. 10: 5–26. Wiegrefe, S. J., K. J. Sytsma, and R. P. Guries. 1994. Phylogeny of elms (*Ulmus*, Ulmaceae): Molecular evidence for a sectional classification. Syst. Bot. 19: 590–612.

Key using reproductive characters

1. Flowers appearing in late summer–fall.
  2. Inflorescences long, to 5 cm, racemose, 8–12-flowered. . . . . 6. *Ulmus serotina*
  2. Inflorescences short, much less than 5 cm, fasciculate, mostly 2–5-flowered.
    3. Calyx lobes 6–9, hairy. . . . . .3. *Ulmus crassifolia*
    3. Calyx lobes (3–)4–5, glabrous. . . . . 10. *Ulmus parvifolia*
1. Flowers appearing in winter–early summer.
  4. Flowers on slender, drooping pedicels, in racemose cymes to 5 cm, long-pendulous; samaras pubescent and margins short-ciliate; seeds not thickened, inflated . . . . 7. *Ulmus thomasii*
  4. Flowers clustered in short racemes or dense fascicles usually less than 2.5 cm; samaras pubescent or marginally ciliate, not both, or glabrous; seeds thickened, not inflated.
    5. Flowers and fruits drooping on elongate pedicels or in short racemes; samaras marginally ciliate.
      6. Inflorescences in short racemes, not pendulous; calyx deeply lobed, symmetric; samaras lanceolate to oblong-elliptic, cilia white, 1–2 mm . . . . 2. *Ulmus alata*
      6. Inflorescences pendulous fascicles; calyx shallowly lobed, slightly asymmetric; samaras ovate, cilia yellow to white, 1 mm or less. . . . . .1. *Ulmus americana*
    5. Flowers and fruits sessile or subsessile, not pendulous, in dense fascicles, not racemes; samaras not marginally ciliate.
      7. Calyx glabrous; samaras glabrous. . . . . 5. *Ulmus pumila*
      7. Calyx pubescent; samaras pubescent, sometimes only on apical margin.
        8. Calyx villous; samaras mostly glabrous, apex marginally pubescent . . . . 9. *Ulmus procera*
        8. Calyx reddish pubescent; samaras pubescent.
          9. Calyx shallowly lobed; samaras pubescent on body only. . . . . 4. *Ulmus rubra*
          9. Calyx lobed at least halfway; samaras pubescent only on central vein of wing. . . . . .8. *Ulmus glabra*

Key using vegetative characters

1. Leaf blade less than 7 cm, margins crenate to serrate, apex obtuse or acute.
  2. Leaf blade adaxially harshly pubescent, margins crenate to doubly serrate, apex obtuse. . . . . .3. *Ulmus crassifolia*
  2. Leaf blade adaxially glabrous, margins singly to doubly serrate, apex acute.
    3. Mature leaf blade 2–3.5 cm wide, lateral veins forking to 3 times per side. . . . . 5. *Ulmus pumila*
    3. Mature leaf blade less than 2 cm wide, lateral veins forking 5 or more times per side . . . . . . 10. *Ulmus parvifolia*

1. Leaf blade greater than 7 cm (except in *U. alata*), margins at least partially doubly serrate, apex acute to acuminate.
   4. Leaf blade abaxially tomentose or villous with tufts of hair in axils of veins, adaxially harshly scabrous or strigose to sparsely scabrous.
      5. Bud scales red, margins red-tomentose; leaf blade marginally ciliate. . . . . 4. *Ulmus rubra*
      5. Bud scales reddish brown to dark brown or brown, margins white- or pale-ciliate; leaf blade not ciliate.
         6. Leaf blade base strongly oblique, one side strongly overlapping, covering petiole; branchlets not corky . . . . 8. *Ulmus glabra*
         6. Leaf blade base oblique, not covering petiole; branchlets with corky wings . . . . 9. *Ulmus procera*
   4. Leaf blade abaxially tomentose to glabrous, pubescence in axils of veins present or absent, adaxially glabrous.
      7. Leaf blade lanceolate to oblanceolate, less than 7 cm, base somewhat cordate to oblique; young and old-growth branches with prominent and regular corky wings . . . . 2. *Ulmus alata*
      7. Leaf blade ovate to obovate, more than 7 cm, bases oblique; old-growth branches smooth or with irregular corky wings or ridges.
         8. Leaf blade abaxially glabrous to slightly pubescent with tufts in axils of veins; branches smooth, not winged . . . . 1. *Ulmus americana*
         8. Leaf blade abaxially soft-pubescent, pubescence absent from or not tufted in axils of veins; branches usually with corky wings.
            9. Leaf blade lanceolate to ovate, abaxially yellow-gold pubescent, pubescence absent from axils, adaxially yellow-green; buds and branches glabrous; leaves mostly 7–8(–14) cm . . . . 6. *Ulmus serotina*
            9. Leaf blade obovate, abaxially white-pubescent, pubescence not tufted in axils, adaxially dark green; buds and young branches pubescent; leaves 9–11(–16) cm. . . . . 7. *Ulmus thomasii*

1. **Ulmus americana** Linnaeus, Sp. Pl. 1: 226. 1753 · American elm, orme d'Amérique E F

Ulmus americana var. *aspera* Chapman; *U. americana* var. *floridana* (Chapman) Little; *U. floridana* Chapman

Trees, 21–35 m; crowns spreading, commonly vase-shaped. **Bark** light brown to gray, deeply fissured or split into plates. **Wood** soft. **Branches** pendulous, old-growth branches smooth, not winged; twigs brown, pubescent to glabrous. **Buds** brown, apex acute, glabrous; scales reddish brown, pubescent. **Leaves:** petiole ca. 5 mm, glabrous to pubescent. **Leaf blade** oval to oblong-obovate, 7–14 × 3–7 cm, base oblique, margins doubly serrate, apex acute to acuminate; surfaces abaxially glabrous to slightly pubescent, tufts in axils of veins, adaxially glabrous to scabrous. **Inflorescences** fascicles, less than 2.5 cm, flowers and fruits drooping on elongate pedicels; pedicel 1–2 cm. **Flowers:** calyx shallowly lobed, slightly asymmetric, lobes 7–9, margins ciliate; stamens 7–9; anthers red; stigmas white-ciliate, deeply divided. **Samaras** yellow-cream when mature, sometimes tinged with reddish purple (s range of species), ovate, ca. 1 cm, narrowly winged, margins ciliate, cilia yellow to white, to 1 mm. **Seeds** thickened, not inflated. $2n = 56$.

Flowering winter–early spring. Alluvial woods, swamp forests, deciduous woodlands, fencerows, pastures, old fields, waste areas; planted as street trees; 0–1400 m; Man., N.B., N.S., Ont., P.E.I., Que., Sask.; Ala., Ark., Conn., Del., D.C., Fla., Ga., Ill., Ind., Iowa, Kans., Ky., La., Maine, Md., Mass., Mich., Minn., Miss., Mo., Mont., Nebr., N.H., N.J., N.Y., N.C., N.Dak., Ohio, Okla., Pa., R.I., S.C., S.Dak., Tenn., Tex., Vt., Va., W.Va., Wis., Wyo.

Ulmus americana is reported as widely escaped in Idaho, outside the natural range of this taxon. It is occasionally cultivated outside its native distribution, and it has escaped sporadically from cultivation. It is also reported as naturalized in Arizona, but I have seen no specimens.

Ulmus americana is the state tree for Massachusetts and for North Dakota.

The American elm is susceptible to numerous diseases, including Dutch elm disease. *Ulmus americana* has been a street and shade tree of choice because of its fast growth and pleasant shape and size. The species still exists in substantial numbers both as shade trees and in nature.

Numerous infraspecific taxa have been recognized in *Ulmus americana* (A. J. Rehder 1949; P. S. Green 1964).

Native American tribes frequently used parts of *Ulmus americana* for a variety of medicinal purposes, including treatment of coughs and colds, sore eyes, dysentery, diarrhea, broken bones, gonorrhea, and pulmonary hemorrhage, as a gynecological aid, as a bath for appendicitis, and as a wash for gun wounds (D. E. Moerman 1986).

2. Ulmus alata Michaux, Fl. Bor.-Amer. 1: 173. 1803 · Winged elm, wahoo E F W1

Ulmus americana Linnaeus var. *alata* (Michaux) Spach

Trees, 10–18 m; crowns open. **Bark** light brown to gray with shallow ridges and plates. **Wood** hard. **Branches:** young and old-growth branches with opposite, prominent, regular corky wings; twigs reddish brown, pubescent to glabrous. **Buds:** apex acute; scales brown to rusty, slightly pubescent. **Leaves:** petiole ca. 2.5 mm, pubescent. **Leaf blade** lanceolate to oblanceolate, 3–6.9 × 0.6–3.2 cm, base somewhat cordate to oblique, margins doubly serrate, apex acute; surfaces abaxially with trichomes on veins, tufts of pubescence in axils of veins, adaxially glabrous to scabrous. **Inflorescences** short racemes, not pendulous, less than 2.5 cm; pedicel 2–7 mm, not fully expanded until fruiting stage. **Flowers:** calyx deeply lobed, symmetric, lobes 5; stamens 5; anthers red. **Samaras** gray-tan, often reddish tinged, lanceolate to oblong-elliptic, ca. 8 mm, narrowly winged, margins ciliate, cilia white, 1–2 mm. **Seeds** slightly thickened, not inflated. $2n = 28$.

Flowering late winter–early spring. Alluvial woods and deciduous woodlands, especially dry, acidic woodlands and glades, along fencerows, waste areas; planted as street trees; 0–600 m; Ala., Ark., Fla., Ga., Ill., Ind., Kans., Ky., La., Miss., Mo., N.C., Ohio, Okla., S.C., Tenn., Tex., Va.

Often planted as a shade tree in the southern United States, *Ulmus alata* is also cultivated outside North America.

The name *Ulmus pumila* was incorrectly applied to this species by Walter in 1788.

3. Ulmus crassifolia Nuttall, Trans. Amer. Philos. Soc., n.s. 5: 169. 1837 · Cedar elm

Trees, 24–27 m; crowns rounded to narrow. **Bark** light brown with shallow ridges and large plates. **Wood** hard. **Branches** often with opposite corky wings; twigs reddish brown, pubescent. **Buds** brown, apex acute, pubescent; scales dark brown, shiny, glabrous. **Leaves:** petiole ca. 1.5 mm, pubescent. **Leaf blade** ovate to elliptic, 2.5–5 × 1.3–2 cm, base oblique or rounded to cuneate, margins crenate to doubly serrate, apex obtuse; surfaces abaxially softly pubescent, adaxially harshly pubescent. **Inflorescences** fascicles, 2–5-flowered, 0.5 cm; pedicel 0.75–1 cm. **Flowers:** calyx deeply lobed, more than 1/2 its length, lobes 6–9, hairy; stamens 5–6, anthers reddish purple; stigmas white, pubescent, exserted and spreading. **Samaras** green to tan, elliptic to oval, ca. 0.75–1. cm, pubescent, margins ciliate, cilia ca. 0.5 mm. **Seeds** somewhat thickened, not inflated. $2n = 28$.

Flowering late summer–early fall. Stream banks, low woods, low hillsides, roadsides, waste places; sometimes shade trees; 0–500 m; Ark., Fla., La., Miss., Okla., Tenn., Tex.; n Mexico.

Except for the Suwanee River Valley in Florida, *Ulmus crassifolia* has not been found east of Webster County, Mississippi. It hybridizes with *U. serotina*.

SELECTED REFERENCES McDaniel, S. 1967. *Ulmus crassifolia* in Florida. Sida 3: 115–116. Sherman-Broyles, S. L., S. B. Broyles, and J. L. Hamrick. 1992. Geographic distribution of allozyme variation in *Ulmus crassifolia* Nutt. Syst. Bot. 17: 33–41.

4. Ulmus rubra Muhlenberg, Trans. Amer. Philos. Soc. 3: 165. 1793 · Slippery elm, orme rouge E F

Ulmus crispa Willdenow; *U. fulva* Michaux; *U. pendula* Willdenow; *U. pubescens* Walter

Trees, 18–35 m; crowns open. **Bark** brown to red, deeply and irregularly furrowed. **Wood** soft. **Branches** spreading; twigs gray, densely pubescent when young, glabrous with age. **Buds** obtuse; scales red, margins red-tomentose. **Leaves:** petiole 5–7 mm, pubescent. **Leaf blade** obovate to ovate, 8–16 × 5–7.5 cm, base oblique, margins doubly serrate in distal 1/2–3/4, singly serrate proximally, basal teeth 6 or fewer, rounded, less distinct, apex acuminate; surfaces abaxially tomentose, dense tufts of white hair in axils of major veins, adaxially harshly scabrous, trichomes pointed toward apex, margins ciliate. **Inflorescences** dense fascicles less than 2.5 cm, 8–20-flowered, flowers and fruits not pendulous, subsessile; pedicel 1–2 mm. **Flowers:** calyx green to reddish, shallowly lobed, lobes 5–9, reddish pubescent; stamens 5–9; anthers reddish; stigmas exserted, pink reddish. **Samaras** yellow to cream, suborbiculate, 12–18 mm diam., broadly winged, samaras pubescent on body only, rusty-tomentose, margins glabrous. **Seeds** thickened, not inflated. $2n = 28$.

Flowering late winter–early spring. Lower slopes, alluvial flood plains, stream banks, riverbanks, and wooded bottom lands; 0–600(–900) m; Ont., Que.; Ala., Ark., Conn., Del., D.C., Fla., Ga., Ill., Ind., Iowa, Kans., Ky., La., Maine, Md., Mass., Mich., Minn., Miss., Mo., Nebr., N.H., N.J., N.Y., N.C., N.Dak., Ohio, Okla., Pa., S.C., S.Dak., Tenn., Tex., Vt., Va., W.Va., Wis.

Scabrous-leaved *Ulmus rubra* is often confused

with *U. americana*. Where ranges coincide, *U. rubra* may freely intergrade with *Ulmus pumila* Linnaeus, a widely introduced species.

The red-rust, mucilaginous inner bark of *Ulmus rubra* is distinctive; its sticky slime gives this tree its common name of slippery elm. Native American tribes used *Ulmus rubra* for a wide variety of medicinal purposes, including inducing labor, soothing stomach and bowels, treating dysentery, coughs, colds, and catarrhs, dressing burns and sores, and as a laxative (D. E. Moerman 1986). Various preparations utilizing it are still marketed.

5. Ulmus pumila Linnaeus, Sp. Pl. 1: 226. 1753 · Siberian elm ⬚W1

Ulmus campestris Linnaeus var. *pumila* Maximowicz; *U. manshurica* Nakai; *U. turkestanica* Requien

Trees, 15 to 30 m; crowns open. **Bark** gray to brown, deeply furrowed with interlacing ridges. **Wood** brittle. **Branches** not winged; twigs gray-brown, pubescent. **Buds** dark brown, ovoid, glabrous; scales light brown, shiny, glabrous to slightly pubescent. **Leaves:** petiole 2–4 mm, glabrous. **Leaf blade** narrowly elliptic to lanceolate, 2–6.5 × 2–3.5 cm, base generally not oblique, margins singly serrate, apex acute; surfaces abaxially with some pubescence in axils of veins, adaxially glabrous; lateral veins forking to 3 times per side. **Inflorescences** tightly clustered fascicles, 6–15-flowered, 0.5 cm, flowers and fruits not pendulous, sessile. **Flowers:** calyx shallowly lobed, lobes 4–5, glabrous; stamens 4–8; anthers brownish red; stigmas green, lobes exserted. **Samaras** yellow-cream, orbiculate, 10–14 mm diam., broadly winged, glabrous, tip notched 1/3–1/2 its length. **Seeds** thickened, not inflated. $2n = 28$.

Flowering late winter–early spring. Commonly escaping from cultivation, waste places, roadsides, fencerows; 0–2200 m; N.B., Ont., Que.; Ala., Ariz., Ark., Calif., Colo., Conn., D.C., Fla., Ga., Idaho, Ill., Ind., Iowa, Kans., Ky., La., Md., Mass., Mich., Minn., Mo., Mont., Nebr., Nev., N.J., N.Mex., N.Y., N.Dak., Ohio, Okla., Pa., S.Dak., Tenn., Tex., Utah, Va., Wis., Wyo.; Asia.

Ulmus pumila probably occurs in Vermont and West Virginia, but it has not been documented for those states.

Planted for quick-growing windbreaks, *Ulmus pumila* has weak wood, and its branches break easily in mature trees. It is easily distinguished from other North American elms by its singly serrate leaf margins. *Ulmus pumila* is similar to *U. parvifolia* Jacquin with its small, singly serrate leaves. *Ulmus parvifolia*, however, has smooth bark that sheds from tan to orange, and it flowers and sets fruit in the fall.

6. Ulmus serotina Sargent, Bot. Gaz. 27: 92. 1899 · September elm, red elm ⬚E⬚F

Trees, to 21 m; crowns spreading, broadly rounded. **Bark** light brown to reddish with shallow fissures. **Wood** hard. **Branches** spreading to pendulous, often developing irregular corky wings with maturity; twigs brown to gray, pubescent to glabrous. **Buds** brown, apex acute, glabrous; scales dark brown, glabrous. **Leaves:** petiole ca. 6 mm, glabrous to pubescent. **Leaf blade** oblong-obovate, 7–10 × 3–4.5 cm, base oblique, margins doubly serrate, apex acuminate; surfaces abaxially yellow-gold soft-pubescent, pubescence absent from axils of veins, adaxially yellow-green, glabrous. **Inflorescences** racemes, 8–12-flowered, long, to 5 cm; pedicel 0.5–1 cm. **Flowers:** calyx lobed almost to base, lobes 5–6; stamens 5–6; anthers yellow-red; stigmas white, pubescent. **Samaras** light brown, ovoid to elliptic, 1–1.5 cm, narrowly winged, pubescent, margins densely ciliate, tip deeply notched. **Seeds** thickened, not inflated. $2n = 28$.

Flowering late summer–fall. Limestone bluffs, stream sides, rich woods; 0–400 m; Ala., Ark., Ga., Ill., Miss., Okla., Tenn., Tex.

Ulmus serotina is infrequent, and few populations are found outside of Tennessee. It reputedly is highly susceptible to Dutch elm disease (W. H. Duncan and M. B. Duncan 1988), and it is sometimes cultivated. *Ulmus serotina* hybridizes with *U. crassifolia*, and plants have been informally designated *U. arkansana*, an unpublished name. In Arkansas and Oklahoma where hybrid swarms are common, specimens are often difficult to assign to either taxon.

SELECTED REFERENCE Duncan, W. H. and M. B. Duncan. 1988. Trees of the Southeastern United States. Athens, Ga. Pp. 234–238.

7. Ulmus thomasii Sargent, Silva 14: 102. 1902 · Rock elm, orme de Thomas ⬚E⬚F

Ulmus racemosa Thomas 1831, not Borkhausen 1800

Trees, to 30 m; crowns oblong. **Bark** gray, deeply fissured with broad, flattened ridges. **Wood** hard. **Branches** short-spreading, young branches pubescent, old-growth with 3–5 prominent, irregular, corky wings; twigs reddish, pubescent. **Buds** brown, ovoid, acute, pubescent; scales brown, pilose on outer surface, ciliate on margins. **Leaves:** petiole ca. 5 mm, pubescent. **Leaf blade** obovate to oblong-oval, (2.5–)9–11(–16) × 2.5–5 cm, base oblique, margins doubly serrate, apex short-acuminate; surfaces abaxially white-pubescent, pubescence not tufted in axils of veins, adaxially dark green, usually glabrous, sometimes scabrous. **Inflorescences** racemose cymes, long-pendulous, (7–)10(–13)-flowered, to 5 cm; pedicel 0.5–1 cm. **Flowers:** calyx deeply lobed, divided nearly to middle, lobes 7–8; stamens 5–8; anthers dark purple; stigmas greenish, pubescent. **Samaras** elliptic to oval, 1.5–2.2 cm, broadly winged, pubescent, margins short-ciliate, apex shallowly notched. **Seeds** inflated, not thickened. $2n = 28$.

Flowering spring. Rocky slopes, limestone outcrops, rich woods, flood plains, stream banks; 30–900 m; Ont.; Ark., Ill., Ind., Iowa, Kans., Ky., Mich., Minn., Mo., Nebr., N.H., N.J., N.Y., Ohio, S.Dak., Tenn., Vt., W.Va., Wis.

8. Ulmus glabra Hudson, Fl. Angl., 95. 1762 · Scotch elm, wych elm, broad-leaved elm

Ulmus montana Withering; *U. scabra* Miller

Trees, to 40 m; trunks often multiple; crowns spreading, broadly rounded or ovate. **Bark** gray, smooth, furrowed with age. **Wood** hard. **Branches** spreading to pendulous, glabrous, branchlets lacking corky wings; twigs ash-gray to red-brown, villous when young. **Buds** obtuse; scales reddish brown, glabrous to marginally white-ciliate. **Leaves:** petiole 2–7 mm, densely villous. **Leaf blade** elliptic to obovate, (4–)7–14(–16) × (3–)4.5–8(–10) cm, base strongly oblique with lowermost lobe strongly overlapping, covering petiole, margins doubly serrate, apex long-acuminate to cuspidate, sometimes with 3 acuminate lobes at broad apex; surfaces abaxially pale green, villous with woolly tufts in vein axils, adaxially dark green, strigose to scabrous, margins not ciliate. **Inflorescences** dense fascicles, 8–20-flowered, less than 2.5 cm, flowers and fruits not pendulous; pedicel short, 0.4–0.8 mm, densely pubescent. **Flowers:** calyx lobed to ca. 1/2 length, lobes 4–8, reddish pubescent; stamens 5–6, purplish; stigmas reddish, with white pubescence. **Samaras** light greenish brown, elliptic to obovate with blunt or rounded tip, 1.5–2.5 × 1–1.8 mm, broadly winged, pubescent only along central vein of wing, apical cleft minute, obscured by persistent, curved styles. **Seeds** thickened, not inflated. $2n = 28$.

Flowering spring–early summer. Along margins of woodlands and disturbed sites; 0–300 m; introduced;

Conn., Maine, Mass., N.Y., R.I., Vt.; native to Europe and Asia.

In the absence of carefully documented naturalized populations, the North American distribution of *Ulmus glabra* is very poorly known. The species is established locally in British Columbia and California, and probably elsewhere. It has been reported from Newfoundland, Nova Scotia, Ontario, District of Columbia, Illinois, Iowa, Michigan, Minnesota, Missouri, Pennsylvania, and Virginia.

Ulmus glabra is similar to *U. rubra* in leaf morphology but may be readily distinguished by its smooth bark and glabrous samaras. Some of the weeping elms found in cultivation are varieties of *U. glabra*. The common name, wych, is derived from Gallic and means "drooping."

9. Ulmus procera Salisbury, Prodr. Stirp. Chap. Allerton, 391. 1796 · English elm, English cork elm

Trees, to 40 m; crowns open. **Bark** grayish brown, deeply ridged, flaking. **Wood** hard. **Branches:** old-growth branchlets with corky ridges; twigs reddish brown, villous to scabrous. **Buds** ovoid; scales dark brown, sparsely pubescent, marginally pale-ciliate. **Leaves:** petiole 3–12 mm, villous to scabrous. **Leaf blade** broadly lanceolate-elliptic to ovate, (3–)7–10 × (3–)4–6(–10) cm, base strongly oblique, not covering petiole, margins doubly serrate, apex acute to acuminate; surfaces abaxially villous with woolly tufts in vein axils, pale in contrast to adaxial surface, adaxially dark green, glabrous to sparsely scabrous, margins not ciliate. **Inflorescences** dense clusters of subsessile flowers borne on lateral shoots resembling short racemes, flowers and fruits not pendulous. **Flowers:** calyx green to reddish purple or tan, shallowly lobed, lobes 5–8, marginally villous; stamens 3–5(–6); anthers dark brown, globose; stigmas white, puberulous, persistent in fruit, slender lobes incurved. **Samaras** light brown, darker brown to red in area covering seed, orbiculate, about as long as broad, 0.9–1.8 × 0.9–1.6 cm, broadly winged, glabrous except for pubescence along margin of apex, apex shallowly notched. **Seeds** thickened, not inflated. $2n = 28$.

Flowering early–late spring. Persisting, sometimes naturalizing locally where species has been planted; 0–400 m; introduced; Ont.; Calif., Conn., Ill., Mass., Mo., N.Y., R.I.; native to Europe.

In the absence of carefully documented naturalized populations, the North American distribution of *Ulmus procera* is very poorly known. It is locally established in British Columbia, Arizona, Louisiana, Maryland, and Michigan. It has been reported from

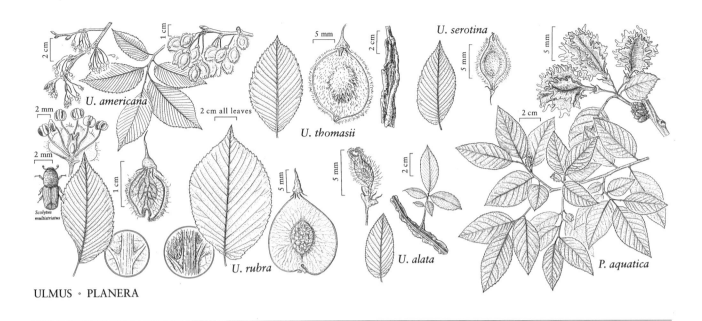

U. americana

U. thomasii

U. serotina

U. rubra

U. alata

P. aquatica

ULMUS · PLANERA

Georgia, North Carolina, Virginia, and West Virginia. Off-shoots from the root systems are often found close to planted trees, forming dense clones, especially in and around urban centers.

Some authors have combined *Ulmus minor* Miller and *U. procera* Salisbury. This treatment follows that of Tutin and colleagues (T. G. Tutin et al. 1964–1980, vol. 1, pp. 65–66), in which the species were regarded as separate. Reports of *Ulmus minor* Miller, in the strict sense, in North America are not confirmed. Hybrids of *Ulmus procera* and *U. glabra* are reported from New York (R. S. Mitchell 1988) and may be spreading. Both *U. procera* and *U. glabra* are involved in the parentage of Dutch elm, *Ulmus* ×*hollandica* Pallas.

SELECTED REFERENCES Melville, R. 1978. On the discrimination of species in hybrid swarms with special reference to *Ulmus* and the nomenclature of *U. minor* Mill. and *U. carpinifolia* Gled. Taxon 27: 345–351. Richens, R. H. 1977. New designations in *Ulmus minor* Mill. Taxon 26: 582–593.

10. **Ulmus parvifolia** Jacquin, Pl. Hort. Schoenbr. 3: 6, plate 262. 1798 · Chinese elm, lacebark elm

Trees, 25 m; crowns rounded, open. **Bark** olive green to gray, shedding in irregular, tan to orange plates. **Branches** long-pendulous, not winged; twigs tan to dark brown, glabrous to pubescent. **Buds** acute to obtuse; scales brown, pubescent. **Leaves:** petiole 2–6(–8) mm, glabrous or sparsely pubescent with short hairs. **Leaf blade** elliptic to ovate-obovate, (3.5–)4–5(–6) × 1.5–2.5 cm, base oblique, margins mostly singly serrate (some doubly serrate), apex acute; surfaces abaxially pale, glabrate, adaxially dark green, lustrous, glabrous; lateral veins forking 5 or more times per side. **Inflorescences** fascicles, (2–)3–4(–8)-flowered; pedicel 8–10 mm. **Flowers:** calyx reddish brown, deeply lobed, lobes (3–)4–5, glabrous; stamens 3–4; anthers reddish; stigma lobes white-pubescent, exserted, recurved and spreading with maturity. **Samaras** green to light brown, elliptic to ovate, ca. 1 cm, not winged, seeds nearly filling samara, notched at apex, glabrous. **Seeds** thickened, not inflated. $2n = 28$.

Flowering late summer–early fall. In woods and in disturbed sites; 0–400 m; introduced; Calif., D.C., Ga., Ky., Maine, Md., Mass., Va.; native to Asia (China and Japan).

Ulmus parvifolia appears to naturalize more readily than *U. procera* or *U. glabra*. It has been reported but not documented from Idaho and West Virginia.

Ulmus parvifolia is valued in cultivation for its pleasing form and ornamental bark. It is ruderal primarily in the southeastern United States.

2. PLANERA J. F. Gmelin, Syst. Nat. 1: 150. 1791 · Water-elm, planer tree [for Johann Jacob Planer, German botanist and physician, 1743–1789]

William T. Barker

Trees, to 12 m; crowns low, broad. **Bark** gray or light brown, scaly and flaky. **Wood** weak, brittle. **Branches** unarmed, spreading; twigs hirtellous when young, glabrate. **Leaves:** stipules falling early; petiole present. **Leaf blade** rhombic-ovate to ovate-oblong, base oblique to rounded or cuneate, margins unequally serrate, apex acute; venation pinnate. **Inflorescences:** staminate fascicles; pistillate solitary flowers or 2–3 together. **Flowers** normally unisexual, pedicellate, staminate and pistillate on same plants, a few bisexual flowers usually present; calyx deeply 4–5-lobed; stamens when present 4–5. **Fruits** nutlike, ellipsoid, with irregular, thickened ribs.

Species 1 (1 in the flora): North America (se United States).

SELECTED REFERENCE Small, J. K. 1903. Flora of the Southeastern United States. . . . New York.

1. Planera aquatica J. F. Gmelin, Syst. Nat. 1: 150. 1791 · Water-elm, planer tree [E] [F]

Leaves: petiole 3–6 mm. **Leaf blade** 3–8 × (2–)2.5(–4) cm. **Flowers:** if bisexual, ovary stalked, 1-locular, ovoid, tubercular; styles 2, reflexed. **Fruits** stalked, compressed, ca. 8 mm, leathery, hirtellous, covered with irregular warty excrescences. **Seeds** ovoid.

Flowering spring (Apr–May). Swamps, streams, lakes, alluvial flood plains, often forming large stands; [0–200 m]; Ala., Ark., Fla., Ga., Ill., Ky., La., Miss., Mo., N.C., Okla., S.C., Tenn., Tex.

Planera aquatica is endemic to southeastern United States.

3. CELTIS Linnaeus, Sp. Pl. 2: 1043. 1753; Gen. Pl. ed. 5, 467. 1754 · Hackberry, sugarberry, bois inconnu [Classical Latin, Pliny's name for *Celtis australis* Linnaeus, the "lotus" of the ancient world]

Trees or rarely shrubs, to 30 m; crowns spreading. **Bark** usually gray, smooth or often fissured and conspicuously warty. **Branches** without or with thorns, slender, glabrous or pubescent. **Leaves:** stipules falling early. **Leaf blade** deltate to ovate to oblong-lanceolate, base oblique or cuneate to rounded, margins entire or serrate-dentate; venation 3(–5)-pinnate. **Inflorescences:** staminate inflorescences cymes or fascicles; pistillate solitary or few-flowered clusters. **Flowers** usually unisexual, staminate and pistillate on same plants, along with a few bisexual flowers, pedicellate on branches of current year, appearing in mid or late spring. **Staminate flowers:** filaments incurved in bud, exserted after anthesis; gynoecium minute, rudimentary. **Pistillate flowers:** calyx slightly to deeply 4(–5)-lobed; stamens 4–5, inserted on pilose receptacle, included, often nonfunctional filaments usually shorter than in staminate flowers, rarely absent; anthers ovate, face to face in bud, extrorse; ovaries sessile, ovoid, 1-locular; styles short, sessile, divided into 2 divergent, elongate, reflexed lobes, lobes entire or 2-cleft. **Fruits** fleshy drupes, ovoid or globose; outer mesocarp thick, firm, inner mesocarp thin, fleshy; stones thick walled, ripening in autumn, persisting after leaves fall. $x = 10$.

Species ca. 60 (6 in the flora): tropical and temperate regions, worldwide.

The hackberries provide important wildlife habitat, forming thickets that give shelter and fleshy drupes that ripen in autumn, persist after leaves fall, and supply winter food for birds and mammals. The treatment presented here is a simplified circumscription of species with no elaboration of infraspecific variation or interspecific hybridization. The group is taxonomically complex and in need of revision.

SELECTED REFERENCE   Correll, D. S. and M. C. Johnston. 1970. Manual of the Vascular Plants of Texas. Renner, Tex.

1. Branches with thorns; leaf blade usually less than 2 cm wide . . . . 6. *Celtis pallida*
1. Branches without thorns; leaf blade usually much more than 2 cm wide.
   2. Leaf blade typically elliptic-lanceolate to ovate-lanceolate, apex sharply acute to acuminate, margins mostly entire. . . . . 1. *Celtis laevigata*
   2. Leaf blade typically broadly to narrowly ovate to oblong-lanceolate, apex blunt or obtuse to abruptly long-acuminate, acute, or short-acuminate, margins variable.
      3. Leaf blade abaxially white tomentose; fruits usually light brown; near San Antonio, Texas . . . 2. *Celtis lindheimeri*
      3. Leaf blade abaxially essentially glabrous or with coarse pubescence mainly on veins; fruits mostly reddish orange to purple; widespread.
         4. Leaf blade typically 4.5 cm or less, margins entire or somewhat serrate above middle. . . . . . . 4. *Celtis reticulata*
         4. Leaf blade mostly 5 cm or more, margins coarsely serrate for at least part of length.
            5. Shrubs or small trees; leaf blade serrate and sparingly toothed toward apex, entire proximally; fruits orange to brown to cherry red . . . . 5. *Celtis tenuifolia*
            5. Trees; leaf blade conspicuously serrate to well below middle; fruits dark orange to purple- or blue-black . . . . 3. *Celtis occidentalis*

1. Celtis laevigata Willdenow, Enum. Pl., suppl: 67. 1814
· Sugarberry, palo blanco [F]

Celtis laevigata var. *anomala* Sargent; *C. laevigata* var. *brachyphylla* Sargent; *C. laevigata* var. *smallii* (Beadle) Sargent; *C. laevigata* var. *texana* Sargent; *C. mississippiensis* Bosc; *C. smallii* Beadle

Trees, to 30 m; trunks to 1 m diam., crowns broad, spreading. **Bark** light gray, smooth or covered with corky warts. **Branches** without thorns, often pendulous, young branches pubescent at first, then glabrous. **Leaves:** petiole 6–10 mm. **Leaf blade** typically elliptic-lanceolate to ovate-lanceolate, (4–)6–8 (–15) × (2–)3–4 cm, thin and membranaceous to leathery, base broadly cuneate to rounded, margins entire or rarely with a few long teeth, apex sharply acute to acuminate; surfaces glabrous or nearly so, margins ciliate. **Inflorescences:** flowers solitary or few-flowered clusters at base of leaves. **Drupes** orange to brown or red when ripe, nearly orbicular, 5–8 mm diam., beakless; pedicel 6–15 mm. **Stones** 4.5–7 × 5–6 mm. $2n$ = 20, 30, and 40.

Flowering late spring–early fall (May–Oct). In rich bottom lands along streams, in flood plains, and on rocky slopes; 0–300 m; Ala., Ark., Fla., Ga., Ill., Ind., Kans., Ky., La., Md., Miss., Mo., N.C., Okla., S.C., Tenn., Tex., Va., W.Va.; n Mexico.

The Houma used preparations from the bark of *Celtis laevigata* to treat sore throats and venereal disease (D. E. Moerman 1986).

2. Celtis lindheimeri Engelmann ex K. Koch, Dendrologie 2: 434. 1872 · Lindheimer hackberry [C]

Celtis helleri Small

Trees, to 12 m; trunks to 15 dm diam; crowns widely spreading, much branched. **Bark** with corky warts. **Branches** without thorns, spreading to pendulous, smooth; young branches and twigs villous-pubescent. **Leaves:** petiole 3–4(–11) mm. **Leaf blade** ovate to ovate-lanceolate or oblong-lanceolate, 4–9 × 2–5 cm, leathery, base rounded to cordate, margins entire or with a few serrations, apex obtuse to acute or shortly acuminate; surfaces abaxially white-tomentose, adaxially dark green, scabrous. **Inflorescences** erect dense clusters, 2–9-flowered, at base of leaves. **Drupes** light brown, globose, 7–9 mm diam., smooth. **Seeds** ovoid, prominently 4-ribbed, reticulate.

Flowering late winter–spring (Mar–May). Ravines and brushlands; of conservation concern; 100–200 m; Tex.; n Mexico.

Celtis lindheimeri warrants further study. It grows near San Antonio, and it is known from the Edwards Plateau of Texas and from northern Mexico. The extent of its range within Texas is uncertain.

3. Celtis occidentalis Linnaeus, Sp. Pl. 2: 1044. 1753
· Hackberry, micocoulier occidental, bois inconnu F

Celtis occidentalis var. *canina*
(Rafinesque) Sargent; *C. occidentalis*
var. *crassifolia* (Lamarck) A. Gray; *C.
occidentalis* var. *pumila* (Pursh)
A. Gray; *C. pumila* Pursh; *C. pumila*
var. *deamii* Sargent

Trees or shrubs, size varying
greatly in response to habitat;
crowns rounded. **Bark** gray,
deeply furrowed, warty with age. **Wood** light yellow,
weak. **Branches** without thorns, spreading, young
branches mostly pubescent. **Leaves:** petiole 0.5–1.2
mm. **Leaf blade** lance-ovate to broadly ovate or deltate, 5–12 × 3–6(–9) cm (on fertile branches), leathery, base oblique or obliquely somewhat acuminate,
margins conspicuously serrate to well below middle,
teeth 10–40, apex acuminate; surfaces scabrous. **Inflorescences** dense pendulous clusters. **Drupes** dark orange to purple- or blue-black when ripe, orbicular, to
7–11(–20) mm diam., commonly with thick beak;
pedicel to 15 mm. **Stones** cream colored, 7–9 × 5–8
mm, reticulate. 2*n* = 20, 30, and 40.

Flowering late winter–spring (Mar–May). In rich
moist soil along streams, on flood plains, on rock, on
wooded hillsides, and in woodlands; 0–1800 m; Man.,
Ont., Que.; Ala., Ark., Colo., Conn., Del., D.C., Ga.,
Ill., Ind., Iowa, Kans., Ky., Maine, Md., Mass., Mich.,
Minn., Miss., Mo., Nebr., N.H., N.J., N.Y., N.C.,
N.Dak., Ohio, Okla., Pa., R.I., S.C., S.Dak., Tenn.,
Tex., Vt., Va., W.Va., Wis., Wyo.

Celtis occidentalis is valued as an ornamental street
tree because of its tolerance to drought.

Native Americans used decoctions prepared from
the bark of *Celtis occidentalis* medicinally as an aid in
menses and to treat sore throat (D. E. Moerman
1986).

This is a highly variable species. Segregates named
as varieties follow an east-west geographic gradient
and are based primarily on leaf size, shape, and pubescence.

4. Celtis reticulata Torrey, Ann. Lyceum Nat. Hist. New
York 2: 247. 1828 · Netleaf hackberry, palo blanco

Celtis brevipes S. Watson; *C. douglasii*
Planchon; *C. laevigata* Willdenow var.
reticulata (Torrey) L. D. Benson; *C.
occidentalis* Linnaeus var. *reticulata*
(Torrey) Sargent; *C. reticulata* var.
vestita Sargent

Trees or shrubs, (1–)7(–16) m;
trunks rarely 6 dm diam.;
crowns ± rounded. **Bark** gray
with corky ridges. **Branches** without thorns, upright,
villous when young. **Leaves:** petiole 3–8 mm. **Leaf
blade** ovate, 2–4.5(–7) × 1.5–3.5 cm, thick, rigid,
base cordate or occasionally oblique, margins entire or

somewhat serrate above middle, apex obtuse to acute
or somewhat acuminate; surfaces pubescent, abaxially
yellow-green, adaxially gray-green, grooved, scabrous
or not. **Inflorescences** of 1–4 flowers in axils of young
leaves. **Drupes** reddish or reddish black when ripe, orbicular, (5–)8–10 mm diam., beaked; pedicel (4–)10–
14 mm.

Flowering late winter–spring. On dry hills, often on
limestone or basalt, ravine banks, rocky outcrops, and
occasionally in sandy soils; 300–2300 m; Ariz., Calif.,
Colo., Idaho, Kans., Nev., N.Mex., Okla., Oreg., Tex.,
Utah, Wash., Wyo.; n Mexico.

The Navajo-Kayenta used *Celtis reticulata* medicinally in the treatment of indigestion (D. E. Moerman
1986).

5. Celtis tenuifolia Nuttall, Gen. N. Amer. Pl. 1: 202. 1818
· Dwarf hackberry, micocoulier de Soper E

Celtis georgiana Small; *C. occidentalis*
Linnaeus var. *georgiana* (Small) Ahles;
C. pumila Pursh var. *georgiana*
(Small) Sargent; *C. tenuifolia* var.
georgiana (Small) Fernald & B. G.
Schubert; *C. tenuifolia* var. *soperi*
B. Boivin

Shrubs or small trees, to 8 m;
trunks to 30 cm; crowns nar-
row. **Bark** light gray, furrowed, warty. **Branches** without thorns, upright to spreading, irregular. **Leaves:**
petiole 6–10 mm. **Leaf blade** ovate to occasionally
ovate-elliptic, (2–)5–8 × (1–)3–4 cm, base unequal, 1
side rounded, margins mostly entire, serrate and sparingly toothed toward apex, apex blunt, acute, or
short-acuminate; surfaces abaxially gray-green, harshly
pubescent, adaxially dark gray-green, scabrous. **Inflorescences:** flowers solitary or few-flowered clusters.
Drupes orange to brown or cherry red, glaucous, orbicular, 5–8 mm diam., beakless; pedicel 3–13 mm.
Stones cream colored, 5–7 × 5–6 mm, reticulate.

Flowering spring (Apr–May). On slopes and along
streams in open woods; 0–500 m; Ont.; Ala., Ark.,
Del., D.C., Fla., Ga., Ill., Ind., Kans., Ky., La., Md.,
Mich., Miss., Mo., N.J., N.C., Ohio, Okla., Pa., S.C.,
Tenn., Tex., Va., W.Va.

6. Celtis pallida Torrey in W. H. Emory, Rep. U.S. Mex.
Bound. 2: 203. 1859 · Desert hackberry F

Celtis spinosa Sprengel var. *pallida*
(Torrey) M. C. Johnston; *Momisia
pallida* (Torrey) Planchon

Shrubs, to 3 m; crowns
rounded. **Bark** gray, smooth.
Branches spreading, flexuous,
whitish gray, with thorns, puberulent; thorns single or in
pairs, 3–25 mm. **Leaf blade**
ovate to ovate-oblong, to 2–3 × 1.5–2 cm, thickish,

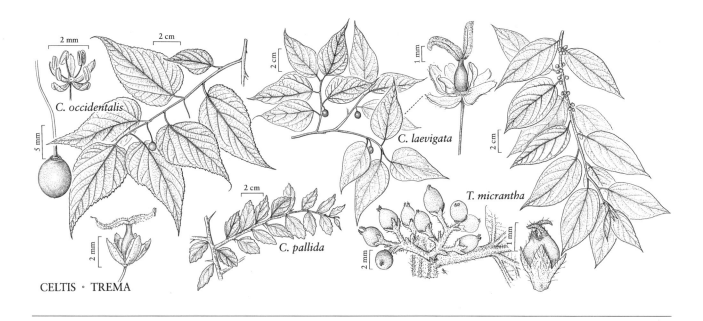

margins entire or crenate-dentate, apex rounded to acute; surfaces scabrous. **Inflorescences** cymes, 3–5-flowered, longer than petiole, flowers mostly staminate on proximal branches, terminal flower bisexual. **Drupes** orange, yellow, or red, ovoid, 6–7 mm; pedicel 1–2 mm.

Flowering late winter–spring (Mar–May). In deserts, canyons, mesas, washes, foothills, thickets, brushland, and grassland near gravelly or well-drained sandy soil; 1000–1300 m; Ariz., Fla., N.Mex., Tex.; Mexico; Central America; South America (to n Argentina).

Celtis pallida is closely related to *C. iguanaea* (Jacquin) Sargent from Mexico. Reports of *C. iganaea* from Florida and Texas are unconfirmed. *Celtis iguanaea* can be identified by its longer leaves (to 4 cm wide), small fruits (4–5 mm), and single thorns. Its fruits have acid, juicy pulp.

4. TREMA Loureiro, Fl. Cochinch. 2: 562. 1790 · Trema, nettletree

Leila M. Shultz

Trees or shrubs, spindly, to 15 m; crowns variable. **Bark** dark brown or gray brown, smooth, shallowly furrowed, sometimes appearing warty with large, raised lenticels. **Branches** unarmed, stout; twigs hoary tomentose. **Leaves:** stipules ephemeral. **Leaf blade:** ovate, base oblique to cordate or truncate, margins crenate to serrate; venation palmate at base, pinnate on remainder of blade. **Inflorescences** cymes, compact to lax, 12–20-flowered. **Flowers** mostly unisexual, usually staminate and pistillate on same plants, appearing after leaves on new stems, in 1 series, pedicellate; calyx 5-parted. **Bisexual flowers,** if present: pedicel present; ovaries ± globose; styles persistent, 2, glabrous; stigmas 2, unbranched. **Staminate flowers:** nearly sessile; pistillodes present. **Pistillate flowers:** pedicel present; staminodes absent. **Fruits** drupes, globose, fleshy. **Stones** thick walled.

Species ca. 15 (2 in the flora): tropical and subtropical regions, North America (Florida), Mexico, West Indies, Central America, South America (to n Argentina), Asia, and Africa.

Trema cannabina Loureiro and *T. orientalis* (Linnaeus) Blume, sometimes reported for North America, are Old World species occasionally planted but not known to have escaped from cultivation.

Trema species are fast-growing pioneer trees with economically important alkaloids.

Trema is a member of the subfamily Celtoideae. Species are locally called nettletrees in reference to their superficial resemblance to members of the Urticaceae. Further studies of variation in this group are needed in the field and in the laboratory, giving special consideration to the morphologic variants within *Trema micrantha*.

SELECTED REFERENCES Baehni, C. 1937. *Trema*. In: J. F. MacBride et al., ed. 1936 + . Flora of Peru. 6 + parts. Chicago. Part 2(2), pp. 269–270. Gardiner, R. C. 1965. Studies in the Leaf Anatomy and Geographic Distribution of *Trema*. M.S. thesis. University of Connecticut. Nevling, L. I. Jr. 1960. *Trema*. In: R. E. Woodson Jr. and R. W. Schery et al., eds. 1943–1981. Flora of Panama. 41 fasc. St. Louis. [Ann. Missouri Bot. Gard. 47: 108–110.] Small, J. K. 1903. Flora of the Southeastern United States. . . . New York. Standley, P. C. 1922–1926. Trees and Shrubs of Mexico. 5 parts. Washington. Part 2, p. 199. [Contr. U.S. Natl. Herb. 23.]

1. Leaf blade mostly longer than 5 cm, abaxially velvety white-pubescent, veins conspicuous but scarcely raised; fruits bright red-orange to yellow. . . . . 1. *Trema micrantha*
1. Leaf blade mostly shorter than 4 cm, harshly pubescent on both surfaces, venation abaxially pronounced, raised; fruits pink. . . . . 2. *Trema lamarckiana*

1. **Trema micrantha** (Linnaeus) Blume, Mus. Bot. 2: 58. 1856 · Florida trema, guacimilla [F] [W1]

Rhamnus micranthus Linnaeus, Syst. Nat. ed. 10, 2: 937. 1759; *Celtis micranthus* (Linnaeus) Swartz; *Sponia micrantha* (Linnaeus) Decaisne; *Trema floridana* Britton; *Trema melinona* Blume; *Trema micrantha* var. *floridana* (Britton) ex Small Standley & Steyermark

Shrubs to small trees, 2–5.5 (–10) m. **Bark** dark brown, smooth when young, developing small, warty projections in maturity. **Branchlets** copiously pubescent. **Leaf blade** ovate to narrowly ovate, 5–6.5(–9) × 2.5–4(–4.5) cm, base oblique to cordate, margins evenly serrate, apex acute to long-acuminate; abaxial surface softly, velvety white-pubescent; venation conspicuous but scarcely raised. **Flowers:** calyx greenish white. **Fruits** bright red-orange to yellow, 1.5–3.5 mm diam. $2n = 20$ (from Costa Rica).

Flowering most of year (Mar–Nov). Hammocks and prairies, often weedy along roadsides, in burned areas, and on calcareous ground; 0–100 m; Fla.; Mexico; West Indies; Central America; South America.

Trema micrantha, as interpreted here, is widespread in tropical regions of the New World. Small-leaved populations may be confused with *T. lamarckiana.*

The soft wood of *Trema micrantha* is suitable for the construction of tea chests and match sticks.

2. **Trema lamarckiana** (Roemer & Schultes) Blume, Mus. Bot. 2: 58. 1856 · West Indian trema

Celtis lamarckiana Roemer & Schultes, Syst. Veg. 6: 311. 1820

Small trees or shrubs, to 3 m. **Bark** gray-brown, smooth. **Branchlets** glabrous to densely pubescent. **Leaf blade** ovate, mostly 2–4(–6) × 2–2.5(–3) cm, base obtuse to nearly truncate, margins crenate, apex acute; surfaces abaxially and adaxially harshly pubescent with conspicuous pustules; venation prominent, raised abaxially, reticulate. **Flowers:** calyx pink to white. **Fruits** pink, 2–2.5 mm diam.

Flowering late winter–spring (Mar–Apr), sometimes summer–fall (Jul–Nov). Disturbed areas along margins of woodlands and roadways; 0–25 m; Fla.; West Indies.

Trema lamarckiana is endemic to the Florida Keys and West Indies.

24. CANNABACEAE Endlicher

• Hemp Family

Ernest Small

Herbs, annual or perennial, taprooted or rhizomatous, erect or twining, aromatic, pubescent with small glands and hairs, hairs with or without cystoliths (structures within cells with basal calcium carbonate concretions or crystals). **Stems** usually branched, usually ridged or furrowed. **Leaves** decussate proximally, often alternate distally, simple to palmately lobed or compound; stipules persistent, triangular. **Leaf blade:** margins serrate. **Inflorescences** axillary or terminal, bracteate; staminate inflorescences compound cymes or panicles, loose, erect or somewhat pendent; pistillate inflorescences spikes, pseudospikes, or racemes. **Flowers** unisexual, staminate and pistillate usually on different plants, when on same plants (in some populations and cultivars), then staminate flowers produced before pistillate flowers, transitional-bisexual flowers sometimes occurring. **Staminate flowers** 20–200+, pedicellate; sepals 5, hypogynous, greenish or whitish; stamens 5, hypogynous, opposite sepals; anthers dehiscing longitudinally and basipetally. **Pistillate flowers** 10–50, often paired, pairs often interpreted as cymes, subsessile, tightly covered or loosely subtended by bract or bracteole; perianth a thin undivided layer adhering to ovary, obscure; pistil 1, usually 2-carpellate; ovary superior, 1-locular; ovule 1 per locule; style short, apical; stigma 2-branched, long, filiform. **Fruits** achenes, crustaceous, covered loosely or tightly by persistent, accrescent perianth. **Seeds** fleshy; endosperm scant; embryo curved or coiled.

Genera 2, species 4 (2 genera, 3 species in the flora): nearly worldwide.

Genera in Cannabaceae have sometimes been included in Moraceae (H. A. Gleason 1952, vol. 2).

Cannabaceae are wind pollinated. They are indigenous to the temperate Northern Hemisphere, widely cultivated, often introduced, and often ruderal.

Cystoliths, including those of *Cannabis sativa,* are frequently used in police laboratories to make positive identification of leaf fragments.

SELECTED REFERENCE   Miller, N. G. 1970. The genera of the Cannabaceae in the southeastern United States. J. Arnold Arbor. 51: 185–203.

1. CANNABIS Linnaeus, Sp. Pl. 2: 1027. 1753; Gen. Pl. ed. 5, 453. 1754 · Hemp, Indian hemp, marijuana, marihuana, chanvre, cannabis [Greek *kannabis,* hemp, said to come from Arabic *kinnab* or Persian *kannab*]

Herbs, annual, erect, taprooted. **Stems** simple to well branched, without 2-branched hairs. **Leaves** palmately compound; petiole not twining, without 2-branched hairs. **Leaf blade:** surfaces abaxially sparsely to densely pubescent. **Inflorescences:** staminate inflorescences compound cymes or panicles, erect; pistillate pseudospikes, congested, erect to spreading. **Flowers:** staminate and pistillate on different plants, sometimes on same plants, especially in cultivars. **Achenes** lenticular, enclosed within enlarged perianth; embryo curved. $x = 10$.

Species 1 (1 in the flora): widespread in temperate regions, nearly worldwide.

Many populations of *Cannabis sativa* have been established largely from escapes from former cultivation and, sporadically, from clandestine cultivation.

One of the oldest cultivated plants, hemp was widely used in Neolithic China in the Yang Shao culture (ca. 4000 B.C.). Many legends understandably surround its origins and popularity. Its tough and durable fiber, excellent for rope, cordage, paper, canvas, sailcloth, and fish nets, prompted its initial spread throughout the world. The seeds are very nutritious; they are an important constituent in birdseed mixes, and the seeds, as well as the edible oil from seeds, are marketed as an excellent food source for human consumption. Oil from the seeds was also used in paints and varnishes and as fuel for lamps (B. B. Simpson and M. Conner-Ogorzaly 1986). Hemp was a major economic crop in the American colonies because of the demand for rope in agricultural, maritime, and military pursuits. Probably best known today for its psychoactive chemicals, it is used legally by physicians in the treatment of glaucoma and to relieve nausea in cancer patients undergoing chemotherapy or radiation (B. B. Simpson and M. Conner-Ogorzaly 1986).

Until 1970 marihuana was legally controlled in the United States by the Marihuana Tax Act of 1937, which levied a transfer tax for which no stamps or licenses were available to private citizens. *Cannabis* is now controlled by the Comprehensive Drug Abuse Act of 1970. In Canada marihuana has been controlled since 1938 by an amendment to the Narcotic Control Act.

The vernacular name hemp refers both to the plant and to its commercially extracted bast fibers. Most other terms refer both to the plant and to drug preparations of it.

SELECTED REFERENCES Schultes, R. E., W. M. Klein, T. Plowman, and T. E. Lockwood. 1974. *Cannabis:* An example of taxonomic neglect. Bot. Mus. Leafl. 23: 337–367. Small, E. 1979. The Species Problem in *Cannabis:* Science & Semantics. 2 vols. Toronto. Small, E. and A. Cronquist. 1976. A practical and natural taxonomy for *Cannabis.* Taxon 25: 405–435.

1. Cannabis sativa Linnaeus, Sp. Pl. 2: 1027. 1753
· Hemp, marihuana (marijuana), pot, grass, maryjane; chanvre, cannabis [F][W1]

Staminate plants usually taller, less robust than pistillate plants. **Stems** 0.2–6 m. **Leaves:** petioles 2–7 cm. **Leaflet blades** mostly 3–9, linear to linear-lanceolate, 3–15 × 0.2–1.7 cm, margins coarsely serrate; surfaces abaxially whitish green with scattered, yellowish brown, resinous dots, strigose, adaxially darker green with large, stiff, bulbous-based conic hairs. **Inflorescences** numerous. **Flowers** unisexual, often transitional flowers and flowers of opposite sex developing later. **Staminate flowers:** pedicels 0.5–3 mm; sepals ovate to lanceolate, 2.5–4 mm, puberulent; stamens caducous after anthesis, somewhat shorter than sepals; filaments 0.5–1 mm. **Pistillate flowers** ± sessile, enclosed by glandular, beaked bracteole and subtended by bract; perianth appressed to and surrounding base of ovary. **Achenes** white or greenish, mottled with purple, ovoid, somewhat compressed, 2–5 mm, with ± persistent perianth that sometimes flakes off. 2*n* = 20.

Flowering early summer–fall; staminate plants generally dying after anthesis, pistillate plants remaining dark green, persisting until frost. Well-manured, moist farmyards, and in open habitats, waste places (roadsides, railways, vacant lots), occasionally in fallow fields and open woods; 0–2000 m; introduced; principal naturalized range (see map) Ont., Que.; Ark., Conn., Del., Ill., Ind., Iowa, Kans., Ky., Maine, Md., Mass., Mich., Minn., Mo., Nebr., N.H., N.J., N.Y., N.Dak., Ohio, Okla., Pa., R.I., S.Dak., Vt., Va., W.Va., Wis.; native to Asia.

Cannabis sativa has been reported as cultivated illegally and as apparently ruderal in all provinces and states except Alaska. It has been collected least frequently in Mississippi and Idaho. It seems to be best established in the prairies and plains of central North America.

Hemp is a short-day plant; flowering depends upon the latitude of origin. Races originating closer to the equator (and generally higher in psychointoxicant) require a longer induction period for flowering than races originating farther north.

The taxonomy of *Cannabis sativa,* a polymorphic species, has been debated in scientific and legal forums.

The name *C. sativa* subsp. *indica* (Lamarck) E. Small & Cronquist has been applied to plants with a mean dry leaf content of the psychotomimetic (hallucinatory) delta-9-tetrahydrocannabinol of at least 0.3%; those with a lesser content fall under *C. sativa* subsp. *sativa.* When separate species are recognized, the name *C. indica* Lamarck has generally been applied to variants with high levels of the intoxicant chemical, whereas the name *C. sativa* Linnaeus, interpreted in a restricted sense, has generally been applied to plants selected for their yield of bast fibers in the stems. (The latter generally have taller, hollow stems with longer internodes and less branching than races selected for drug content.)

Superimposed on this dimension of variation is selection for nonabscising achenes in cultivation and abscising achenes in the wild (i.e., outside of cultivation). This is analagous to selection of nonshattering cereals from wild, shattering grasses. Achenes selected for cultivation tend to be longer than 3.8 mm and lack a basal constricted zone; by contrast, achenes selected for wild existence tend to be shorter than 3.8 mm and to have a basal constricted zone that seems to facilitate disarticulation and a mottled, persistent perianth apparently serving as camouflage.

Within *Cannabis sativa* subsp. *sativa,* the wild phase has been named *C. sativa* var. *spontanea* Vavilov (= *C. ruderalis* Janishevsky), in contrast to the domesticated *C. sativa* var. *sativa.* Within *C. sativa* subsp. *indica,* the wild phase (not to be expected in North America) has been designated *C. sativa* var. *kafiristanica* (Vavilov) E. Small & Cronquist, as distinct from the domesticated *C. sativa* var. *indica.* The chemical and morphologic distinctions by which *Cannabis* has been split into taxa are often not readily discernible, appear to be environmentally modifiable, and vary in a continuous fashion. For most purposes it will suffice to apply the name *Cannabis sativa* to all plants encountered in North America.*

The Iroquois used *Cannabis sativa* medicinally to convince patients that they had recovered. They also found it useful as a stimulant (D. E. Moerman 1986).

*An alternative view of the taxonomic status of variants within *Cannabis* is presented by authors who espouse a polytypic generic concept. For information on this complex and lively debate, see: L. C. Anderson (1974) and W. A. Emboden (1974). The long history of cultivation of this plant, and its selection by humans for particular characteristics, obscures natural patterns of morphologic variations, leaving fertile ground for a debate on species circumscription.–Ed.

2. HUMULUS Linnaeus, Sp. Pl. 2: 1028. 1753; Gen. Pl. ed. 5, 453. 1754 · Hop, houblon

[Latin *humulus,* applied to hop plant; from Latin *humus,* soil, an allusion to the soil-hugging habit if plants are not supported]

Herbs, taprooted annuals or rhizomatous perennials, normally rightward-twining. **Stems** usually branched, armed with rigid 2-branched, stalked hairs that facilitate climbing. **Leaves** simple; petioles often twining, with 2-branched hairs. **Leaf blade** mostly cordate, palmately lobed, sometimes unlobed; surfaces abaxially resin-dotted and/or gland-dotted. **Inflorescences:** staminate inflorescences axillary and terminal, cymose panicles, erect to pendent, (10–)20–100+-flowered, flowers small; pistillate axillary, spikes or racemes, flowers solitary or paired, short-pedicellate, subtended by bracts and bracteoles. **Staminate and pistillate flowers** usually on different plants. **Achenes** lenticular or terete, ensheathed by brownish or sometimes mottled persistent perianth; embryo coiled. $x = 10$.

Species 3 (2 in the flora): temperate Northern Hemisphere.

SELECTED REFERENCES Small, E. 1978. A numerical and nomenclatural analysis of morpho-geographic taxa of *Humulus.* Syst. Bot. 3: 37–76. Small, E. 1980. The relationships of hop cultivars and wild variants of *Humulus lupulus.* Canad. J. Bot. 58: 676–686.

1. Veins on abaxial surfaces of leaf blades pubescent, with stiff hairs; marginal areas of adaxial surfaces of younger leaf blades with stiff cystolithic hairs; petioles usually longer than blades; margins of pistillate bracteole densely ciliate-hairy; anthers without glands. . . . . 1. *Humulus japonicus*
1. Veins on abaxial surfaces of leaf blades glabrous or with soft pubescence, without straight, erect hairs; marginal areas of adaxial surfaces of younger leaf blades with few or no cystolithic hairs; petioles usually shorter than blades; margins of pistillate bracteole not ciliate-hairy; anthers glandular. . . . . 2. *Humulus lupulus*

1. Humulus japonicus Siebold & Zuccarini, Abh. Math.-Phys. Cl. Königl. Bayer. Akad. Wiss. 4(3): 213. 1846 · Japanese hop, houblon japonais W1

Herbs, annual, vining, 0.5–2.5 m. **Stems** usually branched. **Leaves:** petioles usually longer than blades. **Leaf blade** cordate, palmately 5–9-lobed, 5–12 cm, margins of lobes serrulate, apex acuminate; surfaces abaxially with veins pubescent, with stiff hairs, glands yellow, sessile, discoid, adaxially margins of younger leaf blades with stiff cystolithic hairs. **Inflorescences:** staminate inflorescences erect, 15–25 cm, flower anthers without glands; pistillate inflorescences spikes, conelike, ovoid; bracteole ovate-orbiculate, 7–10 mm, pilose, margins densely ciliate-hairy. **Infructescences** pendulous, green, conelike, ovoid to oblong, (1–)1.5–3(–4) cm; bracteoles without yellow glands. **Achenes** yellow-brown, ovoid-orbicular, inflated to lenticular, 4–5 mm, glandless. $2n = 20$, including 6 chromosomes concerned with sex determination.

Flowering early–mid summer. Roadsides, fencerows, waste places, riverbanks; 0–1000 m; introduced; Ont., Que.; Ala., Ark., Conn., Del., Ga., Ill., Ind., Iowa, Kans., Ky., Maine, Md., Mass., Mich., Minn., Mo., Nebr., N.J., N.Y., N.C., N.Dak., Ohio, Pa., R.I., S.C., S.Dak., Tenn., Vt., Va., W.Va., Wis.; Asia.

Although I have no records from New Hampshire, the state is within the geographic range of *Humulus japonicus.*

Variegated forms of *Humulus japonicus,* cultivated as ornamentals, are sometimes spontaneous. The vernacular name Japanese hop is occasionally misapplied to *H. lupulus* var. *cordifolius* (Miquel) Maximowicz, a variety not found in North America.

The disposition of the name *Humulus scandens* (Loureiro) Merrill, based on *Antidesma scandens* Loureiro, is problematic. E. D. Merrill (1935) was convinced that the name *A. scandens* applied to the species *Humulus japonicus.* If Merrill was correct, then the combination *Humulus scandens* would have priority. The material described by Loureiro, however, was not preserved, and his description does not coincide with that of *H. japonicus. Humulus scandens* is not included in synonymy in this treatment.

I. A. Grudzinskaya (1988) segregated *Humulus japonicus* as a new monotypic genus, *Humulopsis,* with the single species *Humulopsis scandens* (Loureiro) Grudzinskaya.

2. Humulus lupulus Linnaeus, Sp. Pl. 2: 1028. 1753 · Hop, common hop(s), houblon

Herbs, perennial, rhizomatous, 1–6(–7) m. **Stems** branched. **Leaves:** petioles usually shorter than blades. **Leaf blade** ± cordate, palmately 3–7-lobed, sometimes unlobed, 3–15 cm, margins dentate-serrate; surfaces

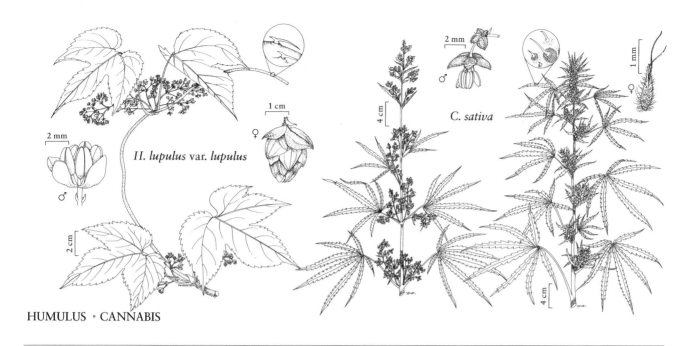

HUMULUS · CANNABIS

abaxially with veins glabrous or with soft pubescence, without straight, erect hairs, glands yellow, adaxially margins of younger leaf blades with few or no cystolithic hairs. **Inflorescences:** staminate with flowers whitish to yellowish, anthers glandular; pistillate usually racemes, 10–20 mm, pedunculate; bracteole margins not ciliate-hairy. **Infructescences** pendulous, pale yellow, conelike, ovoid to oblong, (1–)2–3(–6) cm; bracteoles with yellow glands. **Achenes** yellowish, ovoid, compressed, glandular. $2n = 20$, including 2 or more sex-determining chromosomes.

Varieties 5 (4 in the flora): North America, Eurasia.

Until recently, knowledge of the existence of indigenous kinds of North American *Humulus lupulus* was uncertain, although the appellation American hop was applied sometimes to *H. lupulus* var. *neomexicanus* and sometimes confusingly to other hop varieties. The distinctive Japanese variety *H. lupulus* var. *cordifolius* (Miquel) Maximowicz has not been collected from North America. Hops cultivated commercially in North America for flavoring alcoholic beverages are forms of the European *H. lupulus* var. *lupulus*. The European variety may have been introgressed by the American varieties.

Humulus lupulus has often been transplanted from the wild to homesites as an ornamental. When such sites are abandoned, the plants often persist, and it may appear that they are present naturally. As well, suppliers of ornamental plants may sell hops collected from one site to buyers in a quite distant site. The hop varieties discussed here may therefore be found occasionally beyond the distribution ranges given in this treatment.

Native Americans used *Humulus lupulus* medicinally to induce sleep, for breast and womb problems, for inflamed kidneys, rheumatism, bladder problems, intestinal pain, fever, earaches, pneumonia, coughs, and nervousness, as a tonic and a stimulant, and as a witchcraft medicine (D. E. Moerman 1986).

The measurements mentioned in couplet 1 of the following key are taken in the middle abaxial portion of the central lobe on 4–6 cm leaf blades attached to flowering or fruiting twigs.

1. Leaf blades usually with fewer than 20 hairs per cm on length of midrib, fewer than 25 glands per 10 sq. mm between veins; nodes relatively limited in pubescence, usually fewer than 15 hairs per 0.1 sq. mm at most pubescent portion (excluding angle of petiole with stem) . . . . 2a. *Humulus lupulus* var. *lupulus*
1. Leaf blades usually with more than 20 hairs per cm on length of midrib, more than 25 glands per 10 sq. mm between veins; nodes relatively pubescent, usually more than 15 hairs per 0.1 sq. mm at most pubescent portion (excluding angle of petiole with stem).
    2. Leaf blades 10 cm or more usually having at least 5 lobes; smaller blades (ca. 5 cm) usually with more than 3 easily visible veins branching off midrib (excluding proximal branches). . . 2b. *Humulus lupulus* var. *neomexicanus*
    2. Leaf blades 10 cm or more usually having fewer than 5 lobes; smaller blades (ca. 5 cm)

often with no more than 3 easily visible veins branching off midrib (excluding proximal branches).

    3. Leaf blades conspicuously pubescent abaxially, more than 100 hairs per cm on length of medial midrib, hairs present between veins . . . .
. . . . 2c. *Humulus lupulus* var. *pubescens*

    3. Leaf blades not conspicuously pubescent abaxially, usually fewer than 100 hairs per cm on length of medial midrib, hairs usually absent between veins . . . .
. . . . 2d. *Humulus lupulus* var. *lupuloides*

2a. Humulus lupulus Linnaeus var. lupulus   F

Stems with sparse pubescence at nodes, usually fewer than 15 hairs per 0.1 sq. mm at most pubescent portion (excluding angle of petiole with stem). **Leaf blades** usually with fewer than 20 hairs per cm on length of midrib, fewer than 25 glands per 10 sq. mm between veins.

Roadsides, moist thickets, waste places, edges of woods; 0–2000 m; introduced; Man., N.B., N.S., Ont., P.E.I., Que.; Calif., Conn., Del., Ill., Ind., Maine, Md., Mass., Mich., Mo., N.H., N.J., N.Y., Ohio, Pa., R.I., Vt., Va., W.Va., Wis.

Humulus lupulus var. *lupulus* is indigenous to Europe. It has become established by introduction for ornament and for flavoring beer. Although not yet collected in all parts of the range, it should be expected, particularly as pistillate clones around abandoned homesteads. *Humulus lupulus* var. *lupulus* is reported from Oregon, but no precise locality is known.

2b. Humulus lupulus Linnaeus var. neomexicanus
A. Nelson & Cockerell, Proc. Biol. Soc. Wash. 16: 45. 1903

Stems relatively pubescent at nodes, usually more than 15 hairs per 0.1 sq. mm at most pubescent portion (excluding angle of petiole with stem). **Leaf blades** 10 cm or more usually having at least 5 lobes; smaller blades (ca. 5 cm) usually with more than 3 easily visible veins branching off midrib (excluding proximal branches); surfaces usually more than 20 hairs per cm on length of midrib, more than 25 glands per 10 sq. mm between veins, abaxial glands in exceptionally dense concentration.

On shrubs and trees on slopes, riverbanks, alluvial woods; 300–3000 m; B.C., Man., Sask.; Ariz., Calif.,

Colo., Kans., Mont., Nebr., Nev., N.Mex., N.Dak., S.Dak., Tex., Utah, Wyo.; Mexico.

This taxon includes almost all specimens of *Humulus* originating from south of the Canadian border and west of the 100th meridian. It intergrades with *H. lupulus* var. *lupuloides* in the Prairie Provinces of Canada.

2c. Humulus lupulus Linnaeus var. pubescens E. Small, Syst. Bot. 3: 63, figs. 14, 18, 27, 36. 1978   E

Stems relatively pubescent at nodes, usually more than 15 hairs per 0.1 sq. mm at most pubescent portion (excluding angle of petiole with stem). **Leaf blades** 10 cm or more usually having fewer than 5 lobes; smaller blades (ca. 5 cm) often with no more than 3 easily visible veins branching off midrib (excluding proximal branches); surfaces abaxially conspicuously pubescent, more than 100 hairs per cm on length of medial midrib, more than 25 glands per 10 sq. mm between veins, hairs present between veins.

Moist thickets, woods; 0–1000 m; Ark., Ill., Ind., Iowa, Kans., Md., Minn., Mo., Nebr., N.Y., N.C., Ohio, Okla., Pa., Va.

Humulus lupulus var. *pubescens* intergrades with other varieties. It is found primarily in the midwestern United States.

2d. Humulus lupulus Linnaeus var. lupuloides E. Small, Syst. Bot. 3: 63, figs. 16, 17, 28, 37. 1978   E

Humulus americanus Nuttall

Stems relatively pubescent at nodes, usually more than 15 hairs per 0.1 sq. mm at most pubescent portion (excluding angle of petiole with stem). **Leaf blades** 10 cm or more usually having fewer than 5 lobes; smaller blades (ca. 5 cm) often with no more than 3 easily visible veins branching off midrib (excluding proximal branches); surfaces abaxially not conspicuously pubescent, usually fewer than 100 hairs per cm on length of medial midrib, more than 25 glands per 10 sq. mm between veins, hairs usually absent between veins.

Moist thickets, woods, riverbanks; 0–2000 m; Alta., Man., N.B., Nfld., N.S., Ont., P.E.I., Que., Sask.; Conn., Del., Ill., Ind., Iowa, Kans., Ky., Maine, Md., Mass., Mich., Minn., Mont., Nebr., N.H., N.J., N.Y., N.Dak., Ohio, Pa., R.I., S.Dak., Vt., Va., W.Va., Wis.

Humulus lupulus var. *lupuloides,* defined as an intermediate, encompasses plants not falling into the other three varieties occurring in North America that are much more distinctive. It intergrades where it is sympatric with the introduced *H. lupulus* var. *lupulus* in southern Ontario, southern Quebec, and northeastern United States; with *H. lupulus* var. *pubescens* in midwestern United States; and with *H. lupulus* var. *neomexicanus* in the southern Canadian prairies and the Dakotas.

25. MORACEAE Link

· Mulberry Family

Richard P. Wunderlin

Trees, shrubs, herbs, or vines, deciduous or evergreen, frequently with milky sap. **Leaves** alternate (rarely opposite or whorled), simple; stipules present, persistent or caducous; petiole adaxially grooved. **Leaf blade:** margins entire, toothed, or lobed; venation pinnate or with 3–5 basal palmate veins; cystoliths often present in epidermal cells. **Inflorescences** racemes, cymes, or capitula. **Flowers** unisexual, staminate and pistillate on same or different plants, small, occasionally on flattened torus, more often enclosed within fleshy, flask-shaped receptacle (syconium); sepals 2–6, distinct or partly connate (vestigial in *Brosimum*). **Staminate flowers:** stamens equal in number to sepals or calyx lobes and opposite them, straight or inflexed; anthers 1–2-locular. **Pistillate flowers:** sepals or calyx lobes 4, ± connate; pistils 1, 1–2-carpellate; ovary 1, superior or inferior, 1(–2)-locular; ovules 1 per locule; styles or style branches 1–2; stigmas 1–2, entire. **Fruits** multiple (syncarps); individual achenes or drupelets partly or completely enclosed by enlarged common receptacle or by individual calyces.

Genera ca. 40, species nearly 1100 (7 genera, 18 species in the flora): widespread in tropical and subtropical regions, less common in temperate areas.

Members of the large and diverse mulberry family are mainly woody and tropical; they are most abundant in Asia. The largest genera are *Ficus,* with approximately 750 species, and *Dorstenia,* with about 170 species. The family includes important timber trees, e.g., *Chlorophora excelsa* (Welwitsch) Bentham, iroko, from tropical Africa; *Brosimum guianense* (Aublet) Huber, letterwood, snakewood; and *Ficus* spp. Genera with species bearing edible fruits include the mulberries, *Morus* spp.; breadfruit and jackfruit, e.g., *Artocarpus altilis* (Parkinson) Fosberg and *A. heterophyllus* Lamarck; and figs, *Ficus* spp. Several species of *Ficus* are commonly cultivated in subtropical regions of the United States. These include *F. carica* Linnaeus; *F. elastica* Roxburgh ex Hornemann, India rubber plant; *F. benghalensis* Linnaeus, banyan; *F. benjamina* Linnaeus, weeping fig; *F. pumila* Linnaeus, creeping fig; and *F. microcarpa* Linnaeus f., Indian-laurel.

Rubber plants and weeping figs are commonly sold as houseplants. Economically, the most important species are those associated with the silk trade. *Morus alba* Linnaeus, *M. indica*

Linnaeus, *M. laevigata* Wallis, and *M. serrata* Roxburgh, cultivated in many temperate and tropical countries, provide the natural food source for the silkworm, *Bombyx mori* Linnaeus.

Cudrania tricuspidata (Carrière) Bureau ex Lavallée, used as a food source for silkworms when *Morus* spp. are in short supply, is cultivated in North America as a hedge plant. The fruit is edible. Native to Korea and China, *C. tricuspidata* is known from a collection made in 1956 in McIntosh County, Georgia (S. B. Jones Jr. and N. C. Coile 1988), and it is naturalized in Orange County, North Carolina (R. D. Whetstone, pers. comm.).

SELECTED REFERENCES Engler, H. G. A. 1888b. Moraceae. In: H. G. A. Engler and K. Prantl, eds. 1887–1915. Die natürlichen Pflanzenfamilien. . . . 254 fasc. Leipzig. Fasc. 18[III,1], pp. 66–96. Rohwer, J. G. 1993b. Moraceae. In: K. Kubutzki et al., eds. 1990+. The Families and Genera of Vascular Plants. 2+ vols. Berlin etc. Vol. 2, pp. 438–453. Tomlinson, P. B. 1980. The Biology of Trees Native to Tropical Florida. Allston, Mass.

1. Herbs.
  2. Plants lacking evident aerial stems, rhizomatous, perennial; inflorescences axillary, long-pedunculate . . . . 6. *Dorstenia*, p. 395
  2. Plants caulescent, taprooted, annual; inflorescences axillary, short-pedunculate. . . . . 1. *Fatoua*, p. 389
1. Trees, shrubs, or vines.
  3. Flowers all borne on inside of syconium; terminal vegetative bud surrounded by pair of stipules. . . . . 7. *Ficus*, p. 396
  3. Flowers not borne on inside of syconium or only a solitary female flower immersed in receptacle; terminal vegetative bud scaly, not surrounded by pair of stipules.
    4. Margins of leaf blade toothed, often lobed; venation appearing palmate, or weakly 3-veined from base.
      5. Pistillate inflorescences globose; styles unbranched. . . . . 3. *Broussonetia*, p. 393
      5. Pistillate inflorescences cylindric; styles 2-branched . . . . 2. *Morus*, p. 390
    4. Margins of leaf blade entire, never lobed; venation pinnate.
      6. Leaf blade ovate to lanceolate, not leathery; trees deciduous; syncarps 8–12 cm diam. . . . . 4. *Maclura*, p. 393
      6. Leaf blade oblong, leathery; trees evergreen; syncarps 1.5 cm diam . . . . 5. *Brosimum*, p. 395

1. FATOUA Gaudichaud-Beaupré, Voy. Uranie 12: 509. 1830 · Crabweed

Herbs, annual, caulescent, taprooted; sap not milky. **Leaves** alternate; stipules caducous, free. **Leaf blade** broadly ovate, margins toothed; venation nearly palmate. **Inflorescences** axillary, capitate cymes, short-pedunculate, bracteate. **Flowers:** staminate and pistillate on same plants. **Staminate flowers:** calyx 4-lobed; stamens 4, inflexed. **Pistillate flowers:** calyx green, 4-lobed, pubescent; ovary superior, oblique, 1-locular; style unbranched, nearly lateral. **Syncarps** globose; each achene surrounded by its enlarged, persistent calyx.

Species 2 (1 in the flora): North America, tropical e Asia, Africa (Madagascar), Australia.

SELECTED REFERENCES Thieret, J. W. 1964. *Fatoua villosa* (Moraceae) in Louisiana: New to North America. Sida 1: 248. Vincent, M. A. 1993. *Fatoua villosa* (Moraceae), mulberry weed, in Ohio. Ohio J. Sci. 93: 147–149.

1. **Fatoua villosa** (Thunberg) Nakai, Bot. Mag. (Tokyo) 41: 516. 1927 · Mulberry-weed [F]

Urtica villosa Thunberg, Fl. Jap., 70. 1784

Herbs, to ca. 8 dm. **Stems** erect, branched, pubescent with hooked trichomes. **Leaves:** stipules linear to linear-lanceolate, 1.8–2.5 mm; petiole 1–6 cm, often ± as long as leaf blade. **Leaf blade** to 2.5–10 × 1–7 cm, papery, base cordate to truncate, margins crenate-dentate, apex acute to acuminate; surfaces abaxially and adaxially appressed-hirsute. **Inflorescences** cymes, dense, 4–8 mm wide, subtended by narrow bract; peduncle 1–2 cm. **Flowers** light green, staminate and pistillate in same cyme. **Staminate flowers:** calyx campanulate; stamens exserted. **Pistillate flowers:** calyx boat-shaped; ovary globose, puberulent, somewhat depressed in axis; style reddish purple, filiform. **Achenes** white, oval, 3-angled, ca. 1 mm, minutely muricate, with 2 triangular, membranous appendages. **Seeds** explosively expelled.

Flowering summer–fall. Disturbed sites; 0–300 m; introduced; Ala., Ark., Fla., Ga., Ky., La., Md., Miss., Mo., N.C., Ohio, Okla., S.C., Tenn., Tex., Va., W.Va.; West Indies (Bahamas); native to Asia.

Fatoua villosa was first reported for North America from Louisiana by J. W. Thieret (1964). It has become widespread in the eastern and lower midwestern states where it often occurs as a weed in greenhouses and disturbed sites. Apparently it spreads from the distribution of horticultural materials.

In late 1996, the author reported this species from Indiana.

2. MORUS Linnaeus, Sp. Pl. 2: 986. 1753; Gen. Pl. ed. 5, 424. 1754 · Mulberry, mûrier [Latin *morum*, mulberry]

Trees or shrubs, deciduous; sap milky. **Terminal buds** surrounded by bud scales. **Leaves** alternate; stipules caducous. **Leaf blade** ovate to broadly ovate, margins entire or lobed, dentate; venation nearly palmate. **Inflorescences** pedunculate catkins, erect or pendent, cylindric. **Flowers:** staminate and pistillate on same or different plants. **Staminate flowers:** sepals 4 (4–5 in *M. alba*); stamens 4, inflexed. **Pistillate flowers:** sepals 4, green, of 2 sizes, ciliate; ovary superior, 2-locular; style 2-branched, branches linear. **Syncarps** short-cylindric; each achene enclosed by its enlarged, fleshy calyx. $x = 14$.

Species 10 (3 in the flora): widespread in temperate and tropical regions, North America, Europe, and Asia.

Morus nigra Linnaeus has been reported in floras by various authors (J. K. Small 1903, 1933; R. W. Long and O. Lakela 1971), apparently based on dark-fruited *M. alba.* It is native to Asia, commonly cultivated in Europe for its fruit, and locally naturalized in southern Europe. Occasionally cultivated in North America, it is not known to be naturalized. Because of the similarity to and confusion with *M. alba,* some American authors place it in synonymy with that species.

1. Mature leaf blade less than 7 cm, abaxially harshly scabrous or pubescent, adaxially harshly scabrous; petiole to 1.5 cm . . . . 2. *Morus microphylla*
1. Mature leaf blade usually more than 8 cm, adaxially slightly if at all scabrous; petiole 2 cm or more.
  2. Leaf blade abaxially glabrous or with pubescence only along major veins or in tufts in axils of principal lateral veins and midribs, adaxially glabrous to sparsely pubescent . . . . 1. *Morus alba*
  2. Leaf blade abaxially pubescent or puberulent, adaxially with short, stiff, antrorsely appressed trichomes, usually scabrous. . . . . 3. *Morus rubra*

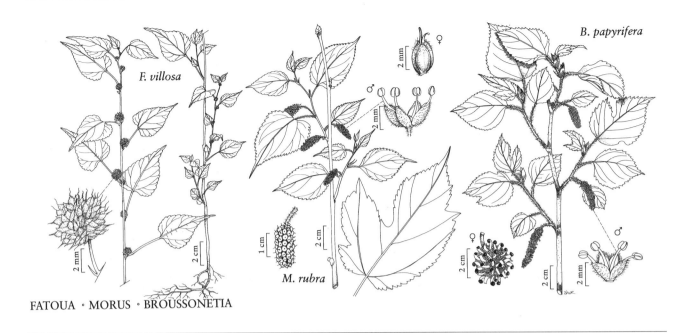

FATOUA · MORUS · BROUSSONETIA

1. Morus alba Linnaeus, Sp. Pl. 2: 986. 1753 · White mulberry, Russian mulberry, silkworm mulberry, mûrier blanc, moral blanco W1

Morus alba var. *tatarica* (Linnaeus) Seringe; *M. tatarica* Linnaeus

Shrubs or trees, to 15 m. **Bark** brown tinged with red or yellow, thin, shallowly furrowed, with long, narrow ridges. **Branchlets** orange-brown or dark green with reddish cast, pubescent or occasionally glabrous; lenticels reddish brown, elliptic, prominent. **Buds** ovoid, 4–6 mm, apex acute to rounded; outer scales yellow-brown with dark margins, glabrous or with a few marginal trichomes; leaf scars half round, bundle scars numerous, in circle. **Leaves:** stipules ovate to lanceolate, 5–9 mm, pubescent; petiole 2.5–5 cm, short-pubescent. **Leaf blade** ovate, often deeply and irregularly lobed, (6–)8–10 × 3–6 cm, base cuneate, truncate, or cordate, margins coarsely serrate to crenate, apex acute to short-acuminate; surfaces abaxially glabrous or sparingly pubescent along major veins or in tufts in axils of principal lateral veins and midribs, adaxially glabrous to sparsely pubescent. **Catkins:** peduncle and axis pubescent; staminate catkins 2.5–4 cm; pistillate catkins 5–8 mm. **Flowers:** staminate and pistillate on same or different plants. **Staminate flowers:** sepals distinct, green with red tip, ca. 1.5 mm, pubescent; filaments ca. 2.7 mm. **Pistillate flowers:** ovary green, ovoid, slightly compressed, ca. 2 mm, glabrous; style branches divergent, red-brown, 0.5–1 mm; stigma papillose. **Syncarps** red when imma-ture, becoming black, purple, or nearly white, cylin-dric, 1.5–2.5 × 1 cm; achenes light brown, ovoid, 2–3 mm.

Flowering spring–summer. Disturbed areas, wood-land margins, fencerows, dry to moist thickets; 0–1500 m; introduced; Ont.; Ala., Ark., Colo., Conn., Del., D.C., Fla., Ga., Ill., Ind., Iowa, Kans., Ky., La., Maine, Md., Mass., Mich., Minn., Miss., Mo., Nebr., N.H., N.J., N.Y., N.C., N.Dak., Ohio, Okla., Pa., R.I., S.C., S.Dak., Tenn., Tex., Vt., Va., W.Va., Wis.; Europe; native to e Asia.

Morus alba is sometimes planted and possibly natu-ralized in Arizona, California, and New Mexico.

Mulberry leaves provide the natural food for silk-worms. Commercially cultivated mulberries are varie-ties of *Morus alba;* they are prized as shade trees with edible fruits.

Morus alba and *M. rubra* are both highly variable and are often confused. Both species have deeply lobed to entire leaves and are variable in pubescence. Some individuals are intermediate in leaf pubescence, suggesting the possibility of hybridization.

Native Americans used infusions made from the bark of *Morus alba* medicinally in various ways: as a laxative, as a treatment for dysentery, and as a purga-tive (D. E. Moerman 1986).

In late 1996, the author reported *Morus alba* from Idaho, Utah, and Washington.

2. Morus microphylla Buckley, Proc. Acad. Nat. Sci. Philadelphia 1862: 8. 1863 · Mountain mulberry, littleleaf mulberry, Texas mulberry

Shrubs or trees, to 7.5 m. **Bark** gray, fissured, scaly. **Branchlets** greenish, pubescent; lenticels light colored, elliptic, prominent. **Buds** ovoid, slightly compressed, 3–4 mm, apex acute; outer scales dark brown, pubescent and minutely ciliate; leaf scars half round to irregularly circular, bundle scars numerous, in circle. **Leaves:** stipules linear-lanceolate, 3–5 mm, papery, pubescent; petiole 0.3–0.6(–1.5) cm, pubescent. **Leaf blade** ovate, sometimes 3–5-lobed, 2–7(–9) × 1–4(–7) cm, base rounded to nearly cordate, margins serrate or crenate-serrate, apex acute to acuminate; surfaces abaxially harshly scabrous or pubescent, somewhat paler than adaxial surface, adaxially harshly scabrous. **Catkins:** staminate, 1–2 cm; pistillate, 8–12 × 5–7 mm, peduncle 3–7 mm, pubescent. **Flowers:** staminate and pistillate on different plants. **Staminate flowers:** calyx lobes green to reddish, rounded, hairy; stamens 4; filiments filiform. **Pistillate flowers:** ovary dark green, broadly ovoid, slightly compressed, 1.5–2 × 1 mm, glabrous; style branches divergent, whitish, sessile, ca. 1.5 mm; stigma papillate. **Syncarps** red, purple, or black, short-cylindric, 1–1.5 cm; achenes yellowish, oval, flattened, ca. 2 mm, smooth.

Flowering spring. In canyons on limestone and igneous slopes, usually along streams; 200–2200 m; Ariz., N.Mex., Okla., Tex.; Mexico.

3. Morus rubra Linnaeus, Sp. Pl. 2: 986. 1753 · Red mulberry, mûrier rouge, moral   E   F   W1

Morus rubra var. *tomentosa* (Rafinesque) Bureau

Shrubs or trees, to 20 m. **Bark** gray-brown with orange tint, furrows shallow, ridges flat, broad. **Branchlets** red-brown to light greenish brown, glabrous or with a few trichomes; lenticels light colored, elliptic, prominent. **Buds** ovoid, slightly compressed, 3–7 mm, apex acute; outer scales dark brown, often pubescent and minutely ciliate; leaf scars oval to irregularly circular, bundle scars numerous, in circle. **Leaves:** stipules linear, 10–13 mm, thin, pubescent; petiole 2–2.5 cm, glabrous or pubescent. **Leaf blade** broadly ovate, sometimes irregularly lobed, 10–18(–36) × 8–12(–15.5) cm, base rounded to nearly cordate, sometimes oblique, margins serrate or crenate, apex abruptly acuminate; surfaces abaxially sparsely to densely pubescent or puberulent, adaxially with short, antrorsely appressed trichomes, usually scabrous. **Catkins:** peduncle pubescent; staminate catkins 3–5 cm; pistillate catkins 8–12 × 5–7 mm. **Flowers:** staminate and pistillate on different plants. **Staminate flowers:** sepals connate at base, green tinged with red, 2–2.5 mm, pubescent outside, ciliate toward tip; stamens 4; filiments 3–3.5 mm. **Pistillate flowers:** calyx tightly surrounding ovary; ovary green, broadly ellipsoid or obovoid, slightly compressed, 1.5–2 × 1 mm, glabrous; style branches divergent, whitish, sessile, ca. 1.5 mm; stigma papillose. **Syncarps** black or deep purple, cylindric, (1.5–)2.5–4(–6) × 1 cm; fleshy calyx surrounding achenes; achenes yellowish, oval, flattened, ca. 2 mm, smooth.

Flowering spring–summer. Moist forests and thickets; 0–300 m; Ont.; Ala., Ark., Conn., D.C., Fla., Ga., Ill., Ind., Iowa, Kans., Ky., La., Md., Mass., Mich., Minn., Miss., Mo., Nebr., N.J., N.Y., N.C., Ohio, Okla., Pa., R.I., S.C., Tenn., Tex., Vt., Va., W.Va., Wis.

Morus rubra is sporadically established along fencerows in southern New Mexico (R. Spellenberg, pers. comm.).

Morus rubra is a common tree of eastern North America. The leaves are highly variable, often with deeply lobed and entire leaves on the same plant. The abaxial surface of the leaf varies from sparsely to densely pubescent.

According to D. E. Moerman (1986), Native American tribes used infusions of the bark of *Morus rubra* medicinally to check dysentery, as a laxative, and as a purgative; infusions of the root for weakness and urinary problems; and tree sap rubbed directly on the skin as treatment for ringworm.

3. BROUSSONETIA L'Héritier ex Ventenat, Tabl. Règn. Vég. 3: 547. 1799, name conserved · Paper-mulberry [for Pierre Marie Auguste Broussonet, 1761–1807, French biologist at Montpellier]

Trees, deciduous; sap milky. **Terminal buds** surrounded by bud scales. **Leaves** alternate, opposite, or whorled; stipules caducous, free. **Leaf blade** ovate, lobed or entire, margins dentate; venation appearing palmate or weakly 3-veined from base. **Staminate inflorescences** pedunculate, cylindric spikes; pistillate inflorescences short-pedunculate, globose capitula. **Flowers:** staminate and pistillate on different plants. **Staminate flowers:** sepals 4, connate at base; stamens 4, inflexed. **Pistillate flowers:** sepals 4, connate, forming tube; ovary superior, stipitate, 1-locular; style unbranched. **Fruits** globose; each drupelet partly protruding from its enlarged calyx. $x = 13$.

Species 7–8 (1 in the flora): North America, Asia, and Pacific Islands (Polynesia).

1. **Broussonetia papyrifera** (Linnaeus) Ventenat, Tabl. Règn. Vég. 3: 547. 1799 · Paper-mulberry [F] [W1]

Morus papyrifera Linnaeus, Sp. Pl. 2: 986. 1753; *Papyrius papyrifera* (Linnaeus) Kuntze

Trees, to 15 m. **Bark** tan, smooth or moderately furrowed. **Branchlets** brown, spreading pubescent. **Terminal bud** absent, axillary buds dark brown, short-pubescent; leaf scars nearly circular, somewhat elevated. **Leaves:** stipules ovate to ovate-oblong, apex attenuate; petiole shorter than or equal to blade. **Leaf blade** entire or 3–5-lobed, 6–20 × 5–15 cm, base shallowly cordate, often oblique, truncate, or broadly rounded, margins serrate, apex acuminate; surfaces abaxially densely gray-pubescent, adaxially scabrous. **Staminate inflorescences** 6–8 cm; peduncle 2–4 cm. **Pistillate inflorescences** ca. 2 cm diam., villous. **Staminate flowers:** sepals pubescent. **Pistillate flowers:** style elongate-filiform. **Syncarps** globose, 2–3 cm diam.; drupes red or orange, oblanceolate, each exserted from its calyx.

Flowering spring. Disturbed thickets; 0–600 m; introduced; Ala., Ark., Conn., Del., D.C., Fla., Ga., Ill., Iowa, Kans., Ky., La., Md., Mass., Miss., Mo., Nebr., N.J., N.Y., N.C., Okla., Pa., R.I., S.C., Tenn., Tex., Va., W.Va.; native to Asia.

Broussonetia papyrifera is now widely naturalized in eastern United States. Frequently planted as a shade tree around dwellings, it is often considered undesirable because of its aggressiveness, shallow root system, and soft, brittle wood. The bark of the tree is used to produce a barkcloth.

In late 1996, the author reported *Broussonetia papyrifera* from Indiana and Ohio.

4. MACLURA Nuttall, Gen. N. Amer. Pl. 2: 233. 1818, name conserved · Osage-orange, bois d'arc [for American geologist William Maclure, 1763–1840]

Trees, deciduous; sap milky. **Branches** with axillary spines. **Terminal buds** surrounded by bud scales. **Leaves** alternate; stipules caducous, free. **Leaf blade** ovate to lanceolate, not leathery, margins entire, never lobed; venation pinnate. **Inflorescences:** flowers borne outside receptacle; staminate inflorescences loose short racemes; pistillate inflorescences dense heads. **Flowers:** staminate and pistillate on different plants. **Staminate flowers:** calyx 4-lobed; stamens 4, inflexed; filaments filiform; anthers introrse, with short connective. **Pistillate flowers:** sepals 4, 2 outer sepals wider than inner ones; ovary 1, superior, 1-locular; style unbranched, filiform. **Syncarps** globose, 8–12 cm or more diam.; each achene completely enclosed by its enlarged, fleshy calyx.

Species 1 (1 in the flora): North America.

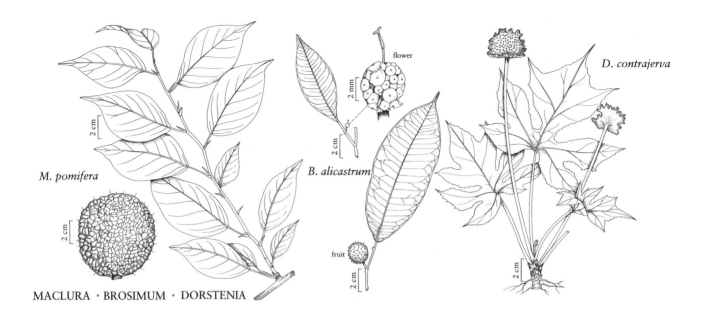

M. pomifera

B. alicastrum

flower

fruit

D. contrajerva

MACLURA · BROSIMUM · DORSTENIA

1. Maclura pomifera (Rafinesque) C. K. Schneider, Ill. Handb. Laubholzk. 1: 806. 1906 · Osage-orange, hedge-apple, bois d'arc E F W1

Ioxylon pomiferum Rafinesque, Amer. Monthly Mag. & Crit. Rev. 2: 118. 1817; *I. aurantiacum* (Nuttall) Rafinesque; *Maclura aurantiaca* Nuttall

Trees, to 20 m. **Bark** dark orange-brown, shallowly furrowed, ridges flat, often peeling into long, thin strips. **Branchlets** greenish yellow, becoming orange-brown; thorns stout, straight, to 1.5 cm, usually lateral to spur branch, spur branches often paired. **Buds** often paired, larger one red-brown, globose, 1.5–2 mm; scales ciliate; leaf scars half round, bundle scars arranged in oval. **Leaves:** stipules lanceolate, 1.5–2 mm, pubescent and long-ciliate; petiole 1–2.5 cm, pubescent. **Leaf blade** 4–12 × 2–6 cm, base rounded, apex acuminate; surfaces abaxially pale, glabrate, midrib and veins pubescent, adaxially lustrous, glabrous, midrib somewhat pubescent. **Staminate inflorescences** clustered on lateral spur branches; peduncle 1–1.5 cm, pubescent; heads globose or cylindric, 1.3–2.3 cm; pedicels 2–10 mm, glabrate. **Pistillate inflorescences:** peduncle 2–2.5 mm, glabrous or pubescent; heads globose, sessile on obconic receptacle, to 1.5 cm diam. **Staminate flowers:** sepals distinct, yellow-green, ca. 1 mm, apex acute, pubescent; filaments ca. 2 mm, closely appressed to sepals, flattened. **Pistillate flowers:** sepals green, obovate, 3 mm, enclosing and closely appressed to ovary, hood-like, ciliate near tip; ovary ovoid, compressed, ca. 1 mm; style base green, ca. 3 mm, branches 4–6 mm, glabrous; stigma yellowish, papillose. **Syncarps** yellow-green to green, spheric, surface irregular, exuding milky sap when broken, peduncle short, glabrous or pubescent; achenes completely covered by accrescent, thickened calyx lobes and deeply embedded in receptacle. **Seeds** cream colored, oval to oblong, 8–12 × 5–6 mm, base truncate or rounded with 1–3 minute points, margins with narrow groove, apex rounded, mucronate; surfaces minutely striated or pitted.

Flowering spring. Thickets; 0–1500 m; Ala., Ark., Calif., Conn., Del., D.C., Fla., Ga., Ill., Ind., Iowa, Kans., Ky., La., Md., Mass., Mich., Miss., Mo., Nebr., N.J., N.Mex., N.Y., N.C., Ohio, Okla., Pa., R.I., S.C., S.Dak., Tenn., Tex., Va., W.Va., Wis.

Maclura pomifera is native to southwestern Arkansas, southeastern Oklahoma, and Texas; it is introduced and naturalized elsewhere in the United States. In late 1996, the author reported *M. pomifera* from Colorado.

Maclura pomifera has been widely used in fence-rows on farms and along roadways in the midwest and eastern states as windbreaks and wildlife shelter.

The Comanches used *Maclura pomifera* as an eye medication (D. E. Moerman 1986).

5. BROSIMUM Swartz, Prodr., 12. 1788, name conserved · Breadnut [Greek *brosimos*, edible]

Trees, evergreen; sap milky. **Terminal buds** surrounded by bud scales. **Leaves** alternate; stipules caducous, free. **Leaf blade** oblong, leathery, margins entire, never lobed; venation pinnate. **Inflorescences** globose heads. **Flowers:** staminate and pistillate on same plants; perianth vestigial. **Staminate flowers** numerous, surrounding 1–2 central pistillate flowers; stamen 1, straight; anther peltate, dehiscence circumscissile. **Pistillate flowers** borne inside receptacle; bracts peltate; ovary 1, inferior, 1-locular; style 2-branched. **Syncarps** somewhat globose; each achene embedded in enlarged, pulpy receptacle. $x = 13$.

Species ca. 24 (1 in the flora): neotropical.

SELECTED REFERENCE Berg, C. C. 1972. Olmedieae, Brosimeae (Moraceae). In: Organization for Flora Neotropica. 1968+. Flora Neotropica. 65+ vols. New York. Vol. 7: 1–229.

1. Brosimum alicastrum Swartz, Prodr., 12. 1788
· Breadnut F

Alicastrum brownei Kuntze

Trees, to 30 m. **Branchlets** gray-brown, glabrous. **Leaves:** stipules clasping, ca. 4 mm; petiole 0.5–0.7 cm. **Leaf blade** 5–15 × 2–6 cm, base broadly obtuse to rounded, apex obtuse to short-acuminate, nearly cuspidate; surfaces abaxially and adaxially glabrous; veins 12–18 pairs. **Inflorescences** nearly globose, 3–6 mm diam.; peduncle slender, equal to or shorter than head. **Staminate flowers:** anther ca. 1 mm diam. **Pistillate flowers:** style 1.5–8.5 mm; stigmas 0.2–8 mm, unequal. **Syncarps** yellow, 1.5–2 cm diam.

Flowering all year. Disturbed areas; 0–50 m; introduced; Fla.; Mexico; West Indies (Cuba and Jamaica); Central America; n South America.

Brosimum alicastrum, native to tropical America, is cultivated in tropical Florida as an ornamental; it rarely escapes.

6. DORSTENIA Linnaeus, Sp. Pl. 1: 121. 1753; Gen. Pl. ed. 5, 56. 1754 · Tusilla [for Theodor Dorsten, d. 1539, German herbalist and professor of medicine at Marburg]

Herbs, perennial, rhizomatous; stems short, fleshy, aerial stems not evident; sap milky. **Leaves** alternate; stipules persistent, free. **Leaf blade** ovate to orbiculate, margins entire or pinnately lobed; venation palmate near base of blade, otherwise pinnate. **Inflorescences** axillary, flowers embedded in long-pedunculate, flat receptacle. **Flowers:** staminate and pistillate on same plant. **Staminate flowers:** calyx minute, 2–3-lobed; stamens 1–3, inflexed in bud. **Pistillate flowers:** calyx tubular, 4-lobed; ovary inferior, embedded in receptacle, 1-locular; style 2-branched. **Syncarps** disc- or cup-shaped; drupes embedded in enlarged, fleshy, common receptacle. **Seeds** explosively expelled.

Species ca. 170 (1 in the flora): tropical.

1. Dorstenia contrajerva Linnaeus, Sp. Pl. 1: 121. 1753
· Contra yerba, tusilla F

Herbs, to 4.5 dm. **Stems** covered with persistent petiole bases. **Leaves:** stipules persistent; petiole 8–25 cm. **Leaf blade** oblong-ovate, deltate-ovate, or orbiculate, entire or deeply pinnately lobed, 6–20 × 7–22 cm, pubescent. **Inflorescences:** peduncle 7–25 cm; receptacle flat, curved, or undulate, quadrangular or lobed, to 3.5 cm square. **Drupes** somewhat globose. **Seeds** yellowish, explosively expelled. $2n = 30$.

Flowering all year. Moist, disturbed sites; 0–20 m; introduced; Fla.; s Mexico; Central America; n South America.

Dorstenia contrajerva is a weed in greenhouses and nurseries; it rarely occurs in the wild in North America. It is sometimes cultivated as a house plant.

7. FICUS Linnaeus, Sp. Pl. 2: 1059. 1753; Gen. Pl. ed. 5, 482. 1754 · Fig, figuier [Latin *ficus*, an old name for edible fig, *Ficus carica*]

Trees, shrubs, or woody vines, evergreen or deciduous, commonly epiphytic or scandent as seedlings; sap milky. **Terminal buds** surrounded by pair of stipules. **Leaves** alternate, monomorphic (dimorphic in *F. pumila*); stipules caducous, fused, enclosing naked buds. **Leaf blade:** margins entire (lobed in *F. carica*), rarely dentate; venation pinnate or nearly palmate. **Inflorescences** small, borne on inner walls of fruitlike and fleshy receptacle (syconium). **Flowers:** staminate and pistillate on same plant. **Staminate flowers** sessile or pedicellate; calyx of 2–6 sepals; stamens 1–2, straight. **Pistillate flowers** sessile; ovary 1-locular; style unbranched, lateral. **Syconia** globose to pyriform; achenes completely embedded in enlarged, fleshy, common receptacle and accessible by apical opening (ostiole) closed by small scales. $x = 13$.

Species ca. 750 (10 in the flora): tropics and subtropics, chiefly Asian.

Worldwide, *Ficus* is one of the largest genera of flowering plants. Members of the genus are usually treated as a separate tribe within Moraceae because of their unique inflorescence and wasp-dependent system of pollination.

The floral characters (especially of the American species, which are quite uniform) are exceedingly difficult to use or of little value in distinguishing species. Therefore they are not used in the species descriptions. The form of the syconium, however, is often significant and taxonomically useful.

Ficus palmata Forsskål was cited by P. A. Munz (1974) as an occasional escape in the Santa Barbara region. It is not cited by other workers, and I have seen no specimens.

Ficus rubiginosa Desfontaines ex Ventenat cultivar 'Florida', a species native to Australia, has recently been reported as naturalized in the Los Angeles area (Michael O'Brien, pers. comm.). It is a small tree with rusty-pubescent branchlets, petiole, and abaxial leaf surfaces; ovate to elliptic-oblong, leathery, 10-cm leaves; and paired axillary, globose, warty, rusty-pubescent syconia 1 cm in diameter. Vernacular names include Port Jackson fig, rusty fig, and littleleaf fig.

1. Plants climbing, attaching by nodal adventitious roots, or trailing; leaves dimorphic. . . . . 1. *Ficus pumila*
1. Plants erect or essentially so; leaves monomorphic.
   2. Leaf blade palmately 3–5-lobed, pubescent. . . . . .2. *Ficus carica*
   2. Leaf blade entire, glabrous (abaxially puberulent in *F. benghalensis*).
      3. Apex of leaf blade abruptly long-caudate or long-acuminate, ca. 1/2 length of blade. . . . . 3. *Ficus religiosa*
      3. Apex of leaf blade obtuse to acute or if caudate, then much shorter in proportion to blade.
         4. Basal leaf veins (2–)3–4 pairs; fruit pubescent . . . . 4. *Ficus benghalensis*
         4. Basal leaf veins 1(–2) pairs; fruit glabrous.
            5. Leaf blade with more than 10 uniform lateral veins, these regularly spaced.
               6. Leaf blade 4–6(–11) cm; stipules 0.8–1.2 cm; syconia nearly globose . . . . 5. *Ficus benjamina*
               6. Leaf blade 9–30 cm; stipules 3–10 cm; syconia oblong-ovoid . . . . 6. *Ficus elastica*
            5. Leaf blade with fewer than 10 lateral veins, or if more than 10, these not uniformly spaced.
               7. Syconia on peduncles (2–)5–10(–15) mm.
                  8. Petiole (0.7–)1.5–6 cm; syconia spotted; base of leaf blade usually cordate or rounded to obtuse. . . . . 7. *Ficus citrifolia*
                  8. Petiole 0.2–1 cm; syconia not spotted; base of leaf blade usually acute or cuneate to obtuse . . . . 8. *Ficus americana*
               7. Syconia sessile or subsessile, rarely with peduncles to 5 mm.
                  9. Leaf blade 6–12(–15) cm; syconia 6–15 mm diam . . . . .9. *Ficus aurea*
                  9. Leaf blade 3–11 cm; syconia 5–6 mm diam . . . . 10. *Ficus microcarpa*

1. Ficus pumila Linnaeus, Sp. Pl. 2: 1060. 1753 · Climbing fig

Woody vines or sprawling shrubs, vines closely appressed to substrate, shrubs loosely ascending, evergreen. **Roots** adventitious, nodal. **Branches** appressed-pubescent when young, glabrous in age. **Leaves** dimorphic; stipules 0.3–0.8 cm; petiole 1.5–2 cm. **Leaf blade** oblong to ovate-elliptic or obovate, 4–10 × 2.5–4.5 cm, those of appressed climbing stems distichous, appressed, smaller (than those of loose, extended, flowering stems), spreading, leathery, base obtuse to rounded, margins recurved, apex obtuse to nearly acute; surfaces abaxially glabrous or puberulent on veins, adaxially glabrous, prominently reticulate; basal pair of veins 1; lateral pairs of veins 3–6, straight; secondary veins prominent. **Syconia** solitary, pedunculate, green, oblong, obovoid, pyriform, or nearly globose, 3–4 × 3–4 cm, slightly pubescent but becoming glabrescent in age; peduncle thick, 8–15 mm; subtending bracts ovate, 5–7 mm; ostiole closed by 3 bracts, umbonate.

Flowering all year. Disturbed thickets; 0–10 m; introduced; Fla.; native to s, se Asia.

Ficus pumila was reported from Georgia and Louisiana in late 1996. It is occasionally cultivated as an ornamental on walls.

Ficus scandens Lamarck is a nomenclaturally illegitimate name.

2. Ficus carica Linnaeus, Sp. Pl. 2: 1059. 1753 · Common fig F W1

Shrubs or small trees, deciduous, to 5 m. **Roots** not adventitious. **Bark** grayish, slightly roughened. **Branchlets** pubescent. **Leaves:** stipules 1–1.2 cm; petiole 8–20 cm. **Leaf blade** obovate, nearly orbiculate, or ovate, palmately 3–5-lobed, 15–30 × 15–30 cm, base cordate, margins undulate or irregularly dentate, apex acute to obtuse; surfaces abaxially and adaxially scabrous-pubescent; basal veins 5 pairs; lateral veins irregularly spaced. **Syconia** solitary, sessile, green, yellow, or red-purple, pyriform, 5–8 cm, pubescent; peduncle ca. 1 cm; subtending bracts ovate, 1–2 mm; ostiole with 3 subtending bracts, umbonate.

Flowering spring–summer. Disturbed sites; 0–300 m; introduced; Calif., Fla., Mass., N.C., S.C.; Mexico; West Indies; native to Asia.

Ficus carica is known to escape in Alabama and West Virginia, although no specific localities are documented. In late 1996, it was reported from Georgia, Kentucky, Louisiana, Maryland, Michigan, Mississippi, New York, Pennsylvania, Tennessee, Texas, Utah, and Virginia.

Ficus carica was first known from Caria in southwestern Asia. It is cultivated for its edible fruit and becomes established outside of cultivation only sporadically in the United States. It can sometimes be found persisting around old habitations and old orchards.

3. Ficus religiosa Linnaeus, Sp. Pl. 2: 1059. 1753 · Bo tree, sacred fig

Urostigma religiosum (Linnaeus) Gasparrini

Trees, evergreen, to 30 m. **Bark** of trunks and older branches brown, smooth. **Branchlets** glabrous. **Leaves:** stipules ovate, to 5 cm; petiole slender, 3.5–13 cm. **Leaf blade** broadly ovate to ovate-orbiculate, 7–25 × 4–16 cm, thinly leathery, base rounded to truncate, margins entire, occasionally wavy, apex abruptly long-caudate or long-acuminate, tip to 2.5–9 cm; surfaces occasionally glaucous, glabrous; basal veins 2(–3) pairs; lateral veins 6–9 pairs, the main veins finely reticulate. **Syconia** paired, sessile, dark purple, nearly globose, 1–1.5 × 1–1.5 cm, glabrous; subtending bracts ovate, 3–5 mm, silky-puberulous; ostiole closed by 3 bracts 2–3 mm wide, umbonate.

Flowering all year. Disturbed thickets; 0–10 m; introduced; Fla.; Asia (native to India and Southeast Asia).

4. Ficus benghalensis Linnaeus, Sp. Pl. 2: 1059. 1753 · Banyan tree F

Trees, evergreen, to 30 m. **Roots** aerial, often descending to ground level and forming pillar-roots **Bark** of trunks and older branches brown, smooth. **Branchlets** puberulent, glabrescent in age. **Leaves:** stipules stout, 1.5–2.5 cm; petiole 1.5–7 cm. **Leaf blade** ovate, 10–30 × 7–20 cm, leathery, base cordate, margins entire, apex obtuse; surfaces abaxially puberulent, adaxially glabrous; basal veins (2–)3–4 pairs, 1/3–1/2 length of blade, reticulations regular; lateral veins 5–6(–7) pairs. **Syconia** paired, sessile, orange or red, depressed-globose, 1.5–2 × 2–2.5 mm, pubescent; subtending bracts ovate, 3–7 mm, puberulous; ostiole closed by 3 flat or nearly umbonate apical bracts 3–4 mm wide.

Flowering all year. Disturbed thickets; 0–10 m; introduced; Fla.; Asia (native to Pakistan and India).

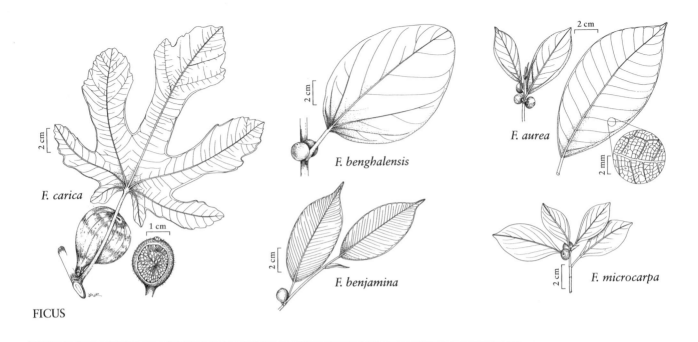

F. carica

F. benghalensis

F. aurea

F. benjamina

F. microcarpa

FICUS

5. Ficus benjamina Linnaeus, Mant. Pl., 129. 1767
· Weeping fig F

Urostigma benjamina (Linnaeus)
Miquel

Trees, evergreen, to 10 m. **Roots**
adventitious, occasionally hanging. **Bark** gray, smooth. **Branchlets** brown, glabrous. **Leaves:**
stipules 0.8–1.2 cm; petiole 0.5–
2(–3) cm. **Leaf blade** oblong, elliptic, lanceolate, or ovate, 4–6
(–11) × 1.5–6 cm, nearly leathery, base rounded or
cuneate, margins entire, apex acuminate or cuspidate;
surfaces abaxially and adaxially glabrous; basal veins
1(–2) pairs, short; lateral veins (6–)12(–14) pairs, regularly spaced, uniform; secondary veins prominent.
Syconia solitary or paired, sessile or subsessile, orange,
yellow, or dark red, nearly globose, 8–12 × 7–10
mm, glabrous; subtending bracts 2–3, crescent-shaped,
0.5–1.5 mm, glabrous; ostiole closed by 3 small, flat,
apical bracts 1.5–2 mm wide, umbonate.

Flowering all year. Disturbed thickets and hammocks; 0–10 m; introduced; Fla.; West Indies (Lesser
Antilles); native to Asia.

Ficus benjamina is commonly cultivated as a
houseplant. The name probably refers to the supposed
relation of the plant to the source of a resin or benzoin procured from the Orient in antiquity.

6. Ficus elastica Roxburgh ex Hornemann, Suppl. Hort.
Bot. Hafn., 7. 1819 · India rubber plant

Macrophthalma elastica (Roxburgh ex
Hornemann) Gasparrini; *Urostigma
elasticum* (Roxburgh ex Hornemann)
Miquel; *Visiania elastica* (Roxburgh
ex Hornemann) Gasparrini

Trees, evergreen, to 12 m, epiphytic when young. **Roots** aerial, abundant. **Bark** gray,
smooth or slightly roughened.
Branchlets greenish brown, glabrous. **Leaves:** stipules
3–10 cm; petiole 2.5–10 cm. **Leaf blade** oblong-elliptic
to obovate, 9–30 × 5–12 cm, leathery, base rounded,
margins entire, apex abruptly short-acuminate or apiculate; surfaces abaxially and adaxially glabrous; basal
veins 1(–2) pairs; lateral veins 10 or more, parallel;
secondary veins inconspicuous. **Syconia** paired, sessile,
greenish yellow, oblong-ovoid, ca. 2 × ca. 1.5 cm,
glabrous; subtending bracts caducous, leaving annual
scar, entirely enclosing young syconia, glabrous; ostiole closed by 3 apical bracts, umbonate.

Flowering all year. Disturbed sites; 0–10 m; introduced; Fla.; West Indies (Lesser Antilles); Asia (native
to India).

Ficus elastica is commonly cultivated. It has not
been collected recently in the area of the flora.

7. **Ficus citrifolia** Miller, Gard. Dict. ed. 8 Ficus no. 10. 1768 · Wild banyan tree

Ficus brevifolia Nuttall; *F. laevigata* Vahl; *F. laevigata* var. *brevifolia* (Nuttall) D'Arcy; *F. populifolia* Desfontaines; *F. populnea* Willdenow subvar. *floridana* Warburg; *F. populnea* var. *brevifolia* (Nuttall) Warburg

Shrubs or trees, evergreen, to 15 m. **Roots** adventitious, aerial, hanging. **Bark** brownish, smooth. **Branchlets** grayish, glabrous or sparingly pubescent. **Leaves:** stipules 0.5–2 cm, glabrous; petiole (0.7–)1.5–6 cm. **Leaf blade** ovate to elliptic or obovate, 3–14 × 1.5–8 cm, nearly leathery, base usually cordate or rounded to obtuse, margins entire, apex obtuse to acute or acuminate; surfaces abaxially and adaxially glabrous; basal veins 1 (–2) pairs; lateral veins fewer than 10, if more than 10, not uniformly spaced. **Syconia** solitary or paired, pedunculate, yellow or red, spotted, globose to globose-ovoid, 8–18 mm diam., glabrous; peduncles to ca. 15 mm; subtending bracts 2, shortly connate, deltate or broadly rounded, 2–3 mm wide, glabrous or puberulent; ostiole subtended by 3 bracts, bracts ovate, ca. 1 × 2–3 mm, slightly umbonate.

Flowering spring–summer. Tropical hammocks; 0–10 m; Fla.; Mexico; West Indies; Central America; South America.

Ficus citrifolia is the large and graceful banyan tree that is planted for shade around verandas.

8. **Ficus americana** Aublet, Hist. Pl. Guiane, 952. 1775 · West Indian laurel fig

Trees, evergreen, to 12 m. **Roots** adventitious, aerial. **Bark** grayish to brown, smooth. **Branchlets** grayish, smooth. **Leaves:** stipules 0.7–0.9 cm; petiole 0.2–1 cm. **Leaf blade** elliptic to obovate, 2–8 × 1–4 cm, base usually acute or cuneate to nearly obtuse, margins entire, apex acute, obtuse, or short-apiculate; surfaces abaxially and adaxially glabrous; basal veins 1(–2) pairs; lateral veins 6–14 pairs, not uniformly spaced. **Syconia** paired, red, not spotted, globose, 3–7 mm diam., glabrous; peduncles 2–5 mm; subtending bracts 2, basally connate, ovate, 1–1.5 mm; ostiole ca. 1 mm wide, subtended by 3 bracts, bracts ca. 1 mm, not umbonate.

Flowering all year. Disturbed thickets; 0–10 m; introduced; Fla.; Mexico; native to West Indies; Central America; South America.

The name *Ficus perforata* Linnaeus (Pl. Surin., 17. 1775) is an illegitimate name, based on the same type collection as *F. americana* Aublet. *Ficus americana* is ocally escaped from cultivation.

9. **Ficus aurea** Nuttall, N. Amer. Sylv. 2: 4, plate 43. 1846 · Strangler fig, golden fig F

Ficus aurea var. *latifolia* Nuttall

Trees, evergreen, to 20 m. **Roots** aerial, sometimes present on branches, pendent, sometimes reaching ground and forming pillar-roots. **Bark** gray, smooth. **Branchlets** yellow. **Leaves:** stipules 1–1.5 cm; petiole 1–6 cm. **Leaf blade** ovate to oblong or obovate, 6–12(–15) × 3.5–6 cm, leathery, base rounded to cuneate, margins entire, apex obtuse or shortly and bluntly acuminate; surfaces abaxially and adaxially glabrous; basal veins 1(–2) pairs; lateral veins fewer than 10, if more, not uniformly spaced. **Syconia** usually paired, usually sessile, rarely with peduncles to 5 mm, red or yellow, obovoid, 6–15 mm diam., glabrous; subtending bracts 2, 3–5 mm, glabrous; ostiole prominent, closed by 3 conspicuous scales.

Flowering spring–summer. Frequent in swamps, tropical hammocks, borders of mangrove swamps; 0–10 m; Fla.; West Indies.

10. **Ficus microcarpa** Linnaeus f., Suppl. Pl., 442. 1782 · Indian laurel F

Urostigma microcarpa (Linnaeus f.) Miquel

Trees, evergreen, to 30 m. **Roots** aerial, abundant, sometimes developing pillar-roots. **Bark** gray. **Branchlets** brown, glabrous. **Leaves:** stipules 0.7–0.9 cm; petiole 0.5–1 cm. **Leaf blade** elliptic, obovate to ovate, 3–11 × 1.5–6 cm, thinly leathery, base obtuse to cuneate, margins entire, apex nearly acute to acuminate; surfaces abaxially and adaxially glabrous; basal veins 1(–2) pairs; lateral veins 5–9 pairs, uniformly spaced. **Syconia** paired, sessile, purple or black, obovoid, pyriform, or nearly globose, 9–11 × 5–6 mm; subtending bracts ovate-lanceolate, 1.5–3.5 mm, apex obtuse to subacute; ostiole closed by 3 flat, apical bracts 2–2.5 mm wide, umbonate.

Flowering all year. Disturbed sites; 0–20 m; introduced; Fla.; West Indies; native to Eastern Hemisphere.

Ficus microcarpa is commonly cultivated in Florida. In late 1996, word had been received (Michael O'Brien, pers. comm.) that *F. microcarpa* was recently found in the Los Angeles area, where the pollinating wasp apparently has been present since 1992.

26. URTICACEAE Jussieu

· Nettle Family

David E. Boufford

Herbs or small shrubs [lianas, trees], herbs annual or rhizomatous perennial, usually pubescent, sometimes with stinging hairs, deciduous. **Leaves** opposite or alternate and spirally arranged, simple; stipules present or absent; petioles present. **Leaf blades** equal in size (except in *Pilea,* which may have unequally paired leaves), dotted with linear or rounded marks formed by cystoliths (variously shaped calcium carbonate crystals inside epidermal cells). **Inflorescences** axillary or terminal, of paniculately or racemosely arranged cymes, or spikelike. **Flowers** bisexual or unisexual (staminate or pistillate), staminate and pistillate flowers on same or different plants; perianth hypogynous. **Staminate flowers** usually pedicellate; tepals 4–5, white or green; stamens 4–5, equaling tepals in number; filaments inflexed in bud, reflexing suddenly as flowers open; anthers basifixed, dehiscing by longitudinal slits; pollen ejected explosively; pistillode 1. **Pistillate flowers** usually sessile; tepals 2–4, hypogynous, greenish or reddish, distinct or connate; staminodes present or absent; pistil 1, 1-locular; placentation basal; ovule 1; style present or stigma sessile; stigma linear [capitate]. **Bisexual flowers:** tepals 4; stamens 4; pistil 1. **Fruits** achenes, free or loosely or tightly surrounded by persistent, accrescent perianth.

Genera ca. 45, species ca. 800 (8 genera, 21 species in the flora): nearly worldwide, primarily tropical and subtropical regions.

Cystoliths cause patterns on epidermal surfaces. Forms of the cystoliths given in descriptions are readily discernible from surface patterns.

Stinging hairs in Urticaceae have a distinct bulbous or cylindric base and a stiff, translucent apex. Nonstinging hairs are soft and flexible and lack a bulbous or cylindric base.

The compounds producing the stinging sensation caused by contact with some members of Urticaceae have been reported to be histamine, acetylcholine, 5-hydroxytryptamine, and, in extracts from which the other three have been removed, an unknown substance that produces pain (E. L. Thurston and N. R. Lersten 1969). E. L. Thurston (1969) was not able to find these compounds in *Urtica chamaedryoides* using analytic techniques, but J. M. Kingsbury (1964, p. 67) reported that the same species "... contains toxicologically significant amounts of acetylcholine and histamine." The tip of the stinging hair breaks off upon slight contact, leaving a sharp point that readily pierces skin and allows fluid contents of the hair to enter flesh through the body of the hair, which acts as a miniature hypodermic needle.

Economically the Urticaceae are most important for their fibers (see D. W. Woodland 1989). They can be troublesome weeds (species of *Urtica* and *Parietaria*), pot herbs (species of *Pilea* in the tropics, of *Urtica* in temperate zones), and frequently cultivated ornamentals (*Pilea*) (I. Friis 1993).

SELECTED REFERENCES Bassett, I. J., C. W. Crompton, and D. W. Woodland. 1974. The family Urticaceae in Canada. Canad. J. Bot. 52: 503–516. I. Friis. 1993. Urticaceae. In: K. Kubitzki et al., eds. 1990+. The Families and Genera of Vascular Plants. 2+ vols. Berlin etc. Vol. 2, pp. 612–630. Miller, N. G. 1971. The genera of the Urticaceae in the southeastern United States. J. Arnold Arbor. 52: 40–68. Weddell, H. A. 1856. Monographie de la Famille des Urticacées. Paris. Woodland, D. W. 1989. Biology of temperate Urticaceae (nettle family). In: P. R. Crane and S. Blackmore, eds. 1989. Evolution, Systematics, and Fossil History of the Hamamelideae. 2 vols. Oxford. Vol. 2, pp. 309–318. [Syst. Assoc. Special Vol. 40A, B.]

1. Plants with stinging hairs (hairs with distinct bulbous or cylindric base and stiff, translucent apex); tepals of pistillate flowers 2 or 4, distinct or connate and completely enclosing ovary and achene.
  2. Leaves alternate; style present, persistent even in fruit . . . . 3. *Laportea*, p. 405
  2. Leaves opposite; style absent.
    3. Tepals of pistillate flowers distinct, inner 2 equal to achene, outer 2 smaller, without hooked hairs . . . . 1. *Urtica*, p. 401
    3. Tepals of pistillate flowers connate, forming saclike structure tightly enclosing mature achene, covered with delicate, hooked hairs. 2. *Hesperocnide*, p. 404
1. Plants without stinging hairs (hairs, if any, soft and flexible and without bulbous or cylindric base); tepals of bisexual and pistillate flowers 3–4, distinct or connate forming tubular perianth, or perianth appearing absent.
  4. Leaves opposite (opposite or alternate in *Boehmeria*).
    5. Plants glabrous; cystoliths linear. . . . . 5. *Pilea*, p. 408
    5. Plants pubescent; cystoliths rounded. . . . . 8. *Boehmeria*, p. 412
  4. Leaves alternate.
    6. Blade margins dentate or serrate. . . . . 8. *Boehmeria*, p. 412
    6. Blade margins entire.
      7. Bracts subtending pistillate flowers developing 3 corky wings covered with fine, hooked hairs in fruit; leaves with elongate, linear cystoliths . . . . 7. *Soleirolia*, p. 411
      7. Bracts subtending pistillate flowers not developing corky wings or hooked hairs in fruit; leaves with rounded cystoliths.
        8. Stipules absent (but axillary involucral bracts appear stipular); tepals distinct, ascending, loosely enclosing achene; achenes acute or mucronate. . . . . 4. *Parietaria*, p. 406
        8. Stipules present; tepals connate, appressed to, and tightly enclosing achene; achenes acute, not mucronate . . . . 6. *Pouzolzia*, p. 411

1. URTICA Linnaeus, Sp. Pl. 2: 983. 1753; Gen. Pl. ed. 5, 423. 1754 · Nettle, ortie [Latin *urtica*, nettle; derived from Latin *uro*, to burn]

Herbs, annual or perennial, with stinging and nonstinging hairs on same plant. **Stems** simple or branched, erect, ascending, or sprawling. **Leaves** opposite; stipules present. **Leaf blades** elliptic, lanceolate, ovate, or orbiculate, margins dentate to serrate; cystoliths rounded or ± elongate. **Inflorescences** axillary, lax, of cymes arranged in racemes or panicles. **Flowers** unisexual, staminate and pistillate flowers in loose to tight clusters intermixed in same inflorescence or in separate inflorescences on same or different plants; bracts narrowly triangular to lanceolate, lacking hooked hairs. **Staminate flowers:** tepals 4, distinct, equal; stamens 4; pistillode cuplike. **Pistillate flowers:** tepals 4, distinct, inner 2 equal to achene, outer 2 smaller, without hooked hairs; staminodes absent; style absent; stigma tufted, persistent or deciduous. **Achenes** sessile, laterally compressed, ovoid or deltoid, loosely enclosed by inner tepals. $x = 12, 13$.

Species 45 (4 in the flora): nearly worldwide.

SELECTED REFERENCES Woodland, D. W. 1982. Biosystematics of the perennial North American taxa of *Urtica*. II. Taxonomy. Syst. Bot. 7: 282–290. Woodland, D. W., I. J. Bassett, and C. W. Crompton. 1976. The annual species of stinging nettle (*Hesperocnide* and *Urtica*) in North America. Canad. J. Bot. 54: 374–383. Woodland, D. W., I. J. Bassett, L. Crompton, and S. Forget. 1982. Biosystematics of the perennial North American taxa of *Urtica*. I. Chromosome number, hybridization, and palynology. Syst. Bot. 7: 269–281.

1. Plants perennial, rhizomatous; inflorescences either staminate or pistillate . . . . 1. *Urtica dioica*
1. Plants annual, with taproot; inflorescences with both staminate and pistillate flowers (or staminate and pistillate flowers in separate inflorescences in *Urtica gracilenta*).
   2. Leaf blades elliptic to broadly elliptic, widest near middle, base cuneate; achenes 1.5–1.8 mm . . . . 4. *Urtica urens*
   2. Leaf blades narrowly ovate to orbiculate, usually widest below middle or near base; base ± cordate to truncate or rounded, occasionally cuneate, sometimes distal leaves ovate-lanceolate to lanceolate; achenes 1.2–1.6 mm.
      3. Inflorescences ± globose; staminate and pistillate flowers intermixed in same inflorescence . . . . 2. *Urtica chamaedryoides*
      3. Inflorescences elongate; staminate and pistillate flowers in separate inflorescences, or a few pistillate flowers at apex of staminate inflorescences . . . . 3. *Urtica gracilenta*

1. Urtica dioica Linnaeus, Sp. Pl. 2: 984. 1753 · Stinging nettle, ortie [W2]

Herbs, perennial, rhizomatous, 5–30 dm. **Stems** simple or branched, erect or sprawling. **Leaf blades** elliptic, lanceolate, or narrowly to broadly ovate, 6–20 × 2–13 cm, base rounded to cordate, margins coarsely serrate, sometimes doubly serrate, apex acute or acuminate; cystoliths rounded. **Inflorescences** paniculate, pedunculate, elongate. **Flowers** unisexual, staminate and pistillate on same or different plants, staminate ascending, the pistillate lax or recurved. **Pistillate flowers:** outer tepals linear to narrowly spatulate or lanceolate, 0.8–1.2 mm, inner tepals ovate to broadly ovate, 1.4–1.8 × 1.1–1.3 mm. **Achenes** ovoid to broadly ovoid, 1–1.3(–1.4) × 0.7–0.9 mm.

Subspecies 3 (3 in the flora): North America, Mexico, Eurasia.

1. Staminate and pistillate flowers on different plants; leaf blades abaxially hispid, both surfaces with stinging hairs . . . . . 1a. *Urtica dioica* subsp. *dioica*
1. Staminate and pistillate flowers mostly on same plants; leaf blades abaxially bearing stinging hairs, otherwise glabrous, puberulent, or tomentose and moderately strigose, adaxially without (rarely with a few) stinging hairs.
   2. Stems glabrous or strigose, with a few stinging hairs; leaf blades abaxially glabrous or puberulent. . . . . 1b. *Urtica dioica* subsp. *gracilis*
   2. Stems softly pubescent, also with stinging hairs; leaf blades abaxially tomentose to moderately strigose. . .1c. *Urtica dioica* subsp. *holosericea*

1a. Urtica dioica Linnaeus subsp. **dioica** [F]

Stems hispid, with stinging hairs. **Leaf blades** abaxially hispid, both surfaces with stinging hairs. **Flowers** unisexual, staminate and pistillate on different plants. $2n = 52$.

Flowering late spring–early fall. Alluvial woods, margins of deciduous woodlands, fencerows, waste places; 0–500 m; introduced; Greenland; B.C., N.B., Nfld., N.S., Ont., P.E.I., Que.; Ala., Alaska, Calif., Conn., Del., D.C., Fla., Ga., Maine, Md., Mass., Mich., Mo., N.H., N.J., N.Y., N.C., Ohio, Okla., Oreg., Pa., Tenn., Va., Wash., W.Va.; native to Eurasia.

I have seen no specimens of *Urtica diocia* subsp. *dioica* from Rhode Island or Vermont, although it most likely occurs there.

1b. Urtica dioica Linnaeus subsp. **gracilis** (Aiton) Selander, Svensk Bot. Tidskr. 41: 271. 1947 [E][F][W1]

Urtica gracilis Aiton, Hort. Kew. 3: 341. 1789; *U. californica* Greene; *U. dioica* var. *angustifolia* Schlechtendal; *U. dioica* var. *californica* (Greene) C. L. Hitchcock; *U. dioica* var. *gracilis* (Aiton) Roy L. Taylor & MacBryde; *U. dioica* var. *lyallii* (S. Watson) C. L. Hitchcock; *U. dioica* var. *procera* (Muhlenberg ex Willdenow) Weddell; *U. lyallii* S. Watson; *U. lyallii* var. *californica* (Greene) Jepson; *U. procera* Muhlenberg ex Willdenow; *U. viridis* Rydberg

Stems glabrous or strigose, with a few stinging hairs. **Leaf blades** abaxially bearing stinging hairs, otherwise glabrous or puberulent, adaxially without or rarely with a few stinging hairs. **Flowers** unisexual, staminate and pistillate mostly on same plant. $2n = 26, 52$.

Flowering late spring–summer. Alluvial woods, margins of deciduous woodlands, fencerows, waste places; 0–3100 m; St. Pierre and Miquelon; Alta., B.C., Man., N.B., Nfld., N.W.T., N.S., Ont., P.E.I., Que., Sask., Yukon; Ala., Alaska, Ariz., Calif., Colo., Conn., Del., Ga., Idaho, Ill., Ind., Iowa, Kans., Ky., La., Maine, Md., Mass., Mich., Minn., Miss., Mo., Mont., Nebr., N.H., N.J., N.Mex., N.Y., N.C., N.Dak., Ohio, Okla., Oreg., Pa., R.I., S.Dak., Tenn., Tex., Utah, Vt., Va., Wash., W.Va., Wis., Wyo.

Native Americans used *Urtica dioica* subsp. *gracilis* medicinally for rheumatism, upset stomach, childbirth, paralysis, fevers, colds, tuberculosis, and as a general tonic, and as a witchcraft medicine (D. E. Moerman 1986).

1c. Urtica dioica Linnaeus subsp. **holosericea** (Nuttall) Thorne, Aliso 6: 68. 1967   F

Urtica holosericea Nuttall, Proc. Acad. Nat. Sci. Philadelphia 4: 25. 1848; *U. breweri* S. Watson; *U. dioica* var. *holosericea* (Nuttall) C. L. Hitchcock; *U. dioica* var. *occidentalis* S. Watson; *U. gracilis* Aiton var. *greenei* (Jepson) Jepson; *U. gracilis* subsp. *holosericea* (Nuttall) W. A. Weber; *U. gracilis* var. *holosericea* (Nuttall) Jepson

Stems softly pubescent, also with stinging hairs. **Leaf blades** abaxially sparsely to densely tomentose to moderately strigose, soft to touch, with stinging hairs, adaxially without or rarely with a few stinging hairs. **Flowers** unisexual, staminate and pistillate mostly on same plant. $2n = 26$.

Flowering late spring–summer. Alluvial woods, margins of deciduous or mixed woodlands, fencerows, waste places; 0–3100 m; Ariz., Calif., Colo., Idaho, Mont., Nev., N.Mex., Oreg., Utah, Wash., Wyo.; Mexico.

Urtica dioica subsp. *holosericea* is highly variable in leaf shape and degree of pubescence. The least pubescent plants appear to grade into *U. dioica* subsp. *gracilis,* and it is sometimes difficult to separate the two.

The name *U. serra* Blume has been misapplied to this taxon.

2. Urtica chamaedryoides Pursh, Fl. Amer. Sept. 1: 113. 1814   F   W1

Urtica chamaedryoides var. *runyonii* Correll

Herbs, annual, with taproot, 1.5–8 dm. **Stems** usually branched from base, erect or reclining. **Leaf blades** narrowly ovate to orbiculate, distal blades sometimes lanceolate, 2–8 × 1–6 cm, base nearly cordate to rounded, sometimes cuneate in distal leaves, margins serrate, apex rounded to acute; cystoliths rounded or ± elongate. **Inflorescences** cymose, sessile to short-pedunculate ± globose. **Flowers** unisexual, staminate and pistillate intermixed in same inflorescence. **Pistillate flowers:** outer tepals linear, 0.4–0.8 mm, inner tepals ovate, 1.4–2 × 1–1.4 mm. **Achenes** ovoid to broadly ellipsoid, 1.2–1.4(–1.6) × 0.8–1 mm. $2n = 26$.

Flowering all year except early winter. Rich, wooded slopes, bluffs, stream banks, swamps, waste places, and fields, often on limestone or nearly neutral soils; 0–600 m; Ala., Ark., Fla., Ga., Ill., Kans., Ky., La., Miss., Mo., N.C., Ohio, Okla., S.C., Tenn., Tex.; n Mexico.

3. Urtica gracilenta Greene, Bull. Torrey Bot. Club 8: 122. 1881

Herbs, annual, with taproot, 3–20 dm. **Stems** simple or branched from base, erect. **Leaf blades** ovate to broadly ovate, distal blades becoming ovate-lanceolate to lanceolate, 7–15 × 5–10 cm, base truncate to cordate, margins coarsely dentate, apex acute to caudate; cystoliths rounded or occasionally elongate. **Inflorescences** racemose, elongate. **Flowers** unisexual, staminate and pistillate in separate inflorescences, or with a few pistillate flowers at apex of staminate inflorescences, subsessile to short-pedunculate. **Pistillate flowers:** outer tepals lanceolate to narrowly ovate, 0.8–1 mm, inner tepals broadly ovate, 1.4–2.2 × 1.3–1.4 mm. **Achenes** ovoid, 1.4–1.6 × 1–1.1 mm. $2n = 26$.

Flowering summer–fall, occasionally all year. Alluvial or calcareous soils, often in moist, shaded places; 1200–2800 m; Ariz., N.Mex., Tex.; n Mexico.

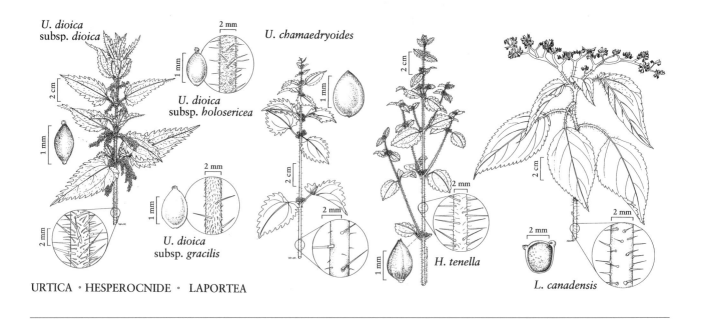

U. dioica subsp. *dioica*

2 mm

1 mm

U. dioica subsp. *holosericea*

U. chamaedryoides

1 mm

2 cm

2 mm

1 mm

2 mm

U. dioica subsp. *gracilis*

2 cm

2 mm

2 mm

1 mm

H. tenella

2 mm

2 mm

2 mm

L. canadensis

URTICA · HESPEROCNIDE · LAPORTEA

4. Urtica urens Linnaeus, Sp. Pl. 2: 984. 1753 · Dog nettle, burning nettle W2

Herbs, annual, with taproot, 1–8 dm. **Stems** simple or branched, erect. **Leaf blades** elliptic to broadly elliptic, widest near middle, 1.8–9 × 1.2–4.5 cm, base cuneate, margins coarsely serrate, serrations often with lateral lobes, apex acute; cystoliths rounded. **Inflorescences** spikelike or paniculate. **Flowers** unisexual, staminate and pistillate in same inflorescence, subsessile to short-pedunculate. **Pistillate flowers:** outer tepals ovate, 0.5–

0.7 mm, inner tepals broadly ovate, 0.6–0.9 × 1.2–1.4 mm. **Achenes** ovoid, 1.5–1.8 × 1.1–1.3 mm. $2n = 24, 26$.

Flowering spring–summer. Waste places, roadsides, pastures, barnyards, cultivated fields, rich woodlands; 0–700 m; introduced; Greenland; St. Pierre and Miquelon; Alta., B.C., Man., N.B., Nfld., N.S., Ont., P.E.I., Que., Sask., Yukon; Ala., Alaska, Ariz., Calif., Conn., Fla., Ill., Maine, Mass., Mich., Mo., Nev., N.H., N.Mex., N.Y., Okla., Oreg., Pa., R.I., S.C., Tex., Vt., Wash.; Eurasia.

Within the flora, *Urtica urens* is most abundant in California and in eastern Canada.

2. HESPEROCNIDE Torrey, Pacif. Railr. Rep. 4(5): 139. 1857 · [Greek *hesperos,* west, and *knide,* nettle]

Herbs, annual, with stinging and nonstinging hairs. **Stems** usually branched, erect, spreading, or reclining. **Leaves** opposite; stipules present. **Leaf blades** ovate to broadly ovate, distal blades sometimes broadly elliptic, margins serrate; cystoliths elongate. **Inflorescences** axillary, globose, nearly globose, or elongate-racemose or paniculate. **Flowers** unisexual, staminate and pistillate in loose to dense clusters in same inflorescence; bracts absent. **Staminate flowers:** tepals 4, distinct, equal; stamens 4; pistillode present. **Pistillate flowers:** tepals 4, connate, forming persistent saclike structure covered with delicate, hooked hairs and completely enclosing mature, flattened achene; staminodes absent; style absent, stigma tufted, persistent. **Achenes** subsessile, laterally compressed, ovoid, tightly enclosed in persistent tepals. $x = 12$.

Species 2 (1 in the flora): 1 in California and Mexico, 1 in Hawaii.

SELECTED REFERENCE Woodland, D. W., I. J. Bassett, and C. W. Crompton. 1976. The annual species of stinging nettle (*Hesperocnide* and *Urtica*) in North America. Canad. J. Bot. 54: 374–383.

1. Hesperocnide tenella Torrey, Pacif. Railr. Rep. 4(5): 139. 1857 · Western-nettle F

Plants with taproot, 0.8–6 dm. **Leaf blades** 1–6 × 1–4 cm, base broadly cuneate, rounded, or cordate, apex acute or short-acuminate to obtuse tip. **Flowers** sessile to short-pedicellate; perianth sac microscopically 2–4-lobed at apex, 0.4–0.8 mm. **Achenes** ovoid to broadly ovoid, 1–1.4 × 0.8–1 mm. $2n = 24$.

Flowering spring–early summer. Moist, shaded places in chaparral, deserts, grasslands, and woodlands, or in open areas; 0–1000 m; Calif.; Mexico (n Baja California).

3. LAPORTEA Gaudichaud-Beaupré, Voy. Uranie 12: 498. 1830, name conserved · [For F. L. de Laporte de Castelnau, leader of expeditions to South America]

Urticastrum Heister ex Fabricius

Herbs, annual or perennial, with stinging and nonstinging hairs on same plant. **Stems** simple, erect. **Leaves** alternate; stipules present. **Leaf blades** narrowly ovate to orbiculate, margins dentate or serrate; cystoliths rounded. **Inflorescences** axillary and terminal, of paniculately arranged cymes. **Flowers** unisexual, proximal panicles staminate and distal pistillate, or staminate and pistillate flowers in same panicle; bracts absent. **Staminate flowers:** tepals 4–5, distinct, equal; stamens 4–5; pistillode knoblike. **Pistillate flowers:** tepals 2–4, distinct, outer pair minute or absent, without hooked hairs; staminodes absent; style persistent even in fruit, hooklike or elongate; stigma extended along style. **Achenes** stipitate, laterally compressed, ± orbicular, free from and not enclosed by perianth. $x = 13$.

Species 22 (2 in the flora): North America, West Indies, Central America, South America, Asia, Africa, Pacific Islands.

Most species of *Laportea* occur in Africa and Madagascar.

SELECTED REFERENCE Chew, W. L. 1969. A monograph of *Laportea* (Urticaceae). Gard. Bull. Straits Settlem. 25: 111–178.

1. Plants with stinging hairs and stipitate-glandular hairs; base of leaf blade rounded or abruptly attenuate, auriculate; achenes less than 1.5 mm . . . . 1. *Laportea aestuans*
1. Plants with stinging hairs but without stipitate-glandular hairs; base of leaf blade rounded, truncate, or broadly cuneate, not auriculate; achenes 2 mm or more . . . . 2. *Laportea canadensis*

1. Laportea aestuans (Linnaeus) Chew, Gard. Bull. Straits Settlem. 21: 200. 1965

Urtica aestuans Linnaeus, Sp. Pl. ed. 2, 2: 1397. 1763

Herbs, annual, 1–10 dm, sparsely to densely pubescent with stinging hairs and stipitate-glandular, nonstinging hairs. **Leaf blades** broadly ovate to nearly orbiculate, 9–20 × 6–16 cm, base rounded or abruptly attenuate, auriculate, margins regularly serrate or dentate, apex short-acuminate. **Inflorescences** with both staminate and pistillate flowers in same panicle, or proximal panicles with staminate flowers. **Staminate flowers** ca. 2 mm across; tepals 4–5, equal in length; stamens 4–5, opposite tepals; filaments longer than tepals. **Pistillate flowers** ca. 0.7 mm; tepals 2–4, appressed, inner pair ca. 1/2 length of ovary; ovary ovoid to ellipsoid; style persistent, hooked and beaklike, ca. 0.2 mm, becoming knoblike in fruit. **Achenes** strongly compressed, ± orbicular, ca. 0.9 × 1.3 mm.

Flowering fall–winter. Waste places, cultivated ground; 0–10 m; possibly introduced; Fla.; Mexico; West Indies; Central America (Costa Rica and Panama).

Plants of *Laportea aestuans* in the flora have slightly auriculate leaf bases; those from outside the flora area frequently have rounded or truncate leaf bases.

2. Laportea canadensis (Linnaeus) Weddell, Ann. Sci. Nat., Bot., sér. 4, 1: 181. 1854 · Wood-nettle, grande ortie F

Urtica canadensis Linnaeus, Sp. Pl. 2: 985. 1753; *Urticastrum divaricatum* (Linnaeus) Kuntze

Herbs, annual or perennial, rhizomatous, with tuberous roots, 3–15 dm, sparsely to densely covered with stinging hairs and nonglandular, nonstinging hairs, stipitate-glandular hairs absent. **Leaf blades** narrowly to broadly ovate, 6–30 × 3–18 cm, base rounded, truncate, or broadly cuneate, not auriculate, margins regularly serrate, apex acuminate. **Inflorescences** with staminate and pistillate flowers in separate panicles, proximal panicles staminate, distal panicles pistillate. **Staminate flowers** ca. 1–1.5 mm across; tepals 5, equal in length; stamens 5, opposite tepals; filaments slightly longer than tepals. **Pistillate flowers** ca. 0.5 mm; tepals 2–4, appressed, inner pair as long as ovary; ovary compressed, nearly orbicular to crescent-shaped; style persistent, feathery, 2–3 mm or more. **Achenes** strongly compressed, ± orbicular, ca. 2–3 mm. $2n = 26$.

Flowering spring–early fall. Rich, moist, deciduous forests, often along seepages and streams; 0–2000 m; St. Pierre and Miquelon; Man., N.B., N.S., Ont., Que., Sask.; Ala., Ark., Conn., Del., D.C., Fla., Ga., Ill., Ind., Iowa, Kans., Ky., La., Maine, Md., Mass., Mich., Minn., Miss., Mo., Nebr., N.H., N.J., N.Y., N.C., N.Dak., Ohio, Okla., Pa., R.I., S.C., S.Dak., Tenn., Vt., Va., W.Va., Wis.; Mexico.

4. PARIETARIA Linnaeus, Sp. Pl. 2: 1052. 1753; Gen. Pl. ed. 5, 471. 1754 · Pellitory

[Latin *paries*, wall, referring to habitat of original species]

Herbs, annual or perennial, sparsely to densely pubescent with hooked and straight, nonstinging hairs on all parts of plant, stinging hairs absent. **Stems** often branched from base, erect, ascending, or decumbent. **Leaves** alternate; stipules absent. **Leaf blades** deltate, orbiculate to narrowly elliptic, or lanceolate, margins entire; cystoliths rounded. **Inflorescences** axillary. **Flowers** bisexual, staminate, or pistillate, proximal flowers usually bisexual and staminate, distal flowers pistillate; involucral bracts linear to lanceolate, without hooked hairs; tepals 4, distinct, ascending, lacking hooked hairs; stamens 4; style persistent or not; stigma tufted, deciduous. **Achenes** stipitate, ovoid, acute or mucronate (style base sometimes persisting as apical or subapical mucro), loosely enclosed by tepals. $x = 7, 8, 10, 13$.

Species 20–30 (5 in the flora): primarily in temperate and subtropical regions.

Mature achenes are necessary for certain determination.

Parietaria nummularifolia (Schwartz) Weddell was collected once in 1992 in Palm Beach County, Florida, in mesic woods bordering a creek (R. P. Wunderlin, pers. comm.). This species is occasionally cultivated, and the Florida collection probably represents an escape.

1. Herbs perennial; achenes dark brown, apex acute, mucro absent or minute. .1. *Parietaria judaica*
1. Herbs annual or short-lived perennial; achenes light brown, apex obtuse, mucro distinct, apical or subapical.
    2. Leaf blades orbiculate to deltate, apex smoothly attenuate or occasionally slightly acuminate; achenes less than 0.9 mm, stipes short-cylindric, abruptly flared basally . . . . 2. *Parietaria floridana*
    2. Leaf blades ovate to narrowly elliptic or lanceolate, or if deltate or orbiculate, then apex acuminate; achenes 0.9 mm or more, stipes basally dilated cylinders.
        3. Mucro subapical; stipe not centered . . . . .5. *Parietaria praetermissa*
        3. Mucro apical; stipe centered.
            4. Leaf blades narrowly to very broadly ovate, oblong, orbiculate, or reniform, base rounded; proximal pair of lateral veins arising at junction of blade and petiole; involucral bracts usually more than 2 times length of achene. . . . . 3. *Parietaria hespera*
            4. Leaf blades narrowly to broadly elliptic, lanceolate, oblong, or ovate, base narrowly cuneate; proximal pair of lateral veins arising distal to junction of blade and petiole; involucral bracts usually less than 2 times length of achene . . . . .4. *Parietaria pensylvanica*

1. Parietaria judaica Linnaeus, Fl. Palaest., 32. 1756

Herbs, perennial from crown, 1–8 dm. **Stems** ascending, erect, or decumbent. **Leaf blades** narrowly to broadly elliptic, lance-elliptic, or ovate, 1.3–9 × 0.8–4.5 cm, base attenuate, cuneate, or broadly rounded, apex abruptly acuminate to long-attenuate. **Flowers:** involucral bracts 1.5–2.5 mm; tepals ca. 2–3.5 mm, longer than bracts. **Achenes** dark brown, symmetric, 1–1.2 × 0.6–0.9 mm, apex acute, mucro absent or minute; stipe centered, on cylindric base.

Flowering all year, with peak in late winter–spring. Cracks in sidewalks, ballast heaps, waste places, frequently about ports and coastal areas; 0–200 m; introduced; Calif., Fla., La., Mich., N.J., N.Y., Pa., Tex.; Eurasia; n Africa.

Parietaria judaica, which, in North America, is most abundant in scattered localities in California, is the only long-lived perennial species of *Parietaria* in the flora. Because of confusion in Europe over the correct name, plants in North America have been called *P. judaica, P. officinalis* of authors, not Linnaeus, *P. officinalis* var. *erecta* (Mertens & Koch) Weddell, and *P. officinalis* var. *diffusa* (Mertens & Koch) Weddell. For a clarification of the nomenclature and taxonomy of this complex, see C. C. Townsend (1968).

Parietaria judaica was first reported from Louisiana as *P. diffusa* Mertens & Koch, another name commonly used on herbarium specimens (J. W. Thieret 1969).

2. Parietaria floridana Nuttall, Gen. N. Amer. Pl. 2: 208. 1818 [F] [W2]

Parietaria nummularia Small

Herbs, annual or short-lived perennial, 1–4 dm. **Stems** 10–20-branched, decumbent to ascending. **Leaf blades** orbiculate to deltate, 0.7–2.7 × 0.5–1.7 cm, base truncate, rounded, or very broadly cuneate, apex smoothly attenuate or occasionally slightly acuminate. **Flowers:** involucral bracts 1.5–2 mm; tepals ca. 1.5 mm, nearly equal to bracts. **Achenes** light brown, symmetric, 0.5–0.8 × 0.3–0.6 mm or less, apex obtuse, mucro ± apical; stipe centered, short-cylindric, abruptly flared basally.

Flowering winter–spring. Weedy places, around masonry, woodland and shrub borders, shell mounds, sandy beaches, roadsides; Atlantic and Gulf coastal plains; 0–30 m; Del., Fla., Ga., La., Miss., N.C., S.C., Tex.; Mexico; West Indies; South America.

Parietaria praetermissa has been misidentified as *P. floridana* by some authors.

3. Parietaria hespera B. D. Hinton, Sida 3: 293, figs. 1, 2. 1969

Herbs, annual, 0.2–5.5 dm. **Stems** simple or freely branched, sometimes densely matted, prostrate, decumbent, ascending, or erect. **Leaf blades** narrowly to very broadly ovate, oblong, orbiculate, or reniform, 0.2–4.5 × 0.2–2.7 cm, base broadly cuneate, rounded, truncate, or nearly cordate, apex acuminate, acute, obtuse, or rounded; proximal pair of lateral veins arising at junction of blade and petiole. **Flowers:** involucral bracts 1–4.5 mm; tepals 2–2.8 mm, shorter or longer than bracts. **Achenes** light brown, symmetric, 0.9–1.2 × 0.6–0.7 mm, apex obtuse, mucro apical; stipe centered, short-cylindric, basally dilated.

Varieties 2 (2 in the flora): North America, Mexico.

1. Tepals erect, loosely connivent, apex acute; leaf blades conspicuously longer than wide, base broadly cuneate, rounded, or truncate. . . . . 3a. *Parietaria hespera* var. *hespera*
1. Tepals spreading or recurved and twisted at maturity, distinct, apex long-acuminate, attenuate, or caudate; leaf blades as long as or slightly longer than wide, base rounded to nearly cordate. . . . . 3b. *Parietaria hespera* var. *californica*

3a. Parietaria hespera B. D. Hinton var. **hespera**

Herbs, 0.3–5.5 dm. **Leaf blades** narrowly to broadly ovate, less frequently oblong or nearly orbiculate, 0.4–4.5 × 0.4–2.7 cm, conspicuously longer than wide, base broadly cuneate, rounded, or truncate, apex distally rounded or acuminate to acute or obtuse. **Flowers:** involucral bracts 1–4.5 mm; tepals erect, loosely connivent at maturity, ca. 2–2.8 mm, apex acute.

Flowering late winter–early summer, rarely at other times. Chaparral, deserts, roadsides, sand dunes, often in shaded and moist places; 0–1400 m; Ariz., Calif., Nev., N.Mex., Utah; nw Mexico.

3b. Parietaria hespera B. D. Hinton var. **californica** B. D. Hinton, Sida 3: 295, figs. 1, 2. 1969

Herbs, 0.2–4 dm. **Leaf blades** oblong, broadly ovate to orbiculate or reniform, 0.2–1.7 × 0.2–1.2 cm, as long as or slightly longer than wide, base rounded to nearly cordate, apex distally short-acuminate to obtuse or rounded. **Flowers:** involucral bracts 1.2–3.2 mm; tepals spreading or recurved and twisted at maturity, distinct, ca. 2–2.7 mm, apex long-acuminate, attenuate, or caudate.

Flowering spring. Moist, often shaded, sandy or rocky places; 0–1200 m; Calif.; Mexico (Baja California).

The two varieties of *Parietaria hespera* are generally distinct, but intermediates occur. All are found in California and have been collected in Los Angeles and Monterey counties and on Santa Barbara Island.

4. Parietaria pensylvanica Muhlenberg ex Willdenow, Sp. Pl. 4(2): 955. 1806 F W2

Parietaria obtusa Rydberg ex Small; *P. occidentalis* Rydberg; *P. pensylvanica* var. *obtusa* (Rydberg ex Small) Shinners

Herbs, annual, 0.4–6 dm. **Stems** simple or freely branched, decumbent, ascending, or erect. **Leaf blades** narrowly to broadly elliptic, lanceolate, oblong, or ovate, (1–)2–9 × 0.4–3 cm, base narrowly cuneate, apex acuminate to long-attenuate or obtuse to rounded; proximal pair of lateral veins arising above junction of blade and petiole. **Flowers:** involucral bracts 1.8–5 mm, usually less than 2 times length of achene; tepals 1.5–2 mm, shorter than bracts. **Achenes** light brown, symmetric, 0.9–1.2 × 0.6–0.9 mm, apex obtuse, mucro apical; stipe straight, short-cylindric, centered, basally dilated. $2n = 14, 16$.

Flowering spring–late fall. Dry ledges, talus slopes, waste and shaded places, primarily in neutral to basic soils, and reported from margins of hot springs in northernmost locations; 0–2400 m; Alta., B.C., Man., Ont., Que., Sask., Yukon; Ala., Ariz., Ark., Calif., Colo., Conn., D.C., Fla., Ga., Idaho, Ill., Ind., Iowa, Kans., Ky., La., Maine, Md., Mass., Mich., Minn., Miss., Mo., Mont., Nebr., Nev., N.H., N.J., N.Mex., N.Y., N.C., N.Dak., Ohio, Okla., Oreg., Pa., S.C., S.Dak., Tenn., Tex., Utah, Vt., Va., Wash., W.Va., Wis., Wyo.; Mexico.

Some extremes of *Parietaria pensylvanica* with short, oblong or ovate leaf blades strongly resemble *P. hespera* var. *hespera*. *Parietaria hespera* is usually more delicate and has thinner leaves with the proximal pair of lateral veins arising at the junction of blade and petiole. Leaf shape and texture tend to overlap in the two species, but in *P. pensylvanica* the proximal pair of lateral veins clearly arise above the junction of blade and petiole. The extremes of *P. pensylvanica* frequently are found where the ranges of the two species approach or overlap. Examples of these intermediates are from Gila, Mohave, and Yuma counties, Arizona. A mixed collection from Rock Springs, Gila County, Arizona, suggests that the two species occasionally grow together.

5. Parietaria praetermissa B. D. Hinton, Sida 3: 192, fig. 1. 1968 E

Herbs, annual, 1–5.5 dm. **Stems** freely branched, decumbent to ascending. **Leaf blades** narrowly to broadly ovate or rarely lanceolate, 1–6.5 × 0.6–4 cm, base rounded or very broadly cuneate, apex short- to long-acuminate or attenuate. **Flowers:** involucral bracts 1.3–4.7 mm; tepals 1.7–2.3 mm, equal to or shorter than bracts. **Achenes** light brown, asymmetric, 1–1.4 × 0.7–1.1 mm, apex obtuse, mucro subapical; stipe slanting, short-cylindric, not centered, basally dilated.

Flowering winter–summer. Shell mounds, calcareous outcrops, hammocks, waste places; 0–10 m; Fla., Ga., La., N.C., S.C.

Parietaria praetermissa is endemic to the Atlantic and Gulf coastal plains of southeastern United States. The name *Parietaria floridana* has been incorrectly applied to this species.

5. PILEA Lindley, Coll. Bot., plate 4 and text on facing page. 1821, name conserved

· [Latin *pileus*, felt cap, because of the calyx covering the achene]

Adicea Rafinesque ex Britton & A. Brown

Herbs, shrubs, or subshrubs, annual or perennial, glabrous. **Stems** simple or branched, erect, ascending, or repent. **Leaves** opposite; stipules present. **Leaf blades** equal or unequal, ovate, margins dentate or entire; cystoliths linear, ± conspicuous. **Inflorescences** axillary, compact to lax cymes. **Flowers** unisexual, staminate and pistillate flowers in same cyme; bracts deltate to linear. **Staminate flowers:** tepals 4; stamens 4; pistillode conic. **Pistillate flowers:** tepals 3, equal or sometimes 1 tepal enlarged and hoodlike; staminodes 3, opposite tepals, under tension and ejecting mature achene; style and tufted stigma deciduous. **Achenes** sessile, laterally compressed, ovoid to teardrop-shaped, free from perianth at maturity, partly covered by hoodlike tepal. $x = 12, 13$.

Species ca. 400 (5 in the flora): mostly tropical and subtropical regions worldwide except Australia and New Zealand.

The genus *Pilea* is in need of worldwide revision.

SELECTED REFERENCES Chen, C. J. 1982. A monograph of *Pilea* (Urticaceae) in China. Bull. Bot. Res., Harbin 2: 1–132. Fernald, M. L. 1936. *Pilea* in eastern North America. Contr. Gray Herb. 113: 169–170. Hermann, F. J. 1940. The geographic distribution of *Pilea fontana*. Torreya 40: 118–120.

1. Leaf margins dentate; paired blades equal.
  2. Achenes uniformly black except for very narrow, pale, marginal band, surface conspicuously pebbled with raised bosses. . . . . 4. *Pilea fontana*
  2. Achenes uniformly light colored or streaked with purple, surface smooth or purple striations sometimes slightly raised . . . . 5. *Pilea pumila*
1. Leaf margins entire; paired blades unequal.
  3. Stems repent or prostrate . . . . 1. *Pilea herniarioides*
  3. Stems erect or diffusely ascending.
    4. Herbs; leaf blades spatulate to obovate . . . . 2. *Pilea microphylla*
    4. Shrubs or subshrubs; leaf blades of 2 types, obovate and orbiculate to orbiculate-cordate . . .3. *Pilea trianthemoides*

1. Pilea herniarioides (Swartz) Lindley, Coll. Bot., index (see also plate 4 and text on facing page). 1821

Urtica herniarioides Swartz, Kongl. Vetensk. Acad. Nya Handl. 8: 64. 1787

Herbs, annual or short-lived perennial, 0.2–1 dm. **Stems** 5–10-branched, repent or prostrate, sometimes mat-forming. **Leaf blades** broadly ovate to orbiculate, paired blades only slightly unequal, 1.5–6 × 1.6–5 mm, margins entire. **Inflorescences** crowded. **Flowers** ca. 0.2–0.3 mm across. **Achenes** uniformly light brown, slightly compressed, ovoid-cylindric, ca. 0.4 × ca. 0.3 mm, pebbled.

Flowering fall–spring (sporadically throughout summer). Hammocks, waste places; 0–10 m; Fla.; West Indies; Central America; South America (Colombia).

2. Pilea microphylla (Linnaeus) Liebmann, Kongel. Vidensk. Selsk. Skr., Naturvidensk. Math. Afd., ser. 5, 2: 296. 1851 · Artillery weed F W1

Parietaria microphylla Linnaeus, Syst. Nat. ed. 10, 2: 1308. 1759

Herbs, annual or short-lived perennial, 0.3–2 dm. **Stems** 10–40-branched, erect. **Leaf blades** spatulate to obovate, paired blades unequal, the larger 3–10 × 1.5–5.5 mm, the smaller 1.5–4 × 0.7–2 mm, margins entire. **Inflorescences** crowded. **Flowers** ca. 0.5 mm across. **Achenes** uniformly light brown, slightly compressed, ovoid-cylindric, ca. 0.5(–1.1) × 0.3 mm, smooth.

Flowering all year. Waste places, hammocks, rocky woods, cultivated plots, on masonry; 0–100 m; Fla., Ga., La., S.C.; Mexico; West Indies; Central America; tropical South America; Pacific Islands (Hawaii); Asia.

Pilea microphylla has been collected once in Tennessee and once in Michigan, but it is unlikely that the species persists so far north. It is widely grown as a houseplant in the north and a border plant in the south. It is a greenhouse weed in various parts of the flora.

3. Pilea trianthemoides (Swartz) Lindley, Coll. Bot., plate 4 and text on facing page. 1821

Urtica trianthemoides Swartz, Kongl. Vetensk. Acad. Nya Handl. 8: 68. 1787

Shrubs or subshrubs, 1.5–5 dm. **Stems** 10–30-branched, erect or diffusely ascending. **Leaf blades** with members of pair unequal, the larger obovate, 5–10 × 4–6 mm, the smaller nearly orbiculate to orbiculate-cordate, 2–5 mm wide, margins entire. **Inflorescences** crowded. **Flowers** ca. 0.3–0.5 mm across. **Achenes** uniformly light brown, slightly compressed, ovoid-cylindric, ca. 0.7 × 0.5 mm, pebbled.

Flowering all year. Waste places, moist thickets, persisting after cultivation; 0–20 m; introduced; Fla.; West Indies.

The group including *Pilea microphylla* and *P. trianthemoides* is in need of taxonomic study. Although in the United States these two species generally can be separated without difficulty, in the West Indies they appear to intergrade. In the West Indies, *P. trianthemoides* is sometimes treated as *P. microphylla* var. *trianthemoides* (Swartz) Grisebach.

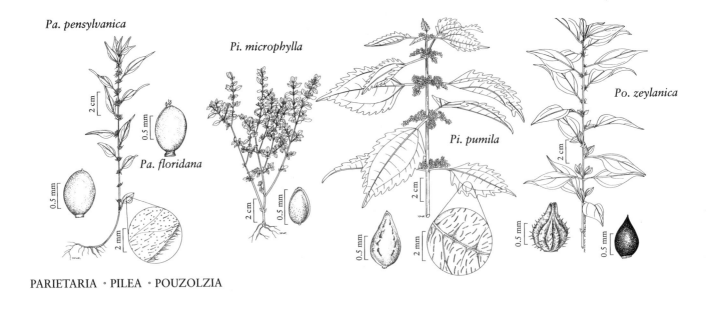

Pa. pensylvanica

Pi. microphylla

Pa. floridana

Po. zeylanica

Pi. pumila

PARIETARIA · PILEA · POUZOLZIA

The name *Pilea serpyllifolia* (Poiret) Weddell was misapplied to this taxon by J. K. Small.

4. Pilea fontana (Lunell) Rydberg, Brittonia 1: 87. 1931 · Clearweed E

Adicea fontana Lunell, Amer. Midl. Naturalist 3: 7. 1913

Herbs, annual, 1–7 dm. **Stems** simple or slightly branched, erect. **Leaf blades** elliptic to ovate, paired blades equal, 1–10 × 0.6–4.5 cm, margins dentate. **Inflorescences** crowded or lax. **Flowers** ca. 1 mm across. **Achenes** uniformly black except for very narrow, pale, often inconspicuous, marginal band, compressed, teardrop-shaped, 1.3–1.7 × 1–1.5 mm, conspicuously pebbled or warty with raised bosses.

Flowering late summer–fall. Mixed woods, along streams, swamps, seepages, and marshes; 0–300 m; Ont.; Ala., Conn., Fla., Ga., Ill., Ind., Iowa, Md., Mass., Mich., Minn., Nebr., N.J., N.Y., N.C., N.Dak., Ohio, Pa., R.I., S.C., S.Dak., Vt., Va., Wis.

Pilea fontana and *P. pumila* are separated primarily by differences in their mature achenes. In addition, leaves of *P. fontana* are often more opaque and less shiny than those of *P. pumila*. A few collections of *P. pumila* from Bourbon, Owen, and Robertson counties, Kentucky, and Macon County, Tennessee, have the black achenes of *P. fontana,* but without the bosses, and show striations on the younger achenes as in *P. pumila*. I have seen only two mixed collections (Chisago County, Minnesota, and Richland-Ransom

county line, South Dakota), which probably indicates that these two very similar species seldom occur together, even though their ranges overlap completely.

5. Pilea pumila (Linnaeus) A. Gray, Manual, 437. 1848 · Clearweed F W1

Urtica pumila Linnaeus, Sp. Pl. 2: 984. 1753

Herbs, 0.7–7 dm. **Stems** simple, erect. **Leaf blades** elliptic to broadly elliptic or ovate, paired blades equal, 2–13 × 1–9 cm, margins dentate. **Inflorescences** crowded to lax. **Flowers** ca. 1 mm across. **Achenes** uniformly light colored or with streaks of purple, compressed, teardrop-shaped, 1.3–1.7 × 0.6–1.1 mm, smooth or purple striations sometimes raised. 2*n* = 24, 26.

Flowering summer–fall. Moist to wet woods, woodland margins, along streams, shaded waste places; 0–2000 m; N.B., Ont., P.E.I., Que.; Ala., Ark., Conn., Del., D.C., Fla., Ga., Ill., Ind., Iowa, Kans., Ky., La., Maine, Md., Mass., Mich., Minn., Miss., Mo., Nebr., N.H., N.J., N.Y., N.C., N.Dak., Ohio, Okla., Pa., R.I., S.C., S.Dak., Tenn., Tex., Vt., Va., W.Va., Wis.; Asia.

Typical plants have leaf blades with cuneate bases and 3–11 rounded teeth on each margin; plants with rounded leaf bases and 11–17 less rounded or acute teeth on each margin have been called *Pilea pumila* var. *deamii* (Lunell) Fernald (M. L. Fernald 1936) [*Adicea deamii* Lunell, Amer. Midl. Naturalist 3: 10. 1913.]. Typical *P. pumila* also is found in eastern Asia, where three infraspecific taxa, *P. pumila* var. *pumila,* *P. pumila* var. *hamaoi* (Makino) C. J. Chen, and *P. pumila* var. *obtusifolia* C. J. Chen are recognized. This complex, which also includes *P. pauciflora* C. J. Chen, has been placed in *Pilea* series *Pumilae* C. J. Chen. Although the Asian plants are often vegetatively and florally indistinguishable from the North American plants, minor differences do occur in the achenes, especially in their markings and sculpturing when mature. Detailed studies are needed to clarify exact relationships.

6. POUZOLZIA Gaudichaud-Beaupré, Voy. Uranie 12: 503. 1830 · [For P. C. M. de Pouzolz, botanist and collector in Corsica, France, and the Pyrénées]

Herbs or subshrubs [and shrubs], perennial, with nonstinging hairs. **Stems** simple, erect. **Leaves** alternate, less often opposite; stipules present. **Leaf blades** elliptic, lanceolate, or ovate, margins entire; cystoliths rounded. **Inflorescences** axillary. **Flowers** unisexual, staminate and pistillate flowers in sessile or subsessile clusters in same inflorescences; bracts oblanceolate, lacking hooked hairs. **Staminate flowers:** tepals 4, equal; stamens 4; pistillode present. **Pistillate flowers:** tepals 4, connate, appressed to and tightly enclosing ovary; staminodes absent; style deciduous, elongate. **Achenes** ovoid, acute, not mucronate, tightly enclosed by persistent tepals. $x = 13$.

Species ca. 50 (1 in the flora): mostly in tropical and subtropical regions worldwide.

1. Pouzolzia zeylanica (Linnaeus) Bennett & R. Brown in J. J. Bennett et al., Pl. Jav. Rar., 67. 1838   [F]

Parietaria zeylanica Linnaeus, Sp. Pl. 2: 1052. 1753

Plants 1–20 dm, sparsely to moderately covered on all parts with appressed and spreading hairs. **Leaf blades** 1–5 × 0.6–2 cm, base rounded to broadly cuneate, apex acute or nearly acuminate. **Staminate flowers** ca. 3 mm across; filaments longer than tepals; pistillode elongate, knoblike. **Pistillate flowers** ca. 1–1.5 mm, tepals ribbed. **Achenes** ca. (0.9–)1.2 × ca. 0.6 mm, base truncate.

Flowering late summer–winter. Roadsides, old fields, waste places, disturbed areas; 0–20 m; introduced; Fla.; Asia.

Pouzolzia zeylanica was misidentified as *Parietaria officinalis* Linnaeus by J. K. Small.

7. SOLEIROLIA Gaudichaud-Beaupré, Voy. Uranie 12: 503. 1830 · [For Captain Soleirol, collector in Corsica]

Herbs, perennial, with nonstinging hairs. **Stems** 1–25-branched, repent, filiform, mat-forming. **Leaves** alternate; stipules absent. **Leaf blades** orbiculate to oblong, margins entire; cystoliths elongate, linear. **Inflorescences** axillary, 1-flowered. **Flowers** unisexual, proximal flowers pistillate, the distal staminate; bracts subtending pistillate flowers developing 3 corky wings covered with fine, hooked hairs in fruit. **Staminate flowers:** tepals 4, distinct, equal; stamens 4; pistillode obovate. **Pistillate flowers:** tepals 4, connate, lacking hooked hairs; staminodes absent; stigma filiform, deciduous. **Achenes** sessile, ovoid, apex acute, tightly enclosed by scarious perianth and surrounded by connivent, winged bracteoles. $x = 10$.

Species 1: North America; Europe (Sardinia and Corsica).

This genus was first reported naturalized in California in 1940.

SELECTED REFERENCE   Robbins, W. W. 1940. Alien plants growing without cultivation in California. Univ. Calif. Agric. Exp. Sta. Bull. 637: 1–128.

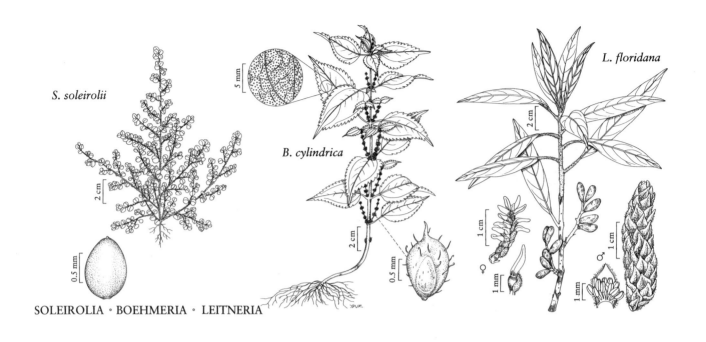

S. soleirolii

B. cylindrica

L. floridana

SOLEIROLIA · BOEHMERIA · LEITNERIA

1. Soleirolia soleirolii (Requien) Dandy, Feddes Repert. 70: 4. 1965 · Baby's-tears [F]

Helxine soleirolii Requien, Ann. Sci. Nat. (Paris) 5: 384. 1825

Herbs, 0.2–2.5 dm, sparsely to densely pubescent with falcate and straight hairs. **Stems** rooting at nodes. **Leaf blades** 3–8 × 2–4 mm, base prominently oblique. **Achenes** light brown, shiny, symmetric, 0.8–0.9 × 0.6 mm, hard.

Flowering late winter–spring. Weed in lawns and gardens, waste places, shaded banks, sidewalks; 0–100 m; introduced; Calif.; Mexico (Baja California); Europe.

8. BOEHMERIA Jacquin, Enum. Syst. Pl., 9. 1760 · False-nettle [for G. R. Böhmer, German botanist]

Herbs, subshrubs, or shrubs, perennial, rhizomatous, without stinging hairs, sparsely to ± densely pubescent or tomentose with hooked, curved, or straight, nonstinging hairs on all parts of plant. **Stems** simple to freely branched, erect. **Leaves** opposite or nearly opposite, or alternate; stipules present. **Leaf blades** lanceolate to broadly ovate, base rounded to cordate, less often cuneate, margins dentate or serrate, apex acuminate; cystoliths rounded. **Inflorescences** axillary, spikelike or paniculate. **Flowers** unisexual, staminate and pistillate flowers on same plant, rarely on different plants, in remote or crowded clusters in same or separate inflorescence; bracts linear. **Staminate flowers:** tepals 4; stamens 4; pistillode globose. **Pistillate flowers:** tepals 4, connate, adnate to ovary; staminodes absent; style persistent, elongate; stigma linear, straight or hooked. **Achenes** sessile, laterally compressed, ovoid, nearly orbicular, or ellipsoid, tightly enclosed by persistent perianth. $x = 14$.

Species 50 (2 in the flora): tropical and subtropical regions worldwide.

1. Leaf blades abaxially glabrous, puberulent, or short-pilose, not white-tomentose; inflorescences spike-like . . . . 1. *Boehmeria cylindrica*
1. Leaf blades abaxially densely white-tomentose; inflorescences paniculate . . . . 2. *Boehmeria nivea*

1. Boehmeria cylindrica (Linnaeus) Swartz, Prodr., 34. 1788 · False-nettle [F]

Urtica cylindrica Linnaeus, Sp. Pl. 2: 984. 1753; *Boehmeria cylindrica* var. *drummondiana* (Weddell) Weddell; *B. decurrens* Small; *B. drummondiana* Weddell; *B. scabra* (Porter) Small

Herbs or subshrubs, 1–16 dm. **Leaves** opposite or nearly opposite, rarely alternate. **Leaf blades** elliptic, lanceolate to broadly ovate, 5–18 × 2–10 cm, almost glabrous on both surfaces or abaxially densely short-pilose or puberulent, adaxially scabrous. **Inflorescences** spikelike, often leafy at apex. **Flowers** in remote or crowded clusters of 1– few staminate and several pistillate flowers or rarely staminate and pistillate flowers on different plants. **Achenes** ovoid to nearly orbicular, 0.9–1.6 × 0.9–1.2 mm, almost glabrous or pubescent with straight and hooked hairs; seeds prominent, conspicuous in outline, surrounded except at base by corky tissue. $2n = 28$.

Flowering summer–fall. Alluvial or moist, deciduous woods, swamps, bogs, marshes, wet meadows, ditches; 0–1800 m; N.B., Ont., Que.; Ala., Ariz., Ark., Conn., Del., D.C., Fla., Ga., Ill., Ind., Iowa, Kans., Ky., La., Maine, Md., Mass., Mich., Minn., Miss., Mo., Nebr., N.H., N.J., N.Mex., N.Y., N.C., Ohio, Okla., Pa., R.I., S.C., S.Dak., Tenn., Tex., Utah, Vt., Va., W.Va., Wis.; Mexico; West Indies; Bermuda; Central America; South America (Argentina, s Brazil, Paraguay, Uruguay, and Venezuela).

Populations of *Boehmeria cylindrica* are disjunct in South America.

Plants with thick, often drooping, lanceolate leaf blades, more or less pilose or puberulent abaxially, scabrous adaxially, with short petioles, pilose or puberulent stems, and densely pubescent achenes have been called *Boehmeria cylindrica* var. *drummondiana.* These plants are found mostly, but not exclusively, in the southeastern United States and are totally sympatric with more typical plants. The above characteristics may or may not occur together, and each grades into the state found in more typical plants through abundant intermediates. Field and experimental studies are needed to clarify the biologic basis of this variation.

2. Boehmeria nivea (Linnaeus) Gaudichaud-Beaupré, Voy. Uranie 12: 499. 1830 · Ramie

Urtica nivea Linnaeus, Sp. Pl. 2: 985. 1753; *Ramium niveum* (Linnaeus) Small

Shrubs or subshrubs, 2(–3) m. **Leaves** alternate. **Leaf blades** broadly ovate to nearly orbiculate, 8–15 × 5–12 cm, abaxial surface densely white-tomentose, adaxial surface slightly scabrous. **Inflorescences** panicles of moniliform (beaded) clusters, branches not leafy at apex; staminate flowers in proximal leaf axils, pistillate flowers in distil axils. **Achenes** compressed or lenticular, ovoid or ellipsoid, ca. 1.5 × ca. 0.9 mm, pubescent with straight or slightly curved hairs, uniformly smooth; seeds not conspicuous in outline, corky tissue absent.

Flowering late summer–fall. Roadsides, waste places, vacant lots, cultivated fields, along Atlantic and Gulf coastal plains; 0–200 m; introduced; Ala., Calif., Fla., Ga., La., S.C., Tex., Va.; Central America; Asia.

Boehmeria nivea, ramie, is an important source of fiber in Asia and was introduced into the United States in 1855 as a commercial crop. The fibers are exceptionally strong but difficult to extract.

27. LEITNERIACEAE Bentham

· Corkwood Family

A. Linn Bogle

Shrubs or small trees. Wood white to yellowish, soft, brittle, even-grained; secretory canals in outer pith; resin yellowish. **Leaves** alternate, not aromatic; stipules absent; petiole and veins with secretory canals. **Inflorescences** axillary catkins, erect or lax. **Flowers** unisexual, staminate and pistillate on different (rarely on same) plants, inconspicuous, sessile; perianth absent or of small, undifferentiated sepals, hypogynous; pistils simple (rarely compound), 1(–2)-carpellate; ovary superior; placentation parietal; ovule 1, pendulous; style 1. **Fruits** drupes, leathery, dry.

Genus 1, species 1: se United States.

Leitneriaceae are a relict family confined to the flora area and occurring on coastal and riverine flood plains of the Atlantic and Gulf coasts. The fossil record of the family extends to the Middle Oligocene (32–30 M.Y.B.P.).

Taxonomic affinities are uncertain. Various alliances have been suggested: Hamamelidales (A. Cronquist 1981), Sapindales (R. M. T. Dahlgren 1980), and Rutales (R. F. Thorne 1983) of the Rutiflorae. Biochemical studies (D. E. Giannasi 1986; F. P. Petersen and D. E. Fairbrothers 1983, 1985) suggest affinities with Simaroubaceae in Rutales.

SELECTED REFERENCES Abbe, E. C. and T. T. Earle. 1940. Inflorescence, floral anatomy and morphology of *Leitneria floridana*. Bull. Torrey Bot. Club 67: 173–193. Channell, R. B. and C. E. Wood Jr. 1962. The Leitneriaceae in the southeastern United States. J. Arnold Arbor. 43: 435–438. Giannasi, D. E. 1986. Phytochemical aspects of phylogeny in Hamamelidae. Ann. Missouri Bot. Gard. 73: 417–437. Jarvis, C. E. 1989. A review of the order Leitneriales Engler. In: P. R. Crane and S. Blackmore, eds. 1989. Evolution, Systematics, and Fossil History of the Hamamelidae. 2 vols. Oxford. Vol. 2, pp. 189–192. [Syst. Assoc. Special Vol. 40B.] Petersen, F. P. and D. E. Fairbrothers. 1983. A serotaxonomic appraisal of *Amphipterygium* and *Leitneria*—two amentiferous taxa of Rutiflorae (Rosidae). Syst. Bot. 8: 134–148.

1. LEITNERIA Chapman, Fl. South. U.S., 427. 1860 · Corkwood [for Dr. Edward Frederick Leitner, 1812–1838, German physician, naturalist, and explorer of southern Florida]

Shrubs or small trees, forming dense thickets. **Leaves** 5-ranked, clustered toward branch tips; leaf scars with 3 bundle scars. **Leaf blade:** margins entire; surfaces densely pubescent, becoming ± glabrous. **Inflorescences** in flower before leaves appear, in axils of previous year's leaves; basal and apical bracts often sterile; staminate catkins compound, bracts 40–50, bracteoles absent; pistillate catkins simple, spikelike, bracts 10–15, each subtending 2 small bracteoles. **Flowers** subtended by bracts; bracts spirally arranged, deltate-ovate, apex acute, surface silky-pubescent abaxially. **Staminate flowers:** cymules 3-flowered, sessile; stamens clustered; pistils absent. **Pistillate flowers** solitary; perianth present; sepals small, scalelike; staminodes absent; ovary sessile, style linear, stigma deciduous, decurrent, abaxially grooved, groove facing bract. **Drupes** subtended by persistent bracts, bracteoles, and sepals. $x = 16$.

Bisexual cymules and 2-locular ovaries may occur sporadically; they merit investigation (R. K. Godfrey and A. F. Clewell 1965).

1. Leitneria floridana Chapman, Fl. South. U.S., 427. 1860 [C][E][F]

Myrica floridana (Chapman) A. W. Wood

Shrubs or small trees, 1–4(–8) m, to 15 cm diam. at base. **Stems** unbranched or sparsely short-branched distally, slender, pubescent, becoming glabrous. **Bark** gray to dark reddish brown; lenticels numerous, gray. **Leaves:** petiole 2–7 cm, densely pubescent, becoming ± glabrous. **Leaf blade** lanceolate to elliptic-lanceolate, 5–20 × 3.5–6 cm, leathery, base acute, margins narrowly revolute, apex acute to acuminate; surfaces silky-pubescent when young, becoming glabrous and glossy adaxially. **Inflorescences:** staminate catkins cylindric in bud, lax and outward curving at anthesis, 2–6 × 1–1.5 cm; pistillate stiffly erect, 1–3 × 1 cm. **Staminate flowers:** stamens (3–)10–12(–15); filaments short; anthers basifixed, broadly oblong, 4-sporangiate, 2-locular at anthesis, dehiscing longitudinally. **Pistillate flowers:** sepals (3–)4(–8), irregularly inserted, unequal; ovary elliptic, ca. 2 mm, finely pubescent; style 4–5 mm; stigma exserted, often recurved or twisted, reddish. **Drupes** erect, green, becoming chestnut brown, elliptic to oblong-elliptic, often compressed on drying, 1–2.5 cm, glabrous, base and apex blunt, apex with conspicuous dark stylar scar, vascular bundles toughening the thin flesh; endocarp brown, bony, surface rough-reticulate. **Seeds** 1, compressed; seed coat membranous, hilum blackish, elongate; endosperm thin, starchy; embryo large, straight; cotyledons ovoid, slightly fleshy. $2n = 32$.

Flowering late winter–early spring; fruiting spring–summer. Open or forested swamps, wet thickets, roadside ditches, saw-grass–palmetto marshes, estuarine tidal shores; of conservation concern; 0–100 m; Ark., Fla., Ga., Mo., Tex.

Vegetative reproduction is predominant, forming large clones from adventitious buds on shallow roots. Large, old plants ("small trees") are apparently rare in the field. Florida plants begin growth about a month before Missouri plants (J. W. Day 1975). *Leitneria floridana* is successfully cultivated as far north as Chicago, Rochester, and Boston.

The wood (specific gravity 0.21) is lighter than cork (sp. gr. 0.24); it is used locally for fishnet floats or bottle stoppers.

28. JUGLANDACEAE A. Richard ex Kunth

• Walnut Family

Donald E. Stone

Trees, rarely shrubs, deciduous [evergreen], with gray or brownish bark. **Terminal buds** larger than lateral buds. **Leaves** alternate [opposite], usually odd-, rarely even-, pinnately compound, glandular-peltate scales often resinous, aromatic; stipules absent; petiole present. **Leaflets** 3–23, margins serrate or wavy, ± entire [entire]. **Inflorescences:** staminate catkins solitary or fasciculate, pendulous [rarely erect], elongate, on reduced shoots arising on branches of previous year or at base of current year's growth; pistillate catkins solitary or few-flowered spikes [many-flowered racemes]. **Flowers** unisexual, staminate and pistillate on same plant; bract 1, bracteoles (0–)2. **Staminate flowers:** calyx 4-lobed or absent; corolla absent; stamens 3–50; filaments very short or absent; anthers usually pubescent. **Pistillate flowers:** calyx 4-lobed or absent; corolla absent; ovary 1, inferior, 2-carpellate, 1-locular distally; ovule 1; stigmas 2, fleshy or plumose. **Fruits** large nuts [winged nutlets], nuts enclosed in dehiscent or indehiscent, fibrous-fleshy or hard involucres (husks), thus ± drupelike, base 2–4-chambered. **Seeds** 1; endosperm absent; cotyledons fleshy and oily, variously lobed.

Genera 9, species 60+ (2 genera, 17 species in the flora): Western Hemisphere and Eurasia.

The fruit in Juglandaceae superficially resembles a drupe, with a hard "stone" surrounded by a soft, often fleshy husk. The husk is not part of the ovary wall (it develops from the involucre and calyx), and the fruit is actually a nut (T. S. Elias 1972; W. E. Manning 1978).

The morphology and classification of the family were discussed by W. E. Manning (1978). The female inflorescences and fruits show considerable variation in Latin American and Eurasian taxa.

SELECTED REFERENCES Elias, T. S. 1972. The genera of Juglandaceae in the southeastern United States. J. Arnold Arbor. 53: 26–51. Manchester, S. R. 1987. The fossil history of the Juglandaceae. Monogr. Syst. Bot. Missouri Bot. Gard. 21: 1–137. Manning, W. E. 1978. The classification within the Juglandaceae. Ann. Missouri Bot. Gard. 65: 1058–1087. Stone, D. E. 1993. Juglandaceae. In: K. Kubitzki et al., eds. 1990+. The Families and Genera of Vascular Plants. 2+ vols. Berlin etc. Vol. 2, pp. 348–359.

1. Branchlets with solid and homogeneous pith; distal leaflets largest; staminate catkins in sessile or pendunculate fascicles of 3, stamens 3–10(–15) per flower; fruits with husks completely or partially dehiscent, nuts smooth, verrucose, or rugulose . . . . .1. *Carya*, 417

1. Branchlets with chambered pith; leaflets uniform in size or median leaflets largest; staminate catkins sessile, solitary, stamens 7–50 per flower; fruits with husks indehiscent, nuts grooved, ridged, furrowed, or smooth . . . . .2. *Juglans,* 425

1. CARYA Nuttall, Gen. N. Amer. Pl. 2: 220. 1818, name conserved · Hickory, caryer, hicorier [Greek *káryon,* nut, kernel]

Donald E. Stone

Hicoria Rafinesque

Trees, rarely shrubs, 3–52 m. **Bark** gray or brownish, smooth with fissures in younger trees, becoming ridged and sometimes deeply furrowed or exfoliating with small platelike scales or long strips or broad plates. **Twigs** greenish, orangish, reddish, or rusty brown, or bronze, terete, slender or stout, pubescent and scaly or glabrous; leaf scars shield-shaped or 3-lobed, large; pith solid and homogeneous. **Bud scales** valvate or imbricate [absent], glabrous or variously pubescent; axillary buds protected by pair of valvate bracteoles (i.e., prophylls) or bracteoles fused into hood. **Leaves** odd-pinnate; petiole pubescent and/or scaly or glabrous. **Leaflets** 3–17(–21), petiolulate, distal leaflets largest, 2–26 × 1–14 cm; surfaces abaxially with nonglandular hairs (unicellular common to all species, fasciculate with 2–8 rays in 1 rank, multiradiate with 8–17 rays in 2 ranks) and glandular scales (capitate-glandular and large peltate scales common to all species; small peltate scales round, irregular, or 2- or 4-lobed), adaxially with scattered hairs and scattered to abundant scales in spring or concentrated along midrib and secondary veins to essentially glabrous in the fall. **Staminate catkins** in fascicles of 3 from 1st-, sometimes 2d-year twigs, sessile or pedunculate; stamens 3–10(–15) per flower, with or without hairs. **Pistillate flowers** in terminal few-flowered spikes. **Fruits** nuts enclosed in husks, compressed or not compressed; husks completely or partially dehiscing, sutures smooth or winged; nuts brown, reddish brown, or tan, sometimes mottled with black or tan, compressed or not compressed, angled or not angled, smooth, rugulose, or verrucose, base 2–4-chambered; shells thin or thick. **Seeds** sweet or bitter. $x = 16$.

Species 18 (11 in the flora): e North America, Mexico, e Asia.

Carya was widespread during the Tertiary; fossils have been reported from the states of Colorado and Washington, and from China, Japan, Europe, and western Siberia. *Carya* sect. *Sinocarya* occurs in southeastern Asia; sect. *Rhamphocarya* (southeastern Asia) is now recognized as the distinct genus *Annamocarya*. *Carya* sect. *Apocarya*, the so-called pecan hickories, and *Carya* sect. *Carya*, the true hickories, occur in North America. Both $2n = 32$ and $2n = 64$ chromosome numbers are known for the genus; tetraploidy is confined to sect. *Carya*.

The commercial use of *Carya* is substantial. The cultivated pecan, *C. illinoinensis,* is the most important nut tree native to North America, and the wood of the true hickories is unequaled for its use in tool handles because of the combined strength and shock resistance. Hickory nuts are also an important, high-quality food source for wildlife because they are high in proteins and fats. *Carya cordiformis, C. glabra,* and *C. ovata* are grown extensively in central Europe for timber.

Characters of the buds and bark are taxonomically important in *Carya,* but shoots with terminal buds and information about bark characteristics are frequently absent on herbarium specimens. Phenotypic variation from tree to tree is often considerable and difficult to quantify. Most of this variation undoubtedly results from adaptation to local and regional conditions; hybridization has probably played a subtle role as well. Sympatry of two or more species is common, and artificial pollinations suggest that even diploid × tetraploid crosses produce viable seed.

SELECTED REFERENCES Hardin, J. W. and D. E. Stone. 1984. Atlas of foliar surface features in woody plants, VI. *Carya* (Juglandaceae) of North America. Brittonia 36: 140–153. Manning, W. E. 1950. A key to the hickories north of Virginia with notes on the two pignuts, *Carya glabra* and *C. ovalis*. Rhodora 52: 188–199. Manning, W. E. 1973. The northern limits of the distributions of hickories in New England. Rhodora 75: 34–51. Stone, D. E. 1962. Affinities of a Mexican endemic, *Carya palmeri*, with American and Asian hickories. Amer. J. Bot. 49: 199–212. Stone, D. E., G. A. Adrouny, and R. H. Flake. 1969. New World Juglandaceae. II. Hickory nut oils, phenetic similarities, and evolutionary implications in the genus *Carya*. Amer. J. Bot. 56: 928–935. Thompson, T. E. and L. J. Grauke. 1990. Pecans and other hickories *(Carya)*. In: J. Moore and J. R. Ballington Jr., eds. 1990. Genetic Resources of Temperate Fruit and Nut Crops. 2 vols. Wageningen. Vol. 2, pp. 837–904. Woodworth, R. H. 1930c. Meiosis of sporogenesis in the Juglandaceae. Amer. J. Bot. 17: 863–869.

1. Scales of terminal buds pseudovalvate, with little or no overlap of opposing bud scales; axillary buds protected by pair of valvate bracteoles or by bracteoles fused into hood; leaflets (5–)7–13(–17), symmetric or falcate; staminate catkins at base of leafy shoots on new wood, and commonly on reduced shoots from old wood; husk sutures winged; shells thin or thick; seeds sweet or bitter (sect. *Apocarya*).

  2. Terminal buds sulfur yellow to brown; axillary buds protected by pair of valvate bracteoles; leaflets (5–)7–9(–11), symmetric, abaxially large peltate scales retained, concentrated near margins of base and apex on fall specimens; fruits not compressed or only slightly so; husks dehiscing to middle or slightly below; nuts not compressed or only slightly so, not angled, rugulose; seeds bitter . . . . 3. *Carya cordiformis*

  2. Terminal buds yellowish brown to reddish brown or black; axillary buds protected by bracteoles fused into hood; leaflets (5–)7–13(–17), symmetric or falcate, abaxially large peltate scales mainly lost by fall or at least not concentrated near margins of base and apex; fruits compressed or not compressed; husks dehiscing to base; nuts compressed or not compressed, angled or not angled, smooth or verrucose; seeds bitter or sweet.

    3. Bark exfoliating in long strips or platelike scales; leaflets (5–)9–11(–13), margins serrate to ± entire; lateral petiolules 0–2 mm; midribs adaxially villous near base; fascicles of male catkins pedunculate; nuts compressed, angled, verrucose; seeds bitter. . . . . 1. *Carya aquatica*

    3. Bark ridged or with appressed scales or exfoliating with small platelike scales; leaflets (7–) 9–13(–17), margins serrate; lateral petiolules 0–7 mm; midribs mostly adaxially glabrous, rarely hirsute near base; fascicles of male catkins sessile or pedunculate; nuts not compressed, not angled, smooth; seeds sweet. . . . . 2. *Carya illinoinensis*

1. Scales of terminal buds imbricate; axillary buds protected by bracteoles fused into hood; leaflets 3–9, symmetric; staminate catkins at base of leafy shoots on new wood, rarely on reduced shoots from old wood (*C. texana*); husk sutures usually without wings, infrequently with narrow wings (*C. floridana, C. glabra, C. texana*); shells thick; seeds sweet (sect. *Carya*).

  4. Leaflets (3–)5(–7), serrations with hairs tufted below apex, and at least some hairs persisting into fall; bark exfoliating in long strips or broad plates; fruits spheric or nearly so; husks thick, dehiscing to base . . . . 5. *Carya ovata*

  4. Leaflets 3–9, serrations sometimes lightly ciliate, subapical tuft absent; bark ridged, often deeply furrowed or exfoliating in long strips or broad plates; fruits spheric to obovoid; husks thin to thick, dehiscing partially or completely to base.

    5. Twigs stout; terminal buds 8–20 mm; leaflets (5–)7–9(–11), abaxially hirsute, with abundant unicellular, fasciculate, and multiradiate hairs; husks 4–13 mm thick; nuts attenuate toward stylar end.

      6. Bark exfoliating in long strips or broad plates; petiole and rachis lightly pubescent; leaflets apically acuminate, tapering, abaxially hirsute with abundant unicellular and fasciculate hairs, occasional multiradiate hairs; husks minutely hirsute. 6. *Carya laciniosa*

      6. Bark ridged; petiole and rachis densely hirsute; leaflets apically acute, abaxially hirsute with low density of unicellular hairs and high density of fasciculate and multiradiate hairs; husks rough, glabrous . . . . 7. *Carya tomentosa*

    5. Twigs slender; terminal buds 3–15 mm; leaflets 3–7(–9), abaxially glabrous except near midrib, occasionally hirsute with unicellular and fasciculate hairs, never with multiradiate hairs; husks 2–5 mm thick; nuts not attenuate toward stylar end.

      7. Terminal buds 4–10 mm; leaflets (5–)7(–9), abaxially with dense covering of small 4-lobed, irregular, and round peltate scales.

        8. Petiole and rachis hirsute with scattered fasciculate hairs and hairs concentrated adaxi-

ally near leaflet insertions; leaflets abaxially with inconspicuous 4-lobed, irregular scales, large silvery tan peltate scales dense; mainly e of Mississippi River .8. *Carya pallida*

8. Petiole and rachis with few hairs, hairs not concentrated adaxially near leaflet insertions; leaflets abaxially with conspicuous small rusty brown scales, large silvery tan peltate scales infrequent; mainly w of Mississippi River . . . . .9. *Carya texana*

7. Terminal buds 3–15 mm; leaflets 3–7(–9), abaxially with sparse to dense covering of small irregular and round peltate scales, 4-lobed scales uncommon.

  9. Bark fissured, usually exfoliating, separating freely into long strips or broad plates; leaflets (5–)7–9, abaxially with dense coating of overlapping peltate scales; husk sutures prominently winged. . . . . .4. *Carya myristiciformis*

  9. Bark ridged, sometimes fissured, or exfoliating with small platelike scales or short narrow strips; leaflets 3–7(–9), abaxially with coarse to moderate covering of peltate scales; husk sutures smooth or slightly winged.

    10. Terminal buds 3–9 mm, densely scaly, golden brown to rusty; leaflets 3–7, margins coarsely serrate, surfaces abaxially without small round, dark brown peltate scales; fruits obovoid to oblong, bronze to dark brown; husk sutures slightly winged, dehiscing to base; c Florida . . . . 10. *Carya floridana*

    10. Terminal buds 5–15 mm, sparsely scaly, reddish brown to tan; leaflets (3–)5–7 (–9), margins finely to coarsely serrate, surfaces abaxially with small round, dark brown peltate scales; fruits obovoid to spheric, tan to reddish brown; husk sutures smooth, dehiscing to base or only partially dehiscent; throughout e United States. . . . 11. *Carya glabra*

1. Carya aquatica (F. Michaux) Nuttall, Gen. N. Amer. Pl. 2: 222. 1818 · Water hickory, bitter pecan E F

Juglans aquatica F. Michaux, Hist. Arbr. Forest. 1(2): 182, plate 5. 1810; *Hicoria aquatica* (F. Michaux) Britton

Trees, to 46 m. **Bark** light gray or brownish, exfoliating, separating freely into long strips or broad plates, less commonly with small platelike scales. **Twigs** brown to reddish brown or black, slender, villous becoming glabrous. **Terminal buds** brown, reddish brown, or black, oblong, 8–10 mm, yellow-scaly, villous; bud scales valvate; axillary buds protected by bracteoles fused into hood. **Leaves** 4–6 dm; petiole 3–8 cm, villous becoming glabrous. **Leaflets** (5–)9–11(–13), lateral petiolules 0–2 mm, terminal petiolules (2–)6–10(–14) mm; blades ovate-lanceolate, often falcate, 2–19 × 1–4 cm, margins finely or coarsely serrate to ± entire and wavy, without tufts of hairs, apex acuminate; surfaces abaxially villous with unicellular and 2–8-rayed fasciculate hairs along midrib and secondary veins, densely scaly in spring with large peltate scales and small round, irregular, and 4-lobed peltate scales, adaxially villous along midrib near base, glabrous between veins. **Staminate catkins** pedunculate, to 21 cm, stalks villous, bracts scaly; anthers sparsely pubescent. **Fruits** brown, bronze, or black, obovoid, compressed, 1.5–3 × 1.5–2.5 cm; husks rough, 1 mm thick, dehiscing to base or nearly so, sutures winged; nuts chocolate brown, broadly obovoid, compressed, 2-angled, verrucose; shells thin. **Seeds** bitter. 2*n* = 32.

Flowering spring. Bayous, river flood plains, bluffs, and levees, temporarily flooded bottomlands; 0–200 m; Ala., Ark., Fla., Ga., Ill., Ky., La., Miss., Mo., N.C., Okla., S.C., Tenn., Tex., Va.

Carya aquatica hybridizes with *C. illinoinensis* (*C.* ×*lecontei* Little [= *Hicoria texana* Le Conte]) and is reported to hybridize with the tetraploid *C. texana* [*C.* ×*ludoviciana* (Ashe) Little].

2. Carya illinoinensis (Wangenheim) K. Koch, Dendrologie 1: 593. 1869 · Pecan, pecanier, nogal morado, nuez encarcelada

Juglans illinoinensis Wangenheim, Beytr. Teut. Forstwiss., 54, plate 18, fig. 43. 1787, excluding fruit; *Hicoria oliviformis* (Michaux) Nuttall; *H. pecan* (Marshall) Britton

Trees, to 44 m. **Bark** light gray or brownish, ridged with appressed scales or exfoliating with small platelike scales. **Twigs** tan to reddish brown, slender, hirsute, conspicuously scaly, sometimes becoming glabrous. **Terminal buds** yellowish brown, oblong, 6–12 mm, hirsute, scaly; bud scales valvate; axillary buds protected by bracteoles fused into hood. **Leaves** 4–7 dm; petiole 4–8 cm, glabrous to scurfy with short single hairs or scattered fascicles. **Leaflets** (7–)9–13(–17), lateral petiolules 0–7 mm, terminal petiolules 5–25 mm; blades ovate-lanceolate, often falcate, 2–16 × 1–7 cm, margins finely to coarsely serrate, without tufts of hairs, apex acuminate; surfaces abaxially hirsute or with scattered unicellular and 2-rayed fasciculate hairs, scaly with large peltate scales and small round peltate scales, adaxially without hairs or rarely hirsute with unicellular hairs along midrib, and with scattered 2–6-rayed fasciculate hairs, moderately scaly in spring. **Staminate catkins** essentially sessile, to 18 cm, stalks with small capitate-glandular trichomes; anthers sparsely pilose. **Fruits** dark brown, ovoid-ellipsoid, not compressed, 2.5–6 × 1.5–3 cm; husks rough, 3–4 mm thick, dehiscing to base or nearly so, sutures winged; nuts tan to brown and mottled with black patches, ovoid-ellipsoid, not compressed, not angled, smooth; shells thin. **Seeds** sweet. $2n = 32$.

Flowering spring. Along stream banks, river flood plains, and on well-drained soils; 0–600(–1000) m; Ark., Ill., Ind., Iowa, Kans., Ky., La., Miss., Mo., Okla., Tenn., Tex.; Mexico.

Extensive cultivation and naturalization have confounded interpretation of the natural range of *Carya illinoinensis*. The pecan hybridizes with *C. aquatica* (*C.* ×*lecontei* Little [= *Hicoria texana* Le Conte]), *C. cordiformis* (*C.* ×*brownii* Sargent), *C. laciniosa* (*C.* ×*nussbaumeri* Sargent), and *C. ovata,* and reputedly with the tetraploid *C. tomentosa* (*C.* ×*schneckii* Sargent). J. W. Thieret (1961) pointed out that *C. illinoinensis* was the spelling Wangenheim used throughout his publication, and there is no valid basis for accepting the deviant spelling "*illinoensis*" that is so widely used today (E. L. Little Jr. 1979).

Carya illinoinensis is the state tree of Texas.

Native Americans used *Carya illinoinensis* medicinally as a dermatological aid and as a remedy for tuberculosis (D. E. Moerman 1986).

3. **Carya cordiformis** (Wangenheim) K. Koch, Dendrologie 1: 597. 1869 · Bitternut hickory, pignut, noyer amer, caryer cordiforme E F

Juglans cordiformis Wangenheim, Beytr. Teut. Forstwiss., 25, plate 10, fig. 25. 1787; *Hicoria cordiformis* (Wangenheim) Britton; *H. minima* (Marshall) Britton

Trees, to 52 m. **Bark** gray or brownish, smooth or ridged or exfoliating with small platelike scales. **Twigs** tan, slender, glabrous except scaly near tip. **Terminal buds** sulfur yellow to tan, oblong, 10(–19) mm, densely scaly with yellow peltate scales, pilose near apex; bud scales valvate; axillary buds protected by pair of valvate bracteoles. **Leaves** 2–4 dm; petiole 3–7 cm, glabrous near base, hirsute near rachis. **Leaflets** (5–)7–9(–13), lateral petiolules 0–1 mm, terminal petiolules 2–8 mm; blades ovate-lanceolate, rarely falcate, 3–19 × 1–7 cm, margins finely to coarsely serrate, without tufts of hairs, apex acuminate; surfaces abaxially villous with unicellular and 2–4-rayed fasciculate hairs along midrib and major veins, densely to sparsely pubescent throughout, and with abundant large peltate scales and small round and 2- or 4-lobed peltate scales in spring, still present near margins at base and apex in fall, adaxially villous along midrib near base, sparsely scaly in spring. **Staminate catkins** pedunculate, to 16 cm, stalks without hairs or hirsute, bracts scaly; anthers hirsute. **Fruits** brown, cylindric, obovoid, or nearly

spheric, not compressed or only slightly compressed, 2–3 × 2–3.2 cm; husks rough, 2–3 mm thick, dehiscing to middle or slightly below, sutures winged; nuts light brown, ellipsoid to ovoid, not compressed or only slightly compressed, not angled, rugulose; shells thin. **Seeds** bitter. $2n = 32$.

Flowering spring. River flood plains, well-drained hillsides, and limestone glades; 0–900 m; Ont., Que.; Ala., Ark., Conn., Del., D.C., Fla., Ga., Ill., Ind., Iowa, Kans., Ky., La., Maine, Md., Mass., Mich., Minn., Miss., Mo., Nebr., N.H., N.J., N.Y., N.C., Ohio, Okla., Pa., R.I., S.C., Tenn., Tex., Vt., Va., W.Va., Wis.

Carya cordiformis hybridizes with *C. illinoinensis* (*C.* ×*brownii* Sargent), *C. ovata* (*C.* ×*laneyi* Sargent), and *C. laciniosa,* and reputedly with the tetraploid *C. glabra* (*C.* ×*demareei* Palmer).

The Fox Indians used *Carya cordiformis* medicinally as a diuretic, a laxative, and a panacea (D. E. Moerman 1986).

4. **Carya myristiciformis** (F. Michaux) Nuttall, Gen. N. Amer. Pl. 2: 222. 1818 · Nutmeg hickory, nogal

Juglans myristiciformis F. Michaux, Hist. Arbr. Forest., 1(2): 211, plate 10. 1810 (as *myristicaeformis*); *Hicoria myristiciformis* (F. Michaux) Britton

Trees, to 35 m. **Bark** gray to brownish, fissured or exfoliating, separating freely into long strips or broad plates. **Twigs** brown to bronze, slender, without hairs, densely scaly. **Terminal buds** bronze, ovoid, 4–6 mm, essentially without hairs, densely scaly; bud scales imbricate; axillary buds protected by bracteoles fused into hood. **Leaves** 3–6 dm; petiole 3–10 cm, densely scaly. **Leaflets** (5–)7–9, lateral petiolules 0–2 mm, terminal petiolules 2–3 mm; blades ovate or obovate to elliptic, not falcate, 3–17 × 1–8 cm, margins finely to coarsely serrate, without tufts of hairs, apex acuminate; surfaces abaxially with unicellular and 2–4-rayed fasciculate hairs along midrib in spring, densely scaly with coating of large peltate scales and small irregular, round, and 4-lobed peltate scales, imparting bronze color, adaxially pubescent along midrib and major veins in spring, with scattered peltate scales. **Staminate catkins** pedunculate, to 6 cm, stalks and bracts scaly; anthers hirsute. **Fruits** light tan to bronze, obovoid to ellipsoid, not compressed, 2–3 × 1.5–2 cm; husks rough, 2 mm thick, dehiscing to base, sutures winged; nuts reddish brown mottled with tan patches, ellipsoid, not compressed, not angled, smooth; shells thick. **Seeds** sweet. $2n = 32$.

Flowering spring. River bottomlands, edges of streams, bluffs, and hillsides, often on calcareous prairie soils and marl ridges; 0–500 m; Ala., Ark., La.,

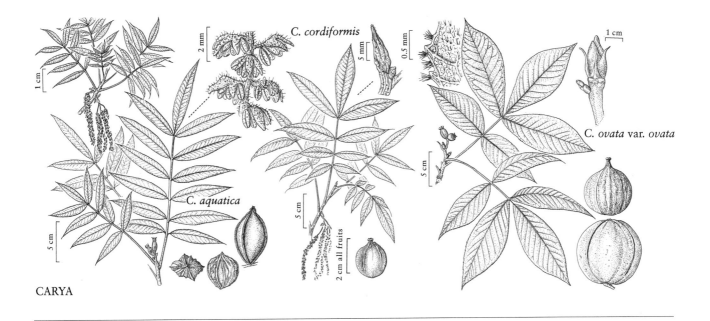

C. cordiformis

C. ovata var. *ovata*

C. aquatica

CARYA

Miss., N.C., Okla., S.C., Tex.; Mexico (Nuevo León).

Carya myristiciformis is the rarest species in the genus. It is patchily distributed from the mountains of northeastern Mexico to the coastal plain of North Carolina.

5. **Carya ovata** (Miller) K. Koch, Dendrologie 1: 598. 1869 · Shagbark hickory, shellbark hickory, noyer tendre, caryer ovale

Juglans ovata Miller, Gard. Dict. ed. 8, Juglans no. 6. 1768; *Hicoria ovata* (Miller) Britton

Trees, to 46 m. **Bark** light gray, fissured or exfoliating, separating freely into long strips or broad plates that persist, ends often curling away from trunk. **Twigs** greenish, reddish, or orangish brown, retaining color or turning black on drying, stout or slender, hirsute or glabrous. **Terminal buds** tan to dark brown to black, ovoid, 6–18 mm, tomentose or nearly glabrous; bud scales imbricate; axillary buds protected by bracteoles fused into hood. **Leaves** 3–6 dm; petiole 4–13 cm, petiole and rachis hirsute or mainly glabrous. **Leaflets** (3–)5(–7), lateral petiolules 0–1 mm, terminal petiolules 3–17 mm; blades ovate, obovate, or elliptic, not falcate, 4–26 × 1–14 cm, margins finely to coarsely serrate, with tufts of hairs in axils of proximal veins of serrations, often weathering to only a few in fall, apex acute to acuminate; surfaces abaxially hirsute with unicellular and 2–4-rayed fasciculate hairs, occasionally restricted to midrib and major veins or essentially without hairs, with few to many large peltate scales and small round, irregular, and 4-lobed peltate scales. **Staminate catkins** pedunculate, to 13 cm, stalks and bracts without hairs; anthers hirsute. **Fruits** brown to reddish brown, spheric to depressed-spheric, not compressed, 2.5–4 × 2.5–4 cm; husks rough, 4–15 mm

thick, dehiscing to base, sutures smooth; nuts tan, ovoid, obovoid, or ellipsoid, compressed, 4-angled, rugulose; shells thick. **Seeds** sweet.

Varieties 3 (2 in the flora): North America.

Carya ovata hybridizes with *C.* × *cordiformis* (*C.* × *laneyi* Sargent), *C. illinoinensis*, and *C. laciniosa* (*C.* × *dunbarii* Sargent).

The Mexican hickory (*Carya ovata* var. *mexicana* (Hemsley) W. E. Manning) appears to be a good variety.

Native Americans used *Carya ovata* medicinally as an antirheumatic, a gynecological aid, a tonic, and an anthelmintic (D. E. Moerman 1986).

1. Twigs stout, hirsute, rarely turning black on drying; terminal buds tan to dark brown, tomentose; leaflets abaxially hirsute, with conspicuous peltate scales . . . . 5a. *Carya ovata* var. *ovata*
1. Twigs slender, glabrous, turning black on drying; terminal buds reddish brown, mainly glabrous; leaflets abaxially essentially glabrous, with few peltate scales . . . . 5b. *Carya ovata* var. *australis*

5a. **Carya ovata** (Miller) K. Koch var. **ovata** [E] [F]

Carya ovata var. *fraxinifolia* Sargent; *C. ovata* var. *nuttallii* Sargent; *C. ovata* var. *pubescens* Sargent;

Trees, to 46 m. **Twigs** greenish, reddish, or grayish brown, normally retaining color on drying, stout, hirsute. **Terminal buds** tan to dark brown, 9–18 mm, tomentose. **Leaves** 3–6 dm; petiole 5–13 cm, hirsute. **Leaflets** 6–26 × 3–14 cm, terminal petiolules 5–17 mm; blades abaxially hirsute, with small irregular, round, and 4-lobed peltate scales. **Staminate catkins** to 13 cm. **Fruits** 3.5–4 × 3.5–4 cm; husks 4–15 mm thick. $2n = 32$.

Flowering spring. Wet bottomlands, rocky hillsides, and limestone outcrops; 0–1400 m; Ont., Que.; Ala., Ark., Conn., Del., D.C., Ga., Ill., Ind., Iowa, Kans., Ky., La., Maine, Md., Mass., Mich., Minn., Miss., Mo., Nebr., N.H., N.J., N.Y., N.C., Ohio, Okla., Pa., R.I., S.C., Tenn., Tex., Vt., Va., W.Va., Wis.

5b. Carya ovata (Miller) K. Koch var. **australis** (Ashe) Little, Phytologia 19: 188. 1969 · Carolina hickory [E]

Carya australis Ashe, Bull. Charleston Mus. 14: 12. 1918; *C. carolinae-septentrionalis* (Ashe) Engler & Graebner

Trees, to 39 m. **Twigs** reddish brown, often turning black by fall or upon drying, slender, glabrous. **Terminal buds** reddish brown to black, ovoid, 6–15 mm, mainly glabrous. **Leaves** 2–3 dm; petiole 4–12 cm, mainly glabrous. **Leaflets** 4–19 × 1–6.5 cm, terminal petiolules 3–12 mm; blades abaxially with a few hairs, and with small 2- and 4-lobed peltate scales. **Staminate catkins** to 6 cm. **Fruits** 2.5–3 × 2.5–3 cm; husks 5–8 mm thick. $2n = 32$.

Flowering spring. Wet bottomlands, rocky hillsides, and limestone outcrops; 150–200 m; Ala., Ga., Miss., N.C., S.C., Tenn.

6. Carya laciniosa (F. Michaux) Loudon, Hort. Brit., 384. 1830 · Shellbark hickory, kingnut [E]

Juglans laciniosa F. Michaux, Hist. Arbr. Forest. 1(2): 199, plate 8. 1810; *Hicoria laciniosa* (F. Michaux) Sargent

Trees, to 41 m. **Bark** light gray, fissured or exfoliating, separating freely into large, thick, broad plates that persist. **Twigs** orange-tan, stout, hirsute, scaly. **Terminal buds** tan, broadly ovoid with apices of outer scales prolonged, 12–20 mm, tomentose; bud scales imbricate; axillary buds protected by bracteoles fused into hood. **Leaves** 6–9 dm; petiole 6–13 cm, minutely hirsute, becoming glabrous at base. **Leaflets** (5–)7–9(–11), lateral petiolules 0–1 mm, terminal petiolules 2–14 mm; blades ovate to obovate or elliptic, not falcate, 9–20 × 3–10 cm, margins coarsely serrate, apex narrowly acuminate; surfaces abaxially hirsute with unicellular, 2–6-rayed fasciculate and occasionally multiradiate hairs, scaly with abundant large peltate scales and small round peltate scales, adaxially hirsute along midrib, puberulent throughout. **Staminate catkins** pedunculate, to 20 cm, stalks and bracts minutely hirsute, capitate-

glandular; anthers hirsute. **Fruits** tan to brown, spheric to ellipsoid, not compressed or slightly so, 4.5–6 × 4–5 cm; husks minutely hirsute, 7–13 mm thick, dehiscing to base, sutures smooth; nuts tan, ellipsoid, compressed, 4-angled, rugulose; shells thick. **Seeds** sweet. $2n = 32$.

Flowering spring. Rich bottomlands, along creeks, and in open cedar glades; 20–300 m; Ont.; Ala., Ark., Del., Ga., Ill., Ind., Iowa, Kans., Ky., Md., Mich., Miss., Mo., N.Y., N.C., Ohio, Okla., Pa., Tenn., Tex., Va., W.Va.

The most southern locality of *Carya laciniosa* is an outlier from Hardin County, Texas. *Carya laciniosa* hybridizes with *C. illinoinensis* (*C.* ×*nussbaumeri* Sargent) and *C. ovata* (*C.* ×*dunbarii* Sargent), and possibly *C. cordiformis.*

Cherokee Indians used *Carya laciniosa* medicinally as an analgesic, a gastrointestinal aid, and a general disease remedy (D. E. Moerman 1986).

7. Carya tomentosa (Poiret) Nuttall, Gen. N. Amer. Pl. 2: 221. 1818 · Mockernut hickory [E] [F]

Juglans tomentosa Poiret in J. Lamarck et al., Encycl. 4: 504. 1798

Trees, to 36 m. **Bark** dark gray, fissured or ridged. **Twigs** reddish brown, stout, hirsute and scaly. **Terminal buds** tan (after early loss of outer scales), broadly ovoid, 8–20 mm, tomentose; bud scales imbricate; axillary buds protected by bracteoles fused into hood. **Leaves** 3–5 dm; petiole 3–12 cm, petiole and rachis hirsute, with conspicuous large and small round peltate scales. **Leaflets** (5–)7–9, lateral petiolules 0–2 mm, terminal petiolules 2–13 mm; blades ovate to elliptic or obovate, not falcate, 4–19 × 2–8 cm, margins finely to coarsely serrate, apex acute, rarely acuminate; surfaces abaxially hirsute with unicellular, 2–8-rayed fasciculate and multiradiate hairs, and with large and small round peltate scales abundant, adaxially hirsute along midrib and major veins, puberulent with fasciculate hairs and scales in spring. **Staminate catkins** pedunculate, to 14 cm, stalks and bracts hirsute, scaly, apex of each bract with coarse hairs; anthers hirsute. **Fruits** reddish brown, finely mottled, spheric to ellipsoid or obovoid, not compressed to compressed, 3–5 × 3–5 cm; husks rough, 4–10 mm thick, dehiscing to middle or nearly to base, sutures smooth; nuts tan, spheric to ellipsoid, compressed, prominently to faintly 4-angled, rugulose; shells thick. **Seeds** sweet. $2n = 64$.

Flowering spring. Well-drained sandy soils, rolling hills and rocky hillsides, occasionally on limestone outcrops, 0–900 m; Ala., Ark., Conn., Del., Fla., Ga.,

Ill., Ind., Iowa, Kans., Ky., La., Md., Mass., Miss., Mo., N.J., N.Y., N.C., Ohio, Okla., Pa., R.I., S.C., Tenn., Tex., Va., W.Va.

Both the mockernut hickory and the shagbark hickory were formerly known as *Carya alba* (Linnaeus) K. Koch [or *Hicoria alba* (Linnaeus) Britton], based on *Juglans alba* of Linnaeus. A. J. Rehder (1945) pointed out that the original circumscription included two taxa, and *C. alba* (*J. alba*) should therefore be rejected as ambiguous in favor of *C. tomentosa* and *C. ovata*, respectively.

Carya tomentosa hybridizes with *C. texana* (*C. ×collina* Laughlin) and is reported to hybridize with the diploid *C. illinoinensis* (*C. ×schneckii* Sargent).

Cherokee Indians used *Carya tomentosa* medicinally as an analgesic, especially as an aid for polio, and as a cold aid; the Delaware, as a gynecological aid and a tonic (D. E. Moerman 1986).

8. **Carya pallida** (Ashe) Engler & Graebner, Notizbl. Königl. Bot. Gart. Berlin, App. 9: 19. 1902 · Sand hickory [E]

Hicoria pallida Ashe, Gard. & Forest 10: 304, fig. 39. 1897

Trees, to 29 m. **Bark** dark gray, ridged, often deeply furrowed. **Twigs** reddish brown, slender, slightly scaly, sometimes pubescent. **Terminal buds** reddish brown, ovoid, 4–11 mm, sparsely to densely scaly; outer bud scales with coarse hairs on midribs, bud scales imbricate; axillary buds protected by bracteoles fused into hood. **Leaves** 3–6 dm; petiole 3–10 cm, rachis sparingly hirsute near base, densely hirsute and scaly distally. **Leaflets** (5–)7(–9), lateral petiolules 0–1 mm, terminal petiolules 2–5 mm; blades ovate to obovate or elliptic, not falcate, 2–15 × 1–6 cm, margins finely to coarsely serrate, apex acuminate; surfaces abaxially hirsute toward base of midrib, otherwise without hairs or rarely hirsute with unicellular and 2–8-rayed fasciculate hairs, abundant large peltate scales and small 4-lobed, irregular, and round peltate scales imparting silvery tan color, adaxially glabrous except for dense fasciculate hairs at base near leaf insertions, moderately to densely scaly in spring. **Staminate catkins** pedunculate, to 13 cm, stalks hirsute, scaly, bracts scaly, hirsute at apex; anthers hirsute. **Fruits** tan to reddish brown, obovoid to spheric or ellipsoid, slightly compressed, 3–4 × 2–3 cm; husks rough, 2–4 mm thick, dehiscing to middle or base, sutures smooth; nuts tan, obovoid to spheric or ellipsoid, slightly compressed, not angled, rugulose; shells thick. **Seeds** sweet.

Flowering spring. Well-drained sandy or rocky soils on bluffs, ridges, rolling hills, and dry woods; 0–500

m; Ala., Ark., Del., Fla., Ga., Ill., Ind., Ky., La., Md., Miss., Mo., N.J., N.C., S.C., Tenn., Va.

Carya pallida occurs principally east of the Mississippi River. It seems to intergrade with *C. texana* in eastern Missouri and southern Illinois, and it may hybridize with *C. glabra*.

9. **Carya texana** Buckley, Proc. Acad. Nat. Sci. Philadelphia 12: 444. 1860 · Black hickory [E]

Carya arkansana Sargent; *C. buckleyi* Durand; *C. glabra* (Miller) Sweet var. *villosa* (Sargent) B. L. Robinson; *C. texana* Buckley var. *villosa* (Sargent) Little

Trees, to 41 m. **Bark** dark gray to black, ridged and deeply furrowed. **Twigs** rusty brown, slender, densely scaly, often pubescent. **Terminal buds** rusty brown, ovoid, 4–9 mm, densely scaly; bud scales imbricate; axillary buds protected by bracteoles fused into hood. **Leaves** 2–5 dm; petiole 3–8 cm, glabrous or with scattered coarse hairs, or rarely pubescent, usually with dense coating of scales imparting rusty brown color. **Leaflets** (5–)7 (–9), lateral petiolules 0–1 mm, terminal petiolules 2–10 mm; blades ovate to obovate, elliptic, or linear-elliptic, not falcate, 3–15 × 1–8 cm, margins finely to coarsely serrate, apex acuminate; surfaces abaxially hirsute along base of midrib, otherwise without hairs or hirsute with unicellular and 2–8-rayed fasciculate hairs, densely covered in spring with a few silvery tan, large, peltate scales and many small, 4-lobed, irregular, and round peltate scales imparting rusty brown color, adaxially without hairs, moderately scaly, becoming glabrous. **Staminate catkins** pedunculate, to 16 cm, stalks with dense coating of rusty brown scales, bracts scaly, with hairs at apex; anthers hirsute. **Fruits** bronze to reddish brown, obovoid to spheric, not compressed, 3–5 × 2.5–3.5 cm; husks 2–4 mm thick, dehiscing to base or nearly so, sutures narrowly winged; nuts tan, obovoid, slightly compressed, usually not angled, occasionally 2–4-angled, rugulose; shells thick. **Seeds** sweet. $2n = 64$.

Flowering spring. Well-drained sandy soils on rolling hills and rocky hillsides, occasionally on low flat lands and marl soils; 0–500 m; Ark., Ill., Ind., Kans., La., Miss., Mo., Okla., Tex.

Carya texana occurs principally west of the Mississippi River. It hybridizes with *C. glabra and C. tomentosa* (*C. ×collina* Laughlin), seemingly intergrades with *C. pallida* in eastern Missouri and southern Illinois, and is reported to hybridize with the diploid *C. aquatica* [*C. ×ludoviciana* (Ashe) Little].

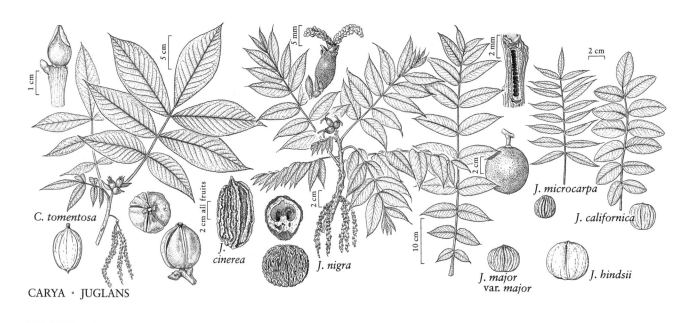

C. tomentosa

J. cinerea

J. nigra

J. microcarpa

J. californica

J. major
var. major

J. hindsii

CARYA · JUGLANS

10. Carya floridana Sargent, Trees & Shrubs 2: 193, plate 177. 1913 · Scrub hickory [E]

Hicoria floridana (Sargent) Sudworth

Trees, to 25 m; more often shrubs, 3–5 m. **Bark** gray, smooth or fissured. **Twigs** reddish brown, slender, glabrous to scaly. **Terminal buds** golden to rusty brown, ovoid, 3–9 mm, hirsute along margins and densely scaly; bud scales imbricate; axillary buds protected by bracteoles fused into hood. **Leaves** 2–3 dm; petiole 4–7 cm, with coarse hairs and scales in spring. **Leaflets** 3–7, lateral petiolules 0–1 mm, terminal petiolules 0–5 mm; blades ovate to elliptic or obovate, not falcate, 4–10 × 2–4 cm, margins often coarsely serrate, apex acuminate; surfaces abaxially with unicellular and 2–8-rayed fasciculate hairs restricted to axils of secondary veins, large peltate scales and small irregular and round peltate scales imparting rusty brown color, adaxially with peltate scales in spring. **Staminate catkins** pedunculate, to 6 cm, stalks with coarse hairs, rusty brown scales, bracts scaly, sparingly hirsute; anthers hirsute. **Fruits** bronze to dark brown, obovoid to oblong, not compressed to slightly compressed, not angled, 3–4 × 2–2.5 cm; husks rough, 2–3 mm thick, dehiscing to base, sutures slightly winged; nuts tan, ellipsoid, slightly compressed, not angled, rugulose; shells thick. **Seeds** sweet. $2n = 64$.

Flowering spring. Sandy ridges of sand-pine scrub; 0–50 m; Fla.

Carya floridana intergrades with *C. glabra* in areas where their ranges overlap.

11. Carya glabra (Miller) Sweet, Hort. Brit., 97. 1826 · Pignut hickory, sweet pignut [E] [W1]

Juglans glabra Miller, Gard. Dict. ed. 8., Juglans no. 5. 1768; *Carya glabra* var. *megacarpa* Sargent; *C. glabra* var. *odorata* (Marshall) Little; *C. leiodermis* Sargent; *C. magnifloridana* Murrill; *C. ovalis* (Wangenheim) Sargent; *C. ovalis* var. *hirsuta* (Ashe) Sargent; *C. ovalis* var. *obcordata* (Muhlenberg & Willdenow) Sargent; *C. ovalis* var. *obovalis* Sargent; *C. ovalis* var. *odorata* (Marshall) Sargent; *Hicoria austrina* Small; *H. microcarpa* (Nuttall) Britton

Trees, to 30 m. **Bark** light gray, smooth or fissured or exfoliating with small platelike scales or narrow strips. **Twigs** reddish brown, slender, essentially glabrous or sparsely scaly. **Terminal buds** reddish brown to tan, ovoid, 5–15 mm; outer scales sparsely scaly, hirsute to glabrous, inner scales finely pubescent, sparsely scaly, bud scales imbricate; axillary buds protected by bracteoles fused into hood. **Leaves** 2–6 dm; petiole 3–14 cm, glabrous to moderately pubescent near rachis, moderately scaly, rachis glabrous or finely puberulent. **Leaflets** (3–)5–7(–9), lateral petiolules 0–2 mm, terminal petiolules 2–18 mm; blades ovate to elliptic or obovate, not falcate 4–21 × 2–10 cm, margins finely to coarsely serrate, apex acuminate to narrowly acuminate; surfaces abaxially glabrous to densely pubescent with unicellular and 2–4-rayed fasciculate hairs, large peltate scales and small irregular, round, and 4-lobed peltate scales in spring, usually becoming glabrous in fall, adaxially scaly in spring. **Staminate catkins** pedunculate, to 13 cm, stalks glabrous or densely pubescent, bracts hirsute at tips; anthers hirsute. **Fruits** tan to reddish brown, obovoid, spheric, or ellipsoid, not compressed to compressed, not angled, 2–4.5 × 2–3.5 cm; husks rough, 2–5 mm thick, partially dehiscent or dehiscing to base, sutures smooth or slightly winged; nuts tan, obovoid to ellipsoid, not compressed to compressed, not angled, rugulose; shells thick. **Seeds** sweet.

Flowering spring. Edge of bayous, deep flood plains, well-drained sandy soils, rolling hills and slopes, dry rocky soils, or thin soils on edge of granite outcrops; 0–800 m; Ont.; Ala., Ark., Conn., Del., D.C., Fla., Ga., Ill., Ind., Iowa, Kans., Ky., La., Md., Mass., Mich., Miss., Mo., N.H., N.J., N.Y., N.C., Ohio, Okla., Pa., R.I., S.C., Tenn., Tex., Vt., Va., W.Va.

Carya glabra is a highly polymorphic species. Tight-barked trees bearing large pear-shaped fruits are common along the Gulf Coast (*C. glabra* var. *megacarpa*, *C. magnifloridana*, and *C. leiodermis*). Trees with exfoliating bark, reddish petioles, and small, compressed, ellipsoid fruits that dehisce to the base (e.g., *C. ovalis*) are more common at higher latitudes. *Carya glabra* intergrades with *C. floridana*, *C. pallida*, and *C. texana*, and it is reported to hybridize with the diploid *C. cordiformis* (*C.* ×*demareei* Palmer). The extreme northern *ovalis* form of the species also appears to hybridize with the typical *glabra* in areas of sympatry.

2. JUGLANS Linnaeus, Sp. Pl. 2: 997. 1753; Gen. Pl. ed. 5, 431. 1754 · Walnut, nogal

Alan T. Whittemore
Donald E. Stone

Shrubs or trees, 3–50 m. **Bark** light to dark gray or gray-brown, smooth or split into ridges or plates. **Twigs** purplish brown, terete, stout, sparsely to densely covered with glands and capitate-glandular hairs, sometimes also with scales or fasciculate hairs, early in season with multiradiate hairs; leaf scars triangular or 3-lobed, large; pith chambered. **Bud scales** valvate, densely hirsute. **Leaves** usually odd-, sometimes even-pinnate; petiole and rachis with indument as twigs. **Leaflets** 5–25, sessile or subsessile, often aromatic, uniform in size or median leaflets largest, (2.5–)4.3–15(–17.5) × 0.8–6.5 cm; surfaces usually with nonglandular hairs (simple and/or fasciculate), glandular hairs, sessile glands, and/or scales, sometimes glabrous. **Staminate catkins** solitary from 2d-year twigs, sessile; stamens 7–50 per flower, glabrous or pilose. **Pistillate flowers** solitary or in terminal racemes. **Fruits** nuts enclosed in husks, not compressed; husks thick, indehiscent; nuts tan, neither compressed nor angled, grooved, ridged, rugulose, or smooth, base 2–4-chambered; shells thick. **Seeds** sweet. $x = 16$.

Species 21 (6 in the flora): North America, Mexico, West Indies, Central America, South America, Eurasia.

Juglans is a very important source of edible nuts, dyes, and wood for cabinet work, furniture, and construction. *Juglans regia* Linnaeus, the walnut of commerce, is widely cultivated in California; it is easily distinguished from native species by its leaves with 5–11 broad, entire leaflets and nuts with thin rugulose shells, not grooved or ridged. Because of its sensitivity to native pathogens, *J. regia* is usually grown as stem-grafts on roots of native or hybrid walnuts (see discussion under *J. hindsii*). Occasional seedlings of *J. regia* have been reported from the vicinity of cultivated plants, but these seldom, if ever, live to maturity.

The growth form, bark, and fruit are important taxonomically in *Juglans*, but these usually are not available on herbarium specimens. As with many woody plants, the first one or two leaves of the season (i.e., the lowermost leaves on the twig) are sometimes atypical in structure, having broader, blunter leaflets. The fasciculate hairs on the veins normally have more rays than those on the blade. In addition to the types of hairs described below, small multiradiate hairs are found on the immature twigs, petioles, rachises, and midribs. They are usually gone when the leaf is fully expanded, but they may persist for a short time afterwards.

SELECTED REFERENCE McGranahan, G. H. and P. B. Catlin. 1987. *Juglans* rootstocks. In: R. C. Rom and R. F. Carlson, eds. 1987. Rootstocks for Fruit Crops. New York. Pp. 411–450.

1. Fruits ellipsoid to ovoid or cylindric, surface of nut with ca. 8 high, narrow, longitudinal main ridges, with narrow, interrupted, longitudinal ridges or lamellae between main ridges, base 2-chambered; terminal buds 12–18 mm; distal edge of leaf scar straight, bordered by well-defined velvety ridge; white walnuts (*Juglans* sect. *Trachyocaryon* Dode ex Manning) . . . . 1. *Juglans cinerea*
1. Fruits globose, rarely ellipsoid, surface of nut smooth or longitudinally grooved, base 4-chambered; terminal buds 3–10 mm; distal edge of leaf scar notched, usually glabrous, sometimes velvety, never forming prominent velvety ridge; black walnuts (*Juglans* sect. *Rhysocaryon* Dode).
  2. Leaflets narrowly elliptic to lance-elliptic, apex rounded to acute; abaxial surface without tufts of hairs in vein axils; s California . . . . 6. *Juglans californica*
  2. Leaflets lanceolate to lance-ovate or narrowly triangular, apex acuminate (sometimes blunter on 1st leaf of season); abaxial surface with conspicuous tufts of fasciculate hairs in vein axils (except sometimes *J. microcarpa*); widespread but not in s California.
    3. Leaflet blades glabrous, hairs confined to major veins; nuts smooth or nearly so; California . . . . . 5. *Juglans hindsii*
    3. Leaflet blades and veins with hairs, at least abaxially (hairs of blade sometimes becoming sparse late in season); nuts longitudinally grooved; widespread but not in California.
      4. Fruits 1.4–2.3 cm; nuts 1.1–1.7 cm; leaflets 0.8–1.1(–2.2) cm wide, with capitate-glandular hairs, nonglandular hairs limited to axils of proximal veins on abaxial surface; shrubs or trees 3–10 m . . . . 4. *Juglans microcarpa*
      4. Fruits 2–8 cm; nuts 1.8–4 cm; leaflets 1.5–5.5 cm wide, with both glandular and nonglandular hairs abaxially; trees 5–50 m.
        5. Fruits 2–3.5 cm, nuts longitudinally grooved, surface between grooves smooth; leaflets 9–15, adaxially with capitate-glandular hairs . . . . 3. *Juglans major*
        5. Fruits 3.5–8 cm, nuts longitudinally grooved, surface between grooves coarsely warty; leaflets (9–)15–19(–23), adaxially glabrous except for scattered hairs on midrib. . . . . 2. *Juglans nigra*

1. Juglans cinerea Linnaeus, Syst. Nat. ed. 10, 1272. 1759 · Butternut, white walnut, noyer cerdré C E F

Wallia cinerea (Linnaeus) Alefeld

Trees, to 20(–30) m. **Bark** light gray or gray-brown, shallowly divided into smooth or scaly plates. **Twigs** with distal edge of leaf scar straight or nearly so, bordered by well-defined, tan-gray, velvety ridge; pith dark brown. **Terminal buds** conic, flattened, 12–18 mm. **Leaves** 30–60 cm; petiole 3.5–12 cm. **Leaflets** (7–)11–17, ovate to lanceolate or oblong-lanceolate, ± symmetric, (2.5–)5–11(–17.5) × 1.5–6.5 cm, margins serrate, apex acuminate; surfaces abaxially with abundant 4–8-rayed fasciculate hairs, scales, and sometimes capitate-glandular hairs, axils of proximal veins with prominent tufts of fasciculate hairs, adaxially with scattered fasciculate hairs or becoming glabrescent; terminal leaflet present, usually large. **Staminate catkins** 6–14 cm; stamens 7–15 per flower; pollen sacs 0.8–1.2 mm. **Fruits** 3–5, ellipsoid to ovoid or cylindric, 4–8 cm, smooth, with dense capitate-glandular hairs; nuts ellipsoid to subcylindric or ovoid, 3–6 cm, surface with ca. 8 high, narrow, longitudinal main ridges, with narrow, interrupted, longitudinal ridges or lamellae between main ridges, base 2-chambered.

Flowering spring (Apr–Jun). Rich woods of river terraces and valleys, also dry rocky slopes; of conservation concern; 0–1000 m; N.B., Ont., Que.; Ala., Ark., Conn., Del., Ga., Ill., Ind., Iowa, Ky., Maine, Md., Mass., Mich., Minn., Miss., Mo., N.H., N.J., N.Y., N.C., Ohio, Pa., R.I., S.C., Tenn., Vt., Va., W.Va., Wis.

The butternut canker is killing *Juglans cinerea* across its range. Because the trees do not root-sprout, the range is contracting.

2. Juglans nigra Linnaeus, Sp. Pl. 2: 997. 1753 · Black walnut, noyer noir E F

Wallia nigra (Linnaeus) Alefeld

Trees, to 40(–50) m. **Bark** medium to dark gray or brownish, deeply split into narrow rough ridges. **Twigs** with distal edge of leaf scar notched, usually deeply, not bordered by well-defined band of pubescence; pith light brown. **Terminal buds** ovoid or subglobose, weakly flattened, 8–10 mm. **Leaves** 20–60 cm; petiole 6.5–14 cm. **Leaflets** (9–)15–19(–23), lanceolate or ovate-lanceolate, symmetric or weakly falcate, (3–)6–15 × 1.5–5.5 cm, margins serrate, apex acuminate; surfaces abaxially with capitate-glandular hairs, simple or 2-rayed fasciculate hairs, and scales scattered over veins and blade, axils of proximal veins with prominent tufts of fasciculate hairs, adaxially glabrous except for scattered capitate-glandular and fasciculate hairs on midrib; terminal leaflet small or often absent. **Staminate catkins** 5–10 cm; stamens 17–50 per flower; pollen sacs 0.8–0.9 mm. **Fruits** 1–2, subglobose to globose, rarely ellipsoid, 3.5–8 cm, warty, with scales and capitate-glandular hairs; nuts subglobose to globose, rarely ellipsoid, 3–4 cm, very deeply longitudinally grooved, surface between grooves coarsely warty, base 4-chambered. $2n = 32$.

Flowering spring (Apr–May). Rich woods; 0–1000 m; Ont.; Ala., Ark., Conn., Del., D.C., Fla., Ga., Ill., Ind., Iowa, Kans., Ky., La., Md., Mass., Mich., Minn., Miss., Mo., Nebr., N.J., N.Y., N.C., Ohio, Okla., Pa., R.I., S.C., S.Dak., Tenn., Tex., Vt., Va., W.Va., Wis.

Variation of *Juglans nigra* in central Texas and south-central Oklahoma should be studied; specimens seemingly intermediate between *J. nigra* and both *J. major* and *J. microcarpa* have been seen from this area. E. C. Twisselmann (1967) incorrectly reported that *J. nigra* was locally naturalized in California; his specimens were all *J. hindsii* (possibly introgressed with *J. nigra*) and *J. californica*.

Juglans nigra is frequently cultivated as an ornamental, and the nuts are prized for their strong, distinctive flavor.

Native Americans used *Juglans nigra* medicinally as a miscellaneous disease remedy, a dermatological aid, and a psychological aid (D. E. Moerman 1986).

3. Juglans major (Torrey) A. Heller, Muhlenbergia 1: 50. 1904

Juglans rupestris Engelmann ex Torrey var. (β) *major* Torrey in L. Sitgreaves, Rep. Exped. Zuni Colorado Rivers, 171, plate 16. 1853; *J. microcarpa* Berlandier var. *major* (Torrey) L. D. Benson

Varieties 2 (1 in the flora): United States, Mexico.

3a. Juglans major (Torrey) A. Heller var. **major** · Arizona walnut, nogal, nogal silvestre $\boxed{\text{F}}$

Trees, 5–18 m. Bark light to medium gray or brownish, divided into narrow checkered plates. Twigs with distal edge of leaf scar notched, glabrous or bordered by poorly defined velvety zone; pith brown. Terminal buds narrowly ovoid or conic, flattened, 4–7 mm. Leaves 18–38 cm; petiole 3–6 cm. Leaflets 9–15, lanceolate to lance-ovate, symmetric or falcate, 6.5–10.5 × 1.5–3.4 cm, margins serrate, apex narrowly acuminate; surfaces abaxially with capitate-glandular hairs, simple or 2–4-rayed fasciculate hairs, and often scales scattered over veins and blade, axils of proximal veins with prominent tufts of fasciculate hairs, adaxially with capitate-glandular hairs, sometimes also scattered fasciculate hairs, becoming glabrate except along major veins; terminal leaflet usually small or none. Staminate catkins 5–8 cm; stamens 20–40 per flower; pollen sacs 1.2–1.4 mm. Fruits 1–3, subglobose or short-ovoid, 2–3.5 cm, smooth, densely covered with capitate-glandular hairs and peltate scales, when very immature also fasciculate

hairs; nuts globose to ovoid, 1.8–2.7 cm, deeply longitudinally grooved, surfaces between grooves, base 4-chambered smooth.

Flowering spring (Apr–May). Along streams and rocky canyon sides; 300–2100 m; Ariz., N.Mex., Okla., Tex.; Mexico (Chihuahua, Durango, Sinaloa, and Sonora).

Specimens intermediate between *Juglans major* and both *J. microcarpa* and *J. nigra* are discussed under the latter species.

4. Juglans microcarpa Berlandier in J. L. Berlandier and R. Choval, Diario Viaje Comis. Limites, 276. 1850 · Little walnut, nogal, nogalito, namboca $\boxed{\text{F}}$

Juglans rupestris Engelmann ex Torrey

Shrubs or small trees, to 10 m. Bark medium gray, split into ± rough ridges. Twigs with distal edge of leaf scar notched, glabrous or bordered by poorly defined velvety zone; pith light to dark brown. Terminal buds globose to short-ovoid, not flattened, 3–5 mm. Leaves 12–29 cm; petiole 1–3(–4) cm. Leaflets 17–25, lanceolate or narrowly lanceolate, weakly to strongly falcate, 5.2–6.3(–9.6) × 0.8–1.1 (–2.2) cm, margins entire or toothed, apex long-acuminate; surfaces abaxially with capitate-glandular hairs (sometimes becoming sparse late in season except along veins), often scattered scales, axils of proximal veins usually, not always, with prominent tufts of fasciculate hairs, adaxially with capitate-glandular hairs (late in season hairs sometimes becoming sparse except along veins); terminal leaflet usually small. Staminate catkins 3–7 cm; stamens 20–25(–35) per flower; pollen sacs 0.8–1 mm. Fruits 1–3, globose, 1.4–2.3 cm, smooth, with capitate-glandular hairs; nuts globose to depressed-globose, 1.1–1.7 cm, grooved, surface between grooves smooth, base 4-chambered.

Flowering spring (Mar–Apr[–Jun]). Along creeks and rivers, 200–2000 m; Kans., N.Mex., Okla., Tex.; Mexico (Chihuahua, Coahuila, and Nuevo León).

Specimens of *Juglans microcarpa* with larger leaflets (parenthetical numbers above) may result from introgression with *J. major*. These have sometimes been treated as *J. microcarpa* var. *stewartii* (I. M. Johnston) W. E. Manning, but W. E. Manning (1978) reported this variety only from Mexico. Intermediates between *J. microcarpa* and *J. nigra* (q.v.) are also known.

5. **Juglans hindsii** Jepson ex R. E. Smith, Univ. Calif. Agric. Exp. Sta. Bull. 203: 27. 1909 · Northern California walnut, Hinds's black walnut [C] [E] [F]

Juglans californica S. Watson var. *hindsii* Jepson

Trees, 6–23 m. **Bark** light or medium gray, split into smooth or ± scaly plates. **Twigs** with distal edge of leaf scar shallowly to deeply notched, not bordered by well-defined band of pubescence; pith light brown. **Terminal buds** ellipsoid to oblong, flattened, 6–8 mm. **Leaves** 22–45 cm; petiole 3–8 cm. **Leaflets** 13–21, narrowly triangular to lanceolate, symmetric or weakly falcate, (5.6–)7.3–13 × (1–)1.9–2.8 cm, margins serrate, apex acuminate; surfaces abaxially glabrous or with sparse glands, sparse glands and few capitate-glandular hairs scattered along major veins, fasciculate hairs conspicuously tufted in axils of proximal veins, sometimes also on adjacent blade and edges of midrib, adaxially glabrous or with scattered scales, major veins glabrous or with sparse scattering of glands and few capitate-glandular hairs, without nonglandular hairs; terminal leaflet well developed. **Staminate catkins** 6–15 cm; stamens 20–40 per flower; pollen sacs 1–1.4 mm. **Fruits** 1–2, globose, 3.5–5 cm; nuts ovoid to ovoid-globose, 2.4–3.2 cm, smooth or nearly so or shallowly and indistinctly ridged or grooved, base 4-chambered.

Flowering spring (Apr–May). Along streams, sometimes on disturbed slopes; of conservation concern; 0–300 m; Calif.

Before 1850, *Juglans hindsii* was restricted to a few locations (R. E. Smith et al. 1912). It has been widely used as a rootstock for grafting *J. regia* and has been planted extensively in many parts of California for this purpose. It is now naturalized in many areas where it apparently did not occur before the introduction of commercial walnut growing. Possibly some of these naturalized populations are introgressed with *J. nigra,* since spontaneous hybridization between *J. hindsii* and *J. nigra* has been reported in areas where both species have been planted. These hybrids are difficult to distinguish from *J. hindsii* unless fruit are present. Currently most commercial walnut orchards use hybrid rootstocks, usually *J. hindsii* × *J. regia* (G. H. McGranahan and P. B. Catlin 1987).

6. **Juglans californica** S. Watson, Proc. Amer. Acad. Arts 10: 349. 1875 · Southern California walnut, California black walnut [E] [F]

Shrubs or small trees, to 6–9 m. **Bark** light or medium gray, divided into rough plates. **Twigs** with distal edge of leaf scar notched, often shallowly so, glabrescent or bordered by poorly defined velvety patch; pith brown. **Terminal buds** ovoid to ellipsoid, somewhat flattened, 5–6 mm. **Leaves** 15–24 cm; petiole 2–5 cm. **Leaflets** (9–)11–15(–17), usually narrowly oblong-elliptic to lance-elliptic, occasionally lanceolate, symmetric or weakly falcate, 4.3–9.5 × 1.6–2.6 cm, margins finely serrate, apex rounded to acute; surfaces abaxially without tufts of hair in vein axils, abaxially and adaxially glabrous with scales but no hairs, main veins glandular, often sparsely so, leaflets without nonglandular hairs (except for multiradiate hairs early in season); terminal leaflet well developed. **Staminate catkins** 5–14 cm; stamens 15–35 per flower; pollen sacs 0.6–1 mm. **Fruits** 1–3, globose, 2.1–3.5 cm, smooth, at first glandular, with scattered scales, soon glabrescent; nuts depressed-globose, 1.8–2.2(–2.5) cm, shallowly grooved, surface between grooves smooth, base 4-chambered.

Flowering spring (Mar–May). Hillsides and canyons; 30–900 m; Calif.

Juglans californica is the most distinctive western walnut, but some care must be taken in identifying it. The distinctive leaflet shape of *J. californica* is occasionally replicated by early-season leaves of other species. Furthermore, *J. californica* is distinctive in lacking simple and fasciculate hairs on the leaves, but like most other walnuts, multiradiate hairs are normally present on the young vegetative growth (stems, petioles, and midribs) in the spring. The hairs are usually deciduous early in the growing season. They have short (0.1–0.2 mm), crisped rays and are never clustered or especially associated with vein axils. The fasciculate hairs found in all of our other species (except sometimes *J. microcarpa*) are persistent, have longer (0.3–0.4 mm), straight rays, and are concentrated in clusters abaxially in the axils of the main lateral veins.

29. MYRICACEAE Blume

· Wax-myrtle Family

Allan J. Bornstein

Shrubs or small trees, evergreen or deciduous, usually aromatic and resinous. **Roots** commonly with nitrogen-fixing nodules. **Leaves** alternate, simple or pinnatifid; stipules absent or present; petiole present. **Leaf blade** commonly with peltate, multicellular, glandular trichomes. **Inflorescences** axillary catkins; bracts present. **Flowers** usually unisexual, occasionally bisexual, staminate and pistillate flowers usually on different plants, occasionally on same plants; perianth absent. **Staminate flowers** subtended by solitary bract; stamens 2–14(–22); filaments filiform, distinct or basally connate; anthers dorsifixed, 2-locular, extrorsely dehiscent by longitudinal slits. **Pistillate flowers** subtended by solitary bract, bracteoles present or absent, usually 2–4(–8); pistils 1, 2-carpellate, 1-locular; ovules 1, basal, erect; styles, if present, short; stigmas 2. **Fruits** drupaceous or nutlike, smooth or often covered with warty protuberances, these commonly with waxy coating; persistent, accrescent bracts and bracteoles sometimes enclosing fruits. **Seeds** with little or no endosperm; embryo straight, with 2 plano-convex cotyledons.

Genera 2–4, species ca. 50 (2 genera, 8 species in the flora): widespread in temperate and subtropical regions.

Significant disagreement exists concerning the number of genera to be recognized in Myricaceae. *Myrica* in the broad sense is sometimes divided into three genera. *Comptonia* L'Héritier ex Aiton is often segregated on the basis of leaf type, presence of stipules, and the burlike fruits with 6–8 accrescent bracts and bracteoles. *Morella* Loureiro sometimes is elevated from its usual rank of subgenus to emphasize differences concerning position of the catkins, size of the staminate bracts, and appearance of the fruits (A. Chevalier 1901; J. R. Baird 1968). The real question is the appropriate rank at which recognition should be made (T. S. Elias 1971; R. L. Wilbur 1994). I follow a traditional approach in recognizing just *Myrica* and *Comptonia* in North America.

SELECTED REFERENCES Baird, J. R. 1968. A Taxonomic Revision of the Plant Family Myricaceae of North America, North of Mexico. Ph.D. thesis. University of North Carolina. Chevalier, A. 1901. Monographie des Myricacées. Mém. Soc. Sci. Nat. Cherbourg 32: 85–340. Elias, T. S. 1971. The genera of Myricaceae in the southeastern United States. J. Arnold Arbor. 52: 305–318. Sheffy, M. V. 1972. A Study of the Myricaceae from Eocene Sediments of Southeastern North America. Ph.D. thesis. Indiana University. Wilbur, R. L. 1994. The Myricaceae of the United States and Canada: Genera, subgenera, and series. Sida

16: 93–107. Youngken, H. W. 1919. The comparative morphology, taxonomy and distribution of the Myricaceae of the eastern United States. Contr. Bot. Lab. Morris Arbor. Univ. Pennsylvania 4: 339–400.

1. M Y R I C A Linnaeus, Sp. Pl. 2: 1024. 1753; Gen. Pl. ed. 5, 449. 1754 · Wax-myrtle, bayberry, sweet gale, myrique [Greek *myrike,* name for tamarisk or another aromatic shrub; possibly from *myrizein,* to perfume]

Cerothamnus Tidestrom; *Gale* Duhamel; *Morella* Loureiro

Shrubs or small trees, often aromatic and resinous. **Branches** spreading, terete, glabrous or pubescent, often gland-dotted. **Leaves** persistent or deciduous; stipules absent. **Leaf blade** aromatic when crushed (except *M. inodora*), oblanceolate, elliptic, obovate, or oblong-ovate, membranous or leathery, margins entire or serrate-denticulate, especially in distal 1/2, pubescent or glabrous, usually gland-dotted. **Inflorescences** ± erect, ellipsoid to short-cylindric or ovoid, appearing before or with leaves; bracts ovate, glabrous or variously pubescent. **Flowers** unisexual, rarely bisexual, staminate and pistillate flowers usually on different plants, infrequently on same plants. **Staminate flowers:** stamens (2–)3–12(–22), shorter or longer than subtending bract; filaments distinct or often connate into branching staminal column, each branch terminated by anther; rudimentary ovary occasionally present. **Pistillate flowers:** ovary subtended by 2–6 broadly ovate bracteoles, these sometimes persistent and accrescent, always shorter than fruit, sometimes completely absent; styles short. **Fruits** globose or ovoid to lenticular, smooth or more commonly with warty protuberances, usually covered with waxy coating that dries white. $x = 8$.

Species ca. 50 (7 in the flora): nearly worldwide.

Myrica is often cultivated. *Myrica* species were used by various tribes of Native Americans for medicinal purposes. Leaves were used for a gynecological aid and an emetic; the bark, as a blood purifier and a kidney aid (D. E. Moerman 1986). Bayberry candles were used by early settlers, and they remain popular household items, both decorative and functional.

SELECTED REFERENCES Davey, A. J. and C. M. Gibson. 1917. Note on the distribution of sexes in *Myrica gale*. New Phytol. 16: 147–151. Houghton, W. M. 1988. The Systematics of Section *Cerophora* of the Genus *Myrica* (Myricaceae) in North America. M.S. thesis. University of Georgia. MacDonald, A. D. 1977. Myricaceae: Floral hypothesis for *Gale* and *Comptonia*. Canad. J. Bot. 55: 2636–2651. MacDonald, A. D. and R. Sattler. 1973. Floral development of *Myrica gale* and the controversy over floral concepts. Canad. J. Bot. 51: 1965–1975. Thieret, J. W. 1966. Habit variation in *Myrica pensylvanica* and *M. cerifera*. Castanea 31: 183–185.

1. Bracts of staminate flowers longer than stamens; bracteoles of pistillate flowers 2, accrescent and adnate to fruit; fruits smooth (without protuberances), lacking waxy deposit (subg. *Myrica*).
  2. Bracteoles of pistillate flowers glabrous; leaf margins serrate, usually minutely so, with 1–4 pairs of teeth restricted to distal 1/3 of blade; not extending s of Oregon in w part of range. . . . . .1. *Myrica gale*
  2. Bracteoles of pistillate flowers densely pilose, especially at apex; leaf margins serrate, often coarsely so, with 4–12 pairs of teeth ± in distal 1/2 of blade; California only. . . . . .2. *Myrica hartwegii*
1. Bracts of staminate flowers shorter than stamens; bracteoles of pistillate flowers 4–6, not accrescent or adnate to fruit; fruits with numerous protuberances, usually covered with waxy coating that dries white or blue-white to gray (subg. *Morella*).

3. Staminate flowers with 6 or more stamens, rarely 2–3, especially in distal flowers; fruit wall, but not warty protuberances, pubescent.

    4. Margins of leaf blade entire, rarely serrate at apex; blade not aromatic when crushed; flowers unisexual, staminate and pistillate on different plants; restricted to Gulf Coast region . . . . 3. *Myrica inodora*

    4. Margins of leaf blade conspicuously serrate almost their entire length; blade fragrant when crushed; flowers bisexual, staminate, and pistillate, all on same plant; Pacific Coast region. . . . 4. *Myrica californica*

3. Staminate flowers with 3–5(–7) stamens; fruit wall usually glabrous, if pubescent, warty protuberances also pubescent.

    5. Leaf blade densely glandular on both surfaces . . . . 5. *Myrica cerifera*

    5. Leaf blade densely glandular only abaxially.

        6. Fruit wall and warty protuberances densely hirsute when young; branches whitish gray in age; leaves deciduous, membranous; fruits 3.5–5.5 mm. . . . . .6. *Myrica pensylvanica*

        6. Fruit wall glabrous or sparsely glandular, warty protuberances ± glandular; branches black; leaves persistent or tardily deciduous, leathery; fruits 3–4.5 mm . . . . .7. *Myrica heterophylla*

---

1. Myrica gale Linnaeus, Sp. Pl. 2: 1024. 1753 · Sweet gale, meadow-fern, myrique baumier, bois-sent-bon ⬚F⬚

Gale palustris (Lamarck) A. Chevalier; *G. palustris* var. *denticulata* A. Chevalier; *G. palustris* var. *lusitanica* A. Chevalier; *G. palustris* var. *subglabra* A. Chevalier; *G. palustris* var. *tomentosa* (C. de Candolle) A. Chevalier; *Myrica gale* var. *subglabra* (A. Chevalier) Fernald; *M. gale* var. *tomentosa* C. de Candolle; *M. palustris* Lamarck

Shrubs, deciduous, much branched, to 1.5(–2) m. **Branchlets** purple-black, gland-dotted, glands brownish yellow. **Leaf blade** oblanceolate to obovate, 1.5–6.5 × 0.5–1.5 cm, ± leathery, base cuneate, margins usually minutely serrate, with 1–4 pairs of teeth usually restricted to distal 1/3 of blade, occasionally entire throughout, apex rounded or obtuse; surfaces abaxially pale green, glabrous to densely pilose, adaxially dark green, glabrous to pilose, both surfaces variously gland-dotted; glands bright yellow to orange. **Inflorescences**: staminate ca. 1–1.5 cm; pistillate to 1.5 cm. **Flowers** unisexual, staminate and pistillate mostly on different plants, occasionally on same plants. **Staminate flowers**: bract of each flower longer than stamens, stamens mostly 3–5. **Pistillate flowers**: bracteoles 2, accrescent and adnate to base of fruit wall, laterally compressed, glabrous but gland-dotted; ovary glabrous. **Fruits** ovoid, flattened, 2.5–3 mm; fruit wall smooth (no protuberances), without waxy deposit, with glandular deposit, enclosed by spongy bracteoles. $2n$ = ca. 96.

Flowering spring–early summer, fruiting in summer. Coastal and inland swamps, bogs, borders of lakes, ponds, and streams; 0–670 m; St. Pierre and Miquelon; Alta., B.C., Man., N.B., Nfld., N.W.T., N.S., Ont., P.E.I., Que., Sask., Yukon; Alaska, Conn., Maine, Mass., Mich., Minn., N.H., N.J., N.Y., N.C., Oreg., Pa., R.I., Vt., Wash., Wis.; Eurasia.

I have seen at least two specimens of *Myrica gale* from Seneca County, Ohio, although they could have been very old collections. They apparently do not represent the current situation in Ohio.

The spongy bracteoles that surround the fruits aid in dispersal by acting as flotation devices in water. A. J. Davey and C. M. Gibson (1917), as well as others, have commented on the sexual distribution in this species. A. D. MacDonald and R. Sattler (1973) and A. D. MacDonald (1977) have used this species to investigate the nature of the flower/inflorescence in Myricaceae.

The pounded branches of *Myrica gale* were utilized by the Bella Coola to prepare decoctions taken as a diuretic or as a treatment for gonorrhea (D. E. Moerman 1986).

2. Myrica hartwegii S. Watson, Proc. Amer. Acad. Arts 10: 350. 1875 · Sierra sweet-bay ⬚E⬚

Gale hartwegii (S. Watson) A. Chevalier

Shrubs, deciduous, to 1.8 m. **Branchlets** purple-black, glanddotted, glands yellow. **Leaf blade** oblanceolate to elliptic, 3.2–10.4 × 1–3.4 cm, membranous, base attenuate to cuneate, margins serrate, often coarsely so, with 4–12 pairs of teeth in ± distal 1/2 of blade, rarely entire, apex acute to occasionally obtuse; surfaces abaxially pale green, pilose, adaxially dark green, pilose, often glabrate, both surfaces gland-dotted, density quite variable; glands yellow to orange. **Inflorescences**: staminate 0.8–2.6 cm; pistillate 3–6 mm at anthesis, enlarging to 2 cm in fruit. **Flowers** unisexual, staminate and pistillate on different plants. **Staminate flowers**: bract of flower longer than stamens, gland-dotted at base, distal margins transparent, occasionally ciliate; stamens 3–5. **Pistillate flowers**: bracteoles 2, accrescent and adnate to base of fruit wall, laterally compressed, densely pilose with some hairs persistent, especially toward apex, glandular; ovary glabrous. **Fruits** ovoid, flattened, 1.5–2.5 mm; fruit wall smooth, without waxy deposits, enclosed by spongy bracteoles.

Flowering spring–summer, fruiting in summer. Borders of streams; 250–1800 m; Calif.

Myrica hartwegii is endemic to the northern and central Sierra Nevada of California.

3. Myrica inodora W. Bartram, Travels Carolina, 403. 1791 · Odorless bayberry, odorless wax-myrtle, waxberry, candleberry, waxtree [E]

Cerothamnus inodorus (W. Bartram) Small; *Morella inodora* (W. Bartram) Small; *Myrica laureola* C. de Candolle; *M. obovata* C. de Candolle

Shrubs or small trees, evergreen, to 7 m. **Branchlets** reddish brown and gland-dotted when young, glands colorless to white. **Leaf blade** lacking odor when crushed, oblong-obovate to elliptic, 3.5–10.5(–11.8) × 1.4–3.7(–4.4) cm, leathery, base attenuate to cuneate, margins entire, rarely serrate distally, slightly revolute, apex acute to rounded; surfaces abaxially pale green, glabrous, sometimes with a few scattered hairs, adaxially dark green, shiny, glabrous, both surfaces gland-dotted, pitted; glands minute, colorless or white. **Inflorescences:** staminate 0.7–2.2 cm; pistillate 0.4–4 (–5) cm. **Flowers** unisexual, staminate and pistillate on different plants. **Staminate flowers:** bract of flower shorter than staminal column, margins opaque, densely ciliate; stamens mostly 6–10, as few as 3 in more distal flowers. **Pistillate flowers:** bracteoles persistent, 4, obscure in fruit, not accrescent or adnate to fruit wall, glabrous except for ciliate margins; ovary densely villous. **Fruits** globose-ellipsoid, 4–8 mm; fruit wall densely pubescent, obscured by enlarged, glandular protuberances and thin (usually) coat of white-gray wax.

Flowering late winter–early spring, fruiting mid summer. Coastal pineland swamps, swamp margins, bogs, pond edges, and stream banks; 0–10 m; Ala., Fla., Ga., La., Miss.

4. Myrica californica Chamisso, Linnaea 6: 535. 1831 · Pacific bayberry, California wax-myrtle [E]

Gale californica (Chamisso) Greene

Shrubs or small trees, evergreen, 2–10 m. **Branchlets** green when young, becoming red-brown, eventually black to gray with age, densely gland-dotted, glands colorless to black, pilose to villous, ultimately glabrous. **Leaf blade** fragrant when crushed, narrowly elliptic to elliptic-oblanceolate, 4–13 × 0.7–3.1 cm, sometimes membranous, more commonly leathery, base cuneate-attenuate, margins variable, from nearly entire (less common) to remotely and coarsely serrate entire length of blade, apex acute;

surfaces abaxially pale green, adaxially dark green, shiny, both surfaces gland-dotted; glands colorless to black, considerably more dense abaxially, midrib pilose to glabrate adaxially. **Inflorescences:** staminate 0.6–1.7(–2.5) cm; bisexual 0.6–1.9(–3) cm; flowers bisexual, staminate, or pistillate within any 1 spike. **Staminate flowers:** bract of flower shorter than staminal column, margins opaque, densely ciliate; stamens (2–)6–12(–22). **Pistillate and bisexual flowers:** bracteoles usually persistent in fruit, 4–6, not accrescent or adnate to fruit wall, margins ciliate; stamens 1–5, in bisexual flowers hypogynous, free or often adnate to ovary, especially near styles; ovary ± villous, especially at apex. **Fruits** globose-ellipsoid, 4–6.5 mm; fruit wall glabrate to sparsely villous, obscured by enlarged, glabrous protuberances, with or without light to very heavy coat of white wax.

Flowering spring–early summer, fruiting summer–early fall. Coastal conifer forests, bogs, sand dunes, stream banks, wet meadows, marshes, low, moist hillsides; 0–1000 m; B.C.; Calif., Oreg., Wash.

On any one branchlet, staminate inflorescences are borne proximal to bisexual inflorescences; the most distal inflorescences may be completely pistillate.

It is quite common for two or three pistillate or bisexual flowers to occur per bract and for the ovaries to fuse to form a syncarp. In the fruiting condition this can usually be detected by counting the number of style branches (two per ovary, therefore four for a syncarp derived from two fused ovaries). Many specimens apparently do not produce any wax, in which case the fruits appear purple-black rather than white.

5. Myrica cerifera Linnaeus, Sp. Pl. 2: 1024. 1753 · Southern bayberry, southern wax-myrtle [F] [W1]

Cerophora lanceolata Rafinesque; *Cerothamnus arborescens* (Castiglioni) Tidestrom; *C. ceriferus* (Linnaeus) Small; *C. pumilus* (Michaux) Small; *Morella cerifera* (Linnaeus) Small; *M. pumila* (Michaux) *Myrica cerifera* var. *angustifolia* Aiton; *M. cerifera* var. *arborescens* Castiglioni; *M. cerifera* var. *dubia* A. Chevalier; *M. cerifera* var. *pumila* Michaux; *M. pumila* (Michaux) Small; *M. pusilla* Rafinesque

Shrubs or small trees, evergreen, often forming large, rhizomatous colonies of much-branched specimens, to 14 m. **Branchlets** reddish brown, densely gland-dotted when young, otherwise glabrous to densely pilose, eventually glabrate; glands yellow. **Leaf blade** aromatic when crushed, linear-oblanceolate to obovate, (1.1–)2–10.5(–13.3) × 0.4–3.3 cm, leathery, base cuneate to attenuate, margins entire or coarsely serrate beyond middle, apex acute to slightly rounded; surfaces abaxially pale yellow-green, glabrous except for pilose midrib, adaxially dark green, glabrous to pilose, both sur-

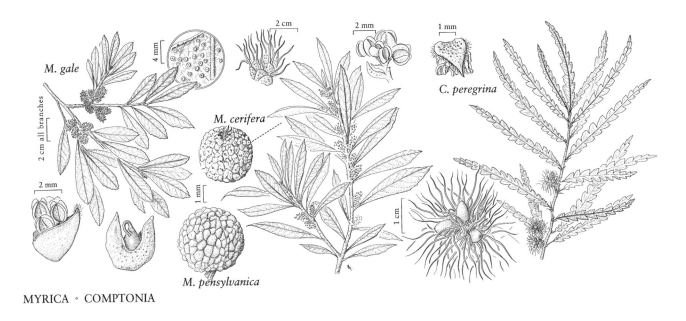

M. gale

M. cerifera

C. peregrina

M. pensylvanica

MYRICA · COMPTONIA

faces densely glandular; glands yellow to orange. **Inflorescences:** staminate 0.4–1.9 cm; pistillate 0.3–1.5 cm. **Flowers** unisexual, staminate and pistillate on different plants. **Staminate flowers:** bract of flower shorter than staminal column, margins opaque, densely ciliate, abaxially densely gland-dotted; stamens mostly 3–4. **Pistillate flowers:** bracteoles persistent in fruit, 4, not accrescent or adnate to fruit wall, margins ciliate, abaxially densely gland-dotted; ovary glandular, especially at apex near style base. **Fruits** globose-ellipsoid, 2–3.5(–4) mm; fruit wall glabrous or sparsely glandular when young, obscured by enlarged protuberances and thick coat of blue-white wax.

Flowering mid winter–spring, fruiting summer–fall. Bogs, edges of marshes, ponds, creeks, and swamps, pine forests, mixed deciduous forests, pine barrens, coastal sand dunes, open fields, sandy hillsides; 0–450 m; Ala., Ark., Del., Fla., Ga., La., Md., Miss., N.J., N.C., Okla., S.C., Tex., Va.; Mexico; West Indies; Bermuda; Central America.

Myrica cerifera is an extremely variable species with respect to habitat and corresponding habit/vegetative morphology. In general, plants that occupy dry, sandy (more xeric) areas tend to be strongly rhizomatous, colonial, and smaller in stature, and to possess smaller leaves (commonly recognized as *M. cerifera* var. *pumila*). In contrast, plants of more mesic areas are seldom rhizomatous, not colonial, and often large and treelike, and they have larger leaves. These "extremes pass insensibly into each other" (J. W. Thieret 1966). I agree with Thieret's contention that these differences do not constitute reliable criteria upon which one should base taxonomic distinctions. Until it can be determined with certainty whether these differences are due to genetics

or environment, the question will remain open. I have chosen the conservative route.

Myrica cerifera has often been confused with *M. pensylvanica* and with *M. heterophylla*. It is distinguished from *M. pensylvanica* on the basis of gland density on the leaves, the presence of glands versus hirsute pubescence on the fruit wall and protuberances (especially visible on young fruits), and less reliably on the size of the fruit (2–3.5 versus 3.5–5.5 mm). *Myrica cerifera* is distinguished from *M. heterophylla* by the density of the glands on the leaves and the glandular versus glabrous (usually) fruit wall.

Native Americans used a decoction of the leaves and stems of *Myrica cerifera* to treat fevers; and roots, to treat inflamed tonsils and stomachaches, and as a stimulant (D. E. Moerman 1986).

6. **Myrica pensylvanica** Mirbel in H. Duhamel du Monceau et al., Traité Arbr. Arbust. Nouv. ed. 2: 190. 1804 · Northern bayberry, waxberry, tallow bayberry, small waxberry, tallowshrub, swamp candleberry, candlewood, candletree, tallowtree, myrique de Pennsylvanie E F W1

Cerothamnus pensylvanica (Mirbel) Moldenke; *Myrica cerifera* Linnaeus var. *frutescens* Castiglioni; *M. macfarlanei* Youngken

Shrubs or rarely small trees, deciduous, rhizomatous, colonial, to 2(–4.5) m. **Branchlets** reddish brown and gland-dotted when young, becoming whitish gray in age, otherwise densely pilose; glands yellow. **Leaf blade** aromatic when crushed, oblanceolate to elliptic, occasionally obovate, 2.5–6.5(–7.8) × 1.5–2.7 cm, usually membranous, less often leathery, base cuneate to attenuate, margins sometimes entire, usually serrate distal to middle, apex obtuse to rounded, sometimes acute, short-apiculate; surfaces abaxially pale green, pilose on veins, moderately to densely glandular, adaxially dark green, pilose (especially along midrib), glandless or sparsely glandular; glands yellow-brown. **Inflorescences:** staminate 0.4–1.8 cm; pistillate 0.3–1.4 cm. **Flowers** unisexual, staminate and pistillate on different plants. **Staminate flowers:** bract of flower shorter than staminal column, margins opaque, apically ciliate or completely glabrous, usually abaxially glabrous, occasionally densely pilose; stamens mostly 3–4. **Pistillate flowers:** bracteoles persistent in fruit, 4, not accrescent or adnate to fruit wall, margins slightly ciliate or glabrous, abaxially usually densely glanddotted; ovary wall densely hirsute near apex, otherwise glabrous. **Fruits** globose-ellipsoid, 3.5–5.5 mm; fruit wall and warty protuberances hirsute, at least when young, hairs usually obscured by thick coat of white wax.

Flowering spring–early summer, fruiting late summer–fall. Coastal dunes, pine barrens, pine-oak forests, old fields, bogs, edges of streams, ponds, and swamps; 0–325 m; St. Pierre and Miquelon; N.B., Nfld., N.S., Ont., P.E.I., Que.; Conn., Del., Maine, Md., Mass., N.H., N.J., N.Y., N.C., Ohio, Pa., R.I., Va.

Where their ranges overlap, *Myrica pensylvanica* hybridizes quite readily with both *M. cerifera* and *M. heterophylla*. This ease of hybridization obviously contributes to an already complicated taxonomic situation; it is a matter for further field-based investigation.

7. Myrica heterophylla Rafinesque, Alsogr. Amer., 9. 1838 · Evergreen bayberry, myrtle, wax-myrtle E

Cerothamnus carolinensis (Miller) Tidestrom; *Myrica cerifera* Linnaeus var. *augustifolia* C. de Candolle; *M. cerifera* var. *latifolia* Aiton; *M. curtissii* A. Chevalier; *M. curtissii* var. *media* (Michaux) A. Chevalier; *M. heterophylla* var. *curtissii* (A. Chevalier) Fernald; *M. sessilifolia* Rafinesque; *M. sessilifolia* var. *latifolia* (Aiton) Rafinesque

Shrubs or small trees, evergreen or tardily deciduous, often forming rhizomatous colonies of much-branched individuals, to 3 m. **Branchlets** appearing black, glabrous to densely pilose; glands sparse or dense, yelloworange. **Leaf blade** aromatic when crushed, oblanceolate to elliptic, occasionally obovate, 3–12.4(–14.2) × 1–5.2 cm, sometimes membranous, more often leathery, base cuneate to attenuate, margins entire or serrate distal to middle, apex rounded to acute, apiculate; surfaces abaxially pilose (especially on major veins) or glabrate, densely glandular, adaxially pilose or glabrous, lacking glands or very sparsely glandular; glands yellow. **Inflorescences:** staminate 0.5–1.8 cm; pistillate 0.3–1.1 cm. **Flowers** unisexual, staminate and pistillate on different plants. **Staminate flowers:** bract of flower shorter than staminal column, margins opaque, ciliate, especially at apex and laterally, abaxially glabrous or with few glands; stamens 3–5(–7). **Pistillate flowers:** bracteoles persistent in fruit, 4, not accrescent or adnate to fruit wall, abaxially pilose, usually along midrib, lacking glands; ovary glabrous or sparsely glandular, not pubescent. **Fruits** globose-ellipsoid, 3–4.5 mm; fruit wall glabrous or sparsely glandular, obscured by enlarged protuberances (± glandular) and thin to thick coat of gray to white wax.

Flowering spring–early summer, fruiting summer–fall. Bogs, stream, pond and lake margins, moist regions of mixed deciduous forests, pine flatlands near pitcher-plant bogs, swamps; 0–250 m; Ala., Ark., D.C., Fla., Ga., La., Md., Miss., N.J., N.C., Pa., S.C., Tex., Va.

I have not seen any specimens of *Myrica heterophylla* from Delaware although it is listed in other floras as occurring there.

To distinguish between *Myrica heterophylla* and *M. pensylvanica* in the vegetative condition is difficult at best. Although *M. heterophylla* tends to have larger, persistent leaves (versus smaller, deciduous ones in *M. pensylvanica*), the difference breaks down, especially in the northern portion of the range of *M. heterophylla*. Male specimens provide little help in resolving this problem because the inflorescences are virtually identical. Female specimens are most useful (essential?) for definitive delimitation because the ovary and young fruit (wall and protuberances) of *M. pensylvanica* are pubescent in contrast to the glabrous or sparsely glandular fruits of *M. heterophylla*. Whether these differences are sufficient to warrant the recognition of distinct species is yet to be satisfactorily resolved. W. M. Houghton (1988) attempted to settle this matter, eventually recognizing two subspecies in *M. pensylvanica*, but he did not examine floral features in his analysis. Again, I have taken the conservative route, leaving the question largely unanswered.

2. COMPTONIA L'Héritier ex Aiton, Hort. Kew. 3: 334. 1789 · Sweet-fern, comptonie [for Henry Compton, amateur horticulturist and Bishop of London]

Shrubs, fragrant. **Branches** spreading-ascending, terete, pubescent to glabrate, glandular when young. **Leaves** mostly deciduous, occasionally persistent, ± deeply pinnatifid; stipules present, deciduous or ± persistent. **Leaf blade** linear-lanceolate, with 2–10 rounded to pointed lobes, membranous, glabrous or densely pubescent and glandular. **Inflorescences** cylindric; staminate eventually flexuous, pistillate globose-ovoid at maturity, appearing before leaves; bracts ovate or cordate, glabrous or variously pubescent. **Flowers** unisexual, staminate and pistillate usually on different plants, occasionally on same plants. **Staminate flowers:** stamens 3–8, shorter than subtending bract, filaments free or slightly fused. **Pistillate flowers:** ovary subtended by persistent bract and 2 linear-subulate bracteoles at anthesis, bracteoles accrescent and developing 4–8 tertiary bracteoles (= scales of other authors), these longer than and concealing fruit; styles 2, elongate. **Fruits** oblong-ovoid, smooth (without protuberances), waxless. $x = 8$.

Species 1 (1 in the flora): North America.

1. Comptonia peregrina (Linnaeus) J. M. Coulter, Mem. Torrey Bot. Club 5: 127. 1894 · Sweet-fern, comptonie voyageuse E F W1

Liquidambar peregrina Linnaeus, Sp. Pl. 2: 999. 1753; Syst. Nat. ed. 10, 2: 1273. 1759; *Comptonia ceterach* Mirbel; *C. peregrina* var. *asplenifolia* (Linnaeus) Fernald; *C. peregrina* var. *tomentosa* A. Chevalier; *Myrica asplenifolia* Linnaeus; *M. comptonia* C. de Candolle; *M. peregrina* (Linnaeus) Kuntze

Shrubs, to 1.5 m, forming rhizomatous colonies. **Branchlets** red-brown to gray, pilose to villose, sometimes puberulent. **Leaves:** stipules nearly cordate, long-acuminate. **Leaf blade** very aromatic when crushed, 3–15.5 × 0.3–2.9 cm, lobes alternate to nearly opposite, base truncate, cuneate to attenuate, or oblique, apex acute; surfaces abaxially pale gray-green, densely pilose to puberulent, adaxially dark green, densely pilose to glabrate, glanddotted, especially adaxially. **Inflorescences:** staminate in clusters at ends of branches, elongating to 5 cm, bracts broadly ovate to trullate, margins ciliate, apex acute to long-acuminate, abaxially gland-dotted; pistillate to 5 mm at anthesis, elongating in fruit to 2 cm, bracteoles to 1.3 cm, pilose, gland-dotted. **Fruits** 2.5–5.5 mm.

Flowering spring, fruiting summer. Dry, sterile, sandy to rocky soils in pinelands or pine barrens, clearings, or edges of woodlots; 0–1800 m; N.B., N.S., Ont., P.E.I., Que.; Conn., Del., Ga., Ill., Ind., Ky., Maine, Md., Mass., Mich., Minn., N.H., N.J., N.Y., N.C., Ohio, Pa., R.I., S.C.,Vt., Va., W.Va., Wis.

Comptonia peregrina has been reported from Tennessee (Scott County), but I have not seen any specimens.

Many Native American tribes used different parts of *Comptonia peregrina* variously: as an incense for ritual ceremonies; for medicinal purposes; as a stimulant or tonic; as a food seasoning; and as a poison (D. E. Moerman 1986).

30. FAGACEAE Dumortier

• Beech Family

Kevin C. Nixon

Trees or shrubs, evergreen or deciduous, shrubs sometimes rhizomatous. **Winter buds** sessile, with few to many imbricate scales (2 valvate scales enclosing imbricate scales in *Castanea*); terminal bud present or absent. **Leaves** alternate, spirally arranged, simple; stipules deciduous (usually), distinct, scarious; petiole present. **Leaf blade** lobed or unlobed, pinnately veined, margins serrate, dentate, or entire; surfaces usually pubescent, at least when young, sometimes with scales. **Inflorescences** unisexual or androgynous catkins; staminate and androgynous catkins spicate or capitate, rigid, flexible, or lax, consisting of few- to many-flowered clusters, bracts present or absent; pistillate catkins rigid or flexible, with 1–several spicately arranged, or rarely solitary terminal cupules bearing 1–3(–15 or more) pistillate flowers. **Staminate flowers** bracteate, bracts often caducous; sepals (3–)4–6(–8); stamens (3–)6–12(–18 or more); petals absent; anthers 2-locular, dehiscing by longitudinal slits, pollen sacs contiguous; pistillode often present and indurate, or vestigial as central tuft of trichomes. **Pistillate flowers:** calyx of 4–6 distinct or connate sepals; petals absent; pistil 1, 3(–6 or more)-carpellate; ovary inferior, locules as many as carpels; placentation axile; ovules pendulous, 2 in each locule, all but 1 in each pistil usually aborting; styles as many as carpels, distinct to base; stigmas dry; staminodes present or absent. **Fruits** nuts, sometimes winged, 1-seeded, subtended or enclosed individually or in groups of 2–3(–15) by scaly or spiny, multibracteate cupule; seed coat membranous; endosperm none; embryo straight, as long as seed; cotyledons fleshy, starchy or oily.

Genera 9, species probably 600–800 (5 genera, 97 species, and numerous hybrids in the flora): widespread, often dominant forest trees in temperate, subtropical, and tropical areas, mostly Northern Hemisphere.

In the Western Hemisphere, Fagaceae are found from southern Canada to Colombia; they are absent or infrequent in most of the northern Great Plains and northern Rocky Mountain region.

Fagaceae are one of the most important families of Northern Hemisphere woody plants in terms of total biomass and economic use. They are widely used for lumber, firewood, and horticultural plantings; the nuts are often used for animal fodder and, in some species, for human food (particularly *Castanea*). As dominants in forests, woodlands, and chaparral, native stands of fagaceous trees and shrubs provide optimal wildlife habitat, often harboring an exceptionally diverse insect fauna. Most of the diversity of the family in the Western Hemisphere is concentrated in the genus *Quercus,* with the greatest number of species in Mexico (at least 125 species), and a secondary area of diversity in the southeastern United States.

Polyploidy has not been reported in any natural populations of species of Fagaceae. Natural interspecific hybridization is common in the family, particularly in *Quercus,* and also in *Castanea* and *Lithocarpus.*

The most important diagnostic feature of Fagaceae is the cupule, which occurs as the cup or cap of the acorn in *Quercus* and *Lithocarpus* and the spiny bur that surrounds the fruits of *Castanea* and *Chrysolepis.* The cupule is sometimes referred to as an involucre. A true involucre, however, is made up of bracts, while the cupule has been shown to be a complex structure that is interpreted as an indurated, condensed, partial inflorescence formed by fusion of stem axes with several orders of branching, bearing bracts that are modified as scales and/ or spines (see B. S. Fey and P. K. Endress 1983).

SELECTED REFERENCES   Elias, T. S. 1971b. The genera of Fagaceae in the southeastern United States. J. Arnold Arbor. 52: 159–195.   Fey, B. S. and P. K. Endress. 1983. Development and morphological interpretation of the cupule in Fagaceae. Flora 173: 451–468.   Forman, L. L. 1966. On the evolution of cupules in the Fagaceae. Kew Bull. 18: 385–419.   Hjelmquist, H. 1948. Studies on the floral morphology and phylogeny of the Amentiferae. Bot. Not., Suppl. 2: 1–171.   Nixon, K. C. and W. L. Crepet. 1989. *Trigonobalanus* (Fagaceae): Taxonomic status and phylogenetic relationships. Amer. J. Bot. 76: 828–841.   Soepadmo, E. 1972. Fagaceae. In: C. G. G. J. Van Steenis, ed. 1950 + . Flora Malesiana. . . . Series I. Spermatophyta. 11 + vols. in parts. Djakarta and Leiden. Vol. 7, part 2, pp. 265–403.

1. Fruits acorns, acorn a solitary nut, circular in cross section, at least partially covered by scaly cup, cup unlobed, without visible sutures or valves; scales not noticeably spinose; nut not completely enclosed by cup at maturity (except in *Quercus lyrata*).
    2. Cup scales strongly reflexed, hooked at tip; staminate and androgynous inflorescences erect or ascending, rigid or flexible, often appearing terminal and branched. . . . . . . 3. *Lithocarpus*, p. 442
    2. Cup scales various, or if somewhat reflexed (rarely), then not noticeably hooked at tip; staminate inflorescences lax, axillary or clustered near base of new growth; androgynous inflorescences absent. . . . . .5. *Quercus*, p. 445
1. Fruits of 1–several nuts, nut usually 3-angled or rounded-angular in cross section, enclosed in spiny or prickled cupule; cupule valves 2–4(–8 +), distinct or marginally connate along sutures, these ± completely enveloping nut(s) until maturity.
    3. Spines/scales of cupule unbranched, stout, not obscuring surface of cupule; inflorescences unisexual (staminate below pistillate on same branch); pistillate flowers (and fruits) typically 2 per cupule; staminate inflorescences lax, loosely capitate; nut sharply angular, slightly winged. . . . . .4. *Fagus*, p. 443
    3. Spines of cupule branched, interlocking and usually obscuring surface of cupule; inflorescences staminate and androgynous (staminate below androgynous on same twig); pistillate flowers (and fruits) 1–3 or many per cupule (rarely but not consistently 2); staminate inflorescences spicate, rigid or flexible; nut angular or rounded, not winged.
        4. Plants evergreen; leaves thick, leathery, margins entire (rarely spinose in sprouts), secondary veins obscure, not strongly parallel; adjacent nuts separated from each other by internal cupule valves; bud scales imbricate; spines of cupule without simple hairs, with large, yellowish, multicellular glands; styles 3. . . . . 1. *Chrysolepis*, p. 438
        4. Plants winter-deciduous; leaves thin, somewhat leathery, secondary veins prominent, parallel, ending in prominent marginal teeth or awns; adjacent nuts not separated by cupule valves within cupule; buds with 2 unequal opposite outer scales that cover several imbricate inner scales; spines of cupule densely or sparely clothed with simple hairs; styles 6 or more . . . . 2. *Castanea*, p. 439

1. CHRYSOLEPIS Hjelmquist, Bot. Not., Suppl. 2(1): 117. 1948 · Western chinkapin [Greek *chrysos*, gold, and *lepis*, scale, referring to yellow glands on various organs of the plant]

Trees or shrubs, evergreen. **Terminal buds** present, ovoid or subglobose, scales imbricate. **Leaves:** stipules prominent on new growth, often persistent around buds. **Leaf blade** thick, leathery, margins entire or obscurely toothed, secondary veins obscure, branching and anastomosing before reaching margin. **Inflorescences** staminate or androgynous, axillary, clustered at ends of branches, spicate, ascending, rigid or flexible; androgynous inflorescences with pistillate cupules/flowers toward base and staminate flowers distally. **Staminate flowers:** sepals distinct; stamens (6–)12(–18), typically surrounding indurate pistillode covered with silky hairs. **Pistillate flowers** (1–)3 or more per cupule; sepals distinct; carpels and styles typically 3. **Fruits:** maturation in 2d year following pollination (termed biennial by many authors); cupule 2–several-valved, valves distinct, completely enclosing nuts, densely spiny, spines irregularly branched, interlocking, without simple hairs, with large, yellowish, multicellular glands; nuts (1–)3–several per cupule, 3-angled to rounded in cross section, not winged, adjacent nuts separated from each other by internal cupule valves. $x = 12$.

Species 2 (2 in the flora): w United States.

Nuts are sweet and edible but difficult to remove from the spiny cupules unless completely ripe. The two species of *Chrysolepis* have sometimes been included in *Castanopsis;* the latter, however, is a related genus of Fagaceae, native to Asia, with very different cupule structure (H. Hjelmquist 1948; L. L. Forman 1966).

1. Leaf apex acute or acuminate; trees or erect shrubs, bark thick, rough . . . . 1. *Chrysolepis chrysophylla*
1. Leaf apex obtuse, occasionally somewhat acute; low rhizomatous shrubs, bark thin, smooth . . . . .
. . . . 2. *Chrysolepis sempervirens*

1. Chrysolepis chrysophylla (Douglas ex Hooker) Hjelmquist, Bot. Not., Suppl. 2(1): 117. 1948 · Giant golden chinkapin [E]

Castanea chrysophylla Douglas ex Hooker, Fl. Bor.-Amer. 2: 159. 1838; *Castanopsis chrysophylla* (Douglas ex Hooker) A. de Candolle

Trees or erect shrubs, to 45 m. **Bark** thick, rough. **Twigs** densely covered with tight, yellowish, peltate trichomes. **Leaves:** petiole 5–8 mm. **Leaf blade** flat or folded upward along midvein, margins entire, apex acute or acuminate, surfaces abaxially usually golden or brownish with dense glandular trichomes. **Fruits:** cupule yellowish, 20–60 mm thick, densely spiny, surface obscured; nut light brown, (6–)8–12(–15) mm, glabrous, completely enclosed by cupule until maturity.

Varieties 2 (2 in the flora): w United States.

The two varieties intergrade considerably where they are in contact.

1. Tree; leaves typically flat. . . . .
. . . . 1a. *Chrysolepis chrysophylla* var. *chrysophylla*
1. Shrub; leaves folded upward along midvein . . . .
. . . . 1b. *Chrysolepis chrysophylla* var. *minor*

1a. Chrysolepis chrysophylla (Douglas ex Hooker) Hjelmquist var. **chrysophylla** [E]

Trees, to 30(–45) m. **Leaves** typically flat.

Flowering summer (Jun–Sep). Redwood forest, mixed evergreen forest, conifer forest; 0–2000 m; Calif., Oreg., Wash.

Chrysolepis chrysophylla var. *chrysophylla* grows mostly near the coast, extending inland to the Sierra Nevada.

1b. Chrysolepis chrysophylla (Douglas ex Hooker) Hjelmquist var. **minor** (Bentham) Munz, Suppl. Calif. Fl., 120. 1968 [E]

Castanea chrysophylla var. *minor* Bentham, Pl. Hartw., 377. 1857

Shrubs, erect, to 2–5(–10) m. **Leaves** strongly folded upward along midvein.

Flowering summer (Jun–Sep). Rocky or gravelly open slopes, conifer forest, closed cone pines and chaparral; 300–1800 m; Calif., Oreg.

2. Chrysolepis sempervirens (Kellogg) Hjelmquist, Bot. Not. 113: 377. 1960 · Bush golden chinquapin, Sierra chinkapin E F

Castanea sempervirens Kellogg, Proc. Calif. Acad. Sci. 1: 71. 1855; *Castanopsis sempervirens* (Kellogg) Dudley ex Merriam

Shrubs, rhizomatous-spreading, 0.2–1(–2.5) m. **Bark** gray or brown, thin, smooth. **Twigs** densely covered with tight, yellowish, peltate trichomes. **Leaves:** petiole 10–15 mm. **Leaf blade** oblong to oblanceolate, to 75 mm, margins entire, apex obtuse, occasionally somewhat acute; surfaces abaxially rusty or golden pubescent, often becoming glabrate and glaucous with age. **Fruits:** cupule yellowish, 20–60 mm thick, densely spiny, surface obscured; nut light brown, 8–13 mm, glabrous, completely enclosed by cupule until maturity.

Flowering summer (Jul–Aug). Rocky slopes, chaparral, conifer forest, mostly at high elevations; 0–3300 m; Calif., Oreg.

2. CASTANEA Miller, Gard. Dict. Abr. ed. 4, vol. 3. 1754 · Chestnut, châtaignier [Classical Latin, from Greek *kastanaion karuon,* nut from Castania, probably referring either to Kastanaia in Pontus or Castana in Thessaly]

Trees or shrubs, winter-deciduous, sometimes rhizomatous. **Terminal buds** absent, pseudoterminal bud (axillary bud of youngest leaf) ovoid, with 2 unequal opposite outer scales enclosing several imbricate inner scales. **Leaves:** stipules prominent on new growth, soon deciduous. **Leaf blade** thin, somewhat leathery, secondary veins unbranched, ± parallel, extending to margin, each vein ending in sharp tooth or well-developed awn. **Inflorescences** staminate or androgynous, axillary, spicate, erect, rigid or flexible; androgynous inflorescences with pistillate cupules/flowers toward base and staminate flowers distally. **Staminate flowers:** sepals distinct; stamens 12(–18), typically surrounding indurate pistillode covered with silky hairs. **Pistillate flowers** 1–3 per cupule; sepals distinct; carpels and styles typically 6(–9). **Fruits:** maturation in 1st year following pollination (termed annual by many authors); cupule 2–4-valved, valves connate marginally until maturity, ± completely enclosing nut(s), spiny, spines irregularly branched, often interlocking, densely or sparsely covered in simple hairs; nuts 1–3 per cupule, plano-convex, or if 3, then central nut often reduced and flattened, or if solitary, then often rounded in cross section, not winged, adjacent nuts not separated by internal cupule valves. $x = 12$.

Species ca. 8–10 (3 in the flora, often interpreted as 2): North America, Europe, Asia.

As evidenced by United States breeding programs, all species are probably interfertile (including American × Asian species). Local morphologic intergradation between species is to be expected.

SELECTED REFERENCES Hardin, J. W. and G. P. Johnson. 1985. Atlas of foliar surface features in woody plants, VIII. *Fagus* and *Castanea* (Fagaceae) of eastern North America. Bull. Torrey Bot. Club 112: 11–20. Johnson, G. P. 1988. Revision of *Castanea* sect. *Balanocastanon* (Fagaceae). J. Arnold Arbor. 69: 25–49. Paillet, F. L. 1993. Growth form and life histories of American chestnut and Allegheny and Ozark chinquapin at various North American sites. Bull. Torrey Bot. Club 120: 257–268. Paillet, F. L. and P. A. Rutter. 1989. Replacement of native oak and hickory tree species by the introduced American chestnut *Castanea dentata* in southwestern Wisconsin, USA. Canad. J. Bot. 67: 3457–3469. Tucker, G. E. 1975. *Castanea pumila* var. *ozarkensis* (Ashe) Tucker, comb. nov. Proc. Arkansas Acad. Sci. 29: 67–69.

1. Cupules 4-valved, enclosing 3 flowers/fruits; leaf blade abaxially without stellate trichomes, appearing glabrous, with minute multicellular glands, these often embedded on blade, and simple trichomes on veins; nut obovate, flattened at least on 1 side, beak thin, flexible, to 8 mm or more excluding styles . . . . 1. *Castanea dentata*
1. Cupules 2-valved, enclosing 1 flower/fruit; leaf blade abaxially bearing stellate trichomes (occasionally visible only with magnification), often with simple trichomes on veins; nut round in cross section, ovoid-conic, beak less than 3 mm excluding styles.

2. Longest spines of cupule often exceeding 10 mm; young twigs glabrous; petiole usually 8–10(–15) mm; bark brownish, moderately to deeply fissured . . . . .2. *Castanea ozarkensis*
2. Longest spines of cupule usually less than 10 mm; young twigs puberulent (sometimes glabrate with age); petiole usually 3–7(–10) mm; bark gray to brown, smooth, not fissured or only shallowly fissured . . . . 3. *Castanea pumila*

1. Castanea dentata (Marshall) Borkhausen, Theor. Prakt. Handb. Forstbot. 1: 741. 1800 · American chestnut, châtaigner d'Amérique ⎡E⎤

Fagus-Castanea dentata Marshall, Arbust. Amer., 46. 1785

Trees, often massive, formerly to 30 m, now persisting mostly as multistemmed resprouts to 5–10 m because of widespread destruction by blight. **Bark** gray, smooth when young, furrowed in age. **Twigs** glabrous. **Leaves:** petiole (8–)10–30(–40) mm. **Leaf blade** narrowly obovate to oblanceolate, 90–300 × 30–100 mm, base cuneate, margins sharply serrate, each tooth triangular, gradually tapering to awn often more than 2 mm, apex acute or acuminate, surfaces abaxially often without stellate trichomes, appearing glabrous but with evenly distributed, minute, multicellular, embedded glands between veins and sparse, straight, simple trichomes concentrated on veins, stellate or tufted trichomes absent. **Staminate flowers** with conspicuous pistillodes, whitish or yellowish straight hairs in center of flower. **Pistillate flowers** 3 per cupule. **Fruits:** cupule 4-valved, enclosing 3 flowers/fruits, valves irregularly dehiscing along 4 sutures at maturity, spines of cupule essentially glabrous, with a few scattered simple trichomes; nuts 3 per cupule, obovate, 18–25 × 18–25 mm, flattened on 1 or both sides, beak to 8 mm excluding styles.

Flowering summer (Jun–Jul). Previously common in rich deciduous and mixed forests, particularly with oak; 0–1200 m; Ont.; Ala., Conn., Del., Fla., Ga., Ill., Ind., Ky., La., Maine, Md., Mass., Mich., Miss., Mo., N.H., N.J., N.Y., N.C., Ohio, Pa., R.I., S.C., Tenn., Vt., Va., W.Va., Wis.

The American chestnut was one of the most important dominant forest trees of eastern North America prior to 1930. The nuts, sweet and edible, were a favorite confection in the eastern United States. The wood is light, strong, and resistant to decay; it was widely used for construction, furniture, and decorative trim. The bark was used for tanning leather.

Native Americans used various parts of the plants of *Castanea dentata* medicinally as a cough syrup and to treat whooping cough, for heart trouble, and as a powder for chafed skin (D. E. Moerman 1986).

After 1930, most populations of *Castanea dentata* were nearly destroyed by the chestnut blight, caused by the introduced fungus *Cryphonectria parasitica* (Murrill) M. E. Barr [= *Endothia parasitica* (Murrill) P. J. Anderson & H. W. Anderson]. While chestnuts persist in many localities, the plants are mostly resprouts that rarely, if ever, produce viable seed.

Virtually all known natural populations remain infected with the blight, and various studies continue in an effort to find ways to improve growth and vitality of infected trees. The species was widely planted outside of its native range (e.g., Illinois, Indiana, Iowa, Michigan, and Wisconsin), and some of these plantings remain blight-free because of their isolation. One particularly large grove was planted near West Salem, Wisconsin, in 1880, and continuing regeneration through seedlings has been documented (F. L. Paillet and P. A. Rutter 1989). Unfortunately, chestnut blight has recently been discovered in this isolated population and probably is extending to other isolated plantings in the west.

As part of the effort to introduce blight-resistant strains of the American chestnut, breeding programs have produced hybrids of *Castanea dentata* in various combinations with exotic species of chestnut. These hybrids are often extremely difficult to identify because they may derive from as many as three parents in complicated backcrosses. When identifying trees suspected of being introduced, one should be aware of the three most commonly cultivated exotic chestnut species, all of which have been collected as sporadic escapes or persistent waifs:

Castanea sativa Miller • Spanish chestnut
Leaf blade abaxially with sparse to dense covering of stellate hairs, also with conspicuous glands as in *C. dentata.* Petiole relatively long (30 mm or more).
Castanea mollissima Blume • Chinese chestnut
Twigs with spreading hairs. Leaf blade abaxially cobwebby-pubescent, without conspicuous foliar glands found in *Castanea dentata.* Resistant to blight and widely cultivated in the United States, where it occasionally escapes.
Castanea crenata Siebold & Zuccarini • Japanese chestnut
Leaf blade abaxially with minute, glandular, peltate scales mixed with dense, tangled tomentum. Vegetatively, this species may be difficult to distinguish from *Castanea pumila; C. crenata* is typically a tree (as opposed to shrub), and it has three nuts per cupule in contrast to the solitary nut found in *C. pumila.*

Putative hybrids between *Castanea dentata* and *C. pumila* are known as *C.* ×*neglecta* Dode. These are rather widespread and difficult to separate from *C. pumila;* they tend to have few stellate trichomes and a greater proportion of glandular-bulbous trichomes on the leaves, along with intermediate leaf shape and size, and 1–2 nuts per cupule (G. P. Johnson 1988). The *C. ×neglecta* hybrids are known from scattered localities in Louisiana (probably not native), Tennessee, North Carolina, Virginia, Pennsylvania, Maryland, and New Jersey.

2. Castanea ozarkensis Ashe, Bull. Torrey Bot. Club 50: 360. 1923 · Ozark chinkapin [E]

Castanea arkansana Ashe; *C. pumila* Miller var. *ozarkensis* (Ashe) G. E. Tucker

Trees, occasionally shrubs, previously often massive, to 20 m, now rarely more than 10 m, mostly resprouting following blight. **Bark** brownish, deeply or moderately fissured. **Twigs** glabrous when young. **Leaves:** petiole usually (8–)10–15 mm. **Leaf blade** narrowly obovate or oblanceolate, (40–)120–200(–260) × 30–100 mm, base rounded to slightly cordate or slightly cuneate, margins sharply serrate, each lateral vein terminating in cuneate, gradually acuminate tooth with awn usually more than 2 mm, apex acute or acuminate; surfaces abaxially densely to sparsely covered with appressed, whitish, minute, stellate trichomes, sometimes essentially glabrate, especially on shade leaves, veins glabrous or with a few simple trichomes. **Pistillate flower** 1 per cupule. **Fruits:** cupule 2-valved, enclosing 1 flower/fruit, valves irregularly dehiscing along 2 sutures, longest spines usually more than 10 mm; nut 1 per cupule, oval-conic, 9–19 × 8–14 mm, round in cross section, not flattened, beak less than 3 mm excluding styles.

Flowering June. Deciduous forest; 150–600 m; Ala., Ark., La., Mo., Okla., Tex.

Castanea ozarkensis is concentrated in the Ozark Mountains, extending into the Ouachita Mountains (Arkansas) as well, where some intermediates with *C. pumila* may be found (G. E. Tucker 1975; G. P. Johnson 1988). Some authors have interpreted the putative hybrids as evidence to support inclusion of the Ozark chinkapin as a subspecies of *C. pumila.* Because virtually all chestnut species are interfertile, the occurrence of hybridization cannot be used as evidence of conspecificity, unless one is willing to accept a single chestnut species worldwide. The nature of the character differences between the Ozark populations and populations of *C. pumila* are substantial, and they are similar to differences seen between closely related species of *Quercus,* that nonetheless may hybridize locally. In many characteristics, *C. ozarkensis* differs from *C.*

pumila in the direction of *C. dentata* (e.g., tree habit, glabrous twigs, leaf shape and size), including its resprouting pattern following chestnut blight (F. L. Paillet 1993). Given the ability of the chestnut to hybridize, and its intermediacy toward *C. dentata,* the origin of the Ozark chinkapin as a mere geographic race of *C. pumila* is questionable, and for the present, *C. ozarkensis* is best treated as a separate species.

Populations from northern Alabama are apparently no longer extant, probably eliminated by the chestnut blight (G. P. Johnson 1988).

3. Castanea pumila Miller, Gard. Dict. ed. 8, Castanea no. 2. 1768 · Chinquapin, chinkapin, dwarf chestnut [E] [F]

Castanea alnifolia Nuttall; *C. alnifolia* var. *floridana* Sargent; *C. floridana* (Sargent) Ashe; *C. pumila* var. *ashei* Sudworth

Shrubs or trees, to 15 m, often rhizomatous. **Bark** gray to brown, smooth to slightly fissured. **Twigs** puberulent with spreading hairs, occasionally glabrate with age. **Leaves:** petiole 3–7(–10) mm. **Leaf blade** narrowly elliptic to narrowly obovate or oblanceolate, 40–210 × 20–80 mm, base rounded to cordate, margins obscurely to sharply serrate, each abruptly acuminate tooth with awn usually less than 2 mm; surfaces abaxially typically densely covered with appressed stellate or erect-woolly, whitish to brown trichomes, sometimes essentially glabrate especially on shade leaves, veins often minutely puberulent. **Pistillate flower** 1 per cupule. **Fruits:** cupule 2-valved, enclosing 1 flower, valves irregularly dehiscing along 2 sutures, longest spines usually less than 10 mm; nut 1 per cupule, ovate-conic, 7–21 × 7–19 mm, round in cross section, not flattened, beak less than 3 mm excluding styles.

Flowering spring (May–Jun). Forest, open woods, forest understory, dry sandy and wet sandy barrens; 0–1000 m; Ala., Ark., Del., Fla., Ga., Ind., Ky., La., Md., Mass., Miss., Mo., N.J., N.Y., N.C., Ohio, Okla., Pa., S.C., Tenn., Tex., Va., W.Va.

Castanea pumila was thought to be extirpated from Long Island, New York, but it was recently recollected in Suffolk County.

Numerous names have been applied to populations included here under *Castanea pumila* (see also *C. ozarkensis*). In general, the pattern of morphologic variation suggests three forms that have some geographic and ecologic continuity, but among which sufficient clinal variation occurs to prevent easy recognition. Plants of higher elevation and northern populations tend to be trees or large shrubs with densely tomentose vesture of the abaxial leaf surface,

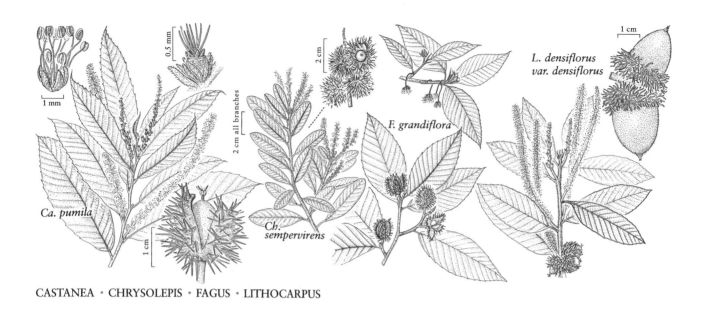

CASTANEA · CHRYSOLEPIS · FAGUS · LITHOCARPUS

while southern coastal populations typically have flat-stellate vestiture, often very sparse or even glabrate (e.g., *C. floridana*). These coastal populations may be separated into forms that are trees or large shrubs with leaves sparsely to not at all glandular versus low rhizomatous shrubs with prominent globose glands on twigs and leaves (centered in northern Florida and adjacent Georgia). The latter form is probably the same as *C. alnifolia*.

Various preparations of the leaves of *Castanea alnifolia* were used by Native Americans medicinally to relieve headaches and as a wash for chills and cold sweats; preparations from unspecified parts of the plants were used to treat fever blisters (D. E. Moerman 1986).

3. LITHOCARPUS Blume, Bijdr. Fl. Ned. Ind. 10: 526. 1826 · [Greek *lithos,* stone, and *carpos,* fruit, referring to the hard fruit wall]

Trees or shrubs, evergreen. **Terminal buds** present, ovate, all scales imbricate. **Leaves:** stipules prominent on new growth, persistent around buds. **Leaf blade** leathery, margins entire or obscurely toothed to serrate, secondary veins unbranched, ± parallel, extending to margin. **Inflorescences** staminate and androgynous, axillary, often appearing terminal and branched by reduction of leaves, spicate, erect or ascending, rigid or flexible; androgynous inflorescences with pistillate cupules/flowers toward base and staminate flowers distally. **Staminate flowers:** sepals distinct; stamens 12(–18 or more) typically surrounding indurated pistillode covered with silky hairs. **Pistillate flower** 1 per cupule; sepals distinct; carpels and styles 3. **Fruits:** maturation in 2d year following pollination; cupule cup-shaped, without any indication of valves, covering proximal portion of nut, scaly, spines absent, scales strongly reflexed, hooked at tip; nut 1 per cupule, round in cross section, not winged. $x = 12$.

Species 100–200 (1 in the flora): North America, e Asia.

Although fruit of *Lithocarpus* closely resembles that of *Quercus,* the two genera differ in characters of the inflorescence, flowers, and pollen. These characters indicate that *Lithocarpus* is more closely related to *Castanea, Chrysolepis,* and other Asian genera of subfamily Castaneoideae than to *Quercus,* and the similarity in fruit is because of convergence.

1. Lithocarpus densiflorus (Hooker & Arnott) Rehder in L. H. Bailey, Stand. Cycl. Hort. 6: 3569. 1917 · Tanoak, tanbark-oak [E]

Quercus densiflora Hooker & Arnott, Bot. Beechey Voy., 391. 1841

Shrubs or trees, to 20(–45) m. **Bark** gray or brown, smooth or deeply furrowed. **Twigs** densely yellowish tomentose. **Leaf blade** adaxially convex, to 60–120 mm, leathery to brittle, margins often revolute, regularly toothed, teeth prominent to obscure; surfaces abaxially prominently and densely woolly, often glabrate at maturity, revealing gray or bluish green waxy surface, veins often distally impressed. **Fruits:** cup scales subulate, spreading to strongly recurved, hooked; nut yellowish brown, globose to cylindric-tapered, to 15–35 mm, extremely hard, densely tomentose, eventually glabrate.

Varieties 2 (2 in the flora): w United States.

Sterile specimens of *Lithocarpus densiflorus* are often confused with *Chrysolepis* and vice versa. Nonfruiting material of *L. densiflorus* is recognizable by the loose tomentose pubescence of the leaves and inflorescences (although the leaves are often glabrate with age). *Chrysolepis* lacks this tomentose pubescence and has only a tight vestiture of glandular-peltate trichomes, except for some stellate and straight simple trichomes associated with the flowers.

The Costanoan used infusions prepared from the bark of *Lithocarpus densiflora* (no varieties specified) as a wash for facial sores and to tighten loose teeth (D. E. Moerman 1986).

1. Trees, 20(–45) m at maturity; leaves to 120 mm, convex adaxially, secondary veins conspicuous and strongly impressed adaxially. . . . .
. . . . 1a. *Lithocarpus densiflorus* var. *densiflorus*

1. Shrubs, 3 m or less at maturity; leaves 60 mm or less, flat, secondary veins inconspicuous and not strongly impressed adaxially. . . . .
. . . . 1b. *Lithocarpus densiflorus* var. *echinoides*

1a. Lithocarpus densiflorus (Hooker & Arnott) Rehder var. **densiflorus** [E] [F]

Trees, to 20(–45) m. **Twigs** densely tomentose. **Leaves** convex adaxially, to 120 mm, veins often impressed adaxially. **Fruits:** cup scales spreading to strongly recurved.

Flowering summer (Jul–Aug). Mixed evergreen forest and redwood forest; 0–1500 m; Calif., Oreg.

1b. Lithocarpus densiflorus (Hooker & Arnott) Rehder var. **echinoides** (R. Brown ter) Abrams in L. Abrams and R. S. Ferris, Ill. Fl. Pacific States 1: 516. 1923 · Dwarf tanbark [E]

Quercus echinoides R. Brown ter, Ann. Mag. Nat. Hist., ser. 4, 7: 251. 1871

Shrubs, ascending, to 2(–3) m. **Twigs** densely tomentose. **Leaf blade** usually flat, to 60 mm, veins not deeply impressed. **Fruits:** cup scales usually strongly recurved.

Flowering summer (Jul–Aug). Open conifer forest, margins of woods, and dry slopes; 600–2200 m; Calif., Oreg.

Lithocarpus densiflorus var. *echinoides* is found in scattered locations in the Siskiyou region and the Sierra Nevada south to Mariposa County, California.

4. FAGUS Linnaeus, Sp. Pl. 2: 997. 1753; Gen. Pl. ed. 5, 432. 1754 · Beech, hêtre, haya [Classical Latin name, from Greek *figos,* an oak with edible acorns, probably from Greek *fagein,* to eat]

Trees, winter-deciduous. **Terminal buds** present, long, tapered in maturity, all scales imbricate. **Leaves:** stipules prominent on new growth, soon deciduous. **Leaf blade** thin, secondary veins unbranched, ± parallel, extending to margin, each vein ending in acute or obscure tooth. **Inflorescences** unisexual, axillary in new growth leaves; staminate inflorescence lax, loosely capitate cluster of flowers; pistillate inflorescence short, stiff, cupule 1, terminal. **Staminate flowers:** sepals connate; stamens 6–16; pistillode typically absent. **Pistillate flowers** 2 per cupule; sepals distinct; carpels and styles 3. **Fruits:** maturation in 1st year following pollination;

cupule 4-valved, valves distinct, ± completely enclosing nuts until maturity, prickly, prickles stout, unbranched, short, not obscuring surface of cupule, internal valves absent; nuts 2 per cupule, sharply 3-angled, slightly winged. $x = 12$.

Species 8–10 (1 in the flora): temperate, subtropical, and montane tropical forests, North America (e United States), Mexico, Europe, Asia.

SELECTED REFERENCES Cooper, A. W. and E. P. Mercer. 1977. Morphological variation in *Fagus grandifolia* Ehrh. in North Carolina. J. Elisha Mitchell Sci. Soc. 93: 136–149. Hardin, J. W. and G. P. Johnson. 1985. Atlas of foliar surface features in woody plants, VIII. *Fagus* and *Castanea* (Fagaceae) of eastern North America. Bull. Torrey Bot. Club 112: 11–20. Rehder, A. J. 1907. Some new or little known forms of New England trees. Rhodora 9: 109–116.

1. Fagus grandifolia Ehrhart, Beitr. Naturk. 3: 22. 1788
· American beech, hêtre américain [E] [F]

Fagus grandifolia var. *caroliniana* (Loudon) Fernald & Rehder

Trees, to 20(–30) m. **Bark** gray, smooth. **Twigs** glabrous at maturity, or with scattered, straight, silky, simple hairs, prominent ringlike bud scale scars at beginning of each year's growth. **Buds** narrowly fusiform, to 15–20 mm, apex acute, scales few, silky light brown or glabrous. **Leaves:** petiole 4–12 mm. **Leaf blade** ovate or narrowly ovate, rarely obovate, 60–120 × 25–75 mm, base cuneate or subacute, apex acuminate; surfaces abaxially with scattered straight silky hairs, these often concentrated on midrib, occasionally glabrous or much more villous. **Fruits:** cupule brown to reddish brown, 15–20(–25) mm, opening at maturity to reveal nuts; nut 15–20 × 10–18 mm wide, glabrous or puberulent, often hollow even when full-sized.

Flowering spring (Apr–Jun). Rich woods, deciduous forest and mixed broadleaf–conifer forest; 0–1000 m; N.B., N.S., Ont., P.E.I., Que.; Ala., Ark., Conn., Del., Fla., Ga., Ill., Ind., Ky., La., Maine, Md., Mass., Mich., Miss., Mo., N.H., N.J., N.Y., N.C., Ohio, Okla., Pa., R.I., S.C., Tenn., Tex., Vt., Va., W.Va., Wis.

A. J. Rehder (1907) argued for recognition of a southern variety (*Fagus grandifolia* var. *caroliniana*) of this somewhat variable species. The northern populations in general are characterized by cupules with denser, longer prickles, somewhat narrower leaves with a greater proportion of cuneate leaf bases, and larger fruits that exceed the cupules slightly. Others (e.g., W. H. Camp 1951) have suggested the existence of three races within United States *F. grandifolia*, often referred to as gray beech, red beech, and white beech. I follow J. W. Hardin and G. P. Johnson (1985) and others in not recognizing subspecific or varietal variation within eastern United States *F. grandifolia*. Examination of material over the geographic range of the species suggests that this variation is broadly clinal and can only be demonstrated statistically, with much variation indicative of the other races within most populations. It seems a matter of taste as to whether such variation be recognized with infraspecific names. In addition, forms with more densely pubescent leaves occur locally in both the north and south; they have been formally recognized by some authors. Clearly, additional taxonomic work on variation in *F. grandifolia* is desirable; it is possible that other characters that could adequately delimit subgeneric/varietal taxa might yet be identified.

Fagus sylvatica Linnaeus, the European Beech, is cultivated in temperate eastern North America and to a lesser extent in western United States and Canada. Escapes are to be expected. Various cultivars are known, particularly purple-leaf, tricolor-leaf, and cut-leaf forms. When encountered, *F. sylvatica* is easily distinguished from *F. grandifolia* by the crenate leaf margin (without distinct teeth) and the softer, less stout, less reflexed spines of the cupule of *F. sylvatica*.

Native Americans used various preparations from plants of *Fagus sylvatica* medicinally for worms, consumption, chancre, and heart trouble, to purify the blood, as a poultice for burns and scalds, and as a wash for poison ivy (D. E. Moerman 1986).

5. QUERCUS Linnaeus, Sp. Pl. 2: 994. 1753; Gen. Pl. ed. 5, 431. 1754 · Oak, chêne

[Classical Latin for the English oak, *Quercus robur,* from some central European language]

Kevin C. Nixon

Trees or shrubs, evergreen or winter-deciduous, sometimes rhizomatous. **Terminal buds** spheric to ovoid, terete or angled, all scales imbricate. **Leaves:** stipules deciduous and inconspicuous (except in *Quercus sadleriana*). **Leaf blade** lobed or unlobed, thin or leathery, margins entire, toothed, or awned-toothed, secondary veins either unbranched, ± parallel, extending to margin, or branching and anastomosing before reaching margin. **Inflorescences** unisexual, in axils of leaves or bud scales, usually clustered at base of new growth; staminate inflorescences lax, spicate; pistillate inflorescences usually stiff, with terminal cupule and sometimes 1–several sessile, lateral cupules. **Staminate flowers:** sepals connate; stamens (2–)6(–12), surrounding tuft of silky hairs (apparently a reduced pistillode). **Pistillate flower** 1 per cupule; sepals connate; carpels and styles 3(–6). **Fruits:** maturation annual or biennial; cup variously shaped (saucer- to cup- or bowl- to goblet-shaped), without indication of valves, covering base of nut (rarely whole nut), scaly, scales imbricate or reduced to tubercles, not or weakly reflexed, never hooked; nut 1 per cup, round in cross section, not winged. $x = 12$.

Species ca. 400 (90 in the flora): North America, Mexico, West Indies, Central America, South America (Colombia only), Eurasia, n Africa.

Quercus is without doubt one of the most important woody genera of the Northern Hemisphere. Historically, oaks have been an important source of fuel, fodder, and building materials throughout their range. Other products include tannins and dyes, and oak bark and leaves were often used for tanning leather. Acorns were historically an important food for indigenous people in North America, Central America, Europe, and Asia. In some areas, acorn consumption is still important, but in general, because of the intense preparation necessary to remove tannins and strong flavor of acorn products, they have fallen out of use as human food in developed areas. They do remain, however, an important mast for wildlife and domesticated animals in many rural areas.

Among the most important diagnostic characters within *Quercus,* and particularly the white oak group (*Quercus* sect. *Quercus*), are features of the foliar trichomes. Often these can be seen with a 10× or 15× hand lens; higher magnifications are sometimes required and are useful particularly when characters for a species or complex are first studied and mastered for later use in the field. Although these microscopic characters may seem intimidating, the alternative characters of leaf shape and dentition, so often used in the field, are unreliable in many cases. The large number of misidentified specimens in herbaria that can be easily identified properly with the use of trichome characters illustrates this point. Additionally, many specimens are encountered, both in field and herbarium, that lack fruit or have only immature fruit. Very few species require mature fruit for proper diagnosis; most can be adequately identified with a representative selection of mature sun leaves attached, if possible, to twigs with mature buds. The combination of leaf vesture, form of the margin (entire, lobed, toothed, spinose), twig vesture, and bud form and vesture constitute the majority of diagnostic features minimally required at species level.

Staminate floral and inflorescence characters have not been used to any significant extent in the taxonomy of *Quercus.* Immature, flowering material is often difficult to identify with certainty, and floral features such as number and form of sepals, number of stamens, and pubescence of flowers or floral rachises seem to vary independently of species affinity within many groups. Because of these problems, descriptions of staminate features are excluded in this treatment as unreliable and of little diagnostic value. When collecting flowering oaks, make a point of gathering fallen fruit and mature leaves carefully from the ground, if available, and revisit such populations again when mature material is available to verify identifications.

The character of acorn maturation in the first year (annual maturation) or second year (biennial maturation) after pollination is commonly used to differentiate major groups within *Quercus*. All North American white oaks have annual maturation; all the *Protobalanus* group have biennial maturation; and the vast majority of red oaks have biennial maturation, with one eastern North American and a few western species having annual maturation. In the field, this character can be observed throughout the growing season by examining a sample of twigs from the same tree. If developing fruits exhibit a single size class and are found only on the current year's growth, maturation is annual; if the developing fruits exhibit two size classes with small pistillate flowers on new growth and larger developing fruit on the previous year's twigs, maturation is biennial. In *Quercus* sect. *Protobalanus,* biennial maturation may be mistaken for annual maturation because all the species are fully evergreen, and often the twigs bearing fruit do not produce new growth in the second year after pollination. In such cases, careful examination of a broad sample of twigs from within one tree and throughout a population will verify biennial maturation. Herbarium specimens are sometimes inadequate for this determination.

Hybridization among species of *Quercus* has been widely documented and even more widely suspected. An astounding number of hybrid combinations have been reported in the literature, and many of these have been given species names, either before or after their hybrid status was known (E. J. Palmer 1948). Hybrids are known to occur in the wild only between members of the same section, and attempts at artificial crosses between species from different sections or subgenera within *Quercus* have failed with very few exceptions (W. P. Cottam et al. 1982).

Hybridization in most cases results in solitary unusual trees or scattered clusters of intermediate individuals (J. W. Hardin 1975). In some cases, however, populations of fairly broad distribution and extreme variability, often with a majority of intermediate types, may occur. Such instances occur in both the red oak and white oak groups, and to a lesser extent in the *Protobalanus* group.

When dealing with a suspected hybrid, therefore, one should first consider the possibility of intraspecific variation or environmental plasticity, and then seek parentage among sympatric members of the same section. Because of the almost infinite number of possible hybrid combinations, and the myriad names applied to them, only those that appear to be prominent either locally or in widespread areas are dealt with here. The interested reader is referred to various discussions of oak hybridization in the literature (e.g., E. J. Palmer 1948; J. W. Hardin 1975).

SELECTED REFERENCES Hunt, D. M. 1989. A Systematic Review of *Quercus* Series *Laurifoliae, Marilandicae* and *Nigrae.* Ph.D. dissertation. University of Georgia. Muller, C. H. 1951. The oaks of Texas. Contr. Texas Res. Found., Bot. Stud. 1: 21–323. Muller, C. H. 1961. The live oaks of the series *Virentes.* Amer. Midl. Naturalist 65: 17–39. Nixon, K. C. 1993. Infrageneric classification of *Quercus* (Fagaceae) and typification of sectional names. Ann. Sci. Forest. 50(suppl.1): 25–34. Nixon, K. C. 1993b. The genus *Quercus* in Mexico. In: T. P. Ramamoorthy et al., eds. 1993. Biological Diversity of Mexico: Origin and Distribution. New York. Pp. 447–458. Palmer, E. J. 1948. Hybrid oaks of North America. J. Arnold Arbor. 29: 1–48. Sargent, C. S. 1918. Notes on North American trees. I. *Quercus.* Bot. Gaz. 65: 423–459. Tillson, A. H. and C. H. Muller. 1942. Anatomical and taxonomic approaches to subgeneric segregation in American *Quercus.* Amer. J. Bot. 29: 523–529. Trelease, W. 1924. The American oaks. Mem. Natl. Acad. Sci. 20: 1–255.

Key to North American sections of *Quercus*

1. Mature bark smooth or deeply furrowed (but see also *Q. montana*), not scaly or papery; cup scales flattened, rarely tuberculate, never embedded in tomentum; leaf blade if lobed then with awned teeth, if entire then often with bristle at apex; endocarp densely tomentose or silky over entire inner surface of nut. . . . . .5a. *Quercus* sect. *Lobatae* (red or black oaks), p. 447
1. Mature bark scaly or papery, rarely deeply furrowed; cup scales thickened, tuberculate (with darkened or hairy callous at base of each scale), or embedded in tomentum; leaf blade if lobed without awned teeth, if unlobed without bristle at apex, merely mucronate; endocarp glabrous or somewhat sparsely and irregularly hairy, not densely tomentose over entire inner surface of nut.
    2. Acorn maturation biennial, nut requiring 2 seasons to mature; cup scales embedded in tawny or glandular tomentum, only scale tips visible; plants evergreen, leaves leathery, abaxially glaucous; California, Oregon, and Arizona. . . . .
    . . . . 5b. *Quercus* sect. *Protobalanus* (intermediate oaks, golden oaks), p. 468
    2. Acorn maturation annual, nut requiring 1 growing season to mature; cup scales not embedded in tawny or glandular tomentum, although scales may be tomentose; plants variously evergreen or deciduous; widespread . . . . 5c. *Quercus* sect. *Quercus* (white oaks), p. 471

5a. Quercus Linnaeus sect. **Lobatae** Loudon, Hort. Brit., 385. 1830 · Red or black oaks

Richard J. Jensen

Trees or shrubs, evergreen or deciduous. **Bark** gray to dark brown or black, smooth or furrowed. **Leaf blade** lobed or unlobed, margins entire or toothed, teeth if present usually bristle-tipped. **Staminate flowers:** calyx 2–6-lobed; anthers usually retuse, rarely apiculate. **Pistillate flowers:** calyx free from ovary, forming free skirt or flange; styles linear-spatulate. **Acorns:** maturation biennial, rarely annual; cup with scales distinct, flat, thin, rarely somewhat keeled or tuberculate; nut with inner wall silky-tomentose, abortive ovules apical (rarely in some species variable in position or subbasal), seed coat adhering to seed at maturity. **Cotyledons** distinct, rarely partially connate.

Species ca. 195 (35 in the flora): North America, Mexico, Central America, South America (Colombia only).

Cup dimensions in *Quercus* sect. *Lobatae* refer to height of the cup as seen in profile view. This measurement, taken from apex of peduncle to cup rim, has been used for those situations (quite common in herbarium specimens and fresh collections) in which the cup cannot be separated from the nut without seriously damaging one or both of these units. The depth of the cup will always be less than its height, but a simple translation of height to depth does not exist.

Key to species in *Quercus* sect. *Lobatae*

1. Fruits annual (maturing at end of 1st season); if fruits not available, key as if biennial.
    2. Leaf blade abaxially glabrous except for tufts of tomentum in vein axils or at base.
        3. Leaf blade distinctly convex, adaxially rugose; California . . . . .2. *Quercus agrifolia*
        3. Leaf blade planar, adaxially not rugose; Arizona to w Texas. . . . . .4. *Quercus emoryi*
    2. Leaf blade abaxially uniformly pubescent or tomentose, sometimes glabrate.
        4. Leaf blade less than 3 times as long as wide; nut ovoid or conic, 15–35 mm; California . . . . . .2. *Quercus agrifolia*
        4. Leaf blade more than 3 times as long as wide; nut globose or oblong, 9–15 mm; Arizona and east.
            5. Leaf blade abaxially tawny or white-tomentose; cup covering 1/3 nut or less; Arizona to w Texas. . . . . .6. *Quercus hypoleucoides*
            5. Leaf blade abaxially glabrate or gray- to brown-pubescent; cup covering 1/3–2/3 nut; Mississippi to North Carolina. . . . . .16. *Quercus pumila*

1. Fruits biennial (maturing at end of 2d season) except in *Q. agrifolia*, *Q. emoryi*, *Q. hypoleucoides*, and *Q. pumila*.
     6. Evergreen trees or shrubs; leaf blade adaxially noticeably rugose, margins entire or spinose, strongly revolute, abaxially densely tawny or white-tomentose. . . . . .6. *Quercus hypoleucoides*
     6. Plants without above combination of characters.
          7. Petiole less than 10 mm.
               8. Leaf blade more than 4.5 times as long as wide.
                    9. Petiole glabrous. . . . . .14. *Quercus phellos*
                    9. Petiole pubescent.
                         10. Low shrubs. . . . . .16. *Quercus pumila*
                         10. Trees.
                              11. Leaf blade narrowly lanceolate, abaxially glabrous or with tufts of tomentum at base; sc Arizona. . . . . 5. *Quercus viminea*
                              11. Leaf blade elliptic or ovate, abaxially uniformly pubescent or tomentose; e Texas to North Carolina . . . . 11. *Quercus incana*
               8. Leaf blade less than 4.5 times as long as wide.
                    12. Leaf margins entire or merely spinose.
                         13. Abaxial leaf surface uniformly pubescent or tomentose.
                              14. Leaf blade convex, adaxially pubescent or tomentose . . . . .2. *Quercus agrifolia*
                              14. Leaf blade planar, adaxially essentially glabrous.
                                   15. Leaf blade more than 2.5 times as long as wide.
                                        16. Low shrubs; leaf margins revolute . . . . .16. *Quercus pumila*
                                        16. Trees; leaf margins not revolute.
                                             17. Petioles and twigs pubescent; inner surface of cup uniformly pubescent . . . . 11. *Quercus incana*
                                             17. Petioles and twigs glabrous; inner surface of cup glabrous or with only ring of hairs around nut scar . . . . 17. *Quercus imbricaria*
                                   15. Leaf blade less than 2.5 times as long as wide.
                                        18. Leaf blade widest distal to middle, apex broadly obtuse to rounded; e Texas to Georgia . . . . 23. *Quercus arkansana*
                                        18. Leaf blade widest at or proximal to middle, apex acute; North Carolina to Maine . . . . .25. *Quercus ilicifolia*
                         13. Abaxial leaf surface glabrous or sparingly pubescent.
                              19. Leaf blade distinctly convex, adaxially rugose.
                                   20. Apex of leaf blade blunt to attenuate; nut more than 15 mm; California. . . 2. *Quercus agrifolia*
                                   20. Apex of leaf blade obtuse or rounded; nut less than 15 mm; Florida . . . . 19. *Quercus inopina*
                              19. Leaf blade planar, not adaxially rugose.
                                   21. Trees or shrubs evergreen (often tardily deciduous in *Q. laurifolia*, *Q. hemisphaerica*, and *Q. nigra*).
                                        22. Twigs pubescent.
                                             23. Nut 21–44 mm; California . . . . .1. *Quercus wislizeni*
                                             23. Nut less than 20 mm; Arizona and east.
                                                  24. Base of leaf blade cuneate to rounded; Mississippi to South Carolina . . . . .18. *Quercus myrtifolia*
                                                  24. Base of leaf blade cordate; Arizona to w Texas. . . . . .4. *Quercus emoryi*
                                        22. Twigs glabrous.
                                             25. Leaf blade widest near apex. . . . . 15. *Quercus nigra*
                                             25. Leaf blade widest at or proximal to middle.
                                                  26. Leaf blade rarely 3 times as long as wide; nut 21–44 mm; California . . . . .1. *Quercus wislizeni*
                                                  26. Leaf blade commonly more than 3 times as long as wide; nut 8–16 mm; e Texas to Virginia.
                                                       27. Leaf blade leathery, base obtuse or rounded, apex acute or acuminate; trees on dry, sandy uplands . . . . . .13. *Quercus hemisphaerica*
                                                       27. Leaf blade thin, base attenuate or cuneate, apex obtuse or rounded; trees on low, wet flood plains and bottoms . . . . 12. *Quercus laurifolia*
                                   21. Trees deciduous (often tardily deciduous in *Q. laurifolia*, *Q. hemisphaerica*, and *Q. nigra*).

28. Twigs pubescent.
    29. Leaf blade widest distal to middle, apex broadly obtuse to
        rounded; e Texas to Georgia . . . . 23. *Quercus arkansana*
    29. Leaf blade widest proximal to middle, apex acute or long-
        attenuate; w Texas . . . . 8. *Quercus robusta*
28. Twigs glabrous.
    30. Leaf blade widest near apex. . . . . 15. *Quercus nigra*
    30. Leaf blade widest at or proximal to middle.
        31. Leaf blade leathery, base obtuse or rounded, apex acute or
            acuminate; trees on dry, sandy uplands. . . . . .13. *Quercus hemisphaerica*
        31. Leaf blade thin, base attenuate or cuneate, apex obtuse or
            rounded; trees on low, wet flood plains and bottoms. . . . . 12. *Quercus laurifolia*
12. Leaf margins shallowly or deeply lobed, not merely spinose.
    32. Terminal buds 6–12 mm.
        33. Twigs glabrous . . . . 15. *Quercus nigra*
        33. Twigs pubescent.
            34. Leaf blade widest near middle, base attenuate to acute, blade decurrent;
                margin of cup involute. . . . . .22. *Quercus laevis*
            34. Leaf blade widest near apex, base rounded to cordate, blade not decur-
                rent; margin of cup not involute. . . . . 24. *Quercus marilandica*
    32. Terminal buds 2–6 mm.
        35. Twigs pubescent.
            36. Apex of leaf broadly obtuse to rounded; e Texas to Long Island.
                37. Terminal buds glabrous or only ciliate on scale margins . . . . 23. *Quercus arkansana*
                37. Terminal buds uniformly pubescent. . . . . 24. *Quercus marilandica*
            36. Apex of leaf acute to attenuate; w Texas.
                38. Base of leaf blade subcordate to broadly obtuse, awns usually fewer
                  than 10 . . . . 8. *Quercus robusta*
                38. Base of leaf blade rounded to cuneate, awns usually more than 10 . . . 10. *Quercus gravesii*
        35. Twigs glabrous.
            39. Leaf blade widest distal to middle.
                40. Apex of leaf blade obtuse to rounded or blunt. . . . . 15. *Quercus nigra*
                40. Apex of leaf blade oblong to acute. . . . . .27. *Quercus georgiana*
            39. Leaf blade widest at or proximal to middle.
                41. Petiole rarely more than 5 mm; margins of leaf blade rarely with
                  more than 4 awns; trees evergreen or tardily deciduous . . . . . .13. *Quercus hemisphaerica*
                41. Petiole commonly more than 5 mm; margins of leaf blade often with
                  more than 4 awns; trees deciduous.
                  42. Mature leaf blade abaxially with conspicuous (readily discernible
                    to naked eye) tufts of tomentum in vein axils; bark gray to light
                    brown, scaly; Alabama to South Carolina. . . . . .27. *Quercus georgiana*
                  42. Mature leaf blade abaxially with minute (often detectable only
                    with magnification) tufts of tomentum in vein axils (or such tufts
                    absent); bark brown to black, roughly furrowed; w Texas.
                    43. Base of leaf blade subcordate to broadly obtuse, awns usu-
                      ally fewer than 10 . . . . 8. *Quercus robusta*
                    43. Base of leaf blade rounded to cuneate, awns usually more
                      than 10. . . . . 10. *Quercus gravesii*
7. Petiole more than 10 mm.
    44. Leaf margins entire or spinose or with 3–7 lobes separated by shallow (less than 1/3
        distance to midrib) sinuses; if lobed, each lobe generally with single awn.
        45. Leaf blade abaxially pubescent to tomentose.
            46. Plants evergreen; California or w Texas.
                47. Leaf blade convex; California . . . . .2. *Quercus agrifolia*
                47. Leaf blade planar; w Texas. . . . . .9. *Quercus tardifolia*
            46. Plants deciduous; e Texas to New England.
                48. Leaf blade 2.5–4 times as long as wide; twigs glabrous . . . . 17. *Quercus imbricaria*
                48. Leaf blade rarely as much as 2.5 times as long as wide; twigs pubescent.
                  49. Terminal buds uniformly pubescent, generally exceeding 5 mm.
                    50. Petiole mostly more than 20 mm; tips of cup scales appressed. . . . . . 20. *Quercus falcata*

50. Petiole rarely as much as 20 mm; tips of cup scales loose, especially
at cup margin... 24. *Quercus marilandica*
49. Terminal buds glabrous or hairy only at apex, rarely as much as 5 mm.
   51. Apex of leaf blade broadly obtuse or rounded, margins entire or
with 2–3 rounded lobes; e Texas to Georgia .... 23. *Quercus arkansana*
   51. Apex of leaf blade acute or obtuse, margins always with 3–4 acute
lobes; North Carolina to Maine........................ 25. *Quercus ilicifolia*
45. Leaf blade abaxially glabrous except for tufts of tomentum in vein axils, especially
along midrib.
  52. Leaf blade commonly more than 3 times as long as wide............... 7. *Quercus graciliformis*
  52. Leaf blade rarely 3 times as long as wide.
    53. Leaf blade convex, adaxially rugose .... 2. *Quercus agrifolia*
    53. Leaf blade planar, not adaxially rugose.
      54. Plants evergreen; California .... 1. *Quercus wislizeni*
      54. Plants deciduous; w Texas and east.
        55. Leaf blade widest distal to middle.
          56. Twigs glabrous.................................. 27. *Quercus georgiana*
          56. Twigs pubescent.
            57. Terminal buds uniformly pubescent .... 24. *Quercus marilandica*
            57. Terminal buds glabrous or at most ciliate on scale margins
.... 23. *Quercus arkansana*
        55. Leaf blade widest at or proximal to middle.
          58. Leaf blade usually with more than 10 awns.
            59. Petiole less than 25 mm; bark black and furrowed; cup less
than 15 mm wide; w Texas...................... 10. *Quercus gravesii*
            59. Petiole more than 25 mm; bark gray or dark gray with
wide shiny ridges separated by shallow fissures; cup more
than 15 mm wide; Oklahoma to Nova Scotia............ 32. *Quercus rubra*
          58. Leaf blade usually with fewer than 10 awns.
            60. Base of leaf blade subcordate to broadly obtuse; cup
deeply cup-shaped, 6–9 mm high; w Texas............. 8. *Quercus robusta*
            60. Base of leaf blade cuneate to obtuse; cup shallowly saucer-
shaped, 4–6 mm high; Alabama to South Carolina ..... 27. *Quercus georgiana*
44. Leaf margins with 3–11 lobes separated by deep (more than 1/3 distance to midrib) sinuses; lobes generally with 2 or more awns.
  61. Leaf blade abaxially uniformly pubescent or tomentose.
    62. Nut oblong to broadly ellipsoid, 21–34 mm; tips of cup scales loose; California
to Oregon .... 3. *Quercus kelloggii*
    62. Nut ovoid or subglobose, 9–16 mm; tips of cup scales appressed; Texas and east.
      63. Base of leaf blade rounded or U-shaped, terminal lobe much longer than
lateral lobes.. 20. *Quercus falcata*
      63. Base of leaf blade cuneate to truncate, terminal lobe rarely exceeding lateral
lobes.
        64. Trees; terminal buds 4–9 mm, light reddish brown, uniformly puberu-
lent .... 21. *Quercus pagoda*
        64. Shrubs; terminal buds 2–4.5 mm, dark reddish brown, puberulent only
at apex.. 25. *Quercus ilicifolia*
  61. Leaf blade abaxially glabrous or with a few scattered hairs, not uniformly pubescent.
    65. Terminal buds uniformly pubescent.
      66. Leaf base attenuate or acute, blade decurrent; margin of cup involute........ 22. *Quercus laevis*
      66. Leaf base obtuse to truncate, blade not decurrent; margin of cup not invo-
lute... 35. *Quercus velutina*
    65. Terminal buds glabrous or pubescent only on distal 1/2.
      67. Terminal buds silvery or reddish pubescent on distal 1/2.
        68. Base of leaf blade attenuate or acute, decurrent; petiole commonly less
than 20 mm; margin of cup involute....................... 22. *Quercus laevis*
        68. Base of leaf blade obtuse to truncate, not decurrent; petiole commonly
more than 20 mm; margin of cup not involute.
          69. Leaf blade adaxially dull green; sinuses usually extending less than
1/2 distance to midrib, lobes acute to oblong................. 32. *Quercus rubra*

69. Leaf blade adaxially glossy green; sinuses usually extending more than 1/2 distance to midrib, lobes distally expanded.

    70. Cup turbinate to hemispheric, scales with broad glossy base, scale margins often strongly concave; nut ovoid to subglobose, with 1 or more concentric rings of pits at apex . . . . .33. *Quercus coccinea*

    70. Cup deeply cup-shaped to turbinate, scales pubescent with straight or slightly concave margins; nut ellipsoid to ovoid, without rings of pits at apex . . . . 34. *Quercus ellipsoidalis*

67. Terminal buds glabrous or with only a few reddish hairs at apex.

  71. Leaf blades mostly wider than long. . . . . 31. *Quercus acerifolia*

  71. Leaf blades mostly longer than wide.

    72. Twigs and/or terminal buds yellowish gray, gray, or grayish brown.

      73. Mature leaf blade abaxially glabrous or with minute (often detectable only with magnification) tufts of tomentum in vein axils . . . . .29. *Quercus buckleyi*

      73. Mature leaf blade abaxially with conspicuous (readily discernible to naked eye) tufts of tomentum in vein axils.

        74. Cup with thin (less than 1.5 mm in cross section) walls, deeply goblet-shaped, covering 1/3–1/2 nut, inner surface pubescent. . . . . 28. *Quercus texana*

        74. Cup with thick (more than 1.5 mm in cross section) walls, saucer- or cup-shaped, covering 1/4–1/3 nut, inner surface glabrous or with ring of hairs around scar . . . . .30. *Quercus shumardii*

    72. Twigs and terminal buds brown to dark reddish brown or red.

      75. Leaf blade abaxially with conspicuous (readily discernible to naked eye) tufts of tomentum in vein axils.

        76. Leaf blade 3–5-lobed with fewer than 10 awns . . . . .27. *Quercus georgiana*

        76. Leaf blade 5–11-lobed with more than 10 awns.

          77. Cup deeply goblet-shaped, covering 1/3–1/2 nut; nut broadly ovoid to broadly ellipsoid, rarely less than 16 mm . . . . .28. *Quercus texana*

          77. Cup shallowly saucer-shaped, covering less than 1/3 nut; nut globose or ovoid, rarely more than 16 mm . . .26. *Quercus palustris*

      75. Leaf blade abaxially glabrous or with minute (often detectable only with magnification) tufts of tomentum in vein axils.

        78. Cup generally more than 12 mm high; cup scales more than 4 mm, attenuate or acuminate to acute with loose tips, especially at cup margin; California to Oregon . . . . . .3. *Quercus kelloggii*

        78. Cup rarely as much as 12 mm high; cup scales less than 4 mm, acute to obtuse with tightly appressed tips; w Texas and east.

          79. Leaf blade adaxially dull green, abaxially pale green or glaucous; cup 18–30 mm wide; Oklahoma to Nova Scotia . . . . .32. *Quercus rubra*

          79. Leaf blade adaxially shiny or glossy, abaxially light green or coppery green; cup 7–18 mm wide; Texas to Oklahoma.

            80. Petiole more than 20 mm; leaf lobes usually distally expanded; c Texas to Oklahoma . . . . .29. *Quercus buckleyi*

            80. Petiole usually less than 20 mm; leaf lobes acute; w Texas . . . . 10. *Quercus gravesii*

1. **Quercus wislizeni** A. de Candolle in A. P. de Candolle and A. L. P. de Candolle, Prodr. 16(2): 67. 1864 · Sierra live oak, interior live oak [E] [F]

Quercus parvula Greene; *Q. parvula* var. *shrevei* (C. H. Muller) Nixon; *Q. parvula* var. *tamalpaisensis* S. K. Langer; *Q. wislizeni* var. *frutescens* Engelmann

Trees or shrubs, evergreen, to 22 m. **Bark** nearly black, deeply furrowed with broad scaly ridges. **Twigs** brown to red-brown, 1.5–3 mm diam., glabrous or sparsely pubescent. **Terminal buds** light chestnut brown to dark reddish brown, ovoid to conic, 3–9 mm, glabrous or with tuft of minute hairs at apex. **Leaves:** petiole 3–20 mm, glabrous or sparsely pubescent. **Leaf blade** circular to oblong, usually ovate, planar, 25–70 × 20–50 mm, base obtuse to cordate, margins entire or spinose with up to 16 awns, apex acute to rounded; surfaces abaxially and adaxially glabrous, veins little raised on either surface. **Acorns** biennial; cup deeply and narrowly cup-shaped or U-shaped, 9–19 mm high × 7–18 mm wide, covering 1/3–1/2(–2/3) nut, outer surface glabrous to sparsely puberulent, inner surface glabrous or pubescent on innermost 1/3, occasionally uniformly pubescent, scales acute, tips loose; nut narrowly conic or ovoid to narrowly oblong, 21–44 × 8–14 mm, glabrous, scar diam. 2.5–7.5 mm. $2n = 24$.

Flowering late spring. Valleys, slopes, and sand chaparral; 300–1900 m; Calif.

Shrubs with oval leaves 25–38 mm and margins entire or deeply lobed-dentate may be treated as *Quercus wislizeni* var. *frutescens*. J. M. Tucker (1993) treated *Q. parvula* as a distinct species, distinguished from *Q. wislizeni* by its larger leaves (30–90 versus 20–50 mm), by the dull, olive-green, abaxial leaf surface (versus shiny, yellow-green), and by nuts that are abruptly tapered proximal to the middle (versus gradually tapered).

Tucker recognized two varieties of *Quercus parvula*: *Q. parvula* var. *parvula* is a shrub of 1–3 m and *Q. parvula* var. *shrevei* is a tree less than 17 m. S. K. Langer (1993) recognized a third variety, *Q. parvula* var. *tamalpaisensis*, based on several small populations on or near Mount Tamalpais, differentiated primarily by having larger leaves (50–160 × 20–60 mm) with attenuate-dentate margins.

Quercus wislizeni reportedly hybridizes with *Q. agrifolia* and *Q. kelloggii* (W. B. Brophy and D. R. Parnell 1974).

2. **Quercus agrifolia** Née, Anales Ci. Nat. 3: 271. 1801 · Coast live oak [F]

Quercus acroglandis Kellogg; *Q. agrifolia* var. *oxyadenia* (Torrey) J. T. Howell; *Q. pricei* Sudworth

Trees, evergreen, to 25 m. **Bark** gray to dark brown or black, ridges broad, rounded. **Twigs** brown to red-brown, 1.5–3 mm diam., with scattered pubescence or uniformly pubescent. **Terminal buds** light chestnut brown, ovoid, occasionally subconic, 3–6(–7) mm, glabrous except for cilia along scale margins. **Leaves:** petiole 4–15(–18) mm, sparsely to densely pubescent. **Leaf blade** broadly elliptic to ovate or oblong, 15–75 × 10–40 mm, base rounded or cordate, margins entire or spinose, with up to 24 awns, apex blunt to attenuate; surfaces abaxially glabrous or with small axillary tufts of tomentum, veins raised, adaxially distinctly convex, rugose, glabrous, occasionally densely uniformly pubescent. **Acorns** annual; cup turbinate to cup- or bowl-shaped, rarely saucer-shaped, 9–13 mm high × 9–15 mm wide, covering 1/4–1/3(–1/2) nut, outer surface glabrous to sparsely puberulent, inner surface pubescent on innermost 1/3 to uniformly pubescent, scales acute, tips loose; nut ovoid to oblong or conic, 15–35 × 10–15 mm, glabrous, scar diam. 3.5–8 mm. $2n = 24$.

Flowering early to mid spring. Moderately dry sites; to 1400 m; Calif.; Mexico (Baja California).

Quercus agrifolia is found in the Coast Ranges from Sonoma County, California, south to Baja California. Plants with densely pubescent leaves, especially abaxially, have been treated as *Q. agrifolia* var. *oxyadenia*.

This species reportedly hybridizes with *Quercus kelloggii* and *Q. wislizeni*.

The Mahuna used *Quercus agrifolia* medicinally to heal the bleeding navel of a newborn (D. E. Moerman 1986).

3. **Quercus kelloggii** Newberry in War Department, Pacif. Railr. Rep. 6: 28, 89, fig. 6. 1859 · California black oak [E] [F]

Quercus californica (Torrey) Cooper; *Q. tinctoria* W. Bartram var. *californica* Torrey

Trees, deciduous, to 25 m. **Bark** dark brown to black, ridges broad, irregular. **Twigs** brown to red-brown, (1.5–)2–3.5 mm diam., glabrate. **Terminal buds** chestnut brown, ovoid, 4–10 mm, glabrous or with scales ciliate on margins. **Leaves:** petiole 10–60 mm, glabrous to densely pubescent. **Leaf blade** ovate or broadly elliptic to obovate, 60–200 × 40–140 mm, base cordate to obtuse, occa-

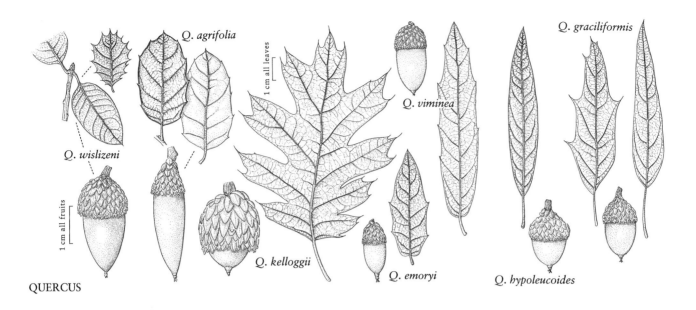

Q. wislizeni
Q. agrifolia
Q. graciliformis
Q. viminea
Q. kelloggii
Q. emoryi
Q. hypoleucoides

1 cm all leaves
1 cm all fruits

QUERCUS

sionally rounded, margins with 7–11 lobes and 13–45 awns, lobes acute to distally expanded, separated by deep sinuses, apex acute; surfaces abaxially glabrous with small axillary tufts of tomentum to densely pubescent, adaxially glabrous to minutely pubescent, veins raised on both surfaces. **Acorns** biennial; cup saucer-shaped to deeply bowl-shaped, 13–27 mm high × 20–28 mm wide, covering 1/2–2/3 nut, outer surface glabrous to sparsely puberulent, inner surface 1/3 to completely pubescent, scales more than 4 mm long, attenuate or acuminate to acute, smooth, occasionally tuberculate near base of cup, tips loose, especially at margin of cup; nut oblong to broadly ellipsoid, 21–34 × 14–22 mm, puberulent, especially at apex, scar diam. 5.5–10.5 mm. $2n = 24$.

Flowering late spring. On slopes and valleys of hills and mountains; 300–2400 m; Calif., Oreg.

The abundant crops of acorns from *Quercus kelloggii* were at one time an important food source for Native Americans.

The species reportedly hybridizes with *Quercus agrifolia* (= *Q.* ×*ganderi* C. B. Wolf) and *Q. wislizeni* (= *Q.* ×*morehus* Kellogg).

4. **Quercus emoryi** Torrey in W. H. Emory, Not. Milit. Reconn., 151, plate 9. 1848 · Emory oak, bellota [F]

Quercus hastata Liebmann

Trees or shrubs, evergreen, to 15 m. **Bark** dark brown to black, deeply fissured. **Twigs** dark reddish brown, 1–3 mm diam., pubescent. **Terminal buds** reddish brown, ovoid to subconic, 2.5–6.5 mm, glabrous except for tuft of hairs at apex, occasionally hairy on distal 1/2. **Leaves:** petiole 3–7(–10) mm, pubescent. **Leaf blade** ovate to narrowly oblong to obovate, planar, 28–95 × 15–45 mm, base cordate, margins entire or spinose, with up to 13 awns, apex blunt to acute; surfaces abaxially glabrous except for tuft of tomentum on each side of midrib at base of blade, rarely completely glabrous, adaxially not rugose, glabrous or with a few hairs along midrib. **Acorns** annual; cup cup-shaped, 5–7.5 mm high × 7–12 mm wide, covering 1/4–1/2 nut, outer surface pubescent to sparsely puberulent, inner surface pubescent to floccose, scale tips appressed, blunt; nut ellipsoid to oblong, 10–18 × 6–10 mm, glabrous to puberulent, especially at apex, scar diam. 3–5.5 mm.

Flowering spring. Foothills and slopes; 1000–2200 m; Ariz., N.Mex., Tex.; n Mexico.

Quercus emoryi reportedly hybridizes with *Q. graciliformis* (= *Q.* ×*tharpii* C. H. Muller).

5. **Quercus viminea** Trelease, Mem. Natl. Acad. Sci. 20: 123, plate 222. 1924 [F]

Trees, evergreen or drought-deciduous, to 10 m. **Twigs** brown to dark reddish brown, 2–3 mm diam., sparsely to uniformly pubescent. **Terminal buds** brown to reddish brown, acutely ovoid, 1.5–2.5 mm, glabrous. **Leaves:** petiole 2–5 mm, pubescent. **Leaf blade** narrowly lanceolate, widest at or proximal to middle, 35–60[–150] × 7–10 mm, base usually rounded or obtuse, often oblique, occasionally somewhat cordate, margins entire with 1 apical awn, apex acute to acuminate; surfaces abaxially glabrous or with prominent tufts of tomentum near base of blade, adaxially glabrous or with pubescence along midrib and scattered on blade. **Acorns** (not known within our range) biennial; cup cupshaped, 5–8 mm high × 8–10 mm wide, covering ca. 1/3 nut, inner surface pubescent, scales with appressed, obtuse tips; nut ovoid, 9–15 × 7–10 mm.

Flowering spring or summer. Habitat not specified; 1500 m; Ariz.; Mexico (Chihuahua, Sonora, and southward).

As with many of the Madrean counterparts of the oak flora, in very dry years trees progressively lose their leaves through the long dry spring and may become virtually leafless by the time the rains come in early summer (R. W. Spellenberg, pers. comm.).

In the flora, *Quercus viminea* is known from a single sterile specimen collected at Red Mountain, Santa Cruz County, Arizona. Recent attempts to find the taxon at this site have been fruitless. The Arizona specimen is similar to specimens from northern Mexico that have small, usually entire, narrowly lanceolate leaves with short petioles. These northern forms appear to grade clinally into larger-leaved southern forms that fit the general description of *Q. bolanyosensis* Trelease.

6. Quercus hypoleucoides A. Camus, Bull. Mus. Natl. Hist. Nat., sér. 2, 4: 124. 1932 · Silverleaf oak [F]

Quercus hypoleuca Engelmann

Trees or shrubs, evergreen, to 10 m. **Bark** black, deeply furrowed. **Twigs** dark reddish brown, 1.5–3 mm diam., pubescent. **Terminal buds** light chestnut brown, ovoid, 2.5–4.5 mm, glabrous except for ciliate scale margins, occasionally with tuft of hairs at apex. **Leaves:** petiole 1.5–13 mm, densely pubescent. **Leaf blade** narrowly ovate to ovate or elliptic, 45–120 × 15–40 mm, base cuneate to rounded, margins strongly revolute, entire or spinose with up to 11 awns, apex acute to attenuate; surfaces abaxially densely tawny- or white-tomentose, adaxially noticeably rugose, glabrous. **Acorns** annual or biennial; cup deeply saucer- or cup-shaped, 4.5–7 mm high × 6–13 mm wide, covering 1/3 nut or less, outer surface pubescent to sparsely puberulent, inner surface pubescent to floccose, scales appressed, blunt; nut ellipsoid to oblong, 8–16 × 5–10 mm, glabrous, scar diam. 2.5–5.5 mm.

Flowering spring. Common in moist canyons and on ridges; 1500–2700 m; Ariz., N.Mex., Tex.; n Mexico.

Quercus hypoleucoides reportedly hybridizes with *Q. gravesii* (*Q.* ×*inconstans* E. J. Palmer [= *Q. livermorensis* C. H. Muller]) (see C. H. Muller 1951). Several specimens from Pima County, Arizona, fall out-

side the range of typical *Q. hypoleucoides,* suggesting hybridization with the Mexican *Q. mcvaughii* Spellenberg (R. Spellenberg 1992).

7. Quercus graciliformis C. H. Muller, Torreya 34: 120. 1934 · Chisos oak [C] [E] [F]

Quercus canbyi Cory & Parks; *Q. graciliformis* var. *parvilobata* C. H. Muller

Trees, tardily deciduous, to 8 m. **Bark** gray, furrowed. **Twigs** straw colored to brown or deep reddish brown, 1–2 mm diam., glabrate or somewhat pubescent at apex. **Terminal buds** glossy brown or red-brown, ovoid, 1.5–3 mm, minutely ciliate. **Leaves:** petiole 10–20 mm, glabrous or glabrate. **Leaf blade** lanceolate to narrowly elliptic, widest proximal to middle, 45–90 × 10–25 mm, base rounded to cuneate, margins entire or with 8–10 teeth or shallow lobes with rounded sinuses, awns 1–9, apex acute to long-attenuate; surfaces abaxially glabrous or occasionally with small axillary tufts of tomentum, adaxially glabrous except for scattered pubescence near base and along midrib. **Acorns** biennial; cup saucer- to shallowly bowl-shaped, 4–6 mm high × 7–10 mm wide, covering 1/4–1/3 nut, outer surface puberulent to glabrate, inner surface uniformly pubescent, scales appressed, acute, occasionally tuberculate at base; nut ovoid to narrowly ellipsoid, 9–18 × 7–10 mm, minutely puberulent, scar diam. 3–4.5 mm.

Flowering spring. Dry rocky canyons of the Chisos Mountains; of conservation concern; to 1650 m; Tex.

Quercus graciliformis reportedly hybridizes with *Q. emoryi*.

8. Quercus robusta C. H. Muller, Torreya 34: 119. 1934 [C] [E] [F]

Trees, deciduous, to 13 m. **Bark** brown or black, roughly furrowed. **Twigs** dark reddish brown, 1.5–2.5 mm diam., densely pubescent or glabrate. **Terminal buds** glossy light brown, acutely ovoid, 4–7 mm, glabrous or pubescent on apical 1/2. **Leaves:** petiole 5–20 mm, pubescent or glabrate. **Leaf blade** acutely ovate to elliptic, widest at or proximal to middle, 55–120 × 20–50 mm, base cuneate to rounded or subcordate, margins with 6–8 teeth or shallow lobes with rounded sinuses, rarely entire, 1–10 awns, apex acute or attenuate; surfaces abaxially glabrous except for small axillary tufts of tomentum or pubescent along midrib, adaxially glabrous or persistently pubescent near base and along midrib. **Acorns** biennial; cup deeply cup-shaped, 6–9 mm high × 8–12 mm wide, covering 1/4–1/3 nut, outer surface puberulent or glabrate, inner surface uniformly pubescent, scales tightly appressed, acute or attenuate; nut oblong to broadly ellipsoid, 10–22 × 7–10 mm, glabrate, scar diam. 3.5–4.5 mm.

Flowering spring. Moist wooded canyons in Chisos Mountains; of conservation concern; 1500 m; Tex.

After describing this species, C. H. Muller later (1951, 1970) concluded that it represented a hybrid between *Quercus emoryi* and *Q. gravesii*. The extreme forms (e.g., the type specimens) of *Q. robusta* and *Q. gravesii* are easily differentiated, but these two taxa appear to occupy the ends of a morphologic continuum. Muller's recent view, however, is that *Q. robusta* deserves species status. The origin of this taxon is still worthy of study.

9. Quercus tardifolia C. H. Muller, Bull. Torrey Bot. Club 63: 154. 1936 · Lateleaf oak [C] [F]

Trees, evergreen. **Bark** gray, furrowed. **Twigs** dark reddish brown, 1.5–2.5 mm diam., densely pubescent. **Terminal buds** brown or reddish brown, ellipsoid or ovoid, 3.5–5.5 mm, apex hairy, scales with ciliate margins. **Leaves:** petiole 10–20 mm, pubescent or glabrate. **Leaf blade** broadly elliptic or obovate, widest at or distal to middle, planar, 50–100 × 20–70 mm, base cordate or occasionally rounded, margins with 3–4 lobes with shallow sinuses, 6–12 awns, apex acute or obtuse; surfaces abaxially conspicuously tomentose, primary and secondary veins raised, adaxially somewhat rugose, glabrate. **Acorns** biennial, immature acorns in pairs, mature acorns not known.

Flowering spring. Wooded arroyos; of conservation concern; 2000 m; Tex.

Quercus tardifolia was reported from Mexico (Coahuila) (A. M. Powell 1988), but I have not seen the specimens. It should be expected in the ranges (e.g., Sierra del Carmen) adjacent to the Big Bend area.

This distinctive species is apparently quite infrequent, only two small clumps being known from the Chisos Mountains (C. H. Muller 1951). Recent efforts to locate *Quercus tardifolia* have not been successful (M. Powell, pers. comm.). Its status is also in question; Muller and K. C. Nixon (pers. comm.) think that it might be a hybrid between *Quercus gravesii* and *Q. hypoxantha* Trelease.

10. Quercus gravesii Sudworth, Check List For. Trees U.S., 86. 1927 · Graves oak, Chisos red oak [F]

Quercus chesosensis (Sargent) C. H. Muller; *Q. shumardii* Buckley var. *microcarpa* (Torrey) Shinners; *Q. stellipila* (Sargent) H. B. Parks ex Cory; *Q. texana* var. *chesosensis* Sargent; *Q. texana* var. *stellapila* Sargent

Trees, deciduous, to 13 m. **Bark** black, roughly furrowed. **Twigs** light brown to dark reddish brown, (1–)1.5–2.5 mm diam., glabrous or glabrate. **Terminal buds** brown or reddish brown, ovoid or ellipsoid to subconic, 2–5 mm, glabrous or with tuft of hairs at apex. **Leaves:** petiole 5–25 mm, glabrate. **Leaf blade** ovate to elliptic, 45–140 × 20–120 mm, base rounded (rarely subcordate) to obtuse or cuneate, margins with 3–5 acute lobes, 9–20 awns, apex broadly obtuse or acute to attenuate, occasionally falcate; surfaces abaxially light green or coppery green, glabrous with small axillary tufts of tomentum or pubescent along midrib and veins, adaxially shiny or glossy, glabrous except for scattered pubescence near base and along midrib. **Acorns** biennial; cup turbinate or deeply cup-shaped, 4.5–8.5 mm high × 7.5–12 mm wide, covering 1/3–1/2 nut, outer surface glabrate, inner surface glabrous to pubescent on inner 2/3, scales less than 4 mm, tips appressed, acute; nut ovoid to ellipsoid, rarely subglobose or oblong, 9–16 × 5.5–11 mm, occasionally striate, glabrous to puberulent, especially at apex, scar diam. 3–6 mm.

Flowering spring. Davis, Glass, and Chisos mountains; above 1200 m; Tex.; n Mexico.

Quercus gravesii reportedly hybridizes with *Q. hypoleucoides*. As noted above, forms of *Q. gravesii* and *Q. robusta* are easily confused and give the impression of belonging to a single morphologic continuum.

11. Quercus incana W. Bartram, Travels Carolina, 378. 1791 · Bluejack oak [E] [F] [W1]

Quercus cinerea Michaux

Trees, deciduous, to 10 m. **Bark** dark brown to black with square plates. **Twigs** brown to reddish brown, 1–2.5 mm diam., tomentose to sparsely pubescent. **Terminal buds** light brown to reddish brown, narrowly ovoid to conic, 3.5–7 mm, distinctly 5-angled in cross section, scales pubescent, often tuft of reddish or silvery hairs at apex. **Leaves:** petiole 2–8(–10) mm, tomentose. **Leaf blade** narrowly ovate or elliptic to obovate, usually widest near middle, planar, 30–100 × 12–35 mm, base acute (rarely attenuate) to rounded, margins entire, with 1 apical awn (leaves on juvenile or 2d-flush growth may have 2–3 shallow lobes and 3–5 awns), apex acute or obtuse, rarely rounded; surfaces abaxially densely tomentose, hairs in vein axils often reddish, easily distinguished from others, adaxially often glossy, sparsely pubescent, especially along midrib and near base, veins often raised. **Acorns** biennial; cup saucer-shaped to bowl-shaped, 4.5–8 mm high × 10–18 mm wide, covering 1/4–1/3(–1/2) nut, outer surface pubescent or puberulent, inner surface uniformly pubescent, scale tips tightly appressed, obtuse or acute; nut ovoid (rarely subglobose) to broadly ellipsoid, 10–17 × 10–16 mm, occasionally striate, glabrate, scar diam. 5.5–10.5 mm.

Flowering spring. Well-drained sandy soils of barrens, hammocks, dunes, and upland ridges; 0–250 m; Ala., Ark., Fla., Ga., La., Miss., N.C., Okla., S.C., Tex., Va.

Quercus incana reportedly hybridizes with *Q. falcata* [= *Q.* ×*subintegra* (Engelmann) Trelease], *Q. hemisphaerica* (D. M. Hunt 1989), *Q. laurifolia* (= *Q.* ×*atlantica* Ashe), *Q. laevis* (= *Q.* ×*asheana* Little), *Q. marilandica* (= *Q.* ×*cravenensis* Little), *Q. nigra* (= *Q.* ×*caduca* Trelease), *Q. phellos* (E. J. Palmer 1948), *Q. pumila* (D. M. Hunt 1989), *Q. velutina* (= *Q.* ×*podophylla* Trelease), and questionably, *Q. myrtifolia*.

12. **Quercus laurifolia** Michaux, Hist. Chênes Amér., no. 10, plate 17. 1801 · Swamp laurel oak [E] [F] [W1]

Quercus obtusa (Willdenow) Ashe; *Q. rhombica* Sargent

Trees, tardily deciduous, to 40 m. **Bark** dark brown to black, ridges flat, furrows deep. **Twigs** red-brown, (1–)1.5–2.5 mm diam., glabrous. **Terminal buds** dark red-brown, ovoid to subconic, 2.5–6 mm, distinctly 5-angled in cross section, glabrous or with tuft of reddish hairs at apex. **Leaves:** petiole 1.5–5 mm, glabrous. **Leaf blade** rhombic or broadly elliptic to obovate, occasionally oblong or spatulate, 30–120 × 15–45 mm, thin, base attenuate or cuneate, rarely obtuse, margins entire with 1 apical awn, apex obtuse or rounded; surfaces abaxially glabrous, adaxially glabrous, veins raised. **Acorns** biennial; cup shallowly saucer-shaped to deeply bowl-shaped, 3.5–9 mm high × 11–17 mm wide, covering 1/4–1/2 nut, outer surface puberulent, inner surface pubescent at least 1/2 distance to rim, scale tips appressed, acute or attenuate; nut globose or ovoid, 8.5–16 × 10–16 mm, glabrate, scar diam. 6.5–11.5 mm.

Flowering spring. Sandy flood plains and bottoms, riverbanks, and terraces, occasionally on poorly drained uplands; 0–150 m; Ala., Ark., Fla., Ga., La., Miss., N.C., S.C., Tex., Va.

Quercus laurifolia apparently flowers two weeks earlier than sympatric *Quercus hemisphaerica* (W. H. Duncan and M. B. Duncan 1988). It reportedly hybridizes with *Q. falcata*, *Q. incana*, and *Q. nigra* (H. A. Fowells 1965); with *Q. hemisphaerica*, *Q. marilandica*, *Q. myrtifolia*, *Q. phellos*, and *Q. shumardii* (D. M. Hunt 1989); and with *Q. velutina*.

13. **Quercus hemisphaerica** W. Bartram ex Willdenow, Sp. Pl. 4(1): 443. 1805 · Laurel oak, Darlington oak [E] [F]

Trees, evergreen or tardily deciduous, to 35 m. **Bark** much like that of *Q. laurifolia.* **Twigs** light brown to dark red-brown, 1–2.5 mm diam., glabrous. **Terminal buds** reddish to purplish brown, ovoid, 2.5–5 mm, glabrous or with ciliate scale margins. **Leaves:** petiole 1–5(–6) mm, glabrous. **Leaf blade** narrowly ovate or elliptic to oblanceolate, 30–120 × 10–40 mm, leathery, base obtuse to rounded, rarely attenuate, margins entire or with a few shallow lobes or teeth near apex, awns 1–4 (rarely as many as 8–10 on 2d-flush growth), apex acute or acuminate, occasionally obtuse; surfaces abaxially glabrous, rarely with minute axillary tufts of tomentum, adaxially glabrous. **Acorns** biennial; cup saucer-shaped to bowl-shaped, rarely turbinate, 3–10 mm high × 11–18 mm wide, covering 1/4–1/3 nut, outer surface puberulent, inner surface pubescent at least 1/2 distance to rim, scales occasionally distinctly tuberculate, tips appressed, acute to obtuse; nut broadly ovoid to hemispheric, 9–16 × 9–16 mm, glabrate, scar diam. 6–9.5 mm.

Flowering spring. Moderately dry sandy soils, scrub sandhills, stream banks, occasionally on hillsides and ravines; 0–150 m; Ala., Ark., Fla., Ga., La., Miss., N.C., S.C., Tex., Va.

Quercus hemisphaerica flowers about two weeks later than sympatric *Q. laurifolia* (W. H. Duncan and M. B. Duncan 1988).

Most authors have treated *Quercus hemisphaerica* as synonymous with *Q. laurifolia*. M. L. Fernald (1946) carefully examined the situation and concluded that *Q. hemisphaerica* is a distinct entity, but C. H. Muller (1951) argued that these two taxa ". . . are now certainly not separable even as varieties of the same species." Later (1970), Muller recanted by recognizing *Q. hemisphaerica* as a common component of stream terraces along the Gulf Coast.

Quercus hemisphaerica reportedly hybridizes with *Q. falcata* (C. H. Muller 1970); with *Q. arkansana, Q.*

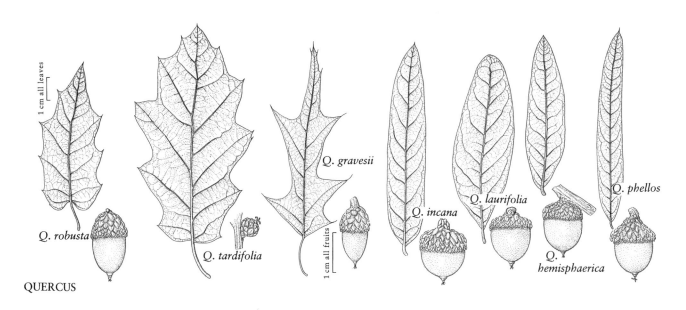

1 cm all leaves

Q. gravesii

Q. phellos

Q. incana

Q. laurifolia

Q. robusta

Q. tardifolia

1 cm all fruits

Q. hemisphaerica

QUERCUS

inopina, Q. marilandica, Q. myrtifolia, Q. nigra, Q. phellos, Q. pumila, and *Q. shumardii* (D. M. Hunt 1989); and with *Q. incana* (= *Q.* × *sublaurifolia* Trelease), and *Q. laevis* (= *Q.* × *mellichampii* Trelease).

14. Quercus phellos Linnaeus, Sp. Pl. 2: 994. 1753 · Willow oak [E] [F] [W1]

Trees, deciduous, to 30 m. **Bark** dark gray and smooth, becoming darker and irregularly fissured with age, inner bark light orange. **Twigs** reddish brown, 1–2 mm diam., glabrous. **Terminal buds** chestnut brown, ovoid, 2–4 mm, apex acute, glabrous. **Leaves:** petiole 2–4(–6) mm, glabrous, rarely sparsely hairy. **Leaf blade** linear to narrowly elliptic, usually widest near middle, 50–120 × 10–25 mm, base acute, margins entire with 1 apical awn, apex acute; surfaces abaxially pale green, glabrous, rarely softly pubescent, adaxially light green, glabrous. **Acorns** biennial; cup shallowly saucershaped, 3–6.5 mm high × 7.5–11 mm wide, covering 1/4–1/3 nut, outer surface puberulent, inner surface light brown, pubescent, scale tips tightly appressed, acute; nut ovoid to hemispheric, 8–12 × 6.5–10 mm, often striate, glabrate, scar diam. 4.5–6 mm.

Flowering spring. Of bottomland flood plains, also on stream banks, dunes, and terraces, and, occasionally, on poorly drained uplands; 0–400 m; Ala., Ark., Del., D.C., Fla., Ga., Ill., Ky., La., Md., Miss., Mo., N.J., N.Y., N.C., Okla., Pa., S.C., Tenn., Tex., Va.

Individual trees with leaves softly pubescent abaxi-

ally may be classified as *Quercus phellos* forma *intonsa* Fernald; however, such leaves are known to occur on second-flush shoots from twigs bearing typical leaves.

Quercus phellos reportedly hybridizes with *Q. coccinea* (W. W. Ashe 1894), *Q. ilicifolia* (= *Q.* × *giffordii* Trelease), *Q. incana* (E. J. Palmer 1948), *Q. marilandica, Q. nigra, Q. pagoda* (= *Q.* × *ludoviciana* Sargent), *Q. palustris, Q. rubra, Q. shumardii,* and *Q. velutina.* D. M. Hunt (1989) cited evidence of hybridization with *Q. hemisphaerica, Q. imbricaria, Q. laurifolia,* and *Q. pumila.*

15. Quercus nigra Linnaeus, Sp. Pl. 2: 995. 1753 · Water oak, chêne gris [E] [F] [W1]

Quercus nana Willdenow; *Q. nigra* var. *tridentifera* Sargent; *Q. uliginosa* Wangenheim

Trees, deciduous or tardily deciduous, to 30 m. **Bark** grayish black, fissures irregular, shallow, inner bark pinkish. **Twigs** dark red-brown, 1.5–2.5 mm diam., glabrous. **Terminal buds** reddish brown, ovoid, 3–6.5 mm, puberulent throughout, occasionally densely pubescent on apical 2/3. **Leaves:** petiole 2–9 mm, glabrous. **Leaf blade** distinctly obtrullate, rarely elliptic or merely obovate, widest near apex, 30–120(–160) × 15–60(–70) mm, base attenuate or cuneate, rarely rounded, margins entire with 1 apical awn or with 2–3 shallow lobes and 2–5 awns (leaves on juvenile or 2d-flush growth may be deeply lobed with more awns), apex obtuse to blunt or rounded; surfaces abaxially glabrous except for minute or conspicuous axillary tufts of tomentum, veins rarely raised, adaxially glabrous with secondary veins somewhat impressed. **Acorns** biennial; cup saucer-shaped, 2.5–5.5 mm high × 10–18 mm wide, covering 1/4 nut or less, outer surface puberulent, inner surface sparsely to uniformly pubescent, scale tips tightly appressed, acute; nut broadly ovoid, 9.5–14 × 9.5–14.5 mm, often faintly striate, glabrate, scar diam. 6–11.5 mm.

Flowering spring. Mesic alluvial and lowland sites, also barrens, dunes, hammocks, and low ridges to steep slopes; 0–450 m; Ala., Ark., Del., Fla., Ga., Ky., La., Md., Miss., Mo., N.J., N.C., Okla., S.C., Tenn., Tex., Va.

Typically on mesic alluvial and lowland sites, *Quercus nigra* also occurs on a wide variety of soil types and in a diversity of habitats.

Trees with 3-lobed leaves with attenuate bases have been recognized as *Quercus nigra* var. *tridentifera* Sargent.

Quercus nigra reportedly hybridizes with *Q. falcata* (= *Q.* ×*garlandensis* E. J. Palmer), *Q. incana*, *Q. laevis* (= *Q.* ×*walteriana* Ashe), *Q. marilandica* (= *Q.* ×*sterilis* Trelease), *Q. phellos* (= *Q.* ×*capesii* W. Wolf), *Q. shumardii* (= *Q.* ×*neopalmeri* Sudworth), and *Q. velutina* (*Q.* ×*demarei* Ashe). In addition, D. M. Hunt (1989) cited evidence of hybridization with *Q. arkansana*, *Q. georgiana*, *Q. hemisphaerica*, *Q. laurifolia*, *Q. myrtifolia*, *Q. palustris*, *Q. rubra*, and *Q. texana*.

16. Quercus pumila Walter, Fl. Carol., 234. 1788 · Runner oak E F W1

Shrubs, deciduous or tardily deciduous, to 1 m. **Bark** gray to dark brown. **Twigs** gray-brown to reddish brown, 1–2 mm diam., sparsely to uniformly pubescent. **Terminal buds** brown to red-brown, ovoid, 2.5–4.5 mm, glabrous or with ciliate scale margins. **Leaves:** petiole 1.5–4.5 mm, pubescent. **Leaf blade** oblong to narrowly obovate, 25–100 × 10–33 mm, base acute to rounded, margins entire, revolute, with 1 apical awn, apex acute or obtuse to rounded; surfaces abaxially uniformly gray-brown pubescent, rarely glabrate, adaxially somewhat convex, rugose, glabrous or with scattered hairs along midrib. **Acorns** annual; cup deeply saucer-shaped to turbinate, 5–12 mm high × 10–15 mm wide, covering 1/3–1/2(–2/3) nut, outer surface pubescent, inner surface densely pubescent, scales rarely involute, often tuberculate, tips tightly appressed, acute; nut globose to ovoid or broadly oblong, 9.5–15 × 9–12 mm, glabrate, scar diam. 5–8 mm.

Flowering spring. Dry sandy soils of savannas, low ridges and oak-pine scrub, occasionally at margins of poorly drained sites; 0–100 m; Ala., Fla., Ga., Miss., N.C., S.C.

Although no hybrid combinations have been formally proposed, D. M. Hunt (1989) has reported evidence of hybridization with *Quercus hemisphaerica*, *Q. incana*, *Q. myrtifolia*, and *Q. phellos*.

17. Quercus imbricaria Michaux, Hist. Chênes Amér., no. 9, plates 15, 16. 1801 · Shingle oak E F

Trees, deciduous, to 20 m. **Bark** grayish brown, fissures and ridges shallow, inner bark pinkish. **Twigs** greenish brown to brown, 1.5–3(–4) mm diam., glabrous or sparsely pubescent. **Terminal buds** brown to reddish brown, ovoid, 3–6 mm, distinctly 5-angled in cross section, scales minutely ciliate on margins. **Leaves:** petiole 10–20 mm, glabrous. **Leaf blade** ovate or elliptic to obovate, usually widest near middle, 80–200 × 15–75 mm, base obtuse to cuneate, occasionally rounded, margins entire with 1 apical awn, apex acute to obtuse; surfaces abaxially uniformly pubescent, adaxially lustrous, glabrous. **Acorns** biennial; cup deeply saucer-shaped to cup-shaped, 5–9 mm high × 10–18 mm wide, covering 1/3–1/2 nut, outer surface puberulent, inner surface light brown to reddish brown and glabrous or with a few hairs around nut scar, scale tips tightly appressed, acute; nut ovoid to subglobose, 9–18 × 10–18 mm, often striate, having 1 or more indistinct rings of minute pits at apex, glabrate, scar diam. 5–9 mm. 2*n* = 24.

Flowering spring. Moderately dry to mesic slopes and uplands, occasionally in ravines and bottoms; 100–700 m; Ark., Del., D.C., Ill., Ind., Iowa, Kans., Ky., La., Md., Mich., Mo., N.J., N.C., Ohio, Pa., Tenn., Tex., Va., W.Va.

The wood of *Quercus imbricaria* was once an important source of shingles, hence its common name.

The Cherokee used the bark of *Quercus imbricaria* to treat indigestion, chronic dysentery, mouth sores, chapped skin, general sores, chills and fevers, lost voice, milky urine, and as an antiseptic and a general tonic (D. E. Moerman 1986).

This species reportedly hybridizes with *Q. coccinea* (W. H. Wagner Jr. and D. J. Schoen 1976), *Q. falcata* (= *Q.* ×*anceps* E. J. Palmer), *Q. ilicifolia* (D. M. Hunt 1989), *Q. marilandica*, *Q. palustris*, *Q. phellos* (H. A. Gleason 1952), *Q. rubra*, *Q. shumardii*, and *Q. velutina*.

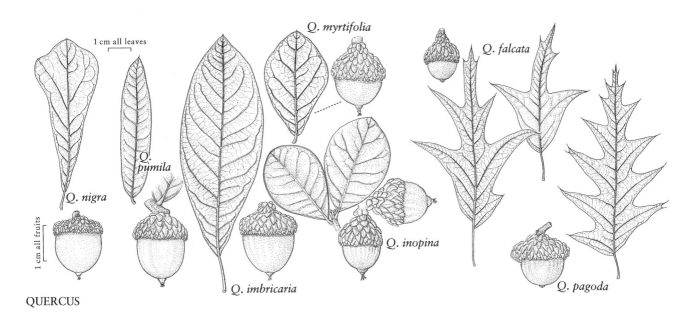

1 cm all leaves

Q. myrtifolia

Q. falcata

Q. nigra

Q. pumila

1 cm all fruits

Q. inopina

Q. imbricaria

Q. pagoda

QUERCUS

18. Quercus myrtifolia Willdenow, Sp. Pl. 4(1): 424. 1805 · Myrtle oak E F W1

Trees or shrubs, evergreen, to 12 m. Bark gray and smooth distally, dark and shallowly furrowed near base. Twigs dark red-brown, 1–2.5(–3) mm diam., persistently pubescent, rarely almost glabrous. Terminal buds reddish to purplish brown, ovoid, 2–5.5 mm, glabrous or with tuft of reddish hairs at apex. Leaves: petiole 1–5 mm, glabrous, occasionally sparsely pubescent. Leaf blade elliptic to narrowly or broadly obovate, occasionally spatulate, 15–50(–70) × 10–25(–35) mm, base cuneate to rounded, margins entire, somewhat revolute, with 1–4 awns, apex obtuse or rounded; surfaces abaxially glabrous except for axillary tufts of tomentum, occasionally yellow-scurfy, adaxial surface planar, glabrous. Acorns biennial; cup saucer-shaped to shallowly goblet-shaped, 4–7 mm high × 8.5–14.5 mm wide, covering 1/4–1/3 nut, outer surface puberulent, inner surface half to fully pubescent, scale tips tightly appressed, acute; nut broadly ovoid to globose, 9.5–14 × 8–13 mm, glabrate, scar diam. 5–8 mm.

Flowering spring. Dunes, hammocks, sandhills, dry sandy ridges, and oak scrub; 0–100 m; Ala., Fla., Ga., Miss., S.C.

This species flowers one to two weeks earlier than *Q. inopina* (A. F. Johnson and W. G. Abrahamson 1982).

Quercus myrtifolia reportedly hybridizes with *Q. incana* (= *Q.* ×*oviedoensis* Sargent), but E. J. Palmer (1948) questioned the identification of the type specimen; the brief description by Sargent suggests that the specimen may be from an individual of *Q. inopina*. D. M. Hunt (1989) cited evidence of hybridization with *Q. arkansana, Q. hemisphaerica, Q. inopina, Q. laurifolia, Q. marilandica, Q. nigra,* and *Q. pumila* (Hunt suggested that the last may give rise to occasional reports of annual fruiting in *Q. myrtifolia*).

19. Quercus inopina Ashe, Rhodora 31: 79. 1929 E F

Shrubs, evergreen, to 5 m. Bark gray. Twigs light to dark purplish brown, (1.5–)2–3(–4) mm diam., glabrate to sparsely pubescent, especially at apex. Terminal buds dark purplish brown, ovoid to subconic, 2–6 mm, noticeably 5-angled in cross section, glabrous to tawny strigose on apical 1/3. Leaves: petiole 1.5–8.5 mm, glabrous, occasionally sparsely pubescent. Leaf blade ovate or elliptic to obovate, occasionally spatulate, (25–)45–85 × (15–)25–45 mm, base acute to rounded or cordate, margins entire, strongly revolute, with 1 apical awn, apex obtuse or rounded; surfaces abaxially yellow-scurfy, occasionally somewhat pubescent, rarely glabrous, adaxially distinctly convex, rugose, glabrous or with scattered hairs, especially along midrib and at base. Acorns biennial; cup cup-shaped to bowl-shaped, 6–8 mm high × 10–15 mm wide, covering 1/3–1/2 nut, outer surface puberulent, inner surface half to fully pubescent, scale tips tightly appressed, acute; nut ovoid to broadly ellipsoid, 10–14 × 9–13 mm, glabrate, scar diam. 4.5–8 mm.

Flowering spring. On deep white "sugar" sands of low sandhill ridges, scrub communities, and flat upland terraces; 0–50 m; Fla.

Quercus inopina occurs from Orange County, Florida, southwest to Manatee County and south to Martin County. It flowers one to two weeks later than *Q. myrtifolia* (A. F. Johnson and W. G. Abrahamson 1982).

The leaves of this species often have numerous small black dots on the adaxial surface. These are ascocarps (the sexual fruiting bodies of ascomycete fungi) of the genus *Asterina* (D. M. Hunt, pers. comm).

Although no hybrids have been formally described, evidence of hybridization of *Quercus inopina* with *Q. hemisphaerica*, *Q. laevis*, and *Q. myrtifolia* has been reported (D. M. Hunt 1989).

20. Quercus falcata Michaux, Hist. Chênes Amér., no. 16, plate 28. 1801 · Southern red oak, Spanish oak, chêne rouge E F W1

Quercus digitata Sudworth; *Q. falcata* Michaux var. *triloba* (Michaux) Nuttall

Trees, deciduous, to 30 m. **Bark** dark brown to black, narrowly fissured with scaly ridges, inner bark orange. **Twigs** reddish brown, (1–)1.5–3.5(–4.5) mm diam., pubescent. **Terminal buds** light reddish brown, ovoid, 4–8 mm, puberulent throughout. **Leaves:** petiole 20–60 mm, glabrous to sparsely pubescent. **Leaf blade** ovate to elliptic or obovate, 100–300 × 60–160 mm, base rounded or U-shaped, margins with 3–7 deep lobes and 6–20 awns, terminal lobe often long-acuminate, much longer than lateral lobes, apex acute; surfaces abaxially sparsely to uniformly tawny-pubescent, adaxially glossy and glabrous or puberulent along midrib, secondary veins raised on both surfaces. **Acorns** biennial; cup saucer-shaped to cup-shaped, 3–7 mm high × 9–18 mm wide, covering 1/3–1/2 nut, outer surface puberulent, inner surface pubescent, scale tips tightly appressed, acute; nut subglobose, 9–16 × 8–15 mm, often striate, puberulent, scar diam. 5–10 mm.

Flowering spring. Dry or sandy upland sites; 0–800 m; Ala., Ark., Del., D.C., Fla., Ga., Ill., Ind., Ky., La., Md., Miss., Mo., N.J., N.Y., N.C., Ohio, Okla., Pa., S.C., Tenn., Tex., Va., W.Va.

Native Americans used *Quercus falcata* in various ways to treat indigestion, chronic dysentery, sores, chapped skin, chills and fevers, lost voice, asthma, milky urine, and as an antiseptic, a tonic, and an emetic (D. E. Moerman 1986).

Quercus falcata reportedly hybridizes with *Q. ilici-*folia (= *Q. ×caesariensis* Moldenke), *Q. imbricaria*, *Q. incana*, *Q. laevis*, *Q. laurifolia* (= *Q. ×beaumontiana* Sargent), *Q. marilandica* (E. J. Palmer 1948); *Q. nigra*, *Q. pagoda* (S. A. Ware 1967; R. J. Jensen 1989), and *Q. phellos*, *Q. shumardii*, *Q. hemisphaerica*, and *Q. velutina*.

21. Quercus pagoda Rafinesque, Alsogr. Amer., 23. 1838 · Cherrybark oak E F

Quercus falcata Michaux var. *leucophylla* (Ashe) E. J. Palmer & Steyermark; *Q. falcata* var. *pagodifolia* Elliott; *Q. leucophylla* Ashe; *Q. pagodifolia* (Elliott) Ashe

Trees, deciduous, to 40 m. **Bark** nearly black with narrow and noticeably flaky ridges, often resembling that of wild black cherry, inner bark orange. **Twigs** yellowish brown, 2–3.5 mm diam., pubescent. **Terminal buds** light reddish brown, ovoid, 4–9 mm, strongly 5-angled in cross section, puberulent throughout. **Leaves:** petiole 20–50 mm, glabrate or pubescent. **Leaf blade** ovate to elliptic or obovate, 90–300 × 60–160 mm, base cuneate to rounded or truncate, margins with 5–11 lobes and 10–25 awns, lobes acute or oblong, rarely falcate, terminal lobe rarely exceeding lateral lobes in length, apex acute; surfaces abaxially pale, tomentose, adaxially glossy, glabrous, secondary veins raised on both surfaces. **Acorns** biennial; cup saucer-shaped to cup-shaped, 3–7 mm high × 10–18 mm wide, covering 1/3–1/2 nut, outer surface puberulent, inner surface pubescent, scale tips tightly appressed, acute; nut subglobose, 9–15 × 8–16 mm, often striate, puberulent, scar diam. 5–9 mm.

Flowering spring. Poorly drained bottoms and mesic slopes; 0–300 m; Ala., Ark., Fla., Ga., Ill., Ind., Ky., La., Miss., Mo., N.C., Okla., S.C., Tenn., Tex., Va.

Quercus pagoda is often treated as a variety of *Q. falcata*; it is quite distinctive, however, both morphologically and ecologically (S. A. Ware 1967; R. J. Jensen 1989).

This species reportedly hybridizes with *Q. falcata* and *Q. phellos* (D. M. Hunt 1989).

22. Quercus laevis Walter, Fl. Carol., 234. 1788 · Turkey oak E F W1

Quercus catesbaei Michaux

Trees or shrubs, deciduous, to 20 m. **Bark** bluish gray, deeply furrowed, inner bark orangish or reddish. **Twigs** dark reddish brown with distinct grayish cast, (1.5–)2–3.5(–4) mm diam., sparsely pubescent to almost glabrous. **Terminal buds** light

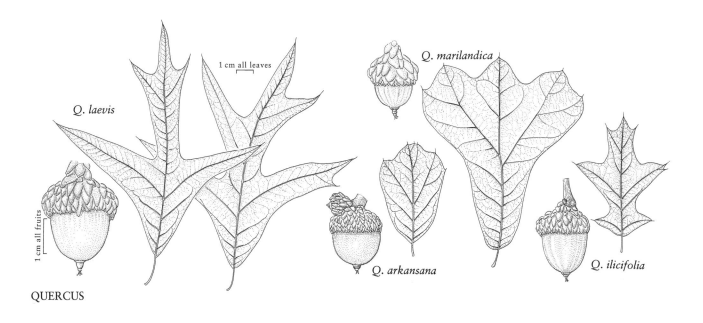

Q. laevis

1 cm all leaves

Q. marilandica

1 cm all fruits

Q. arkansana

Q. ilicifolia

QUERCUS

brown to reddish brown, conic or narrowly ovoid-ellipsoid, 5.5–12 mm, pubescent. **Leaves:** petiole 5–25 mm, glabrous. **Leaf blade** circular or broadly ovate-elliptic, widest near or proximal to middle, 100–200 × 80–150 mm, base attenuate to acute, occasionally obtuse or rounded, blade decurrent on petiole, margins with 3–7(–9) lobes and 7–20 awns, lobes attenuate to falcate, occasionally oblong or distally expanded, apex acute to acuminate; surfaces abaxially occasionally orange-scurfy, usually glabrous except for conspicuous axillary tufts of tomentum, adaxially glabrous, secondary veins raised on both surfaces. **Acorns** biennial; cup somewhat goblet-shaped, 9–14 mm high × 16–24 mm wide, covering 1/3 nut, outer surface puberulent, inner surface pubescent, scales occasionally tuberculate, tips loose, especially at margin of cup, acute, margin conspicuously involute; nut ovoid to broadly ellipsoid, 17–28 × 12–18 mm, often faintly striate, glabrate, scar diam. 6–10 mm.

Flowering early to mid spring. Dry sandy soils of barrens, sandhills, and well-drained ridges; 0–150 m; Ala., Fla., Ga., La., Miss., N.C., S.C., Va.

Quercus laevis reportedly hybridizes with *Q. falcata* (= *Q.* ×*blufftonensis* Trelease), *Q. hemisphaerica*, *Q. incana*, *Q. marilandica* (C. S. Sargent 1918), *Q. nigra*, and (as cited by D. M. Hunt 1989) with *Q. arkansana*, *Q. coccinea*, *Q. myrtifolia*, *Q. phellos*, *Q. shumardii*, and *Q. velutina*.

23. **Quercus arkansana** Sargent, Trees & Shrubs 2: 121. 1911 · Arkansas oak [E] [F]

Quercus caput-rivuli Ashe

Trees, deciduous, to 15 m. **Bark** black with long rough ridges separated by deep furrows. **Twigs** 1.5–3 mm diam., gray-pubescent, rarely glabrate. **Terminal buds** red-brown, ovoid, 2–5 mm, glabrous or with scales somewhat ciliate on margins, especially at apex. **Leaves:** petiole 5–25 mm, pubescent, rarely glabrate. **Leaf blade** rhombic to obovate or obtrullate, 50–150 × 35–100 mm, base acute to cordate, margins entire or with 2–3 shallow lobes and up to 10 awns, apex broadly obtuse to rounded; surfaces abaxially uniformly pubescent to glabrous except for conspicuous axillary tufts of tomentum, veins prominent, adaxially planar or somewhat rugose with a few persistent hairs near base. **Acorns** biennial; cup thin, shallow goblet- to almost saucer-shaped, 5–9 mm high × 10–16 mm wide, covering 1/4–1/2 nut, outer surface puberulent, inner surface sparsely to uniformly pubescent, scale tips appressed, acute; nut broadly ellipsoid to subglobose, 10–15 × 9–15 mm, puberulent, scar diam. 5–10 mm.

Flowering spring. An understory tree of well-drained, sandy soils, on ravine heads (pocosins, steepheads); 50–150 m; Ala., Ark., Fla., Ga., La., Tex.

Quercus arkansana reportedly hybridizes with *Q. incana* (= *Q.* ×*venulosa* Ashe) and *Q. nigra* (D. M. Hunt 1986; W. H. Duncan and M. B. Duncan 1988). While agreeing that an isotype of the former clearly indicated a relationship to *Q. arkansana*, E. J. Palmer

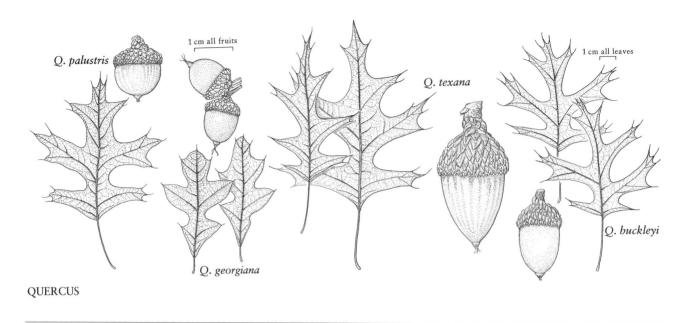

Q. palustris

1 cm all fruits

Q. texana

1 cm all leaves

Q. georgiana

Q. buckleyi

QUERCUS

(1948) questioned the identity of the second parent and noted that *venulosa* is a name of a fossil species. In addition, D. M. Hunt (1989) cited evidence of hybridization with *Q. hemisphaerica, Q. laevis, Q. marilandica, Q. myrtifolia,* and either *Q. falcata* or *Q. velutina.*

24. **Quercus marilandica** Münchhausen, Hausvater 5(1): 253. 1770 · Blackjack oak [E] [F] [W1]

Trees, deciduous, to 15 m. **Bark** almost black, with irregular or rectangular blocks, inner bark orangish. **Twigs** ashy brown, (1–)2–4(–5) mm diam., pubescent to tomentose. **Terminal buds** conic or narrowly ovoid-ellipsoid, 5–10 mm, noticeably 5-angled in cross section, tawny pubescent. **Leaves:** petiole 5–20 mm, densely to sparsely pubescent. **Leaf blade** obovate to obtrullate, (50–)70–200 × (40–)70–200 mm, base rounded or cordate, blade not decurrent, margins with 3–5 shallow, often very broad lobes and 3–10 awns, apex acute to obtuse, rarely rounded; surfaces abaxially scurfy or with scattered pubescence, adaxially glossy, glabrous, secondary veins raised on both surfaces. **Acorns** biennial; cup turbinate, 6–10 mm high × 13–18 mm wide, covering 1/3 nut, outer surface puberulent, inner surface pubescent, scale tips loose, especially at margin of cup, acute or acuminate; nut broadly ovoid or ellipsoid, 12–20 × 10–18 mm, often striate, glabrate, scar diam. 5–8 mm. $2n = 24$. [R. C. Friesner 1930, as $2n = 12$. J. W. Duffield (1940) suggested that Friesner

was counting bivalents; if so, then $2n = 24$, consistent with all other reports for *Quercus.*]

Flowering spring. Poor shallow soils of glades, barrens and flatwoods, disturbed fields, rocky outcrops, and dry ridges; 0–900 m; Ala., Ark., Del., D.C., Fla., Ga., Ill., Ind., Iowa., Kans., Ky., La., Md., Miss., Mo., Nebr., N.J., N.Y., N.C., Ohio, Okla., Pa., S.C., Tenn., Tex., Va., W.Va.

Evidence suggests that small trees from the western portion of the range (primarily Texas and Oklahoma) should be recognized as *Quercus marilandica* Münchhausen var. *ashei* Sudworth (D. M. Hunt 1989). These trees are characterized by 50–70 × 40–60 mm leaf blades with abaxial surfaces gray-tomentose in vein axils.

Quercus marilandica reportedly hybridizes with *Q. buckleyi, Q. falcata* (E. J. Palmer 1948), *Q. georgiana, Q. ilicifolia* (= *Q.* ×*brittonii* W. T. Davis), *Q. imbricaria* [= *Q.* ×*tridentata* (A. de Candolle) Engelmann], *Q. incana, Q. nigra, Q. phellos* (= *Q.* ×*rudkinii* Britton), and *Q. velutina* (= *Q.* ×*bushii* Sargent). D. M. Hunt (1989) cited evidence of hybridization with *Q. arkansana, Q. hemisphaerica, Q. laevis, Q. laurifolia, Q. myrtifolia, Q. palustris,* and *Q. rubra.*

The Choctaw used *Quercus marilandica* to ease cramps and to aid in childbirth (D. E. Moerman 1986).

25. Quercus ilicifolia Wangenheim, Beytr. Teut. Forstwiss., 79, plate 6, fig. 17. 1787 · Bear oak, scrub oak E F

Quercus nana (Marshall) Sargent

Trees or shrubs, deciduous, to 6 m. **Bark** dark gray, becoming shallowly fissured and scaly, inner bark pinkish. **Twigs** brown or yellowish brown, 1.5–3 mm diam., pubescent. **Terminal buds** dark reddish brown, ovoid, 2–4.5 mm, apex puberulent. **Leaves:** petiole (8–)10–25 mm, pubescent. **Leaf blade** ovate to elliptic or obovate, 50–120 × 30–90 mm, base cuneate to obtuse, margins with 3–7 acute lobes separated by shallow sinuses and 5–14 awns, apex acute; surfaces abaxially pale green to gray, tomentose, adaxially glossy dark green, glabrous, secondary veins raised on both surfaces. **Acorns** biennial; cup saucer-shaped to cup-shaped, 5–9 mm high × 10.5–17 mm wide, covering 1/4–1/3(–1/2) nut, outer surface puberulent, inner surface pubescent, scale tips tightly appressed, acute; nut ovoid to subglobose, 9.5–16 × 8–11 mm, striate, puberulent, scar diam. 4.8–7 mm. $2n = 24$.

Flowering spring. Dry, sandy soils and open rocky outcrops; 0–1500 m; Ont.; Conn., Del., Maine, Md., Mass., N.H., N.J., N.Y., N.C., Pa., R.I., Vt., Va., W.Va.

Quercus ilicifolia reportedly hybridizes with *Q. coccinea, Q. falcata, Q. imbricaria, Q. marilandica, Q. phellos, Q. rubra,* and *Q. velutina.*

The Iroquois considered *Quercus ilicifolia* very helpful in treating gynecological problems (D. E. Moerman 1986).

26. Quercus palustris Münchhausen, Hausvater 5(1): 253. 1770 · Pin oak E F W1

Trees, deciduous, to 25 m. **Bark** grayish brown, fissures broad, shallow, inner bark pinkish. **Twigs** reddish brown, 1.5–3(–4) mm diam., soon becoming glabrous. **Terminal buds** brown to reddish brown, ovoid, 3–5 mm, glabrous or with a few fine hairs at apex. **Leaves:** petiole 20–60 mm, glabrous. **Leaf blade** elliptic to oblong, 50–160 × 50–120 mm, base cuneate to broadly obtuse or truncate with basal pair of lobes often somewhat recurved, margins with 5–7 lobes and 10–30 awns, lobes acute or attenuate or distally expanded, apex acute to acuminate; surfaces abaxially glabrous except for conspicuous axillary tufts of tomentum, veins raised, adaxially planar, glabrous. **Acorns** biennial; cup thin, saucer-shaped, 3–6 mm high × 9.5–16 mm wide, covering

1/4 nut, outer surface glabrous or puberulent, inner surface glabrous or with a few hairs around scar, scale tips tightly appressed, acute to obtuse; nut globose or ovoid, 10–16 × 9–15 mm, often conspicuously striate, glabrous, scar diam. 5.5–9 mm. $2n = 24$.

Flowering spring. Bottoms and poorly drained upland clay soils; 0–350 m; Ont.; Ark., Conn., Del., D.C., Ill., Ind., Iowa, Kans., Ky., Md., Mass., Mich., Mo., N.J., N.Y., N.C., Ohio, Okla., Pa., R.I., Tenn., Va., W.Va., Wis.

Quercus palustris is especially common in landscape and street plantings. Its persistent dead branchlets (pins) and branching pattern (drooping lower branches, horizontal middle branches, ascending upper branches) are quite distinctive.

This species reportedly hybridizes with *Quercus coccinea* (E. J. Palmer 1948), *Q. imbricaria* (= *Q. ×exacta* Trelease), *Q. marilandica, Q. nigra, Q. phellos* (= *Q. ×schochiana* Dieck), *Q. rubra, Q. shumardii,* and *Q. velutina.*

Some Native American tribes used infusions prepared from the bark of *Quercus palustris* to alleviate intestinal pains (D. E. Moerman 1986).

27. Quercus georgiana M. A. Curtis, Amer. J. Sci. Arts, ser. 2, 7: 406. 1849 · Georgia oak E F

Trees, deciduous, to 15 m. **Bark** gray to light brown, scaly. **Twigs** deep red, 1–2 mm diam., glabrous. **Terminal buds** red-brown, ovoid to subconic, 2.5–5 mm, glabrous or scales somewhat ciliate. **Leaves:** petiole 6–23 mm, glabrous or with a few persistent hairs. **Leaf blade** broadly ovate to elliptic or obovate, 40–130 × 20–90 mm, base cuneate to obtuse, margins with 3–5(–7) acute or oblong lobes and up to 10 awns, apex oblong to acute; surfaces abaxially glabrous except for conspicuous axillary tufts of tomentum, veins raised, adaxially planar, glabrous. **Acorns** biennial; cup thin, saucer-shaped, 4–6 mm high × 9–14 mm wide, covering 1/3 nut, outer surface puberulent, inner surface glabrous or with a few hairs around scar, scale tips appressed, acute; nut globose or ovoid, 9–14 × 9–14 mm, glabrous, scar diam. 4–7.5 mm.

Flowering spring. Granitic outcrops and dry slopes and knolls; 50–500 m; Ala., Ga., S.C.

Quercus georgiana reportedly hybridizes with *Q. marilandica* (= *Q. ×smallii* Trelease) and *Q. nigra,* although D. M. Hunt (1989) has questioned the validity of the former report.

28. **Quercus texana** Buckley, Proc. Acad. Nat. Sci. Philadelphia 12: 444. 1860 · Texas red oak, Nuttall's oak E F

Quercus nuttallii E. J. Palmer; *Q. rubra* Linnaeus var. *texana* (Buckley) Buckley; *Q. shumardii* Buckley var. *texana* (Buckley) Ashe

Trees, deciduous, to 25 m. **Bark** dark brown with flat ridges divided by shallow fissures. **Twigs** red-brown to gray, 1.5–3(–3.5) mm diam., glabrous. **Terminal buds** gray to gray-brown, ovoid, 3–7 mm, glabrous or with scales somewhat ciliate at apex. **Leaves:** petiole 20–50 mm, glabrous. **Leaf blade** ovate to elliptic or obovate, 75–200 × 55–130 mm, base cuneate to almost truncate, often inequilateral, margins with 6–11 lobes and 9–24 awns, lobes acute to distally expanded, rarely falcate, apex acute; surfaces abaxially glabrous except for conspicuous axillary tufts of tomentum, veins raised, adaxially planar, glabrous. **Acorns** biennial; cup thin (scale bases visible on inner surface), deeply goblet-shaped with pronounced constriction at base, 10–16 mm high × 15–22 mm wide, covering 1/3–1/2 nut, outer surface glabrous to sparsely puberulent, inner surface sparsely to uniformly pubescent, scale tips appressed, acute; nut broadly ovoid to broadly ellipsoid, 15–26 × 13–18 mm, glabrous or sparsely puberulent, scar diam. 8–13 mm, scar often orangish.

Flowering spring. Flood plains and bottomlands; 0–200 m; Ala., Ark., Ill., Ky., La., Miss., Mo., Tenn., Tex.

For many years the name *Quercus texana* was erroneously used for *Q. buckleyi* (L. J. Dorr and K. C. Nixon 1985). A few authors have also used the name for *Q. gravesii*.

Quercus nuttallii E. J. Palmer var. *cachensis* E. J. Palmer was described as a small-fruited form (nuts 16–18 × 12–16 mm) from specimens collected in east-central Arkansas (E. J. Palmer 1937). Noting the similarity between *Q. nuttallii* var. *cachensis* and *Q. palustris,* Palmer discounted the possibility of the former being of hybrid origin because (1) he had not observed *Q. palustris* in the type locality, and (2) the leaves and buds of the former were essentially the same as in *Q. nuttallii* var. *nuttallii.*

C. H. Muller (1942), on the other hand, argued that *Quercus nuttallii* was nothing more than a form [forma *nuttallii* (E. J. Palmer) C. H. Muller] of *Q. palustris.* This is a puzzling conclusion because it was based largely on the premise that *Q. nuttallii* occurred "... with the parent species throughout a large part of the latter's southern range (Mississippi to eastern Texas and southeastern Missouri)." The range of *Q.*

palustris does not extend into Mississippi or eastern Texas, although its range does overlap that of *Q. texana* in eastern Arkansas and southeastern Missouri. E. J. Palmer (1948) and D. M. Hunt (1989) have suggested hybridization with *Q. shumardii* and *Q. nigra,* respectively. See L. J. Dorr and K. C. Nixon (1985) for an explanation of the nomenclatural confusion regarding this taxon.

29. **Quercus buckleyi** Nixon & Dorr, Taxon 34: 225. 1985 · Buckley's oak E F

Trees, deciduous, to 15 m. **Bark** gray and smooth or black and furrowed. **Twigs** grayish brown to red-brown, rarely somewhat yellowish or gray, 1.5–3 mm diam., glabrous. **Terminal buds** grayish brown to reddish brown, ovoid to subfusiform, (2.5–)3–7 mm, scales on apical 1/2 distinctly ciliate. **Leaves:** petiole 20–45 mm, glabrous. **Leaf blade** circular or broadly elliptic to obovate, 55–100 × 51–112 mm, base cuneate to truncate, often inequilateral, margins with 7–9 lobes and 12–35 awns, lobes acute to distally expanded, apex acute to acuminate; surfaces abaxially light green or coppery green, glabrous or with small, axillary tufts of tomentum, adaxially shiny or glossy, glabrous, veins raised on both surfaces. **Acorns** biennial; cup goblet- to cup-shaped, rarely saucer-shaped, 5–11.5 mm high × 10–18 mm wide, covering 1/3–1/2 nut, outer surface glabrous to sparsely puberulent, inner surface glabrous except for a few hairs around scar, scales acute, less than 4 mm long, occasionally tuberculate, especially at base of cup, tips appressed; nut broadly ovoid to broadly ellipsoid, rarely oblong, 12–18.5 × 8–14 mm, occasionally with faint rings of pits at apex, glabrous or sparsely puberulent, scar diam. 3.5–8 mm.

Flowering spring. Limestone ridges and slopes, creek bottoms, occasionally along larger streams; 150–500 m; Okla., Tex.

For many years the names *Quercus texana, Q. rubra* var. *texana,* and *Q. shumardii* var. *texana* were erroneously used for *Q. buckleyi* (L. J. Dorr and K. C. Nixon 1985).

Quercus buckleyi reportedly hybridizes with *Q. marilandica* (= *Q.* ×*hastingsii* Sargent). *Quercus* ×*hastingsii* may be derived from *Q. marilandica* var. *ashei* (D. M. Hunt 1989). Hybridization with *Q. shumardii* may also occur (L. J. Dorr and K. C. Nixon 1985).

30. Quercus shumardii Buckley, Proc. Acad. Nat. Sci. Philadelphia 12: 444. 1860 · Shumard oak E F

Quercus schneckii Britton

Trees, deciduous, to 35 m. **Bark** gray-brown to dark brown, shallowly fissured with scaly or light-colored flat ridges, inner bark pinkish. **Twigs** gray to light brown, (1.5–)2–3.5(–4.5) mm diam., glabrous. **Terminal buds** gray to grayish brown, ovoid or broadly ellipsoid, 4–8 mm, often noticeably 5-angled in cross section, glabrous. **Leaves:** petiole 20–60 mm, glabrous. **Leaf blade** broadly elliptic to obovate, 100–200 × 60–150 mm, base obtuse to truncate, occasionally acute, margins with 5–9 lobes and 15–50 awns, lobes oblong or distally expanded, apex acute; surfaces abaxially glabrous except for prominent axillary tufts of tomentum, adaxially glossy, glabrous, secondary veins raised on both surfaces. **Acorns** biennial; cup saucer-shaped to cup-shaped, 7–12 mm high × 15–30 mm wide, covering 1/4–1/3 nut, outer surface glabrous or puberulent, inner surface light-brown to red-brown, glabrous or with ring of pubescence around scar, scales often with pale margins, tips tightly appressed, obtuse or acute; nut ovoid to oblong, occasionally subglobose, 14–30 × 10–20 mm, glabrous, scar diam. 6.5–12 mm.

Flowering spring. Mesic slopes and bottoms, stream banks and poorly drained uplands; 0–500 m; Ont.; Ala., Ark., Fla., Ga., Ill., Ind., Kans., Ky., La., Md., Mich., Miss., Mo., Nebr., N.C., Ohio, Okla., Pa., S.C., Tenn., Tex., Va., W.Va.

Trees with shallow cups covering ca. one-fourth of the nut are treated as *Quercus shumardii* var. *shumardii;* those with more deeply rounded cups covering ca. one-third of the nut are treated as *Q. shumardii* var. *schneckii* (Britton) Sargent. *Quercus shumardii* var. *stenocarpa* Laughlin was described from several trees in Missouri and Illinois having ellipsoid acorns that were covered less than one-third their length by very small (5.5–7 mm high × 12.5–18 mm wide), shallow cups (K. Laughlin 1969).

Quercus shumardii reportedly hybridizes with *Q. buckleyi, Q. falcata* (= *Q.* ×*joorii* Trelease), *Q. hemisphaerica, Q. imbricaria* (= *Q.* ×*egglestonii* Trelease), *Q. laevis, Q. laurifolia, Q. marilandica* (see *Q. buckleyi*), *Q. nigra, Q. palustris* (= *Q.* ×*mutabilis* E. J. Palmer & Steyermark), *Q. phellos* (= *Q.* × *moultonensis* Ashe), *Q. rubra,* and *Q. velutina* (= *Q.* × *discreta* Laughlin).

31. Quercus acerifolia (E. J. Palmer) Stoynoff & W. J. Hess, Sida 14: 268. 1990 · Maple-leaf oak C E F

Quercus shumardii Buckley var. *acerifolia* E. J. Palmer, J. Arnold Arbor, 8: 54. 1927

Trees or shrubs, deciduous, to 15 m. **Bark** dark gray to almost black, sometimes becoming rough and furrowed. **Twigs** grayish brown to reddish brown, 1.5–3(–3.5) mm diam., glabrous or sparsely pubescent. **Terminal buds** gray to grayish brown, ovoid or broadly ellipsoid, 3.5–5.5 mm, glabrous. **Leaves:** petiole 20–45 mm, glabrous. **Leaf blade** oblate to broadly elliptic, 70–140 × (60–)100–150(–180) mm, base cordate-truncate to obtuse, margins with 5–7(–9) lobes and 11–48 awns, lobes ovate-oblong or markedly distally expanded, the middle or apical pairs often overlapping, apex acute; surfaces abaxially glabrous or with prominent axillary tufts of tomentum, occasionally with scattered pubescence, adaxially glabrous, secondary veins raised on both surfaces. **Acorns** biennial; cup saucer- to cup-shaped, 4–7 mm high × 10–20 mm wide, covering 1/4–1/3 nut, outer surface glabrous or puberulent, inner surface light brown to red-brown, glabrous or with ring of pubescence around scar, scales often with pale margins, tips tightly appressed, obtuse or acute; nut ovoid to oblong, 10.5–20 × 9–15 mm, glabrous or pubescent, scar diam. 5–9 mm.

Flowering spring. Dry glades, slopes, and ridge tops; of conservation concern; 500–800 m; Ark.

Quercus acerifolia is known only from four localities in Arkansas: Magazine Mountain, Logan County; Porter Mountain, Polk County; Pryor Mountain, Montgomery County; and Sugarloaf Mountain, Sebastian County (N. Stoynoff and W. J. Hess 1990; G. P. Johnson 1992, 1994). Some specimens suggest hybridization with *Q. marilandica* and/or *Q. velutina,* but no hybrids have been reported.

32. Quercus rubra Linnaeus, Sp. Pl. 2: 996. 1753 · Northern red oak, chêne rouge E F

Quercus borealis Michaux; *Q. maxima* Ashe

Trees, deciduous, to 30 m. **Bark** gray or dark gray, ridges wide, shiny, separated by shallow fissures, inner bark pinkish. **Twigs** reddish brown, 2–3.5(–4.5) mm diam., glabrous. **Terminal buds** dark reddish brown, ovoid to ellipsoid, 4–7 mm, glabrous or with tuft of reddish hairs at apex. **Leaves:** petiole 25–50 mm, glabrous, often red tinged. **Leaf blade** ovate to elliptic or obovate,

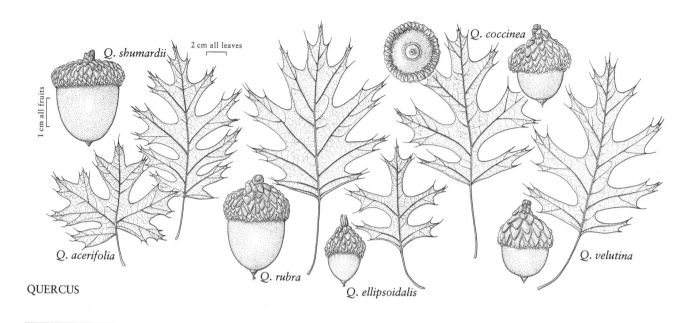

Q. shumardii   2 cm all leaves

1 cm all fruits

Q. coccinea

Q. acerifolia

Q. rubra

Q. ellipsoidalis

Q. velutina

QUERCUS

120–200 × 60–120 mm, base broadly cuneate to almost truncate, margins with 7–11 lobes and 12–50 awns, lobes acute to oblong, occasionally distally expanded, separated by shallow sinuses, sinuses usually extending less than 1/2 distance to midrib, apex acute; surfaces abaxially pale green, often glaucous, glabrous except for minute axillary tufts of tomentum, adaxially dull green, glabrous, secondary veins raised on both surfaces. **Acorns** biennial; cup saucer- to cup-shaped, 5–12 mm high × 18–30 mm wide, covering 1/4–1/3 nut, outer surface puberulent, inner surface light brown to red-brown, glabrous or with ring of pubescence around scar, scales less than 4 mm long, often with dark margins, tips tightly appressed, obtuse; nut ovoid to oblong, 15–30 × 10–21 mm, glabrous, scar diam. 6.5–12.5 mm. $2n = 24$.

Flowering spring. Commonly on mesic slopes and well-drained uplands, occasionally on dry slopes or poorly drained uplands; 0–1800 m; N.B., N.S., Ont., P.E.I., Que.; Ala., Ark., Conn., Del., D.C., Ga., Ill., Ind., Iowa, Kans., Ky., Maine, Md., Mass., Mich., Minn., Miss., Mo., Nebr., N.H., N.J., N.Y., N.C., Ohio, Okla., Pa., R.I., S.C., Tenn., Vt., Va., W.Va., Wis.

Trees with large nuts only one-fourth covered by flat saucer-shaped cups often are treated as *Quercus rubra* var. *rubra*; those with smaller nuts one-third covered by cup- or bowl-shaped cups are treated as *Q. rubra* var. *borealis* (F. Michaux) Farwell. While E. J. Palmer (1942) suggested that these two varieties do not breed true, K. M. McDougal and C. R. Parks (1986) found evidence of correspondence between morphologic types and flavonoid chemotypes. This is one of the most important ornamental and timber trees in the genus.

Native Americans used *Quercus rubra* for a number of medicinal purposes, including the treatment of sores, weakness, lung problems, sore throat, dysentery, indigestion, chapped skin, chills and fevers, lost voice, asthma, cough, milky urine, heart trouble, and blood diseases, and as an appetizer (D. E. Moerman 1986).

Quercus rubra reportedly hybridizes with *Q. coccinea* (= *Q.* ×*benderi* Baenitz), *Q. ellipsoidalis* (P. C. Swain 1972; R. J. Jensen et al. 1993), *Q. ilicifolia* (= *Q.* ×*fernaldii* Trelease), *Q. imbricaria* [= *Q.* ×*runcinata* (A. de Candolle) Engelmann], *Q. marilandica* (E. J. Palmer 1948; D. M. Hunt 1989), *Q. nigra* (D. M. Hunt 1989), *Q. palustris* (= *Q.* ×*columnaris* Laughlin), *Q. phellos* (= *Q.* ×*heterophylla* F. Michaux), *Q. shumardii* (= *Q.* ×*riparia* Laughlin), and *Q. velutina* (= *Q.* ×*hawkinsii* Sudworth).

33. Quercus coccinea Münchhausen, Hausvater 5(1): 254. 1770 · Scarlet oak [E] [F]

Quercus coccinea var. *tuberculata* Sargent

Trees, deciduous, to 30 m; lower trunk without stubs of dead branches. **Bark** dark gray to dark brown, irregularly fissured with scaly ridges, inner bark orangish pink. **Twigs** reddish brown, (1–)2–3.5 mm diam., glabrous. **Terminal buds** dark reddish brown, conic to ovoid, 4–7 mm, noticeably 5-angled in cross section, usually silvery- or tawny-pubescent distal to middle. **Leaves:** petiole 25–60 mm, glabrous. **Leaf blade** elliptic to ovate or obovate, 70–160 × 80–130 mm, base obtuse to truncate, margins with 5–9 deep lobes and 18–50 awns, lobes distally expanded, sinuses usually extending more than 1/2 distance to midrib, apex acute; surfaces abaxially glabrous except for minute axillary tufts of tomentum, adaxially glossy light green, glabrous, secondary veins raised on both surfaces. **Acorns** biennial; cup turbinate to hemispheric, 7–13 mm high × 16.5–31.5 mm wide, covering 1/3–1/2 nut, outer surface light to dark reddish brown, glossy, glabrous to puberulent, inner surface light brown, glabrous, occasionally with ring of pubescence around scar, scales often tuberculate, base broad, glossy, margins strongly concave with tips tightly appressed, acute to attenuate; nut oblong to subglobose, 12–22 × 10–21 mm, glabrous, with 1 or more rings of fine pits at apex, scar diam. 6.5–13.5 mm. $2n = 24$.

Flowering spring. Poor soils, well-drained uplands, dry slopes, and ridges, occasionally on poorly drained sites; 0–1500 m; Ala., Ark., Conn., Del., D.C., Ga., Ill., Ind., Ky., Maine, Md., Mass., Mich., Miss., Mo., N.H., N.J., N.Y., N.C., Ohio, Pa., R.I., S.C., Tenn., Vt., Va., W.Va., Wis.

Trees having acorns with broad, distinctly warty cups are sometimes classified as *Quercus coccinea* var. *tuberculata* Sargent.

Quercus coccinea reportedly hybridizes with *Q. imbricaria, Q. ilicifolia* (= *Q.* ×*robbinsii* Trelease), *Q. laevis, Q. palustris* (E. J. Palmer 1948), *Q. phellos, Q. rubra,* and *Q. velutina* (= *Q.* ×*fontana* Laughlin).

34. Quercus ellipsoidalis E. J. Hill, Bot. Gaz. 27: 204, plates 2, 3. 1899 · Northern pin oak, jack oak, Hill's oak E F

Quercus ellipsoidalis var. *kaposianensis* J. W. Moore

Trees, deciduous, to 20 m; lower trunk often with stubs of dead branches. **Bark** dark gray-brown, shallowly fissured, inner bark orangish. **Twigs** dark reddish brown, (1–)1.5–3 mm diam., glabrous. **Terminal buds** dark reddish brown, ovoid, 3–5 mm, often conspicuously 5-angled in cross section, usually silvery- or tawny-pubescent toward apex. **Leaves:** petiole 20–50 mm, glabrous. **Leaf blade** elliptic, 70–130 × 50–100 mm, base obtuse to truncate, margins with 5–7 deep lobes and 15–55 awns, lobes distally expanded, sinuses usually extending more than 1/2 distance to midrib, apex acute; surfaces abaxially glabrous except for minute axillary tufts of tomentum, adaxially glossy light green, glabrous, secondary veins raised on both surfaces. **Acorns** biennial; cup narrowly turbinate to deeply cup-shaped, 6–11 mm high × 10–19 mm wide, covering 1/3–1/2 nut, outer surface reddish brown, puberulent, inner surface light brown, glabrous, rarely with ring of pubescence around scar, scales with straight or slightly concave margins, tips tightly appressed, obtuse or acute; nut ellipsoid to ovoid, rarely subglobose, 10–20 × 9–15 mm, occasionally striate, glabrous, occasionally with 1 or more faint rings of fine pits at apex, scar diam. 4–8 mm.

Flowering spring. Dry sandy sites, rarely on moderately mesic slopes or uplands; 150–500 m; Ont.; Ill., Ind., Iowa, Mich., Minn., Ohio, Wis.

In many treatments (e.g., E. G. Voss 1972+, vol. 2), *Quercus ellipsoidalis* is included in *Q. coccinea.* Variation in fruit morphology has led to recognition of several formae (W. Trelease 1919; see also R. J. Jensen 1986) and one variety (*Q. ellipsoidalis* var. *kaposianensis,* based on specimens from St. Paul, Minnesota, in which the cup tightly encloses the nut for two-thirds its length at maturity).

Quercus ellipsoidalis reportedly hybridizes with *Q. rubra* and *Q. velutina.*

The Menominee used *Quercus ellipsoidalis* medicinally to treat suppressed menses caused by cold (D. E. Moerman 1986).

35. Quercus velutina Lamarck in J. Lamarck et al., Encycl. 1: 721. 1785 · Black oak E F

Quercus tinctoria W. Bartram

Trees, deciduous, to 25 m. **Bark** dark brown to black, deeply furrowed, ridges often broken into irregular blocks, inner bark yellow or orange. **Twigs** dark reddish brown, (1.5–)2.5–4.5(–5) mm diam., glabrous or sparsely pubescent. **Terminal buds** ovoid or ellipsoid to subconic, 6–12 mm, noticeably 5-angled in cross section, tawny- or gray-pubescent. **Leaves:** petiole 25–70 mm, glabrous to sparsely pubescent. **Leaf blade** ovate to obovate, (80–)100–300 × 80–150 mm, base obtuse to truncate, inequilateral, margins with 5–9 lobes and 15–50 awns, lobes oblong or distally expanded, separated by deep sinuses, apex acute to obtuse; surfaces abaxially pale green, glabrous except for small axillary tufts of tomentum or with scattered pubescence, especially along veins, adaxially glossy, dark green, glabrous, secondary veins raised on both surfaces. **Acorns** biennial; cup cup-shaped or turbinate, 7–14 mm high × 12–22 mm wide, covering 1/2 nut, cup margins not involute, outer surface puberulent, inner surface pubescent, scale tips loose, especially at margin of cup, acute to acuminate; nut subglobose to ovoid, 10–20 × 10–18 mm, glabrate, scar diam. 5.5–12 mm. $n = 12 \pm 1$; $2n = 24$.

Flowering spring. Commonly on dry slopes and upland areas, occasionally on sandy lowlands (especially in north) and poorly drained uplands and terraces; 0–1500 m; Ont.; Ala., Ark., Conn., Del., D.C., Fla., Ga., Ill., Ind., Iowa, Kans., Ky., La., Maine, Md., Mass., Mich., Minn., Miss., Mo., Nebr., N.H., N.J., N.Y., N.C., Ohio, Okla., Pa., R.I., S.C., Tenn., Tex., Vt., Va., W.Va., Wis.

The bark of this species (quercitron) is rich in tannins and was once an important source of these chemicals used for tanning leather. (The yellow dye obtained from the bark is also called quercitron.)

Native Americans used *Quercus velutina* medicinally for indigestion, chronic dysentery, mouth sores, chills and fevers, chapped skin, hoarseness, milky urine, lung trouble, sore eyes, and as a tonic, an antiseptic, and an emetic (D. E. Moerman 1986).

Quercus velutina reportedly hybridizes with *Q. coccinea*, *Q. ellipsoidalis* (= *Q.* ×*palaeolithicola* Trelease), *Q. falcata* [= *Q.* ×*willdenowiana* (Dippel) Zabel] (= *Q.* ×*pinetorum* Moldenke)], *Q. ilicifolia* (= *Q.* ×*rehderi* Trelease), *Q. imbricaria* (= *Q.* ×*leana* Nuttall), *Q. incana*, *Q. laevis*, *Q. laurifolia* (= *Q.* ×*cocksii* Sargent, although E. J. Palmer [1948] challenged the validity of this claim), *Q. marilandica*, *Q. nigra*, *Q. palustris* (= *Q.* ×*vaga* E. J. Palmer & Steyermark), *Q. phellos* (= *Q.* ×*filialis* Little), *Q. rubra*, *Q. shumardii*, and possibly *Q. arkansana* (D. M. Hunt 1989).

5b. Quercus Linnaeus sect. Protobalanus (Trelease) A. Camus, Chênes 1: 157. 1938 · Intermediate oaks, golden-cup oaks

Paul S. Manos

Quercus Linnaeus subg. *Protobalanus* Trelease in P. C. Standley, Trees Shrubs Mexico 2: 176. 1922

Trees or shrubs, evergreen. **Bark** grayish white to reddish brown, scaly to smooth with furrows. **Leaf blade** never lobed, margins entire or toothed, teeth if present usually spinose. **Staminate flowers:** calyx 5–6-lobed; anthers attenuate-apiculate. **Pistillate flowers:** calyx adnate to ovary, not forming flange; styles short and dilated to long and abruptly enlarged. **Acorns:** maturation biennial; cup pedunculate, cup scales distinct or laterally connate, base thickened and corky (tuberculate); nut with inner wall densely to sparsely tomentose, abortive ovules apical to lateral or basal, sometimes variable within individual plants, seed coats adhering to seed, sometimes to fruit wall at maturity. **Cotyledons** distinct.

Measurements for the cup of the acorn in the next two sections is for depth of cup, not height.

Species 5 (4 in the flora): sw North America and nw Mexico.

SELECTED REFERENCES Manos, P. S. 1993. Foliar trichome variation in *Quercus* section *Protobalanus* (Fagaceae). Sida 15: 391–403. Tucker, J. M. and H. S. Haskell. 1960. *Quercus dunnii* and *Q. chrysolepis* in Arizona. Brittonia 12: 196–219.

Key to species in *Quercus* sect. *Protobalanus*

1. Twigs rigid, 1.5–3 mm diam., branching at 65–90° angles; cup scales obscured, laterally connate into concentric rings; nut oblong to fusiform . . . . 38. *Quercus palmeri*
1. Twigs flexible, if stiff 3–4 mm diam., branching at less than 60° angles; cup scales often laterally connate and embedded in tomentum, not in noticeable concentric rings; nut ovoid.
  2. Leaf blade glossy dark green, brittle, margins entire to mucronately toothed, often strongly revolute; surfaces abaxially densely tomentose . . . . 39. *Quercus tomentella*
  2. Leaf blade dull gray-green to yellowish green, leathery, margins entire to spinescent or mucronately toothed, usually flat; surfaces abaxially glabrous or pubescent, not tomentose.
    3. Shrubs; leaf margins entire to mucronately toothed, surfaces glabrous or pubescent with sparse cover of minute stellate hairs, glandular hairs absent; nut 8–17 mm, cup thin, 10–15 mm wide; nut scar to 3 mm diam . . . . 37. *Quercus vaccinifolia*
    3. Trees or large shrubs; leaf margins entire to spinescent, surfaces glabrate to pubescent with golden glandular and multiradiate hairs; nut 15–30 mm, cup thick, 15–40 mm wide; nut scar 4–10 mm diam . . . . 36. *Quercus chrysolepis*

36. Quercus chrysolepis Liebmann, Overs. Kongel. Danske Vidensk. Selsk. Forh. Medlemmers Arbeider 1854: 173. 1854 · Canyon live oak, maul oak F

Quercus chrysolepis var. *nana* (Jepson) Jepson; *Q. wilcoxii* Rydberg

Trees or shrubs, trees small to medium-sized, to 25 m, shrubs of variable size. **Twigs** branching at 60° angles or less, golden brown, 1–2 mm diam., flexible, densely pubescent 1st year, moderately so 2d year. **Terminal buds** conic, 2–8 mm, scales brown with ciliate margins. **Leaves:** petiole 3–14 mm, rusty-pubescent, adaxially flattened. **Leaf blade** oblong, acuminate, usually flat to slightly concave, 20–70 × 10–35 mm, thick, leathery, base obtuse to rounded, secondary veins 12 or more pairs, branching at ca. 50° angles, slightly raised abaxially, margins often slightly revolute with moderately thickened cell walls, entire to spinulose-dentate (especially on juvenile growth), regularly toothed, teeth terminating with mucronate to spinescent tip, apex acute or obtuse, mucronate to spinescent; surfaces abaxially glabrate to pubescent with bluish white wax layer, often obscured by golden glandular and multiradiate hairs, adaxially yellowish green, scurfy with multiradiate hairs, later in season slightly pubescent. **Acorns** solitary or paired, rarely in 3s or 4s; cup saucer-shaped, 4–10 mm deep × 15–40 mm wide, rims often corky and thickened, scales appressed, deeply embedded in tomentum, often appearing swollen and keeled, tuberculate; nut ovoid, 15–30 × 10–20 mm, apex blunt, glabrous; nut scar 4–10 mm diam.

Flowering usually in spring, occasionally in fall. Mountain ridges, canyons, and moist slopes; 200–2600 m; Ariz, Calif., Nev., N.Mex., Oreg.; Mexico (Baja California and Chihuahua).

Quercus chrysolepis is one of the most variable North American oaks. Historically, individuals with extreme variation in fruit and leaf characteristics led to the recognition of several varieties and forms; for example, shrubs with small leaves have been called *Q. chrysolepis* var. *nana* (Jepson) Jepson. Studies of quantitative and qualitative variation in these characteristics do not support the recognition of infraspecific taxa; geographic variation is apparent, however, based on populations with nearly stabilized character combinations that loosely define widespread variants. This oak is distinguished consistently from other species of the complex by the presence of multiradiate trichomes on both leaf surfaces (P. S. Manos 1993). The leaf morphology and branching habit present on juvenile growth, suckers, and shade forms may approach typical *Q. palmeri*. Similarly, various extreme forms often resemble other species of the complex.

Putative hybrids have been reported from narrow zones of range overlap with the three other North American species of *Quercus* sect. *Protobalanus*.

The Mendocin Indians considered the nuts of *Quercus chrysolepis* poisonous (D. E. Moerman 1986).

37. Quercus vaccinifolia Kellogg ex Curran, Bull. Calif. Acad. Sci. 1: 146. 1885 · Huckleberry oak E F

Shrubs, low spreading to often prostrate, to 1.5 m. **Twigs** branching at 45° angles or less, reddish brown, 1–1.5 mm diam., flexible, glabrous to sparsely pubescent. **Terminal buds** conic, 2.5 mm, scales brown with ciliate margins. **Leaves:** petiole 5–8 mm, sparsely pubescent, flattened adaxially. **Leaf blade** oblong-ovate, 10–35 × 7–15 mm, flat, thin, leathery, base slightly rounded to acute, secondary veins inconspicuous, 6–8 pairs, branching at 45–60° angles, with weakly thickened cell walls, margins entire or indistinctly and irregularly mucronately toothed, apex acute or rarely obtuse; surfaces abaxially whitish green with waxy layer, glabrous or slightly pubescent with stellate hairs, adaxially dull gray-green, glabrous or sparsely pubescent with stellate hairs. **Acorns** solitary or rarely paired; cup shallowly saucer-shaped to slightly turbinate, 3–4 mm deep × 10–15 mm wide, scales appressed, slightly embedded, moderately silvery brown-pubescent; nut ovoid, 8–17 × 5–10 mm, apex acute; nut scar to 3 mm diam.

Flowering in early summer. Dry ridges, steep slopes, and rocky areas from montane coniferous zone to near treeline; 900–2800 m; Calif., Nev., Oreg.

Typical high-elevation populations in the Sierra Nevada of California can be distinguished from all shrubby forms of *Quercus chrysolepis* by the absence of glandular trichomes and by thin cups with small nut-attachment scars. At lower elevations in northern California and southwestern Oregon, secondary contact with *Q. chrysolepis* has resulted in the formation of hybrids.

38. Quercus palmeri Engelmann in S. Watson, Bot. California 2: 97. 1880 · Palmer oak F

Quercus dunnii Kellogg ex Curran

Small trees and shrubs, to 2–3 m. **Twigs** rigid, divaricately branched at 65–90° angles, reddish brown, 1.5–3 mm diam., pubescent, sparsely so in 2d year. **Terminal buds** ovoid, 1–1.5 mm, apex rounded, glabrous. **Leaves:** petiole 2–5 mm, round in cross section, glabrous to sparsely

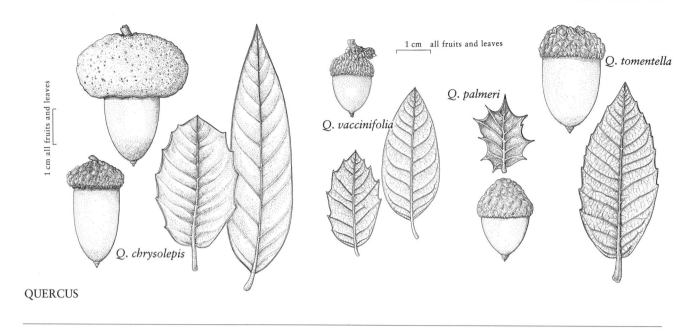

QUERCUS

fasciculate-pubescent. **Leaf blade** suborbiculate, elliptic to round-ovate, 20–30(–50) × 20–40 mm, crisped, leathery and brittle, base obtuse to strongly subcordate, secondary veins 5–8(–12) pairs, each terminating in spine, basal pairs recurving, others branching at 45° angles, raised abaxially, margins spinose-dentate to occasionally entire, with highly thickened cell walls, spines cartilaginous, (1–)1.5–2 mm, apex broadly rounded or subacute, spinose; surfaces abaxially glaucous with waxy layer, often obscured by golden brown glandular hairs, adaxially grayish dark green, scurfy with fasciculate erect and twisting hairs. **Acorns** solitary or rarely paired; cup turbinate to saucer-shaped, margins involute, often irregular, 7–10 mm deep × 10–25(–35) mm wide, scales appressed, embedded, often appearing laterally connate into concentric rings with only tip of scale visible, tuberculate, densely golden-tomentose throughout; nut oblong to fusiform, 20–30 × 10–15 mm, apex acute.

Flowering in spring. Disjunct in canyons, mountain washes, dry thickets, and margins of chapparal communities; 700–1800 m; Ariz, Calif.; Mexico (n Baja California).

Populations of *Quercus palmeri* are often small and may exist as single clones. The disjunct populations of California and Baja California are consistent morphologically. In Arizona populations, individuals tend to have flatter leaves bearing fewer teeth; this distinction is not constant, however. Morphologically aberrant populations identified as *Q. palmeri* in eastern Arizona (Chiracahua, Huachuca, and Santa Catalina mountains) and southwestern New Mexico are most likely the result of introgression from *Q. palmeri* to *Q. chrysolepis* (J. M. Tucker and H. S. Haskell 1960).

Those populations tend to be intermediate in overall morphology, but all lack the diagnostic trichomes and biochemical markers of *Q. palmeri*; they are best classified as *Q. chrysolepis* affinity *Q. palmeri*.

39. Quercus tomentella Engelmann, Trans. Acad. Sci. St. Louis, 3: 393. 1877 · Channel Island oak [F]

Trees, to 20 m. **Twigs** branching at 45° angles, reddish brown, 3–4 mm diam., somewhat rigid, densely tomentose, persistent into 2d year. **Terminal buds** conic, 7–10 mm, scales brown with ciliate margins. **Leaves:** petiole 3–10 mm, rusty-villous, flattened adaxially. **Leaf blade** wavy or distinctly concave, oblong-lanceolate or elliptic, acuminate, (30–)70–10(–120) × 25–40 mm, leathery and brittle, base obtuse to cordate, secondary veins 8–10(–12) pairs, branching at 45–50° angles, strongly pinnate, raised abaxially, often sunken adaxially, margins often strongly revolute, with slightly thickened cell walls, entire or crenate to dentate with mucronate teeth, apex rounded or acute, mucronate; surfaces abaxially densely tomentose with whitish nonglandular hairs, midrib pilose, adaxially glossy dark green, sparsely pubescent. **Acorns** solitary or rarely paired; cup shallowly cup-shaped, 4–8 mm deep × 15–30 mm wide, scales laterally connate, appressed, deeply imbedded in tomentum, with only thin, brown, elongated apices visible, tuberculate, densely whitish brown tomentose throughout; nut ovoid, 20–30 × 15–20 mm, apex rounded.

Flowering in spring, occasionally in fall. Lower portions of steep canyons and occasionally ridge tops; 100–650 m; Calif.; Mexico (Baja California and on Guadalupe Island).

The insular endemic *Quercus tomentella* is a relict as evidenced by its widespread representation in mainland late Tertiary fossil floras. Hybridization with *Q. chrysolepis* is apparent on the Channel Islands: Santa Catalina, Santa Cruz, and possibly San Clemente and Anacapa. Putative hybrids have been observed in narrow zones of contact on the islands of Santa Cruz and Santa Catalina. On those islands, taxonomically distinct individuals of *Q. chrysolepis* occur at the highest elevations, whereas *Q. tomentella* generally is found in moist canyons at lower elevations. Populations of Channel Island oak are in decline because of overgrazing and poor seedling recruitment. The greatest number of populations occur on Santa Rosa Island, and those are taxonomically and genetically noteworthy because *Q. chrysolepis* apparently is absent from the island.

5c. Quercus Linnaeus sect. Quercus · White oaks

Kevin C. Nixon

Cornelius H. Muller

Trees or shrubs, evergreen or deciduous. **Bark** nearly white, gray, brown, or black, smooth, scaly, flaky, or rarely furrowed. **Leaf blade** lobed or unlobed, margins entire or toothed, teeth if present acute or spinose, never bristle-tipped. **Staminate flowers:** calyx 2–6-lobed; anthers usually somewhat apiculate, occasionally retuse. **Pistillate flowers:** calyx adnate to ovary, not forming flange; styles usually abruptly enlarged or dilated, sometimes gradually enlarged to subulate. **Acorns:** maturation annual; cup with scales distinct or laterally connate, keeled or with bases thickened and corky (tuberculate); nut with inner wall glabrate or minutely tomentulose near base and apex, abortive ovules basal, seed coats adhering to fruit wall at maturity. **Cotyledons** distinct or connate.

Species ca. 200 (51 in the flora): North America, Mexico, West Indies, Central America, Eurasia, n Africa.

Key to species in *Quercus* sect. *Quercus*

1. Leaf blade regularly or irregularly lobed, at least some sinuses extending more than 1/3 distance from margin to midrib, or distally 3-lobed (3-dentate).
  2. Mature leaves glabrous or at least appearing so macroscopically on abaxial surface; twigs glabrous.
    3. Leaf blade blue-green, glaucous . . . . .51. *Quercus laceyi*
    3. Leaf blade green or grayish green.
      4. Leaf blade moderately to deeply lobed, lobe apex sharply angular; acorn cup burlike, usually enclosing nut or only apex of nut visible, orifice smaller than nut diameter, often splitting irregularly at maturity . . . . .48. *Quercus lyrata*
      4. Leaf blade sinuate or shallowly to deeply lobed, lobes rounded, sometimes retuse, without acute teeth; acorn cup not burlike, enclosing less than 1/2 nut, orifice ± equaling largest nut diameter, not splitting.
        5. Leaf base strongly cordate; acorn on axillary peduncle 25–100 mm . . . . .50. *Quercus robur*
        5. Leaf base cuneate or attenuate; acorn subsessile or on axillary peduncle 15–25(–50) mm.
          6. Leaf blade moderately to deeply lobed its entire length . . . . .49. *Quercus alba*
          6. Leaf blade merely sinuately lobed or narrowly obovate, distally lobed or 3-dentate, margins proximally 1/3 or more entire or at most sinuate.
            7. Cup shallow, saucer-shaped, enclosing 1/4 nut or less; leaf blade abaxially with minute, appressed-stellate hairs, often silvery, shade leaves sometimes glabrate . . . .77. *Quercus sinuata*
            7. Cup hemispheric or goblet-shaped, often enclosing 1/3–1/2 nut; leaf blade glabrate, abaxially without appressed-stellate hairs, with deciduous scattered, erect, stalked, multiradiate hairs, these often persisting in sheltered spots near midrib. . . 78. *Quercus austrina*
  2. Leaf blade variously, conspicuously hairy on 1 or both surfaces at maturity, or abaxially strongly glandular or waxy; twigs glabrous or hairy, stellate or tomentose.
    8. Twigs initially glabrous or eventually glabrate in 1st season.
      9. Leaf blade abaxially sparsely to densely covered with persistent erect, velvety hairs.

10. Leaf blade abaxially covered with appressed-stellate hairs intermixed with erect velvety hairs; acorn pedunculate, peduncle to 60 mm or more; scales near rim of cup often with stout recurved awns; large trees of wet areas .46. *Quercus bicolor*

10. Leaf blade abaxially with erect velvety hairs but without appressed-stellate hairs; acorn subsessile in leaf axil; scales of cup without awns; shrubs and small trees of sand areas
. . . . .53. *Quercus margaretta*

9. Leaf blade abaxially silvery with minute appressed-stellate hairs, or often glabrate in shade leaves.

11. Axillary peduncle (20–)40 mm or more . . . . .46. *Quercus bicolor*

11. Axillary peduncle to 8 mm (acorn usually subsessile).

12. Cup shallow, enclosing less than 1/2 nut; leaf blade shallowly lobed (sinuses extending less than 1/2 distance to midrib) or undivided, lobes if present rounded . . .77. *Quercus sinuata*

12. Cup deeply hemispheric or goblet-shaped, enclosing 2/3 nut or more; leaf blade deeply lobed (some sinuses extending more than 1/2 distance to midrib), lobes often blunt-acute . . . . .48. *Quercus lyrata*

8. Twigs hairy, stellate or tomentose, sometimes finely so.

13. Leaf blade with a few deep sinuses (extending more than 1/2 distance to midrib) in proximal 1/2 or middle, with shallower sinuses or merely toothed in distal 1/3; older twigs often with flat, radiating, corky wings; cup with soft awns on scales forming fringe along rim . . . . 47. *Quercus macrocarpa*

13. Leaf blade ± regularly lobed, or sinuses deeper toward middle and apex; twigs without corky wings; cup without soft awns along rim, although sometimes (*Q. bicolor*) with scattered short, stout awns, these not forming fringe on rim.

14. Leaf blade abaxially with tight, usually interlocking, appressed or semi-appressed, stellate hairs (these may be absent in shade leaves), without erect straight hairs.

15. Nut fusiform, 30–60 mm; California . . . . 62. *Quercus lobata*

15. Nut oblong, ovoid, or globose, less than 30 mm; c, e United States.

16. Leaf blade shallowly to deeply lobed, often strongly cruciform, 2 larger lobes in distal 1/2 extending at right angles to midrib . . . . .52. *Quercus stellata*

16. Leaf blade narrowly obovate or cuneiform, shallowly lobed, often 3-lobed, lobes ovate or oblong-narrow, conspicuously concentrated toward apex of leaf.

17. Low rhizomatous shrubs, 0.2–2(–3) m, rarely small trees. . . . . 54. *Quercus boyntonii*

17. Forest trees with single straight trunks to 25 m, not rhizomatous. . . . . . . 55. *Quercus similis*

14. Leaf blade abaxially with erect or nearly erect hairs, often felty or velvety to touch.

18. Axillary peduncle (20–)40 mm or more . . . . .46. *Quercus bicolor*

18. Axillary peduncle less than 10 mm, acorn subsessile in leaf axil.

19. Leaf blade shallowly lobed (sinuses extending less than 1/2 distance to midrib), or merely toothed.

20. Leaf blade and petiole blue-green; California . . . . 63. *Quercus douglasii*

20. Leaf blade green or yellow-green; petiole reddish; w Texas. . . . .
. . . . 61. *Quercus carmenensis* (see also hybrids of *Q. gambelii*)

19. Leaf blade with sinuses extending 1/2 or more distance from margin to midrib.

21. Terminal buds 2–12 mm; nut (12–)25–30(–40) mm; Pacific slope; California, Oregon, Washington, British Columbia. . . . . 60. *Quercus garryana*

21. Terminal buds ca. 3 mm; nut (8–)12–15(–33) mm; Rocky Mountain region. . . . . .59. *Quercus gambelii*

1. Leaf blade unlobed or sometimes shallowly lobed or sinuate-lobed, if lobed, sinuses less than 1/3 distance to midrib; leaf margins never 3-dentate.

22. Leaf blade with 6–20 ± parallel secondary veins on each side, these ± straight and not branching, margins dentate, regularly toothed with single tooth for each secondary (at least in distal 1/2 leaf), teeth never spinose.

23. Leaf blade abaxially with erect hairs with 1–4 rays, these evenly distributed, velvety to touch.

24. Leaf blade abaxially with appressed-stellate hairs in addition to erect velvety hairs; peduncle of acorn 20–70 mm; leaf blade sometimes with irregular deep sinuses near base; mature nut 12–25 mm; some cup scales often with a few irregularly distributed, stout, spinose awns near rim of cup. . . . . .46. *Quercus bicolor*

24. Leaf blade abaxially without flat appressed-stellate hairs, with only erect hairs (may be

glabrate in shade leaves); peduncle of acorn to 20 mm; leaf blade regularly toothed, without deep sinuses; mature nut 25–35 mm; cup scales always without awns . . . . . .44. *Quercus michauxii*

23. Leaf blade abaxially without evenly distributed erect hairs (1 species with spreading hairs in patches along midvein), glabrous or with appressed-stellate hairs, not velvety to touch.

    25. Axillary peduncle of acorn 20–70 mm; leaf blade sometimes with irregular deep sinuses near base; cup scales often with stout spinose awns near rim of cup . . . . .46. *Quercus bicolor*

    25. Axillary peduncle of acorn to 20 mm; leaf blade always dentate, serrate, or toothed, without irregular deep sinuses near base; cup scales without stout spinose awns near rim of cup.

        26. Plants evergreen, leaves persistent on stems 2–3 years old; stipules golden-silky, persistent around buds; California, Oregon . . . . .45. *Quercus sadleriana*

        26. Plants deciduous, leaves only on current year's twigs; stipules quickly deciduous, glabrate or ciliate; e North America.

            27. Abaxial leaf blade usually with clearly visible tufts of spreading hairs along midvein, with scattered, irregular, 2–4-rayed, microscopic stellate hairs; bark deeply furrowed . . . . 40. *Quercus montana*

            27. Abaxial leaf blade without tufts of erect (spreading) hairs along midvein, with scattered to dense appressed-stellate, 6–10-rayed hairs; bark scaly.

                28. Secondary veins 10 or more on each side; trees; usually on limestone or calcareous soils . . . . 42. *Quercus muhlenbergii*

                28. Secondary veins 5–8(–9), occasionally fewer, on each side; shrubs; dry sand and ridges . . . . 43. *Quercus prinoides*

22. Leaf blade with 2–12 irregular secondary veins on each side, these curved, crooked, and/or branching, or obscure, not noticeably parallel, margins entire or irregularly toothed or spinose, if toothed then not regularly dentate.

  29. Leaves and twigs glabrous at maturity, leaf blade glaucous and blue-green to gray-green.

    30. Leaf margins with spinose teeth; abaxial leaf epidermis papillose under strong magnification.

      31. Leaf blade subrotund, sometimes wider than long, to 15 mm, teeth 2–3 on each side; petiole less than 2 mm; axillary peduncle to 4 mm; low rhizomatous shrubs, usually less than 1 m; w Texas . . . . 69. *Quercus hinckleyi*

      31. Leaf blade ovate or oblong, (10–)15–35(–50) mm, teeth 4–6(–8) on each side; petiole 3–4 mm; axillary peduncle (5–)30–50 mm; erect shrubs to 2–3 m; Arizona, New Mexico . . . . 70. *Quercus ajoensis*

    30. Leaf blade unlobed or sinuately lobed, margins entire or obscurely toothed, not sharply toothed or spinose.

      32. Leaf blade unlobed to sinuately lobed, rarely deeply lobed; petiole (3–)5–25 mm; cotyledons distinct.

        33. Leaf blade never lobed, veinlets raised abaxially, forming raised reticulum; acorn cup 10–13 mm deep, enclosing ca. 1/2 nut . . . . 41. *Quercus polymorpha*

        33. Leaf blade lobed or unlobed, veinlets not raised abaxially; acorn cup 4–7 mm deep, enclosing 1/3 nut or less . . . . .51. *Quercus laceyi*

      32. Leaf blade unlobed, margins entire to obscurely toothed; petiole 1–5(–8) mm; cotyledons connate.

        34. Low rhizomatous shrubs less than 1 m; most or all mature leaves less than 25 mm . . . . .86. *Quercus depressipes*

        34. Trees or occasionally erect shrubs, not rhizomatous; most or all mature leaves greater than 25 mm.

            35. Largest cup scales 1–1.5 mm wide, moderately, regularly tuberculate; cup usually 6–8(–13) mm deep; Arizona, New Mexico, Texas, Mexico. . . 80. *Quercus oblongifolia*

            35. Largest cup scales 1.5–3 mm wide, strongly, irregularly tuberculate; cup usually 8–10 mm deep; California . . . . 81. *Quercus engelmannii*

  29. Mature leaves pubescent, hairy or glandular on abaxial surface (visible with 10× lens), twigs variously hairy or sometimes glabrate, leaf blade greenish to grayish or blue-green.

    36. Mature leaf blade abaxially yellowish, glandular or glabrate, without stellate or erect fasciculate hairs, or with spreading hairs only along midevin.

      37. Leaf blade adaxially dull or glossy green or yellowish green, not strikingly reflective; mature leaf abaxially with erect or spreading straight hairs unevenly distributed, concentrated in tufts along midvein and near base of blade; Arizona, New Mexico, Texas, Mexico . . . . .71. *Quercus toumeyi*

37. Leaf blade adaxially glossy green, strikingly reflective; mature leaf abaxially merely glandular, without straight hairs in tufts; se United States. . . . . 56. *Quercus chapmanii*
36. Mature leaf blade abaxially with stellate or erect fasciculate hairs distributed ± evenly (hairs sometimes minute or obscured by glandular hairs or wax).
  38. Leaf blade usually convexly cupped, abaxially with prominent raised reticulum formed by ultimate venation; secondary veins often adaxially impressed.
    39. Leaf blade abaxially with minute appressed-stellate hairs visible under magnification, often with scattered erect (spreading), straight, felty, fasciculate hairs as well; se United States . . . . 88. *Quercus geminata*
    39. Leaf blade abaxially without minute appressed-stellate hairs, with scattered to dense erect, usually curly, fasciculate hairs.
      40. Acorn on axillary peduncle 30–60 mm, usually 3–several acorns per peduncle; leaf blade broadly obovate or panduriform to orbiculate . . . . 74. *Quercus rugosa*
      40. Acorn subsessile or on axillary peduncle to 30 mm, usually 1–2(–3) acorns per peduncle; leaf blade elliptic to lanceolate, ovate, or obovate.
        41. Leaf blade 10–25 × 5–13 mm; petiole 2–3 mm; blade margins very coarsely revolute, often undulate-crisped . . . . .85. *Quercus intricata*
        41. Leaf blade (30–)40–100(–150) × 15–60(–80) mm; petiole 3–25 mm; blade margins plane or less strongly revolute.
          42. Leaf blade 15–30 mm wide, adaxially sparsely and minutely stellate-pubescent; nut 8–12 mm; cotyledons connate . . . . 79. *Quercus arizonica*
          42. Leaf blade 30–60(–80) mm wide, adaxially floccose or tomentose when immature, soon glabrate; nut 14–20(–25) mm; cotyledons distinct. . . . . 41. *Quercus polymorpha*
  38. Leaf blade abaxially even, without prominent raised reticulum formed by ultimate venation; secondary veins not strongly impressed adaxially.
    43. Leaf blade abaxially with obvious or minute appressed-stellate hairs, sometimes appearing glabrous or glaucous or yellowish glandular to naked eye, not felty or velvety to touch.
      44. Acorn on axillary peduncle 3–30 mm; cotyledons distinct or connate.
        45. Leaf margins regularly toothed, teeth often spinose, leaf blade grayish glaucous or yellowish glandular, base cordate; cotyledons distinct; se California to w Texas. . . . . .68. *Quercus turbinella*
        45. Leaf margins entire or irregularly toothed, teeth mucronate (rarely spinose in suckers or juveniles), leaf blade green, never glandular, base cuneate or rounded, rarely if ever cordate; cotyledons connate; east of Pecos River.
          46. Low rhizomatous shrubs, with several ± straight stems 0.2–0.7 (–2) m, emerging from ground, bearing acorns at this size, leaves dimorphic (usually some leaves on proximal parts of stems, blades asymmetrically obovate, margins with 1–5 irregularly spaced teeth, distal leaf blades usually narrowly elliptic, margins entire); twigs ± glabrous. . . . . 89. *Quercus minima*
          46. Trees or nonrhizomatous shrubs, or if spreading rhizomatously, without multiple straight, short (less than 0.7 m), erect stems emerging from ground, or if so then branches mixed with other larger branches, infertile; leaves not dimorphic, none asymmetric; twigs minutely puberulent or stellate.
            47. Leaf blade convex-cupped, margins strongly revolute, secondary veins impressed adaxially; blade abaxially usually with scattered erect fasciculate (sometimes deciduous) hairs in addition to appressed-stellate hairs . . . . 88. *Quercus geminata*
            47. Leaf blade ± planar, margins not strongly revolute, secondary veins not impressed adaxially; blade abaxially only with appressed-stellate hairs, without erect fasciculate hairs.
              48. Nut 15–20(–25) mm, apex rounded or blunt; e Texas eastward . . . . .87. *Quercus virginiana*
              48. Nut (17–)20–30(–33) mm, apex acute; c, w Texas, Oklahoma . . . . .90. *Quercus fusiformis*

44. Acorn subsessile or on axillary peduncle to 5 mm; cotyledons distinct.
  49. Leaf blade adaxially dull, grayish because of conspicuous stellate hairs.
    50. Leaf blade strongly bicolored, abaxially densely covered with compact appressed-stellate hairs (8–)10–14(–16)-rayed, appearing whitish; cup strongly tuberculate. . . . . 67. *Quercus cornelius-mulleri*
    50. Leaf blade ± unicolored, abaxially sparsely, rarely densely, covered with appressed-stellate hairs (8–)10–12-rayed, appearing grayish or yellowish; cup not strongly tuberculate. . . . . 64. *Quercus john-tuckeri*
  49. Leaf blade adaxially green, glabrous, or if with scattered stellate hairs then lustrous.
    51. Leaf margins strongly undulate, ± regularly toothed or shallowly lobed, apex acute.
      52. Leaf blade adaxially sandpapery with harsh hairs; nut to 10 mm . . . . .76. *Quercus pungens*
      52. Leaf blade adaxially not sandpapery, without harsh hairs; nut 12–25 mm . . . . .58. *Quercus havardii*
    51. Leaf margins ± planar, not strongly undulate, entire, irregularly toothed or sinuate, apex rounded, obtuse, or acute.
      53. Cup shallow, saucer-shaped, less than 5 mm deep, to 15 mm wide; Arizona eastward.
        54. Leaves deciduous, secondary veins 7–11 on each side; bark light brown, papery-scaly; buds 2–3 mm . . . . .77. *Quercus sinuata*
        54. Leaves deciduous, secondary veins 4–6 on each side; bark dark brown, furrowed and exfoliating in long strips; buds 1–1.5 mm. . . . . 75. *Quercus vaseyana*
      53. Cup hemispheric or deeper, 8–20 mm deep × 15–25 mm wide; California.
        55. Leaf blade abaxially densely covered with compact appressed-stellate hairs (8–)10–14(–16)-rayed, rays imparting whitish color, not strongly waxy, glandular hairs absent . . . . .67. *Quercus cornelius-mulleri*
        55. Leaf blade abaxially sparsely covered with appressed-stellate hairs (4–)8(–10)-rayed, often waxy or glandular.
          56. Nut tapered, apex acute; leaf base consistently cuneate or narrowly attenuate; Channel Islands of California . . . . 66. *Quercus pacifica*
          56. Nut apex rounded or blunt; leaf base usually truncate, rounded, or broadly attenuate, rarely cuneate; mainland California . . . . 65. *Quercus berberidifolia*
43. Leaf blade abaxially with erect fasciculate or (nonappressed) erect or semi-erect stellate hairs, these often felty or velvety to touch.
  57. Hairs of abaxial leaf surface with 1–4 rays.
    58. Leaf blade 50–150 mm, ± planar, not strongly convexly cupped; se United States. . . . . .57. *Quercus oglethorpensis*
    58. Leaf blade less than 50 mm, moderately to strongly convex-cupped.
      59. Leaf base rounded, attenuate, or cuneate; twigs with prominent, yellowish, spreading hairs; nut cylindric or ovoid, apex rounded; serpentine soils n of Los Angeles County, and gneiss soils at base of San Gabriel Mountains, Los Angeles County . . . . 73. *Quercus durata*
      59. Leaf base cordate or angular-cordate; twigs glabrate or with sparse stellate hairs; nut fusiform, apex acute; various soils. . . . . 72. *Quercus dumosa*
  57. Hairs of abaxial leaf surface with (4–)6–many rays.
    60. Leaf blade blue-green adaxially and abaxially; trees; California . . . . 63. *Quercus douglasii*
    60. Leaf blade adaxially green or gray-green, abaxially whitish, grayish, or yellowish.
      61. At least some leaf margins undulate, with 2–3 rounded teeth on each side; low rhizomatous shrubs on stabilized sand dunes; cotyledons distinct. . . . . .58. *Quercus havardii*

61. Leaf margins not undulate (flat or undulate in *Q. mohriana*), entire or with sharp teeth; trees or shrubs, on various substrates; cotyledons connate.
    62. Bud dark red-brown, glabrous or occasionally puberulent on outer scales, not subtended by persistent, hairy, subulate stipules; acorn subsessile or on axillary peduncle sometimes 10–15 mm; limestone and calcareous substrates . . . . 82. *Quercus mohriana*
    62. Bud yellowish because of stellate hairs, at least on outer scales, terminal bud usually subtended by 1–4 persistent, subulate, hairy stipules; acorn often on axillary peduncle to 20–30 mm; igneous substrates.
        63. Leaves usually less than 40 mm; leaf blade adaxially not felty to touch, with scattered to dense, semi-erect stellate hairs . . . . .83. *Quercus grisea*
        63. Leaves usually greater than 40 mm; leaf blade adaxially densely felty to touch, with stiff erect hairs almost obscuring surface . . . . .84. *Quercus chihuahuensis*

40. Quercus montana Willdenow, Sp. Pl. 4(1): 440. 1805
· Mountain chestnut oak, rock chestnut oak   [E] [F]

Trees, deciduous, to 30 m. **Bark** dark gray or brown, hard, with deep V-shaped furrows. **Twigs** light brown, 2–3(–4) mm diam., glabrous. **Buds** light brown, ovoid, (3–)4–6 mm, occasionally apex acute, glabrous. **Leaves:** petiole (3–)10–30 mm. **Leaf blade** obovate to narrowly elliptic or narrowly obovate, (100–)120–200(–220) × 60–100(–120) mm, base subacute or rounded-acuminate, often unequal, margins regularly toothed, teeth rounded or rarely somewhat acute, secondary veins ± parallel, straight or moderately curved, 10–14(–16) on each side, apex broadly acuminate; surfaces abaxially light green, appearing glabrous but with scattered minute, asymmetric, appressed-stellate hairs and usually visible, larger, simple or fascicled erect hairs along veins, adaxially dark green, glossy, glabrous or with minute, scattered, simple hairs. **Acorns** 1–3, subsessile or on peduncle 8–20(–25) mm; cup shallowly cup-shaped to hemispheric or deeply goblet-shaped, rim thin, often flared and undulate, helmetlike, 9–15 mm deep × 18–25 mm wide, scales often in concentric or transverse rows, laterally connate, gray, broadly ovate, tips reddish, glabrous; nut light brown, ovoid-ellipsoid, 15–30 × 10–20(–25) mm, glabrous. **Cotyledons** distinct. $2n = 24$.

Flowering mid–late spring. Rocky upland forest, dry ridges, mixed deciduous forests on shallow soils; 0–1400 m; Ala., Conn., Del., Ga., Ill., Ind., Ky., Maine, Md., Mass., Mich., Miss., N.H., N.J., N.Y., N.C., Ohio, Pa., R.I., S.C., Tenn., Vt., Va., W.Va.

The name *Quercus prinus* Linnaeus is often applied to this species, particularly in the forestry literature, and in many regional floras. In a number of works, however, *Q. prinus* has been applied to the species here treated as *Q. michauxii.* Following the recommendations of J. W. Hardin (1979), because of the persistent confusion in the application of the name *Q. prinus* and uncertainty regarding the identity of the Linnaean type materials, the names *Q. montana* and *Q. michauxii* should be used for the two species that have been variously called *Q. prinus. Quercus prinus* under this interpretation is a name of uncertain position.

The four species of the chestnut oak group in eastern North America (*Quercus montana, Q. michauxii, Q. muhlenbergii,* and *Q. prinoides*) are somewhat difficult to distinguish unless careful attention is paid to features of leaf vesture and fruit and cup morphology. Attempts to identify these species mostly or solely on basis of leaf shape and dentition (as in many other oak species complexes) have resulted in a plethora of misidentified material in herbaria and erroneous reports in the literature. The closely appressed, asymmetric trichomes on the abaxial surface of the mature leaf, in combination with longer simple hairs along the midvein, are unique to *Q. montana* among North American species of *Quercus.* Immature leaves and densely shaded leaves sometimes exhibit a more erect trichome that could be confused with the longer, felty hairs of *Q. michauxii,* so it is important to evaluate mature sun leaves when possible.

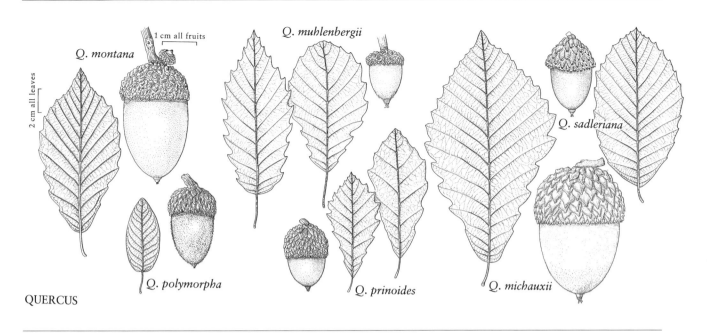

Q. montana
Q. muhlenbergii
Q. sadleriana
Q. polymorpha
Q. prinoides
Q. michauxii
QUERCUS

1 cm all fruits
2 cm all leaves

41. Quercus polymorpha Schlechtendal & Chamisso, Linnaea 5: 78. 1830 · Net-leaf white oak [F]

Trees, subevergreen, to 20 m. **Bark** gray to brown, scaly. **Twigs** reddish brown, 2–3 mm diam., tomentose, soon glabrate. **Buds** reddish brown, ovoid, 3–10 mm, apex acute, pubescent or glabrate. **Leaves:** petiole (6–)15–25 mm. **Leaf blade** elliptic or ovate or lance-ovate, sometimes obovate, 50–100(–150) × 30–60(–80) mm, base rounded or cordate, margins entire or obscurely or prominently serrate-toothed in distal 1/3 blade, revolute, secondary veins moderately curved, 10–12(–14) on each side, apex rounded, acuminate or retuse, sometimes with prominent drip-tip; surfaces abaxially light green, sometimes rather glaucous, veinlets raised, forming raised reticulum, floccose or tomentose with erect, golden hairs, soon glabrate, adaxially dark or light green, glossy, floccose or tomentose when immature, soon glabrate, secondary and tertiary veins impressed. **Acorns** 1–2 on peduncle 5–30 mm; cup hemispheric or funnel-shaped, 10–13 mm deep × 12–20 mm wide, including ca. 1/2 nut, scales appressed, thickened basally, gray-canescent; nut light brown, ovoid-ellipsoid or barrel-shaped, 14–20(–25) × 8–13 mm, glabrous. **Cotyledons** distinct.

Flowering in spring. Riparian forest gallery, margins of thorn scrub, dry tropical forest, lower margins of oak-pine woodland, and cloud forest; 400–2100 m; Tex.; Mexico (Chiapas, Hidalgo, Nuevo León, Oaxaca, Tamaulipas, San Luis Potosí, and Veracruz); Central America (Guatemala).

This widespread species of Mexico and Central America has only recently been discovered in the United States as a small grove of trees about 30 km from the international border in Texas (B. J. Simpson et al. 1992). *Quercus polymorpha* is becoming available in the nursery trade in Texas and the southeastern United States. It is distinct from the superficially similar *Q. splendens* Née (= *Q. sororia* Liebmann) of western Mexico, with which it is sometimes placed in synonymy, in that *Q. splendens* has connate cotyledons instead of distinct cotyledons, as in *Q. polymorpha*.

42. Quercus muhlenbergii Engelmann, Trans. Acad. Sci. St. Louis 3: 391. 1887 · Chinkapin oak, chinquapin oak, yellow chestnut oak [F]

Quercus acuminata (Michaux) Sargent; *Q. brayi* Small; *Q. prinus* Linnaeus var. *acuminata* Michaux

Trees, deciduous, moderate to large, to 30 m, occasionally large shrubs (ca. 3 m) on drier sites. **Bark** gray, thin, flaky to papery. **Twigs** brownish, 1.5–3 (–4) mm diam., sparsely fine-pubescent, soon becoming glabrate, graying in 2d year. **Buds** brown to red-brown, subrotund to broadly ovoid, 20–40 × (10–)15–25 mm, apex rounded, very sparsely pubescent. **Leaves:** petiole (7–)10–30(–37) mm. **Leaf blade** usually obovate, sometimes lanceolate to oblanceolate, (32–)50–150(–210) × (10–)40–80 (–106) mm, leathery, base truncate to cuneate, margins regularly undulate, toothed or shallow-lobed, teeth or lobes rounded, or acute-acuminate, often strongly antrorse, secondary veins usually (9–)10–14

(–16) on each side, ± parallel, apex short-acute to acuminate or apiculate; surfaces abaxially glaucous or light green, appearing glabrate but with scattered or crowded minute, appressed, symmetric, 6–10-rayed stellate hairs, adaxially lustrous dark green, glabrate. **Acorns** 1–2, subsessile or on axillary peduncle to 8 mm; cup hemispheric or shallowly cupped, 4–12 mm deep × 8–22 mm wide, enclosing 1/4–1/2 nut, base rounded, margin usually thin, scales closely appressed, moderately to prominently tuberculate, uniformly short gray-pubescent; nut light brown, oblong to ovoid, (13–)15–20(–28) × 10–13(–16) mm. **Cotyledons** distinct. $2n = 24$.

Flowering late winter–spring. Mixed deciduous forest, woodlands and thickets, sometimes restricted to n slopes and riparian habitats in w parts of range, limestone and calcareous soils, rarely on other substrates; 0–2300 m; Ont.; Ala., Ark., Conn., Fla., Ga., Ill., Ind., Iowa, Kans., Ky., La., Md., Mass., Mich., Minn., Miss., Mo., Nebr., N.J., N.Mex., N.Y., N.C., Ohio, Okla., Pa., S.C., Tenn., Tex., Vt., Va., W.Va., Wis.; Mexico (Coahuila, Nuevo León, Hidalgo, and Tamaulipas).

Shrubby forms of *Quercus muhlenbergii* are difficult to distinguish from *Quercus prinoides,* but *Q. muhlenbergii* does not spread clonally or produce acorns on small shrubs as does *Q. prinoides.* The edaphic preferences of these two species are distinctive, with *Q. muhlenbergii* never far from limestone substrates and *Q. prinoides* occurring mostly on dry shales and deep sands. Populations of *Q. muhlenbergii* from the southwest part of its range, on the Edwards Plateau of Texas and westward, sometimes are segregated as *Q. brayi* Small, but the variation appears to be clinal with inconsistent differences. Distributed from Hidalgo, Mexico to Maine, *Q. muhlenbergii* is one of the most widespread species of temperate North American trees.

The Delaware-Ontario prepared infusions from the bark of *Quercus muhlenbergii* to stop vomiting (D. E. Moerman 1986).

43. **Quercus prinoides** Willdenow, Ges. Naturf. Freunde Berlin Neue Schriften 3: 397. 1801 · Dwarf chinkapin oak, scrub chestnut oak E F

Quercus prinoides var. *rufescens* Rehder

Shrubs, deciduous, (0.5–)1–3(–5, 10?) m, sometimes spreading-rhizomatous. **Bark** gray, thin, flaky to papery. **Twigs** brownish, 1.5–3(–4) mm diam., sparsely fine-pubescent, soon becoming glabrate, graying in 2d year. **Buds** brown to red-brown, subrotund to broadly

ovoid, 1–3 mm, apex rounded, very sparsely pubescent. **Leaves:** petiole (7–)8–15(–25) mm. **Leaf blade** lanceolate to oblanceolate or usually obovate, 40–140 × 20–60(–80) mm, leathery, base truncate to cuneate, margins regularly undulate, toothed or shallow-lobed, teeth usually acute, sometimes rounded, or acute-acuminate, often strongly antrorse, secondary veins usually 5–8(–9) on each side, ± parallel, apex short-acute to acuminate; surfaces abaxially glaucous or light green, appearing glabrate, with scattered or crowded minute, appressed, symmetric, 6–10-rayed, stellate hairs, adaxially lustrous dark green, glabrate. **Acorns** solitary or paired, subsessile or on axillary peduncle to 3–8 mm; cup deeply or shallowly cup-shaped, 9–12 mm deep × 13–17(–22) mm wide, enclosing 1/4–1/3 nut, base rounded, margin usually thin, scales rather tightly appressed, moderately tuberculate, uniformly short gray-pubescent; nut light brown, oblong to ovoid, (13–)15–20 × 10–13 mm. **Cotyledons** distinct.

Flowering in spring. Pine barrens, scrublands, forest margins, prairies, and exposed ridges, on deep sands or dry shale, rarely reported on calcareous soils; 0–500 m; Ont.; Ala., Ark., Conn., Ga., Ill., Ind., Iowa, Ky., Mass., Mich., Mo., N.H., N.J., N.Y., N.C., Ohio, Okla., Pa., R.I., S.C., Tenn., Vt., Va., W.Va.

Some reports of *Quercus prinoides* growing in calcareous soils are probably based on shrub forms of *Q. muhlenbergii.*

The debate over whether *Quercus prinoides* is distinct from *Q. muhlenbergii* has continued for most of the last century. Little doubt can exist that strong genetic differences, as expressed by characteristics of habit, leaf form, and habitat preference, separate the two taxa; the question is merely whether they are best treated as subspecies or varieties or as separate species. Over most of the eastern United States, the two taxa occur sympatrically over broad areas with little immediate contact (syntopy), because *Q. muhlenbergii* is found on calcareous soils and *Q. prinoides* occurs on sands (often acidic) and dry shales. Seedlings of *Q. prinoides* can flower and produce acorns in as few as 3–5 years from planting, when only 20–50 cm, and maintain their dwarf, clonal habit in cultivation. *Quercus muhlenbergii* begins fruiting as a small tree of 3 m or more in height. Because of these differences, and interdigitating geographic distributions, the two taxa seem to be similar in pattern of variation and interaction to other closely related oak species of eastern North America, and dissimilar in pattern to infraspecific taxa such as we see in *Q. sinuata* var. *sinuata* and *Q. sinuata* var. *breviloba.* The populations that are difficult to determine are usually small scrubby trees, probably mostly *Q. muhlenbergii,* stunted because of less than favorable moisture conditions, with or without indications of introgression from *Q. prinoides*.

Material of *Quercus prinoides* from Long Island, coastal Massachusetts, Nantucket Island, and Martha's Vineyard has been segregated as *Q. prinoides* Willdenow var. *rufescens* Rehder on the basis of vesture and minor differences in leaf form. These populations have appressed-stellate leaf pubescence abaxially, as throughout the range of the species; in addition they have reddish, erect, fasiculate hairs similar to those found in *Q. michauxii.* The hairs make the abaxial leaf surface somewhat felty to the touch. Variability in this characteristic and lack of other consistently correlated features preclude taxonomic recognition of *Q. prinoides* var. *rufescens,* but this problem is worthy of further investigation.

44. Quercus michauxii Nuttall, Gen. N. Amer. Pl. 2: 215. 1818 · Basket oak, cow oak, swamp chestnut oak   E F

Quercus houstoniana C. H. Muller

Trees, deciduous, to 20 m. **Bark** light brown or gray, scaly. **Twigs** brown or reddish brown, 2–3 mm diam., with sparse spreading hairs or glabrate. **Buds** reddish brown, ovoid, apex rounded or acute, glabrous or minutely puberulent. **Leaves:** petiole 5–20 mm. **Leaf blade** broadly obovate or broadly elliptic, (60–)100–280 × 50–180 mm, base rounded-acuminate or broadly cuneate, margins regularly toothed, teeth rounded, dentate, or acuminate, secondary veins 15–20 on each side, parallel, straight or somewhat curved, apex broadly rounded or acuminate; surfaces abaxially light green or yellowish, felty to touch because of conspicuous or minute, erect, 1–4-rayed hairs, adaxially glabrous or with minute simple or fascicled hairs. **Acorns** 1–3, subsessile or more often on axillary peduncle to 20–30 mm; cup hemispheric, broadly hemispheric or even short-cylindric, 15–25 mm deep × 25–40 mm wide, enclosing 1/2 nut or more, scales very loosely appressed, distinct to base, gray or light brown, moderately to heavily tuberculate, tips silky-tomentose; nut light brown, ovoid or cylindric, 25–35 × 20–25 mm, glabrous. **Cotyledons** distinct.

Flowering early–late spring. Bottomlands, rich sandy woods and swamps, on variety of soils; 0–600 m; Ala., Ark., Del., Fla., Ga., Ill., Ind., Ky., La., Md., Miss., Mo., N.J., N.C., Pa., S.C., Tenn., Tex., Va.

Quercus michauxii is easily distinguished from other chestnut-leaved oaks by the felty hairs of the abaxial leaf surface and rather large acorn cups with attenuate-acute, loose scales. This species is no longer extant in Oklahoma. Historical reports from Connect-

icut, Massachusetts, and New York have not been confirmed; possibly populations are no longer extant. (See *Quercus montana* for a discussion of nomenclature and the uncertain application of the name *Q. prinus*).

45. Quercus sadleriana R. Brown ter, Ann. Mag. Nat. Hist., ser. 4, 7: 249. 1871   E F

Shrubs, evergreen, to 1–3 m, rhizomatous. **Bark** gray, smooth. **Twigs** reddish or brown, pruinose, 3–4 mm diam., glabrous, sometimes sparsely puberulent around buds. **Buds** yellowish or tan, broadly ovoid or globose, 8–10 mm; scales loose, acute-ovate, silky. **Leaves** persisting on branches 2–3 years old; stipules often persistent, to 15 mm, golden-silky; petiole 15–25 mm. **Leaf blade** obovate or elliptic, 70–140 × 35–80 mm, base rounded or rounded-acuminate, rarely subcordate, margins serrate, not deeply lobed, teeth antrorse-acuminate, sharply mucronate, secondary veins prominent, parallel, straight, (8–)10–15 on each side, apex acuminate or acute; surfaces abaxially light green, waxy, with sparse or scattered, minute, asymmetric, appressed, 4–8-rayed stellate hairs 0.1–0.2 mm diam., and prominent raised stomates giving surface minutely granular appearance, adaxially dark green, glabrous. **Acorns** solitary or paired, subsessile; cup hemispheric or funnel-shaped, 7–9 mm deep × 10–18 mm wide, scales gray, moderately tuberculate, tips reddish brown; nut light brown, ovoid or subglobose, 15–20 × 10–15 mm. **Cotyledons** distinct.

Flowering spring. Open slopes in coniferous forest; 600–2200 m; Calif., Oreg.

Quercus sadleriana is one of the most distinctive western oaks, with strong similarities to certain eastern North American and Asian species of *Quercus* with "chestnut" leaves. Its restricted distribution in the Siskiyou region and uncertain relationships suggest it is a relictual species. It hybridizes occasionally with *Q. garryana* var. *breweri* (see treatment).

46. Quercus bicolor Willdenow, Ges. Naturf. Freunde Berlin Neue Schriften 3: 396. 1801 · Swamp white oak, chêne bicolore   E F

Quercus bicolor var. *angustifolia* Dippel; *Q. bicolor* var. *cuneiformis* Dippel; *Q. bicolor* var. *platanoides* A. de Candolle; *Q. platanoides* (Lamarck) Sudworth

Trees, deciduous, to 30 m. **Bark** dark gray, scaly or flat-ridged. **Twigs** light brown or tan, 2–3 (–4) mm diam., glabrous. **Buds**

light or dark brown, globose to ovoid, 2–3 mm, glabrous. **Leaves:** petiole (4–)10–25(–30) mm. **Leaf blade** obovate to narrowly elliptic or narrowly obovate, (79–)120–180(–215) × (40–)70–110(–160) mm, base narrowly cuneate to acute, margins regularly toothed, or entire with teeth in distal 1/2 only, or moderately to deeply lobed, or sometimes lobed proximally and toothed distally, secondary veins arched, divergent, (3–)5–7 on each side, apex broadly rounded or ovate; surfaces abaxially light green or whitish, with minute, flat, appressed-stellate hairs and erect, 1–4-rayed hairs, velvety to touch, adaxially dark green, glossy, glabrous. **Acorns** 1–3(–5) mm, on thin axillary peduncle (20–)40–70 mm; cup hemispheric or turbinate, 10–15 mm deep × 15–25 mm wide, enclosing 1/2–3/4 nut, scales closely appressed, finely grayish tomentose, those near rim of cup often with short, stout, irregularly recurved and sometimes branched, spinose awns emerging from tubercle; nut light brown, ovoid-ellipsoid or oblong, (12–)15–21(–25) × 9–18 mm, glabrous. **Cotyledons** distinct. $2n = 24$.

Flowering in spring. Low swamp forests, moist slopes, poorly drained uplands; 0–1000 m; Ont., Que.; Ala., Conn., Del., Ill., Ind., Iowa, Ky., Maine, Md., Mass., Mich., Minn., Mo., N.H., N.J., N.Y., N.C., Ohio, Pa., R.I., Tenn., Vt., Va., W.Va., Wis.

Putative hybrids between *Quercus bicolor* and *Q. macrocarpa* are common in areas of contact. The hybrids tend to have more deeply lobed leaves and varying degrees of development of awns as a fringe along the margin of the acorn cup. Such characteristics occur sporadically throughout many populations of *Q. bicolor*; in some cases they may occur because of subtle introgression.

The Iroquois used *Quercus bicolor* in the treatment of cholera, broken bones, consumption, and as a witchcraft medicine (D. E. Moerman 1986).

47. **Quercus macrocarpa** Michaux, Hist. Chênes Amér., plates 2, 3. 1801 · Bur oak, mossy-cup oak, chêne à gros fruits E F

Quercus macrocarpa var. *depressa* (Nuttall) Engelmann; *Q. mandanensis* Rydberg

Trees, deciduous, to 30(–50) m. **Bark** dark gray, scaly or flat-ridged. **Twigs** grayish or reddish, 2–4 mm diam., often forming extensive flat, radiating, corky wings, finely pubescent. **Buds** 2–5(–6) mm, glabrous. **Leaves:** petiole (6–)15–25(–30) mm. **Leaf blade** obovate to narrowly elliptic or narrowly obovate, often fiddle-shaped, (50–)70–150 (–310) × (40–)50–130(–160) mm, base rounded to cuneate, margins moderately to deeply lobed, toothed, deepest sinuses near midleaf (at least in proximal 2/3),

sinuses reaching nearly to midrib, longer lobes grading into shallow lobes or merely simple teeth distally, shallower, compound lobes proximally, secondary veins arched, divergent, 4–5(–10) on each side, apex broadly rounded or ovate; surfaces abaxially light green or whitish, with minute appressed-stellate hairs forming dense, rarely sparse, tomentum, erect felty hairs absent, adaxially dark green or dull gray, sparsely puberulent to glabrate. **Acorns** 1–3 on stout peduncle (0–)6–20(–25) mm; cup hemispheric or turbinate, (8–)15–50 mm deep × (10–)20–60 mm wide, enclosing 1/2–7/8 nut or more, scales closely appressed, laterally connate, broadly triangular, keeled, tuberculate, finely grayish tomentose, those near margins often with soft awns to 5–10 mm or more, forming fringe around nut; nut light brown or grayish, ovoid-ellipsoid or oblong, (15–)25–50 × (10–)20–40 mm, finely puberulent or floccose. **Cotyledons** distinct. $2n = 24$.

Flowering in spring. Bottomlands, riparian slopes, poorly drained areas, prairies, usually on limestone or calcareous clays (in nw part of range on dry slopes and ridges, prairies); 0–1000 m; Man., N.B., Ont., Que., Sask.; Ala., Ark., Conn., Ill., Ind., Iowa, Kans., Ky., La., Maine, Md., Mass., Mich., Minn., Mo., Mont., Nebr., N.J., N.Y., N.Dak., Ohio, Okla., Pa., S.Dak., Tenn., Tex., Vt., Va., W.Va., Wis., Wyo.

Quercus macrocarpa is one of our most cold-tolerant oak species; it also endures a wide variety of other harsh conditions including poor dry soils and wet, poorly drained, and inundated locations. Putative hybrids with *Q. bicolor* are common in the northeastern part of its range, where the two species often occur together in wet, poorly drained habitats. The effect of this contact may be partially responsible for morphologic differences across the range of *Q. macrocarpa*. The large acorns are best developed in the southern part of the range, and a clinal decrease in acorn size and extent of the mossy fringe on the acorn cup seems to occur as one travels from south to north. In the northwest part of its range, *Q. macrocarpa* varies clinally to smaller, shrubbier forms on bluffs and hillsides, with smaller, less fringed cups, that are the basis of *Q. macrocarpa* var. *depressa* (Nuttall) Engelmann and *Q. mandanensis* Rydberg. These scrubby forms may merit formal recognition after more thorough study; they are treated here as clinal variants of the species. *Quercus macrocarpa* forms putative hybrids also with *Q. alba* in the savanna-type regions of the midwest. Putative hybrids with *Q. gambelii* occur out of the range of *Q. macrocarpa*.

Quercus macrocarpa is the only oak species native to Montana (in the southeast corner). Wood of *Q. macrocarpa* is similar to that of *Q. alba* and produces one of the best and most durable oak lumbers.

Native Americans used *Quercus macrocarpa* medic-

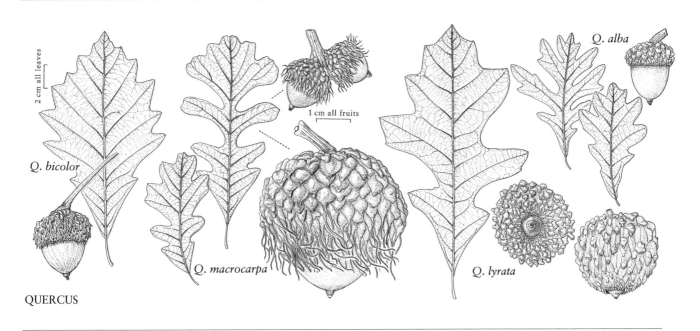

Q. bicolor

2 cm all leaves

Q. macrocarpa

1 cm all fruits

Q. lyrata

Q. alba

QUERCUS

---

inally to treat heart troubles, cramps, diarrhea, and broken bones, to expel pinworms, and as an astringent (D. E. Moerman 1986).

48. **Quercus lyrata** Walter, Fl. Carol., 235. 1788 · Overcup oak E F

Quercus bicolor Willdenow var. *lyrata* (Walter) Dippel

Trees, deciduous, to 20 m. **Bark** light gray, tinged with red, with thick plates underlying scales. **Twigs** grayish or reddish, (2–)3(–4) mm diam., villous, soon glabrate. **Buds** 3 mm, gray-puberulent. **Leaves:** petiole 8–20(–25) mm. **Leaf blade** obovate or broadly obovate, 100–160(–200) × 50–100(–120) mm, base narrowly cuneate to acute, margins moderately to deeply lobed, lobes somewhat to sharply angular or spatulate, often with 2–3 teeth, sinuses nearly to midrib, secondary veins arched, divergent, (3–)5–7 on each side, apex broadly rounded or ovate; surfaces abaxially light green or somewhat glaucous, tomentose, tomentum persisting or soon falling, adaxially dark green or dull gray, sparsely puberulent to glabrate. **Acorns** 1–2 on axillary peduncles to 40 mm; cup goblet-shaped, bur-like, or spheroid, 15–20 mm deep × 20–30 mm wide, usually completely enclosing nut or merely apex visible, rarely enclosing only 1/2 nut, orifice smaller than nut diameter, often splitting irregularly at maturity, scales closely appressed, especially about margin, laterally connate, broadly triangular, keeled-tuberculate, finely grayish tomentose; nut light brown or grayish, ovoid-ellipsoid or oblong, (15–)25–50 × (10–)20–40 mm, finely puberulent or floccose. **Cotyledons** distinct.

Flowering spring. Bottoms, lowlands, wet forest, streamside forests, swamp forests, periodically inundated areas; 0–200 m; Ala., Ark., Del., Fla., Ga., Ill., Ind., Ky., La., Md., Miss., Mo., N.J., N.C., Okla., S.C., Tenn., Tex., Va.

The large acorns with hardened cups that enclose all or most of the nut are diagnostic.

49. **Quercus alba** Linnaeus, Sp. Pl. 2: 996. 1753 · White oak, eastern white oak, chêne blanc E F W1

Trees, deciduous, to 25 m. **Bark** light gray, scaly. **Twigs** green or reddish, becoming gray, 2–3(–4) mm diam., initially pubescent, soon glabrous. **Buds** dark reddish brown, ovoid, ca. 3 mm, apex obtuse, glabrous. **Leaves:** petiole (4–)10–25(–30) mm. **Leaf blade** obovate to narrowly elliptic or narrowly obovate, (79–)120–180(–230) × (40–)70–110(–165) mm, base narrowly cuneate to acute, margins moderately to deeply lobed, lobes often narrow, rounded distally, sinuses extending 1/3–7/8 distance to midrib, secondary veins arched, divergent, (3–)5–7 on each side, apex broadly rounded or ovate; surfaces abaxially light green, with numerous whitish or reddish erect hairs, these quickly shed as leaf expands, adaxially light gray-green, dull or glossy. **Acorns** 1–3, subsessile or on peduncle to 25(–50) mm; cup hemispheric, enclosing 1/4 nut, scales closely appressed, finely grayish tomentose; nut light brown, ovoid-ellipsoid or oblong, (12–)15–21(–25) × 9–18 mm, glabrous. **Cotyledons** distinct. $2n = 24$.

Flowering in spring. Moist to fairly dry, deciduous forests usually on deeper, well-drained loams, also on thin soils on dry upland slopes, sometimes on barrens; 0–1600 m; Ont., Que.; Ala., Ark., Conn., Del., Fla., Ga., Ill., Ind., Iowa, Kans., Ky., La., Maine, Md., Mass., Mich., Minn., Miss., Mo., Nebr., N.H., N.J., N.Y., N.C., Ohio, Okla., Pa., R.I., S.C., Tenn., Tex., Vt., Va., W.Va., Wis.

Considerable variation in depth of lobing occurs in the leaves of *Quercus alba* (M. J. Baranski 1975; J. W. Hardin 1975); the species is easily distinguished from others, however, by the light gray-green, glabrous mature leaves and cuneate leaf bases.

In the past *Quercus alba* was considered to be the source of the finest and most durable oak lumber in America for furniture and shipbuilding. Now it has been replaced almost entirely in commerce by various species of eastern red oak (e.g., *Q. rubra*, *Q. velutina*, and *Q. falcata*) that are more common and have faster growth and greater yields. These red oaks also lack tyloses and therefore are more suited to pressure treating with preservatives, even though they are less decay-resistant without treatment.

Medicinally, *Quercus alba* was used by Native Americans to treat diarrhea, indigestion, chronic dysentery, mouth sores, chapped skin, asthma, milky urine, rheumatism, coughs, sore throat, consumption, bleeding piles, and muscle aches, as an antiseptic, and emetic, and a wash for chills and fevers, to bring up phlegm, as a witchcraft medicine, and as a psychological aid (D. E. Moerman 1986).

Numerous hybrids between *Quercus alba* and other species of white oak have been reported, and some have been named. J. W. Hardin (1975) reviewed the hybrids of *Quercus alba*. Nothospecies names based on putative hybrids involving *Q. alba* include: *Q.* ×*beadlei* Trelease (= *Q. alba* × *prinus*), *Q.* ×*bebbiana* C. K. Schneider (= *Q. alba* × *macrocarpa*), *Q.* ×*bimundorum* E. J. Palmer (= *Q. alba* × *robur*), *Q.* ×*deamii* Trelease (= *Q. alba* × *muhlenbergii*), *Q.* ×*faxonii* Trelease (= *Q. alba* × *prinoides*), *Q.* ×*jackiana* Schneider (= *Q. alba* × *bicolor*), and *Q.* ×*saulei* Schneider (= *Q. alba* × *montana*).

SELECTED REFERENCES Baranski, M. J. 1975. An Analysis of Variation within White Oak (*Quercus alba* L.). Raleigh. [North Carolina Agric. Exp. Sta. Techn. Bull. 236.] Hardin, J. W. 1975. Hybridization and introgression in *Quercus alba*. J. Arnold Arbor. 56: 336–363.

50. **Quercus robur** Linnaeus, Sp. Pl. 2: 996. 1753 · English oak, pedunculate oak, chêne pédoncule

Quercus pedunculata Ehrhart

Trees, deciduous, to 30 m. **Bark** light gray, scaly. **Twigs** brown, 2–3 mm diam., glabrous. **Buds** dark brown, ovoid, distally obtuse, 2–3 mm, glabrous. **Leaves:** petiole 3–6 mm. **Leaf blade** obovate to narrowly elliptic or narrowly obovate (some cultivars oblanceolate), (50–)70–150(–200) × (20–)35–85(–100) mm, base strongly cordate, often minutely revolute or folded, margins moderately to deeply lobed, lobes rounded or retuse distally, sinuses extending 1/3–7/8 distance to midrib, secondary veins arched, divergent, (3–)5–7 on each side, apex broadly rounded; surfaces abaxially light green, glabrous or sparsely pubescent, glabrous at maturity, adaxially deep green to light green or gray, dull or glossy. **Acorns** 1–3, on very thin (1–2 mm diam.), flexuous peduncle (25–)35–65(–100) mm; cup hemispheric to deeply goblet-shaped, enclosing 1/4–1/2 nut or more, scales closely appressed, often in concentric rows, finely grayish tomentose; nut brown, ovoid, oblong, or cylindric, 15–30(–35) × 12–20 mm, glabrous. **Cotyledons** distinct. $2n = 24$.

Flowering spring. Roadsides, pastures, forest margins and woodlands; 0–1000 m; introduced from Europe; B.C., N.B., N.S., P.E.I.

Quercus robur is one of the oaks most commonly cultivated in temperate and subtropical parts of the world. In North America it is most commonly seen in the eastern and northwestern parts of the United States and in southeastern and southwestern Canada, where it tolerates a wide array of conditions and is extremely hardy. In a few areas, apparently reproducing populations persist in the wild. Elsewhere, although actual naturalization appears to be rare, *Q. robur* should be expected to persist around old homesites and other places of cultivation.

Quercus robur most closely resembles our native species *Q. alba* in leaf form. In contrast with *Q. alba*, which has relatively long petioles (longer than 10 mm), acute leaf bases, and subsessile fruit (rarely on peduncles to 25 mm), *Q. robur* is easily distinguished by its shorter petioles (less than 10 mm), cordate, almost clasping, leaf bases, and fruit on long (more than 35 mm), thin peduncles.

Quercus robur is one of the oaks most widely celebrated in literature; it has wood of exceptionally high quality for the manufacture of furniture, and it previously was the most important wood used in the manufacture of wooden sailing vessels in Europe.

51. Quercus laceyi Small, Bull. Torrey Bot. Club 28: 358. 1901 [F]

Quercus breviloba (Torrey) Sargent subsp. *laceyi* (Small) A. Camus

Trees, deciduous, to 5–8(–10) m. **Bark** light colored, papery or scaly. **Twigs** gray, 1.5–2 mm diam., pubescent with erect stellate hairs, these soon shed, at maturity reddish and pruinose to tan and glabrous. **Buds** brown, ovoid to ovoid-lanceoloid, 1.5–3 × 1–2 mm, apex acute, glabrous. **Leaves:** petiole (3–)5–9(–12) mm. **Leaf blade** blue-green, glaucous, obovate or elliptic, (20–)40–90(–210) × (20–)30–60(–110) mm, thin, base cuneate and decurrent on petiole to rounded or rarely somewhat cordate, margins thin, flat, entire to shallowly lobed or (rarely in shade forms) deeply lobed, lobes if present oblong, squarish, often retuse, secondary veins 6–9 on each side, each terminating in tooth or arching near margins, apex broadly rounded, retuse; surfaces abaxially whitish, with erect stellate hairs, hairs shed as leaves expand, becoming glabrous, glaucous, adaxially glabrous, glaucous. **Acorns** annual, solitary or paired, subsessile or on short peduncle to 10(–20) mm in leaf axil; cup saucer-shaped or shallowly cup-shaped, 4–7 mm deep × 10–12(–18) mm wide, enclosing 1/3 nut or less, scales moderately tuberculate, finely tomentose; nut oblong or barrel-shaped, often flattened at both ends, (11–)13–15(–20) × 9–11(–14) mm. **Cotyledons** distinct.

Flowering in spring. Limestone hills, woodlands and riparian forests, canyons and streamsides; 350–2200 m; Tex.; Mexico (Coahuila and Nuevo León).

Material from Texas and northeastern Mexico, excluding the type, has been incorrectly referred to *Quercus glaucoides* M. Martens & Galeotti by some authors (K. C. Nixon and C. H. Muller 1992).

On the Edwards Plateau of Texas, *Quercus laceyi* occurs mostly at 350–600 m elevation; in Coahuila and Nuevo León, it occurs at 1500–2200 m. This species is sometimes associated with remnant mesic forests, which include *Acer grandidentatum* Nuttall, *Tilia* species, *Quercus muhlenbergii* Engelmann, and various pine and other oak species. The leaves are shallowly lobed or entire, although occasional specimens on moist sites are deeply lobed and resemble the leaves of *Q. alba* in outline.

SELECTED REFERENCE Nixon, K. C. and C. H. Muller. 1992. The taxonomic resurrection of *Quercus laceyi* Small (Fagaceae). Sida 15: 57–69.

52. Quercus stellata Wangenheim, Beytr. Teut. Forstwiss., 78, plate 6, fig. 15. 1787 · Post oak [E] [F] [W1]

Quercus minor (Marshall) Sargent; *Q. obtusiloba* Michaux

Trees, deciduous, to 20(–30) m. **Bark** light gray, scaly. **Twigs** yellowish or grayish, (2–)3–5 mm diam., densely stellate-pubescent. **Buds** reddish brown, ovoid, to 4 mm, apex obtuse or acute, sparsely pubescent. **Leaves:** petiole 3–15(–30) mm. **Leaf blade** obovate to narrowly obovate, elliptic or obtriangular, 40–150 (–200) × 20–100(–120) mm, rather stiff and hard, base rounded-attenuate to cordate, sometimes cuneate, margins shallowly to deeply lobed, lobes rounded or spatulate, usually distal 2 lobes divergent at right angles to midrib in cruciform pattern, secondary veins 3–5 on each side, apex broadly rounded; surfaces abaxially yellowish green, with crowded yellowish glandular hairs and scattered minute, 6–8-rayed, appressed or semi-appressed stellate hairs, not velvety to touch, adaxially dark or yellowish green, dull or glossy, sparsely stellate, often somewhat sandpapery with harsh hairs. **Acorns** 1–3, subsessile or on peduncle to 6(–40) mm; cup deeply saucer-shaped, proximally rounded or constricted, 7–12(–18) mm deep × (7–)10–15(–25) mm wide, enclosing 1/4–2/3 nut, scales tightly appressed, finely grayish pubescent; nut light brown, ovoid or globose, 10–20 × 8–12(–20) mm, glabrous or finely puberulent. **Cotyledons** distinct.

Flowering spring. Usually on xeric sites, dry gravelly and sandy ridges and uplands, dry clays, prairies and limestone hills, woodlands and deciduous forests; 0–750 m; Ala., Ark., Conn., Del., Fla., Ga., Ill., Ind., Iowa, Kans., Ky., La., Md., Mass., Miss., Mo., N.J., N.Y., N.C., Ohio, Okla., Pa., R.I., S.C., Tenn., Tex., Va., W.Va.

Quercus stellata is often identified by its commonly cross-shaped leaf form, particularly in the eastern part of its range. All individuals and populations do not express this characteristic, however. Moreover, *Q. stellata* has broad overlap with *Q. margaretta* and even with some forms of the blackjack oak, *Q. marilandica,* one of its most common associates. The thick yellowish twigs with indument of stellate hairs and the dense harsh stellate hairs on the abaxial leaf surface are better diagnostic characteristics when variation includes leaf forms that are not obviously cruciform.

Native Americans used *Quercus stellata* medicinally for indigestion, chronic dysentery, mouth sores, chapped skin, hoarseness, and milky urine, as an antiseptic, and as a wash for fever and chills (D. E. Moerman 1986).

Putative hybrids are known with *Quercus marga-*

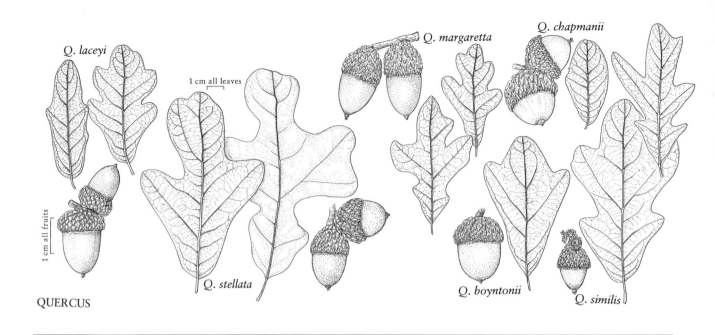

Q. laceyi

1 cm all leaves

Q. margaretta

Q. chapmanii

1 cm all fruits

Q. stellata

Q. boyntonii

Q. similis

QUERCUS

retta, Q. alba, and various other white oaks. *Quercus stellata* is also one of the few oaks that appears to produce hybrids with species in the live oak group, although obvious intermediates are rarely encountered. Nothospecies names based on putative hybrids involving *Q. stellata* include: *Q.* ×*stelloides* E. J. Palmer (= *Q. prinoides* × *Q. stellata*), *Q.* ×*mahonii* E. J. Palmer (as *Q. sinuata* var. *breviloba* × *Q. stellata*), *Q.* ×*pseudomargaretta* Trelease (= *Q. margaretta* × *Q. stellata*), *Q.* ×*sterrettii* Trelease (= *Q. lyrata* × *Q. stellata*), *Q.* ×*macnabiana* Sudworth (= *Q. sinuata* × *Q. stellata*), *Q.* ×*guadalupensis* Sargent (= *Q. sinuata* × *Q. stellata*), *Q.* ×*fernowii* Trelease (= *Q. alba* × *Q. stellata*), and *Q.* ×*bernardensis* W. Wolf (= *Q. montana* × *Q. stellata*).

53. Quercus margaretta (Ashe) Ashe in J. K. Small, Fl. S.E. U.S., 355. 1903 · Sand post oak, dwarf post oak E F

Quercus minor (Marshall) Sargent var. *margaretta* Ashe, J. Elisha Mitchell Sci. Soc. 11: 94. 1894; *Q. stellata* Wangenheim var. *margaretta* (Ashe) Sargent

Small trees or shrubs, deciduous, to 12 m, sometimes rhizomatous. **Bark** light gray, scaly. **Twigs** green or reddish, becoming gray, 1–2(–3) mm diam., glabrous. **Buds** reddish brown, ovoid, 2–3(–6) mm, apex obtuse, sparsely pubescent to glabrate. **Leaves:** petiole 3–10(–15) mm. **Leaf blade** obovate to narrowly obovate, (25–)40–80(–135) × 20–40(–80) mm, base cuneate to rounded-attenuate, margins moderately to deeply lobed, lobes rounded or spatulate, sometimes middle or distal 2 lobes compound, with secondary lobe on proximal side, divergent at right angles, forming cruciform pattern, secondary veins 3–5 on each side, apex broadly rounded; surfaces abaxially light green, with interlocking, erect, 2–4(–6)-rayed stellate hairs, velvety to touch, adaxially dark green, glossy, glabrous or sparsely stellate, not harsh to touch. **Acorns** 1–3, subsessile or on peduncle to 20 mm; cup deeply cup-shaped, basally rounded or constricted, (7–)9–12 mm deep × 12–20 mm wide, enclosing to 3/4 nut, scales loosely appressed, finely grayish pubescent; nut light brown, ovoid, 15–25(–30) × 9–13(–16) mm, glabrous. **Cotyledons** distinct.

Deep sands and gravels, often in dense woods as understory or in open scrubland and pine barrens; 0–600 m; Ala., Ark., Fla., Ga., La., Miss., Mo., N.C., Okla., S.C., Tex., Va.

Historical records for *Quercus margaretta* exist for New York, but no current population is known there.

Populations of post oak in east Texas (the Cross Timbers region) on sands and gravels exhibit characteristics somewhat intermediate between *Quercus stellata* and *Q. margaretta;* most of the trees at those localities have leaves with abaxial surface similar to *Q. margaretta,* leaf shape more similar to *Q. stellata,* and twigs somewhat intermediate between the two species in diameter and varying from tightly pubescent to glabrate. Acorn characters tend toward *Q. margaretta* as well. These populations have been treated as *Q. drummondii* Liebmann, the Drummond post oak. Similar intermediates occur sporadically throughout the range of the post oaks in southeastern United

States, but they do not form such continuous and morphologically stable populations; perhaps the Texas material is best treated as a nothospecies, *Q.* ×*drummondii*.

54. Quercus boyntonii Beadle, Biltmore Bot. Stud. 1: 47.

1901 · Boynton oak [C][E][F]

Quercus stellata Wangenheim var. *boyntonii* (Beadle) Sargent

Shrubs, rarely small trees, deciduous or subevergreen, shrubs low, under 2 m, often trailing, rhizomatous, trees to 6 m. **Bark** brown, scaly. **Twigs** light brown, 1.5–3 mm diam., densely tomentulose. **Buds** reddish brown, ovoid, 2–3(–4) mm, apex acute or rounded, sparsely pubescent. **Leaves:** petiole (4–)5–10(–15) mm. **Leaf blade** obovate or oblanceolate, (39–)50–100(–125) × 20–60(–91) mm, base cuneate, margins minutely revolute, broadly 3-lobed distally or with 3–5 rounded, irregular lobes in distal 1/2, secondary veins curved, 6–8 on each side, apex broadly ovate or triangular-lobed; surfaces abaxially grayish or silvery, densely tomentulose-glandular with minute, appressed-stellate hairs, adaxially dark green, glossy, glabrous or with minute, scattered, simple hairs. **Acorns** 1–2, on peduncle 2–10(–35) mm; cup deeply or shallowly cup-shaped, 5–10 mm deep × 10–13 mm wide, including 1/3–1/2 nut, scales closely appressed, gray, tomentulose; nut light brown, ovoid, 10–17 × 7–13 mm, apex rounded, glabrous. **Cotyledons** distinct.

Flowering spring. Deep sands and crevices in pine forests, along streams; of conservation concern; 0–200 m; Ala., Tex.

Quercus boyntonii is a rare and poorly known species of somewhat uncertain distribution; probably it is often overlooked. Some intermediates between *Q. boyntonii* and *Q. margaretta* are known. These tend to be larger shrubs, to 2 m with felty hairs proximally but with the rhizomatous habit of *Q. boyntonii*.

SELECTED REFERENCE Muller, C. H. 1956. The distribution of *Quercus boyntoni*. Madroño 13: 221–225.

55. Quercus similis Ashe, J. Elisha Mitchell Sci. Soc. 40:

43. 1924 · Swamp post oak [E][F]

Quercus ashei Sterret; *Q. stellata* Wangenheim var. *paludosa* Sargent

Trees, deciduous, to 25 m, with single straight trunk. **Bark** brown, scaly. **Twigs** grayish, 2–3 mm diam., persistently tomentulose. **Buds** brown, ovoid, 2–3 mm, apex acute or rounded, proximally pubescent. **Leaves:**

petiole 3–10 mm. **Leaf blade** obovate or narrowly obovate, (50–)75–120(–150) × 50–65(–80) mm, base rounded-attenuate or acute, margins flat, shallowly 2–3-lobed on each side, lobes usually simple, oblong or rounded, rarely spatulate, not cruciform, secondary veins 3–5 on each side, apex broadly ovate or acute; surfaces abaxially grayish, sparsely glandular and sparsely appressed-stellate, adaxially dark green, glossy, sparsely stellate. **Acorns** 1–3, subsessile; cup 6–7 mm deep × 10–13 mm wide, scales closely appressed, grayish, finely tomentulose; nut light brown or dark reddish brown, ovoid or oblong, 12–16 × 8–12 mm, puberulent or glabrate. **Cotyledons** distinct.

Flowering spring. Forests in wet stream bottoms, flatwoods, river valleys; 0–300 m; Ala., Ark., Ga., La., Miss., S.C., Tex.

56. Quercus chapmanii Sargent, Gard. & Forest 8: 93.

1895 · Chapman oak [E][F]

Shrubs, deciduous or subevergreen, 0.5–3(–6) m, often rhizomatous. **Bark** brown, scaly. **Twigs** yellowish, 1–2 mm diam., densely fine-tomentulose. **Buds** reddish brown, globose, 1–2(–3) mm, proximal scales densely tomentulose, distal scales glabrous. **Leaves:** petiole 1–3(–5) mm. **Leaf blade** obovate or oblanceolate, 30–70(–85) × 14–30 (–45) mm, base cuneate or attenuate, margins minutely revolute, entire or sinuately lobed, sometimes obscurely 3-lobed distally or with 3–5 rounded, irregular lobes in distal 1/2, secondary veins curved, 8–9 on each side, apex ovate or triangular-lobed, often retuse; surfaces abaxially grayish or yellowish, with yellowish, erect branched hairs, these soon shed, leaving matted glandular and waxy hairs except on ± glabrate yellowish veins, adaxially bright glossy, very reflective, glabrous or with minute, scattered, stellate hairs. **Acorns** 1–2, on peduncle 1–6(–35) mm; cup hemispheric, 5–11 m deep × 10–15 mm wide, including 1/3–1/2 nut, scales closely appressed, gray, tomentulose; nut light brown, ovoid to barrel-shaped, 15–20 × 9–13 mm, apex rounded, glabrous or puberulent. **Cotyledons** distinct.

Flowering late winter–early spring. Open pine forests, scrublands, xerophytic scrub oak, on sand near coast; 0–100 m; Fla., Ga., S.C.

57. Quercus oglethorpensis Duncan, Amer. Midl.
Naturalist 24: 755. 1940 · Oglethorpe oak [C] [E] [F]

Trees, deciduous, to 18(–25) m.
Bark light gray or whitish, scaly.
Twigs brownish red, ca. 1 mm
diam., sparsely pubescent, glabrate with age. **Buds** reddish
brown, globose, to 2–2.5 mm,
sparsely pubescent or glabrous.
Leaves: petiole 2–7 mm. **Leaf
blade** narrowly elliptic or oblan-
ceolate, ± planar, not strongly convexly cupped, 50–
150 × 20–45 mm, base cuneate to cordate, margins
entire or on vigorous shoots sometimes sinuate near
apex, secondary veins 3–5 on each side, apex rounded,
obtuse or broadly acute; surfaces abaxially yellowish
green, covered with persistent velvety branched hairs,
adaxially dark green, dull or glossy, sparsely stellate,
often somewhat sandpapery with harsh hairs. **Acorns**
1–2, subsessile or on peduncle to 7 mm; cup turbinate,
somewhat constricted proximally, 8 mm deep × 10
mm wide, enclosing 1/3 nut or more, scales closely appressed, finely tan-pubescent; nut gray brown, ovoid,
9–11 × (5–)7–9 mm, finely puberulent. **Cotyledons**
distinct.

Flowering spring. Alluvial flatwoods and streamsides in rich woods, low pastures, and edge of bottomland forests; of conservation concern; 0–200 m;
Ga., La., Miss., S.C.

Since its original discovery in 1940 in Oglethorpe
County, Georgia, *Quercus oglethorpensis* has been
found to be more common locally near the type site in
Georgia and South Carolina than originally thought.
It remains one of the least-known oak species of the
southeastern United States. *Quercus oglethorpensis* is
one of our most distinctive eastern oaks, easily recognized by its narrow, entire, abaxially felty leaves.

SELECTED REFERENCE   Haenhle, G. G. and S. M. Jones. 1985. Geographical distribution of *Quercus oglethorpensis*. Castanea 50: 26–
31.

58. Quercus havardii Rydberg, Bull. New York Bot. Gard.
2: 213, plate 29, fig. 2. 1901 (as **havardi**) · Havard
oak [E] [F] [W1]

Shrubs, deciduous, low, forming
clones 0.3–1.5 × 10 m, rhizomatous. **Bark** light gray, scaly-
papery. **Twigs** brown or grayish,
1–2.5 mm diam., glabrous or
densely short grayish or yellowish tomentulose, glabrate in age.
Buds dark red-brown, subglobose, ca. 2 mm, sparsely pubes-
cent. **Leaves:** petiole to 7 mm. **Leaf blade** green, often
turning brownish with age, polymorphic, oblong or elliptic or sometimes lanceolate to oblanceolate or ovate

to obovate, (30–)50–100 × (10–)20–50 mm, rather
thick and hard, base rounded to cuneate, margins flat
to revolute, at least some undulate, 2–3 rounded teeth
on each side, secondary veins 5–8 on each side, much
branched, apex broadly rounded, rarely acute; surfaces
abaxially densely grayish or yellowish tomentulose or
stellate-pubescent, sometimes only sparsely pubescent,
secondary veins quite prominent, adaxially lustrous,
very sparsely stellate-pubescent or glabrate, secondary
veins very slightly if at all raised. **Acorns** solitary or
paired, subsessile or on peduncle to 10(–18) mm; cup
from deeply cup-shaped to goblet-shaped, 10–12 mm
deep × 15–25 mm wide, enclosing 1/3–1/2 nut, base
rounded or slightly constricted, margin very thin and
smooth, scales reddish brown, triangular-ovate to long-
acute, proximally moderately to markedly tuberculate,
pubescent, often canescent, tips loosely appressed; nut
brown, ovoid, 12–25 × 14–18 mm. **Cotyledons** distinct.

Flowering spring. Deep, shifting or stabilized sand
dunes, off deep sands in putative hybrid populations;
500–1500 m; N.Mex., Okla., Tex.

Individual clones emerging to heights of 2–3 m from
thickets occur sporadically across the Texas range of
Quercus havardii and express some characteristics of
Q. stellata, such as more deeply lobed leaves and
smaller acorns. Such putative hybrids increase in frequency in the eastern part of the range of the species.

Material of *Quercus havardii* from the Navajo Basin
of Utah and adjacent Arizona has been treated as *Q.
havardii* var. *tuckeri* Welsh. Welsh followed J. M.
Tucker (1970) and interpreted these intermediate populations as putative hybrids between *Q. havardii* and
both *Q. turbinella* and *Q. gambelii.* Giving varietal
rank, instead of nothospecies status, to such populations seems arbitrary, and it certainly is inconsistent
with their putative hybrid origins.

SELECTED REFERENCE   Tucker, J. M. 1970. Studies in the *Quercus
undulata* complex. IV. The contribution of *Q. havardii*. Amer. J. Bot.
57: 71–84.

59. Quercus gambelii Nuttall, J. Acad. Nat. Sci.
Philadelphia, ser. 2, 1(2): 179. 1848 · Gambel oak
[F] [W2]

Quercus douglasii Hooker & Arnott
var. *gambelii* (Nuttall) A. de Candolle;
Q. gambelii var. *gunnisonii* Wenzig;
Q. lesueuri C. H. Muller; *Q. marshii*
C. H. Muller; *Q. novomexicana*
Rydberg; *Q. undulata* Torrey var.
gambelii (Nuttall) Engelmann; *Q.
utahensis* Rydberg

Shrubs   or   trees, deciduous,
shrubs sometimes clumped and spreading, trees small
or moderately large. **Bark** gray or brown, scaly. **Twigs**
brown or reddish brown with few, inconspicuous lenticels, 1.5–2.5 mm diam., glabrous or stellate-
pubescent. **Buds** brown, ovoid, ca. 3 mm, apex acute

or obtuse, sparsely pubescent, becoming glabrate. **Leaves:** petiole 10–20 mm. **Leaf blade** elliptic to obovate or oblong, deeply to shallowly 4–6-lobed, (40–)80–120(–160) × (25–)40–60(–100) mm, membranous, base truncate to cuneate, margins entire or coarsely toothed, lobes oblong, rounded or subacute, sinuses acute or narrowly rounded at base, reaching more than 1/2 distance to midrib, secondary veins 4–6 on each side, each passing into lobe, branched, apex broadly rounded; surfaces abaxially dull green, sometimes glaucous, densely velvety with erect 4–6-rayed hairs, sometimes glabrate or persistently villous only near midribs, secondary veins prominent, adaxially lustrous dark green, appearing glabrate, microscopically pubescent, secondary veins slightly raised. **Acorns** solitary or paired, subsessile or on peduncle to 10(–30) mm; cup deeply cup-shaped, 5–8(–17) mm deep × 7–15(–25) mm wide, enclosing 1/4–1/2 nut, base round, margin thin, scales closely appressed, ovate, markedly tuberculate, proximally gray-tomentulose; nut light brown, ovoid to ellipsoid, (8–)12–15(–33) × 7–12(–18) mm. **Cotyledons** distinct.

Flowering mid–late spring. Montane conifer, oak-maple, and higher margins of pinyon-juniper woodlands; 1000–3030 m; Ariz., Colo., Nev., N.Mex., Okla., Tex., Utah, Wyo.; Mexico (Chihuahua, Coahuila, and Sonora).

Numerous hybrids of *Quercus gambelii* with various scrub oaks have been reported, including *Q. grisea* and *Q. turbinella*. Such hybrids in general have shallowly lobed or dentate, semipersistent leaves and intermediate characteristics of pubescence and fruit between parental types. Such hybrids are usually referred to as the *Quercus* ×*undulata* complex because of widespread application of the latter name to various populations.

One population from San Juan County, Utah, with larger fruit but otherwise not differing from typical *Quercus gambelii*, has been recognized as *Q. gambelii* var. *bonina* Welsh. Unless other characters are found to support this segregation, the plants are best not treated as a formal taxon, particularly considering the extensive variation and hybridization associated with *Q. gambelii* throughout its range.

Numerous putative hybrid swarms occur throughout the range of *Quercus gambelii* that involve a number of suspected parental species. Most of these populations have, at one time or another, been referred to *Quercus undulata* Torrey. The putative hybrids have serrate or shallowly lobed leaves and considerable variation in habit, leaf pubescence, and acorn morphology. J. M. Tucker (1961, 1969, 1971) and J. M. Tucker et al. (1961) have identified the major components of the *Q. undulata* complex as *Q. turbinella* (western Utah and northwestern Arizona, and central Colorado), *Q. grisea* (New Mexico and southern Col-

orado), *Q. havardii* (southeastern Utah and northwestern Arizona), *Q. mohriana* (northeastern and southern New Mexico), *Q. arizonica* (central Arizona), and *Q. muhlenbergii* (eastern and central New Mexico). *Quercus macrocarpa* has been implicated as a parent of variable populations in New Mexico (J. M. Tucker and J. R. Maze 1966). Because of the complex variability in these populations, no effort has been made to treat them separately here; indeed, it would be impossible to produce usable keys if these were included as formal taxa.

Hybrids derived from *Quercus gambelii* and an evergreen species are often semideciduous, retaining a variable portion of green or brownish leaves over the winter.

Quercus gambelii was used medicinally by the Navajo-Ramah to alleviate postpartum pain, as a cathartic, as a ceremonial emetic, and as a life medicine (D. E. Moerman 1986).

60. **Quercus garryana** Douglas ex Hooker, Fl. Bor.-Amer. 2: 159. 1840 · Oregon white oak, Garry oak [E] [F]

Quercus douglasii Hooker & Arnott var. *neaei* (Liebmann) A. de Candolle; *Q. garryana* var. *jacobi* (R. Brown ter) Zabel; *Q. jacobi* R. Brown ter; *Q. lobata* Née var. *breweri* (Engelmann) Wenzig; *Quercus neaei* Liebmann

Trees or shrubs, deciduous, trees to 15(–20) m, with solitary trunks, shrubs to 0.1–3 m, multitrunked. **Bark** light gray or almost white, scaly. **Twigs** brown, red, or yellowish, 2–4 mm diam., densely puberulent with spreading hairs or glabrate. **Buds** brown or yellowish, ovoid or fusiform and apex acute, 2–12 mm, glandular-puberulent or densely pubescent. **Leaves:** petiole 4–10 mm. **Leaf blade** obovate, elliptic or subrotund, moderately to deeply lobed, 25–120(–140) × 15–85 mm, base rounded-attenuate or cuneate, rarely subcordate, often unequal, margins with sinuses usually reaching more than 1/2 distance to midrib, lobes oblong or spatulate, obtuse, rounded or blunt, larger lobes usually with 2–3 sublobes or teeth, veins often ending in retuse teeth, secondary veins yellowish, 4–7 on each side, the more distal veins often branching within distal lobes, apex broadly rounded; surfaces abaxially light green or waxy yellowish, often felty to touch, densely to sparsely covered with semi-erect or erect, simple and (2–)4–8-rayed, fasciculate hairs 0.1–1 mm, secondary veins raised, adaxially bright or dark green, glossy or somewhat scurfy because of sparse stellate hairs. **Acorns** 1–3, subsessile, rarely on peduncle to 10(–20) mm; cup saucer-shaped, cup-shaped, or hemispheric, 4–10 mm deep × 12–22 mm wide; scales yellowish or reddish brown, often long-acute near rim of cup, moderately or scarcely tuberculate, canescent or tomentulose; nut light brown, oblong to globose, (12–)25–30(–40) × (10–)14–20(–22) mm, apex blunt or rounded, glabrous or often persistently puberulent. **Cotyledons** distinct. 2*n* = 24.

Varieties 3 (3 in the flora): w North America.

Quercus garryana (no varieties specified) was used medicinally by Native Americans to treat tuberculosis and as a drink and a rub for mothers before childbirth (D. E. Moerman 1986).

1. Trees to 15 m or more, trunk usually solitary; buds yellowish or cream, usually fusiform, 6–12 mm, apex acute, densely pubescent; twigs persistently puberulent, with spreading hairs . . . . .60a. *Quercus garryana* var. *garryana*
1. Shrubs or small trees usually less than 5 m, multitrunked, spreading and clonal; buds reddish brown, ovoid, 2–5 mm, sparsely glandular-puberulent; twigs sparsely puberulent or glabrate, without spreading hairs.
  2. Leaf blade abaxially velvety to touch, hairs usually 4–6-rayed, rays 0.25–0.5 mm . . . . 60b. *Quercus garryana* var. *breweri*
  2. Leaf blade abaxially not velvety but sometimes felty, hairs 6–8-rayed, rays less than 0.3 mm. . . . . 60c. *Quercus garryana* var. *semota*

60a. Quercus garryana Douglas ex Hooker var. garryana E F

Trees, to 20 m, trunks usually solitary. **Twigs** yellowish or brown, persistently puberulent, with spreading hairs. **Buds** yellowish or cream, usually fusiform, 6–12 mm, apex acute, densely pubescent. **Leaf blade** abaxially light green, velvety to touch, sparsely to densely covered with erect simple and (2–)4–5(–6)-rayed hairs 0.3–1 mm.

Flowering spring. Oak woodlands, margins of redwood forests, mixed evergreen forests; 0–800 m; B.C.; Calif., Oreg., Wash.

Quercus garryana var. *garryana* possibly extends to Los Angeles County as isolated trees in riparian situations.

60b. Quercus garryana Douglas ex Hooker var. breweri (Engelmann) Jepson, Fl. Calif. 1(2): 354. 1909 E F

Quercus breweri Engelmann in S. Watson, Bot. Calif. 2: 96. 1880

Shrubs, spreading and clonal, to 2–3 m, multitrunked. **Twigs** reddish, sparsely puberulent, often glabrate, without spreading hairs. **Buds** reddish brown, ovoid, 2–5 mm, glandular-puberulent. **Leaf blade** abaxially light green, velvety to touch, sparsely to densely covered with erect (2–)4–6-rayed hairs 0.2–0.5 mm.

Flowering late spring–early summer. Montane conifer forests and chaparral; 1400–1900 m; Calif., Oreg.

Quercus garryana var. *breweri* appears to be endemic to the Siskiyou region of California and Oregon; it may extend into the northern Sierra Nevada of California. Specimens sometimes placed here from the Coast Ranges of northern California are probably shrubby forms of *Q. garryana* var. *garryana* or hybrids between the latter and *Q. durata* (see treatment). It should be noted that key characteristics separating *Q. garryana* vars. *breweri* and *semota* from var. *garryana* (clonal habit, smaller, glabrate, brown buds, montane habitat of the former two varieties) suggest a relationship of these two varieties with the Rocky Mountain *Q. gambelii*. The latter species has smaller fruit than *Q. garryana*, but the extent of variation in this characteristic in *Q. garryana* var. *breweri* is unknown.

60c. Quercus garryana Douglas ex Hooker var. semota Jepson, Fl. Calif. 1(2): 354. 1909 E F

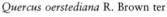

Quercus oerstediana R. Brown ter

Shrubs or small multitrunked trees, spreading and clonal, to 1–5 m. **Twigs** reddish brown or tan, sparsely puberulent, without spreading hairs. **Buds** brown or light brown, ovoid, 2–5 mm, sparsely glandular-puberulent. **Leaf blade** abaxially light or yellowish green, waxy, not conspicuously velvety, but sometimes tightly felty, sparsely to densely covered with semi-erect 6–8-rayed hairs 0.1–0.3 mm.

Flowering spring. Dry slopes in open montane conifer forests and chaparral; 1250–1800 m; Calif., Oreg.

Quercus garryana var. *semota* is common on the west slope of the Sierra Nevada and north slope of the Tehachapi Mountains. Its northern limit appears to be on dry volcanics in southern Oregon. The fruit of this variety is often rather large, falling in the range of that seen in *Q. garryana* var. *garryana*. Some material with elongate-oblong acorns suggests introgression from *Q. lobata* may occur in isolated localities at the lower reaches of *Q. garryana* var. *semota* and higher limits of *Q. lobata*.

61. Quercus carmenensis C. H. Muller, Amer. Midl. Naturalist 18: 847. 1937 C F

Shrubs or trees, deciduous, shrubs 0.5–2 m, rhizomatous, trees (on better sites) to 12 m, trunk 0.75 m diam. **Bark** light gray, checkered or furrowed. **Twigs** often strikingly red, 1–1.5 mm diam., sparingly (rarely densely) stellate-pubescent, somewhat glabrescent and gray 2d year. **Buds** light brown, nearly round, 1–1.5 mm, indu-

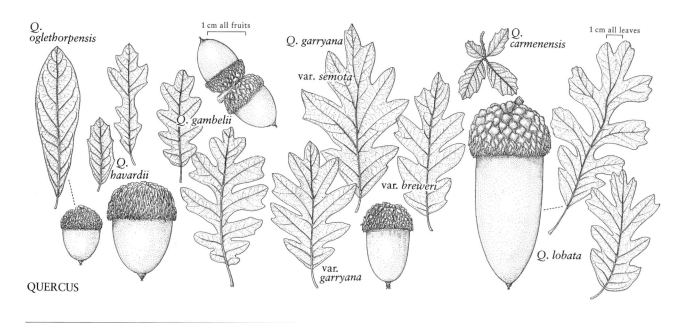

Q.
oglethorpensis

Q.
havardii

Q. gambelii

1 cm all fruits

Q. garryana

var. *semota*

var. *breweri*

var.
garryana

QUERCUS

Q.
carmenensis

1 cm all leaves

Q. lobata

mentum similar to twigs. **Leaves:** petiole usually strikingly red, (2–)5–10 mm, ca. 1 mm diam. **Leaf blade** obovate or narrowly obovate, (20–)30–50 × 10–30 mm, thin to moderately leathery, base cuneate to rounded, margins shallowly and irregularly lobed or coarsely toothed in distal 1/2, rarely subentire, teeth mucronate, secondary veins 9–12 on each side, branching or passing directly to teeth, apex acute, sometimes broadly rounded; surfaces abaxially light green or yellow-green, prominently pubescent with minute, erect velvety hairs, adaxially surfaces dark green, sparsely and minutely stellate-pubescent. **Acorns** solitary or paired, subsessile or short-pedunculate (immature); cup (mature) unknown; scales (immature) light brown, tip acute, canescent. **Nut** unknown. **Cotyledons** unknown.

Shrublands and woodlands on limestone; of conservation concern; 2200–2500 m; Tex.; Mexico (Coahuila).

Quercus carmenensis is known in the United States from only one collection from the Chisos Mountains, Texas; otherwise, it is known in the Sierra del Carmen region, Coahuila, Mexico.

62. Quercus lobata Née, Anales Ci. Nat. 3: 277. 1801 · Valley oak E F

Quercus hindsii Bentham; *Q. lobata* var. *hindsii* (Bentham) Wenzig; *Q. longiglanda* Frémont

Trees, deciduous, to 25(–35) m, usually with solitary trunks. **Bark** gray, scaly, deeply checkered in age. **Twigs** yellowish, gray, occasionally reddish, 2–4 mm diam., densely or sparsely

tomentulose. **Buds** yellowish or light brown, ovoid, (2–)3–5(–6) mm, apex occasionally acute, densely pubescent. **Leaves:** petiole 5–12 mm. **Leaf blade** broadly obovate or elliptic, moderately to deeply lobed, (40–)50–100(–120) × 30–60(–75) mm, base rounded-attenuate, cuneate, or truncate, rarely subcordate, margins with sinuses usually reaching more than 1/2 distance to midrib, lobes oblong or spatulate, obtuse, rounded, or blunt, secondary veins 5–10 on each side, apex broadly rounded; surfaces abaxially whitish or light green, densely to sparsely covered with interlocking appressed or semi-erect, 8–10(–14)-rayed stellate hairs, adaxially dark green or grayish, glossy or somewhat scurfy because of sparse stellate hairs. **Acorns** solitary or paired, subsessile; cup deeply cup-shaped, hemispheric or turbinate, rim thick, 10–30 mm deep × 14–30 mm wide, scales grayish or cream, more acute near rim, strongly and irregularly tuberculate, especially toward base of cup; nut light brown, oblong or fusiform, 30–60 × (12–)15–25 mm, tapering to acute or rounded apex. **Cotyledons** distinct.

Flowering late winter–early spring. Valley floors and moderate slopes, open grasslands, savanna and oak woodlands, riparian areas in chaparral; 0–1700 m; Calif.

Mature trees of *Quercus lobata* are among the largest oaks of the United States. The species hybridizes with numerous other species, but the hybrids are not common in most parts of its range. On Santa Cruz and Santa Catalina islands, however, occur extensive and relatively stable populations that show intermediate characteristics with *Q. pacifica* (see treatment). The hybrids have been given the name *Q.* ×*macdonaldii*, and they differ from *Q. lobata* in the following: leaf sinuses reaching less than half the distance to the midrib; leaves usually smaller, the lobes often more acute and brighter green; and acorns smaller, with more acute apices.

The Yuri used *Quercus lobata* in the treatment of diarrhea (D. E. Moerman 1986).

63. Quercus douglasii Hooker & Arnott, Bot. Beechey Voy., 391. 1840 · Blue oak E F

Quercus douglasii var. *ransomii* (Kellogg) Beissner; *Q. ransomii* Kellogg

Trees, deciduous, with single trunks, sometimes with few to several trunks. **Bark** gray, scaly. **Twigs** reddish or yellowish, ca. 2 mm, densely or sparsely puberulent, occasionally glabrate with age. **Buds** reddish brown, rarely yellowish, broadly ovoid to rarely subglobose, (2–)3–5 mm; scales glabrous except for ciliate margins, sometimes sparsely or densely pubescent. **Leaves:** petiole blue-green, 2–6 mm. **Leaf blade** obovate or elliptic, oblong or oblanceolate, (20–)40–60(–80) × (15–)20–30(–40) mm, base rounded-attenuate or rounded, rarely cuneate, margins shallowly lobed or irregularly toothed, sometimes entire, lobes mucronate or rounded, secondary veins 6–10 on each side, apex rounded, rarely moderately acute; surfaces abaxially light green or blue-green, waxy, with scattered to crowded, semierect, (2–)4–6(–8)-rayed stellate hairs usually 0.2–0.6 mm diam. or larger, adaxially blue-green, glaucous or grayish, vestiture similar to abaxial surface. **Acorns** subsessile, solitary; cup hemispheric or cup-shaped, rarely deeper, 5–10 mm deep × 10–15 mm wide, enclosing only base of nut, scales thin and not tuberculate to strongly and irregularly tuberculate, particularly toward base of cup; nut thin-walled, fusiform or subcylindric, 20–30 × 10–16 mm. **Cotyledons** distinct. $2n = 24$.

Flowering late winter–spring. Oak woodlands, margins of chaparral and grasslands; 0–1200 m; Calif.

Populations of *Quercus douglasii* in coastal southern California and on the Channel Islands consist of small stands or solitary individuals often associated with *Q. lobata* and scrub oaks. Some doubt exists as to whether some or all of those populations are natural stands or are historical introductions near Native American settlements. Along the canyons of Santa Barbara County, putative hybrids between *Q. douglasii* and *Q. dumosa* are referred to *Q.* ×*kinseliae* (C. H. Muller) Nixon. In the interior Coast Ranges of California are found numerous populations that are intermediate in form between *Q. douglasii* and *Q. john-tuckeri* (= *Q. turbinella* var. *californica*). This appears to be an area of secondary contact, and the

two species remain distinct in nearby populations. Because of the widespread nature of the intermediates, following Tucker's extensive studies they can be conveniently referred to as the nothospecies *Q.* ×*alvordiana* Eastwood. The plants tend to be shrubs to small trees, with somewhat more spinose leaves than *Q. douglasii* and fruit similar to those of the latter species.

64. Quercus john-tuckeri Nixon & C. H. Muller, Novon 4: 391. 1994 · Tucker oak, desert scrub oak E F

Quercus turbinella Greene subsp. *californica* J. M. Tucker

Shrubs, subevergreen or evergreen, 1–3(–5) m. **Bark** light gray or brown, scaly. **Twigs** yellowish or dingy gray, 1–1.5(–2) mm diam., densely tomentulose. **Buds** brown, ovoid or globose, 1.5–2(–3) mm, glabrous except for ciliate margins of scales; proximal scales often yellowish puberulent. **Leaves:** petiole 1–4 mm. **Leaf blade** unicolored, elliptic or obovate, (10–)15–30(–40) × (8–)10–15(–20) mm, thick and leathery, often brittle, base truncate or rounded-attenuate, rarely subcordate, margins irregularly spinose-toothed, occasionally shallowly lobate, secondary veins (3–)4–7, often some veins branching near margin and passing into more than 1 tooth, apex acute or rounded; surfaces abaxially waxy grayish, light green, or yellowish, sparse to moderately dense (8–)10–12-rayed, (loosely) appressed-stellate hairs, often 0.2–0.5 mm diam., and sparse to crowded, yellowish, glandular hairs, adaxially dull grayish, with stellate hairs, similar to abaxial surface. **Acorns** solitary or paired, subsessile; cup cup-shaped or obconic to hemispheric, 5–7 mm deep × 10–15 mm wide, thin, scales whitish or yellowish, moderately or scarcely tuberculate, puberulent; nut fusiform, ovoid, or conic, 20–30 mm, apex acute. **Cotyledons** distinct.

Dry slopes, chaparral, pinyon and juniper woodlands, margins of oak woodlands and sagebrush; 900–2000 m; Calif.

Endemic to California, *Quercus john-tuckeri* occurs from Los Angeles County northward in the interior Coast Ranges and Sierra Foothills to the northern edge of Sacramento Valley.

Quercus john-tuckeri bears some resemblance to both *Q. turbinella* and *Q. berberidifolia*. *Quercus turbinella* has pedunculate fruit and cordate leaf bases, however, and *Q. berberidifolia* has a glabrate adaxial leaf surface, substantially smaller stellate trichomes with fewer rays on the abaxial leaf surface, heavier tuberculate acorn cups, and blunt or rounded (instead of acute) acorns.

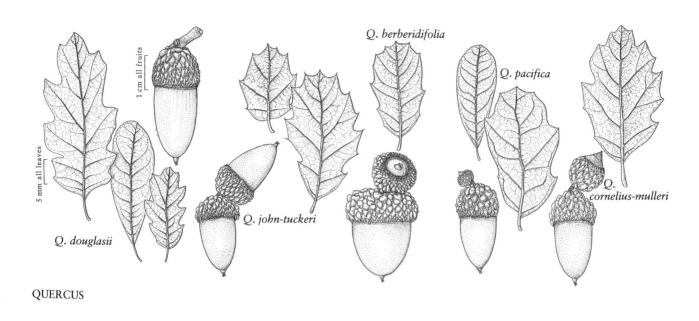

Q. berberidifolia

Q. pacifica

Q. john-tuckeri

Q. douglasii

Q. cornelius-mulleri

QUERCUS

65. Quercus berberidifolia Liebmann, Overs. Kongel. Danske Vidensk. Selsk. Forh. Medlemmers Arbeider 1854: 172. 1854 · California scrub oak [E] [F]

Quercus agrifolia Née var. *berberidifolia* (Liebmann) Wenzig; *Q. dumosa* Nuttall var. *munita* Greene

Shrubs, subevergreen, 1–2(–4) m. **Bark** gray, scaly. **Twigs** gray, yellowish, rarely reddish, 1–3 mm diam. **Buds** reddish brown, globose or ovoid, 2–3 mm, minutely puberulent. **Leaves:** petiole 2–4 mm. **Leaf blade** bicolored, obovate, elliptic, occasionally subrotund, planar or moderately convex, (10–)15–30 × (8–)10–20 mm, base truncate or rounded-attenuate, rarely cuneate, margins irregularly toothed and spinose, often sublobate, rarely entire, secondary veins (3–)4–7 on each side, apex broadly rounded or acute; surfaces abaxially waxy, light green or glaucous, with scattered minute, appressed, (4–)8 (–10)-rayed hairs less than 0.2 mm diam. and sparse to dense yellowish, glandular hairs, adaxially glossy or dull green, glabrous or glabrate. **Acorns** solitary or paired, subsessile; cup hemispheric or turbinate, rarely shallowly cup-shaped, rim thick, 8–15 mm deep × 15–20(–22) mm wide, enclosing to 1/2 nut, scales reddish or yellowish, usually strongly, irregularly tuberculate, puberulent or canescent; nut light to dark brown, ovoid, ellipsoid, or barrel-shaped, (10–)15–30 × (8–) 10–20 mm, apex rounded, glabrous at maturity. **Cotyledons** distinct.

Flowering spring. Chaparral, margins of coastal sage scrub; 100–1800 m; Calif.

The name *Quercus dumosa* (see species treatment no. 72) has often been applied to this species.

Quercus berberidifolia is the most common scrub oak of central and southern California, mostly at midelevations in the Coast Ranges. In central California it is replaced in drier interior habitats by *Q. john-tuckeri*, and south of the transverse ranges by *Q. cornelius-mulleri*. From Santa Barbara south, it does not descend to the low elevation coastal sites typical of *Q. dumosa* in the strict sense. *Quercus berberidifolia* hybridizes with numerous other white oaks of California. In southern California, putative hybrids with *Q. john-tuckeri* are noticeable in the mountains above Ventura and on the north slope of the Tehachapi Mountains.

66. Quercus pacifica Nixon & C. H. Muller, Novon 4: 391. 1994 [C] [E] [F]

Quercus dumosa Nuttall var. *polycarpa* Greene

Shrubs, rarely small trees, subevergreen, shrubs to 2 m, trees to 5 m or taller. **Bark** scaly on older branches and trunk. **Twigs** brownish or reddish, minutely puberulent, becoming glabrate and gray with age. **Buds** light or chestnut brown, ovate or globose, 2–3 × 1–2 mm. **Leaves:** petiole 2–5 mm. **Leaf blade** obovate or oblong, planar to moderately convex or undulate, 15–40 × 7–20(–40) mm, base cuneate or rounded, attenuate-decurrent along petiole, margins minutely cartilaginous, entire or with 1–5 irregular teeth on each side, secondary veins obscure, 1–5 on each side, apex blunt or rounded, occasionally subacute with mucronate tip; surfaces abaxially waxy, glandular, with scattered minute, flat, appressed, ± 8-rayed stellate hairs, not obscuring surface, adaxially green, glossy, glabrate or with minute, scattered, stellate hairs. **Acorns** paired or solitary in leaf axil, subsessile, rarely pedunculate in teratological forms; cup hemispheric to turbinate, to 15 mm deep × 20 mm wide, enclosing only 1/4–1/2 nut, scales moderately to heavily tuberculate, irregularly formed; nut light brown, acute-cylindric or fusiform, tapered, (15–)20–30 × (6–)9–15 m, apex acute, glabrate. **Cotyledons** distinct.

Flowering spring. Chaparral, oak woodlands, margins of grasslands, understory in closed-cone pine stands; of conservation concern; 0–300 m; Calif.

Quercus pacifica is endemic to three of the California Channel Islands: Santa Cruz, Santa Catalina, and Santa Rosa. It is not known from the mainland, but it bears a superficial similarity to some of the tree forms that are putative hybrids between *Q. engelmannii* and *Q. cornelius-mulleri* in San Diego County. The latter populations, sometimes treated as *Q.* ×*acutidens*, differ in having much greater variability in leaf shape; thicker, more leathery leaves; denser abaxial leaf vestiture; much smaller hairs, having more than 10 rays; and variable levels of connation of cotyledons (always distinct in *Q. pacifica*). *Quercus pacifica* appears to be most closely related to *Q. douglasii*, whether by direct descent or by introgression with another species no longer extant on the islands.

Quercus ×*macdonaldii* Greene (as a species) [= *Quercus dumosa* var. *macdonaldii* (Greene) Jepson] is a stabilized hybrid complex between *Quercus pacifica* and *Q. lobata* Née. The plants tend to be small to moderate trees with leaves that resemble those of *Q. lobata*; the leaves are much more shallowly lobed and always less than two-thirds the distance from the margin to the midrib. *Quercus* ×*macdonaldii* is known from Santa Rosa, Santa Cruz, and Santa Catalina islands.

67. Quercus cornelius-mulleri Nixon & K. P. Steele, Madroño 28: 210, figs. 1–6. 1981 · Muller oak [F]

Shrubs, evergreen or subevergreen, densely branched, 1–2(–3) m. **Bark** gray, scaly. **Twigs** gray, yellowish, or brownish, 1–1.5 mm diam., densely tomentulose, rarely glabrate. **Buds** dull brown, ovoid, 2.5–3 mm, glabrous except for margins of scales. **Leaves:** petiole 2–5 mm. **Leaf blade** strongly bicolored, ovate to oblong or narrowly

obovate, 15–35 × 10–20 mm, rather thick and leathery, base cuneate or attenuate-rounded, margins entire or irregularly, shallowly toothed, teeth mucronate, rarely spinose, secondary veins 6–7 on each side, apex rounded or acute; surfaces abaxially whitish, densely covered with minute, compact, appressed, (8–)10–14 (–16)-rayed stellate hairs less than 0.2 mm diam. (lateral fusion of rays visible under high magnification), without glandular seriate hairs, adaxially dull, light green, grayish, or yellowish green, with scattered, appressed-stellate hairs to 0.2 mm diam. **Acorns** solitary or clustered, subsessile; cup deeply cup-shaped or turbinate, to 5–13 mm deep × 12–20 mm wide, scales whitish or cream, strongly tuberculate especially near base of cup; nut dark brown, fusiform or cylindric, 20–30 × 10–30 mm. **Cotyledons** distinct.

Flowering early spring. Open chaparral, pinyon and juniper woodlands, desert margins, often on loose granitic soils; 1000–1800 m; Calif.; Mexico (Baja California).

Quercus cornelius-mulleri is easily distinguished from other California scrub oaks by the strongly bicolored leaves, dense minute tomentum of the abaxial leaf surface, and large acute acorns in deep tuberculate cups. Of greater difficulty are swarms of putative hybrids with *Q. engelmannii*, sometimes referred to as *Q.* ×*acutidens*.

In Joshua Tree National Monument a lone tree and several shrubs appear to be hybrids and backcrosses between *Quercus cornelius-mulleri* and *Q. lobata*. This tree is the basis of *Quercus* ×*munzii* J. M. Tucker.

68. Quercus turbinella Greene, Ill. W. Amer. Oaks 1: 37. 1889 · Sonoran scrub oak [F] [W1]

Quercus dumosa Nuttall var. *turbinella* (Greene) Jepson; *Q. subturbinella* Trelease

Shrubs or small trees, evergreen or subevergreen, to 4 m. **Bark** light gray or brown, scaly. **Twigs** brown to gray, 1–3 mm diam., usually tomentulose, sometimes glabrous, becoming glabrate. **Buds** brown, round to ovoid, 1–2 mm, minutely pubescent. **Leaves:** petiole 1–4 mm. **Leaf blade** elliptic or ovate, (1.5–)20–30 × (5–)10–15(–20) mm, thick, leathery, base cordate or rounded, margins planar or slightly crisped-undulate, coarsely 3–5-toothed or very shallowly lobed on each side, teeth spinose with spines 1–1.5 mm, secondary veins 4–8 on each side, apex acute or obtuse; surfaces abaxially yellow or reddish, usually glaucous, minutely stellate-puberulent, adaxially grayish, glaucous, or yellowish glandular, glabrous or sparsely and minutely stellate-pubescent. **Acorns** solitary or several, on axillary peduncle 10–40 mm; cup hemispheric or shallowly cup-shaped, 4–6 mm deep × 8–12 mm wide, covering 1/4–1/2 nut, scales tightly appressed, ovate, moderately tuberculate, grayish or yellowish puberulent; nut light brown, ovoid, to 20 × 11 mm, minutely puberulent or glabrate. **Cotyledons** distinct.

Flowering spring. Dry desert slopes, often in juniper and pinyon woodlands; 800–2000 m; Ariz., Calif., Colo., N.Mex., Nev., Tex., Utah; Mexico (Baja California, Sonora, and probably n Chihuahua).

Formerly, California populations of what here is referred to as *Quercus john-tuckeri* have been included in the concept of *Q. turbinella*. *Quercus john-tuckeri* has subsessile fruit and noncordate leaf bases as opposed to the consistently pedunculate fruit and strongly cordate leaf bases of *Q. turbinella*. The two species seem to be no more closely related to each other than each might be to other southwestern oaks, and *Q. john-tuckeri* shares at least as many characteristics with *Q. berberidifolia* as with *Q. turbinella*. Thus, treatment of these two taxa as varieties of the same species is inappropriate.

Quercus turbinella forms putative hybrid swarms with *Q. gambelii* (see treatment), as well as with *Q. grisea*.

SELECTED REFERENCE Tucker, J. M. 1961b. Studies in the *Quercus undulata* complex. II. The contribution of *Q. turbinella*. Amer. J. Bot. 48: 329–339.

69. **Quercus hinckleyi** C. H. Muller, Contr. Texas Res. Found., Bot. Stud. 1: 40. 1951 · Hinckley oak [C] [F]

Shrubs, evergreen, low, to 0.75 (–1.5) m, spreading rhizomatously in thickets, intricately branched. **Bark** gray, scaly. **Twigs** light brown, pruinose, becoming waxy-glaucous in 2d season, 1–1.5 mm diam., glabrous or sparsely and minutely stellate-pubescent. **Buds** minute, subrotund, 0.5–1 mm; scales reddish brown, glabrous except for ciliate margins. **Leaves:** petiole to 2 mm. **Leaf blade** subrotund or rotund, to 15 × 15 mm, thick, leathery, base cordate or auriculate, margins strongly crisped with 2–3 coarse, spinescent teeth on each side, cartilaginous-thickened secondary veins obscure, apex acute or obtuse, spine-tipped; surfaces abaxially blue-green, glaucous, glabrous, microscopically markedly papillose, adaxially blue-green, glaucous, glabrous, secondary veins slightly raised on both surfaces. **Acorns** solitary, subsessile or on axillary peduncle to 4 mm; cup shallow, saucer-shaped, 1–3 mm deep × 10–15 mm wide, enclosing base of nut only, margin irregularly undulate, scales closely appressed, minute, basally tuberculate-thickened, glabrous except for thin ciliate

margins; nut ovoid, 10–20 × 8–12 mm, glabrous. **Cotyledons** distinct.

Flowering spring. On dry desert slopes; of conservation concern; 1150–1400 m; Tex.; Mexico.

This species is known only from two sites in the United States, El Solitario and near Shafter, Texas. The Shafter population includes some individuals with characteristics that suggest hybridization with *Quercus pungens* Liebmann. These plants are larger, with more pubescent twigs and leaves, and hemispheric acorn cups to 10 mm deep. Such plants have recently been collected in adjacent Mexico. Fossil evidence from packrat middens indicates *Q. hinckleyi* probably had a broader distribution and was a dominant shrub between 19,000 and 9500 years ago.

70. **Quercus ajoensis** C. H. Muller, Madroño 12: 140, fig. 1. 1954 · Ajo Mountain scrub oak [F]

Quercus turbinella Greene subsp. *ajoensis* (C. H. Muller) Felger & C. H. Lowe; *Q. turbinella* var. *ajoensis* (C. H. Muller) Little

Shrubs, rarely trees, evergreen, to 2–3 m. **Bark** gray, scaly or furrowed. **Twigs** light brown, 1–2 mm diam., inconspicuously short stellate-pubescent or glabrate. **Buds** brown or reddish brown, ovoid or globose, 1–1.5 mm, variously short stellate-pubescent, tomentose, or glabrate. **Leaves:** petiole (2–)3–4 m. **Leaf blade** ovate to narrowly ovate or oblong, (10–)15–35(–50) × (5–)10–20(–30) mm, rather leathery, base cordate, rarely rounded, margins crispate, sometimes flat, cartilaginous, with 4–6(–8) long-attenuate, spinose-awned teeth on each side, secondary veins 5–8 on each side, whitish, apex acute or obtuse with bristly distal teeth; surfaces abaxially blue-green, waxy-glaucous, microscopically papillose, glabrous, sometimes sparsely stellate-pubescent along midrib, adaxially blue-green, glaucous, glabrous or sparingly stellate-pubescent along midrib, secondary veins raised on both surfaces. **Acorns** solitary or paired on thin axillary peduncle (5–)30–50 mm; cup shallowly cup-shaped, thin, 3–4 mm deep × 6–8(–10) mm wide, enclosing only base of nut, scales brownish, moderately tuberculate, pubescent; nut oblong to narrowly ovoid, 12–15 × 5–8 mm. **Cotyledons** distinct.

Flowering in spring. Rare to locally abundant on igneous slopes; 500–1500 m; Ariz.; Mexico (Baja California).

Populations of *Quercus ajoensis* in southern New Mexico show characteristics suggesting introgression from hybridization with *Quercus toumeyi*, such as increased twig and leaf pubescence and sometimes the prominent golden puberulum of the abaxial leaf surfaces. Hybrids between *Q. ajoensis* and both *Q. turbinella* and *Q. gambelii* (Utah) are also known.

71. Quercus toumeyi Sargent, Gard. & Forest 8: 92, figs. 13, 14. 1895 · Toumey oak [F]

Quercus hartmanii Trelease

Shrubs or small trees, deciduous or subevergreen. **Bark** dark gray to almost black, scaly. **Twigs** brownish, 1–2 mm, usually persistently pubescent. **Buds** reddish brown, ovoid, ca. 1 mm. **Leaves:** petiole 2–3.5 mm. **Leaf blade** oblong-elliptic or lanceolate, 15–25(–30) × (6–)8–12(–15) mm, base obtuse or cuneate, rarely subcordate, margins strongly cartilaginous, entire, sometimes sparsely mucronate-dentate toward apex, secondary veins 7–8 on each side, apex acute, sometimes rounded; surfaces abaxially dull gray, microscopically pubescent with long, soft, white or yellow hairs concentrated in tufts along midvein and base, adaxially glossy green, sparsely minutely stellate-pubescent or glabrate. **Acorns** solitary or paired, subsessile or on peduncle 2 mm; cup cup-shaped, 6 mm deep × ca. 8–9 mm wide, enclosing ca. 1/3 nut, scales moderately tuberculate; nut light brown, narrowly ovoid or elliptic, 8–15 × 6–8 mm. **Cotyledons** distinct.

Flowering spring. Rocky slopes, oak woodlands, and open chaparral; 1500–1800 m; Ariz., N.Mex., Tex.; Mexico (Chihuahua and Sonora).

Quercus toumeyi, particularly the more spinescent-leaved form, is often confused with *Q. turbinella.* The latter species has acorns on peduncles greater than 10 mm, and more or less evenly distributed minute, flat, stellate trichomes on the abaxial leaf surface, in contrast to the subsessile acorns and longer straight hairs along the midvein of the abaxial leaf surface in *Q. toumeyi.*

72. Quercus dumosa Nuttall, N. Amer. Sylv. 1: 7. 1842
Coastal sage scrub oak [C][F]

Shrubs, subevergreen, 1–2(–2.5) m, dense, divaricately branching, leaves brittle, often falling when branches disturbed. **Bark** smooth when young, eventually scaly. **Twigs** reddish or grayish, 1–1.5(–2) mm diam., glabrous or sparsely stellate-pubescent, soon glabrous. **Buds** reddish brown, globose or ovoid, 1–2 mm, glabrous, rarely puberulent near apex. **Leaves:** petiole 1–2(–3) mm. **Leaf blade** undulate or strongly to moderately cupped, occasionally subplanar, 10–20(–25) × 6–15(–20) mm, base cordate or angular-cordate, margins irregularly spinose-toothed or shallowly lobed, rarely entire, often somewhat revolute, secondary veins 3–5(–6) on each side, irregularly branched, apex rounded or spinose-acute; surfaces abaxially sparsely to densely covered with erect, curly, (2–) 4(–6)-rayed fasciculate hairs to 0.5 mm, felty to touch in young leaves, adaxially glossy green, glabrate or with scattered stellate hairs, secondary veins somewhat impressed, puberulent. **Acorns** solitary or paired, subsessile; cup reddish, deeply cup-shaped, 5–8 mm deep × 8–15 mm wide, enclosing 1/3 nut or less, scales long-acute, moderately or scarcely tuberculate at base; nut fusiform or subcylindric, 10–20(–30) × 5–10(–12) mm, apex acute. **Cotyledons** distinct.

Flowering spring. Open chaparral, coastal sage scrub; of conservation concern; 0–300 m; Calif.; Mexico (Baja California).

The name *Quercus dumosa* has been applied to virtually all scrub oaks in the white oak group of central and southern California and adjacent Baja California. Through the years, and following independent studies by various authors, the concept of this species has gradually narrowed from the original, which included plants here segregated as *Q. turbinella, Q. johntuckeri, Q. cornelius-mulleri, Q. berberidifolia,* and *Q. pacifica.* In degree and constancy, the differences among these species are similar to those separating other commonly recognized tree species of the western United States. The majority of populations referred to *Q. dumosa* in recent treatments are now included in *Q. berberidifolia* (see treatment). All the scrub oaks have a striking superficial similarity because of their shrubby habit and small, often spiny leaves; they differ dramatically in leaf and twig vestiture and acorn form. The concept of *Q. dumosa* presented here limits it to populations of scraggly shrubs with short petioles, cordate leaf bases, erect curly trichomes on the abaxial leaf surface, and narrow acute acorns that occur at low elevations almost always within sight of the ocean. Because these locations are typically prime real estate, the species, which probably never was common, is highly at risk. It rarely comes into contact with other white oaks because of its low elevation and dry habitat preference; putative hybrids are known, however, with *Q. engelmannii* and *Q. lobata.* Some populations of *Q. berberidifolia* from higher elevations near populations of *Q. dumosa* show signs of introgression.

Named hybrids include *Quercus* ×*kinseliae* (C. H. Muller) Nixon & C. H. Muller (= *Q. dumosa* Nuttall × *Q. lobata* Née) and *Q. dumosa* Nuttall var. *kinselae* C. H. Muller (= *Q. dumosa* × *Q. engelmannii*).

The Luisenos used gall nuts from *Quercus dumosa* medicinally for sores and wounds and as an astringent (D. E. Moerman 1986).

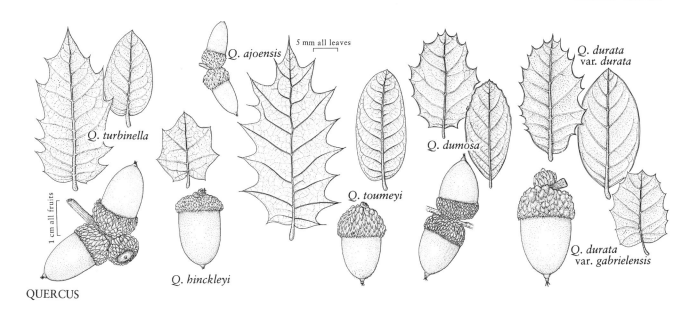

Q. ajoensis

5 mm all leaves

Q. turbinella

Q. dumosa

Q. durata var. durata

1 cm all fruits

Q. toumeyi

Q. hinckleyi

QUERCUS

Q. durata var. gabrielensis

73. Quercus durata Jepson, Fl. Calif. 1(2): 356. 1909 · Leather oak E

Quercus dumosa Nuttall var. *bullata* Engelmann; *Q. dumosa* var. *revoluta* Sargent

Shrubs, evergreen, 1–2(–3) m. **Bark** scaly. **Twigs** gray or yellowish, 1–3 mm diam., densely or sparsely to-mentulose, often with prominent, yellowish, spreading hairs. **Buds** brown or reddish brown, ovoid or globose, 1–2 mm, glabrous or puberulent. **Leaves:** petiole 1–5 mm. **Leaf blade** cupped or convex, rarely somewhat planar, (10–)15–40 × 7–15(–20) mm, base cuneate, rounded-attenuate, or truncate, margins entire or irreg-ularly toothed, sometimes spinose, usually unevenly revolute, secondary veins 4–6 on each side, apex rounded or subacute; surfaces abaxially densely to sparsely covered with erect, stipitate, (1–)2–4(–6)-rayed hairs 1–4 mm, felty to touch, secondary veins promi-nent, adaxially grayish or yellowish, with dense or scat-tered, semi-erect or appressed hairs, secondary veins obscure or somewhat impressed. **Acorns** solitary or paired, subsessile; cup reddish, hemispheric, deeply cup-shaped or turbinate, 4–6 mm deep × 12–18 mm wide, enclosing to 1/2 nut or more, scales reddish or yellowish, weakly to strongly tuberculate, often some-what glandular; nut globose, ovoid or cylindric, 15–25 × 10–25 mm, apex rounded or obtuse, persistently minute-puberulent. **Cotyledons** distinct.

Chaparral, oak woodlands, open pine forests, on serpentine and nonserpentine soils; 150–1500 m.

Varieties 2 (2 in the flora): Calif.

1. Leaf blade adaxially grayish stellate, usually deeply cupped, growth compact; leaves densely

crowded on twigs; serpentine soils, Santa Bar-bara County and northward. . . . .
. . . . 73a. *Quercus durata* var. *durata*
1. Leaf blade adaxially greenish, glabrate or essen-tially so, rarely deeply cupped, usually moder-ately cupped or subplanar, growth open, scrag-gly; leaves not densely crowded; nonserpentine soils, Los Angeles County. . . . .
. . . . 73b. *Quercus durata* var. *gabrielensis*

73a. Quercus durata Jepson var. **durata** E F

Shrubs, densely branched. **Leaf blade** usually deeply cupped, margins revolute; surfaces abaxi-ally with hairs usually 1–2 mm, rays curly, adaxially grayish or yellowish, with persistent stellate pubescence.

Flowering spring. Serpentine soils; 150–1500 m; Calif.

Quercus durata var. *durata* is found from Santa Bar-bara County north to Shasta County.

73b. Quercus durata Jepson var. **gabrielensis** Nixon &
C. H. Muller, Novon 4: 392. 1994 [C] [E] [F]

Shrubs, openly branched, scrag-
gly. Leaf blade moderately
cupped, sometimes almost planar; surfaces abaxially with hairs
usually 2–4 mm, rays arched,
curved or straight, not curly, adaxially green, glossy, usually glabrate or sometimes sparsely
stellate-pubescent.

Flowering spring. Chaparral on dry, exposed, loose
slopes in nonserpentine soils; of conservation concern;
450–1000 m; Calif.

Quercus durata var. *gabrielensis* occurs only in Los
Angeles County, on the southern slope of the San Ga-
briel Mountains from La Canada to Pomona. In this
area along the lower elevational limits of this variety,
occasional intermediates with *Q. engelmannii* occur
[= *Q.* ×*grandidentata* Ewan (as species)]. These putative hybrids are large shrubs or small trees with leaves
that are persistently woolly on the abaxial surface and
have coarse regular teeth. Unfortunately, most of these
hybrids have been eliminated by the same uncontrolled
growth that has largely extirpated *Q. engelmannii* in
this area.

74. Quercus rugosa Née, Anales Ci. Nat. 3: 275. 1801
· Netleaf oak [F]

Quercus ariifolia Trelease; *Q.
diversicolor* Trelease; *Q. durangensis*
Trelease; *Q. reticulata* Humboldt &
Bonpland; *Q. rhodophlebia* Trelease;
Q. vellifera Trelease

Shrubs or trees, evergreen, usually moderate-sized, rarely large.
Bark light or dark brown, scaly.
Twigs brown, turning gray with
age, 1–2 mm diam., tomentose to tomentulose, variously glabrate or persistently pubescent. Buds brown,
ovoid, 2–4 mm, apex obtuse, sparsely pubescent or
eventually glabrate. Leaves: petiole to 7 mm. Leaf
blade broadly obovate or panduriform to orbiculate
or elliptic, rarely narrowly obovate, usually cupped,
strongly concave proximally, sometimes planar, to 100
× 70 mm, stiff, leathery, base deeply or shallowly
cordate, margins usually somewhat revolute, cartilaginously thickened, undulately crisped or flat with inconspicuous or coarse mucronate teeth near apex, secondary veins 8–10(–12) on each side, branched, apex
broadly rounded, rarely subacute; surfaces abaxially
dull, glaucous, or densely brownish tomentose, becoming nearly glabrate or pubescence persistent, especially about midribs, secondary veins very prominently
raised, reticulate, adaxially dark green, lustrous,
sparsely stellate-pubescent especially about base of
midrib, secondary veins impressed. Acorns 1–3 or

more on slender axillary peduncle 30–60 mm; cup
deeply cup-shaped to saucer-shaped, to 9 mm deep ×
15 mm wide, enclosing to 1/2 nut, scales loosely appressed, characteristically somewhat spreading, brown,
ovate, tuberculate-thickened or only slightly so, tomentose or obscurely tomentulose; nut light brown,
ovoid to elliptic, to 20 × 15 mm, glabrous or minutely villous. Cotyledons distinct, often reddish or
purple.

Flowering early–late spring. Wooded slopes; 2000–
2500 m; Ariz., N.Mex., Tex.; Mexico; Central
America (Guatemala).

Quercus rugosa occurs on wooded slopes at high
elevations in trans-Pecos Texas, southern New Mexico, and Arizona, and throughout most of the mesic
montane parts of Mexico, south to Guatemala.

75. Quercus vaseyana Buckley, Bull. Torrey Bot. Club 10:
91. 1883 · Vasey oak [F]

Quercus pungens Liebmann var.
vaseyana (Buckley) C. H. Muller; *Q.
undulata* Torrey var. *vaseyana*
(Buckley) Rydberg

Shrubs or small trees, evergreen
or subevergreen, to 10 m. Bark
dark brown, furrowed and exfoliating in long strips. Twigs reddish or grayish brown, 1–1.5
mm diam., short stellate-tomentose or tomentulose,
later glabrate or persistently pubescent, rarely glabrous. Buds dark red-brown or gray, round-ovoid, 1–
1.5 mm, apex obtuse, sparsely pubescent or glabrate.
Leaves: petiole to 5 mm. Leaf blade narrowly lanceolate to usually oblong, mostly planar or slightly convex, 20–60(–90) × 10–20 mm, often rather leathery,
base cuneate to rounded, margins coarsely 3–5-
toothed on each side or shallowly lobed or entire,
with teeth or lobes acute or obtuse, mucronate-tipped,
secondary veins 4–6 on each side, usually branched,
apex acute, rarely obtuse; surfaces abaxially densely
stellate with minute appressed hairs, rarely glabrate
and lustrous green, adaxially dark green, lustrous, glabrous or very sparsely stellate-puberulent. Acorns
subsessile or on peduncle 2–3 mm; cup saucer-shaped
to cup-shaped, 3–4 mm deep × 10 mm wide, margin
thin, scales reddish brown, strongly, regularly tuberculate; nut light brown, ovoid to oblong or subcylindric,
to 12 × 12 mm, glabrous. Cotyledons distinct.

Flowering spring. Dry limestone slopes, oak and
mesquite woodlands, juniper woodlands, and canyons
and ravines in otherwise dry, open grasslands, sometimes descending into margins of dry scrub; 300–600
m; Tex.; Mexico (Chihuahua, Coahuila, and Nuevo
León).

Apparent hybridization between *Quercus vaseyana*
and *Q. pungens* is discussed under the latter species.

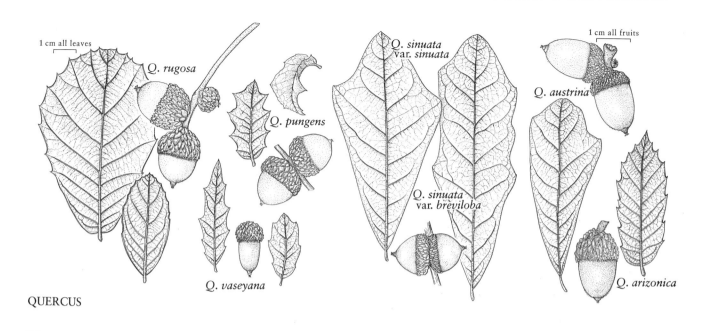

QUERCUS

76. Quercus pungens Liebmann, Overs. Kongel. Danske Vidensk. Selsk. Forh. Medlemmers Arbeider 1854: 171. 1854; not Gandoger 1890 · Pungent oak · F

Quercus undulata Torrey var. *pungens* (Liebmann) Engelmann

Shrubs or moderate-sized trees, evergreen or subevergreen. **Bark** light brown, papery. **Twigs** gray, 1–2 mm diam., short velvety-tomentose, glabrate with age. **Buds** dark red-brown, ca. 2 mm, sparsely pubescent. **Leaves:** petiole to 10 mm. **Leaf blade** elliptic to oblong, 10–40 (–90) × 10–20(–40) mm, rather thick, leathery, stiff, base rounded or minutely cordate, very rarely cuneate, margins regularly undulate-crisped, not revolute, coarsely toothed or incised with acute teeth or spinose lobes, secondary veins 5–8(–14) on each side, usually branched before passing into teeth, apex acute or obtuse, rarely rounded, spine-tipped; surfaces abaxially canescent, usually densely stellate-pubescent, and mixed with stiff, harsh, stellate hairs, often sandpapery to touch, rarely glabrate, adaxially yellowish green, glossy, usually rough and sandpapery because of minute, persistent hair bases, rarely glabrate. **Acorns** subsessile or on peduncle to 3 mm; cup shallowly to deeply cup-shaped or turbinate, to 8 mm deep × 13 mm wide, covering ca. 1/4 nut, margin thin, scales reddish brown, moderately tuberculate or keeled, densely gray-tomentose; nut light brown, broadly ovoid to subcylindric, to 10 × 10 mm, apex rounded to subacute, glabrous. **Cotyledons** distinct.

Flowering spring. On dry limestone or igneous slopes, usually in oak, pinyon, and juniper woodlands, chaparral, and sometimes descending into desert vegetation; 800–2000 m; Ariz., N.Mex., Tex.; Mexico (Chihuahua and Coahuila).

Numerous populations appear to be hybrid swarms between *Quercus pungens* and *Q. vaseyana*, which is sometimes treated as a variety of *Q. pungens.* No other evidence for a close relationship exists for these two species, and such a treatment risks erecting a polyphyletic assemblage. To the west and south within the range of *Q. pungens* no indication of introgression exists, and the two species are strikingly different and easily separable. I interpret the contact as secondary.

77. Quercus sinuata Walter, Fl. Carol., 235. 1788 · Bastard oak, bastard white oak, Durand oak

Quercus durandii Buckley

Trees or shrubs, deciduous, to 15(–20) m, with solitary or multiple trunks. **Bark** gray to light brown, flaky to papery and exfoliating. **Twigs** light gray or gray, 1–2 (–3) mm diam., glabrous, rarely minutely puberulent. **Buds** brown or reddish brown, broadly ovoid, 2–3 mm, essentially glabrous. **Leaves:** petiole 2–5(–8) mm. **Leaf blade** oblong to oblanceolate, or narrowly rhomboid, or cuneiform, or rounded–3-dentate, (25–)30–120(–140) × (15–)25–60 mm, base acute, cuneate, attenuate-rounded, or obtuse, margins entire to irregularly toothed or moderately, sinuately lobed, flat, secondary veins ca. 7–11 on each side, apex broadly rounded, rarely attenuately narrowed or obscurely 3-lobed; surfaces abaxially silvery or dull green, with scattered to crowded, minute, appressed-stellate, 8–10-rayed hairs, or glabrate or glabrous, especially in shade forms, adaxially green or dull green, glabrous. **Acorns**

solitary or paired, subsessile or on axillary peduncle to 1–7 mm; cup saucer-shaped to shallowly cup-shaped, rarely deeper, 2–8 mm deep × 8–15(–20) mm wide, enclosing 1/8–1/4 nut, rarely more, base flat, rounded, or constricted, margin thin, scales closely appressed, grayish with reddish margins, ovate, flat, obtuse, not tuberculate; nut light brown, depressed-ovoid to oblong, 7–15 × 7–12(–17) mm, glabrous. **Cotyledons** distinct.

Varieties 2 (2 in the flora): North America.

The question of the correct name for this species has persisted, with some authors rejecting the usage here in favor of *Quercus durandii*. Although no type material is extant, the original description of *Q. sinuata* is consistent with the concept presented here, as by W. W. Ashe (1916) and W. Trelease (1924), and inconsistent with any other oak from the broad area covered by Thomas Walter's *Flora Caroliniana* (1788).

The two varieties differ in habit, habitat, leaf size and lobing, and geographic range, and considerable variability exists within both varieties as to the degree and density of silvery stellate-pubescence on the abaxial surface of the leaf. Sun leaves of both tend to have a higher proportion of silvery pubescence, and shade leaves and some individual trees tend to have more glabrate leaves, although evidence of flat-stellate trichomes is usually apparent. Plants with young, expanding leaves sometimes are mistaken for *Quercus nigra*, a member of the red oak group.

1. Trees to 20 m, trunk usually solitary; moist bottomlands and riparian habitats. . . . .
. . . . .77a. *Quercus sinuata* var. *sinuata*
1. Shrubs or small trees to 3(–5) m, often forming thickets, trunks multiple from near ground; dry limestone hills. . . . .
. . . . 77b. *Quercus sinuata* var. *breviloba*

77a. Quercus sinuata Walter var. sinuata E F

Trees, to 15(–20) m, with solitary trunks. **Leaves:** petiole (2–)3–5(–8) mm. **Leaf blade** 50–120(–140) × (20–)25–60 mm. **Acorns:** cup very shallowly saucer-shaped and flat abaxially and adaxially to shallowly cup-shaped with constricted bases, 2–5 mm deep × (9–)10–15(–20) mm wide, enclosing 1/4 nut or less; nut subrotund to ovoid or elliptic, 12–15(–18) × 8–12(–17) mm.

Flowering spring. Low wet areas, hummocks, riparian and limestone bluffs and prairies; 0–400 m; Ala., Ark., Fla., Ga., La., Miss., N.C., S.C., Tex.

77b. Quercus sinuata Walter var. breviloba (Torrey) C. H. Muller, J. Arnold Arbor. 25: 439. 1944   F

Quercus obtusifolia D. Don var. *breviloba* Torrey in W. H. Emory, Rep. U.S. Mex. Bound. 2(1): 206. 1859; *Q. annulata* Buckley 1860[1861], not J. E. Smith 1819, not Korthals 1839–1842[1844]; *Q. breviloba* (Torrey) Sargent; *Q. durandii* Buckley var. *breviloba* (Torrey) E. J. Palmer; *Q. durandii* var. *san-sabeana* (Buckley ex M. J. Young) Buckley; *Q. san-sabeana* Buckley ex M. J. Young

Shrubs or trees, rarely small, usually moderate, (0.5–)1–3(–5) m, often clonal with multiple trunks. **Leaves:** petiole 2–3 mm. **Leaf blade** (25–)30–60(–100) × (15–) 20–40(–60) mm. **Acorns:** cup saucer-shaped to shallowly cup-shaped, rarely deeper, to 3–8 mm deep × 8–12 mm wide, enclosing 1/4 nut, rarely more, base flat, rounded, or constricted, margin thin; nut depressed-ovoid to oblong, 7–12(–15) × 7–10 mm, glabrous.

Flowering spring. Open oak woodlands, dry scrublands, margins of grasslands, and along streams and arroyos, on limestone, rarely on granitics; 200–600 m; Okla., Tex.; Mexico (Coahuila, Nuevo León, and Tamaulipas).

Quercus sinuata var. *breviloba* replaces var. *sinuata* on the Edwards Plateau of Texas and extends south at lower elevations along the eastern side of Sierra Madre Oriental in northern Nuevo León and Tamaulipas. Although habitats of these two varieties are very different, along streams through limestone hills in central Texas the two varieties are in contact, and numerous problematic, morphologically intermediate forms may be found. The lack of broad geographic sympatry and intergradation argue for the treatment of these two taxa at varietal rank.

78. Quercus austrina Small, Fl. S.E. U.S., 353. 1903 · Bastard white oak E F

Quercus durandii Buckley var. *austrina* (Small) E. J. Palmer

Trees, deciduous, to 20(–25) m. **Bark** pale gray, scaly, eventually divided into broad ridges. **Twigs** dark brown to somewhat reddish, 2–2.5 mm diam., often with prominent, corky, white lenticels. **Buds** dark reddish brown, ovoid, distally acute, 3–5 × 2–2.5 mm, puberulent. **Leaves:** petiole 3–5 mm. **Leaf blade** green or grayish green, narrowly obovate or elliptic, (40–)70–100(–200) × (13–)30–50(–115) mm, base cuneate or attenuate, margins sinuately and irregularly shallowly lobed, lobes rounded, sometimes obtuse, secondary veins 4–6(–8) on each side, apex narrowly or broadly rounded; surfaces abaxially loosely covered with semierect stellate hairs to 0.5 mm diam., glabrous at maturity, often with a few hairs remaining along veins near midrib, adaxially glabrous, glossy. **Acorns** subsessile or on stout axillary peduncle to 15 mm; cup hemispheric or deeply goblet- or cup-shaped, 9–10 mm deep × 10–13 mm wide, enclosing 1/3–1/2 nut, scales loosely appressed, gray, narrowly ovate, sometimes thickened near base, not tuberculate, canescent; nut ovoid or elliptic, 17 × 12 mm. **Cotyledons** distinct.

Flowering spring. River bottoms, wet forests, flatwoods; 0–200 m; Ala., Fla., Ga., Miss., N.C., S.C.

Quercus austrina is probably the most misunderstood oak of the southeastern United States. Although the species is fairly widespread, it is apparently abundant only in local areas and is poorly represented in herbaria. This may be partly because *Q. austrina* is often misidentified as *Q. sinuata,* which it superficially resembles, or as *Q. nigra,* a red oak with similarly shaped leaves. It is easily distinguished from *Q. sinuata* by its larger, more acute buds; darker twigs; deeper, turbinate acorn cups; and absence of minute, appressed, stellate hairs on the abaxial leaf surface. Instead, *Q. austrina* has a tomentum of soft erect hairs on young leaves, and glabrate mature leaves.

79. Quercus arizonica Sargent, Gard. & Forest 8: 92. 1895 · Arizona oak F

Quercus sacame Trelease

Trees, evergreen or subevergreen, small to moderate-sized trees, rarely to 18 m. **Bark** scaly. **Twigs** yellowish, 1.5–2.5 mm diam., persistently felty-tomentose, eventually dingy gray. **Buds** dull russet-brown, ovoid, distally subacute or rounded, 3 mm, sparsely pubescent or glabrate. **Leaves:** petiole to 3–10 mm. **Leaf blade** elliptic or oblong to narrowly obovate or oblanceolate, planar or moderately convex, to (30–)40–80(–90) × 15–30 mm, thick and leathery, usually stiff, base cordate or rounded and weakly cordate, margins entire or coarsely toothed especially near apex, cartilaginously revolute, teeth mucronatetipped, obscure or prominent, secondary veins ca. 7–11 on each side, branching, passing into teeth when present, apex acute to usually obtuse or broadly rounded; surfaces abaxially dull, sparsely pubescent or subtomentose with curly branched hairs, reticulate from prominent, raised secondary veins, usually glaucous where exposed, adaxially dark or bluish green, moderately lustrous, sparsely and minutely stellatepubescent, secondary veins slightly raised or prominent within depressions or impressed. **Acorns** solitary or paired, subsessile, occasionally on peduncle to 15 mm; cup hemispheric or cup-shaped, 5–10(–15) mm deep × 10–15 mm wide, enclosing ca. 1/2 nut, base rounded, margin rather coarse, scales cream to brown, broadly ovate, evenly and strongly tuberculate, tomentose, tips closely appressed; nut light brown, ovoid or oblong, 8–12 mm, nearly glabrous. **Cotyledons** connate.

Flowering spring. Oak and pinyon woodlands, margins of chaparral, arroyos; 1300–2500(–3000) m; Ariz., N.Mex., Tex.; Mexico (Chihuahua, Coahuila, Durango, and Sonora).

Some of the specimens previously referred to *Quercus endemica* by C. H. Muller belong here instead.

Putative hybrids between *Quercus arizonica* and *Q. grisea* (= *Q.* ×*organensis* Trelease) are problematic in local areas of contact from southeastern Arizona to western Texas. These intermediates tend to have narrower leaves than *Q. arizonica,* with moderately reticulate patterns of venation, and more densely hairy leaves. *Quercus arizonica* and *Q. grisea* are amply distinct elsewhere, including large areas in northern Mexico, and they appear to be more closely related to other species than to one another (e.g., *Q. arizonica* with *Q. oblongifolia* and *Q. laeta* Liebmann, and *Q. grisea* with *Q. mohriana* and *Q. microphylla* Née). Thus, *Q. arizonica* and *Q. grisea* are best treated as distinct species that hybridize, and not as conspecific populations.

SELECTED REFERENCE Tucker, J. M. 1963. Studies in the *Quercus undulata* complex. III. The contribution of *Q. arizonica.* Amer. J. Bot. 50: 699–708.

80. Quercus oblongifolia Torrey in L. Sitgreaves, Rep. Exped. Zuni Colorado Rivers, 173, plate 19. 1853 · Sonoran blue oak F

Trees, evergreen, to 10 m. **Bark** gray or whitish, closely furrowed. **Twigs** light brown, 1–1.5 mm diam., densely or sparsely stellate-tomentose, soon glabrate. **Buds** reddish brown, subspheric to broadly ovoid, 1–2 mm, glabrous or basal scales pubescent; stipules persistent about terminal buds. **Leaves:** petiole 2–5(–8) mm. **Leaf blade** oblong to elliptic, occasionally lanceolate or ovate, (20–)30–60(–80) × (5–)10–25(–30) mm, base cuneate to cordate, margins entire, undulate, sometimes irregularly toothed especially toward apex, secondary veins 7–8(–10) on each side, branched, apex acute or broadly rounded; surfaces abaxially densely and loosely glandular-tomentose, quickly glabrate or persistently floccose, especially about base of midrib, at maturity strongly glaucous, adaxially dull pale green, bluish green, or glaucous, sparsely stellate-tomentose, quickly glabrate. **Acorns** solitary or paired, subsessile or on peduncle 4–12 mm; cup cup-shaped, about 6–8 (–13) mm deep × 10–13 mm wide, enclosing ca. 1/3 nut, scales to 1–1.5 mm wide, moderately, regularly tuberculate near base of cup, gray-pubescent; nut light brown, ovoid or oblong, 12–17(–19) × (7–)10–12 mm, glabrate or puberulent about apex. **Cotyledons** connate.

Flowering in spring. Common in high grasslands and midelevation woodlands, mesas, and canyons; 1300–1650 m; Ariz., N.Mex., Tex.; Mexico (Baja California South, Sonora, Chihuahua, and Coahuila).

81. Quercus engelmannii Greene, Ill. W. Amer. Oaks 1: 33, plate 17. 1889 · Engelmann oak F

Trees, subevergreen, to 10 m. **Bark** gray or whitish, closely furrowed. **Twigs** light brown, 1–1.5 mm diam., densely or sparsely stellate-tomentose, soon glabrate. **Buds** reddish brown, subspheric to broadly ovoid, 1–2 mm, glabrous or basal scales pubescent; stipules persistent about terminal buds. **Leaves:** petiole (2–)3–4(–6) mm. **Leaf blade** oblong to elliptic, occasionally lanceolate or ovate, (20–)30–60(–80) × (5–)10–20(–25) mm, base cuneate to cordate, margins entire, undulate, sometimes irregularly toothed, especially toward apex, secondary veins 7–8(–10) on each side, branched, apex acute or broadly rounded; surfaces abaxially blue-green or pale green, densely and loosely glandular-tomentose, quickly glabrate or persistently floccose, especially about base of midrib, at maturity strongly glaucous, adaxially gray-green or pale green, bluish green or glaucous. **Acorns** solitary or paired, subsessile or on peduncle to 5–6 mm; cup cup-shaped or shallowly cup-shaped, 8–10 mm deep × 10–15 mm wide, enclosing 1/3 nut, scales 1.5–3 mm wide, strongly and regularly tuberculate near base of cup, gray-pubescent; nut light brown, ovoid or oblong, 15–25 × 12–14 mm, glabrate or puberulent about apex. **Cotyledons** connate. $2n = 24$.

Flowering in spring. Oak woodlands, margins of chaparral, arroyos, slopes and bajadas; 50–1200 m; Calif.; Mexico (Baja California).

Quercus engelmannii is closely related to and possibly conspecific with *Q. oblongifolia*. The cups of *Q. engelmannii* are larger, deeper, and generally more tuberculate than those of *Q. oblongifolia*, and the scales are usually larger. Based on available samples, the nuts of *Q. engelmannii* are consistently larger than those of *Q. oblongifolia*, apparently with little, if any, overlap in diameter. Considerably more variation occurs within

Q. engelmannii in leaf form, possibly reflecting introgression from other white oak species such as *Q. cornelius-mulleri, Q. dumosa,* and *Q. durata* (see treatment).

On Catalina Island, *Quercus engelmannii* is known only from a small grove of trees. Putative hybrids between *Q. engelmannii* and *Q. cornelius-mulleri* are common in areas of contact between the two species in Riverside and San Diego counties in southern California. Such a population was the basis for *Q. acutidens* Torrey [*Q. dumosa* var. *acutidens* (Torrey) Wenzig]. Other names applied to those populations are *Q. macdonaldii* var. *elegantula* Greene and *Q. dumosa* var. *elegantula* (Greene) Jepson. Variable in leaf form and stature, those intermediates form extensive populations and are probably best disposed of under the name *Q.* ×*acutidens*.

82. Quercus mohriana Buckley, Bull. New York Bot. Gard. 2: 219, plate 31, figs. 1, 2. 1901 · Mohr oak F

Shrubs or trees, evergreen or deciduous, shrubs erect, rhizomatous, trees small, 0.5–3 m. **Bark** pale, rough and deeply furrowed. **Twigs** yellowish or whitish, 1–2 mm diam., felty-tomentose. **Buds** dark red-brown, round-ovoid, 2 mm, glabrous, occasionally puberulent on outer scales, not subtended by persistent, hairy, subulate stipules. **Leaves:** petiole 2–5 mm. **Leaf blade** usually strongly bicolored, oblong or elliptic, (15–)30–50(–80) × (10–)20–30 (–35) mm, leathery, base rounded, rarely cuneate or cordulate, margins entire or toothed or denticulate, undulate or flat, secondary veins 8–9 on each side, apex rounded or acute; surfaces abaxially densely gray- or white-tomentose with semi-erect curly, stellate hairs, secondary veins rather prominently raised, adaxially dark or dull green, lustrous or somewhat glaucous, with minute, scattered, semi-erect or appressed-stellate, (4–)6- or many-rayed hairs, not felty to touch, secondary veins slightly raised or prominent within depressions. **Acorns** solitary or paired, subsessile or peduncle sometimes 10–15 mm, tomentose like twigs; cup shallowly to very deeply cup-shaped, 5–12 mm deep × 8–18 mm wide, enclosing 1/2 nut, base rounded or flat, margin thin, scales triangular-ovate to oblong, proximal scales coarsely tuberculate and canescent-tomentose, distal ones usually elongate and narrowed, tips appressed, reddish, thin, nearly glabrous; nut light brown, ellipsoid to ovoid, 8–15 × 5–12 mm. **Cotyledons** connate.

Flowering spring. Limestone hills and slopes, calcareous substrates; 600–2500 m; N.Mex., Okla., Tex.; Mexico (Coahuila).

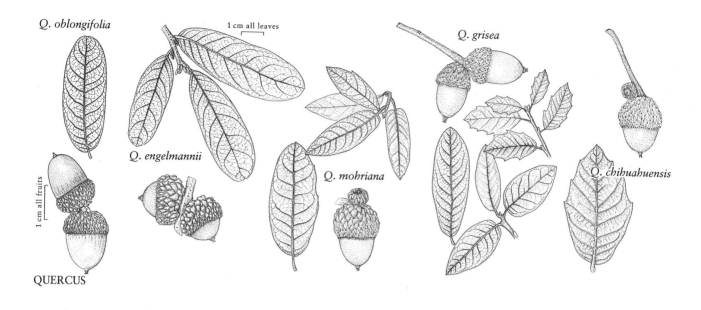

Q. oblongifolia

1 cm all leaves

Q. engelmannii

Q. mohriana

Q. grisea

Q. chihuahuensis

1 cm all fruits

QUERCUS

Putative hybrids between *Quercus mohriana* Buckley and *Q. grisea* Liebmann are problematic and highly polymorphic. They are restricted to zones of contact between limestone, the preferred habitat of *Q. mohriana,* and igneous substrates, the preferred habitat of *Q. grisea,* or sometimes on dolomite, in western Texas.

83. Quercus grisea Liebmann, Overs. Kongel. Danske Vidensk. Selsk. Forh. Medlemmers Arbeider 1854: 171. 1854 · Gray oak F

Quercus undulata Torrey var. *grisea* (Liebmann) Engelmann

Large shrubs or moderate trees, deciduous or subevergreen, to 10 m. **Bark** gray, fissured. **Twigs** gray, 1–2 mm diam., sparsely or densely stellate-tomentulose or tomentose when young. **Buds** dark red-brown, ovoid to sub-globose, 1–2 mm, stellate hairs causing yellowish color, at least on outer scales; stipules persistent, 1–4, subulate, pubescent, at base of terminal buds. **Leaves:** petiole 3–10 mm. **Leaf blade** oblong to elliptic or ovate, (15–)25–35(–80) × (7–)15–30(–40) mm, thick and leathery, base cordate or rounded, margins minutely revolute, entire or dentate with mucronate teeth, secondary veins 6–10 on each side, branched, apex acute, sometimes obtuse, rarely rounded; surfaces abaxially dull gray-green or yellowish, minutely stellate-pubescent with interlocking hairs, secondary veins very prominent, adaxially dull green, very sparsely and minutely stellate-pubescent, secondary veins slightly raised. **Acorns** solitary or paired, subsessile or on peduncle 0–30 mm; cup from deeply goblet- to deeply cup-shaped, 4–10 mm deep × 8–15 mm wide, enclosing to 1/2 nut, scales broadly ovate to oblong, proximal scales slightly or markedly tuberculate and whitish canescent, tips closely appressed, red-brown, thin, glabrate; nut light brown, ovoid to narrowly ovoid or ellipsoid, 12–18 × 8–12 mm. **Cotyledons** connate.

Flowering spring. Igneous or dolomitic slopes, oak woodlands, juniper woodlands, desert chaparral; usually above 1500 m; Ariz., N.Mex., Tex.; Mexico (Sonora, Chihuahua, and Durango).

Some of the specimens referred to *Quercus endemica* by C. H. Muller should be placed in *Q. grisea.*

Numerous hybrids between *Quercus grisea* and other white oaks, including *Q. gambelii, Q. mohriana, Q. arizonica,* and numerous species in northern Mexico have been reported. In the Hueco and Quitman mountains of trans-Pecos Texas, putative hybrids of *Q. grisea* × *Q. turbinella* Greene occur.

84. Quercus chihuahuensis Trelease, Mem. Natl. Acad. Sci. 20: 85. 1924 · Chihuahua oak, felt oak F

Quercus infralutea Trelease; *Q. jaliscensis* Trelease; *Q. santaclarensis* C. H. Muller

Shrubs or trees, deciduous, to 10 m. **Bark** gray, furrowed, checkered, or scaly. **Twigs** gray, 2–3(–4) mm diam., densely tomentose. **Buds** reddish brown, broadly ovoid, distally rounded, 2–2.5 mm, densely yellowish pubescent; scales gray-puberulent; stipules persistent, 1–4, subulate, pubescent, at base of terminal buds. **Leaves:** petiole 3–5(–8) mm. **Leaf blade** elliptic or oblong to ovate or obovate, (25–)40–50(–85) × (18–)20–30(–50) mm, base rounded or shallowly cordate, margins entire or toothed to sublobate, secondary veins 8 to 10 on each side, somewhat branching, apex broadly rounded to acute; surfaces abaxially yellowish or grayish, densely stellate with velvety hairs, adaxially green, sparsely soft-pubescent with prominent, spreading, stellate hairs, felty to touch, secondary veins somewhat prominent on both surfaces, even under dense tomentum. **Acorns** 1–3 on tomentose peduncle 15–35(–60) mm; cup hemispheric, 7–10 mm deep × 10–15 mm wide, enclosing 1/2 nut, scales proximally thickened, distally appressed, densely gray-puberulent, tips reddish, ultimately glabrate; nut ovoid, 14–18 × 10–12 mm, puberulent, eventually glabrate. **Cotyledons** connate.

Flowering spring. Oak and pinyon-juniper woodlands, grassy hills, sometimes extending into dry thorn scrub and bursera woodland (Mexico); 400–2000 m; Tex.; Mexico (Chihuahua, Sonora, Zacatecas, and San Luis Potosí).

Quercus chihuahuensis is a distinctive species throughout its range, mostly in dry montane western Mexico; it occurs in the United States only as putative hybrids with *Q. grisea* (the Eagle and Quitman mountains) and *Q. arizonica* (Hueco Tanks) in Texas.

85. **Quercus intricata** Trelease, Mem. Natl. Acad. Sci. 20: 84. 1924 · Intricate oak [F]

Shrubs, evergreen, clonal, intricately branched. **Bark** gray, scaly. **Twigs** gray- or yellow-tomentose, darkened, 1–1.5 mm diam., persistently pubescent for several seasons. **Buds** dark reddish brown, 1–1.5 mm, apex round, sparsely pubescent to glabrate. **Leaves:** petiole 2–3 mm. **Leaf blade** oblong, sometimes ovate, often strongly cupped, 10–25 × 5–13 mm, extremely thick, leathery, base cuneate to cordate, margins very coarsely revolute, often undulate-crisped, entire, rarely with a few teeth, secondary veins 8 or 9 on each side, apex acute or obtuse; surfaces abaxially brownish or buff, persistently tomentose with erect curly hairs, rarely glabrate in 2d season, midribs (and sometimes principal veins) glabrous and brown against tomentum, secondary veins sometimes prominently raised, usually obscured by tomentum, adaxially dark or gray-green, lustrous, sparsely or moderately stellate-pubescent, secondary veins impressed. **Acorns** solitary or paired, subsessile or on peduncle to 15 mm; cup deeply cup-shaped, 7–8 mm deep × ca. 10 mm wide, base round, margin thin,

scales ovate or narrower, proximally canescent-tomentose, moderately or markedly tuberculate, tips closely appressed, reddish, thin, nearly glabrous; nut light brown, ovoid, 9–12 × 8–10 mm. **Cotyledons** connate.

Flowering spring. Open chaparral and pinyon-oak woodlands, on dry, rocky, limestone slopes (in Mexico also on gypsophilous soils); 1500–2500 m; Tex.; Mexico (Coahuila, Nuevo León, Durango, and Zacatecas).

Quercus intricata, a fairly common element of the mountains of the Chihuahuan Desert region, is known in the United States only from two localities: a population in the Chisos Mountains and another in the Eagle Mountains of west Texas.

86. **Quercus depressipes** Trelease, Mem. Natl. Acad. Sci. 20: 90. 1924 · Depressed oak [F]

Quercus bocoynensis C. H. Muller

Shrubs, evergreen or subevergreen, low, to 1 m, often forming dense thickets, rhizomatous. **Bark** gray, scaly. **Twigs** tan-brown, becoming reddish gray, 1–1.5 mm diam., glabrate or hairy. **Buds** tan or brown, subglobose, 1–1.5 mm, glabrate or scales inconspicuously ciliate. **Leaves:** petiole 1–2 mm, rarely longer. **Leaf blade** oblong to elliptic, 10–25 (–60) × 8–25 mm, thick, leathery, base moderately to deeply cordate, petiole strongly depressed in basal sinus, margins inconspicuously toothed in distal 1/2, rarely entire, sometimes sublobate, somewhat revolute, secondary veins 5 or 6 on each side with few intermediates, branching, apex broadly rounded to subacute; surfaces abaxially dull gray-green or glaucous, completely glabrous or with a few stellate hairs on midrib, adaxially similar to abaxial surface, secondary veins somewhat raised on both surfaces. **Acorns** paired on peduncle 7–15 mm; cup 4–7 mm deep × 8–13 mm wide, goblet-shaped, enclosing 1/4–1/2 nut, base somewhat constricted or rounded, scales moderately tuberculate, proximally densely gray-tomentose, tips rather closely appressed, reddish brown, abaxially glabrous, ciliate; nut tan-brown, elliptic to ovoid or globose, to 10–15 × 10–11 mm, apex rounded, glabrous. **Cotyledons** connate.

Flowering spring. Grassland and open wooded slopes; 2100–2600 m; Tex.; Mexico (Chihuahua, Durango, and Zacatecas).

Quercus depressipes enters the United States in only one population on the highest portion of Mt. Livermore in trans-Pecos Texas; it has a wider distribution in the dry altiplano of northern Mexico. Its most distinctive characteristics are the combination of

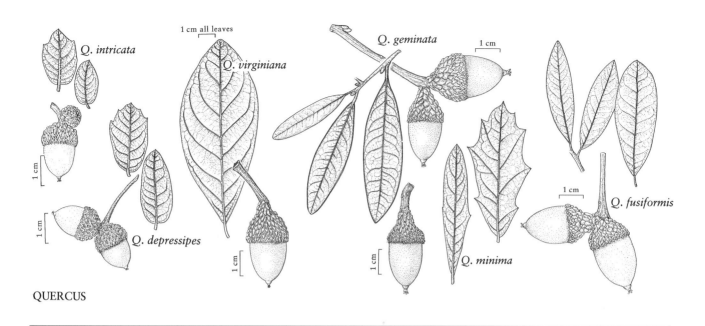

1 cm all leaves

Q. intricata

Q. virginiana

Q. geminata

1 cm

1 cm

1 cm

Q. depressipes

1 cm

Q. fusiformis

1 cm

Q. minima

QUERCUS

dwarf clonal habit, small glaucous leaves without spinose teeth, and connate cotyledons. In northern Mexico, it hybridizes locally with *Q. rugosa.*

87. Quercus virginiana Miller, Gard. Dict. ed. 8, Quercus no. 16. 1768 · Southern live oak E F W1

Quercus virginiana Miller var. *eximea* Sargent

Trees, sometimes shrubs, subevergreen, trees to 35 m, shrubs sometimes rhizomatous. **Bark** dark brown or black, scaly. **Twigs** yellowish to light gray, 1–3 mm diam., minutely puberulent or stellate-pubescent, glabrate in 2d year. **Buds** reddish or dark brown, subglobose or ovate, 1–2 mm; scale margins glabrous or puberulent. **Leaves:** petiole 1–10(–20) mm. **Leaf blade** obovate to oblanceolate, sometimes orbiculate or lance-ovate, ± planar, (10–)35–90(–150) × (15–)20–40(–85) mm, base cuneate to rounded, rarely truncate or cordate, margins minutely revolute or flat, entire or irregularly 1–3-toothed on each side, teeth mucronate, secondary veins obscure, 6–9(–12) on each side, apex obtuse-rounded or acute; surfaces abaxially whitish or glaucous, densely covered with minute, appressed, fused-stellate hairs, light green and glabrate in shade leaves, adaxially dark or light green, glossy, glabrous or with minute, scattered, stellate hairs. **Acorns** 1–3, on peduncle (3–)10–20 mm; cup hemispheric or deeply goblet-shaped, 8–15 mm deep × 8–15 mm wide, base often constricted; scales whitish or grayish,

proximally thickened, keeled, tomentulose, tips reddish, acute-attenuate, glabrous or puberulent; nut dark brown, barrel-shaped, ovoid, or cylindric, 15–20 (–25) × 8–15 mm, apex rounded or blunt, glabrous. **Cotyledons** connate.

Flowering late winter–early spring. Coastal plain, open evergreen woodlands, scrublands, and hummocks on loam, clay, and rarely on sand on immediate coast; 0–200 m; Ala., Fla., Ga., La., Miss., N.C., S.C., Tex., Va.

Quercus virginiana is one of the commonest and best known species in the coastal region of the southeastern United States. In the past, it was widely used for structural pieces in the manufacture of wooden ships, and large groves were actually considered a strategic resource by the federal government. Historically oil pressed from the acorns was utilized. Like other members of the live oak group (*Q. minima, Q. geminata,* and *Q. fusiformis*), *Q. virginiana* seedlings form swollen hypocotyls that may develop into large, starchy, underground tubers. In the past, the tubers were gathered, sliced, and fried like potatoes for human consumption. The tendency for the tree members of this group to produce rhizomatous growth and clonal shrubs in juvenile stages, and in response to damage, fire, and poor soil conditions, has led to considerable confusion in delimiting the species. This is exacerbated by considerable plasticity in leaf form. When evaluating specimens an effort should be made to sample broadly within a population. The tuberous condition mentioned above suggests that live oaks have different phases in their life history that may persist depending on the environmental conditions. This is not uncommon in other woody plants that occur in seasonally dry, fire-prone habitats of the southeastern United States.

The Houma used *Quercus virginiana* medicinally for healing dysentery (D. E. Moerman 1986).

Putative hybrids between *Quercus virginiana* and *Q. minima* are known, but care should be taken to avoid assigning hybrid status to clonal phases of *Q. virginiana* solely on the basis of habit. Hybrids with *Q. fusiformis* and *Q. geminata* are discussed under those species. Occasional putative hybrids with *Q. stellata* are also found, and those tend to be semi-evergreen with shallowly lobed leaves.

Some named putative hybrids are: *Q. ×burnetensis* Little (= *Q. macrocarpa* × *Q. virginiana*); *Q. ×comptoniae* Sargent (= *Q. lyrata* × *Q. virginiana*); and the artificially produced hybrid, *Q. ×nessiana* E. J. Palmer (= *Q. bicolor* × *Q. virginiana*).

88. Quercus geminata Small, Bull. Torrey Bot. Club 24: 438. 1897 · Sand live oak [F][W1]

Quercus virginiana Miller var. *geminata* (Small) Sargent

Trees, sometimes shrubs, subevergreen, trees to 25 m, shrubs sometimes rhizomatous (if spreading rhizomatously, without numerous straight, short, erect stems emerging from gound, or if so, mixed with other larger branches, infertile, and without dimorphic or asymmetric leaf form). **Bark** dark brown or black, scaly. **Twigs** yellowish, becoming light gray, 1.5–3 mm diam., tomentulose, glabrate in 2d year. **Buds** reddish or dark brown, globose or ovoid, 1–2.5(–3) mm; scale margins glabrous or puberulent. **Leaves:** petiole 3–10(–20) mm. **Leaf blade** narrowly lanceolate or elliptic, rarely orbiculate, convex-cupped, (10–)35–60(–120) × (7–)10–30(–45) mm, base narrowly cuneate, rarely truncate or rounded, margins strongly revolute, entire, secondary veins 8–10(–12) on each side, apex acute, sometimes obtuse; surfaces abaxially whitish or glaucous, densely covered with minute, appressed, fused-stellate hairs (visible under magnification), and with additional scattered, erect, felty or spreading hairs (sometimes deciduous), or light green and glabrate in shade leaves, adaxially dark or light green, glossy, glabrous or with minute, scattered, stellate hairs, secondary veins moderately to deeply impressed. **Acorns** 1–3, on peduncle 10–100 mm; cup hemispheric or deeply goblet-shaped, sometimes saucer-shaped, 8–15 mm deep × 5–15 mm wide, base often constricted, scales whitish or grayish, thickened basally, keeled, acute-attenuate, tomentulose, tips red-

dish, glabrous or puberulent; nut dark brown, ovoid, barrel-shaped, or acute, (13–)15–20(–25) × (8–)9–12(–15) mm, glabrous. **Cotyledons** connate.

Flowering spring. Coastal plain, open evergreen woodlands and scrublands on deep sandy soils, often with pines; 0–200 m; Ala., Fla., Ga., La., Miss., N.C., S.C.

Quercus geminata occurs in Cuba as putative hybrids.

Although some recent authors prefer to treat *Quercus geminata* as a variety of *Q. virginiana*, the two species are easily separable and rarely intergrade through most of the broad range in which they are sympatric. Apparently this is primarily because of habitat separation, but additionally *Q. geminata* flowers much later than *Q. virginiana* in any given geographic area. At the northern extreme of the range of *Q. geminata,* apparent intermediates with *Q. virginiana* are more common, possibly because flowering times of the two species overlap to a greater extent because of slower warming in the spring. Scattered intermediates also occur where the two species are sympatric on sands in coastal Mississippi.

89. Quercus minima (Sargent) Small, Bull. Torrey Bot. Club 24: 438. 1897 · Minimal oak [E][F]

Quercus virginiana Miller var. *minima* Sargent, Silva 8: 101. 1895; *Q. virens* Aiton var. *dentata* Chapman; *Q. virginiana* var. *dentata* (Chapman) Sargent

Shrubs, subevergreen or evergreen, low, rhizomatous, forming clonal patches of straight, often unbranched stems 0.2–0.7(–2) m. **Bark** light gray or brown, smooth. **Twigs** light gray, 1.5–3 mm diam., glabrous or very finely tomentulose and glabrate in 2d year. **Buds** dark gray-brown, globose, 0.5–1(–1.5) mm; scale margins glabrous or puberulent. **Leaves** dimorphic; petiole 1–4(–10) mm. **Leaf blade** obovate to oblanceolate, sometimes orbiculate or lance-ovate, often asymmetric, 40–120 × 20–50 mm proximally on shoots, narrowly elliptic, 20–60 × 5–20 mm distally on shoots, base cuneate, rarely truncate or cordate, margins minutely revolute or flat, coarsely and irregularly 1–5-toothed on each side (proximal leaves of shoots) or entire (distal leaves on shoots), teeth mucronate (rarely spinose in suckers or juveniles), secondary veins obscure, 4–10(–12) on each side, usually 4–6 in proximal leaves, apex obtuse-rounded or acute; surfaces abaxially whitish or glaucous, densely covered with minute, appressed, fused-stellate hairs, light green and glabrate in shade leaves, adaxially dark or light green, glossy, glabrous or with minute, scattered, stellate hairs. **Acorns** 1–3, on peduncle 3–30 mm; cup hemispheric or deeply goblet-shaped, (8–)10–15 mm deep × 6–15 mm wide, base often constricted, scales whitish or grayish, thickened basally, keeled, acute-attenuate, tomentulose, tips reddish, glabrous or puberulent; nut dark brown, narrowly ovoid or subcylindric, 13–22 (–25) × 8–15 mm, glabrous. **Cotyledons** connate.

Flowering spring. Coastal plain, open evergreen woodlands and scrublands on deep sandy soils, often as understory with pines; 0–200 m; Ala., Fla., Ga., Miss., N.C., S.C.

Quercus minima is one of the most distinctive oaks of southeastern United States in morphology, habit, and habitat. It is one of our most strongly rhizomatous species, and it flowers and fruits as early as three years from seed, on vertical stems as small as 0.2–0.3 m. Other related live oaks of *Quercus* series *Virentes* (see C. H. Muller 1961) can produce rhizomatous clonal patches as well, but these typically do not bear fruit, are usually associated with large shrub or tree forms, and probably represent response to fire and/or drought conditions. Because of the juvenile phases in other live oaks that may resemble *Q. minima,* it is necessary to evaluate form and reproductive capacity within populations to identify those species accurately.

Putative hybrid swarms between *Quercus minima* and *Q. chapmanii* occur along the eastern part of Florida. These are rhizomatous shrubs with leaves somewhat intermediate between the two parents. The leaves are broadly rounded apically and basally, and sparsely stellate abaxially. The fruit resembles that of *Q. minima.* Material of such affinity, but with larger nuts (to 25 mm), was the basis for *Q. rolfsii* Small. More data on variation in cotyledon fusion in the putative hybrid swarms would be useful because *Q. minima* has connate cotyledons and those of *Q. chapmanii* are distinct.

Quercus minima has been reported from coastal Texas (C. H. Muller 1970), but that material is probably best referred to *Q. fusiformis* and possible introgressants from *Q. virginiana* and *Q. oleoides.* Typical *Q. minima* is not known from west of Mississippi. The Texas populations that have been referred here lack leaf dimorphism, leaf shape, and venation of *Q. minima.* Instead, they seem to be part of an extremely variable complex of erect shrubs, rhizomatous shrubs, and trees that have leaves varying in form between those of *Q. fusiformis* and *Q. oleoides* (e.g., the type of *Q. oleoides* var. *quaterna* C. H. Muller). If *Q. minima* occurred on the Texas coast in the past, its presence there has been so diluted by introgression that it is not now recognizable by the very distinctive and reliable characteristics that are used to identify it.

90. Quercus fusiformis Small, Bull. Torrey Bot. Club 28: 357. 1901 · Texas live oak F W1

Quercus virginiana Miller var. *fusiformis* (Small) Sargent

Trees, sometimes shrubs, subevergreen, trees to 25 m, shrubs often forming large clonal stands. **Bark** dark brown or black, scaly. **Twigs** light gray, 1.5–3 mm diam., tomentulose, tomentulum often persistent in age. **Buds** reddish or dark brown, subglobose or ovate, 1.5–3 mm; scale margins glabrous or puberulent. **Leaves:** petiole 2–8(–12) mm. **Leaf blade** oblong-elliptic to narrowly ovate or lanceolate, sometimes obovate, ± planar, (10–)35–90(–150) × (15–)20–40 (–85) mm, base rounded to truncate or cordate, rarely cuneate, margins minutely revolute or flat, entire or irregularly 1–3-toothed on each side, teeth mucronate (rarely spinose in suckers or juveniles), secondary veins obscure, 8–10 on each side, apex obtuse-rounded or acute; surfaces abaxially whitish or glaucous, densely covered with minute, appressed, fused-stellate hairs, light green and glabrate in shade leaves, adaxially dark or light green, glossy, glabrous or with minute, scattered, stellate hairs. **Acorns** 1–3, on peduncle 3–30 mm; cup funnel-shaped, hemispheric, or deeply goblet-shaped, 8–15 mm deep × 6–12(–15) mm wide, base often constricted, scales whitish or grayish, thickened basally, keeled, acute-attenuate, tomentulose, tips reddish, glabrous or puberulent; nut dark brown, often with light brown longitudinal stripes, subfusiform and acute to narrowly barrel-shaped, rarely distally rounded, (17–)20–30(–33) × 8–15 mm, glabrous. **Cotyledons** connate.

Flowering spring. Hills, grasslands, scrublands, open woodlands, oak-juniper woodlands, and margins of thorn scrub, often on limestone or deep calcareous loams, sometimes on granular sand or gravel; 0–1200 m; Okla., Tex.; Mexico (Coahuila, Nuevo León, and Tamaulipas).

The difficulty in distinguishing Texas populations of *Quercus fusiformis* from *Q. virginiana* is reflected in a variety of taxonomic treatments, including reducing *Q. fusiformis* to varietal rank under *Q. virginiana.* The latter disposition is problematic, however, because *Q. fusiformis* in northeastern Mexico is amply distinct from *Q. virginiana* and appears to be more closely related to *Q. brandegei* Goldmann, an endemic of Baja California, Mexico. Thus, here we assume that the intergradation of *Q. virginiana* and *Q. fusiformis* is a result of secondary contact, and is not primary clinal variation. Under this interpretation, *Q. virginiana* in typical form extends into Texas only as far west as the Brazos River drainage along the coast from there to the escarpment of the Edwards Plateau; most populations elsewhere are either intermediate between the two species or show greater affinity with *Q. fusiformis*. On the Edwards Plateau, the live oak populations are small trees forming rhizomatous copses (shinneries) and having mostly acute acorns.

Populations of live oak on deep sands in south Texas differ from typical *Quercus fusiformis* in having broader, more rounded leaves, often with the second-ary veins somewhat impressed abaxially, and relatively blunt, barrel-shaped acorns. These characteristics suggest introgression from the Mexican–Central American species *Q. oleoides* Schlechtendal & Chamisso, which in its typical form reaches north only as far as southern Tamaulipas, Mexico. The name *Q. oleoides* var. *quaterna* C. H. Muller has been applied to what is apparently a shrub form of one of these *Q. fusiformis* × *Q. oleoides* hybrids.

31. BETULACEAE Gray

• Birch Family

John J. Furlow

Trees and shrubs, deciduous. **Winter buds** stipitate or sessile, with either 2–3 valvate scales (stipules) or few to many imbricate scales (or occasionally naked); terminal bud absent. **Leaves** alternate, spirally arranged, 2–3-ranked, simple; stipules deciduous, distinct; petioles present. **Leaf blade** sometimes lobed, pinnately veined, margins toothed, serrate to nearly entire; surfaces glabrous to tomentose, abaxially often with resinous glands. **Inflorescences** unisexual; staminate catkins pendulous, elongate, cylindric, conspicuously bracteate, consisting of crowded, reduced, 1–3-flowered clusters; pistillate inflorescences either of erect to pendulous bracteate catkins, or of compact 2–3-flowered clusters subtended by leafy involucres; bracts often nearly foliaceous or woody in infructescences. **Staminate flowers** bracteate; stamens (1–)4–6; anthers 2-locular, dehiscing by longitudinal slits, pollen sacs often ± distinct; pistillode sometimes present. **Pistillate flowers** small, highly reduced; pistil 1, 2(–3)-carpellate; ovary inferior, usually 2-locular proximally, 1-locular distally; placentation axile; ovules 1–2 per locule, pendulous; styles 2, distinct or nearly so; stigmas dry; staminodes usually absent. **Fruits** nuts, nutlets, or 2-winged samaras, 1-seeded, without multibracteate cupule, often subtended or enclosed by foliaceous hull developed from 2–3 bracts; seed coat membranous; endosperm present, thin at maturity; embryo straight, as long as seed; cotyledons flat or greatly thickened, oily.

Genera 6, species ca. 125 (5 genera, 33 species in the flora): primarily boreal and cool temperate zones of Northern Hemisphere; North America, Central America, South America, Asia.

Betulaceae belong to an ancient lineage, traceable in the fossil record to the upper Cretaceous. They are easily distinguished by their woody habit; simple, pinnately veined, usually ovate, sharp-toothed leaves; long staminate catkins that often develop the season before anthesis; and (except in *Corylus* and *Ostryopsis*) strobiluslike infructescences. The family is held together on the basis of many features, including habit, leaf structure and arrangement, trichome morphology, wood anatomy, inflorescence morphology, ovary and ovule structure, pollen morphology, embryo structure, and fertilization and germination patterns. Five of the six constituent genera inhabit the boreal and cool temperate zones of Eurasia, North America, and the mountains of Mexico and Central America, and two genera also grow in the Andes to northern Argentina in South America. The remaining genus, *Ostryopsis* Decaisne (most closely related to *Corylus* Linnaeus), consists of two species of shrubs restricted to northern and western China.

The group is sometimes divided into two families, Betulaceae (*Alnus* and *Betula*) and Corylaceae (*Carpinus, Ostrya, Corylus,* and *Ostryopsis*), especially in Europe (e.g., H. K. Airy Shaw in J. C. Willis 1973; R. M. T. Dahlgren 1980; J. Hutchinson 1959). In America, this treatment has been followed by A. J. Rehder (1940), J. K. Small (1903), and a few others. Some of those writers have based their recognition of two families in part on the belief that a fundamental difference exists in the staminate inflorescences of the two groups. This view is no longer widely accepted, and most modern authors maintain the family as a single group composed of two subfamilies, Betuloideae and Coryloideae.

Betulaceae may have been derived from hamamelidaceous stock (O. Tippo 1938); this idea has been advanced by others (e.g., J. A. Doyle 1969; P. K. Endress 1977; M. W. Chase et al. 1993). The ancestor of more advanced Hamamelidae (including Betulaceae) may have been a member of the Normapolles complex, a Cretaceous fossil pollen flora that includes grains similar in structure to those of the modern Betulaceae (J. A. Doyle 1969). The pollen of modern Betulaceae is much more specialized than that of modern Hamamelidales, however, and the relationships of the Normapolles complex to Juglandaceae, Fagaceae, and other amentiferous groups, as well as to Betulaceae, are not certain. Additional fossil information will be needed if botanists are to gain stronger insights into the phylogeny of these families (P. R. Crane 1989; K. C. Nixon 1989).

The most obvious evolutionary trends within the family are those related to fruit dispersal (winged fruits in the Betuloideae, nutlets with accessory winglike bracts in *Carpinus* and *Ostrya,* and dispersal by small mammals in *Corylus*), and with apparent adaptations for survival in cold climates (including shrubby growth forms, small leaves, and the protection of pistillate catkins during winter by bud scales in boreal species). Some radiation has also occurred with respect to the pollination system (E. J. Jäger 1980). Please see P. R. Crane (1989) and J. J. Furlow (1990) for reviews of literature related to the origin and diversification of the family.

Members of Betulaceae are economically important as timber trees, as the source of hazelnuts and filberts (*Corylus*), as ornamental trees and shrubs, and as aids in soil nitrification and stabilization (*Alnus*). Some are important as causes of pollen allergies in regions where they grow abundantly.

SELECTED REFERENCES   Abbe, E. C. 1935. Studies in the phylogeny of the Betulaceae. I. Floral and inflorescence anatomy and morphology. Bot. Gaz. 97: 1–67.   Abbe, E. C. 1938. Studies in the phylogeny of the Betulaceae. II. Extremes in the range of variation of floral and inflorescence morphology. Bot. Gaz. 99: 431–469.   Bousquet, J., S. H. Strauss, and Li P. 1992. Complete congruence between morphological and *rbc*L-based molecular phylogenies in birches and related species (Betulaceae). Molec. Biol. Evol. 9: 1076–1088.   Brunner, F. and D. E. Fairbrothers. 1979. Serological investigation of the Corylaceae. Bull. Torrey Bot. Club 106: 97–103.   Endress, P. K. 1977. Evolutionary trends in the Hamamelidales-Fagales group. Plant Syst. Evol. Suppl. 1: 321–347.   Furlow, J. J. 1983. The phylogenetic relationships of the genera and infrageneric taxa of the Betulaceae. [Abstract.] Amer. J. Bot. 70(suppl.): 114.   Furlow, J. J. 1990. The genera of Betulaceae in the southeastern United States. J. Arnold Arbor. 71: 1–67.   Hall, J. W. 1952. The comparative anatomy and phylogeny of the Betulaceae. Bot. Gaz. 113: 235–270.   Hardin, J. W. and J. M. Bell. 1986. Atlas of foliar surface features in woody plants, IX. Betulaceae of eastern United States. Brittonia 38: 133–144.   Hjelmquist, H. 1948. Studies in the floral morphology and phylogeny of the Amentiferae. Bot. Not., Suppl. 2: 1–171.   Jäger, E. J. 1980. Progressionen im Synfloreszenzbau und in der Verbreitung bei den Betulaceae. Flora 170: 91–113.   Petersen, F. P. and D. E. Fairbrothers. 1985. A serotaxonomic appraisal of the "Amentiferae." Bull. Torrey Bot. Club 112: 43–52.   Regel, E. 1861. Monographische Bearbeitung der Betulaceen. Nouv. Mém. Soc. Imp. Naturalistes Moscou 13(2): 59–187.   Winkler, H. 1904. Betulaceae. In: H. G. A. Engler, ed. 1900–1953. Das Pflanzenreich. . . . 107 vols. Berlin. Vol. 19[IV,61], pp. 1–149.

1. Infructescences 1–4 cm, conelike, of small, crowded, woody or leathery scales, these deciduous with fruits or persistent; fruits tiny samaras, laterally winged, wings sometimes reduced to ridges; pericarp thin, leathery (subfamily Betuloideae).
    2. Winter buds stipitate (nearly sessile in some species); stamens (3–)4(–6), anthers and filaments undivided; pistillate flowers 2 per scale; infructescence scales with 5 lobes, greatly thickened, woody, persistent long after release of fruits. . . . . .1. *Alnus,* p. 509
    2. Winter buds sessile; stamens usually (1–)2–3(–4), filaments divided below anthers; pistillate flowers (1–)3 per scale; infructescence scales with (1–)3 lobes, thickened or leathery but not woody, usually deciduous with release of fruits. . . . . 2. *Betula,* p. 516
1. Infructescences mostly longer than 4 cm, of relatively uncrowded clusters with large, nearly foliaceous bracts, these deciduous with fruits; fruits tiny to moderately large nuts, hard-shelled, not winged (subfamily Coryloideae).
    3. Leaf blade broadly ovate to nearly orbiculate, lateral veins 8 or fewer pairs; infructescence irregular cluster of several nuts, each nut usually 1 cm or more, surrounded by involucre of several coarsely toothed, leaflike bracts, involucre sometimes long and tubular (tribe Coryleae).. . . . . 5. *Corylus,* p. 535
    3. Leaf blade narrowly ovate to elliptic, lateral veins 10 or more pairs; infructescence elongate, loosely arranged racemose cluster of 3 or more pairs of leafy bracts, each bract either subtending or enclosing single nutlet, nutlet less than 0.8 cm (tribe Carpineae).
        4. Infructescence bracts flat, open, (1–)3-lobed, variously toothed. . . . . 3. *Carpinus,* p. 531
        4. Infructescence bracts forming inflated bladders, these completely enclosing fruits.. . . . 4. *Ostrya,* p. 533

31a. BETULACEAE Gray subfam. BETULOIDEAE Koehne, Deut. Dendrol., 106, 1893 (as Betuleae)

Trunks and branches terete. **Bark** thin, close or exfoliating in thin sheets, becoming thicker and frequently furrowed or broken in age; lenticels often present, prominent, sometimes becoming greatly expanded horizontally. **Bark and wood** strongly tanniferous. **Young twigs and buds** often covered with small to large, resinous glands; pith triangular in cross section. **Leaves** 3-ranked, occasionally nearly 2-ranked. **Staminate flowers:** perianth of 4(–6) sepals, well defined, minute, membranaceous. **Pistillate flowers** 2–3 per scale, scales arranged in conelike catkins; perianth not obvious; ovules with 1 integument. **Infructescences** 1–4 cm, conelike, composed of many scales; scales either persistent or deciduous with fruits, crowded, small, woody or leathery. **Fruits** tiny samaras, lateral wings 2, membranous, sometimes reduced to ridges; pericarp thin, leathery.

Genera 2, species 60 (2 genera, 26 species in the flora): primarily boreal and cool temperate zones of Northern Hemisphere; North America, Central America, South America.

1. ALNUS Miller, Gard. Dict. Abr. ed. 4, vol. 1. 1754 · Alder, aulne, aune [Latin *alnus,* alder]

Trees or shrubs, to 35 m; trunks usually several, branching excurrent to deliquescent. **Bark** of trunks and branches light gray to dark brown, thin, smooth, close; lenticels often present, pale, prominent, sometimes horizontally expanded. **Wood** nearly white, turning reddish upon exposure to air, moderately light and soft, texture fine. **Branches, branchlets, and twigs** nearly

2-ranked to diffuse; young twigs uniform or (*Alnus* subg. *Alnobetula*) differentiated into long and short shoots. **Winter buds** stipitate (nearly sessile in *Alnus* subg. *Alnobetula*), narrowly to broadly ovoid or ellipsoid, terete, apex acute to rounded; scales 2–3, valvate, or (*Alnus* subg. *Alnobetula*) several, imbricate, smooth, or (*Alnus* subg. *Clethropsis*) sometimes none. **Leaves** borne on long or short shoots, 3-ranked to nearly 2-ranked. **Leaf blade** ovate to elliptic or obovate, thin to leathery, base variable, cuneate to rounded, margins doubly serrate, serrate, serrulate, or nearly entire, apex variable, acute to obtuse or acuminate to rounded; surfaces glabrous to tomentose, abaxially sometimes resinous-glandular. **Inflorescences:** staminate catkins lateral, in racemose clusters or (*Alnus* subg. *Clethropsis*) solitary, formed (*Alnus* subg. *Alnus* and *Clethropsis*) during previous growing season and exposed or enclosed in buds during winter, or (*Alnus* subg. *Clethropsis*) formed and expanding during same growing season, expanding before or with leaves; pistillate catkins proximal to staminate catkins, solitary or in relatively small racemose clusters, erect to nearly pendulous, ovoid to ellipsoid, firm; scales and flowers crowded, developing and maturing at same time as staminate catkins. **Staminate flowers** in catkins, 3 per scale; stamens (3–)4(–6); anthers and filaments undivided. **Pistillate flowers** usually 2 per scale. **Infructescences** erect or pendulous; scales persistent long after release of fruits, with 5 lobes, greatly thickened, woody. **Fruits** tiny samaras, lateral wings 2, leathery or membranaceous, reduced or essentially absent in some species. $x = 7$.

Species ca. 25 (8 species in the flora): forested temperate and boreal Northern Hemisphere; North America; at higher elevations, Central America, South America; Asia.

Alders resemble birches but are easily distinguished from them by the infructescences, which consist of persistent, 5-lobed, woody scales (versus deciduous, 3-lobed, thin scales). Except in members of *Alnus* subg. *Alnobetula* Petermann (which have nearly sessile buds with several imbricate scales), alders are also distinctive in their stipitate buds bearing two stipular scales. The fruits, borne two to a scale, are laterally winged, although the wings are sometimes reduced or absent.

The genus is diverse, including several very distinct lines of specialization. The shrubby or arborescent *Alnus* subg. *Alnus* is characterized by winter buds with long stalks and two valvate scales, inflorescences borne in racemose clusters, and development of both pistillate and staminate inflorescences during the growing season prior to anthesis, with these fully exposed during winter. It includes the common *A. rubra*, *A. incana*, *A. oblongifolia*, and *A. serrulata*. *Alnus* subg. *Alnobetula* (represented in North America by three subspecies of *A. viridis*) consists of shrubby species of cold-climate regions. In this group, the buds are nearly sessile and covered by several imbricate scales. Both staminate and pistillate catkins are formed the season before anthesis, but only the staminate ones are exposed during winter. The predominantly Asian *Alnus* subg. *Clethropsis* (Spach) Regel is represented in America by a single species, *A. maritima*, a small tree or large shrub of stream banks, marshes, and the shores of shallow lakes. Members of this group are unique in that they bloom in autumn rather than spring. They also differ from other native species in *Alnus* in having essentially naked buds, leaves with semicraspedodromous venation (i.e., with the secondary veins branching and anastomosing with each other near the margin before reaching the teeth), and solitary pistillate inflorescences borne in the axils of foliage leaves. All the alders associate symbiotically with species of the actinomycete *Frankia,* leading to the formation of nodules on the roots of the plants and the fixation of atmospheric nitrogen.

SELECTED REFERENCES Furlow, J. J. 1979. The systematics of the American species of *Alnus* (Betulaceae). Rhodora 81: 1–121, 151–248. Hylander, N. 1957. On cut-leaved and small-leaved forms of *Alnus glutinosa* and *A. incana*. Svensk Bot. Tidskr. 51:

437–453. Murai, S. 1964. Phytotaxonomical and geobotanical studies on gen. *Alnus* in Japan (III). Taxonomy of whole world species and distribution of each sect. Bull. Gov. Forest Exp. Sta. 171: 1–107. Trappe, J. M., J. F. Franklin, R. F. Tarrant, and G. M. Hansen, eds. 1968. Biology of Alder. . . . Portland.

1. Winter buds nearly sessile (stalks usually not over 1 mm), covered by 4–6 unequal, imbricate scales; staminate inflorescences formed late in growing season before blooming, exposed during winter; pistillate inflorescences enclosed within buds during winter, exposed with first new growth in spring (subg. *Alnobetula*) . . . . 7. *Alnus viridis*
1. Winter buds distinctly stalked, covered, sometimes incompletely, by 2–3 nearly equal, valvate scales; staminate and pistillate inflorescences both formed mid to late in growing season, not with first new growth in spring.
   2. Pistillate inflorescences and infructescences solitary in leaf axils along main stems; flowering near end of growing season (subg. *Clethropsis*) . . . . 8. *Alnus maritima*
   2. Pistillate inflorescences (and later infructescences) on short branchlets in racemose clusters; flowering at beginning of growing season (subg. *Alnus*).
      3. Leaf blade margins serrulate or finely serrate, without noticeably larger secondary teeth (although sometimes slightly lobulate).
         4. Leaf blade broadly elliptic to obovate, apex obtuse to rounded; staminate flowers with 4 stamens; large shrubs of e North America. . . . . 6. *Alnus serrulata*
         4. Leaf blade narrowly elliptic to rhombic, apex acute or obtuse, usually not rounded; stamens 2, or 4 with 2 reduced in size; trees of mountainous w United States. . . . . 3. *Alnus rhombifolia*
      3. Leaf blade margins doubly serrate or crenate, with distinctly larger secondary teeth, or coarsely serrate or serrate-dentate.
         5. Leaf blade margins strongly revolute; large trees of nw North America . . . . .1. *Alnus rubra*
         5. Leaf blade margins flat or only slightly revolute; trees and shrubs.
            6. Leaf blade narrowly ovate or lanceolate to narrowly elliptic; major teeth sharp, acuminate; trees of mountainous s Arizona and New Mexico, adjacent nw Mexico . . . . 2. *Alnus oblongifolia*
            6. Leaf blade ovate, elliptic, obovate, or nearly orbiculate; major teeth acute to obtuse or rounded.
               7. Leaf blade obovate to ± orbiculate, apex rounded to retuse or obcordate; moderately large introduced trees naturalized in ne United States, adjacent Canada. . . . . 4. *Alnus glutinosa*
               7. Leaf blade ovate to elliptic, apex acute to obtuse; native shrubs or shrubby trees. . . . . 5. *Alnus incana*

1. **Alnus rubra** Bongard, Mém. Acad. Sci. St.-Pétersbourg, Sér. 6, Sci. Math. 2: 162. 1833 · Red alder, Oregon alder [E] [F] [W1]

Alnus oregona Nuttall; *A. rubra* var. *pinnatisecta* Starker

Trees, to 28 m; trunks often several, crowns narrow or pyramidal. **Bark** gray, smooth, darkening and breaking into shallow rectangular plates in age; lenticels inconspicuous. **Winter buds** stipitate, ellipsoid, 6–10 mm, apex rounded, long; stalks 2–8 mm; scales 2–3, outer 2 equal and valvate, usually heavily resin-coated. **Leaf blade** ovate to elliptic, 6–16 × 3–11 cm, leathery, base broadly cuneate to rounded, margins strongly revolute, deeply doubly serrate or crenate, with distinctly larger secondary teeth, apex acute to obtuse; surfaces abaxially glabrous to sparsely pubescent. **Inflorescences** formed season before flowering and exposed during winter; staminate catkins in 1 or more clusters of 2–6, 3.5–14 cm; pistillate catkins in 1 or more clusters of 3–8. **Flowering** before new growth in spring. **Infructescences** ovoid to nearly globose, 1–3.5 × 0.6–1.5 cm; peduncles 1–10 mm. **Samaras** ovate or elliptic, wings much narrower than body, irregularly elliptic to obovate, leathery. $2n = 28$.

Flowering early spring. Stream banks, moist flood plains, lake shores, wet slopes, and sandy, open coasts; 0–300 m; B.C., Yukon; Alaska, Calif., Idaho, Oreg., Wash.

Alnus rubra is the largest alder in North America north of Mexico; it often forms extensive stands along streams and on low-lying flood plains in the Pacific Northwest. The strongly revolute margins of its leaf blades make it easily distinguished from all the other alders in the flora. It is an important commercial tree; the wood is used to make inexpensive furniture, small wooden items, and paper pulp.

Native Americans used various parts of plants of *Alnus rubra* medicinally as a purgative, an emetic, for aching bones, headaches, coughs, biliousness, stomach problems, scrofula sores, tuberculosis, asthma, and eczema, and as a general panacea (D. E. Moerman 1986).

2. Alnus oblongifolia Torrey in W. H. Emory, Rep. U.S. Mex. Bound. 2: 204. 1859 · Arizona alder, New Mexican alder, aliso (Mexico)

Trees, to 30 m; trunks often several, crowns spreading. **Bark** dark gray, smooth, becoming blackish and breaking into shallow vertical plates in age; lenticels inconspicuous. **Winter buds** stipitate, ovoid, 4–8 mm, apex rounded; stalks 1.5–4 mm; scales 2, equal, valvate, sometimes incompletely covering underlying leaves, moderately resin-coated. **Leaf blade** narrowly ovate or lanceolate to narrowly elliptic, 5–9 × 3–6 cm, leathery, base narrowly to broadly cuneate or narrowly rounded, margins flat, sharply and coarsely doubly serrate, rarely evenly and densely short-serrate, major teeth sharp, acuminate, secondary teeth distinctly larger, apex long- to short-acuminate, rarely acute; surfaces abaxially glabrous to sparsely pubescent or infrequently villous, moderately resin-coated. **Inflorescences** formed season before flowering and exposed during winter; staminate catkins in 1 or more clusters of 3–6, 3.5–10 cm; pistillate catkins in 1 or more clusters of 2–7. **Flowering** before new growth in spring. **Infructescences** ovoid, ellipsoid, or nearly cylindric, 1–2.5 × 0.8–1.5 cm; peduncles 5–10 mm. **Samaras** elliptic to obovate, wings narrower than body, irregular in shape, leathery.

Flowering early spring. Sandy or rocky stream banks and moist slopes, often in mountain canyons; 1000–2300 m; Ariz., N.Mex.; Mexico (n Chihuahua and n Sonora).

Alnus oblongifolia is closely related to the Mexican and Central American *A. acuminata,* with which it has sometimes been confused. It is found only in scattered populations in the temperate deciduous forest vegetation zone of high mountains in the arid Southwest.

3. Alnus rhombifolia Nuttall, N. Amer. Sylv. 1: 49. 1842 · White alder, California alder [E]

Alnus rhombifolia var. *bernardina* Munz & I. M. Johnston

Trees, to 25 m; trunks often several, crowns spreading, open. **Bark** light gray, smooth, becoming darker and breaking into scales in age; lenticels inconspicuous. **Winter buds** stipitate, ellipsoid to obovoid, 3–9 mm, apex rounded; stalks 3–5 mm; scales 2, equal, valvate, sometimes incompletely covering underlying leaves, moderately to heavily resin-coated. **Leaf blade** narrowly elliptic to rhombic, rarely ovate, 4–9 × 2–5 cm, base cuneate to rounded, margins flat, finely ser-

rate or serrulate, sometimes slightly lobed, without noticeably larger secondary teeth, apex acute or obtuse to rounded; surfaces abaxially sparsely pubescent to villous. **Inflorescences** formed season before flowering and exposed during winter; staminate catkins in 1 or more clusters of 3–7, 3–10 cm, stamens 2, or 4 with 2 reduced in size; pistillate catkins in 1 or more clusters of 2–6. **Flowering** before new growth in spring. **Infructescences** ovoid to nearly cylindric, 1–2.2 × 0.7–1 cm; peduncles 1–10 mm. **Samaras** broadly elliptic, wings narrower than body, irregular in shape, leathery.

Flowering early spring. Open, rocky stream banks and adjacent (often rather dry) slopes; 100–2400 m; Calif., Idaho, Mont., Oreg., Wash.

Alnus rhombifolia is the common alder throughout the dry Mediterranean climatic zone of coastal western United States. Mexican populations are not known, but because *A. rhombifolia* has been collected as far south as San Diego, California, it should be expected in adjacent Baja California.

Native Americans used various parts of *Alnus rhombifolia* medicinally for diarrhea, consumption, and burns, as a blood purifier, an emetic, and a wash for babies with skin diseases, and to facilitate childbirth (D. E. Moerman 1986).

4. Alnus glutinosa (Linnaeus) Gaertner, Fruct. Sem. Pl. 2: 54. 1790 · Black alder, European alder [W1]

Betula alnus Linnaeus var. (α) *glutinosa* Linnaeus, Sp. Pl. 2: 983. 1753

Trees, to 20 m; trunks often several, crowns narrow. **Bark** dark brown, smooth, becoming darker and breaking into shallow fissures in age; lenticels pale, horizontal. **Winter buds** stipitate, ellipsoid to obovoid, 6–10 mm, apex obtuse; stalks 2–5 mm; scales 2–3, outer 2 equal, valvate, usually heavily resin-coated. **Leaf blade** obovate to nearly orbiculate, 3–9 × 3–8 cm, leathery, base obtuse to broadly cuneate, margins flat, coarsely and often irregularly doubly serrate to nearly dentate, major teeth acute to obtuse or rounded, apex often retuse or obcordate, or occasionally rounded; surfaces abaxially glabrous to sparsely pubescent, often more heavily on veins, both surfaces heavily resin-coated. **Inflorescences** formed season before flowering and exposed during winter; staminate catkins in 1 or more clusters of 2–5, 4–13 cm; pistillate catkins in 1 or more clusters of 2–5. **Flowering** before new growth in spring. **Infructescences** ovoid to nearly globose, 1.2–2.5 × 1–1.5 cm; peduncles 1–10(–20) mm. **Samaras** obovate, wings reduced to narrow, thickened ridges. $2n = 28$.

Flowering early spring. Stream banks, moist flood plains, damp depressions, borders of wetlands; 0–200 m; Ont.; Conn., Ill., Ind., Iowa, Mass., Mich., Minn., N.J., N.Y., Ohio, Pa., R.I., Wis.; Europe.

Alnus glutinosa is cultivated as an ornamental tree throughout eastern North America and is available in a variety of cultivars, including cut-leafed and compact-branching forms. This species has also been used extensively to control erosion and improve the soil on recently cleared or unstable substrates, such as sand dunes and mine spoils. It has escaped and become widely naturalized throughout the temperate Northeast, occasionally becoming a weedy pest. In Europe the black alder has served for many centuries as an important source of hardwood for timbers and carved items, including wooden shoes.

Alnus glutinosa has been called *A. vulgaris* Hill in some older literature; that name was not validly published.

5. Alnus incana (Linnaeus) Moench, Methodus, 424. 1794

Betula alnus Linnaeus var. (β) *incana* Linnaeus, Sp. Pl. 2: 983. 1753

Trees and shrubs, to 25 m; crowns open. **Bark** light to dark gray, reddish, or brown, smooth, or in age broken into irregular plates; lenticels present or absent, conspicuous, enlarged or unexpanded. **Winter buds** stipitate, ellipsoid, 4–7 mm, apex rounded to nearly acute; stalk 1–3 mm; scales 2–3, equal, valvate, resin-coated. **Leaf blade** narrowly ovate to elliptic, base cuneate to narrowly rounded, margins doubly serrate, with distinctly larger secondary teeth, apex acute or short-acuminate to obtuse. **Inflorescences** formed season before flowering and exposed during winter. **Flowering** before new growth in spring. **Infructescences** ovoid to nearly cylindric; peduncles relatively short and stout. **Samaras** elliptic to obovate, wings narrower than body, irregular in shape.

Subspecies 4 (2 in the flora): United States, Canada.

Native Americans used *Alnus incana* medicinally to treat anemia, as an emetic, a compress or wash for sore eyes, and a diaphoretic, for internal bleeding, urinary problems, sprains, bruises or backaches, itches, flux, and piles, to cure saddle gall in horses, and when mixed with powdered bumblebees, as an aid for difficult labor (D. E. Moerman 1986).

1. Leaf blade thick, major teeth sharp; large shrubs of ne United States and e Canada . . . . 5a. *Alnus incana* subsp. *rugosa*
1. Leaf blade thin and papery, secondary teeth rounded or blunt; large shrubs or small trees of w United States, Canada . . . . 5b. *Alnus incana* subsp. *tenuifolia*

5a. Alnus incana (Linnaeus) Moench subsp. **rugosa** (Du Roi) R. T. Clausen, Cornell Univ. Agric. Exp. Sta. Mem. 291: 8. 1949 · Speckled alder, tag alder, swamp alder, aulne blanchâtre E W1

Betula alnus Linnaeus var. *rugosa* Du Roi, Diss. Observ. Bot., 31. 1771; *Alnus glauca* Michaux; *A. incana* var. *americana* Regel; *A. rugosa* (Du Roi) Sprengel var. *americana* (Regel) Fernald; *A. rugosa* var. *tomophylla* (Fernald) Fernald

Shrubs, open, spreading, to 9 m. **Bark** dark grayish to reddish brown, smooth; lenticels whitish, prominent, horizontal. **Winter buds** parallel to twig, ellipsoid, 3–7 mm, apex acute or obtuse; stalks 2–4 mm. **Leaf blade** ovate to elliptic, 4–11 × 3–8 cm, firm, base cuneate to narrowly rounded, margins usually coarsely doubly serrate, toothed to base, apex acute or short-acuminate to slightly obtuse; surfaces abaxially often glaucous, glabrous to sparsely pubescent, often more densely so on veins, slightly to not noticeably resin-coated. **Inflorescences:** staminate catkins in 1 or more clusters of 2–4, 2–7 cm; pistillate catkins in 1 or more clusters of 2–6. **Infructescences** ovoid, 1–1.7 × 0.8–1.2 cm; peduncles 1–5 mm. $2n = 28$.

Flowering early spring. Stream banks, lake shores, bogs, swamps, margins of wet fields, swales, and roadsides, often forming dense thickets; 0–800 m; St. Pierre and Miquelon; Man., N.B., Nfld., N.S., Ont., P.E.I., Que., Sask.; Conn., Ill., Ind., Iowa, Maine, Md., Mass., Mich., Minn., N.H., N.J., N.Y., N.Dak., Ohio, Pa., R.I., Vt., Va., W.Va., Wis.

Alnus incana subsp. *rugosa* is an important shoreline and meadow colonizer in boreal and north temperate areas of the Canadian Shield, and a weedy successional species in damp areas along roadsides throughout its range. It overlaps in range and intergrades with *A. incana* subsp. *tenuifolia* to the west (in Saskatchewan and Manitoba) and with *A. serrulata* to the south. It is only slightly differentiated from the more treelike European *A. incana* subsp. *incana*.

5b. Alnus incana (Linnaeus) Moench subsp. **tenuifolia** (Nuttall) Breitung, Amer. Midl. Naturalist 58: 25. 1957 · Thinleaf alder, mountain alder E

Alnus tenuifolia Nuttall, N. Amer. Sylv. 1: 48. 1842; *A. incana* var. *occidentalis* (Dippel) C. L. Hitchcock; *A. incana* var. *virescens* S. Watson; *A. occidentalis* Dippel; *A. rugosa* (Du Roi) Sprengel var. *occidentalis* (Dippel) C. L. Hitchcock

Shrubs or trees, to 12 m; shrubs ascending, open, spreading, trees small, shrubby. **Bark** light gray to dark brown, smooth; lenticels pale, orbiculate to elliptic. **Winter**

buds nearly divergent, ellipsoid, 4–7 mm, apex obtuse; stalks 1–3 mm; scales 2, equal, valvate. **Leaf blade** ovate to elliptic, 4–10 × 2.5–8 cm, thin, base broadly cuneate to rounded, margins distinctly doubly serrate to nearly crenate or lobulate, teeth relatively blunt or rounded, apex acute to obtuse; surfaces abaxially glabrous to sparsely pubescent, slightly to not noticeably resin-coated. **Inflorescences:** staminate catkins in 1 or more clusters of 3–5, 4–10 cm; pistillate catkins in 1 or more clusters of 2–5. **Infructescences** ovoid, 1–2 × 0.8–1.3 cm; peduncles 1–5 mm. $2n = 28$.

Flowering early spring. Stream banks, lake shores, margins of wet fields and meadows, bog margins, and muskegs; 100–3000 m; Alta., B.C., N.W.T., Sask., Yukon; Alaska, Ariz., Calif., Colo., Idaho, Mont., Nev., N.Mex., Oreg., Utah, Wash., Wyo.

Alnus incana subsp. *tenuifolia* is somewhat more treelike than the eastern *A. incana* subsp. *rugosa*, from which it also differs in leaf shape, leaf margins, and other characters. It is a frequent component of streamside vegetation throughout the Rocky Mountains and other mountainous parts of western North America.

Native Americans used *alnus incana* subsp. *tenuifolia* medicinally for pains in the lungs or hips, for scrofula, as a laxative, and as a diuretic for gonorrhea (D. E. Moerman 1986).

6. **Alnus serrulata** (Aiton) Willdenow, Sp. Pl. 4(1): 336. 1805 · Smooth alder, hazel alder [E]

Betula serrulata Aiton, Hort. Kew. 1: 338. 1789; *Alnus noveboracensis* Britton; *A. rubra* Desfontaines ex Spach; *A. rugosa* (Du Roi) Sprengel var. *serrulata* (Aiton) H. J. P. Winkler

Shrubs, open to rather densely ascending, to 10 m. **Bark** light gray, smooth; lenticels small, inconspicuous. **Winter buds** stipitate, ellipsoid to obovoid, 3–6 mm, apex mostly rounded; stalks 2–4 mm; scales 2, equal, valvate, moderately to heavily resin-coated. **Leaf blade** broadly elliptic to obovate, 5–14 × 3.5–8 cm, leathery, base broadly to narrowly cuneate, margins flat, serrulate, without noticeably larger secondary teeth, apex obtuse to rounded; surfaces abaxially glabrous to moderately villous, slightly to moderately resin-coated. **Inflorescences** formed season before flowering and exposed during winter; staminate catkins in 1 or more clusters of 2–5, 3–8.5 cm, stamens 4; pistillate catkins in 1 or more clusters of 3–5. **Flowering** before new growth in spring. **Infructescences** ovoid-ellipsoid, 1–2.2 × 0.6–1.2 cm; peduncles 1–3(–5) mm. **Samaras** obovate, wings narrower than body, irregularly elliptic or obovate, leathery. $2n = 28$.

Flowering early spring. Stream banks, ditches,

edges of sloughs, swampy fields and bogs, and lakeshores; 0–800 m; N.S., Que.; Ala., Ark., Conn., Del., D.C., Fla., Ga., Ill., Ind., Ky., La., Maine, Md., Mass., Miss., Mo., N.H., N.J., N.Y., N.C., Ohio, Okla., Pa., R.I., S.C., Tenn., Tex., Vt., Va., W.Va.

Primarily an Atlantic coastal species, *Alnus serrulata* also grows along the St. Lawrence river system and the lower Great Lakes westward to the dunes of southern Lake Michigan, and across the southern states to the Gulf Coast and east Texas. *Alnus serrulata* was erroneously called *A. rugosa* in a number of earlier floristic works (J. K. Small 1903, 1933; N. L. Britton and A. Brown 1896, 1913; and B. L. Robinson and M. L. Fernald 1908), and the mistake was perpetuated in both editions of *Flora Europaea* (T. G. Tutin et al. 1964–1980, vol. 1; 1993+, vol. 1).

Alnus incana subsp. *rugosa* hybridizes with *A. serrulata* (= *Alnus serrulata* var. *subelliptica* Fernald). Extensive hybrid swarms occur where the ranges of these species overlap, including the area along the St. Lawrence River and the southern edge of the Great Lakes (F. L. Steele 1961). R. H. Woodworth's conclusion (1929, 1930) that apomixis occurs in *A. serrulata* resulted from his use of material selected from a hybrid swarm. The remainder of the species appears to reproduce normally. The two species and their hybrids are usually easily distinguished by leaf shape and margin characters.

Various preparations of *Alnus serrulata* were used medicinally by Native Americans to alleviate pain of childbirth, as a blood tonic, an emetic and purgative, for coughs and fevers, to stimulate kidneys, to bathe hives or piles, for eye troubles, indigestion, biliousness, jaundice, heart trouble, mouth soreness in babies, and toothaches, to lower blood pressure, and to clear milky urine (D. E. Moerman 1986).

7. **Alnus viridis** (Villars) de Candolle in J. Lamarck and A. P. de Candolle, Fl. Franç. ed. 3, 3: 304. 1805

Betula viridis Villars, Hist. Pl. Dauphiné 3(1): 789. 1789; *Alnus alnobetula* (Ehrhart) K. Koch; *A. ovata* (Schrank) Loddiges

Shrubs, spreading to compact, to 10 m. **Bark** smooth; lenticels scattered, conspicuous to inconspicuous, small, mostly unenlarged. **Winter buds** nearly sessile, ovoid, apex acuminate; stalks usually not over 1 mm; scales 4–6, unequal, imbricate. **Leaf blade** broadly to narrowly ovate or elliptic, 3–11 × 3–8 cm, base rounded, obtuse, or cuneate, sometimes nearly cordate, margins serrulate to coarsely doubly serrate, apex acute to rounded; surfaces abaxially glabrous to tomentose, lightly to heavily resin-coated. **Inflorescences:** staminate catkins in 1 cluster of 2–4, formed late in growing season before flowering and exposed during winter; pistillate catkins in 1 or more clusters of 2–10, formed season before blooming, enclosed in buds during winter, exposed with new growth in spring. **Flowering** with new growth in spring. **Infructescences** ovoid to ellipsoid or nearly cylindric; peduncles relatively long, thin. **Samaras** elliptic to obovate, wings wider than body, membranaceous.

Subspecies 4 (3 in the flora): southern arctic, subarctic, and n mountainous regions, North America and Asia.

Alnus viridis is distinctive among the alders in its essentially sessile buds with several imbricate scales and in its relatively long, thin, infructescence peduncles. Like the birches, only the staminate catkins are exposed during the winter prior to blooming.

1. Leaf blade coarsely doubly serrate, thin, light or yellowish green, glabrous to sparsely pubescent; mountainous nw United States, Alaska, and Canada . . . . 7c. *Alnus viridis* subsp. *sinuata*
1. Leaf blade serrulate to finely and densely serrate or doubly serrate, firm, dark green, sometimes abaxially glabrous or sparsely to densely pubescent.
   2. Leaf blade broadly to narrowly ovate or elliptic, margins serrulate or finely serrate, apex obtuse to acute; e United States, n Canada, Alaska, and s Greenland . . . .
   . . . . .7a. *Alnus viridis* subsp. *crispa*
   2. Leaf blade broadly ovate, margins sharply and densely doubly serrate, apex acute to short-acuminate; w coastal North America, adjacent subarctic Asia . . . .
   . . . . .7b. *Alnus viridis* subsp. *fruticosa*

7a. Alnus viridis (Villars) de Candolle subsp. **crispa** (Aiton) Turrill, Bot. Mag. 173: 382. 1962 · Green alder, mountain alder, aulne vert, aulne crispé ⸂E⸣ ⸂F⸣

Betula crispa Aiton, Hort. Kew. 2: 339. 1789; *Alnus alnobetula* (Ehrhart) K. Koch var. *crispa* (Aiton) H. J. P. Winkler; *A. crispa* var. *elongata* Raup; *A. crispa* var. *harricanensis* Lepage; *A. crispa* var. *mollis* (Fernald) Fernald; *A. crispa* var. *stragula* Fernald; *A. mitchelliana* M. A. Curtis ex A. Gray; *A. mollis* Fernald; *A. viridis* var. *crispa* (Michaux) House

Shrubs, spreading or rather compact, to 3(–4) m. **Bark** grayish brown; lenticels pale. **Leaf blade** dark green, broadly to narrowly ovate or elliptic, 3.5–6(–10) × 3–5(–7) cm, leathery, base rounded, obtuse, or cuneate, sometimes nearly cordate, margins serrulate or finely serrate, apex obtuse to acute; surfaces abaxially glabrous to velutinous or occasionally tomentose, moderately to heavily resin-coated. **Inflorescences:** staminate catkins 2.5–9 cm. **Infructescences** 1.2–2 × 0.5–1.2 cm; peduncles 1–5 cm. $2n = 28$.

Flowering spring. Singly or in thickets along streams, lakeshores, coasts, and bog or muskeg margins, or on sandy or gravelly slopes or flats; 0–2000 m; St. Pierre and Miquelon; Greenland; Alta., Man., N.B., Nfld., N.W.T., N.S., Ont., P.E.I., Que., Sask.; Maine, Mass., Mich., Minn., N.H., N.Y., N.C., Pa., Tenn., Vt., Wis.

Alnus viridis subsp. *crispa* grows across much of the continent in the far North; widely disjunct populations occur in the Appalachians in Pennsylvania and on the summit of Roan Mountain on the North Carolina–Tennessee border (R. B. Clarkson 1960; E. T. Wherry 1960).

The Cree used *Alnus viridis* subsp. *crispa* medicinally for the astringent qualities of the bark and to treat dropsy (D. E. Moerman 1986).

7b. Alnus viridis (Villars) de Candolle subsp. **fruticosa** (Ruprecht) Nyman, Consp. Fl. Eur., 672. 1881 · Siberian alder

Alnus fruticosa Ruprecht, Distr. Crypt. Vasc. Ross., 53. 1845; *A. viridis* var. *fruticosa* (Ruprecht) Regel

Shrubs, spreading, to 3(–6) m. **Bark** gray-brown; lenticels pale. **Leaf blade** dark green, broadly ovate, 5–8(–10) × 3–6(–7) cm, base rounded to nearly truncate or nearly cordate, margins flat, sharply and densely doubly serrate, apex acute to short-acuminate; surfaces abaxially glabrous to sparsely pubescent, especially on veins, moderately to heavily resin-coated. **Inflorescences:** staminate catkins 3.5–6 cm. **Infructescences** 1.2–2 × 0.5–1.2 cm; peduncles 1–3 cm. $2n = 28$.

Flowering spring. Rocky or sandy coasts, stream banks, lakeshores, and damp, open areas; 0–500 m; Alta., B.C., N.W.T., Sask., Yukon; Alaska, Calif., Oreg., Wash.; n Asia.

This primarily subarctic Asian subspecies has long been mistaken in western North America for *Alnus viridis* subsp. *crispa,* which it closely resembles, or for subsp. *sinuata* (J. J. Furlow 1983b). It can be separated from the former by its larger and more coarsely toothed leaves, and from the latter by its much thicker, mostly single-toothed leaf blades.

7c. Alnus viridis (Villars) de Candolle subsp. **sinuata** (Regel) Á. Löve & D. Löve, Univ. Colorado Stud., Ser. Biol. 17: 20. 1965 · Sitka alder, mountain alder ⸂E⸣

Alnus viridis var. (δ) *sinuata* Regel, Bull. Soc. Imp. Naturalistes Moscou 38(3): 422. 1865; *A. crispa* (Aiton) Pursh subsp. *sinuata* (Regel) Hultén; *A. sinuata* (Regel) Rydberg; *A. sitchensis* (Regel) Sargent

Shrubs, spreading, to 5(–10) m. **Bark** light gray to reddish brown; lenticels inconspicuous.

Leaf blade light or yellowish green, narrowly to broadly ovate, 4–10 × 3–8 cm, thin, papery, base rounded to cordate, margins flat, sharply and coarsely doubly serrate, apex acuminate; surfaces abaxially glabrous to sparsely pubescent, lightly to moderately resin-coated. **Inflorescences:** staminate catkins 2.5–13 cm. **Infructescences** 1.5–2.5 × 0.8–1.3 cm; peduncles 1–3 cm. $2n = 28$.

Flowering spring. Along gravelly or rocky stream banks, lakeshores, and coasts, on moist rocky slopes, outcrops, in open coniferous woodlands; 0–2500 m; Alta., B.C., N.W.T., Yukon; Alaska, Calif., Idaho, Mont., Oreg., Wash., Wyo.

Alnus viridis subsp. *sinuata* is one of the first successional taxa to appear in the northwestern mountains following disruption of the mature vegetation. It often forms dense thickets on avalanche and talus slopes. Sitka alder differs from the two previous subspecies in its paper-thin, light or yellowish green, doubly serrate leaves.

The Bella Coola used *Alnus viridis* subsp. *sinuata* medicinally although D. E. Moerman (1986) did not specify the nature of the remedies.

8. **Alnus maritima** (Marshall) Muhlenburg ex Nuttall, N. Amer. Sylv. 1: 50. 1842 · Seaside alder, brook alder [E] [F]

Betula-alnus maritima Marshall, Arbust. Amer., 20. 1785

Shrubs or trees, to 10 m; crowns narrow. **Bark** light gray, smooth; lenticels small, inconspicuous. **Winter buds** stipitate, ovoid to ellipsoid, 2.5–5 mm, apex rounded; stalks 1–3 mm; scales 2–3, subequal, often poorly developed, heavily resin-coated. **Leaf blade** narrowly elliptic, oblong, or narrowly obovate, 4.5–9 × 2–5 cm, leathery, base acute to cuneate, margins flat, teeth low, single, relatively distant, apex acute, obtuse, or rounded; surfaces abaxially mostly glabrous, resin-coated when young. **Inflorescences:** catkins formed during same season as flowering; staminate catkins in 1 terminal cluster of 2–4, 2–6 cm; pistillate catkins solitary in leaf axils proximal to staminate catkins. **Infructescences** ovoid, 1.2–2.8 × 1.2–2.2 cm; peduncles 5–10 mm. **Samaras** elliptic, wings reduced to narrow, leathery ridges. $2n = 28$.

Flowering late summer–early fall. Along edges of ponds and small streams, often in standing water; 0–100 m; Del., Md., Okla.

Alnus maritima consists of widely disjunct populations in Delaware, Maryland, and southern Oklahoma. The populations probably represent remnants of Pleistocene and post-Pleistocene distributions and migrations. It is our only member of the predominantly Asian fall-blooming *Alnus* subg. *Clethropsis*.

2. BETULA Linnaeus, Sp. Pl. 2: 982. 1753; Gen. Pl. ed. 5, 433. 1754 · Birch [Latin *betula*, birch]

Trees or shrubs, to 30 m; trunks often several, branching excurrent, becoming deliquescent. **Bark** of trunks and branches dark brown to chalky white, smooth, often exfoliating; lenticels dark, prominent, sometimes horizontally expanded. **Wood** nearly white to reddish brown, light and soft to moderately heavy and hard, texture fine. **Branches, branchlets, and twigs** nearly 2-ranked; young twigs differentiated into long and short shoots, sometimes with taste and odor of wintergreen. **Winter buds** sessile, slender, terete, apex acute; scales several, imbricate, smooth. **Leaves** mostly on short shoots, nearly 2-ranked. **Leaf blade** ovate to deltate, elliptic, or nearly orbiculate, 0.5–10(–14) × 0.5–8 cm, thin, margins doubly serrate or serrate (or crenate to shallowly round-lobed in dwarf northern species); surfaces glabrous to tomentose, sometimes abaxially resinous-glandular. **Inflorescences:** staminate catkins mostly terminal on branchlets, solitary or in small racemose clusters, formed previous growing season and often exposed during winter, expanding with leaves; pistillate catkins proximal to staminate catkins, mostly solitary, erect, ovoid to cylindric, firm; scales and flowers crowded, enclosed

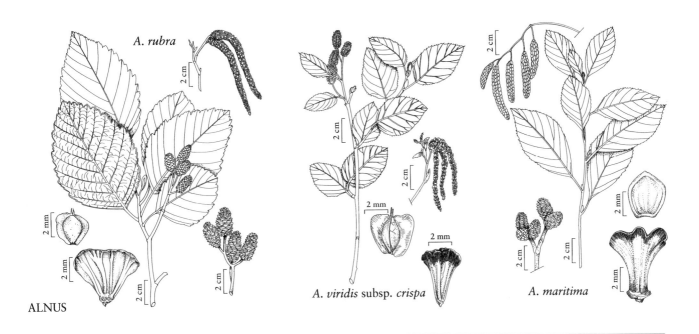

A. rubra

ALNUS

A. viridis subsp. *crispa*

A. maritima

within buds during winter, expanding with leaves. **Staminate flowers** in catkins, 3 per scale; stamens (1–)2–3(–4), filaments divided below anthers, nearly to base. **Pistillate flowers** (1–)3 per scale. **Infructescences** erect or pendulous; scales usually deciduous with release of fruits (although persisting into winter in a few species), (1–)3-lobed, thickened or leathery but not woody. **Fruits** samaras, lateral wings 2, moderately wide to broad, membranaceous. $x = 14$.

Species ca. 35 (18 species in the flora): throughout n temperate, boreal, and arctic zones of the Northern Hemisphere; North America, Asia.

Birches, like alders, are common trees and shrubs of northern temperate and boreal zones of the Northern Hemisphere. Like *Alnus,* the group is highly diversified, especially in the Old World. The species are well known for their free hybridization, and specimens are therefore frequently difficult to identify. Birches occupy habitats in cool, moist regions, including peatlands, stream banks, and lakeshores, cool, damp woods, and moist slopes in cool coves. The wood of species that grow to a large size (including especially *B. alleghaniensis*) has many uses, including the manufacture of doors and windows, flooring, cabinetry, interior molding, wood paneling, furniture, and plywood.

Betula sect. *Costatae* (Regel) Koehne consists of large, mesophytic trees, often with dark, close or exfoliating bark, large thin leaves, infructescence scales with long narrow lobes, and fruits with narrow wings. North American representatives of this group include *Betula alleghaniensis, B. lenta,* and *B. nigra.* The mostly circumboreal *Betula* sect. *Betula* consists of small to medium trees with rather large thin leaves and fruits with wide wings (wider than the fruit body). A characteristic feature of trees in this group is their white bark that often peels apart in sheets. These include the familiar paper birch, *B. papyrifera,* and its European counterpart, *B. pubescens,* as well as the common eastern *B. neoalaskana.* A third line, *Betula* sect. *Humiles* W. D. Koch, consists of dwarf shrubby species of the cold circumpolar region. In North America this section is represented by *B. glandulosa, B. pumila,* and *B. nana.*

Birches are a difficult group taxonomically because of their high vegetative variability and frequent hybridization. Many morphologic and cytologic studies have attempted to deal with variation within the genus or its subgroups. Species of *Betula* form a polyploid series, with chromosome numbers of $2n = 28, 56, 70, 84$, and 112, plus additional numbers in some hybrids. This and other research in the genus has been reviewed by J. J. Furlow (1990).

SELECTED REFERENCES Alam, M. T. and W. F. Grant. 1972. Interspecific hybridization in birch (*Betula*). Naturaliste Canad. 99: 33–40. Brittain, W. H. and W. F. Grant. 1965. Observations on Canadian birch (*Betula*) collections at the Morgan Arboretum. I. *B. papyrifera* in eastern Canada. Canad. Field-Naturalist 79: 189–197. Brittain, W. H. and W. F. Grant. 1965b. Observations on Canadian birch (*Betula*) collections at the Morgan Arboretum. II. *B. papyrifera* var. *cordifolia*. Canad. Field-Naturalist 79: 253–257. Brittain, W. H. and W. F. Grant. 1966. Observations on Canadian birch (*Betula*) collections at the Morgan Arboretum. III. *B. papyrifera* from British Columbia. Canad. Field-Naturalist 80: 147–157. Brittain, W. H. and W. F. Grant. 1967. Observations on Canadian birch (*Betula*) collections at the Morgan Arboretum. IV. *B. caerulea-grandis* and hybrids. Canad. Field-Naturalist 81: 116–127. Brittain, W. H. and W. F. Grant. 1967b. Observations on Canadian birch (*Betula*) collections at the Morgan Arboretum. V. *B. papyrifera* and *B. cordifolia* from eastern Canada. Canad. Field-Naturalist 81: 251–262. Brittain, W. H. and W. F. Grant. 1968. Observations on Canadian birch (*Betula*) collections at the Morgan Arboretum. VI. *B. papyrifera* from the Rocky Mountains. Canad. Field-Naturalist 82: 44–48. Brittain, W. H. and W. F. Grant. 1969. Observations on Canadian birch (*Betula*) collections at the Morgan Arboretum. VII. *B. papyrifera* and *B. resinifera* from northwestern Canada. Canad. Field-Naturalist 83: 185–202. Brittain, W. H. and W. F. Grant. 1969b. Observations on Canadian birch (*Betula*) collections at the Morgan Arboretum. VIII. *Betula* from Grand Manon Island, New Brunswick. Canad. Field-Naturalist 83: 361–383. Dugle, J. R. 1966. A taxonomic study of western Canadian species in the genus *Betula*. Canad. J. Bot. 44: 929–1007. Fernald, M. L. 1902. The relationships of some American and Old World birches. Amer. J. Sci. 169: 167–194. Fernald, M. L. 1945. Some North American Corylaceae (Betulaceae). 1. Notes on *Betula* in eastern North America. Rhodora 47: 303–329. Fredskild, B. 1991. The genus *Betula* in Greenland—Holocene history, present distribution, and synecology. Nordic J. Bot. 11: 393–412. Grant, W. F. and B. K. Thompson. 1975. Observations on Canadian birches, *Betula cordifolia*, *B. populifolia*, *B. papyrifera* and *B. ×caerulea*. Canad. J. Bot. 53: 1478–1490. Johnsson, H. 1945. Interspecific hybridization within the genus *Betula*. Hereditas (Lund) 31: 163–176. Lepage, E. 1976. Les bouleaux arbustifs du Canada et de Alaska. Naturaliste Canad. 103: 215–233. Sulkinoja, M. 1990. Hybridization, introgression and taxonomy of the mountain birch in SW Greenland compared with related results from Iceland and Finnish Lapland. Meddel. Grønland, Biosci. 33: 21–29.

1. Larger leaf blades generally more than 5 cm, with 5–18 pairs of lateral veins; medium to large trees, 10–30 m.
  2. Samaras with wings narrower than body; bark of mature trunks and branches mostly dark to medium red or reddish or yellowish brown, sometimes partly (but not completely) creamy or grayish white, with or without horizontally expanded lenticels; infructescences erect, conic or nearly globose, scales often persistent into early winter.
    3. Leaf blade rhombic-ovate, base broadly cuneate to truncate, apex acute; scales of infructescences lobed distal to middle, with 3 narrow, ascending lobes, ± equal to somewhat unequal in length and breadth; twigs without wintergreen taste or odor . . . . .5. *Betula nigra*
    3. Leaf blade ovate to elliptic or oblong, base rounded, cordate, or nearly cordate, apex acute to acuminate; scales of infructescences lobed at or proximal to middle, lobes dissimilar in size and shape; twigs with distinctive wintergreen taste and odor.
      4. Apex of leaf blade acute or only slightly acuminate; lateral veins 7–10 pairs; swampy habitats . . . . 2. *Betula murrayana*
      4. Apex of leaf blade distinctly acuminate; lateral veins 12–18 pairs; mesic, often streamside habitats.
        5. Margins of leaf blade with coarse, rather irregular teeth; bark of mature trunks and branches yellowish, irregularly exfoliating or sometimes darkening and remaining close; scales of infructescences sparsely to moderately pubescent . . . . 1. *Betula alleghaniensis*
        5. Margins of leaf blade with fine, sharp teeth; bark of mature trunks and branches light grayish brown, close, not freely exfoliating; scales of infructescences mostly glabrous . . . . .3. *Betula lenta*
  2. Samaras with wings as broad as or broader than body; bark of mature trunks and branches bright chalky to creamy, reddish, pinkish, or brownish white to reddish tan or bronze, often with greatly expanded, dark, horizontal lenticels; infructescences pendulous (erect to nearly pendulous in *B. populifolia* and *B. pendula*), elongate, cylindric, readily shattering when mature.
    6. Leaf blade ovate or rhombic-ovate to elliptic or oblong, abaxially pubescent at least along veins, apex acute or short-acuminate; central lobe of infructescence scales equal to or longer than basal and lateral lobes; bark of mature trunks exfoliating.
      7. Base of larger leaf blade cuneate, rounded, or truncate; lateral veins 9 or fewer pairs; lateral lobes of infructescence scales held nearly at right angles to axis; mature bark creamy to chalky white or pale to (infrequently) dark brown . . . . .6. *Betula papyrifera*
      7. Base of larger leaf blade mostly cordate (rarely rounded); lateral veins 9–12 pairs; lateral lobes of fruiting catkin scales turned toward apex; mature bark pinkish or brownish white to reddish tan or bronze . . . . 7. *Betula cordifolia*
    6. Leaf blade broadly ovate to ± rhombic, abaxially glabrous to somewhat pubescent on veins,

apex acuminate to long-acuminate; central lobe of infructescence scales shorter than basal and lateral lobes; bark of mature trunks close or exfoliating.

 8. Apex of leaf blade acuminate, but not extended into long tapering tip; central lobe of infructescence scales nearly as long as lateral lobes; twigs densely covered with large resinous glands . . . . .14. *Betula neoalaskana*

 8. Apex of leaf blade drawn out into long tapering tip; central lobe of infructescence scales much shorter than lateral lobes; twigs sometimes glandular, glands small, inconspicuous.

  9. Mature bark grayish white; leaf apex long-caudate; infructescence scales densely pubescent on adaxial surface; native trees of ne North America . . . . .12. *Betula populifolia*

  9. Mature bark creamy to silvery white, exfoliating as long strands; leaf apex acuminate; infructescence scales sparsely pubescent on adaxial surface; domesticated trees adventive or persisting after cultivation. . . . . .13. *Betula pendula*

1. Larger leaf blades mostly less than 5 cm, with 2–6 pairs of lateral veins; shrubs and small to medium trees, 0.5–12 m (*B. pubescens* to 20 m).

 10. Leaf blade 0.5–2 cm, orbiculate, orbiculate-obovate, or reniform, margins simple-toothed; depressed or low upright shrubs.

  11. Scales of infructescences 3-lobed; samaras with narrow but definite lateral wings. . . . . .17. *Betula nana*

  11. Scales of infructescences unlobed; samaras with wings reduced to narrow ridges or without apparent wings . . . . 18. *Betula michauxii*

 10. Leaf blade 2.5–5(–7) cm, ovate to nearly orbiculate, margins distinctly to obscurely doubly serrate, dentate, or crenate; trees and small to large shrubs.

  12. Leaf blade nearly orbiculate to broadly elliptic, base rounded to cordate or truncate, apex broadly obtuse to rounded; twigs with taste and odor of wintergreen. 4. *Betula uber*

  12. Leaf blade ovate to elliptic, base rounded to cuneate or truncate, apex acute to obtuse or rounded; twigs without taste and odor of wintergreen.

   13. Margins of leaf blade crenate to blunt-dentate; shrubs with close bark.

    14. Leaf blade 2.5–5(–7) cm, broadly ovate, obovate, or elliptic to nearly orbiculate, base cuneate, apex broadly acute to obtuse or rounded, surfaces abaxially glabrous to tomentose; twigs pubescent or glabrous, sometimes inconspicuously small-glandular, without large resinous glands . . . . 15. *Betula pumila*

    14. Leaf blade 1–2(–4) cm, mostly obovate to nearly orbiculate, base cuneate to rounded, apex obtuse to rounded, surfaces abaxially mostly glabrous to moderately pubescent; twigs essentially glabrous and warty with large resinous glands. . . . . . .16. *Betula glandulosa*

   13. Leaf blade margins simply or doubly serrate to dentate, teeth obtuse to relatively sharp; trees and shrubs with close or exfoliating bark.

    15. Young twigs covered with scattered short, stiff, erect hairs; infructescences 1–6 cm; bark light brownish or tannish white, not readily exfoliating; native shrubby trees of sw Greenland, and introduced trees adventive or persisting after cultivation in United States and Canada. . . . . 9. *Betula pubescens*

    15. Young twigs glabrous to pubescent, without short, stiff, erect hairs; infructescences mostly (1–)2–3(–4) cm; bark light to dark, close or exfoliating; native trees and shrubs.

     16. Central lobe of infructescence scales much shorter than lateral lobes; bark dark reddish brown to bronze, not readily exfoliating; large shrubs and small trees of Rocky Mountains, nw Great Plains, w, c northern Canada, and Alaska. . . 11. *Betula occidentalis*

     16. Central lobe of infructescence scales equal in length to longer than lateral lobes; bark dark brown to grayish white, exfoliating in thin sheets or close; large shrubs or small trees, nw, boreal, and subalpine ne North America.

      17. Bark brown to pinkish or grayish white, exfoliating in thin sheets; small trees of nw North America . . . . .10. *Betula kenaica*

      17. Bark dark reddish brown, not readily exfoliating; shrubs of boreal and subalpine ne North America . . . . 8. *Betula minor*

1. Betula alleghaniensis Britton, Bull. Torrey Bot. Club 31: 166. 1904 · Yellow birch, merisier, bouleau jaune [E] [F]

Betula alleghaniensis var. *fallax* (Fassett) Brayshaw; *B. alleghaniensis* var. *macrolepis* (Fernald) Brayshaw; *B. lutea* F. Michaux; *B. lutea* var. *macrolepis* Fernald

Trees, to 30 m; trunks straight, crowns narrowly round. **Bark** of young trunks and branches dark reddish brown, in maturity tan, yellowish, or grayish, lustrous, smooth, irregularly exfoliating, or sometimes darkening and remaining close; lenticels dark, horizontally expanded. **Twigs** with odor and taste of wintergreen when crushed, glabrous to sparsely pubescent, usually covered with small resinous glands. **Leaf blade** narrowly ovate to ovate-oblong with (9–)12–18 pairs of lateral veins, 6–10 × 3–5.5 cm, base rounded to cuneate or cordate, margins sharply doubly serrate, teeth coarse, rather irregular, apex acuminate; surfaces abaxially usually moderately pubescent, especially along major veins and in vein axils, often with scattered, minute, resinous glands. **Infructescences** erect, ovoid, 1.5–3 × 1–2.5 cm, generally remaining intact after release of fruits in late fall; scales sparsely to moderately pubescent, lobes diverging proximal to middle, central lobe tapering to narrow tip, lateral lobes ascending or partially extended, broader, rounded. **Samaras** with wings narrower than body, broadest near summit, not or only slightly extended beyond body apically. $2n = 84$.

Flowering late spring. Stream banks, swampy woods, and rich, moist, forested slopes; 0–500 m; St. Pierre and Miquelon; N.B., Nfld., N.S., Ont., P.E.I., Que.; Ala., Conn., Ga., Ill., Ind., Iowa, Ky., Maine, Md., Mass., Mich., Minn., N.H., N.J., N.Y., N.C., Ohio, Pa., R.I., S.C., Tenn., Vt., Va., W.Va., Wis.

Betula alleghaniensis is a characteristic tree of the northern Appalachians and the hemlock hardwoods forest of the Great Lakes region. It was formerly widely known by the illegitimate (superfluous) name *B. lutea* F. Michaux.

Native Americans used *Betula alleghaniensis* medicinally as an emetic or cathartic, to remove bile from intestines, as a blood purifier, and as a diuretic (D. E. Moerman 1986, as *Betula lutea*).

Betula alleghaniensis is very closely related to *B. lenta,* which it resembles in many features (T. L. Sharik and R. H. Ford 1984). A distinctive feature is usually its freely exfoliating bark, although in certain populations the bark remains close and dark (B. P. Dancik 1969; B. P. Dancik and B. V. Barnes 1971).

Betula alleghaniensis Britton × *B. papyrifera* Marshall has seldom been reported, but it may actually be more common than realized in the northeastern states.

In most features it is intermediate between the parents (B. V. Barnes et al. 1974).

Betula ×*purpusii* Schneider (= *Betula alleghaniensis* Britton × *B. pumila* Linnaeus, $2n = 70$) is a rather common hybrid wherever the parent species occur together. The large shrubby plants show strikingly intermediate leaf characteristics.

2. Betula murrayana B. V. Barnes & Dancik, Canad. J. Bot. 63: 226, figs. 1D, 2, 3. 1985 · Murray's birch [C] [E]

Trees, to 15 m; trunks usually several. **Bark** of mature trunk and branches dark red to reddish brown, smooth, close; lenticels pale, conspicuous, horizontally expanded. **Twigs** with taste and odor of wintergreen when crushed, glabrous to sparsely pubescent, covered with small resinous glands. **Leaf blade** ovate with 7–10 pairs of lateral veins, 5–11 × 3–6 cm, base cuneate, margins sharply and obscurely doubly serrate, apex acute or only slightly acuminate; surfaces abaxially sparsely pubescent to glabrous. **Infructescences** erect, ovoid, 2–4 × 1.5–3 cm, remaining intact for a period after release of fruits in late fall; scales sparsely pubescent to glabrous, lobes ascending, branching at middle, slightly unequal in length. **Samaras** with wings narrower than body, broadest near summit, not extended beyond body apically. $2n = 112$.

Flowering late spring. Wet, swampy forests containing *Betula pumila;* of conservation concern; 0–300 m; Mich.

Betula murrayana is an octoploid derivative of *Betula* ×*purpusii* (= *B. alleghaniensis* Britton × *B. pumila* Linnaeus) (B. V. Barnes and B. P. Dancik 1985). It is intermediate between *B. alleghaniensis* and *B. pumila* in most vegetative features, but in characters such as leaf size, it approaches *B. alleghaniensis.*

3. Betula lenta Linnaeus, Sp. Pl. 2: 983. 1753 · Sweet birch, cherry birch [E]

Trees, to 20 m; trunks tall, straight, crowns narrow. **Bark** of mature trunks and branches light grayish brown to dark brown or nearly black, smooth, close, furrowed and broken into shallow scales with age. **Twigs** with taste and odor of wintergreen when crushed, glabrous to sparsely pubescent, usually covered with small resinous glands. **Leaf blade** ovate to oblong-ovate with 12–18 pairs of lateral veins, 5–10 × 3–6 cm, base rounded to cordate, margins finely and sharply serrate or obscurely doubly serrate, teeth fine, sharp, apex acuminate; surfaces abaxially mostly glabrous, except sparsely pubescent along major veins and in vein axils, often with scattered, minute, resinous glands. **Infructescences** erect, ovoid to nearly globose, 1.5–4 × 1.5–2.5 cm, usually remaining intact for a period after release of fruits in fall; scales mostly glabrous, lobes diverging at or proximal to middle, central lobe short, cuneate, lateral lobes extended to slightly ascending, longer and broader than central lobe. **Samaras** with wings narrower than body, broadest near center, not extended beyond body apically. $2n = 28$.

Flowering late spring. Rich, moist, cool forests, especially on protected slopes, to rockier, more exposed sites; 0–1500 m; Ont.; Ala., Conn., Ga., Ky., Maine, Md., Mass., Miss., N.H., N.J., N.Y., N.C., Ohio, Pa., R.I., S.C., Tenn., Vt., Va., W.Va.

Betula lenta is a dominant tree in the northern hardwood forests of the northern Appalachians and a valuable source of timber. It was formerly the chief commercial source of wintergreen oil (methyl salicylate), which is distilled from its wood. *Betula lenta* is most easily separated from *B. alleghaniensis* by its close bark and the glabrous scales of infructescences.

Native Americans used *Betula lenta* medicinally to treat dysentery, colds, diarrhea, fevers, soreness, and milky urine, and as a spring tonic (D. E. Moerman 1986).

4. **Betula uber** (Ashe) Fernald, Rhodora 47: 325. 1945 · Virginia roundleaf birch [C][E]

Betula lenta Linnaeus var. *uber* Ashe, Rhodora 20: 64. 1918

Trees, slender, to 10 m. **Bark** dark brown, smooth, close. **Twigs** with taste and odor of wintergreen when crushed, glabrous, covered with small resinous glands. **Leaf blade** nearly orbiculate to broadly elliptic with 2–6 pairs of lateral veins, 2–5 × 2–4 cm, base rounded to cordate or truncate, margins irregularly serrate or dentate, apex broadly obtuse to rounded; surfaces abaxially glabrous to sparsely pubescent, especially along major veins and in vein axils, often with scattered resinous glands. **Infructescences** erect, ellipsoid-cylindric, 1–2 × 1–1.5 cm, shattering with fruits in fall; scales glabrous, lobes diverging distal to middle, central lobe ascending, shorter than lateral lobes. **Samaras** with wings narrower than to as wide as body, broadest near summit, extended beyond body apically.

Flowering late spring. Stream banks and adjacent flood plains in rich mesic forest; of conservation concern; 500 m; Va.

Betula uber, described in 1918, was not seen again until its widely celebrated rediscovery in 1974 (P. M. Mazzeo 1974; C. F. Reed 1975; D. W. Ogle and P. M. Mazzeo 1976; D. J. Preston 1976). It is apparently allied to *B. lenta* (W. J. Hayden and S. M. Hayden 1984; T. L. Sharik and R. H. Ford 1984); whether it constitutes a separate species or simply mutant individuals of *B. lenta* is a matter of controversy. Seeds obtained from the original single extant population of 17 trees and grown at the U.S. National Arboretum have produced an apparent hybrid swarm of offspring varying in leaf characteristics from those of *B. uber* to those of *B. lenta* (with which it occurs).

5. **Betula nigra** Linnaeus, Sp. Pl. 2: 982. 1753 · River birch, red birch [E][F][W1]

Betula rubra F. Michaux

Trees, to 25 m; trunks often several, crowns round. **Bark** of mature trunks and branches grayish brown, yellowish, reddish, or creamy white, smooth, irregularly shredding and exfoliating in shaggy sheets when mature; lenticels dark, horizontally expanded. **Twigs** without wintergreen taste or odor, glabrous to sparsely pubescent, often with scattered, tiny, resinous glands. **Leaf blade** rhombic-ovate, with 5–12 pairs of lateral veins, 4–8 × 3–6 cm, base broadly cuneate to truncate, margins coarsely doubly serrate to dentate, apex acuminate; surfaces abaxially moderately pubescent to velutinous, especially along major veins and in vein axils, often with scattered, minute, resinous glands. **Infructescences** erect, conic or nearly globose, 1.5–3 × 1–2.5 cm, shattering with fruits in late spring or early summer; scales often persistent into early winter, lobes 3, ascending, branching distal to middle, narrow, elongate, equal to somewhat unequal in length, apex acute. **Samaras** with wings narrower than body, usually broadest near summit, not extended beyond body apically. $2n = 28$.

Flowering late spring. Riverbanks and flood plains, often where land is periodically inundated; 0–300 m; Ala., Ark., Conn., Del., D.C., Fla., Ga., Ill., Ind., Iowa, Kans., Ky., La., Md., Mass., Minn., Miss., Mo., N.J., N.Y., N.C., Ohio, Okla., Pa., S.C., Tenn., Tex., Vt., Va., W.Va., Wis.

Betula nigra is a large and characteristic floodplain tree. Like several other species of this habitat (e.g., *Acer saccharinum* Marshall and *Ulmus americana* Linnaeus), it releases its fruits in early summer; the seeds germinate immediately (at a time when the surrounding land is unlikely to be flooded). The wood of *Betula nigra* is not in high demand for timber because of its generally poor quality. Cultivars with freely ex-

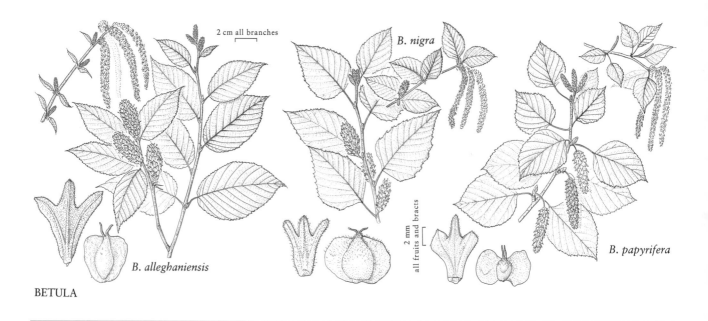

2 cm all branches

B. nigra

2 mm all fruits and bracts

B. alleghaniensis

B. papyrifera

BETULA

foliating bark are commonly cultivated in the Northeast and Midwest.

Native Americans used *Betula nigra* medicinally to treat dysentery, colds, and milky urine (D. E. Moerman 1986).

6. Betula papyrifera Marshall, Arbust. Amer., 19. 1785 · Paper birch, canoe birch, white birch, bouleau blanc, bouleau à papier [E] [F]

Betula alba Linnaeus var. *papyrifera* (Marshall) Spach; *B. papyracea* Aiton; *B. papyrifera* var. *commutata* (Regel) Fernald; *B. papyrifera* var. *elobata* (Fernald) Sargent; *B. papyrifera* var. *macrostachya* Fernald; *B. papyrifera* var. *pensilis* Fernald; *B. papyrifera* var. *subcordata* (Rydberg) Sargent

Trees, to 30 m, usually 20 m or shorter; trunks often single, sometimes 2 or more, mature crowns narrowly round. **Bark** of young trunks and branches dark reddish brown, smooth, in maturity creamy to chalky white or pale to (infrequently) dark brown, readily exfoliating in paper-thin sheets; lenticels pale, horizontal, in maturity dark, much expanded, horizontal. **Twigs** without strong odor and taste of wintergreen, slightly to moderately pubescent, infrequently with scattered, small, resinous glands. **Leaf blade** ovate with 9 or fewer pairs of lateral veins, 5–9(–12) × 4–7 cm, base rounded, cuneate, or truncate, margins sharply to coarsely or irregularly doubly serrate or serrate-dentate, apex acute to short-acuminate; surfaces abaxially sparsely to moderately pubescent, often velutinous along major veins and in

vein axils, covered with minute, resinous glands. **Infructescences** pendulous, cylindric, 2.5–5 × 0.6–1.2 cm, readily shattering with fruits in late fall; scales pubescent to glabrous, lobes diverging at or proximal to middle, central lobe narrowly elongate, obtuse, lateral lobes about equal in length to central lobe but several times broader, strongly divergent, held nearly at right angles to axis. **Samaras** with wings as broad as or slightly broader than body, extended nearly beyond body apically. $2n = 56, 70, 84$.

Flowering late spring. Moist, ± open, upland forest, especially on rocky slopes, also sometimes in swampy woods; 300–900 m; St. Pierre and Miquelon; Alta., B.C., Man., N.B., Nfld., N.W.T., N.S., Ont., P.E.I., Que., Sask., Yukon; Alaska, Colo., Conn., Idaho, Ill., Ind., Iowa, Maine, Mass., Mich., Minn., Mont., Nebr., N.H., N.J., N.Y., N.Dak., Oreg., Pa., R.I., S.Dak., Vt., Va., Wash., Wis., Wyo.

Betula papyrifera is a well-known tree of the northern forest with its paper-thin, white, peeling bark. The bark, which has a high oil content and is consequently waterproof, was used for a wide variety of building and clothing purposes by the American Indians, including the covering of the familiar birch bark canoe. It is still used for various purposes, including basketmaking, in Canada and Alaska. Variants having more or less close, dark brown bark (*B. papyrifera* var. *commutata*) occur locally throughout the wide range of this species; this characteristic appears to be largely environmentally caused. The species is an important successional tree, coming up readily after fires, logging, or the abandonment of cultivated land. The relatively soft, whitish wood is used extensively for such items as clothespins, spools, ice cream sticks, and toothpicks, as well as for pulpwood for paper.

Betula papyrifera is the state tree of New Hampshire.

Native Americans used *Betula papyrifera* medicinally in enemas, to shrivel the womb, to alleviate stomach cramps and pain, and as a tonic (D. E. Moerman 1986).

Betula ×*sandbergii* Britton is a fairly common hybrid, occurring where the ranges of the parents (*B. papyrifera* Marshall and *B. pumila* Linnaeus) come into contact. In most vegetative features it is intermediate between the parental conditions (K. E. Clausen 1963; C. O. Rosendahl 1928).

7. **Betula cordifolia** Regel, Nouv. Mém. Soc. Imp. Naturalistes Moscou 13(2): 86. 1861 · Heartleaf birch, mountain white birch, bouleau à feuilles cordées, bouleau blanc E

Betula alba Linnaeus var. *cordifolia* (Regel) Fernald; *B. papyrifera* Marshall var. *cordifolia* (Regel) Fernald

Trees or shrubs, large, irregular, to 20 m; trunks often several, trees with narrow crowns. **Bark** of young trunks and branches dark reddish brown, close, in maturity reddish white to reddish tan or bronze, exfoliating in paper-thin sheets; lenticels dark, horizontally expanded. **Twigs** glabrous to sparsely pubescent, often covered with conspicuous, warty, resinous glands. **Leaf blade** narrowly ovate to ovate with 9–12 pairs of lateral veins, 6–10(–14) × 4–8 cm, base usually cordate, rarely rounded, margins coarsely or irregularly doubly serrate, apex short-acuminate, abaxially sparsely to moderately pubescent, sometimes velutinous or tomentose along major veins and in vein axils, covered with minute, resinous glands. **Infructescences** pendulous or nearly pendulous, cylindric, 2.5–5.5 × 0.6–1 cm, shattering with fruits in fall; scales glabrous to moderately pubescent, lobes diverging proximal to middle, central lobe elongate, obtuse, lateral lobes ascending, shorter and slightly broader than central lobe. **Samaras** with wings broader than body, broadest near summit, extended beyond body apically. $2n = 28, 56$.

Flowering late spring. Moist, rocky slopes or rich, open forest; 800–2000 m; St. Pierre and Miquelon; N.B., Nfld., N.S., Ont., P.E.I., Que.; Maine, Mass., Mich., Minn., N.H., N.Y., N.C., Pa., Vt., Va., W.Va., Wis.

Betula cordifolia has been reported from Connecticut; I have not seen specimens.

In recent years *Betula cordifolia* has usually been treated as a variety of *B. papyrifera,* and perhaps it should be considered an ecological race of that species. It differs from *B. papyrifera* in polyploid level (diploid and tetraploid in *B. cordifolia* versus tetraploid, pentaploid, and hexaploid in *B. papyrifera*) and in vegetative characters, including the number of lateral veins of leaves and the color of bark (W. H. Brittain and W. F. Grant 1967; P. E. DeHond and C. S. Campbell 1989). In the Adirondacks, *B. cordifolia* and *B. papyrifera* occur in rather distinct ecological zones (*B. cordifolia* mostly above 800 m and *B. papyrifera* generally below this elevation). The species does not appear to occur as far west (Iowa) as stated by M. L. Fernald (1950).

8. **Betula minor** (Tuckerman) Fernald, Rhodora 47: 306. 1945 · Dwarf birch, bouleau mineur E

Betula papyracea Aiton var. *minor* Tuckerman, Amer. J. Sci. Arts, 45: 31. 1843; *B. pubescens* Ehrhart subsp. *minor* (Tuckerman) Á. Löve & D. Löve; *B. saxophila* Lepage

Shrubs, erect, irregularly spreading, or depressed, to 5 m. **Bark** dark, reddish brown, smooth, close, not readily exfoliating; lenticels pale, horizontally expanded. **Twigs** without odor and taste of wintergreen, glabrous to sparsely pubescent, often dotted with resinous glands. **Leaf blade** ovate with 2–6 pairs of lateral veins, 1.5–5.5 (–8) × 1.5–3(–5) cm, base rounded or cuneate to truncate, margins coarsely doubly serrate, teeth obtuse to rather sharp, toothed nearly to base, apex acute to obtuse; surfaces abaxially glabrous to moderately pubescent, usually more densely pubescent along major veins, often covered with small resinous glands. **Infructescences** erect, cylindric 1–3 × 0.5–1 cm, shattering with fruits in fall; scales glabrous to moderately pubescent, lobes diverging at middle, central lobe elongate, apex obtuse, lateral lobes ascending, as long as to nearly shorter and broader than central lobe. **Samaras** with wings equal to or broader than body, broadest near summit, extending beyond body apically. $2n = 56$.

Flowering late spring. Rocky slopes, barrens, and subalpine summits; (0–)1000–2000 m; N.B., Nfld., N.S., Ont., Que.; Maine, N.H., N.Y.

The origin and relationships of this small birch have not been adequately determined. *Betula minor* resembles *B. pubescens* (as *B. odorata* Bechstein) of Greenland and northern Europe (M. L. Fernald 1950), and it has been combined into that species (Á. Löve and D. Löve 1966). Northern and maritime populations of the complex have often been segregated as a separate species (*B. borealis* Spach sensu M. L. Fernald 1950; *B. saxophila* of E. Lepage 1976); the name *B. minor* has been mostly restricted to the subalpine form of northern Appalachian peaks. These two taxa actually constitute a single, somewhat variable, morphologic entity; they are indistinguishable by the minor character differences that have been used to separate them in the past. Because Spach's type of *B. borealis* consists of material of *B. pumila* (B. Boivin 1967b), that name must be rejected for this species.

Further complicating matters, E. Lepage (1976) concluded that the type of *Betula minor* represents a hybrid between individuals of the dwarf species and *B. papyrifera,* and on that basis, following nomenclatural rules, he renamed the dwarf species *B. saxophila,* retaining the name *B. minor* for the hybrid. Leaf shapes and other visible characters of the type fall easily within the limits of variation of *B. saxophila,* however, and the group is considered here to consist of a single entity, designated by the older name *B. minor.*

At least in the Adirondacks, *Betula minor* usually occurs near populations of *B. cordifolia* and *B. glandulosa,* and it has frequently been suggested (e.g., E. Hultén 1968; E. Lepage 1976; J. J. Furlow 1990) that it may have originated through hybridization between these species (perhaps followed by polyploidy). The northern populations may similarly consist of a hybrid swarm involving *B. papyrifera* or *B. cordifolia* and *B. glandulosa.* Critical examination of the entire complex, including experimental studies of the patterns of hybridization present, are necessary to unravel its problems satisfactorily.

9. **Betula pubescens** Ehrhart, Beitr. Naturk. 5: 160. 1790

Betula alba Linnaeus var. *pubescens* (Ehrhart) Spach

Trees and shrubs; trunks 1–many. **Bark** when young dark reddish brown, in maturity light reddish brown to tan or brownish or grayish white, smooth, rather close or readily exfoliating in paper-thin sheets; lenticels pale, horizontal, in maturity dark, horizontally expanded. **Twigs** without taste and odor of wintergreen, usually covered with short bristly hairs. **Leaf blade** ovate or rhombic-ovate, margins serrate, apex acute; surfaces abaxially sparsely pubescent to velutinous, especially along major veins and in vein axils, without prominent resinous glands. **Fruiting catkins** pendulous or subpendulous, cylindric, shattering with fruits in late fall; scales puberulent to glabrous, often ciliate, lobes diverging at middle. **Samaras** with wings equal to or somewhat broader than body, broadest near summit, extended beyond body apically.

Subspecies 3 (2 in the flora): Greenland; introduced elsewhere in North America; Iceland; Eurasia.

Betula pubescens was used medicinally by the Cree for chafed skin, and by the Ojibwa as a seasoner in medicines and a component in a maple syrup mixture used to relieve stomach cramps (D. E. Moerman 1986, as *B. alba*).

Betula alba Linnaeus is a long-standing nomen ambiguum that had not been in use (until recently) because it included two taxa whose names had been widely adopted long ago. At this time a proposal to reject *Betula alba* is in press, and possibly a decision will be made before the end of 1996 (R. Brummitt, pers. comm.; Fred Barrie, pers. comm.)

1. Leaf blade 3–4(–6) cm; twigs usually without conspicuous resinous glands; wing of samara 1–1.5 times as wide as body; trees usually with single trunk, persisting or escaped from cultivation . . . . .9a. *Betula pubescens* subsp. *pubescens*
1. Leaf blade 1–2.5(–3.5) cm; twigs ± glandular; wing of samara about as wide as body; native shrubs of sw Greenland . . . . .9b. *Betula pubescens* subsp. *tortuosa*

9a. **Betula pubescens** Ehrhart subsp. **pubescens**
· European white birch, downy birch

Trees, narrow, to 20 m; trunk usually 1, branches ascending or spreading. **Twigs** usually without conspicuous resinous glands. **Leaf blade** broadly ovate to rhombic-ovate, 3–4(–6) × 2–4 (–6) cm, base rounded, truncate, or cuneate, margins finely to coarsely toothed or dentate, apex acute; surfaces abaxially moderately pubescent to velutinous, especially along major veins and in vein axils. **Fruiting catkins** 2–3 × 0.8–1.2 cm; scales puberulent and ciliate, lobes diverging at middle, central lobes ovate to deltate, apex acute to obtuse, lateral lobes divergent, about equal in length, several times broader. **Samaras** with wings somewhat broader than body, broadest near summit, extended beyond body apically. $2n = 56$.

Flowering late spring. Rocky slopes, scrubs, heaths, and open woods where native, elsewhere in abandoned plantings, moist open roadsides, swales, swampy thickets; 0–200 m; introduced; B.C.; Conn., Ind., Maine, Mass., Ohio, Pa., Vt.; native to Europe.

Betula pubescens subsp. *pubescens* is commonly cultivated in northeastern North America, where it has sometimes escaped and persisted, or become adventive but not widely naturalized. It is distinguished from other light-barked species in the East by its relatively small leaves, pubescent twigs, and brownish, mostly unpeeling bark.

9b. Betula pubescens Ehrhart subsp. **tortuosa** (Ledebour) Nyman, Consp. Fl. Eur., 672. 1881

Betula tortuosa Ledebour, Fl. Ross. 3: 652. 1851; *B. odorata* Bechstein

Shrubs, 1–12 m; trunks usually many, often interlacing, branches ascending. **Twigs** with numerous small resinous glands. **Leaf blade** ovate, rhombic-ovate, or suborbiculate-rhombic, 1–2.5(–3.5) × 1–2(–3) cm, base cuneate to truncate, margins coarsely serrate or dentate, apex acute; surfaces abaxially moderately pubescent. **Infructescences** 1–1.5(–2) × 0.4–0.8 cm; scales pubescent to glabrous, often ciliate, central lobe oblong or narrowly triangular, apex acute to obtuse, lateral lobes divergent and ascending, about equal in length but somewhat broader. **Samaras** with wings about equal in diameter to body, broadest near summit, usually extended beyond body apically. $2n = 56$.

Flowering late spring. Protected inland valleys; 0–200 m; native, Greenland; Europe.

Betula pubescens subsp. *tortuosa,* which also occurs in Iceland, the mountains of Scandinavia, and arctic Europe, is native to southwestern Greenland; it has also been reported as well from Baffin Island, although I have not seen specimens. Based on pollen analyses (B. Fredskild 1991), this subspecies arrived in Greenland from Europe ca. 3500 years ago. Hybridization of this taxon with other birches in Greenland, Iceland, and Scandinavia has been studied by M. Sulkinoja (1990).

10. Betula kenaica W. H. Evans, Bot. Gaz. 27: 481. 1899 · Kenai birch [E]

Betula kamtschatica (Regel) V. N. Vassiljev var. *kenaica* (W. H. Evans) C.-A. Jansson; *B. neoalaskana* Sargent var. *kenaica* (W. H. Evans) B. Boivin; *B. papyrifera* Marshall var. *kenaica* (W. H. Evans) A. Henry

Trees, to 12 m; crowns narrow. **Bark** dark reddish brown, sometimes becoming pinkish or grayish white, smooth, in maturity exfoliating in thin sheets; lenticels dark, horizontally expanded. **Twigs** without taste and odor of wintergreen, slightly to moderately pubescent, often with scattered resinous glands. **Leaf blade** ovate to nearly deltate with 2–6 pairs of lateral veins, 4–5(–7.5) × 2.5–4.5 cm, base rounded to cuneate, margins coarsely doubly serrate to dentate, teeth relatively sharp, apex acute to short-acuminate; surfaces abaxially sparsely to moderately pubescent, especially along major veins and in vein axils, often with scattered resinous glands. **Infructes-**cences erect to nearly pendulous, cylindric, 2–5 × 0.5–1 cm, shattering with fruits in fall; scales ciliate, lobes diverging at middle, nearly equal in length, strongly divergent. **Samaras** with wings as broad as to somewhat narrower than body, broadest near middle, not extended beyond body apically. $2n = 70$.

Flowering late spring. Rocky slopes in the subalpine zone; 0–300 m; Yukon; Alaska.

The relationship of *Betula kenaica* to other white-barked birches is not well understood, although it and the following species are evidently closely allied to *B. papyrifera,* from which they have likely been derived. *Betula kenaica* differs from *B. papyrifera* primarily in its smaller stature and in its smaller, blunter-tipped, more coarsely and regularly serrate leaves.

Betula ×*hornei* Butler (= *Betula kenaica* W. H. Evans × *B. nana* Linnaeus), variously intermediate between its parents, is common throughout the range of *B. kenaica* (which is mostly overlapped by that of *B. nana*).

11. Betula occidentalis Hooker, Fl. Bor.-Amer. 2: 155. 1838 · Water birch, river birch [E] [F]

Betula fontinalis Sargent; *B. fontinalis* var. *inopina* (Jepson) Jepson; *B. microphylla* Bunge var. *fontinalis* (Sargent) M. E. Jones; *B. occidentalis* var. *fecunda* (Britton) Fernald; *B. occidentalis* var. *inopina* (Jepson) C. L. Hitchcock; *B. papyrifera* Marshall var. *occidentalis* (Hooker) Sargent; *B. papyrifera* subsp. *occidentalis* (Hooker) Hultén

Shrubs, spreading, to 10 m. **Bark** dark reddish brown to bronze, smooth, close, not readily exfoliating; lenticels pale, horizontally expanded. **Twigs** without the odor or taste of wintergreen, glabrous to sparsely pubescent, covered with conspicuous, reddish, resinous glands. **Leaf blade** broadly ovate to rhombic-ovate with 2–6 pairs of lateral veins, 2–5.8 × 1–4.5 cm, base truncate to rounded or cuneate, margins sharply and coarsely serrate or irregularly doubly serrate, teeth mostly long and sharp, basal portion untoothed, apex acute to occasionally short-acuminate; surfaces abaxially sparsely to moderately pubescent, covered with minute, resinous glands. **Infructescences** erect to nearly pendulous, cylindric, 2–3(–3.9) × 0.8–1.5 cm, shattering with fruits in fall; scales glabrous, ciliate, lobes diverging at middle, central lobe narrower and longer than ascending lateral lobes. **Samaras** with wings broader than body, broadest near summit, extended beyond body apically. $2n = 28$.

Flowering late spring. Montane stream banks, slopes, and ridges, also in moist open woods, at edges of marshes, along lakeshores, and in wet swales; 100–

3000 m; Alta., B.C., Man., N.W.T., Ont., Sask., Yukon; Alaska, Ariz., Calif., Colo., Idaho, Mont., Nebr., Nev., N.Mex., N.Dak., Oreg., S.Dak., Utah, Wash., Wyo.

Betula occidentalis is a common, streamside, shrubby birch throughout much of the Rocky Mountains, extending eastward to northwestern Ontario. It has been widely known by the later name *B. fontinalis* because of questions concerning the legitimacy of Hooker's epithet (J. R. Dugle 1966). Recent changes to the *International Code of Botanical Nomenclature* (W. Greuter et al. 1994) have clarified the situation, however, and the consensus now is that the earlier name is correct. E. Hultén (1968) believed that the species in Alaska that has been called *B. occidentalis* consists of an extensive hybrid swarm between *B. neoalaskana* (as *B. resinifera*) and *B. glandulosa*. The studies of J. R. Dugle (1966) do not support a hybrid origin of *B. occidentalis* in other parts of its range. Additional study will be needed to resolve this problem, both in Alaska and southward.

Betula ×*utahensis* Britton (= *B. occidentalis* Hooker × *B. papyrifera* Marshall) is a common hybrid marked by intermediate characteristics.

Betula papyrifera Marshall var. *subcordata* (Rydberg) Sargent, formerly recognized in several state, provincial, and regional floras, consists of introgressants of *B. occidentalis* into *B. papyrifera* (J. R. Dugle 1966).

12. **Betula populifolia** Marshall, Arbust. Amer., 19. 1785 · Gray birch, white birch, fire birch, bouleau à feuilles de peuplier, bouleau gris [E] [F] [W1]

Betula alba Linnaeus subsp. *populifolia* (Marshall) Regel; *B. alba* var. *populifolia* (Marshall) Spach

Trees, broadly pyramidal, to 10 m; trunks usually several. **Bark** when young dark reddish brown, in maturity becoming grayish white, smooth, close; lenticels dark, horizontally expanded. **Twigs** without taste and odor of wintergreen, glabrous to sparsely pubescent, dotted with small, inconspicuous, resinous glands. **Leaf blade** broadly ovate to deltate or rhombic with 5–18 pairs of lateral veins, 3–10 × 3–8 cm, base truncate to cuneate, marginally coarsely, irregularly, or sometimes obscurely doubly serrate, apex abruptly long-acuminate; surfaces abaxially glabrous or sparsely pubescent, often covered with minute, resinous glands. **Infructescences** erect to nearly pendulous, nearly cylindric, 1–2.5(–3) × 0.8–1 cm, shattering with fruits in early fall; scales adaxially densely pubescent, lobes diverging distal to middle, central lobe cuneate, acute, much shorter than lateral lobes, lateral lobes divergent, broad, irregularly

angular. **Samaras** with wings much broader than body, broadest near middle, often extended beyond body both apically and basally. $2n = 28$.

Flowering late spring. Rocky or sandy open woods, moist to dryish slopes, old fields, and waste places; 100–600 m; N.B., N.S., Ont., P.E.I., Que.; Conn., Del., Ill., Ind., Maine, Md., Mass., N.H., N.J., N.Y., N.C., Ohio, Pa., R.I., S.C., Vt., Va.

Betula populifolia is an important successional tree on burned, cleared, or abandoned land in the Northeast. It is closely related to *Betula pendula* Roth of Europe, *B. neoalaskana* of the Northwest, and several Asian taxa. This species is easily distinguished from the paper birch, with which it is often sympatric, by the long tapering apices of its leaves, its nonpeeling bark, and the characteristic expanded, black triangular patches on the trunks below the branches.

The Iroquois used *Betula populifolia* medicinally to treat bleeding piles; the Micmac, to treat infected cuts and as an emetic (D. E. Moerman 1986).

The blue birches (*Betula* ×*caerulea* Blanchard) have been variously considered to represent a true species or a hybrid between *B. papyrifera* Marshall and *B. populifolia* Marshall (T. C. Brayshaw 1966) or *B. papyrifera* and the big blue birch *B. caerulea-grandis* (M. L. Fernald 1922). Both *B.* ×*caerulea* and *B. caerulea-grandis* have been shown in more recent experimental studies to be of hybrid origin between *B. cordifolia* Regel and *B. populifolia* (A. G. Guerriero et al. 1970; W. F. Grant and B. K. Thompson 1975; P. E. DeHond and C. S. Campbell 1989). Individuals of these hybrids combine characteristics of the parents, the infructescence scales and leaves somewhat resembling those of *B. populifolia*, and the habit and exfoliating reddish bark that of *B. cordifolia*.

SELECTED REFERENCE   Catling, P. M. and K. W. Spicer. 1988. The separation of *Betula populifolia* and *Betula pendula* and their status in Ontario. Canad. J. Forest Res. 18: 1017–1026.

13. **Betula pendula** Roth, Tent. Fl. Germ. 1: 405. 1788 · Weeping birch, European white birch, silver birch, bouleau pleureur

Betula verrucosa Ehrhart

Trees, to 25 m; trunks usually several, crowns spreading. **Bark** of mature trunks and branches creamy to silvery white, smooth, exfoliating as long strands; lenticels dark, horizontally expanded. **Branches** pendulous; twigs glabrous, usually dotted with small resinous glands. **Leaf blade** broadly ovate to rhombic with 5–18 pairs of lateral veins, 3–7 × 2.5–5 cm, base cuneate, rarely truncate, margins coarsely and sharply doubly serrate, apex acuminate;

surfaces abaxially glabrous to sparsely pubescent, covered with minute, resinous glands. **Infructescences** erect to nearly pendulous, cylindric, 2–3.5 × 0.6–1 cm, shattering with fruits in fall; scales adaxially sparsely pubescent, lobes diverging at middle, central lobe obtuse, much shorter than lateral lobes, lateral lobes broad, rounded, extended. **Samaras** with wings much broader than body, broadest near center, extended beyond body apically. $2n = 28, 56$.

Flowering late spring. Abandoned plantings, roadsides, edges of bogs, waste places; 0–350 m; B.C., Man., Ont.; Conn., Mass., N.H., N.Y., Ohio, Pa., Vt., Wash.; Europe; Asia.

The Eurasian weeping birch (*Betula pendula*) is extensively cultivated throughout the temperate range of the flora, and it has been known to persist or to become locally naturalized in several areas, particularly in the Northeast. In vegetative features it resembles *B. populifolia* Marshall, to which it is closely allied; it can easily be distinguished from the latter by its peeling bark, as well as by its mostly pubescent leaves with somewhat shorter, acuminate apices.

14. Betula neoalaskana Sargent, J. Arnold Arbor. 3: 206. 1922 · Resin birch, paper birch [E]

Betula alaskana Sargent, Bot. Gaz. 31: 236. 1901, not Lesquereux 1883; *B. papyrifera* Marshall subsp. *humilis* (Regel) Hultén; *B. papyrifera* var. *humilis* (Regel) Fernald & Raup; *B. papyrifera* var. *neoalaskana* (Sargent) Raup; *B. resinifera* (Regel) Britton

Trees, to 25 m, crowns narrow. **Bark** of young trunks and branches reddish brown, when mature becoming pinkish white to light red (starkly white in interior Alaska; D. F. Murray, pers. comm.), smooth, exfoliating in thin sheets. **Twigs** glabrous, covered with conspicuous resinous glands. **Leaf blade** deltate-ovate with 5–18 pairs of lateral veins, 3–8 × 2–6 cm, base rounded or broadly cuneate, margins coarsely doubly serrate, apex long-acuminate; surfaces abaxially glabrous to sparsely pubescent, pubescent along major veins and in vein axils, covered with small resinous glands. **Infructescences** pendulous, cylindric, 2–4 × 0.8–1.2 cm, shattering with fruits in fall; scales glabrous, margins ciliate, lobes diverging distal to middle, central lobe narrower and equal to or slightly shorter than lateral lobes, lateral lobes broadly angular, extended. **Samaras** with wings broader than body, broadest near summit, extended beyond body apically. $2n = 28$.

Flowering late spring. Rocky or peaty slopes, bog margins, sandhills, open woods; 100–1200 m; Alta., B.C., Man., N.W.T., Ont., Sask., Yukon; Alaska.

Betula neoalaskana belongs to a circumpolar complex including *B. pendula* Roth and *B. populifolia* Marshall (but not *B. papyrifera* Marshall, with which it has sometimes erroneously been merged). It is most closely related to the Asian members of this group, including *B. japonica* Siebold, *B. mandshurica* (Regel) Nakai, and *B. platyphylla* Sukaczev (T. C. Brayshaw 1976). The species was formerly widely known by the name *B. resinifera* (Regel) Britton, but that name has been shown to be illegitimate (B. Boivin 1967–1979, 15: 414–418; J. R. Dugle 1969). The name was based on a mixture of Siberian and North American material and has never been lectotypified.

Betula neoalaskana Sargent is known to hybridize with *B. papyrifera* Marshall, producing *B.* ×*winteri* Dugle, and with *B. glandulosa,* producing *B.* ×*uliginosa* Dugle.

15. Betula pumila Linnaeus, Mant. Pl., 124. 1767 · Bog birch, dwarf birch, bouleau nain [E]

Betula borealis Spach; *B. glandulifera* (Regel) B. T. Butler; *B. glandulosa* Michaux var. *glandulifera* (Regel) Gleason; *B. glandulosa* var. *hallii* (Howell) C. L. Hitchcock; *B. hallii* Howell; *B. nana* Linnaeus var. *glandulifera* (Regel) B. Boivin; *B. pubescens* Ehrhart subsp. *borealis* (Spach) Á. Löve & D. Löve; *B. pumila* Linnaeus var. *glabra* Regel; *B. pumila* var. *glandulifera* Regel; *B. pumila* var. *renifolia* Fernald

Shrubs, coarse, irregular, or spreading, to 4 m. **Bark** dark reddish brown, smooth, close; lenticels pale, inconspicuous. **Twigs** without taste and odor of wintergreen, glabrous to moderately pubescent, with scattered small resinous glands, especially near nodes. **Leaf blade** elliptic, obovate, or nearly orbiculate (to sometimes reniform) with 2–6 pairs of lateral veins, 2.5–5 (–7) × 1–5 cm, base cuneate to rounded, margins crenate to dentate, apex usually broadly acute or obtuse to rounded; surfaces abaxially glabrous or slightly pubescent to heavily velutinous or tomentose, often with scattered resinous glands. **Infructescences** erect, cylindric, 0.8–1.5(–2) × 0.8–1 cm, shattering with fruits in fall; scales glabrous to pubescent, lobes diverging slightly distal to middle, central lobe narrow, elongate, lateral lobes shorter and broader, extended. **Samaras** with wings slightly narrower than body, broadest near center, not extended beyond body apically. $2n = 56$.

Flowering late spring. Bogs, calcareous fens, wooded swamps, muskegs, lake shores; 0–1000 m; St. Pierre and Miquelon; Alta., B.C., Man., N.B., Nfld., N.W.T., N.S., Ont., P.E.I., Que., Sask., Yukon; Calif., Colo., Conn., Idaho, Ill., Ind., Iowa, Kans., Maine, Mass., Mich., Minn., Mont., Nebr., N.J., N.Y., N.Dak., Ohio, Oreg., Pa., S.Dak., Vt., Wash., Wis., Wyo.

Betula pumila is sometimes treated (in part) as a variety of *B. glandulosa* Michaux, to which it is related

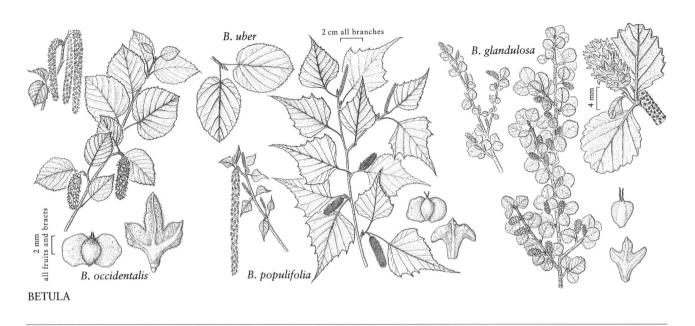

BETULA

at a subgeneric or sectional level. On the basis of morphology, however, it forms a cohesive and distinct entity (J. J. Furlow 1984). The two main varieties into which *B. pumila* is often divided (a more southern *B. pumila* var. *pumila,* with mostly pubescent, glandless leaves, and a more northern *B. pumila* var. *glandulifera,* with less pubescent, gland-bearing leaves) may represent geographic races; these are not well marked, however, and they do not hold up well when the complex is examined as a whole.

The Ojibwa used *Betula pumila* medicinally as a gynecological aid and as a respiratory aid (D. E. Moerman 1986).

16. Betula glandulosa Michaux, Fl. Bor.-Amer. 2: 180. 1803 · Dwarf birch, resin birch, bouleau glanduleux E F

Shrubs, spreading or ascending, to 3 m. **Bark** dark brown, smooth, close; lenticels pale, inconspicuous, unexpanded. **Twigs** without taste or odor of wintergreen, essentially glabrous to sparsely pubescent, usually conspicuously covered with large, warty, resinous glands. **Leaf blade** mostly obovate to nearly orbiculate with 2–6 pairs of lateral veins, 0.5–3 × 1–2.5 cm, base cuneate to rounded, margins dentate-crenate, teeth obtuse to rounded, apex obtuse to rounded; surfaces abaxially glabrous to moderately pubescent, especially along major veins and in vein axils, often covered with resinous glands. **Infructescences** erect, cylindric, 1–2.5 × 0.5–

1.2 cm, shattering with fruits in fall; scales glabrous, lobes diverging distal to middle, central lobe elongate, lateral lobes ascending, somewhat shorter and broader than central lobe. **Samaras** with wings narrower than body, broadest near summit, extended slightly beyond body apically. $2n = 28$.

Flowering late spring. Arctic and alpine tundra, acidic rocky slopes and barrens, muskegs, peat bogs, stream banks, open subalpine summits; 0–3400 m; Greenland; Alta., B.C., Man., N.B., Nfld., N.W.T., N.S., Ont., P.E.I., Que., Sask., Yukon; Alaska, Calif., Colo., Idaho, Maine, Mont., N.H., N.Y., Oreg., S.Dak., Utah, Wash., Wyo.

Betula glandulosa is the characteristic dwarf birch of upland habitats throughout much of the mountainous west, occurring as well in dry open areas across the north. Where their ranges meet, *B. glandulosa* intergrades with both *B. pumila* Linnaeus and *B. nana* Linnaeus subsp. *exilis* (Sukaczev) Hultén, creating a confusing complex of intermediate forms. In the east, it reaches its southernmost limit on the subalpine slopes of high Adirondack peaks, including Mt. Washington, where it forms low sprawling thickets and scrubs.

Specimens of *Betula glandulosa* have been reported from the St. Lawrence Valley, but I have not seen them.

Wherever *Betula glandulosa* comes in contact with *B. pumila,* it forms a bewildering swarm of plants, known as *B.* ×*sargentii* Dugle, having intermediate states of most vegetative characters.

Plants intermediate between *Betula glandulosa* and *B. nana* subsp. *exilis* make up a continuum of forms linking the typical forms of *Betula nana* and *B. glandu-*

losa in parts of Alaska where the ranges of these species overlap. Wherever they occur in isolation, the species remain reasonably distinct and easy to identify. In southern Greenland, *Betula glandulosa* hybridizes with *B. nana* subsp. *nana* and with *B. pubescens.*

Betula ×*eastwoodiae* Sargent (= *B. glandulosa* × *occidentalis*) occurs in montane meadows and marshes in Alberta, British Columbia, Northwest Territories, Saskatchewan, Yukon, Alaska, Colorado, and Wyoming, where the range of the parents overlap.

Betula ×*dugleana* Lepage (= *Betula glandulosa* Michaux × *B. neoalaskana* Sargent) is common throughout Alaska and the Yukon, where the parent species frequently come into contact (E. Hultén 1941–1950, vol. 4; E. Lepage 1976).

Betula ×*dutillyi* Lepage [= *Betula glandulosa* Michaux × *B. minor* (Tuckerman) Fernald] is a putative hybrid that occupies the same general range as *Betula minor*. Like that species, however, it has not been studied experimentally. Careful examination of the entire complex to which this taxon belongs will be necessary before any of its parts can be truly understood. *Betula* ×*dutillyi* exhibits many of the same characteristics as *B. minor*, but it is slightly smaller in habit, and its leaves are smaller with somewhat blunter tips and more cuneate bases (E. Lepage 1976).

17. Betula nana Linnaeus, Sp. Pl. 2: 983. 1753 · Arctic dwarf birch, bouleau nain

Shrubs, sprawling, creeping, or upright, to 1 m. **Bark** gray to dark brown, smooth, close; lenticels inconspicuous, unexpanded. **Twigs** without taste and odor of wintergreen, glabrous to sparsely or moderately pubescent, with or without heavy resinous coating, sometimes covered with warty resinous glands. **Leaf blade** broadly orbiculate or obovate-orbiculate to reniform, with 2–6 pairs of lateral veins, often broader than long, base rounded to nearly cordate, margins deeply crenate, apex rounded; surfaces abaxially glabrous to sparsely or moderately pubescent. **Staminate and pistillate catkins** produced season before flowering but retained in buds during winter, expanding along with new growth in spring. **Infructescences** erect, nearly cylindric, shattering with fruits in fall. **Samaras** with wings much narrower than body, broadest near center, not extended beyond body apically.

Subspecies 3 (2 in the flora): subarctic and arctic of North America, Europe, and Asia.

1. Young twigs pubescent, not covered with conspicuous resinous layer; subarctic and arctic ne Canada, s Greenland. . . . . 17a. *Betula nana* subsp. *nana*
1. Young twigs glabrous or only puberulent, covered with thick resinous coating; Alaska, Yukon, n Asia . . . . .17b. *Betula nana* subsp. *exilis*

17a. Betula nana Linnaeus subsp. **nana** · Arctic dwarf birch

Shrubs, procumbent, widely spreading, or upright, to 1 m. **Bark** dark brown. **Twigs** sparsely to densely pubescent, dull, without heavy resinous coating, without large, warty, resinous glands. **Leaf blade** broadly orbiculate or obovate-orbiculate to reniform, 0.8–1.5 × 0.8–2 cm, base rounded or cuneate to nearly cordate; surfaces abaxially sparsely pubescent to glabrous. **Infructescences** 0.5–1.5 × 0.4–1 cm; scales glabrous, lobes diverging distal to middle, ascending, about equal in length. $2n = 28$.

Flowering late spring. Tundra, open rocky barrens, and scrubs, in Europe subalpine mountain moors; 0–600 m; Greenland; N.W.T.; n Europe.

Betula nana subsp. *nana* occurs in both eastern and western Greenland at latitudes north of 63 degrees. In Europe, the range of this subspecies extends across the subarctic zone and southward in the Alps and other ranges.

17b. Betula nana Linnaeus subsp. **exilis** (Sukaczev) Hultén, Fl. Alaska Yukon 4: 579. 1944 · Arctic dwarf birch

Betula exilis Sukaczev, Trudy Bot. Muz. Imp. Akad. Nauk 8: 213. 1911; *B. glandulosa* Michaux var. *sibirica* (Ledebour) C. K. Schneider

Shrubs, spreading, depressed, to 1 m. **Bark** reddish brown. **Twigs** glabrous to very sparsely pubescent, conspicuously resinous, with relatively few small, resinous glands. **Leaf blade** broadly orbiculate to reniform, 0.5–1.2 × 0.5–1.6 cm, base rounded to nearly cordate or cuneate; surfaces abaxially glabrous, occasionally with only a few small, resinous glands. **Infructescences** 0.5–1.2 × 0.5–0.7 cm; scales glabrous, lobes diverging distal to middle, held nearly parallel, center lobe ascending, slightly longer than lateral lobes. $2n = 28$.

Flowering late spring. Wet tundra and margins of bogs, terraces and open slopes; 0–2400 m; Alta., B.C., Man., N.W.T., Sask., Yukon; Alaska; n Asia.

The range of *Betula nana* subsp. *exilis* stretches through Siberia and across northwestern North America. It is sometimes confused with *Betula glandulosa* Michaux; it can be distinguished from that species by its usually much smaller, rounded to truncate-tipped leaves that have teeth all the way to the base (the base being mostly toothless in *B. glandulosa*). *Betula nana* subsp. *exilis* has been combined into *B.*

glandulosa by several authors (E. H. Moss and J. G. Packer 1983; A. E. Porsild and W. J. Cody 1980). The two taxa hybridize and form a sometimes bewildering assortment of intermediates in the extensive region where their ranges overlap; they constitute well-defined entities away from the area of contact. The current treatment follows the view of E. Hultén (1968) that two distinct species of dwarf arctic birches inhabit northern North America. Little is understood, however, about the evolutionary or phytogeographic history of this complex, and many problems remain to be unraveled before a truly satisfactory classification can be attained.

Western Eskimos used decoctions prepared from the leaves of *Betula nana* subsp. *exilis* to alleviate intestinal discomfort (D. E. Moerman 1986).

18. **Betula michauxii** Spach, Ann. Sci. Nat., Bot., sér. 2, 15: 195. 1841 · Newfoundland dwarf birch, Michaux's birch, bouleau de Michaux E

Betula terra-novae Fernald

Shrubs, spreading, dwarfed, to ca. 0.5 m. **Bark** dark brown, smooth, close; lenticels pale, inconspicuous, circular. **Twigs** without taste and odor of wintergreen, moderately to densely pubescent, not conspicuously resin-coated, without large, warty, resinous glands. **Leaf blade** obovate-reniform, with 2–3 pairs of lateral veins, 0.5–1 × 0.5–1.2 cm, base cuneate, margins deeply crenate-dentate, apex broadly rounded to nearly truncate; surfaces abaxially usually glabrous. **Infructescences** erect, short-cylindric, 0.5–1 × 0.5–0.8 cm, shattering with fruits in fall; scales unlobed (lateral lobes sometimes present but greatly reduced), glabrous. **Samaras** with wings not apparent or reduced to narrow ridges.

Flowering late spring. Sphagnum bogs, around pools, and wet peaty meadows; 0–700 m; St. Pierre and Miquelon; Nfld., N.S., Que.

This infrequent dwarf birch is distinguished from *Betula nana* mostly on the basis of its reduced infructescence scales and wetter habitat (J. J. Furlow 1984), characteristics that are also occasionally noted in *B. nana*. It perhaps might better be treated as a race of that species; in the absence of thorough study of this complex, however, it seems best to follow the traditional treatment (M. L. Fernald 1950c; J. Rousseau and M. Raymond 1950).

31b. BETULACEAE Gray subfam. CORYLOIDEAE Koehne, Deut. Dendrol., 106, 116. 1893

Trunk and branches terete. **Bark** thin, close or becoming furrowed or broken into plates; lenticels not conspicuous. **Bark and wood** tanniferous. **Young twigs and buds** usually without prominent, large, resinous glands; pith circular to remotely triangular in cross section. **Leaves** 2-ranked. **Staminate flowers:** perianth absent. **Pistillate flowers** 2 per bract; perianth adnate to ovary, often visible as membranaceous fringe at summit; ovules with 2 integuments. **Infructescences** usually longer than 4 cm, consisting of relatively uncrowded clusters with large, nearly foliaceous bracts; bracts deciduous with fruits. **Fruits** tiny to moderately large nuts, not winged; pericarp thick and bony.

Genera 4, species ca. 45 (3 genera, 7 species in the flora): primarily boreal and cool temperate zones of Northern Hemisphere; in mountains of Mexico and Central America, not extending into South America.

3. CARPINUS Linnaeus, Sp. Pl. 2: 998. 1753; Gen. Pl. ed. 5, 432. 1754 · Hornbeam

[Latin *carpinus,* hornbeam, possibly from *carpentum,* a Roman horse-drawn vehicle with wheels made from its hard wood]

Trees, 8–25 m; trunks usually 1, branching mostly deliquescent, trunk and branches irregularly longitudinally ridged, fluted. **Bark** of trunk and branches bluish to brownish gray, thin, smooth, close [thicker, broken or shredded]; lenticels generally inconspicuous. **Wood** nearly white to light brown, very hard and heavy, texture fine. **Branches, branchlets, and twigs** conspicuously 2-ranked; young twigs differentiated into long and short shoots. **Winter buds** sessile, ovoid, 4-angled in cross section, apex acute; scales many, imbricate, smooth. **Leaves** on long and short shoots, 2-ranked. **Leaf blade** narrowly ovate to ovate, elliptic, or obovate with 10 or more pairs of lateral veins, 3–12 × 3–6 cm, thin, margins doubly serrate to serrulate; surfaces abaxially glabrous to tomentose, sometimes covered with small glands. **Inflorescences:** staminate catkins solitary or in small racemose clusters, lateral, formed previous growing season and enclosed [exposed] in buds during winter, expanding with leaves; pistillate catkins distal to staminate on short, leafy new growth, solitary, ± erect, elongate; bracts and flowers uncrowded. **Staminate flowers** in catkins 3 per scale, crowded together on pilose receptacle; stamens 3(–6), short; filaments often distinct part way to base; anthers divided into 2 parts, each 1-locular, apex pilose. **Pistillate flowers** 2 per bract. **Infructescences** loose racemose clusters of paired bracts, clusters pendulous, elongate; paired bracts deciduous with fruit, expanded, (1–)3-lobed, variously toothed, foliaceous, each bract subtending 1 fruit. **Fruits** small nutlets, deltoid, longitudinally ribbed, often crowned with persistent sepals and styles. $x = 8$.

Species ca. 25 (1 in the flora): mostly north temperate zone; mountainous regions of Mexico, Central America; Europe, Asia (s to India, Iran).

In the flora, *Carpinus* consists of a single native species, *C. caroliniana,* which is composed of two fairly distinctive geographic races (J. J. Furlow 1987, 1987b), treated here as subspecies. Worldwide it includes about 25 species, some of which become large trees. The European *C. betulus* is frequently planted in North America and persists long after other signs of human development have vanished. It seldom escapes, however, and it has not become naturalized. In the mountains of Mexico and Central America, the larger *C. tropicalis* (Donnell Smith) Lundell is widespread in the temperate forest zone.

Closely related to *Ostrya, Carpinus* is easily recognized by its smooth, gray, often fluted stems and racemose infructescences consisting of pairs of uncrowded, foliaceous, 3-lobed bracts, each subtending a small triangular nutlet. The staminate (but not the pistillate) catkins develop in the autumn and are enclosed within buds throughout the winter prior to anthesis (in *Ostrya,* these are exposed during the winter). The pistillate catkins are produced on the first new growth in the spring.

Of relatively minor economic importance, *Carpinus* has limited use for its very hard wood, especially in Europe, where it is used for making mallet heads, tool handles, levers, and other small, hard, wooden objects.

SELECTED REFERENCES Furlow, J. J. 1987. The *Carpinus caroliniana* complex in North America. I. A multivariate analysis of geographical variation. Syst. Bot. 12: 21–40. Furlow, J. J. 1987b. The *Carpinus caroliniana* complex in North America. II. Systematics. Syst. Bot. 12: 416–434. Winkler, H. 1914. Neue Revision der Gattung *Carpinus.* Bot. Jahrb. Syst. 15(suppl.): 488–508. Winstead, J. E., B. J. Smith, and G. I. Wardell. 1977. Fruit weight clines in populations of ash, ironwood, cherry, dogwood and maple. Castanea 42: 56–60.

1. Carpinus caroliniana Walter, Fl. Carol., 236. 1788

Carpinus americana Michaux

Trees, to 12 m; trunks short, often crooked, longitudinally or transversely fluted, crowns spreading. **Bark** gray, smooth to somewhat roughened. **Wood** whitish, extremely hard, heavy. **Winter buds** containing inflorescences squarish in cross section, somewhat divergent, 3–4 mm. **Leaf blade** ovate to elliptic, 3–12 × 3–6 cm, margins doubly serrate, teeth typically obtuse and evenly arranged, primary teeth often not much longer than secondary; surfaces abaxially slightly to moderately pubescent, especially on major veins, with or without conspicuous dark glands. **Inflorescences:** staminate inflorescences 2–6 cm; pistillate inflorescences 1–2.5 cm. **Infructescences** 2.5–12 cm; bracts relatively uncrowded, 2–3.5 × 1.4–2.8 cm, lobes narrow, elongate, apex nearly acute, obtuse, or rounded, central lobe (1–)2–3 cm.

Subspecies 2 (2 in the flora).

Carpinus caroliniana consists of two rather well-marked geographical races, treated here as subspecies. These hybridize or intergrade in a band extending from Long Island along the Atlantic coast through coastal Virginia and North Carolina, and then westward in northern South Carolina, Georgia, Alabama, Mississippi, and Louisiana. Plants with intermediate features are also found throughout the highlands of Missouri and Arkansas. J. J. Furlow (1987b) has described the variation of this complex in detail.

Native Americans used *Carpinus caroliniana* medicinally to treat flux, navel yellowness, cloudy urine, consumption, diarrhea, and constipation, as an astringent, a tonic, and a wash, and to facilitate childbirth (D. E. Moerman 1986; no subspecies specified).

1. Leaf blade narrowly ovate to oblong-ovate, 3–8.5(–12) cm, apex acute to obtuse; secondary teeth small and blunt; surfaces abaxially without small dark glands . . . .
. . . . 1a. *Carpinus caroliniana* subsp. *caroliniana*
1. Leaf blade ovate to elliptic, mostly 8–12 cm, apex usually abruptly narrowing, nearly caudate, sometimes long, gradually tapered, long-acuminate; secondary teeth often almost as large as primary teeth, sharp-tipped; surfaces abaxially covered with tiny, dark brown glands . . . .
. . . . 1b. *Carpinus caroliniana* subsp. *virginiana*

1a. Carpinus caroliniana Walter subsp. **caroliniana**
· American hornbeam, blue-beech [E]

Trees, to 8 m; trunks short, crooked, shallowly fluted, crowns open, spreading. **Bark** steel gray or brownish gray, smooth to somewhat roughened. **Winter buds** containing inflorescences squarish in cross section, 3–4 mm. **Leaf blade** narrowly ovate to oblong-ovate, 3–8.5 (–12) × 3–6 cm, base narrowly rounded to cordate, margins doubly serrate, apex acute or obtuse; surfaces abaxially slightly to moderately pubescent, especially on major veins, without small dark glands. **Infructescences** 2.5–7 cm; bracts relatively uncrowded, 2–3 × 1.4–2.3 cm, lobes narrow, elongate, apex acute, obtuse, or rounded.

Flowering late spring. Stream banks, flood plains, and moist slopes, frequently in understory; 0–200 m; Ala., Ark., Del., Fla., Ga., Ill., Ky., La., Md., Miss., Mo., N.J., N.C., Okla., S.C., Tenn., Tex., Va.

Carpinus caroliniana subsp. *caroliniana* is characteristic of the southern Atlantic and Gulf of Mexico coastal plains and also extends northward in the Mississippi Embayment. It differs from *C. caroliniana* subsp. *virginiana* in its smaller, more ovate, more acute-tipped leaves that have finer secondary teeth and lack conspicuous dark glands abaxially.

1b. Carpinus caroliniana Walter subsp. **virginiana** (Marshall) Furlow, Syst. Bot. 12: 429. 1987 · American hornbeam, blue beech, musclewood, ironwood, charme de Caroline, bois de fer [E] [F]

Carpinus betulus Linnaeus [var.] *virginiana* Marshall, Arbust. Amer., 25. 1785; *C. caroliniana* var. *virginiana* (Marshall) Fernald; *C. virginiana* (Marshall) Sudworth 1893, not Miller 1768

Trees, to 12 m; trunks short, crooked, shallowly to deeply and often irregularly fluted, crowns broadly spreading. **Bark** bluish gray, smooth to somewhat roughened. **Leaf blade** ovate or elliptic to narrowly elliptic, (6–)8–12 × 3.5–6 cm, base narrowly rounded to cordate, margins coarsely and unevenly doubly serrate, teeth sharp and slender, secondary teeth almost as large as primary teeth, apex usually abruptly nearly caudate, but sometimes long, gradually tapered; surfaces abaxially usually moderately pubescent, especially on major veins, covered with numerous tiny, dark brown glands. **Inflorescences:** staminate inflorescences 2–6 cm; pistillate inflorescences 1–3 cm. **Infructescences** 4.5–12 cm; bracts

2.5–3.5 × 1.5–2.8 cm, lobes narrowly triangular, sharp-tipped. $2n = 16$.

Flowering late spring. In understory stratum in rich deciduous forest along stream banks, on flood plains, and on moist hillsides; 0–300 m; Ont., Que.; Ala., Ark., Conn., Del., D.C., Ga., Ill., Ind., Iowa, Ky., Maine, Md., Mass., Mich., Minn., Miss., Mo., N.H., N.J., N.Y., N.C., Ohio, Okla., Pa., R.I., S.C., Tenn., Vt., Va., W.Va., Wis.

Carpinus caroliniana subsp. *virginiana* is the familiar hornbeam of the Appalachians and interior forested northeastern North America. The leaves are distinctive in that they bear scattered dark glands on the abaxial surface. This subspecies hybridizes and intergrades with subsp. *caroliniana* where their ranges overlap in a broad band running from the Carolinas south to northern Georgia and westward to Missouri, Arkansas, and southeastern Oklahoma (J. J. Furlow 1987).

4. OSTRYA Scopoli, Fl. Carniol., 414. 1760, name conserved · Hop-hornbeam [Latin *ostrya*, hop-hornbeam, from Greek *ostryos*, scale, in reference to the scaly infructescences]

Trees, 9–18 m; trunks usually 1, branching mostly deliquescent, trunk and branches terete. **Bark** of trunk and branches brownish gray to light brown, thin, smooth, breaking and shredding into shaggy vertical strips and scales; lenticels generally inconspicuous. **Wood** nearly white to light brown, very hard and heavy, texture fine. **Branches, branchlets, and twigs** conspicuously 2-ranked; young twigs differentiated into long and short shoots. **Winter buds** sessile, ovoid, somewhat laterally compressed, apex acute; scales many, imbricate, longitudinally striate. **Leaves** on long and short shoots, 2-ranked. **Leaf blade** narrowly ovate to ovate, elliptic, or obovate with 10 or more pairs of lateral veins, 2.5–13 × 1.5–6 cm, thin, margins doubly serrate to serrulate; surfaces abaxially glabrous to tomentose. **Inflorescences:** staminate catkins terminal on branches, mostly in small, racemose clusters, formed previous growing season and exposed during winter, expanding with leaves; pistillate catkins proximal to staminate on short, lateral, leafy new growth, solitary, ± erect, elongate, bracts and flowers uncrowded. **Staminate flowers** in catkins, 3 per bract, crowded together on pilose receptacle; stamens 3 (–6), short; filaments often divided part way to base; anthers divided into 2 parts, each 1-locular, apex pilose. **Pistillate flowers** 2 per bract. **Infructescences** loosely imbricate, strobiloid clusters of closed inflated bracts; clusters pendulous, elongate; bracts deciduous with fruit, inflated, bladderlike, each bract enclosing 1 fruit. **Fruits** small nutlets, ovoid, longitudinally ribbed, often crowned with persistent sepals and styles. $x = 8$.

Species ca. 5 (3 in the flora): mostly north temperate zones; North America, c Mexico, Europe, Asia (n, w China).

In North America *Ostrya* consists of small trees in the northern temperate deciduous forest zone and in the mountains of southwestern United States and adjacent Mexico. Mexican populations have generally been treated as conspecific with *O. virginiana* of eastern United States and Canada. They differ in various respects, however, including leaf shape and indumentum; the morphologic variation and phytogeography of the complex as a whole should be carefully examined. *Ostrya carpinifolia* Scopoli is a common and important forest tree of southern Europe.

Ostrya shares many features with *Carpinus*. The staminate catkins in most species of *Ostrya* are produced the season before anthesis but, unlike *Carpinus*, they are exposed during the winter. Dispersal occurs as it does in *Carpinus*, except that the bracts form closed, bladderlike structures rather than flat wings.

The wood of *Ostrya* is used for fuel, fence posts, and various other purposes. It was formerly utilized for manufacturing items subject to prolonged friction, including sleigh runners, wheel rims, and airplane propellers. Because of its hardness, it has been used for tool handles, mallet heads, and other hard wooden objects.

SELECTED REFERENCE Fernald, M. L. 1936b. Plants from the outer coastal plain of Virginia. Rhodora 38: 376–404, 414–452.

1. Leaf blade (5–)8–10(–13) cm, apex usually abruptly acuminate; infructescences 3.5–6.5 cm; e, nc United States, adjacent Canada . . . . .1. *Ostrya virginiana*
1. Leaf blade 2.5–6.5 cm, apex acute, obtuse, or rounded; infructescences 2–4 cm; sw United States.
  2. Leaf blade ovate or broadly ovate-elliptic to broadly elliptic to nearly orbiculate; petiole and young twigs often bearing stipitate glands; staminate catkins 2–3 cm; w Texas, New Mexico, Arizona, Utah . . . . 2. *Ostrya knowltonii*
  2. Leaf blade elliptic to elliptic-lanceolate; petiole and young twigs without stipitate glands; staminate catkins 3.5–5 cm; endemic to Chisos Mountains in Big Bend National Park, Texas. . . . . 3. *Ostrya chisosensis*

1. Ostrya virginiana (Miller) K. Koch, Dendrologie 2(2): 8. 1873 · Eastern hop hornbeam, ironwood, ostryer de Virginie, bois de fer [E] [F]

Carpinus virginiana Miller, Gard. Dict. ed. 8, Carpinus no. 4. 1768; *Ostrya virginiana* subsp. *lasia* (Fernald) E. Murray; *O. virginiana* var. *glandulosa* Sargent; *O. virginiana* var. *lasia* Fernald

Trees, to 18 m; trunks short, crowns open, narrow to broadly rounded. **Bark** grayish brown or steel gray, shredding into narrow, sometimes rather ragged, vertical strips. **Twigs** sparsely pubescent to densely velutinous. **Leaves:** petiole glabrous to pubescent, without stipitate glands. **Leaf blade** narrowly ovate or elliptic to oblong-lanceolate, (5–)8–10(–13) × 4–5(–6) cm, base narrowly rounded to cordate or cuneate, margins sharply and unevenly doubly serrate, apex usually abruptly acuminate, sometimes acute or gradually tapering; surfaces abaxially sparsely to moderately pubescent (or sometimes densely villous), especially on major veins. **Inflorescences:** staminate catkins 2–5 cm; pistillate catkins 0.8–1.5 cm. **Flowering** with leaves in late spring. **Infructescences** 3.5–6.5 × 2–2.5 cm; bracts 1–1.8 × 0.8–1 cm. $2n = 16$.

Flowering late spring. Moist, open to forested hillsides to dry upland slopes and ridges, sometimes also on moist, well-drained flood plains; 0–300 m; Man., N.B., N.S., Ont., P.E.I., Que.; Ala., Ark., Conn., D.C., Fla., Ga., Ill., Ind., Iowa, Kans., Ky., La., Maine, Md., Mass., Mich., Minn., Miss., Mo., Nebr., N.H., N.J., N.Y., N.C., N.Dak., Ohio, Okla., Pa., R.I., S.C., S.Dak., Tenn., Tex., Vt., Va., W.Va., Wis., Wyo.

The shaggy bark and winter-exposed terminal staminate catkins of *Ostrya virginiana* permit easy recognition of this characteristic tree of dryish eastern forests. Along the Atlantic Coastal Plain, *Ostrya virginiana,* like *Carpinus caroliniana,* has smaller, blunter, often more pubescent leaves (*O. virginiana* var. *lasia* Fernald). This variety has not been studied carefully; from the available material, however, it does not seem as distinct as the coastal subspecies in *C. caroliniana.*

Native Americans used *Ostrya virginiana* medicinally to treat toothache, to bathe sore muscles, for hemorrhages from lungs, for coughs, kidney trouble, female weakness, cancer of the rectum, consumption, and flux (D. E. Moerman 1986).

2. Ostrya knowltonii Coville, Gard. & Forest 7: 114. 1894 · Knowlton's hop-hornbeam, western hop-hornbeam, ironwood [E]

Ostrya baileyi Rose

Trees, to 9 m; crowns open, narrowly rounded. **Bark** brownish gray, broken into narrow vertical scales or rather ragged strips. **Twigs** sparsely to moderately pubescent, often with stipitate glands. **Leaves:** petiole covered with stipitate glands. **Leaf blade** ovate or broadly ovate-elliptic to broadly elliptic or nearly orbiculate, 2.5–6.5 × 1.5–5 cm, base narrowly rounded to cordate or cuneate, margins sharply and unevenly doubly serrate, apex acute; surfaces abaxially pubescent, especially on veins. **Inflorescences:** staminate catkins 2–3 cm; pistillate catkins 0.6–1 cm. **Infructescences** 2.5–4 × 1.8–2.5 cm; bracts 1–1.8 × 0.5–1 cm.

Flowering late spring. Streamsides and rocky slopes in moist canyons; 1200–2400 m; Ariz., N.Mex., Tex., Utah.

Ostrya knowltonii occurs sporadically throughout the arid Southwest, including both rims of the Grand Canyon. On the basis of morphology and phytogeography, it appears to be more closely allied with *Ostrya* in mountainous western Mexico than with the eastern *O. virginiana.*

3. Ostrya chisosensis Correll, Wrightia 3: 128. 1965 · Chisos hop-hornbeam, Big Bend hop-hornbeam [C] [E]

Ostrya knowltonii Coville subsp. *chisosensis* (Correll) E. Murray

Trees, to 12 m; crowns open, cylindric. **Bark** brownish gray, broken into narrow vertical strips. **Twigs** sparsely to moderately pubescent, without stipitate glands. **Leaves:** petiole glabrous or pubescent, without stipitate glands. **Leaf blade** broadly elliptic to elliptic-lanceolate, 3.5–5 × 2–3 cm, base narrowly rounded to cordate or cuneate, margins finely doubly serrate, apex obtuse to acute; surfaces abaxially sparsely pubescent, especially along veins. **Inflorescences:** staminate catkins 3–4 cm; pistillate catkins 0.8–1.5 cm. **Infructescences** 2–4 × 1.5–2.5 cm; bracts 1–1.8 × 0.5–1 cm.

Flowering late spring. Streamsides and moist slopes in canyons; of conservation concern; 1500–2300 m; Tex.

Ostrya chisosensis is endemic to Big Bend National Park, Texas. It appears to be related to *O. knowltonii* and to populations of hop-hornbeams in the mountains of western Mexico; the complex has not been studied in detail.

5. **CORYLUS** Linnaeus, Sp. Pl. 2: 998. 1753; Gen. Pl. ed. 5, 433. 1754 · Hazel [Latin *corylus,* hazel, from Greek *korus,* helmet, for shape and hardness of nut shells]

Shrubs and trees, 3–15 m; tree trunks usually 1, branching mostly deliquescent, trunks and branches terete. **Bark** grayish brown, thin, smooth, close, breaking into vertical strips and scales in age; prominent lenticels absent. **Wood** nearly white to light brown, moderately hard, heavy, texture fine. **Branches, branchlets, and twigs** nearly 2-ranked to diffuse; young twigs differentiated into long and short shoots. **Winter buds** sessile, broadly ovoid, apex acute; scales several, imbricate, smooth. **Leaves** on long and short shoots, 2-ranked. **Leaf blade** broadly ovate with 8 or fewer pairs of lateral veins, 4–12 × 3.5–12 cm, thin, bases often cordate, margins doubly serrate, apex occasionally nearly lobed; surfaces abaxially usually pubescent, sometimes glandular. **Inflorescences:** staminate catkins on short shoots lateral on branchlets, in numerous racemose clusters, formed previous growing season and exposed during winter, expanding well before leaves; pistillate catkins distal to staminate catkins, in small clusters of flowers and bracts, reduced, only styles protruding from buds containing them at anthesis, expanding at same time as staminate. **Staminate flowers** in catkins, 3 per scale, congested; stamens 4, divided nearly to base to form 8 half-stamens; filaments very short, adnate with 2 bractlets to bract. **Pistillate flowers** 2 per bract. **Infructescences** compact clusters of several fruits, each subtended and surrounded by involucre of bracts, bracts 2, hairy [spiny], expanded, foliaceous, sometimes connate into short to elongate tube. **Fruits** relatively thin-walled nuts, nearly globose to ovoid, somewhat laterally compressed, longitudinally ribbed. $x = 11$.

Species ca. 15 (3 in the flora): throughout north temperate zone; North America, Europe, Asia.

Corylus differs from other Betulaceae in various features, most notably in the infructescences, which consist of small clusters of well-developed nuts, each enclosed by a loose involucre of leaflike bracts. As in *Ostrya,* the staminate catkins are formed during the summer and are exposed through the winter prior to anthesis. In *Corylus,* however, pistillate catkins develop at the same time as the staminate, and they consist of only a few flowers, protected by the scales of special buds rather than being arranged in elongate pistillate catkins. The staminate flowers are unique in the family in that well-developed sepals are occasionally present, clearly defining the three individual flowers that make up each cymule.

A longstanding disparity occurs in the literature regarding the diploid chromosome number found in *Corylus* species, with both $2n = 22$ and $2n = 28$ being cited. J. G. Packer (pers.

comm.) believes that the $2n = 28$ for several species (R. H. Woodworth 1929c) was in error because of a misinterpretation of Woodworth's meiotic preparations, a number of which actually indicate eleven haploid chromosomes. Woodworth's count may be largely, if not entirely responsible for the persistence of this number in the literature.

The genus consists of three major subgroups, the first composed of shrubby plants having a short, open involucre of two bracts surrounding the fruits (*Corylus* sect. *Corylus*). Members of *Corylus* sect. *Tuboavellana* Spach are of similar habit but have the involucre modified into a tubular beak, and *Corylus* sect. *Acanthochlamnys* Spach is characterized by densely spiny bracts. Recent treatments have avoided applying sectional names. The genus as a whole should be considered for taxonomic revision.

Corylus is the source of hazelnuts and filberts. Commercial filberts (*C. colurna* Linnaeus and *C. maxima* Miller) are cultivated in various parts of the world, particularly Turkey, Italy, Spain, China, and the United States. Wild hazelnuts (*C. americana* and *C. cornuta*) are smaller but similar in flavor to those of the cultivated species.

SELECTED REFERENCES Drumke, J. S. 1965. A systematic survey of *Corylus* in North America. Diss. Abstr. 25: 4925–4926. Kasapligil, B. 1964. A contribution to the histotaxonomy of *Corylus* (Betulaceae). Adansonia, n. s. 4: 43–90. Rose, J. N. 1895. Notes upon *Corylus rostrata* and *C. californica*. Gard. & Forest 8: 263. Wiegand, K. M. 1909. Recognition of *Corylus rostrata* and *Corylus americana*. Rhodora 11: 107.

1. Fruit surrounded by soft bristly involucre connate to summit into narrow tube 2–5 times length of fruit; branchlets and petioles glabrous to pubescent, with or without glandular hairs . . . . .3. *Corylus cornuta*
1. Fruit surrounded by involucre of 2 downy, expanded, foliaceous bracts, distinct nearly to base; branchlets and petioles covered with bristly glandular hairs.
  2. Involucre slightly longer than to 2 times length of fruit; staminate catkins mostly in groups of 1–2, peduncles 5 mm or shorter; slender native shrubs to ca. 3 m . . . . .1. *Corylus americana*
  2. Involucre shorter than to only slightly longer than fruit; staminate catkins mostly in groups of 2–4, peduncles more than 5 mm; broad, spreading, introduced shrubs to ca. 5 m. . . . . 2. *Corylus avellana*

1. Corylus americana Walter, Fl. Carol., 236. 1788 · American hazel or hazelnut, noisetier d'Amérique [E] [F]

Corylus americana var. *altior* Farwell; *C. americana* var. *indehiscens* E. J. Palmer & Steyermark; *C. americana* var. *missourensis* A. de Candolle

Shrubs, open, upright, rounded, to 3(–5) m. **Bark** light gray, smooth. **Branches** ascending; twigs pubescent, covered with bristly glandular hairs. **Winter buds containing inflorescences** broadly ovoid, 3–4 × 3–4 mm, apex obtuse to rounded. **Leaves:** petiole pubescent, densely glandular-bristly. **Leaf blade** broadly ovate, often with straight sides and slight lobes near apex, giving them squarish appearance, 5–16 × 4–12 cm, moderately thin, base narrowly cordate to narrowly rounded, margins sharply serrate or obscurely doubly serrate, apex abruptly acuminate to long-acuminate; surfaces abaxially sparsely to moderately pubescent, velutinous to tomentose along major veins and in vein axils. **Inflorescences:** staminate catkins lateral along branchlets on very short shoots, usually in clusters of 1–2, 4–8 × 0.5–0.8 cm; peduncles mostly 1–5 mm. **Nuts** in clusters of 2–5, sometimes partially visible; bracts much enlarged, leaflike, distinct nearly to base, slightly longer than to 2 times length of nuts, apex deeply and irregularly laciniate; bract surfaces downy-pubescent, abaxially stipitate-glandular. $2n = 22, 28$.

Flowering very early spring. Moist to dry open woods and thickets, hillsides, roadsides, fencerows, and waste places; 0–750 m; Man., Ont., Sask.; Ala., Ark., Conn., Del., D.C., Ga., Ill., Ind., Iowa, Kans., Ky., La., Maine, Md., Mass., Mich., Minn., Miss., Mo., Nebr., N.H., N.J., N.Y., N.C., N.Dak., Ohio, Okla., Pa., R.I., S.C., S.Dak., Tenn., Vt., Va., W.Va., Wis.

Corylus americana is a weedy species, sometimes considered a pest in carefully managed forests. The nuts are smaller but of the same general quality and flavor as commercial filberts (*Corylus maxima* Miller and *C. colurna* Linnaeus).

Native Americans used *Corylus americana* medicinally for hives, biliousness, diarrhea, cramps, hay fever, childbirth, hemorrhages, prenatal strength, and teething, to induce vomiting, and to heal cuts (D. E. Moerman 1986).

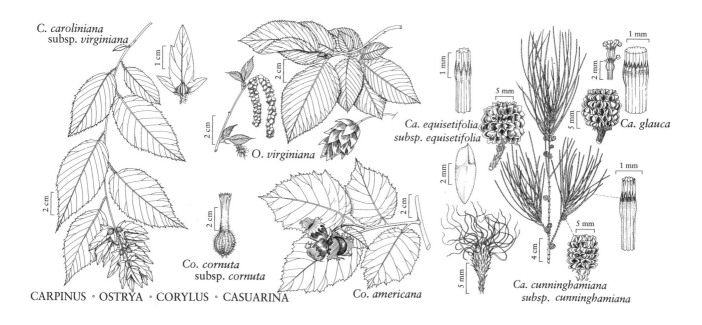

C. caroliniana subsp. *virginiana*

O. virginiana

Ca. equisetifolia subsp. *equisetifolia*

Ca. glauca

Co. cornuta subsp. *cornuta*

Co. americana

Ca. cunninghamiana subsp. *cunninghamiana*

CARPINUS · OSTRYA · CORYLUS · CASUARINA

2. Corylus avellana Linnaeus, Sp. Pl. 2: 998. 1753

· European hazel or hazelnut, avelline

Shrubs, broadly spreading, to 5 (–8) m. **Bark** coppery brown, smooth, sometimes exfoliating in thin papery strips. **Branches** ascending; twigs pubescent, covered with bristly glandular hairs. **Winter buds containing inflorescences** ovoid, 5–6 × 3–4 mm, apex obtuse. **Leaves:** petiole pubescent, covered with bristly glandular hairs. **Leaf blade** broadly ovate to broadly elliptic, often nearly angled to lobulate near apex, 5–12 × 4–12 cm, moderately thin, base narrowly cordate to narrowly rounded, margins coarsely and doubly serrate, apex abruptly acuminate, abaxially sparsely to moderately pubescent, velutinous to tomentose along major veins and in vein axils. **Inflorescences:** staminate catkins lateral along branchlets on relatively long short shoots, usually in clusters of 2–4, 3–8 × 0.7–1 cm; peduncles mostly 5–12 mm. **Nuts** in clusters of 2–4; bracts much enlarged, distinct nearly to base, expanded, shorter than to only slightly longer than nuts, apex deeply lobed; bract surfaces pubescent. $2n = 22, 28$.

Flowering very early spring. Abandoned plantings, roadsides, thickets, waste places; 0–700 m; introduced; B.C.

Corylus avellana is widely grown as an ornamental shrub in temperate North America, and it sometimes persists following cultivation, although it seldom becomes established.

Corylus avellana is similar to *C. americana* in habit, leaves, and fruit characteristics, although it becomes much larger. If fruits are present, the two species can be distinguished by the involucre, which is shorter than the nut in *C. avellana*. The best technical character for separating these species in the absence of fruits is the length of the peduncles of the staminate catkins (which are formed during the summer prior to the season of blooming).

3. Corylus cornuta Marshall, Arbust. Amer., 37. 1785

· Beaked hazel or hazelnut [E]

Shrubs or trees, open spreading, 4–8(–15) m. **Bark** light to dark brown, smooth. **Branches** ascending; twigs glabrous to sparsely pubescent, sometimes with glandular hairs. **Winter buds containing inflorescences** ovoid, 3–5 × 3–4 mm, acute. **Leaves:** petiole glabrous to moderately pubescent, with or without glandular hairs. **Leaf blade** nearly orbiculate to narrowly ovate or ovate-oblong, often nearly angular and slightly lobulate near apex, 4–10 × 3.5–12 cm, thin to leathery, base narrowly cordate to narrowly rounded, margins coarsely and often irregularly doubly serrate, apex obtuse to acute or acuminate; surfaces abaxially glabrous to moderately pubescent, usually pubescent on major veins and in vein axils. **Inflorescences:** staminate catkins lateral along branchlets on short shoots, usually in clusters of 2–3, 4–6 × 0.5–0.8 cm; peduncles 0.5–10 mm. **Nuts** in clusters of 2–6, completely concealed; bracts bristly, connate at summit, lengthened into extended tubular beak.

Subspecies 2 (2 in the flora).

Corylus cornuta was used medicinally by Native Americans as an emetic, for teething, to expel worms, to heal cuts, and as an astringent (D. E. Moerman 1986).

1. Leaf blade ovate to narrowly elliptic, apex distinctly acuminate; twigs and petioles without glandular hairs; involucral tubular beak 2 times or more length of fruit; small to large shrubs of e, c, n North America . . . .
. . . . .3a. *Corylus cornuta* subsp. *cornuta*
1. Leaf blade nearly orbiculate or broadly elliptic, apex broadly acute to obtuse; twigs and petioles usually bearing glandular hairs; involucral tubular beak less than 2 times length of fruit; large shrubs or small trees of Pacific coastal region of North America . . .3b. *Corylus cornuta* subsp. *californica*

3a. Corylus cornuta Marshall subsp. **cornuta** · Beaked hazel, noisetier à long bec E F

Corylus cornuta var. *megaphylla* Victorin & J. Rousseau; *C. rostrata* Aiton

Shrubs, open spreading, to (4–)6 m. **Bark** light brown, smooth. **Branches** ascending; twigs glabrous to sparsely pubescent, without glandular hairs. **Winter buds containing inflorescences** ovoid, 3–5 × 3–4 mm, apex acute. **Leaves:** petiole glabrous to moderately pubescent, without glandular hairs. **Leaf blade** ovate to obovate or narrowly elliptic, often nearly angular and lobulate near apex, 5–12 × 3.5–9 cm, base narrowly cordate to narrowly rounded, margins coarsely and often irregularly doubly serrate, apex usually distinctly acuminate; surfaces abaxially glabrous to moderately pubescent, usually pubescent on major veins and in vein axils. **Inflorescences:** staminate catkins usually in clusters of 2–3, 4.5–6 × 0.5–0.8 cm; peduncles mostly 0.5–2 mm. **Nuts** in clusters of 2–6; involucral tubular beak long, narrow, 2–3(–4) times length of nuts, densely bristly. $2n = 22, 28$.

Flowering very early spring. Moist to dry roadsides, edges of woods, fencerows, waste places, and thickets, or as understory in open woodlands; 100–500 m; St. Pierre and Miquelon; Alta., B.C., Man., N.B., Nfld., N.S., Ont., P.E.I., Que., Sask.; Ala., Colo., Conn., Ga., Idaho, Iowa, Ky., Maine, Md., Mass., Mich., Minn., Mont., N.H., N.J., N.Y., N.C., N.Dak., Ohio, Pa., R.I., S.C., S.Dak., Tenn., Vt., Va., Wash., W.Va., Wis., Wyo.

Like *Corylus americana* Walter, the beaked hazel (*C. cornuta* subsp. *cornuta*) is a weedy shrub and is sometimes considered a pest in carefully managed northern forests. The fruits are similar to those of *C. americana,* except that the surrounding bracts are connate into a long, narrow, tubular beak. Vegetative individuals of *C. cornuta* subsp. *cornuta* can be distinguished from *C. americana* by the absence of glandular hairs on the petioles and young twigs.

3b. Corylus cornuta Marshall subsp. **californica** (A. de Candolle) E. Murray, Kalmia 12: 19. 1982 · California hazel E

Corylus rostrata Aiton var. *californica* A. de Candolle, in A. P. de Candolle and A. L. P. de Candolle, Prodr. 16(2): 133. 1864; *C. californica* (A. de Candolle) Rose; *C. cornuta* var. *californica* (A. de Candolle) Sharp; *C. cornuta* var. *glandulosa* B. Boivin; *C. rostrata* var. *tracyi* Jepson

Shrubs or trees, open spreading, to 8(–15) m; trunks usually several. **Bark** dark brown to blackish. **Branches** ascending; twigs sparsely to moderately pubescent, bearing glandular hairs. **Winter buds** containing inflorescences broadly ovoid, 3–5 × 3–5 mm, apex acute. **Leaves:** petiole pubescent, often bearing well-developed glandular hairs. **Leaf blade** nearly orbiculate or broadly elliptic, 4–7 × 3.5–7 cm, leathery, base nearly cordate, margins coarsely doubly serrate, apex obtuse to acute, abaxially moderately pubescent, villous to tomentose on major veins and in vein axils. **Inflorescences:** staminate catkins usually in clusters of 2–3, 4–6 × 0.5–0.8 cm; peduncles mostly 5–10 mm. **Nuts** in clusters of 2–4; involucral tubular beak less than 2 times length of nuts, hispid. $2n = 22$.

Flowering very early spring. Damp rocky slopes and stream banks in coastal mountain ranges; 1000–2500 m; B.C.; Calif., Oreg., Wash.

The California hazel (*Corylus cornuta* subsp. *californica*) is most often treated as a variey of the northern *C. cornuta*. The two may not be very closely related, however, differing conspicuously in habit, leaf shape, pubescence, the presence of glandular hairs, form and size of the involucre, habitat, phytogeography, and various other features (J. N. Rose 1895; J. S. Drumke 1965). A thorough taxonomic study of this group should be undertaken.

32. CASUARINACEAE R. Brown

· She-oak or Casuarina Family

Karen L. Wilson

Trees [or shrubs], evergreen. **Photosynthetic branchlets** slender, wiry, with several very short, basal segments and 1–numerous elongate segments; segments terete [quadrangular], with as many longitudinal ridges as leaves; ridges separated by furrows containing stomates. **Leaves** reduced to small teeth in whorls of [4–]6–17 at apex of each segment of photosynthetic branchlets. **Inflorescences** of alternating whorls of flowers, each flower subtended by toothlike bract and 2 bracteoles, bracteoles usually persistent, lateral, scalelike; staminate inflorescences catkinlike spikes, short to elongate; pistillate inflorescences heads, globular to ovoid. **Flowers** unisexual, staminate and pistillate on same or different plants. **Staminate flowers:** sepals deciduous at anthesis, 1–2, hooded, scalelike; stamen 1; anthers basifixed, 2-locular. **Pistillate flowers:** perianth absent; pistil 1, compound, 2-carpellate, 1 fertile, the other usually reduced or absent; ovules 2, an additional 2 abortive ovules in reduced carpel; styles 2-branched, reddish. **Infructescences** ± woody, cylindric, conelike; floral bracteoles 2, enlarged as lateral valves. **Fruits** compressed, winged nuts (samaras). **Seeds** 1 in each samara.

Genera 4, species 90 (1 genus, 3 species in the flora): tropical and subtropical dry regions, warm temperate areas; naturalized, North America; native Southeast Asia, s Pacific Islands (to Tahiti and Samoa), and Australia.

Casuarinaceae are occasionally referred to as the beefwood family or the Australian-pine family.

Species have been cultivated in the warmest parts of the flora area as ornamentals and shelterbelts, and for sand binding. Their suitability for such uses is partly because of the presence in root nodules of actinomycetes (*Frankia*); such actinomycetes are effective in fixation of atmospheric nitrogen. Vesicular-arbuscular, endotrophic mycorrhizae have also been reported. In addition to the species described here, the following have all been recorded as cultivated in the flora, but they are not known to be naturalized: *Allocasuarina decussata* (Bentham) L. A. S. Johnson, *A. helmsii* (Ewart & M. Gordon) L. A. S. Johnson, *A. littoralis* (Salisbury) L. A. S. Johnson (*Casuarina suberosa* Otto & Dietrich), *A. torulosa* (Aiton) L. A. S. Johnson, *A. verticillata* (Lamarck) L. A. S. Johnson (*C. stricta* Aiton), *C. cristata* Miquel (*C. lepidophloia* F. Mueller), and *Gymnostoma sumatranum* (Junghuhn ex de Vriese) L. A. S.

Johnson (*C. sumatrana* Junghuhn ex de Vriese). See K. L. Wilson and L. A. S. Johnson (1989) for distinguishing features of the Australian species.

Dried specimens differ significantly from fresh material. When fresh, fruiting bracteoles of the infructescence are nearly always appressed to each other, enclosing the samara; when the infructescence dries out, the bracteoles separate. Measurements given here for infructescence body diameter do not include any portion of the bracteoles extending beyond the main body of the infructescence. The softer tissues of branchlets contract when dried, so that features such as angularity or convexity of longitudinal ridges are emphasized in dried specimens. The key and descriptions are generally based on dried specimens.

SELECTED REFERENCES Johnson, L. A. S. and K. L. Wilson. 1989. Casuarinaceae: A synopsis. In: P. R. Crane and S. Blackmore, eds. 1989. Evolution, Systematics and Fossil History of the Hamamelidae. 2 vols. Oxford. Vol. 2, pp. 167–188. [Syst. Assoc. Special Vol. 40A,B.] Johnson, L. A. S. and K. L. Wilson. 1993. Casuarinaceae. In: K. Kubitzki et al., eds. 1990+. The Families and Genera of Vascular Plants. 2+ vols. Berlin etc. Vol. 2, pp. 237–242. Rogers, G. K. 1982b. The Casuarinaceae in the southeastern United States. J. Arnold Arbor. 63: 357–373. Wilson, K. L. and L. A. S. Johnson. 1989. Casuarinaceae. In: R. Robertson et al., eds. 1981+. Flora of Australia. 14+ vols. Canberra. Vol. 3, pp. 100–174. Woodall, S. L. and T. F. Geary. 1985. Identity of Florida casuarinas. Res. Notes S. E., U.S. Forest Serv. 332: 1–10.

1. CASUARINA Linnaeus, Amoen. Acad. 4: 143. 1759 · She-oak [Neo-Latin *casuarius*, cassowary, from resemblance of drooping branchlets to feathers of the cassowary]

Young persistent branchlets distinguished from deciduous branchlets by shorter segments and differences in shape or size of leaves; furrows deep and closed, concealing stomates. **Infructescences** pedunculate, pubescent at least when immature; bracts thin in exposed portion, not vertically expanded; bracteoles ± protruding from surface of infructescence, never greatly thickened, always lacking dorsal protuberance. **Samaras** pale yellow-brown or grayish, dull, glabrous. $x = 9$.

Species 17 (3 in the flora): almost throughout range of family.

Hybrids are frequent in cultivation; in the flora, hybrids are known between all combinations of the three species.

1. Longitudinal ridges of branchlets flat or slightly rounded-convex; leaf teeth 12–17; teeth on young permanent shoots long-recurved. . . . . .3. *Casuarina glauca*
1. Longitudinal ridges of branchlets prominently angular; leaf teeth (6–)7–10; teeth on young permanent shoots erect to spreading.
    2. Branchlets ± densely and obviously pubescent; teeth (6–)7–8, not marcescent; infructescence body 12–24 × 9–11 mm, bracteoles acute; samaras 6–8 mm. . . . . 1a. *Casuarina equisetifolia* subsp. *equisetifolia*
    2. Branchlets sparsely and minutely pubescent; teeth 8–10, marcescent; infructescence body 7–14 × 4–6 mm, bracteoles broad-acute; samaras 3–4 mm. . . . . .2a. *Casuarina cunninghamiana* subsp. *cunninghamiana*

1. Casuarina equisetifolia Linnaeus, Amoen. Acad. 4: 143. 1759 · Beach she-oak, coast she-oak, casuarina, Australian-pine, horsetail casuarina W1

Subspecies 2 (1 in the flora): North America; Southeast Asia; s Pacific Islands (e to Tahiti and Samoa); Australia.

1a. Casuarina equisetifolia Linnaeus subsp. **equisetifolia** F W1

Trees, 7–35 m, not suckering. Bark gray-brown to black, scaly. Branchlets drooping; segments 5–8[–13] × 0.5–0.7[–1] mm, usually densely pubescent at least in furrows, not waxy; longitudinal ridges angular with median rib; teeth not marcescent, (6–)7–8, erect, 0.3–0.8 mm. **Young per-**

manent shoots with erect to spreading teeth. **Flowers** unisexual, staminate and pistillate on same plant. **Staminate spikes** 0.7–4 cm, 7–11.5 whorls per cm; anthers 0.6–0.8 mm. **Infructescences** sparsely pubescent [tomentose]; peduncles 3–10 mm; infructescence body 12–24 × 9–11 mm; bracteoles acute. **Samaras** 6–8 mm.

Flowering spring–summer. Sandy seasides; native to tropical and subtropical coastlines; 0–20 m; introduced, Fla.; native, Southeast Asia; native, s Pacific Islands (e to Tahiti and Samoa); native, Australia.

Casuarina litorea Rumphius ex Stickman is not a valid name.

Casuarina equisetifolia is widely cultivated in many parts of the world because of its salt tolerance; it is now considered an invasive pest in parts of Florida. Only *C. equisetifolia* subsp. *equisetifolia* is known from the flora area. *Casuarina equisetifolia* subsp. *incana* (Bentham) L. A. S. Johnson, from Australia and the Pacific Islands region, is cultivated elsewhere in the world and possibly has been introduced to (although not yet discovered in) the flora area. That subspecies is a smaller tree with a more rounded crown, longer and thicker branchlets, and more pubescent branchlets and infructescences.

2. **Casuarina cunninghamiana** Miquel, Rev. Crit. Casuarinarum, 56, plate 6A. 1848 · River she-oak, Cunningham's beefwood W1

Subspecies 2 (1 in the flora): North America; native to ne, e Australia.

2a. **Casuarina cunninghamiana** Miquel subsp. **cunninghamiana** F

Trees, 15–35 m, usually not suckering. **Bark** gray-brown, finely fissured and scaly. **Branchlets** drooping in vigorous specimens, erect in depauperate specimens; segments 6–9 × 0.4–0.6 mm, sparsely and minutely pubescent in furrows, not waxy, edges of furrows often marked (when dry) by slight ridge; longitudinal ridges angular with median rib; teeth marcescent, [6–]8–10, erect,

0.3–0.5 mm. **Young permanent shoots** with erect to spreading teeth. **Flowers** unisexual, staminate and pistillate on different plants. **Staminate spikes** 0.4–4 cm, 11–13 whorls per cm; anthers 0.4–0.7 mm. **Infructescences** sparsely pubescent; peduncles 2–9 mm; infructescence body 7–14 × 4–6 mm; bracteoles broadly acute [to acute]. **Samaras** 3–4 mm.

Flowering summer. Naturally on riverbanks, in somewhat drier sites in the flora; 0–500 m; introduced, Fla.; native, ne, e Australia.

Casuarina cunninghamiana subsp. *cunninghamiana* is less common in Florida than the other two species.

The type subspecies is widely cultivated in many parts of the world. *Casuarina cunninghamiana* subsp. *miodon* L. A. S. Johnson is not known to be cultivated.

3. **Casuarina glauca** Sieber ex Sprengel, Syst. Veg. 3: 803. 1826 · Swamp she-oak, gray she-oak, suckering Australian-pine, scaly-bark beefwood, Brazilian beefwood F W1

Trees, 8–20 m, frequently producing root suckers. **Bark** gray-brown, finely fissured and scaly. **Branchlets** drooping; segments 8–20 × 0.9–1.2 mm, glabrous, occasionally waxy; longitudinal ridges flat to slightly rounded-convex; teeth usually marcescent, 12–17, erect, 0.6–0.9 mm. **Young permanent shoots** with long-recurved teeth. **Flowers** unisexual, staminate and pistillate on different plants. **Staminate spikes** 1.2–4 cm, 7–10 whorls per cm; anthers ca. 0.8 mm. **Infructescences** rust-colored pubescent to white-pubescent, becoming glabrous; peduncles 3–12 mm; infructescence body 9–18 × 7–9 mm; bracteoles broadly acute. **Samaras** 3.5–5 mm.

Commonly near brackish water; 0–50 m; introduced, Fla.; native, e coast Australia.

Casuarina glauca is widely cultivated in many parts of the world. Pistillate trees are very infrequent in the flora.

It is now considered a pest species in Florida because of root suckering. Its identification may be confused by the practice of some Florida nurserymen of grafting scions of *Casuarina glauca* onto rootstocks from the other two species.

Literature Cited

Robert W. Kiger, Editor

This is a consolidated list of all works cited in volume 3, whether as selected references, in text, or in nomenclatural contexts. In citations of articles, both here and in the taxonomic treatments, and also in nomenclatural citations, the titles of serials are rendered in the abbreviated forms recommended in G. D. R. Bridson and E. R. Smith (1991). Cross references to the corresponding full serial titles are interpolated here alphabetically by abbreviated form. In nomenclatural citations (only), book titles are rendered in the abbreviated forms recommended in F. A. Stafleu and R. S. Cowan (1976–1988) and F. A. Stafleu and E. A. Mennega (1992 +). Here, those abbreviated forms are indicated parenthetically following the full citations of the corresponding works, and cross references to the full citations are interpolated in the list alphabetically by abbreviated form. Two or more works published in the same year by the same author or group of coauthors will be distinguished uniquely and consistently throughout all volumes of *Flora of North America* by lower-case letters (b, c, d, . . .) suffixed to the date for the second and subsequent works in the set. The suffixes are assigned in order of editorial encounter and do not reflect chronological sequence of publication. The first work by any particular author or group from any given year carries the implicit date suffix "a"; thus, the sequence of explicit suffixes begins with "b". Some citations in this list have dates suffixed "b" but are not preceded by citations of "[a]" works for the same year, or have dates suffixed "c" but are not preceded by citations of "[a]" and/or "b" works for that year. In these cases, the missing "[a]" and "b" works are ones cited (and encountered first from) elsewhere in the *Flora* that are not pertinent here.

Abbe, E. C. 1935. Studies in the phylogeny of the Betulaceae. I. Floral and inflorescence anatomy and morphology. Bot. Gaz. 97: 1–67.

Abbe, E. C. 1938. Studies in the phylogeny of the Betulaceae. II. Extremes in the range of variation of floral and inflorescence morphology. Bot. Gaz. 99: 431–469.

Abbe, E. C. and T. T. Earle. 1940. Inflorescence, floral anatomy and morphology of *Leitneria floridana*. Bull. Torrey Bot. Club 67: 173–193.

Abh. K. K. Zool.-Bot. Ges. Wien = Abhandlungen der Kaiserlich-königlichen zoologisch-botanischen Gesellschaft in Wien.

Abh. Math.-Phys. Cl. Königl. Bayer. Akad. Wiss. = Abhandlungen der Mathematisch-physikalischen Classe der Königlich bayerischen Akademie der Wissenschaften.

Abrams, L. 1934. The mahonias of the Pacific states. Phytologia 1: 89–94.

Abrams, L. and R. S. Ferris. 1923–1960. Illustrated Flora of the Pacific States: Washington, Oregon, and California. 4 vols. Stanford. (Ill. Fl. Pacific States)

Academia Sinica. 1959+. Flora Reipublicae Popularis Sinicae. 80+ vols. Beijing.

Account Exped. Pittsburgh—See: E. James 1823.

Acta Bot. Sin. = Acta Botanica Sinica. [Zhiwu Xuebao.]

Acta Horti Berg. = Acta Horti Bergiani.

Acta Lit. Univ. Hafn. = Acta Literaria Universitatis Hafniensis.

Acta Phytotax. Geobot. = Acta Phytotaxonomica et Geobotanica. [Shokubutsu Bunrui Chiri.]

Acta Univ. Lund. = Acta Universitatis Lundensis. Nova Series. Sectio 2, Medica, Mathematica, Scientiae Rerum Naturalium. [Lunds Universitets Årsskrift N.F., Avd. 2.]

Adanson, M. 1763[–1764]. Familles des Plantes. 2 vols. Paris. [Vol. 1, 1764; vol. 2, 1763.] (Fam. Pl.)

Adansonia = Adansonia; Recueil Périodique d'Observations Botaniques.

Addisonia = Addisonia; Colored Illustrations and Popular Descriptions of Plants.

Aiton, W. 1789. Hortus Kewensis; or, a Catalogue of the Plants Cultivated in the Royal Botanic Garden at Kew. 3 vols. London. (Hort. Kew.)

Aiton, W. and W. T. Aiton. 1810–1813. Hortus Kewensis; or a Catalogue of the Plants Cultivated in the Royal Botanic Garden at Kew. 5 vols. London. (Hortus Kew.)

Alan, M. T. and W. F. Grant. 1972. Interspecific hybridization in birch (*Betula*). Naturaliste Canad. 99: 33–40.

Alsogr. Amer.—See: C. S. Rafinesque 1838.

Amer. Forests = American Forests [title varies].

Amer. Hort. Mag. = American Horticultural Magazine.

Amer. J. Bot. = American Journal of Botany.

Amer. J. Sci. Arts = American Journal of Science, and Arts.

Amer. Midl. Naturalist = American Midland Naturalist; Devoted to Natural History, Primarily That of the Prairie States.

Amer. Monthly Mag. & Crit. Rev. = American Monthly Magazine and Critical Review.

Amer. Naturalist = American Naturalist. . . .

Amer. Nurseryman = American Nurseryman.

American Forestry Association 1994. National Register of Big Trees. Washington.

Amoen. Acad.—See: C. Linnaeus 1749[–1769].

Anales Ci. Nat. = Anales de Ciencias Naturales.

Anderson, E. 1934. The genus *Akebia*. Bull. Popular Inform. Arnold Arbor., ser. 4, 2: 17–20.

Anderson, J. P. 1959. Flora of Alaska and Adjacent Parts of Canada: An Illustrated Descriptive Text of All Vascular Plants Known to Occur within the Region Covered. Ames. (Fl. Alaska)

Anderson, L. C. 1974. A study of systematic wood anatomy in *Cannabis*. Bot. Mus. Leafl. 23: 61–69.

Animadv. Bot. Ranunc. Cand.—See: D. F. L. von Schlechtendal 1819–1820.

Ann. Bot. Fenn. = Annales Botanici Fennici.

Ann. Bot. (König & Sims) = Annals of Botany. [Edited by König & Sims.]

Ann. Bot. (Usteri) = Annalen der Botanick. . . . [Edited by P. Usteri.]

Ann. Lyceum Nat. Hist. New York = Annals of the Lyceum of Natural History of New York.

Ann. Mag. Nat. Hist. = Annals and Magazine of Natural History, Including Zoology, Botany, and Geology.

Ann. Missouri Bot. Gard. = Annals of the Missouri Botanical Garden.

Ann. New York Acad. Sci. = Annals of the New York Academy of Sciences.

Ann. Sci. Forest. = Annales des Sciences Forestières.

Ann. Sci. Nat., Bot. = Annales des Sciences Naturelles. Botanique.

Ann. Sci. Nat. (Paris) = Annales des Sciences Naturelles.

Aquatic Bot. = Aquatic Botany; International Scientific Journal Dealing with Applied and Fundamental Research on Submerged, Floating and Emergent Plants in Marine and Freshwater Ecosystems.

Arbust. Amer.—See: H. Marshall 1785.

Arch. Mus. Hist. Nat. = Archives du Muséum d'Histoire Naturelle.

Argus, G. W. and K. M. Pryer. 1990. Rare Vascular Plants in Canada—Our Natural Heritage. Ottawa.

Argus, G. W. and D. J. White, eds. 1982–1983. Atlas of the Rare Vascular Plants of Ontario. 2 vols. in 4 parts. Ottawa.

Arnoldia (Jamaica Plain) = Arnoldia; a Continuation of the Bulletin of Popular Information.

Arr. Brit. Pl. ed. 3—See: W. Withering 1796.

Ashe, W. W. 1894. A new post and hybrid oaks. J. Elisha Mitchell Sci. Soc. 11: 87–95.

Asia Life Sci. = Asia Life Sciences; the Asian International Journal of Life Sciences.

Aublet, J. B. 1775. Histoire des Plantes de la Guiane Française. . . . 4 vols. Paris. [Vols. 1 and 2: text, paged consecutively; vols. 3 and 4: plates.] (Hist. Pl. Guiane)

Baehni, C. 1937. *Trema*. In: J. F. Macbride et al. 1936+. Flora of Peru. 6+ parts. Chicago. Part 2(2), pp. 269–270.

Bailey, I. W. and C. G. Nast. 1948. Morphology and relationships of *Illicium*, *Schisandra* and *Kadsura*. J. Arnold Arbor. 29: 77–89.

Bailey, L. H. 1914–1917. The Standard Cyclopedia of Horticulture. . . . 6 vols. New York. [Vols. paged consecutively.] (Stand. Cycl. Hort.)

Baileya = Baileya; a Quarterly Journal of Horticultural Taxonomy.

Baird, J. R. 1968. A Taxonomic Revision of the Plant Family Myricaceae of North America, North of Mexico. Ph.D. thesis. University of North Carolina.

Baldwin, J. T. and B. M. Speese. 1949. Cytogeography of *Saururus cernuus*. Bull. Torrey Bot. Club 76: 213–216.

Baranski, M. J. 1975. An Analysis of Variation within White

Oak (*Quercus alba* L.). Raleigh. [North Carolina Agric. Exp. Sta. Techn. Bull. 236.]

Barker, W. T. 1986. Ulmaceae. In: Great Plains Flora Association. 1986. Flora of the Great Plains. Lawrence, Kans. Pp. 119–123.

Barnes, B. V. and B. P. Dancik. 1985. Characteristics and origin of a new birch species, *Betula murrayana,* from southeastern Michigan. Canad. J. Bot. 63: 223–226.

Barnes, B. V., B. P. Dancik, and T. L. Sharik. 1974. Natural hybridization of yellow birch and paper birch. Forest Sci. 20: 215–221.

Barringer, K. 1993. New combinations in North American *Asarum.* Novon 3: 225–227.

Bartonia = Bartonia; a Botanical Annual.

Bartram, W. 1791. Travels through North and South Carolina, Georgia, East and West Florida, the Cherokee Country, the Extensive Territories of the Muscogulges, or Creek Confederacy, and the Country of the Chactaws. . . . Philadelphia. (Travels Carolina)

Bassett, I. J., C. W. Crompton, and D. W. Woodland. 1974. The family Urticaceae in Canada. Canad. J. Bot. 52: 503–516.

Bayly, I. L. and K. Jongejan. 1982. A morphological and ecological variant of the tuberous water lily, *Nymphaea tuberosa* Paine, from the Jock River, Ottawa, Ontario. Canad. Field-Naturalist 96: 301–306.

Beal, E. O. 1956. Taxonomic revision of the genus *Nuphar* Sm. of North America and Europe. J. Elisha Mitchell Sci. Soc. 72: 317–346.

Beal, E. O. and R. M. Southall. 1977. The taxonomic significance of experimental selection by vernalization in *Nuphar* (Nymphaeaceae). Syst. Bot. 2: 49–60.

Behnke, H.-D. 1988. Sieve-element plastids, phloem protein, and evolution of flowering plants. III. Magnoliidae. Taxon 37: 699–732.

Beitr. Biol. Pflanzen = Beiträge zur Biologie der Pflanzen.

Beitr. Naturk.—See: J. F. Ehrhart 1787–1792.

Bemerk. Reise Russ. Reich—See: J. G. Georgi 1775.

Bennett, J. J., R. Brown, and T. Horsfield. 1838–1852. Plantae Javanicae Rariores, Descriptae Iconibus Illustratae, Quas in Insula Java, Annis 1802–1818. . . . 4 parts. London. (Pl. Jav. Rar.)

Benson, L. D. 1943. Revisions of status of southwestern trees and shrubs. Amer. J. Bot. 30: 230–240.

Benson, L. D. 1948. A treatise on the North American *Ranunculi.* Amer. Midl. Naturalist 40: 1–261.

Benson, L. D. 1954. Supplement to a treatise on the North American *Ranunculi.* Amer. Midl. Naturalist 52: 328–369.

Bentham, G. 1839[–1857]. Plantas Hartwegianas Imprimis Mexicanas. . . . London. [Issued by gatherings with sequential signatures and consecutive pagination.] (Pl. Hartw.)

Berchtold, F. and J. S. Presl. 1823–1835. O Přirozenosti Rostlin aneb Rostlinář. . . . 3 vols. Prague. [Vols. 2 and 3 by Presl only.] (Přir. Rostlin)

Berg, C. C. 1972. Olmedieae, Brosimeae (Moraceae). In: Organization for Flora Neotropica. 1968 + . Flora Neotropica. 65 + vols. New York. Vol. 7, pp. 1–229.

Berg, R. Y. 1969. Adaptation and evolution in *Dicentra* (Fumariaceae), with special reference to seed, fruit, and dispersal mechanism. Nytt Mag. Bot. 16: 49–75.

Berg, R. Y. 1972. Dispersal ecology of *Vancouveria* (Berberidaceae). Amer. J. Bot. 59: 109–122.

Bergens Mus. Årbok = Bergens Museums Årbok; Afhandlinger og Aarsberetning.

Berlandier, J. L. and R. Chovel. 1850. Diario de Viaje de la Comisión de Límites Que Puso el Gobierno de la República baja el Dirección del Exmo. Sr. Gral. D. Manuel Mier y Terán. Mexico City. (Diario Viaje Comis. Limites)

Berry, E. W. 1901. The origin of stipules in *Liriodendron.* Bull. Torrey Bot. Club 28: 493–498.

Beytr. Teut. Forstwiss.—See: F. A. J. von Wangenheim 1787.

Bibl. Index N. Amer. Bot.—See: S. Watson 1878.

Bigelow, J. 1814. Florula Bostoniensis. A Collection of Plants of Boston and Its Environs. . . . Boston. (Fl. Boston.)

Bigelow, J. 1824. Florula Bostoniensis A Collection of Plants of Boston and Its Vicinity . . . , ed. 2. Boston. (Fl. Boston. ed. 2)

Bijdr. Fl. Ned. Ind.—See: C. L. Blume 1825–1826.

Biltmore Bot. Stud. = Biltmore Botanical Studies; a Journal of Botany Embracing Papers by the Director and Associates of the Biltmore Herbarium.

Biochem. Syst. & Ecol. = Biochemical Systematics and Ecology.

Blomquist, H. L. 1957. A revision of *Hexastylis* of North America. Brittonia 8: 255–281.

Blomquist, H. L. and H. J. Oosting. 1948. A Guide to the Spring and Early Summer Flora of the Piedmont, North Carolina. [Durham, N.C.] (Spring Fl. Piedmont)

Blume, C. L. 1825–1826. Bijdragen tot de Flora van Nederlandsch Indië. 10 parts. Batavia. [Parts paged consecutively.] (Bijdr. Fl. Ned. Ind.)

Blume, C. L. 1849–1851[–1857]. Museum Botanicum Lugduno-Batavum. . . . 2 vols. Leiden. (Mus. Bot.)

Blumea = Blumea; Tijdschrift voor die Systematiek en die Geografie der Planten (A Journal of Plant Taxonomy and Plant Geography).

Böcher, T. W., K. Holmen, and K. Jakobsen. 1968. The Flora of Greenland, ed. 2. Copenhagen.

Bogle, A. L. 1970. Floral morphology and vascular anatomy of the Hamamelidaceae: The apetalous genera of Hamamelidoideae. J. Arnold Arbor. 51: 310–366.

Bogle, A. L. 1986. The floral morphology and vascular anatomy of the Hamamelidaceae: Subfamily Liquidambaroideae. Ann. Missouri Bot. Gard. 73: 325–347.

Boivin, B. 1944. American Thalictra and their Old World allies. Rhodora 46: 337–377.

Boivin, B. 1948. Key to Canadian species of Thalictra. Canad. Field-Naturalist 62: 169–170.

Boivin, B. 1953. Notes on *Aquilegia.* Amer. Midl. Naturalist 50: 509–510.

Boivin, B. 1966. Énumération des plantes du Canada. Naturaliste Canad. 93: 583–646.

Boivin, B. 1967b. Notes sur les *Betula.* Naturaliste Canad. 94: 229–231.

Boivin, B. 1967–1979. Flora of the prairie provinces. Part I (–IV). Phytologia 15–18, 22–23, 42–43: passim. [Reprinted as Provancheria 2–5 with auxiliary pagination.]

Bojo, A. C., E. Albano-Garcia, and G. N. Pocsidio. 1994. The antibacterial activity of *Peperomia pellucida* (L.) HBK. (Piperaceae). Asia Life Sci. 3: 35–44.

Boothroyd, L. E. 1930. The morphology and anatomy of the inflorescence and flower of the Platanaceae. Amer. J. Bot. 17: 678–693.

Boraiah, G. and M. Heimburger. 1964. Cytotaxonomic studies on New World *Anemone* (section *Eriocephalus*) with woody rootstocks. Canad. J. Bot. 42: 891–922.

Borkhausen, M. B. 1800–1803. Theoretisches-praktisches Handbuch der Forstbotanik und Forsttechnologie. 2 vols. Giessen and Darmstadt. [Vols. paged consecutively.] (Theor. Prakt. Handb. Forstbot.)

Borsch, T. and W. Barthlott. 1994 [1996]. Classification and distribution of the genus *Nelumbo* Adans. (Nelumbonaceae). Beitr. Biol. Pflanzen. 68: 421–450.

Bot. Beechey Voy.—See: W. J. Hooker and G. A. W. Arnott [1830–]1841.

Bot. Cab. = Botanical Cabinet; Consisting of Coloured Delineations of Plants from All Countries. . . . [Edited by G. Loddiges.]

Bot. California—See: S. Watson 1876–1880.

Bot. Exped.—See: J. C. Schaeffer 1760.

Bot. Gaz. = Botanical Gazette; Paper of Botanical Notes.

Bot. J. Linn. Soc. = Botanical Journal of the Linnean Society.

Bot. Jahrb. Syst. = Botanische Jahrbücher für Systematik, Pflanzengeschichte und Pflanzengeographie.

Bot. Mag. = Botanical Magazine; or, Flower-garden Displayed. . . . [Edited by Wm. Curtis. With vol. 15, 1801, title became Curtis's Botanical Magazine; or. . . .]

Bot. Mag. (Tokyo) = Botanical Magazine. [Shokubutsu-gaku Zasshi.]

Bot. Mater. Gerb. Glavn. Bot. Sada RSFSR = Botanicheskie Materialy Gerbariya Glavnogo Botanicheskogo Sada R S F S R.

Bot. Misc. = Botanical Miscellany.

Bot. Mus. Leafl. = Botanical Museum Leaflets. [Harvard University.]

Bot. Not. = Botaniska Notiser.

Bot. Not., Suppl. = Botisker Notiser. Supplement.

Bot. Reg. = Botanical Register. . . .

Bot. Repos. = Botanists' Repository, for New, and Rare Plants.

Bot. Rev. (Lancaster) = Botanical Review, Interpreting Botanical Progress.

Bot. Surv. Nebraska = Botanical Survey of Nebraska.

Bot. Voy. Herald—See: B. Seemann 1852–1857.

Bot. Zhurn. (Moscow & Leningrad) = Botanicheskii Zhurnal.

Botany (Fortieth Parallel)—See: S. Watson 1871.

Boufford, D. E. 1982. Notes on *Peperomia* (Piperaceae) in the southeastern United States. J. Arnold Arbor. 63: 317–325.

Boufford, D. E. and S. A. Spongberg. 1981. *Calycanthus floridus* (Calycanthaceae)—A nomenclatural note. J. Arnold Arbor. 62: 265–266.

Bousquet, J., S. H. Strauss, and Li P. 1992. Complete congruence between morphological and *rbc*L-based molecular phylogenies in birches and related species (Betulaceae). Molec. Biol. Evol. 9: 1076–1088.

Bradford, J. L. and D. L. March. 1977. Comparative studies of the witch hazels *Hamamelis virginiana* and *H. vernalis*. Proc. Arkansas Acad. Sci. 31: 29–31.

Brayshaw, T. C. 1966. What are the blue birches? Canad. Field-Naturalist 80: 187–194.

Brayshaw, T. C. 1976. Catkin Bearing Plants (Amentiferae) of British Columbia. Victoria. [Brit. Columbia Prov. Mus., Occas. Pap. 18.]

Brayshaw, T. C. 1989. Buttercups, Waterlilies, and Their Relatives (the Order Ranales) in British Columbia. Victoria. [Roy. Brit. Columbia Mus. Mem. 1.]

Brett, J. F. 1981. The Morphology and Taxonomy of *Caulophyllum thalictroides* (L.) Michx. (Berberidaceae) in North America. M.S. thesis. University of Guelph.

Bridson, G. D. R. and E. R. Smith. 1991. B-P-H/S. Botanico-Periodicum-Huntianum/Supplementum. Pittsburgh.

Brink, D. E. 1980. Reproduction and variation in *Aconitum columbianum* (Ranunculaceae), with emphasis on California populations. Amer. J. Bot. 67: 263–273.

Brink, D. E. 1982. Tuberous *Aconitum* (Ranunculaceae) of the continental United States: Morphological variation, taxonomy and disjunction. Bull. Torrey Bot. Club 109: 13–23.

Brink, D. E. and J. M. J. de Wet. 1980. Interpopulation variation in nectar production in *Aconitum columbianum* (Ranunculaceae). Oecologia 47: 160–163.

Brink, D. E., J. A. Woodson, and K. R. Stern. 1994. Bulbiferous *Aconitum* (Ranunculaceae) of the western United States. Sida 16: 9–15.

Brit. Fl. Gard.—See: R. Sweet 1823–1837.

Brittain, W. H. and W. F. Grant. 1965. Observations on Canadian birch (*Betula*) collections at the Morgan Arboretum. I. *B. papyrifera* in eastern Canada. Canad. Field-Naturalist 79: 189–197.

Brittain, W. H. and W. F. Grant. 1965b. Observations on Canadian birch (*Betula*) collections at the Morgan Arboretum. II. *B. papyrifera* var. *cordifolia*. Canad. Field-Naturalist 79: 253–257.

Brittain, W. H. and W. F. Grant. 1966. Observations on Canadian birch (*Betula*) collections at the Morgan Arboretum. III. *B. papyrifera* from British Columbia. Canad. Field-Naturalist 80: 147–157.

Brittain, W. H. and W. F. Grant. 1967. Observations on Canadian birch (*Betula*) collections at the Morgan Arboretum. IV. *B. caerulea-grandis* and hybrids. Canad. Field-Naturalist 81: 116–127.

Brittain, W. H. and W. F. Grant. 1967b. Observations on Canadian birch (*Betula*) collections at the Morgan Arboretum. V. *B. papyrifera* and *B. cordifolia* from eastern Canada. Canad. Field-Naturalist 81: 251–262.

Brittain, W. H. and W. F. Grant. 1968. Observations on Canadian birch (*Betula*) collections at the Morgan Arboretum. VI. *B. papyrifera* from the Rocky Mountains. Canad. Field-Naturalist 82: 44–48.

Brittain, W. H. and W. F. Grant. 1969. Observations on Canadian birch (*Betula*) collections at the Morgan Arboretum. VII. *B. papyrifera* and *B. resinifera* from northwestern Canada. Canad. Field-Naturalist 83: 185–202.

Brittain, W. H. and W. F. Grant. 1969b. Observations on Canadian birch (*Betula*) collections at the Morgan Arboretum. VIII. *Betula* from Grand Manon Island, New Brunswick. Canad. Field-Naturalist 83: 361–383.

Britton, N. L. 1891. The American species of the genus *Anemone* and the genera which have been referred to it. Ann. New York Acad. Sci. 6: 215–238.

Britton, N. L. 1901. Manual of the Flora of the Northern States and Canada. New York. (Man. Fl. N. States)

Britton, N. L. 1905. Hamamelidaceae. In: N. L. Britton et al., eds. 1905+. North American Flora. . . . 47+ vols. New York. Vol. 22, pp. 185–187.

Britton, N. L. et al., eds. 1905+. North American Flora. . . . 47+ vols. New York. [Vols. 1–34, 1905–1957; ser. 2, vols. 1–13, 1954+.]

Britton, N. L. and A. Brown. 1896. An Illustrated Flora of the Northern United States, Canada and the British Possessions from Newfoundland to the Parallel of the Southern Boundary of Virginia, and from the Atlantic Ocean Westward to the 102d Meridian. . . . 3 vols. New York.

Britton, N. L. and A. Brown. 1913. An Illustrated Flora of the Northern United States, Canada and the British Possessions from Newfoundland to the Parallel of the Southern Boundary of Virginia, and from the Atlantic Ocean Westward to the 102d Meridian . . . , ed. 2. 3 vols. New York.

Britton, N. L., E. E. Sterns, J. F. Poggenburg, A. Brown, T. C. Porter, and C. A. Hollick. 1888. Preliminary Catalogue of Anthophyta and Pteridophyta Reported As Growing Spontaneously within One Hundred Miles of New York City. New York. [Authorship often attributed as B.S.P. in nomenclatural contexts.] (Prelim. Cat.)

Brittonia = Brittonia; a Journal of Systematic Botany. . . .

Brophy, W. B. and D. R. Parnell. 1974. Hybridization between *Quercus agrifolia* and *Quercus wislizenii* (Fagaceae). Madroño 22: 290–302.

Brown, R. 1824. Journal of a Voyage for the Discovery of a North-west Passage. . . . Supplement to the Appendix. XI. Botany. A List of Plants Collected in Melville Island. London. (J. Voy. N.-W. Passage, Bot.)

Browne, P. 1756. The Civil and Natural History of Jamaica. . . . London. (Civ. Nat. Hist. Jamaica)

Brunner, F. and D. E. Fairbrothers. 1979. Serological investigation of the Corylaceae. Bull. Torrey Bot. Club 106: 97–103.

Buker, W. E. and S. A. Thompson. 1986. Is *Stylophorum diphyllum* (Papaveraceae) native to Pennsylvania? Castanea 51: 66–67.

Bull. Bot. Res., Harbin = Bulletin of Botanical Research. [Zhiwi Yanjiu.]

Bull. Calif. Acad. Sci. = Bulletin of the California Academy of Sciences.

Bull. Charleston Mus. = Bulletin of the Charleston Museum.

Bull. Gov. Forest Exp. Sta. = Bulletin of the Government Forest Experiment Station. [Ringyo Shikenjo Kenkyu Hokoku.]

Bull. Herb. Boissier = Bulletin de l'Herbier Boissier.

Bull. Misc. Inform. Kew = Bulletin of Miscellaneous Information, Royal Gardens, Kew.

Bull. Mus. Natl. Hist. Nat. = Bulletin du Muséum National d'Histoire Naturelle.

Bull. New York Bot. Gard. = Bulletin of the New York Botanical Garden.

Bull. New York State Mus. Sci. Serv. = Bulletin of the New York State Museum and Science Service.

Bull. Popular Inform. Arnold Arbor. = Bulletin of Popular Information: Arnold Arboretum Harvard University.

Bull. S. Calif. Acad. Sci. = Bulletin of the Southern California Academy of Sciences.

Bull. Soc. Imp. Naturalistes Moscou = Bulletin de la Société Impériale des Naturalistes de Moscou.

Bull. Soc. Roy. Bot. Belgique = Bulletin de la Société Royale de Botanique de Belgique.

Bull. Torrey Bot. Club = Bulletin of the Torrey Botanical Club.

Bunge, A. A. [1836.] Verzeichniss der im Jahre 1832, im östlichen Theile des Altai-Gebirges gesammelten Pflanzen. Ein Supplement zur Flora altaica. [St. Petersburg.] (Verz. Altai Pfl.)

Butterwick, M., B. D. Parfitt, and D. Hillyard. 1991. Vascular plants of the northern Hualapai Mountains, Arizona. J. Arizona-Nevada Acad. Sci. 24/25: 31–49.

Calder, J. A. and R. L. Taylor. 1963. A new species of *Isopyrum* endemic to the Queen Charlotte Islands of British Columbia and its relation to other species in the genus. Madroño 17: 69–76.

Calder, J. A. and R. L. Taylor. 1968. Flora of the Queen Charlotte Islands. 2 vols. Ottawa.

Callaway, D. J. 1994. The World of Magnolias. Portland.

Camp, W. H. 1951. A biogeographic and paragenetic analysis of the American beech. Yearb. Amer. Philos. Soc. 1960: 166–169.

Camus, A. A. 1934–1954. Les Chênes. Monographie du Genre *Quercus*. 3 vols. + 3 vols. atlas. Paris. (Chênes)

Canad. Field-Naturalist = Canadian Field-Naturalist.

Canad. J. Bot. = Canadian Journal of Botany.

Canad. J. Forest Res. = Canadian Journal of Forest Research.

Canad. J. Genet. Cytol. = Canadian Journal of Genetics and Cytology.

Candolle, A. P. de. [1817]1818–1821. Regni Vegetabilis Systema Naturale. . . . 2 vols. Paris, Strasbourg, and London. (Syst. Nat.)

Candolle, A. P. de and A. L. P. de Candolle, eds. 1823–1873. Prodromus Systematis Naturalis Regni Vegetabilis. . . . 17 vols. Paris etc. [Vols. 1–7 edited by A. P. de Candolle, vols. 8–17 by A. L. P. de Candolle.] (Prodr.)

Candolle, C. de 1869. Piperaceae. In: A. P. de Candolle and A. L. P. de Candolle, eds. 1823–1873. Prodromus Systematis Naturalis Regni Vegetabilis. . . . 17 vols. Paris etc. Vol. 16, part 1, pp. 235–471.

Canright, J. E. 1960. The comparative morphology and relationships of the Magnoliaceae. III. Carpels. Amer. J. Bot. 47: 145–155.

Capperino, M. E. and E. L. Schneider. 1985. Floral biology of *Nymphaea mexicana* Zucc. (Nymphaeaceae). Aquatic Bot. 23: 83–93.

Carlquist, S. 1982. Wood anatomy of *Illicium* (Illiciaceae): Phylogenetic, ecological, and functional interpretations. Amer. J. Bot. 69: 1587–1598.

Caron, D. M., ed. 1979. Increasing Production of Agricultural Crops through Increased Insect Pollination: Proceedings of the IVth International Symposium on Pollination. College Park, Md. [Maryland Agric. Exp. Sta. Special Misc. Publ. 1.]

Castanea = Castanea; Journal of the Southern Appalachian Botanical Club.

Cat. Pl. Oneida Co.—See: J. A. Paine 1865.

Cat. Rais. Pl. Dordogne—See: C. R. A. Des Moulins 1840.

Catling, P. M. and K. W. Spicer. 1988. The separation of *Betula populifolia* and *Betula pendula* and their status in Ontario. Canad. J. Forest Res. 18: 1017–1026.

Cely, J. E. 1979. The ecology and distribution of banana waterlily and its utilization by canvasback ducks. Proc. Annual Conf. SouthE. Assoc. Fish Wildlife Agencies 33: 43–47.

Cent. Pl. I—See: C. Linnaeus 1755.

Channell, R. B. and C. E. Wood Jr. 1962. The Leitneriaceae in the southeastern United States. J. Arnold Arbor. 43: 435–438.

Chapman, A. W. 1860. Flora of the Southern United States. . . . New York. (Fl. South. U.S.)

Chapman, A. W. 1883. Flora of the Southern United States . . . , ed. 2. New York. (Fl. South. U.S. ed. 2)

Chase, M. W. et al. 1993. Phylogenetics of seed plants: An analysis of nucleotide sequences from the plastid gene *rbc*L. Ann. Missouri Bot. Gard. 80: 528–580.

Chater, A. O. 1993. *Consolida*. In: T. G. Tutin et al., eds. 1993+. Flora Europaea, ed. 2. 1+ vol. Cambridge and New York. Vol. 1, pp. 260–262.

Check List For. Trees U.S.—See: G. B. Sudworth 1927.

Chen, C. J. 1982. A monograph of *Pilea* (Urticaceae) in China. Bull. Bot. Res., Harbin 2: 1–132.

Chênes—See: A. A. Camus 1934–1954.

Chevalier, A. 1901. Monographie des Myricacées. Mém. Soc. Sci. Nat. Cherbourg 32: 85–340.

Chew, W. L. 1969. A monograph of *Laportea* (Urticaceae). Gard. Bull. Straits Settlem. 25: 111–178.

Civ. Nat. Hist. Jamaica—See: P. Browne 1756.

Clark, C. 1978. Systematic studies of *Eschscholzia* (Papaveraceae). I. The origin and affinities of *E. mexicana*. Syst. Bot. 3: 374–385.

Clark, C. and M. Faull. 1991. A new subspecies and a combination in *Eschscholzia minutiflora* (Papaveraceae). Madroño 38: 73–79.

Clark, C. and J. A. Jernstedt. 1978. Systematic studies of *Eschscholzia* (Papaveraceae). II. Seed coat microsculpturing. Syst. Bot. 3: 386–402.

Clarkson, R. B. 1960. Notes on the distribution of *Alnus crispa* in eastern North America. Castanea 25: 83–86.

Class-book Bot. ed. 2(a)—See: A. Wood 1847.

Clausen, K. E. 1963. Characteristics of a hybrid birch and its parent species. Canad. J. Bot. 41: 441–458.

Coker, W. C. 1943. *Magnolia cordata* Michx. J. Elisha Mitchell Sci. Soc. 59: 81–88.

Coll. Bot.—See: J. Lindley 1821[–1826].

Collect. Bot. (Barcelona) = Collectanea Botanica; a Barcionensi Botanico Instituto Edita.

Collectanea—See: N. J. Jacquin 1786[1787]–1796[1797].

Compton, J. 1992. *Cimicifuga* L.: Ranunculaceae. Plantsman 14: 99–115.

Comun. Inst. Trop. Invest. Ci. Univ. El Salvador = Comunicaciones. Instituto Tropical de Investigaciones Científicas, Universidad de El Salvador.

Conard, H. S. 1905. The waterlilies: A monograph of the genus *Nymphaea*. Publ. Carnegie Inst. Wash. 4: 1–279.

Consp. Fl. Eur.—See: C. F. Nyman 1878–1890.

Contr. Bot. Lab. Morris Arbor. Univ. Pennsylvania = Contributions from the Botanical Laboratory and the Morris Arboretum of the University of Pennsylvania.

Contr. Dudley Herb. = Contributions from the Dudley Herbarium of Stanford University.

Contr. Gray Herb. = Contributions from the Gray Herbarium of Harvard University. [Some numbers reprinted from other periodicals, e.g. Rhodora.]

Contr. Texas Res. Found., Bot. Stud. = Contributions from the Texas Research Foundation, Botanical Studies.

Contr. U.S. Natl. Herb. = Contributions from the United States National Herbarium.

Contr. W. Bot. = Contributions to Western Botany.

Cook, C. D. K. 1966. A monographic study of *Ranunculus* subgenus *Batrachium* (DC.) A. Gray. Mitt. Bot. Staatssamml. München 6: 47–237.

Cooper, A. W. and E. P. Mercer. 1977. Morphological variation in *Fagus grandifolia* Ehrh. in North Carolina. J. Elisha Mitchell Sci. Soc. 93: 136–149.

Cornell Univ. Agric. Exp. Sta. Mem. = Cornell University Agricultural Experiment Station Memoir.

Corner, E. J. H. 1976. The Seeds of Dicotyledons. 2 vols. Cambridge and New York.

Correll, D. S. and H. B. Correll. 1982. Flora of the Bahama Archipelago. Vaduz.

Correll, D. S. and M. C. Johnston. 1970. Manual of the Vascular Plants of Texas. Renner, Tex.

Cottam, W. P., J. M. Tucker, and F. S. Santamour Jr. 1982. Oak Hybridization at the University of Utah. Salt Lake City. [State Arbor. Utah Publ. 1.]

Crane, P. R. 1989. Early fossil history and evolution of the Betulaceae. In: P. R. Crane and S. Blackmore, eds. 1989. Evolution, Systematics, and Fossil History of the Hamamelidae. 2 vols. Oxford. Vol. 2, pp. 87–116.

Crane, P. R. and S. Blackmore, eds. 1989. Evolution, Systematics, and Fossil History of the Hamamelidae. 2 vols. Oxford. [Syst. Assoc. Special Vol. 40A,B.]

Crantz, H. J. N. 1762–1767. Stirpium Austriarum Fasciculus

I [–III]. 3 fasc. Vienna and Leipzig. (Stirp. Austr. Fasc.)

Cronquist, A. 1981. An Integrated System of Classification of Flowering Plants. New York.

Cycl.—See: A. Rees [1802–]1819–1820.

Dahlgren, R. M. T. 1980. A revised system of classification of the angiosperms. Bot. J. Linn. Soc. 80: 91–124.

Dancik, B. P. 1969. Dark-barked birches of southern Michigan. Michigan Bot. 8: 38–41.

Dancik, B. P. and B. V. Barnes. 1971. Variability in bark morphology of yellow birch in an even-aged stand. Michigan Bot. 10: 34–38.

Danert, S. 1958. Zur Systematik von *Papaver somniferum* L. Kulturpflanze 6: 61–88.

Darby, J. 1841. A Manual of Botany Adapted to the Productions of the Southern States. 2 parts. Macon. [Parts paged separately.] (Man. Bot.)

Darlington, C. D. and A. P. Wylie. 1955. Chromosome Atlas of Flowering Plants. London.

Darwiniana = Darwiniana; Carpeta del "Darwinion."

Davey, A. J. and C. M. Gibson. 1917. Note on the distribution of sexes in *Myrica gale*. New Phytol. 16: 147–151.

Day, J. W. 1975. The Autecology of *Leitneria floridana*. Ph.D. thesis. Mississippi State University.

DeHond, P. E. and C. S. Campbell. 1989. Multivariate analyses of hybridization between *Betula cordifolia* and *B. populifolia* (Betulaceae). Canad. J. Bot. 67: 2252–2260.

Demuth, P. and F. S. Santamour Jr. 1978. Carotenoid flower pigments in *Liriodendron* and *Magnolia*. Bull. Torrey Bot. Club 105: 65–66.

Dendrologie—See: K. H. E. Koch 1873.

Denham, D. and H. Clapperton. 1826. Narrative of Travels and Discoveries in Northern and Central Africa in the Years 1822, 1823 and 1824. . . . London. [Botanical appendix by R. Brown, pp. 208–246.] (Narr. Travels Africa)

Dennis, W. M. 1976. A Biosystematic Study of *Clematis* Section *Viorna* Subsection *Viornae*. Ph.D. dissertation. University of Tennessee.

Dennis, W. M. 1979. *Clematis pitcheri* T. & G. var. *dictyota* (Greene) Dennis, comb. nov. (Ranunculaceae). Sida 8: 194–195.

Denton, M. F. 1978. *Ranunculus californicus,* a new record for the state of Washington. Madroño 25: 132.

DePoe, C. E. and E. O. Beal. 1969. Origin and maintenance of clinal variation in *Nuphar* (Nymphaeaceae). Brittonia 21: 15–28.

De Steven, D. 1983. Floral ecology of witch-hazel (*Hamamelis virginiana*). Michigan Bot. 22: 163–171.

Des Moulins, C. R. A. 1840. Catalogue Raisonné des Plantes Qui Croissent Spontanément dans le Département de la Dordogne. . . . 1 part (only). Bordeaux. (Cat. Rais. Pl. Dordogne)

Desfontaines, R. L. [1798–1799.] Flora Atlantica sive Historia Plantarum, Quae in Atlante, Agro Tunetano et Algeriensi Crescunt. 2 vols. in 9 parts. Paris. (Fl. Atlant.)

Deut. Dendrol.—See: B. A. E. Koehne 1893.

Dewick, P. M. 1983. Tumour inhibitors from plants. In: G. E. Trease and W. C. Evans. 1983. Pharmacognosy, ed. 12. London. Pp. 629–647.

Diario Viaje Comis. Limites—See: J. L. Berlandier and R. Chovel 1850.

Diels, F. L. E. 1910. Menispermaceae. In: H. G. A. Engler, ed. 1900–1953. Das Pflanzenreich. . . . 107 vols. Berlin. Vol. 46[IV,94], pp. 334–345.

Dietrich, A. G. 1831–1832. Caroli a Linné Species Plantarum. . . . Olim Curante Carolo Ludovico Willdenow. Editio Sexta. 2 vols. Berlin. (Sp. Pl.)

Dilcher, D. L. 1990. The occurrence of fruits with affinities to Ceratophyllaceae in lower and mid-Cretaceous sediments. [Abstract.] Amer. J. Bot. 76: 162.

Dirr, M. A. 1986. Hardy *Illicium* species display commendable attributes. Amer. Nurseryman 163: 92–100.

Diss. Abstr. = Dissertation Abstracts; Abstracts of Dissertations and Monographs in Microfilm.

Diss. Observ. Bot.—See: J. P. Du Roi 1771.

Distr. Crypt. Vasc. Ross.—See: F. Ruprecht 1845.

Don, G. 1831–1838. A General History of the Dichlamydeous Plants. . . . 4 vols. London. (Gen. Hist.)

Doroszewska, A. 1974. The genus *Trollius* L., a taxonomical study. Monogr. Bot. 41: 1–167, plates 1–16.

Dorr, L. J. and K. C. Nixon. 1985. Typification of the oak (*Quercus*) taxa described by S. B. Buckley (1809–1884). Taxon 34: 211–228.

Douglas, G. W., G. B. Straley, and D. Meidinger. 1989. The Vascular Plants of British Columbia. 3 vols. Victoria. [B.C. Minist. Forests, Special Rep. 1–3.]

Doyle, J. A. 1969. Cretaceous angiosperm pollen of the Atlantic coastal plain and its evolutionary significance. J. Arnold Arbor. 50: 1–35.

Drumke, J. S. 1965. A systematic survey of *Corylus* in North America. Diss. Abstr. 25: 4925–4926.

Drummond, J. R. and J. Hutchinson. 1920. A revision of *Isopyrum* (Ranunculaceae) and its nearer allies. Bull. Misc. Inform. Kew 1920: 145–169.

Duchartre, P. 1864. Aristolochiaceae. In: A. P. de Candolle and A. L. P. de Candolle, eds. 1823–1873. Prodromus Systematis Naturalis Regni Vegetabilis. . . . 17 vols. Paris etc. Vol. 15, pp. 421–498.

Dugle, J. R. 1966. A taxonomic study of western Canadian species in the genus *Betula*. Canad. J. Bot. 44: 929–1007.

Dugle, J. R. 1969. Some nomenclatural problems in North American *Betula*. Canad. Field-Naturalist 83: 251–253.

Duhamel du Monceau, H. et al. [1800–]1801–1819. Traité des Arbres et Arbustes, Que l'On Cultivé in France. . . . Nouvelle Édition. . . . 7 vols. Paris. (Traité Arbr. Arbust. nouv. ed.)

Duke, J. A. 1985. CRC Handbook of Medicinal Herbs. Boca Raton.

Dunal, M. F. 1817. Monographie de la Famille des Anonacées. Paris, London, Strasbourg, and Montpellier. (Monogr. Anonac.)

Duncan, T. 1980. A taxonomic study of the *Ranunculus hispidus* Michaux complex in the Western Hemisphere. Univ. Calif. Publ. Bot. 77: 1–125.

Duncan, T. and C. S. Keener. 1991. A classification of the Ranunculaceae with special reference to the Western Hemisphere. Phytologia 70: 24–27.

Duncan, W. H. 1959. Leaf variation in *Liquidambar styraciflua* L. Castanea 24: 99–111.

Duncan, W. H. and M. B. Duncan. 1988. Trees of the Southeastern United States. Athens, Ga.

Du Roi, J. P. 1771. Dissertatio Inauguralis Observationes Botanicas Sistens. Helmstad. (Diss. Observ. Bot.)

Econ. Bot. = Economic Botany; Devoted to Applied Botany and Plant Utilization.

Edinburgh J. Sci. = Edinburgh Journal of Science.

Edinburgh Philos. J. = Edinburgh Philosophical Journal.

Edwards's Bot. Reg. = Edwards's Botanical Register.

Ehrhart, J. F. 1787–1792. Beiträge zur Naturkunde. . . . 7 vols. Hannover and Osnabrück. (Beitr. Naturk.)

Eichenberger, M. D. and G. R. Parker. 1976. Goldenseal (*Hydrastis canadensis* L.) distribution, phenology and biomass in an oak-hickory forest. Ohio J. Sci. 76: 204–210.

Elench. Pl.—See: M. Lagasca y Segura 1816.

Elias, T. S. 1970. The genera of Ulmaceae in the southeastern United States. J. Arnold Arbor. 51: 18–40.

Elias, T. S. 1971. The genera of Myricaceae in the southeastern United States. J. Arnold Arbor. 52: 305–318.

Elias, T. S. 1971b. The genera of Fagaceae in the southeastern United States. J. Arnold Arbor. 52: 159–195.

Elias, T. S. 1972. The genera of Juglandaceae in the southeastern United States. J. Arnold Arbor. 53: 26–51.

Elkan, L. 1839. Tentamen Monographiae Generis *Papaver*. Königsberg. (Tent. Monogr. Papaver)

Elliott, S. [1816–]1821–1824. A Sketch of the Botany of South-Carolina and Georgia. 2 vols. in 13 parts. Charleston. (Sketch Bot. S. Carolina)

Emboden, W. A. 1974. *Cannabis*—a polytypic genus. Econ. Bot. 28 304–310.

Emory, W. H. 1848. Notes of a Military Reconnoissance, from Fort Leavenworth, in Missouri, to San Diego, in California, Including Part of the Arkansas, Del Norte, and Gila Rivers. . . . Made in 1846–7, with the Advanced Guard of the "Army of the West." Washington. (Not. Milit. Reconn.)

Emory, W. H. 1857–1859. Report on the United States and Mexican Boundary Survey, Made under the Direction of the Secretary of the Interior. 2 vols. in parts. Washington. (Rep. U.S. Mex. Bound.)

Encycl.—See: J. Lamarck et al. 1783–1817.

Endress, P. K. 1977. Evolutionary trends in the Hamamelidales-Fagales group. Pl. Syst. Evol., Suppl. 1: 321–347.

Endress, P. K. 1987. Aspects of evolutionary differentiation of the Hamamelidaceae and the lower Hamamelidae. Pl. Syst. Evol. 162: 193–211.

Endress, P. K. 1989. A suprageneric taxonomic classification of the Hamamelidaceae. Taxon 38: 371–376.

Endress, P. K. 1989b. Phylogenetic relationships in the Hamamelidoideae. In: P. R. Crane and S. Blackmore, eds. 1989. Evolution, Systematics, and Fossil History of the Hamamelidae. 2 vols. Oxford. Vol. 1, pp. 227–248.

Engler, H. G. A. 1888. Saururaceae. In: H. G. A. Engler and K. Prantl, eds. 1887–1915. Die natürlichen Pflanzenfamilien. . . . 254 fasc. Leipzig. Fasc. 14[III,1], pp. 1–3.

Engler, H. G. A. 1888b. Moraceae. In: H. G. A. Engler and K. Prantl, eds. 1887–1915. Die natürlichen Pflanzenfamilien. . . . 254 fasc. Leipzig. Fasc. 18[III,1], pp. 66–96.

Engler, H. G. A., ed. 1900–1953. Das Pflanzenreich. . . . 107 vols. Berlin. [Sequence of volume (Heft) numbers (order of publication) is independent of the sequence of series and family (Roman and Arabic) numbers (taxonomic order).] (Pflanzenr.)

Engler, H. G. A., H. Harms, J. Mattfeld, H. Melchior, and E. Werdermann, eds. 1924+. Die natürlichen Pflanzenfamilien . . . , ed. 2. 26+ vols. Leipzig and Berlin. (Nat. Pflanzenfam. ed. 2)

Engler, H. G. A. and K. Prantl, eds. 1887–1915. Die natürlichen Pflanzenfamilien. . . . 254 fasc. Leipzig. [Sequence of fascicle (Lieferung) numbers (order of publication) is independent of the sequence of division (Teil) and subdivision (Abteilung) numbers (taxonomic order).]

Enum. Pl.—See: M. Vahl 1804–1805; C. L. Willdenow 1809–1813[–1814].

Enum. Syst. Pl.—See: N. J. Jacquin 1760.

Erickson, R. O. 1943. Taxonomy of *Clematis* section *Viorna*. Ann. Missouri Bot. Gard. 30: 1–62, plate 1.

Ernst, W. R. 1962. A Comparative Morphology of the Papaveraceae. Ph.D. dissertation. Stanford University.

Ernst, W. R. 1962b. The genera of Papaveraceae and Fumariaceae in the southeastern United States. J. Arnold Arbor. 43: 315–343.

Ernst, W. R. 1963. The genera of Hamamelidaceae and Platanaceae in the southeastern United States. J. Arnold Arbor. 44: 193–210.

Ernst, W. R. 1964. The genera of Berberidaceae, Lardizabalaceae, and Menispermaceae in the southeastern United States. J. Arnold Arbor. 45: 1–35.

Ernst, W. R. 1964b. The genus *Eschscholzia* in the south Coast Ranges of California. Madroño 17: 281–294.

Ernst, W. R. 1967. Floral morphology and systematics of *Platystemon* and its allies *Hesperomecon* and *Meconella* (Papaveraceae: Platystemonoideae). Univ. Kansas Sci. Bull. 47: 25–70.

Erythea = Erythea; a Journal of Botany, West American and General.

Essig, F. B. 1990. The *Clematis virginiana* (Ranunculaceae) complex in the southeastern United States. Sida 14: 49–68.

Essig, F. B. 1992. Seedling morphology in *Clematis* (Ranunculaceae) and its taxonomic implications. Sida 14: 377–390.

Evol. Biol. = Evolutionary Biology.

Evolution = Evolution, International Journal of Organic Evolution.

Ewan, J. 1945. A synopsis of the North American species of *Delphinium*. Univ. Colorado Stud., Ser. D, Phys. Sci. 2: 55–244.

Exell, A. W. 1927. William Bartram and the genus *Asimina* in North America. J. Bot. 65: 65–70.

Fabergé, A. C. 1944. Genetics of the *Scapiflora* section of *Papaver*. III. Interspecific hybrids and genetic homology. J. Genet. 46: 125–149.

Fagerström, L. and G. Kvist. 1983. Vier neue arktische und subarktische *Ranunculus auricomus*-Sippen. Ann. Bot. Fenn. 20: 237–243.

Fahselt, D. 1970. The anthocyanins of *Dicentra* (Fumariaceae). Canad. J. Bot. 48: 49–53.

Fahselt, D. 1971. Flavonoid components of *Dicentra canadensis* (Fumariaceae). Canad. J. Bot. 49: 1559–1563.

Fahselt, D. and M. Ownbey. 1968. Chromatographic comparison of *Dicentra* species and hybrids. Amer. J. Bot. 55: 334–345.

Fam. Pl.—See: M. Adanson 1763[–1764].

Famil. Lessons Bot.—See: M. J. Young 1873.

Fassett, N. C. 1953. North American *Ceratophyllum*. Comun. Inst. Trop. Invest. Ci. Univ. El Salvador 2: 25–45.

Fassett, N. C. 1953b. A monograph of *Cabomba*. Castanea 13: 116–128.

Fedde, F. 1909. Papaveraceae–Hypecoideae et Papaveraceae–Papaveroideae. In: H. G. A. Engler, ed. 1900–1953. Das Pflanzenreich. . . . 107 vols. Berlin. Vol. 40[IV,104], pp. 1–430.

Fedde, F. 1936. Papaveraceae. In: H. G. A. Engler et al., eds. 1924+. Die natürlichen Pflanzenfamilien . . . , ed. 2. 26+ vols. Leipzig and Berlin. Vol. 17b, pp. 5–145.

Feddes Repert. = Feddes Repertorium.

Fernald, M. L. 1922. The relationships of some American and Old World birches. Amer. J. Sci. 169: 167–194.

Fernald, M. L. 1928b. The North American species of *Anemone* § *Anemonanthea*. Rhodora 30: 180–188.

Fernald, M. L. 1929. *Coptis trifolia* and its eastern American representative. Rhodora 31: 136–142.

Fernald, M. L. 1936. *Pilea* in eastern North America. Contr. Gray Herb. 113: 169–170.

Fernald, M. L. 1936b. Plants from the outer coastal plain of Virginia. Rhodora 38: 376–404, 414–452.

Fernald, M. L. 1940. What is *Actaea alba?* Rhodora 42: 260–265.

Fernald, M. L. 1943. Morphological differentiation of *Clematis ochroleuca* and its allies. Rhodora 45: 401–412, figs. 776–782.

Fernald, M. L. 1945. Some North American Corylaceae (Betulaceae). 1. Notes on *Betula* in eastern North America. Rhodora 47: 303–329.

Fernald, M. L. 1950. Gray's Manual of Botany, ed. 8. New York.

Fernald, M. L. 1950c. *Betula Michauxii*, a brief symposium. 1. Introductory note. Rhodora 52: 25–27.

Ferry, R. J. Sr. and R. J. Ferry Jr. 1987. *Calycanthus brockiana* (Calycanthaceae), a new spicebush from north central Georgia. Sida 12: 339–341.

Fey, B. S. and P. K. Endress. 1983. Development and morphological interpretation of the cupule in Fagaceae. Flora 173: 451–468.

Field Mus. Nat. Hist., Bot. Ser. = Field Museum of Natural History. Botanical Series.

Figlar, R. B. 1981. A last stand in Arkansas [*Magnolia macrophylla*]. J. Amer. Magnolia Soc. 17(1): 17–20.

Fischer, F. E. L. von, C. A. Meyer, E. R. von Trautvetter, and J. L. É. Avé-Lallemant. 1835–1846. Index Seminum, Quae Hortus Botanicus Imperialis Petropolitanus pro Mutua Commutatione Offert. 11 vols. St. Petersburg. (Index Sem. Hort. Petrop.)

Fl. Alaska—See: J. P. Anderson 1959.

Fl. Alaska Yukon—See: E. Hultén 1941–1950.

Fl. Aleut. Isl.—See: E. Hultén 1937.

Fl. Amer. Sept.—See: F. Pursh 1814.

Fl. Angl.—See: W. Hudson 1762.

Fl. Atlant.—See: R. L. Desfontaines [1798–1799].

Fl. Bor.-Amer.—See: W. J. Hooker [1829–]1833–1840; A. Michaux 1803.

Fl. Boston.—See: J. Bigelow 1814.

Fl. Boston. ed. 2—See: J. Bigelow 1824.

Fl. Brit. W. I.—See: A. H. R. Grisebach [1859–]1864.

Fl. Calif.—See: W. L. Jepson 1909–1943.

Fl. Carniol.—See: J. A. Scopoli 1760.

Fl. Carol.—See: T. Walter 1788.

Fl. Chil.—See: C. Gay 1845–1854.

Fl. Cochinch.—See: J. de Loureiro 1790.

Fl. Franç. ed. 3—See: J. Lamarck and A. P. de Candolle 1805[–1815].

Fl. France—See: G. Rouy et al. 1893–1913.

Fl. Francisc.—See: E. L. Greene 1891–1897.

Fl. Graec. Prodr.—See: J. Sibthorp and J. E. Smith 1806–1813[–1816].

Fl. Jap.—See: C. P. Thunberg 1784.

Fl. Jen.—See: J. H. Rudolph [1781].

Fl. Kamtchatka—See: E. Hultén 1927–1930.

Fl. Lapp.—See: G. Wahlenberg 1812.

Fl. Mexic. ed. 2—See: M. de Sessé y Lacasta and J. M. Moçiño 1894.

Fl. N. Amer.—See: J. Torrey and A. Gray 1838–1843.

Fl. New York—See: J. Torrey 1843.

Fl. Palaest.—See: C. Linnaeus 1756.

Fl. Peruv.—See: H. Ruiz Lopez and J. A. Pavon 1798–1802.

Fl. Peruv. Prodr.—See: H. Ruiz Lopez and J. A. Pavon 1794.

Fl. Pl. Ferns Ariz.—See: T. H. Kearney and R. H. Peebles 1942.

Fl. Rocky Mts.—See: P. A. Rydberg 1917.

Fl. Ross.—See: C. F. von Ledebour [1841]1842–1853.

Fl. S.E. U.S.—See: J. K. Small 1903.

Fl. Serres Jard. Eur. = Flore des Serres et des Jardins de l'Europe.

Fl. South. U.S.—See: A. W. Chapman 1860.

Fl. South. U.S. ed. 2—See: A. W. Chapman 1883.

Fl. Tangut.—See: C. J. Maximowicz 1889.

Flora = Flora; oder (allgemeine) botanische Zeitung. [Vols. 1–16, 1818–33, include "Beilage" and "Ergänzungsblätter"; vols. 17–25, 1834–42, include "Beiblatt" and "Intelligenzblatt."]

Flora of North America Editorial Committee, eds. 1993+. Flora of North America North of Mexico. 3+ vols. New York and Oxford.

Florida Sci. = Florida Scientist.

Forest Sci. = Forest Science.

Forman, L. L. 1966. On the evolution of cupules in the Fagaceae. Kew Bull. 18: 385–419.

Foster, S. 1989. Phytogeographic and botanical considerations of medicinal plants in eastern Asia and eastern North America. Herbs Spices Med. Pl. 4: 115–144.

Fowells, H. A. 1965. Sylvics of Forest Trees of the United States. Washington. [Agric. Handb. 271.]

Fredskild, B. 1991. The genus *Betula* in Greenland—Holocene history, present distribution, and synecology. Nordic J. Bot. 11: 393–412.

Frémont, J. C., J. Torrey, and J. Hall. 1843–1845. Report of the Exploring Expedition to the Rocky Mountains in the Year 1842, and to Oregon and North California in the Year 1843–44. 2 parts. Washington. [Parts paged consecutively.] (Rep. Exped. Rocky Mts.)

Fries, R. E. 1931. Revision der Arten einiger Annonaceen-Gattungen. Acta Horti Berg. 10: 1–341.

Fries, R. E. 1934. Revision der Arten einiger Annonaceen-Gattungen. Acta Horti Berg. 12: 1–220.

Fries, R. E. 1939. Revision der Arten einiger Annonaceen-Gattungen. Acta Horti Berg. 12: 289–577.

Friis, I. 1993. Urticaceae. In: K. Kubitzki et al., eds. 1990+. The Families and Genera of Vascular Plants. 2+ vols. Berlin etc. Vol. 2, pp. 612–630.

Frodin, D. G. 1964. A Preliminary Revision of the Section *Anemonanthea* of *Anemone* in Eastern North America, with Special Reference to the Southern Appalachian Mountains. M.S. thesis. University of Tennessee.

Fruct. Sem. Pl.—See: J. Gaertner 1788–1791[–1792].

Fu, D.-Z. 1990. Phylogenetic considerations on the subfamily Thalictroideae (Ranunculaceae). Cathaya 2: 181–190.

Fukuda, I. 1967. The biosystematics of *Achlys*. Taxon 16: 308–316.

Fukuda, I. and H. G. Baker. 1970. *Achlys californica*—(Berberidaceae)—A new species. Taxon 19: 341–344.

Fulling, F. H. 1953. American witch hazel—History, nomenclature, and modern utilization. Econ. Bot. 7: 359–389.

Furlow, J. J. 1979. The systematics of the American species of *Alnus* (Betulaceae). Rhodora 81: 1–121, 151–248.

Furlow, J. J. 1983. The phylogenetic relationships of the genera and infrageneric taxa of the Betulaceae. [Abstract.] Amer. J. Bot. 70(suppl.): 114.

Furlow, J. J. 1983b. Evolutionary divergence and classification of the *Alnus viridis* complex (Betulaceae). [Abstract.] Amer. J. Bot. 70(suppl.): 114.

Furlow, J. J. 1984. The evolution and classification of the *Betula nana* complex. [Abstract.] Amer. J. Bot. 71(suppl.): 166.

Furlow, J. J. 1987. The *Carpinus caroliniana* complex in North America. I. A multivariate analysis of geographical variation. Syst. Bot. 12: 21–40.

Furlow, J. J. 1987b. The *Carpinus caroliniana* complex in North America. II. Systematics. Syst. Bot. 12: 416–434.

Furlow, J. J. 1990. The genera of Betulaceae in the southeastern United States. J. Arnold Arbor. 71: 1–67.

Gaddy, L. L. 1986. A new *Hexastylis* (Aristolochiaceae) from Transylvania County, North Carolina. Brittonia 38: 82–85.

Gaddy, L. L. 1987. A review of the taxonomy and biogeography of *Hexastylis* (Aristolochiaceae). Castanea 52: 186–196.

Gaertner, J. 1788–1791[–1792]. De Fructibus et Seminibus Plantarum. . . . 2 vols. Stuttgart and Tübingen. [Vol. 1 in 1 part, vol. 2 in 4 parts.] (Fruct. Sem. Pl.)

Gard. & Forest = Garden and Forest; a Journal of Horticulture, Landscape Art and Forestry.

Gard. Bull. Straits Settlem. = Gardens' Bulletin. Straits Settlements.

Gard. Chron. = Gardener's Chronicle.

Gard. Dict. Abr. ed. 4—See: P. Miller 1754.

Gard. Dict. ed. 8—See: P. Miller 1768.

Gardiner, R. C. 1965. Studies in the Leaf Anatomy and Geographic Distribution of *Trema*. M.S. thesis. University of Connecticut.

Gaudichaud-Beaupré, C. 1826[–1830]. Voyage Autour du Monde . . . Exécuté sur les Corvettes de S.M. l'Uranie et la Physicienne . . . Publié . . . par M. Louis Freycinet. Botanique par M. Charles Gaudichaud. . . . 12 parts, atlas. Paris. [Botanical portion of larger work by H. L. C. Freycinet.] (Voy. Uranie)

Gay, C. 1845–1854. Historia Física y Política de Chile. . . . Botánica [Flora Chilena]. 8 vols., atlas. Paris. (Fl. Chil.)

Gen. Amer. Bor.—See: A. Gray 1848–1849.

Gen. Hist.—See: G. Don 1831–1838.

Gen. N. Amer. Pl.—See: T. Nuttall 1818.

Gen. Pl.—See: J. C. Schreber 1789–1791; J. Wernischeck 1763.

Gen. Pl. ed. 5—See: C. Linnaeus 1754.

Gentes Herb. = Gentes Herbarum; Occasional Papers on the Kinds of Plants.

Georgi, J. G. 1775. Bemerkungen einer Reise im Russischen Reich im Jahre 1772. 2 vols. St. Petersburg. [Vols. paged consecutively.] (Bemerk. Reise Russ. Reich)

Ges. Naturf. Freunde Berlin Neue Schriften = Der Gesellschaft naturforschender Freunde zu Berlin, neue Schriften.

Giannasi, D. E. 1978. Generic relationships in the Ulmaceae based on flavonoid chemistry. Taxon 27: 331–344.

Giannasi, D. E. 1986. Phytochemical aspects of phylogeny in Hamamelidae. Ann. Missouri Bot. Gard. 73: 417–437.

Giannasi, D. E. and K. J. Niklas. 1977. Flavonoids and other constituents of fossil Miocene *Celtis* and *Ulmus* (Succor Creek Flora). Science 197: 765–767.

Gilg, E. 1925. Canellaceae. In: H. G. A. Engler et al., eds. 1924+. Die natürlichen Pflanzenfamilien . . . , ed. 2. 26+ vols. Leipzig and Berlin. Vol. 21, pp. 323–328.

Gleason, H. A. 1952. The New Britton and Brown Illustrated Flora of the Northeastern United States and Adjacent Canada. 3 vols. New York.

Gleason, H. A. and A. Cronquist. 1991. Manual of Vascular Plants of Northeastern United States and Adjacent Canada, ed. 2. Bronx.

Gmelin, J. F. 1791[–1792]. Caroli à Linné . . . Systema Naturae per Regna Tria Naturae. . . . Tomus II. Editio Decima Tertia, Aucta, Reformata. 2 parts. Leipzig. (Syst. Nat.)

Godfrey, R. K. 1988. Trees, Shrubs, and Woody Vines of Northern Florida and Adjacent Georgia and Alabama. Athens, Ga.

Godfrey, R. K. and A. F. Clewell. 1965. Polygamodioecious *Leitneria floridana* (Leitneriaceae). Sida 2: 172–173.

Goldblatt, P. 1974. Biosystematic studies in *Papaver* section *Oxytona*. Ann. Missouri Bot. Gard. 61: 264–296.

Goldblatt, P. and P. K. Endress. 1977. Cytology and evolution in Hamamelidaceae. J. Arnold Arbor. 58: 67–71.

Grana = Grana; an International Journal of Palynology Including World Pollen and Spore Flora.

Grant, W. F. and B. K. Thompson. 1975. Observations on Canadian birches, *Betula cordifolia, B. populifolia, B. papyrifera* and *B.* ×*caerulea.* Canad. J. Bot. 53: 1478–1490.

Gray, A. 1848. A Manual of the Botany of the Northern United States. . . . Boston, Cambridge, and London. (Manual)

Gray, A. 1848–1849. Genera Florae Americae Boreali-orientalis Illustrata. The Genera of the Plants of the United States. . . . 2 vols. Boston, New York, and London. (Gen. Amer. Bor.)

Gray, A. 1867. A Manual of the Botany of the Northern United States . . . , ed. 5. New York and Chicago. [Pteridophytes by D. C. Eaton.] (Manual ed. 5)

Gray, A., S. Watson, B. L. Robinson, et al. 1878–1897. Synoptical Flora of North America. 2 vols. in parts and fasc. New York etc. [Vol. 1(1,1), 1895; vol. 1(1,2), 1897; vol. 1(2), 1884; vol. 2(1), 1878.] (Syn. Fl. N. Amer.)

Gray, S. F. 1821. A Natural Arrangement of British Plants. . . . 2 vols. London. (Nat. Arr. Brit. Pl.)

Great Plains Flora Association. 1986. Flora of the Great Plains. Lawrence, Kans.

Green, P. S. 1964. Registration of cultivar names in *Ulmus.* Arnoldia (Jamaica Plain) 24: 41–80.

Greene, E. L. 1889–1890. Illustrations of West American Oaks. 2 parts. San Francisco. [Parts paged consecutively.] (Ill. W. Amer. Oaks)

Greene, E. L. 1891–1897. Flora Franciscana. An Attempt to Classify and Describe the Vascular Plants of Middle California. 4 parts. San Francisco. [Parts paged consecutively.] (Fl. Francisc.)

Greene, E. L. 1905. Suggestions regarding *Sanguinaria.* Pittonia 5: 306–308.

Greene, E. L. 1905b. Revision of *Eschscholtzia.* Pittonia 5: 205–308.

Gregory, M. P. 1956. A phyletic rearrangement of the Aristolochiaceae. Amer. J. Bot. 43: 110–122.

Gregory, W. C. 1941. Phylogenetic and cytological studies in the Ranunculaceae Juss. Trans. Amer. Philos. Soc., n. s. 31: 443–[521].

Greuter, W., F. R. Barrie, H. M. Burdet, W. G. Chaloner, V. Demoulin, D. L. Hawksworth, P. M. Jørgensen, D. H. Nicolson, P. C. Silva, P. Trehane, and J. McNeill, eds. 1994. International Code of Botanical Nomenclature (Tokyo Code): Adopted by the Fifteenth International Botanical Congress, Yokohama, August–September 1993. Königstein. [Regnum Veg. 131.]

Grey-Wilson, C. 1989. *Clematis orientalis* (Ranunculaceae) and its allies. Kew Bull. 44: 33–60.

Grey-Wilson, C. 1993. Poppies: A Guide to the Poppy Family in the Wild and in Cultivation. Portland.

Griffin, J. R. and W. B. Critchfield. 1972. The Distribution of Forest Trees in California. Berkeley. [U.S.D.A. Forest Serv., Res. Pap. PSW-82.]

Grisebach, A. H. R. [1859–]1864. Flora of the British West Indian Islands. 7 parts. London. [Parts paged consecutively.] (Fl. Brit. W. I.)

Grisebach, A. H. R. 1879. Symbolae ad Floram Argentinam. Zweite Bearbeitung argentinischer Pflanzen. Göttingen. (Symb. Fl. Argent.)

Groh, H. 1949. Plants of clearing and trail between Peace River and Fort Vermillion, Alberta. Canad. Field-Naturalist 63: 119–134.

Grudzinskaya, I. A. 1965. The Ulmaceae and reasons for distinguishing the Celtidoideae as a separate family Celtidaceae Link. Bot. Zhurn. (Moscow & Leningrad) 52: 1723–1749.

Grudzinskaya, I. A. 1988. K sistematike semeistva Cannabaceae. [The taxonomy of the family Cannabaceae.] Bot. Zhurn. (Moscow & Leningrad) 73: 589–593.

Guerriero, A. G., W. F. Grant, and W. H. Brittain. 1970. Interspecific hybridization between *Betula cordifolia* and *B. populifolia* at Valcartier, Quebec. Canad. J. Bot. 48: 2241–2247.

Gunn, C. R. 1980. Seeds and fruits of Papaveraceae and Fumariaceae. Seed Sci. Techn. 8: 3–58.

Gunn, C. R. and M. J. Seldin. 1976. Seeds and Fruits of North American Papaveraceae. Washington. [U.S.D.A. Agric. Res. Serv., Techn. Bull. 1517.]

Haenhle, G. G. and S. M. Jones. 1985. Geographical distribution of *Quercus oglethorpensis.* Castanea 50: 26–31.

Hall, J. W. 1952. The comparative anatomy and phylogeny of the Betulaceae. Bot. Gaz. 113: 235–270.

Hall, T. F. 1940. The biology of *Saururus cernuus.* Amer. Midl. Naturalist 24: 253–260.

Hall, T. F. and W. T. Penfound. 1944. The biology of the American lotus *Nelumbo lutea* (Willd.) Pers. Amer. Midl. Naturalist 31: 744–758.

Hannan, G. L. 1979. Floral Polymorphism and the Reproductive Biology of *Platystemon californicus* Benth. (Papaveraceae). Ph.D. dissertation. University of California, Berkeley.

Hannan, G. L. 1982. Correlation of morphological variation in *Platystemon californicus* (Papaveraceae) with flower color and geography. Syst. Bot. 7: 35–47.

Hara, H. 1975. The identity of *Clematis terniflora* DC. J. Jap. Bot. 50: 155–158.

Hardin, J. W. 1954. An analysis of variation within *Magnolia acuminata* L. J. Elisha Mitchell Sci. Soc. 70: 298–312.

Hardin, J. W. 1964. Variation in *Aconitum* of eastern United States. Brittonia 16: 80–94.

Hardin, J. W. 1972. Studies of the southeastern United States flora. III. Magnoliaceae and Illiciaceae. J. Elisha Mitchell Sci. Soc. 88: 30–32.

Hardin, J. W. 1975. Hybridization and introgression in *Quercus alba.* J. Arnold Arbor. 56: 336–363.

Hardin, J. W. 1979. *Quercus prinus* L.—nomen ambiguum. Taxon 28: 355–357.

Hardin, J. W. 1984. The Demoulin Rule in nomenclature. Sida 10: 252.

Hardin, J. W. and J. M. Bell. 1986. Atlas of foliar surface features in woody plants, IX. Betulaceae of eastern United States. Brittonia 38: 133–144.

Hardin, J. W. and G. P. Johnson. 1985. Atlas of foliar surface features in woody plants, VIII. *Fagus* and *Castanea*

(Fagaceae) of eastern North America. Bull. Torrey Bot. Club 112: 11–20.

Hardin, J. W. and K. A. Jones. 1989. Atlas of foliar surface features in woody plants, X. Magnoliaceae of the United States. Bull. Torrey Bot. Club 116: 164–173.

Hardin, J. W. and D. E. Stone. 1984. Atlas of foliar surface features in woody plants, VI. *Carya* (Juglandaceae) of North America. Brittonia 36: 140–153.

Harms, H. 1936. Reihe Rhoeadales. In: H. G. A. Engler et al., eds. 1924+. Die natürlichen Pflanzenfamilien . . . , ed. 2. 26+ vols. Leipzig and Berlin. Vol. 17b, pp. 1–4.

Harshberger, J. W. 1903. Juvenile and adult forms of bloodroot. Pl. World 6: 106–108.

Hausvater—See: O. von Münchhausen 1765–1773.

Hayden, W. J. and S. M. Hayden. 1984. Wood anatomy and relationships of *Betula uber*. Castanea 49: 26–30.

Heimburger, M. 1959. Cytotaxonomic studies in the genus *Anemone*. Canad. J. Bot. 37: 587–612.

Heiser, C. B. 1962. Some observations on pollination and compatibility in *Magnolia*. Proc. Indiana Acad. Sci. 72: 259–266.

Herbs Spices Med. Pl. = Herbs, Spices, and Medicinal Plants; Recent Advances in Botany, Horticulture, and Pharmacology.

Herendeen, P. S., D. H. Les, and D. L. Dilcher. 1990. Fossil *Ceratophyllum* (Ceratophyllaceae) from the Tertiary of North America. Amer. J. Bot. 77: 7–16.

Hermann, F. J. 1940. The geographic distribution of *Pilea fontana*. Torreya 40: 118–120.

Heyn, C. C. and B. Pazy. 1989. The annual species of *Adonis* (Ranunculaceae)—A polyploid complex. Pl. Syst. Evol. 168: 181–193.

Hickey, L. J. 1971. Evolutionary significance of leaf architectural features in the woody dicots. [Abstract.] Amer. J. Bot. 58: 469.

Hickman, J. C., ed. 1993. The Jepson Manual. Higher Plants of California. Berkeley, Los Angeles, and London.

Hikobia, Suppl. = Hikobia. Supplement.

Hill, A. F. 1952. Economic Botany. New York.

Hist. Arbr. Forest.—See: F. Michaux 1810–1813.

Hist. Chênes Amér.—See: A. Michaux 1801.

Hist. Nat. Vég.—See: E. Spach 1834–1848.

Hist. Pl. Dauphiné—See: D. Villars 1786–1789.

Hist. Pl. Guiane—See: J. B. Aublet 1775.

Hitchcock, C. L., A. Cronquist, M. Ownbey, and J. W. Thompson. 1955–1969. Vascular Plants of the Pacific Northwest. 5 vols. Seattle. (Vasc. Pl. Pacif. N.W.)

Hjelmquist, H. 1948. Studies in the floral morphology and phylogeny of the Amentiferae. Bot. Not., Suppl. 2: 1–171.

Holm, T. 1905. *Anemiopsis* [sic] *californica* (Nutt.) H. and A. An anatomical study. Amer. J. Sci., ser. 4, 19: 76–82.

Holm, T. 1926. *Saururus cernuus* L. A morphological study. Amer. J. Sci., ser. 5, 12: 162–168.

Holm, T. 1930. Leaf-variation in *Liquidambar styraciflua* L. Rhodora 32: 95–105.

Hooker, W. J. [1829–]1833–1840. Flora Boreali-Americana; or, the Botany of the Northern Parts of British America. . . . 2 vols. in 12 parts. London, Paris, and Strasbourg. (Fl. Bor.-Amer.)

Hooker, W. J. and G. A. W. Arnott. [1830–]1841. The Botany of Captain Beechey's Voyage; Comprising an Account of the Plants Collected by Messrs Lay and Collie, and Other Officers of the Expedition, during the Voyage to the Pacific and Bering's Strait, Performed in His Majesty's Ship Blossom, under the Command of Captain F. W. Becchcy . . . in the Years 1825, 26, 27, and 28. 10 parts. London. [Parts paged and plates numbered consecutively.] (Bot. Beechey Voy.)

Hooker, W. J., J. D. Hooker, D. Oliver, W. T. Thiselton-Dyer, D. Prain, A. W. Hill, E. J. Salisbury, G. Taylor, and J. P. M. Brenan, eds. [1836–]1837–1975+. Icones Plantarum; or Figures with Brief Descriptive Characters and Remarks, of New or Rare Plants. . . . 38+ vols. in 5 series. London, Oxford, and Kew. (Icon. Pl.)

Hoot, S. B. 1991. Phylogeny of the Ranunculaceae based on epidermal microcharacters and macromorphology. Syst. Bot. 16: 741–755.

Hoot, S. B., A. A. Reznicek, and J. D. Palmer. 1994. Phylogenetic relationships in *Anemone* (Ranunculaceae) based on morphology and chloroplast DNA. Syst. Bot. 19: 169–200.

Horae Phys. Berol.—See: C. G. D. Nees 1820.

Horn, K. 1938. Chromosome numbers in Scandinavian *Papaver* species. Norske Vidensk.-Akad., Mat.-Naturvidensk. Kl., Avh. 5: 1–13.

Hornemann, J. W. 1813–1815. Hortus Regius Botanicus Hafniensis. . . . 2 vols. Copenhagen. (Hort. Bot. Hafn.)

Hornemann, J. W. 1819. Supplementum Horti Botanici Hafniensis. . . . Copenhagen. (Suppl. Hort. Bot. Hafn.)

Hort. Bot. Hafn.—See: J. W. Hornemann 1813–1815.

Hort. Brit.—See: J. C. Loudon 1830; R. Sweet 1826.

Hort. Brit. ed. 2—See: R. Sweet 1830.

Hort. Kew.—See: W. Aiton 1789.

Hortus Kew.—See: W. Aiton and W. T. Aiton 1810–1813.

Houghton, W. M. 1988. The Systematics of Section *Cerophora* of the Genus *Myrica* (Myricaceae) in North America. M.S. thesis. University of Georgia.

Houttuyn, M. 1773–1783. Natuurlijke Historie of Uitvoerige Beschrijving der Dieren, Planten en Mineraalen, Volgens het Samenstel van den Heer Linnaeus. Met Naauwkeurige Afbeeldingen. Tweede Deels [Planten]. . . . 14 fasc. Amsterdam. (Nat. Hist.)

Howe, W. H. 1975. The Butterflies of North America. Garden City, N.Y.

Hudson, W. 1762. Flora Anglica. . . . London. (Fl. Angl.)

Hultén, E. 1927–1930. Flora of Kamtchatka and Adjacent Islands. . . . 4 vols. Stockholm. [Vols. designated as numbers of Kongl. Svenska Vetenskapsakad. Handl., ser. 3.] (Fl. Kamtchatka)

Hultén, E. 1937. Flora of the Aleutian Islands and Westernmost Alaska Peninsula with Notes on the Flora of Commander Islands. . . . Stockholm. (Fl. Aleut. Isl.)

Hultén, E. 1941–1950. Flora of Alaska and Yukon. 10 vols. Lund and Leipzig. [Vols. paged sequentially and designated as simultaneous numbers of Lunds Univ. Årsskr. (= Acta Univ. Lund.) and Kungl. Fysiogr. Sällsk. Handl.] (Fl. Alaska Yukon)

Hultén, E. 1968. Flora of Alaska and Neighboring Territories: A Manual of the Vascular Plants. Stanford.

Hultén, E. 1971. The circumpolar plants. II. Dicotyledons. Kongl. Svenska Vetenskapsakad. Handl., ser. 4, 13: 1–463.

Humboldt, A. von, A. J. Bonpland, and C. S. Kunth. 1815[1816]–1825. Nova Genera et Species Plantarum Quas in Peregrinatione Orbis Novi Collegerunt, Descripserunt. . . . 7 vols. in 36 parts. Paris. (Nov. Gen. Sp.)

Hunt, D. M. 1986. Distribution of *Quercus arkansana* in Georgia. Castanea 51: 183–187.

Hunt, D. M. 1989. A Systematic Review of *Quercus* Series *Laurifoliae, Marilandicae* and *Nigrae*. Ph.D. dissertation. University of Georgia.

Hutchinson, J. 1920. *Bocconia* and *Macleaya*. Bull. Misc. Inform. Kew 1920: 275–282.

Hutchinson, J. 1921. The genera of the Fumariaceae and their distribution. Bull. Misc. Inform. Kew 1921: 97–115.

Hutchinson, J. 1923. Contributions toward a phylogenetic classification of flowering plants. II. The genera of Annonaceae. Bull. Misc. Inform. Kew 1923: 241–261.

Hutchinson, J. 1925. Contributions towards a phylogenetic classification of flowering plants: V. The genera of Papaveraceae. Bull. Misc. Inform. Kew 1925: 161–168.

Hutchinson, J. 1959. The Families of Flowering Plants, ed. 2. 2 vols. Oxford.

Hutchinson, J. 1964–1967. The Genera of Flowering Plants (Angiospermae). . . . 2 vols. Oxford.

Hutchinson, J. 1973. The Families of Flowering Plants, ed. 3. Oxford.

Hylander, N. 1957. On cut-leaved and small-leaved forms of *Alnus glutinosa* and *A. incana*. Svensk Bot. Tidskr. 51: 437–453.

I. A. W. A. Bull., N.S. = I A W A Bulletin. New Series. Quarterly Periodical of the International Association of Wood Anatomists.

Icon. Pl.—See: W. J. Hooker et al. [1836–]1837–1975+.

Ill. Fl. Pacific States—See: L. Abrams and R. S. Ferris 1923–1960.

Ill. Handb. Laubholzk.—See: C. K. Schneider [1904–]1906–1912.

Ill. W. Amer. Oaks—See: E. L. Greene 1889–1890.

Index Sem. Hort. Petrop.—See: F. E. L. von Fischer et al. 1835–1846.

Israel J. Bot. = Israel Journal of Botany.

Izv. Imp. Akad. Nauk = Izvestiya Imperatorskoi Akademii Nauk.

J. Acad. Nat. Sci. Philadelphia = Journal of the Academy of Natural Sciences of Philadelphia.

J. Adelaide Bot. Gard. = Journal of the Adelaide Botanic Gardens.

J. Amer. Magnolia Soc. = Journal, American Magnolia Society.

J. Arboric. = Journal of Arboriculture.

J. Arizona-Nevada Acad. Sci. = Journal of the Arizona-Nevada Academy of Science.

J. Arnold Arbor. = Journal of the Arnold Arboretum.

J. Bot. = Journal of Botany, British and Foreign.

J. Bot. (Desvaux) = Journal de Botanique. [Edited by Desvaux.]

J. Cincinnati Soc. Nat. Hist. = Journal of the Cincinnati Society of Natural History.

J. Elisha Mitchell Sci. Soc. = Journal of the Elisha Mitchell Scientific Society.

J. Fac. Sci. Univ. Tokyo, Sect. 3, Bot. = Journal of the Faculty of Science, University of Tokyo. Section III. Botany.

J. Gén. Hort. = Journal Général d'Horticulture.

J. Genet. = Journal of Genetics.

J. Heredity = Journal of Heredity.

J. Jap. Bot. = Journal of Japanese Botany.

J. Korean Pl. Taxon. = Journal of Korean Plant Taxonomy.

J. Linn. Soc., Bot. = Journal of the Linnean Society. Botany.

J. New York Bot. Gard. = Journal of the New York Botanical Garden.

J. Phys. Chim. Hist. Nat. Arts = Journal de Physique, de Chimie, d'Histoire Naturelle et des Arts.

J. Tennessee Acad. Sci. = Journal of the Tennessee Academy of Science.

J. Voy. N.-W. Passage, Bot.—See: R. Brown 1824.

J. Wash. Acad. Sci. = Journal of the Washington Academy of Sciences.

Jacquin, N. J. 1760. Enumeratio Systematica Plantarum, Quas in Insulis Caribaeis Vicinaque Americes Continente Detexit Novas. . . . Leiden. (Enum. Syst. Pl.)

Jacquin, N. J. 1786[1787]–1796[1797]. Collectanea ad Botanicam, Chemiam, et Historiam Naturalem Spectantia. . . . 5 vols. Vienna. (Collectanea)

Jacquin, N. J. 1797–1804. Plantarum Rariorum Horti Caesarei Schoenbrunnensis Descriptiones et Icones. 4 vols. Vienna, London, and Leiden. (Pl. Hort. Schoenbr.)

Jäger, E. J. 1980. Progressionen im Synfloreszenbau und in der Verbreitung bei den Betulaceae. Flora 170: 91–113.

James, E. 1823. Account of an Expedition from Pittsburgh to the Rocky Mountains, Performed in the Years 1819 and '20 . . . under the Command of Major Stephen H. Long. 2 vols. + atlas. Philadelphia. (Account Exped. Pittsburgh)

James, J. F. 1883. Revision of the genus *Clematis* of the United States. J. Cincinnati Soc. Nat. Hist. 6: 118–135.

Janish, J. R. 1977. Nevada's vanishing bear-poppies. Mentzelia 3: 2–5.

Jarvis, C. E. 1989. A review of the order Leitneriales Engler. In: P. R. Crane and S. Blackmore, eds. 1989. Evolution, Systematics, and Fossil History of the Hamamelidae. 2 vols. Oxford. Vol. 2, pp. 189–192.

Jenne, G. E. 1966. A Study of Variation in North American *Hamamelis* L. (Hamamelidaceae). Masters thesis. Vanderbilt University.

Jepson, W. L. 1909–1943. A Flora of California. . . . 3 vols. in 12 parts. San Francisco etc. [Pagination sequential within each vol.; vol. 1 page sequence independent of part number sequence (chronological); part 8 of vol. 1 (pp. 1–32, 579–index) never published.] (Fl. Calif.)

Jensen, R. J. 1986. Geographic spatial autocorrelation in *Quercus ellipsoidalis*. Bull. Torrey Bot. Club 113: 431–439.

Jensen, R. J. 1989. The *Quercus falcata* Michx. complex in

Land Between The Lakes, Kentucky and Tennessee: A study of morphological variation. Amer. Midl. Naturalist 121: 245–255.

Jensen, R. J., S. C. Hokanson, J. G. Isebrands, and J. F. Hancock. 1993. Morphometric variation in oaks of the Apostle Islands in Wisconsin: Evidence of hybridization between *Quercus rubra* and *Quercus ellipsoidalis* (Fagaceae). Amer. J. Bot. 80: 1358–1366.

Johnson, A. F. and W. G. Abrahamson. 1982. *Quercus inopina*: A species to be recognized from south-central Florida. Bull. Torrey Bot. Club 109: 392–395.

Johnson, D. L. 1989. Species and Cultivars of the Genus *Magnolia* (Magnoliaceae) Cultivated in the United States. M.S. thesis. Cornell University.

Johnson, D. L. 1989b. Nomenclatural changes in *Magnolia*. Baileya 23: 55–56.

Johnson, D. S. 1900. On the development of *Saururus cernuus* L. Bull. Torrey Bot. Club 27: 365–372.

Johnson, G. P. 1988. Revision of *Castanea* sect. *Balanocastanon* (Fagaceae). J. Arnold Arbor. 69: 25–49.

Johnson, G. P. 1992. *Quercus shumardii* Buckl. var. *acerifolia* Palmer (Fagaceae). Castanea 57: 150–151.

Johnson, G. P. 1994. *Quercus shumardii* Buckl. var. *acerifolia* Palmer (Fagaceae). Castanea 59: 78.

Johnson, L. A. S. and K. L. Wilson. 1989. Casuarinaceae: A synopsis. In: P. R. Crane and S. Blackmore, eds. 1989. Evolution, Systematics, and Fossil History of the Hamamelidae. 2 vols. Oxford. Vol. 2, pp. 167–186.

Johnson, L. A. S. and K. L. Wilson. 1993. Casuarinaceae. In: K. Kubitzki et al., eds. 1990+. The Families and Genera of Vascular Plants. 2+ vols. Berlin etc. Vol. 2, pp. 237–242.

Johnsson, H. 1945. Interspecific hybridization within the genus *Betula*. Hereditas (Lund) 31: 163–176.

Jones, E. N. 1931. The morphology and biology of *Ceratophyllum demersum*. Stud. Nat. Hist. Iowa Univ. 13: 11–55.

Jones, S. B. and N. C. Coile. 1988. The Distribution of the Vascular Flora of Georgia. Athens, Ga.

Joseph, C. and M. Heimburger. 1966. Cytotaxonomic studies on New World species of *Anemone* (section *Eriocephalus*) with tuberous rootstocks. Canad. J. Bot. 44: 899–928.

Juzepczuk, S. V. 1970. *Anemone*. In: V. L. Komarov et al., eds. 1963+. Flora of the U.S.S.R. (Flora SSSR). Translated from Russian. 22+ vols. Jerusalem. Vol. 7, pp. 184–218.

Kadereit, J. W. 1986. A revision of *Papaver* section *Argemonidium*. Notes Roy. Bot. Gard. Edinburgh 44: 25–43.

Kadereit, J. W. 1986b. A revision of *Papaver* L. sect. *Papaver* (Papaveraceae). Bot. Jahrb. Syst. 108: 1–16.

Kadereit, J. W. 1988. Sectional affinities and geographical distribution in the genus *Papaver* L. (Papaveraceae). Beitr. Biol. Pflanzen 63: 139–156.

Kadereit, J. W. 1988b. *Papaver* L. sect. *Californicum* Kadereit, a new section of the genus. Rhodora 90: 7–13.

Kadereit, J. W. 1989. A revision of *Papaver* L. section *Rhoeadium* Spach. Notes Roy. Bot. Gard. Edinburgh 45: 225–286.

Kadereit, J. W. 1990. Some suggestions on the geographical origin of the central, west and north European synanthropic species of *Papaver* L. Bot. J. Linn. Soc. 103: 221–231.

Kadereit, J. W. 1993. Papaveraceae. In: K. Kubitzki et al., eds. 1990+. The Families and Genera of Vascular Plants. 2+ vols. Berlin etc. Vol. 2, pp. 494–506.

Kadota, Y. 1987. A Revision of *Aconitum* Subgenus *Aconitum* (Ranunculaceae) of East Asia. Utsunomiya.

Kalmia = Kalmia; Botanic Journal.

Kapil, R. N. and S. Jalan. 1964. *Schisandra* Michaux—its embryology and systematic position. Bot. Not. 117: 285–306.

Kasapligil, B. 1964. A contribution to the histotaxonomy of *Corylus* (Betulaceae). Adansonia, n. s. 4: 43–90.

Kearney, T. H. and R. H. Peebles. 1942. Flowering Plants and Ferns of Arizona. . . . Washington. (Fl. Pl. Ferns Ariz.)

Keener, C. S. 1967. A biosystematic study of *Clematis* subsection *Integrifoliae* (Ranunculaceae). J. Elisha Mitchell Sci. Soc. 83: 1–41.

Keener, C. S. 1975. Studies in the Ranunculaceae of the southeastern United States. III. *Clematis* L. Sida 6: 33–47.

Keener, C. S. 1975b. Studies in the Ranunculaceae of the southeastern United States. I. *Anemone* L. Castanea 40: 36–44.

Keener, C. S. 1976. Studies in the Ranunculaceae of the southeastern United States. IV. Genera with zygomorphic flowers. Castanea 41: 12–20.

Keener, C. S. 1977. Studies in the Ranunculaceae of the southeastern United States. VI. Miscellaneous genera. Sida 7: 1–12.

Keener, C. S. 1993. A review of the classification of the genus *Hydrastis* (Ranunculaceae). Aliso 13: 551–558.

Keener, C. S. and W. M. Dennis. 1982. The subgeneric classification of *Clematis* (Ranunculaceae) in temperate North America north of Mexico. Taxon 31: 37–44.

Keener, C. S., E. T. Dix, and B. E. Dutton. 1996. The identity of *Anemone riparia* (Ranunculaceae). Bartonia 59: 37–47.

Keener, C. S. and B. E. Dutton. 1994. A new species of *Anemone* (Ranunculaceae) from central Texas. Sida 16: 191–202.

Kelso, L. 1932. A note on *Anemopsis californica*. Amer. Midl. Naturalist 13: 110–113.

Kew Bull. = Kew Bulletin.

Kieckhefer, B. J. 1964. Phenology and Energy Relationships of *Dicentra canadensis* and *D. cucullaria*. Ph.D. thesis. University of Illinois.

Kiger, R. W. 1973. Sectional nomenclature in *Papaver* L. Taxon 22: 579–582.

Kiger, R. W. 1975. *Papaver* in North America north of Mexico. Rhodora 77: 410–422.

Kiger, R. W. 1985. Revised sectional nomenclature in *Papaver* L. Taxon 34: 150–152.

Kingsbury, J. M. 1964. Poisonous Plants of the United States and Canada. Englewood Cliffs.

Klokov, M. V. 1978. The genus *Ceratocephala* Moench., on aspects of its overall differentiation. Novosti Sist. Vyssh. Rast. 15: 7–73. [In Russian.]

Knaben, G. 1958. *Papaver*-studier, med et forsvar for *P. radi-*

catum Røttb. som en Islandsk-Skandinavisk art. Blyttia 16: 61–80.

Knaben, G. 1959. On the evolution of the *radicatum*-group of the *Scapiflora* Papavers as studied in 70 and 56 chromosome species. Part. A. Cytotaxonomical aspects. Part. B. Experimental studies. Opera Bot. 2(3), 3(3).

Knaben, G. and N. Hylander. 1970. On the typification of *Papaver radicatum* Røttb. and its nomenclatural consequences. Bot. Not. 123: 338–345.

Koch, K. H. E. 1869–1873. Dendrologie. Bäume, Sträucher und Halbsträucher, welche in Mittel- und Nord-Europa im Freien kultivirt werden. 2 vols. in 3. Erlangen. (Dendrologie)

Koehne, B. A. E. 1893. Deutsche Dendrologie. Stuttgart. (Deut. Dendrol.)

Komarov, V. L. et al., eds. 1963+. Flora of the U.S.S.R. (Flora SSSR). Translated from Russian. 22+ vols. Jerusalem.

Kongel. Danske Vidensk. Selsk. Skr., Naturvidensk. Math. Afd. = Kongelige Danske Videnskabernes Selskabs Skrifter. Naturvidenskabelig og Mathematisk Afdeling.

Kongl. Svenska Vetenskapsakad. Handl. = Kongl[iga]. Svenska Vetenskapsakademiens Handlingar.

Kongl. Vetensk. Acad. Nya Handl. = Kongl[iga]. Vetenskaps Academiens Nya Handlingar.

Kral, R. 1960. A revision of *Asimina* and *Deeringothamnus* (Annonaceae). Brittonia 12: 233–278.

Kral, R., ed. 1983. A Report on Some Rare, Threatened, or Endangered Forest-related Vascular Plants of the South. Washington. [U.S.D.A., Techn. Publ. R8-TP 2.]

Kral, R. 1983c. *Deeringothamnus.* In: R. Kral, ed. 1983. A Report on Some Rare, Threatened, or Endangered Forest-related Vascular Plants of the South. Washington. Pp. 452–456.

Kubitzki, K., K. U. Kramer, P. S. Green, J. G. Rohwer, and V. Bittrich, eds. 1990+. The Families and Genera of Vascular Plants. 2+ vols. Berlin etc.

Kulturpflanze = Kulturpflanze. Berichte und Mitteilungen aus dem Institut für Kulturpflanzenforschung der Deutschen Akademie der Wissenschaften zu Berlin in Gatersleben Krs. Aschersleben.

Kuntze, O. 1885. Monographie der Gattung *Clematis*. Verh. Bot. Vereins Prov. Brandenburg 26: 6–202.

Kurz, H. 1983. Fortpflanzungsbiologie einiger Gattungen neotropischer Lauraceen und Revision der Gattung *Licaria* (Lauraceae). Ph.D. thesis. University of Hamburg.

Lagasca y Segura, M. 1816. Elenchus Plantarum, Quae in Horto Regio Botanico Matritensi Colebantur Anno MDCCCXV. Madrid. (Elench. Pl.)

Lamarck, J. and A. P. de Candolle. 1805[–1815]. Flore Française, ou Descriptions Succinctes de Toutes les Plantes Qui Croissent Naturellement en France . . . , ed. 3. 5 tomes in 6 vols. Paris. [Tomes 1–4(2), 1805; tome 5, 1815.] (Fl. Franç. ed. 3)

Lamarck, J., J. Poiret, A. P. de Candolle, L. Desrousseaux, and M. Savigny. 1783–1817. Encyclopédie Méthodique. Botanique. . . . 13 vols. Paris and Liège. [Vols. 1–8, suppls. 1–5.] (Encycl.)

Langer, S. K. 1993. A new oak on Mount Tamalpais. Four Seasons 9: 21–30.

Langlet, O. F. 1932. Über Chromosomenverhältnisse und Systematik der Ranunculaceae. Svensk Bot. Tidskr. 26: 381–401.

Larson, G. E. 1986. *Ranunculus.* In: Great Plains Flora Association. 1986. Flora of the Great Plains. Lawrence, Kans. Pp. 95–106.

Laughlin, K. 1969. *Quercus shumardii* var. *stenocarpa* Laughlin, stenocarp shumard oak. Phytologia 19: 57–64.

Leafl. Bot. Observ. Crit. = Leaflets of Botanical Observation and Criticism.

Leafl. W. Bot. = Leaflets of Western Botany.

Lecoyer, J. C. 1885. Monographie du genre *Thalictrum*. Bull. Soc. Roy. Bot. Belgique 24: 78–324.

Ledebour, C. F. von. [1841]1842–1853. Flora Rossica sive Enumeratio Plantarum in Totius Imperii Rossici Provinciis Europaeis, Asiaticis, et Americanis Hucusque Observatarum. . . . 4 vols. Stuttgart. (Fl. Ross.)

Leonard, S. W. 1987. Inventory of populations of *Thalictrum cooleyi* and Its Occurrence Sites in North Carolina. Report to the North Carolina Natural Heritage Program. Raleigh.

Lepage, E. 1976. Les bouleaux arbustifs du Canada et de l'Alaska. Naturaliste Canad. 103: 215–233.

Les, D. H. 1985. The taxonomic significance of plumule morphology in *Ceratophyllum* (Ceratophyllaceae). Syst. Bot. 10: 338–346.

Les, D. H. 1986. The phytogeography of *Ceratophyllum demersum* and *C. echinatum* in glaciated North America. Canad. J. Bot. 64: 498–509.

Les, D. H. 1986b. The evolution of achene morphology in *Ceratophyllum* (Ceratophyllaceae), I. Fruit spine variation and relationships of *C. demersum, C. submersum,* and *C. apiculatum.* Syst. Bot. 11: 549–558.

Les, D. H. 1988. The origin and affinities of the Ceratophyllaceae. Taxon 37: 326–345.

Les, D. H. 1988b. The evolution of achene morphology in *Ceratophyllum* (Ceratophyllaceae), II. Fruit variation and systematics of the "spiny-margined" group. Syst. Bot. 13: 73–86.

Les, D. H. 1988c. The evolution of achene morphology in *Ceratophyllum* (Ceratophyllaceae), III. Relationships of the "facially-spined" group. Syst. Bot. 13: 509–518.

Les, D. H. 1989. The evolution of achene morphology in *Ceratophyllum* (Ceratophyllaceae), IV. Summary of proposed relationships and evolutionary trends. Syst. Bot. 14: 254–262.

Les, D. H. 1993. Ceratophyllaceae. In: K. Kubitzki et al., eds. 1990+. The Families and Genera of Vascular Plants. 2+ vols. Berlin etc. Vol. 2, pp. 246–250.

Les, D. H., D. K. Garvin, and C. F. Wimpee. 1991. Molecular evolutionary history of ancient aquatic angiosperms. Proc. Natl. Acad. Sci. U.S.A. 88: 10119–10123.

Lewis, H. and C. Epling. 1954. A taxonomic study of Californian delphiniums. Brittonia 8: 1–22.

Lewis, H. and C. Epling. 1959. *Delphinium gypsophilum*, a diploid species of hybrid origin. Evolution 13: 511–525.

Lewis, H., C. Epling, G. A. L. Mehlquist, and C. C. Wyckoff. 1951. Chromosome numbers of Californian *Delphinium* and their geographical occurrence. Ann. Missouri Bot. Gard. 38: 101–117.

Lewis, H. and R. Snow. 1951. A cytotaxonomic approach to *Eschscholzia*. Madroño 11: 141–143.

Lewis, W. H. and M. P. F. Elvin-Lewis. 1977. Medical Botany: Plants Affecting Man's Health. New York.

Lidén, M. 1986. Synopsis of Fumarioideae (Papaveraceae) with a monograph of the tribe Fumarieae. Opera Bot. 88: 1–133.

Lindley, J. 1821[–1826]. Collectanea Botanica; or, Figures and Botanical Illustrations of Rare and Curious Exotic Plants. . . . 8 parts. London. (Coll. Bot.)

Linnaea = Linnaea. Ein Journal für die Botanik in ihrem ganzen Umfange.

Linnaeus, C. 1749[–1769]. Amoenitates Academicae seu Dissertationes Variae Physicae, Medicae Botanicae. . . . 7 vols. Stockholm and Leipzig. (Amoen. Acad.)

Linnaeus, C. 1753. Species Plantarum. . . . 2 vols. Stockholm. (Sp. Pl.)

Linnaeus, C. 1754. Genera Plantarum, ed. 5. Stockholm. (Gen. Pl. ed. 5)

Linnaeus, C. 1755. Centuria I. Plantarum. . . . Uppsala. (Cent. Pl. I)

Linnaeus, C. 1756. Flora Palaestina. . . . Uppsala. [Respondent: J. Strand.] (Fl. Palaest.)

Linnaeus, C. 1758[–1759]. Systema Naturae per Regna Tria Naturae . . . , ed. 10. 2 vols. Stockholm. (Syst. Nat. ed. 10)

Linnaeus, C. 1762–1763. Species Plantarum . . . , ed. 2. 2 vols. Stockholm. (Sp. Pl. ed. 2)

Linnaeus, C. 1766[–1768]. Systema Naturae per Regna Tria Naturae . . . , ed. 12. 3 vols. Stockholm. (Syst. Nat. ed. 12)

Linnaeus, C. 1767[–1771]. Mantissa Plantarum. 2 parts. Stockholm. [Mantissa [1] and Mantissa [2] Altera paged consecutively.] (Mant. Pl.)

Linnaeus, C. 1774. Systema Vegetabilium. . . . Editio Decima Tertia. . . . Göttingen and Gotha. (Syst. Veg. ed. 13)

Linnaeus, C. 1775. Plantae Surinamenses. . . . Uppsala. (Pl. Surin.)

Linnaeus, C. f. 1781[1782]. Supplementum Plantarum Systematis Vegetabilium Editionis Decimae Tertiae, Generum Plantarum Editionis Sextae, et Specierum Plantarum Editionis Secundae. Braunschweig. (Suppl. Pl.)

List Pl. Nevada Utah—See: S. Watson [1871]b.

Little, E. L. Jr. 1979. Checklist of United States Trees (Native and Naturalized). Washington. [Agric. Handb. 541.]

Loconte, H. 1993. Berberidaceae. In: K. Kubitzki et al., eds. 1990+. The Families and Genera of Vascular Plants. 2+ vols. Berlin etc. Vol. 2, pp. 147–152.

Loconte, H. and W. H. Blackwell. 1985. Intrageneric taxonomy of *Caulophyllum* (Berberidaceae). Rhodora 87: 463–469.

Loconte, H. and J. R. Estes. 1989. Generic relationships within Leonticeae (Berberidaceae). Canad. J. Bot. 67: 2310–2316.

Loconte, H. and J. R. Estes. 1989b. Phylogenetic systematics of Berberidaceae and Ranunculales (Magnoliidae). Syst. Bot. 14: 565–579.

London J. Bot. = London Journal of Botany.

Long, R. W. and O. Lakela. 1971. A Flora of Tropical Florida: A Manual of the Seed Plants and Ferns of Southern Peninsular Florida. Coral Gables.

Lott, E. J. 1979. Variation and Interrelationships of *Aquilegia* Populations of Trans-Pecos Texas. M.S. thesis. Sul Ross State University.

Lott, E. J. 1985. New combinations in Chihuahuan Desert *Aquilegia* (Ranunculaceae). Phytologia 58: 488.

Loudon, J. C. 1830. Hortus Brittanicus. A Catalogue of All the Plants Indigenous, Cultivated in, or Introduced to Britain. London. (Hort. Brit.)

Loureiro, J. de 1790. Flora Cochinchinensis. . . . 2 vols. Lisbon. [Vols. paged consecutively.] (Fl. Cochinch.)

Löve, Á. 1962b. Typification of *Papaver radicatum*—A nomenclatural detective story. Bot. Not. 115: 113–136.

Löve, Á. 1962c. Nomenclature of North Atlantic Papavers. Taxon 11: 132–138.

Löve, Á. and D. Löve. 1966. Cytotaxonomy of the alpine vascular plants of Mount Washington. Univ. Colorado Stud., Ser. Biol. 24: 1–74.

Löve, D. 1969. *Papaver* at high altitudes in the Rocky Mountains. Brittonia 21: 1–10.

Löve, D. and N. J. Freedman. 1956. A plant collection from SW Yukon. Bot. Not. 109: 153–211.

Lowden, R. M. 1978. Studies in the submerged genus *Ceratophyllum* L. in the neotropics. Aquatic Bot. 4: 127–142.

Lu, K. L. 1982. Pollination biology of *Asarum caudatum* (Aristolochiaceae) in northern California. Syst. Bot. 7: 150–157.

Lu, K. L. and M. R. Mesler. 1983. A re-evaluation of a green-flowered *Asarum* (Aristolochiaceae) from southern Oregon. Brittonia 35: 331–334.

Macbride, J. F. et al. 1936+. Flora of Peru. 6+ parts. Chicago. [Published in numerous fascicles constituting 6 nominal parts (together designated as vol. 13 of Field Mus. Nat. Hist., Bot. Ser.) plus later unnumbered increments (designated as individual issues of Fieldiana, Bot.).]

MacDonald, A. D. 1977. Myricaceae: Floral hypothesis for *Gale* and *Comptonia*. Canad. J. Bot. 55: 2636–2651.

MacDonald, A. D. and R. Sattler. 1973. Floral development of *Myrica gale* and the controversy over floral concepts. Canad. J. Bot. 51: 1965–1975.

Macior, L. W. 1970. The pollination ecology of *Dicentra cucullaria*. Amer. J. Bot. 57: 1–5.

Macior, L. W. 1978. Pollination interactions in sympatric *Dicentra* species. Amer. J. Bot. 65: 57–62.

Madroño = Madroño; Journal of the California Botanical Society [from vol. 3: a West American Journal of Botany].

Maekawa, F. and M. Ono. 1965. Karyotype analysis of the genus *Hexastylis* (Aristolochiaceae). J. Fac. Sci. Univ. Tokyo, Sect. 3, Bot. 9: 151–159.

Mag. Hort. Bot. = Magazine of Horticulture, Botany and All Useful Discoveries and Improvements in Rural Affairs.

Malpighia = Malpighia; Rassegna Mensile di Botanica.

Man. Bot.—See: J. Darby 1841.

Man. Calif. Shrubs—See: H. McMinn 1939.

Man. Fl. N. States—See: N. L. Britton 1901.

Manchester, S. R. 1986. Vegetative and reproductive morphology of an extinct plane tree (Platanaceae) from the Eocene of western North America. Bot. Gaz. 147: 200–226.

Manchester, S. R. 1987. The fossil history of the Juglandaceae. Monogr. Syst. Bot. Missouri Bot. Gard. 21: 1–137.

Manning, W. E. 1950. A key to the hickories north of Virginia with notes on the two pignuts, *Carya glabra* and *C. ovalis*. Rhodora 52: 188–199.

Manning, W. E. 1973. The northern limits of the distributions of hickories in New England. Rhodora 75: 34–51.

Manning, W. E. 1978. The classification within the Juglandaceae. Ann. Missouri Bot. Gard. 65: 1058–1087.

Manos, P. S. 1993. Foliar trichome variation in *Quercus* section *Protobalanus* (Fagaceae). Sida 15: 391–403.

Mant. Pl.—See: C. Linnaeus 1767[–1771].

Manual—See: A. Gray 1848.

Manual ed. 5—See: A. Gray 1867.

Marshall, H. 1785. Arbustrum Americanum: The American Grove. . . . Philadelphia. (Arbust. Amer.)

Mason, H. L. 1957. A Flora of the Marshes of California. Berkeley.

Mason, H. L. and D. E. Stone. 1957. *Myosurus*. In: H. L. Mason. 1957. A Flora of the Marshes of California. Berkeley. Pp. 497–505.

Massey, J. R., D. K. S. Otte, T. A. Atkinson, and R. D. Whetstone. 1983. An Atlas and Illustrated Guide to Threatened and Endangered Vascular Plants of the Mountains of North Carolina and Virginia. Washington. [U.S.D.A. Forest Serv., Gen. Techn. Rep. SE-20.]

Matthews, V. A. 1989. Lardizabalaceae. In: S. M. Walters et al., eds. 1984+. The European Garden Flora. 4+ vols. Cambridge etc. Vol. 3, pp. 396–398.

Maximowicz, C. J. 1889. Flora Tangutica sive Enumeratio Plantarum Regionis Tangut (Amdo) Provinciae Kansu, nec non Tibetiae Praesertim Orientaliborealis atque Tsaidam. 1 fasc. only. St. Petersburg. [Constitutes vol. 1, part 1 of Historia Naturalis Itinerum N. M. Przewalskii per Asiam Centralem . . . Pars Botanica.] (Fl. Tangut.)

Mazzeo, P. M. 1974. *Betula uber*—What is it and where is it? Castanea 39: 273–278.

McCain, J. W. and J. F. Hennen. 1982. Is the taxonomy of *Berberis* and *Mahonia* (Berberidaceae) supported by their rust pathogens *Cumminsiella santa* sp. nov. and other *Cumminsiella* species (Uredinales)? Syst. Bot. 7: 48–59.

McCartney, R. D., K. Wurdack, and J. Moore. 1989. The genus *Lindera* in Florida. Palmetto 9: 3–8.

McDaniel, J. C. 1966. Variations in the sweet bay magnolias. Morris Arbor. Bull. 17: 7–12.

McDaniel, S. 1967. *Ulmus crassifolia* in Florida. Sida 3: 115–116.

McDougal, K. M. and C. R. Parks. 1986. Environmental and genetic components of flavonoid variation in red oak, *Quercus rubra*. Biochem. Syst. & Ecol. 14: 291–298.

McGranahan, G. H. and P. B. Catlin. 1987. *Juglans* rootstocks. In: R. C. Rom and R. F. Carlson, eds. 1987. Rootstocks for Fruit Crops. New York. Pp. 411–450.

McIntosh, F. 1989. *Eranthis*. In: S. M. Walters et al., eds. 1984+. The European Garden Flora. 4+ vols. Cambridge etc. Vol. 3, p. 331.

McIntosh, F. 1989b. In: S. M. Walters et al., eds. 1984+. The European Garden Flora. 4+ vols. Cambridge etc. Vol. 3, pp. 368–369.

McMinn, H. 1939. An Illustrated Manual of California Shrubs. . . . San Francisco. (Man. Calif. Shrubs)

Meacham, C. A. 1980. Phylogeny of the Berberidaceae with an evaluation of classifications. Syst. Bot. 5: 149–172.

Med. Repos. = Medical Repository.

Meddel. Grønland, Biosci. = Meddelelser om Grønland. Bioscience.

Meeuse, B. J. D. and E. L. Schneider. 1980. *Nymphaea* revisited: A preliminary communication. Israel J. Bot. 28: 65–79.

Melchior, H. and W. Schultze-Motel. 1959. Canellaceae. (Supplement to Vol. 21.) In: H. G. A. Engler et al., eds. 1924+. Die natürlichen Pflanzenfamilien . . . , ed. 2. 26+ vols. Leipzig and Berlin. Vol. 17a(2), pp. 221–224.

Melville, R. 1978. On the discrimination of species in hybrid swarms with special reference to *Ulmus* and the nomenclature of *U. minor* Mill. and *U. carpinifolia* Gled. Taxon 27: 345–351.

Mém. Acad. Imp. Sci. St.-Pétersbourg, Sér. 6, Sci. Math. = Mémoires de l'Académie Impériale des Sciences de St.-Pétersbourg. Sixième Série. Sciences Mathématiques, Physiques et Naturelles.

Mem. Amer. Acad. Arts = Memoirs of the American Academy of Arts and Science.

Mem. Fac. Sci. Kyoto Univ., Ser. Biol. = Memoirs of the Faculty of Science, Kyoto University. Series of Biology.

Mem. Natl. Acad. Sci. = Memoirs of the National Academy of Sciences.

Mem. New York Bot. Gard. = Memoirs of the New York Botanical Garden.

Mém. Soc. Linn. Paris = Mémoires de la Société Linnéenne de Paris, Précédés de Son Histoire.

Mém. Soc. Sci. Nat. Cherbourg = Mémoires de la Société des Sciences Naturelles de Cherbourg.

Mem. Torrey Bot. Club = Memoirs of the Torrey Botanical Club.

Mentzelia = Mentzelia; Journal of the Northern Nevada Native Plant Society.

Mesler, M. R. and Lu K. L. 1990. The status of *Asarum marmoratum* (Aristolochiaceae). Brittonia 42: 33–37.

Methodus—See: C. Moench 1794.

Meyer, E. 1830. De Plantis Labradoricis Libri Tres. Leipzig. (Pl. Labrador.)

Michaux, A. 1801. Histoire des Chênes de l'Amérique, ou Descriptions et Figures de Toutes les Espèces et Variétés de Chênes de l'Amérique Septentrionale. . . . Paris. (Hist. Chênes Amér.)

Michaux, A. 1803. Flora Boreali-Americana. . . . 2 vols. Paris and Strasbourg. (Fl. Bor.-Amer.)

Michaux, F. 1810–1813. Histoire des Arbres Forestiers de l'Amérique Septentrionale. . . . 3 vols. in parts and fasc. Paris. [Pagination independent by volumes, plate numbering independent by parts.] (Hist. Arbr. Forest.)

Michigan Bot. = Michigan Botanist.

Millais, J. G. 1927. Magnolias. London.

Miller, G. S. Jr. and P. C. Standley. 1912. The North American species of *Nymphaea*. Contr. U.S. Natl. Herb. 16: 63–108.

Miller, N. G. 1970. The genera of the Cannabaceae in the southeastern United States. J. Arnold Arbor. 51: 185–203.

Miller, N. G. 1971. The genera of the Urticaceae in the southeastern United States. J. Arnold Arbor. 52: 40–68.

Miller, N. G. 1995. *Consolida regalis* Gray naturalized in Genessee County, New York. Newslett. New York Fl. Assoc. 6(1): 6–7.

Miller, P. 1754. The Gardeners Dictionary. . . . Abridged . . . , ed. 4. 3 vols. London. (Gard. Dict. Abr. ed. 4)

Miller, P. 1768. The Gardeners Dictionary . . . , ed. 8. London. (Gard. Dict. ed. 8)

Miller, R. B. 1981. Hawkmoths and the geographic patterns of floral variation in *Aquilegia caerulea*. Evolution 35: 763–774.

Miller, R. B. 1985. Hawkmoth pollination of *Aquilegia chrysantha* (Ranunculaceae) in southern Arizona. SouthW. Naturalist 30: 69–76.

Minnesota Bot. Stud. = Minnesota Botanical Studies.

Miquel, F. A. W. 1848. Revisio Critica Casuarinum. Amsterdam. [Also published in Nieuwe Verh. Eerste Kl. Kon. Ned. Inst. Wetensch. Amsterdam 13: 267–350, plates 1–12. 1848.] (Rev. Crit. Casuarinum)

Mitchell, R. S. 1986. A checklist of New York State plants. Bull. New York State Mus. Sci. Serv. 458.

Mitchell, R. S. 1988. Contributions to a flora of New York State, ser. 6, 7–12. Bull. New York State Mus. Sci. Serv. 464.

Mitchell, R. S. and J. K. Dean. 1982. Ranunculaceae (Crowfoot Family) of New York State. Bull. New York State Mus. Sci. Serv. 446.

Mitt. Bot. Staatssaml. München = Mitteilungen (aus) der Botanischen Staatssammlung München.

Mitt. Deutsch. Dendrol. Ges. = Mitteilungen der Deutschen dendrologischen Gesellschaft.

Moench, C. 1794. Methodus Plantas Horti Botanici et Agri Marburgensis. . . . Marburg. (Methodus)

Moerman, D. E. 1986. Medicinal Plants of Native America. 2 vols. Ann Arbor. [Univ. Michigan, Mus. Anthropol., Techn. Rep. 19.]

Molec. Biol. Evol. = Molecular Biology and Evolution.

Monogr. Anonac.—See: M. F. Dunal 1817.

Monogr. Bot. = Monographiae Botanicae.

Monogr. Syst. Bot. Missouri Bot. Gard. = Monographs in Systematic Botany from the Missouri Botanical Garden.

Monson, P. H. 1960. Variation in *Nymphaea*, the white waterlily, in the Itasca State Park region. Proc. Minnesota Acad. Sci. 25/26: 26–39.

Moore, J. and J. R. Ballington Jr., eds. 1990. Genetic Resources of Temperate Fruit and Nut Crops. 2 vols. Wageningen.

Moore, R. J. 1963. Karyotype evolution in *Caulophyllum*. Canad. J. Genet. Cytol. 5: 384–388.

Moran, R. V. 1982. *Berberis claireae,* a new species from Baja California; and why not *Mahonia*. Phytologia 52: 221–226.

Moricand, M. E. 1833–1846[–1847]. Plantes Nouvelles d'Amérique. . . . 9 fasc. Geneva. [Fascicles paged consecutively.] (Pl. Nouv. Amér.)

Morris, M. I. 1973. A biosystematic analysis of the *Caltha leptosepala* (Ranunculaceae) complex in the Rocky Mountains. III. Variability in seed and gross morphological characters. Canad. J. Bot. 51: 2259–2268.

Morris Arbor. Bull. = Morris Arboretum Bulletin.

Moseley, M. F. Jr. 1961. Morphological studies of the Nymphaeaceae. II. The flower of *Nymphaea*. Bot. Gaz. 122: 233–259.

Mosquin, T. 1961. *Eschscholzia covillei* Greene, a tetraploid species from the Mojave Desert. Madroño 16: 91–96.

Moss, E. H. and J. G. Packer. 1983. Flora of Alberta, ed. 2. Toronto.

Mozingo, H. N. and Margaret Williams. 1980. Threatened and Endangered Plants of Nevada: An Illustrated Manual. [Washington.]

Muenscher, W. C. 1940. Fruits and seedlings of *Ceratophyllum*. Amer. J. Bot. 27: 231–233.

Muenscher, W. C. 1980. Weeds, ed. 2, [new] foreword and appendixes by Peter A. Hyypio. Ithaca, N.Y., and London.

Muhlenbergia = Muhlenbergia; a Journal of Botany.

Muller, C. H. 1942. Notes on the American flora, chiefly Mexican. Amer. Midl. Naturalist 27: 470–490.

Muller, C. H. 1951. The oaks of Texas. Contr. Texas Res. Found., Bot. Stud. 1: 21–323.

Muller, C. H. 1956. The distribution of *Quercus boyntoni*. Madroño 13: 221–225.

Muller, C. H. 1961. The live oaks of the series *Virentes*. Amer. Midl. Naturalist 65: 17–39.

Muller, C. H. 1970. *Quercus*. In: D. S. Correll and M. C. Johnston. 1970. Manual of the Vascular Plants of Texas. Renner, Tex. Pp. 467–492.

Münchhausen, O. von. 1765–1773. Der Hausvater. 6 vols. in parts. Hannover. (Hausvater)

Munz, P. A. 1945. The cultivated aconites. Gentes Herb. 6: 462–505.

Munz, P. A. 1946. The cultivated and wild columbines. Gentes Herb. 7: 1–150.

Munz, P. A. 1959. A California Flora. Berkeley and Los Angeles.

Munz, P. A. 1967. A synopsis of African species of *Delphinium* and *Consolida*. J. Arnold Arbor. 48: 30–55.

Munz, P. A. 1967b. A synopsis of the Asian species of *Consolida* (Ranunculaceae). J. Arnold Arbor. 48: 159–202.

Munz, P. A. 1968. Supplement to A California Flora. Berkeley and Los Angeles. (Suppl. Calif. Fl.)

Munz, P. A. 1974. A Flora of Southern California. Berkeley.

Murai, S. 1964. Phytotaxonomical and geobotanical studies on gen. *Alnus* in Japan (III). Taxonomy of whole world species and distribution of each sect. Bull. Gov. Forest Exp. Sta. 171: 1–107.

Murray, D. F. 1995. New names in *Papaver* section *Meconella* (Papaveraceae). Novon 5: 294–295.

Mus. Bot.—See: C. L. Blume 1849–1851[–1857].

N. Amer. Sylv.—See: T. Nuttall 1842–1849.

Nakai, T. 1915–1936. Flora Sylvatica Koreana. . . . 22 parts. Seoul and Keijyo.

Narr. Travels Africa—See: D. Denham and H. Clapperton 1826.

Nash, G. V. 1896. Revision of the genus *Asimina* in North America. Bull. Torrey Bot. Club 23: 234–242.

Nat. Arr. Brit. Pl.—See: S. F. Gray 1821.

Nat. Hist.—See: M. Houttuyn 1773–1783.

Nat. Pflanzenfam. ed. 2—See: H. G. A. Engler et al. 1924 +.

Naturaliste Canad. = Naturaliste Canadien. Bulletin de Recherches, Observations et Découvertes se Rapportant à l'Histoire Naturelle du Canada.

Nature = Nature; a Weekly Illustrated Journal of Science.

Nees, C. G. D., ed. 1820. Horae Physicae Berolinenses. . . . Bonn. (Horae Phys. Berol.)

Nees, C. G. D. 1836. Systema Laurinarum. Berlin. (Syst. Laur.)

Nelson, D. R. and K. T. Harper. 1991. Site characteristics and habitat requirements of the endangered dwarf bear-claw poppy (*Arctomecon humilis* Coville, Papaveraceae). Great Basin Naturalist 51: 167–175.

Nelson, D. R. and S. L. Welsh. 1993. Taxonomic revision of *Arctomecon* Torr. & Frem. Rhodora 95: 197–213.

Neogenyton—See: C. S. Rafinesque 1825.

Nesom, G. L. 1993. *Ranunculus* (Ranunculaceae) in Nuevo León, with comments on the *R. petiolaris* group. Phytologia 75: 391–398.

Nevling, L. I. Jr. 1960. *Trema*. In: R. E. Woodson Jr. and R. W. Schery et al., eds. 1943–1981. Flora of Panama. 41 fasc. St. Louis. [Ann. Missouri Bot. Gard. 47: 108–110.]

New Phytol. = New Phytologist; a British Botanical Journal.

Newslett. New York Fl. Assoc. = Newsletter, New York Flora Association.

Newton, H. and W. P. Jenney. 1880. Report on the Geology and Resources of the Black Hills of Dakota: With Atlas. Washington. (Rep. Geol. Resources Black Hills)

Nicely, K. A. 1965. A monographic study of the Calycanthaceae. Castanea 30: 38–81.

Nieuwland, J. A. 1910. Notes on the seedlings of bloodroot. Amer. Midl. Naturalist 1: 199–203.

Nixon, K. C. 1989. Origins of Fagaceae. In: P. R. Crane and S. Blackmore, eds. 1989. Evolution, Systematics, and Fossil History of the Hamamelidae. 2 vols. Oxford. Vol. 2, pp. 23–43.

Nixon, K. C. 1993. Infrageneric classification of *Quercus* (Fagaceae) and typification of sectional names. Ann. Sci. Forest. 50(suppl. 1): 25–34.

Nixon, K. C. 1993b. The genus *Quercus* in Mexico. In: T. P. Ramamoorthy et al., eds. 1993. Biological Diversity of Mexico: Origin and Distribution. New York. Pp. 447–458.

Nixon, K. C. and W. L. Crepet. 1989. *Trigonobalanus* (Fagaceae): Taxonomic status and phylogenetic relationships. Amer. J. Bot. 76: 828–841.

Nixon, K. C. and C. H. Muller. 1992. The taxonomic resurrection of *Quercus laceyi* Small (Fagaceae). Sida 15: 57–69.

Nooteboom, J. P. 1985. Notes on Magnoliaceae. Blumea 31: 65–121.

Nordic J. Bot. = Nordic Journal of Botany.

Norske Vidensk.-Akad., Mat.-Naturvidensk. Kl., Avh. = Norske Videnskaps-Akademi, Matematisk-Naturvidenskapelig Klasse. Avhandlinger.

Not. Milit. Reconn.—See: W. H. Emory 1848.

Notes Roy. Bot. Gard. Edinburgh = Notes from the Royal Botanic Garden, Edinburgh.

Notizbl. Königl. Bot. Gart. Berlin = Notizblatt des Königlichen botanischen Gartens und Museums zu Berlin.

Nouv. Mém. Soc. Imp. Naturalistes Moscou = Nouveaux Mémoires de la Société Impériale des Naturalistes de Moscou.

Nov. Gen. Pl.—See: C. P. Thunberg 1781–1801.

Nov. Gen. Sp.—See: A. von Humboldt et al. 1815[1816]–1825.

Novák, J. and V. Preininger. 1987. Chemotaxonomic review of the genus *Papaver*. Preslia 59: 1–13.

Novon = Novon; a Journal for Botanical Nomenclature.

Nowicke, J. W. and J. J. Skvarla. 1981. Pollen morphology and phylogenetic relationships of the Berberidaceae. Smithsonian Contr. Bot. 50: 1–83.

Nozeran, R. 1955. Contribution à l'étude de quelques structures florales. (Essai de morphologie florale comparée). Ann. Sci. Nat., Bot., sér. 11, 16: 1–224.

Nuttall, T. 1818. The Genera of North American Plants, and Catalogue of the Species, to the Year 1817. . . . 2 vols. Philadelphia. (Gen. N. Amer. Pl.)

Nuttall, T. 1842–1849. The North American Sylva. . . . 3 vols. Philadelphia. (N. Amer. Sylv.)

Nyman, C. F. 1878–1890. Conspectus Florae Europaeae. . . . 4 parts and 2 suppls. Örebro. [Parts 1–4 and suppl. 1 paged consecutively; suppl. 2 paged separately.] (Consp. Fl. Eur.)

Nytt Mag. Bot. = Nytt Magasin for Botanikk.

Oesterr. Bot. Z. = Oesterreichische botanische Zeitschrift. Gemeinütziges Organ für Botanik.

Ogle, D. W. and P. M. Mazzeo. 1976. *Betula uber,* the Virginia round-leaf birch, rediscovered in southwest Virginia. Castanea 41: 248–256.

Ohio J. Sci. = Ohio Journal of Science.

Ohwi, J. 1965. Flora of Japan (in English). . . . A Combined, Much Revised, and Extended Translation by the Author of His Flora of Japan (1953) and Flora of Japan—Pteridophyta (1957). Edited by Frederick G. Meyer . . . and Egbert H. Walker. . . . Washington.

Olsen, J. D., G. Manners, and S. W. Pelletier. 1990. Poisonous properties of larkspur (*Delphinium* spp.). Collect. Bot. (Barcelona) 19: 141–151.

Opera Bot. = Opera Botanica a Societate Botanice Lundensi.

Ørgaard, M. 1991. The genus *Cabomba* (Cabombaceae)—A taxonomic study. Nordic J. Bot. 11: 179–203.

Organization for Flora Neotropica. 1968+. Flora Neotropica. 65+ vols. New York.

Osborn, J. M. and E. L. Schneider. 1988. Morphological studies of the Nymphaeaceae sensu lato. XVI. The floral biology of *Brasenia schreberi*. Ann. Missouri Bot. Gard. 75: 778–794.

Osol, A., R. Pratt, and M. D. Altschule. 1960. The United States Dispensatory and Physicians' Pharmacology, ed. 26. Philadelphia.

Ovchinnikov, P. N. 1970. *Ranunculus*. In: V. L. Komarov et al., eds. 1963+. Flora of the U.S.S.R. (Flora SSSR). Translated from Russian. 22+ vols. Jerusalem. Vol. 7, pp. 271–388.

Overs. Kongel. Danske Vidensk. Selsk. Forh. Medlemmers Arbeider = Oversigt over det Kongelige Danske Videnskabernes Selskabs Forhandlinger og dets Medlemmers Arbeider.

Ownbey, G. B. 1947. Monograph of the North American species of *Corydalis*. Ann. Missouri Bot. Gard. 34: 187–260.

Ownbey, G. B. 1951. On the cytotaxonomy of the genus *Corydalis*, section *Eucorydalis*. Amer. Midl. Naturalist 45: 184–186.

Ownbey, G. B. 1958. Monograph of the genus *Argemone* for North America and the West Indies. Mem. Torrey Bot. Club 21(1): 1–159.

Ownbey, G. B. 1961. The genus *Argemone* in South America and Hawaii. Brittonia 13: 91–109.

Pacif. Railr. Rep.—See: War Department 1855–1860.

Pacif. Railr. Rep. 4(5)—See: J. Torrey 1857.

Paillet, F. L. 1993. Growth form and life histories of American chestnut and Allegheny and Ozark chinquapin at various North American sites. Bull. Torrey Bot. Club 120: 257–268.

Paillet, F. L. and P. A. Rutter. 1989. Replacement of native oak and hickory tree species by the introduced American chestnut *Castanea dentata* in southwestern Wisconsin, USA. Canad. J. Bot. 67: 3457–3469.

Paine, J. A. 1865. Catalogue of Plants Found in Oneida County and Vicinity. [Albany.] (Cat. Pl. Oneida Co.)

Pallas, P. S. 1771–1776. Reise durch verschiedene Provinzen des russischen Reichs. . . . 3 vols. St. Petersburg. (Reise Russ. Reich.)

Palmer, E. J. 1937. Notes on North American trees and shrubs. J. Arnold Arbor. 18: 136.

Palmer, E. J. 1942. The red oak complex in the United States. Amer. Midl. Naturalist 24: 732–740.

Palmer, E. J. 1948. Hybrid oaks of North America. J. Arnold Arbor. 29: 1–48.

Park, M. M. 1992. A Biosystematic Study of *Thalictrum* Section *Leucocoma* (Ranunculaceae). Ph.D. dissertation. Pennsylvania State University.

Payne, W. W. and J. L. Seago. 1968. The open conduplicate carpel of *Akebia quinata* (Berberidales: Lardizabalaceae). Amer. J. Bot. 55: 575–581.

Payson, E. B. 1918. The North American species of *Aquilegia*. Contr. U.S. Natl. Herb. 20: 133–157.

Pellmyr, O. 1985. The pollination biology of *Actaea pachypoda* and *A. rubra* (including *A. erythrocarpa*) in northern Michigan and Finland. Bull. Torrey Bot. Club 112: 265–273.

Persoon, C. H. 1805–1807. Synopsis Plantarum. . . . 2 vols. Paris and Tubingen. (Syn. Pl.)

Petersen, F. P. and D. E. Fairbrothers. 1983. A serotaxonomic appraisal of *Amphipterygium* and *Leitneria*—two amentiferous taxa of Rutiflorae (Rosidae). Syst. Bot. 8: 134–148.

Petersen, F. P. and D. E. Fairbrothers. 1985. A serotaxonomic appraisal of the "Amentiferae." Bull. Torrey Bot. Club 112: 43–52.

Pfeifer, H. W. 1966. Revision of the North and Central American hexandrous species of *Aristolochia*. Ann. Missouri Bot. Gard. 53: 1–114.

Pfeifer, H. W. 1970. A Taxonomic Revision of the Pentandrous Species of *Aristolochia*. [Storrs.]

Philos. Trans. = Philosophical Transactions: Giving Some Account of the Present Undertakings, Studies, and Labours of the Ingenious in Many Parts of the World.

Phipps, C. J. 1774. A Voyage to the North Pole Undertaken by His Majesty's Command 1773. . . . London. (Voy. North Pole)

Phytologia = Phytologia; Designed to Expedite Botanical Publication.

Pl. Dis. = Plant Disease; International Journal of Applied Plant Pathology.

Pl. Hartw.—See: G. Bentham 1839[–1857].

Pl. Hort. Schoenbr.—See: N. J. Jacquin 1797–1804.

Pl. Jav. Rar.—See: J. J. Bennett et al. 1838–1852.

Pl. Labrador.—See: E. Meyer 1830.

Pl. Nouv. Amér.—See: M. E. Moricand 1833–1846[–1847].

Pl. Surin.—See: C. Linnaeus 1775.

Pl. Syst. Evol. = Plant Systematics and Evolution.

Pl. Syst. Evol., Suppl. = Plant Systematics and Evolution. Supplementum.

Pl. World = Plant World.

Porsild, A. E. and W. J. Cody. 1980. Vascular Plants of Continental Northwest Territories, Canada. Ottawa.

Porter, T. C. and J. M. Coulter. 1874. Synopsis of the Flora of Colorado. . . . Washington. (Syn. Fl. Colorado)

Praglowski, J. 1974. Magnoliaceae Juss. Taxonomy by J. E. Dandy. World Pollen Spore Fl. 3: 1–48.

Prain, D. 1895. An account of the genus *Argemone*. J. Bot. 33: 129–135, 176–178, 207–209, 307–312, 325–333, 363–371.

Prantl, K. 1888. Lardizabalaceae. In: H. G. A. Engler and K. Prantl, eds. 1887–1915. Die natürlichen Pflanzenfamilien. . . . 254 fasc. Leipzig. Fasc. 19[III,2], pp. 67–70.

Prelim. Cat.—See: N. L. Britton et al. 1888.

Preslia = Preslia. Věstník (Časopis) Československé Botanické Společnosti.

Preston, D. J. 1976. The rediscovery of *Betula uber*. Amer. Forests 82(8): 16–20.

Priestley, D. A. and M. A. Posthumus. 1982. Extreme longevity of lotus seeds from Pulantien. Nature 299: 148–149.

Pringle, J. S. 1971. Taxonomy and distribution of *Clematis,* sect. *Atragene* (Ranunculaceae), in North America. Brittonia 23: 361–393.

Prir. Rostlin—See: F. Berchtold and J. S. Presl 1823–1835.

Proc. Acad. Nat. Sci. Philadelphia = Proceedings of the Academy of Natural Sciences of Philadelphia.

Proc. Amer. Acad. Arts = Proceedings of the American Academy of Arts and Sciences.

Proc. & Trans. Roy. Soc. Canada = Proceedings and Transactions of the Royal Society of Canada.

Proc. Annual Conf. SouthE. Assoc. Fish Wildlife Agencies = Proceedings of the Annual Conference Southeastern Association of Fish and Wildlife Agencies.

Proc. Arkansas Acad. Sci. = Proceedings of the Arkansas Academy of Science.

Proc. Biol. Soc. Wash. = Proceedings of the Biological Society of Washington.

Proc. Boston Soc. Nat. Hist. = Proceedings of the Boston Society of Natural History.

Proc. Calif. Acad. Sci. = Proceedings of the California Academy of Science.

Proc. Indiana Acad. Sci. = Proceedings of the Indiana Academy of Science.

Proc. Kon. Ned. Akad. Wetensch. C. = Proceedings, Koninklijke Nederlandse Akademie van Wetenschappen. Series C, Biological and Medical Sciences.

Proc. Louisiana Acad. Sci. = Proceedings of the Louisiana Academy of Sciences.

Proc. Minnesota Acad. Sci. = Proceedings of the Minnesota Academy of Science.

Proc. Natl. Acad. Sci. U.S.A. = Proceedings of the National Academy of Sciences of the United States of America.

Prodr.—See: A. P. de Candolle and A. L. P. de Candolle 1823–1873; O. P. Swartz 1788.

Prodr. Pl. Cap.—See: C. P. Thunberg 1794–1800.

Prodr. Stirp. Chap. Allerton—See: R. A. Salisbury 1796.

Publ. Carnegie Inst. Wash. = Publications of the Carnegie Institution of Washington.

Pursh, F. 1814. Flora Americae Septentrionalis; or, a Systematic Arrangement and Description of the Plants of North America. 2 vols. London. (Fl. Amer. Sept.)

Qiu, Y. L., M. W. Chase, D. H. Les, and C. R. Parks. 1993. Molecular phylogenetics of the Magnoliidae: Cladistic analyses of nucleotide sequences of the plastid gene *rbc*L. Ann. Missouri Bot. Gard. 80: 587–606.

Quibell, C. H. 1941. Floral anatomy and morphology of *Anemopsis californica.* Bot. Gaz. 102: 749–758.

Rachelle, L. D. 1974. Pollen morphology of the Papaveraceae of the northeastern U.S. and Canada. Bull. Torrey Bot. Club 101: 152–159.

Rafinesque, C. S. 1825. Neogenyton, or Indication of Sixty-six New Genera of Plants of North America. [Lexington, Ky.] (Neogenyton)

Rafinesque, C. S. 1838. Alsographia Americana, or an American Grove. . . . Philadelphia. (Alsogr. Amer.)

Rafinesque, C. S. 1838b. Sylva Telluriana. Mantis. Synopt.

. . . Being a Supplement to the Flora Telluriana. Philadelphia. (Sylva Tellur.)

Raju, M. V. S. 1961. Morphology and anatomy of the Saururaceae. I. Floral anatomy and embryology. Ann. Missouri Bot. Gard. 48: 107–124.

Ramamoorthy, T. P. et al., eds. 1993. Biological Diversity of Mexico: Origin and Distribution. New York.

Ramsey, G. W. 1965. A Biosystematic Study of the Genus *Cimicifuga* (Ranunculaceae). Ph.D. thesis. University of Tennessee.

Ramsey, G. W. 1987. Morphological considerations in North American *Cimicifuga.* Castanea 52: 129–141.

Rändel, U. 1974. Beitrage zur Kenntnis der Sippenstruktur der Gattung *Papaver* Sectio *Scapiflora* Reichenb. (Papaveraceae). Feddes Repert. 84: 655–732.

Rändel, U. 1975. Die Beziehungen von *Papaver pygmaeum* Rydb. aus den Rocky Mountains zum nordamerikanischen *P. kluanense* D. Löve sowie zu einigen nordostasiatischen Vertretern der Sektion *Scapiflorae* Reichenb. im Vergleich mit *P. alpinum* L. (Papaveraceae). Feddes Repert. 86: 19–37.

Rändel, U. 1977. Über Sippen des subarktisch-arktischen Nordamerikas, des Beringia-Gebietes und Nordost-Asiens der Sektion *Lasiotrachyphylla* Bernh. (Papaveraceae) und deren Beziehungen zu einander und zu Sippen anderer Arealteile der Sektion. Feddes Repert. 88: 421–450.

Rändel, U. 1977b. Über die grönländischen Vertreter der Sektion *Lasiotrachyphylla* Bernh. (Papaveraceae) und deren Beziehungen zu Vertretern anderer arktischer Arealteile der Sektion. Feddes Repert. 88: 451–464.

Raynie, D. E., D. R. Nelson, and K. T. Harper, K. T. 1991. Alkaloidal relationships in the genus *Arctomecon* (Papaveraceae) and herbivory in *A. humilis.* Great Basin Naturalist 51: 397–403.

Recent Advances Phytochem. = Recent Advances in Phytochemistry.

Recueil Trav. Bot. Néerl. = Recueil des Travaux Botaniques Néerlandais.

Reed, C. F. 1975. *Betula uber* (Ashe) Fernald rediscovered in Virginia. Phytologia 32: 305–311.

Rees, A. [1802–]1819–1820. The Cyclopaedia; or, Universal Dictionary of Arts, Sciences, and Literature. . . . 39 vols. in 79 parts. London. [Pages unnumbered.] (Cycl.)

Regel, E. 1861. Monographische Bearbeitung der Betulaceen. Nouv. Mém. Soc. Imp. Naturalistes Moscou 13(2): 59–187.

Rehder, A. J. 1907. Some new or little known forms of New England trees. Rhodora 9: 109–116.

Rehder, A. J. 1940. Manual of Cultivated Trees and Shrubs Hardy in North America . . . , ed. 2. New York.

Rehder, A. J. 1944. *Schisandra* Michaux, nomen genericum conservandum. J. Arnold Arbor. 25: 129–131.

Rehder, A. J. 1945. *Carya alba* proposed as nomen ambiguum. J. Arnold Arbor. 26: 482–483.

Rehder, A. J. 1949. Bibliography of Cultivated Trees and Shrubs Hardy in the Cooler Temperate Regions of the Northern Hemisphere. Jamaica Plain.

Rehder, A. J. and W. A. Dayton. 1944. A new combination in *Asimina.* J. Arnold Arbor. 25: 84.

Reise Russ. Reich.—See: P. S. Pallas 1771–1776.

Rep. (Annual) Michigan Acad. Sci. = Report (Annual) of the Michigan Academy of Science, (Arts, and Letters).

Rep. (Annual) Regents Univ. State New York State Cab. Nat. Hist. = Report (Annual) of the Regents of the University of the State of New York on the Condition of the State Cabinet of Natural History [etc.].

Rep. Exped. Rocky Mts.—See: J. C. Frémont 1843–1845.

Rep. Exped. Zuni Colorado Rivers—See: L. Sitgreaves 1853.

Rep. Geol. Resources Black Hills—See: H. Newton and W. P. Jenney 1880.

Rep. U.S. Mex. Bound.—See: W. H. Emory 1857–1859.

Repert. Bot. Syst.—See: W. G. Walpers 1842–1847.

Repert. Spec. Nov. Regni Veg. = Repertorium Specierum Novarum Regni Vegetabilis.

Res. Notes S. E., U.S. Forest Serv. = Research Notes S E, United States Forest Service.

Rev. Crit. Casuarinum—See: F. A. W. Miquel 1848.

Rhodes, D. G. 1975. A revision of the genus Cissampelos. Phytologia 30: 415–484.

Rhodora = Rhodora; Journal of the New England Botanical Club.

Richens, R. H. 1977. New designations in Ulmus minor Mill. Taxon 26: 582–593.

Rickett, H. W. and F. A. Stafleu. 1959. Nomina generica conservanda et rejicienda spermatophytorum. Taxon 8: 256–274.

Risser, P. G. and G. Cottam. 1968. Carbohydrate cycles in the bulbs of some spring ephemerals. Bull. Torrey Bot. Club 95: 359–369.

Robbins, W. W. 1940. Alien plants growing without cultivation in California. Univ. Calif. Agric. Exp. Sta. Bull. 637: 1–128.

Roberts, M. L. and R. R. Haynes. 1983. Ballistic seed dispersal in Illicium (Illiciaceae). Pl. Syst. Evol. 143: 227–232.

Robertson, R. et al., eds. 1981+. Flora of Australia. 14+ vols. Canberra.

Robinson, B. L. and M. L. Fernald. 1908. Gray's New Manual of Botany: A Handbook of the Flowering Plants and Ferns of the Central and Northeastern United States and Adjacent Canada, ed. 7. New York, Cincinnati, and Chicago.

Rockwell, H. C. 1966. The Genus Magnolia in the United States. M.S. thesis. West Virginia University.

Roelfs, A. P. 1982. Effects of barberry eradication on stem rust in the United States. Pl. Dis. 66: 177–181.

Roemer, J. J. and J. A. Schultes. 1817[–1830]. Caroli a Linné . . . Systema Vegetabilium . . . editione XV. . . . 7 vols. Stuttgart. (Syst. Veg.)

Rogers, G. K. 1982b. The Casuarinaceae in the southeastern United States. J. Arnold Arbor. 63: 357–373.

Rohwer, J. G. 1993. Lauraceae: Nectandra. In: Organization for Flora Neotropica. 1968+. Flora Neotropica. 65+ vols. New York. Monogr. 60.

Rohwer, J. G. 1993b. Moraceae. In: K. Kubitzki et al., eds. 1990+. The Families and Genera of Vascular Plants. 2+ vols. Berlin etc. Vol. 2, pp. 438–453.

Rom, R. C. and R. F. Carlson, eds. 1987. Rootstocks for Fruit Crops. New York.

Rose, J. N. 1895. Notes upon Corylus rostrata and C. californica. Gard. & Forest 8: 263.

Rosendahl, C. O. 1928. Evidence of the hybrid nature of Betula sandbergi. Rhodora 30: 125–129.

Roth, A. W. 1788–1800. Tentamen Florae Germanicae. . . . 3 vols. in 5 parts. Leipzig. (Tent. Fl. Germ.)

Rousseau, J. and M. Raymond. 1950. Betula michauxii, a brief symposium. 2. Betula michauxii in northeastern America. Rhodora 52: 27–32.

Rouy, G., J. Foucaud, and E. G. Camus. 1893–1913. Flore de France. . . . 14 vols. Asnières. (Fl. France)

Rudolph, J. H. [1781.] Florae Jenensis Plantas ad Polyandriam Monogyniam Linnaei Pertinentes. Jena. (Fl. Jen.)

Ruiz Lopez, H. and J. A. Pavon. 1794. Flora Peruvianae, et Chilensis Prodromus. . . . Madrid. (Fl. Peruv. Prodr.)

Ruiz Lopez, H. and J. A. Pavon. 1798–1802. Flora Peruviana, et Chilensis, sive Descriptiones, et Icones Plantarum Peruvianarum, et Chilensium. . . . 3 vols. [Madrid.] (Fl. Peruv.)

Ruprecht, F. 1845. Distributio Cryptogamarum Vascularium in Imperio Rossico. St. Petersburg. [Alt. title: Beiträge zur Pflanzenkunde des Russischen Reiches. . . . Dritte Lieferung.] (Distr. Crypt. Vasc. Ross.)

Rusby, H. H. 1935. The custard-apple family in Florida. J. New York Bot. Gard. 36: 233–239.

Ryberg, M. 1955. A taxonomical survey of the genus Corydalis Ventenat, with reference to cultivated species. Acta Horti Berg. 17: 155–175.

Ryberg, M. 1960. Studies in the Morphology and Taxonomy of the Fumariaceae. Uppsala.

Ryberg, M. 1960b. A morphological study of the Fumariaceae and the taxonomic significance of the characters examined. Acta Horti Berg. 19: 121–248.

Rydberg, P. A. 1917. Flora of the Rocky Mountains and Adjacent Plains. New York. (Fl. Rocky Mts.)

Safford, W. E. 1914. Classification of the genus Annona, with descriptions of the new and imperfectly known species. Contr. U.S. Natl. Herb. 18: i–xii, 1–68.

Salisbury, R. A. 1796. Prodromus Stirpium in Horto ad Chapel Allerton Vigentium. . . . London. (Prodr. Stirp. Chap. Allerton)

Santamour, F. S. Jr. 1969b. Cytology of Magnolia hybrids. I. Morris Arbor. Bull. 20: 63–65.

Santamour, F. S. Jr. 1972b. Interspecific hybridization in Platanus. Forest Sci. 18: 236–239.

Santamour, F. S. Jr. 1986. Checklist of cultivated Platanus (planetree). J. Arboric. 12: 78–83.

Santamour, F. S. Jr. and F. G. Meyer. 1971. The two tuliptrees. Amer. Hort. Mag. 50(2): 87–89.

Sargent, C. S. 1890–1902. The Silva of North America. . . . 14 vols. Boston and New York. (Silva)

Sargent, C. S. 1891. The fruit of Akebia quinata. Gard. & Forest 4: 136–137.

Sargent, C. S. [1902–]1905–1913. Trees and Shrubs. . . . 2 vols. in parts. Boston and New York. (Trees & Shrubs)

Sargent, C. S. 1918. Notes on North American trees. I. Quercus. Bot. Gaz. 65: 423–459.

Sargent, C. S. 1922. Manual of the Trees of North America

(Exclusive of Mexico), ed. 2. Boston and New York. [Facsimile edition in 2 vols. 1961, reprinted 1965, New York.]

Sargentia = Sargentia; Continuation of the Contributions from the Arnold Arboretum of Harvard University.

Schaeffer, J. C. 1760. Botanica Expeditior. Regensburg. (Bot. Exped.)

Schery, R. W. 1952. Plants for Man. New York.

Schlechtendal, D. F. L. von. 1819–1820. Animadversiones Botanicae in Ranunculeas Candollii. . . . 2 parts. Berlin. [Parts paged separately.] (Animadv. Bot. Ranunc. Cand.)

Schmidt, O. C. 1935. Aristolochiaceae. In: H. G. A. Engler et al., eds. 1924+. Die natürlichen Pflanzenfamilien . . . , ed. 2. 26+ vols. Leipzig and Berlin. Vol. 16b, pp. 202–242.

Schmitt, D. 1966. Pistillate inflorescence of sweetgum (*Liquidambar styraciflua* L.). Silvae Genet. 15: 33–35.

Schneider, C. K. [1904–]1906–1912. Illustriertes Handbuch der Laubholzkunde. . . . 2 vols. in 12 fasc. Jena. (Ill. Handb. Laubholzk.)

Schneider, C. K. 1916. Beiträge zur Kenntnis der Gattung *Ulmus*. Oesterr. Bot. Z. 66: 21–34, 65–82.

Schneider, E. L. 1979. Pollination biology of the Nymphaeaceae. In: D. M. Caron, ed. 1979. Increasing Production of Agricultural Crops through Increased Insect Pollination: Proceedings of the IVth International Symposium on Pollination. College Park, Md. Pp. 419–429.

Schneider, E. L. 1982. Notes on the floral biology of *Nymphaea elegans* (Nymphaeaceae) in Texas. Aquatic Bot. 12: 197–200.

Schneider, E. L. and J. D. Buchanan. 1980. Morphological studies of the Nymphaeaceae. XI. The floral biology of *Nelumbo pentapetala*. Amer. J. Bot. 67: 182–193.

Schneider, E. L. and J. M. Jeter. 1982. Morphological studies of the Nymphaeaceae. XII. The floral biology of *Cabomba caroliniana*. Amer. J. Bot. 69: 1410–1419.

Schneider, E. L. and L. A. Moore. 1977. Morphological studies of the Nymphaeaceae. VII. The floral biology of *Nuphar lutea* subsp. *macrophylla*. Brittonia 29: 88–99.

Schreber, J. C. 1789–1791. Caroli a Linné . . . Genera Plantarum. . . . 2 vols. Frankfurt am Main. (Gen. Pl.)

Schultes, R. E., W. M. Klein, T. Plowman, and T. E. Lockwood. 1974. *Cannabis:* An example of taxonomic neglect. Bot. Mus. Leafl. 23: 337–367.

Schwarzwalder, R. N. Jr. and D. L. Dilcher. 1991. Systematic placement of the Platanaceae in the Hamamelidae. Ann. Missouri Bot. Gard. 78: 962–969.

Sci. Rep. Coll. Gen. Educ. Osaka Univ. = Science Reports, College of General Education, Osaka University.

Science = Science; an Illustrated Journal [later: a Weekly Journal Devoted to the Advancement of Science]. [American Association for the Advancement of Science.]

Scoggan, H. J. 1978–1979. The Flora of Canada. 4 parts. Ottawa. [Natl. Mus. Nat. Sci. Publ. Bot. 7.]

Scopoli, J. A. 1760. Flora Carniolica. . . . Vienna. (Fl. Carniol.)

Seed Sci. Techn. = Seed Science and Technology; Proceedings of the International Seed Testing Association.

Seemann, B. 1852–1857. The Botany of the Voyage of H.M.S. Herald . . . during the Years 1845–51. 10 parts. London. [Parts paged consecutively.] (Bot. Voy. Herald)

Sell, P. D. 1994. *Ranunculus ficaria* L. *sensu lato.* Watsonia 20: 41–50.

Sessé y Lacasta, M. de and J. M. Moçiño. 1894. Flora Mexicana, ed. 2. Mexico City. (Fl. Mexic. ed. 2)

Sharik, T. L. and R. H. Ford. 1984. Variation and taxonomy of *Betula uber, B. lenta* and *B. alleghaniensis*. Brittonia 36: 307–316.

Sheffy, M. V. 1972. A Study of the Myricaceae from Eocene Sediments of Southeastern North America. Ph.D. thesis. Indiana University.

Shen, Y. 1954. Phylogeny and wood anatomy of *Nandina*. Taiwania 5: 89–91.

Sherman, S. L. 1987. Flavonoid Systematics of *Ulmus* L. in the United States. Masters thesis. University of Georgia.

Sherman, S. L. and D. E. Giannasi. 1988. Foliar flavonoids of *Ulmus* in eastern North America. Biochem. Syst. & Ecol. 16: 51–56.

Sherman-Broyles, S. L., S. B. Broyles, and J. L. Hamrick. 1992. Geographic distribution of allozyme variation in *Ulmus crassifolia* Nutt. Syst. Bot. 17: 33–41.

Shimizu, T. 1981. Miscellaneous notes from my Appalachian trip. Hikobia, Suppl. 1: 445–453.

Shteinberg, E. I. 1970. *Aconitum* L. In: V. L. Komarov et al., eds. 1963+. Flora of the U.S.S.R. (Flora SSSR). Translated from Russian. 22+ vols. Jerusalem. Vol. 7, pp. 143–184.

Sibthorp, J. and J. E. Smith. 1806–1813[–1816]. Florae Graecae Prodromus: Sive Plantarum Omnium Enumeratio, Quas in Provinciis aut Insulis Graeciae Invenit. . . . 2 vols. London. (Fl. Graec. Prodr.)

Sida = Sida; Contributions to Botany.

Silva—See: C. S. Sargent 1890–1902.

Silvae Genet. = Silvae Genetica.

Simpson, B. B. and M. Conner-Ogorzaly. 1986. Economic Botany: Plants in Our World. New York.

Simpson, B. J., P. Karges, and J. M. Carpenter. 1992. *Quercus polymorpha* (Fagaceae) new to Texas and the United States. Sida 15: 153.

Sitgreaves, L. 1853. Report of an Expedition Down to the Zuni and Colorado Rivers. . . . Washington. [Botany by J. Torrey, pp. 153–178, plates 1–21.] (Rep. Exped. Zuni Colorado Rivers)

Sketch Bot. S. Carolina—See: S. Elliott [1816–]1821–1824.

Skr. Kiøbenhavnske Selsk. Laerd. Elsk. = Skrifter, Som Udi det Kiøbenhavnske Selskab af Saerdoms og Videnskabers Elskere ere Fremlagte og Oplaeste.

Small, E. 1978. A numerical and nomenclatural analysis of morpho-geographic taxa of *Humulus*. Syst. Bot. 3: 37–76.

Small, E. 1979. The Species Problem in *Cannabis:* Science & Semantics. 2 vols. Toronto.

Small, E. 1980. The relationships of hop cultivars and wild variants of *Humulus lupulus*. Canad. J. Bot. 58: 676–686.

Small, E. and A. Cronquist. 1976. A practical and natural taxonomy for *Cannabis*. Taxon 25: 405–435.

Small, J. K. 1903. Flora of the Southeastern United States. . . . New York. (Fl. S.E. U.S.)

Small, J. K. 1926. *Deeringothamnus pulchellus.* Addisonia 11: 33–34.

Small, J. K. 1930. *Deeringothamnus rugelii.* Addisonia 15: 17–18.

Small, J. K. 1933. Manual of the Southeastern Flora, Being Descriptions of the Seed Plants Growing Naturally in Florida, Alabama, Mississippi, Eastern Louisiana, Tennessee, North Carolina, South Carolina and Georgia. New York.

Smit, P. G. 1967. Taxonomical and ecological studies in *Caltha palustris* L. (preliminary report). Proc. Kon. Ned. Akad. Wetensch. C. 70: 500–510.

Smit, P. G. 1968. Taxonomical and ecological studies in *Caltha palustris* L. II. Proc. Kon. Ned. Akad. Wetensch. C. 71: 280–292.

Smit, P. G. 1973. A revision of *Caltha* (Ranunculaceae). Blumea 21: 119–150.

Smit, P. G. and W. Punt. 1969. Taxonomy and pollen morphology of the *Caltha leptosepala* complex. Proc. Kon. Ned. Akad. Wetensch. C. 72: 16–27.

Smith, A. C. 1947. The families Illiciaceae and Schisandraceae. Sargentia 7: 1–224.

Smith, L. E. 1812. *Leontice triphylla.* In: A. Rees. [1802–]1819–1820. The Cyclopaedia; or, Universal Dictionary of Arts, Sciences, and Literature. . . . 39 vols. in 79 parts. London. Vol. 20.

Smith, R. E., C. O. Smith, and H. J. Ramsey. 1912. Walnut culture in California. Walnut blight. Univ. Calif. Agric. Exp. Sta. Bull. 231: 119–398.

Smith, R. R., ed. 1986. Proceedings of the First Symposium on the Botany of the Bahamas. . . . San Salvador, Bahamas.

Smithsonian Contr. Bot. = Smithsonian Contributions to Botany.

Smithsonian Contr. Knowl. = Smithsonian Contributions to Knowledge.

Soepadmo, E. 1972. Fagaceae. In: C. G. G. J. van Steenis, ed. 1950+. Flora Malesiana. . . . Series I. Spermatophyta. 11+ vols. in parts. Djakarta and Leiden. Vol. 7, part 2, pp. 265–403.

Solereder, H. 1889. Aristolochiaceae. In: H. G. A. Engler and K. Prantl, eds. 1887–1915. Die natürlichen Pflanzenfamilien. . . . 254 fasc. Leipzig. Fasc. 35[III,1], pp. 264–273.

Solereder, H. 1889b. Beiträge zur vergleichenden Anatomie der Aristolochiaceen. Bot. Jahrb. Syst. 10: 410–523, tables 12–14.

Soltis, D. E. 1984. Karyotypes of species of *Asarum* and *Hexastylis* (Aristolochiaceae). Syst. Bot. 9: 490–493.

SouthW. Naturalist = Southwestern Naturalist.

Sp. Pl.—See: A. G. Dietrich 1831–1832; C. Linnaeus 1753; C. L. Willdenow 1797–1830.

Sp. Pl. ed. 2—See: C. Linnaeus 1762–1763.

Spach, E. 1834–1848. Histoire Naturelle des Végétaux. Phanérogames. . . . 14 vols., atlas. Paris. (Hist. Nat. Vég.)

Spellenberg, R. 1992. A new species of black oak (*Quercus* subg. *Erythrobalanus,* Fagaceae) from the Sierra Madre Occidental, Mexico. Amer. J. Bot. 79: 1200–1206.

Spongberg, S. A. 1976. Magnoliaceae hardy in temperate North America. J. Arnold Arbor. 57: 250–312.

Spongberg, S. A. and I. H. Burch. 1979. Lardizabalaceae hardy in temperate North America. J. Arnold Arbor. 60: 302–315.

Sprengel, K. [1824–]1825–1828. Caroli Linnaei . . . Systema Vegetabilium. Editio Decima Sexta. . . . 5 vols. Göttingen. [Vol. 4 in 2 parts, each paged separately; vol. 5 by A. Sprengel.] (Syst. Veg.)

Spring Fl. Piedmont—See: H. L. Blomquist and H. J. Oosting 1940.

Stafleu, F. A. and R. S. Cowan. 1976–1988. Taxonomic Literature: A Selective Guide to Botanical Publications and Collections with Dates, Commentaries and Types, ed. 2. 7 vols. Utrecht, Antwerp, The Hague, and Boston.

Stafleu, F. A. and E. A. Mennega. 1992+. Taxonomic Literature: A Selective Guide to Botanical Publications and Collections with Dates, Commentaries and Types. Supplement. 3+ vols. Königstein.

Stand. Cycl. Hort.—See: L. H. Bailey 1914–1917.

Standley, P. C. 1920–1926. Trees and Shrubs of Mexico. 5 parts. Washington. [Contr. U.S. Natl. Herb. 23.] (Trees Shrubs Mexico)

Stearn, W. T. 1938. *Epimedium* and *Vancouveria,* a monograph. J. Linn. Soc., Bot. 51: 409–535.

Stearn, W. T. 1944. Some dwarf dicentras. Gard. Chron. 116: 40.

Stearn, W. T. 1956. Gender of generic names. In: P. M. Synge, ed. 1956. Supplement to the Dictionary of Gardening: A Practical and Scientific Encyclopaedia of Horticulture. Oxford. Pp. 222–223.

Stebbins, G. L. 1993. Concepts of species and genera. In: Flora of North America Editorial Committee, eds. 1993+. Flora of North America North of Mexico. 3+ vols. New York and Oxford. Vol. 1, pp. 229–246.

Steele, F. L. 1961. Introgression of *Alnus serrulata* and *Alnus rugosa.* Rhodora 63: 297–302.

Steenis, C. G. G. J. van, ed. 1950+. Flora Malesiana. . . . Series I. Spermatophyta. 11+ vols. in parts. Djakarta and Leiden.

Stermitz, F. R. 1968. Alkaloid chemistry and the systematics of *Papaver* and *Argemone.* Recent Advances Phytochem. 1: 161–183.

Stermitz, F. R., D. E. Nicodem, Wei C. C., and K. D. McMurtrey. 1969. Alkaloids of *Argemone polyanthemos, A. corymbosa, A. chisosensis, A. sanguinea, A. aurantiaca* and general *Argemone* systematics. Phytochemistry 8: 615–620.

Stern, K. R. 1961. Revision of *Dicentra* (Fumariaceae). Brittonia 13: 1–57.

Stern, K. R. 1962. The use of pollen morphology in the taxonomy of *Dicentra.* Amer. J. Bot. 49: 362–368.

Stern, K. R. 1968. Cytogeographic studies in *Dicentra.* I. *Dicentra formosa* and *D. nevadensis.* Amer. J. Bot. 55: 626–628.

Stern, K. R. 1970. Pollen aperture variation and phylogeny in *Dicentra.* Madroño 20: 354–359.

Stern, K. R. and M. Ownbey. 1971. Hybridization and cytotaxonomy of *Dicentra.* Amer. J. Bot. 58: 861–866.

Steyermark, J. A. 1934. *Hamamelis virginiana* in Missouri. Rhodora 36: 97–100.

Steyermark, J. A. 1949. *Lindera melissaefolia*. Rhodora 51: 153–162.

Steyermark, J. A. 1963. Flora of Missouri. Ames.

Steyermark, J. A. and C. S. Steyermark. 1960. *Hepatica* in North America. Rhodora 62: 223–232.

Stipes, R. J. and R. J. Campana, eds. 1981. Compendium of Elm Diseases. St. Paul.

Stirp. Austr. Fasc.—See: H. J. N. Crantz 1762–1767.

Stockmarr, J. 1974. SEM studies on pollen grains of North European *Ulmus* species. Grana 14: 103–107.

Stone, D. E. 1959. A unique balanced breeding system in the vernal pool mouse-tails. Evolution 13: 151–174.

Stone, D. E. 1962. Affinities of a Mexican endemic, *Carya palmeri*, with American and Asian hickories. Amer. J. Bot. 49: 199–212.

Stone, D. E. 1968. Cytological and morphological notes on the southeastern endemic *Schisandra glabra* (Schisandraceae). J. Elisha Mitchell Sci. Soc. 84: 351–356.

Stone, D. E. 1993. Juglandaceae. In: K. Kubitzki et al., eds. 1990+. The Families and Genera of Vascular Plants. 2+ vols. Berlin etc. Vol. 2, pp. 348–359.

Stone, D. E., G. A. Adrouny, and R. H. Flake. 1969. New World Juglandaceae. II. Hickory nut oils, phenetic similarities, and evolutionary implications in the genus *Carya*. Amer. J. Bot. 56: 928–935.

Stoynoff, N. and W. J. Hess. 1990. A new status for *Quercus shumardii* var. *acerifolia* (Fagaceae). Sida 14: 267–271.

Stud. Nat. Hist. Iowa Univ. = Studies in Natural History, Iowa University.

Sudworth, G. B. 1927. Check List of the Forest Trees of the United States, Their Names and Ranges. . . . Washington. [U.S.D.A. Div. Forestry, Bull. 17.] (Check List For. Trees U.S.)

Sugawara, T. 1982. Taxonomic studies of *Asarum* sensu lato. II. Karyotype and C-banding in two species of *Hexastylis* and *Asarum epigynum*. Bot. Mag. (Tokyo) 95: 295–302.

Sulkinoja, M. 1990. Hybridization, introgression and taxonomy of the mountain birch in SW Greenland compared with related results from Iceland and Finnish Lapland. Meddel. Grønland, Biosci. 33: 21–29.

Suppl. Calif. Fl.—See: P. A. Munz 1968.

Suppl. Hort. Bot. Hafn.—See: J. W. Hornemann 1819.

Suppl. Pl.—See: C. Linnaeus f. 1781[1782].

Sutherland, D. M. 1967. A Taxonomic Revision of the Low Larkspurs of the Pacific Northwest. Ph.D. dissertation. University of Washington, Seattle.

Svensk Bot. Tidskr. = Svensk Botanisk Tidskrift Utgifven af Svenska Botaniska Föreningen.

Swain, P. C. 1972. An Analysis of Morphological Differences among Oaks in Selected Minnesota Stands of the *Quercus borealis*–*Q. ellipsoidalis* Complex. M.S. thesis. University of Minnesota.

Swanson, E. E., H. W. Youngken, C. J. Zufall, W. J. Husa, J. C. Munch, and J. B. Wolffe. 1938. Aconite. Washington. [Amer. Pharm. Assoc. Monogr. 1.]

Swartz, O. P. 1788. Nova Genera & Species Plantarum seu Prodromus. . . . Stockholm, Uppsala, and Åbo. (Prodr.)

Sweet, R. 1823–1837. The British Flower Garden. . . . 7 vols. London. [Vols. 5–7 also issued as vols. 1–3 of series 2.] (Brit. Fl. Gard.)

Sweet, R. 1826. Hortus Britannicus. . . . 2 parts. London. [Parts paged consecutively.] (Hort. Brit.)

Sweet, R. 1830. Hortus Britannicus . . . , ed. 2. London. (Hort. Brit. ed. 2)

Sweitzer, E. M. 1971. The comparative anatomy of Ulmaceae. J. Arnold Arbor. 52: 523–585.

Sylva Tellur.—See: C. S. Rafinesque 1838b.

Symb. Fl. Argent.—See: A. H. R. Grisebach 1879.

Syn. Fl. Colorado—See: T. C. Porter and J. M. Coulter 1874.

Syn. Fl. N. Amer.—See: A. Gray et al. 1878–1897.

Syn. Pl.—See: C. H. Persoon 1805–1807.

Synge, P. M., ed. 1956. Supplement to the Dictionary of Gardening: A Practical and Scientific Encyclopaedia of Horticulture. Oxford.

Syst. Bot. = Systematic Botany; Quarterly Journal of the American Society of Plant Taxonomists.

Syst. Bot. Monogr. = Systematic Botany Monographs; Monographic Series of the American Society of Plant Taxonomists.

Syst. Laur.—See: C. G. D. Nees 1836.

Syst. Nat.—See: A. P. de Candolle [1817]1818–1821; J. F. Gmelin 1791[–1792].

Syst. Nat. ed. 10—See: C. Linnaeus 1758[–1759].

Syst. Nat. ed. 12—See: C. Linnaeus 1766[–1768].

Syst. Veg.—See: J. J. Roemer and J. A. Schultes 1817[–1830]; K. Sprengel [1824–]1825–1828.

Syst. Veg. ed. 13—See: C. Linnaeus 1774.

Tabl. Règn. Vég.—See: E. P. Ventenat [1799].

Takhtajan, A. L. 1986. Floristic Regions of the World. Berkeley and Los Angeles.

Takhtajan, A. L. 1987. Systema Magnoliophytorum. Leningrad.

Tamura, M. 1963. Morphology, ecology and phylogeny of the Ranunculaceae I. Sci. Rep. Coll. Gen. Educ. Osaka Univ. 11: 115–126.

Tamura, M. 1968. Morphology, ecology and phylogeny of the Ranunculaceae. VIII. Sci. Rep. Coll. Gen. Educ. Osaka Univ. 17: 41–56.

Tamura, M. 1984. Phylogenetical considerations on the Ranunculaceae. J. Korean Pl. Taxon. 14: 33–42.

Tamura, M. 1991. A new classification of the Ranunculaceae. 2. Acta Phytotax. Geobot. 42: 177–187.

Tamura, M. 1992. A new classification of the family Ranunculaceae. Acta Phytotax. Geobot. 43: 53–58.

Tamura, M. 1993. Ranunculaceae. In: K. Kubitzki et al., eds. 1990+. The Families and Genera of Vascular Plants. 2+ vols. Berlin etc. Vol. 2, pp. 563–583.

Tamura, M. and L. A. Lauener. 1968. A revision of *Isopyrum*, *Dichocarpum* and their allies. Notes Roy. Bot. Gard. Edinburgh 28: 267–273.

Taxon = Taxon; Journal of the International Association for Plant Taxonomy.

Taylor, R. J. 1960. The genus *Delphinium* in Wyoming. Univ. Wyoming Publ. 24: 9–21.

Taylor, R. L. and G. A. Mulligan. 1968. Flora of the Queen Charlotte Islands. Part 2. Cytological Aspects of the Vascular Plants. Ottawa.

Tent. Fl. Germ.—See: A. W. Roth 1788–1800.

Tent. Monogr. Papaver—See: L. Elkan 1839.

Terebayashi, S. 1985. The comparative floral anatomy and systematics of the Berberidaceae. I. Morphology. Mem. Fac. Sci. Kyoto Univ., Ser. Biol. 10: 73–90.

Terebayashi, S. 1985b. The comparative floral anatomy and systematics of the Berberidaceae. II. Systematic considerations. Acta Phytotax. Geobot. 36: 1–13.

Theor. Prakt. Handb. Forstbot.—See: M. B. Borkhausen 1800–1803.

Thien, L. B. 1974. Floral biology of *Magnolia*. Amer. J. Bot. 61: 1037–1045.

Thien, L. B., W. H. Heimermann, and R. T. Holman. 1975. Floral odors and quantitative taxonomy of *Magnolia* and *Liriodendron*. Taxon 24: 557–568.

Thien, L. B., D. A. White, and L. Y. Yatsu. 1983. The reproductive biology of a relict—*Illicium floridanum* Ellis. Amer. J. Bot. 70: 719–727.

Thieret, J. W. 1961. The specific epithet of the pecan. Rhodora 63: 296.

Thieret, J. W. 1964. *Fatoua villosa* (Moraceae) in Louisiana: New to North America. Sida 1: 248.

Thieret, J. W. 1966. Habit variation in *Myrica pensylvanica* and *M. cerifera*. Castanea 31: 183–185.

Thieret, J. W. 1969. Twenty-five species of vascular plants new to Louisiana. Proc. Louisiana Acad. Sci. 32: 79.

Thorne, R. F. 1976. A phylogenetic classification of the Angiospermae. Evol. Biol. 9: 35–106.

Thorne, R. F. 1983. Proposed new realignments in the angiosperms. Nordic J. Bot. 3: 85–117.

Thunberg, C. P. 1781–1801. Nova Genera Plantarum. . . . 16 diss. Uppsala. [Paged consecutively.] (Nov. Gen. Pl.)

Thunberg, C. P. 1784. Flora Japonica Sistens Plantas Insularum Japonicarum Secundum Systema Sexuale. . . . Leipzig. (Fl. Jap.)

Thunberg, C. P. 1794–1800. Prodromus Plantarum Capensium, Quas in Promontorio Bonae Spei Africes, Annis 1772–1775, Collegit. 2 parts. Upsala. (Prodr. Pl. Cap.)

Thurston, E. L. 1969. An Anatomical and Fine Structure Study of Stinging Hairs in Some Members of the Urticaceae, Euphorbiaceae, and Loasaceae. Ph.D. thesis. Iowa State University.

Thurston, E. L. and N. R. Lersten. 1969. The morphology and toxicology of plant stinging hairs. Bot. Rev. (Lancaster) 35: 393–412.

Tillson, A. H. and C. H. Muller. 1942. Anatomical and taxonomic approaches to subgeneric segregation in American *Quercus*. Amer. J. Bot. 29: 523–529.

Tippo, O. 1938. Comparative anatomy of the Moraceae and their presumed allies. Bot. Gaz. 100: 1–99.

Tobe, H. and R. C. Keating. 1985. The morphology and anatomy of *Hydrastis* (Ranunculales): Systematic re-evaluation of the genus. Bot. Mag. (Tokyo) 98: 291–316.

Tobe, J. D. 1993. A Molecular Systematic Study of Eastern North American Species of *Magnolia* L. Ph.D. thesis. Clemson University.

Tolmatchew, A. I., ed. 1960+. Flora Arctica U.R.S.S. 10+ vols. Moscow and Leningrad.

Tolmatchew, A. I. 1971. *Ranunculus*. In: A. I. Tolmatchew, ed. 1960+. Flora Arctica U.R.S.S. 10+ vols. Moscow and Leningrad. Vol. 6, pp. 123–231.

Tolmatchew, A. I. and V. V. Petrovsky. 1975. *Papaver*. In: A. I. Tolmatchew, ed. 1960+. Flora Arctica U.R.S.S. 10+ vols. Moscow and Leningrad. Vol. 7, pp. 7–32.

Tomlinson, P. B. 1980. The Biology of Trees Native to Tropical Florida. Allston, Mass.

Torrey, J. 1843. A Flora of the State of New York. . . . 2 vols. Albany. (Fl. New York)

Torrey, J. 1857. Explorations and Surveys for a Railroad Route from the Mississippi River to the Pacific Ocean. War Department. Route Near the Thirty-fifth Parallel, Explored by Lieutenant A. W. Whipple, Topographical Engineers, in 1853 and 1854. Report on the Botany of the Expedition. Washington. [Pacif. Railr. Rep. 4(5)]

Torrey, J. and A. Gray. 1838–1843. A Flora of North America. . . . 2 vols. in 7 parts. New York, London, and Paris. (Fl. N. Amer.)

Torreya = Torreya; a Monthly Journal of Botanical Notes and News.

Townsend, C. C. 1968. *Parietaria officinalis* and *P. judaica*. Watsonia 6: 365–370.

Traité Arbr. Arbust. nouv. ed.—See: H. Duhamel du Monceau et al. [1800–]1801–1819.

Trans. Acad. Sci. St. Louis = Transactions of the Academy of Science of St. Louis.

Trans. Amer. Philos. Soc. = Transactions of the American Philosophical Society Held at Philadelphia for Promoting Useful Knowledge.

Trans. Hort. Soc. London = Transactions, of the Horticultural Society of London.

Trans. Illinois State Acad. Sci. = Transactions of the Illinois State Academy of Science.

Trans. Linn. Soc. London = Transactions of the Linnean Society of London.

Trans. New York Acad. Sci. = Transactions of the New York Academy of Sciences.

Trappe, J. M., J. F. Franklin, R. F. Tarrant, and G. M. Hansen, eds. 1968. Biology of Alder. . . . Portland.

Travels Carolina—See: W. Bartram 1791.

Trease, G. E. and W. C. Evans. 1983. Pharmacognosy, ed. 12. London.

Trees & Shrubs—See: C. S. Sargent [1902–]1905–1913.

Trelease, W. 1919. The jack oak. Trans. Illinois State Acad. Sci. 12: 108–118, 7 plates.

Trelease, W. 1924. The American oaks. Mem. Natl. Acad. Sci. 20: 1–255.

Trelease, W. and T. G. Yuncker. 1950. The Piperaceae of Northern South America. 2 vols. Urbana.

Treseder, N. G. 1978. Magnolias. Boston.

Trop. Woods = Tropical Woods. . . .

Trudy Bot. Muz. Imp. Akad. Nauk = Trudy Botanicheskogo Muzeya Imperatorskoi Akademii Nauk.

Tucker, G. E. 1975. *Castanea pumila* var. *ozarkensis* (Ashe) Tucker, comb. nov. Proc. Arkansas Acad. Sci. 29: 67–69.

Tucker, J. M. 1961. Studies in the *Quercus undulata* complex. I. A preliminary statement. Amer. J. Bot. 48: 202–208.

Tucker, J. M. 1961b. Studies in the *Quercus undulata* complex. II. The contribution of *Q. turbinella*. Amer. J. Bot. 48: 329–339.

Tucker, J. M. 1963. Studies in the *Quercus undulata* complex. III. The contribution of *Q. arizonica*. Amer. J. Bot. 50: 699–708.

Tucker, J. M. 1970. Studies in the *Quercus undulata* complex. IV. The contribution of *Q. havardii*. Amer. J. Bot. 57: 71–84.

Tucker, J. M. 1971. Studies in the *Quercus undulata* complex. V. The type of *Q. undulata*. Amer. J. Bot. 58: 329–341.

Tucker, J. M. 1993. Fagaceae. In: J. C. Hickman, ed. 1993. The Jepson Manual. Higher Plants of California. Berkeley, Los Angeles, and London. Pp. 657–663, 665.

Tucker, J. M. and H. S. Haskell. 1960. *Quercus dunnii* and *Q. chrysolepis* in Arizona. Brittonia 12: 196–219.

Tucker, J. M. and J. Maze. 1966. Bur oak (*Quercus macrocarpa*) in New Mexico. SouthW. Naturalist 11: 402–405.

Tucker, S. C. 1975. Floral development in *Saururus cernuus* (Saururaceae). 1. Floral initiation and stamen development. Amer. J. Bot. 62: 993–1007.

Tucker, S. C. 1976. Floral development in *Saururus cernuus* (Saururaceae). 2. Carpel initiation and floral vasculature. Amer. J. Bot. 63: 289–301.

Tucker, S. C. 1979. Ontogeny of the inflorescence of *Saururus cernuus* (Saururaceae). Amer. J. Bot. 66: 227–236.

Turner, N. J. and A. F. Szczawinski. 1991. Common Poisonous Plants and Mushrooms of North America. Portland.

Turrill, W. B., E. Milne-Redhead, C. E. Hubbard, and R. M. Polhill, eds. 1952+. Flora of Tropical East Africa. 152+ vols. London and Rotterdam. [Unnumbered volumes by family, some in parts.]

Tutin, T. G. and J. R. Akeroyd. 1993. *Ranunculus*. In: T. G. Tutin et al., eds. 1993+. Flora Europaea, ed. 2. 1+ vol. Cambridge and New York. Vol. 1, pp. 269–286.

Tutin, T. G., V. H. Heywood, N. A. Burges, D. H. Valentine, S. M. Walters, D. A. Webb, et al., eds. 1964–1980. Flora Europaea. 5 vols. Cambridge.

Tutin, T. G. et al., eds. 1993+. Flora Europaea, ed. 2. 1+ vol. Cambridge and New York.

Twisselman, E. C. 1965. A flora of Kern County, California. Wasmann J. Biol. 25: 1–395.

Univ. Calif. Agric. Exp. Sta. Bull. = University of California Agricultural Experiment Station, Bulletin.

Univ. Calif. Publ. Bot. = University of California Publications in Botany.

Univ. Colorado Stud., Ser. Biol. = University of Colorado Studies. Series in Biology.

Univ. Colorado Stud., Ser. D, Phys. Sci. = University of Colorado Studies. Series D, Physical and Biological Sciences.

Univ. Kansas Sci. Bull. = University of Kansas Science Bulletin.

Univ. Wash. Publ. Biol. = University of Washington Publications in Biology.

Univ. Wyoming Publ. = University of Wyoming Publications.

Uphof, J. C. T. 1933. Die nordamerikanischen Arten der Gattung *Asimina*. Mitt. Deutsch. Dendrol. Ges. 45: 61–76.

Vahl, M. 1804–1805. Enumeratio Plantarum. . . . 2 vols. Copenhagen. (Enum. Pl.)

Van Buren, R., K. T. Harper, W. R. Andersen, D. J. Stanton, S. Seyoum, and J. L. England. 1994. Evaluating the relationship of autumn buttercup (*Ranunculus acriformis* var. *aestivalis*) to some close congeners using random amplified polymorphic DNA. Amer. J. Bot. 81: 514–519.

Vasc. Pl. Pacif. N.W.—See: C. L. Hitchcock et al. 1955–1969.

Vázquez-G., J. A. 1990. Taxonomy of the Genus *Magnolia* in Mexico and Central America. M.S. thesis. University of Wisconsin.

Vázquez-G., J. A. 1994. *Magnolia* (Magnoliaceae) in Mexico and Central America: A synopsis. Brittonia 46: 1–23.

Ventenat, E. P. [1799.] Tableau du Règne Végétal. . . . 4 vols. Paris. (Tabl. Règn. Vég.)

Verdcourt, B. 1989. Nymphaeaceae. In: W. B. Turrill et al., eds. 1952+. Flora of Tropical East Africa. 152+ vols. London and Rotterdam. Unnumbered vol.

Verh. Bot. Vereins Prov. Brandenburg = Verhandlungen des Botanischen Vereins der Provinz Brandenburg.

Verh. Mitth. Siebenbürg. Vereins Naturwiss. Hermannstadt = Verhandlungen und Mittheilungen des siebenbürgischen Vereins für Naturwissenschaften zu Hermannstadt.

Verz. Altai Pfl.—See: A. A. Bunge [1836].

Villars, D. 1786–1789. Histoire des Plantes de Dauphiné. 3 vols. Grenoble, Lyon, and Paris. (Hist. Pl. Dauphiné)

Vincent, M. A. 1993. *Fatoua villosa* (Moraceae), mulberry weed, in Ohio. Ohio J. Sci. 93: 147–149.

Voss, E. G. 1965. On citing the names of publishing authors. Taxon 14: 154–160.

Voss, E. G. 1972+. Michigan Flora. 2+ vols. Ann Arbor.

Voy. North Pole—See: C. J. Phipps 1774.

Voy. Uranie—See: C. Gaudichaud-Beaupré 1826[–1830].

W. Amer. Sci. = West American Scientist.

Wagner, W. H. Jr. and D. J. Schoen. 1976. Shingle oak (*Quercus imbricaria*) and its hybrids in Michigan. Michigan Bot. 15: 141–155.

Wahlenberg, G. 1812. Flora Lapponica Exhibens Plantas Geographice et Botanice Consideratas. . . . Berlin. (Fl. Lapp.)

Walker, J. W. and A. G. Walker. 1984. Ultrastructure of lower Cretaceous angiosperm pollen and the origin and early evolution of flowering plants. Ann. Missouri Bot. Gard. 71: 464–521.

Walker, R. I. 1932. Chromosome numbers in *Ulmus*. Science 75: 107.

Walpers, W. G. 1842–1847. Repertorium Botanices Systematicae. . . . 6 vols. Leipzig. (Repert. Bot. Syst.)

Walter, T. 1788. Flora Caroliniana, Secundum Systema Vegetabilium Perillustris Linnaei Digesta. . . . London. (Fl. Carol.)

Walters, S. M. et al., eds. 1984+. The European Garden Flora. 4+ vols. Cambridge etc.

Wang, W.-T. 1980. *Anemone*. In: Academia Sinicae. 1959+.

Flora Reipublicae Popularis Sinicae. 80+ vols. Beijing. Vol. 28, pp. 1–56.

Wangenheim, F. A. J. von. 1787. Beytrag zur teutschen holzgerechten Forstwissenschaft, die Anpflanzung nordamericanischer Holzarten. . . . Göttingen. (Beytr. Teut. Forstwiss.)

War Department [U.S.]. 1855–1860. Reports of Explorations and Surveys, to Ascertain the Most Practicable and Economical Route for a Railroad from the Mississippi River to the Pacific Ocean. Made under the Direction of the Secretary of War, in 1853–[6]. . . . 12 vols. in 13. Washington. (Pacif. Railr. Rep.)

Ward, D. B. 1977. Keys to the flora of Florida. 4. *Nymphaea* (Nymphaeaceae). Phytologia 37: 443–448.

Ward, D. B. 1977b. Night-flowering waterlilies in Florida. Florida Sci. 40: 155–159.

Ware, S. A. 1967. The morphological varieties of southern red oak. J. Tennessee Acad. Sci. 42: 29–36.

Warnock, M. J. 1981. Biosystematics of the *Delphinium carolinianum* complex (Ranunculaceae). Syst. Bot. 6: 38–54.

Warnock, M. J. 1993. *Delphinium*. In: J. C. Hickman, ed. 1993. The Jepson Manual. Higher Plants of California. Berkeley, Los Angeles, and London. Pp. 916–922.

Warnock, M. J. 1995. A taxonomic conspectus of North American *Delphinium*. Phytologia 78: 73–101.

Wasmann J. Biol. = Wasmann Journal of Biology.

Watson, S. [1871]b. United States Geological Exploration of the 40th Parallel. Clarence King, U.S. Geologist, in Charge. List of Plants Collected in Nevada and Utah 1867–'69; Numbered as Distributed. Sereno Watson, Collector. [Washington.] (List Pl. Nevada Utah)

Watson, S. 1878. Bibliographical Index to North American Botany; or Citations of Authorities for All the Recorded Indigenous and Naturalized Species of the Flora of North America. . . . Part 1, Polypetalae. Washington. [Smithsonian Misc. Collect. 258.] (Bibl. Index N. Amer. Bot.)

Watson, S., W. H. Brewer, and A. Gray. 1876–1880. Geological Survey of California. . . . Botany. . . . 2 vols. Cambridge, Mass. (Bot. California)

Watson, S., D. C. Eaton, et al. 1871. United States Geological Exploration [sic] of the Fortieth Parallel. Clarence King, Geologist-in-charge. [Vol. 5] Botany. By Sereno Watson. . . . Washington. [Botanical portion of larger work by C. King.] [Botany (Fortieth Parallel)]

Watsonia = Watsonia; Journal of the Botanical Society of the British Isles.

Weaver, R. E. 1969. Studies in the North American genus *Fothergilla* (Hamamelidaceae). J. Arnold Arbor. 50: 599–619.

Weber, J. L. 1981. A taxonomic revision of *Cassytha* (Lauraceae) in Australia. J. Adelaide Bot. Gard. 3: 187–262.

Weddell, H. A. 1856. Monographie de la Famille des Urticacées. Paris.

Welsh, S. L. 1974. Anderson's Flora of Alaska and Adjacent Parts of Canada. Provo.

Welsh, S. L. and K. L. Thorne. 1979. Illustrated Manual of Proposed Endangered and Threatened Plants of Utah. [Washington.]

Wernischeck, J. 1763. Genera Plantarum Cum Characteribus Suis Essentialibus et Naturalibus. . . . Vienna. (Gen. Pl.)

Wheeler, E., C. A. LaPasha, and Regis B. Miller. 1988. Wood anatomy of elm (*Ulmus*) and hackberry (*Celtis*) species native to the United States. I. A. W. A. Bull., N.S. 10: 5–26.

Wherry, E. T. 1960. Intermediate occurrences of *Alnus crispa*. Castanea 25: 135.

White, D. J. and H. L. Dickson. 1983. *Hydrastis canadensis*. In: G. W. Argus and D. J. White, eds. 1982–1983. Atlas of the Rare Vascular Plants of Ontario. 2 vols. in 4 parts. Ottawa. Vol. 2.

Wiegand, K. M. 1909. Recognition of *Corylus rostrata* and *Corylus americana*. Rhodora 11: 107.

Wiegrefe, S. J., K. J. Sytsma, and R. P. Guries. 1994. Phylogeny of elms (*Ulmus*, Ulmaceae): Molecular evidence for a sectional classification. Syst. Bot. 19: 590–612.

Wiersema, J. H. 1982. Distributional records for *Nymphaea lotus* (Nymphaeaceae) in the Western Hemisphere. Sida 9: 230–234.

Wiersema, J. H. 1987. A monograph of *Nymphaea* subgenus *Hydrocallis* (Nymphaeaceae). Syst. Bot. Monogr. 16: 1–112.

Wiersema, J. H. 1988. Reproductive biology of *Nymphaea* (Nymphaeaceae). Ann. Missouri Bot. Gard. 75: 795–804.

Wiersema, J. H. 1996. *Nymphaea tetragona* and *Nymphaea leibergii* (Nymphaeaceae): Two species of diminutive water-lilies in North America. Brittonia 48: 520–531.

Wiersema, J. H. and C. B. Hellquist. 1994. Nomenclatural notes in Nymphaeaceae for the North American flora. Rhodora 96: 170–178.

Wiersema, J. H. and J. L. Reveal. 1991. Proposals to reject two 1788 Thomas Walter names of American waterlilies (Nymphaeaceae). Taxon 40: 509–516.

Wilbur, R. L. 1970. Taxonomic and nomenclatural observations on the eastern North American genus *Asimina* (Annonaceae). J. Elisha Mitchell Sci. Soc. 86: 88–96.

Wilbur, R. L. 1994. The Myricaceae of the United States and Canada: Genera, subgenera, and series. Sida 16: 93–107.

Willdenow, C. L. 1809–1813[–1814]. Enumeratio Plantarum Horti Regii Botanici Berolinensis. . . . 2 parts + suppl. Berlin. (Enum. Pl.)

Willdenow, C. L., C. F. Schwägrichen, and J. H. F. Link. 1797–1830. Caroli a Linné Species Plantarum. . . . Editio Quarta. . . . 6 vols. Berlin. [Vols. 1–5(1), 1797–1810, by Willdenow; vol. 5(2), 1830, by Schwägrichen; vol. 6, 1824–1825, by Link.] (Sp. Pl.)

Williams, G. R. 1970. Investigations in the white waterlilies (*Nymphaea*) of Michigan. Michigan Bot. 9: 72–86.

Willis, J. C. 1973. A Dictionary of the Flowering Plants and Ferns, ed. 8, revised by H. K. Airy Shaw. Cambridge.

Wilson, K. L. and L. A. S. Johnson. 1989. Casuarinaceae. In: R. Robertson et al. eds. 1981+. Flora of Australia. 14+ vols. Canberra. Vol. 3, pp. 100–174.

Wilson, P. 1905. Altingiaceae: *Liquidambar*. In: N. L. Britton et al., eds. 1905+. North American Flora. . . . 47+ vols. New York. Vol. 22, p. 189.

Wilson, T. K. 1960. The comparative morphology of the Canellaceae. I. Synopsis of genera and wood anatomy. Trop. Woods 112: 1–27.

Wilson, T. K. 1964. Comparative morphology of the Canellaceae. III. Pollen. Bot. Gaz. 125: 192–197.

Wilson, T. K. 1966. Comparative morphology of the Canellaceae. IV. Floral morphology and conclusions. Amer. J. Bot. 53: 336–343.

Wilson, T. K. 1986. The natural history of *Canella alba* (Canellaceae). In: R. R. Smith, ed. 1986. Proceedings of the First Symposium on the Botany of the Bahamas. . . . San Salvador, Bahamas. Pp. 101–115.

Winkler, H. 1904. Betulaceae. In: H. G. A. Engler, ed. 1900–1953. Das Pflanzenreich. . . . 107 vols. Berlin. Vol. 19[IV,61], pp. 1–149.

Winkler, H. 1914. Neue Revision der Gattung *Carpinus*. Bot. Jahrb. Syst. 15(suppl.): 488–508.

Winstead, J. E., B. J. Smith, and G. I. Wardell. 1977. Fruit weight clines in populations of ash, ironwood, cherry, dogwood and maple. Castanea 42: 56–60.

Withering, W. 1796. An Arrangement of British Plants . . . , ed. 3. 4 vols. London. (Arr. Brit. Pl. ed. 3)

Wofford, B. E. 1983. A new *Lindera* from North America. J. Arnold Arbor. 64: 325–331.

Wood, A. 1847. A Class-book of Botany. . . . Boston and Claremont, N.H. [Class-book Bot. ed. 2(a)]

Wood, C. E. Jr. 1958. The genera of the woody Ranales in the southeastern United States. J. Arnold Arbor. 39: 296–346.

Wood, C. E. Jr. 1959. The genera of the Nymphaeaceae and Ceratophyllaceae in the southeastern United States. J. Arnold Arbor. 40: 94–112.

Wood, C. E. Jr. 1971b. The Saururaceae in the southeastern United States. J. Arnold Arbor. 52: 479–485.

Wood, H. C. and A. Osol. 1943. The Dispensatory of the United States of America, ed. 23. Philadelphia.

Woodall, S. L. and T. F. Geary. 1985. Identity of Florida casuarinas. Res. Notes S. E., U.S. Forest Serv. 332: 1–10.

Woodell, S. R. J. and M. Kootin-Sanwu. 1971. Intraspecific variation in *Caltha palustris*. New Phytol. 70: 173–186.

Woodland, D. W. 1982. Biosystematics of the perennial North American taxa of *Urtica*. II. Taxonomy. Syst. Bot. 7: 282–290.

Woodland, D. W. 1989. Biology of temperate Urticaceae (nettle family). In: P. R. Crane and S. Blackmore, eds. 1989. Evolution, Systematics, and Fossil History of the Hamamelidae. 2 vols. Oxford. Vol. 2, pp. 309–318.

Woodland, D. W., I. J. Bassett, and C. W. Crompton. 1976. The annual species of stinging nettle (*Hesperocnide* and *Urtica*) in North America. Canad. J. Bot. 54: 374–383.

Woodland, D. W., I. J. Bassett, L. Crompton, and S. Forget. 1982. Biosystematics of the perennial North American taxa of *Urtica*. I. Chromosome number, hybridization, and palynology. Syst. Bot. 7: 269–281.

Woodson, R. E. Jr., R. W. Schery, et al., eds. 1943–1981. Flora of Panama. 41 fasc. St. Louis. [Fascicles published as individual issues of Ann. Missouri Bot. Gard. and aggregating 8 nominal parts + introduction and indexes.]

Woodworth, R. H. 1929. Parthenogenesis and polyploidy in *Alnus rugosa* (Du Roi) Spreng. Science 70: 192–193.

Woodworth, R. H. 1929c. Cytological studies in the Betulaceae. II. *Corylus* and *Alnus*. Bot. Gaz. 88: 383–399.

Woodworth, R. H. 1930. Cytological studies in the Betulaceae. III. Parthenogenesis and polyembryony in *Alnus rugosa*. Bot. Gaz. 90: 108–115.

Woodworth, R. H. 1930c. Meiosis of sporogenesis in the Juglandaceae. Amer. J. Bot. 17: 863–869.

World Pollen Spore Fl. = World Pollen and Spore Flora.

Wrightia = Wrightia; a Botanical Journal.

Wunderlin, R. P. and D. H. Les. 1980. *Nymphaea ampla* (Nymphaeaceae), a waterlily new to Florida. Phytologia 45: 83–84.

Wyoming Agric. Exp. Sta. Bull. = Wyoming Agricultural Experiment Station Bulletin.

Xiao, P. G. 1989. Excerpts of the Chinese pharmacopoeia. Herbs Spices Med. Pl. 4: 42–114.

Yearb. Amer. Philos. Soc. = Yearbook of the American Philosophical Society.

Ying, T. S., S. Terabayashi, and D. E. Boufford. 1984. A monograph of *Diphylleia* (Berberidaceae). J. Arnold Arbor. 65: 57–94.

Young, M. J. 1873. Familiar Lessons in Botany, with Flora of Texas, Adapted to General Use in the Southern States. New York. (Famil. Lessons Bot.)

Youngken, H. W. 1919. The comparative morphology, taxonomy and distribution of the Myricaceae of the eastern United States. Contr. Bot. Lab. Morris Arbor. Univ. Pennsylvania 4: 339–400.

Zavada, M. 1983. Pollen morphology of Ulmaceae. Grana 22: 23–30.

Ziman, S. N. and C. S. Keener. 1989. A geographical analysis of the family Ranunculaceae. Ann. Missouri Bot. Gard. 76: 1012–1049.

Zimmerman, G. A. 1941. Hybrids of the American pawpaw. J. Heredity 32: 83–91.

Zohary, M. 1983. The genus *Nigella* (Ranunculaceae)—A taxonomic revision. Pl. Syst. Evol. 142: 71–107.

Index

Names in *italics* are synonyms, casually mentioned hybrids, or plants not established in the flora. Page numbers in **boldface** indicate the primary entry for a taxon. Page numbers in *italics* indicate an illustration. Roman type is used for all other entries, including author names, vernacular names, and accepted scientific names for plants treated as established members of the flora.

Political Map of North America North of Mexico

Canadian Provinces				
Alta.	Alberta	N.S.	Nova Scotia
B.C.	British Columbia	Ont.	Ontario
Man.	Manitoba	P.E.I.	Prince Edward Island
N.B.	New Brunswick	Que.	Quebec
Nfld.	Newfoundland (incl. Labrador)	Sask.	Saskatchewan
N.W.T.	Northwest Territories	Yukon	

United States				
Ala.	Alabama	Mont.	Montana
Alaska		Nebr.	Nebraska
Ariz.	Arizona	Nev.	Nevada
Ark.	Arkansas	N.H.	New Hampshire
Calif.	California	N.J.	New Jersey
Colo.	Colorado	N.Mex.	New Mexico
Conn.	Connecticut	N.Y.	New York
Del.	Delaware	N.C.	North Carolina
D.C.	District of Columbia	N.Dak.	North Dakota
Fla.	Florida	Ohio	
Ga.	Georgia	Okla.	Oklahoma
Idaho		Oreg.	Oregon
Ill.	Illinois	Pa.	Pennsylvania
Ind.	Indiana	R.I.	Rhode Island
Iowa		S.C.	South Carolina
Kans.	Kansas	S.Dak.	South Dakota
Ky.	Kentucky	Tenn.	Tennessee
La.	Louisiana	Tex.	Texas
Maine		Utah	
Md.	Maryland	Vt.	Vermont
Mass.	Massachusetts	Va.	Virginia
Mich.	Michigan	Wash.	Washington
Minn.	Minnesota	W.Va.	West Virginia
Miss.	Mississippi	Wis.	Wisconsin
Mo.	Missouri	Wyo.	Wyoming